Reveal MATH®
Integrated II

Mc
Graw
Hill

mheducation.com/prek-12

Cover: (t to b, l to r) donskarpo/Shutterstock, Ingram Publishing

Send all inquiries to:
McGraw-Hill Education
8787 Orion Place
Columbus, OH 43240

ISBN: 978-0-07-700688-4
MHID: 0-07-700688-7

Printed in the United States of America.

4 5 6 7 8 9 10 1112 13 14 LWI 28 27 26 25 24 23 22 21

Contents in Brief

Reveal AGA® Makes Math Meaningful...

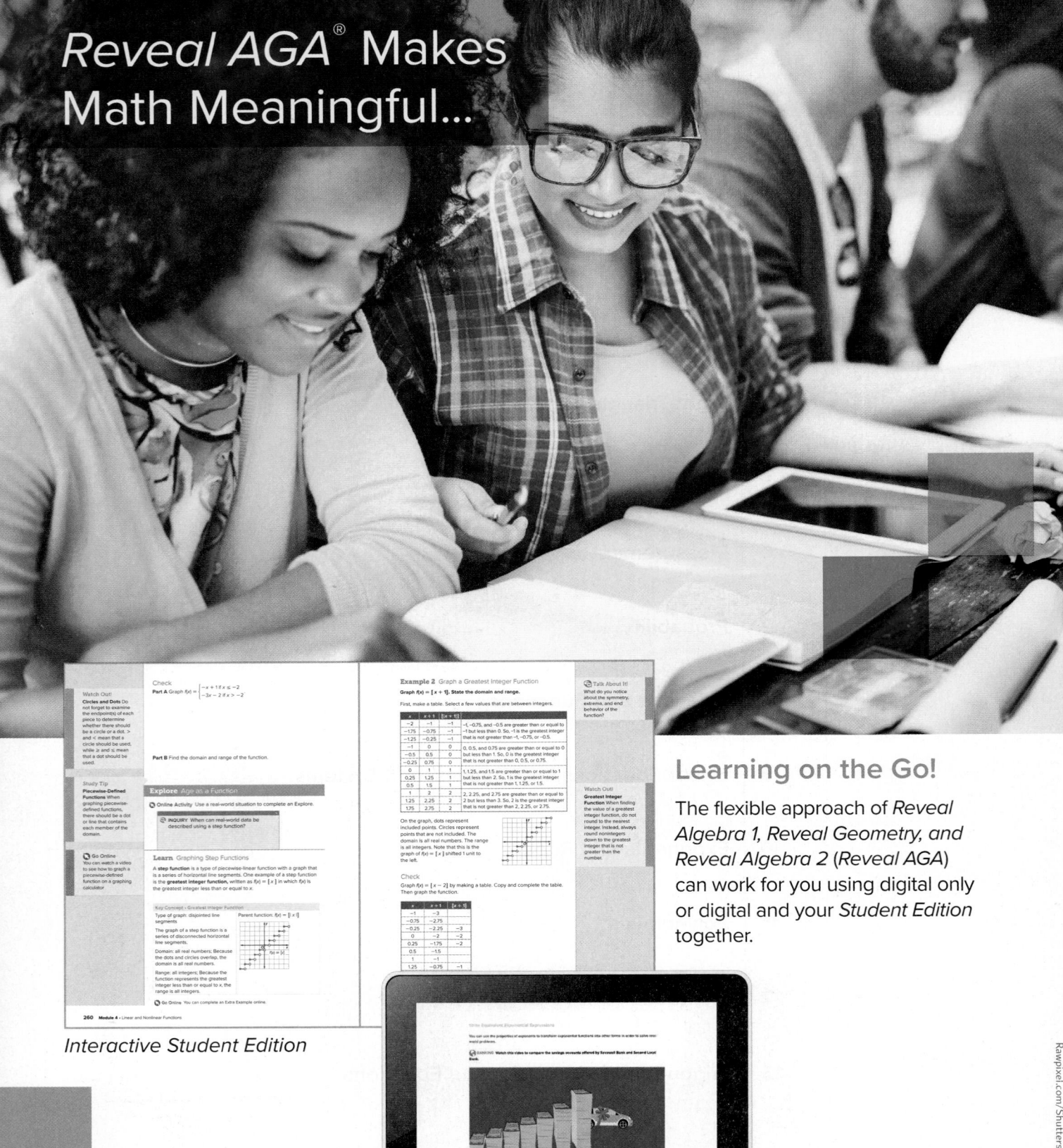

Interactive Student Edition

Student Digital Center

Learning on the Go!

The flexible approach of *Reveal Algebra 1, Reveal Geometry,* and *Reveal Algebra 2 (Reveal AGA)* can work for you using digital only or digital and your *Student Edition* together.

...to Reveal YOUR Full Potential!

Reveal AGA® Brings Math to Life in Every Lesson

Reveal AGA is a blended print and digital program that supports access on the go. Use your student edition as a reference as you work through assignments in the Student Digital Center, access interactive content, animations, videos, eTools, and technology-enhanced practice questions.

Go Online!
my.mheducation.com

Web Sketchpad® Powered by The Geometer's Sketchpad®- Dynamic, exploratory, visual activities embedded at point of use within the lesson.

Animations and Videos – Learn by seeing mathematics in action.

Interactive Tools – Get involved in the content by dragging and dropping, selecting, highlighting, and completing tables.

Personal Tutors – See and hear a teacher explain how to solve problems.

eTools – Math tools are available to help you solve problems and develop concepts.

dolgachov/123RF

Module 1
Relationships in Triangles

Module 3
Similarity

Module 4

Right Triangles and Trigonometry

Module 5
Circles

Module 6
Measurement

Module 7
Probability

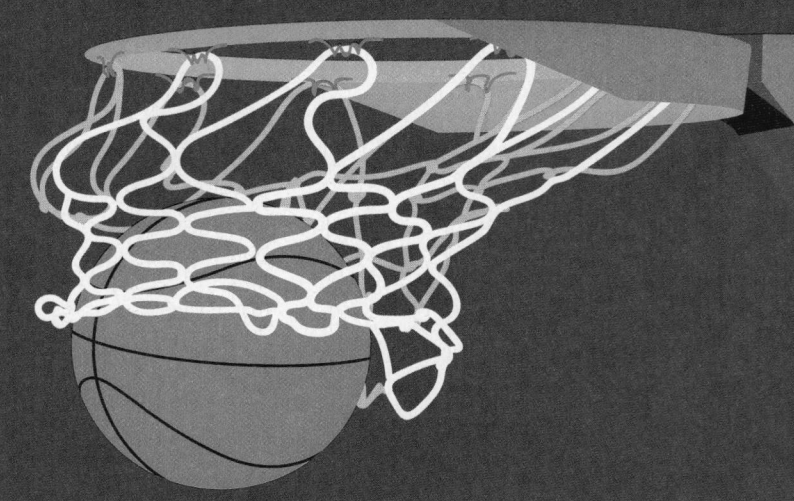

Relations and Functions

Module 9
Linear Equations, Inequalities, and Systems

Module 10

Exponents and Roots

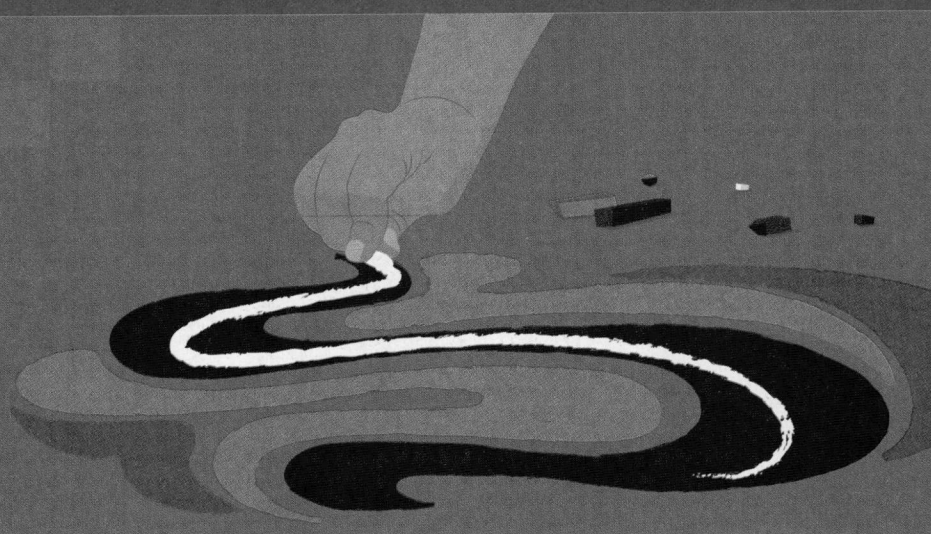

Module 11
Polynomials

Module 12
Quadratic Functions

Module 13
Trigonometric Identities and Equations

Relationships in Triangles

e Essential Question
How can relationships in triangles be used in real-world situations?

What Will You Learn?

How much do you already know about each topic **before** starting this module?

KEY

👎 — I don't know. 👍 — I've heard of it. 👍 — I know it!

	Before			After		
	👎	👍	👍	👎	👍	👍
solve problems using perpendicular bisectors in triangles						
solve problems using angle bisectors						
solve problems using medians in triangles						
solve problems using altitudes in triangles						
solve problems using inequalities in the angles in a triangle						
solve problems using inequalities in the angles and sides in a triangle						
prove algebraic and geometric relationships by using indirect proof						
apply the Triangle Inequality Theorem						
apply the Hinge Theorem and its converse						

📙 Foldables Make this Foldable to help you organize your notes about relationships in triangles. Begin with seven sheets of grid paper.

1. **Stack** the sheets. Fold the top right corner to the bottom edge to form an isosceles triangle.

2. **Fold** the rectangular part in half.

3. **Staple** the sheets along the rectangular fold in four places.

4. **Label** each sheet with a lesson number and the rectangular tab with the module title.

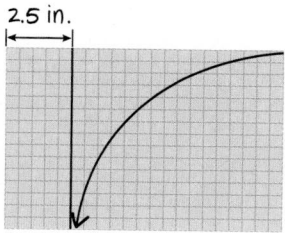
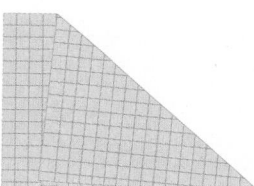
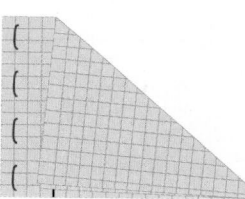
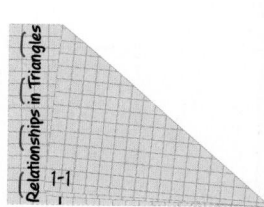

What Vocabulary Will You Learn?

- altitude of a triangle
- centroid
- circumcenter
- concurrent lines
- incenter
- indirect proof
- indirect reasoning
- median
- orthocenter
- perpendicular bisector
- point of concurrency
- proof by contradiction

Are You Ready?

Complete the Quick Review to see if you are ready to start this module.
Then complete the Quick Check.

Quick Review

Example 1

Given that $m\angle DBF = 52°$, find $m\angle DBA$.

$\angle DBF$ and $\angle DBA$ form a right angle, so their sum is 90°.

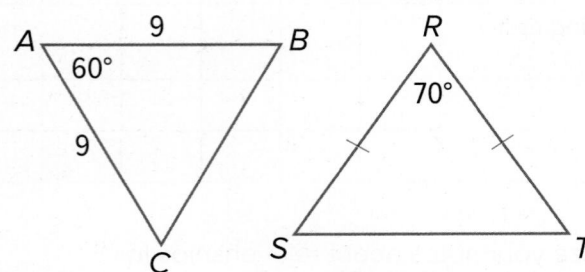

$52 + x = 90$ Let $x = m\angle DBA$.

$x = 38$ Subtract 52 from

$m\angle DBA = 38°$ each side.

Example 2

Solve $3x + 5 > 2x$.

$3x + 5 > 2x$	Original inequality
$3x - 3x + 5 > 2x - 3x$	Subtract $3x$ from each side.
$5 > -x$	Simplify.
$-5 < x$	Divide each side by -1. Reverse the inequality symbol.

Quick Check

Find each measure.

1. BC

2. $m\angle RST$

3. Two sides of a right triangular flower bed are 7 feet long each. What is the length of the third side to the nearest foot?

Solve each inequality.

4. $x + 13 < 41$

5. $6x + 9 < 7x$

6. $x - 6 > 2x$

7. $8x + 15 > 9x - 26$

How Did You Do?

Which exercises did you answer correctly in the Quick Check?

Perpendicular Bisectors of Segments

Learn Perpendicular Bisectors of Segments

A **perpendicular bisector** is a line, segment, or ray that passes through the midpoint of a segment and is perpendicular to that segment.

Theorem 1.1: Perpendicular Bisector Theorem

Words	If a point is on the perpendicular bisector of a segment, then it is equidistant from the endpoints of the segment.
Example	If $\overline{CD}$ is a $\perp$ bisector of $\overline{AB}$, then $AC = BC$.

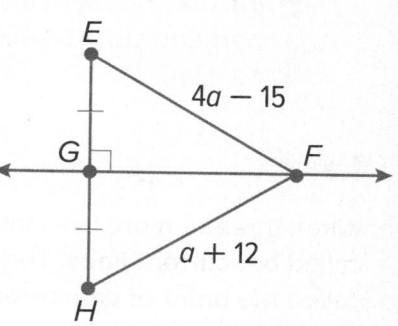

Theorem 1.2: Converse of the Perpendicular Bisector Theorem

Words	If a point is equidistant from the endpoints of a segment, then it is on the perpendicular bisector of the segment.
Example	In the triangle above, if $AC = BC$, then C lies on the $\perp$ bisector of $\overline{AB}$.

You will prove Theorems 1.1 and 1.2 in Exercises 15 and 16, respectively.

Example 1 Use the Perpendicular Bisector Theorem

Find EF.

$\overleftrightarrow{FG}$ is the perpendicular bisector of $\overline{EH}$.

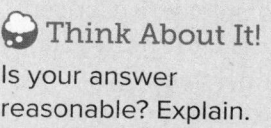

$EF = HF$	Perpendicular Bisector Theorem
$4a - 15 = a + 12$	Substitution
$a = 9$	Solve.

So, $EF = 4(9) - 15$ or 21.

Check

Find *RT*.

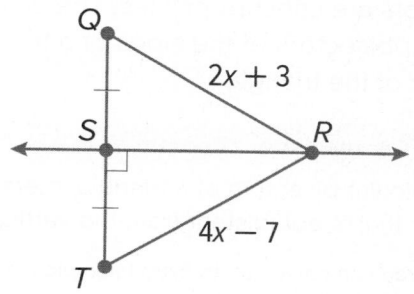

$RT = \underline{\ \ ?\ \ }$

Today's Goals
- Prove theorems and solve problems about perpendicular bisectors of line segments.
- Prove theorems and apply geometric methods to solve design problems using the perpendicular bisectors of triangles.

Today's Vocabulary
perpendicular bisector

concurrent lines

point of concurrency

circumcenter

💭 **Think About It!**

Is your answer reasonable? Explain.

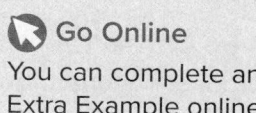

Go Online

You can complete an Extra Example online.

Example 2 Use the Converse of the Perpendicular Bisector Theorem

Find XY.

Because $WX = WZ$ and $\overleftrightarrow{WY} \perp \overline{XZ}$, $\overrightarrow{WY}$ is the perpendicular bisector of $\overline{XZ}$ by the Converse of the Perpendicular Bisector Theorem. By the definition of segment bisector, $XY = ZY$. Because $ZY = 22.4$, $XY = 22.4$.

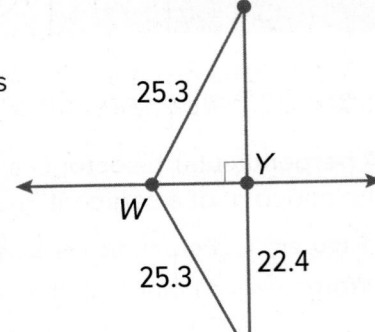

Check

Find *WY*.

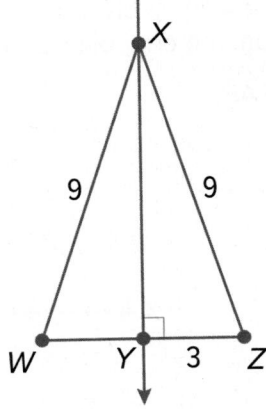

$WY = \underline{\;?\;}$

Explore Relationships Formed by Perpendicular Bisectors

Go Online
You may want to complete the Concept Check to check your understanding.

Online Activity Use dynamic geometry software to complete the Explore.

> **INQUIRY** What relationships exist among the perpendicular bisectors in triangles?

Study Tip

Circum – The prefix *circum* – means about or around. The circumcenter is the center of a circle around a triangle that contains the vertices of the triangle.

Learn Perpendicular Bisectors of Triangles

When three or more lines intersect at a common point, the lines are called **concurrent lines**. The point of intersection of concurrent lines is called the **point of concurrency**.

A triangle has three sides, so it also has three perpendicular bisectors. These bisectors are concurrent lines. The point of concurrency of the perpendicular bisectors of the sides of a triangle is called the **circumcenter** of the triangle.

Go Online
A proof of Theorem 1.3 is available.

> **Theorem 1.3: Circumcenter Theorem**
>
> The perpendicular bisectors of a triangle intersect at a point called the *circumcenter* that is equidistant from the vertices of the triangle.

Go Online You can complete an Extra Example online.

Example 3 Use the Circumcenter Theorem

Find BF if D is the circumcenter of $\triangle ABC$, $AC = 9$, $DE = 1.83$, and $DF = 1.53$.

D is the circumcenter of $\triangle ABC$, so $\overline{DE}$, $\overline{DF}$, and $\overline{DG}$ are the perpendicular bisectors of the triangle. Because $\overline{DE}$ bisects $\overline{AC}$, $EC = \frac{1}{2}AC$. So, $EC = \frac{1}{2}(9)$ or 4.5.

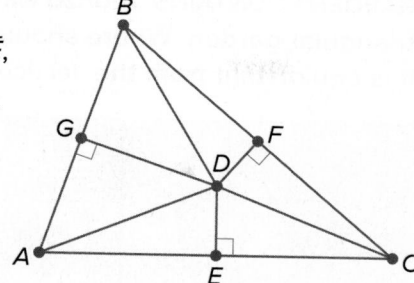

Use the Pythagorean Theorem to find DC.

$c^2 = a^2 + b^2$	Pythagorean Theorem
$DC^2 = 1.83^2 + 4.5^2$	Substitution
$DC^2 = 3.35 + 20.25$	Simplify.
$DC^2 = 23.60$	Add.
$DC \approx \pm 4.86$	Take the square root of each side.

Because length cannot be negative, use the positive square root, 4.86.

By the Circumcenter Theorem, $DC = DB$. So, $DB \approx 4.86$.

Because $DB = 4.86$ and $DF = 1.53$, we can use the Pythagorean Theorem to find BF.

$a^2 + b^2 = c^2$	Pythagorean Theorem
$1.53^2 + BF^2 = 4.86^2$	Substitution
$2.34 + BF^2 = 23.62$	Simplify.
$BF^2 = 21.28$	Subtract.
$BF \approx \pm 4.61$	Take the square root of each side.

The length of the segment must be positive, so $BF = 4.61$.

Check

Find SV if S is the circumcenter of $\triangle QPR$, $QR = 7.79$, and $PS = 4.25$. If necessary, round your answer to the nearest tenth.

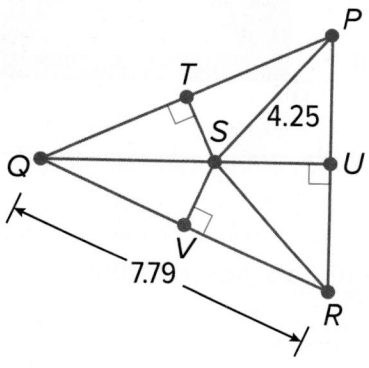

$SV = \underline{\quad ? \quad}$

🔲 **Go Online** You can complete an Extra Example online.

Problem-Solving Tip
Make a Plan Before solving for unknown measures, analyze the information you are given, develop a plan, and determine the theorems you will need to apply to find a specific measure.

💬 **Talk About It!**
Determine whether the statement is *sometimes*, *always*, or *never* true. Justify your argument.

The perpendicular bisectors of a triangle intersect at a point that is equidistant from the sides of the triangle.

Think About It!

If Alonzo is installing the fountain in his backyard garden, what unit of measure is most appropriate for the dimensions of the garden?

🌐 Example 4 Use Perpendicular Bisectors in Design Problems

GARDEN FOUNTAINS **Alonzo wants to install a fountain in his triangular garden. Where should Alonzo place the fountain so it is equidistant from the vertices of the garden?**

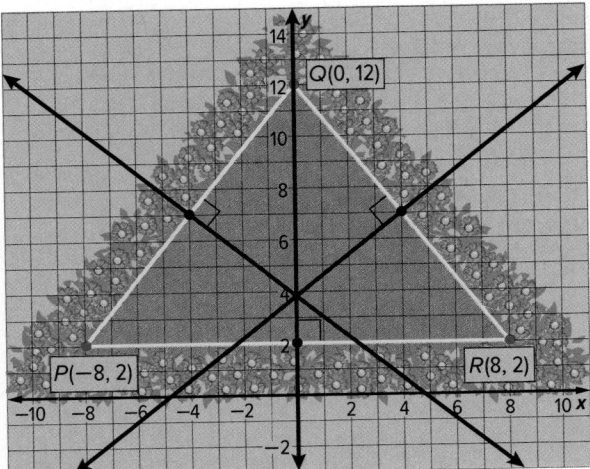

The circumcenter of a triangle is equidistant from the vertices of the triangle. Use a compass and straightedge or dynamic geometry software to construct the perpendicular bisectors of △PQR. Find the intersection point of the perpendicular bisectors to determine where Alonzo should install the fountain.

The intersection of the perpendicular bisectors appears to occur at (0, 3.8). So, Alonzo should install the fountain at (0, 3.8).

Check

HOME IMPROVEMENT Lana wants to install a skylight in her bedroom. The section of roof where she plans to install the skylight has an incline and is triangular. The vertices of the roof are at X(2, 1), Y(7, 10), and Z(12, 1). Where should the center of the skylight be located so that it is equidistant from the vertices of the roof? If necessary, round your answer to the nearest whole number.

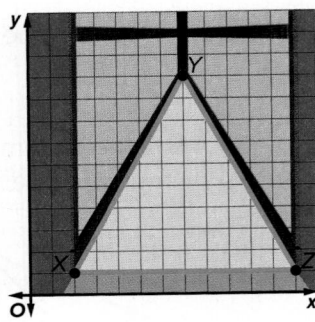

Pause and Reflect

Did you struggle with anything in this lesson? If so, how did you deal with it?

🌐 Go Online

You may want to complete the construction activities for this lesson.

Practice

Go Online You can complete your homework online.

Examples 1 and 2

Find each measure.

1. *FM*

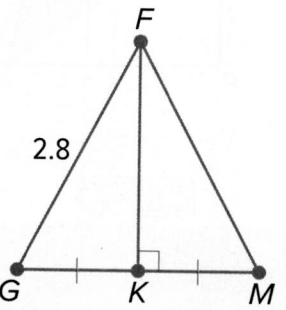

2. *XW*

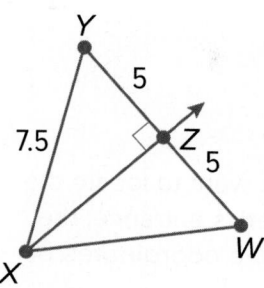

3. *BF*

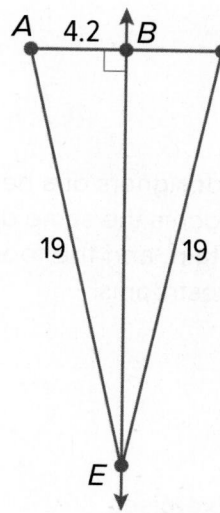

4. *KL*

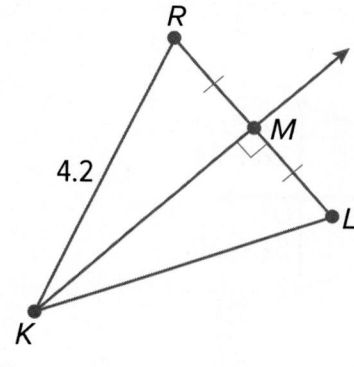

5. *TP*

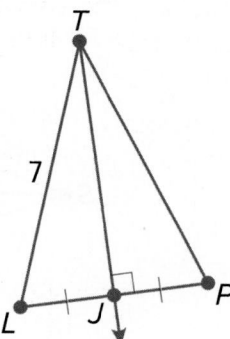

6. *KL*

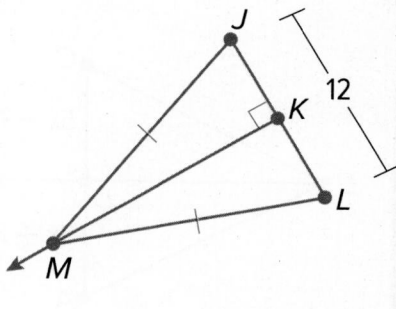

Example 3

7. Find *BR* if *D* is the circumcenter of △*ABC*, *CD* = 11.1, and *RD* = 5.2. Round to the nearest tenth.

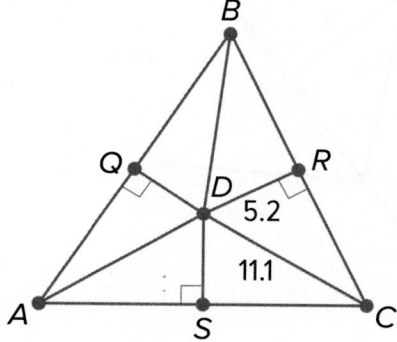

8. Find *FL* if *H* is the circumcenter of △*EFG*, *EH* = 5.06, and *LH* = 2.74. Round to the nearest hundredth.

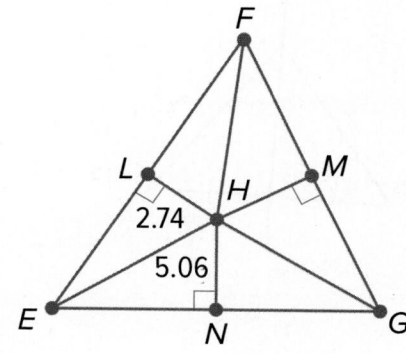

Example 4

9. A monument will be at the circumcenter of a triangular plot of land at the state capital. On the coordinate grid, the vertices of the triangular plot of land are $A(1, 5)$, $B(7, 5)$, and $C(8, 0)$. Find the coordinates of the location where the monument will be built.

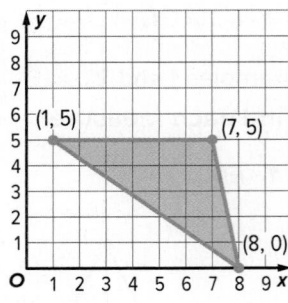

10. The designers of a new amusement park want to locate the restrooms the same distance from the park's entrance, the fountain, and the food court. Estimate the coordinates of the restrooms.

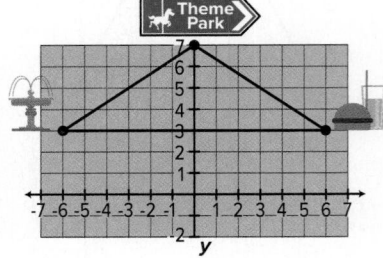

Mixed Exercises

Find the value of x.

11.

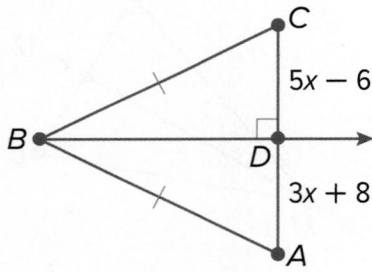

12.

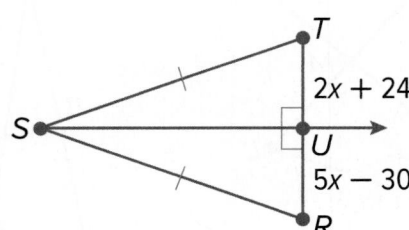

Determine whether there is enough information given in each diagram to find the value of x. If there is, find the value of x. If there is not, explain what needs to be given.

13.

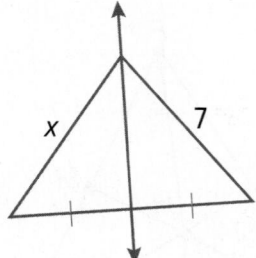

14.

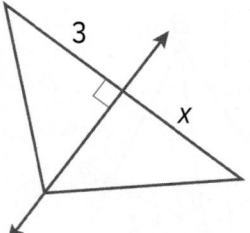

PROOF Use the figure to complete the following proofs.

15. Write a paragraph proof of the Perpendicular Bisector Theorem (Theorem 6.1).

    Given: $\overline{CD}$ is the perpendicular bisector of $\overline{AB}$.

    Prove: C is equidistant from A and B.

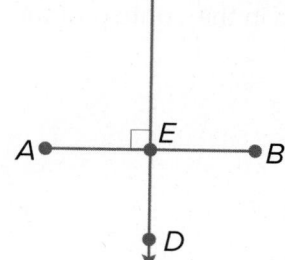

16. Write a two-column proof of the Converse of the Perpendicular Bisector Theorem (Theorem 6.2).

    Given: $\overline{CA} \cong \overline{CB}, \overline{AD} \cong \overline{BD}$

    Prove: C and D are on the perpendicular bisector of $\overline{AB}$.

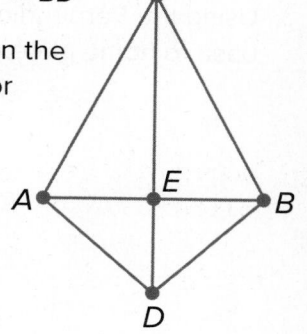

Find the coordinates of the circumcenter of the triangle with the given vertices. Explain.

17. A(0, 0), B(0, 6), C(10, 0)

18. J(5, 0), K(5, −8), L(0, 0)

19. Consider $\overleftrightarrow{CD}$. Describe the set of all points in space that are equidistant from C and D.

20. **YARDWORK** Martina has a front yard with three trees. The figure shows the locations of the trees on a coordinate plane. Martina would like to place an inground sprinkler at a location that is the same distance from all three trees. At what point on the coordinate grid should Martina place the sprinkler?

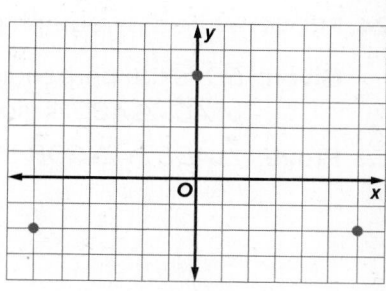

21. CREATE On a baseball diamond, home plate and second base lie on the perpendicular bisector of the line segment that joins first and third base. First base is 90 feet from home plate. How far is it from third base to home plate? Sketch a baseball diamond, labeling home plate as point *A*, first base as *B*, second base as *C*, and third base as *D*. Label the intersection of $\overline{AC}$ and $\overline{BD}$ as *E*. Using the Perpendicular Bisector Theorem, determine how far it is from third base to home plate. Describe your conclusion in the context of the situation.

22. FIND THE ERROR Thiago says that from the information supplied in the diagram, he can conclude that *K* is on the perpendicular bisector of $\overline{LM}$. Caitlyn disagrees. Is either of them correct? Explain your reasoning.

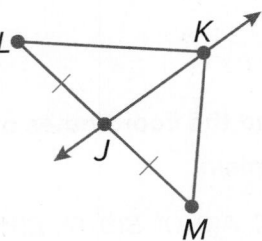

23. PROOF Write a two-column proof.

Given: Plane *Y* is a perpendicular bisector of $\overline{DC}$.

Prove: $\angle ADB \cong \angle ACB$

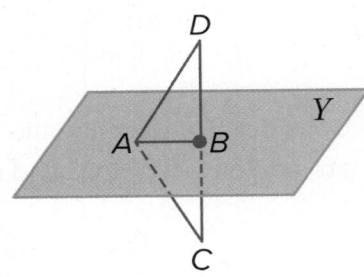

24. PROOF Write a paragraph proof.

Given: $\overline{BD}$ is the perpendicular bisector of $\overline{AC}$. $\triangle ABC$ is isosceles with base $\overline{AC}$.

Prove: $\triangle ADB \cong \triangle CDB$

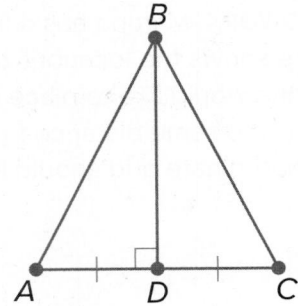

Angle Bisectors

Learn Angle Bisectors

You may recall that an angle bisector divides an angle into two congruent angles. The angle bisector can be a line, segment, or ray.

Theorem 1.4: Angle Bisector Theorem	
Words	If a point is on the bisector of an angle, then it is equidistant from the sides of the angle.
Example	If $\overrightarrow{BF}$ bisects $\angle DBE$, $\overline{FD} \perp \overrightarrow{BD}$, and $\overline{FE} \perp \overrightarrow{BE}$, then $DF = FE$.

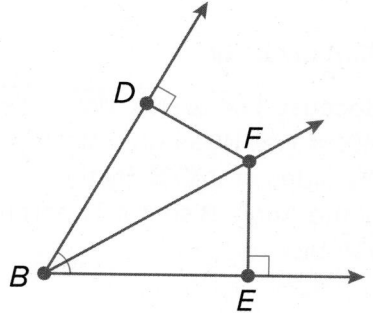

Theorem 1.5: Converse of the Angle Bisector Theorem	
Words	If a point in the interior of an angle is equidistant from the sides of the angle, then it is on the bisector of the angle.
Example	If $\overline{FD} \perp \overrightarrow{BD}$, $\overline{FE} \perp \overrightarrow{BE}$, and $DF = FE$, then $\overrightarrow{BF}$ bisects $\angle DBE$.

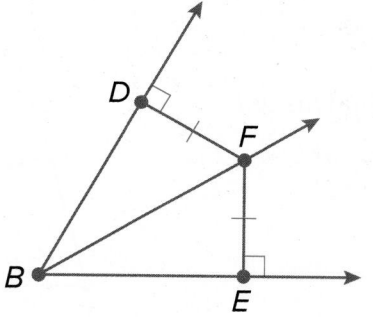

You will prove Theorems 1.4 and 1.5 in Exercises 17 and 18, respectively.

Example 1 Use the Angle Bisector Theorem

Find QT.

$\overrightarrow{RT}$ is the angle bisector of $\angle QRS$.

$$QT = ST \quad \text{Angle Bisector Theorem}$$
$$4x + 8 = 9x - 7 \quad \text{Substitution}$$
$$-5x = -15 \quad \text{Simplify.}$$
$$x = 3 \quad \text{Simplify.}$$

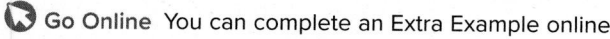

So, $QT = 4(3) + 8$ or 20.

 Go Online You can complete an Extra Example online.

Today's Goals
• Prove theorems and solve problems about angle bisectors.

• Prove theorems and apply geometric methods to solve design problems using the angle bisectors of triangles.

Today's Vocabulary
incenter

Study Tip
Angle Bisector In the figure below, there is not enough information to conclude that $\overrightarrow{BD}$ bisects $\angle ABC$. You must also know that $\overline{AD} \perp \overrightarrow{BA}$ and $\overline{CD} \perp \overrightarrow{BC}$.

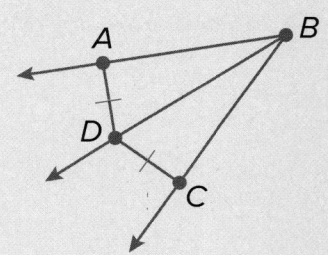

Check

Find *SP*.

SP = ___?___

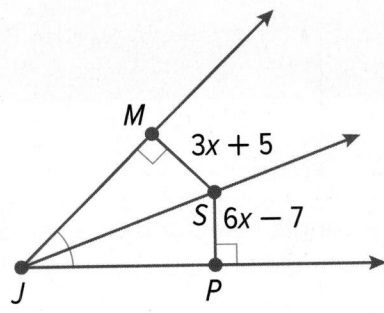

Example 2 Use the Converse of the Angle Bisector Theorem

Find *m∠ZYW*.

Because $\overrightarrow{WX} \perp \overrightarrow{YX}$, $\overline{WZ} \perp \overrightarrow{YZ}$, and *WX* = *WZ*, *W* is equidistant from the sides of ∠*XYZ*. By the Converse of the Angle Bisector Theorem, $\overrightarrow{YW}$ bisects ∠*XYZ*.

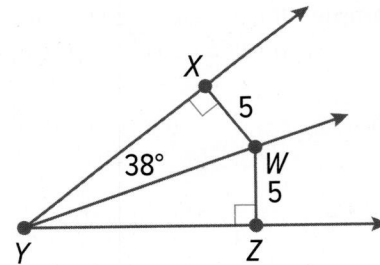

∠*ZYW* ≅ ∠*XYW*   Definition of angle bisector

m∠ZYW = *m∠XYW*   Definition of congruent angles

m∠ZYW = 38°   Substitution

Check

Find *m∠JKL*.

m∠JKL = ___?___

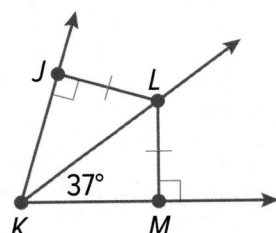

Explore Relationships Formed by Angle Bisectors

Online Activity Use dynamic geometry software to complete the Explore.

> ✎ **INQUIRY** What relationships exist among the angle bisectors in triangles?

⊗ **Go Online** You can complete an Extra Example online.

Learn Angle Bisectors of Triangles

Because a triangle has three angles, it also has three angle bisectors. The angle bisectors of a triangle are concurrent, and their point of concurrency is called the **incenter** of a triangle.

Theorem 1.6: Incenter Theorem	
Words	The angle bisectors of a triangle intersect at a point called the incenter that is equidistant from the sides of the triangle.
Example	If P is the incenter of $\triangle ABC$, then $PD = PE = PF$. 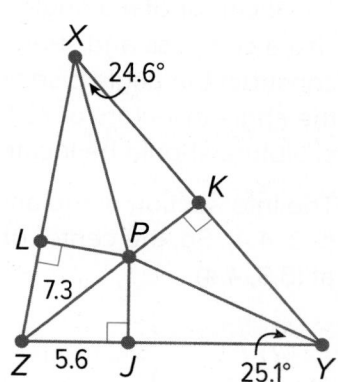

You will prove Theorem 1.6 in Exercise 19.

💭 **Think About It!**

Determine whether the statement is *sometimes*, *always*, or *never* true. Justify your argument.

The angle bisectors of a triangle intersect at a point that is equidistant from the vertices of the triangle.

Example 3 Use the Incenter Theorem

P is the incenter of $\triangle XYZ$. Find $m\angle LZP$.

P is the incenter of $\triangle XYZ$, so $\overline{XP}$ bisects $\angle LXK$, $\overline{YP}$ bisects $\angle KYJ$, and $\overline{ZP}$ bisects $\angle LZJ$. Thus, $m\angle LXK = 2m\angle PXK$. So, $m\angle LXK = 2(24.6)$ or $49.2°$.

Likewise, $m\angle KYJ = 2m\angle PYJ$, so $m\angle KYJ = 2(25.1)$ or $50.2°$.

Find $m\angle LZJ$ using the Triangle Angle-Sum Theorem.

$m\angle LXK + m\angle KYJ + m\angle LZJ = 180°$ Triangle Angle-Sum Theorem

$\qquad 49.2 + 50.2 + m\angle LZJ = 180°$ Substitution

$\qquad\qquad\qquad m\angle LZJ = 80.6°$ Simplify.

Because $\overline{ZP}$ bisects $\angle LZJ$, $2m\angle LZP = m\angle LZJ$. This means that $m\angle LZP = \frac{1}{2}m\angle LZJ$, so $m\angle LZP = \frac{1}{2}(80.6)$ or $40.3°$.

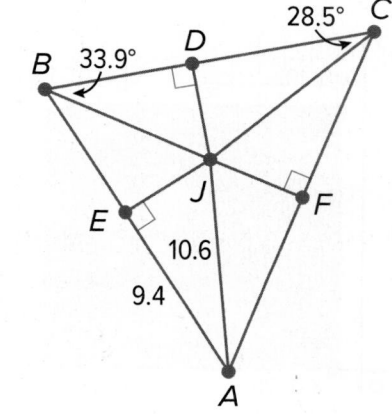

Check

Find each measure if J is the incenter of $\triangle ABC$.

$JF = $ ___?___

$m\angle JAC = $ ___?___

🧭 **Go Online** You can complete an Extra Example online.

🌐 **Example 4** Using Angle Bisectors in Design Problems

COURTYARD **A new art sculpture will be placed in the courtyard of a high school. The entrances to the courtyard are located at points *A*, *B*, and *C*. Find the location of the center of the sculpture so that it is equidistant from the sides of the courtyard.**

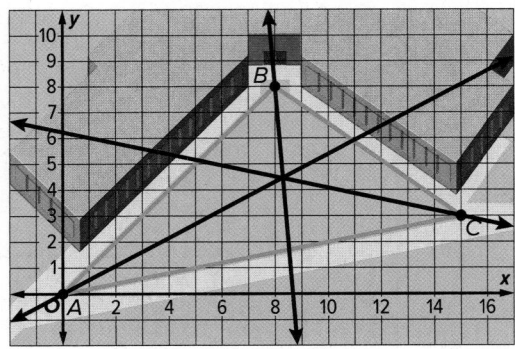

The incenter of a triangle is equidistant from the sides of the triangle. Use a compass and straightedge or dynamic geometry software to construct the angle bisectors of △*ABC*. Find the intersection point of the angle bisectors of △*ABC* to determine where the center of the sculpture should be located.

The intersection of the angle bisectors appears to be at about (8.3, 4.4). So, the center of the sculpture should be located at (8.3, 4.4).

Check

HOME IMPROVEMENT Tyrice's parents want to install a hot tub on the back deck of their house. The vertices of the deck are located at points *X*, *Y*, and *Z*. Find the location of the center of the hot tub so it is equidistant from the edges of the deck. If necessary, round your answer to the nearest whole number.

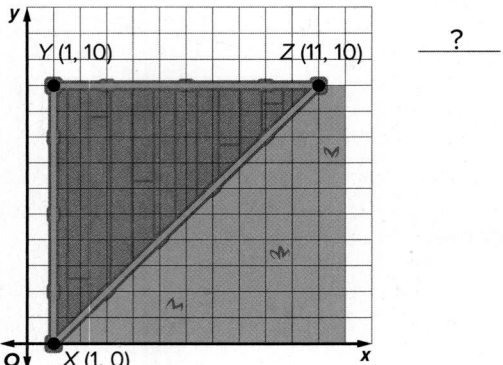

🔵 **Go Online**
You may want to complete the construction activities for this lesson.

🔵 **Go Online** You can complete an Extra Example online.

Practice

Go Online You can complete your homework online.

Examples 1 and 2

Find each measure.

1. m∠ABE

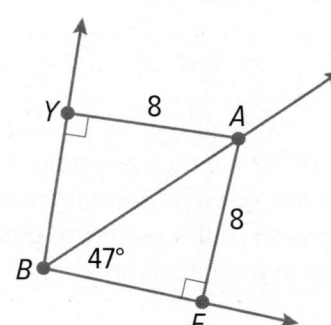

2. m∠YBA

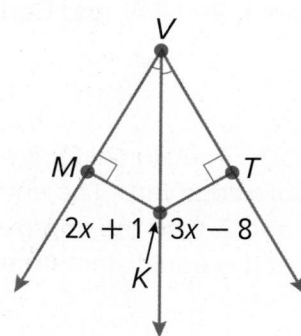

3. MK

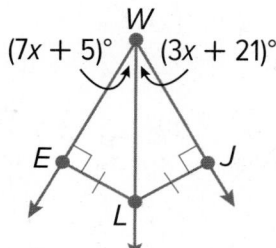

4. m∠EWL

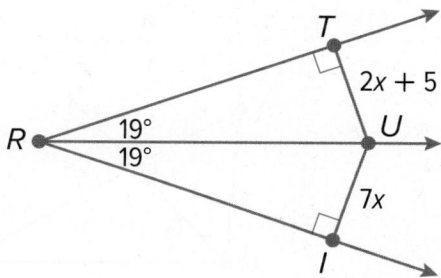

5. IU

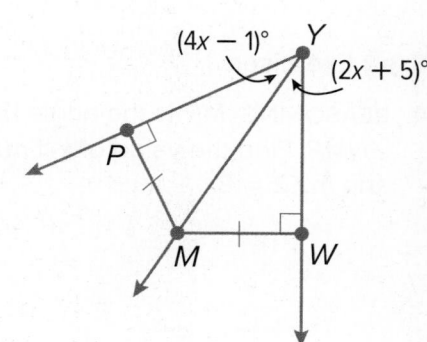

6. m∠MYW

Example 3

A is the incenter of △PQR. Find each measure.

7. m∠ARU

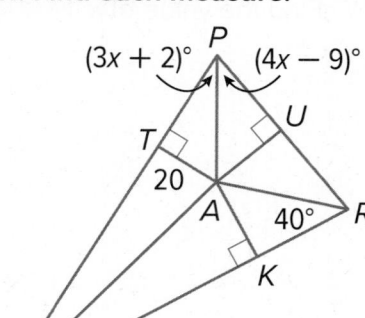

8. AU

U is the incenter of △GHY. Find each measure.

9. m∠UGM

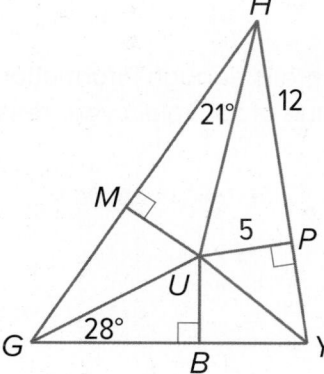

10. m∠PHU

11. HU

Example 4

12. CITY PLANNING City planning officials want the location of a new electric car charging station to be equidistant from the three townships shown on the coordinate plane. Find the approximate location of the charging station so that it is equidistant from the roads connecting the townships of Fairmont, Bethany, and Cedarview.

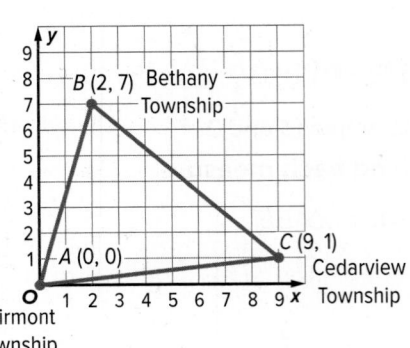

13. SCHOOL The alumni foundation will donate a new fountain for the high school's courtyard. The entrances to the courtyard are located at points X, Y, and Z. Find the approximate location of the center of the fountain so that it is equidistant from the sides of the courtyard.

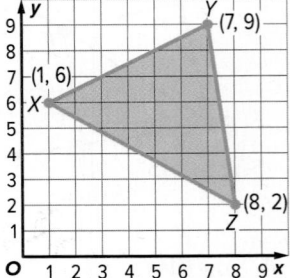

Mixed Exercises

14. REASONING $\overrightarrow{MR}$ is the angle bisector of $\angle NMP$. Find the value of x if $m\angle 1 = 5x + 8$ and $m\angle 2 = 8x - 16$.

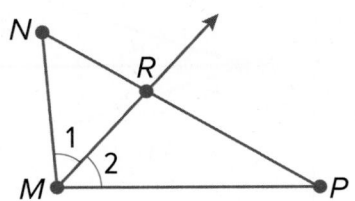

15. REASONING $\overrightarrow{YL}$ is the angle bisector of $\angle JYF$. Find the value of x if $m\angle 1 = 8x + 10$ and $m\angle 2 = 11x - 8$.

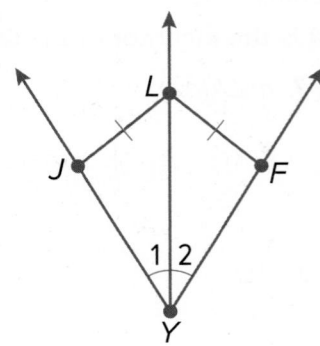

16. Determine whether there is enough information given in the diagram to find the value of x. Explain your reasoning.

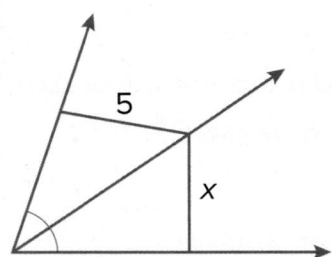

17. **PROOF** Write a paragraph proof of the Angle Bisector Theorem.

   Given: $\overline{BD}$ is the angle bisector of $\angle ABC$.

   Prove: D is equidistant from $\overline{AB}$ and $\overline{BC}$.

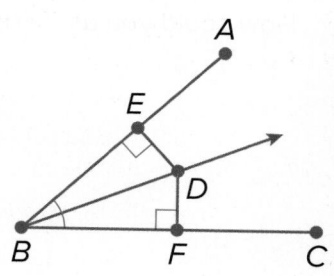

18. **PROOF** Write a paragraph proof of the Converse of the Angle Bisector Theorem.

   Given: P is in the interior of $\angle BAC$, and P is equidistant from $\overline{AB}$ and $\overline{AC}$ at D and E, respectively.

   Prove: $\overline{AP}$ is the angle bisector of $\angle BAC$.

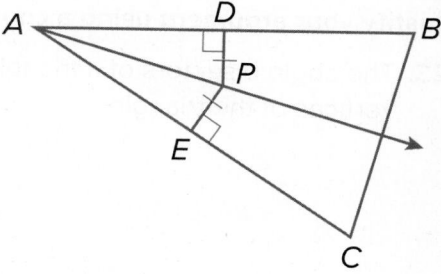

19. **PROOF** Write a two-column proof of the Incenter Theorem.

   Given: $\triangle ABC$ has angle bisectors $\overline{AD}$, $\overline{BE}$, and $\overline{CF}$.
   $\overline{KP} \perp \overline{AB}, \overline{KQ} \perp \overline{BC}, \overline{KR} \perp \overline{AC}$

   Prove: $KP = KQ = KR$

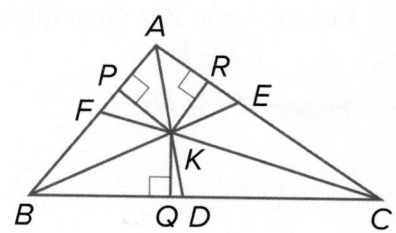

20. **CONSTRUCT ARGUMENTS** State whether the following sentence is *sometimes*, *always*, or *never* true. Justify your argument.

   The three angle bisectors of a triangle intersect at a point in the exterior of the triangle.

21. USE TOOLS Copy the triangle shown. Construct the incenter. How could you use a ruler to verify your construction?

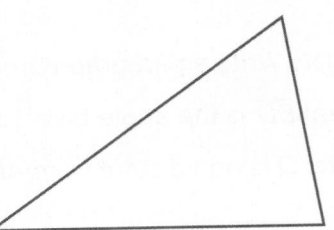

22. CREATE Draw a triangle with an incenter located inside the triangle and a circumcenter located outside. Justify your drawing using a straightedge and a compass to find both points of concurrency.

ANALYZE Determine whether each statement is *sometimes*, *always*, or *never* true. Justify your argument using a counterexample or proof.

23. The angle bisectors of a triangle intersect at a point that is equidistant from the vertices of the triangle.

24. In an isosceles triangle, the perpendicular bisector of the base is also the angle bisector of the opposite vertex.

25. PROOF Write a two-column proof.

Given: Plane Z is an angle bisector of $\angle KJH$.
$\overline{KJ} \cong \overline{HJ}$

Prove: $\overline{MH} \cong \overline{MK}$

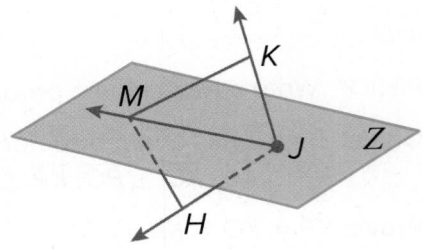

26. WRITE Compare and contrast the perpendicular bisectors and angle bisectors of a triangle. How are they alike? How are they different? Be sure to compare their points of concurrency.

27. WRITE Write a biconditional statement for Theorem 6.4 and its converse.

Medians and Altitudes of Triangles

Today's Goals
- Solve problems by applying the Centroid Theorem.
- Use altitudes and the slope criteria for perpendicular lines to determine the coordinates of the orthocenters of triangles on the coordinate plane.

Today's Vocabulary
median
centroid
altitude of a triangle
orthocenter

Explore Centroid of a Triangle

🢒 **Online Activity** Use dynamic geometry software to complete the Explore.

> ⓠ **INQUIRY** How is the location of the centroid related to the medians of a triangle? ×

Learn Medians of Triangles

In a triangle, a **median** is a line segment with endpoints that are a vertex of the triangle and the midpoint of the side opposite the vertex.

Every triangle has three medians that are concurrent. The point of concurrency of the medians of a triangle is called the **centroid**, and it is always inside the triangle.

> **Theorem 1.7: Centroid Theorem**
>
> The medians of a triangle intersect at a point called the centroid that is two-thirds of the distance from each vertex to the midpoint of the opposite side.

You will prove Theorem 1.7 in Exercise 22.

All polygons have a balancing point or *center of gravity*. This is the point at which the weight of a region is evenly dispersed and all sides of the region are balanced. The centroid is the center of gravity for a triangular region.

Example 1 Use the Centroid Theorem

In △ABC, P is the centroid and BL = 6. Find BP and PL.

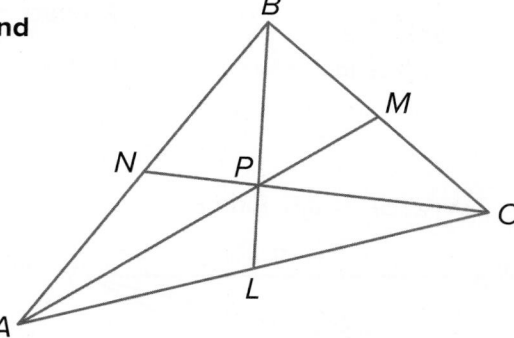

$BP = \frac{2}{3} BL$ Centroid Theorem

$ = \frac{2}{3} (6)$ or 4 $BL = 6$

$BP + PL = 6$ Segment Addition Postulate

$4 + PL = 6$ $BP = 4$

$PL = 2$ Subtract.

🢒 **Go Online** You can complete an Extra Example online.

💭 **Think About It!**

How could you find the coordinates of the centroid of △PQR?

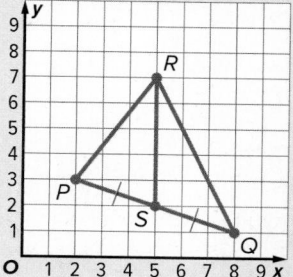

Check

In △ABC, Q is the centroid and BE = 9. Find BQ and QE.

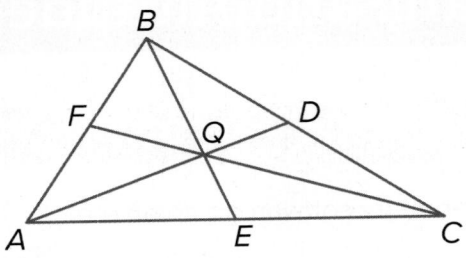

BQ = __?__ and QE = __?__

Example 2 Apply the Centroid Theorem

In △LMN, PY = 7. Find LP.

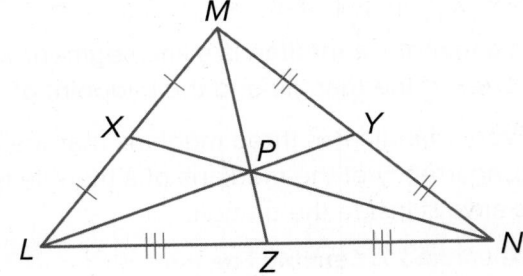

Because $\overline{LX} \cong \overline{XM}$, X is the midpoint of $\overline{LM}$, and $\overline{NX}$ is a median of △LMN. Likewise, Y and Z are the midpoints of $\overline{MN}$ and $\overline{LN}$ respectively, so $\overline{LY}$ and $\overline{MZ}$ are also medians of △LMN. Therefore, point P is the centroid of △LMN.

$LP = \frac{2}{3}LY$	Centroid Theorem
$LP = \frac{2}{3}(LP + PY)$	Segment Addition and Substitution
$LP = \frac{2}{3}(LP + 7)$	$PY = 7$
$LP = \frac{2}{3}LP + \frac{14}{3}$	Distributive Property
$\frac{1}{3}LP = \frac{14}{3}$	Subtract $\frac{2}{3}LP$ from each side.
$LP = 14$	Multiply each side by 3.

Talk About It!

How can you find LP without solving an equation? Justify your argument.

Check

In △XYZ, SP = 3.5. Find PZ.

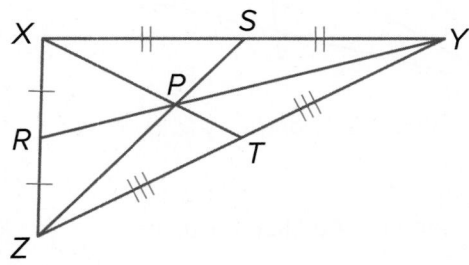

PZ = __?__

Go Online You can complete an Extra Example online.

🌐 Apply Example 3 Find a Centroid on the Coordinate Plane

CHIMES Lashaya needs to hang a wind chime with a single piece of cord. The pipes of the wind chime are attached to a triangular platform. When the platform is placed on a coordinate plane, the vertices of the triangle are located at (1, 1), (11, 5), and (7, 10). What are the coordinates of the point where the cord should be attached to the platform so the wind chime stays balanced?

1 What is the task?

Describe the task in your own words. Then list any questions that you may have. How can you find answers to your questions?

Sample answer: I need to find the balancing point of the triangular platform. The balancing point of a triangular region is the centroid, so I need to find the centroid of the triangle that is described.

2 How will you approach the task? What have you learned that you can use to help you complete the task?

Sample answer: I will graph the triangle on the coordinate plane. Then, I will use the Midpoint Formula to calculate the midpoint of one side of the triangle. Then, I will use the Centroid Theorem and what I have learned about calculating fractional distance to find a point that is $\frac{2}{3}$ of the distance from the vertex opposite the midpoint that I found to the midpoint.

3 What is your solution?

Use your strategy to solve the problem.

Graph the triangular platform and the medians of $\triangle ABC$.

The midpoint of $\overline{AB}$ is (6, 3).

The centroid of $\triangle ABC$ is $\left(\frac{19}{3}, \frac{16}{3}\right)$.

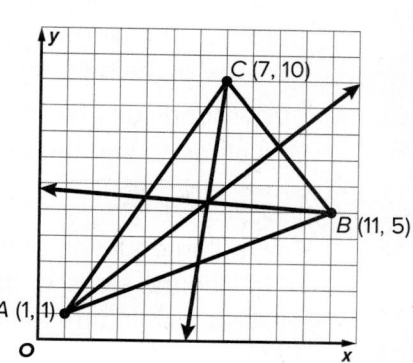

4 How can you know that your solution is reasonable?

🔺 **Write About It!** Write an argument that can be used to defend your solution.

Sample answer: When I graph the medians of $\triangle ABC$, the intersection appears to be at about (6.5, 5.5). I calculated that the centroid is located at $\left(\frac{19}{3}, \frac{16}{3}\right)$ which is about (6.33, 5.33). So, my answer is reasonable.

🌐 Go Online You can complete an Extra Example online.

💭 **Think About It!**
What assumption did you make while solving this problem?

🔖 **Go Online**
You may want to complete the construction activities for this lesson.

🔖 **Go Online** to practice what you've learned about points of concurrency in triangles in the Put It All Together over Lessons 1-1 through 1-3.

Learn Altitudes of Triangles

An **altitude of a triangle** is a segment from a vertex of the triangle to the line containing the opposite side and perpendicular to that side. An altitude can lie in the interior, in the exterior, or on the side of the triangle.

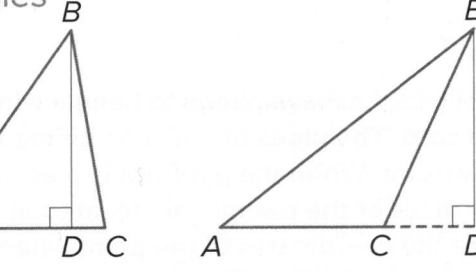

$\overline{BD}$ is an altitude from B to the line containing $\overline{AC}$.

Every triangle has three altitudes. If extended, the altitudes of a triangle are concurrent. The point of concurrency is called the **orthocenter**.

The lines containing altitudes $\overline{AF}$, $\overline{CD}$, and $\overline{BG}$ intersect at P, the orthocenter of $\triangle ABC$.

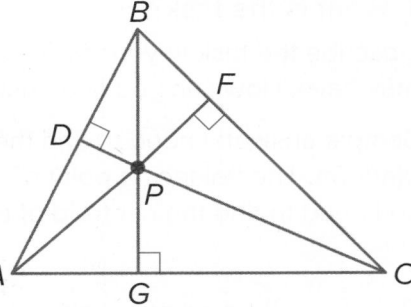

Example 4 Find an Orthocenter on the Coordinate Plane

The vertices of $\triangle ABC$ are $A(4, 0)$, $B(-2, 4)$, and $C(0, 6)$. Find the coordinates of the orthocenter of $\triangle ABC$.

Step 1 Graph $\triangle ABC$.

To find the orthocenter of $\triangle ABC$, find the point where two of the three altitudes intersect.

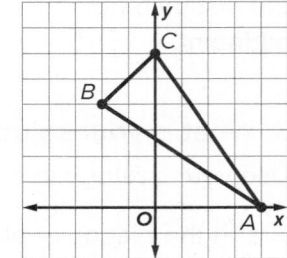

Step 2 Find equations of the altitudes.

Find an equation of the altitude from B to $\overline{AC}$. The slope of $\overline{AC}$ is $-\frac{3}{2}$, so the slope of the altitude, which is perpendicular to $\overline{AC}$, is $\frac{2}{3}$. Use the slope and point B on the altitude to find the equation of the line.

$y = \frac{2}{3}x + \frac{16}{3}$

Find an equation of the altitude from A to $\overline{BC}$. The slope of $\overline{BC}$ is 1, so the slope of the altitude is -1. Use the slope and point A on the altitude to find the equation of the line.

$y = -x + 4$

Step 3 Solve the system of equations.

$x = -\frac{4}{5}$ and $y = \frac{24}{5}$

The coordinates of the orthocenter of $\triangle ABC$ are $\left(-\frac{4}{5}, \frac{24}{5}\right)$.

Check

The vertices of $\triangle FGH$ are $F(-2, 4)$, $G(4, 4)$, and $H(1, -2)$. What are the coordinates of the orthocenter of $\triangle FGH$? If necessary, round your answer to the nearest tenth.

🧭 **Go Online** You can complete an Extra Example online.

Practice

🔵 **Go Online** You can complete your homework online.

Examples 1 and 2

In △*CDE*, *U* is the centroid, *UK* = 12, *EM* = 21, and *UD* = 9. Find each measure.

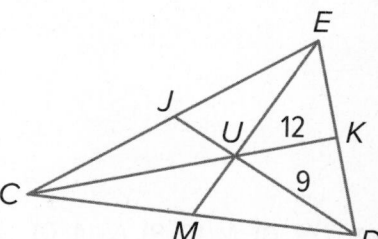

1. *CU*

2. *MU*

3. *EU*

4. *JD*

In △*ABC*, *AU* = 16, *BU* = 12, and *CF* = 18. Find each measure.

5. *CU*

6. *AD*

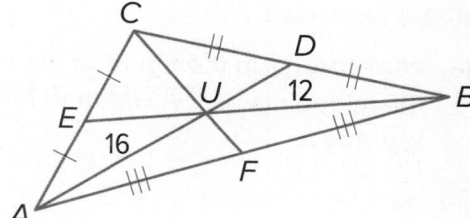

7. *UF*

8. *BE*

Example 3

Find the coordinates of the centroid of each triangle with the given vertices.

9. *X*(−3, 15) *Y*(1, 5), *Z*(5, 10)

10. *S*(2, 5), *T*(6, 5), *R*(10, 0)

11. **DECORATING** Camilla made a collage with pictures of her trip to Europe. She wants to hang the collage from the ceiling in her room so that it is parallel to the ceiling. She draws a model of the collage on a coordinate plane. At what point should she place the string?

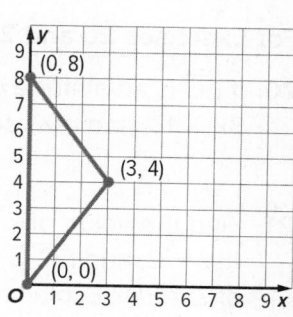

Example 4

Find the coordinates of the orthocenter of the triangle with the given vertices.

12. $J(1, 0)$, $H(6, 0)$, $I(3, 6)$

13. $S(1, 0)$, $T(4, 7)$, $U(8, -3)$

14. $L(8, 0)$, $M(10, 8)$, $N(14, 0)$

15. $D(-9, 9)$, $E(-6, 6)$, $F(0, 6)$

Mixed Exercises

16. REASONING In the figure at the right, if J, P, and L are the midpoints of $\overline{KH}$, $\overline{HM}$, and $\overline{MK}$, respectively, find x, y, and z.

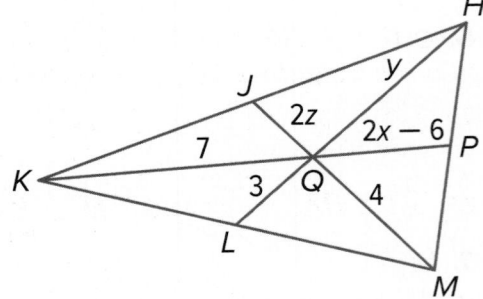

Given $\triangle RST$ with medians $\overline{RM}$, $\overline{SL}$, and $\overline{TK}$, and centroid J, find each value of x.

17. $SL = x(JL)$

18. $JT = x(TK)$

19. $JM = x(RJ)$

For Exercises 20 and 21, refer to the figure at the right.

20. If $\overline{EC}$ is an altitude of $\triangle AED$, $m\angle 1 = 2x + 7$, and $m\angle 2 = 3x + 13$, find $m\angle 1$ and $m\angle 2$.

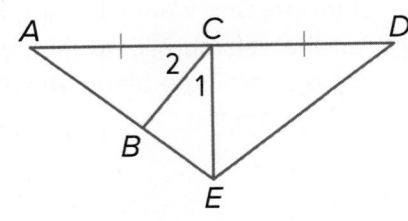

21. Find the value of x if $AC = 4x - 3$, $DC = 2x + 9$, $m\angle ECA = 15x + 2$, and $\overline{EC}$ is a median of $\triangle AED$. Is $\overline{EC}$ also an altitude of $\triangle AED$? Explain.

22. PROOF Write a coordinate proof to prove the Centroid Theorem.

Given: $\triangle ABC$, medians $\overline{AR}$, $\overline{BS}$, and $\overline{CQ}$

Prove: The medians intersect at point P, and P is two thirds of the distance from each vertex to the midpoint of the opposite side.

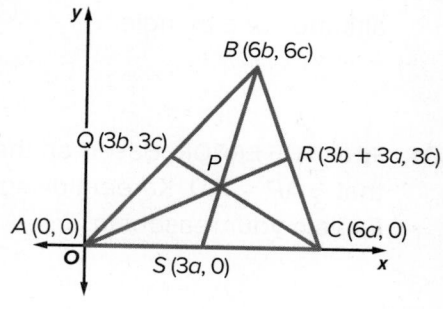

(*Hint*: First, find the equations of the lines containing the medians. Then find the coordinates of point P and show that all three medians intersect at point P. Next, use the Distance Formula and multiplication to show

$$AP = \frac{2}{3}\,AR,\ BP = \frac{2}{3}\,BS,\text{ and } CP = \frac{2}{3}\,CQ.)$$

CONSTRUCT ARGUMENTS **Use the given information to determine whether $\overline{LM}$ is a *perpendicular bisector*, *median*, and/or *altitude* of $\triangle JKL$.**

23. $\overline{LM} \perp \overline{JK}$

24. $\triangle JLM \cong \triangle KLM$

25. $\overline{JM} \cong \overline{KM}$

26. $\overline{LM} \perp \overline{JK}$ and $\overline{JL} \cong \overline{KL}$

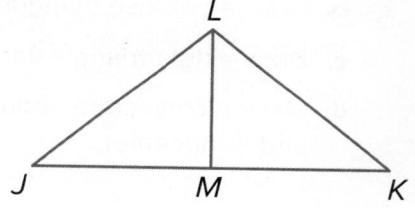

REASONING **In $\triangle JLP$, $m\angle JMP = 3x - 6$, $JK = 3y - 2$, and $LK = 5y - 8$.**

27. If $\overline{JM}$ is an altitude of $\triangle JLP$, find the value of x.

28. Find LK if $\overline{PK}$ is a median.

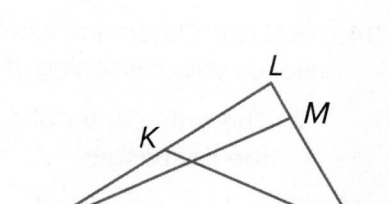

29. PROOF Write a paragraph proof.

    Given: $\triangle XYZ$ Is isosceles.

        $\overline{WY}$ bisects $\angle Y$.

    Prove: $\overline{WY}$ is a median.

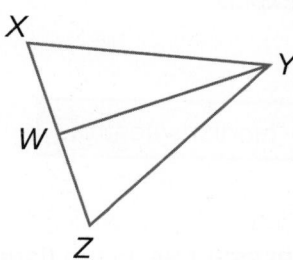

30. WRITE Compare and contrast the perpendicular bisectors, medians, and altitudes of a triangle.

31. FIND THE ERROR Based on the figure at the right, Laura says that $\frac{2}{3}AP = AD$. Kareem disagrees. Is either of them correct? Explain your reasoning.

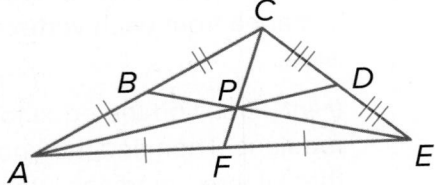

32. PERSEVERE △ABC has vertices $A(-3, 3)$, $B(2, 5)$, and $C(4, -3)$. What are the coordinates of the centroid of △ABC? Explain the process you used to reach your conclusion.

33. CREATE In this problem, you will investigate the relationships among three points of concurrency in a triangle.

    a. Draw an acute triangle and find the circumcenter, centroid, and orthocenter.

    b. Draw an obtuse triangle and find the circumcenter, centroid, and orthocenter.

    c. Draw a right triangle and find the circumcenter, centroid, and orthocenter.

    d. Make a conjecture about the relationships among the circumcenter, centroid, and orthocenter.

34. ANALYZE Determine whether the following statement is *true* or *false*. If true, explain your reasoning. If false, provide a counterexample.

    The orthocenter of a right triangle is always located at the vertex of the right angle.

35. WHICH ONE DOESN'T BELONG? Choose the term pairing that is not correct. Explain.

medians/centroid	perpendicular bisectors/incenter	altitudes/orthocenter

36. PERSEVERE In the figure at the right, segments $\overline{AD}$ and $\overline{CE}$ are medians of △ACB, $\overline{AD} \perp \overline{CE}$, $AB = 10$, and $CE = 9$. Find CA.

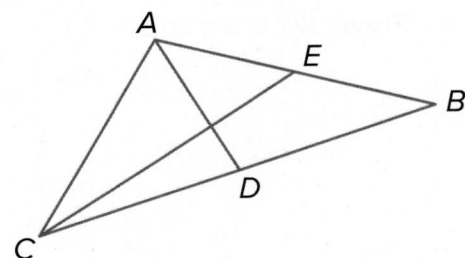

Inequalities in One Triangle

Explore Angle and Angle-Side Inequalities in Triangles

📡 **Online Activity** Use dynamic geometry software to complete the Explore.

> ⊘ **INQUIRY** What relationship exists between the sides and angles of a triangle?

Learn Angle Inequalities in One Triangle

The inequality relationship between two real numbers is often applied in proofs. Recall that for any real numbers a and b, $a > b$ if and only if there is a positive number c such that $a = b + c$.

Key Concept • Properties of Inequality

Comparison Property of Inequality

Words	The value of a must be less than, greater than, or equal to the value of b.
Symbols	For all real numbers a, b, and c, the following is true: $a < b$, $a > b$, or $a = b$.

Transitive Property of Inequality

Words	If a is less than b and b is less than c, then a is less than c. If a is greater than b and b is greater than c, then a is greater than c.
Symbols	For all real numbers a, b, and c, the following are true. If $a < b$ and $b < c$, then $a < c$. If $a > b$ and $b > c$, then $a > c$.

Addition Property of Inequality

Words	If the same number is added to each side of a true inequality, then the resulting inequality is also true.
Symbols	For all real numbers a, b, and c, the following are true. If $a > b$, then $a + c > b + c$. If $a < b$, then $a + c < b + c$.

Subtraction Property of Inequality

Words	If the same number is subtracted from each side of a true inequality, then the resulting inequality is also true.
Symbols	For all real numbers a, b, and c, the following are true. If $a > b$, then $a - c > b - c$. If $a < b$, then $a - c < b - c$.

The definition of inequality and the properties of inequalities can be applied to the measures of angles and segments, because these are real numbers.

Today's Goals
- Solve problems by applying the Exterior Angle Inequality Theorem.
- Prove and apply theorems about inequalities in one triangle.

Theorem 1.8: Exterior Angle Inequality Theorem

The measure of an exterior angle of a triangle is greater than the measure of either of its corresponding remote interior angles.

▶ **Go Online** A proof of Theorem 1.8 is available.

Remember, each exterior angle of a triangle has two *remote interior angles* that are not adjacent to the exterior angle.

Example 1 Use the Exterior Angle Inequality Theorem

List the angles that satisfy the stated condition. Justify your reasoning using the Exterior Angle Inequality Theorem.

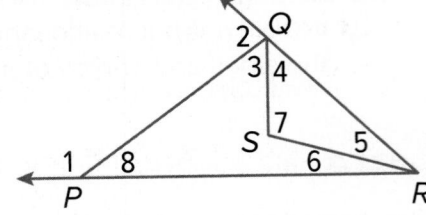

a. measures $< m\angle 1$

$\angle 1$ is an exterior angle of $\triangle PQR$, with $\angle PQR$ and $\angle PRQ$ as corresponding remote interior angles. By the Exterior Angle Inequality Theorem, $m\angle 1 > m\angle PQR$ and $m\angle 1 > m\angle PRQ$. Because $m\angle PQR = m\angle 3 + m\angle 4$ and $m\angle PRQ = m\angle 5 + m\angle 6$, $m\angle 1 > m\angle 3 + m\angle 4$ and $m\angle 1 > m\angle 5 + m\angle 6$ by substitution. So, the angles with measures less than $m\angle 1$ are $\angle 3, \angle 4, \angle 5, \angle 6$, $\angle PQR$, and $\angle PRQ$.

b. measures $> m\angle 8$

$\angle 2$ is an exterior angle of $\triangle PQR$. So, by the Exterior Angle Inequality Theorem, $m\angle 2 > m\angle 8$.

Check

List the angles that satisfy the stated condition. Justify your reasoning using the Exterior Angle Inequality Theorem.

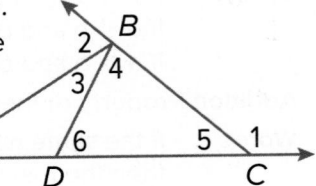

a. measures $< m\angle 1$

b. measures $> m\angle 7$

Sidebar

▶ **Go Online**

An alternate method is available for this example.

💭 **Think About It!**

Why must the markings be incorrect in the given diagram? Justify your argument.

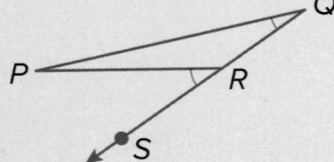

Learn Angle-Side Inequalities in One Triangle

We know that if two sides of a triangle are congruent, or the triangle is isosceles, then the angles opposite those sides are congruent.

Angle-Side Relationships in Triangles
Theorem 1.9
If one side of a triangle is longer than another side, then the angle opposite the longer side has a greater measure than the angle opposite the shorter side.
Theorem 1.10
If one angle of a triangle has a greater measure than another angle, then the side opposite the angle with the greater measure is longer than the side opposite the angle with the lesser measure.

You will prove Theorem 1.10 in Lesson 1-5, Exercise 19.

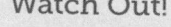 **Go Online** A proof of Theorem 1.9 is available.

Example 2 Order Triangle Angle Measures

List the angles of △LMN in order from smallest to largest.

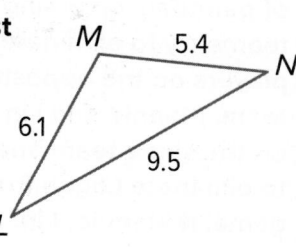

The sides from shortest to longest are $\overline{MN}$, $\overline{LM}$, and $\overline{LN}$.

So, the angles from smallest to largest are

∠L, ∠N, and ∠M.

Check

List the angles of △ABC in order from smallest to largest.

The angles from smallest to largest are
_____?_____.

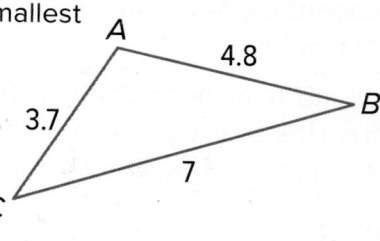

Example 3 Order Triangle Side Lengths

List the sides of △WXY in order from shortest to longest.

First find the missing angle measure using the Triangle Angle-Sum Theorem.
$m\angle X = 180 - (51 + 90)$ or 39°

So, the angles from smallest to largest are ∠X, ∠W, and ∠Y. So, the sides from shortest to longest are
$\overline{WY}$, $\overline{YX}$, and $\overline{WX}$.

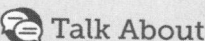 **Go Online** You can complete an Extra Example online.

Check

List the sides of △FGH in order from shortest to longest.

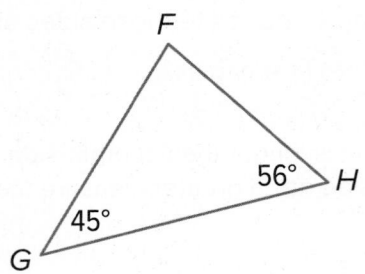

🌐 **Example 4** Use Angle-Side Relationships

PAINTBALL During a game of paintball, opposing teams try to eliminate players on the opposite team. Mannie and Lin are on the same team and want to eliminate Logan from the game. If Mannie, Lin, and Logan are located at the positions shown on the diagram, who is closer to Logan? Explain your reasoning.

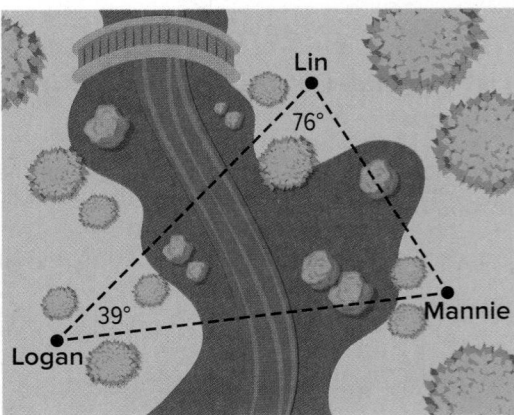

By the Triangle Angle-Sum Theorem, the measure of the angle across from the segment between Logan and Lin is 65°. Because 65 < 76, according to Theorem 1.10, Lin is closer to Logan.

Check

SPORTS Gabrielle, Diego, and Lucy are passing a football. Lucy wants to practice throwing the ball long distances. Which player should she throw the ball to next if she wants to pass the football the farthest distance?

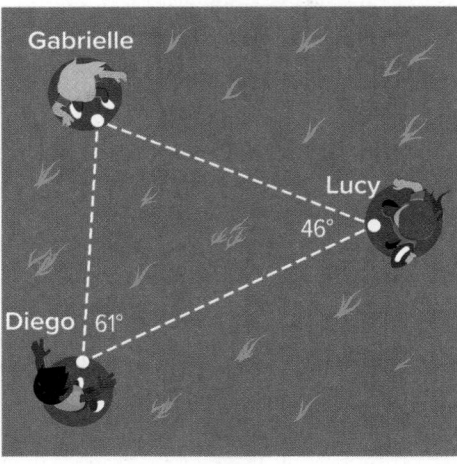

🅑 **Go Online** You can complete an Extra Example online.

Practice

 **Go Online** You can complete your homework online.

Example 1

List the angles that satisfy the stated condition.

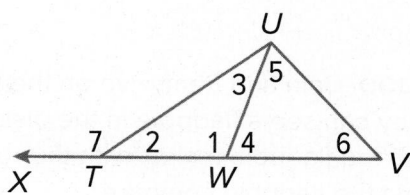

1. measures are greater than $m\angle 3$

2. measures are less than $m\angle 1$

3. measures are greater than $m\angle 1$

4. measures are less than $m\angle 7$

5. measures are greater than $m\angle 2$

Examples 2 and 3

List the angles and sides of each triangle in order from smallest to largest.

6.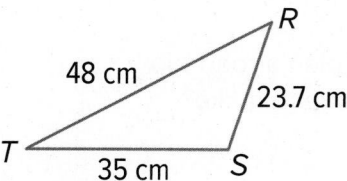

7.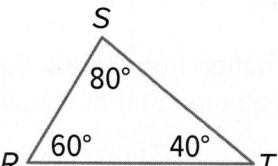

8.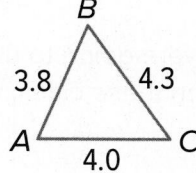

9.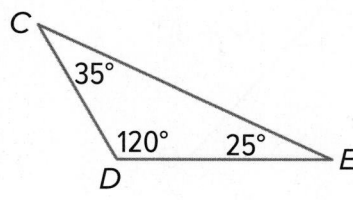

10.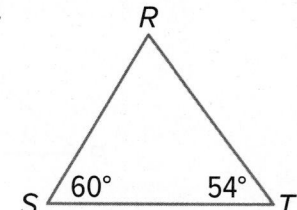

11.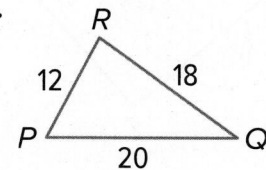

Lesson 1-4 • Inequalities in One Triangle 31

Example 4

12. **SPORTS** The figure shows the position of three trees on one part of a disc golf course. At which tree position is the angle between the trees the greatest?

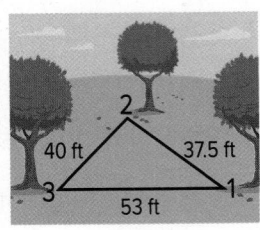

13. **NEIGHBORHOOD** Cain and Remy live on the same straight road. From their balconies, they can see a flagpole in the distance. The angle that each person's line of sight to the flagpole makes with the road is the same. How do their distances from the flagpole compare?

Mixed Exercises

14. **MAPS** Sata is going to Texas to visit a friend. As she looked at a map to see where she might want to go, she noticed that Austin, Dallas, and Abilene form a triangle. She wanted to determine how the distances between the cities were related, so she used a protractor to measure two angles.

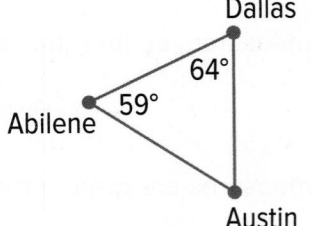

a. Based on the information in the figure, which of the two cities are nearest to each other?

b. Based on the information in the figure, which of the two cities are farthest apart from each other?

c. If you were going to use the information from Sata's sketch to plan a road trip between these cities, what is an assumption that you would have to make?

REASONING List the angles and sides of each triangle in order from smallest to largest.

15.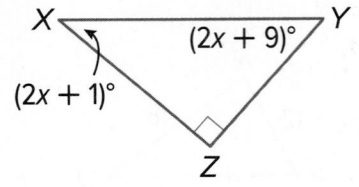

16.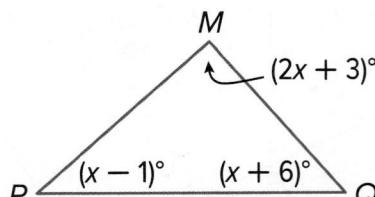

In △ABC, $\overline{AY}$ bisects ∠A, $\overline{BX}$ bisects ∠B, and Q is the intersection of $\overline{AY}$ and $\overline{BX}$. Use the figure for Exercises 17–19.

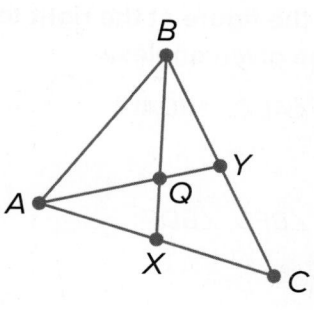

17. Suppose BC > AC. Compare m∠BAY and m∠ABX. Justify your reasoning.

18. Cynda claims that $\overline{XC} \cong \overline{YC}$ if BC > AC. Can Cynda make this conclusion? Justify your argument.

19. If m∠AQB > m∠BQY, which side of △ABQ has the greatest length?

20. PROOF Write a paragraph proof.
Given: WY > YX
Prove: m∠ZWY > m∠YWX

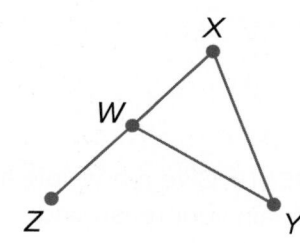

21. SQUARES Mahlik has three different squares. He arranges the squares to form a triangle as shown. Based on the information, list the squares in order from the one with the least perimeter to the one with the greatest perimeter.

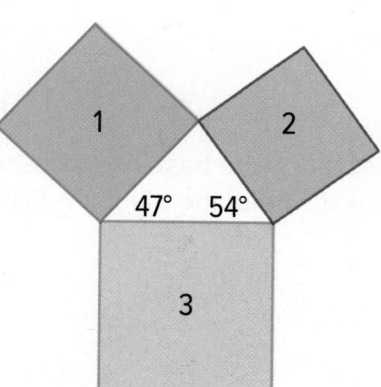

Use the figure at the right to determine which angle has the greatest measure.

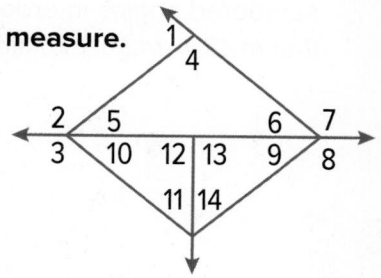

22. ∠1, ∠5, ∠6

23. ∠2, ∠4, ∠6

24. ∠7, ∠4, ∠5

25. ∠3, ∠11, ∠12

26. ∠3, ∠9, ∠14

27. ∠8, ∠10, ∠11

Use the figure at the right to determine the relationship between the measures of the given angles.

28. ∠ABD, ∠BDA

29. ∠BCF, ∠CFB

30. ∠BFD, ∠BDF

31. ∠DBF, ∠BFD

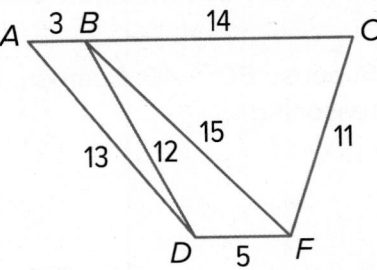

Use the figure at the right to determine the relationship between the given lengths.

32. SM, MR

33. RP, MP

34. RQ, PQ

35. RM, RQ

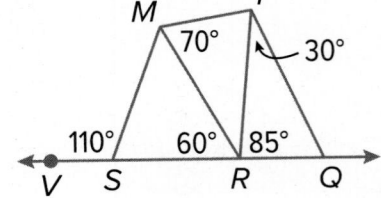

36. PERSEVERE Using only a ruler, draw △ABC such that m∠A > m∠B > m∠C. Justify your drawing.

37. CREATE Give a possible measure for $\overline{AB}$ in △ABC shown. Explain your reasoning.

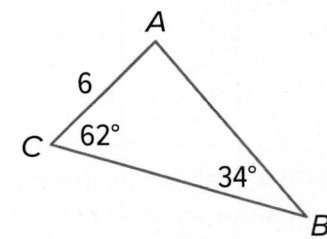

38. ANALYZE Is the base of an isosceles triangle *sometimes, always,* or *never* the longest side of the triangle? Justify your argument.

39. PERSEVERE Use the side lengths in the figure to list the numbered angles in order from smallest to largest given that m∠2 = m∠5. Explain your reasoning.

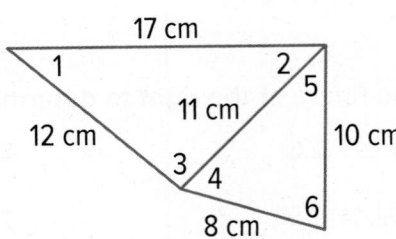

Indirect Proof

Explore Applying Indirect Reasoning

🔾 **Online Activity** Use the video to complete the Explore.

❓ **INQUIRY** How can you use a contradiction to prove a conclusion? ✕

Learn Indirect Proof

A direct proof is one that starts with a true hypothesis, and the conclusion is proved to be true. **Indirect reasoning** eliminates all possible conclusions but one, so the one remaining conclusion must be true. In an **indirect proof**, or **proof by contradiction**, one assumes that the statement to be proved is false and then uses logical reasoning to deduce that a statement contradicts a postulate, theorem, or one of the assumptions. Once a contradiction is obtained, one concludes that the statement assumed false must in fact be true.

Key Concept • How to Write an Indirect Proof

Step 1 Identify the conclusion that you are asked to prove. Make the assumption that this conclusion is false by assuming that the negation is true.

Step 2 Use logical reasoning to show that this assumption leads to a contradiction of the hypothesis or some other fact such as a definition, postulate, theorem, or corollary.

Step 3 State that because the assumption leads to a contradiction, the original conclusion, what you were asked to prove, must be true.

In indirect proofs, you should assume that the conclusion you are trying to prove is false. If, in the proof, you prove that the hypothesis is then false, this is a *proof by contrapositive*. If, in the proof, you assume that the hypothesis is true and prove that some other known fact is false, this is a *proof by contradiction*.

Example 1 Write an Indirect Algebraic Proof

Write an indirect proof to show that if $-4x - 4 < 12$, then $x > -4$.

Given: $-4x - 4 < 12$

Prove: $x > -4$

Indirect proof:

Step 1 Make an assumption.

The negation of $x > -4$ is $x \leq -4$. So, assume that $x < -4$ or $x = -4$ is true.

(continued on the next page)

Today's Goals
• Prove theorems about triangles by using indirect proof.

Today's Vocabulary
indirect reasoning
indirect proof
proof by contradiction

💬 **Talk About It!**

Consider the statement: *If 4 is a factor of x, then 2 is a factor of x*. The conclusion of the conditional statement is that *2 is a factor of x*. What assumption is necessary to start an indirect proof?

💭 **Think About It!**

Why do we have to consider both cases: that $x < -4$ or $x = -4$?

🔾 **Go Online**
You can complete an Extra Example online.

Step 2 Contradict the hypothesis.

Case 1: $x = -4$
When $x = -4$, $-4x - 4 = 12$. Because $12 \not< 12$, the assumption contradicts the given information for $x = -4$.

Case 2: $x < -4$
For all values of $x < -4$, $-4x - 4 > 12$, so the assumption contradicts the given information for $x < -4$.

Step 3 Reason indirectly.

In both cases, the assumption leads to a contradiction of the given information that $-4x - 4 < 12$. Therefore, the assumption that $x \le -4$ must be false, so the original conclusion that $x > -4$ must be true.

🌐 **Example 2** Apply Indirect Reasoning

HIKING Marco hiked more than 10.5 miles on a path, making just two stops along the way. Use indirect reasoning to prove that he hiked more than 3.5 miles on at least one leg of his hike.

Let x equal the distance traveled on the first leg of the hike, y equal the distance traveled on the second leg of the hike, and z equal the distance traveled on the third leg of the hike.

Given: $x + y + z > 10.5$

Prove: $x > 3.5$ or $y > 3.5$ or $z > 3.5$

Indirect proof:
Step 1 Make an assumption.
Assume that no leg of the hike is more than 3.5 miles. That is, $x \le 3.5$, $y \le 3.5$, and $z \le 3.5$.

Step 2 Contradict the hypothesis.
If $x \le 3.5$, $y \le 3.5$, and $z \le 3.5$, then $x + y + z \le 3.5 + 3.5 + 3.5$ or $x + y + z \le 10.5$.

Step 3 Reason indirectly.
This is a contradiction of the given statement. Therefore, the assumption is false and $x > 3.5$ or $y > 3.5$ or $z > 3.5$. Marco traveled more than 3.5 miles on at least one leg of the hike.

Check

FUNDRAISING The senior class is holding a dinner to raise funds for the school's music and arts program. The cost of a nonstudent ticket is $7, and the cost of a student ticket is $3.50. If 256 total tickets were sold and the revenue was more than $1246, prove that at least 100 nonstudent tickets were sold.

Let x equal the number of nonstudent tickets sold and let y equal the number of student tickets sold.

Given: $x + y = 256$; $7x + 3.5y > 1246$

Prove: $x \ge 100$

Indirect Proof:

Assume that ____?____. If $0 \le x < 100$, then using the given $x + y = 256$, we know that $157 \le y \le 256$. For all values of x, the revenue earned from the sale of nonstudent tickets would be $0 \le 7x < 700$. For all values of y, the revenue earned from the sale of student tickets would be ____?____ $\le 3.5y \le$ ____?____. Thus, the total revenue would be ____?____ $\le 7x + 3.5y <$ ____?____, which contradicts the given that the revenue was more than $1246. So, the assumption that $0 \le x < 100$ is ____?____. So, $x \ge 100$ must be ____?____.

Example 3 Indirect Proofs in Number Theory

Write an indirect proof to show that if x^2 is an odd integer, then x is an odd integer.

Given: x^2 is an odd integer.

Prove: x is an odd integer.

Indirect proof: Assume that x is an even integer. To assume that x is an even integer means that $x = 2k$ for some integer k. Rewrite x^2 in terms of k.

$x^2 = (2k)^2$	Substitution
$= 4k^2$	Simplify.
$= (2 \cdot 2)k^2$	Multiplication Property of Equality
$= 2(2k^2)$	Associative Property

Because k is an integer, $2k^2$ is also an integer. Let n represent the integer $2k^2$. So, x^2 can be represented by $2n$, where n is an integer. This means that x^2 is an even integer, which contradicts the given statement that x^2 is an odd integer. Because the assumption that x is even leads to a contradiction of the given, the original conclusion that x is odd must be true.

Check

Write an indirect proof to show that if xy is an even integer, then either x or y is an even integer.

Given: xy is an even integer.

Prove: x or y is an even integer.

Indirect proof: Assume that x and y are odd integers.
Let $x = 2n + 1$ and $y = 2k + 1$, for some integers n and k.

$xy = (2n + 1)(2k + 1)$	Substitution
$= 4nk + 2n + 2k + 1$	Distributive Property
$= 2(2nk + n + k) + 1$	Distributive Property

Because k and n are integers, ____?____ is also an integer. Let p represent the integer $2nk + n + k$. So, xy can be represented by ____?____, where p is an integer. This means that xy is an ____?____ integer, but this contradicts the given that xy is an ____?____ integer. Because the assumption that x and y are ____?____ integers leads to a contradiction of the given, the original conclusion that x or y is an ____?____ integer must be true.

Study Tip

Algebraic Proofs
When working with algebraic proofs, it is helpful to remember that you can represent an even number with the expression $2k$ and an odd number with the expression $2k + 1$ for any integer k.

Watch Out!

Counterexamples
Proof by contradiction and using a counterexample are not the same. A counterexample helps you disprove a conjecture. It cannot be used to prove a conjecture.

Example 4 Write an Indirect Geometric Proof

If an angle is an exterior angle of a triangle, prove that its measure is greater than the measure of either of its corresponding remote interior angles.

Given: $\angle 1$ is an exterior angle of $\triangle MNO$.

Prove: $m\angle 1 > m\angle 4$ and $m\angle 1 > m\angle 3$

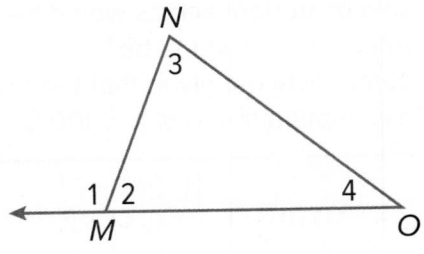

Indirect proof: Assume that $m\angle 1 \leq m\angle 4$ or $m\angle 1 \leq m\angle 3$. $m\angle 1 \leq m\angle 4$ means that either $m\angle 1 = m\angle 4$ or $m\angle 1 < m\angle 4$.

Case 1: $m\angle 1 = m\angle 4$

$m\angle 1 = m\angle 4 + m\angle 3$	Exterior Angle Theorem
$m\angle 1 = m\angle 1 + m\angle 3$	Substitution
$0 = m\angle 3$	Subtract $m\angle 1$ from each side.

This contradicts the fact that the measure of any angle of a triangle is greater than 0. So, $m\angle 1 \neq m\angle 4$.

Case 2: $m\angle 1 < m\angle 4$

By the Exterior Angle Theorem, $m\angle 1 = m\angle 4 + m\angle 3$. Because angle measures in triangles are positive, the definition of inequality implies that $m\angle 1 > m\angle 4$. This contradicts the assumption that $m\angle 1 < m\angle 4$.

The argument for $m\angle 1 \leq m\angle 3$ follows the same reasoning. In both cases, the assumption leads to a contradiction. Therefore, the original conclusion that $m\angle 1 > m\angle 4$ and $m\angle 1 > m\angle 3$ must be true.

Watch Out!

Recognizing Contradictions
Remember that the contradiction in an indirect proof does not always contradict the given information or the assumption. It can contradict a known fact or definition, such as in **Case 1**, that the measure of an angle must be greater than 0.

Check

If a triangle is equilateral, prove that it is also equiangular.

Given: $\triangle JKL$ is equilateral.

Prove: $\triangle JKL$ is equiangular.

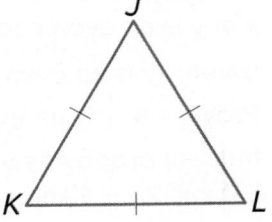

Indirect proof: Assume that $\triangle JKL$ is ___?___. If $\triangle JKL$ is ___?___, then the measure of one angle is greater than the measure of another. Assume that $m\angle K > m\angle L$. Then, by Theorem 6.10, $JL > JK$. This contradicts the given information that $\triangle JKL$ is ___?___. Therefore, the assumption that $\triangle JKL$ is ___?___ must be false, so the original conclusion that $\triangle JKL$ is ___?___ must be true.

Practice

Example 1

PROOF **Write an indirect proof of each statement.**

1. If $x^2 + 8 \leq 12$, then $x \leq 2$.

2. If $-4x + 2 < -10$, then $x > 3$.

3. If $2x - 7 > -11$, then $x > -2$.

4. If $5x + 12 < -33$, then $x < -9$.

5. If $-3x + 4 < 7$, then $x > -1$.

6. If $-2x - 6 > 12$, then $x < -9$.

Example 2

7. SHOPPING Desiree buys two bracelets for a little more than $40 before tax. When she gets home, her mother asks her how much each bracelet costs. Desiree cannot remember the individual prices. Use indirect reasoning to show that at least one of the bracelets cost more than $20.

8. CONCESSIONS Kala worked at the concession stand during his school's basketball game. The cost of a small soft drink is $2, and the cost of a large soft drink is $3. If 120 soft drinks were sold and the total soft drink sales was more than $320, prove that at least 81 large soft drinks were sold.

Examples 3 and 4

PROOF **For 9–16, write an indirect proof of each statement. Assume that x and y are integers.**

9. **Given:** xy is an odd integer.
Prove: x and y are both odd integers.

10. **Given:** x^2 is even.
Prove: x^2 is divisible by 4.

11. **Given:** x is an odd number.
Prove: x is not divisible by 4.

12. **Given:** xy is an even integer.
Prove: x or y is an even integer.

13. In an isosceles triangle, neither of the base angles can be a right angle.

14. A triangle can have only one right angle.

15. Given: $m\angle C = 100°$
Prove: $\angle A$ is not a right angle.

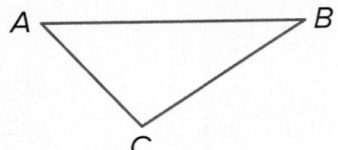

16. Given: $\angle D \not\cong \angle F$
Prove: $DE \neq EF$

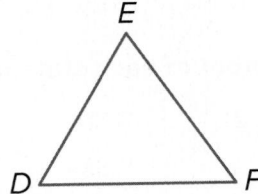

Mixed Exercises

17. PHYSICS Sound travels through air at about 344 meters per second when the temperature is 20°C. If Enrique lives 2 kilometers from the fire station and it takes 5 seconds for the sound of the fire station siren to reach him, how can you prove indirectly that it is not 20°C when Enrique hears the siren?

18. PROOF Write an indirect proof to show that if $\frac{1}{b} < 0$, then b is negative.

19. PROOF Write an indirect proof. (Theorem 6.10)

If one angle of a triangle has a greater measure than another angle, then the side opposite the angle with the greater measure is longer than the side opposite the angle with the lesser measure.

20. WORDS The words *accomplishment, counterexample,* and *extemporaneous* all have 14 letters. Use an indirect proof to show that any word with 14 letters must use a repeated letter or have two letters that are consecutive in the alphabet.

21. WRITE Explain the procedure for writing an indirect proof.

22. CREATE Write a statement that can be proved using an indirect proof. Include the indirect proof of your statement.

23. PERSEVERE If x is a rational number, then it can be represented by a quotient $\frac{a}{b}$ for some integers a and b, if $b \neq 0$. An irrational number cannot be represented by the quotient of two integers. Write an indirect proof to show that the product of a nonzero rational number and an irrational number is an irrational number.

The Triangle Inequality

Explore Relationships Among Triangle Side Lengths

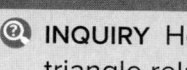 **Online Activity** Use dynamic geometry software to complete the Explore.

> ⊘ **INQUIRY** How are the three side lengths of a triangle related?

Learn The Triangle Inequality

In order for three segments to form a triangle, a special relationship must exist among their lengths.

> **Theorem 1.11: Triangle Inequality Theorem**
>
> The sum of the lengths of any two sides of a triangle must be greater than the length of the third side.

You will prove Theorem 1.11 in Exercise 22.

Example 1 Identify Possible Triangles Given Side Lengths

Is it possible to form a triangle with the given side lengths? If not, explain why not.

a. 9 cm, 12 cm, 18 cm

Check each inequality.

$9 + 12 > 18$ $12 + 18 > 9$ $9 + 18 > 12$

Because the sum of each pair of side lengths is greater than the third side length, sides with lengths 9, 12, and 18 centimeters will form a triangle.

b. 3 in., 5 in., 8 in.

$3 + 5 \not> 8$

Because the sum of one pair of side lengths is not greater than the third side length, sides with lengths 3, 5, and 8 inches will not form a triangle.

Check

Is it possible to form a triangle with the given side lengths? If not, explain why not.

a. 2 mm, 5 mm, 6 mm

b. 3 yd, 4 yd, 8 yd

 Go Online You can complete an Extra Example online.

Today's Goals
- Prove and apply the Triangle Inequality Theorem.

Study Tip

Make a Model You can use interactive geometry software or a ruler and paper to model a triangle for any given side lengths. The scale of the ruler will not matter as long as the side lengths are given in the same unit of measure and you establish a 1-to-1 conversion factor, for example 1 in. = 1 cm.

💭 Think About It!
How can you eliminate triangle side length possibilities by looking at only one inequality?

When the lengths of two sides of a triangle are known, the third side can be any length in a range of values.

🌐 **Example 2** Find Possible Side Lengths

DRONES **A delivery company uses drones to make speedy deliveries around the city. A drone leaves the home office and flies 8 miles east to its first delivery and then 4 more miles southwest to a second delivery. What is the *least* possible whole-number distance the drone will fly to return to the home office?**

Let x represent the length of the third side. Set up and solve each of the three triangle inequalities.

$$4 + 8 > x \qquad\qquad 4 + x > 8 \qquad\qquad x + 8 > 4$$

$$12 > x \qquad\qquad\qquad x > 4 \qquad\qquad\qquad x > -4$$

The least whole-number value between 4 and 12 is 5. The drone has to fly at least 5 miles to the home office after the two deliveries.

Check

HOME IMPROVEMENT To install a smart thermostat, Kelvin is cutting a triangular hole in the wall. He marks side lengths of $2\frac{1}{8}$ inches and $2\frac{3}{16}$ inches.

Part A

Which assumption do you need to make to determine the range for the measure of the third side of the hole?

A The triangular hole is not equiangular.

B The smart thermostat requires a triangular hole.

C Kelvin measured the side lengths accurately.

D Two sides of the hole are longer than the third side.

Part B

What is the possible range for the third side length?

Example 3 Use the Triangle Inequality Theorem

$\triangle HAB$ **and** $\triangle HCB$ **share side** HB, **and** $HC = HB$. **Prove that** $HA + AB > HC$.

Given: $HC = HB$
Prove: $HA + AB > HC$

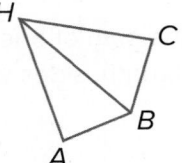

Proof:

Statements	Reasons
1. $HC = HB$	**1.** Given
2. $HA + AB > HB$	**2.** Triangle Inequality Theorem
3. $HA + AB > HC$	**3.** Substitution

 Go Online You can complete an Extra Example online.

Study Tip

Multiple Inequality Symbols
The compound inequality $4 < x < 12$ is read *x is between 4 and 12.*

💭 **Think About It!**

What is the greatest possible whole-number measure for the third side?

Practice

Go Online You can complete your homework online.

Example 1

REASONING **Is it possible to form a triangle with the given side lengths? If not, explain why not.**

1. 9, 12, 18

2. 8, 9, 17

3. 14, 14, 19

4. 23, 26, 50

5. 32, 41, 63

6. 2.7, 3.1, 4.3

7. 0.7, 1.4, 2.1

8. 12.3, 13.9, 25.2

Example 2

Find the range for the measure of the third side of a triangle given the measures of two sides.

9. 6 ft and 19 ft

10. 7 km and 29 km

11. 13 in. and 27 in.

12. 18 ft and 23 ft

13. **CITIES** The distance between New York City, New York, and Boston, Massachusetts, is 187 miles, and the distance between New York City and Hartford, Connecticut, is 97 miles. Hartford, Boston, and New York City form a triangle on a map. What must the distance between Boston and Hartford be greater than?

14. **FENCING** Capria is planning to fence a triangular plot of land. Two of the sides of the plot measure 230 yards and 490 yards.

    a. Find the range for the measure of the third side of the triangular plot of land.

    b. What are the maximum and minimum lengths of fencing Capria will need?

Example 3

PROOF **Write a two-column proof.**

15. **Given:** $\overline{PL} \parallel \overline{MT}$; K is the midpoint of $\overline{PT}$.

    Prove: $PK + KM > PL$

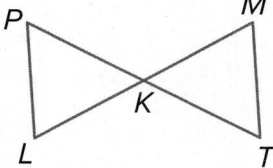

16. **Given:** $\triangle ABC \cong \triangle DEC$

    Prove: $AB + DE > AD - BE$

Mixed Exercises

Find the range of possible measures of *x* if each set of expressions represents measures of the sides of a triangle.

17. *x*, 4, 6

18. *x* − 2, 10, 12

REGULARITY **Determine whether the given coordinates are the vertices of a triangle. Explain.**

19. *X*(1, −3), *Y*(6, 1), *Z*(2, 2)

20. *J*(−7, −1), *K*(9, −5), *L*(21, −8)

21. GARDENING Ennis has 4 lengths of wood from which he plans to make a border for a triangular-shaped herb garden. The lengths of the wood borders are 8 inches, 10 inches, 12 inches, and 18 inches. How many different triangular borders can Ennis make?

22. PROOF Write a two-column proof. (Theorem 6.11)

Given: △*ABC*

Prove: *AC* + *BC* > *AB*

(*Hint:* Draw an auxiliary segment $\overline{CD}$, so that *C* is between *B* and *D* and $\overline{CD} \cong \overline{AC}$.)

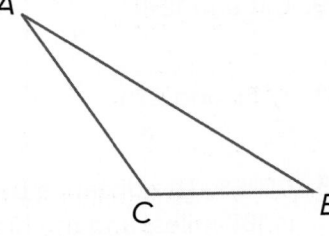

23. PERSEVERE The sides of an isosceles triangle are whole numbers, and its perimeter is 30 units. What is the probability that the triangle is equilateral?

24. CREATE The length of one side of a triangle is 2 inches. Draw a triangle in which the 2-inch side is the shortest side and one in which the 2-inch side is the longest side. Include side and angle measures on your drawing.

25. WRITE What can you tell about a triangle when given three side lengths? Include at least two items.

26. ANALYZE What is the range of lengths of each leg of an isosceles triangle if the measure of the base is 6 inches? Explain.

Inequalities in Two Triangles

Explore Analyzing Inequalities in Two Triangles

Online Activity Use dynamic geometry software to complete the Explore.

INQUIRY How do the included angle measures of two triangles with two pairs of congruent sides compare? ×

Learn Hinge Theorem

Theorem 1.12: Hinge Theorem

If two sides of a triangle are congruent to two sides of another triangle, and the included angle of the first is larger than the included angle of the second triangle, then the third side of the first triangle is longer than the third side of the second triangle.

You will prove Theorem 1.12 in Exercise 18.

🌐 Example 1 Use the Hinge Theorem

BOATING **Two families set sail on their boats from the same dock. The Nguyens sail 3.5 nautical miles north, turn 85° east of north, and then sail 2 nautical miles. The Griffins sail 3.5 nautical miles south, turn 95° east of south, and then sail 2 nautical miles. At this point, which boat is farther from the dock? Explain your reasoning.**

Step 1 Draw a diagram of the situation. The courses of each boat and the straight-line distance from each stopping point back to the boat dock form two triangles. Each boat sails 3.5 nautical miles, turns, and then sails another 2 nautical miles.

Step 2 Determine the interior angle measures.
Use linear pairs to find the measures of the included angles. The measure of the included angle for the Nguyens is 180 − 85 or 95°. The measure of the included angle for the Griffins is 180 − 95 or 85°.

Step 3 Compare the distance each boat is from the boat launch.
Use the Hinge Theorem to compare the distance each boat is from the boat launch.

Because 95 > 85, $NL > GL$ by the Hinge Theorem. So, the Nguyens are farther from the boat launch.

Go Online You can complete an Extra Example online.

Today's Goals
- Prove and apply the Hinge Theorem.
- Prove and apply the Converse of the Hinge Theorem.

🤔 Think About It!
In $\triangle ADB$ and $\triangle CDB$, side $\overline{BD}$ is a common side, $\overline{AB} \cong \overline{BC}$, and $m\angle ABD > m\angle CBD$. Because $m\angle ABD > m\angle CBD$, what is the relationship between AD and DC? Explain your reasoning.

Check

DRONES Esperanza and Landon each fly a drone from the same place in a park and at the same altitude. Esperanza flies her drone 440 feet east, then turns it 122° south of east and flies 300 more feet. Landon flies his drone 440 feet east, then turns it 140° north of east and flies 300 more feet. Whose drone is closer to its original position?

You can use the Hinge Theorem to prove relationships in two triangles.

Example 2 Prove Triangle Relationships by Using the Hinge Theorem

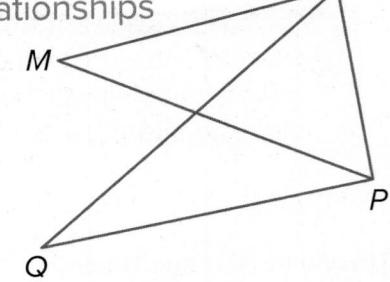

Complete the two-column proof.

Given: $\overline{PQ} \cong \overline{PM}$

Prove: $NQ > NM$

Proof:

Statements	Reasons
1. $\overline{PQ} \cong \overline{PM}$	1. Given
2. $\overline{PN} \cong \overline{PN}$	2. Reflexive Property
3. $m\angle NPQ = m\angle NPM + m\angle MPQ$	3. Angle Addition Postulate
4. $m\angle NPQ > m\angle NPM$	4. Definition of inequality
5. $NQ > NM$	5. Hinge Theorem

Check

Copy and complete the two-column proof.

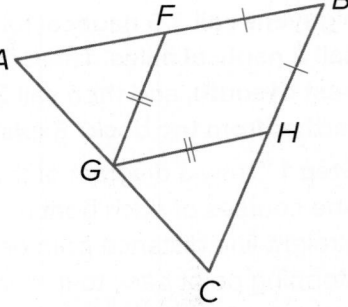

Given: $\overline{HG} \cong \overline{FG}$, $\overline{BH} \cong \overline{BF}$
　　　　G is the midpoint of $\overline{AC}$.
　　　　$m\angle CGH > m\angle AGF$

Prove: $BC > AB$

Statements	Reasons
1. $\overline{HG} \cong \overline{FG}$, $\overline{BH} \cong \overline{BF}$	1. Given
2. G is the midpoint of $\overline{AC}$.	2. Given
3. $m\angle CGH > m\angle AGF$	3. Given
4. $CG = AG$	4. _?_
5. $CH > AF$	5. _?_
6. $BH = BF$	6. _?_
7. $CH + BH > AF + BH$	7. Addition Property
8. $CH + BH > AF + BF$	8. Substitution
9. $BC = CH + BH$, $AB = AF + BF$	9. _?_
10. $BC > AB$	10. _?_

🧭 **Go Online** You can complete an Extra Example online.

Study Tip

Inequalities in Triangles Because we know that $m\angle NPQ$ is the sum of $m\angle NPM$ and $m\angle MPQ$ and neither angle has a degree measure of 0, $m\angle NPQ$ has to be greater than both $m\angle NPM$ and $m\angle MPQ$.

Statements/ Reasons:

Hinge Theorem

Reflexive Property

$m\angle NPQ = m\angle NPM + m\angle MPQ$

$m\angle NPQ > m\angle NPM$

Statements/ Reasons:

Definition of midpoint

Hinge Theorem

Substitution

Segment Addition Postulate

Definition of congruent segments

Learn Converse of the Hinge Theorem

Theorem 1.13: Converse of the Hinge Theorem

If two sides of a triangle are congruent to two sides of another triangle and the third side of the first triangle is longer than the third side of the second triangle, then the included angle measure of the first triangle is greater than the included angle measure of the second triangle.

You will prove Theorem 1.13 in Exercise 19.

Example 3 Apply Algebra to Relationships in Triangles

Find the range of possible values for x.

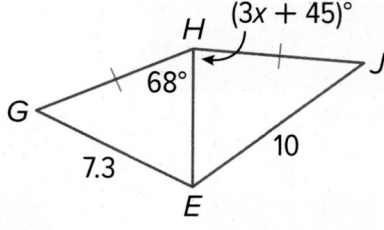

Step 1 Use the Converse of the Hinge Theorem.

From the diagram, we know that $\overline{GH} \cong \overline{JH}$, $\overline{EH} \cong \overline{EH}$, and $JE > GE$.

$m\angle JHE > m\angle GHE$	Converse of the Hinge Theorem
$3x + 45 > 68$	Substitution
$x > \frac{23}{3}$	Solve for x.

Step 2 Use your knowledge of the interior angles of a triangle.

Use the fact that the measure of any angle in a triangle is less than 180° to write a second inequality.

$m\angle JHE < 180°$	
$3x + 45 < 180°$	Substitution
$x < 45°$	Solve for x.

Step 3 Complete the compound inequality.

$$\frac{23}{3} < x < 45$$

Check

Find the range of possible values for x.

_____?_____

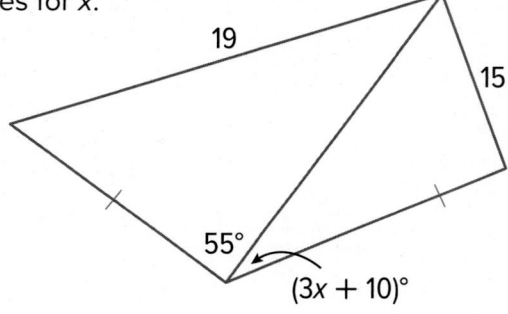

Go Online You can complete an Extra Example online.

Think About It!

In $\triangle XYW$ and $\triangle ZWY$, $\overline{YW}$ is a common side, $\overline{WZ} \cong \overline{YX}$, and $XW > ZY$. Because $XW > ZY$, what is the relationship between $m\angle XYW$ and $m\angle ZWY$? Explain your reasoning.

Think About It!

Do you think that the value of x is likely to be closer to $\frac{23}{3}$ or 45? Justify your argument.

Example 4 Prove Relationships by Using the Converse of the Hinge Theorem

Complete the flow proof.

Given: T is the midpoint of $\overline{ZX}$.

$\overline{ST} \cong \overline{WT}$, $SZ > WX$

Prove: $m\angle XTR > m\angle ZTY$

Proof:

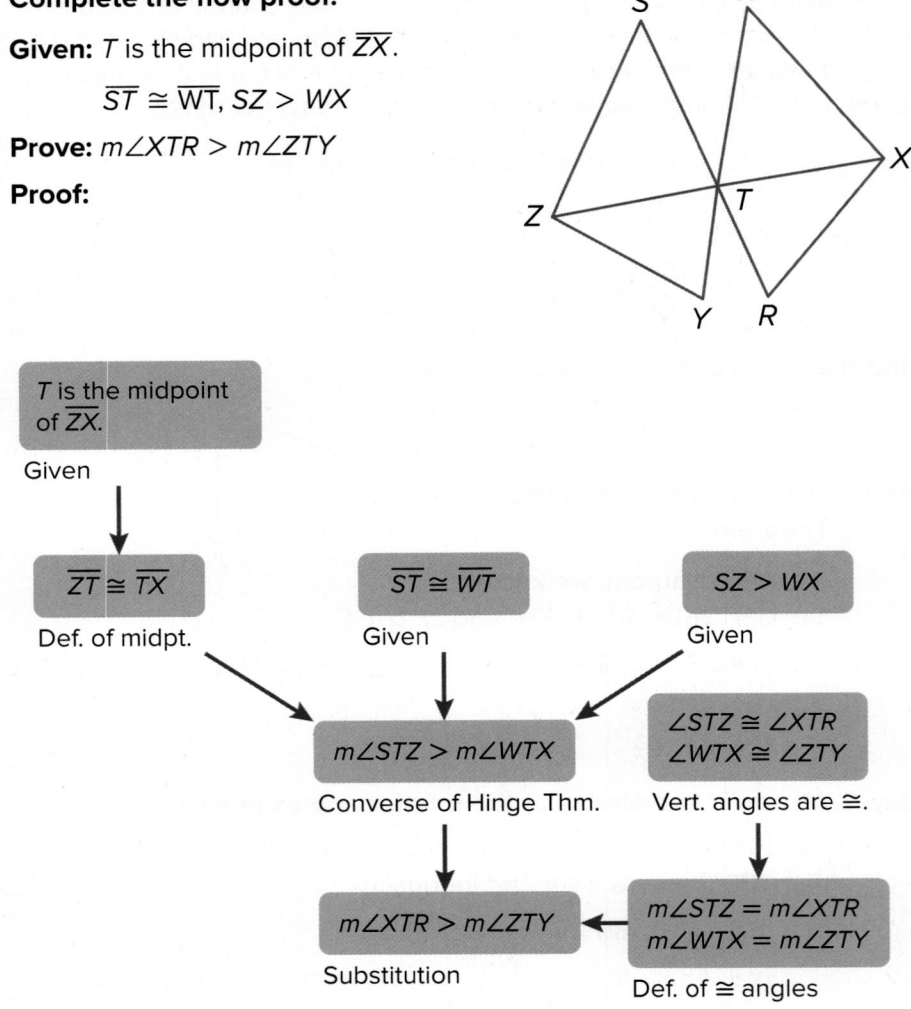

T is the midpoint of $\overline{ZX}$.
Given

$\overline{ZT} \cong \overline{TX}$
Def. of midpt.

$\overline{ST} \cong \overline{WT}$
Given

$SZ > WX$
Given

$m\angle STZ > m\angle WTX$
Converse of Hinge Thm.

$\angle STZ \cong \angle XTR$
$\angle WTX \cong \angle ZTY$
Vert. angles are $\cong$.

$m\angle XTR > m\angle ZTY$
Substitution

$m\angle STZ = m\angle XTR$
$m\angle WTX = m\angle ZTY$
Def. of $\cong$ angles

Pause and Reflect

Did you struggle with anything in this lesson? If so, how did you deal with it?

Go Online You can complete an Extra Example online.

Practice

🔵 **Go Online** You can complete your homework online.

Example 1

1. **FLIGHT** Two planes take off from the same airstrip. The first plane flies west for 150 miles and then flies 30° south of west for 220 miles. The second plane flies east for 220 miles and then flies $x°$ south of east for 150 miles. If $x < 30$, which plane is farther from the airstrip after the second leg? Justify your answer.

2. **HIKING** Gen and Ari start hiking from the same point. Gen hikes 5 miles due east and turns to hike 4.5 miles 30° south of east. Ari hikes 5 miles due west and turns to hike 4.5 miles 15° north of west.

   a. Draw a model to represent the situation.

   b. Who is closer to the starting point? Explain your reasoning.

Example 2

PROOF **Write a two-column proof.**

3. **Given:** $RX = XS$; $m\angle SXT = 97$

   Prove: $ST > RT$

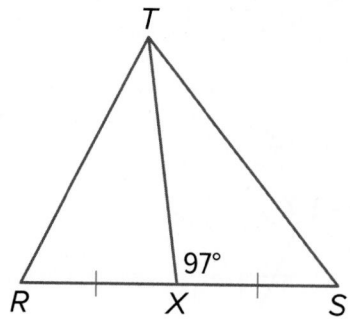

4. **Given:** $\overline{LK} \cong \overline{JK}$, $\overline{RL} \cong \overline{RJ}$, K is the midpoint of $\overline{QS}$, and $m\angle SKL > m\angle QKJ$.

   Prove: $RS > QR$

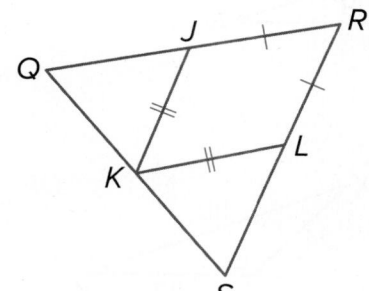

Example 3

Find the range of possible values for x.

5.

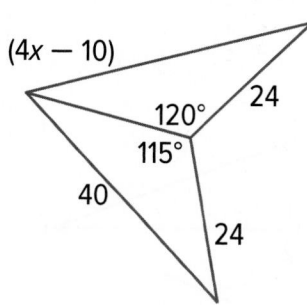

6.

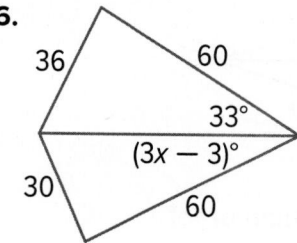

Example 4

PROOF **Write a two-column proof.**

7. Given: $\overline{XU} \cong \overline{VW}$, $VW > XW$
$\overline{XU} \parallel \overline{VW}$

Prove: $m\angle XZU > m\angle UZV$

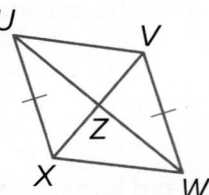

8. Given: $\overline{AF} \cong \overline{DJ}$, $\overline{FC} \cong \overline{JB}$
$AB > DC$

Prove: $m\angle AFC > m\angle DJB$

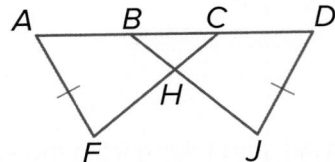

Mixed Exercises

Compare the given measures.

9. MR and RP

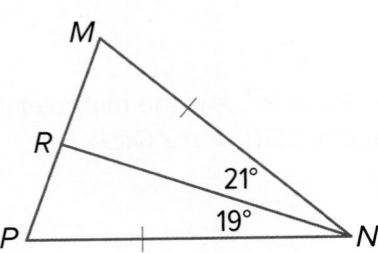

10. AD and CD

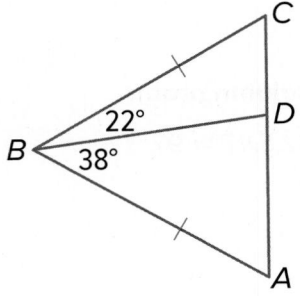

11. $m\angle C$ and $m\angle Z$

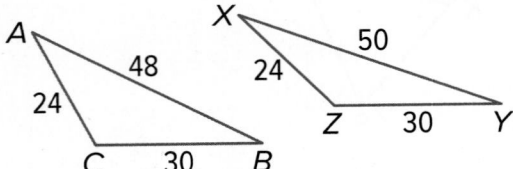

12. $m\angle XYW$ and $m\angle WYZ$

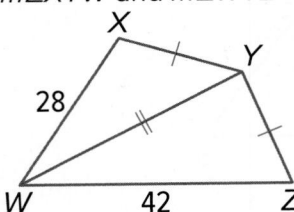

13. $m\angle BXA$ and $m\angle DXA$

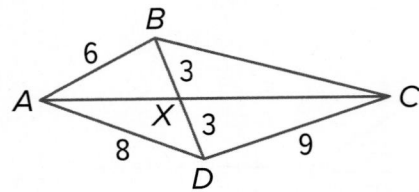

14. PROOF **Write a two-column proof.**

Given: $\overline{BA} \cong \overline{DA}$, $BC > DC$

Prove: $m\angle 1 > m\angle 2$

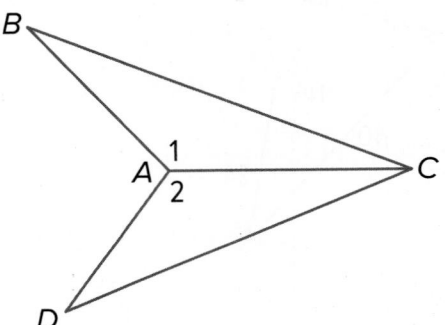

15. PROOF Write a paragraph proof.

Given: $\overline{EF} \cong \overline{GH}$, $m\angle F > m\angle G$.

Prove: $EG > FH$

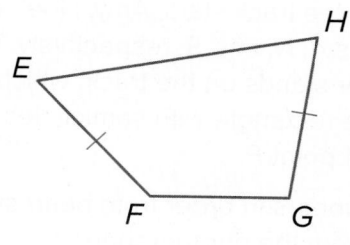

16. REASONING In the figure, $\overline{BA}$, $\overline{BD}$, $\overline{BC}$, and $\overline{BE}$ are congruent and $AC < DE$. How does $m\angle 1$ compare with $m\angle 3$? Explain your thinking.

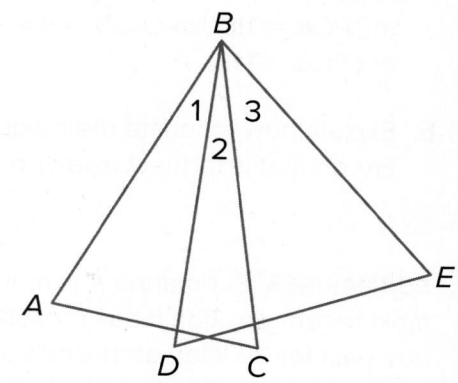

17. CLOCKS The minute hand of a grandfather clock is 3 feet long, and the hour hand is 2 feet long. Is the distance between their ends greater at 3:00 or at 8:00?

18. PROOF Write a paragraph proof to prove the Hinge Theorem. (Theorem 6.12)

Given: $\triangle ABC$ and $\triangle DEF$
$\overline{AC} \cong \overline{DF}$, $\overline{BC} \cong \overline{EF}$
$m\angle F > m\angle C$

Prove: $DE > AB$

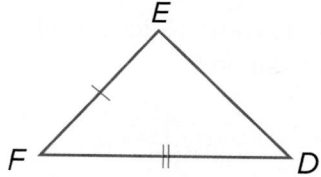
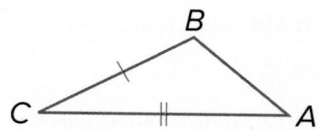

19. PROOF Use an indirect proof to prove the Converse of the Hinge Theorem. (Theorem 6.13)

Given: $\overline{RS} \cong \overline{UW}$
$\overline{ST} \cong \overline{WV}$
$RT > UV$

Prove: $m\angle S > m\angle W$

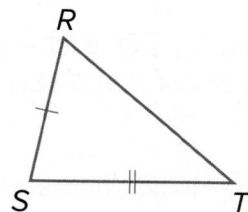
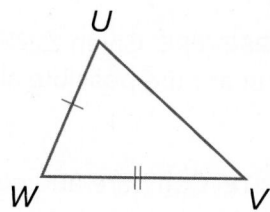

20. **PHOTOGRAPHY** A photographer is taking pictures of three track stars, Amy, Noel, and Beth at points *A*, *N*, and *B*, respectively. The photographer stands on the track, which is shaped like a rectangle with semicircles on both ends, at point *P*.

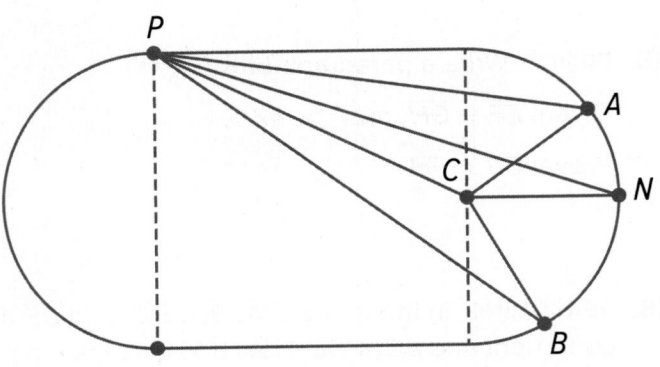

a. List the runners in order from nearest to farthest from the photographer if $m\angle PCA = 118°$, $m\angle ACN = 36°$, and $m\angle PCB = 146°$.

b. Explain how to locate the point along the semicircular curve that the runners are on that is farthest away from the photographer.

21. **SUBMARINES** Submarine A is moving toward the most recent position it has for submarine B. It sails due east for 38 kilometers and then sails 52° north of east for 25 kilometers, arriving at submarine B's starting position. Meanwhile, submarine B sails 38° south of west for 25 kilometers and then due west for 38 kilometers.

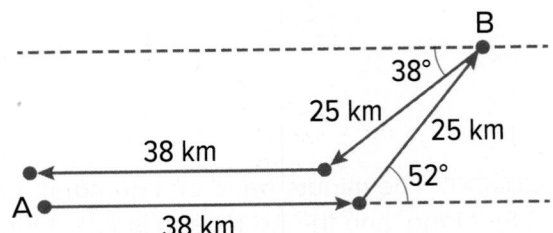

a. Does submarine B end up northeast, northwest, southeast, or southwest of submarine A's starting point?

b. If the lengths of submarine B's legs were switched, what conclusions could you make about submarine B's final position?

🧠 Higher-Order Thinking Skills

22. **CREATE** Give a real-world example of an object that uses a hinge. Draw two sketches in which the hinge on your object is adjusted to two different positions. Use your sketches to explain why Theorem 6.12 is called the Hinge Theorem.

23. **PERSEVERE** Given △*RST* with median $\overline{RQ}$, if *RT* is greater than or equal to *RS*, what are the possible classifications of △*RQT*? Explain your reasoning.

24. **WRITE** Compare and contrast the Hinge Theorem to the SAS Postulate for triangle congruence.

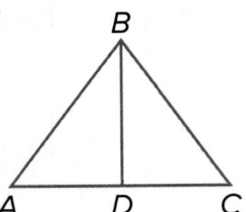

25. **ANALYZE** If $\overline{BD}$ is a median and *AB* < *BC*, then is ∠*BDC* *sometimes*, *always*, or *never* an acute angle? Justify your argument.

Review

 Essential Question

How can relationships in triangles be used in real-world situations?

The relationships between the different parts of a triangle can provide information about the triangle in a real-world context. For example, special segments can model optimal choices in design problems.

Module Summary

Lessons 1-1 and 1-2

Perpendicular Bisectors and Angle Bisectors

- If a point is on the perpendicular bisector of a segment, then it is equidistant from the endpoints of the segment.

- The perpendicular bisectors of a triangle intersect at the circumcenter that is equidistant from the vertices of the triangle.

- If a point is on the bisector of an angle, then it is equidistant from the sides of the angle.

- The angle bisectors of a triangle intersect at the incenter, which is equidistant from the sides of the triangle.

Lesson 1-3

Medians and Altitudes

- A median of a triangle is a line segment with endpoints that are a vertex of the triangle and the midpoint of the side opposite the vertex.

- The medians of a triangle intersect at the centroid, which is two-thirds of the distance from each vertex to the midpoint of the opposite side.

- An altitude of a triangle is a segment from a vertex of the triangle to the line that contains the opposite side and is perpendicular to that side.

- The altitudes of a triangle intersect at a point called the orthocenter.

Lessons 1-4, 1-6, and 1-7

Inequalities in Triangles

- If one side of a triangle is longer than another side, then the angle that is opposite the longer side has a greater measure than the angle that is opposite the shorter side.

- The sum of the lengths of any two sides of a triangle must be greater than the length of the third side.

- If two sides of a triangle are congruent to two sides of another triangle, and the included angle of the first is larger than the included angle of the second triangle, then the third side of the first triangle is longer than the third side of the second triangle.

Lesson 1-5

Indirect Proof

- To write an indirect proof:

  Identify the conclusion you are asked to prove. Make the assumption that this conclusion is false by assuming that the opposite is true.

  Use logical reasoning to show that this assumption leads to a contradiction of the hypothesis or some other fact such as a definition, postulate, theorem, or corollary.

  State that because the assumption leads to a contradiction, the original conclusion, what you were asked to prove, must be true.

Study Organizer

 Foldables

Use your Foldable to review this module. Working with a partner can be helpful. Ask for clarification of concepts as needed.

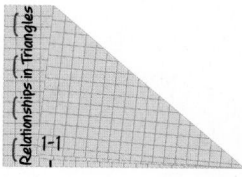

Test Practice

1. **OPEN RESPONSE** In triangle *WXZ*, what is the length of segment *ZY*? (Lesson 1-1)

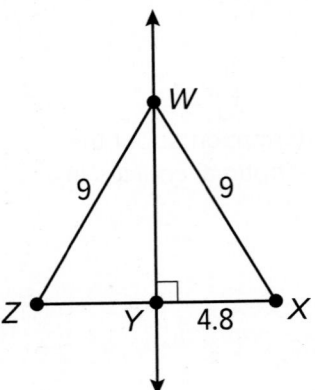

2. **MULTIPLE CHOICE** Macha is a day care provider and has three tables with children eating lunch, as shown on the grid.

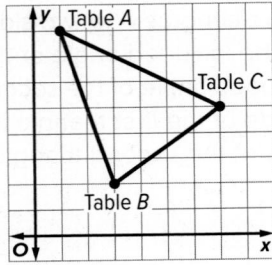

Where should Macha stand to be equidistant from all three tables? (Lesson 1-1)

A. (3.5, 5.5)

B. $\left(\frac{11}{3}, 5\right)$

C. (3.85, 4.72)

D. (4, 4)

3. **MULTIPLE CHOICE** What is the length of side *AB*? (Lesson 1-1)

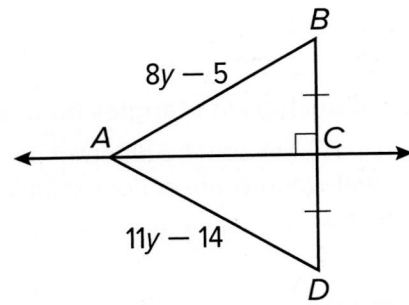

A. 1

B. 3

C. 19

D. 29

4. **OPEN RESPONSE** Given the following, where *D* is the incenter of △*ABC*, *DF* = 6, and *DB* = 10, find the measure of *EB*. (Lesson 1-2)

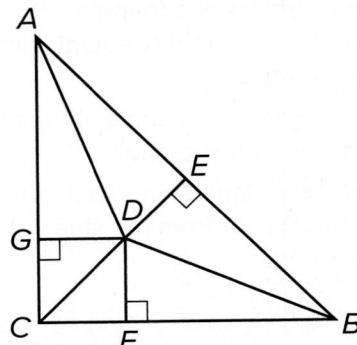

5. **MULTIPLE CHOICE** Use the figure to find the length of segment *RS*. (Lesson 1-2)

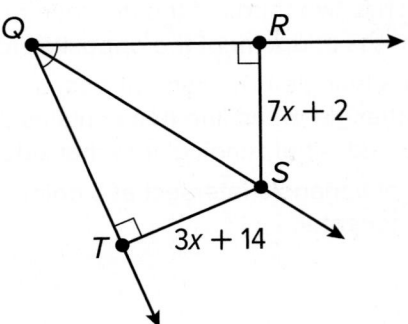

A. 9

B. 16

C. 23

D. 30

6. OPEN RESPONSE Find $m\angle EFH$. (Lesson 1-2)

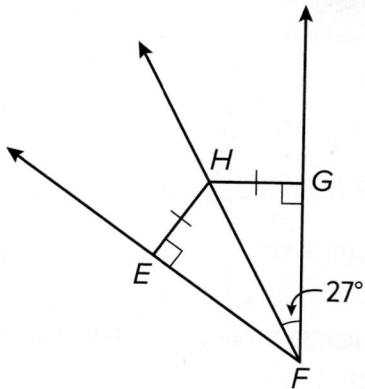

7. OPEN RESPONSE Find the coordinates of the orthocenter of triangle ABC. (Lesson 1-3)

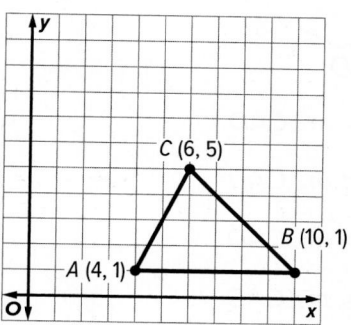

8. OPEN RESPONSE In $\triangle ABC$, G is the centroid and $GE = 4$. Find AG. (Lesson 1-3)

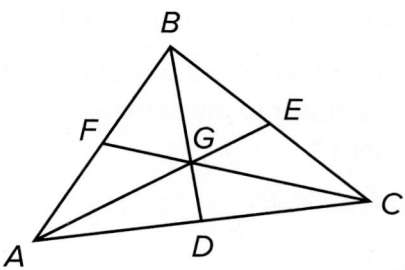

9. OPEN RESPONSE In $\triangle ABC$, G is the centroid and $DC = 36$. Find DG and GC. (Lesson 1-3)

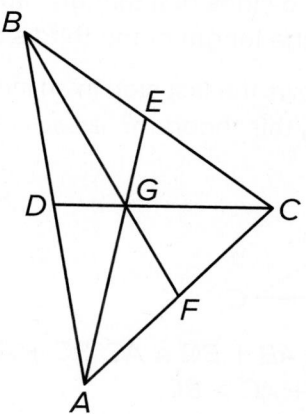

10. MULTIPLE CHOICE Order the angles of the triangle from smallest to largest. (Lesson 1-4)

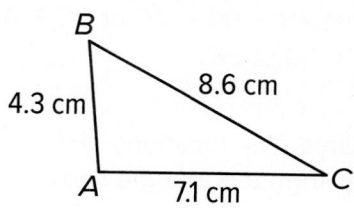

A. $\angle C, \angle B, \angle A$

B. $\angle A, \angle B, \angle C$

C. $\angle C, \angle A, \angle B$

D. $\angle A, \angle C, \angle B$

11. OPEN RESPONSE Using an indirect proof, Antonia must prove that if a triangle is a right triangle, then none of the angles have measures greater than 90°. What assumption must she start with? (Lesson 1-5)

12. MULTIPLE CHOICE Using an indirect proof, Amelia wants to prove that the sum of the lengths of any two sides of a triangle must be greater than the length of the third side.

Which choice shows the first step in an indirect proof for proving this theorem? (Lesson 1-5)

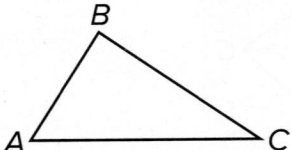

A. Assume that $AB + BC \leq AC$, $BC + AC > AB$, and $AB + AC > BC$.

B. Assume that $AB + BC < AC$ or $BC + AC < AB$ or $AB + AC < BC$.

C. Assume that $AB + BC < AC$ and $BC + AC < AB$ and $AB + AC < BC$.

D. Assume that $AB + BC \leq AC$ or $BC + AC \leq AB$ or $AB + AC \leq BC$.

13. MULTIPLE CHOICE The locations of three friends' homes form the triangle shown.

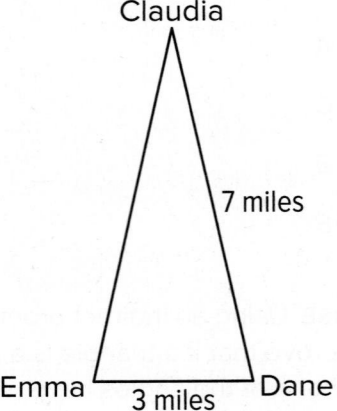

What is the possible range for the distances between Emma's house and Claudia's house? (Lesson 1-6)

A. $x > 4$

B. $x < 10$

C. $4 < x < 21$

D. $4 < x < 10$

14. MULTI-SELECT Select all the sets of lengths that could be the lengths of the sides of a triangle. (Lesson 1-6)

A. 4 in., 5 in., 9 in.

B. 7 mm, 8 mm, 10 mm

C. 4.2 ft, 4.2 ft, 9.1 ft

D. 2 cm, 2 cm, 2 cm

15. MULTIPLE CHOICE Which of the following is a true statement? (Lesson 1-7)

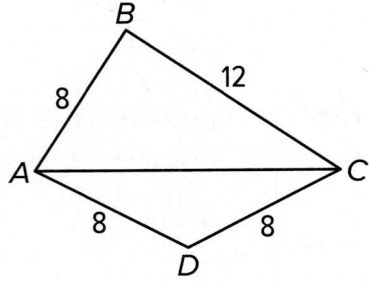

A. $m\angle BAC = m\angle CAD$

B. $m\angle BAC < m\angle CAD$

C. $m\angle BAC > m\angle CAD$

D. $m\angle BAC \leq m\angle CAD$

16. MULTIPLE CHOICE Two groups of cyclists leave from the same starting point.
- Group A rides 12 miles east, turns 45° due north of east, and then rides 8 miles.
- Group B rides 12 miles due west, turns 30° north of west, and then rides 8 miles.

Which of the following is a true statement about who is closer to the starting point? (Lesson 1-7)

A. Group A is closer.

B. Group B is closer.

C. The groups are the same distance.

D. Not enough information to tell

Quadrilaterals

e Essential Question

What are the different types of quadrilaterals, and how can their characteristics be used to model real-world situations?

What Will You Learn?

How much do you already know about each topic **before** starting this module?

KEY

👎 — I don't know.   👊 — I've heard of it.   👍 — I know it!

	Before			After		
	👎	👊	👍	👎	👊	👍
solve problems involving the interior angles of polygons						
solve problems involving the exterior angles of polygons						
solve problems using the properties of parallelograms						
solve problems involving the diagonals of parallelograms						
solve problems using the properties of rectangles						
solve problems using the properties of rhombi						
solve problems using the properties of squares						
solve problems using the properties of trapezoids						
solve problems using the properties of kites						

📖 **Foldables** Make this Foldable to help you organize your notes about quadrilaterals. Begin with one sheet of notebook paper.

1. **Fold** lengthwise.

2. **Fold** along the width of the paper twice and unfold the paper.

3. **Cut** along the fold marks on the left side of the paper to the center.

4. **Label** as shown.

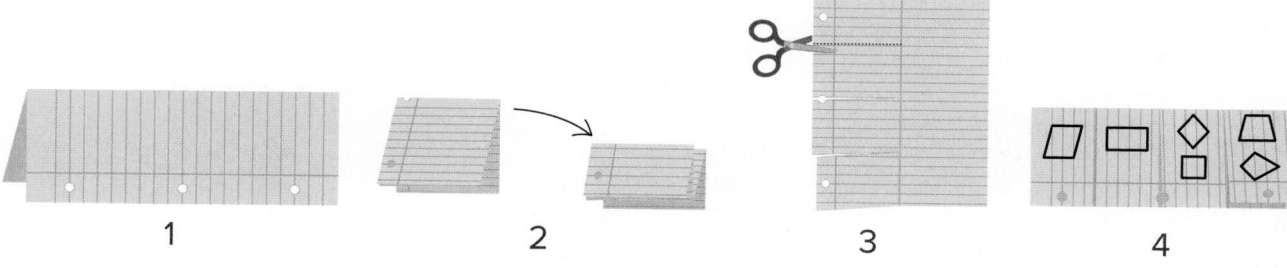

1	2	3	4

What Vocabulary Will You Learn?

- base angle of a trapezoid
- bases of a trapezoid
- diagonal
- isosceles trapezoid
- kite
- legs of a trapezoid
- midsegment of a trapezoid
- parallelogram
- rectangle
- rhombus
- square
- trapezoid

Are You Ready?

Complete the Quick Review to see if you are ready to start this module.
Then complete the Quick Check.

Quick Review

Example 1

Find the measure of each numbered angle.

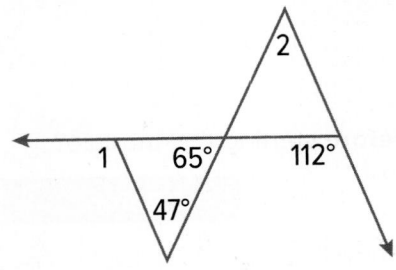

a. $m\angle 1$

$m\angle 1 = 65 + 47$ Exterior Angle Theorem

$m\angle 1 = 112°$ Add.

b. $m\angle 2$

$180 = m\angle 2 + 68 + 65$ Triangle Angle-Sum Thm

$180 = m\angle 2 + 133$ Simplify.

$47° = m\angle 2$ Subtract.

Example 2

Find the lengths of the sides of isosceles $\triangle XYZ$.

$XY = YZ$	Given
$2x + 3 = 4x - 1$	Substitution
$-2x = -4$	Subtract.
$x = 2$	Simplify.
$XY = 2x + 3$	Given
$= 2(2) + 3$, or 7	Substitute $x = 2$.
$YZ = 4x - 1$	Given
$= 4(2) - 1$, or 7	Substitute $x = 2$.
$XZ = 8x - 4$	Given
$= 8(2) - 4$, or 12	Substitute $x = 2$.

Quick Check

Find the value of x to the nearest tenth.

1.

2.

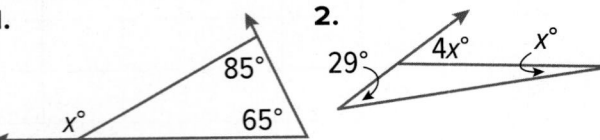

Find the value of x to the nearest tenth.

3.

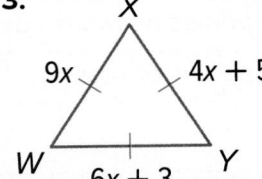

4.

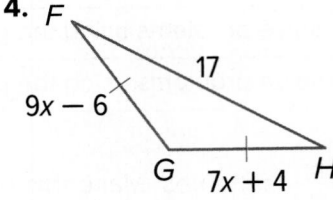

How did you do?

Which exercises did you answer correctly in the Quick Check?

Angles of Polygons

Explore Angles of Polygons

Online Activity Use dynamic geometry software to complete the Explore.

@ **INQUIRY** How can you find the sum of the interior angle measures of a polygon?

Learn Interior Angles of Polygons

A **diagonal** of a polygon is a segment that connects any two nonconsecutive vertices within a polygon. The vertices of polygon *PQRST* that are not consecutive with vertex *P* are vertices *R* and *S*. Notice that the diagonals from vertex *P* separate the polygon into three triangles. The sum of the angle measures of a polygon is the sum of the angle measures of the triangles formed by drawing all of the possible diagonals from one vertex.

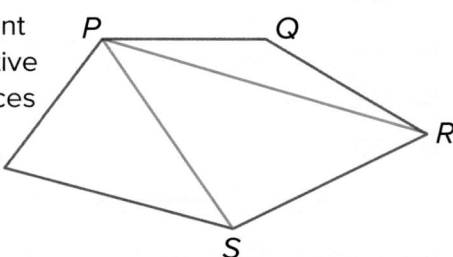

By generalizing this observation for a convex polygon with *n* sides, you can develop the Polygon Interior Angles Sum Theorem.

> **Theorem 2.1: Polygon Interior Angles Sum Theorem**
>
> The sum of the interior angle measures of an *n*-sided convex polygon is $(n - 2) \cdot 180°$.

You will prove Theorem 2.1 in Exercise 34.

Example 1 Find the Interior Angles Sum of a Polygon

Find the measure of each interior angle of pentagon *HJKLM*.

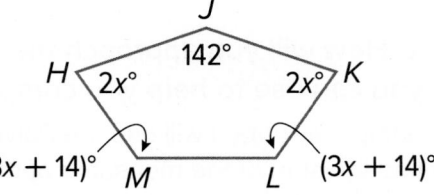

Step 1 Find the sum.
A pentagon has 5 sides. Use the Polygon Interior Angles Sum Theorem to find the sum of its interior angle measures.

$m\angle H + m\angle J + m\angle K + m\angle L + m\angle M$

$= (n - 2) \cdot 180°$ Polygon Interior Angles Sum Thm

$= (5 - 2) \cdot 180°$ Substitute.

$= 540°$ Solve.

(continued on the next page)

Today's Goals
- Prove and use the Polygon Interior Angles Sum Theorem.
- Prove and use the Polygon Exterior Angles Sum Theorem.

Today's Vocabulary
diagonal

Go Online You can complete an Extra Example online.

Step 2 Find the value of x.

Use the sum of the interior angle measures to determine the value of x.

$2x + 2x + (3x + 14) + (3x + 14) + 142 = 540°$ Write an equation.

$x = 37$ Solve.

Step 3 Find the measure of each angle.

Use the value of x to find the measure of each angle.

$m\angle J = 142°$ $m\angle K = 2(37)$ or $74°$ $m\angle L = [3(37) + 14]$ or $125°$

$m\angle M = [3(37) + 14]$ or $125°$ $m\angle H = 2x° = 2(37)$ or $74°$

Check

Find the measure of $\angle E$.

A. 108°

B. 120°

C. 122°

D. 126°

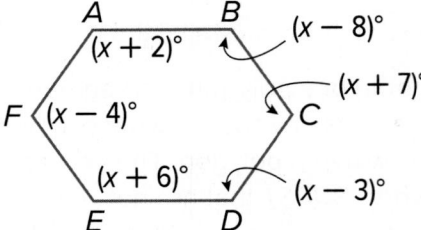

🌐 **Apply Example 2** Interior Angle Measures of a Regular Polygon

FLOOR PLANS Penny is building a house using a floor plan that she designed. What is the measure of $\angle ABC$?

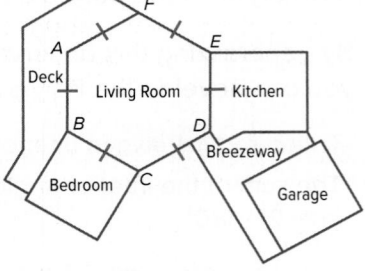

1 What is the task?

Describe the task in your own words. Then list any questions that you may have. How can you find answers to your questions?

Sample answer: I need to find the measure of $\angle ABC$. What is the relationship between the interior angle measures of the regular hexagon?

2 How will you approach the task? What have you learned that you can use to help you complete the task?

Sample answer: I will use the Polygon Interior Angle Sum Theorem to find the sum of the measures of the interior angles of the regular hexagon. Then, I will find the measure of $\angle ABC$ by dividing the sum by the total number of angles. I have learned that the interior angles of a regular polygon are congruent.

🌀 **Go Online** You can complete an Extra Example online.

3 What is your solution?

Use your strategy to solve the problem.

Write the equation that you will use to find the sum of the interior angles of the regular hexagon.

$$m\angle FAB + m\angle ABC + m\angle BCD + m\angle CDE + m\angle DEF + m\angle EFA = (n - 2) \cdot 180°$$

The sum of the interior angles of the regular hexagon is 720°.

$$m\angle ABC = 120°$$

4 How can you know that your solution is reasonable?

✎ **Write About It!** Write an argument that can be used to defend your solution.

Sample answer: 6(120) = 720°, which is the sum of the measures of the interior angles of the hexagon.

Check

PONDS Miguel has commissioned a pentagonal koi pond to be built in his backyard. He wants the pond to have a deck of equal width around it. The lengths of the interior deck sides are the same length, and the lengths of the exterior sides are the same.

The measure of the angle of the pond formed by two sides of the deck is __?__.

Study Tip

Naming Polygons
Remember, a polygon with n-sides is an n-gon, but several polygons have special names.

Number of Sides	Polygon
3	triangle
4	quadrilateral
5	pentagon
6	hexagon
7	heptagon
8	octagon
9	nonagon
10	decagon
11	hendecagon
12	dodecagon
n	n-gon

Example 3 Identify the Polygon Given the Interior Angle Measure

The measure of an interior angle of a regular polygon is 144°. Find the number of sides in the polygon.

Let n = the number of sides in the polygon. Because all angles of a regular polygon are congruent, the sum of the interior angle measures is $144n°$. By the Polygon Interior Angles Sum Theorem, the sum of the interior angle measures can also be expressed as $(n - 2) \cdot 180°$.

$$144n° = (n - 2) \cdot 180° \qquad \text{Write an equation.}$$
$$n = 10 \qquad \text{Solve.}$$

The polygon has 10 sides, so it is a regular decagon.

Check

The measure of an interior angle of a regular polygon is 150°. Find the number of sides in the polygon.

The polygon has __?__ sides.

🧭 **Go Online** You can complete an Extra Example online.

Learn Exterior Angles of Polygons

Theorem 2.2 Polygon Exterior Angles Sum Theorem	
Words	The sum of the exterior angle measures of a convex polygon, one angle at each vertex, is 360°.
Example	$m\angle 1 + m\angle 2 + m\angle 3 + m\angle 4 + m\angle 5 + m\angle 6 = 360°$

You will prove Theorem 2.2 in Exercise 35.

Example 4 Find Missing Values

Find the value of x.

Use the Polygon Exterior Angles Sum Theorem to write an equation. Then solve for x.

$6x + 9x + 2x + 139 = 360$ Write an equation.

$x = 13$ Solve.

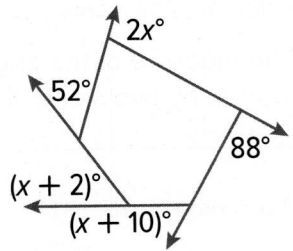

Check

Find the value of x.

A. 45   **B.** 52   **C.** 93   **D.** 97

Example 5 Find Exterior Angle Measures of a Polygon

Find the measure of each exterior angle of a regular dodecagon.

A regular dodecagon has 12 congruent sides and 12 congruent interior angles. The exterior angles are also congruent, because angles supplementary to congruent angles are congruent.

Let n = the measure of each exterior angle and write and solve an equation.

$12n = 360°$ Polygon Exterior Angles Sum Theorem

$n = 30°$ Solve.

The measure of each exterior angle of a regular dodecagon is 30°.

Check

The measure of each exterior angle of a regular octagon is __?__

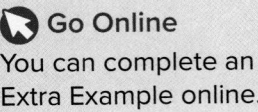

Go Online
An alternate method is available for this example.

Go Online
You can complete an Extra Example online.

Practice

Go Online You can complete your homework online.

Example 1

Find the measure of each interior angle.

1.

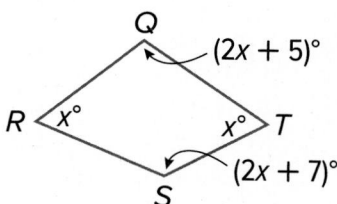

2.

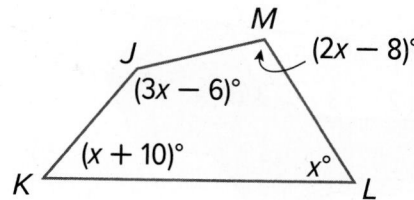

3.

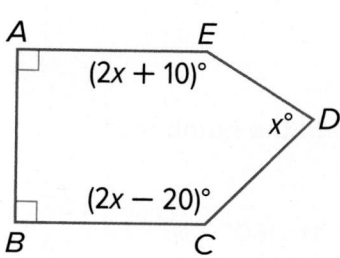

4.

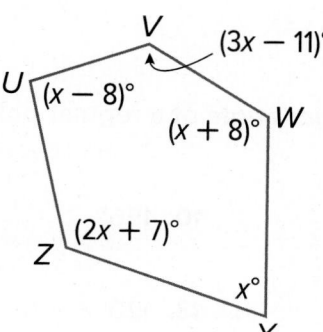

Example 2

5. ARCHITECTURE In the Uffizi gallery in Florence, Italy, there is a room built by Buontalenti called the Tribune (*La Tribuna* in Italian). This room is shaped like a regular octagon. What is the measure of the angle formed by two consecutive walls of the Tribune?

La Tribuna

6. THEATER A theater floor plan is shown. The upper five sides are part of a regular dodecagon. Find $m\angle 1$.

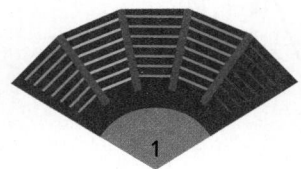

7. FARM An animal pen is in the shape of a regular heptagon. What is the measure of each interior angle of the animal pen? Round to the nearest tenth.

8. POLYGON PATH In Ms. Rickets' math class, students made a "polygon path" that consists of regular polygons of 3, 4, 5, and 6 sides joined together as shown.

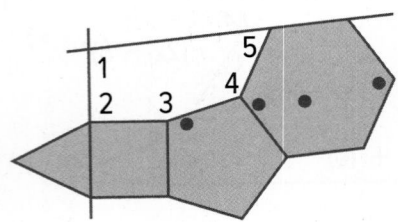

    a. Find $m\angle 2$ and $m\angle 5$.

    b. Find $m\angle 3$ and $m\angle 4$.

    c. What is $m\angle 1$?

Example 3

The measure of an interior angle of a regular polygon is given. Find the number of sides in the polygon.

9. 144°

10. 156°

11. 160°

12. 108°

13. 120°

14. 150°

Example 4

Find the value of x in each diagram.

15.

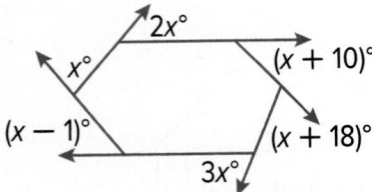

16.

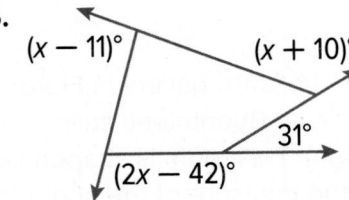

17.

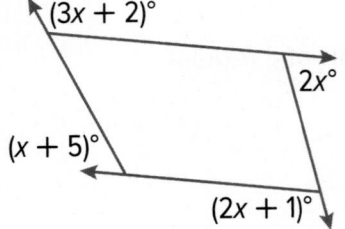

18.

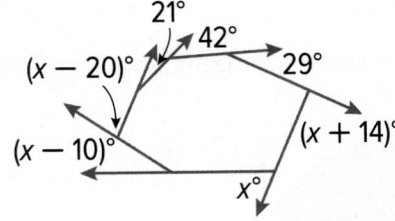

Example 5

Find the measure of each exterior angle of each regular polygon.

19. pentagon

20. 15-gon

21. hexagon

22. octagon

23. nonagon

24. 12-gon

Mixed Exercises

Find the measures of an exterior angle and an interior angle given the number of sides of each regular polygon. Round to the nearest tenth, if necessary.

25. 7

26. 13

27. 14

For Exercises 28 and 29, find the value of x.

28. A convex octagon has interior angles with measures $(x + 55)°$, $(3x + 20)°$, $4x°$, $(4x - 10)°$, $(6x - 55)°$, $(3x + 52)°$, $3x°$, and $(2x + 30)°$.

29. A convex hexagon has interior angles with measures $x°$, $(5x - 103)°$, $(2x + 60)°$, $(7x - 31)°$, $(6x - 6)°$, and $(9x - 100)°$.

For Exercises 30 and 31, find the measure of each interior angle in the given polygon.

30. A decagon in which the measures of the interior angles are $(x + 5)°$, $(x + 10)°$, $(x + 20)°$, $(x + 30)°$, $(x + 35)°$, $(x + 40)°$, $(x + 60)°$, $(x + 70)°$, $(x + 80)°$, and $(x + 90)°$.

31. A polygon $ABCDE$ in which the measures of the interior angles are $(6x)°$, $(4x + 13)°$, $(x + 9)°$, $(2x - 8)°$, and $(4x - 1)°$.

32. Find the measure of each exterior angle of a regular $2x$-gon.

33. Find the sum of the measures of the exterior angles of a convex 65-gon.

34. PROOF Write a paragraph proof to prove the Polygon Interior Angles Sum Theorem.

35. PROOF Use algebra to prove the Polygon Exterior Angles Sum Theorem.

36. REASONING The measure of each interior angle of a regular polygon is 24 more than 38 times the measure of each exterior angle. Find the number of sides of the polygon.

37. **ARCHAEOLOGY** Archaeologists unearthed parts of two adjacent walls of an ancient castle. Before it was unearthed, they knew from ancient texts that the castle was shaped like a regular polygon, but nobody knew how many sides it had. Some said 6, others 8, and some even said 100. From the information in the figure, how many sides did the castle really have?

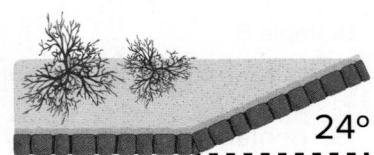

38. **DESIGN** Ronella is designing boxes she will use to ship her jewelry. She wants to shape the box like a regular polygon. For the boxes to pack tightly, she decides to use a regular polygon in which the measure of its interior angles is half the measure of its exterior angles. What regular polygon should she use?

39. **CRYSTALLOGRAPHY** Crystals are classified according to seven crystal systems. The basis of classification is the shape of the faces of the crystal. Turquoise belongs to the triclinic system. Each of the six faces of turquoise is in the shape of a quadrilateral. Find the sum of the measures of the interior angles of one such face.

40. **STRUCTURE** If three of the interior angles of a convex hexagon each measure 140°, a fourth angle measures 84°, and the measure of the fifth angle is 3 times the measure of the sixth angle, find the measure of the sixth angle.

🧠 Higher-Order Thinking Skills

41. **FIND THE ERROR** Marshawn says that the sum of the exterior angles of a decagon is greater than that of a heptagon because a decagon has more sides. Liang says that the sum of the exterior angles for both polygons is the same. Who is correct? Explain your reasoning.

42. **WRITE** Explain how triangles are related to the Polygon Interior Angles Sum Theorem.

43. **CREATE** Sketch a polygon and find the sum of its interior angles. How many sides does a polygon with twice this interior angles sum have? Justify your answer.

44. **PERSEVERE** Find the values of a, b, and c if $QRSTVX$ is a regular hexagon. Justify your answer.

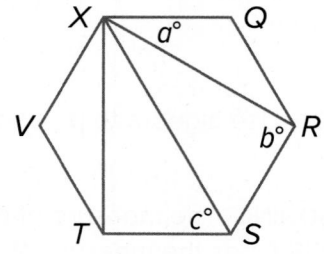

45. **ANALYZE** If two sides of a regular hexagon are extended to meet at a point in the exterior of the polygon, will the triangle formed *sometimes*, *always*, or *never* be equilateral? Justify your argument.

Parallelograms

Explore Properties of Parallelograms

Online Activity Use dynamic geometry software to complete the Explore.

> **INQUIRY** What special properties do parallelograms have? ×

Learn Parallelograms

A **parallelogram** is a quadrilateral with both pairs of opposite sides parallel. To name a parallelogram, use the symbol $\square$. In $\square ABCD$, $\overline{BC} \parallel \overline{AD}$ and $\overline{AB} \parallel \overline{DC}$ by definition.

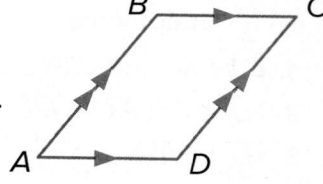

Other properties of parallelograms are given in the theorems below.

Theorems: Properties of Parallelograms

Theorem 2.3

If a quadrilateral is a parallelogram, then its opposite sides are congruent.

Theorem 2.4

If a quadrilateral is a parallelogram, then its opposite angles are congruent.

Theorem 2.5

If a quadrilateral is a parallelogram, then its consecutive angles are supplementary.

Theorem 2.6

If a parallelogram has one right angle, then it has four right angles.

You will prove Theorems 2.3 and 2.5 in Exercises 15 and 17, respectively.

Example 1 Use Properties of Parallelograms

Find CD.

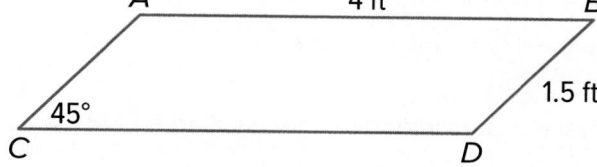

$\overline{CD} \cong \overline{AB}$ Opposite sides of a $\square$ are $\cong$.

$CD = AB$ Definition of congruent

$\quad = 4\ ft$ Substitution

Today's Goals
- Prove and use theorems about the properties of parallelograms.
- Prove and use theorems about the diagonals of parallelograms.

Today's Vocabulary
parallelogram

Watch Out!

Parallelograms
Theorems 2.3 through 2.6 apply only if you already know that the figure is a parallelogram.

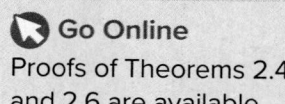 **Go Online**
Proofs of Theorems 2.4 and 2.6 are available.

Talk About It!
Thiago states that because all parallelograms are quadrilaterals, all quadrilaterals are parallelograms. Do you agree? Justify your answer.

Check

Find each measure.

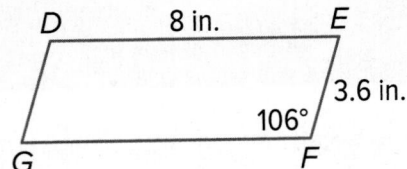

a. $m\angle D = $ ___?___

b. $FG = $ ___?___ in.

You can use the properties of parallelograms to write proofs.

Example 2 Proofs Using the Properties of Parallelograms

Write the correct statements and reasons to complete the two-column proof.

Given: ▱HJKP and ▱PKLM

Prove: $\overline{HJ} \cong \overline{ML}$

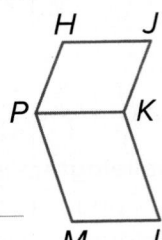

Check

Copy and complete the two-column proof.

Given: ▱JKLM, $\overline{KN} \cong \overline{KL}$

Prove: $\angle J \cong \angle KNL$

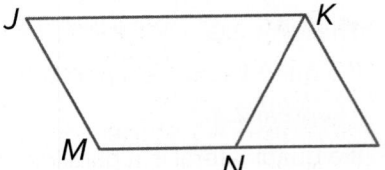

Statements	Reasons
1. ▱JKLM, $\overline{KN} \cong \overline{KL}$	1. ___?___
2. ___?___	2. If a quad. is a ▱, its opp. ∠s are ≅.
3. ___?___	3. Isosceles Triangle Theorem
4. $\angle J \cong \angle KNL$	4. ___?___

Learn Diagonals of Parallelograms

The diagonals of parallelograms have special properties.

Theorems: Diagonals of Parallelograms
Theorem 2.7
If a quadrilateral is a parallelogram, then its diagonals bisect each other.
Theorem 2.8
If a quadrilateral is a parallelogram, then each diagonal separates the parallelogram into two congruent triangles.

You will prove Theorems 2.7 and 2.8 in Exercises 16 and 18, respectively.

🔎 **Go Online** You can complete an Extra Example online.

Example 3 Use Properties of Parallelograms and Algebra

Find the values of *x* and *z* in ☐ABCD.

$m\angle ADC = 4x°$ and $m\angle DAB = (2x − 6)°$.

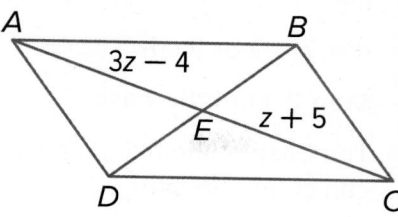

Part A Find the value of *x*.

$180 = m\angle ADC + m\angle DAB$	Consec. $\angle$s in a ☐ are supplementary.
$180 = 4x + (2x − 6)$	Substitution
$x = 31$	Solve.

Part B Find the value of *z*.

$\overline{AE} \cong \overline{CE}$	Diagonals of a ☐ bisect each other.
$AE = CE$	Definition of congruent
$3z − 4 = z + 5$	Substitution
$z = 4.5$	Solve.

🌐 Example 4 Parallelograms and Coordinate Geometry

SCRAPBOOKING Tomas is making envelopes to sell with handmade cards. He uses a different style of paper to create the flap of the envelope, and he edges the envelopes with washi tape. The envelopes are parallelograms, and the edges of the flaps lie along the diagonals of the parallelograms. Find the area of the flap and the perimeter of the envelope.

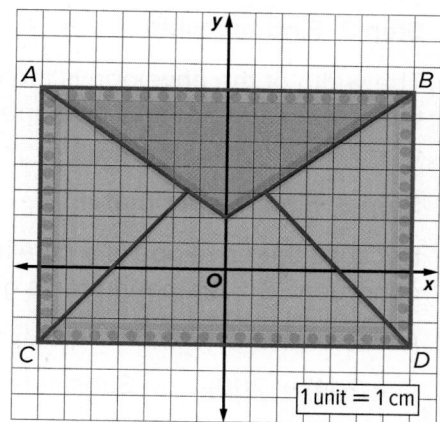

Problem-Solving Tip

Make a Plan To find the area of the paper needed for the envelope flap, you need to calculate the point of intersection of the diagonals of the envelope. Before solving for the area, analyze the information you are given, develop a plan, and determine the theorems you will need to apply.

Part A Find the amount of paper needed to create the flap.

You can approximate the area of the flap with a triangle, so the area is $A = \frac{1}{2}bh$.

Step 1 Find the height.

To find the height, determine the coordinates of the intersection of the diagonals of the envelope, which has vertices at $A(−7, 7)$, $B(7, 7)$, $C(−7, −3)$, and $D(7, −3)$. Because the diagonals of a parallelogram bisect each other, the intersection point is the midpoint of $\overline{AD}$ and $\overline{BC}$. Find the midpoint of $\overline{AD}$.

$$\left(\frac{x_1 + x_2}{2}, \frac{y_1 + y_2}{2}\right) = \left(\frac{−7 + 7}{2}, \frac{7 + (−3)}{2}\right) \qquad \text{Midpoint Formula}$$
$$= (0, 2) \qquad \text{Simplify.}$$

(continued on the next page)

🔵 **Go Online** You can complete an Extra Example online.

The height is the difference in the *y*-coordinates of the intersection of the diagonals and the vertices of the top edge of the envelope.

$h = 7 - 2$ or 5 cm

Step 2 Find the base.

The base of the flap is the length of the top edge of the envelope. You can count the units to determine the base.

$b = 14$ cm

Step 3 Find the area of the flap.

$A = \frac{1}{2}bh$ Area of a triangle

$\quad = \frac{1}{2}(14)(5)$ Substitute.

$\quad = 35 \text{ cm}^2$ Solve.

Part B Find the length of washi tape needed to create the border.

Step 1 Find the length.

The length of the envelope is the same as the base of the triangle determined above.

$\ell = 14$ cm

Step 2 Find the width.

The width of the envelope is the distance between the top edge and the bottom edge of the envelope.

$w = 7 - (-3) = 10$ cm

Step 3 Find the perimeter.

Because the envelope is a parallelogram, opposite sides are congruent. So, the perimeter is given by $P = 2\ell + 2w$.

$P = 2\ell + 2w$ Perimeter formula

$\quad = 2(14) + 2(10)$ Substitute.

$\quad = 48$ cm   Solve.

Check

QUILTING Jimena is making a quilt. Each block is a parallelogram made of a single piece of patterned fabric and is trimmed with a gray border. Find the area of the fabric used to make the block. Find the length of fabric used to make the border of the block. Round to the nearest tenth if necessary.

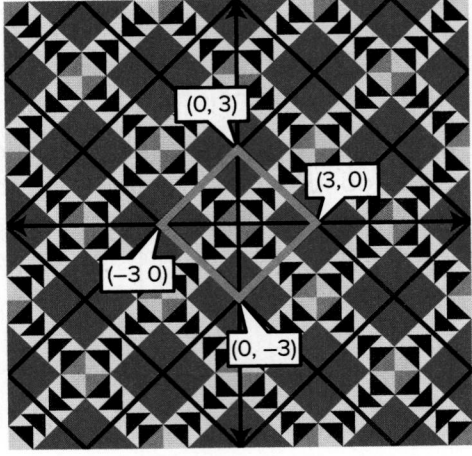

(0, 3)

(3, 0)

(−3 0)

(0, −3)

$A = \underline{\quad ? \quad} \text{ cm}^2$

$P \approx \underline{\quad ? \quad} \text{ cm}$

🔵 **Go Online** You can complete an Extra Example online.

Think About It!

What assumptions did you make when calculating the area of the paper and the length of washi tape?

Practice

Go Online You can complete your homework online.

Example 1

Use □PQRS to find each measure.

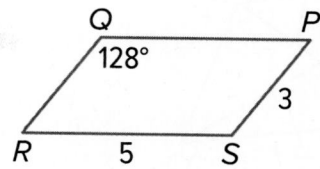

1. $m\angle R$

2. QR

3. QP

4. $m\angle S$

5. **SOCCER** Four soccer players are practicing a drill. Goalie A is facing Player B to receive the ball. Goalie A then turns $x°$ to face Player A to pass her the ball. If Goalie B is facing Player A to receive the ball, then through what angle measure should Goalie B turn to pass the ball to Player B?

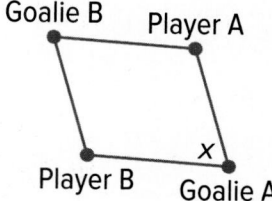

Example 2

PROOF For 6–7, write a two-column proof.

6. **Given:** □BDHA, $\overline{CA} \cong \overline{CG}$

   Prove: $\angle BDH \cong \angle G$

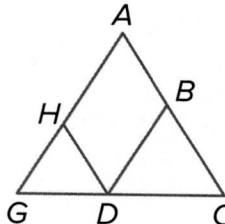

7. **Given:** WXTV and YZTV are parallelograms.

   Prove: $\overline{WX} \cong \overline{YZ}$

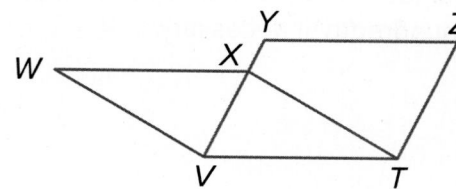

8. Write a paragraph proof.

   Given: □PRST and □PQVU

   Prove: $\angle V \cong \angle S$

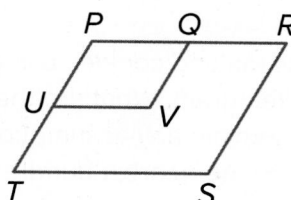

Example 3

Find the value of each variable in each parallelogram.

9.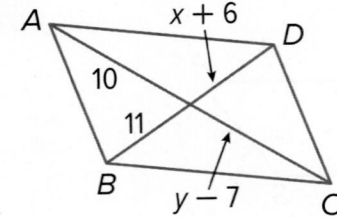

10.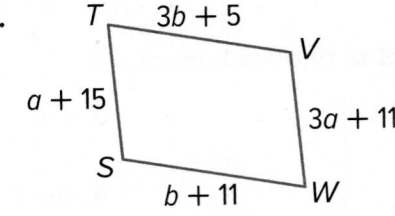

11.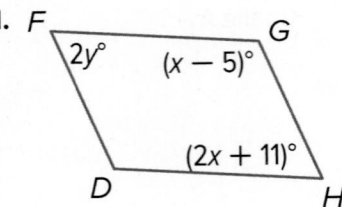

12.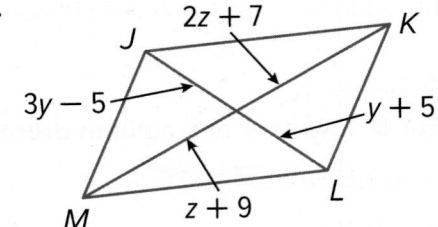

Example 4

13. PARK A new dog park is being designed by a city planner. The park is enclosed by a fence and shaped like a parallelogram. What is the area and perimeter of the dog park? Round your answers to the nearest hundredth, if necessary.

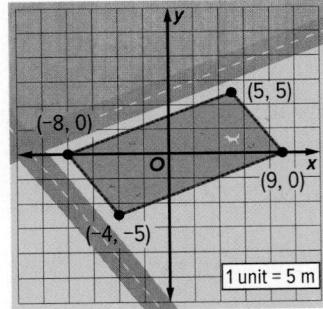

14. STATE YOUR ASSUMPTION Breelyn is making cookies using a cookie cutter in the shape of a parallelogram. What are the perimeter and area of each cookie? Explain any assumptions that you make. Round your answers to the nearest hundredth, if necessary.

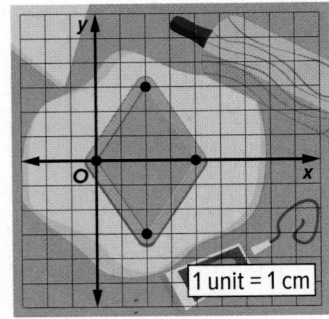

PROOF **Write a two-column proof for each theorem.**

15. Theorem 7.3

    Given: □*PQRS*

    Prove: $\overline{PQ} \cong \overline{RS}$, $\overline{QR} \cong \overline{SP}$

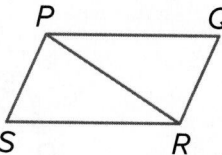

16. Theorem 7.7

    Given: □*ACDE*

    Prove: $\overline{EC}$ bisects $\overline{AD}$.

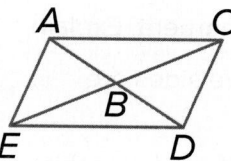

17. Theorem 7.5

    Given: □*GKLM*

    Prove: ∠*G* and ∠*K*, ∠*K* and ∠*L*,

    ∠*L* and ∠*M*, and ∠*M* and ∠*G*

    are supplementary.

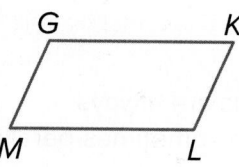

18. Theorem 7.8

    Given: □*WXYZ*

    Prove: △*WXZ* ≅ △*YZX*

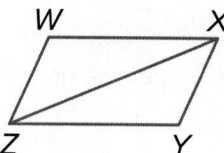

Use ▱ABCD to find each measure or value.

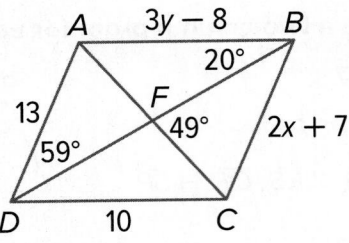

19. x

20. y

21. $m\angle AFB$

22. $m\angle DAC$

23. $m\angle ACD$

24. $m\angle DAB$

25. **REGULARITY** Use the graph shown.

   a. Use the Distance Formula to determine if the diagonals of *JKLM* bisect each other. Explain.

   b. Determine whether the diagonals are congruent. Explain.

   c. Use slopes to determine if the consecutive sides are perpendicular. Explain.

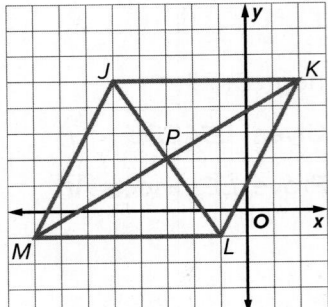

26. **USE TOOLS** Make a Venn diagram showing the relationship between squares, rectangles, and parallelograms.

🍩 **Higher-Order Thinking Skills**

27. **PERSEVERE** *ABCD* is a parallelogram with side lengths as indicated in the figure at the right. The perimeter of *ABCD* is 22. Find *AB*.

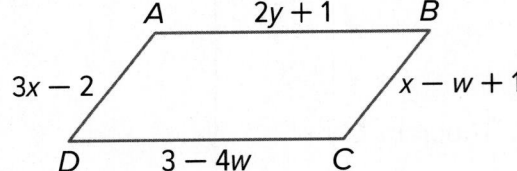

28. **ANALYZE** Explain why parallelograms are always quadrilaterals, but quadrilaterals are sometimes parallelograms.

29. **WRITE** Summarize the properties of the sides, angles, and diagonals of a parallelogram.

30. **CREATE** Provide an example to show that parallelograms are not always congruent if their corresponding sides are congruent.

31. **PERSEVERE** Find $m\angle 1$ and $m\angle 10$ in the figure. Explain your reasoning.

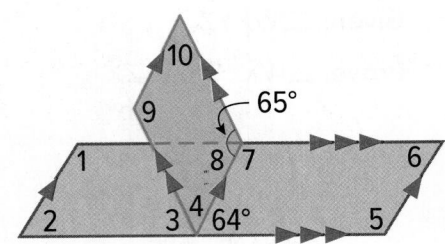

Tests for Parallelograms

Explore Constructing Parallelograms

Online Activity Use dynamic geometry software to complete the Explore.

> **INQUIRY** How can you use the properties of parallelograms to construct parallelograms? ✕

Learn Tests for Parallelograms

If a quadrilateral has each pair of opposite sides parallel, it is a parallelogram by definition. This is not the only test, however, that can be used to determine if a quadrilateral is a parallelogram.

Theorems: Conditions for Parallelograms
Theorem 2.9
If both pairs of opposite sides of a quadrilateral are congruent, then the quadrilateral is a parallelogram.
Theorem 2.10
If both pairs of opposite angles of a quadrilateral are congruent, then the quadrilateral is a parallelogram.
Theorem 2.11
If the diagonals of a quadrilateral bisect each other, then the quadrilateral is a parallelogram.
Theorem 2.12
If one pair of opposite sides of a quadrilateral is both parallel and congruent, then the quadrilateral is a parallelogram.

You will prove Theorems 2.10 and 2.11 in Exercises 18 and 19, respectively.

Example 1 Identify Parallelograms

Determine whether the quadrilateral is a parallelogram. Justify your answer.

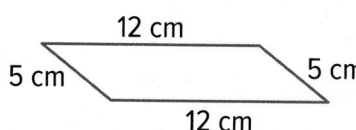

Is the quadrilateral a parallelogram? Yes

What theorem can you use to justify your answer?

If both pairs of opposite sides are congruent, then the quadrilateral is a parallelogram.

Go Online You can complete an Extra Example online.

Today's Goals
- Use the tests for parallelograms to determine whether quadrilaterals are parallelograms.

Go Online
Proofs of Theorems 2.9 and 2.12 are available.

Talk About It!
Jude says that by Theorem 2.12, you only need to show that one pair of opposite sides are congruent to show that the quadrilateral is a parallelogram. Do you agree? Justify your answer.

You can use the conditions of parallelograms to find missing values that make a quadrilateral a parallelogram.

🌐 Example 2   Use Parallelograms to Find Values

SCHOOL SUPPLIES **The top of the eraser appears to be a parallelogram. Find the values of *x* and *y* so that the side of the eraser is a parallelogram.**

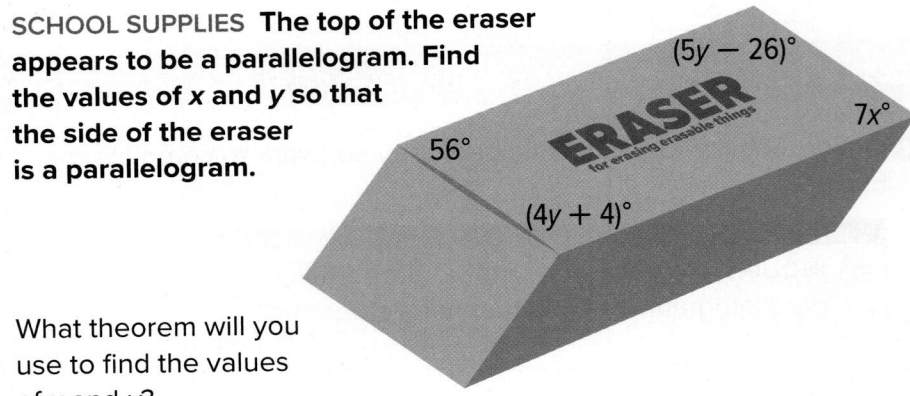

What theorem will you use to find the values of *x* and *y*?

If both pairs of opposite angles are congruent, then the quadrilateral is a parallelogram.

Step 1 Find *x*.

Find *x* such that $7x = 56$.

$$7x = 56$$ Opp. angles of a ▱ are ≅.

$$x = 8$$ Solve.

Step 2 Find *y*.

Find *y* such that $4y + 4 = 5y - 26$.

$$4y + 4 = 5y - 26.$$ Opp. angles of a ▱ are ≅.

$$y = 30$$ Solve.

So, when *x* is 8 and *y* is 30, the quadrilateral is a parallelogram.

Check

MOSAICS The mosaic pattern of the floor is made up of different tiles.

Part A

Find the values of *x* and *y* so that the tile is a parallelogram.

$x =$ __?__

$y =$ __?__

Part B

Select the theorem you used to find the values of *x* and *y*.

A. If both pairs of opp. sides are ≅, then quad. is a ▱.
B. If both pairs of opp. ∠s are ≅, then quad. is a ▱.
C. If diag. bisect each other, then quad. is a ▱.
D. If one pair of opp. sides is ≅ and ∥, then quad. is a ▱.

🔎 **Go Online** You can complete an Extra Example online.

We can use the Distance, Slope, and Midpoint Formulas to determine whether a quadrilateral in the coordinate plane is a parallelogram.

Example 3 Identify Parallelograms on the Coordinate Plane

Determine whether quadrilateral *FGHJ* is a parallelogram. Justify your answer using the Midpoint Formula.

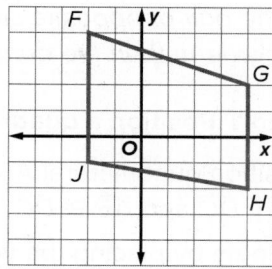

What theorem will you use to determine whether quadrilateral *FGHJ* is a parallelogram?

If the diagonals bisect each other, then the quadrilateral is a parallelogram.

Step 1 Calculate the midpoint of $\overline{GJ}$.

$$M\left(\frac{x_1 + x_2}{2}, \frac{y_1 + y_2}{2}\right) \qquad \text{Midpoint Formula}$$

$$M\left(\frac{4 + (-2)}{2}, \frac{2 + (-1)}{2}\right) \qquad \text{Substitute.}$$

$$M\left(1, \frac{1}{2}\right) \qquad \text{Solve.}$$

Step 2 Calculate the midpoint of $\overline{FH}$.

$$M\left(\frac{x_1 + x_2}{2}, \frac{y_1 + y_2}{2}\right) \qquad \text{Midpoint Formula}$$

$$M\left(\frac{-2 + 4}{2}, \frac{4 + (-2)}{2}\right) \qquad \text{Substitute.}$$

$$M\left(1, 1\right) \qquad \text{Solve.}$$

Step 3 Determine whether *FGHJ* is a ▱.

If the diagonals of a quadrilateral bisect each other, then it is a parallelogram. The diagonals of a quadrilateral bisect each other if the midpoints coincide. Because the midpoints of diagonals $\overline{FH}$ and $\overline{GJ}$ do not have the same coordinates, quadrilateral *FGHJ* is not a parallelogram.

Check

Determine whether quadrilateral *ABCD* is a parallelogram. Justify your answer.

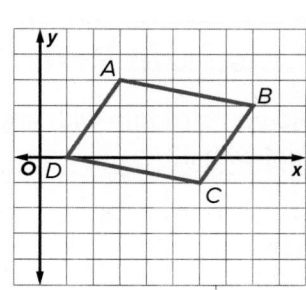

Go Online You can complete an Extra Example online.

> **Think About It!**
> Describe another method you could use to determine whether quadrilateral *FGHJ* is a parallelogram.

You can assign variable coordinates to the vertices of quadrilaterals. Then, you can use the Distance, Slope, and Midpoint Formulas to write coordinate proofs of theorems.

Example 4 Parallelograms and Coordinate Proofs

Write a coordinate proof for the following statement.
If one pair of opposite sides of a quadrilateral is both parallel and congruent, then the quadrilateral is a parallelogram.

Step 1 Position a quadrilateral on the coordinate plane.

Position quadrilateral $ABCD$ on the coordinate plane such that $\overline{AB} \parallel \overline{DC}$ and $\overline{AB} \cong \overline{DC}$.

- Begin by placing the vertex A at the origin.
- Let $\overline{AB}$ have a length of a units. Then B has coordinates $(a, 0)$.
- Because horizontal segments are parallel, position the endpoints of $\overline{DC}$ so that they have the same y-coordinate, c.
- So that the distance from D to C is also a units, let the x-coordinate of D be b and the x-coordinate of C be $b + a$.

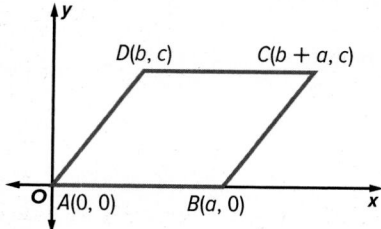

Step 2 Use your figure to write a proof.

Given: quadrilateral $ABCD$, $\overline{AB} \parallel \overline{DC}$, $\overline{AB} \cong \overline{DC}$

Prove: $ABCD$ is a parallelogram.

Proof:
By definition, a quadrilateral is a parallelogram if opposite sides are parallel. We are given that $\overline{AB} \parallel \overline{DC}$, so we need to show that $\overline{AD} \parallel \overline{BC}$.

Use the Slope Formula.

$$\text{slope of } \overline{AD} = \frac{c - 0}{b - 0} = \frac{c}{b} \qquad \text{slope of } \overline{BC} = \frac{c - 0}{b + a - a} = \frac{c}{b}$$

Because $\overline{AD}$ and $\overline{BC}$ have the same slope, $\overline{AD} \parallel \overline{BC}$. So, quadrilateral $ABCD$ is a parallelogram because opposite sides are parallel.

🖱 **Go Online**
You may want to complete the construction activities for this lesson.

Pause and Reflect

Did you struggle with anything in this lesson? If so, how did you deal with it?

🖱 **Go Online** You can complete an Extra Example online.

Practice

🔺 **Go Online** You can complete your homework online.

Example 1

Determine whether each quadrilateral is a parallelogram. Justify your answer.

1.
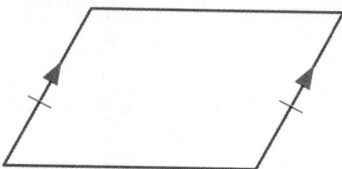

2.
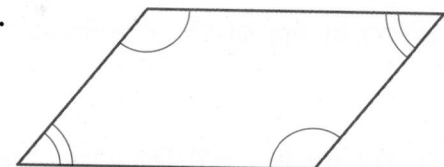

3.
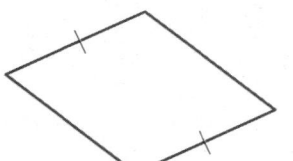

4.
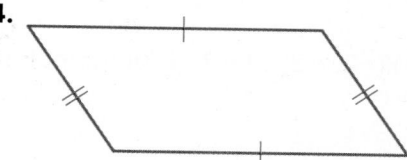

5.
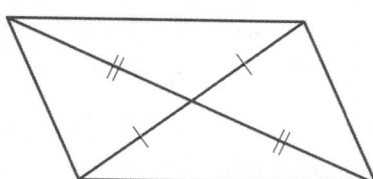

6.
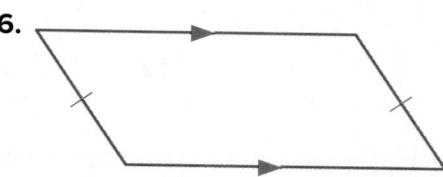

Example 2

7. **ORGANIZATION** The space between the hinges and trays of a collapsible tray organizer appears to be a parallelogram. Find the values of x and y so that the trays and hinges of the organizer form a parallelogram.

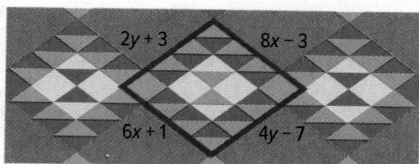

8. **PATTERNS** Many Native American rugs and blankets incorporate parallelograms into the designs. Find the values of x and y so that the quadrilateral shown is a parallelogram.

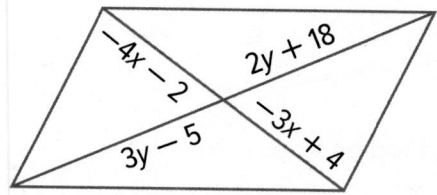

Find the values of x and y so that each quadrilateral is a parallelogram.

9.
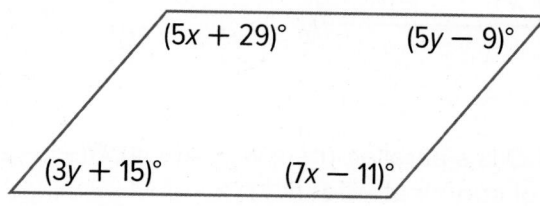

10.

Example 3

11. Determine whether *ABCD* is a parallelogram. Justify your answer.

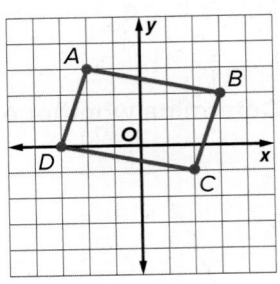

CONSTRUCT ARGUMENTS For Exercises 12–15, graph each quadrilateral with the given vertices. Determine whether the figure is a parallelogram. Justify your argument with the method indicated.

12. *P*(0, 0), *Q*(3, 4), *S*(7, 4), *Y*(4, 0); Slope Formula

13. *S*(−2, 1), *R*(1, 3), *T*(2, 0), *Z*(−1, −2); Distance and Slope Formulas

14. *W*(2, 5), *R*(3, 3), *Y*(−2, −3), *N*(−3, 1); Midpoint Formula

15. *W*(1, −4), *X*(−4, 2), *Y*(1, −1), and *Z*(−2, −3); Slope Formula

Example 4

16. Write a coordinate proof for the statement: *If both pairs of opposite sides of a quadrilateral are congruent, then the quadrilateral is a parallelogram.*

17. Write a coordinate proof for the statement: *If a parallelogram has one right angle, it has four right angles.*

Mixed Exercises

PROOF **Write the specified type of proof for each theorem.**

18. paragraph proof of Theorem 2.10
    Given: ∠*K* ≅ ∠*M*, ∠*N* ≅ ∠*L*
    Prove: *KLMN* is a parallelogram.

19. two-column proof of Theorem 2.11
    Given: $\overline{PR}$ bisects $\overline{TQ}$.
    $\overline{TQ}$ bisects $\overline{PR}$.
    Prove: *PQRT* is a parallelogram.

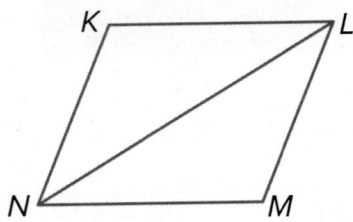

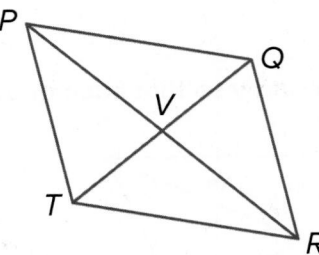

20. *ABCD* is a parallelogram with *A*(5, 4), *B*(−1, −2), and *C*(8, −2). Find one possible set of coordinates for *D*.

21. STRUCTURE A parallelogram has vertices $R(-2, -1)$, $S(2, 1)$, and $T(0, -3)$. Find all possible coordinates for the fourth vertex.

22. If the slope of $\overline{PQ}$ is $\frac{2}{3}$ and the slope of $\overline{QR}$ is $-\frac{1}{2}$, find the slope of $\overline{SR}$ so that $PQRS$ is a parallelogram.

23. If the slope of $\overline{AB}$ is $\frac{1}{2}$, the slope of $\overline{BC}$ is -4, and the slope of $\overline{CD}$ is $\frac{1}{2}$, find the slope of $\overline{DA}$ so that $ABCD$ is a parallelogram.

24. REASONING The pattern shown in the figure is to consist of congruent parallelograms. How can the designer be certain that the shapes are parallelograms?

25. Refer to parallelogram $ABCD$. If $AB = 8$ cm, what is the perimeter of the parallelogram?

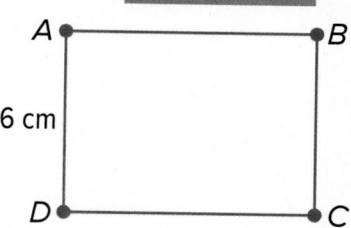

26. PICTURE FRAME Aston is making a wooden picture frame in the shape of a parallelogram. He has two pieces of wood that are 3 feet long and two that are 4 feet long.

a. If he connects the pieces of wood at their ends to each other, in what order must he connect them to make a parallelogram?

b. How many different parallelograms could he make with these four lengths of wood?

c. Explain something Aston might do to specify precisely the shape of the parallelogram.

27. STATE YOUR ASSUMPTION When a coordinate plane is placed over the Harrisville town map, the four street lamps in the center are located as shown. Do the four lamps form the vertices of a parallelogram? Justify your reasoning. Explain any assumptions that you make regarding the coordinate plane and the map.

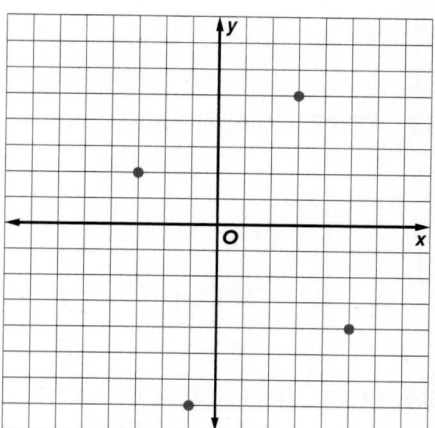

28. USE TOOLS Explain how you can use Theorem 2.11 to construct a parallelogram. Then construct a parallelogram using your method.

29. BALANCING Nikia, Madison, Angela, and Shelby are balancing on an X-shaped floating object. To balance, they want to stand so they are at the vertices of a parallelogram. To achieve this, do all four of them have to be the same distance from the center of the object? Explain.

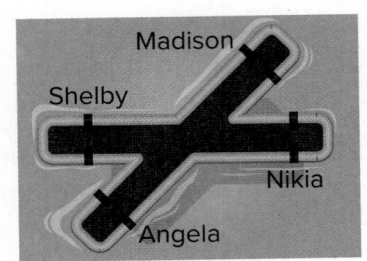

30. FORMATION Four jets are flying in formation. Three of the jets are shown in the graph. If the four jets are located at the vertices of a parallelogram, what are the three possible locations of the missing jet?

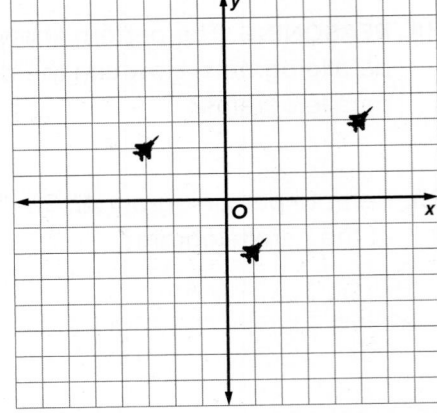

31. PROOF Write a coordinate proof to prove that the segments joining the midpoints of the sides of any quadrilateral form a parallelogram.

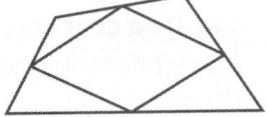

32. ANALYZE If two parallelograms have four congruent corresponding angles, are the parallelograms *sometimes, always,* or *never* congruent? Justify your argument.

33. WRITE Compare and contrast Theorem 2.9 and Theorem 2.3.

34. PERSEVERE If *ABCD* is a parallelogram and $\overline{AJ} \cong \overline{KC}$, show that quadrilateral *JBKD* is a parallelogram.

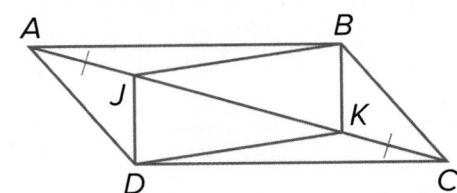

35. ANALYZE The diagonals of a parallelogram meet at the point (0, 1). One vertex of the parallelogram is located at (2, 4), and a second vertex is located at (3, 1). Find the locations of the remaining vertices.

Rectangles

Explore Properties of Rectangles

 Online Activity Use dynamic geometry software to complete the Explore.

@ **INQUIRY** What special properties do rectangles have? ×

Learn Properties of Rectangles

A **rectangle** is a parallelogram with four right angles. From this definition, you know that a rectangle has the following properties:

- All four angles are right angles.
- Opposite sides are parallel and congruent.
- Opposite angles are congruent.
- Consecutive angles are supplementary.
- Diagonals bisect each other.

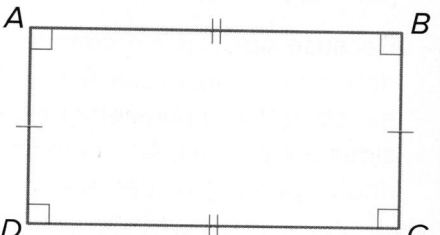

In addition, the diagonals of a rectangle are congruent.

Theorem 2.13: Diagonals of a Rectangle

If a parallelogram is a rectangle, then its diagonals are congruent.

You will prove Theorem 2.13 in Exercise 23.

🌐 Example 1 Use Properties of Rectangles

BASKETBALL The coach is making the basketball team run a new drill along the diagonals of the court, as shown. If $BC = 94$ feet and $FC = 106.5$ feet, find DG.

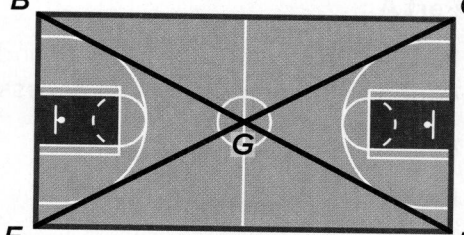

$\overline{FC} \cong \overline{BD}$ If a ▱ is a rectangle, diag. are ≅.

$FC = BD$ Definition of congruence

$BD = 106.5$ Substitution

Because *BCDF* is a rectangle, it is a parallelogram. The diagonals of a parallelogram bisect each other, so $DG = BG$.

$DG + BG = BD$ Segment Addition Postulate

$DG + DG = BD$ Substitution

$2DG = BD$ Simplify.

$DG = \frac{1}{2}BD$ Divide each side by 2.

$DG = \frac{1}{2}(106.5)$ or 53.25 ft   Substitution

Today's Goals
- Recognize and apply the properties of rectangles.
- Determine whether parallelograms are rectangles.

Today's Vocabulary
rectangle

🗯 **Think About It!**

What does point *G* represent in the context of the problem?

Check

FRAMING Jay is framing a barn door with an X-brace as shown. If $RT = 3\frac{9}{16}$ feet, $QR = 7$ feet, and $m\angle RTS = 65°$, find each measure. If a measure is not a whole number, write it as a decimal.

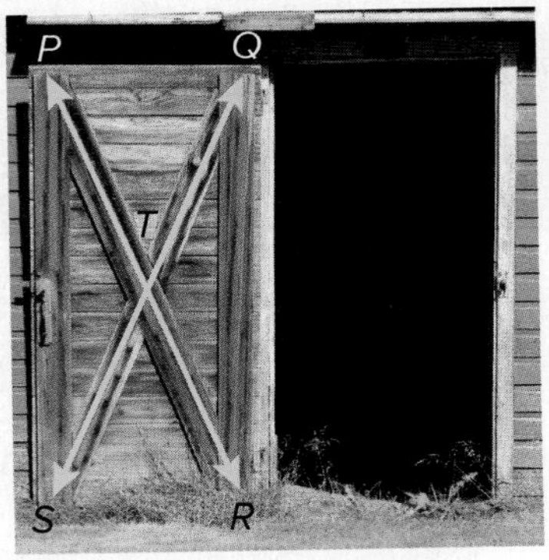

$PS =$ __?__ ft $SQ =$ __?__ ft

$m\angle QTR =$ __?__ $m\angle TQR =$ __?__

Study Tip

Right Angles Recall from Theorem 2.6 that if a parallelogram has one right angle, then it has four right angles.

Think About It!

There are four congruent right triangles formed by the diagonals of a rectangle. How many pairs of congruent triangles are there in all?

Example 2 Use Properties of Rectangles and Algebra

Quadrilateral *ABCD* is a rectangle. If $m\angle BAC = (3x + 3)°$ and $m\angle ACB = (5x - 1)°$, find the value of *x*.

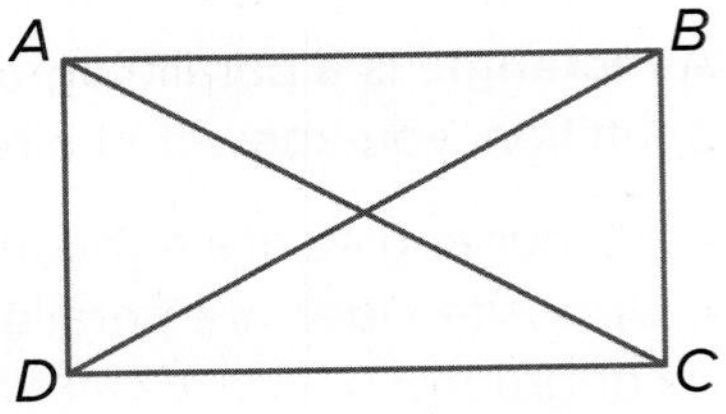

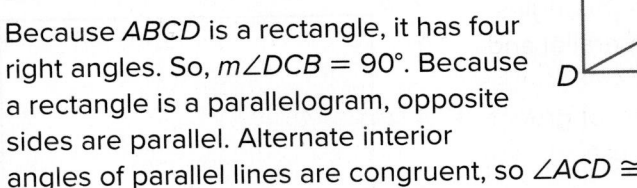

Because *ABCD* is a rectangle, it has four right angles. So, $m\angle DCB = 90°$. Because a rectangle is a parallelogram, opposite sides are parallel. Alternate interior angles of parallel lines are congruent, so $\angle ACD \cong \angle BAC$.

$m\angle ACD + m\angle ACB = 90°$	Angle Addition Postulate
$m\angle BAC + m\angle ACB = 90°$	Substitution
$(3x + 3)° + (5x - 1)° = 90$	Substitution
$x = 11$	Solve.

Check

Quadrilateral *JKLM* is a rectangle.

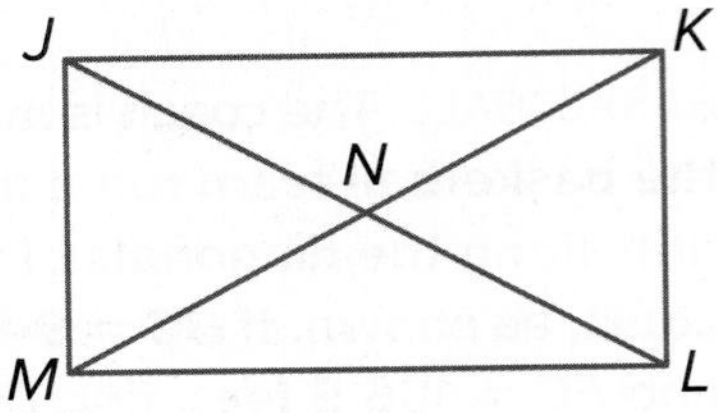

Part A

If $MN = 3x + 1$ and $JL = 2x + 9$, find *MK*. Round to the nearest tenth if necessary.

$MK =$ __?__

Part B

If $m\angle JNK = (5x + 2)°$ and $m\angle JNM = (3x - 6)°$, find each measure.

$m\angle JNK =$ __?__

$m\angle JNM =$ __?__

Go Online You can complete an Extra Example online.

Learn Proving that Parallelograms are Rectangles

You have learned that if a parallelogram is a rectangle, then the diagonals of the parallelogram are congruent. The converse is also true.

Theorem 2.14: Diagonals of a Rectangle

If the diagonals of a parallelogram are congruent, then the parallelogram is a rectangle.

You will prove Theorem 2.14 in Exercise 24.

Example 3 Prove Rectangular Relationships

If $AB = 50$ feet,
$BC = 20$ feet, $CD = 50$ feet,
$AD = 20$ feet, $AC = 54$ feet,
and $BD = 54$ feet, prove
that quadrilateral $ABCD$ is
a rectangle.

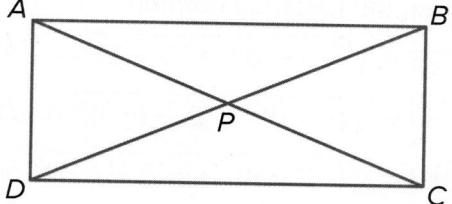

Because $AB = CD$, $BC = AD$, and
$AC = BD$, $\overline{AB} \cong \overline{CD}$, $\overline{BC} \cong \overline{AD}$,
and $\overline{AC} \cong \overline{BD}$. Because $\overline{AB} \cong \overline{CD}$ and $\overline{BC} \cong \overline{AD}$, $ABCD$ is
a parallelogram. Because $\overline{AC}$ and $\overline{BD}$ are congruent diagonals,
$ABCD$ is a rectangle.

Check

Copy and complete the proof with the
correct statements.

Given: $PQRS$ is a rectangle; $\overline{PT} \cong \overline{ST}$.

Prove: $\overline{QT} \cong \overline{RT}$

Proof:

Statements	Reasons
1. $PQRS$ is a rectangle; $\overline{PT} \cong \overline{ST}$.	1. Given
2. $PQRS$ is a parallelogram.	2. Definition of rectangle
3. ___?___	3. Opp. sides of a ▱ are ≅.
4. ___?___	4. Definition of rectangle
5. $\angle S \cong \angle P$	5. All right angles are congruent.
6. ___?___	6. SAS
7. ___?___	7. CPCTC

🔵 **Go Online** You can complete an Extra Example online.

Use a Source

In 1853, the New York State legislature enacted a law to set aside more than 750 acres of land in central Manhattan. This area is now known as Central Park, America's first major landscaped public park. Use available resources to find and use the dimensions of Central Park to prove that it is rectangular.

Statements:
$\angle S$ and $\angle P$ are right $\angle$s.
$\overline{RS} \cong \overline{QP}$
$\overline{QT} \cong \overline{RT}$
$\triangle RST \cong \triangle QPT$

😮 **Think About It!**

Is there another way to show that *GHJK* is a rectangle? If yes, explain.

You can also use the properties of rectangles to prove that a quadrilateral positioned on a coordinate plane is a rectangle given the coordinates of the vertices.

Example 4 Identify Rectangles on the Coordinate Plane

Quadrilateral *GHJK* has vertices *G*(−3, 0), *H*(3, 2), *J*(4, −1), and *K*(−2, −3). Determine whether *GHJK* is a rectangle by using the Distance Formula.

Step 1 Determine whether opposite sides are congruent.
Use the Distance Formula.

$$GH = \sqrt{(-3 - 3)^2 + (0 - 2)^2} \text{ or } \sqrt{40}$$

$$HJ = \sqrt{(3 - 4)^2 + [2 - (-1)]^2} \text{ or } \sqrt{10}$$

$$KJ = \sqrt{(-2 - 4)^2 + [-3 - (-1)]^2} \text{ or } \sqrt{40}$$

$$GK = \sqrt{[-3 - (-2)]^2 + [0 - (-3)]^2} \text{ or } \sqrt{10}$$

Because opposite sides of the quadrilateral have the same measure, they are congruent. So, quadrilateral *GHJK* is a parallelogram.

Step 2 Determine whether diagonals are congruent.
Use the Distance Formula.

$$GJ = \sqrt{(-3 - 4)^2 + [0 - (-1)]^2} \text{ or } \sqrt{50}$$

$$KH = \sqrt{(-2 - 3)^2 + (-3 - 2)^2} \text{ or } \sqrt{50}$$

Because the diagonals have the same measure, they are congruent. So, ▱*GHJK* is a rectangle.

Check

A quadrilateral has vertices *A*(2, 6), *B*(3, 7), and *C*(6, 4). Which of the following points would make *ABCD* a rectangle?

A. *D*(5, 3)

B. *D*(5, 2)

C. *D*(4, 3)

D. *D*(6, 3)

Pause and Reflect

Did you struggle with anything in this lesson? If so, how did you deal with it?

🅑 **Go Online**
You may want to complete the construction activities for this lesson.

🅑 **Go Online** You can complete an Extra Example online.

Practice

Go Online You can complete your homework online.

Example 1

FENCING **X-braces are also used to provide support in rectangular fencing. If _AB_ = 6 feet, _AD_ = 2 feet, and _m∠DAE_ = 65°, find each measure. Round to the nearest tenth, if necessary.**

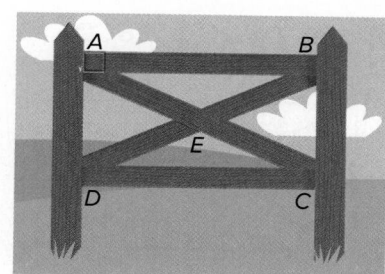

1. _BC_

2. _DB_

3. _m∠CEB_

4. _m∠EDC_

PROM The prom committee is decorating the venue for prom and wants to hang lights above the diagonals of the rectangular room. If _DH_ = 44.5 feet, _EF_ = 39 feet, and _m∠GHF_ = 128°, find each measure.

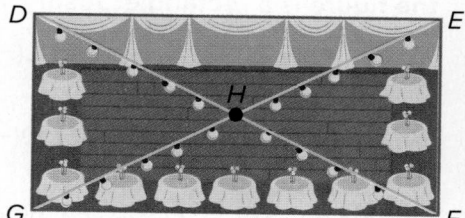

5. _DG_

6. _GE_

7. _m∠EHF_

8. _m∠HEF_

Example 2

9. Quadrilateral _ABCD_ is a rectangle. If _m∠ADB_ = (4x + 8)° and _m∠DBA_ = (6x + 12)°, find the value of _x_.

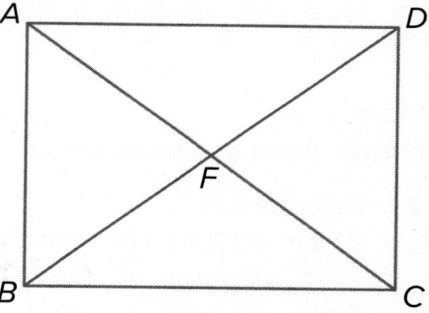

Quadrilateral _EFGH_ is a rectangle. Use the given information to find each measure.

10. If _m∠FEG_ = 57°, find _m∠GEH_.

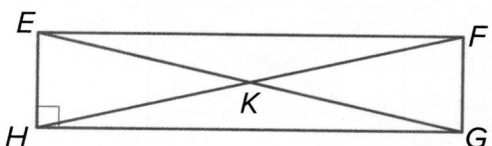

11. If _m∠HGE_ = 13°, find _m∠FGE_.

12. If _FK_ = 32 feet, find _EG_.

13. Find _m∠HEF_ + _m∠EFG_.

14. If _EF_ = 4x − 6 and _HG_ = x + 3, find _EF_.

Example 3

PROOF **Write a two-column proof.**

15. Given: *ABCD* is a rectangle.

  Prove: $\triangle ADC \cong \triangle BCD$

16. Given: *QTVW* is a rectangle, $\overline{QR} \cong \overline{ST}$

  Prove: $\triangle SWQ \cong \triangle RVT$

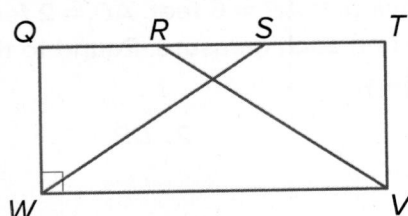

Example 4

PRECISION **Graph each quadrilateral with the given vertices. Determine whether the figure is a rectangle. Justify your answer using the indicated formula.**

17. *B*(−4, 3), *G*(−2, 4), *H*(1, −2), *L*(−1, −3); Slope Formula

18. *N*(−4, 5), *O*(6, 0), *P*(3, −6), *Q*(−7, −1); Distance Formula

19. *C*(0, 5), *D*(4, 7), *E*(5, 4), *F*(1, 2); Slope Formula

20. *P*(−3, −2), *Q*(−4, 2), *R*(2, 4), *S*(3, 0); Slope Formula

21. *J*(−6, 3), *K*(0, 6), *L*(2, 2), *M*(−4, −1); Distance Formula

22. *T*(4, 1), *U*(3, −1), *X*(−3, 2), *Y* (−2, 4); Distance Formula

Mixed Exercises

PROOF **Write a two-column proof to prove each theorem.**

23. Theorem 2.13

  Given: *ABCD* is a rectangle with
    diagonals $\overline{AC}$ and $\overline{BD}$.
  Prove: $\overline{AC} \cong \overline{BD}$

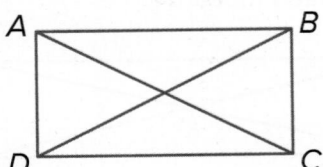

24. Theorem 2.14
  Given: $\overline{PR} \cong \overline{QT}$; *PQRT* is a parallelogram.
  Prove: *PQRT* is a rectangle.

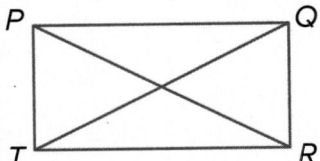

25. LANDSCAPING Huntington Park officials approved a rectangular plot of land for a Japanese Zen garden. Is it sufficient to know that opposite sides of the garden plot are congruent and parallel to determine that the garden plot is rectangular? Explain.

26. Name a property that is true for a rectangle and not always true for a parallelogram.

27. USE TOOLS Construct a rectangle using the construction for congruent segments and the construction for a line perpendicular to another line through a point on the line. Justify each step of the construction.

28. SIGNS The sign is attached to the front of Jackie's lemonade stand. Based on the dimensions given, can Jackie be sure that the sign is a rectangle? Explain your reasoning.

For Exercises 29–30, refer to rectangle WXYZ.

29. If $XW = 3$, and $WZ = 4$, find YW.

30. If $ZY = 6$, and $XY = 8$, find WY.

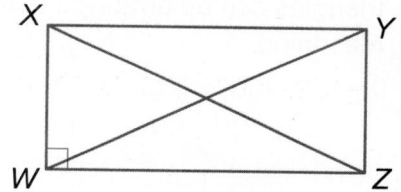

31. FRAMES Jalen makes the rectangular frame shown. Jalen measures the distances BD and AC. How should these two distances compare if the frame is a rectangle?

32. BOOKSHELVES A bookshelf consists of two vertical planks with five horizontal shelves. Are each of the four sections for books rectangles? Explain.

33. REASONING A landscaper is marking off the corners of a rectangular plot of land. Three of the corners are in the place as shown. What are the coordinates of the fourth corner?

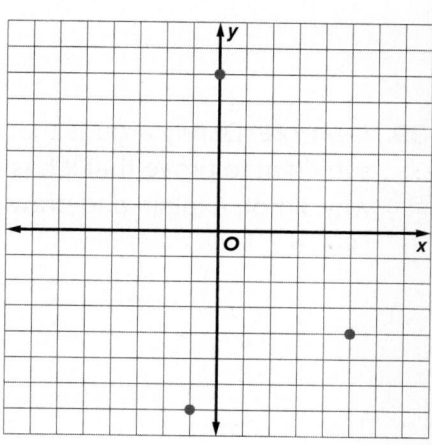

34. STRUCTURE Veronica made the pattern shown out of 7 rectangles with four equal sides. The side length of each rectangle is written inside the rectangle.

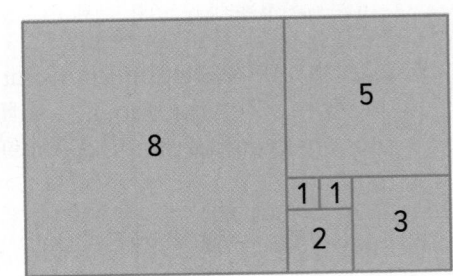

a. How many rectangles can be formed using the lines in this figure?

b. If Veronica wanted to extend her pattern by adding another rectangle with 4 equal sides to make a larger rectangle, what are the possible side lengths of rectangles that she can add?

35. PERSEVERE In rectangle $ABCD$, $m\angle EAB = (4x + 6)°$, $m\angle DEC = (10 - 11y)°$, and $m\angle EBC = 60°$. Find the values of x and y.

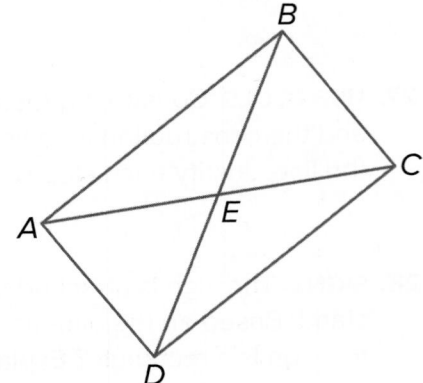

36. FIND THE ERROR Parker says that any two congruent acute triangles can be arranged to make a rectangle. Takeisha says that only two congruent right triangles can be arranged to make a rectangle. Who is correct? Explain your reasoning.

37. WRITE Why are all rectangles parallelograms, but all parallelograms are not rectangles? Explain.

38. CREATE Write the equations of four lines having intersections that form the vertices of a rectangle. Verify your answer using coordinate geometry.

39. ANALYZE Danny argues that to prove a parallelogram is a rectangle, it is sufficient to prove that it has one right angle. Do you agree? If so, explain why. If not, explain and draw a counterexample.

Rhombi and Squares

Explore Properties of Rhombi and Squares

Online Activity Use dynamic geometry software to complete the Explore.

> **INQUIRY** What special properties do rhombi and squares have? ✕

Learn Properties of Rhombi and Squares

A **rhombus** is a parallelogram with all four sides congruent. All of the properties of a parallelogram hold true for a rhombus, in addition to the following two theorems.

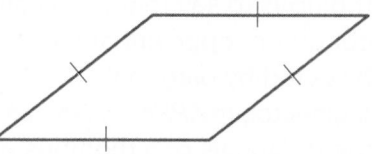

Theorems: Diagonals of a Rhombus
Theorem 2.15
If a parallelogram is a rhombus, then its diagonals are perpendicular.
Theorem 2.16
If a parallelogram is a rhombus, then each diagonal bisects a pair of opposite angles.

You will prove Theorem 2.16 in Exercise 31.

A **square** is a parallelogram with all four sides and all four angles congruent. All of the properties of parallelograms, rectangles, and rhombi apply to squares. For example, the diagonals of a square bisect each other (parallelogram), are congruent (rectangle), and are perpendicular (rhombus).

Example 1 Use the Definition of a Rhombus

If **LM = 2x − 9** and **KN = x + 15** in rhombus **KLMN**, find the value of **x**.

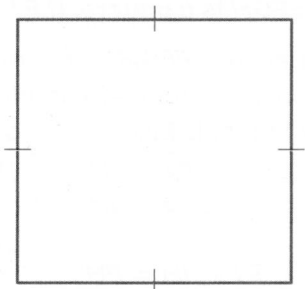

$\overline{LM} \cong \overline{KN}$	Definition of rhombus
$LM = KN$	Definition of congruence
$2x - 9 = x + 15$	Substitution
$x = 24$	Solve.

Go Online You can complete an Extra Example online.

Today's Goals
- Recognize and apply the properties of rhombi and squares.
- Determine whether quadrilaterals are rhombi or squares.

Today's Vocabulary
rhombus
square

Go Online
A proof of Theorem 2.15 is available.

Think About It!
How are the definition of a rhombus and the definition of congruence used to justify the first and second steps?

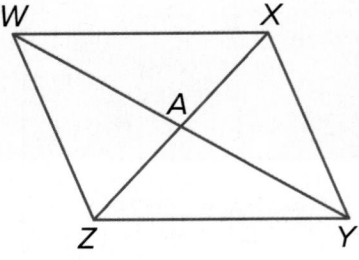

Check

Quadrilateral *WXYZ*
is a rhombus. If *AZ* = 14,
ZY = 22, and *m∠WYZ* = 35°,
find each measure.

XZ = __?__   *m∠XYZ* = __?__   *m∠WXZ* = __?__

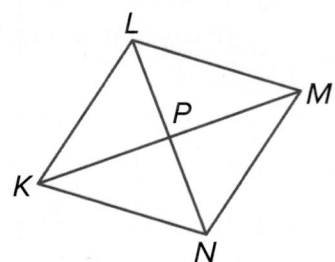

Talk About It!

Compare all of the
properties of the
following quadrilaterals:
parallelograms,
rectangles, rhombi,
and squares.

Example 2 Use the Diagonals of a Rhombus

**The diagonals of rhombus *KLMN*
intersect at *P*. If *m∠LMN* = 75°, find
m∠KNP.**

Because we know that *KLMN* is a
rhombus, we can use the definition of a
rhombus to say that ∠*LKN* and ∠*LMN* are
congruent opposite angles that are
bisected by diagonal $\overline{KM}$. Because $\overline{KM}$ is
a bisector, *m∠PKN* = $\frac{1}{2}$*m∠LKN*. So *m∠PKN* = $\frac{1}{2}$(75°) or 37.5°. Because
the diagonals of a rhombus are perpendicular,
m∠KPN = 90° by the definition of perpendicular lines.

m∠PKN + *m∠KPN* + *m∠KNP* = 180	Triangle Angle-Sum Theorem
37.5 + 90 + *m∠KNP* = 180	Substitution
m∠KNP = 52.5°	Solve.

Example 3 Use the Definition of a Square

EFGH is a square. If *FJ* = 19, find *FH*.

Because *EFGH* is square, it is both a parallelogram
and a rectangle. Therefore, we know that its
diagonals bisect each other and are congruent.

$\overline{FJ} \cong \overline{JH}$	Definition of a square
FJ = *JH*	Definition of congruence
19 = *JH*	Substitution
FJ + *JH* = *FH*	Definition of bisector
19 + 19 = *FH*	Substitution
38 = *FH*	Simplify.

Check

In rhombus *PQRS*, *PQ* = 4*x* + 3, *QR* = 41, and
m∠PQT = (2*x* + 4*y*)°. What must the value of *y*
be for rhombus *PQRS* to be a square?

A. 6.5   **B.** 9.5   **C.** 45   **D.** 90

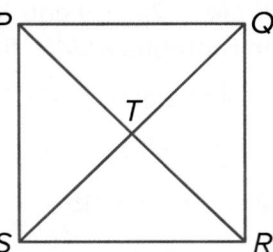

🔁 **Go Online** You can complete an Extra Example online.

Learn Tests for Rhombi and Squares

If a parallelogram meets certain conditions, you can conclude that it is a rhombus or a square.

Theorems: Conditions for Rhombi and Squares
Theorem 2.17
If the diagonals of a parallelogram are perpendicular, then the parallelogram is a rhombus.
Theorem 2.18
If one diagonal of a parallelogram bisects a pair of opposite angles, then the parallelogram is a rhombus.
Theorem 2.19
If two consecutive sides of a parallelogram are congruent, then the parallelogram is a rhombus.
Theorem 2.20
If a quadrilateral is both a rectangle and a rhombus, then it is a square.

You will prove Theorems 2.17, 2.19, and 2.20 in Exercises 32–34.

Go Online A proof of Theorem 2.18 is available.

You can use the properties of rhombi and squares to write proofs.

Example 4 Use Conditions for Rhombi and Squares

Write a paragraph proof.

Given: *TUVW* is a parallelogram.
$\triangle TSW \cong \triangle TSU$

Prove: *TUVW* is a rhombus.

Proof:
Because it is given that $\triangle TSW \cong \triangle TSU$, it must be true that $\overline{WT} \cong \overline{UT}$. Because $\overline{WT}$ and $\overline{UT}$ are congruent, consecutive sides of the given parallelogram, we can prove that *TUVW* is a rhombus by using Theorem 7.19.

🌐 Example 5 Use Properties of a Rhombus

GARMENT DESIGN **Ananya is designing a sweater using an argyle pattern. All four sides of quadrilateral *ABCD* are 2 inches long. How can Ananya be sure that the argyle pattern is a square?**

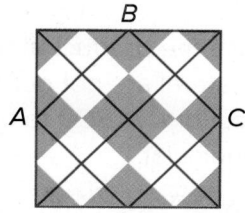

A square has all of the properties of a parallelogram, a rhombus, and a rectangle.
To prove that quadrilateral *ABCD* is a square, prove that it is a parallelogram, a rhombus, and a rectangle.

(continued on the next page)

Go Online You can complete an Extra Example online.

Go Online You may want to complete the Concept Check to check your understanding.

Math History Minute

Robert Ammann (1946–1994) was a programmer who considered himself an amateur mathematician. Although he did not study mathematics in college, Ammann discovered new ways to tile a plane by using quadrilaterals including rhombi. One of the tilings, the Ammann-Beenker tiling, is named for him.

Study Tip

Common Misconceptions
The conditions for rhombi and squares only apply if you already know that a quadrilateral is a parallelogram.

Because both pairs of opposite sides are congruent, *ABCD* is a parallelogram.

Because consecutive sides of ▱*ABCD* are congruent, it is a rhombus.

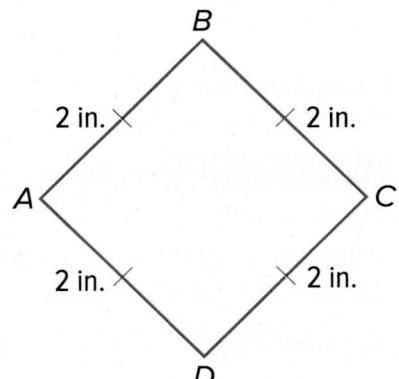

If the diagonals of a parallelogram are congruent, then the parallelogram is a rectangle. So, if Ananya measures the length of each diagonal and finds that they are equal, then *ABCD* is a square.

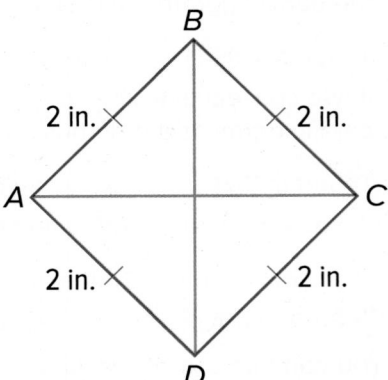

Example 6 Classify Parallelograms by Using Coordinate Geometry

Determine whether ▱*ABCD* with vertices *A*(−3, 2), *B*(−2, 6), *C*(2, 7), and *D*(1, 3) is a *rhombus*, a *rectangle*, a *square*, or *none*. List all that apply. Explain.

Plot and connect the vertices on a coordinate plane.

For ▱*ABCD* to be a rectangle or square, $\overline{AB} \perp \overline{AD}$ must be true.

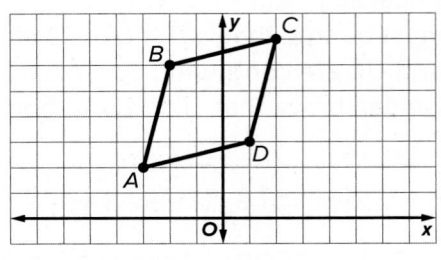

slope of $\overline{AB} = \dfrac{6-2}{-2-(-3)} = \dfrac{4}{1}$ or 4

slope of $\overline{AD} = \dfrac{3-2}{1-(-3)} = \dfrac{1}{4}$

Because the product of the slopes is not −1, the sides of the quadrilateral are not perpendicular. So, ▱*ABCD* is not a rectangle or square. For ▱*ABCD* to be a rhombus, the diagonals must be perpendicular.

slope of $\overline{AC} = \dfrac{7-2}{2-(-3)} = \dfrac{5}{5}$ or 1

slope of $\overline{BD} = \dfrac{3-6}{1-(-2)} = \dfrac{-3}{3}$ or −1

Because the product of the slopes of the diagonals is −1, the diagonals are perpendicular, so ▱*ABCD* is a rhombus.

Go Online You can complete an Extra Example online.

Practice

Go Online You can complete your homework online.

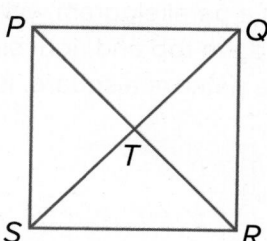

Examples 1 and 2

Quadrilateral *ABCD* is a rhombus. Find each value or measure.

1. If $m\angle ABD = 60°$, find $m\angle BDC$.

2. If $AE = 8$, find AC.

3. If $AB = 26$ and $BD = 20$, find AE.

4. Find $m\angle CEB$.

5. If $m\angle CBD = 58°$, find $m\angle ACB$.

6. If $AE = 3x - 1$ and $AC = 16$, find x.

7. If $m\angle CDB = 6y°$ and $m\angle ACB = (2y + 10)°$, find the value of y.

8. If $AD = 2x + 4$ and $CD = 4x - 4$, find the value of x.

Example 3

9. *PQRS* is a square. If $PR = 42$, find *TR*.

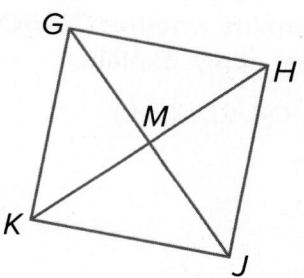

10. *GHJK* is a square. If $KM = 26.5$, find *KH*.

Example 4

Write a two-column proof.

11. Given: *ACDH* and *BCDF* are parallelograms; $\overline{BF} \cong \overline{AB}$.
Prove: *ABFH* is a rhombus.

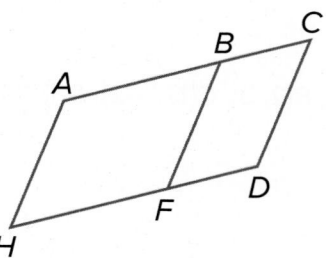

12. Given: *QRST* is a parallelogram; $\overline{TR} \cong \overline{QS}$; $m\angle QPR = 90°$.
Prove: *QRST* is a square.

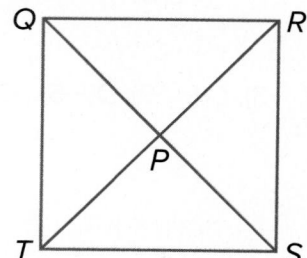

13. Given: $\overline{WZ} \parallel \overline{XY}$, $\overline{WX} \parallel \overline{ZY}$, $\overline{WX} \cong \overline{XY}$
Prove: *WXYZ* is a rhombus.

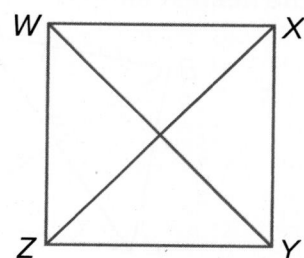

14. Given: *JKQP* is a square.
$\overline{ML}$ bisects $\overline{JP}$ and $\overline{KQ}$.
Prove: *JKLM* is a parallelogram.

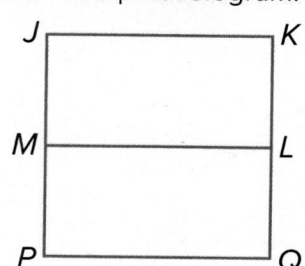

Lesson 2-5 • Rhombi and Squares 95

Example 5

15. PRECISION Jorge is using this box garden to plant his vegetables this year. He knows that the opposite sides of the garden are parallel, so what else does he need to know to ensure that the box garden is a square? Explain.

4 ft

4 ft

16. PRECISION Ingrid is designing a quilt with patches like the one shown. The patch is a parallelogram with all four angles having the same measure and the top and right sides having the same measure. Ingrid says that the patch is a square. Is she correct? Explain.

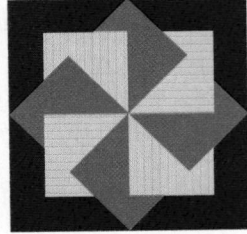

Example 6

REGULARITY Determine whether ▱*ABCD* is a *rhombus*, a *rectangle*, a *square*, or *none*. List all that apply. Explain.

17. *A*(0, 2), *B*(2, 4), *C*(4, 2), *D*(2, 0)

18. *A*(−2, −1), *B*(0, 2), *C*(2, −1), *D*(0, −4)

19. *A*(−6, −1), *B*(4, −6), *C*(2, 5), *D*(−8, 10)

20. *A*(2, −4), *B*(−6, −8), *C*(−10, 2), *D*(−2, 6)

21. *A*(1, 3), *B*(7, −3), *C*(1, −9), *D*(−5, −3)

22. *A*(−9, 1), *B*(2, 3), *C*(12, −2), *D*(1, −4)

Mixed Exercises

BCDF is a square with FD = 55. Find each measure. Round to the nearest tenth, if necessary.

23. *BC*

24. *CD*

25. *GD*

26. *BD*

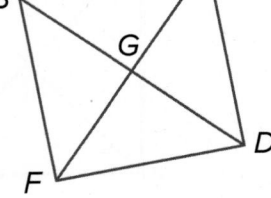

WXYZ is a square. If WT = 3, find each measure.

27. ZX

28. XY

29. m∠WTZ

30. m∠WYX

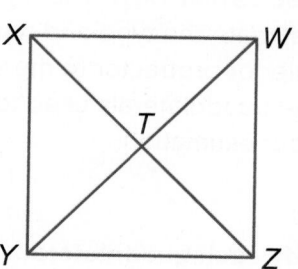

PROOF Write a two-column proof to prove each theorem.

31. Theorem 2.16

Given: *ABCD* is a rhombus.
Prove: $\overline{AC}$ bisects ∠*DAB* and ∠*DCB*.
   $\overline{BD}$ bisects ∠*ABC* and ∠*ADC*.

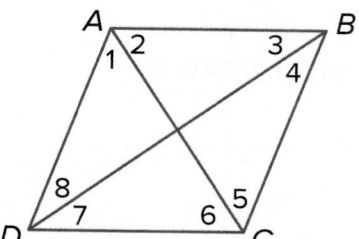

32. Theorem 2.17

Given: *PQRS* is a parallelogram; $\overline{PR} \perp \overline{QS}$.
Prove: *PQRS* is a rhombus.

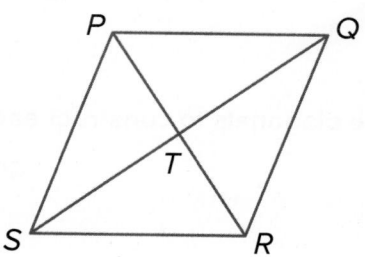

33. Theorem 2.19

Given: *RSTU* is a parallelogram; $\overline{RS} \cong \overline{ST}$.
Prove: *RSTU* is a rhombus.

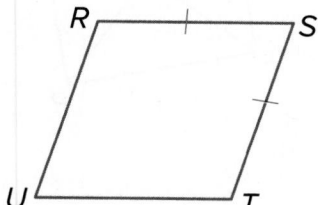

34. Theorem 2.20

Given: *WXYZ* is a rectangle and a rhombus.
Prove: *WXYZ* is a square.

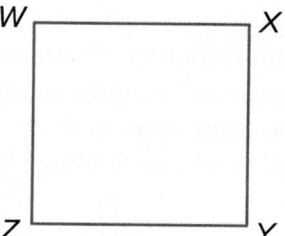

35. USE ESTIMATION The figure is an example of a quilt pattern. Estimate the type and number of shapes in the figure. Use a ruler or protractor to measure the shapes and then name the quadrilaterals used to form the figure. Compare this to your estimation.

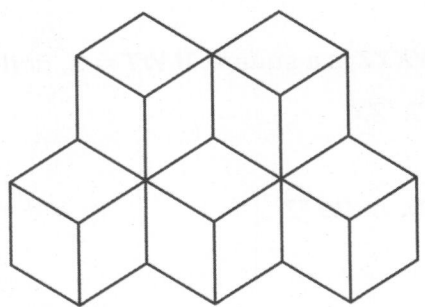

Classify each quadrilateral.

36.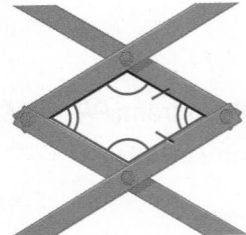

37.

38.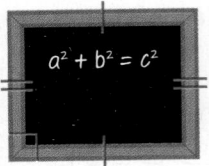

USE TOOLS Use diagonals to construct each figure. Justify each construction.

39. rhombus

40. square

41. CAKE Douglas cuts a rhombus-shaped piece of cake along both diagonals. He ends up with four congruent triangles. Classify these triangles as *acute*, *obtuse*, or *right*.

🧁 **Higher-Order Thinking Skills**

42. PERSEVERE The area of square *ABCD* is 36 square units, and the area of △*EBF* is 20 square units. If $\overline{EB} \perp \overline{BF}$ and *AE* = 2, find *CF*.

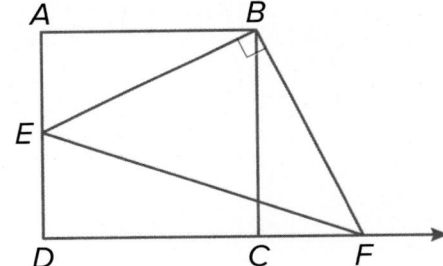

43. WRITE Compare all of the properties of the following quadrilaterals: parallelograms, rectangles, rhombi, and squares.

44. FIND THE ERROR In parallelogram *PQRS*, $\overline{PR} \cong \overline{QS}$. Graciela thinks that the parallelogram is a square, and Xavier thinks that it is a rhombus. Is either of them correct? Explain your reasoning.

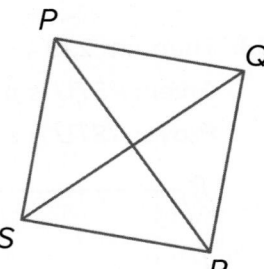

45. ANALYZE Determine whether the statement is *true* or *false*. Then write the converse, inverse, and contrapositive of the statement and determine the truth value of each. Justify your argument.
    If a quadrilateral is a square, then it is a rectangle.

46. CREATE Find the vertices of a square with diagonals that are contained in the graphs of $y = x$ and $y = -x + 6$. Justify your reasoning.

Trapezoids and Kites

Explore Properties of Trapezoids and Kites

🔾 **Online Activity** Use dynamic geometry software to complete the Explore.

❓ **INQUIRY** What special properties do trapezoids and kites have?

Learn Trapezoids

A **trapezoid** is a quadrilateral with exactly one pair of parallel sides. The parallel sides are called **bases of a trapezoid**. The nonparallel sides are called **legs of a trapezoid**. A **base angle of a trapezoid** is formed by a base and one of the legs of the trapezoid. In trapezoid *ABCD*, ∠*A* and ∠*B* are one pair of base angles, and ∠*C* and ∠*D* are the other pair. If the legs of a trapezoid are congruent, then it is an **isosceles trapezoid**.

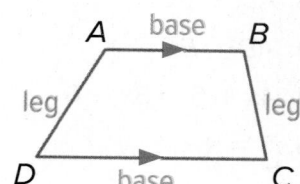

Theorems: Isosceles Trapezoids		
Theorem 2.21		
Words	If a trapezoid is isosceles, then each pair of base angles is congruent.	
Example	If trapezoid *FGHJ* is isosceles with bases $\overline{GH}$ and $\overline{FJ}$, then ∠*G* ≅ ∠*H* and ∠*F* ≅ ∠*J*.	
Theorem 2.22		
Words	If a trapezoid has one pair of congruent base angles, then it is an isosceles trapezoid.	
Example	If ∠*L* ≅ ∠*M*, then trapezoid *KLMP* is isosceles.	
Theorem 2.23		
Words	A trapezoid is isosceles if and only if its diagonals are congruent.	
Example	If trapezoid *QRST* is isosceles, then $\overline{QS} \cong \overline{RT}$. Likewise, if $\overline{QS} \cong \overline{RT}$, then trapezoid *QRST* is isosceles.	

You will prove Theorem 2.22 in Exercise 23.

Today's Goals
• Apply the properties of trapezoids to solve real-world and mathematical problems.
• Apply the properties of trapezoids and use coordinate geometry to find the lengths and endpoints of midsegments.
• Apply the properties of kites to solve real-world and mathematical problems.

Today's Vocabulary
trapezoid
bases of a trapezoid
legs of a trapezoid
base angle of a trapezoid
isosceles trapezoid
midsegment of a trapezoid
kite

🔾 **Go Online**
Proofs of Theorems 2.21 and 2.23 are available.

🌐 Example 1 Use Properties of Isosceles Trapezoids

MUSIC The body of the guitar shown is a trapezoidal prism. The front face of the guitar is an isosceles trapezoid. $AB = 3x - 2$, $CD = 3x + 9$, $AD = 4x + 5$, and $BC = 5x - 6$.

Part A Prove $x = 11$.

Statements	Reasons
1. $ABCD$ is an isosceles trapezoid.	1. Given
2. $\overline{AD} \cong \overline{BC}$	2. Def. of isosceles trapezoid
3. $AD = BC$	3. Def. of congruent segments
4. $4x + 5 = 5x - 6$	4. Substitution
5. $5 = x - 6$	5. Subtraction Prop. of Equality
6. $11 = x$	6. Addition Prop. of Equality
7. $x = 11$	7. Symmetric Prop. of Equality

Part B Find $m\angle A$ if $m\angle C = 72°$.

Because $ABCD$ is an isosceles trapezoid, $\angle C$ and $\angle D$ are congruent base angles. So, $m\angle C = m\angle D = 72°$.

Because $ABCD$ is a trapezoid, $\overline{AB} \parallel \overline{CD}$.

$m\angle A + m\angle D = 180°$ Consecutive Interior Angles Theorem

$m\angle A + 72 = 180$ Substitution

$m\angle A = 108°$ Solve.

Part C Find the perimeter of the front face of the guitar in centimeters.

$P = AB + BC + CD + AD$ Perimeter of trapezoid $ABCD$

$= 3x - 2 + 5x - 6 + 3x + 9 + 4x + 5$ Substitution

$= 15x + 6$ Combine like terms.

$= 15(11) + 6$ $x = 11$

$= 171$ Simplify.

So, the perimeter of the front face of the guitar is 171 centimeters.

Example 2 Isosceles Trapezoids and Coordinate Geometry

Quadrilateral *QRST* has vertices *Q*(−8, −4), *R*(0, 8), *S*(6, 8), and *T*(−6, −10). Show that *QRST* is a trapezoid, and determine whether *QRST* is an isosceles trapezoid.

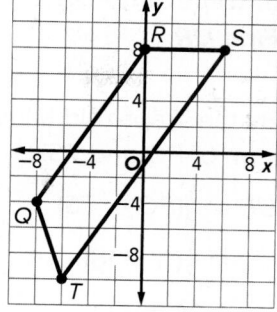

Step 1 Graph quadrilateral *QRST*.

Graph and connect the vertices of *QRST*.

Step 2 Compare the slopes of the opposite sides.

Use the Slope Formula to compare the slopes of opposite sides $\overline{QR}$ and $\overline{ST}$ and opposite sides $\overline{QT}$ and $\overline{RS}$. A quadrilateral is a trapezoid if exactly one pair of opposite sides is parallel.

Opposite sides $\overline{QR}$ and $\overline{ST}$:

slope of $\overline{QR} = \dfrac{8 - (-4)}{0 - (-8)} = \dfrac{12}{8}$ or $\dfrac{3}{2}$

slope of $\overline{ST} = \dfrac{-10 - 8}{-6 - 6} = \dfrac{-18}{-12}$ or $\dfrac{3}{2}$

Because the slopes of $\overline{QR}$ and $\overline{ST}$ are equal, $\overline{QR} \parallel \overline{ST}$.

Opposite sides $\overline{QT}$ and $\overline{RS}$:

slope of $\overline{QT} = \dfrac{-10 - (-4)}{-6 - (-8)} = \dfrac{-6}{2}$ or -3

slope of $\overline{RS} = \dfrac{8 - 8}{6 - 0} = \dfrac{0}{6}$ or 0

Because the slopes of $\overline{QT}$ and $\overline{RS}$ are not equal, $\overline{QT} \nparallel \overline{RS}$. Because quadrilateral *QRST* has only one pair of opposite sides that are parallel, quadrilateral *QRST* is a trapezoid.

Step 3 Compare the lengths of the legs.

Use the Distance Formula to compare the lengths of the legs $\overline{QT}$ and $\overline{RS}$. A trapezoid is isosceles if its legs are congruent.

$QT = \sqrt{[-6 - (-8)]^2 + [-10 - (-4)]^2}$ or $\sqrt{40}$

$RS = \sqrt{(6 - 0)^2 + (8 - 8)^2} = \sqrt{36}$ or 6

Because $QT \neq RS$, legs $\overline{QT}$ and $\overline{RS}$ are *not* congruent. Therefore, trapezoid *QRST* is not isosceles.

Think About It!

What other method could you have used to show that trapezoid *QRST* is not isosceles?

Study Tip

Midsegment The midsegment of a trapezoid can also be called a *median*.

Learn Midsegments of Trapezoids

The **midsegment of a trapezoid** is the segment that connects the midpoints of the legs of the trapezoid.

Theorem 2.24: Trapezoid Midsegment Theorem

The midsegment of a trapezoid is parallel to each base and its length is one half the sum of the lengths of the bases.

Go Online A proof of Theorem 2.24 is available.

Go Online You can complete an Extra Example online.

Example 3 Midsegments of Trapezoids

In the figure, $\overline{UR}$ is the midsegment of trapezoid PQST. Find UR.

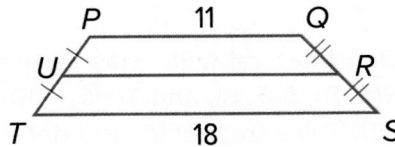

By the Trapezoid Midsegment Theorem, UR is equal to one half the sum of PQ and TS.

$UR = \frac{1}{2}(PQ + TS)$	Trapezoid Midsegment Theorem
$= \frac{1}{2}(11 + 18)$	Substitution
$= 14.5$	Solve.

Example 4 Find Missing Values in Trapezoids

In the figure, $\overline{RN}$ is the midsegment of trapezoid LMPQ. What is the value of x?

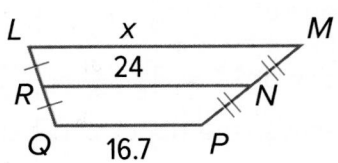

You can use the Trapezoid Midsegment Theorem to write an equation and find the value of x.

$RN = \frac{1}{2}(LM + QP)$	Trapezoid Midsegment Theorem
$24 = \frac{1}{2}(x + 16.7)$	Substitution
$48 = x + 16.7$	Multiply each side by 2.
$31.3 = x$	Solve.

💭 **Think About It!**

If the parallel sides of a trapezoid are contained by the lines $y = x + 4$ and $y = x - 8$, what equation represents the line containing the midsegment?

Example 5 Midsegments and Coordinate Geometry

In trapezoid ABCD, $\overline{AD} \parallel \overline{BC}$. Find the endpoints of the midsegment.

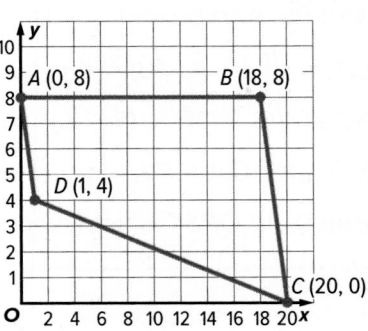

You can use the Midpoint Formula to find the midpoints of $\overline{AB}$ and $\overline{DC}$. These midpoints are the endpoints of the midsegment of trapezoid ABCD.

midpoint of $\overline{AB} = \left(\frac{0 + 18}{2}, \frac{8 + 8}{2}\right) = (9, 8)$

midpoint of $\overline{DC} = \left(\frac{1 + 20}{2}, \frac{4 + 0}{2}\right) = (10.5, 2)$

So, the endpoints of the midsegments are (9, 8) and (10.5, 2).

🅖 **Go Online** You can complete an Extra Example online.

Learn Kites

A **kite** is a convex quadrilateral with exactly two distinct pairs of adjacent congruent sides. Unlike a parallelogram, the opposite sides of a kite are not congruent or parallel.

Theorems: Kites
Theorem 2.25
If a quadrilateral is a kite, then its diagonals are perpendicular.
Theorem 2.26
If a quadrilateral is a kite, then exactly one pair of opposite angles is congruent.

You will prove Theorems 2.25 and 2.26 in Exercises 24 and 25, respectively.

Example 6 Find Angle Measures in Kites

If *KLMN* is a kite, find *m∠N*.

Because a kite can only have one pair of opposite congruent angles and $\angle K \not\cong \angle M$, then $\angle N \cong \angle L$. So, $m\angle N = m\angle L$. You can write and solve an equation to find $m\angle N$.

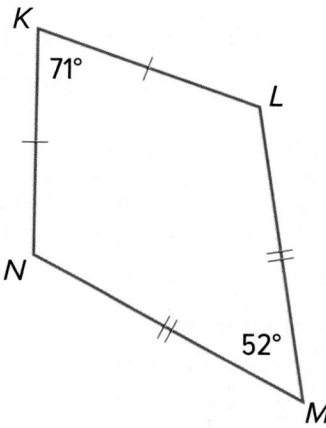

$m\angle K + m\angle L + m\angle M + m\angle N = 360$	Polygon Interior Angles Sum Theorem
$71 + m\angle N + 52 + m\angle N = 360$	Substitution
$2m\angle N + 123 = 360$	Simplify.
$2m\angle N = 237$	Subtract.
$m\angle N = 118.5°$	Divide each side by 2.

Check

If *FGHJ* is a kite, find *m∠F*.

$m\angle F =$ ___?___

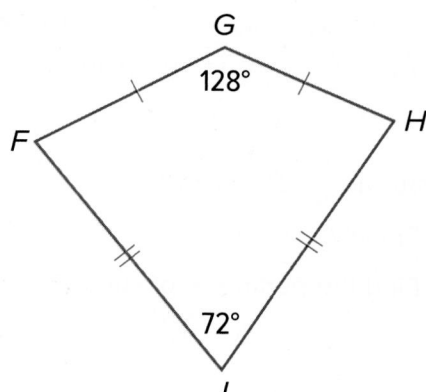

🔵 **Go Online** You can complete an Extra Example online.

Example 7 Find Lengths in Kites

Quadrilateral *ABCD* is a kite.

Part A Find *AD*.

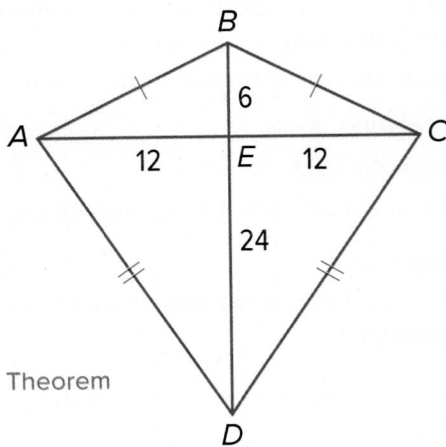

Because the diagonals of a kite are perpendicular, they divide *ABCD* into four right triangles. You can use the Pythagorean Theorem to find *AD*, the length of the hypotenuse of right △*AED*.

$AE^2 + ED^2 = AD^2$ Pythagorean Theorem

$12^2 + 24^2 = AD^2$ Substitution

$144 + 576 = AD^2$ Simplify.

$720 = AD^2$ Simplify.

$\sqrt{720} = AD$ Take the square root of each side.

$12\sqrt{5} = AD$ Simplify.

Part B Find the perimeter of kite *ABCD*.

From the figure, we know $\overline{AB} \cong \overline{BC}$ and $\overline{AD} \cong \overline{CD}$. So, $AB = BC$ and $AD = CD$. We know $AD = 12\sqrt{5}$. So, we can use the Pythagorean Theorem to find *AB*.

$AE^2 + EB^2 = AB^2$ Pythagorean Theorem

$12^2 + 6^2 = AB^2$ Substitution

$144 + 36 = AB^2$ Simplify.

$180 = AB^2$ Simplify.

$\sqrt{180} = AB$ Take the square root of each side.

$6\sqrt{5} = AB$ Simplify.

Use the values of *AB* and *AD* to find the perimeter of kite *ABCD*.

$P = AB + BC + CD + AD$ Perimeter of kite

$= AB + AB + AD + AD$ $AB = BC$ and $AD = CD$

$= 2AB + 2AD$ Simplify.

$= 2(6\sqrt{5}) + 2(12\sqrt{5})$ $AB = 6\sqrt{5}$ and $AD = 12\sqrt{5}$

$= 36\sqrt{5}$ Simplify.

Check

Quadrilateral *ABCD* is a kite.

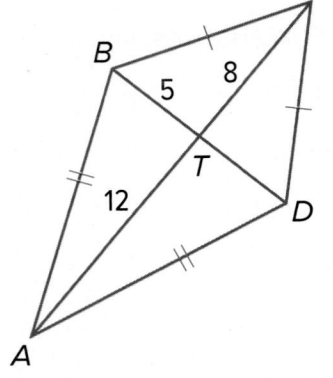

Part A Find *CD*.

Part B Find the perimeter of kite *ABCD*.

 Go Online You can complete an Extra Example online.

💭 Think About It!

How can you find the area of kite *ABCD*? Justify your argument.

🖱 Go Online

to practice what you've learned about types of quadrilaterals in the Put It All Together over Lessons 2-2 through 2-6.

Practice

🔵 **Go Online** You can complete your homework online.

Example 1

1. **SIGNS** The medical sign shown is a trapezoidal prism. The front face of the sign is an isosceles trapezoid. $WX = 2x - 2$, $YZ = 2x + 6$, $WZ = 4x + 5$, and $XY = 5x - 3$.

   a. Prove $x = 8$.

   b. Find $m\angle Z$ if $m\angle W = 106°$.

   c. Find the perimeter of the front face of the sign in inches.

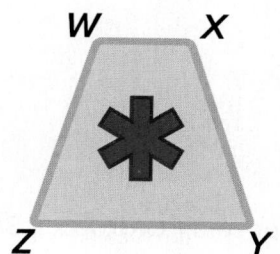

Find each measure.

2. $m\angle T$

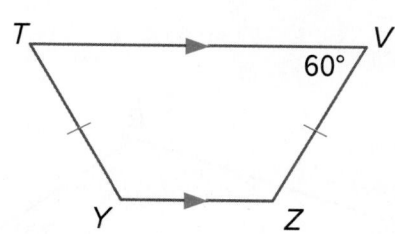

3. $m\angle Y$

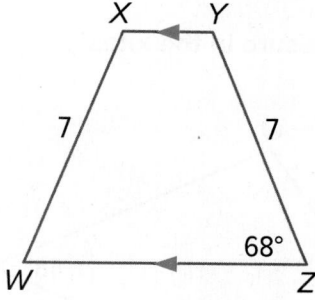

Example 2

4. *RSTU* is a quadrilateral with vertices $R(-3, -3)$, $S(5, 1)$, $T(10, -2)$, and $U(-4, -9)$.

   a. Verify that *RSTU* is a trapezoid.

   b. Is *RSTU* an isosceles trapezoid? Explain.

5. *ABCD* is a quadrilateral with vertices $A(-1, 5)$, $B(3, 2)$, $C(-8, 2)$, and $D(-4, 5)$.

   a. Verify that *ABCD* is a trapezoid.

   b. Is *ABCD* an isosceles trapezoid? Explain.

Examples 3 and 4

$\overline{TS}$ **is the midsegment of trapezoid** *HJKL*.

6. If $HJ = 14$ and $LK = 42$, find TS.

7. If $LK = 19$ and $TS = 15$, find HJ.

8. If $HJ = 7$ and $TS = 10$, find LK.

9. If $KL = 17$ and $JH = 9$, find ST.

10. If $TS = 24$ and $LK = 27.4$, find HJ.

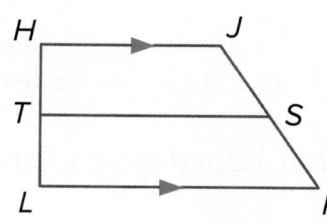

Example 5

11. In trapezoid *ABCD*, $\overline{AD} \parallel \overline{BC}$. Find the endpoints of the midsegment.

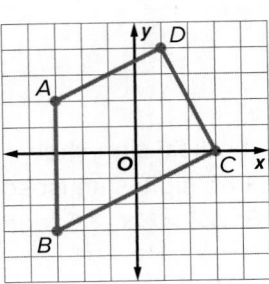

12. In trapezoid *PQRS*, $\overline{PQ} \parallel \overline{SR}$. Find the endpoints of the midsegment.

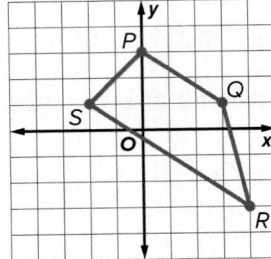

Example 6

Find each measure in the kites.

13. $m\angle Q$

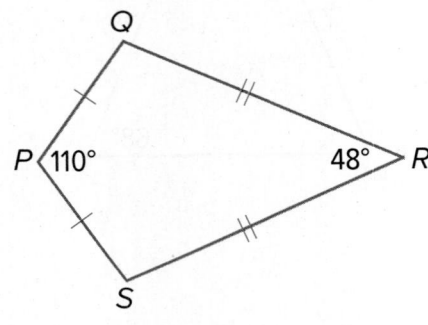

14. $m\angle D$

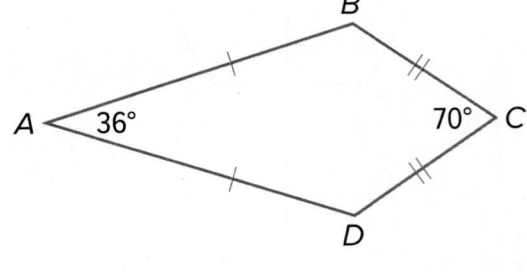

Example 7

15. REASONING Quadrilateral *ABCD* is a kite.

  a. Find *BC*. Write your answer in simplest radical form.

  b. Find the perimeter of kite *ABCD*. Round your answer to the nearest tenth, if necessary.

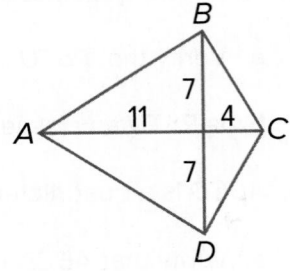

16. REASONING Quadrilateral *HRSE* is a kite.

  a. Find *RH*. Write your answer in simplest radical form.

  b. Find the perimeter of kite *HRSE*. Round your answer to the nearest tenth, if necessary.

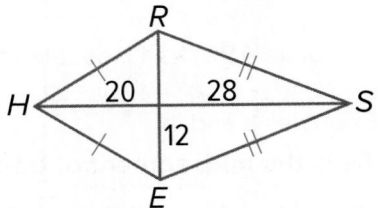

Mixed Exercises

***ABCD* is a trapezoid.**

17. If $AC = 3x - 7$ and $BD = 2x + 8$, find the value of *x* so that *ABCD* is isosceles.

18. If $m\angle ABC = (4x + 11)°$ and $m\angle DAB = (2x + 33)°$, find the value of *x* so that *ABCD* is isosceles.

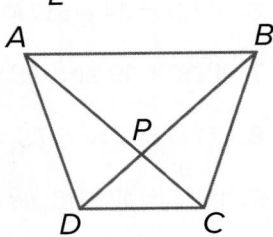

WXYZ is a kite.

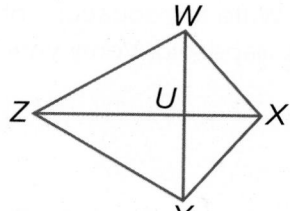

19. If $m\angle WXY = 120°$, $m\angle WZY = (4x)°$, and $m\angle ZWX = (10x)°$, find $m\angle ZYX$.

20. If $m\angle WXY = (13x + 24)°$, $m\angle WZY = 35°$, and $m\angle ZYX = (13x + 14)°$, find $m\angle ZWX$.

21. USE A MODEL A set of stairs leading to the entrance of a building is designed in the shape of an isosceles trapezoid with the longer base at the bottom of the stairs and the shorter base at the top. If the bottom of the stairs is 21 feet wide and the top is 14 feet wide, find the width of the stairs halfway to the top.

22. DESK TOPS A carpenter needs to replace several trapezoid-shaped desktops in a classroom. The carpenter knows the lengths of both bases of the desktop. What other measurements, if any, does the carpenter need?

PROOF **Write a two-column proof to prove each theorem.**

23. Theorem 2.22
  Given: *TUVW* is a trapezoid; $\angle W \cong \angle V$.
  Prove: Trapezoid *TUVW* is isosceles.

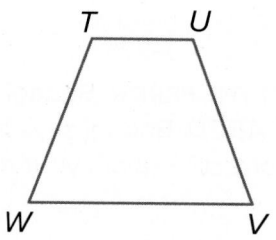

24. Theorem 2.25
  Given: *DEFG* is a kite.
  Prove: $\overline{DF} \perp \overline{EG}$

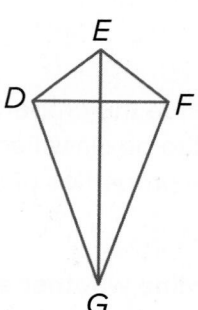

25. PROOF Write a paragraph proof to prove Theorem 2.26.
  Given: *LMNP* is a kite.
  Prove: $\angle M \cong \angle P$; $\angle MLP \not\cong \angle MNP$

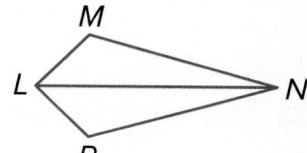

26. USE A SOURCE Go online to research diamond kites.

  a. Find the perimeter of a traditional diamond kite.

  b. Find the area of the kite.

27. **CREATE** Write the equations of four lines that intersect to form the vertices of an isosceles trapezoid. Verify your answer using coordinate geometry.

28. **WHICH ONE DOESN'T BELONG?** The following three characteristics describe all but which of the following quadrilaterals? Justify your conclusion.

- At least one pair of opposite sides is parallel.
- Diagonals are not perpendicular.
- At least one pair of opposite sides is congruent.

| trapezoid | rectangle | square | kite | isosceles trapezoid |

29. **FIND THE ERROR** Bedagi and Belinda are trying to determine $m\angle A$ in kite $ABCD$. Bedagi says $m\angle A = 45°$, and Belinda says $m\angle A = 115°$. Who is correct? Explain your reasoning.

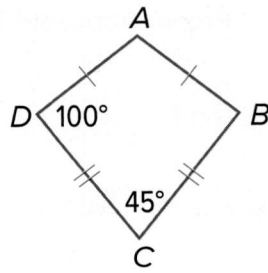

30. **PERSEVERE** $JKLM$ is a kite. $JM = 4y^2 - 4y + 2$, $JK = 3y^2 - 5y + 8$, $ML = 3x^2 - 6x - 10$, and $KL = 2x^2 - 2x - 5$. If $KL > JK$, find the perimeter of $JKLM$.

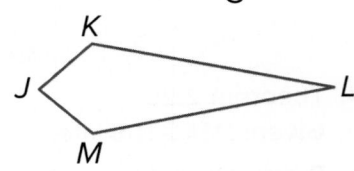

31. **WRITE** Describe the properties that a quadrilateral must possess for the quadrilateral to be classified as a trapezoid, an isosceles trapezoid, or a kite. Compare the properties of all three quadrilaterals.

ANALYZE Determine whether each statement is *sometimes*, *always*, or *never* true. Justify your argument.

32. A square is also a kite.

33. One pair of opposite sides are congruent in a kite.

34. The opposite angles of a trapezoid are supplementary.

Essential Question

What are the different types of quadrilaterals, and how can their characteristics be used to model real-world situations?

Parallelograms, rectangles, rhombi, squares, trapezoids, and kites; You can use these quadrilaterals to model real-world objects, and then you can use what you know about the properties of these shapes to approximate the measures of the real-world objects.

Module Summary

Lesson 2-1

Angles of Polygons

- The sum of the interior angle measures of an n-sided convex polygon is $(n - 2) \cdot 180°$.
- The sum of the exterior angle measures of a convex polygon, one angle at each vertex, is 360°.

Lessons 2-2 and 2-3

Parallelograms

- A parallelogram is a quadrilateral with both pairs of opposite sides parallel.
- In a parallelogram, opposite sides and opposite angles are congruent.
- If the diagonals of a quadrilateral bisect each other, then the quadrilateral is a parallelogram.
- If one pair of opposite sides of a quadrilateral is both parallel and congruent, then the quadrilateral is a parallelogram.

Lesson 2-4

Rectangles

A rectangle has the following properties:

- All four angles are right angles.
- Opposite sides are parallel and congruent.
- Opposite angles are congruent.
- Consecutive angles are supplementary.
- Diagonals bisect each other.

Lesson 2-5

Rhombi and Squares

- A rhombus is a special type of parallelogram with all four sides congruent.
- A square is a special type of parallelogram with all four sides and all four angles congruent.
- If a quadrilateral is both a rectangle and a rhombus, then it is a square.

Lesson 2-6

Trapezoids and Kites

- A trapezoid is a quadrilateral with exactly one pair of parallel sides.
- If a trapezoid is isosceles, then each pair of base angles is congruent.
- The midsegment of a trapezoid is the segment that connects the midpoints of the legs of the trapezoid.
- The midsegment of a trapezoid is parallel to each base and its length is one half the sum of the lengths of the bases.
- A kite is a quadrilateral with exactly two distinct pairs of adjacent congruent sides.

Study Organizer

 **Foldables**

Use your Foldable to review this module. Working with a partner can be helpful. Ask for clarification of concepts as needed.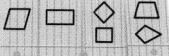

Test Practice

1. MULTIPLE CHOICE A home plate from a baseball field is modeled by the diagram. What is the value of x in degrees? (Lesson 2-1)

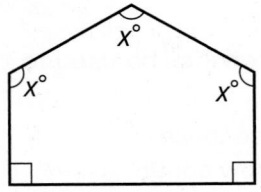

A. 45

B. 90

C. 120

D. 180

2. OPEN RESPONSE Find the measure of $\angle RKL$. (Lesson 2-1)

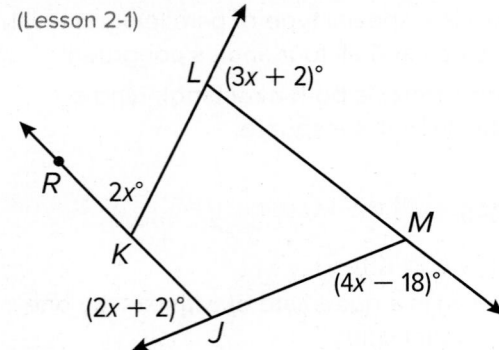

3. MULTIPLE CHOICE A paper fan is made by folding the pattern shown in the diagram.

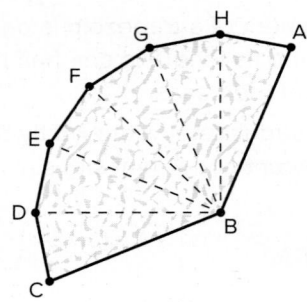

Angles A and C measure 80° and angle B measures 135°. If the remaining angles are congruent to each other, what is the measure of each angle? (Lesson 2-1)

A. 135°   B. 143°   C. 157°   D. 173°

4. OPEN RESPONSE Describe three different methods that you could use to prove that quadrilateral ABCD is a parallelogram.

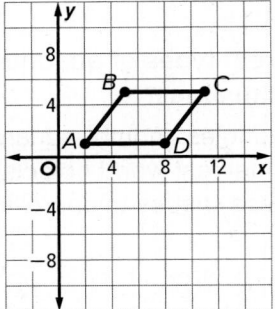

5. OPEN RESPONSE Quadrilateral PQRS is a parallelogram. If $m\angle P = 72°$, then find $m\angle Q$ and $m\angle R$. (Lesson 2-2)

6. OPEN RESPONSE A repeating tile design is made from a rhombus and four congruent parallelograms. (Lesson 2-2)

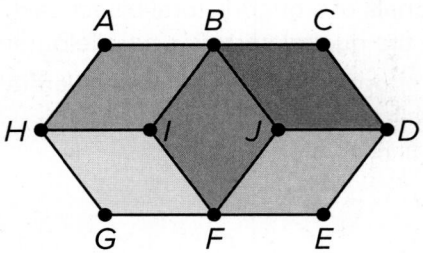

If $m\angle IBJ = 54°$, find each angle measure.

$m\angle BIF =$ ___?___ °

$m\angle JBC =$ ___?___ °

$m\angle BJD =$ ___?___ °

7. MULTIPLE CHOICE Identify which quadrilateral cannot be proven to be a parallelogram. (Lesson 2-3)

A.
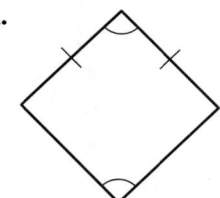

B.
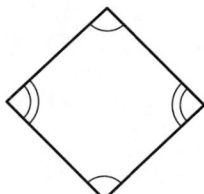

C.
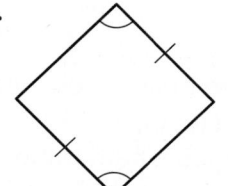

D.
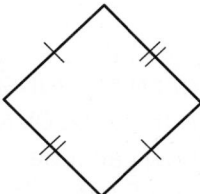

8. OPEN RESPONSE If $\overline{AB}$ and $\overline{AC}$ are two sides of a figure, at which coordinates in Quadrant I should point D be placed so that $ABDC$ is a parallelogram? (Lesson 2-3)

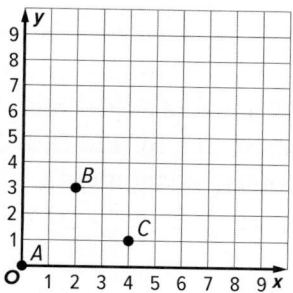

9. MULTIPLE CHOICE Which measurements will ensure that $PQRS$ is a parallelogram? (Lesson 2-3)

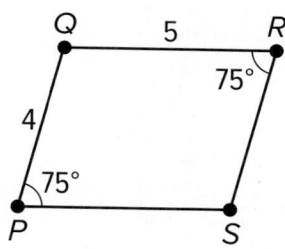

A. $PS = 5$ and $RS = 4$

B. $PS = 4$ and $RS = 5$

C. $PS = 5$ and $m\angle Q = 75°$

D. $PS = 4$ and $m\angle Q = 75°$

10. OPEN RESPONSE Given rectangle $WXYZ$, if $m\angle XZY = 27°$, then find $m\angle WYX$ and $m\angle WVZ$. (Lesson 2-4)

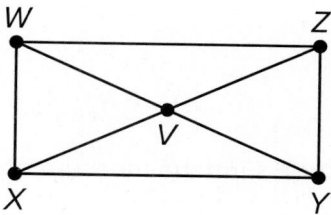

11. MULTIPLE CHOICE A carpenter builds a frame from two 6-foot-long boards and two 8-foot-long boards. Given these side lengths, how can the carpenter ensure that the frame is a rectangle? (Lesson 2-4)

A. If the diagonals are congruent, the frame must be a rectangle.

B. If the opposite sides are congruent, the frame must be a rectangle.

C. If the opposite angles are congruent, the frame must be a rectangle.

D. If the diagonals are perpendicular, the frame must be a rectangle.

12. MULTIPLE CHOICE In parallelogram $ABCD$, $AB = 2x$, $AC = 3x - 2$, and $AD = x + 2$. If the length of segment BD is 10, what value of x will ensure that $ABCD$ is a rectangle? (Lesson 2-4)

A. 2

B. 4

C. 5

D. 8

13. MULTI-SELECT If four bars of equal length are joined at their endpoints, select all the shapes that can be created. (Lesson 2-5)

A. kite

B. parallelogram

C. rectangle

D. rhombus

E. square

F. trapezoid

14. MULTIPLE CHOICE Find the measure of $\angle B$. (Lesson 2-5)

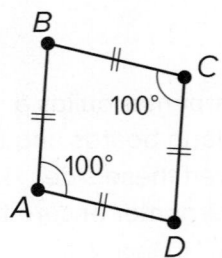

A. 40°

B. 50°

C. 80°

D. 100°

15. MULTI-SELECT If $ABCD$ is a rhombus that is not a square, select all of the true statements. (Lesson 2-5)

A. $\overline{AB} \cong \overline{CD}$

B. $\overline{AC} \perp \overline{BD}$

C. $\angle A \cong \angle C$

D. $\overline{AC} \cong \overline{BD}$

E. $\overline{BC} \cong \overline{DA}$

16. OPEN RESPONSE Given quadrilateral $JKLM$ with $J(-12, 0)$, $K(0, 5)$, $L(6, 0)$, and $M(0, -5)$, how can it be determined whether the quadrilateral is a kite?

17. MULTIPLE CHOICE Trapezoid $ABCD$ has vertices $A(0, 0)$, $B(2, 5)$, $C(3, 5)$, and $D(8, 0)$. What is the length of its midsegment? (Lesson 2-6)

A. 4

B. 4.5

C. 5

D. 5.5

Module 3

Similarity

e Essential Question

What does it mean for objects to be similar, and how is similarity useful for modeling in the real world?

What Will You Learn?

How much do you already know about each topic **before** starting this module?

KEY		Before			After		
— I don't know. — I've heard of it. — I know it!	👎	✊	👍	👎	✊	👍
draw and analyze dilated figures using tools or functions						
solve problems using the definition of similar polygons						
solve problems involving identifying the corresponding parts of similar polygons						
solve problems involving identifying similar polygons based on corresponding sides and angles						
solve problems using the AA Postulate of triangle similarity						
solve problems involving parts of similar triangles						
solve problems using the SSS and SAS Theorems of triangle similarity						
prove geometric theorems using triangle similarity						
use the Converse of the Triangle Proportionality Theorem to determine if lines are parallel						
solve problems and prove relationships using the Triangle Midsegment Theorem and its corollaries						

📖 Foldables Make this Foldable to help you organize your notes about similarity. Begin with four sheets of notebook paper.

1. **Fold** the four sheets of paper in half.

2. **Cut** along the top fold of the papers. Staple along the side to form a book.

3. **Cut** the right side of each paper to create a tab for each lesson.

4. **Label** each tab with a lesson number as shown.

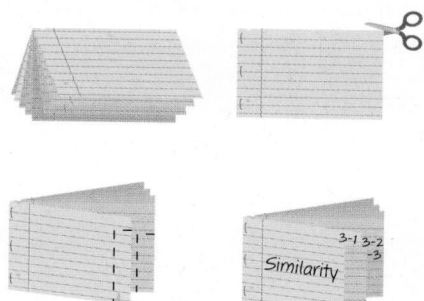

What Vocabulary Will You Learn?

- center of dilation
- dilation
- enlargement
- midsegment of a triangle
- nonrigid motion
- reduction

- scale factor of a dilation
- similar polygons
- similarity ratio
- similarity transformation
- similar triangles

Are You Ready?

Complete the Quick Review to see if you are ready to start this module.
Then complete the Quick Check.

Quick Review

Example 1

Simplify the fraction.

$$\frac{6}{27}$$

$$= \frac{6 \div 3}{27 \div 3}$$ Divide the numerator and denominator by the GCF.

$$= \frac{2}{9}$$ Simplify.

Example 2

Use the scale drawing to find the actual base length and height of the triangle.

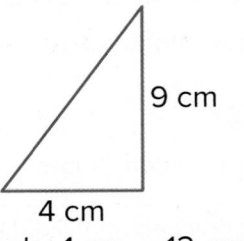

9 cm

4 cm

scale: 1 cm = 12 cm

Multiply the base length in the scale drawing by 12.
actual base length =
$4 \times 12 = 48$ cm
Multiply the height in the scale drawing by 12.
actual height = $9 \times 12 = 108$ cm

Quick Check

Simplify each fraction.

1. $\frac{4}{16}$ 2. $\frac{8}{24}$

3. $\frac{15}{25}$ 4. $\frac{12}{18}$

5. $\frac{36}{45}$ 6. $\frac{10}{12}$

Use the scale drawing to find the actual length and width of the rectangle.

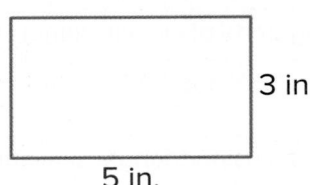

3 in.

5 in.

scale: 1 in. = 8 in.

7. actual length

8. actual width

How Did You Do?

Which exercises did you answer correctly in the Quick Check?

Dilations

Today's Goals
- Use scale factors to calculate the dimensions of dilated images.
- Represent dilations as functions and find the scale factors of dilations.

Today's Vocabulary
nonrigid motion
dilation
center of dilation
scale factor of a dilation
enlargement
reduction

Explore Verifying the Properties of Dilations

 Online Activity Use dynamic geometry software to complete the Explore.

INQUIRY What special properties do dilations have?

Learn Dilations

Recall that a *rigid motion* is an operation that maps an original figure, the *preimage*, onto a new figure called the *image*.

A **nonrigid motion** is a transformation that changes the dimensions of a given figure. A **dilation** is a nonrigid motion that enlarges or reduces a geometric figure. When a figure is enlarged or reduced, the sides of the image are proportional to the sides of the original figure.

The **center of dilation** is the center point from which dilations are performed. The **scale factor of a dilation** is the ratio of a length on an image to a corresponding length on the preimage. The letter k usually represents the scale factor of a dilation.

The value of k determines whether the dilation is an enlargement or a reduction. A dilation with a scale factor greater than 1 produces an **enlargement**, or an image that is larger than the original figure. A dilation with a scale factor between 0 and 1 produces a **reduction**, or an image that is smaller than the original figure.

 Go Online
You may want to complete the Concept Check to check your understanding.

Talk About It!
Does your answer seem reasonable? Explain.

Example 1 Identify a Dilation to Find Scale Factor

Determine whether the dilation from △ABC to △DEF is an *enlargement* or a *reduction*. Then find the scale factor of the dilation.

△DEF is smaller than △ABC, so the dilation is a reduction. The scale factor is equal to the side length of △DEF divided by the corresponding side length of △ABC.

So, the scale factor is $\frac{2}{6}$ or $\frac{1}{3}$.

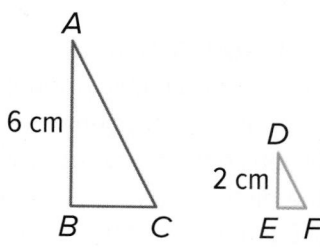

Check

Determine whether the dilation from *WXYZ* to *JKLM* is an *enlargement* or a *reduction*. Then find the scale factor of the dilation.

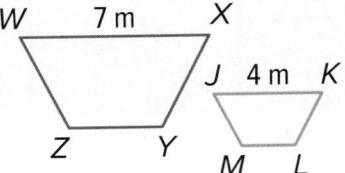

The dilation of *WXYZ* to *JKLM* is a(n) _____?_____ .

The scale factor of the dilation is _____?_____ .

🌐 Example 2 Find and Use a Scale Factor

SCHOOL SPIRIT
Jalal is printing a banner for his school's wheelchair tennis team based on the design shown. By what percent should Jalal enlarge his design so that the dimensions of the banner are 5 times that of the original design? What will be the dimensions of the banner?

Jalal wants to create an enlarged image of his banner design using a commercial printer. The scale factor of his enlargement is 5. Written as a percent, the scale factor is (5 × 100)% or 500%.

Now we can find the dimensions of the enlarged image using the scale factor.

width: 6 in. × 500% = 30 in.   length: 14 in. × 500% = 70 in.

The banner will be 30 inches by 70 inches.

Check

PORTRAITS Natalia wants to print an enlarged family portrait for her mother. By what percent should Natalia enlarge the portrait so that the dimensions of its image are 2.5 times that of the original? What will be the dimensions of the enlarged portrait?

A. 125%; 22.5 cm by 30 cm

B. 125%; 45 cm by 60 cm

C. 250%; 22.5 cm by 30 cm

D. 250%; 45 cm by 60 cm

🔎 **Go Online** You can complete an Extra Example online.

Learn Dilations on the Coordinate Plane

On the coordinate plane, a dilation is a function in which the coordinates of the vertices of a figure are multiplied by the same ratio k.

Key Concept • Dilations on the Coordinate Plane	
Words	To dilate a figure by a scale factor of k with respect to the center of dilation $(0, 0)$, multiply the x- and y-coordinate of each vertex by k.
Symbols	$(x, y) \longrightarrow (kx, ky)$
Example	The image of $\triangle PQR$ dilated by $k = 3$ is $\triangle P'Q'R'$.

🗨 **Think About It!**

What is the relationship between PQ and $P'Q'$? How does this relate to the scale factor?

Study Tip

Center of Dilation
Unless otherwise stated, all dilations on the coordinate plane use the origin as their center of dilation.

Example 3 Dilate a Figure

$\triangle TRS$ has vertices $T(-4, -5)$, $R(0, 6)$, and $S(4, 3)$. Find the coordinates of the vertices of $\triangle T'R'S'$ after a dilation of $\triangle TRS$ by a scale factor of $\frac{1}{2}$.

Because the scale factor is $\frac{1}{2}$, the coordinates of the vertices of $\triangle T'R'S'$ should be half of the value of the coordinates of the vertices of $\triangle TRS$.

Complete the calculations for the dilation when $k = \frac{1}{2}$.

$(x, y) \qquad \longrightarrow \qquad (kx, ky)$

$T(-4, -5) \qquad \longrightarrow \qquad T'\left(\frac{1}{2}(-4), \frac{1}{2}(-5)\right)$ or $T'(-2, -2.5)$

$R(0, 6) \qquad \longrightarrow \qquad R'\left(\frac{1}{2}(0), \frac{1}{2}(6)\right)$ or $R'(0, 3)$

$S(4, 3) \qquad \longrightarrow \qquad S'\left(\frac{1}{2}(4), \frac{1}{2}(3)\right)$ or $S'(2, 1.5)$

Check

$\triangle XYZ$ has vertices $X(3, -4)$, $Y(6, 5)$, and $Z(8, -2)$. Find the coordinates of the vertices of $\triangle X'Y'Z'$ after a dilation of $\triangle XYZ$ by a scale factor of 4.

$X'(\underline{\ ?\ }, \underline{\ ?\ })$ $\qquad Y'(\underline{\ ?\ }, \underline{\ ?\ })$ $\qquad Z'(\underline{\ ?\ }, \underline{\ ?\ })$

🧭 **Go Online** You can complete an Extra Example online.

Example 4 Find the Scale Factor of a Dilation

△A′B′C′ is the image of △ABC after a dilation. Find the scale factor of the dilation.

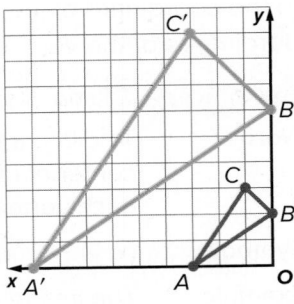

To find the scale factor of the dilation, you must compare the lengths of corresponding sides in △ABC and △A′B′C′.

Step 1 Identify two corresponding sides and their endpoints.

$\overline{AB}$ and $\overline{A'B'}$ are corresponding sides. The endpoints of $\overline{AB}$ are A(−3, 0) and B(0, 2). The endpoints of $\overline{A'B'}$ are A′(−9, 0) and B′(0, 6).

Step 2 Find the lengths of the corresponding sides.

Use the Distance Formula to find AB and A′B′.

A(−3, 0) and B(0, 2)

$$AB = \sqrt{[0 - (-3)]^2 + (2 - 0)^2}$$

$$= \sqrt{3^2 + 2^2}$$

$$= \sqrt{13}$$

A′(−9, 0) and B′(0, 6)

$$A'B' = \sqrt{[0 - (-9)]^2 + (6 - 0)^2}$$

$$= \sqrt{9^2 + 6^2}$$

$$= \sqrt{117}$$

Step 3 Calculate the scale factor.

To find the scale factor, find the ratio of A′B′ to AB.

$AB = \sqrt{13}$ and $A'B' = \sqrt{117}$

$$\frac{A'B'}{AB} = \frac{\sqrt{117}}{\sqrt{13}} = \sqrt{\frac{117}{13}} = \sqrt{9} = 3$$

So, the scale factor of the dilation is 3.

Check

△D′E′F′ is the image of △DEF after a dilation. Find the scale factor of the dilation.

$k = $ _____?_____

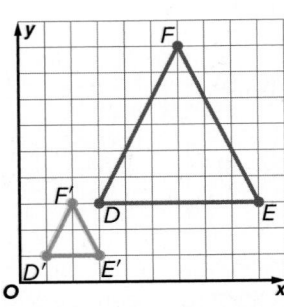

◐ Go Online You can complete an Extra Example online.

Practice

◆ **Go Online** You can complete your homework online.

Example 1

Determine whether the dilation from the figure on the left to the figure on the right is an *enlargement* or a *reduction*. Then find the scale factor of the dilation.

1.

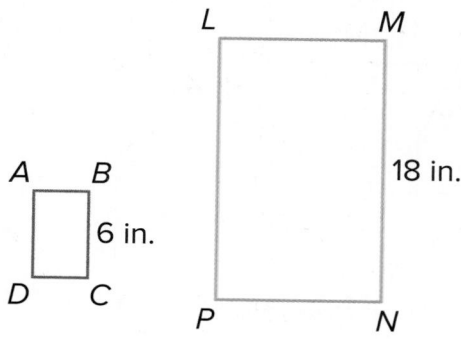

2.

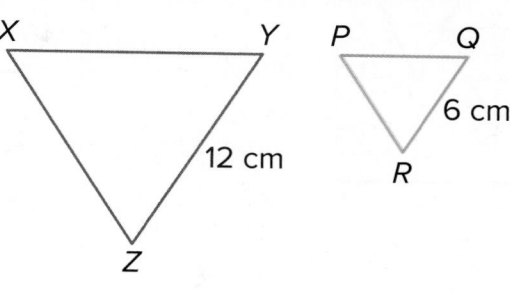

3.

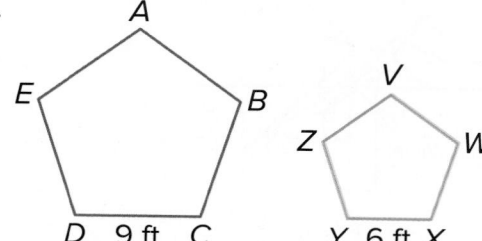

4.

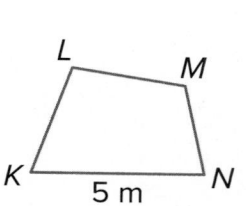 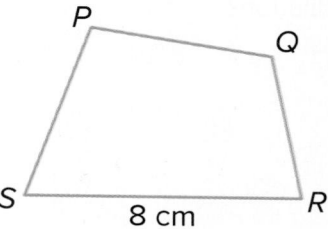

Example 2

5. BLUEPRINTS Ezra is redrawing the blueprint shown of a stage he is planning to build for his band. By what percentage should he multiply the dimensions of the stage so that the dimensions of the image are $\frac{1}{2}$ the size of the original blueprint? What will be the perimeter of the updated blueprint?

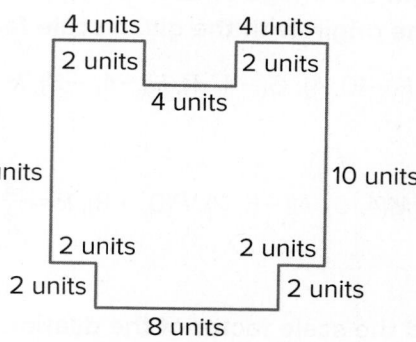

Example 3

For each set of triangle vertices, find and graph the coordinates of the vertices of the image after a dilation of the triangle by the given scale factor.

6. $J(-8, 0)$, $K(-4, 4)$, $L(-2, 0)$, $k = 0.5$

7. $S(0, 0)$, $T(-4, 0)$, $V(-8, -8)$, $k = 1.25$

8. $A(9, 9)$, $B(3, 3)$, $C(6, 0)$, $k = \frac{1}{3}$

9. $D(4, 4)$, $F(0, 0)$, $G(8, 0)$, $k = 0.75$

Example 4

Find the scale factor of the dilation.

10. △J'K'P' is the image of △JKP.

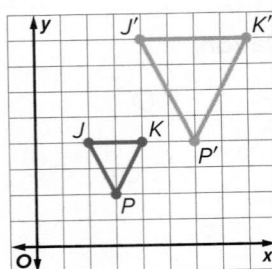

11. △D'F'G' is the image of △DFG.

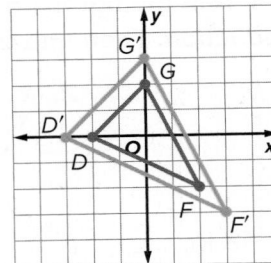

12. Tyrone drew a logo and a dilation of the same logo on the coordinate plane. What is the scale factor of the dilation?

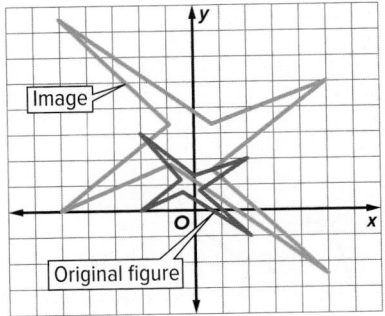

Mixed Exercises

Graph the image of each polygon with the given vertices after a dilation centered at the origin with the given scale factor.

13. F(−10, 4), G(−4, 4), H(−4, −8), k = 0.25

14. X(2, −1), Y(−6, 4), Z(−2, −5), $k = \frac{5}{4}$

15. M(4, 6), N(−6, 2), P(0, −8), $k = \frac{3}{4}$

16. R(−2, 6), S(0, −1), T(−5, 3), k = 1.5

Find the scale factor of the dilation.

17. A'B'C'D' is the image of ABCD.

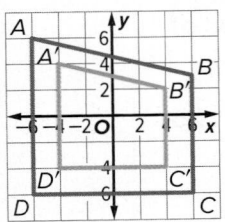

18. △P'Q'R' is the image of △PQR.

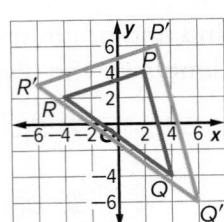

19. Determine whether the dilation from Figure N to N' is an *enlargement* or a *reduction*. Find the scale factor of the dilation.

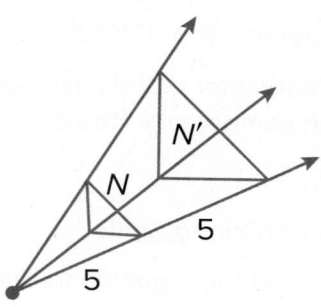

20. $\triangle ABC$ has vertices $A(2, 2)$, $B(3, 4)$, and $C(5, 2)$. What are the coordinates of point C of the image of the triangle after a dilation centered at the origin with a scale factor of 2.5?

21. **USE TOOLS** Use a ruler to draw the image of the figure under a dilation with center M and a scale factor of $\frac{1}{5}$.

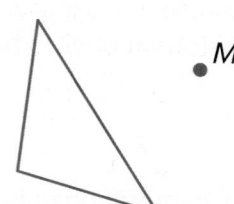

22. Davion is using a coordinate plane to experiment with quadrilaterals, as shown in the figure. Davion creates $M'N'P'Q'$ by enlarging $MNPQ$ with a dilation with a scale factor of 2. Then he creates $M''N''P''Q''$ by dilating $M'N'P'Q'$ with a scale factor of $\frac{1}{3}$. The center of dilation for each dilation is the origin.

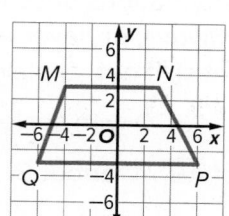

   a. Draw and label the final image, $M''N''P''Q''$.

   b. Can Davion map $MNPQ$ directly to $M''N''P''Q''$ with a single transformation? If so, what transformation should he use?

   c. In general, what can you say about a dilation with scale factor k_1 that is followed by a dilation with scale factor k_2?

23. P' is the image of $P(a, b)$ under a dilation centered at the origin with scale factor $k \neq 1$.

   a. Assuming that P does not lie on the y-axis, what is the slope of $\overleftrightarrow{PP'}$? Explain how you know.

   b. In part **a**, why is it important that P does not lie on the y-axis?

24. *WXYZ* has vertices *W*(6, 2), *X*(3, 7), *Y*(−1, 4), and *Z*(4, −2).

    a. Find the perimeter of *WXYZ*.

    b. Find the perimeter of the image of *WXYZ* after a dilation of $\frac{1}{2}$ centered at the origin and compare it to the perimeter of *WXYZ*.

🌑 Higher-Order Thinking Skills

25. PERSEVERE Find the equation for the dilated image of the line $y = 4x - 2$ if the dilation is centered at the origin with a scale factor of 1.5.

26. WRITE Are parallel lines (parallelism) and collinear points (collinearity) preserved under all transformations? Explain.

27. ANALYZE An *invariant point* is a point that remains fixed under a transformation. For the transformations described below, determine whether there will *sometimes, always,* or *never* exist at least one invariant point. If so, describe the invariant point(s). If not, explain why invariant points are not possible.

    a. dilation of *ABCD* with scale factor of 1

    b. rotation of $\overline{AB}$ 74° about *B*

    c. reflection of △*MNP* in the *x*-axis

    d. translation of *PQRS* along ⟨7, 3⟩

    e. dilation of △*XYZ* centered at the origin with scale factor of 2

28. CREATE Graph a triangle. Dilate the triangle so that its area is four times the area of the original triangle. State the scale factor and center of your dilation.

29. WRITE Can you use transformations to create congruent figures? Explain.

30. ANALYZE Determine whether each statement is *sometimes, always,* or *never* true. Justify your argument.

    a. If *c* is a real number, then a dilation centered at the origin maps the line $y = cx$ to itself.

    b. If $k > 1$, then a dilation with scale factor *k* maps $\overline{AB}$ to a segment that is congruent to $\overline{AB}$.

Similar Polygons

Explore Similarity in Polygons

Online Activity Use graphing technology to complete the Explore.

> ⊘ **INQUIRY** How can you identify whether two polygons are similar?

Learn Similar Polygons

A dilation is a type of similarity transformation. A **similarity transformation** is a transformation composed of a dilation or a dilation and a rigid motion. Two figures are **similar polygons** if one can be obtained from the other by a dilation or a dilation with one or more rigid motions. When two polygons are similar, their angles and sides are related.

Theorem 3.1: Similar Polygons

Two polygons are similar if and only if their corresponding angles are congruent and their corresponding side lengths are proportional.

Like congruence, similarity is reflexive, symmetric, and transitive.

Theorem 3.2: Properties of Similarity

Reflexive Property of Similarity
$\triangle ABC \sim \triangle ABC$

Symmetric Property of Similarity
If $\triangle ABC \sim \triangle DEF$, then $\triangle DEF \sim \triangle ABC$.

Transitive Property of Similarity
If $\triangle ABC \sim \triangle DEF$ and $\triangle DEF \sim \triangle XYZ$, then $\triangle ABC \sim \triangle XYZ$.

Go Online Proofs of Theorems 3.1 and 3.2 are available.

Example 1 Use a Similarity Statement

ABCD ~ PQRS. List all pairs of congruent angles, and write a proportion that relates the corresponding sides.

Use the similarity statement.

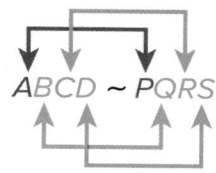

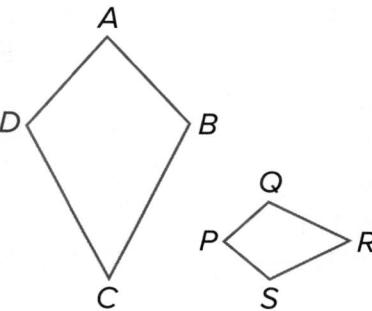

Congruent angles: $\angle A \cong \angle P$, $\angle B \cong \angle Q$, $\angle C \cong \angle R$, $\angle D \cong \angle S$

Proportion: $\dfrac{AB}{PQ} = \dfrac{BC}{QR} = \dfrac{CD}{RS} = \dfrac{DA}{SP}$

Today's Goals
- Determine whether two figures are similar.
- Visualize, describe, and solve problems using the perimeters of similar polygons.

Today's Vocabulary
similarity transformation

similar polygons

similarity ratio

> 💬 **Talk About It!**
> If two polygons are congruent, are they also similar? Justify your argument.

Check

NPQR ~ UVST. List all pairs of congruent angles, and write a proportion that relates the corresponding sides.

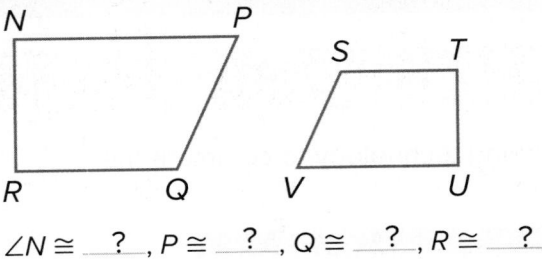

$\angle N \cong$ ___?___ , $P \cong$ ___?___ , $Q \cong$ ___?___ , $R \cong$ ___?___

Example 2 Identify Similar Polygons

Determine whether △*NQP* is similar to △*RST*. If so, find the scale factor. Explain your reasoning.

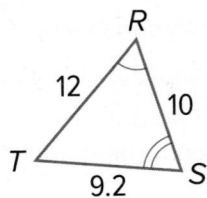

Step 1 Compare the corresponding angles.

Because $\angle N \cong \angle R$ and $\angle Q \cong \angle S$, by the Third Angles Theorem, $\angle P \cong \angle T$. So, the corresponding angles are congruent.

Step 2 Compare the corresponding sides.

$$\frac{NQ}{RS} = \frac{12.5}{10} \text{ or } \frac{5}{4} \qquad \frac{QP}{ST} = \frac{11.5}{9.2} \text{ or } \frac{5}{4} \qquad \frac{PN}{TR} = \frac{15}{12} \text{ or } \frac{5}{4}$$

Because the corresponding angles are congruent and the corresponding sides are proportional, △*NQP* ~ △*RST*. So, the triangles are similar with a scale factor from △*RST* to △*NQP* of $\frac{5}{4}$.

Check

Determine whether quadrilateral *ABCD* is similar to quadrilateral *EFGH*. If so, find the scale factor. Explain your reasoning.

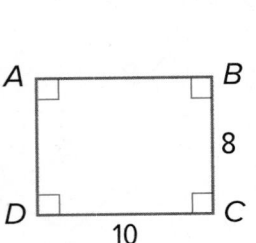
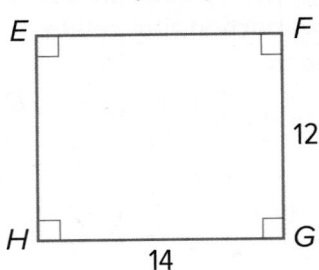

Go Online You can complete an Extra Example online.

Think About It!
What transformations can be used to create △*NQP* from △*RST*?

Example 3 Use Similar Figures to Find Missing Measures

In the diagram, *WXYZ* ~ *PQRS*. Find the value of *y*.

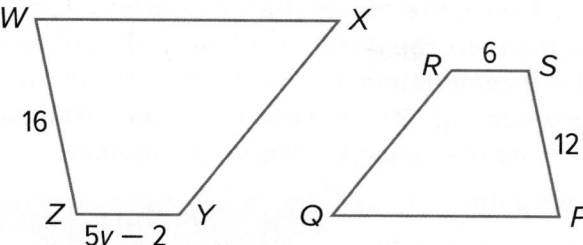

Use the corresponding side lengths to write a proportion.

$\dfrac{WZ}{PS} = \dfrac{YZ}{RS}$	Similarity proportion
$\dfrac{16}{12} = \dfrac{5y - 2}{6}$	Substitute.
$16(6) = 12(5y - 2)$	Multiplication Property of Equality
$96 = 60y - 24$	Multiply.
$120 = 60y$	Add 24 to each side.
$2 = y$	Divide each side by 60.

Check

In the diagram, $\triangle JLM \sim \triangle QST$. Find the value of *x*. Round to the nearest tenth, if necessary.

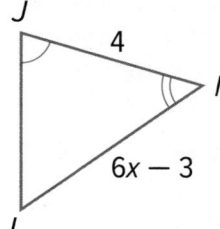 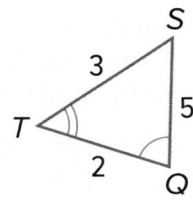

$x =$ _____?_____

🐦 **Go Online** You can complete an Extra Example online.

Learn Perimeters of Similar Polygons

In similar polygons, the ratio of any two corresponding lengths is equal to the scale factor or **similarity ratio** between them. So, you can write a proportion to relate them. This leads to the following theorem about the perimeters of two similar polygons.

> **Theorem 3.3: Perimeters of Similar Polygons Theorem**
>
> If two polygons are similar, then their perimeters are proportional in the same ratio as the scale factor between them.

You will prove Theorem 3.3 in Exercise 21.

Problem-Solving Tip

Redraw Diagrams
When solving problems that use similar figures, sometimes it is difficult to identify corresponding parts. Redraw the given diagram so the similar figures have the same orientation. This will allow you to easily compare corresponding parts and set up a similarity proportion.

👁 **Think About It!**

Will any two regular polygons with the same number of sides be similar? Justify your argument.

🌐 Example 4 Use Similar Polygons to Find Perimeter

STATE FAIR **Geoffrey plans on going to the state fair this summer. He has downloaded a map of the fairgrounds that shows all of the attractions. Geoffrey plans to visit the concert hall, the Ferris wheel, and the sports center. On the map, the distance between the concert hall and the Ferris wheel is 9 centimeters, the distance between the Ferris wheel and the sports center is 8 centimeters, and the distance between the sports center and the concert hall is 4 centimeters.**

Part A Describe Geoffrey's path.

Geoffrey wants to visit the concert hall, the Ferris wheel, and the sports center. If Geoffrey returns to the concert hall after visiting the sports center, what polygon can be used to model Geoffrey's path between the three attractions?

Geoffrey's path can be modeled by a triangle.

On the map, draw line segments between the three attractions. Then label the points that represent the three attractions. Label the concert hall as C, the sports center as S, and the Ferris wheel as F.

Part B Find the total distance.

If the actual distance between the concert hall and the sports center is 20 meters, how far will Geoffrey have to travel to visit all three attractions and then return to his starting point?

$\triangle CSF$ will be similar to the triangle formed by Geoffrey as he walks to the three attractions. Let's call the figure formed by Geoffrey's path $\triangle C'S'F'$. So, $\triangle CSF \sim \triangle C'S'F'$. To find how far Geoffrey will have to travel to visit all three attractions, find the perimeter of $\triangle C'S'F'$.

Because 1 meter = 100 centimeters, 20 meters is equal to 2000 centimeters. So, $C'S' = 2000$ centimeters.

The scale factor of $\triangle CSF$ to $\triangle C'S'F'$ is $\frac{C'S'}{CS} = \frac{2000}{4}$ or 500.

The perimeter of $\triangle CSF$ is $4 + 8 + 9$ or 21.

Use the perimeter of $\triangle CSF$ and the scale factor to write a proportion. Let w represent the perimeter of $\triangle C'S'F'$.

$$\frac{1}{500} = \frac{\text{perimeter of } \triangle CSF}{\text{perimeter of } \triangle C'S'F'}$$

$$\frac{1}{500} = \frac{21}{w}$$

$$w = 500(21)$$

$$w = 10{,}500$$

So, the perimeter of $\triangle C'S'F'$ is 10,500 centimeters or 105 meters.

🌑 **Go Online** You can complete an Extra Example online.

🍩 Think About It!

What assumption did you make while solving this problem?

Study Tip

Units of Measure
When finding a scale factor, the measurements being compared must have the same unit of measure. If they do not have the same unit of measure, you will need to convert one of the measurements.

Practice

Go Online You can complete your homework online.

Example 1

List all pairs of congruent angles, and write a proportion that relates the corresponding sides for each pair of similar polygons.

1. ABCD ~ WXYZ

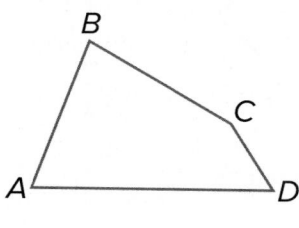 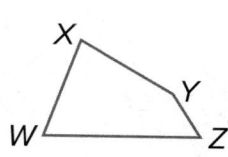

2. MNPQ ~ RSTU

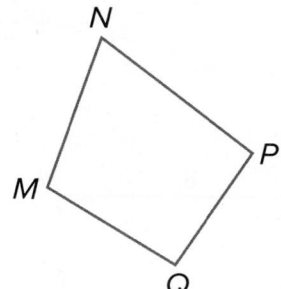 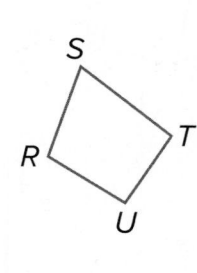

3. △FGH ~ △JKL

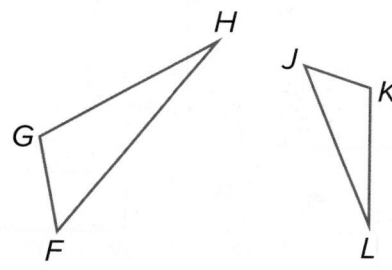

4. △DEF ~ △VWX

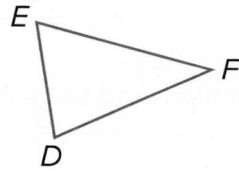 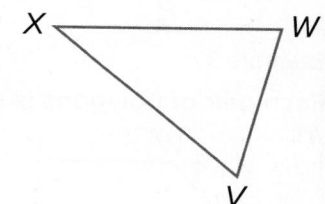

5. ABCD ~ FGHJ

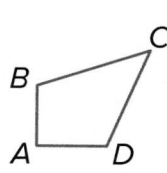 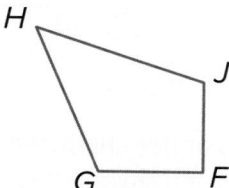

6. △MNP ~ △QRP

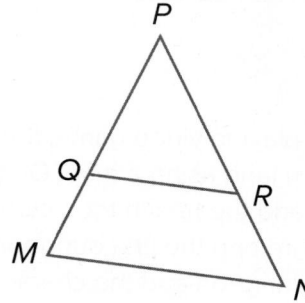

Example 2

Determine whether each pair of figures is similar. If so, find the scale factor. Explain your reasoning.

7.

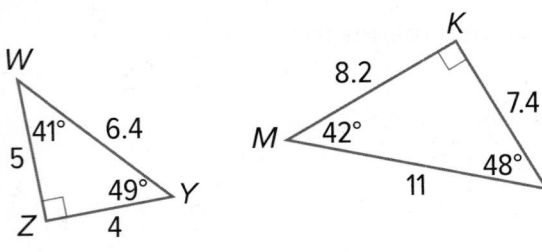

8.

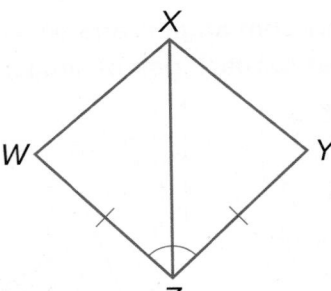

9.

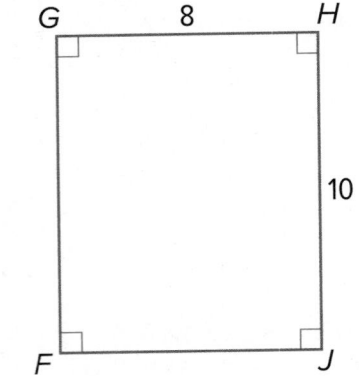

10.

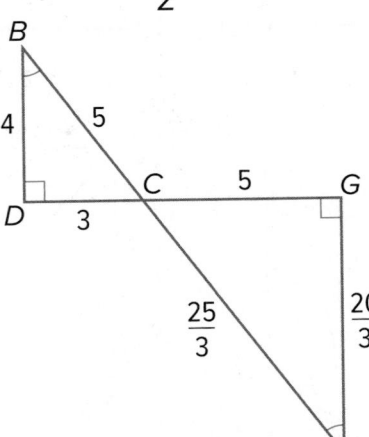

Example 3

Each pair of polygons is similar. Find the value of x.

11.

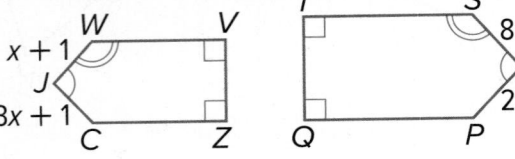

12.

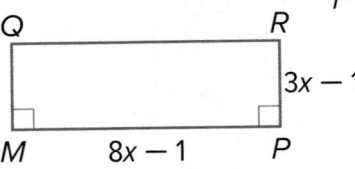

13.

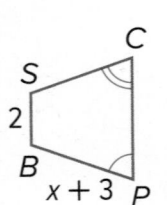 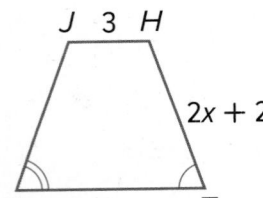

14.

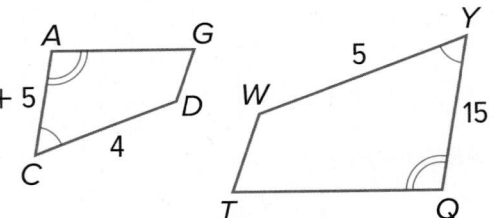

Example 4

15. GAMING In a role-playing video game, the user navigates his or her character across an unknown land using a map. On the map, the distance between the character's home and the health food store is 2 centimeters, the distance between the health food store and the first dungeon is 8 centimeters, and the distance between the first dungeon and the character's home is 7 centimeters. If the actual distance between the health food store and the first dungeon is 4 kilometers, how far will the character have to travel in the game to visit all three destinations and return to their starting point?

16. LAWN CARE Nayeli's rectangular lawn has a perimeter of 126 meters. The ratio of the length of the lawn to the width of the lawn is 5:2. What is the width of Nayeli's lawn?

Mixed Exercises

Find the perimeter of each triangle.

17. △CBH, if △CBH ∼ △FEH, ADEG is a parallelogram, CH = 7, FH = 10, FE = 11, and EH = 6

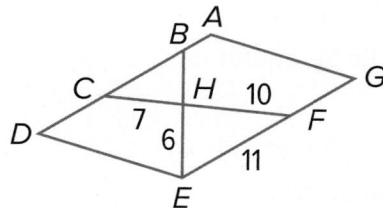

18. △DEF, if △DEF ∼ △CBF, perimeter of △CBF = 27, DF = 6, FC = 8

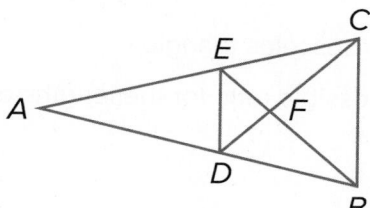

Find the value of x and y for each pair of polygons.

19. ABCD ∼ QSRP

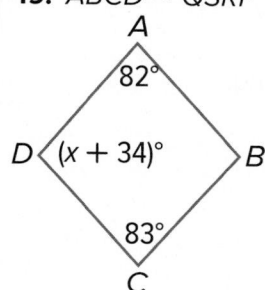

20. △JKL ∼ △WYZ

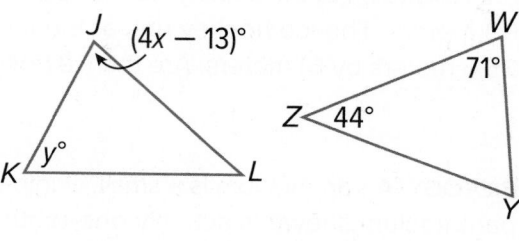

21. PROOF Write a paragraph proof of Theorem 8.3.

Given: △ABC ∼ △DEF and $\frac{AB}{DE} = \frac{m}{n}$

Prove: $\frac{\text{perimeter of } \triangle ABC}{\text{perimeter of } \triangle DEF} = \frac{m}{n}$

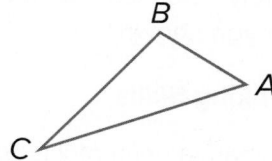
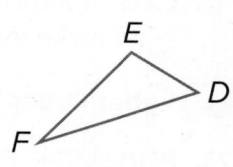

22. If $\triangle ABC \sim \triangle DEC$, find the value of x and the scale factor from $\triangle DEC$ to $\triangle ABC$.

23. Rectangle $ABCD \sim$ rectangle $EFGH$, the perimeter of $ABCD$ is 54 centimeters, and the perimeter of $EFGH$ is 36 centimeters. What is the scale factor from $EFGH$ to $ABCD$?

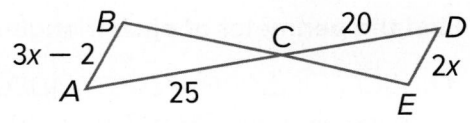

24. $\triangle ABC$ is an isosceles triangle.

  a. Write a possible ratio for the lengths of the sides of $\triangle ABC$ if its perimeter is 42 inches.

  b. Name possible measures for the sides of $\triangle ABC$ using your answer to part **a.**

  c. If $\triangle WXY$ is also isosceles and has a perimeter of 28 and $\triangle ABC$ has sides with the measures you gave in part **b**, what must be the measure of the sides of $\triangle WXY$ so that $\triangle WXY \sim \triangle ABC$?

25. ICE HOCKEY An official Olympic-sized ice hockey rink measures 30 meters by 60 meters. The ice hockey rink at the local community college measures 25.5 meters by 51 meters. Are the ice hockey rinks similar? Explain your reasoning.

26. BIOLOGY A paramecium is a small, single-cell organism. The magnified paramecium shown is actually one-tenth of a millimeter long.

  a. If you want to make a photograph of the original paramecium so that its image is 1 centimeter long, by what scale factor should you magnify it?

  b. If you want to make a photograph of the original paramecium so that its image is 15 centimeters long, by what scale factor should you magnify it?

  c. By approximately what scale factor has the paramecium been enlarged to make the image shown?

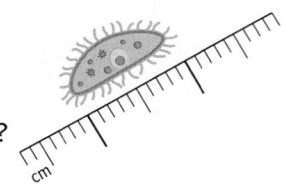

Higher-Order Thinking Skills

27. PERSEVERE For what value(s) of x is $BEFA \sim EDCB$?

28. ANALYZE Recall that an *equivalence relation* is any relationship that satisfies the Reflexive, Symmetric, and Transitive Properties. Is similarity an equivalence relation? Explain.

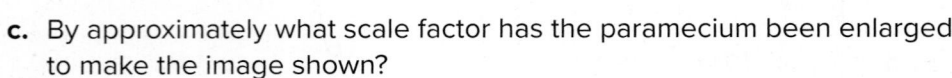

29. CREATE Find a counterexample of the following statement.
  All rectangles are similar.

30. ANALYZE Draw two regular pentagons of different sizes. Are the pentagons similar? Will any two regular polygons with the same number of sides be similar? Explain.

31. WRITE How can you describe the relationship between two figures?

Similar Triangles: AA Similarity

Explore Similarity Transformations and Triangles

Online Activity Use dynamic geometry software to complete the Explore.

> **INQUIRY** How can you determine whether two triangles are similar?

Explore Conditions that Prove Triangles Similar

Online Activity Use dynamic geometry software to complete the Explore.

> **INQUIRY** What shortcut can be used to determine whether two triangles are similar?

Learn Similar Triangles: AA Similarity

In **similar triangles**, all of the corresponding angles are congruent and all of the corresponding sides are proportional. However, you don't need to show that all of the criteria are met to show that two triangles are similar. Angle-Angle Similarity is one of several shortcuts.

Postulate 3.1: Angle-Angle (AA) Similarity

If two angles of one triangle are congruent to two angles of another triangle, then the triangles are similar.

Example 1 Use the AA Similarity Postulate

Determine whether the triangles are similar. Explain your reasoning.

$\angle L \cong \angle L$ by the Reflexive Property of Congruence.

$\angle LPQ \cong \angle LJK$ by the Corresponding Angles Theorem. By AA Similarity, $\triangle KLJ \sim \triangle QLP$.

Check

Determine whether the triangles are similar. Explain your reasoning.

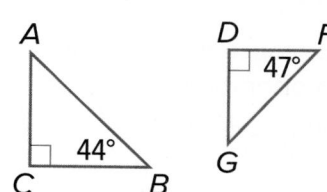

Go Online You can complete an Extra Example online.

🌐 Example 2 Use Parts of Similar Triangles

HANDBALL Demarco is teaching Taye how to play handball. Taye prefers to return the ball on a serve when it bounces to a height of 42 inches. When Demarco serves the ball, where should he aim for the ball to hit the front wall to ensure that it bounces at the short line and up to Taye standing 11 feet behind the short line? Assume that the angles formed by the path of the bouncing handball are congruent.

Create and Describe the Model

- The ball should bounce at the short line, 25 feet from the front wall.

- The ball should bounce to Taye, who is standing 11 feet from the short line.

- Taye will hit the ball when it bounces to a height of 42 inches.

- Find the point on the front wall x where the ball bounces.

- You can model the path of the handball using two triangles.

- You are given that the angles formed by the bouncing handball are congruent. The front wall forms a 90° angle with the ground, and the height to which the ball bounces forms a 90° angle with the ground. Therefore, the two triangles are similar by the AA Similarity Postulate.

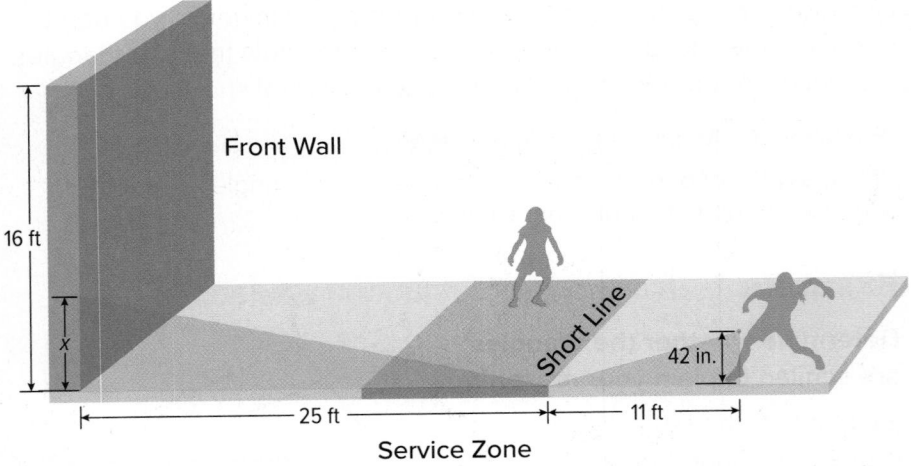

Solve

Because the two triangles are similar, the corresponding sides are proportional.

$$\frac{11 \text{ ft}}{25 \text{ ft}} = \frac{42 \text{ in.}}{x} \qquad\qquad \text{AA Similarity Postulate}$$

$$\frac{11 \text{ ft}}{25 \text{ ft}} = \frac{3.5 \text{ ft}}{x} \qquad\qquad \text{12 inches} = 1 \text{ foot}$$

$$x = 7.95 \text{ ft} \qquad\qquad \text{Solve.}$$

 Go Online You can complete an Extra Example online.

Think About It!

What assumptions did you make when creating your model for the path of the handball?

Watch Out!

Units Remember to convert inches to feet when you solve for x in the example.

Practice

Go Online You can complete your homework online.

Example 1

Determine whether each pair of triangles is similar. Explain your reasoning.

1.
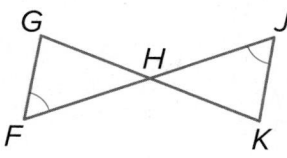

2.
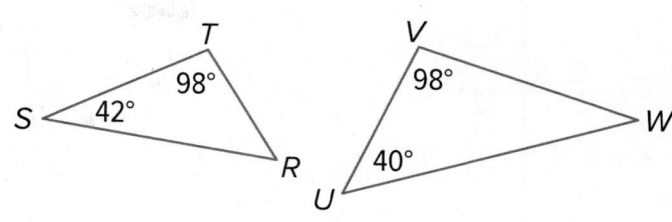

3.
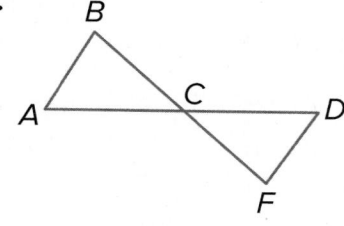

4.
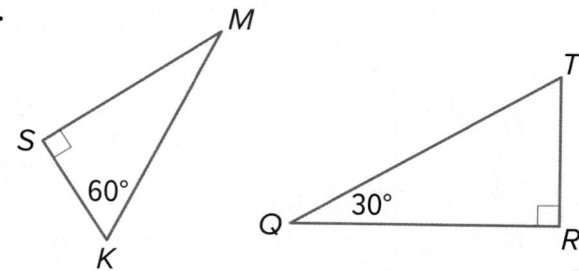

5.
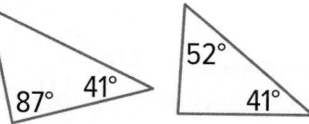

6.
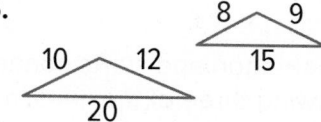

Example 2

7. **CELL TOWERS** A cell phone tower casts a shadow that is 100 feet long. At the same time, Lia stands near the tower and casts a shadow that is 3 feet 4 inches long. If Lia is 4 feet 6 inches tall, how tall is the cell phone tower?

8. **LIGHTHOUSE** Maya wants to know how far she is standing from a lighthouse. The end of Maya's shadow coincides with the end of the lighthouse's shadow.
   a. What is the distance from the lighthouse to the end of the lighthouse's shadow, *x*?
   b. What is the distance from Maya to the lighthouse, *y*?

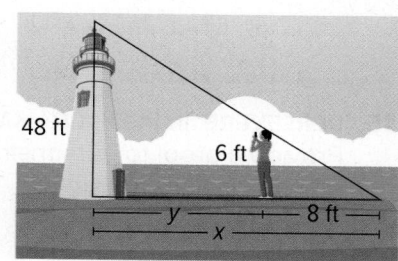

Mixed Exercises

Identify the similar triangles. Then find each measure.

9. *AC*

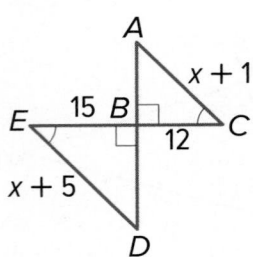

10. *JL*

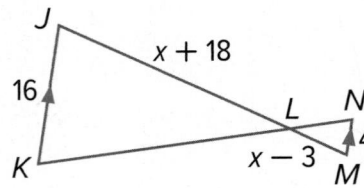

11. *EH*

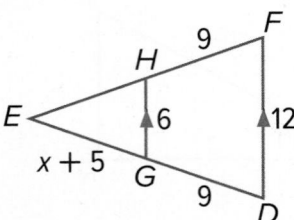

12. *VT*

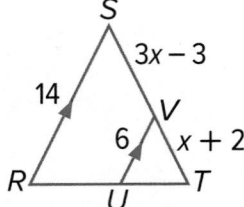

13. Olivia draws a regular pentagon and starts connecting its vertices to make a 5-pointed star. After drawing three of the lines in the star, she becomes curious about two triangles that appear in the figure, △*ABC* and △*CEB*. They look similar to her. Prove that this is the case.

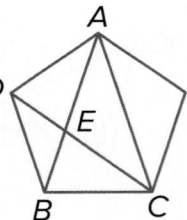

🫧 **Higher-Order Thinking Skills**

14. ANALYZE Write as many triangle similarity statements as possible for the figure shown. How do you know that these triangles are similar?

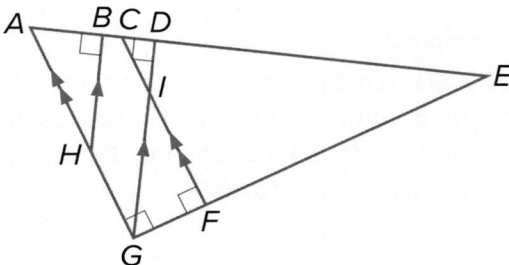

15. PERSEVERE In the figure, $\overline{KM} \perp \overline{JL}$ and $\overline{JK} \perp \overline{KL}$. Is △*JKL* ~ △*JMK*? Provide a proof to demonstrate their similarity or give an explanation of why they are not similar.

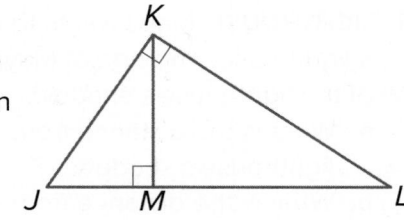

Similar Triangles: SSS and SAS Similarity

Explore Similarity Criteria: SSS, SAS, and SSA

🔵 **Online Activity** Use dynamic geometry software to complete the Explore.

> ❓ **INQUIRY** What shortcuts can be used to determine whether two triangles are similar?   ×

Learn Similar Triangles: SSS and SAS Similarity

You can use the AA Similarity Postulate to prove the following two theorems.

Theorem 3.4: Side-Side-Side (SSS) Similarity	
Words	If the corresponding side lengths of two triangles are proportional, then the triangles are similar.
Example	If $\frac{JK}{MP} = \frac{KL}{PQ} = \frac{LJ}{QM}$, then $\triangle JKL \sim \triangle MPQ$. 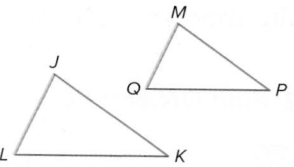

Theorem 3.5: Side-Angle-Side (SAS) Similarity	
Words	If the lengths of two sides of one triangle are proportional to the lengths of two corresponding sides of another triangle and the included angles are congruent, then the triangles are similar.
Example	If $\frac{RS}{XY} = \frac{ST}{YZ}$ and $\angle S \cong \angle Y$, then $\triangle RST \sim \triangle XYZ$. 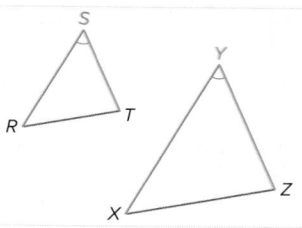

You will prove Theorem 3.5 in Exercise 14.

Example 1 Use the SSS and SAS Similarity Theorems

Determine whether the triangles are similar. Explain your reasoning.

$\frac{JL}{QM} = \frac{12}{9}$ or $\frac{4}{3}$

$\frac{LK}{MP} = \frac{8}{6}$ or $\frac{4}{3}$

$\frac{JK}{QP} = \frac{16}{12}$ or $\frac{4}{3}$

By the SSS Similarity Theorem, $\triangle JLK \sim \triangle QMP$.

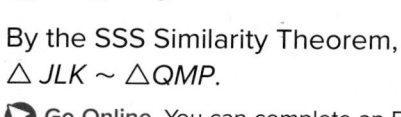 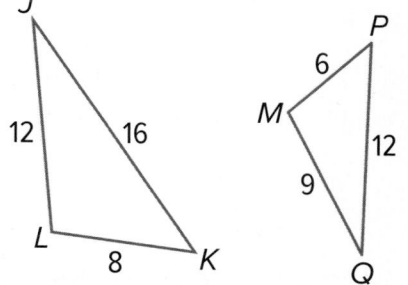

🔵 **Go Online** You can complete an Extra Example online.

Today's Goals
- Use the SSS and SAS Similarity criteria to solve problems and prove triangles similar.

🔵 **Go Online**
A proof of Theorem 3.4 is available.

Study Tip

Corresponding Sides To determine which sides of two triangles correspond, begin by comparing the longest sides, then the next longest sides, and finish by comparing the shortest sides.

Check

Determine whether the triangles are similar. Explain your reasoning.

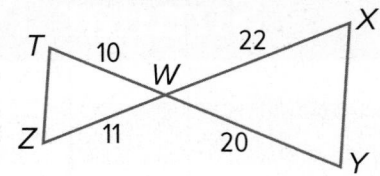

Example 2 Parts of Similar Triangles

Find QN and PO. Justify your answer.

Step 1 Show that △NQM ~ △OPM.

Because $\frac{MP}{MQ} = \frac{8}{5}$ and $\frac{MO}{MN} = \frac{9\frac{3}{5}}{6}$ or $\frac{8}{5}$, the lengths of two sides of △NQM are proportional to the lengths of the corresponding sides of △OPM. By the Reflexive Property of Congruence, $\angle M \cong \angle M$. So, by the SAS Similarity Theorem, △NQM ~ △OPM.

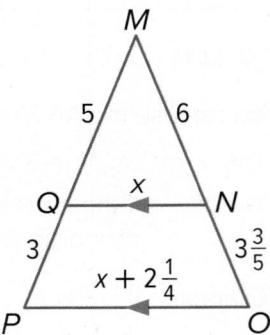

Step 2 Find QN and PO.

$\dfrac{MP}{MQ} = \dfrac{PO}{QN}$	Definition of similar polygons
$\dfrac{8}{5} = \dfrac{x + 2\frac{1}{4}}{x}$	Substitution
$8x = 5x + 11\frac{1}{4}$	Multiplication Property of Equality
$x = 3\frac{3}{4}$	Solve for x.
$QN = 3\frac{3}{4}$	Substitution
$PO = 3\frac{3}{4} + 2\frac{1}{4}$ or 6	Solve.

Check

Find *WR* and *RT*.

WR = _____?_____

RT = _____?_____

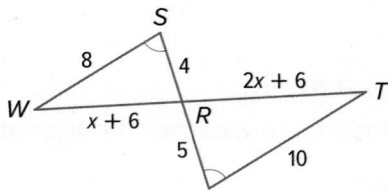

Go Online An alternate method is available for this example.

🖱 **Go Online** You can complete an Extra Example online.

Example 3 Use Similar Triangles to Solve Problems

ARCHITECTURAL DESIGN Julia is designing an A-frame house. The entire house will be 40 feet tall and the base of the house will be 60 feet long. She will build a second-floor balcony along the front of the house, 15 feet above the ground. The left side of the house will be 50 feet long and the balcony will intersect the side 18.75 feet from the bottom. The peak of the house is directly above the midpoints of the base of the house and the balcony. Calculate the total length of the balcony.

Step 1 Draw a diagram.

Step 2 Create a model.

You can model the side of the house with two triangles. Use the information that you know to label the triangles.

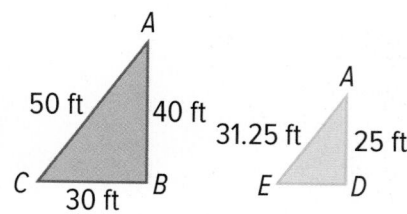

Step 3 Describe the model.

$\frac{AC}{AE} = \frac{50}{31.25}$ or 1.6

$\frac{AB}{AD} = \frac{40}{25}$ or 1.6

By the Reflexive Property of Congruence, $\angle A \cong \angle A$. Therefore, the two triangles are similar by the SAS Similarity Theorem.

Step 4 Solve.

Because the two triangles are similar, the corresponding sides are proportional.

$\frac{50 \text{ ft}}{31.25 \text{ ft}} = \frac{30 \text{ ft}}{ED}$ Definition of similar polygons

$ED = 18.75$ ft   Solve.

D is the midpoint of the balcony, so the total length of the balcony is 2(18.75) or 37.5 feet.

Go Online
to practice what you've learned about similar triangles in the Put It All Together over Lessons 3-3 through 3-4.

Go Online
to learn how to prove the slope criteria in Expand 3-4.

Check

TENNIS Justin is playing tennis. When serving, he stands 12 feet away from the net, which is 3 feet tall. The ball is served from a height of 7.5 feet. Justin thinks the ball travels about 21.4 feet before it hits the ground 8 feet from the net on the opposite side.

Part A

How far does the ball travel before it reaches the net? Round your answer to the nearest tenth, if necessary.

_____?_____ ft

Part B

What assumptions did you make to solve for the distance that the ball travels?

Pause and Reflect

Did you struggle with anything in this lesson? If so, how did you deal with it?

Practice

Go Online You can complete your homework online.

Example 1

Determine whether each pair of triangles is similar. Explain your reasoning.

1.

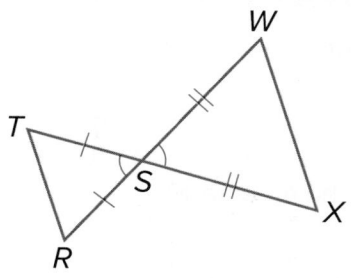

2.

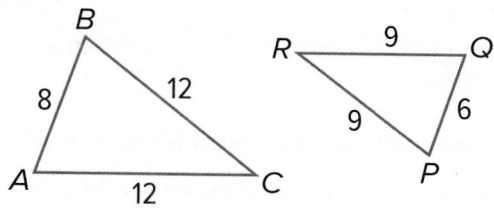

3.

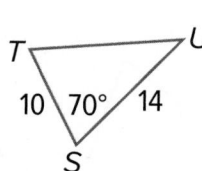

4.

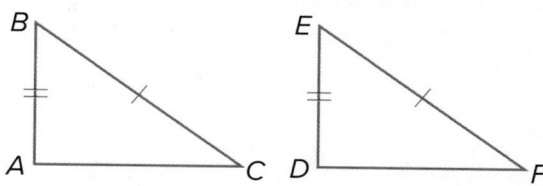

Example 2

Identify the similar triangles. Then find the value of x.

5.

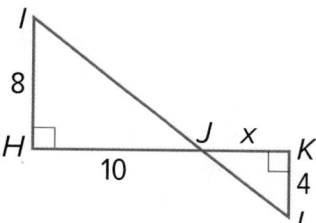

6.

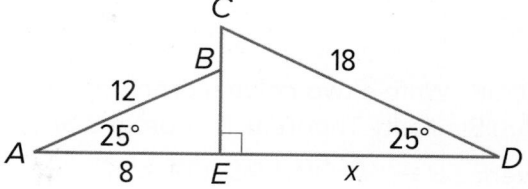

7.

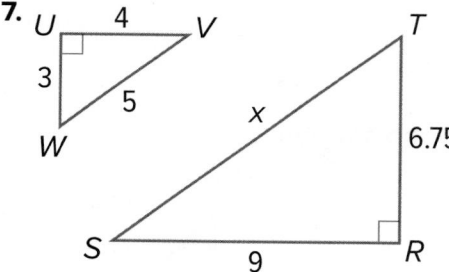

8.

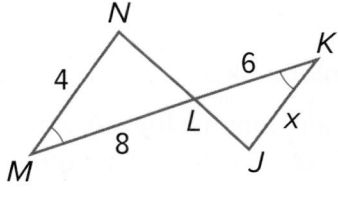

Example 3

9. ROOFING The skeleton of a roof is shown. Find the value of x such that triangles *DEF* and *FBC* in the outline of the roof are similar.

10. RADIO A radio tower casts an 8-foot-long shadow at the same time that a vertical yardstick casts a shadow one half inch long. If the triangles formed by the objects and their shadows are similar, how tall is the radio tower?

11. SAILING The two sailboats shown are participating in a regatta. If the sails are similar, what is the value of x?

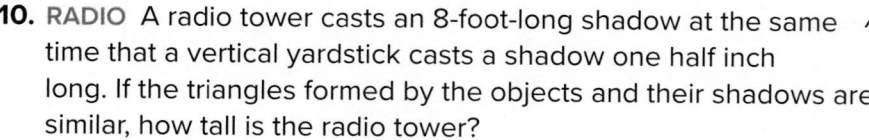

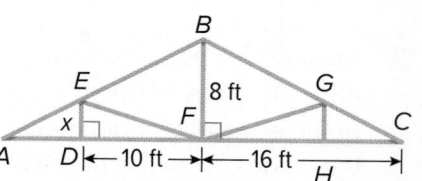

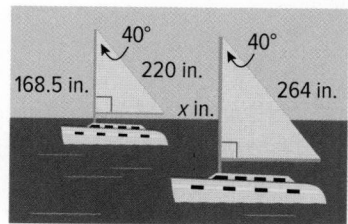

12. MOUNTAIN PEAKS Marcus and Skye want to estimate how far a mountain peak is from their houses. After taking some measurements, they construct a diagram.

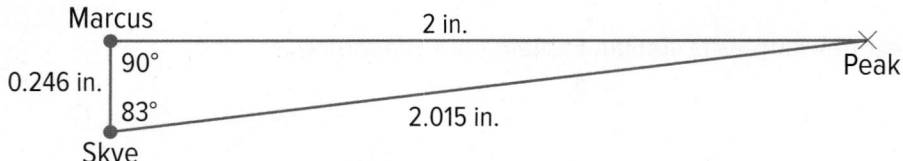

The actual distance between Marcus and Skye's houses is $1\frac{1}{2}$ miles.

a. What is the actual distance from Marcus's house to the peak of the mountain? Round your answer to the nearest tenth of a mile.

b. What is the actual distance from Skye's house to the peak of the mountain? Round your answer to the nearest tenth of a mile.

Mixed Exercises

13. Mia drew triangles *STU* and *SQR*. She claims that △*STU* and △*SQR* are similar. Is she correct? Justify your reasoning.

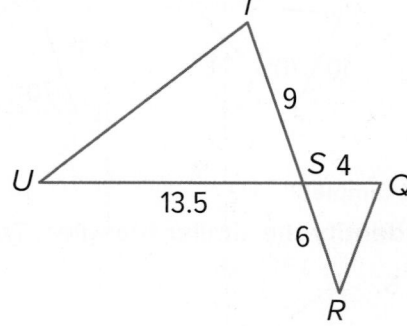

14. PROOF Write a two-column proof for the Side-Angle-Side (SAS) Similarity Theorem. (Theorem 8.5)

Given: $\angle B \cong \angle E$, $\overline{QP} \parallel \overline{BC}$, $\overline{QP} \cong \overline{EF}$, $\frac{AB}{DE} = \frac{BC}{EF}$

Prove: △*ABC* ~ △*DEF*

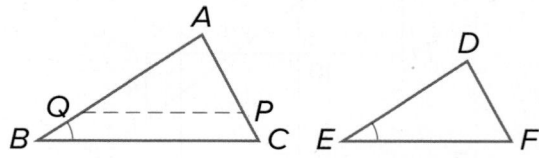

🐛 **Higher-Order Thinking Skills**

15. WRITE Compare and contrast the AA Similarity Postulate, the SSS Similarity Theorem, and the SAS Similarity Theorem.

16. PERSEVERE $\overline{YW}$ is an altitude of △*XYZ*. Find *YW*.

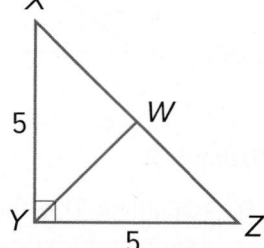

17. ANALYZE A pair of similar triangles has angles that measure 50°, 85°, and 45°. The sides of one triangle measure 3, 3.25, and 4.23 units, and the sides of the second triangle measure $x - 0.46$, x, and $x + 1.81$ units, respectively. Find the value of *x* to the nearest integer.

18. CREATE Draw a triangle that is similar to △*ABC* shown. Explain how you know that it is similar.

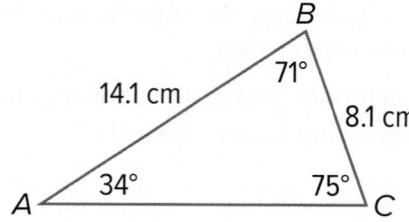

Triangle Proportionality

Explore Proportions in Triangles

 Online Activity Use dynamic geometry software to complete the Explore.

> **? INQUIRY** How do the midsegments of a triangle compare to its sides?

Learn Triangle Proportionality

When a triangle contains a line that is parallel to one of its sides, the two triangles formed can be proven similar using the Angle-Angle Similarity Postulate. Because the triangles are similar, their sides are proportional.

Theorem 3.8: Triangle Proportionality Theorem

If a line is parallel to one side of a triangle and intersects the other two sides, then it divides the sides into segments of proportional lengths.

Theorem 3.9: Converse of Triangle Proportionality Theorem

If a line intersects two sides of a triangle and separates the sides into proportional corresponding segments, then the line is parallel to the third side of the triangle.

You will prove Theorems 3.8 and 3.9 in Exercises 18 and 19, respectively.

Example 1 Use Triangle Proportions to Find the Length of a Side

In $\triangle BCD$, $\overline{PQ} \parallel \overline{CD}$. If $QD = 14.5$, $BP = 9$, and $PC = 15$, find BQ.

$$\frac{BP}{PC} = \frac{BQ}{QD}$$ Triangle Proportionality Theorem

$$\frac{9}{15} = \frac{BQ}{14.5}$$ Substitution

$$15BQ = 130.5$$ Multiplication Property of Equality

$$BQ = 8.7$$ Solve.

Check

In $\triangle FGH$, $\overline{FH} \parallel \overline{BC}$. If $GB = 14.4$, $BF = 4.8$, and $GH = 21$, find GC and CH.

$GC = $ ___?___ $CH = $ ___?___

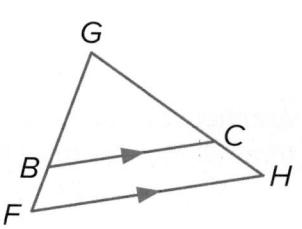

Today's Goals
- Solve problems and prove theorems by using triangle proportionality.
- Solve problems and prove theorems by using the Triangle Midsegment Theorem and its corollaries.

Today's Vocabulary
midsegment of a triangle

🫧 Think About It!
If $BP = 9$, $PC = 15$, and $PQ = 7.5$, could you use $\frac{9}{15} = \frac{7.5}{CD}$ to find CD? Justify your reasoning.

Example 2 Use Triangle Proportions to Determine if Lines are Parallel

In △WXY, YL = 5, LX = 20, and $\overline{JX}$ is four times as long as $\overline{WJ}$. Is $\overline{JL} \parallel \overline{WY}$? Explain your reasoning.

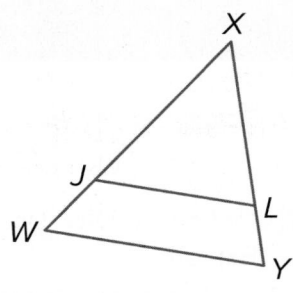

To show that $\overline{JL} \parallel \overline{WY}$, we must show that $\frac{WJ}{JX} = \frac{YL}{LX}$ using the Converse of the Triangle Proportionality Theorem. Find and simplify each ratio. Because $\overline{JX}$ is four times as long as $\overline{WJ}$, you can represent their lengths with x and $4x$, respectively.

$$\frac{WJ}{JX} = \frac{x}{4x} \text{ or } \frac{1}{4} \qquad\qquad \frac{YL}{LX} = \frac{5}{20} \text{ or } \frac{1}{4}$$

Because $\frac{1}{4} = \frac{1}{4}$, the sides are proportional. Therefore, $\overline{JL} \parallel \overline{WY}$.

Check

In △PQR, PK = 34, KQ = 20, and PJ = 1.7JR. Is $\overline{QR} \parallel \overline{KJ}$? Explain your reasoning.

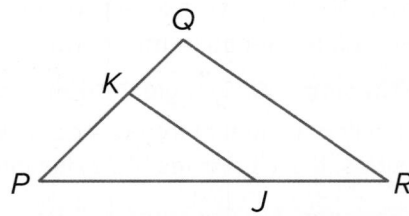

Learn Midsegments and Parallel Lines

A **midsegment of a triangle** is a segment that connects the midpoints of the legs of the triangle. Every triangle has three midsegments. The midsegments of △ABC are $\overline{RP}$, $\overline{PQ}$, and $\overline{RQ}$.

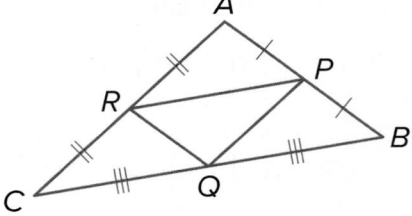

One special case of the Triangle Proportionality Theorem is the Triangle Midsegment Theorem.

Theorem 3.10: Triangle Midsegment Theorem	
Words	A midsegment of a triangle is parallel to one side of the triangle, and its length is one half of the length of that side.
Example	If J and K are midpoints of $\overline{FH}$ and $\overline{HG}$, respectively, then $\overline{JK} \parallel \overline{FG}$ and $JK = \frac{1}{2}FG$.

You will prove Theorem 3.10 in Exercise 20.

Another special case of the Triangle Proportionality Theorem involves three or more parallel lines cut by two transversals.

Corollary 3.1: Proportional Parts of Parallel Lines

Words	If three or more parallel lines intersect two transversals, then they cut off the transversals proportionally.
Example	If $\overline{AE} \parallel \overline{BF} \parallel \overline{CG}$, then $\frac{AB}{BC} = \frac{EF}{FG}$.

You will prove Corollary 3.1 in Exercise 21.

If the scale factor of the proportional segments is 1, they separate the transversals into congruent parts.

Corollary 3.2: Congruent Parts of Parallel Lines

Words	If three or more parallel lines cut off congruent segments on one transversal, then they cut off congruent segments on every transversal.
Example	If $\overline{AE} \parallel \overline{BF} \parallel \overline{CG}$ and $\overline{AB} \cong \overline{BC}$, then $\overline{EF} \cong \overline{FG}$.

Go Online
A proof of Corollary 3.2 is available.

Example 3 Use the Triangle Midsegment Theorem

In the figure, $\overline{EF}$ and $\overline{DE}$ are midsegments of $\triangle ABC$. Find DE.

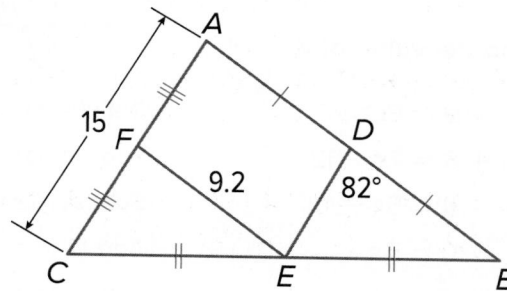

DE

$DE = \frac{1}{2} AC$

$DE = \frac{1}{2} (15)$

$DE = 7.5$

Study Tip

Midsegments The Triangle Midsegment Theorem is similar to the Trapezoid Midsegment Theorem, which states that the midsegment of a trapezoid is parallel to the bases and its length is one half the sum of the measures of the bases.

Check

In the figure, $\overline{FG}$, $\overline{GH}$, and $\overline{FH}$ are midsegments of $\triangle ABC$. Find each measure.

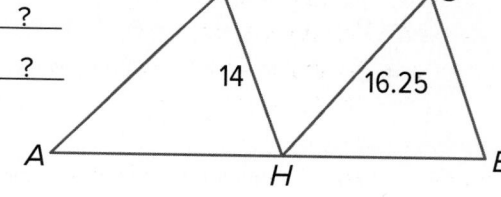

$AC =$ ___?___ $AB =$ ___?___

$CB =$ ___?___ $CG =$ ___?___

$AH =$ ___?___

Think About It!

If $BC = 24$, what is DF?

🌐 Example 4 Use Proportional Segments of Transversals

REAL ESTATE A developer is looking to purchase lots 18 and 19 on the lake and wants to determine the length of the property's boundary that runs along the lake, a measurement known as frontage. Find the lake frontage for Lot 18 to the nearest tenth of a foot.

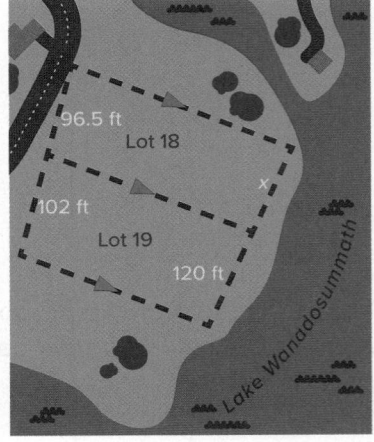

By Corollary 3.1, because the three boundaries are parallel, the segments formed by the front and back property lines are divided into proportional parts. Let x represent the missing length.

$$\frac{x}{120} = \frac{96.5}{102}$$ Corollary 3.1

$$x = 120\left(\frac{96.5}{102}\right)$$ Multiplication Property of Equality

$$x \approx 113.5$$ Solve.

Example 5 Use Congruent Segments of Transversals

Find the values of x and y.

Because $\overleftrightarrow{AJ} \parallel \overleftrightarrow{BK} \parallel \overleftrightarrow{CL}$ and $\overline{JK} \cong \overline{KL}$, then $\overline{AB} \cong \overline{BC}$.

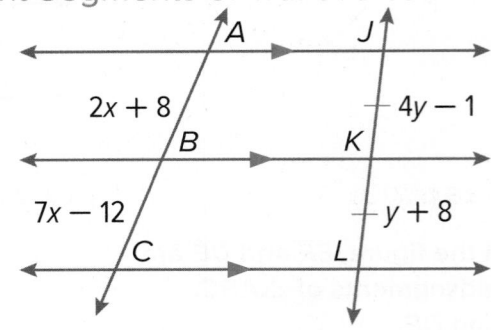

Find the value of x.

$AB = BC$	Definition of congruence
$2x + 8 = 7x - 12$	Substitution
$8 = 5x - 12$	Subtract $2x$ from each side.
$20 = 5x$	Add 12 to each side.
$4 = x$	Divide each side by 5.

Find the value of y.

$JK = KL$	Definition of congruence
$4y - 1 = y + 8$	Substitution
$3y - 1 = 8$	Subtract y from each side.
$3y = 9$	Add 1 to each side.
$y = 3$	Divide each side by 3.

🔵 **Go Online** You can complete an Extra Example online.

Practice

Example 1

Use the figure at the right.

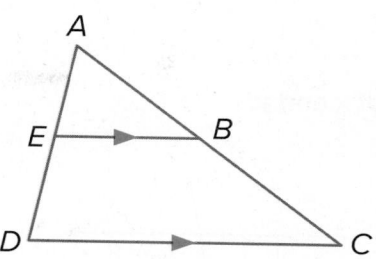

1. If $AB = 6$, $BC = 4$, and $AE = 9$, find ED.

2. If $AB = 12$, $AC = 16$, and $ED = 5$, find AE.

Example 2

Determine whether $\overline{NR} \parallel \overline{PQ}$. Justify your answer.

3. $PM = 18$, $PN = 6$, $QM = 24$, and $RM = 16$

4. $QM = 31$, $RM = 21$, and $PM = 4PN$

Example 3

$\overline{VR}$, $\overline{VZ}$, and $\overline{ZR}$ are midsegments of $\triangle UWY$. Find the value of x.

5.

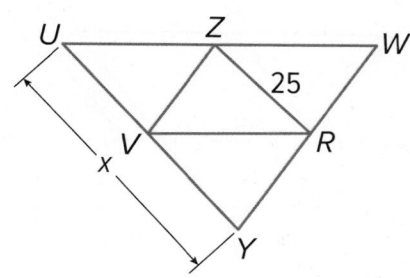

6.

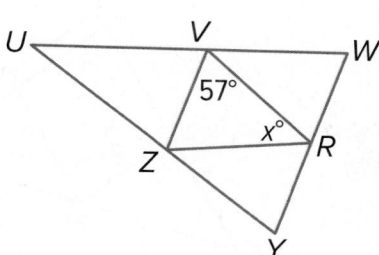

7.

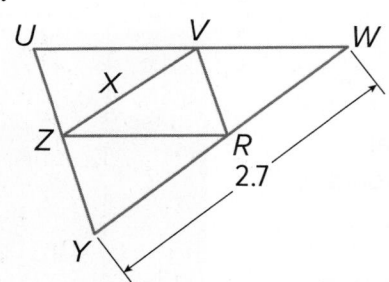

8.

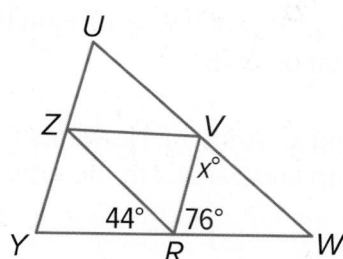

Example 4

9. MAPS In Mika's town, Cay Street and Bay Street are parallel. Find the value of x, the distance from Cay Street to Bay Street along Earl Street.

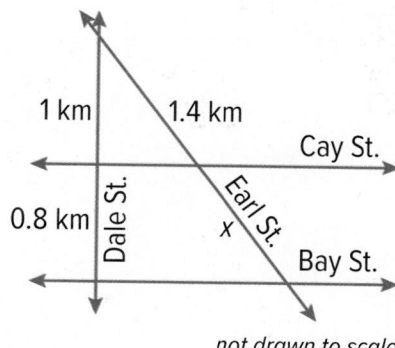

not drawn to scale

10. PLAYSCAPES Prassad is building a two-story playscape using the plans shown. Find the value of x.

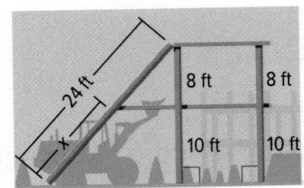

Example 5

Find the values of x and y.

11.

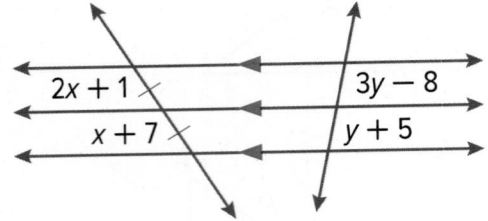

$2x + 1$ $3y - 8$
$x + 7$ $y + 5$

12.

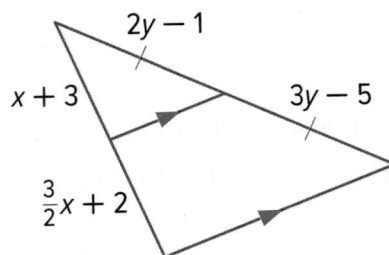

$2y - 1$
$x + 3$ $3y - 5$
$\frac{3}{2}x + 2$

Mixed Exercises

$\overline{JH}$ **is a midsegment of** $\triangle KLM$. **Find the value of** x.

13.

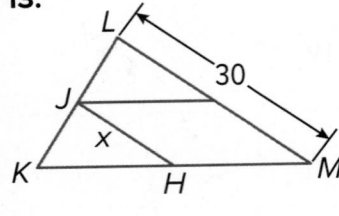

30
x

14.

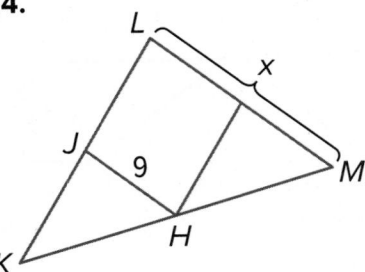

x
9

15.

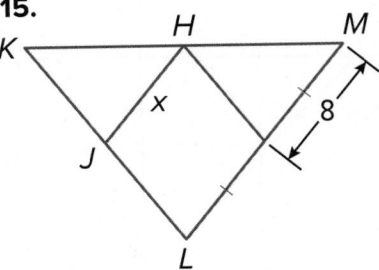

x
8

16. In $\triangle ABC$, $\overline{DE}$ is parallel to $\overline{AC}$ and $DE = 10$. Find the length of $\overline{AC}$ if $\overline{DE}$ is a midsegment of $\triangle ABC$.

17. CARPENTRY Jake is fixing an A-frame. He wants to add a horizontal support beam halfway up and parallel to the ground. How long should this beam be?

18. PROOF Write a paragraph proof of the Triangle Proportionality Theorem. (Theorem 8.8)

Given: $\overline{BD} \parallel \overline{AE}$

Prove: $\frac{BA}{CB} = \frac{DE}{CD}$

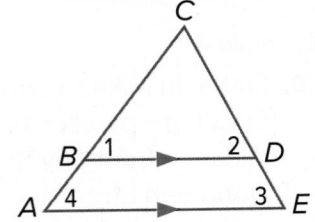

19. **PROOF** Write a paragraph proof of the Converse of the Triangle Proportionality Theorem. (Theorem 8.9).

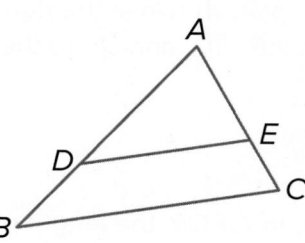

Given: $\frac{AD}{DB} = \frac{AE}{EC}$

Prove: $\overline{DE} \parallel \overline{BC}$

(*Hint: Explain how you can use the given proportion to show that $\frac{AB}{AD} = \frac{AC}{EC}$. Then, use this proportion to complete the proof.*)

20. **PROOF** Write a two-column proof of the Triangle Midsegment Theorem. (Theorem 8.10)

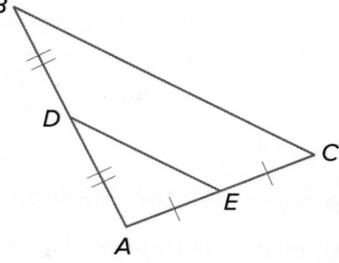

Given: $\triangle ABC$; D is the midpoint of $\overline{AB}$; E is the midpoint of $\overline{AC}$.

Prove: $\overline{DE} \parallel \overline{BC}$, $DE = \frac{1}{2} BC$

21. **PROOF** Write a paragraph proof of Corollary 8.1.

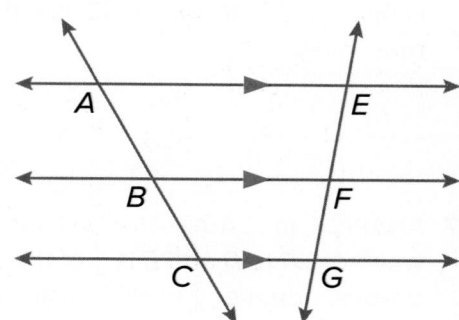

Given: $\overleftrightarrow{AE} \parallel \overleftrightarrow{BF} \parallel \overleftrightarrow{CG}$

Prove: $\frac{AB}{BC} = \frac{EF}{FG}$

22. **REGULARITY** In the figure, $\overline{DE} \parallel \overline{BC}$, $BD = 12$, $EC = 10$, and $AE = 15$. Explain how to find the length of $\overline{AD}$.

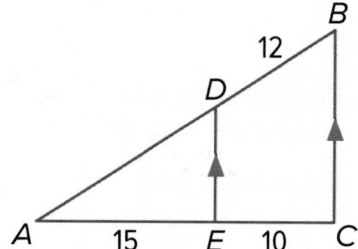

23. **ART** The divider bars between the pieces of colored glass in a stained glass window are called *cames*. In the stained window at the right, the total length of the cames for $\triangle PQR$ is 78 centimeters. What is the total length of the cames for $\triangle JKL$? Give an argument to support your answer.

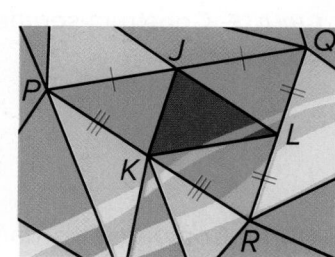

24. SHUFFLEBOARD A crew is laying out a shuffleboard court using the plan shown at the right. Explain how they can find the lengths of $\overline{AB}$, $\overline{BD}$, and $\overline{DF}$ to the nearest tenth of a foot.

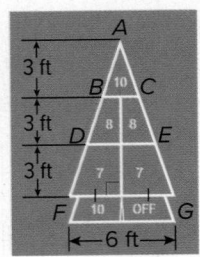

25. In $\triangle PQR$, the length of $\overline{PQ}$ is 16 units. A series of midsegments are drawn such that $\overline{ST}$ is the midsegment of $\triangle PQR$, $\overline{UV}$ is the midsegment of $\triangle STR$, and $\overline{WX}$ is the midsegment of $\triangle UVR$.

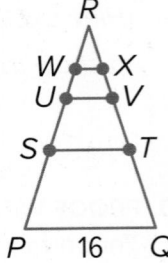

a. What is the length of each midsegment?

$ST =$ $UV =$ $WX =$

b. What would be the measure of midsegment $\overline{YZ}$ of $\triangle WXR$?

🧠 **Higher-Order Thinking Skills**

26. FIND THE ERROR Jacinda and Elaine are finding the value of x in $\triangle JHL$. Jacinda says that MP is one half of JL, so x is 4.5. Elaine says that JL is one half of MP, so x is 18. Is either of them correct? Explain your reasoning.

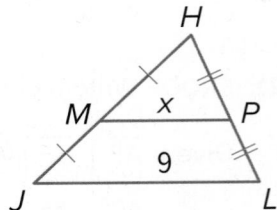

27. ANALYZE In $\triangle ABC$, $AF = FB$ and $AH = HC$. If D is $\frac{3}{4}$ of the way from A to B and E is $\frac{3}{4}$ of the way from A to C, is DE *sometimes, always,* or *never* $\frac{3}{4}$ of BC? Justify your argument.

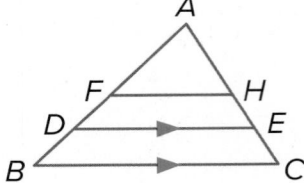

28. PROOF Write a two-column proof.
Given: $AB = 4$, $BC = 4$, and $CD = DE$

Prove: $\overline{BD} \parallel \overline{AE}$

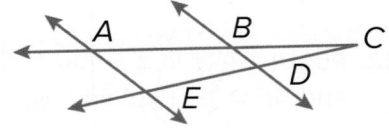

29. CREATE Construct segments a, b, c, and d of all different lengths, such that $\frac{a}{b} = \frac{c}{d}$.

30. WRITE Compare the Triangle Proportionality Theorem and the Triangle Midsegment Theorem.

Parts of Similar Triangles

Explore Special Segments in Triangles

⊙ **Online Activity** Use dynamic geometry software to complete the Explore.

×

ⓘ **INQUIRY** What relationships exist among special segments in similar triangles?

Learn Parts of Similar Triangles

You have learned that the corresponding side lengths of similar triangles are proportional. This concept can be extended to other segments in similar triangles.

⊙ **Go Online**
A proof of Theorem 3.11 is available.

Theorems: Parts of Similar Triangles

Theorem 3.11

Words	If two triangles are similar, the lengths of the corresponding altitudes are proportional to the lengths of corresponding sides.
Symbols	~△s have corr. altitudes proportional to corr. sides.
Example	If $\triangle XYZ \sim \triangle RST$, then $\dfrac{YW}{SQ} = \dfrac{YZ}{ST}$.

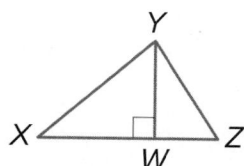 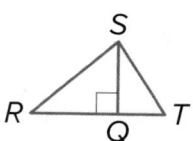

Theorem 3.12

Words	If two triangles are similar, the lengths of corresponding angle bisectors are proportional to the lengths of the corresponding sides.
Symbols	~△s have corr. ∠ bisectors proportional to corr. sides.
Example	If $\triangle DEF \sim \triangle TUV$, then $\dfrac{EW}{UQ} = \dfrac{DE}{TU}$.

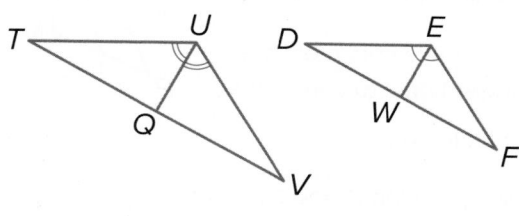

Theorem 3.13

Words	If two triangles are similar, the lengths of corresponding medians are proportional to the lengths of corresponding sides.
Symbols	~△s have corr. medians proportional to corr. sides.
Example	If △PQR ~ △FGH, then $\frac{RS}{HJ} = \frac{PQ}{FG}$.

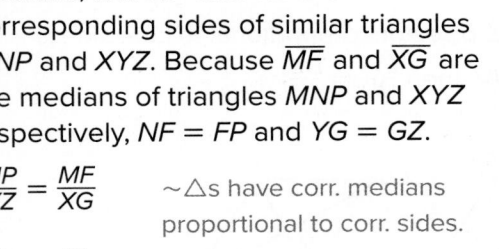

You will prove Theorems 3.12 and 3.13 in Exercises 12 and 13, respectively.

An angle bisector of a triangle also divides the side opposite the angle proportionally.

Theorem 3.14: Triangle Angle Bisector

Words	An angle bisector in a triangle separates the opposite side into two segments that are proportional to the lengths of the other two sides.
Example	If $\overline{AP}$ is an angle bisector of △ABC, then $\frac{BP}{CP} = \frac{BA}{CA}$.

You will prove Theorem 3.14 in Exercise 14.

Study Tip

Proportions Another proportion that could be written using the Triangle Angle Bisector Theorem is $\frac{BP}{BA} = \frac{CP}{CA}$.

Example 1 Use Special Segments in Similar Triangles

In the figure, △MNP ~ △XYZ. Find the value of x.

$\overline{MF}$ and $\overline{XG}$ are corresponding medians, and $\overline{NP}$ and $\overline{YZ}$ are corresponding sides of similar triangles MNP and XYZ. Because $\overline{MF}$ and $\overline{XG}$ are the medians of triangles MNP and XYZ respectively, NF = FP and YG = GZ.

$\frac{NP}{YZ} = \frac{MF}{XG}$ ~△s have corr. medians proportional to corr. sides.

$\frac{2x}{24} = \frac{18}{27}$ Substitution

$54x = 432$ Multiplication Property of Equality

$x = 8$ Solve for x.

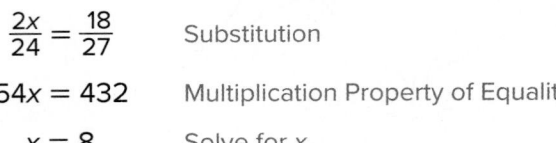

🔲 **Go Online**
An alternate method is available for this example.

🌐 Apply Example 2 Use Similar Triangles to Solve Problems

PHOTOGRAPHY **A digital camera projects an image through its lens and onto its sensor, where it is converted into a digital image. The distance between the camera's lens and its sensor is known as the focal length and is adjusted depending on the size of the object being photographed and its distance from the camera lens. Ms. Elgin sets her camera up 3 meters away from her subject, who is 1.6 meters tall. If the sensor on her camera is 4.8 millimeters tall, what is the optimal focal length?**

1 What is the task?

Describe the task in your own words. Then list any questions that you may have. How can you find answers to your questions?

Sample answer: I need to find the optimal focal length for the digital camera. Are the triangles that model the situation similar? What theorem relating the measures within similar triangles can I use to find the focal length? To answer these questions, I will draw a diagram and use what I know about triangle similarity.

2 How will you approach the task? What have you learned that you can use to help you complete the task?

Sample answer: I will draw a diagram. Then I will identify congruent angles and corresponding sides. I will determine whether the triangles are similar. Then I will write a proportion and solve for the optimal focal length.

3 What is your solution?

Use your strategy to solve the problem.

Complete the diagram. Assume that $\overline{CK}$ and $\overline{CL}$ are altitudes of △ACB and △DCF, respectively, and that $\overline{AB} \parallel \overline{DF}$.

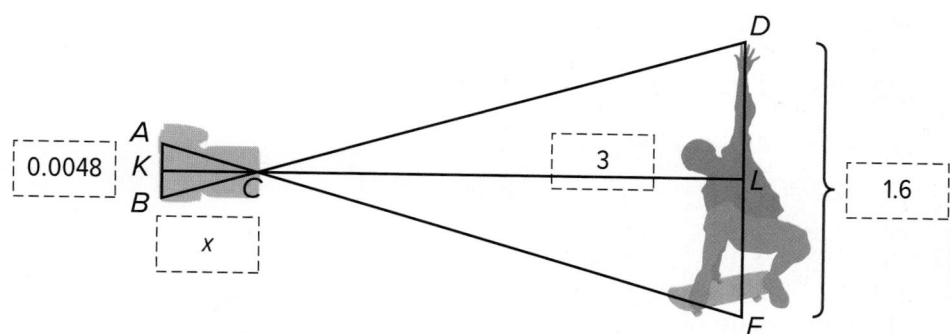

Prove that △ABC is similar to △FDC.

Sample answer: Because $\overline{AB} \parallel \overline{DF}$. $\angle BAC \cong \angle DFC$ and $\angle CBA \cong \angle CDF$ by the Alternate Interior Angles Theorem. △$ABC \sim$ △FDC by AA Similarity.

What is the measure of the optimal focal length of the digital camera?

9 mm

(continued on the next page)

> **Study Tip**
>
> **Assumptions** In this example, we assume that the sensor on the camera is pointing straight, creating altitudes in triangles *ABC* and *DCF*.

> **Watch Out!**
>
> **Altitudes** Not all altitudes separate a triangle proportionally. The altitudes of an obtuse triangle are outside of the triangle and don't separate the triangle at all.

4 How can you know that your solution is reasonable?

🖊 **Write About It!** Write an argument that can be used to defend your solution.
Sample answer: The distance from the object to the camera is 1.875 times as long as the height of the object, so it would follow that the focal length would be 1.875 times as long as the height of the sensor.

Check

TRAFFIC ENFORCEMENT A police officer is determining where to park his vehicle to observe traffic at the red light. If $AC = 512$ feet, $RP = 384$ feet, and Y is 201 feet from B, how far is Z from Q?

_____ ft

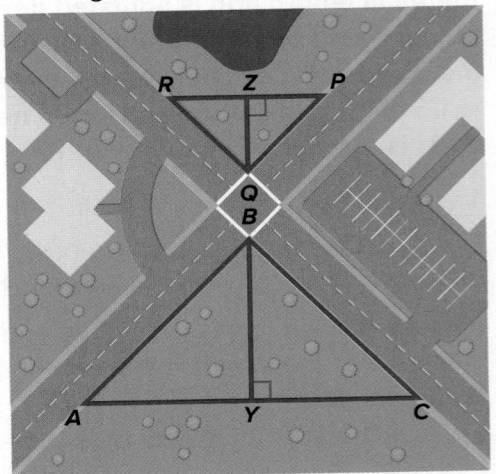

Example 3 Use the Triangle Angle Bisector Theorem

Find the value of x.
Because $\overline{FJ}$ is an angle bisector of $\triangle FGH$, you can use the Triangle Angle Bisector Theorem to write a proportion.

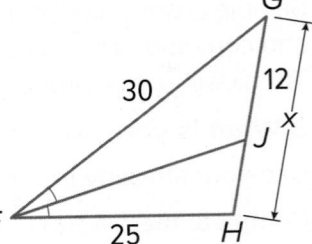

$$\frac{GJ}{JH} = \frac{FG}{FH}$$ Triangle Angle Bisector Theorem

$$\frac{12}{x-12} = \frac{30}{25}$$ Substitution

$12 \cdot 25 = 30(x-12)$ Multiplication Property of Equality

$300 = 30x - 360$ Simplify.

$22 = x$ Solve.

Check

Find the value of x.

$x =$ _____ ?

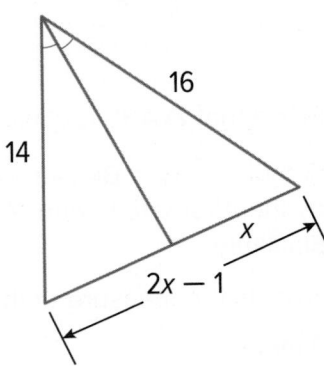

🔎 **Go Online** You can complete an Extra Example online.

Practice

Go Online You can complete your homework online.

Example 1

Each pair of triangles is similar. Find the value of x.

1.
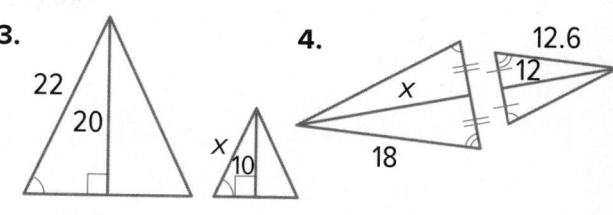
7, 22, 7, x, 5.25, 5.25

2.
33, 15, x, 10

3.
22, 20, x, 10

4.
12.6, 12, x, 18

Example 2

5. If △RST ~ △EFG, $\overline{SH}$ is an altitude of △RST, $\overline{FJ}$ is an altitude of △EFG, ST = 6, SH = 5, and FJ = 7, find FG.

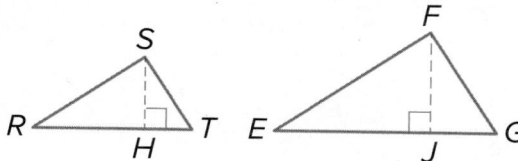

6. If △ABC ~ △MNP, $\overline{AD}$ is an altitude of △ABC, $\overline{MQ}$ is an altitude of △MNP, AB = 24, AD = 14, and MQ = 10.5, find MN.

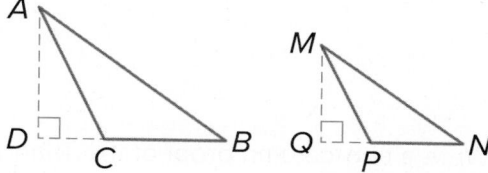

7. **SCENERY** The intersection of the two paths shown forms two similar triangles. If AC is 50 yards, MP is 35 yards, and the fountain is 5 yards from the intersection, about how far from the intersection is the stadium entrance? Round to the nearest yard.

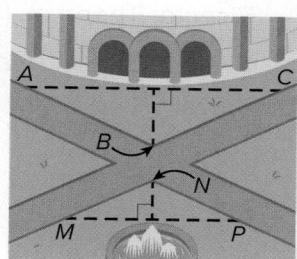

Example 3

Find the value of each variable to the nearest tenth.

8.
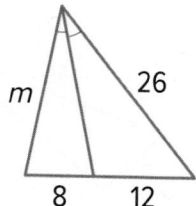
m, 26, 8, 12

9.
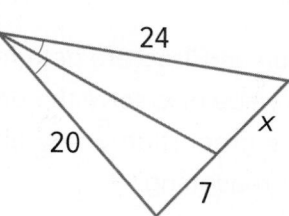
24, 20, x, 7

Mixed Exercises

For Exercises 10 and 11, △ABC ~ △DEF.

10. Find the length of $\overline{XC}$ to the nearest tenth.

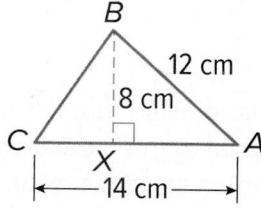

11. Find the length of $\overline{EY}$ to the nearest tenth.

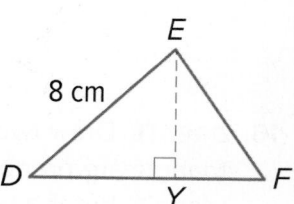

12. PROOF Write a paragraph proof of Theorem 8.12.

Given: △RTS ~ △EGF; $\overline{TA}$ and $\overline{GB}$ are angle bisectors.

Prove: $\frac{TA}{GB} = \frac{RT}{EG}$

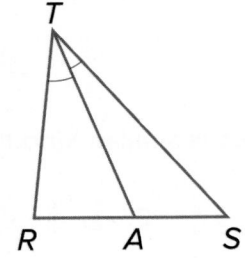
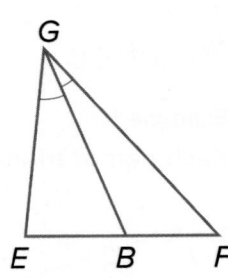

13. PROOF Write a two-column proof of Theorem 8.13.

Given: △ABC ~ △RST; $\overline{AD}$ is a median of △ABC and $\overline{RU}$ is a median of △RST.

Prove: $\frac{AD}{RU} = \frac{AB}{RS}$

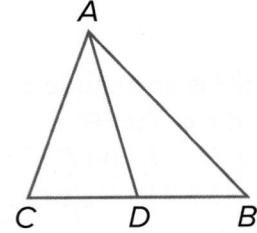
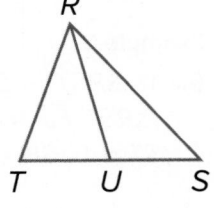

14. PROOF Write a two-column proof of the Triangle Angle Bisector Theorem. (Theorem 8.14)

Given: $\overline{CD}$ bisects ∠ACB. By construction, $\overline{AE} \parallel \overline{CD}$.

Prove: $\frac{AD}{DB} = \frac{AC}{BC}$

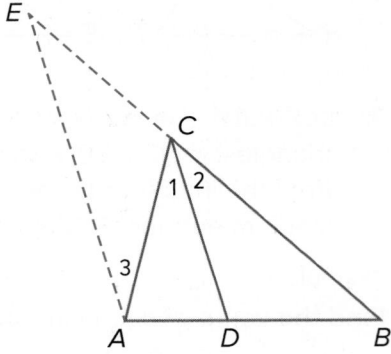

15. FIND THE ERROR Chun and Traci are determining the value of x in the figure. Chun says to find the value of x, solve the proportion $\frac{5}{8} = \frac{15}{x}$, but Traci says to find the value of x, the proportion $\frac{5}{x} = \frac{8}{15}$ should be solved. Is either of them correct? Explain your reasoning.

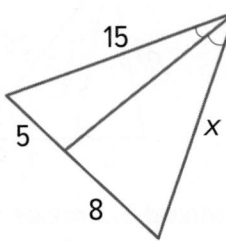

16. CREATE Draw two triangles so that the measures of an altitude and side of one triangle are proportional to the measures of an altitude and side of another triangle, but the triangles are not similar.

17. PERSEVERE The perimeter of △PQR is 94 units. $\overline{QS}$ bisects ∠PQR. Find PS and RS. Round to the nearest tenth, if necessary.

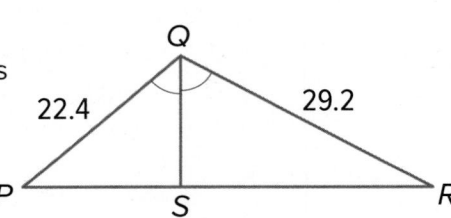

Essential Question

What does it mean for objects to be similar, and how is similarity useful for modeling in the real world?

Objects are similar when they are the same shape but not necessarily the same size. Their measures are in proportion to one another. Similar shapes are used when making scale drawings, such as blueprints, and scale models, such as replicas.

Module Summary

Lessons 3-1 and 3-2

Dilations and Similar Polygons

- A dilation is a nonrigid transformation that enlarges or reduces a geometric figure.

- When a figure is enlarged or reduced, the sides of the image are proportional to the sides of the original figure.

- To dilate a figure by a scale factor of k with respect to (0, 0), multiply the x- and y-coordinate of each vertex by k.

- Two figures are similar if one can be obtained from the other by a dilation or a dilation with one or more rigid motions.

Lessons 3-3 and 3-4

Criteria for Similar Triangles

- Angle-Angle (AA) Similarity: If two angles of one triangle are congruent to two angles of another triangle, then the triangles are similar.

- Side-Side-Side (SSS) Similarity: If the corresponding side lengths of two triangles are proportional, then the triangles are similar.

- Side-Angle-Side (SAS) Similarity: If the lengths of two sides of one triangle are proportional to the lengths of two corresponding sides of another triangle and the included angles are congruent, then the triangles are similar.

Lesson 3-5

Triangle Proportionality

- If a line is parallel to one side of a triangle and intersects the other two sides, then it divides the sides into segments of proportional lengths.

- If three or more parallel lines intersect two transversals, then they cut off the transversals proportionally.

Lesson 3-6

Parts of Similar Triangles

- If two triangles are similar, the lengths of the corresponding altitudes are proportional to the lengths of corresponding sides.

- If two triangles are similar, the lengths of corresponding angle bisectors are proportional to the lengths of the corresponding sides.

Study Organizer

 Foldables

Use your Foldable to review this module. Working with a partner can be helpful. Ask for clarification of concepts as needed.

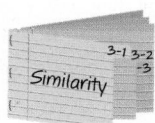

Test Practice

1. MULTIPLE CHOICE Which kind of dilation is the transformation from trapezoid *EFGH* to trapezoid *JKLM*? (Lesson 3-1)

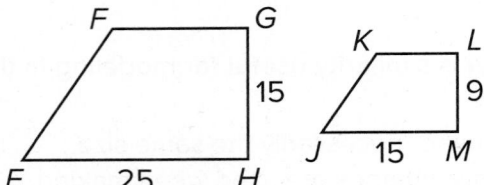

A. Enlargement with a scale factor of $\frac{5}{3}$

B. Enlargement with a scale factor of $\frac{3}{5}$

C. Reduction with a scale factor of $\frac{5}{3}$

D. Reduction with a scale factor of $\frac{3}{5}$

2. MULTIPLE CHOICE Which effect does a dilation with a scale factor from –1 to 0 have on a figure?

A. reduction

B. rotation and reduction

C. enlargement

D. rotation and enlargement

3. OPEN RESPONSE The vertices of △*ABC* are *A*(5, 4), *B*(10, 8), and *C*(20, 0). A dilation centered at the origin with a scale factor of $\frac{4}{5}$ maps △*ABC* onto △*A'B'C'*. What are the coordinates of the vertices of △*A'B'C'*?
(Lesson 3-1)

4. OPEN RESPONSE Given parallelogram *WXYZ*, what are the coordinates of *W'X'Y'Z'* after a dilation centered at the origin with a scale factor of $\frac{3}{2}$? (Lesson 3-1)

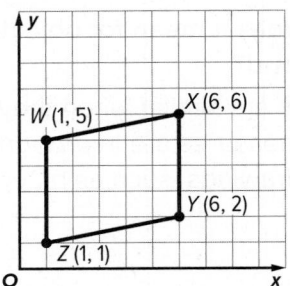

5. OPEN RESPONSE Using the definition of similarity, show that the two quadrilaterals are similar. (Lesson 3-2)

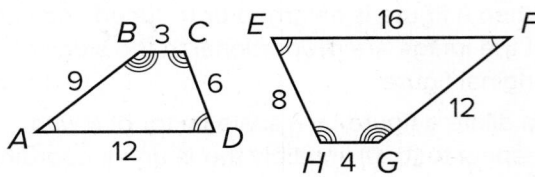

6. OPEN RESPONSE If a similarity transformation maps △*FGH* onto △*LKJ*, what is the similarity ratio? (Lesson 3-2)

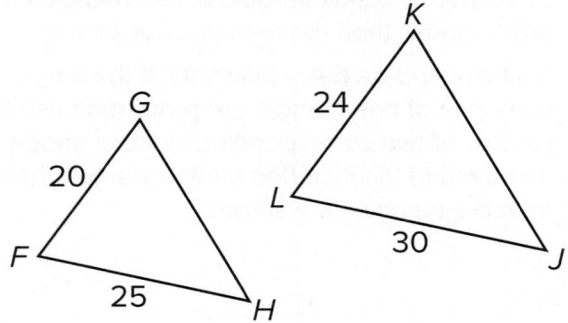

7. MULTIPLE CHOICE On a blueprint, a rectangular kitchen has a length of 4 inches and a width of 3 inches. If the length of the kitchen is 6 feet, what is the perimeter of the kitchen in feet? (Lesson 3-2)

A. 18 feet

B. 20 feet

C. 21 feet

D. 24 feet

8. OPEN RESPONSE Given any two equilateral triangles. Are the triangles similar? Justify your answer. (Lesson 3-3)

9. MULTIPLE CHOICE Given: ∠B ≅ ∠E

Prove: △ADE ~ △CDB

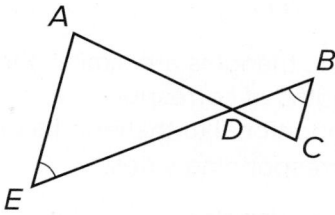

Complete the paragraph proof.

It is given that ∠B ≅ ∠E. ∠ADE ≅ ∠CDB by the ___. Therefore, △ADE is similar to △CDB by the ___. (Lesson 3-3)

A. Corresponding Angles Theorem, Angle-Angle Similarity Postulate

B. Corresponding Angles Theorem, Third Angles Theorem

C. Vertical Angles Theorem, Angle-Angle Similarity Postulate

D. Vertical Angles Theorem, Third Angles Theorem

10. OPEN RESPONSE In △ABC, m∠A = 44° and m∠B = 56°. In △DEF, m∠D = 44° and m∠F = 80°. Is △ABC similar to △DEF? Justify your answer. (Lesson 3-3)

11. MULTIPLE CHOICE What can be used to prove that △ABC is similar to △DBE? (Lesson 3-4)

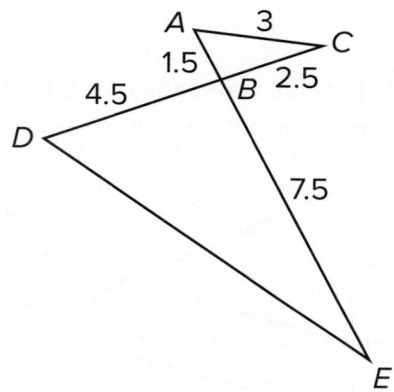

A. AA Similarity Postulate

B. SAS Similarity Theorem

C. SSS Similarity Theorem

D. AAS Similarity Theorem

12. OPEN RESPONSE What proves that △CDE is not similar to △FGH? (Lesson 3-4)

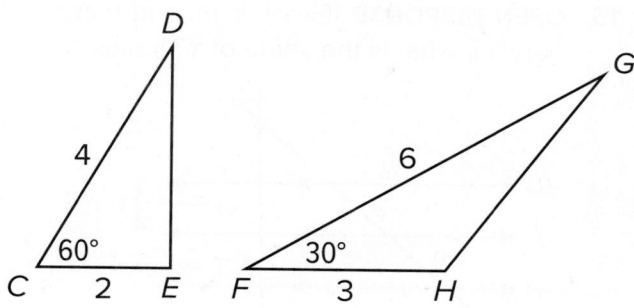

13. OPEN RESPONSE If $\overline{BD} \parallel \overline{AE}$ and $AC = 4$, identify the three unknown lengths. (Lesson 3-5)

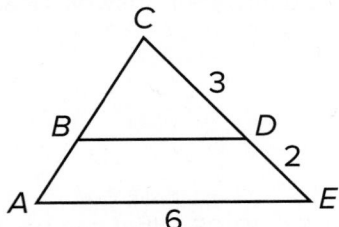

14. MULTI-SELECT Select all of the true statements about the figure. (Lesson 3-5)

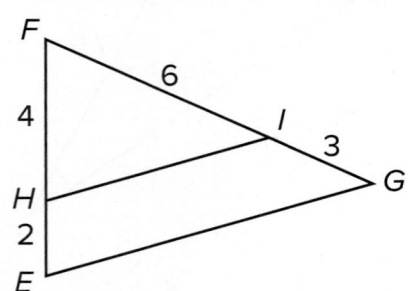

A. $\overline{HI} \parallel \overline{EG}$

B. $\dfrac{FH}{HE} = \dfrac{FI}{IG}$

C. $\dfrac{FH}{HE} = \dfrac{HI}{EG}$

D. $\dfrac{FH}{GI} = \dfrac{FI}{EH}$

E. $\dfrac{HE}{FH} = \dfrac{FI}{IG}$

15. OPEN RESPONSE If lines k, m, and n are parallel, what is the value of x? (Lesson 3-5)

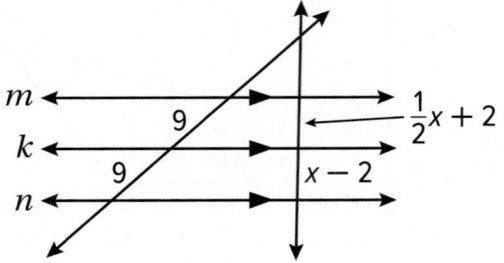

16. OPEN RESPONSE Suppose $AD = 7$. What is the length of $\overline{AB}$? Round your answer to the nearest thousandth. (Lesson 3-6)

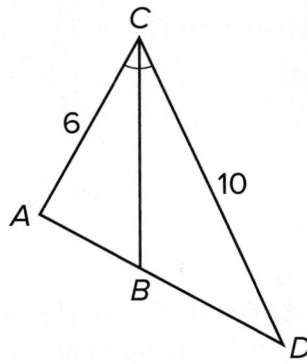

17. MULTIPLE CHOICE If $\triangle ABC \sim \triangle GHE$, which theorem proves $\dfrac{FH}{BD} = \dfrac{GH}{AB}$? (Lesson 3-6)

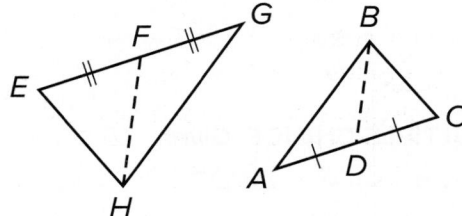

A. If two triangles are similar, then the lengths of corresponding medians are proportional to the lengths of the corresponding sides.

B. If two triangles are similar, then the lengths of corresponding altitudes are proportional to the lengths of the corresponding sides.

C. If two triangles are similar, then the lengths of corresponding angle bisectors are proportional to the lengths of the corresponding sides.

D. An angle bisector in a triangle separates the opposite side into two segments that are proportional to the lengths of the other two sides.

Right Triangles and Trigonometry

What Will You Learn?

How much do you already know about each topic **before** starting this module?

KEY

👎 — I don't know.　　👈 — I've heard of it.　　👍 — I know it!

	Before			After		
	👎	👈	👍	👎	👈	👍
solve problems using geometric mean and relationships between parts of a right triangle when an altitude is drawn to the hypotenuse						
solve problems using the Pythagorean Theorem and its converse						
graph points and find distances using the Distance Formula in three dimensions						
solve problems using the properties of 45°-45°-90° and 30°-60°-90° right triangles						
solve problems using the trigonometric ratios for acute angles						
solve problems using the inverse trigonometric ratios for acute angles						
derive and use a formula for the area of a triangle using trigonometry						
solve problems using the Law of Sines and the Law of Cosines						
determine whether three given measures of a triangle define 0, 1, or 2 triangles using the Law of Sines						

📙 **Foldables** Make this Foldable to help you organize your notes about right triangles and trigonometry. Begin with four sheets of notebook paper and one sheet of construction paper.

1. **Stack** the notebook paper on the construction paper.

2. **Fold** the paper diagonally to form a triangle and cut off the excess.

3. **Open** the paper and staple the inside fold to form a booklet.

4. **Label** each tab with a lesson number and title.

1 　　2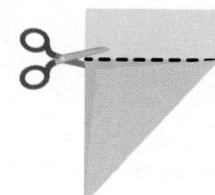

3 　　4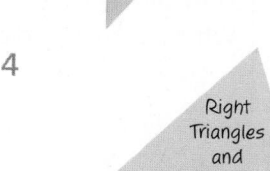

Right Triangles and Trigonometry

What Vocabulary Will You Learn?

- 30°-60°-90° triangle
- 45°-45°-90° triangle
- ambiguous case
- angle of depression
- angle of elevation
- cosine
- geometric mean

- indirect measurement
- inverse cosine
- inverse sine
- inverse tangent
- octant
- ordered triple
- Pythagorean triple

- sine
- solving a triangle
- tangent
- trigonometric ratio
- trigonometry

Are You Ready?

Complete the Quick Review to see if you are ready to start this module.
Then complete the Quick Check.

Quick Review

Example 1

The two triangles are similar. Solve for x.

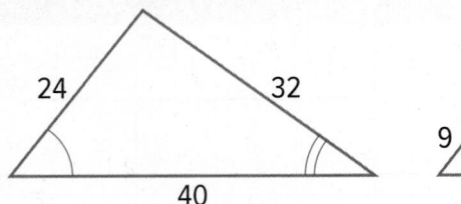

$\frac{x}{40} = \frac{9}{24}$ Similar triangles

$24x = 360$ Multiplication Property of Equality

$x = 15$ Simplify.

Example 2

Find $\sqrt{144}$.

What number multiplied by itself equals 144?

$12 \cdot 12 = 144$

So, $\sqrt{144} = 12$.

Quick Check

Solve for x.

1. △PQR ~ △JKL

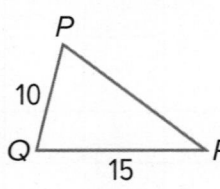

2. △ABC ~ △DEF

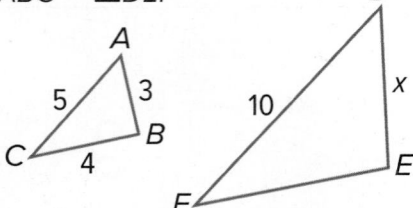

Find each square root.

3. $\sqrt{100}$ **4.** $\sqrt{64}$

5. $\sqrt{196}$ **6.** $\sqrt{81}$

7. $\sqrt{289}$ **8.** $\sqrt{625}$

How Did You Do?

Which exercises did you answer correctly in the Quick Check?

Geometric Mean

Explore Relationships in Right Triangles

Online Activity Use dynamic geometry software to complete the Explore.

> **INQUIRY** In a right triangle, what relationship exists when an altitude is drawn from the vertex of the right angle to the hypotenuse?

Learn Geometric Mean

The **geometric mean** of a set of numbers is the nth root, where n is the number of elements in a set of numbers, of the product of the numbers. So, the geometric mean of two numbers is the positive square root of their product.

Key Concept • Geometric Mean
The geometric mean of two positive numbers a and b is x such that $\frac{a}{x} = \frac{x}{b}$. So, $x^2 = ab$ and $x = \sqrt{ab}$.

The geometric mean can be found in the parts of a right triangle. An altitude drawn from the vertex of the right angle to the hypotenuse of a right triangle forms two additional right triangles. These three right triangles share a special relationship.

Theorem 4.1
If the altitude is drawn to the hypotenuse of a right triangle, then the two triangles formed are similar to the original triangle and to each other.

Theorem 4.2: Geometric Mean (Altitude) Theorem

Words	The altitude drawn to the hypotenuse of a right triangle separates the hypotenuse into two segments. The length of this altitude is the geometric mean between the lengths of these two segments.
Example	If $\overline{CD}$ is the altitude to hypotenuse $\overline{AB}$ of right $\triangle ABC$, then $\frac{x}{h} = \frac{h}{y}$, or $h = \sqrt{xy}$.

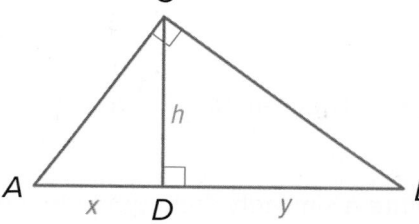

You will prove Theorem 4.2 in Exercise 19.

Today's Goals
- Use similarity criteria for triangles and geometric means to solve problems and to prove relationships in geometric figures.

Today's Vocabulary
geometric mean

> **Study Tip**
>
> **Altitudes** Remember, the altitude of a triangle is a segment from a vertex to the line containing the opposite side and perpendicular to the line containing that side.

Go Online
A proof of Theorem 4.1 is available.

Theorem 4.3: Geometric Mean (Leg) Theorem

Words	The altitude drawn to the hypotenuse of a right triangle separates the hypotenuse into two segments. The length of a leg of this triangle is the geometric mean between the length of the hypotenuse and the segment of the hypotenuse adjacent to that leg.
Example	If $\overline{CD}$ is the altitude to hypotenuse $\overline{AB}$ of right $\triangle ABC$, then $\frac{c}{b} = \frac{b}{x}$, or $b = \sqrt{xc}$ and $\frac{c}{a} = \frac{a}{y}$ or $a = \sqrt{yc}$.

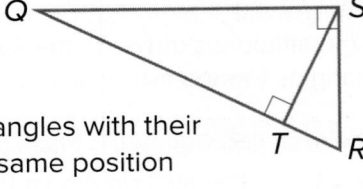

You will prove Theorem 4.3 in Exercise 20.

Example 1 Find a Geometric Mean

Find the geometric mean between 5 and 45.

$x = \sqrt{ab}$ Definition of geometric mean

$ = \sqrt{5 \cdot 45}$ $a = 5$ and $b = 45$

$ = 15$ Simplify.

The geometric mean between 5 and 45 is 15.

Check

Find the geometric mean between 12 and 15. Write your answer in simplest radical form, if necessary.

Example 2 Identify Similar Right Triangles

Write a similarity statement identifying the three similar right triangles in the figure.

The diagram below shows the three triangles with their corresponding angles and sides in the same position as the largest triangle.

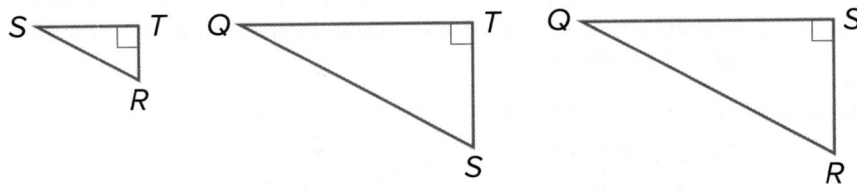

So by Theorem 9.1, $\triangle STR \sim \triangle QTS \sim \triangle QSR$.

Check

Write a similarity statement identifying the three similar right triangles in the figure.

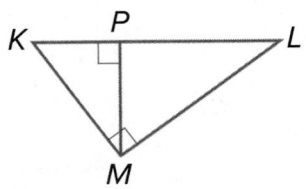

🔵 **Go Online** You can complete an Extra Example online.

Example 3 Use the Geometric Mean with Right Triangles

Find the values of x, y, and z.

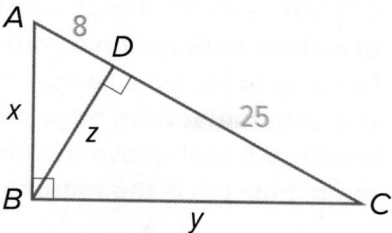

x:

Because x is the measure of leg $\overline{AB}$, x is the geometric mean of AD, the measure of the segment adjacent to $\overline{AB}$, and AC, the measure of the hypotenuse.

$x = \sqrt{AD \cdot AC}$	Geometric Mean (Leg) Theorem
$= \sqrt{8 \cdot (8 + 25)}$	Substitution
$= \sqrt{264}$	Simplify.
$= 2\sqrt{66}$ or about 16.2	Use a calculator to simplify.

y:

Because y is the measure of leg $\overline{BC}$, y is the geometric mean of DC, the measure of the segment adjacent to $\overline{BC}$, and AC, the measure of the hypotenuse.

$y = \sqrt{DC \cdot AC}$	Geometric Mean (Leg) Theorem
$= \sqrt{25 \cdot (8 + 25)}$	Substitution
$= \sqrt{825}$	Simplify.
$= 5\sqrt{33}$ or about 28.7	Use a calculator to simplify.

z:

Because z is the measure of the altitude drawn to the hypotenuse of right $\triangle ABC$, z is the geometric mean of the lengths of the two segments that make up the hypotenuse, AD and DC.

$z = \sqrt{AD \cdot DC}$	Geometric Mean (Altitude) Theorem
$= \sqrt{8 \cdot 25}$	Substitution
$= \sqrt{200}$	Simplify.
$= 10\sqrt{2}$ or about 14.1	Use a calculator to simplify.

Check

Find the values of x, y, and z.

$x = \underline{\ ?\ }$, $y = \underline{\ ?\ }$, and $z = \underline{\ ?\ }$

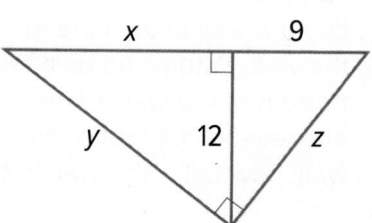

Go Online You can complete an Extra Example online.

🌐 Example 4 Use Indirect Measurement

SKATEBOARDING **Diego wants to find the height of a half pipe ramp at a skate park near his house. To find this height, Diego holds a book up to his eyes so that the top and bottom of the ramp are in line with the bottom edge and binding of the book. If Diego's eye level is 5.5 feet above the ground and he stands 6 feet from the ramp, how tall is the ramp to the nearest foot?**

Step 1 Visualize and describe the situation.

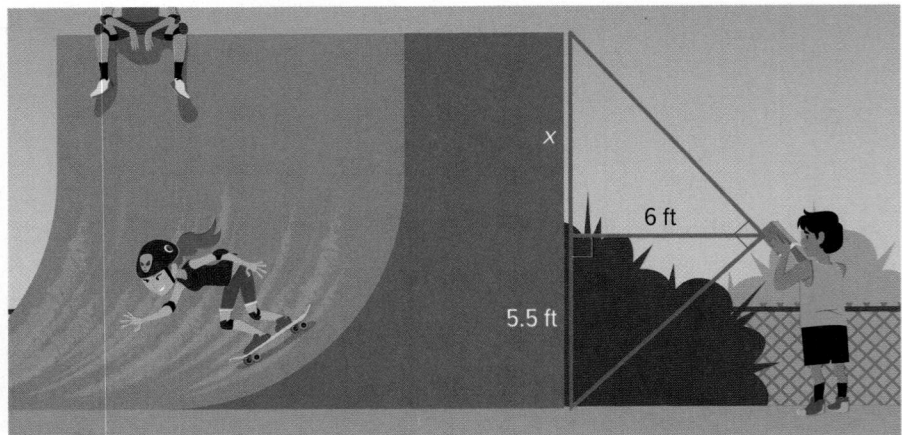

Draw a diagram to model the situation. The distance from Diego to the ramp is the altitude to the hypotenuse of a right triangle. The length of this altitude is the geometric mean between the lengths of the two segments that make up the hypotenuse. The shorter segment is equal to 5.5 feet, the height of Diego's eye level. Let the unknown measure be x feet.

Step 2 Find the height.

You can use the Geometric Mean (Altitude) Theorem to find x.

$$6 = \sqrt{5.5 \cdot x} \qquad \text{Geometric Mean (Altitude) Theorem}$$
$$36 = 5.5x \qquad \text{Square each side.}$$
$$6.55 \approx x \qquad \text{Divide each side by 5.5.}$$

The height of the ramp is the total length of the hypotenuse, $6.55 + 5.5$, or about 12 feet.

Check

EVENTS Katelyn wants to make a banner for homecoming that will cover a wall in the cafeteria of her high school. To find the height of the wall, Katelyn holds a folder up to her eyes so that the top and bottom of the wall are in line with the edges of the folder. If Katelyn's eye level is 5.4 feet above the ground and she stands 11 feet from the wall, how tall is the wall to the nearest foot? ___?___ ft

🅡 **Go Online** You can complete an Extra Example online.

Practice

Go Online You can complete your homework online.

Example 1

Find the geometric mean between each pair of numbers.

1. 4 and 6

2. $\frac{1}{2}$ and 2

3. 4 and 25

4. 12 and 20

5. 17 and 3

6. 3 and 24

Example 2

REGULARITY Write a similarity statement identifying the three similar right triangles in each figure.

7.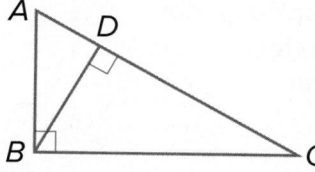

8.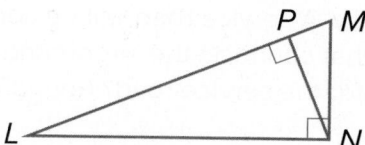

9.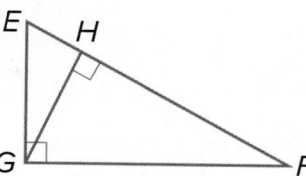

10.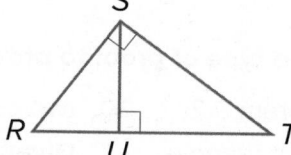

Example 3

Find the values of *x*, *y*, and *z*.

11.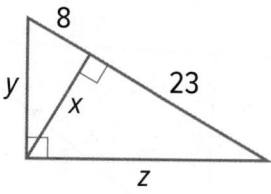

12.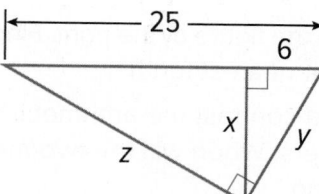

13.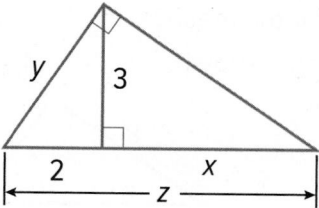

14.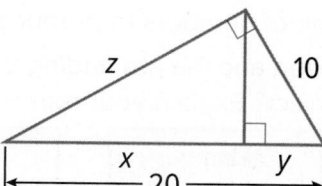

Example 4

15. USE A MODEL A museum has a famous statue on display. The curator places the statue in the corner of a rectangular room and builds a 15-foot-long railing in front of the statue. The railing forms a right triangle with the corner of the room. The legs of the triangle are 12 feet and 9 feet long. Approximate how close visitors will be able to get to the statue. Draw a diagram to model the situation.

16. **USE A MODEL** Noah wants to take a picture of a beach front. He wants to make sure two palm trees located at points *A* and *B* are just inside the edges of the photograph. He walks out on a walkway that goes over the ocean to get the shot. Point *A* is 90 feet from the entrance of the walkway, and point *B* is 40 feet from the entrance. If the walkway is perpendicular to *AB*, and Noah's camera has a viewing angle of 90°, at what distance down the walkway should Noah stop to take his photograph? Draw a diagram to model the situation.

17. **CIVIL ENGINEERING** An airport, a factory, and a shopping center are at the vertices of a right triangle formed by three highways. The airport and factory are 6.0 miles apart. Their distances from the shopping center are 3.6 miles and 4.8 miles, respectively. A service road will be constructed from the shopping center to the highway that connects the airport and factory. What is the shortest possible length for the service road? Round to the nearest hundredth.

Mixed Exercises

18. **REASONING** The geometric mean of a number and four times the number is 22. What is the number?

PROOF Write the specified type of proof to prove each theorem.

19. paragraph proof Theorem 9.2

   Given: △*ADC* is a right triangle.

   $\overline{DB}$ is an altitude of △*ADC*.

   Prove: $\frac{AB}{DB} = \frac{DB}{CB}$

20. two-column proof Theorem 9.3

   Given: ∠*ADC* is a right angle.

   $\overline{DB}$ is an altitude of △*ADC*.

   Prove: $\frac{AB}{AD} = \frac{AD}{AC}, \frac{BC}{DC} = \frac{DC}{AC}$

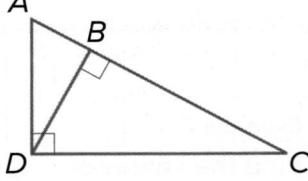

🧠 **Higher-Order Thinking Skills**

21. **PERSEVERE** Refer to the figure at the right. Find the values of *x*, *y*, and *z*. Round to the nearest tenth.

22. **WRITE** Compare and contrast the arithmetic and geometric means of two numbers. When will the two means be equal? Justify your reasoning.

23. **CREATE** Find two pairs of whole numbers with a geometric mean that is also a whole number. What condition must be met in order for a pair of numbers to produce a whole-number geometric mean?

24. **FIND THE ERROR** Aiden and Tia are finding the value of *x* in the triangle shown. Is either of them correct? Explain your reasoning.

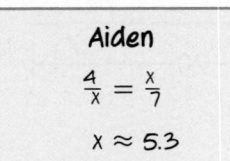

Aiden

$\frac{4}{x} = \frac{x}{7}$

$x \approx 5.3$

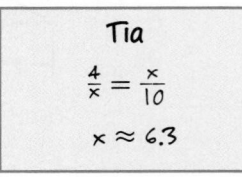

Tia

$\frac{4}{x} = \frac{x}{10}$

$x \approx 6.3$

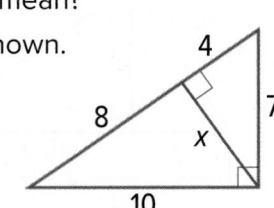

25. **ANALYZE** Determine whether each statement is *sometimes*, *always*, or *never* true. Justify your argument.
    a. The geometric mean for consecutive positive integers is the mean of the two numbers.
    b. The geometric mean for two perfect squares is a positive integer.
    c. The geometric mean for two positive integers is another integer.

Pythagorean Theorem and Its Converse

Explore Proving the Pythagorean Theorem

 Online Activity Use dynamic geometry software to complete the Explore.

> ⓠ **INQUIRY** How can you use triangle similarity to prove the Pythagorean Theorem? ×

Learn The Pythagorean Theorem

The Pythagorean Theorem relates the lengths of the hypotenuse and legs of a right triangle.

Theorem 4.4: Pythagorean Theorem

In a right triangle, the sum of the squares of the lengths of the legs is equal to the square of the length of the hypotenuse.

Go Online A proof of the Pythagorean Theorem is available.

A **Pythagorean triple** is a set of three nonzero whole numbers a, b, and c, such that $a^2 + b^2 = c^2$. The most common Pythagorean triples are shown below in the first row. The triples below them are found by multiplying each number in the triple by the same factor.

Common Pythagorean Triples			
3, 4, 5	**5, 12, 13**	**8, 15, 17**	**7, 24, 25**
6, 8, 10	10, 24, 26	16, 30, 34	14, 48, 50
9, 12, 15	15, 36, 39	24, 45, 51	21, 72, 75
$3x, 4x, 5x$	$5x, 12x, 13x$	$8x, 15x, 17x$	$7x, 24x, 25x$

Example 1 Find Missing Measures by Using the Pythagorean Theorem

Find the value of x.

The side opposite the right angle is the hypotenuse, so $c = x$.

$$a^2 + b^2 = c^2 \quad \text{Pythagorean Theorem}$$
$$17^2 + 7^2 = x^2 \quad a = 17, b = 7, \text{ and } c = x$$
$$289 + 49 = x^2 \quad \text{Simplify.}$$
$$338 = x^2 \quad \text{Add.}$$
$$\sqrt{338} = x \quad \text{Take the positive square root of each side.}$$
$$13\sqrt{2} = x \quad \text{Simplify.}$$

Go Online You can complete an Extra Example online.

Today's Goals
• Use the Pythagorean Theorem to solve problems involving right triangles.
• Classify triangles using the converse of the Pythagorean Theorem.

Today's Vocabulary
Pythagorean triple

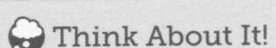 **Go Online**
An alternate method is available for this example.

☁ Talk About It!

Draw a right triangle with side lengths that form a Pythagorean triple. If you double the length of each side, is the resulting triangle *acute*, *right*, or *obtuse*? if you halve the length of each side? Justify your argument.

Example 2 Use a Pythagorean Triple

Use a Pythagorean triple to find the value of x.

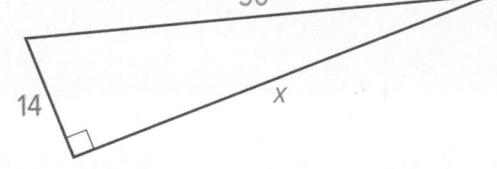

Notice that 14 and 50 are both multiples of 2, because 14 = 2 · 7 and 50 = 2 · 25. Because 7, 24, 25 is a Pythagorean triple, the missing leg length x is 2 · 24 or 48.

Check

Use a Pythagorean triple to find the value of x.

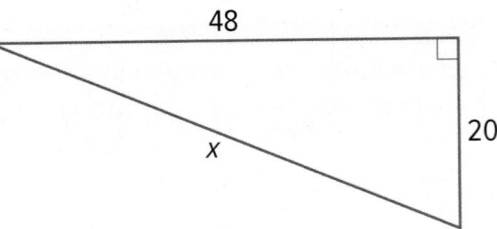

$x = $ ___?___

___

🌐 Example 3 Use the Pythagorean Theorem

ZIP LINING **A summer camp is building a new zip lining course. The designer of the course wants the last zip line to start at a platform 450 meters above the ground and end 775 meters away from the base of the platform. How long must the zip line be to meet the designer's specifications?**

Step 1 Visualize and describe the situation.

The base of the platform and the ground should be approximately perpendicular. We need to find the length of the zip line, which is the hypotenuse of the right triangle. Draw a diagram that models the situation.

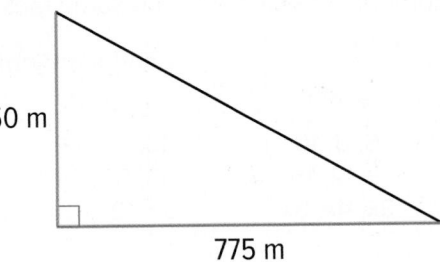

☁ Think About It!

Is your answer reasonable? Use estimation to justify your reasoning.

Step 2 Find the length of the zip line.

Use the Pythagorean Theorem to find the length x of the zip line.

$$450^2 + 775^2 = x^2 \quad \text{Pythagorean Theorem}$$
$$202{,}500 + 600{,}625 = x^2 \quad \text{Simplify.}$$
$$803{,}125 = x^2 \quad \text{Add.}$$
$$\sqrt{803{,}125} = x \quad \text{Take the positive square root of each side.}$$
$$25\sqrt{1285} = x \quad \text{Simplify.}$$

So, the length of the zip line is $25\sqrt{1285}$ or about 896.17 meters.

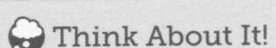 **Go Online** You can complete an Extra Example online.

Check

RAMPS Lincoln High School is installing more wheelchair ramps around the school. One of the new ramps needs to have a base that is 12 feet long and reaches a height of 1 foot. If the side of the ramp forms a right triangle, how long should the inclined surface of the ramp be?

Find the exact length of the inclined surface of the ramp.

____?____ ft

Learn Converse of the Pythagorean Theorem

The converse of the Pythagorean Theorem also holds. You can use this theorem to determine whether a triangle is a right triangle given the measures of all three sides.

> **Theorem 4.5: Converse of the Pythagorean Theorem**
>
> If the sum of the squares of the lengths of the shortest sides of a triangle is equal to the square of the length of the longest side, then the triangle is a right triangle.

Go Online
A proof of Theorem 4.5 is available.

You can also use side lengths to classify a triangle as acute or obtuse.

> **Theorem 4.6**
>
> If the square of the length of the longest side of a triangle is less than the sum of the squares of the lengths of the other two sides, then the triangle is an acute triangle.
>
>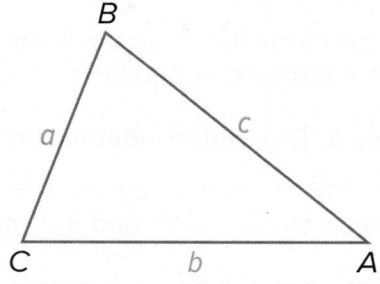
>
> **Example**
>
> If $c^2 < a^2 + b^2$, then $\triangle ABC$ is acute.

Study Tip

Determining the Longest Side If the measures of any of the sides of a triangle are expressed as radicals, you may wish to use a calculator to determine which side is the longest.

> **Theorem 4.7**
>
> If the square of the length of the longest side of a triangle is greater than the sum of the squares of the lengths of the other two sides, then the triangle is an obtuse triangle.
>
>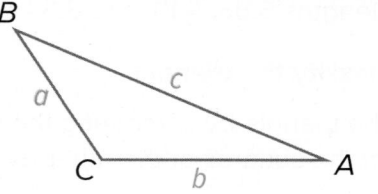
>
> **Example**
>
> If $c^2 > a^2 + b^2$, then $\triangle ABC$ is obtuse.

You will prove Theorems 4.6 and 4.7 in Exercises 23 and 24, respectively.

Example 4 Classify Triangles

Determine whether the points $A(2, 2)$, $B(5, 7)$, and $C(10, 6)$ can be the vertices of a triangle. If so, classify the triangle as *acute*, *right*, or *obtuse*. Justify your answer.

Step 1 Calculate the measures of the sides.

Use the Distance Formula to calculate the measures of $\overline{AB}$, $\overline{BC}$, and $\overline{AC}$.

$$AB = \sqrt{(5-2)^2 + (7-2)^2}$$
$$= \sqrt{3^2 + 5^2}$$
$$= \sqrt{34}$$

$$BC = \sqrt{(10-5)^2 + (6-7)^2}$$
$$= \sqrt{5^2 + (-1)^2}$$
$$= \sqrt{26}$$

$$AC = \sqrt{(10-2)^2 + (6-2)^2}$$
$$= \sqrt{8^2 + 4^2}$$
$$= \sqrt{80}$$
$$= 4\sqrt{5}$$

Use a calculator to approximate the measure of each side. So, $AB \approx 5.83$, $BC \approx 5.10$, and $AC \approx 8.94$.

Step 2 Determine whether the measures can form a triangle.

Use the Triangle Inequality Theorem to determine whether the measures 5.83, 5.10, and 8.94 can form a triangle.

$$5.83 + 5.10 > 8.94 ✓ \qquad 5.83 + 8.94 > 5.10 ✓ \qquad 5.10 + 8.94 > 5.83 ✓$$

The side lengths 5.83, 5.10, and 8.94 can form a triangle.

Step 3 Classify the triangle.

Classify the triangle by comparing the square of the longest side to the sum of the squares of the other two sides.

$c^2 \overset{?}{=} a^2 + b^2$	Compare c^2 and $a^2 + b^2$.
$8.94^2 \overset{?}{=} 5.83^2 + 5.10^2$	$c = 8.94$, $a = 5.83$, and $b = 5.10$
$79.9 > 60.0$	Simplify and compare.

Because $c^2 > a^2 + b^2$, the triangle is obtuse.

Check

Determine whether the points $J(1, 6)$, $K(3, 2)$, and $L(5, 3)$ can be the vertices of a triangle. If so, classify the triangle as *acute*, *right*, or *obtuse*.

 **Go Online** You can complete an Extra Example online.

Study Tip

Approximations
When finding the side lengths of a triangle using the Distance Formula, it may be easier to work with the side lengths after using a calculator to approximate their measures. However, when classifying a triangle, your final calculations will be more accurate if you keep the side lengths in radical form.

⊙ **Think About It!**
Does your conclusion seem reasonable? Explain.

Practice

Go Online You can complete your homework online.

Example 1

Find the value of x.

1.
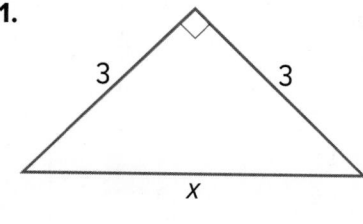

2.
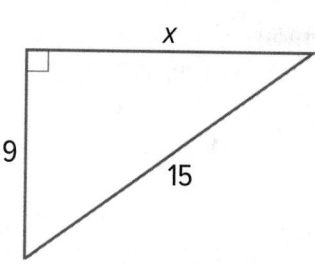

3.
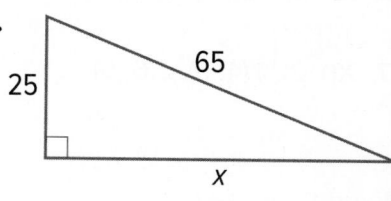

4.
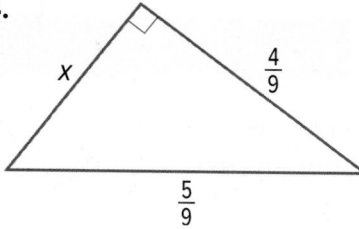

5.
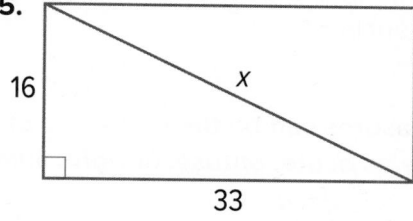

6.
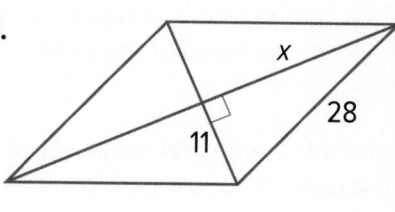

Example 2

Use a Pythagorean Triple to find the value of x.

7.
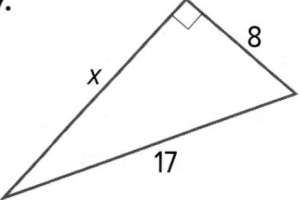

8.
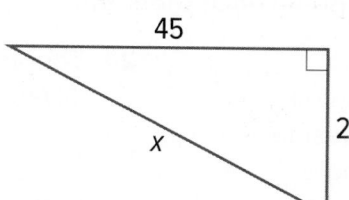

9.
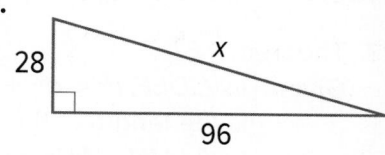

10.
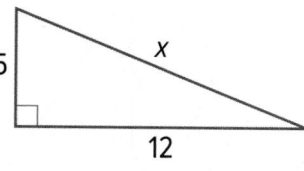

11.
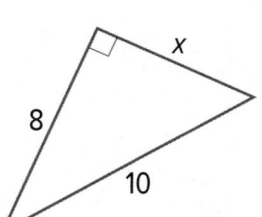

12.
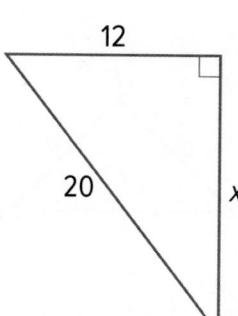

Example 3

13. **CONSTRUCTION** The bottom end of a ramp at a warehouse is 10 feet from the base of the main dock, and the ramp is 11 feet long. How high is the dock?

14. **FLIGHT** An airplane lands at an airport 60 miles east and 25 miles north of where it took off. How far apart are the two airports?

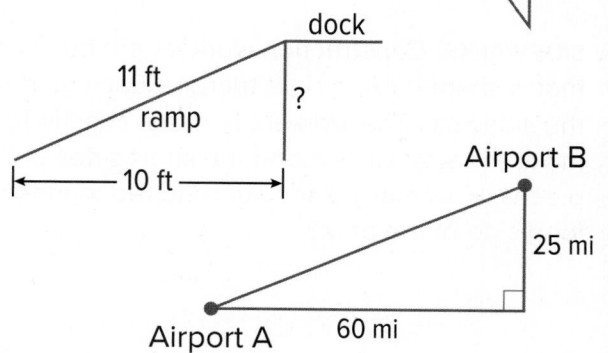

Example 4

Determine whether the points X, Y, and Z can be the vertices of a triangle. If so, classify the triangle as *acute*, *right*, or *obtuse*. Justify your answer.

15. X(−3, −2), Y(−1, 0), Z(0, −1)

16. X(−7, −3), Y(−2, −5), Z(−4, −1)

17. X(1, 2), Y(4, 6), Z(6, 6)

18. X(3, 1), Y(3, 7), Z(11, 1)

Mixed Exercises

19. TETHERS To help support a flag pole, a 50-foot-long tether is tied to the pole at a point 40 feet above the ground. The tether is pulled taut and tied to an anchor in the ground. How far away from the base of the pole is the anchor?

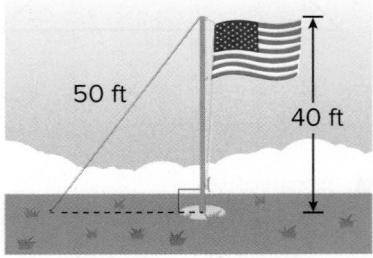

Determine whether each set of measures can be the measures of the sides of a triangle. If so, classify the triangle as *acute*, *obtuse*, or *right*. Justify your answer.

20. $\sqrt{5}, \sqrt{12}, \sqrt{13}$

21. $2, \sqrt{8}, \sqrt{12}$

22. 9, 40, 41

PROOF Write a two-column proof to prove each theorem.

23. Theorem 4.6
   Given: In △DEF, $f^2 < d^2 + e^2$ where f is the length of the longest side. In △LMN, ∠M is a right angle.
   Prove: △DEF is an acute triangle.

24. Theorem 4.7
   Given: In △ABC, $c^2 > a^2 + b^2$ where c is the length of the longest side. In △TUV, ∠V is a right angle.
   Prove: △ABC is an obtuse triangle.

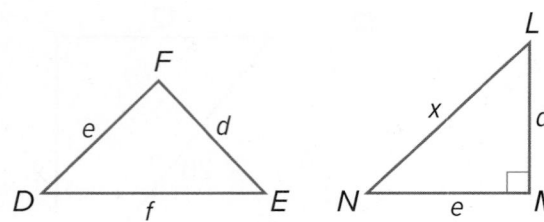

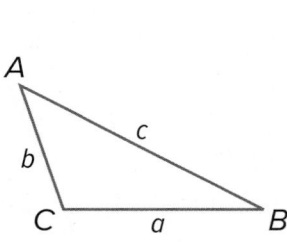

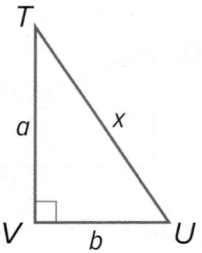

25. SIDEWALKS Construction workers are building a marble sidewalk around a park that is shaped like a right triangle. Each marble slab adds 2 feet to the length of the sidewalk. The workers find that exactly 1071 and 1840 slabs are required to make the sidewalks along the short sides of the park, not counting corner pieces. How many slabs are required to make the sidewalk that runs along the long side of the park?

Find the perimeter and area of each figure.

26.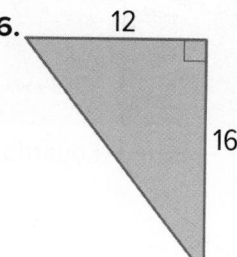
12
16

27.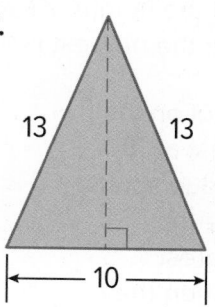
13   13
10

28.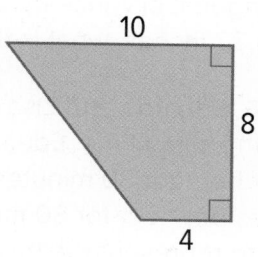
10
8
4

29. The sides of a triangle have measures of x, $x + 5$, and 25. If the measure of the longest side is 25, what value of x makes the triangle a right triangle?

30. **PRECISION** The sides of a triangle have measures of $2x$, 8, and 12. If the measure of the longest side is $2x$, what values of x make the triangle acute?

31. **REASONING** A redwood tree in a national park is 20 meters tall. After it is struck by lightning, the tree breaks and falls over, as shown in the figure. The top of the tree lands at a point 16 feet from the centerline of the tree. A park ranger wants to know the height of the remaining stump of the tree.

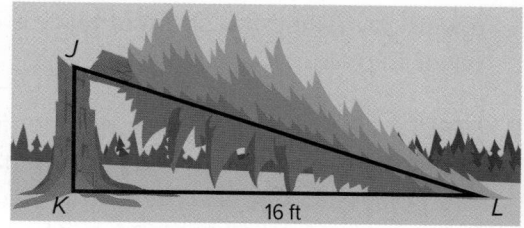

   a. The ranger lets x represent the height of the stump, $\overline{JK}$. Explain how the ranger can write an expression for the length of $\overline{JL}$. Then write an equation that can be used to solve the problem.

   b. Show how to solve the equation from part **a** to find the height of the stump.

32. **CONSTRUCT ARGUMENTS** Valeria and Sanjia are staking out a garden that has one pair of opposite sides measuring 30 feet and the other pair of sides measuring 40 feet. Using only a 60-foot-long tape measure, how can they be sure that their garden is a rectangle?

   a. Draw a model of the garden with diagonal t. Let $p = 30$ and $q = 40$.

   b. If the garden is a rectangle, what must be true about p, q, and t? Why?

   c. Sanjia measures the diagonal and finds that it is 50 feet long. Is there enough information to determine whether their garden is a rectangle? Explain.

Find the value of x.

33.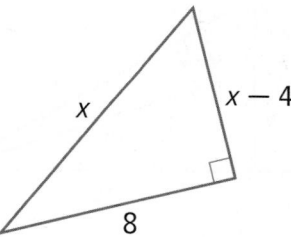
x
$x - 4$
8

34.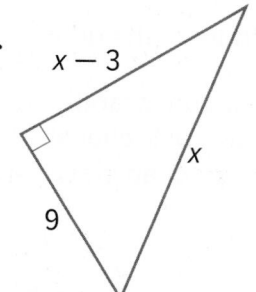
$x - 3$
x
9

35.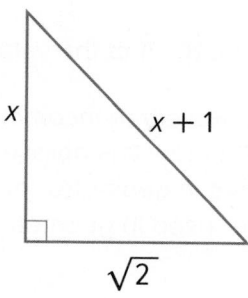
x
$x + 1$
$\sqrt{2}$

36. HDTV The screen aspect ratio, or the ratio of the width to the height, of a high-definition television is 16:9. The size of the television is given by the diagonal distance across the screen. If the height of Raj's HDTV screen is 32 inches, what is the screen size to the nearest inch?

37. DISTANCE Eduardo and Lisa both leave school on their bikes at the same time. Eduardo rides due east at 18 miles per hour for 30 minutes and Lisa rides due south at 16 miles per hour for 30 minutes. Complete the diagram to represent the problem. To the nearest hundredth of a mile, how far apart are they when they stop riding their bikes?

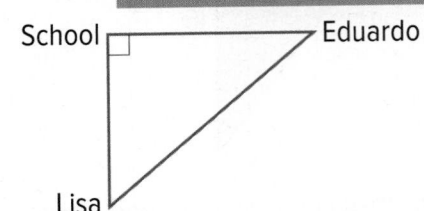

38. OFFICE PARK An office park has a rectangular lawn with the dimensions shown. Employees often take a shortcut by walking from *P* to *R*, rather than from *P* to *S* to *R*. What is the total distance an employee saves in a week by taking the shortcut twice a day for five days? Explain.

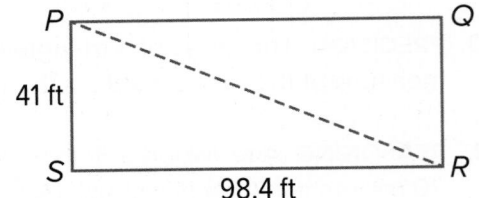

39. STRUCTURE Ms. Jones assigned her fifth-period geometry class the following problem. Let *m* and *n* be two positive integers with $m > n$.

Let $a = m^2 - n^2$, $b = 2mn$, and $c = m^2 + n^2$.

a. Show that there is a right triangle with side lengths *a*, *b*, and *c*.

b. Copy and complete the table shown at the right.

c. Find a Pythagorean triple that corresponds to a right triangle with a hypotenuse $25^2 = 625$ units long. (*Hint:* Use the table you completed for part **b** to find two positive integers *m* and *n* with $m > n$ and $m^2 + n^2 = 625$.)

m	n	a	b	c
2	1	3	4	5
3	1			
3	2			
4	1			
4	2			
4	3			
5	1			

40. ANALYZE *True* or *false*? Any two right triangles with the same hypotenuse have the same area. Explain your reasoning.

41. CREATE Draw a right triangle with side lengths that form a Pythagorean triple. If you double the length of each side, is the resulting triangle *acute*, *right*, or *obtuse*? if you halve the length of each side? Explain.

42. PERSEVERE Find the value of *x* in the figure at the right.

43. WRITE Research *incommensurable magnitudes*, and describe how this phrase relates to the use of irrational numbers in geometry. Include one example of an irrational number used in geometry.

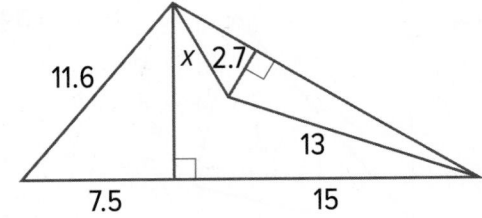

Coordinates in Space

Explore Proving the Distance Formula in Space

🏵 **Online Activity** Use the guiding exercises to complete the Explore.

> ⓠ **INQUIRY** How can you prove the Distance Formula in Space?

Learn Coordinates in Space

You have used an ordered pair of two coordinates to describe the location of a point on the coordinate plane. Because space has three dimensions, an **ordered triple** is required to locate a point in space. An ordered triple is three numbers given in a specific order. A point in space is represented by (x, y, z) where x, y, and z are real numbers. A three-dimensional coordinate system has three axes: the x-, y-, and z-axes. The x- and y-axes lie on a horizontal plane, and the z-axis is vertical. The three axes divide space into eight **octants**. In octant 1, all of the coordinates of a point are positive.

Key Concept · Distance and Midpoint Formulas in Space

A has coordinates $A(x_1, y_1, z_1)$, and B has coordinates $B(x_2, y_2, z_2)$.

Distance Formula:

$$AB = \sqrt{(x_2 - x_1)^2 + (y_2 - y_1)^2 + (z_2 - z_1)^2}$$

Midpoint Formula: $M\left(\dfrac{x_1 + x_2}{2}, \dfrac{y_1 + y_2}{2}, \dfrac{z_1 + z_2}{2}\right)$

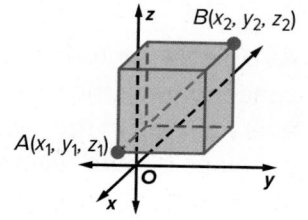

Example 1 Graph a Rectangular Solid

Graph a rectangular solid that contains G(−3, 4, 2) and the origin as vertices. Label the coordinates of each vertex.

Step 1 Plot the x-coordinate.

Plot the x-coordinate. Then draw a segment from the origin three units in the negative direction.

Step 2 Plot the y-coordinate.

To plot the y-coordinate, draw a segment four units in the positive direction from $(−3, 0, 0)$.

Step 3 Plot the z-coordinate.

To plot the z-coordinate, draw a segment two units in the positive direction from $(−3, 4, 0)$. Label the final point G.

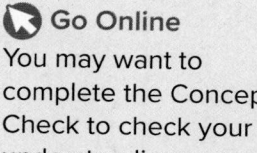

(continued on the next page)

Today's Goals
• Graph points and find distances between points on a three-dimensional coordinate plane.

Today's Vocabulary
ordered triple

octant

🏵 **Go Online**
You may want to complete the Concept Check to check your understanding.

Step 4 Draw the rectangular prism.

Draw the rectangular prism and label each vertex: $G(-3, 4, 2)$, $F(-3, 0, 2)$, $E(0, 0, 2)$, $D(0, 4, 2)$, $H(0, 4, 0)$, $I(0, 0, 0)$, $J(-3, 0, 0)$, and $K(-3, 4, 0)$.

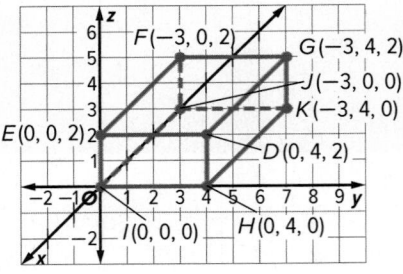

🌐 Example 2 Distance Formula in Space

MEDICINE Doctors use three-dimensional coordinate systems for medical imaging. Medical imaging and positioning systems allow doctors to analyze a person's anatomy from a three-dimensional perspective. On an anatomical coordinate system, the top of a man's spine is located at (5, 0, 65), and the bottom of his spine is located at (3, 0, −6). If each unit on the coordinate system represents a centimeter, what is the length of the man's spine?

Write the coordinates to find the length of the man's spine.

$D = \sqrt{(x_2 - x_1)^2 + (y_2 - y_1)^2 + (z_2 - z_1)^2}$ Distance Formula in Space

$= \sqrt{(5 - 3)^2 + (0 - 0)^2 + [65 - (-6)]^2}$ Substitution

$= \sqrt{2^2 + 0^2 + 71^2}$ Subtract.

$= \sqrt{5045}$ or about 71.0   Use a calculator.

So, the length of the man's spine is about 71.0 centimeters.

Check

AVIATION Air traffic controllers use three-dimensional coordinate space to track the locations of aircraft. By assigning coordinates to every aircraft in the sky, air traffic controllers can describe the positions of other aircraft to pilots to prevent accidents. An airplane is at (17, −14, 23), and the air traffic control tower is at (0, 0, 0). If each unit on the coordinate system represents a kilometer, what is the distance between the airplane and the tower? Round your answer to the nearest tenth, if necessary. ___?___ km

Example 3 Midpoint Formula in Space

Determine the coordinates of the midpoint M of $\overline{DE}$ with endpoints $D(-4, -3, 5)$ and $E(6, 1, -9)$.

$M = \left(\dfrac{x_1 + x_2}{2}, \dfrac{y_1 + y_2}{2}, \dfrac{z_1 + z_2}{2} \right)$ Midpoint Formula in Space

$= \left(\dfrac{-4 + 6}{2}, \dfrac{-3 + 1}{2}, \dfrac{5 + (-9)}{2} \right)$ Substitution

$= (1, -1, -2)$ Simplify.

Check

Determine the coordinates of the midpoint M of $\overline{AB}$ with endpoints $A(-7, 9, 4)$ and $B(5, -3, -4)$.

M ___?___ ___?___ ___?___

Practice

Go Online You can complete your homework online.

Example 1

Graph a rectangular solid that contains the given point and the origin as vertices. Label the coordinates of each vertex.

1. $A(2, 1, 5)$

2. $P(-1, 4, 2)$

3. $C(-2, 2, 2)$

4. $R(3, -4, 1)$

5. $H(4, 5, -3)$

6. $G(4, 1, -3)$

Example 2

Determine the distance between each pair of points.

7. $F(0, 0, 0)$ and $G(2, 4, 3)$

8. $X(-2, 5, -1)$ and $Y(9, 0, 4)$

9. $A(4, -6, 0)$ and $B(1, 0, 1)$

10. $C(8, 7, -2)$ and $D(0, 0, 0)$

11. AIR TRAFFIC CONTROLLERS An air traffic controller knows that the most recent position of an aircraft was at the coordinates shown on the three-dimensional coordinate system to the right. If the control tower is at $(0, 0, 0)$, then what is the distance between the aircraft and the tower if each unit on the coordinate system represents one mile? Round your answer to the nearest tenth, if necessary.

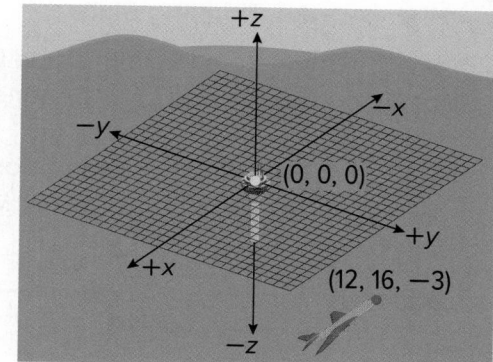

12. ANIMATORS An animator is using three-dimensional software to create the character shown. She labels the coordinates of point R and point T. Point R represents the nose of the dog, and point T represents the tip of the dog's tail. What is the distance between these two points in the animation? Round your answer to the nearest tenth, if necessary.

$R(-9, 2, -1)$ $T(9, -2, 2)$

Example 3

Determine the coordinates of the midpoint M of the segment joining each pair of points.

13. $K(-2, -4, -4)$ and $L(4, 2, 0)$

14. $W(-1, -3, -6)$ and $Z(-1, 5, 10)$

15. $R(3, 3, 4)$ and $V(5, 4, 13)$

16. $A(4, 6, -8)$ and $B(0, 0, 0)$

17. $C(8, 7, 11)$ and $D(2, 1, 8)$

18. $T(-1, -7, 9)$ and $U(5, -1, -6)$

Mixed Exercises

REGULARITY Determine the distance between each pair of points. Then determine the coordinates of the midpoint M of the segment joining the pair of points.

19. $P(-5, -2, -1)$ and $Q(-1, 0, 3)$

20. $J(1, 1, 1)$ and $K(-1, -1, -1)$

21. $F\left(\frac{3}{5}, 0, \frac{4}{5}\right)$ and $G(0, 3, 0)$

22. $G(1, -1, 6)$ and $H\left(\frac{1}{5}, -\frac{2}{5}, 2\right)$

23. $B(\sqrt{3}, 2, 2\sqrt{2})$ and $C(-2\sqrt{3}, 4, 4\sqrt{2})$

24. $S(6\sqrt{3}, 4, 4\sqrt{2})$ and $T(4\sqrt{3}, 5, \sqrt{2})$

25. PROOF Write a coordinate proof of the Midpoint Formula in Space.

    Given: Points $A(x_1, y_1, z_1)$ and $B(x_2, y_2, z_2)$; M is the midpoint of $\overline{AB}$.

    Prove: The coordinates of point M are $\left(\dfrac{x_1 + x_2}{2}, \dfrac{y_1 + y_2}{2}, \dfrac{z_1 + z_2}{2}\right)$.

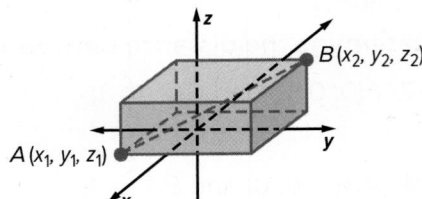

 Higher-Order Thinking Skills

26. WRITE Compare and contrast the Distance and Midpoint Formulas on the coordinate plane and in three-dimensional coordinate space.

27. FIND THE ERROR Camilla and Teion were asked to find the distance between the points $A(2, 5, -8)$ and $B(3, -1, 0)$. Who is correct? Explain your reasoning.

Camilla	Teion
$AB = \sqrt{(x_2 - x_1)^2 + (y_2 - y_1)^2 + (z_2 - z_1)^2}$	$AB = \sqrt{(x_2 - x_1)^2 + (y_2 - y_1)^2 + (z_2 - z_1)^2}$
$= \sqrt{(2 - 3)^2 + (5 - 1)^2 + (-8 - 0)^2}$	$= \sqrt{(2 - 3)^2 + (5 - (-1))^2 + (-8 - 0)^2}$
$= \sqrt{(-1)^2 + (4)^2 + (-8)^2}$	$= \sqrt{(2 - 3)^2 + (5 + 1)^2 + (-8 - 0)^2}$
$= \sqrt{1 + 16 + 64}$	$= \sqrt{(-1)^2 + (6)^2 + (-8)^2}$
$= \sqrt{81}$	$= \sqrt{1 + 36 + 64}$
$= 9$	$= \sqrt{101}$
	≈ 10.05

28. CREATE Graph a cube that has the following characteristics:
- the origin is one of the vertices,
- one of the edges lies on the negative y-axis, and
- one of the faces lies in the negative xz-plane.

29. PERSEVERE Suppose the sphere has a radius of 9 units and passes through point P. Find the missing z-coordinate of point P.

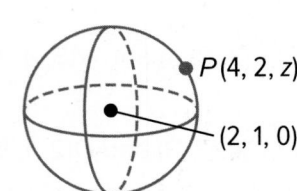

Special Right Triangles

Explore Properties of Special Right Triangles

Online Activity Use dynamic geometry software to complete the Explore.

> **INQUIRY** What is the relationship between side lengths in 45°-45°-90° and 30°-60°-90° triangles? ×

Learn 45°-45°-90° Triangles

The diagonal of a square forms two congruent isosceles right triangles. Because the base angles of an isosceles triangle are congruent, the measure of each acute angle is 90° ÷ 2 or 45°. Such a special right triangle is known as a **45°-45°-90° triangle**.

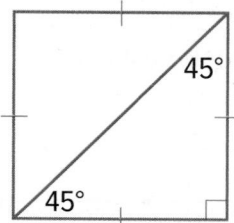

Theorem 4.8: 45°-45°-90° Triangle Theorem

In a 45°-45°-90° triangle, the legs ℓ are congruent and the length of the hypotenuse h is $\sqrt{2}$ times the length of a leg.

You will prove Theorem 4.8 in Exercise 39.

Example 1 Find the Hypotenuse Length Given an Angle Measure

Find the value of x.

The acute angles of a right triangle are complementary, so the measure of the third angle is 90 − 45 or 45°. Because this is a 45°-45°-90° triangle, use the 45°-45°-90° Triangle Theorem.

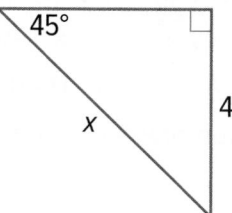

$h = \ell\sqrt{2}$ 45°-45°-90° Triangle Theorem

$x = 4\sqrt{2}$ Substitution

Check

Find the value of x.

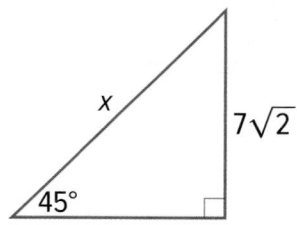

Go Online You can complete an Extra Example online.

Today's Goals
- Understand that by similarity, side ratios in 45-45-90 right triangles are related to the angles in the triangles.
- Understand that by similarity, side ratios in 30-60-90 right triangles are related to the angles in the triangles.

Today's Vocabulary
45°-45°-90° triangle

30°-60°-90° triangle

Talk About It!
Why do all 45°-45°-90° triangles have the same side length ratios? Use similarity to justify your reasoning.

Think About It!
How can you remember the ratios of the side lengths of a 45°-45°-90° triangle?

Example 2 Find the Hypotenuse Length Given a Side Measure

Find the value of *x*.

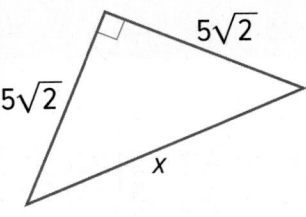

The legs of this right triangle have the same measure, so it is isosceles. Because this is a 45°-45°-90° triangle, use the 45°-45°-90° Triangle Theorem.

$$h = \ell\sqrt{2}$$ 45°-45°-90° Triangle Theorem

$$x = 5\sqrt{2} \cdot \sqrt{2}$$ Substitution

$$x = 5 \cdot 2 \text{ or } 10$$ Solve.

Check

Find the value of *x*.

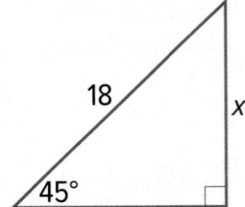

Study Tip

Rationalizing the Denominator You can use the properties of square roots to rationalize the denominator of a fraction with a radical. This involves multiplying the numerator and denominator by a factor that eliminates radicals in the denominator. In this example, $\frac{18}{\sqrt{2}}$ can be simplified by multiplying the numerator and denominator by $\sqrt{2}$.

Example 3 Find Leg Lengths in a 45°-45°-90° Triangle

Find the value of *x*.

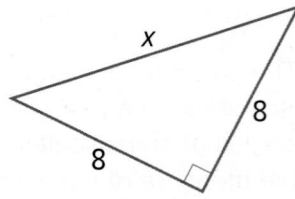

The acute angles of a right triangle are complementary, so the measure of the third angle is 90 − 45 or 45°. So, the triangle is a 45°-45°-90° triangle. Use the 45°-45°-90° Triangle Theorem to find the value of *x*.

$$h = \ell\sqrt{2}$$ 45°-45°-90° Triangle Theorem

$$18 = x\sqrt{2}$$ Substitution

$$\frac{18}{\sqrt{2}} = x$$ Divide each side by $\sqrt{2}$.

The value of *x* is $\frac{18}{\sqrt{2}}$ or $9\sqrt{2}$.

Check

Find the value of *x*.

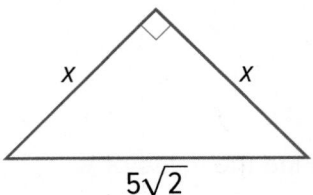

 Go Online You can complete an Extra Example online.

Learn 30°-60°-90° Triangles

A **30°-60°-90° triangle** is a special right triangle or right triangle with side lengths that share a special relationship. You can use an equilateral triangle to find this relationship.

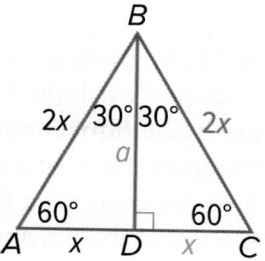

When an altitude is drawn from any vertex of an equilateral triangle, two congruent 30°-60°-90° triangles are formed. In the figure, $\triangle ABD \cong \triangle CBD$, so $\overline{AD} \cong \overline{CD}$. If $AD = x$, then $CD = x$ and $AC = 2x$. Because $\triangle ABC$ is equilateral, $AB = 2x$ and $BC = 2x$.

Use the Pythagorean Theorem to find a, the length of the altitude $\overline{BD}$, which is also the longer leg of $\triangle BDC$.

$a^2 + x^2 = (2x)^2$	Pythagorean Theorem
$a^2 + x^2 = 4x^2$	Simplify.
$a^2 = 3x^2$	Subtract x^2 from each side.
$a = x\sqrt{3}$	Simplify.

Theorem 4.9: 30°-60°-90° Triangle Theorem

In a 30°-60°-90° triangle, the length of the hypotenuse h is 2 times the length of the shorter leg s, and the longer leg ℓ is $\sqrt{3}$ times the length of the shorter leg.

You will prove Theorem 4.9 in Exercise 40.

Example 4 Find Leg Lengths in a 30°-60°-90° Triangle

Find the values of x and y.

Use the 30°-60°-90° Triangle Theorem to find the value of x, the length of the shorter side.

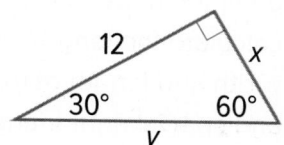

$s\sqrt{3} = \ell$	30°-60°-90° Triangle Theorem
$x\sqrt{3} = 12$	Substitution
$x = \dfrac{12}{\sqrt{3}}$ or $4\sqrt{3}$	Divide each side by $\sqrt{3}$.

Now use the 30°-60°-90° Triangle Theorem to find y, the length of the hypotenuse.

$h = 2s$	30°-60°-90° Triangle Theorem
$y = 2\left(\dfrac{12}{\sqrt{3}}\right)$	Substitution
$y = \dfrac{24}{\sqrt{3}}$ or $8\sqrt{3}$	Simplify.

Check

Find the values of x and y.

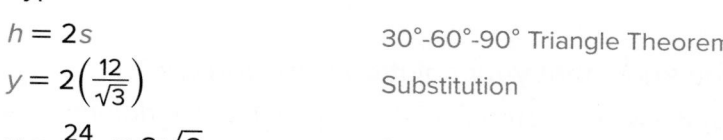

$x =$ ___?___ $y =$ ___?___

Go Online You can complete an Extra Example online.

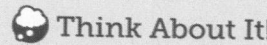

Think About It!

Ian states that in a 30°-60°- 90° triangle, sometimes the 30° angle is opposite the longer leg and the 60° angle is opposite the shorter leg. Do you *agree* or *disagree* with Ian? Justify your answer.

Study Tip

Use Ratios The lengths of the sides of a 30°-60°-90° triangle are in a ratio of 1 to $\sqrt{3}$ to 2 or 1: $\sqrt{3}$: 2.

Apply Example 5 Use Properties of 30°-60°-90° Triangles

JEWELRY **Destiny makes and sells upcycled earrings. The earrings shown are made from congruent equilateral triangles. Each triangle has a height of 2 centimeters. The hooks attached to the top of the earrings are 1 centimeter tall. Destiny needs to mail this pair of earrings to a customer. If she mails the earrings in a rectangular box, what width and length must the base of the box have so the earrings will fit if they are placed side by side in the bottom of the box?**

1 What is the task?

Describe the task in your own words. Then list any questions that you may have. How can you find answers to your questions?

Sample answer: To find the width and length of the box, I should calculate the height and width of the earrings. How does the information provided in the problem relate to the earrings? I can draw and label a diagram to represent the earrings.

2 How will you approach the task? What have you learned that you can use to help you complete the task?

Sample answer: I will draw and label the altitude of the triangle, calculate the length of the base of the triangle, and then calculate the width and length of the box. I will use what I know about equilateral and special right triangles to solve for missing measures.

3 What is your solution?

The altitude of the triangle measures 2 centimeters.

The length of the base of one earring is $\frac{4}{\sqrt{3}}$ or $\frac{4\sqrt{3}}{3}$.

The width of the box must be at least $\frac{8}{\sqrt{3}}$ or about 4.62 centimeters.

The length of the box must be at least 5 centimeters.

4 How can you know that your solution is reasonable?

Write About It! Write an argument that can be used to defend your solution.

Sample answer: Because the equilateral triangles are congruent, it makes sense that the width and length of the box are twice the length of the base and the height of one equilateral triangle, respectively.

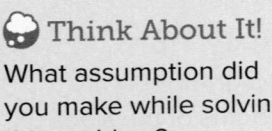 **Think About It!**
What assumption did you make while solving this problem?

 Go Online You can complete an Extra Example online.

Practice

🔵 **Go Online** You can complete your homework online.

Example 1

REGULARITY **Find the value of x.**

1.
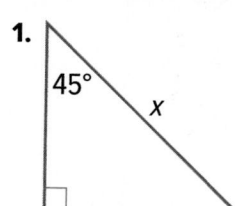
45°
x
7

2.
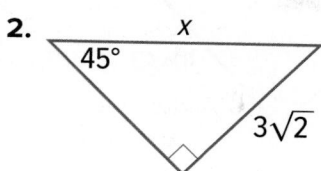
x
45°
$3\sqrt{2}$

3.
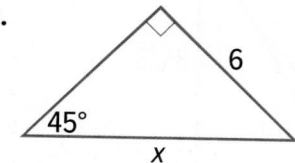
6
45°
x

4.
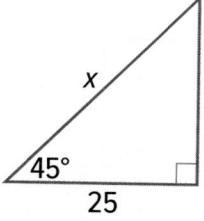
x
45°
25

5.
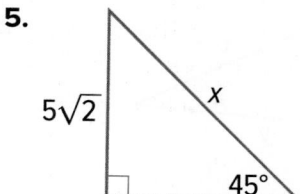
$5\sqrt{2}$
x
45°

6.
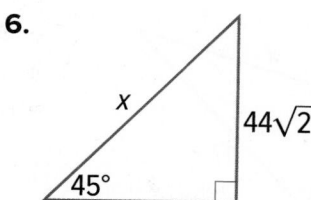
x
$44\sqrt{2}$
45°

Example 2

Find the value of x.

7.
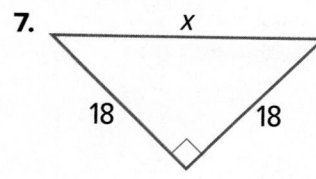
x
18   18

8.
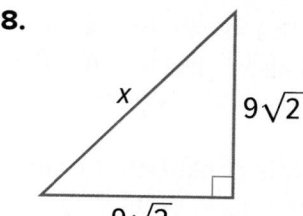
x
$9\sqrt{2}$
$9\sqrt{2}$

9.
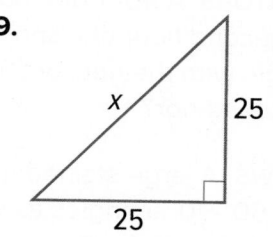
x
25
25

10.
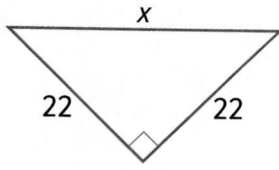
x
22   22

11.
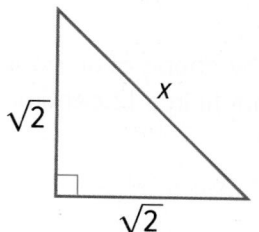
$\sqrt{2}$
x
$\sqrt{2}$

12.
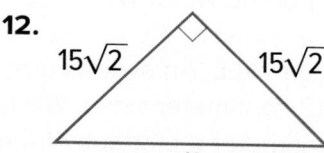
$15\sqrt{2}$ $15\sqrt{2}$
x

Example 3

Find the value of x.

13.
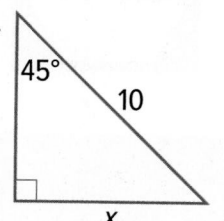
45°
10
x

14.
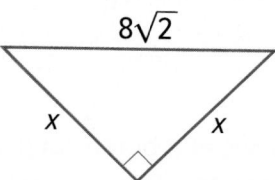
$8\sqrt{2}$
x   x

15.
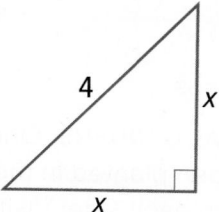
4
x
x

16.
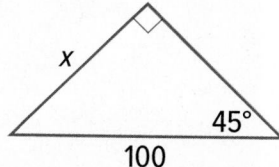
x
45°
100

17.
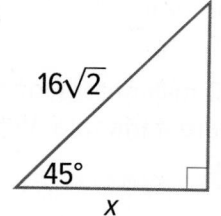
$16\sqrt{2}$
45°
x

18.
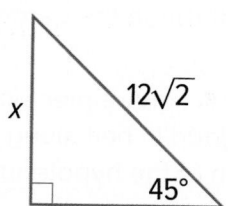
x
$12\sqrt{2}$
45°

Lesson 4-4 • Special Right Triangles   **183**

Example 4

Find the values of *x* and *y*.

19.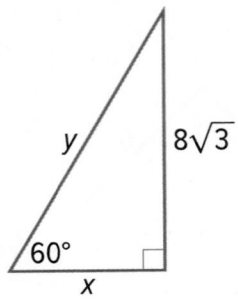

20.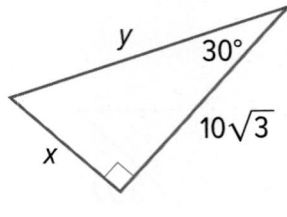

21.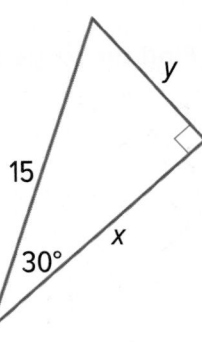

22.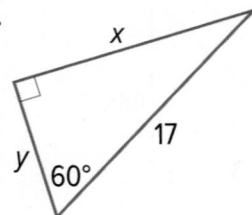

23.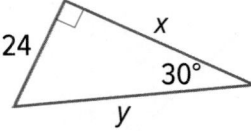

24.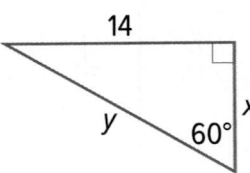

Example 5

25. ESCALATORS A 40-foot-long escalator rises from the first floor to the second floor of a shopping mall. The escalator makes a 30° angle with the horizontal. How high above the first floor is the second floor?

26. WINDOWS A large stained glass window is constructed from six 30°-60°-90° triangles as shown in the figure. What is the height of the window?

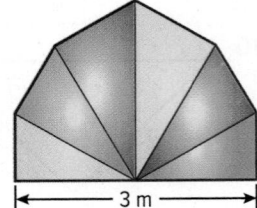

27. USE A MODEL An award certificate is in the shape of an equilateral triangle with 12-centimeter sides. Will the certificate fit in a 12-centimeter by 10-centimeter rectangular frame? Explain.

28. PRECISION A box of chocolates shaped like a regular hexagon is placed snugly inside a rectangular box as shown in the figure. If each side length of the hexagon is 3 inches, what are the dimensions of the rectangular box?

Mixed Exercises

29. BOTANICAL GARDENS One of the displays at a botanical garden is an herb garden planted in the shape of a square. The square measures 6 yards on each side. Visitors can view the herbs from a diagonal pathway through the garden. How long is the pathway?

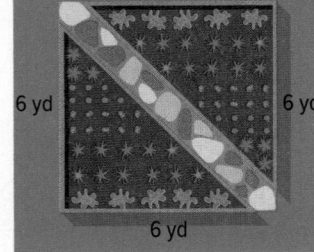

30. ORIGAMI A square piece of paper that is 150 millimeters long on each side is folded in half along a diagonal to create a triangle. What is the length of the hypotenuse of this triangle?

31. REASONING Kim and Yanika are watching a movie in a movie theater. Yanika is sitting x feet from the screen and Kim is 15 feet behind Yanika. The angle that Kim's line of sight to the top of the screen makes with the horizontal is 30°. The angle that Yanika's line of sight to the top of the screen makes with the horizontal is 45°.

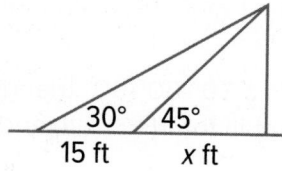

a. How high is the top of the screen in terms of x?

b. What is $\frac{x + 15}{x}$?

c. How far is Yanika from the screen? Round your answer to the nearest tenth.

32. STRUCTURE Each triangle in the figure is a 45°-45°-90° triangle. Find the value of x.

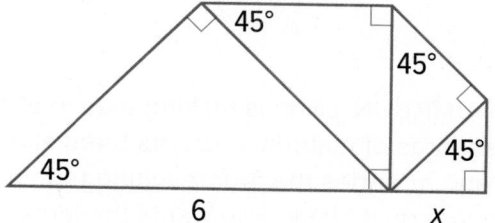

Find the values of x and y.

33.

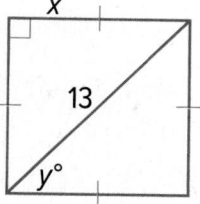

34.

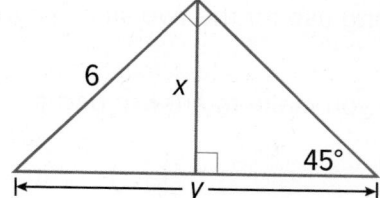

35.

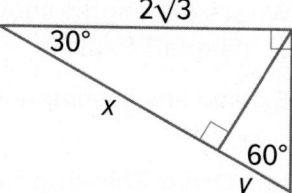

36.

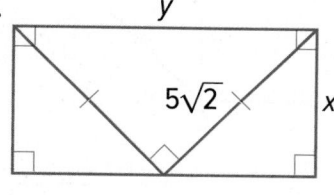

37.

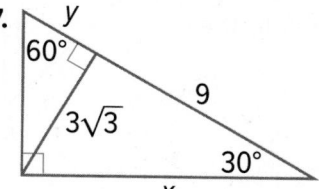

38.

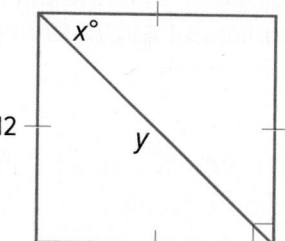

PROOF **Write a paragraph proof to prove each theorem.**

39. 45°-45°-90° Triangle Theorem
  Given: 45°-45°-90° triangle with a leg of length ℓ and a hypotenuse of length h
  Prove: The legs are congruent, and $h = \ell\sqrt{2}$.

40. 30°-60°-90° Triangle Theorem
  Given: equilateral $\triangle DEF$ with sides of length $2x$ and an altitude of length d
  Prove: $DG = x$; $d = x\sqrt{3}$

41. △XYZ is a 45°-45°-90° triangle with right angle Z. Find the coordinates of X in Quadrant I for Y(−1, 2) and Z(6, 2).

42. △EFG is a 30°-60°-90° triangle with m∠F = 90°. Find the coordinates of E in Quadrant III for F(−3, −4) and G(−3, 2). $\overline{FG}$ is the longer leg.

43. USE TOOLS Melody is in charge of building a ramp for a loading dock. According to the plan, the ramp makes a 30° angle with the ground. The plan also states that $\overline{ST}$ is 4 feet longer than $\overline{RS}$. Use a calculator to find the lengths of the three sides of the ramp to the nearest thousandth.

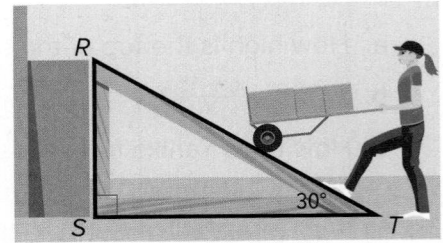

44. STATE YOUR ASSUMPTION Liling is making a quilt. She starts with two small squares of material and cuts them along the diagonal. Then she arranges the four resulting triangles to make a large square quilt block. She wants the large quilt block to have an area of 36 square inches.

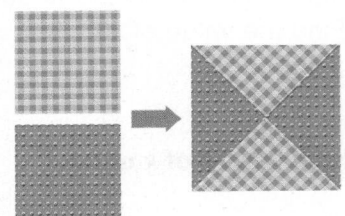

    a. What side lengths should Liling use for the two small squares of material? Explain.

    b. Explain any assumption that you make to answer part **a**.

🧠 **Higher-Order Thinking Skills**

45. PERSEVERE Find the perimeter of quadrilateral ABCD. Round your answer to the nearest tenth.

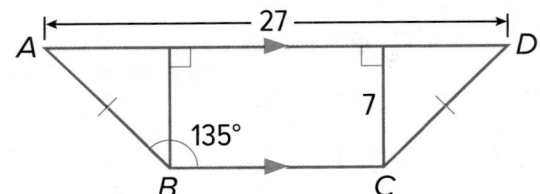

46. WRITE Why are some right triangles considered *special*?

47. FIND THE ERROR Carmen and Audrey want to find the value of x in the triangle shown. Who is correct? Explain your reasoning.

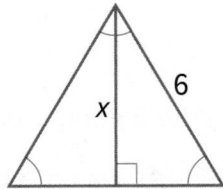

Carmen	Audrey
$x = \dfrac{6\sqrt{3}}{2}$	$x = \dfrac{6\sqrt{2}}{2}$
$x = 3\sqrt{3}$	$x = 3\sqrt{2}$

48. ANALYZE The ratio of the measure of the angles of a triangle is 1:2:3. The length of the shortest side is 8. What is the perimeter of the triangle? Round your answer to the nearest tenth.

49. CREATE Draw a rectangle that has a diagonal twice as long as its width. Then write an equation to find the length of the rectangle.

Trigonometry

Explore Sines and Cosines of Complementary Angles

 Online Activity Use dynamic geometry software to complete the Explore.

> **ⓠ INQUIRY** What can trigonometric ratios tell you about the relationship between the complementary angles in a right triangle?

Explore Trigonometry and Similarity

 Online Activity Use dynamic geometry software to complete the Explore.

> **ⓠ INQUIRY** If two right triangles have the same angle measure, what do you know about the trigonometric ratios of the angle?

Learn Trigonometry

The word **trigonometry** comes from the Greek terms *trigon*, meaning triangle, and *metron*, meaning measure. So the study of trigonometry involves triangle measurement. A **trigonometric ratio** is a ratio of the lengths of two sides of a right triangle.

The names of the three most common trigonometric ratios are given below.

Key Concept • Trigonometric Ratios

Sine: If △ABC is a right triangle, then the sine of each acute angle in △ABC is the ratio of the length of the leg opposite that angle to the length of the hypotenuse.

$\sin A = \frac{\text{opp}}{\text{hyp}}$ or $\frac{a}{c}$; $\sin B = \frac{\text{opp}}{\text{hyp}}$ or $\frac{b}{c}$

Cosine: If △ABC is a right triangle, then the cosine of each acute angle in △ABC is the ratio of the length of the leg adjacent to that angle to the length of the hypotenuse.

$\cos A = \frac{\text{adj}}{\text{hyp}}$ or $\frac{b}{c}$; $\cos B = \frac{\text{adj}}{\text{hyp}}$ or $\frac{a}{c}$

Tangent: If △ABC is a right triangle, then the tangent of each acute angle in △ABC is the ratio of the length of the leg opposite that angle to the length of the leg adjacent to that angle.

$\tan A = \frac{\text{opp}}{\text{adj}}$ or $\frac{a}{b}$; $\tan B = \frac{\text{opp}}{\text{adj}}$ or $\frac{b}{a}$

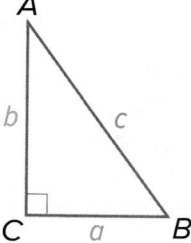

 Go Online You can complete an Extra Example online.

Today's Goal
- Solve problems by using the trigonometric ratios for acute angles.
- Solve problems by using the inverse trigonometric ratios for acute angles.

Today's Vocabulary
trigonometry
trigonometric ratio
sine
cosine
tangent
inverse sine
inverse cosine
inverse tangent
solving a triangle

Study Tip

Memorizing Trigonometric Ratios SOH-CAH-TOA is a mnemonic device for learning the ratios for sine, cosine, and tangent using the first letter of each word in the ratios.

$\sin A = \frac{\text{opp}}{\text{hyp}}$ $\cos A = \frac{\text{adj}}{\text{hyp}}$
$\tan A = \frac{\text{opp}}{\text{adj}}$

Example 1 Find Trigonometric Ratios

Find sin *J*, cos *J*, tan *J*, sin *K*, cos *K*, and tan *K*. Express each ratio as a fraction and as a decimal to the nearest hundredth.

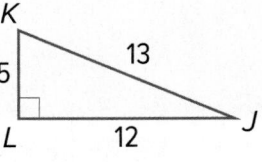

$$\sin J = \frac{opp}{hyp} = \frac{5}{13} \approx 0.38 \qquad \sin K = \frac{opp}{hyp} = \frac{12}{13} \approx 0.92$$

$$\cos J = \frac{adj}{hyp} = \frac{12}{13} \approx 0.92 \qquad \cos K = \frac{adj}{hyp} = \frac{5}{13} \approx 0.38$$

$$\tan J = \frac{opp}{adj} = \frac{5}{12} \approx 0.42 \qquad \tan K = \frac{opp}{adj} = \frac{12}{5} \approx 2.40$$

Special right triangles can be used to find the sine, cosine, and tangent of 30°, 45°, and 60° angles.

Example 2 Use a Special Right Triangle to Find Trigonometric Ratios

Use a special right triangle to express the sine of 60° as a fraction and as a decimal to the nearest hundredth.

Using the 30°-60°-90° Triangle Theorem, write the correct side lengths for each leg of the right triangle with *x* as the length of the shorter leg.

$$\sin 60° = \frac{opp}{hyp} \qquad \text{Definition of sine ratio}$$

$$= \frac{x\sqrt{3}}{2x} \qquad \text{Substitution}$$

$$\approx 0.87 \qquad \text{Use a calculator.}$$

🌐 Example 3 Estimate Measures by Using Trigonometry

ACCESSIBILITY Mathias builds a ramp so his sister can access the back door of their house. The 12-foot ramp to the house slopes upward from the ground at a 4° angle. What is the horizontal distance between the foot of the ramp and the house? What is the height of the ramp? Find the horizontal distance.

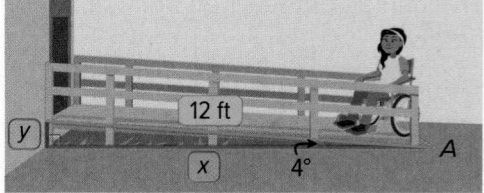

Let $m\angle A = 4°$. The horizontal distance between the foot of the ramp and the house is *x*, the measure of the leg adjacent to ∠*A*. The length of the ramp is the measure of the hypotenuse, 12 feet. Because the lengths of the leg adjacent to a given angle and the hypotenuse are involved, write an equation using the cosine ratio.

$$\cos A = \frac{adj}{hyp} \qquad \text{Definition of cosine ratio}$$

$$\cos 4° = \frac{x}{12} \qquad \text{Substitution}$$

$$12 \cos 4° = x \qquad \text{Multiply each side by 12.}$$

$$x \approx 11.97 \qquad \text{Use a calculator.}$$

The horizontal distance between the foot of the ramp and the house is about 11.97 feet.

Find the height.

The height of the ramp is y, the measure of the leg opposite from $\angle A$. Because the lengths of the leg opposite to a given angle and the hypotenuse are involved, write an equation using a sine ratio.

$$\sin A = \frac{\text{opp}}{\text{hyp}} \qquad \text{Definition of sine ratio}$$

$$\sin 4° = \frac{y}{12} \qquad \text{Substitution}$$

$$12 \cdot \sin 4° = y \qquad \text{Multiply each side by 12.}$$

$$y \approx 0.84 \qquad \text{Use a calculator.}$$

The height y of the ramp is about 0.84 feet or about 10 inches.

Learn Inverse Trigonometric Ratios

If you know the value of a trigonometric ratio for an acute angle, you can use a calculator to find the measure of the angle, which is the inverse of the trigonometric ratio.

Key Concept · Inverse Trigonometric Ratios		
Inverse Sine	Inverse Cosine	Inverse Tangent
Words		
If $\angle A$ is an acute angle and the sine of A is x, then the **inverse sine** of x is the measure of $\angle A$.	If $\angle A$ is an acute angle and the cosine of A is x, then the **inverse cosine** of x is the measure of $\angle A$.	If $\angle A$ is an acute angle and the tangent of A is x, then the **inverse tangent** of x is the measure of $\angle A$.
Symbols		
If $\sin A = x$, then $\sin^{-1}x = m\angle A$.	If $\cos A = x$, then $\cos^{-1}x = m\angle A$.	If $\tan A = x$, then $\tan^{-1}x = m\angle A$.

Example 4 Find Angle Measures by Using Inverse Trigonometric Ratios

Use a calculator to find $m\angle A$ to the nearest tenth.

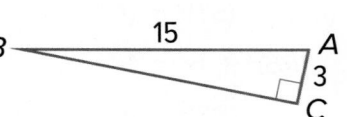

The measures given are those of the leg adjacent to $\angle A$ and the hypotenuse, so write an equation using the cosine ratio.

$$\cos A = \frac{3}{15} \text{ or } \frac{1}{5} \qquad \cos A = \frac{\text{adj}}{\text{hyp}}$$

$$\cos^{-1}\left(\frac{1}{5}\right) = m\angle A \qquad \text{Use the inverse cosine function.}$$

$$\text{So, } m\angle A \approx 78.5°. \qquad \text{Use a calculator.}$$

Go Online You can complete an Extra Example online.

Study Tip

Inverse Trigonometric Ratios The expression $\sin^{-1}x$ is read *the inverse sine of x* and is interpreted as the angle with sine x. Be careful not to confuse this notation with the notation for negative exponents. That is, $\sin^{-1}x \neq \frac{1}{\sin x}$. Instead, this notation is similar to the notation for an inverse function, $f^{-1}(x)$.

Go Online
You can watch a video to see how to solve a right triangle using trigonometry.

Study Tip

Graphing Calculators The second functions of the [sin] [cos] and [tan] keys are usually their inverses.

Talk About It!

What other method could you use to find $m\angle A$? Explain.

Check

Use a calculator to find $m\angle Z$ to the nearest tenth.

$m\angle Z = $ ___?___

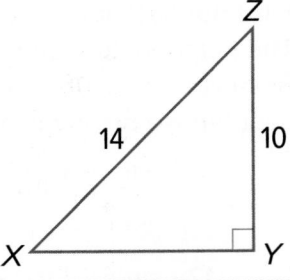

When you are given measurements to find the unknown angle and side measures of a triangle, this is known as **solving a triangle**. To solve a right triangle, you need to know:

- two side lengths or
- one side length and the measure of one acute angle.

Example 5 Solve a Right Triangle

Solve the right triangle. Round side and angle measures to the nearest tenth.

Use a sine ratio to find $m\angle R$.

$$\sin R = \frac{6}{9} \qquad \sin X = \frac{\text{opp}}{\text{hyp}}$$

$$\sin^{-1}\frac{6}{9} = m\angle R \qquad \text{Use the inverse sine function}$$

$$41.8 \approx m\angle R \qquad \text{Use a calculator.}$$

So, $m\angle R \approx 41.8°$.

Use known angles to find $m\angle T$.

$$m\angle R + m\angle T = 90° \qquad \text{Acute } \angle\text{s of rt } \triangle\text{s are comp.}$$

$$41.8° + m\angle T \approx 90° \qquad m\angle R \approx 41.8$$

$$m\angle T \approx 48.2° \qquad \text{Subtract 41.8 from each side.}$$

So, $m\angle T \approx 48.2°$.

Use the Pythagorean Theorem to find RS.

$$(RS)^2 + (ST)^2 = (RT)^2 \qquad \text{Pythagorean Theorem}$$

$$(RS)^2 + 6^2 = 9^2 \qquad \text{Substitution}$$

$$(RS)^2 = 45 \qquad \text{Simplify.}$$

$$RS \approx 6.7 \qquad \text{Use a calculator.}$$

So, $RS \approx 6.7$.

Check

Solve the right triangle. Round side and angle measures to the nearest tenth.

$m\angle C \approx$ ___?___ ; $AB \approx$ ___?___ ; $BC \approx$ ___?___

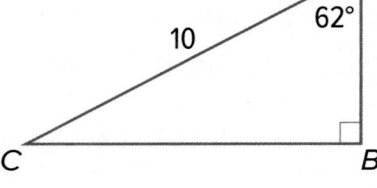

 Go Online You can complete an Extra Example online.

Study Tip

Alternative Methods
Right triangles can often be solved by using different methods. In the example, $m\angle T$ could have been found first using a cosine ratio, and $m\angle T$ and a tangent ratio could have been used to find RS.

Study Tip

Approximation If you are using calculated measures to find other measures in a right triangle, be careful not to round until the last step.

Practice

Go Online You can complete your homework online.

Example 1

Find sin L, cos L, tan L, sin M, cos M, and tan M. Express each ratio as a fraction and as a decimal to the nearest hundredth.

1. $\ell = 15, m = 36, n = 39$

2. $\ell = 12, m = 12\sqrt{3}, n = 24$

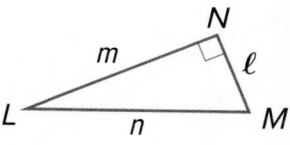

3. Find sin R, cos R, tan R, sin S, cos S, and tan S. Express each ratio as a fraction and as a decimal to the nearest hundredth.

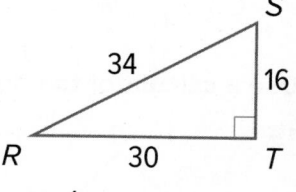

4. Find sin J, cos J, tan J, sin L, cos L, and tan L. Express each ratio as a fraction and as a decimal to the nearest hundredth if necessary.

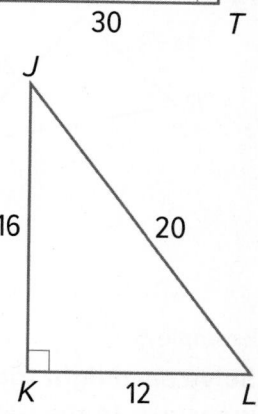

Example 2

Use a special right triangle to express each trigonometric ratio as a fraction and as a decimal to the nearest hundredth if necessary.

5. sin 30°

6. tan 45°

7. cos 60°

8. sin 60°

9. tan 30°

10. cos 45°

Example 3

Find the value of x. Round to the nearest hundredth.

11.

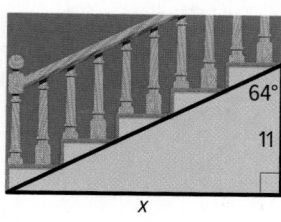

12.

13.

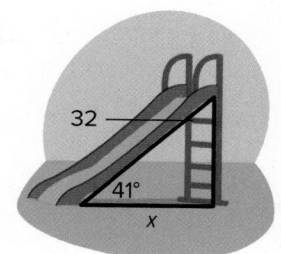

14. **GEOLOGY** Shan used a surveying tool to map a region of land for his science class. To determine the height of a vertical rock formation, he measured the distance from the base of the formation to his position and the angle between the ground and the line of sight to the top of the formation. The distance was 43 meters, and the angle was 36°. What is the height of the formation to the nearest meter?

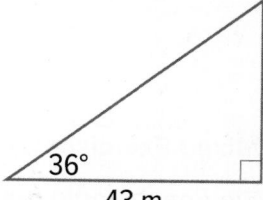

15. **RAMPS** A 60-foot ramp rises from the first floor to the second floor of a parking garage. The ramp makes a 15° angle with the ground. How high above the first floor is the second floor? Express your answer to the nearest tenth of a foot.

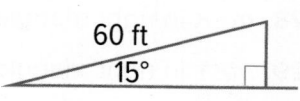

Example 4

Use a calculator to find $m\angle B$ to the nearest tenth.

16.
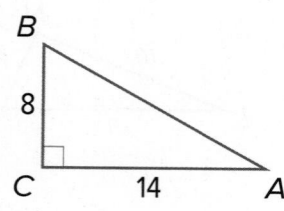

17.
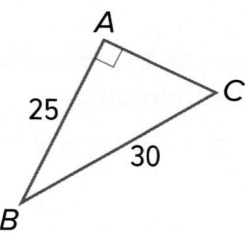

18.
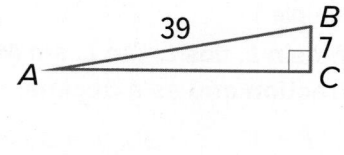

Use a calculator to find $m\angle T$ to the nearest tenth.

19.
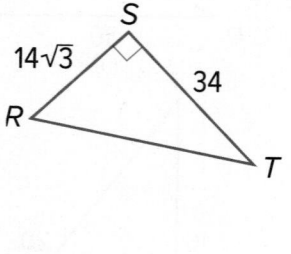

20.
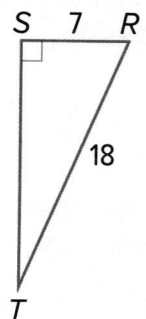

21.
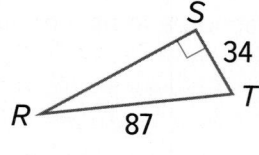

Example 5

Solve each right triangle. Round side measures to the nearest tenth and angle measures to the nearest degree.

22.
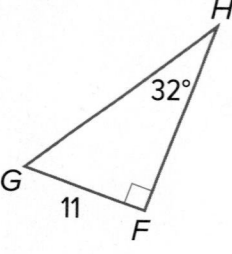

23.
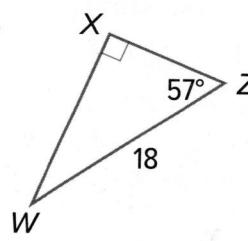

24.
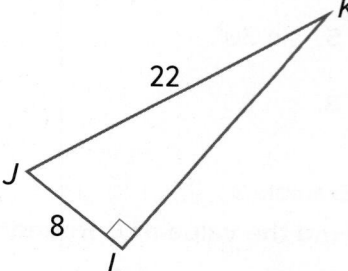

25.
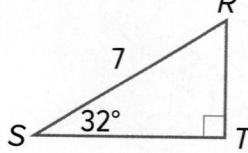

26.
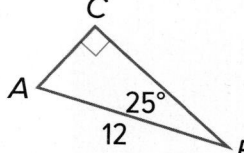

27.
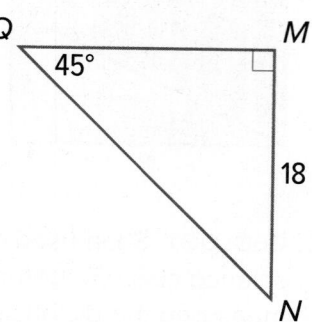

Mixed Exercises

Find each angle measure to the nearest tenth of a degree using the Distance Formula and an inverse trigonometric ratio.

28. $m\angle K$ in right triangle JKL with vertices $J(-2, -3)$, $K(-7, -3)$, and $L(-2, 4)$

29. $m\angle Y$ in right triangle XYZ with vertices $X(4, 1)$, $Y(-6, 3)$, and $Z(-2, 7)$

REASONING Find the perimeter and area of each triangle. Round to the nearest hundredth.

30.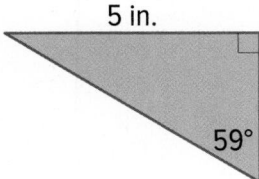
5 in.

59°

31.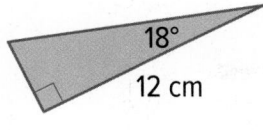
18°

12 cm

32.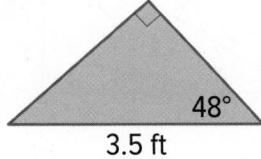
48°

3.5 ft

33. NEIGHBORS Amy, Barry, and Chris live in the same neighborhood. Chris lives up the street and around the corner from Amy, and Barry lives at the corner between Amy and Chris. The three homes are the vertices of a right triangle.

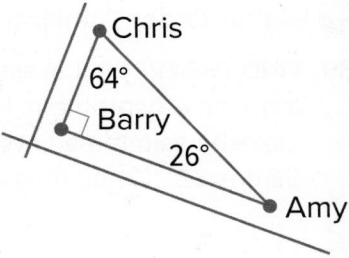

Chris

64°

Barry

26°

Amy

a. Give two trigonometric expressions for the ratio of Barry's distance from Amy to Chris's distance from Amy.

b. Give two trigonometric expressions for the ratio of Barry's distance from Chris to Amy's distance from Chris.

c. Give a trigonometric expression for the ratio of Amy's distance from Barry to Chris's distance from Barry.

34. CONSTRUCT ARGUMENTS A cell phone tower is supported by a guy wire as shown. Chilam wants to determine the height of the tower. She finds that the guy wire makes an angle of 53° with the ground and it is attached to the ground 65 feet from the base of the tower.

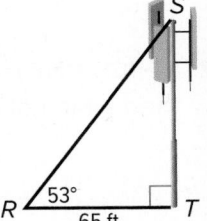

S

R 53°

65 ft

T

a. Let the height of the tower be x. Which trigonometric ratio should you use to write an equation that you can solve for x? Justify your choice.

b. Write and solve an equation for the height of the tower. Round to the nearest tenth of a foot.

c. Suppose Chilam had wanted to find the length of the guy wire. What would you have done differently to solve the problem? Justify your argument.

35. COMPLEMENTARY ANGLES In the right triangle shown, $\sin \alpha = 0.6428$ and $\cos \alpha = 0.7660$. Find $\sin \beta$ and $\cos \beta$ and explain your reasoning.

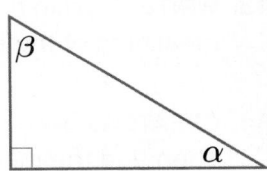

β

α

Find the values of x and y. Round to the nearest tenth.

36.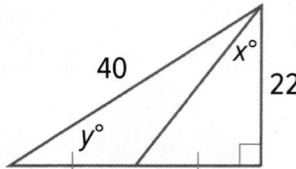

37.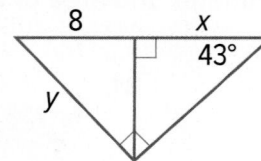

38. STRUCTURE Explain how you can use only the table at the right to find the value of cos 20°.

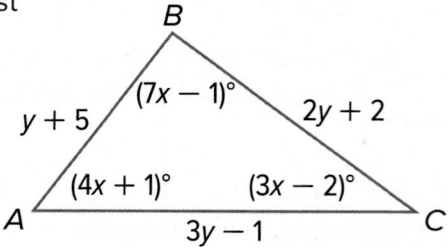

$m\angle A$	sin A
65°	0.9063
70°	0.9397
75°	0.9659
80°	0.9848
85°	0.9962

🧠 **Higher-Order Thinking Skills**

39. FIND THE ERROR Lakasha and Treyvon were both solving the same trigonometry problem. However, after they finished their computations, Lakasha said the answer was 52 sin 27° and Treyvon said the answer was 52 cos 63°. Could they both be correct? Explain your reasoning.

40. PERSEVERE Solve △ABC. Round each measure to the nearest whole number.

41. ANALYZE Are the values of sine and cosine for an acute angle of a right triangle always less than 1? Explain.

42. WHICH ONE DOESN'T BELONG? If the directions say to *Solve the right triangle,* then which of the triangles shown does not belong? Justify your conclusion.

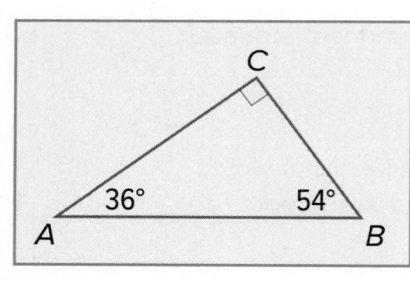

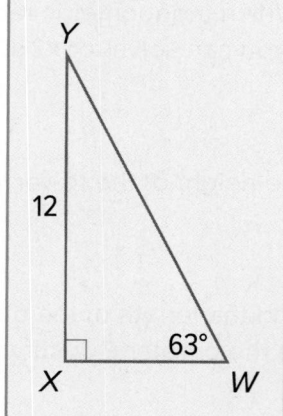

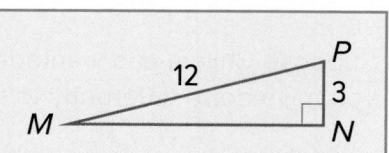

43. WRITE Explain how you can use ratios of the side lengths to find the angle measures of the acute angles in a right triangle.

44. CREATE Draw a right triangle with a tangent ratio of $\frac{3}{2}$ for one of the acute angles. Then find the measure of the other acute angle to the nearest tenth of a degree.

Applying Trigonometry

Explore Angles of Elevation and Depression

Online Activity Use guiding exercises to complete the Explore.

> **@ INQUIRY** How can angles of elevation and depression be used to find measurements? ×

Explore Measuring Angles of Elevation

Online Activity Use a spreadsheet to complete the Explore.

> **@ INQUIRY** When sighting an object from a given distance, how does the tangent of the angle of elevation relate to the height of the object? ×

Learn Angles of Elevation and Depression

Often the only way to measure an object is through the use of **indirect measurement**, which involves using similar figures and proportions to measure an object. In these cases, you measure the object by measuring something else. Indirect measurements use special types of angles. An **angle of elevation** is the angle formed by a horizontal line and an observer's line of sight to an object above the horizontal line. An **angle of depression** is the angle formed by a horizontal line and an observer's line of sight to an object below the horizontal line.

🌐 Example 1 Angle of Elevation

DRONES Rakeem is flying his drone at the park. He spots the drone at an angle of elevation that he estimates to be 30°. The remote control tells Rakeem that his drone is 102 feet above the ground. If Rakeem is 6 feet tall, how far is he from the drone to the nearest foot?

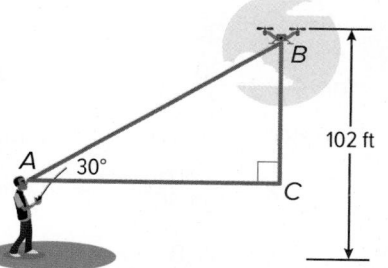

102 ft

Because Rakeem is 6 feet tall, $BC = 102 - 6$ or 96 feet. Let x represent the distance from Rakeem to the drone AB.

$$\sin A = \frac{BC}{AB} \qquad \sin = \frac{\text{opposite}}{\text{hypotenuse}}$$

$$\sin 30° = \frac{96}{x} \qquad m\angle A = 30°, BC = 96, \text{ and } AB = x$$

$$x = \frac{96}{\sin 30°} \qquad \text{Solve for } x.$$

$$x = 192 \qquad \text{Use a calculator.}$$

Rakeem is 192 feet from his drone.

Go Online You can complete an Extra Example online.

Today's Goals
- Solve real-world problems by using the trigonometric ratios and their inverses.
- Derive and use a formula for the area of a triangle by using trigonometry.

Today's Vocabulary
indirect measurement

angle of elevation

angle of depression

🗨 Talk About It!
What determines whether an angle is an angle of depression or an angle of elevation?

💭 Think About It!
If Rakeem set the drone's camera to view himself, what would be the angle of depression? Explain.

Check

SEARCH AND RESCUE A flare is shot vertically into the air approximately 200 meters from base camp. The angle of elevation to the maximum height of the flare is 35°. The group at base camp needs to know the altitude of the flare.

Part A Write the equation that represents the situation if *a* represents the height of the flare.

Part B What is the maximum height of the flare to the nearest meter?

Study Tip

Angles of Elevation and Depression To avoid mislabeling, remember that angles of elevation and depression are always formed with a horizontal line, never with a vertical line.

Use a Source

Chaz hikes to the top of Mount Elbert in Colorado. He uses binoculars to sight his car in the parking lot at the trailhead, which is at an altitude of 10,040 feet. The angle of depression at which he sites his car is about 23.6°. Use available resources to find the total height of Mount Elbert. Then, calculate the distance Chaz is from his car to the nearest foot.

Chaz is about ___?___ feet from his car.

🌐 **Example 2** Angle of Depression

SIGHTSEEING Cottonwood, Idaho's Dog Bark Park Inn is a popular tourist attraction featuring a hotel in the shape of a 30-foot wood-carved beagle. Pedro looks out the window 30 feet from the ground and spots a fire hydrant on the ground at an estimated angle of depression of 40°. What is the horizontal distance from Pedro to the hydrant to the nearest foot?

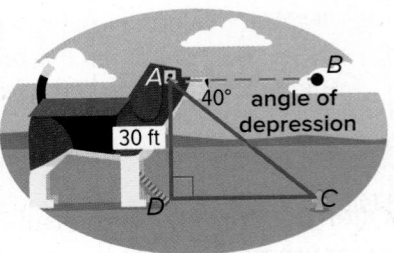

Because $\overline{AB}$ and $\overline{DC}$ are parallel, $m\angle BAC = m\angle ACD$ by the Alternate Interior Angles Theorem and the definition of congruent angles.

Let *x* represent the horizontal distance from the base of the hotel to the hydrant.

$$\tan C = \frac{AD}{DC} \qquad\qquad \tan\theta = \frac{\text{opposite}}{\text{adjacent}}$$

$$\tan 40° = \frac{30}{x} \qquad\qquad m\angle C = 40°, AD = 30, \text{ and } DC = x$$

$$x\tan 40° = 30 \qquad\qquad \text{Multiply each side by } x.$$

$$x = \frac{30}{\tan 40°} \qquad\qquad \text{Divide each side by } \tan 40°.$$

$$x \approx 35.8 \qquad\qquad \text{Use a calculator.}$$

The horizontal distance from Pedro to the hydrant is about 36 feet.

Check

LIFEGUARDING Braylen stands on an 8-foot platform and sights a swimmer at an angle of depression of 5°. If Braylen is 6 feet tall, how far away is the swimmer from the base of the platform to the nearest foot? ___?___ ft

🐾 **Go Online** You can complete an Extra Example online.

Example 3 Use Two Angles of Elevation or Depression

MALL Wei is estimating the height of the second floor in the mall. She sights the second floor at a 10° angle of elevation. She then steps forward 50 feet, until she is 5.5 feet from the wall and sights the second floor again. If Wei's line of sight is 66 inches above the ground, at what angle of elevation does she sight the second floor?

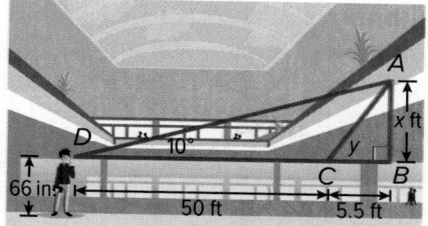

Study Tip

Units of Measure
When solving real-world problems using trigonometric ratios and inverses, be sure to convert all measurements to the same units to avoid unnecessary errors.

$\triangle ABC$ and $\triangle ABD$ are right triangles. To find the angle of elevation, first find the height of the second floor of the mall. This height is the sum of Wei's height and AB.

Use $\triangle ABD$ to write an equation for AB.

$\tan 10° = \dfrac{AB}{BD}$ $\qquad \tan \theta = \dfrac{\text{opposite}}{\text{adjacent}}$ and $m\angle ADB = 10$

$\tan 10° = \dfrac{AB}{55.5}$ $\qquad BD = 50 + 5.5$

$55.5 \tan 10° = AB$ $\qquad$ Multiply each side by 55.5.

Use $\triangle ABC$ to write an equation for AB.

$\tan y° = \dfrac{AB}{BC}$ $\qquad \tan \theta = \dfrac{\text{opposite}}{\text{adjacent}}$

$\tan y° = \dfrac{AB}{5.5}$ $\qquad BC = 5.5$

$5.5 \tan y° = AB$ $\qquad$ Multiply each side by 5.5.

Use the equation for AB from $\triangle ABD$ in the equation for $\triangle ABC$ and solve for y.

$5.5 \tan y° = AB$

$5.5 \tan y° = 55.5 \tan 10°$

$\tan y° = \dfrac{55.5 \tan 10°}{5.5}$

$y \approx 60.66$

Using the equation from $\triangle ABC$, $AB = 5.5 \tan 60.7°$ or about 9.8. The height of the second floor of the mall is about $9.8 + 5.5$ or 15.3, which is about 15 feet.

Check

SIGHTSEEING Looking north, two skyscrapers are sighted from the viewing deck of the Empire State Building at 1250 feet up. One skyscraper is sighted at a 20° angle of depression and a second skyscraper is sighted at a 30° angle of depression. How far apart are the two skyscrapers to the nearest foot?

___?___ ft

 Go Online You can complete an Extra Example online.

Think About It!

Describe the relationship between *a*, *b*, and *C* in a triangle when finding the area using Area = $\frac{1}{2}ab$ sin *C*.

Learn Trigonometry and Areas of Triangles

Key Concept • Area of a Triangle

To find the area of a triangle when the height is not known, you can use Area = $\frac{1}{2}ab$ sin *C*, where *a* and *b* are side lengths and *C* is the included angle.

Example 4 Find the Area of a Triangle When Given the Included Angle

Use trigonometry to find the area of △ABC to the nearest tenth.

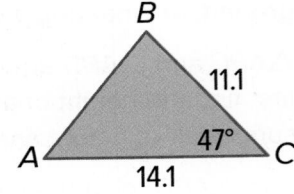

Area = $\frac{1}{2}ab$ sin *C*   Area of a triangle

$\quad= \frac{1}{2}$ (11.1)(14.1) sin 47°   Substitute.

$\quad\approx 57.23$ Simplify.

The area of △ABC is about 57.2 units².

Check

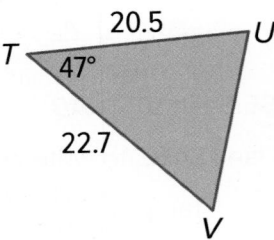

Use trigonometry to find the area of △TUV to the nearest tenth.

____?____ units²

Example 5 Find the Area of Any Triangle

Use trigonometry to find the area of △DEF to the nearest tenth.

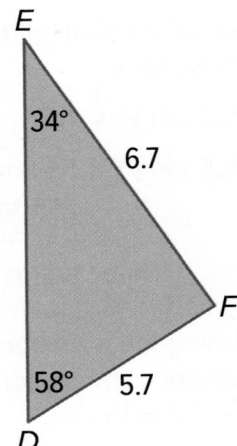

Because you do not know the measure of the included angle *F*, add the measures of angles *D* and *E* and subtract the total from 180.

m∠F = 88°

Area = $\frac{1}{2}de$ sin *F*

$\quad= \frac{1}{2}$(6.7)(5.7) sin 88°

$\quad\approx 19.08$

The area of △DEF is about 19.1 units².

Check

Use trigonometry to find the area of △JKL to the nearest tenth.

____?____ units²

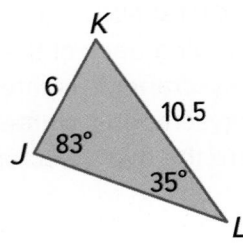

🅡 **Go Online** You can complete an Extra Example online.

Practice

🔖 **Go Online** You can complete your homework online.

Example 1

1. **LIGHTHOUSES** Sailors on a ship at sea spot the light from a lighthouse at an angle of elevation of 25°. The light of the lighthouse is 30 meters above sea level. How far from the shore is the ship? Round your answer to the nearest meter.

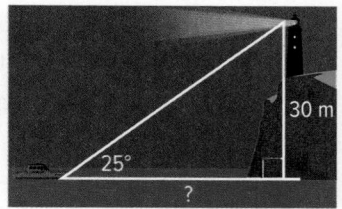

2. **WATER TOWERS** A student can see a water tower from the edge of a soccer field at San Lobos High School. The edge of the field is about 110 feet from the water tower, and the water tower stands at a height of 32.5 feet. What is the angle of elevation if the eye level of the student viewing the tower is 6 feet above the ground? Round your answer to the nearest tenth.

3. **CONSTRUCTION** A roofer props a ladder against a wall so the top of the ladder just reaches a 30-foot roof that needs repair. If the angle of elevation from the bottom of the ladder to the roof is 55°, how far is the ladder from the base of the wall? Round your answer to the nearest foot.

4. **MOUNTAIN BIKING** On a mountain bike trip along the Gemini Bridges Trail in Moab, Utah, Nabuko stopped on the canyon floor to get a good view of the twin sandstone bridges. Nabuko is standing about 60 meters from the base of the canyon cliff, and the natural arch bridges are about 100 meters up the canyon wall. If her line of sight is 5 meters above the ground, what is the angle of elevation to the top of the bridges? Round to the nearest tenth of a degree.

Example 2

5. **ROOFTOP** Lucia is 5.5 feet tall. She is standing on the roof of a building that is 80 feet tall. She spots a fountain at ground level that she knows to be 122 feet away from the base of the building. What is the measure of the angle of depression formed by Lucia's horizontal line of sight and her line of sight to the fountain? Round your answer to the nearest degree.

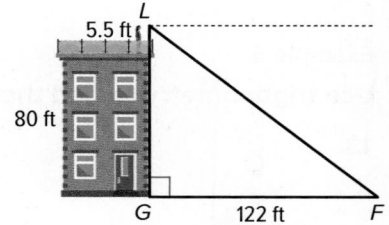

6. **AIR TRAFFIC** From the top of the 120-foot-high tower, an air traffic controller observes an airplane on the runway at an angle of depression of 19°. How far from the base of the tower is the airplane? Round to the nearest tenth of a foot.

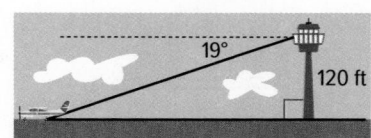

7. **AVIATION** Due to a storm, a pilot flying at an altitude of 528 feet has to land. If he has a horizontal distance of 2000 feet until he reaches the landing strip, what angle of depression should he use to land? Round to the nearest tenth of a degree.

8. **INDIRECT MEASUREMENT** Kenneth is sitting at the end of a pier and using binoculars to watch a whale surface. The pier is 30 feet above the water, and Kenneth's eye level is 3 feet above the pier. If the angle of depression to the whale is 20°, how far is the whale from Kenneth's binoculars? Round your answer to the nearest tenth of a foot.

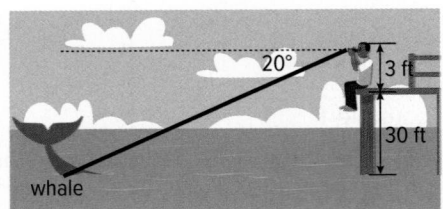

Example 3

9. **GARAGE** To estimate the height of a garage, Carlos sights the top of the garage at a 42° angle of elevation. He then steps back 20 feet and sights the top of the garage at a 10° angle. If Carlos is 6 feet tall, how tall is the garage to the nearest foot?

10. **CLIFF** Sarah stands on the ground and sights the top of a steep cliff at a 60° angle of elevation. She then steps back 50 meters and sights the top of the cliff at a 30° angle. If Sarah is 1.8 meters tall, how tall is the cliff to the nearest meter?

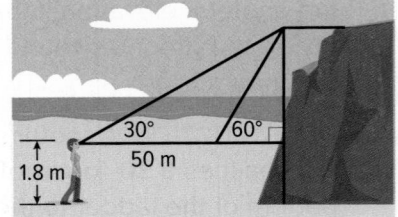

11. **BALLOON** The angle of depression from a hot air balloon to a person on the ground is 36°. When the person steps back 10 feet, the new angle of depression is 25°. If the person is 6 feet tall, how far above the ground is the hot air balloon to the nearest foot?

12. **INDIRECT MEASUREMENT** Mr. Dominguez is standing on a 40-foot ocean bluff near his home. He can see his two friends on the beach below. If his line of sight is 6 feet above the ground and the angles of depression to his friends are 34° and 48°, how far apart are his friends to the nearest foot?

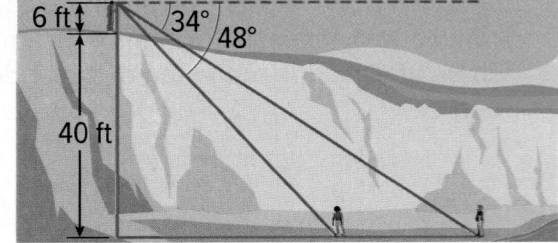

Example 4

Use trigonometry to find the area of △ABC to the nearest tenth.

13.

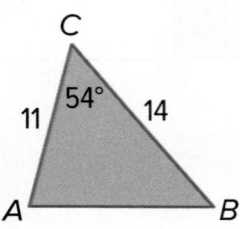

14.

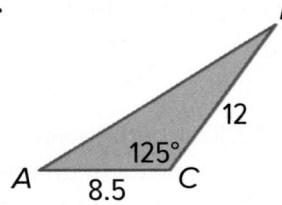

15.

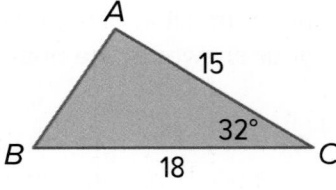

16.

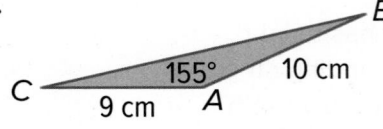

17.

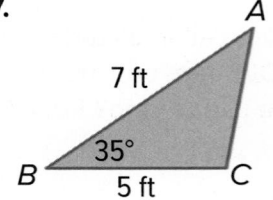

18.

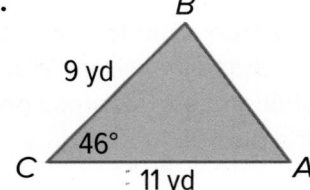

Example 5

Use trigonometry to find the area of △ABC to the nearest tenth.

19.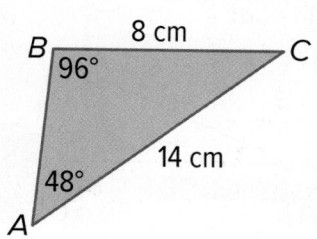

20.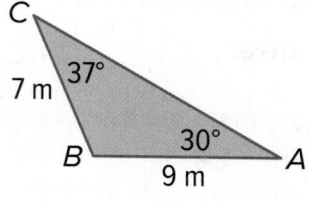

21.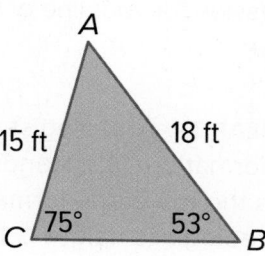

Mixed Exercises

22. USE ESTIMATION The angle of elevation to an airplane viewed from the air traffic control tower at an airport is 7°. The tower is 200 feet tall, and the pilot reports that the altitude of the airplane is 5127 feet.

    a. Explain how you can use angles of elevation and depression to estimate the distance from the air traffic control tower to the airplane.

    b. About how far away is the airplane from the air traffic control tower to the nearest foot?

23. USE ESTIMATION A hiker dropped his backpack over one side of a canyon onto a ledge below. Because of the shape of the cliff, he could not see exactly where it landed. A park ranger is located on the other side of the canyon, at the same height, 113 feet away from the hiker. The ranger sights the backpack at an angle of depression of 32°.

    a. Explain how you can use angles of elevation and depression to estimate the distance that the backpack fell.

    b. About how far down did the backpack fall to the nearest foot?

24. USE A MODEL Jermaine and John are standing 10 meters apart watching a helicopter hover above the ground.

    a. Find two different expressions that can be used to find h, the height of the helicopter.

    b. Equate the two expressions you found for part **a** to solve for x. Round your answer to the nearest hundredth.

    c. How high above the ground is the helicopter? Round your answer to the nearest hundredth.

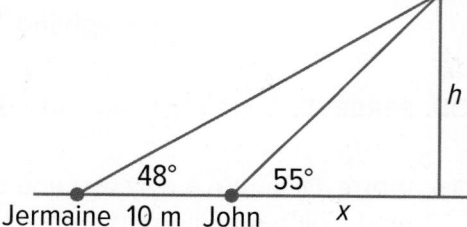

25. USE A SOURCE Go online to research the Sandia Peak Tramway in New Mexico. If you were to stand at the top terminal of Sandia Peak Tramway and look at the base of the second tower along the tramway route, what would be the angle of depression for your line of sight? Round your answer to the nearest tenth of a degree.

26. REGULARITY A geologist wants to determine the height of a rock formation. She stands d meters from the formation and sights the top of the formation at an angle of $x°$, as shown. The geologist's height is 1.8 m. Write a general formula that the geologist can use to find the height h of the rock formation if she knows the values of d and x.

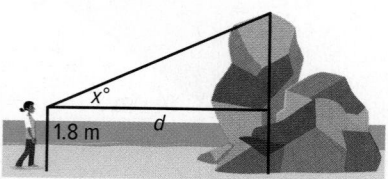

Find the area of $\triangle ABC$ to the nearest tenth.

27. $m\angle A = 20°, c = 4$ cm, $b = 7$ cm

28. $m\angle C = 55°, a = 10$ m, $b = 15$ m

29. $a = 5.6$ ft, $c = 3.7$ ft, $m\angle A = 37°, m\angle C = 24°$

30. $a = 6.3$ in., $c = 7$ in., $m\angle A = 42°, m\angle C = 49°$

31. THEATER Albert is helping to build the set for a play. One piece of scenery is a large triangle that will be constructed out of wood and be painted to represent a mountain. Albert would like to know the area of the piece of scenery so that he can buy the right amount of paint. What is the area of this triangle? Round your answer to the nearest tenth of a foot.

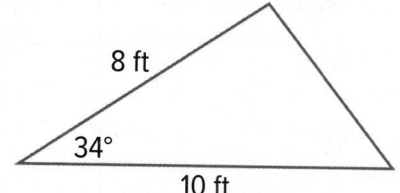

32. FIND THE ERROR Terrence and Rodrigo are trying to determine the relationship between angles of elevation and depression. Terrence says that if you are looking up at someone with an angle of elevation of 35°, then they are looking down at you with an angle of depression of 55°, which is the complement of 35°. Rodrigo disagrees and says that the other person would be looking down at you with an angle of depression equal to your angle of elevation or 35°. Who is correct? Explain your reasoning.

33. CREATE A classmate finds the angle of elevation of an object, but she is trying to find the angle of depression. Write a question to help her solve the problem.

34. ANALYZE Classify the statement below as *true* or *false*. Explain your reasoning.

> As a person moves closer to an object he or she is sighting, the angle of elevation increases.

35. PERSEVERE Find the value of x. Round to the nearest tenth.

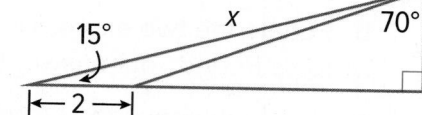

36. WRITE Describe a way that you can estimate the height of an object without using trigonometry. Explain your reasoning.

The Law of Sines

Explore Trigonometric Ratios in Nonright Triangles

Online Activity Use dynamic geometry software to complete the Explore.

×

⊚ INQUIRY How can you use trigonometric ratios to solve for missing side lengths in nonright triangles?

Learn The Law of Sines

The Law of Sines can be used to find side lengths and angle measurements for any triangle.

Theorem 4.10: Law of Sines

If $\triangle ABC$ has lengths a, b, and c, representing the lengths of the sides opposite the angles with measures A, B, and C, then

$$\frac{\sin A}{a} = \frac{\sin B}{b} = \frac{\sin C}{c}.$$

You will prove Theorem 4.10 in Exercise 27.

You can use the Law of Sines to solve a triangle if you know the measures of two angles and any side (AAS or ASA).

Example 1 The Law of Sines (AAS)

Find the value of x to the nearest tenth.

Because we are given the measures of two angles and a nonincluded side, use the Law of Sines to write a proportion.

$\dfrac{\sin A}{a} = \dfrac{\sin C}{c}$ Law of Sines

$\dfrac{\sin 58°}{9} = \dfrac{\sin 35°}{x}$ $m\angle A = 58°$, $a = 9$, $m\angle C = 35°$, and $c = x$

$x \sin 58° = 9 \sin 35°$ Multiplication Property of Equality

$x = \dfrac{9 \sin 35°}{\sin 58°}$ Divide each side by sin 58°.

≈ 6.1 Use a calculator.

Check

Find the value of d to the nearest tenth.

$d = $ _____?_____

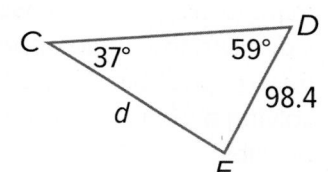

 Go Online You can complete an Extra Example online.

Today's Goals
• Understand and apply the Law of Sines to find unknown measurements in right and nonright triangles.

• Determine whether three given measures of a triangle define 0, 1, or 2 triangles by using the Law of Sines.

Today's Vocabulary
ambiguous case

⊙ Go Online
You may want to complete the Concept Check to check your understanding.

☁ Think About It!
How could you find the value of b?

Example 2 The Law of Sines (ASA)

Find the value of x to the nearest tenth.

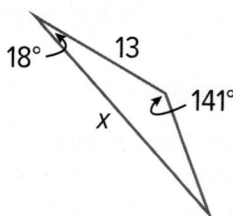

By the Triangle Angle-Sum Theorem,

$$m\angle G = 180 - (60 + 55) \text{ or } 65°.$$

$\dfrac{\sin D}{d} = \dfrac{\sin G}{g}$ Law of Sines

$\dfrac{\sin 60°}{x} = \dfrac{\sin 65°}{73}$ $m\angle D = 60°$, $g = 73$, $m\angle G = 65°$, and $d = x$

$73 \sin 60° = x \sin 65°$ Multiplication Property of Equality

$\dfrac{73 \sin 60°}{\sin 65°} = x$ Divide each side by sin 65°.

$69.8 \approx x$ Use a calculator.

Think About It!

Could you use $\dfrac{\sin D}{d} = \dfrac{\sin F}{f}$ to find the value of x? Justify your reasoning.

Check

Find the value of x to the nearest tenth.

A. 6.4   **B.** 7.4   **C.** 8.2   **D.** 22.8

🌐 Example 3 Indirect Measurement with the Law of Sines

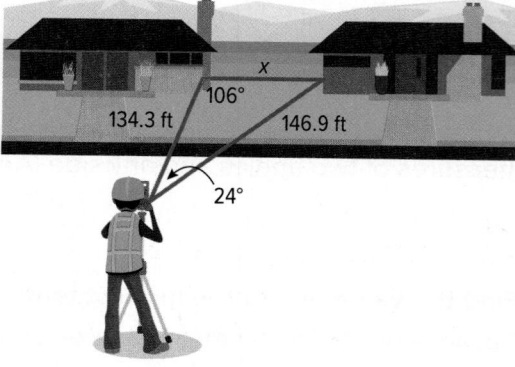

SURVEYING Mr. Fortunado is having a boundary survey done on his property. What is the distance between Mr. Fortunado's home and his neighbor's?

Because we know two angles of a triangle and one nonincluded side, use the Law of Sines.

$\dfrac{\sin 106°}{146.9} = \dfrac{\sin 24°}{x}$ Law of Sines

$x \sin 106° = 146.9 \sin 24°$ Multiplication Property of Equality

$x \approx 62.2$ Use a calculator.

Learn The Ambiguous Case

If you are given the measures of two angles and a side, exactly one triangle is possible. However, if you are given the measures of two sides and the angle opposite one of them, zero, one, or two triangles may be possible. This is known as the **ambiguous case**. So, when solving a triangle using the SSA case, zero, one or two solutions are possible.

 Go Online You can complete an Extra Example online.

Key Concept • Possible Triangles in SSA Case

Consider a triangle in which a, b, and $m\angle A$ are given and h is the altitude of the triangle. Shown below are the triangles that are possible when $\angle A$ is acute and when $\angle A$ is right or obtuse.

Angle A is acute.

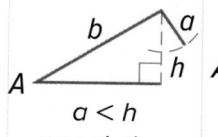 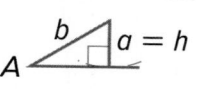 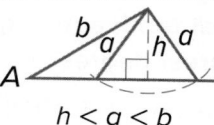 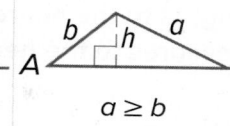

$a < h$	$a = h$	$h < a < b$	$a \geq b$
no solution	one solution	two solutions	one solution

Angle A is right or obtuse.

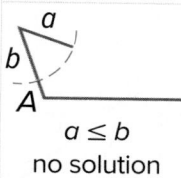 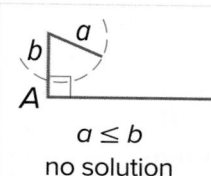 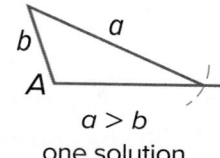 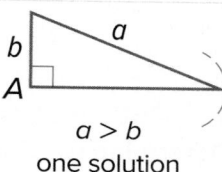

$a \leq b$	$a \leq b$	$a > b$	$a > b$
no solution	no solution	one solution	one solution

Solving a triangle with an obtuse angle sometimes requires finding sine ratios for measures greater than 90°. The sine ratios for obtuse angles are defined based on their supplementary angles.

Postulate 9.1

Words	The sine of an obtuse angle is defined to be the sine of its supplement.
Symbols	If $m\angle A > 90°$ and $x = \sin A$, then $x = \sin(180° - A)$.

Example 4 The Ambiguous Case with One Solution

Because $\sin A = \frac{h}{b}$, you can use $h = b \sin A$ to find h in acute triangles.

In $\triangle MNP$, $m\angle N = 32°$, $n = 7$, and $p = 4$. Determine whether $\triangle MNP$ has no solution, one solution, or two solutions. Then, solve the triangle. Round side lengths to the nearest tenth and angle measures to the nearest degree.

Because $\angle N$ is acute, and $n > p$, you know that one solution exists.

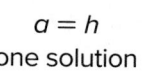

Step 1 Use the Law of Sines to find $m\angle P$.

$$\frac{\sin 32°}{7} = \frac{\sin P}{4} \qquad \text{Law of Sines}$$
$$\frac{4 \sin 32°}{7} = \sin P \qquad \text{Multiply each side by 4.}$$

Use the $\sin^{-1}$ function to find the exact value of $m\angle P$,
$\sin^{-1}\left(\frac{4 \sin 32°}{7}\right)$, or about 17.627°.

Step 2 Use the Triangle Angle-Sum Theorem to find $m\angle M$.

$m\angle M = 180° - (32° + m\angle P)$ or about 130.373°

Step 3 Use the Law of Sines to find m.

$$\frac{\sin M}{m} \approx \frac{\sin 32°}{7} \qquad \text{Law of Sines}$$
$$m \approx \frac{7 \sin 130°}{\sin 32°} \qquad \text{Solve for } m.$$
$$\approx 10.1 \qquad \text{Use a calculator.}$$

So, $m\angle P \approx 18°$, $m\angle M \approx 130°$, and $m \approx 10.1$.

Study Tip

$\angle A$ is Acute In the figures, the altitude h is compared to a because h is the minimum distance from $\angle C$ to $\overline{AB}$ when A is acute.

$\sin A = \dfrac{opposite}{hypotenuse}$

$\sin A = \dfrac{h}{b}$

Talk About It!

If the given angle is a right angle and there is one solution to the triangle, how can you find the third side?

Example 5 The Ambiguous Case with No Solution

In $\triangle RST$, $m\angle R = 95°$, $r = 10$, and $s = 12$. Determine whether $\triangle RST$ has no solution, one solution, or two solutions. Then, solve the triangle. Round side lengths to the nearest tenth and angle measures to the nearest degree.

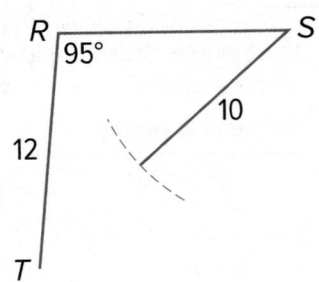

Because $\angle R$ is obtuse, and $10 < 12$, there is no solution.

Example 6 The Ambiguous Case with More than One Solution

In $\triangle ABC$, $m\angle A = 32°$, $a = 15$, and $b = 18$. Determine whether $\triangle ABC$ has no solution, one solution, or two solutions. Then, solve the triangle. Round side lengths to the nearest tenth and angle measures to the nearest degree.

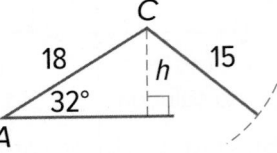

Because $\angle A$ is acute, and $15 < 18$, find h and compare it to a.

$b \sin A = 18 \sin 32°$ $b = 18$ and $m\angle A = 32°$

$ \approx 9.5$ Use a calculator.

Because $9.5 < 15 < 18$, or $h < a < b$, there are two solutions.

$\angle B$ is acute.	$\angle B$ is obtuse.
Find $m\angle B$.	**Find $m\angle B$.**
$\dfrac{\sin B}{18} = \dfrac{\sin 32°}{15}$ Law of Sines	Find an obtuse angle B for which $\sin B \approx 0.6359$.
$\sin B = \dfrac{18 \sin 32°}{15}$ Solve for sin B.	$m\angle B \approx 180° - 39°$ Postulate 9.1
$\sin B = 0.6359$ Use a calculator.	$ $ or $141°$
$m\angle B \approx 39°$ Use the $\sin^{-1}$ function.	
Find $m\angle C$.	**Find $m\angle C$.**
$m\angle C \approx 180° - (32° + 39°)$ or $109°$	$m\angle C \approx 180° - (32° + 141°)$ or $7°$
Find c.	**Find c.**
$\dfrac{\sin 109°}{c} = \dfrac{\sin 32°}{15}$ Law of Sines	$\dfrac{\sin 7°}{c} = \dfrac{\sin 32°}{15}$ Law of Sines
$c = \dfrac{15 \sin 109°}{\sin 32°}$ Solve for c.	$c = \dfrac{15 \sin 7°}{\sin 32°}$ Solve for c.
$c \approx 26.8$ Use a calculator.	$c \approx 3.4$ Use a calculator.

So, one solution is $m\angle B \approx 39°$, $m\angle C \approx 109°$, and $c \approx 26.8$, and another solution is $m\angle B \approx 141°$, $m\angle C \approx 7°$, and $c \approx 3.4$.

Go Online You can complete an Extra Example online.

Practice

Go Online You can complete your homework online.

Examples 1 and 2

Find the value of x to the nearest tenth.

1.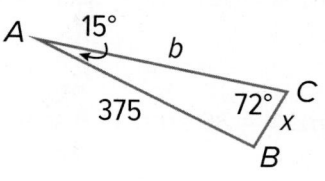

2.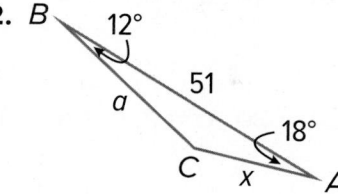

3.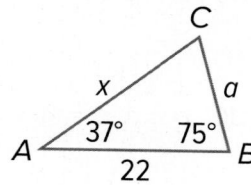

4.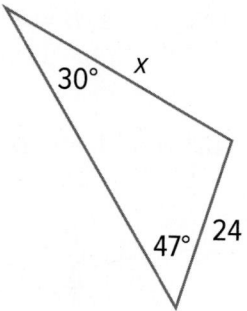

5.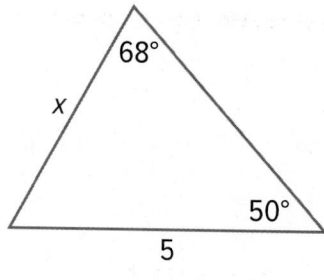

6.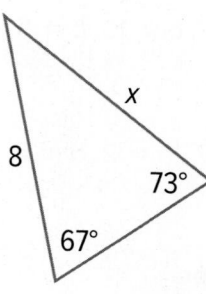

7.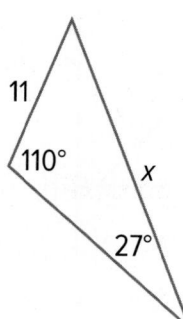

8.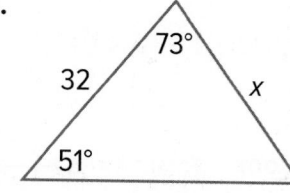

9.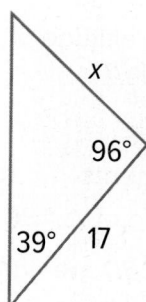

10.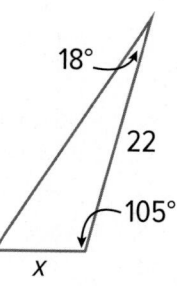

11.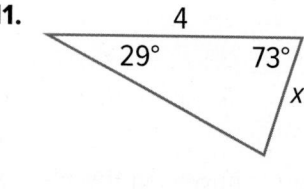

12.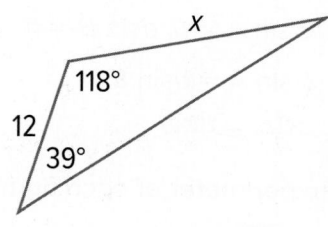

Example 3

13. WILDLIFE An officer for the Department of Fisheries and Wildlife, checks boaters on a lake to make sure they do not disturb two osprey nesting sites. She leaves a dock and heads due north in her boat to the first nesting site. From here, she turns 5° north of due west and travels an additional 2.14 miles to the second nesting site. She then travels 6.7 miles directly back to the dock. How far from the dock is the first osprey nesting site? Round to the nearest tenth.

14. SAILING Observers at two shoreline towers 100 feet apart measure the angles to an incoming sailboat. Using the diagram, find the distance *d* that the sailboat is from Tower A to the nearest tenth of a foot.

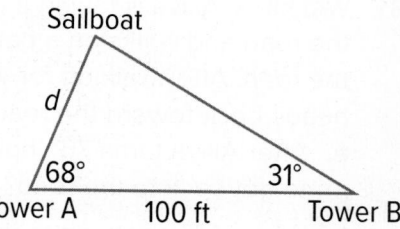

Examples 4–6

Determine whether each triangle has *no solution*, *one solution*, or *two solutions*. Then, solve the triangle. Round side lengths to the nearest tenth and angle measures to the nearest degree.

15. $m\angle A = 50°, a = 34, b = 40$ **16.** $m\angle A = 24°, a = 3, b = 8$ **17.** $m\angle A = 125°, a = 22, b = 15$

18. $m\angle A = 30°, a = 1, b = 4$ **19.** $m\angle A = 30°, a = 2, b = 4$ **20.** $m\angle A = 30°, a = 3, b = 4$

21. $m\angle A = 38°, a = 10, b = 9$ **22.** $m\angle A = 78°, a = 8, b = 5$ **23.** $m\angle A = 133°, a = 9, b = 7$

24. $m\angle A = 127°, a = 2, b = 6$ **25.** $m\angle A = 109°, a = 24, b = 13$ **26.** $m\angle A = 48°, a = 11, \text{ and } b = 16$

Mixed Exercises

27. PROOF Justify each statement for the proof of the Law of Sines.

Given: $\overline{CD}$ is an altitude of $\triangle ABC$.

Prove: $\dfrac{\sin A}{a} = \dfrac{\sin B}{b}$

Proof:

Statements	Reasons
$\overline{CD}$ is an altitude of $\triangle ABC$	Given
$\triangle ACD$ and $\triangle CBD$ are right	Def. of altitude
a. $\sin A = \dfrac{h}{b}, \sin B = \dfrac{h}{a}$	**a.** ___?___
b. $b\sin A = h, a\sin B = h$	**b.** ___?___
c. $b\sin A = a\sin B$	**c.** ___?___
d. $\dfrac{\sin A}{a} = \dfrac{\sin B}{b}$	**d.** ___?___

Find the perimeter of each figure. Round your answer to the nearest tenth.

28.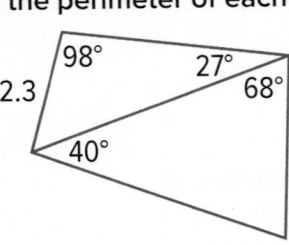

29.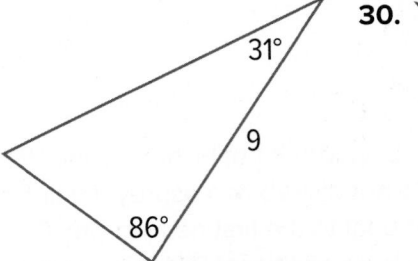

30.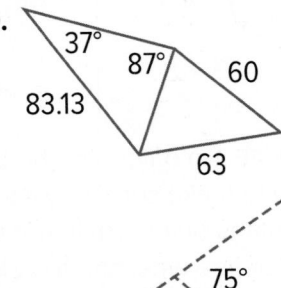

31. WALKING Alliya is taking a walk along a straight road. She leaves the road and walks on a path that makes an angle of 35° with the road. After walking for 450 meters, she turns 75° and heads back toward the road.

a. After Alliya turns 75°, how far does Alliya walk to get back to the road?

b. When Alliya returns to the road, how far along the road is she from where she started?

32. CAMERAS A security camera is located on top of a building at a certain distance from the sidewalk. The camera revolves counterclockwise at a steady rate of one revolution per minute. At one point in the revolution it directly faces a point on the sidewalk that is 20 meters from the camera. Four seconds later, it directly faces a point 10 meters down the sidewalk.

   a. How many degrees does the camera rotate in 4 seconds?

   b. To the nearest tenth of a meter, how far is the security camera from the sidewalk?

33. FISHING A fishing pole is resting against the railing of a boat making an angle of 22° with the deck. The fishing pole is 5 feet long, and the hook hangs 3 feet from the tip of the pole. The movement of the boat causes the hook to sway back and forth. Determine which angles the fishing line must make with the pole in order for the hook to be level with the boat's deck.

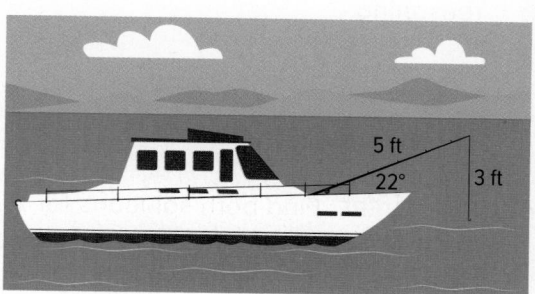

34. REASONING How many triangles can be formed if $a = b$? if $a < b$? if $a > b$?

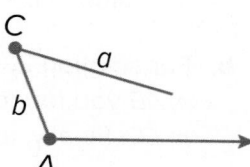

REGULARITY Determine whether the given measures define *0, 1,* or *2* triangles. Justify your answers.

35. $a = 14, b = 16, m\angle A = 55°$ **36.** $a = 7, b = 11, m\angle A = 68°$ **37.** $a = 22, b = 25, m\angle A = 39°$

38. $a = 13, b = 12, m\angle A = 81°$ **39.** $a = 10, b = 10, m\angle A = 45°$ **40.** $a = 17, b = 15, m\angle A = 128°$

41. $a = 13, b = 17, m\angle A = 52°$ **42.** $a = 5, b = 9, c = 6$ **43.** $a = 10, b = 15, m\angle A = 33°$

44. TOWERS Cell towers *A*, *B*, and *C* form a triangular region in one of the suburban districts of Fairfield County. Towers *A* and *B* are 8 miles apart. The angle formed at tower *A* is 112°, and the angle formed at tower *B* is 40°. How far apart are towers *B* and *C*?

45. FIND THE ERROR In $\triangle RST$, $m\angle R = 56°$, $r = 24$, and $t = 12$. Cameron and Gabriela are using the Law of Sines to find $m\angle T$. Who is correct? Explain your reasoning.

Cameron	Gabriela
$\dfrac{\sin T}{12} = \dfrac{\sin 56°}{24}$ $\sin T = 0.4145$ $m\angle T = 24.5$	Because $r > t$, there is no solution.

46. WRITE What two methods can be used to find the value of x in $\triangle ABC$? Write two equations using the different methods and explain your reasoning.

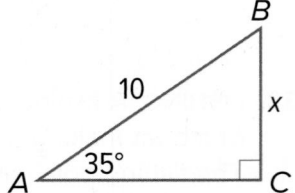

47. PERSEVERE Find both solutions for $\triangle ABC$ if $a = 15$, $b = 21$, and $m\angle A = 42°$. Round angle measures to the nearest degree and side measures to the nearest tenth.

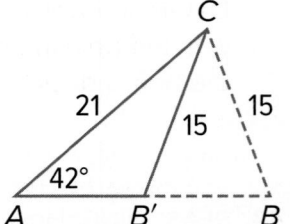

   a. For solution 1, assume that $\angle B$ is acute, and use the Law of Sines to find $m\angle B$. Then find $m\angle C$. Finally, use the Law of Sines again to find the value of c.

   b. For solution 2, assume that $\angle B$ is obtuse. Let this obtuse angle be $\angle B'$. Use $m\angle B$ you found in solution 1 and the diagram shown to find $m\angle B'$. Then find $m\angle C$. Finally, use the Law of Sines to find the value of c.

48. ANALYZE Determine whether the statement below is *true* or *false*. Explain your reasoning.

   The Law of Sines can always be used to solve a triangle if the measures of two sides and their included angle are known.

49. CREATE Give measures for a, b, and an acute $\angle A$ that define the given number of triangles.

   a. 0 triangles   **b.** exactly one triangle   **c.** two triangles

The Law of Cosines

Explore Trigonometric Relationships in Nonright Triangles

 Online Activity Use the guiding exercises to complete the Explore.

> ×
>
> @ **INQUIRY** When can the Law of Cosines be used to solve triangles?

Learn The Law of Cosines

When the Law of Sines cannot be used to solve a triangle, the Law of Cosines may apply. You can use the Law of Cosines to find the length of the third side of a triangle when the measures of two sides and the included angle are known or to find the angle measures of a triangle if the lengths of all three sides are known.

> **Theorem 4.11: Law of Cosines**
>
> If $\triangle ABC$ has lengths a, b, and c, representing the lengths of the sides opposite the angles with measures A, B, and C, then
>
> $a^2 = b^2 + c^2 - 2bc \cos A$,
>
> $b^2 = a^2 + c^2 - 2ac \cos B$, and
>
> $c^2 = a^2 + b^2 - 2ab \cos C$

You will prove Theorem 4.11 in Exercise 20.

You can use the Law of Cosines to solve a triangle if you know the measures of two sides and the included angle (SAS).

Example 1 The Law of Cosines (SAS)

Find the value of x to the nearest tenth.

We are given the measures of two sides and their included angle. Use the Law of Cosines to write an equation.

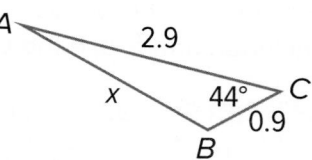

$c^2 = a^2 + b^2 - 2ab \cos C$ Law of Cosines

$x^2 = 0.9^2 + 2.9^2 - 2(0.9)(2.9) \cos 44°$ Substitution

$x^2 = 9.22 - 5.22 \cos 44°$ Simplify.

$x = \sqrt{9.22 - 5.22 \cos 44°}$ Take the positive square root of each side.

$x \approx 2.3$ Use a calculator.

 Go Online You can complete an Extra Example online.

Today's Goals
- Understand and apply the Law of Cosines to find unknown measurements in right and nonright triangles.

🗨 Talk About It!
What happens when you apply the Law of Cosines to find the missing measures of a right triangle?

💭 Think About It!
Why can't you use the Law of Sines to find the value of x?

You can also use the Law of Cosines if you know the three side lengths.

Example 2 The Law of Cosines (SSS)

Find the value of *x* to the nearest whole number.

$p^2 = m^2 + n^2 - 2mn \cos P$	Law of Cosines
$9^2 = 5^2 + 11^2 - 2(5)(11) \cos x°$	Substitution
$81 = 146 - 110 \cos x°$	Simplify.
$-65 = -110 \cos x°$	Subtract 146 from each side.
$\frac{-65}{-110} = \cos x°$	Divide each side by -110.
$x = \cos^{-1}\left(\frac{-65}{-110}\right)$	Use the inverse cosine function.
$x \approx 54°$	Use a calculator.

Check

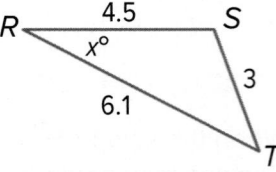

Find the value of *x* to the nearest whole number.

$x = \underline{\quad ? \quad}$

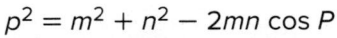

🌐 **Example 3** Indirect Measurement with the Law of Cosines

GOLF Nhat is golfing and uses a distance measuring tool to determine that the tee box where he is standing is 378 yards from the hole. To avoid a water hazard, Nhat turns 32° and hits a shot 261.5 yards up the fairway. Complete the diagram with the correct values. Then find the distance between Nhat's ball and the hole to the nearest yard.

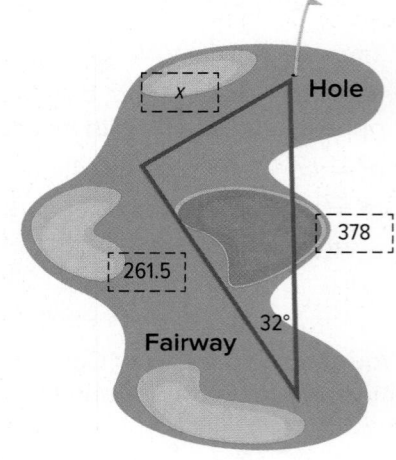

Because we know the measures of two sides of the triangle and the included angle, use the Law of Cosines to find the remaining distance.

$x^2 = 261.5^2 + 378^2 - 2(261.5)(378) \cos 32°$	Law of Cosines
$x^2 = 211{,}266.25 - 197{,}316 \cos 32°$	Simplify.
$x = \sqrt{211{,}266.25 - 197{,}316 \cos 32°}$	Take the positive square root of each side.
$x \approx 210$	Use a calculator.

Nhat's ball is about 210 yards from the hole.

🅑 **Go Online** You can complete an Extra Example online.

Check

CELL PHONE TOWERS A cell phone company builds two towers that are 2 miles apart. They choose a random location 1.1 miles from tower A and 1.5 miles from tower B to test the towers' signal strengths. Find the value of x to the nearest degree.

$x =$ _____?_____

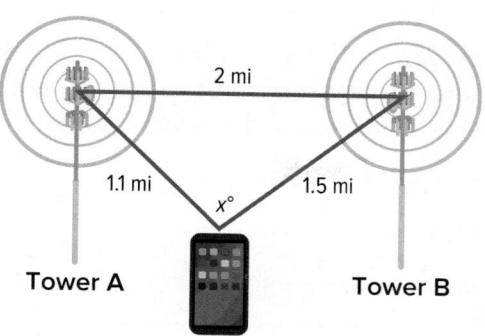

When solving right triangles, you can use sine, cosine, or tangent. When solving any triangle, you can use the Law of Sines or the Law of Cosines, depending on what information is given.

Example 4 Solve a Nonright Triangle with the Law of Cosines

Solve $\triangle ABC$. Round to the nearest degree.

Because $7^2 + 8^2 \neq 11^2$, this is not a right triangle. The measures of all three sides are given (SSS), so decide which angle measure you want to find. Then use the Law of Cosines.

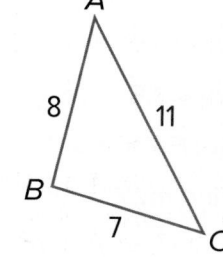

$a^2 = b^2 + c^2 - 2bc \cos A$	Law of Cosines
$7^2 = 11^2 + 8^2 - 2(11)(8) \cos A$	Substitute.
$49 = 185 - 176 \cos A$	Simplify.
$-136 = -176 \cos A$	Subtract 185 from each side.
$\frac{-136}{-176} = \cos A$	Divide each side by -176.
$m\angle A = \cos^{-1}\left(\frac{-136}{-176}\right)$	Use the inverse cosine function.
$m\angle A = 39°$	Use a calculator.

Use the Law of Sines to find $m\angle B$.

$\frac{\sin A}{a} = \frac{\sin B}{b}$	Law of Sines
$\frac{\sin 39°}{7} = \frac{\sin B}{11}$	$m\angle A \approx 39°$, $a = 7$, and $b = 11$
$11 \sin 39° = 7 \sin B$	Multiplication Property of Equality
$\left(\frac{11 \sin 39°}{7}\right) = \sin B$	Divide each side by 7.
$m\angle B = \sin^{-1}\left(\frac{11 \sin 39°}{7}\right)$	Use the inverse sine function.
$m\angle B \approx 81°$	Use a calculator.

By the Triangle Angle-Sum Theorem, $m\angle C \approx 180 - (39 + 81)$ or $60°$.

 **Go Online** You can complete an Extra Example online.

> **Study Tip**
>
> **Rounding** When you round a numerical solution and then use it in later calculations, your answers may be inaccurate. Wait until after you have completed all of your calculations to round.

Check

Solve △ABC when $b = 10.2$, $c = 9.3$, and $m\angle A = 26°$. Round angle measures to the nearest degree and side measures to the nearest tenth.

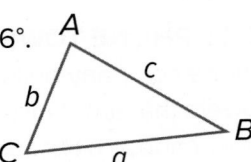

$a = \underline{\quad ? \quad}$

$m\angle B = \underline{\quad ? \quad}$

$m\angle C = \underline{\quad ? \quad}$

Example 5 Solve a Right Triangle with the Law of Cosines

Solve △FGH. Round angle measures to the nearest degree and side measures to the nearest tenth.

Find FG.

$h^2 = f^2 + g^2$	Pythagorean Theorem
$h^2 = 2^2 + 3^2$	Substitution
$h^2 = 13$	Simplify.
$h = \sqrt{13}$	Take the positive square root of each side.
$h \approx 3.6$	Use a calculator.

So, $FG \approx 3.6$.

Find $m\angle F$.

$f^2 = g^2 + h^2 - 2gh \cos F$	Law of Cosines
$2^2 = 3^2 + (\sqrt{13})^2 - 2(3)(\sqrt{13}) \cos F$	Substitution
$4 = 22 - 6\sqrt{13} \cos F$	Simplify.
$-18 = -6\sqrt{13} \cos F$	Subtract 22 from each side.
$\dfrac{-18}{-6\sqrt{3}} = \cos F$	Divide each side by $-6\sqrt{13}$.
$34 \approx m\angle F$	Use a calculator.

So, $m\angle F \approx 34°$.

Find $m\angle G$.

Because we know that $m\angle H = 90°$ and $m\angle F \approx 34°$, find $m\angle G$.

$m\angle G = 90° - 34°$ or $56°$

Study Tip

Obtuse Angles There are also values for sin A, cos A, and tan A when $m\angle A \geq 90°$. Values of the ratios for these angles can be found by using the trigonometric functions on your calculator. It is good practice to solve for smaller angles first, and then use the Triangle Angle Sum Theorem to find the measure of the largest, third angle.

Go Online
to practice what you've learned about solving triangles in the Put It All Together over Lessons 9-7 through 9-8.

Check

Solve △EFG. Round angle measures to the nearest degree and side measures to the nearest tenth.

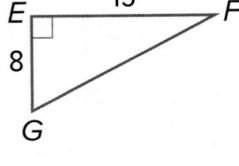

$e = \underline{\quad ? \quad}$

$m\angle F = \underline{\quad ? \quad}$

$m\angle G = \underline{\quad ? \quad}$

Go Online You can complete an Extra Example online.

Practice

⟲ **Go Online** You can complete your homework online.

Examples 1 and 2

Find the value of *x* to the nearest tenth for side lengths and nearest degree for angle measures.

1.

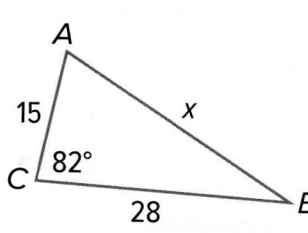

2.

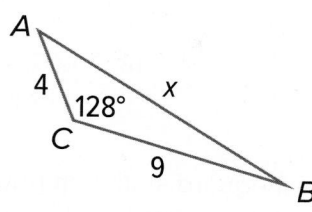

3.

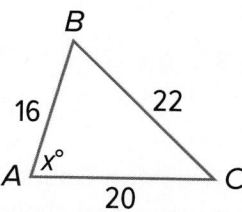

4.

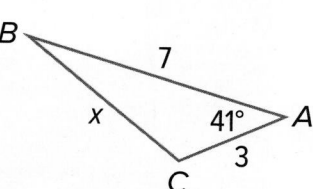

5.

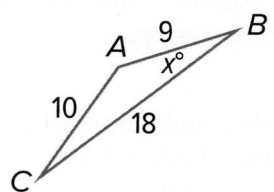

6.

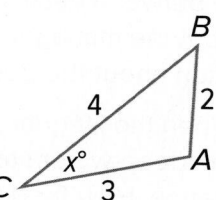

Example 3

7. RADAR Two radar stations 2.4 miles apart are tracking an airplane. The straight-line distance between Station A and the plane is 7.4 miles. The straight-line distance between Station B and the plane is 6.9 miles. What is the angle of elevation from Station A to the plane? Round to the nearest degree.

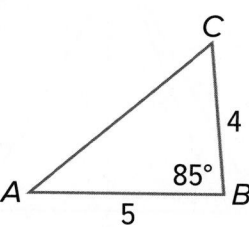

8. DRAFTING Marion is using a drafting program to produce a drawing for a client. She wants to make a triangle first. She begins by drawing a segment 4.2 inches long from point *A* to point *B*. From point *B*, she draws a second segment that forms a 42° angle with $\overline{AB}$, is 6.4 inches long, and ends at point *C*. To the nearest tenth of an inch, how long is the segment from *C* to *A*?

Examples 4 and 5

REASONING Solve each triangle. Round side lengths to the nearest tenth and angle measures to the nearest degree.

9.

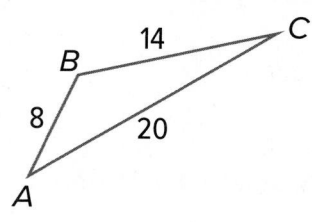

10.

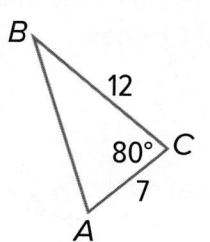

11.

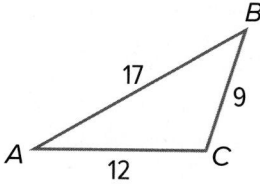

12.

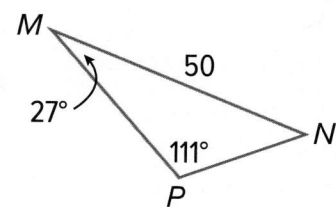

13.

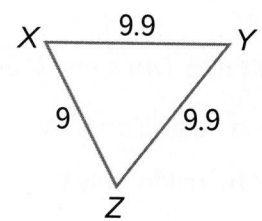

14.

Mixed Exercises

REGULARITY **Determine whether each triangle should be solved by beginning with the *Law of Sines* or the *Law of Cosines*. Then solve the triangle.**

15. $m\angle A = 11°$, $m\angle C = 27°$, $c = 50$

16. $m\angle B = 47°$, $a = 20$, $c = 24$

17. $m\angle A = 37°$, $a = 20$, $b = 18$

18. $m\angle C = 35°$, $a = 18$, $b = 24$

19. POOLS The Perth County pool has a lifeguard station in both the deep-water and shallow-water sections of the pool. The distance between each station and the bottom of the slide is known, but the manager would like to calculate more information about the pool setup.

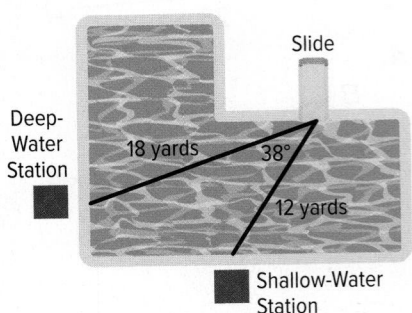

a. When the lifeguards switch positions, the lifeguard at the deep-water station swims to the shallow-water station. How far does the lifeguard swim?

b. If the lifeguard at the deep-water station is directly facing the bottom of the slide, what angle does she need to turn in order to face the lifeguard at the shallow-water station?

20. PROOF Write a two-column proof to prove the Law of Cosines.

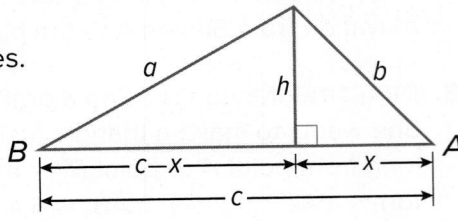

Given: The altitude of $\triangle ABC$ is h.

Prove: $a^2 = b^2 + c^2 - 2bc \cos A$

🧠 **Higher-Order Thinking Skills**

21. PERSEVERE Find the value of x in the figure at the right.

22. ANALYZE Explain why the Pythagorean Theorem is a specific case of the Law of Cosines.

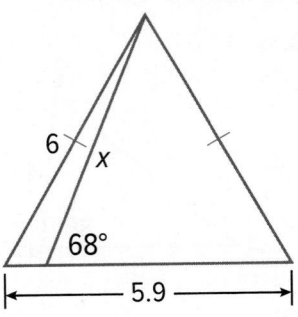

23. WRITE What methods can you use to solve a triangle?

24. CREATE Draw and label a triangle that can be solved:

a. using only the Law of Sines.

b. using only the Law of Cosines.

@ Essential Question

How are right triangle relationships useful in solving real-world problems?

The Pythagorean Theorem can be used to find missing side lengths of right triangles that occur in the real world.

Module Summary

Lesson 4-1

Geometric Mean

- The geometric mean of two positive numbers a and b is x such that $\frac{a}{x} = \frac{x}{b}$.

  So, $x^2 = ab$ and $x = \sqrt{ab}$.

Lesson 4-2

Pythagorean Theorem

- If $\triangle ABC$ is a right triangle with right angle C, then $a^2 + b^2 = c^2$.

Lesson 4-3

Coordinates in Space

- A point in space is represented by (x, y, z) where x, y, and z are real numbers.
- A three-dimensional coordinate system has three axes: the x-, y-, and z-axes. The x- and y-axes lie on a horizontal plane, and the z-axis is vertical.

Lesson 4-4

Special Right Triangles

- In a 45°-45°-90° triangle, the legs ℓ are congruent and the length of the hypotenuse h is $\sqrt{2}$ times the length of a leg.
- In a 30°-60°-90° triangle, the length of the hypotenuse h is 2 times the length of the shorter leg s, and the longer leg ℓ is $\sqrt{3}$ times the length of the shorter leg.

Lesson 4-5

Trigonometry

- $\sin A = \frac{\text{opp}}{\text{hyp}}$, $\cos A = \frac{\text{adj}}{\text{hyp}}$, and $\tan A = \frac{\text{opp}}{\text{adj}}$.
- If $\sin A = x$, then $\sin^{-1} x = m\angle A$. If $\cos A = x$, then $\cos^{-1} x = m\angle A$. If $\tan A = x$, then $\tan^{-1} x = m\angle A$.

Lessons 4-6 through 4-8

Applications of Trigonometry

- Area $= \frac{1}{2}ab \sin C$, where a and b are side lengths and C is the included angle.
- Law of Sines: If $\triangle ABC$ has lengths a, b, and c, representing the lengths of the sides opposite the angles with measures A, B, and C, then $\frac{\sin A}{a} = \frac{\sin B}{b} = \frac{\sin C}{c}$.
- Law of Cosines: If $\triangle ABC$ has lengths a, b, and c, representing the lengths of the sides opposite the angles with measures A, B, and C, then $a^2 = b^2 + c^2 - 2bc \cos A$, $b^2 = a^2 + c^2 - 2ac \cos B$, and $c^2 = a^2 + b^2 - 2ab \cos C$.

Study Organizer

Foldables

Use your Foldable to review this module. Working with a partner can be helpful. Ask for clarification of concepts as needed.

Right Triangles and Trigonometry

Test Practice

1. MULTIPLE CHOICE What is the geometric mean between 6 and 12? (Lesson 4-1)

A. $6\sqrt{2}$ B. $3\sqrt{2}$

C. 9   D. 36

2. MULTIPLE CHOICE What is the geometric mean between 9 and 15? (Lesson 4-1)

A. $9\sqrt{15}$ B. 12

C. $3\sqrt{15}$ D. 135

3. MULTI-SELECT Which of the following triangles is similar to $\triangle JKL$? Select all that apply. (Lesson 4-1)

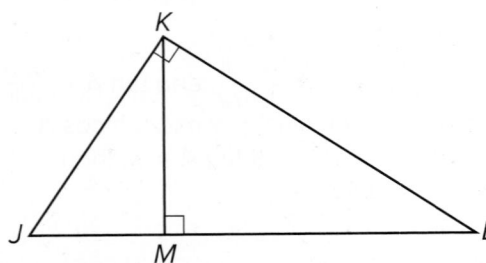

A. $\triangle JMK$ B. $\triangle MKL$

C. $\triangle JKM$ D. $\triangle KML$

E. $\triangle LJK$

4. OPEN RESPONSE A path forms a diagonal of a rectangular city park. A bench is placed along the path at a point that is closest to the fountain. This point is 30 yards from one end of the path and 50 yards from the other end of the path. What is the distance from the bench to the fountain, to the nearest yard? (Lesson 4-1)

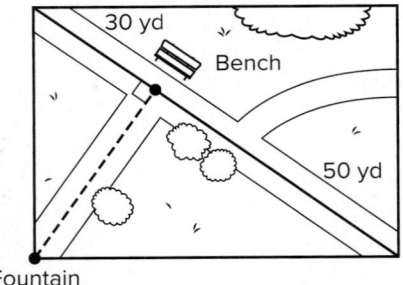

5. OPEN RESPONSE Given right $\triangle ABC$ with point D such that $\overleftrightarrow{CD}$ is perpendicular to $\overleftrightarrow{AB}$ and $\triangle ACB$ is similar to $\triangle ADC$ and to $\triangle CDB$, complete the proof to show $AC^2 + BC^2 = AB^2$. (Lesson 4-1)

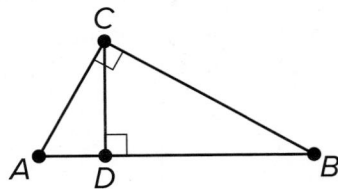

Statements	Reasons
1. $\triangle ABC$ is a right triangle with point D such that $\overleftrightarrow{CD}$ is perpendicular to $\overleftrightarrow{AB}$ and $\triangle ACB$ is similar to $\triangle ADC$ and to $\triangle CDB$.	1. Given
2. $\dfrac{BC}{BA} = \dfrac{BD}{BC}$	2. Definition of similar triangles
3. $\dfrac{AC}{AB} = \dfrac{AD}{AC}$	3. Definition of similar triangles
4. $BC^2 = (BA)(BD)$ and $AC^2 = (AB)(AD)$	4. ___?___
5. $AC^2 + BC^2 = (BA)(BD) + (AB)(AD)$	5. Addition Property
6. $AC^2 + BC^2 = (AB)(BD + AD)$	6. ___?___
7. $BD + AD = AB$	7. Segment Addition Postulate
8. $AC^2 + BC^2 = AB^2$	8. Substitution

6. MULTIPLCE CHOICE In △*DEF*, *DE* = 8 and *EF* = 15. Which length of the third side would create an obtuse triangle? (Lesson 4-2)

A. 15

B. 16

C. 17

D. 18

7. OPEN RESPONSE Use the figure. If *AB* = 3√15, what is *BD*? Write the answer in simplest radical form. (Lesson 4-4)

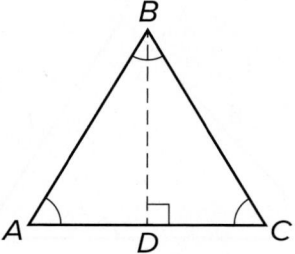

8. MULTIPLE CHOICE What is the value of *y*?
(Lesson 4-4)

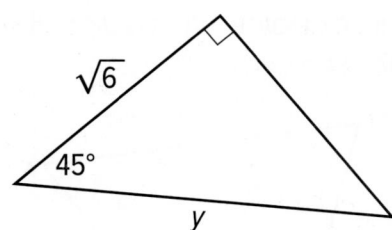

A. $\sqrt{2}$

B. $\sqrt{3}$

C. $2\sqrt{3}$

D. $3\sqrt{2}$

9. OPEN RESPONSE A 10-foot-long ladder leans against a wall so that the ladder and the wall form a 30° angle. What is the distance to the nearest tenth of a foot from the ground to the point where the ladder touches the wall? (Lesson 4-4)

10. MULTPLE CHOICE Find the sine, cosine, and tangent of B written as a decimal value.
(Lesson 4-5)

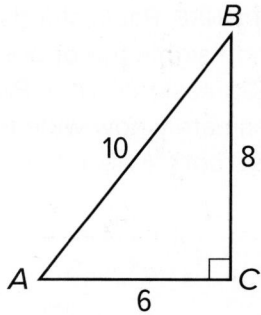

11. OPEN RESPONSE Use special right triangles to evaluate sin 45°. Give an exact answer.
(Lesson 4-5)

12. **MULTIPLE CHOICE** A support wire for an electric pole makes a 50° angle with the ground. If the bottom of the support wire is 8 feet from the base of the pole, which of the following equations can be used to find the height at which the support wire is attached to the pole? (Lesson 4-6)

A. $\frac{x}{8} = \tan 50°$

B. $\frac{8}{x} = \tan 50°$

C. $\frac{x}{8} = \sin 50°$

D. $\frac{8}{x} = \sin 50°$

13. **OPEN RESPONSE** Paula stands on the side of a river and sights the opposite bank at an angle of depression of 7°. If Paula is 5.5 feet tall, approximately how wide is the river, to the nearest foot? (Lesson 4-6)

14. **MULTI-SELECT** In $\triangle DEF$, $e = 7$, $f = 5$, and $m\angle F = 44°$. What are the possible measures of $\angle E$, to the nearest degree? Select all that apply. (Lesson 4-7)

A. 32°   B. 44°

C. 60°   D. 77°

E. 103°   F. 120°

15. **OPEN RESPONSE** What is the value of a, to the nearest tenth? (Lesson 4-7)

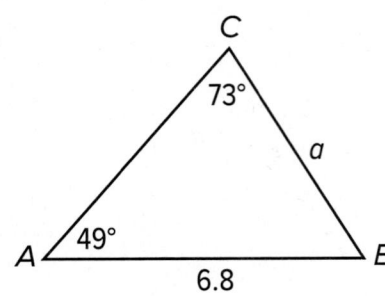

16. **MULTIPLE CHOICE** Two ships leave from the same location. The first ship travels due south for 13 miles. The second ship travels 50° east of due south for 11 miles. What is the distance between the ships? (Lesson 4-8)

A. 6.9 miles

B. 9.2 miles

C. 10.3 miles

D. 10.9 miles

17. **MULTIPLE CHOICE** What is $m\angle E$, to the nearest degree? (Lesson 4-8)

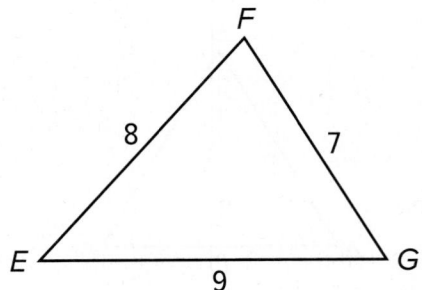

A. 40°

B. 42°

C. 45°

D. 48°

18. **MULTIPLE CHOICE** What is JK to the nearest tenth? (Lesson 4-8)

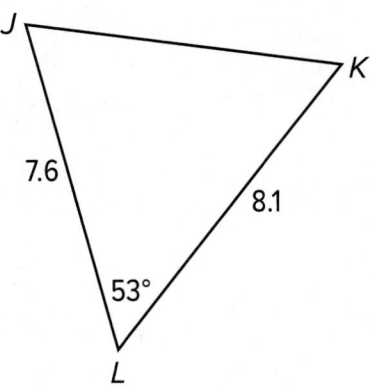

A. 7.0 units   B. 15.4 units

C. 49.3 units   D. 74.1 units

e Essential Question

How can circles and parts of circles be used to model situations in the real world?

What Will You Learn?

How much do you already know about each topic **before** starting this module?

KEY

👎 — I don't know.   👍➤ — I've heard of it.   👍 — I know it!

	Before			After		
	👎	👍➤	👍	👎	👍➤	👍
use the formula for circumference of a circle						
prove all circles are similar						
find measures of angles and arcs using the properties of circles						
solve problems using the relationships between arcs, chords, and diameters						
solve problems using inscribed angles						
solve problems using inscribed polygons						
solve problems using relationships between circles, tangents, and secants						
construct inscribed and circumscribed circles						
use equations of circles to solve problems						
graph equations of parabolas						

📖 **Foldables** Make this Foldable to help you organize your notes about circles. Begin with nine sheets of notebook paper.

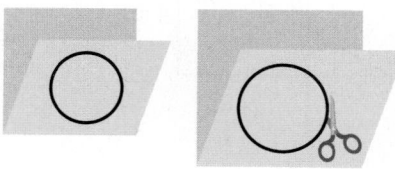

1. **Trace** an 8-inch circle on each paper using a compass.

2. **Cut** out each of the circles.

3. **Staple** an inch from the left side of the papers.

4. **Label** each tab with a lesson number and title.

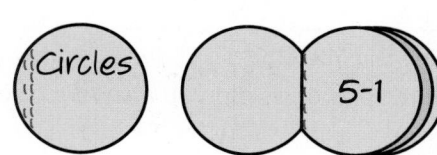

What Vocabulary Will You Learn?

- adjacent arcs
- arc
- arc length
- center of a circle
- central angle of a circle
- chord of a circle
- circle

- circumscribed angle
- circumscribed polygon
- common tangent
- concentric circles
- congruent arcs
- degree
- diameter of a circle
- directrix

- focus
- inscribed angle
- inscribed polygon
- intercepted arc
- major arc
- minor arc
- parabola
- pi

- point of tangency
- radian
- radius of a circle
- secant
- semicircle
- tangent to a circle

Are You Ready?

Complete the Quick Review to see if you are ready to start this module.
Then complete the Quick Check.

Quick Review

Example 1

Find the value of *h*.

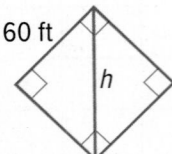 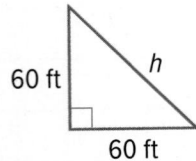

60 ft

60 ft

60 ft

Because *h* is the hypotenuse of the triangle, the triangle can be redrawn as shown.

In a 45°-45°-90° triangle, the hypotenuse is $\sqrt{2}$ times the length of a leg.

$h = 60\sqrt{2}$

≈ 84.85

So, *h* is approximately 84.85 feet.

Example 2

Solve $x^2 + 3x - 40 = 0$ by using the Quadratic Formula. Round to the nearest tenth.

$x = \dfrac{-b \pm \sqrt{b^2 - 4ac}}{2a}$ Quadratic Formula

$= \dfrac{-3 \pm \sqrt{3^2 - 4(1)(-40)}}{2(1)}$ Substitution

$= \dfrac{-3 \pm \sqrt{169}}{2}$ Simplify.

$= 5 \text{ or } -8$ Simplify.

Quick Check

Find the value of *h* in each triangle.

1.

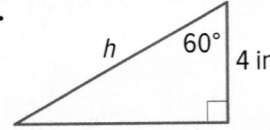

h   60°   4 in.

2.

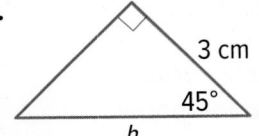

3 cm   45°   *h*

Solve each equation by using the Quadratic Formula. Round to the nearest tenth if necessary.

3. $5x^2 + 4x - 20 = 0$

4. $x^2 = x + 12$

5. $3x^2 - x - 12 = 0$

6. $4x^2 - 16x - 18 = 0$

How Did You Do?

Which exercises did you answer correctly in the Quick Check?

Circles and Circumference

Explore Discovering the Formula for Circumference

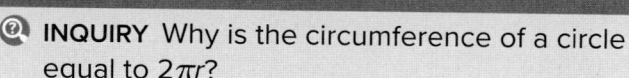 **Online Activity** Use dynamic geometry software to complete the Explore.

> **INQUIRY** Why is the circumference of a circle equal to $2\pi r$? ✕

Learn Parts of Circles

A **circle** is the set of all points in a plane that are the same distance from a given point called the **center of a circle**. The center of the circle below is C.

A **radius of a circle** (plural radii) is a line segment from the center to a point on a circle. $\overline{CD}$, $\overline{CE}$, and $\overline{CF}$ are radii of circle C.

A **chord of a circle** is a segment with endpoints on the circle. $\overline{AB}$ and $\overline{DE}$ are chords of circle C.

A **diameter of a circle** is a chord that passes through the center of a circle. $\overline{DE}$ is a diameter of circle C.

All radii r of a circle are congruent. Because a diameter d is composed of two radii, all diameters of a circle are also congruent.

The words *radius* and *diameter* are used to describe lengths as well as segments.

Key Concept • Radius and Diameter Relationships

If a circle has radius r and diameter d, then the following relationships are true.

Radius Formula $r = \frac{d}{2}$ or $r = \frac{1}{2}d$	**Diameter Formula** $d = 2r$

The circumference of a circle is the distance around the circle. By definition, the ratio $\frac{C}{d}$ is an irrational number called **pi (π)**. Two formulas for circumference can be derived by using this definition.

$\frac{C}{d} = \pi$	Definition of pi
$C = \pi d$	Multiply each side by d.
$C = \pi(2r)$	$d = 2r$
$C = 2\pi r$	Simplify.

Key Concept • Circumference Formula

Words	If a circle has diameter d or radius r, the circumference C equals the diameter times pi or twice the radius times pi.
Symbols	$C = \pi d$ or $C = 2\pi r$

Today's Goals
- Know the precise definition of circle and find the circumferences of circles.
- Find measures in intersecting circles and prove relationships between circles.

Today's Vocabulary
circle
center of a circle
radius of a circle
chord of a circle
diameter of a circle
pi
concentric circles

Math History Minute

Chinese mathematician **Tsu Ch'ung-chih (429 AD-500 AD)** was particularly interested in finding the value of π. He approximated its value to be about the same as the ratio 355:113, which is actually correct to about 6 decimal places.

Example 1 Identify Segments in a Circle

Name the circle and identify a radius, a chord, and a diameter of the circle.

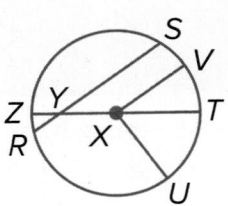

The circle has a center at X, so it is named circle X or $\odot X$.

Four radii are shown: $\overline{XV}$, $\overline{XT}$, $\overline{XU}$, and $\overline{XZ}$.

Two chords are shown: $\overline{RS}$ and $\overline{TZ}$.

$\overline{TZ}$ contains the center, so $\overline{TZ}$ is the diameter.

Example 2 Use Radius and Diameter Relationships

If $TU = 14$ feet, what is the radius of $\odot Q$?

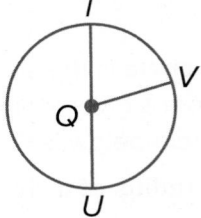

$r = \dfrac{d}{2}$ Radius Formula

$r = \dfrac{14}{2}$ or 7   Substitute and simplify.

The radius of $\odot Q$ is 7 feet.

Check

If $LM = 11$ inches, what is the diameter $\odot L$?

___?___ in.

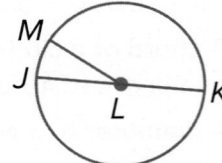

😀 **Think About It!**

What assumption did you make while solving this problem?

Use a Source

Find the diameter of a famous traffic circle. Then calculate the circumference of the traffic circle.

🌎 Example 3 Find Circumference

TRAFFIC CIRCLES **Traffic circles, also known as roundabouts, are circular roadways that reflow traffic in one direction around an island. A car enters a traffic circle and is 18 meters from the center of the island. If the car drives around the traffic circle until it is back to its original position, what is the circumference of the car's path?**

Because the car is 18 meters from the center of the island, the radius of the car's path is 18 meters.

$C = 2\pi r$ Circumference Formula

$\quad = 2\pi(18)$ Substitution

$\quad = 36\pi$ Simplify.

$\quad \approx 113.1$ Use a calculator.

The circumference of the car's path is 36π or about 113.1 meters.

🔘 **Go Online** You can complete an Extra Example online.

Check

AUTOMOBILES Many automobiles have customized rims that are attached to wheels and align with the inner edges of the tires. The rim of a tire has a radius of 7.5 inches, and the width of the tire is 6 inches.

Part A Find the circumference of the rim.

A. $\frac{15\pi}{2}$ in.   **B.** 13.5π in.   **C.** 15π in.   **D.** 27π in.

Part B What assumptions did you make while solving this problem? Select all that apply.

A. I assumed the radius of the tire rim.

B. I assumed that the tire rim was a perfect circle.

C. I assumed the width of the tire.

D. I assumed there was no space between the tire and the rim.

E. I assumed the tire and rim were not the same shape.

Example 4 Find Diameter and Radius

Find the diameter and radius of a circle to the nearest hundredth if the circumference of the circle is 77.8 centimeters.

Find the diameter.

$C = \pi d$ Circumference Formula

$77.8 = \pi d$ Substitution

$\frac{77.8}{\pi} = d$ Divide each side by π.

$24.76 \approx d$ Use a calculator.

The diameter of the circle is about 24.76 centimeters. Use the diameter of the circle to find the radius.

$r = \frac{1}{2} d$ Radius Formula

$\approx \frac{1}{2} (24.76)$ $d \approx 24.76$

≈ 12.38 Use a calculator.

So, the radius of the circle is about 12.38 centimeters.

Check

Find the diameter and radius of a circle to the nearest hundredth if the circumference of the circle is 94.2 yards.

Part A Select the most appropriate estimates of the diameter and radius.

A. 15.7 yd; 7.85 yd.   **B.** 26.91 yd; 13.46 yd

C. 18.84 yd; 9.42 yd   **D.** 31.4 yd; 15.7 yd

Part B Find the diameter and radius of the circle to the nearest hundredth.

diameter = ___?___ yd

radius = ___?___ yd

🔾 **Go Online** You can complete an Extra Example online.

Learn Pairs of Circles

As with other figures, pairs of circles can be congruent, similar, or share other special relationships.

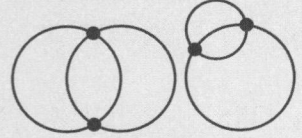

Postulate 5.1	
Words	Two circles are congruent if and only if they have congruent radii.
Example	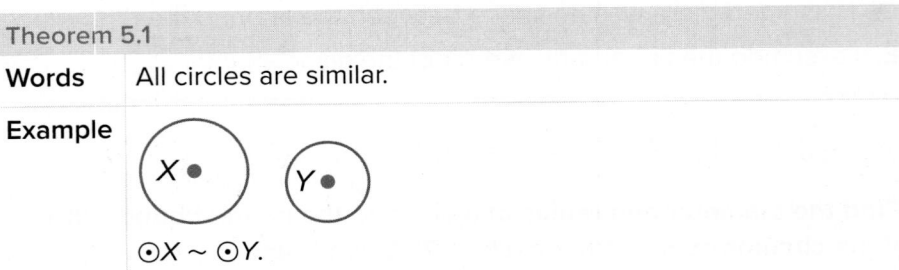 $\overline{GH} \cong \overline{JK}$, so $\odot G \cong \odot J$.

Theorem 5.1	
Words	All circles are similar.
Example	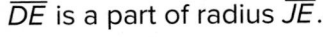 $\odot X \sim \odot Y$.

You will prove Theorem 5.1 in Exercise 42.

Concentric circles are coplanar circles that have the same center.

$\odot A$ with radius $\overline{AB}$ and $\odot A$ with radius $\overline{AC}$ are concentric.

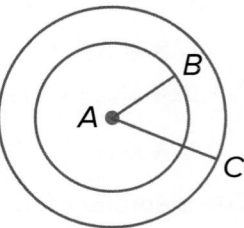

Example 5 Find Measures in Intersecting Circles

The diameter of $\odot K$ is 12 units, the diameter of $\odot J$ is 20 units, and $JD = 8$ units. Find EK.

Because the diameter of $\odot J$ is 20, $JE = 10$.

$\overline{DE}$ is a part of radius $\overline{JE}$.

$JD + DE = JE$	Segment Addition Postulate
$8 + DE = 10$	Substitution
$DE = 2$	Solve.

Because the diameter of $\odot K$ is 12, $DK = 6$. $\overline{DE}$ and $\overline{EK}$ form radius $\overline{DK}$.

$DE + EK = DK$	Segment Addition Postulate
$2 + EK = 6$	Substitution
$EK = 4$	Solve.

So, EK is 4 units.

 **Go Online** You can complete an Extra Example online.

Practice

◤ Go Online You can complete your homework online.

Example 1

For Exercises 1–3, refer to the circle at the right.

1. Name the circle.

2. Name the radii of the circle.

3. Name the chords of the circle.

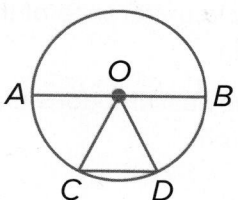

For Exercises 4–8, refer to the circle at the right.

4. Name the circle.

5. Name the radii of the circle.

6. Name the chords of the circle.

7. Name a diameter of the circle.

8. Name a radius not drawn as part of a diameter.

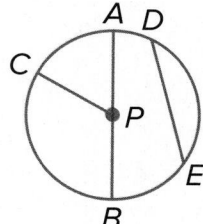

Example 2

For Exercises 9–11, refer to ⊙R.

9. If $AB = 18$ millimeters, find AR.

10. If $RY = 10$ inches, find AR and AB.

11. Is $\overline{AB} \cong \overline{XY}$? Explain.

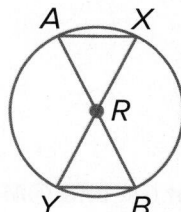

For Exercises 12–14, refer to ⊙L.

12. Suppose the radius of the circle is 3.5 yards. Find the diameter.

13. If $RT = 19$ meters, find LW.

14. If $LT = 4.2$ inches, what is the diameter of ⊙L?

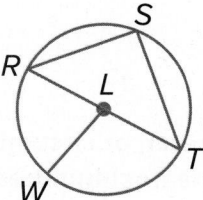

Example 3

15. TIRES A bicycle has tires with a diameter of 26 inches. Find the radius and circumference of each tire. Round your answer to the nearest hundredth, if necessary.

16. STATE YOUR ASSUMPTION Herman purchased a sundial to use as the centerpiece for a garden. The diameter of the sundial is 9.5 inches.

    a. Find the radius of the sundial.

    b. Find the circumference of the sundial to the nearest hundredth.

    c. Explain any assumptions that you make while solving this problem.

Example 4

Find the diameter and radius of a circle to the nearest hundredth with the given circumference.

17. $C = 40$ in.

18. $C = 256$ ft

19. $C = 15.62$ m

20. $C = 9$ cm

21. $C = 79.5$ yd

22. $C = 204.16$ m

Example 5

The diameters of $\odot F$ and $\odot G$ are 5 and 6 units, respectively. Find each measure.

23. BF

24. AB

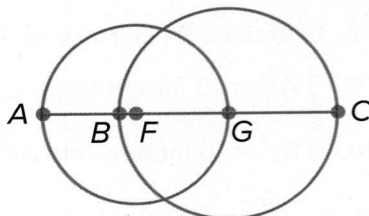

The diameters of $\odot L$ and $\odot M$ are 20 and 13 units, respectively, and $QR = 4$. Find each measure.

25. LQ

26. RM

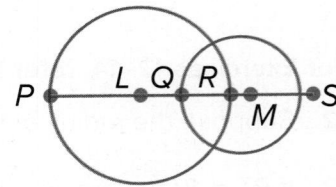

Mixed Exercises

The radius, diameter, or circumference of a circle is given. Find each missing measure to the nearest hundredth.

27. $d = 8\frac{1}{2}$ in., $r =$ _?_ , $C =$ _?_

28. $r = 11\frac{2}{5}$ ft, $d =$ _?_ , $C =$ _?_

29. $C = 628$ m, $d =$ _?_ , $r =$ _?_

30. $d = \frac{3}{4}$ yd, $r =$ _?_ , $C =$ _?_

31. $C = 35x$ cm, $d =$ _?_ , $r =$ _?_

32. $r = \frac{x}{8}$, $d =$ _?_ , $C =$ _?_

Determine whether the circles in the figures below appear to be *congruent,* *concentric,* **or** *neither.*

33.
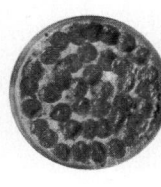 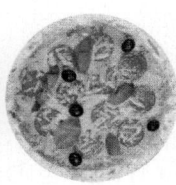

34.

35.

For each circle, find the exact circumference in terms of π.

36.
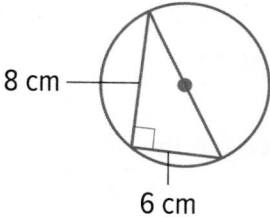

8 cm

6 cm

37.
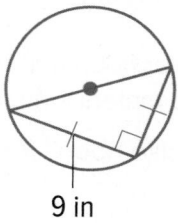

9 in

38.
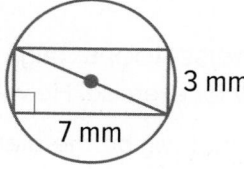

3 mm

7 mm

39.
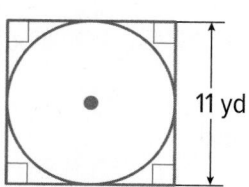

11 yd

40.
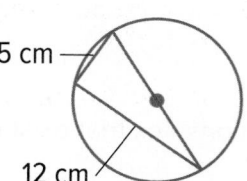

5 cm

12 cm

41.
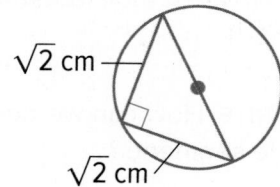

$\sqrt{2}$ cm

$\sqrt{2}$ cm

42. PROOF Write a paragraph proof to prove Theorem 5.1.

    Given: ⊙D and ⊙E

    Prove: ⊙D ∼ ⊙E

43. USE A SOURCE Go online to research a famous clock face. Then use the diameter of the clock face to find the circumference. Round your answer to the nearest hundredth.

44. WHEELS Zack is designing wheels for a concept car. The diameter of the wheel is 18 inches. Zack wants to make spokes in the wheel that run from the center of the wheel to the rim. In other words, each spoke is a radius of the wheel. How long are these spokes?

45. PRECISION Kathy slices through a circular cake. The cake has a diameter of 14 inches. The slice that Kathy made is straight and has a length of 11 inches. Did Kathy cut along a *radius*, a *diameter*, or a *chord* of the circle?

46. REASONING Three identical circular coins are lined up in a row as shown. The distance between the centers of the first and third coins is 3.2 centimeters. What is the radius of one of these coins?

|←3.2 cm→|

47. EXERCISE HOOPS Taiga wants to make a circular hoop that he can twirl around his body for exercise. He will use a tube that is 2.5 meters long.

a. What will be the diameter of Taiga's exercise hoop? Round your answer to the nearest thousandth of a meter.

b. What will be the radius of Taiga's exercise hoop? Round your answer to the nearest thousandth of a meter.

48. WRITE How can we describe the relationships that exist between circles and line segments?

49. PERSEVERE The sum of the circumferences of circles H, J, and K shown at the right is 56π units. Find KJ.

50. ANALYZE Is the distance from the center of a circle to a point in the interior of a circle *sometimes*, *always*, or *never* less than the radius of the circle? Justify your argument.

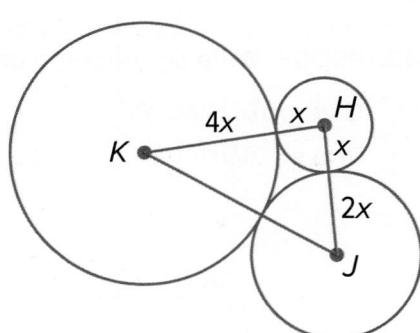

51. CREATE Design a sequence of transformations that can be used to prove that $\odot D$ is similar to $\odot E$.

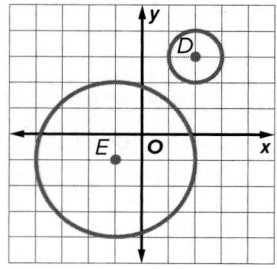

Measuring Angles and Arcs

Learn Measuring Angles and Arcs

A **central angle of a circle** is an angle with a vertex at the center of a circle and sides that are radii. $\angle ABC$ is a central angle of $\odot B$.

A **degree** is $\frac{1}{360}$ of the circular rotation about a point. This leads to the following relationship.

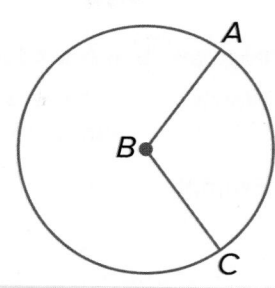

Key Concept • Sum of Central Angles	
Words	The sum of the measures of the central angles of a circle with no interior points in common is 360°.
Example	$m\angle 1 + m\angle 2 + m\angle 3 = 360°$

An **arc** is part of a circle that is defined by two endpoints. A central angle separates the circle into two arcs with measures related to the measure of the central angle.

A **minor arc** has a measure less than 180°. The measure of a minor arc is equal to the measure of its related central angle. $m\,\overarc{PR} = m\angle PQR$.

A **major arc** has a measure greater than 180°. The measure of a major arc is equal to 360° minus the measure of the minor arc with the same endpoints. $m\,\overarc{PSR} = 360° - m\,\overarc{PR}$.

A **semicircle** is an arc that measures exactly 180°. The endpoints of a semicircle lie on a diameter. $m\,\overarc{RST} = 180°$.

Congruent arcs are arcs in the same or congruent circles that have the same measure.

Theorem 5.2	
Words	In the same circle or in congruent circles, two minor arcs are congruent if and only if their central angles are congruent.
Example	If $\angle 1 \cong \angle 2$, then $\overarc{FG} \cong \overarc{HJ}$. If $\overarc{FG} \cong \overarc{HJ}$, then $\angle 1 \cong \angle 2$.

You will prove Theorem 5.2 in Exercise 47.

Today's Goals
- Find measures of angles and arcs using the properties of circles.
- Find arc lengths and convert between degrees and radians.

Today's Vocabulary
central angle of a circle
degree
arc
minor arc
major arc
semicircle
congruent arcs
adjacent arcs
arc length
radian

Study Tip
Naming Arcs Minor arcs can be named by just their endpoints. Major arcs and semicircles are named by their endpoints and another point on the arc that lies between these endpoints.

Go Online

You can watch a video to see the characteristics of arcs.

Go Online

You may want to complete the Concept Check to check your understanding.

Think About It!

What part of a circle is represented by the sides of a central angle?

Arcs in a circle that have exactly one point in common are called **adjacent arcs**. In ⊙M, $\widehat{HJ}$ and $\widehat{JK}$ are adjacent arcs. As with adjacent angles, you can add the measures of adjacent arcs.

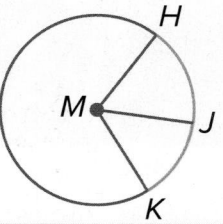

Postulate 5.2: Arc Addition Postulate	
Words	The measure of an arc formed by two adjacent arcs is the sum of the measures of the two arcs.
Example	$m\widehat{XZ} = m\widehat{XY} + m\widehat{YZ}$ 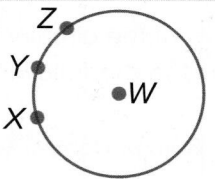

Example 1 Find Measures of Central Angles

Find the value of x.

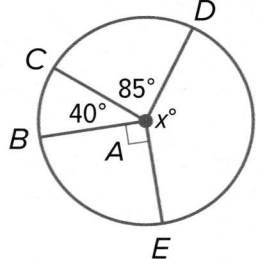

$$m\angle EAB + m\angle BAC + m\angle CAD + m\angle DAE = 360 \quad \text{Sum of central angles}$$

$$90 + 40 + 85 + x = 360 \quad \text{Substitution}$$

$$215 + x = 360 \quad \text{Simplify.}$$

$$x = 145° \quad \text{Solve.}$$

Check

Find the value of x.

$x = $ _____?_____

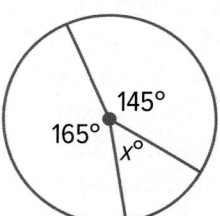

Example 2 Classify Arcs and Find Arc Measures

$\overline{PM}$ **is a diameter of ⊙R. Identify each arc as a** *major arc,* *minor arc,* **or** *semicircle.* **Then find its measure.**

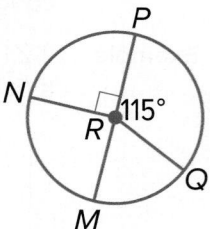

a. $\widehat{MQ}$

$\widehat{MQ}$ is a minor arc, so $m\widehat{MQ} = m\angle MRQ$. Because $\angle MRQ$ and $\angle PRQ$ are linear pairs, $m\angle MRQ = 180 - 115$ or 65°. So, $m\widehat{MQ} = 65°$.

Go Online You can complete an Extra Example online.

b. $\overarc{MNP}$

$\overarc{MNP}$ is a semicircle, so $m\overarc{MNP} = 180°$.

c. $\overarc{MNQ}$

$\overarc{MNQ}$ is a major arc that shares the same endpoints as minor arc $\overarc{MQ}$.

$m\overarc{MNQ} = 360° - m\overarc{MQ}$

$\quad\quad\quad = 360° - 65°$ or $295°$

Check

$\overline{GJ}$ is a diameter of $\odot K$. Identify each arc as a *major arc, minor arc,* or *semicircle.* Then find its measure.

a. $\overarc{GH}$

b. $\overarc{GLH}$

c. $\overarc{GLJ}$

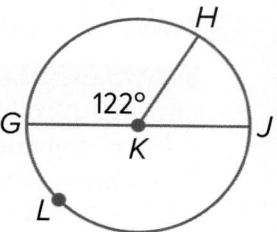

🌐 Example 3 Find Arc Measures in Circle Graphs

ONLINE **The circle graph shows how teenagers spend their time online. Find $m\overarc{DE}$.**

$\overarc{DE}$ is a minor arc.

So, $m\overarc{DE} = m\angle DPE$.

$\angle DPE$ represents 12% of the whole, or 12% of the circle.

$m\angle DPE = 0.12(360°)$ Find 12% of 360°.

$\quad\quad\quad = 43.2°$ Simplify.

So, $m\overarc{DE}$ is 43.2°.

Check

SPORTS The circle graph shows the percent of male athletes that are involved in each type of sport at a high school. Find the measure of $\overarc{CD}$. Round to the nearest tenth if necessary.

$m\overarc{CD} = \underline{\quad ? \quad}$

Example 4 Use Arc Addition to Find Measures of Arcs

Find $m\overarc{WY}$ in $\odot V$.

$m\overarc{WY} = m\overarc{WX} + m\overarc{XY}$ Arc Addition Postulate

$\quad\quad\quad = m\angle WVX + m\angle XVY$ $m\overarc{WX} = m\angle WVX, m°\overarc{XY} = m\angle XVY$

$\quad\quad\quad = 90° + 62°$ or $152°$ Substitution

💬 **Talk About It!**

Are any arcs in $\odot P$ congruent to $\overarc{DE}$? Explain.

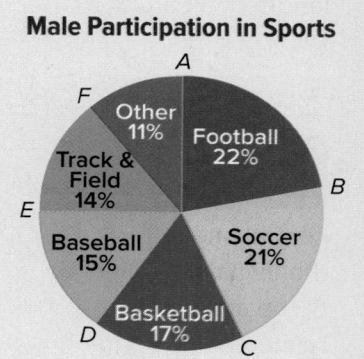

Teenagers' Online Activities

Games 7%
Other 3%
Search Engines 12%
Social Media 44%
Music 14%
Videos 20%

Male Participation in Sports

Other 11%
Football 22%
Track & Field 14%
Soccer 21%
Baseball 15%
Basketball 17%

Check

Find $m\overarc{AD}$ in $\odot F$.

$m\overarc{AD} =$ ____?____

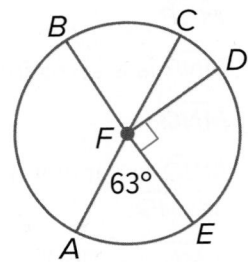

Explore Relationships Between Arc Lengths and Radii

🔗 **Online Activity** Use dynamic geometry software to complete the Explore.

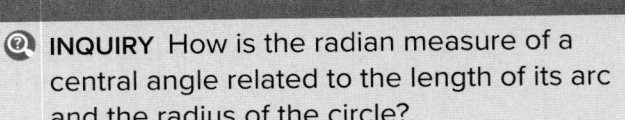

❓ **INQUIRY** How is the radian measure of a central angle related to the length of its arc and the radius of the circle?

Learn Arc Length and Radian Measure

Arc length is the distance between the endpoints of an arc measured along the arc in linear units. Because an arc is a portion of a circle, its length is a fraction of the circumference.

Key Concept • Arc Length in Degrees	
Words	The ratio of the length of an arc ℓ to the circumference of the circle is equal to the ratio of the degree measure of the arc to 360°.
Proportion	$\frac{\ell}{2\pi r} = \frac{x}{360}$
Equation	$\ell = \frac{x}{360} \cdot 2\pi r$

Angles can be measured in radians. A **radian** is a unit of angular measurement equal to $\frac{180°}{\pi}$ or about 57.296°.

Key Concept • Radian Measure	
The radian measure θ of a central angle is the ratio of the arc length to the radius of the circle; $\theta = \frac{\ell}{r}$ radians.	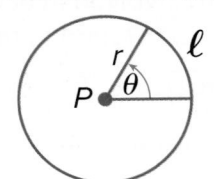

Key Concept • Degree and Radian Conversion Rules
1. To convert a degree measure to radians, multiply by $\frac{\pi \text{ radians}}{180°}$.
2. To convert a radian measure to degrees, multiply by $\frac{180°}{\pi \text{ radians}}$.

🔗 **Go Online** You can complete an Extra Example online.

Watch Out!

Arc Length The *length* of an arc is given in linear units, such as centimeters. The *measure* of an arc is given in degrees.

🔗 **Go Online** You can watch a video to see how to find the length of an arc of a circle.

💭 **Think About It!**

If the measure of a central angle $\theta = \frac{\ell}{r}$, then what is the length of the arc with the related central angle?

Example 5 Find Arc Length by Using Degrees

Find the length of $\overarc{AB}$ to the nearest hundredth.

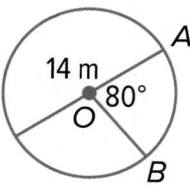

$$\ell = \frac{x}{360} \cdot 2\pi r \qquad \text{Arc Length Equation}$$

$$= \frac{80}{360} \cdot 2\pi(7) \qquad \text{Substitution}$$

$$\approx 9.77 \; m \qquad \text{Use a calculator.}$$

Check

Find the length of $\overarc{PQ}$ to the nearest hundredth.

___?___ cm

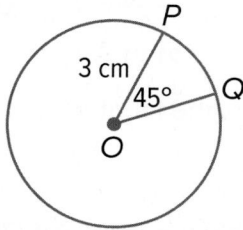

Example 6 Convert from Degrees to Radian Measure

Write 135° in radians as a multiple of π.

$$135° = 135° \times \frac{\pi \, \text{radians}}{180°} \qquad \text{Multiply by } \frac{\pi \, \text{radians}}{180°}.$$

$$= \frac{135\pi}{180} \text{ or } \frac{3\pi}{4} \text{ radians} \qquad \text{Simplify.}$$

Check
Write 240° in radians as a multiple of π.

Example 7 Convert from Radian Measure to Degrees

Write $\frac{11\pi}{6}$ radians in degrees.

$$\frac{11\pi}{6} \text{ radians} = \frac{11\pi}{6} \text{ radians} \times \frac{180°}{\pi \, \text{radians}} \qquad \text{Multiply by } \frac{180°}{\pi \, \text{radians}}.$$

$$= \frac{1980}{6} \text{ or } 330° \qquad \text{Simplify.}$$

Check
Write $\frac{2\pi}{3}$ radians in degrees.

Example 8 Find Arc Length by Using Radian Measure

Find the length of $\widehat{ZY}$ to the nearest hundredth.

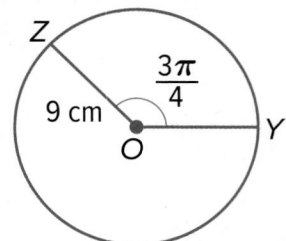

$$\theta = \frac{\ell}{r} \qquad \text{Arc length Equation}$$

$$\frac{3\pi}{4} = \frac{\ell}{9} \qquad \theta = \frac{3\pi}{4}, r = 9$$

$$9\left(\frac{3\pi}{4}\right) = \ell \qquad \text{Multiply each side by 9.}$$

$$\frac{27\pi}{4} = \ell \qquad \text{Use a calculator.}$$

$$21.21 \approx \ell$$

So, the length of $\widehat{ZY}$ is about 21.21 centimeters.

Check

Find the length of $\widehat{MN}$ to the nearest hundredth.

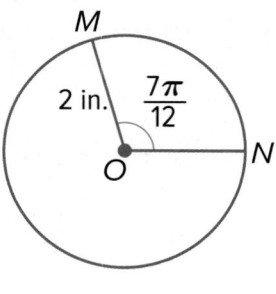

_____?_____ in.

Pause and Reflect

Did you struggle with anything in this lesson? If so, how did you deal with it?

Go Online You can complete an Extra Example online.

Practice

🔍 **Go Online** You can complete your homework online.

Example 1

Find the value of *x*.

1.

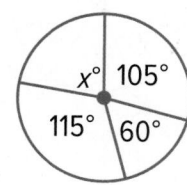

2.

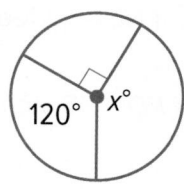

3.

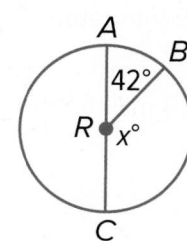

Example 2

$\overline{AC}$ and $\overline{EB}$ are diameters of ⊙R. Identify each arc as a *major arc, minor arc,* or *semicircle*. **Then find its measure.**

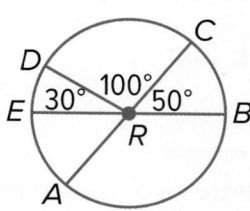

4. $m\widehat{EA}$

5. $m\widehat{CB}$

6. $m\widehat{DC}$

7. $m\widehat{DEB}$

8. $m\widehat{AB}$

9. $m\widehat{CDA}$

Example 3

10. SURVEYS A survey asked students at Westwood High School their preferences for the new school mascot. The results are shown in the circle graph. Find $m\widehat{AB}$.

11. SPORTS The circle graph shows the favorite spectator sport among a group of teens at a local high school. Find $m\widehat{AD}$.

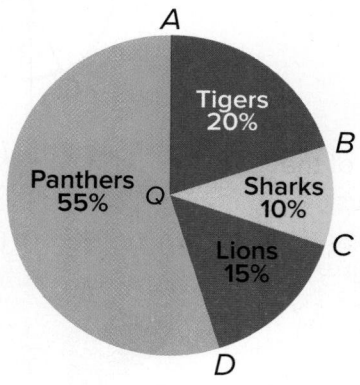

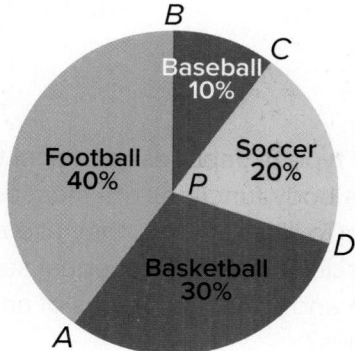

Example 4

$\overline{PR}$ and $\overline{QT}$ are diameters of ⊙A. Find each measure.

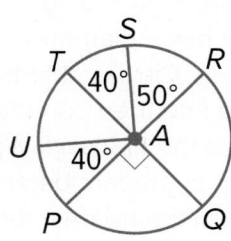

12. $m\widehat{UPQ}$

13. $m\widehat{PQR}$

14. $m\widehat{UTS}$

15. $m\widehat{RS}$

16. $m\widehat{RSU}$

17. $m\widehat{STP}$

18. $m\widehat{PQS}$

19. $m\widehat{PRU}$

Example 5

Use ⊙D to find the length of each arc to the nearest hundredth. $\overline{NL}$ is a diameter.

20. $\widehat{LM}$ if the radius is 5 inches

21. $\widehat{MN}$ if the diameter is 3 yards

22. $\widehat{KL}$ if $JD = 7$ centimeters

23. $\widehat{NJK}$ if $NL = 12$ feet

24. $\widehat{KLM}$ if $DM = 9$ millimeters

25. $\widehat{JK}$ if $KD = 15$ inches

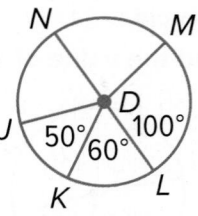

Example 6

Write each degree measure in radians as a multiple of π.

26. 120°   **27.** 45°   **28.** 30°

29. 90°   **30.** 180°   **31.** 225°

Example 7

Write each radian measure in degrees.

32. $\frac{3\pi}{4}$ radians   **33.** $\frac{3\pi}{2}$ radians   **34.** $\frac{\pi}{3}$ radians

35. $\frac{5\pi}{6}$ radians   **36.** 2π radians   **37.** $\frac{\pi}{12}$ radians

Example 8

Use ⊙Z to find the length of each arc to the nearest hundredth.

38. $\widehat{QR}$, if $PZ = 12$ feet

39. $\widehat{ST}$, if $SZ = 8$ inches

40. $\widehat{PQ}$, if $TZ = 14$ centimeters

41. $\widehat{PT}$, if $TR = 20$ inches

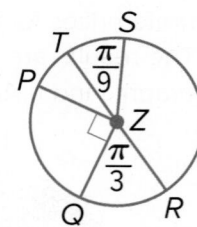

Mixed Exercises

42. CLOCKS Shiatsu is a type of Japanese physical therapy. One of the beliefs is that various body functions are most active at various times during the day. To illustrate this, they use a Chinese clock that is based on a circle divided into 12 equal sections by radii. What are the degree and radian measures of any one of the 12 equal central angles?

43. RIBBONS Cora is wrapping a ribbon around a cylinder-shaped gift box. The box has a diameter of 15 inches, and the ribbon is 60 inches long. Cora is able to wrap the ribbon all the way around the box once and then continue so that the second end of the ribbon passes the first end. What is the measure of the central angle formed by the arc between the ends of the ribbon? Round your answer to the nearest tenth of a degree.

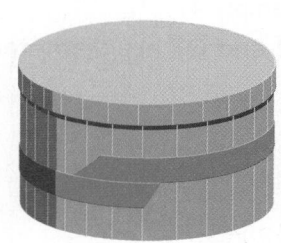

44. PIES Shekeia has divided a circular apple pie into 4 slices by cutting the pie along 4 radii. If the measures of the central angles of the 4 slices are represented by $3x°$, $(6x - 10)°$, $(4x + 10)°$, and $5x°$, what are the measures of the central angles?

45. BIKE WHEELS Louis has to buy a new wheel for his bike. The bike wheel has a diameter of 20 inches.

    a. If Louis rolls the wheel one complete rotation along the ground, how far will the wheel travel? Round your answer to the nearest hundredth of an inch.

    b. If the bike wheel is rolled along the ground so that it rotates 45°, how far will the wheel travel? Round your answer to the nearest hundredth of an inch.

    c. If the bike wheel is rolled along the ground for 10 inches, through what angle does the wheel rotate? Round your answer to the nearest tenth of a degree.

46. USE TOOLS The table on the right shows the number of hours students at Leland High School say they spend on homework each night.

    a. If you were to construct a circle graph of the data, how many degrees would be allotted to each category?

    b. Describe the types of arcs associated with each category.

Homework	
< 1	8
1–2	29
2–3	58
3–4	3
> 4	2

47. PROOF Write a two-column proof of Theorem 10.2.

    Given: $\angle BAC \cong \angle DAE$

    Prove: $\overarc{BC} \cong \overarc{DE}$

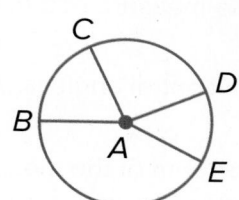

REASONING Find each measure. Round each linear measure to the nearest hundredth and each arc measure to the nearest degree.

48. circumference of ⊙S

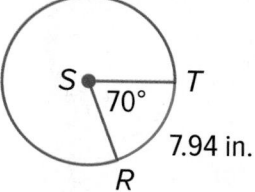

49. $m\overarc{CD}$

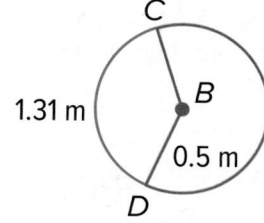

50. radius of ⊙K

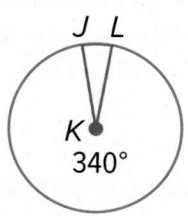

In ⊙C, $m\angle HCG = 2x°$ and $m\angle HCD = (6x + 28)°$. Find each measure. $\overline{HE}$ and $\overline{GD}$ are diameters of the circle.

51. $m\overarc{EF}$ **52.** $m\overarc{HD}$ **53.** $m\overarc{HGF}$

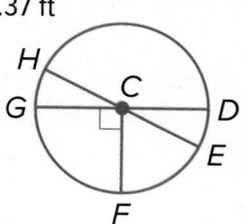

54. STRUCTURE An arc has a related central angle of 45°. Copy and complete the table by finding the length of the arc in terms of π for each given radius.

Radius of Circle, r	3	5	11	15	r
Length of Arc, ℓ					

55. REGULARITY An architect is designing the seating area for a theater. The seating area is formed by a region that lies between two circles. The architect is planning to place a brass rail in front of the first row of seats.

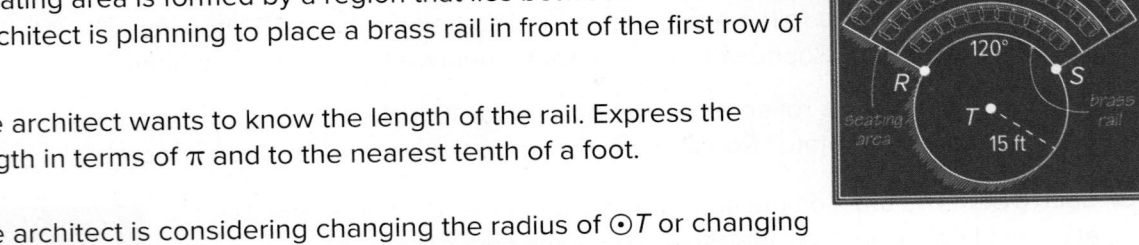

a. The architect wants to know the length of the rail. Express the length in terms of π and to the nearest tenth of a foot.

b. The architect is considering changing the radius of ⊙T or changing the measure of $\overset{\frown}{RS}$. Describe a general method she can use to find the length of $\overset{\frown}{RS}$.

ANALYZE For Exercises 56–58, determine whether each statement is *sometimes*, *always*, or *never* true. Justify your argument.

56. The measure of a minor arc is less than 180°.

57. If a central angle is obtuse, its corresponding arc is a major arc.

58. The sum of the measures of adjacent arcs of a circle depends on the measure of the radius.

59. FIND THE ERROR Brody says that $\overset{\frown}{WX}$ and $\overset{\frown}{YZ}$ are congruent because their central angles have the same measure. Selena says they are not congruent. Who is correct? Explain your reasoning.

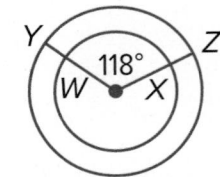

60. PERSEVERE The time shown on an analog clock is 8:10. What is the measure of the angle formed by the hands of the clock?

61. CREATE Draw a circle and locate three points on the circle. Estimate the measures of the three nonoverlapping arcs that are formed. Then use a protractor to find the measure of each arc. Label your circle with the arc measures.

62. WRITE Describe the three different types of arcs in a circle and a method for finding the measure of each one.

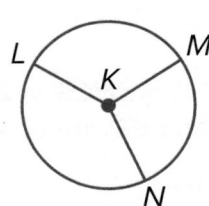

63. PERSEVERE The measures of $\overset{\frown}{LM}$, $\overset{\frown}{MN}$, and $\overset{\frown}{NL}$ are in the ratio of 5:3:4. Find the measure of each arc.

Arcs and Chords

Explore Chords in Circles

▶ **Online Activity** Use dynamic geometry software to complete the Explore.

> ⓠ **INQUIRY** What relationships exist between chords and arcs in circles? ×

Learn Arcs and Chords

A chord is a segment with endpoints on a circle. If a chord is not a diameter, then its endpoints divide the circle into a major and minor arc.

Theorem 5.3

Words	In the same circle or in congruent circles, two minor arcs are congruent if and only if their corresponding chords are congruent.
Example	$\widehat{FG} \cong \widehat{HJ}$ if and only if $\overline{FG} \cong \overline{HJ}$. 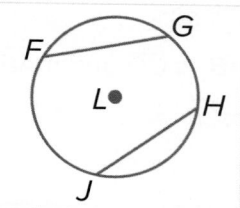

If a line, segment, or ray divides an arc into two congruent arcs, then it bisects the arc.

Theorem 5.4

Words	If a diameter (or radius) of a circle is perpendicular to a chord, then it bisects the chord and its arc.
Example	If diameter $\overline{AB}$ is perpendicular to chord $\overline{XY}$, then $\overline{XZ} \cong \overline{ZY}$ and $\widehat{XB} \cong \widehat{BY}$.

Theorem 5.5

Words	The perpendicular bisector of a chord contains a diameter (or radius) of the circle.
Example	If $\overline{AB}$ is a perpendicular bisector to chord $\overline{XY}$, then $\overline{AB}$ is a diameter of $\odot C$.

(continued on the next page)

Today's Goals
• Solve problems using the relationships between arcs, chords, and diameters.

▶ **Go Online**
Proofs of Theorems 5.3 through 5.5 are available.

Study Tip

Arc Bisectors In the figure, $\overline{FH}$ is an arc bisector of $\widehat{JG}$.

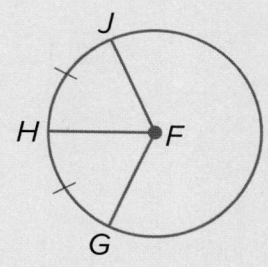

🍫 **Think About It!**

What is the relationship between a chord and a radius perpendicular to the chord?

In addition to Theorem 10.3, you can use the following theorem to determine whether two chords in a circle are congruent.

Theorem 5.6	
Words	In the same circle or in congruent circles, chords are congruent if and only if they are equidistant from the center.
Example	$\overline{FG} \cong \overline{JH}$ if and only if $LX = LY$.

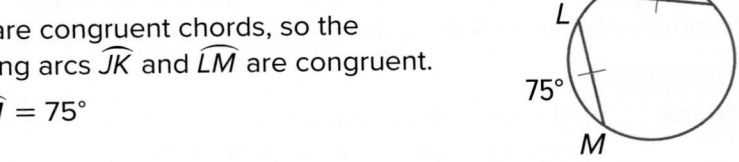

You will prove both cases of Theorem 5.6 in Exercises 21–22.

Talk About It!

In $\odot P$, chords $\overline{DE}$ and $\overline{FG}$ are congruent. What do you know about $\angle DPE$ and $\angle FPG$? Explain.

Example 1 Use Congruent Arcs to Find Arc Measures

$\overline{JK} \cong \overline{LM}$ and $m\widehat{LM} = 75°$. Find $m\widehat{JK}$.

$\overline{JK}$ and $\overline{LM}$ are congruent chords, so the corresponding arcs $\widehat{JK}$ and $\widehat{LM}$ are congruent.

$m\widehat{JK} = m\widehat{LM} = 75°$

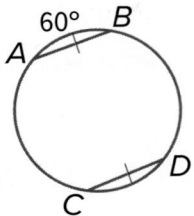

Check

$\overline{AB} \cong \overline{CD}$ and $m\widehat{AB} = 60°$. Find $m\widehat{CD}$.

$m\widehat{CD} = $ ____?____

Example 2 Use Congruent Arcs to Find Chord Length

In $\odot C$, $\widehat{DE} \cong \widehat{FG}$. Find FG.

$\widehat{DE}$ and $\widehat{FG}$ are congruent arcs in the same circle, the corresponding chords $\overline{DE}$ and $\overline{FG}$ are congruent.

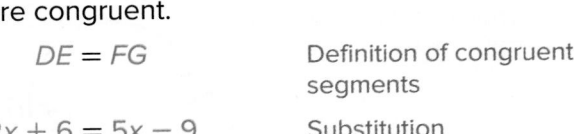

$DE = FG$	Definition of congruent segments
$2x + 6 = 5x - 9$	Substitution
$6 = 3x - 9$	Subtract $2x$ from each side.
$15 = 3x$	Add 9 to each side.
$5 = x$	Divide each side by 3.

So, $FG = 5(5) - 9$ or 16.

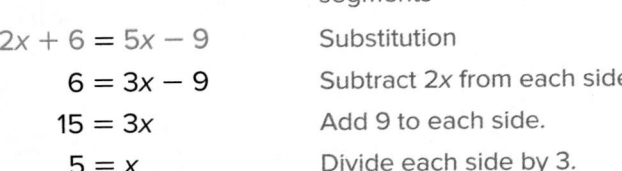

Check

In $\odot W$, $\widehat{RS} \cong \widehat{TV}$. Find RS.

$RS = $ ___?___ units

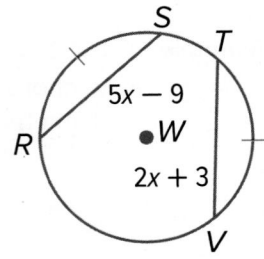

Example 3 Use a Radius Perpendicular to a Chord

In ⊙D, $m\widehat{EFG} = 120°$. Find $m\widehat{FG}$ and EG.

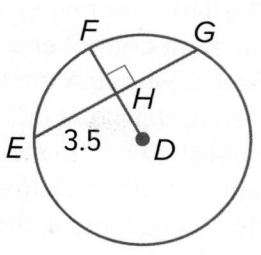

Radius $\overline{DF}$ is perpendicular to chord $\overline{EG}$. So, by Theorem 10.4, $\overline{DF}$ bisects $\widehat{EFG}$ and $\overline{EG}$.

Therefore, $m\widehat{EF} = m\widehat{FG}$ and $EH = HG$.

By substitution, $m\widehat{FG} = \frac{120}{2}$ or 60°.

By the Segment Addition Postulate and substitution, $EG = 2 \cdot EH$. So, $EG = 2 \cdot 3.5$ or 7 units.

Think About It!

In a circle, $\overline{AB}$ is a diameter and $\overline{QR}$ is a chord that intersects $\overline{AB}$ at point X. Is it *sometimes*, *always*, or *never* true that $QX = XR$? Explain.

Check

In ⊙S, $m\widehat{PQR} = 90°$. Find $m\widehat{PQ}$ and PR.

$m\widehat{PQ} = $ __?__

$PR = $ __?__ units

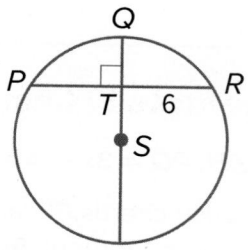

🌐 Example 4 Use a Diameter Perpendicular to a Chord

RECORDS The record shown can be modeled by a circle. Diameter $\overline{CD}$ is 12 inches long, and chord $\overline{FH}$ is 10 inches long. Find EG.

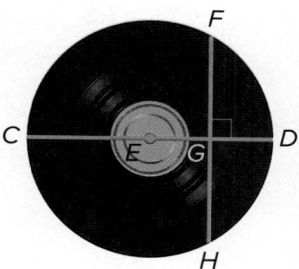

Step 1 Draw radius $\overline{EF}$.
Radius $\overline{EF}$ forms right △EFG.

Step 2 Find EF and FG.
Because $CD = 12$ inches, $ED = 6$ inches. All radii of a circle are congruent, so $EF = 6$ inches. Because diameter $\overline{CD}$ is perpendicular to $\overline{FH}$, $\overline{CD}$ bisects chord $\overline{FH}$ by Theorem 10.4.
So $FG = \frac{1}{2}(10)$ or 5 inches.

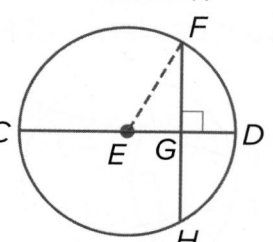

Study Tip

Assumptions
Assuming that objects can be modeled by perfect circles allows you to find reasonable measures in objects.

Step 3 Use the Pythagorean Theorem to find EG.

$EG^2 + FG^2 = EF^2$	Pythagorean Theorem
$EG^2 + 5^2 = 6^2$	$FG = 5$ and $EF = 6$
$EG^2 + 25 = 36$	Simplify.
$EG^2 = 11$	Subtract 25 from each side.
$EG = \sqrt{11}$	Take the positive square root of each side.

So, EG is $\sqrt{11}$ or about 3.32 inches long.

Check

DRIVING Steering devices provide drivers with more control and strength when turning a steering wheel. A technician is installing a steering device at point *P* on the steering wheel shown. The steering device extends to point *T*, and the diameter of the steering wheel is 15 inches long. If chord $\overline{SV}$ is 12 inches long and perpendicular to the diameter of the steering wheel, what is the length of the steering device?

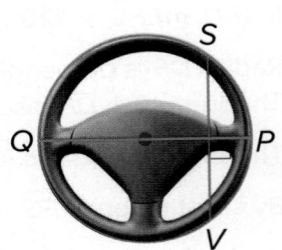

🐦 **Go Online**
You may want to complete the construction activities for this lesson.

Example 5 Chords Equidistant from the Center

In ⊙*H*, *PQ* = 3*x* − 4 and *RS* = 14. Find *x*.

Because chords $\overline{PQ}$ and $\overline{RS}$ are equidistant from *H*, they are congruent. So, *PQ* = *RS*.

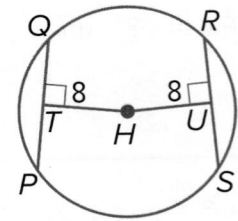

$PQ = RS$	Definition of congruent segments
$3x - 4 = 14$	Substitution
$3x = 18$	Add 4 to each side.
$x = 6$	Divide each side by 3.

Check

In ⊙*P*, *CE* = *FH* = 48. Find *PG*.

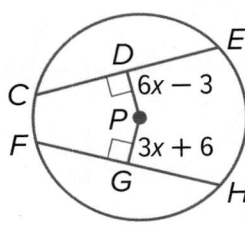

PG = ___?___ units

🐦 **Go Online** You can complete an Extra Example online.

luminis/iStock/Getty Images

Practice

Go Online You can complete your homework online.

Examples 1 and 2

REGULARITY Find the value of *x*.

1.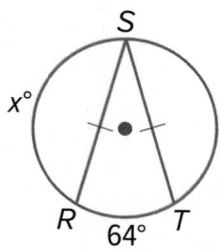

2.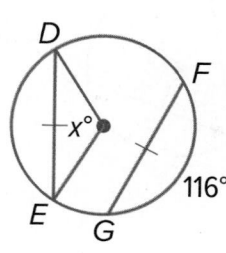

3.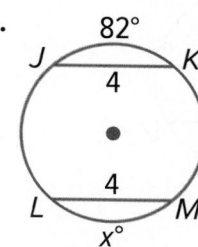

4.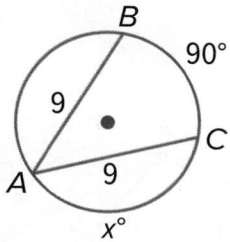

5.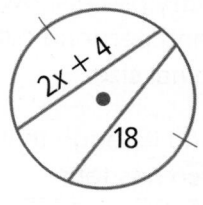

6.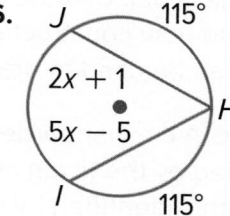

7.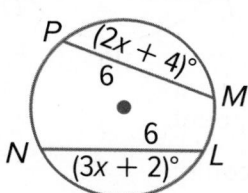

8. ⊙*M* ≅ ⊙*P*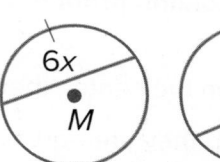

9. ⊙*V* ≅ ⊙*W*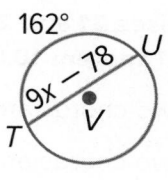

Examples 3 and 4

In ⊙*P*, *PQ* = 13 and *RS* = 24. Find each measure.

10. *RT*

11. *PT*

12. *TQ*

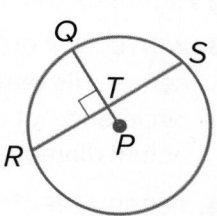

In ⊙*A*, *EB* = 12, *CD* = 8, and $m\widehat{CD}$ = 90°. Find each measure.
Round to the nearest hundredth, if necessary.

13. $m\widehat{DE}$

14. *FD*

15. *AF*

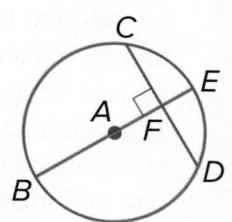

16. **USE A MODEL** For security purposes a jewelry company prints a hidden
watermark on the logo of its official documents. The watermark is a chord
located 0.7 cm from the center of a circular ring that has a 2.5 cm radius. To the
nearest tenth, what is the length of the chord?

Example 5

17. In ⊙R, TS = 21 and
UV = 3x. What is the value of x?

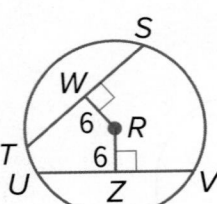

18. In ⊙Q, $\overline{CD} \cong \overline{CB}$, GQ = x + 5, and
EQ = 3x − 6. What is the value of x?

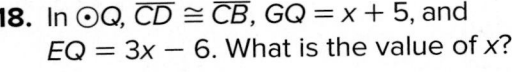

Mixed Exercises

19. USE TOOLS A piece of a broken plate is found during an archaeological dig. Use the sketch of the pottery piece shown to demonstrate how constructions with chords and perpendicular bisectors can be used to draw the plate's original size.

20. REASONING A circular garden has paths around its edge that are identified by the given arc measures. It also has four straight paths, identified by segments $\overline{AC}$, $\overline{AD}$, $\overline{BE}$, and $\overline{DE}$, that cut through the garden's interior. Which two straight paths have the same length?

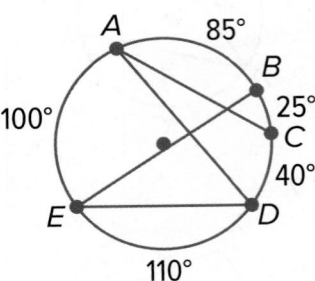

PROOF For Exercises 21 and 22, write a two-column proof of the indicated part of Theorem 10.6.

21. In a circle, if two chords are equidistant from the center, then they are congruent.

22. In a circle, if two chords are congruent, then they are equidistant from the center.

🌥 Higher-Order Thinking Skills

23. WRITE Neil draws a diameter of a circle and then marks its midpoint as the center. His teacher asks Neil how he knows that he drew the diameter of the circle and not a shorter chord. How can Neil determine whether he drew an actual diameter?

24. PERSEVERE Toshelle is following directions for a quilt pattern. The directions are intended to make a rectangle. They say "In a 10-inch diameter circle, measure 3 inches from the center of the circle and mark a chord $\overline{AB}$ perpendicular to the radius of the circle. Then cut along the chord." Toshelle is to repeat this for another chord, $\overline{CD}$. Finally, she is to cut along chord $\overline{DB}$ and $\overline{AC}$. The result should be four curved pieces and one quadrilateral. If Toshelle follows the directions, why might the resulting quadrilateral not be a rectangle? Explain how to adjust the directions.

25. CREATE Construct a circle and draw a chord. Measure the chord and the distance that the chord is from the center. Find the length of the radius.

26. ANALYZE In a circle, $\overline{AB}$ is a diameter, and $\overline{HG}$ is a chord that intersects $\overline{AB}$ at point X. Is it *sometimes*, *always*, or *never* true that HX = GX? Justify your argument.

Inscribed Angles

Explore Angles Inscribed in Circles

⊗ **Online Activity** Use dynamic geometry software to complete the Explore.

> ⊘ **INQUIRY** What is the relationship between an inscribed angle and the arc it intercepts? ✕

Learn Inscribed Angles

An **inscribed angle** has its vertex on a circle and sides that contain chords of the circle. In ⊙C, ∠QRS is an inscribed angle. An **intercepted arc** is the part of a circle that lies between the two lines intersecting it. In ⊙C, minor arc $\overset{\frown}{QS}$ is intercepted by ∠QRS. There are three ways that an angle can be inscribed in a circle.

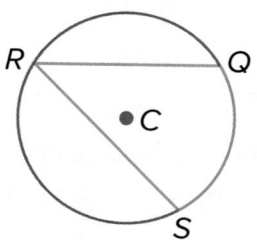

Case 1	Case 2	Case 3
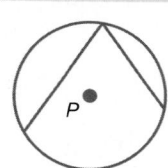	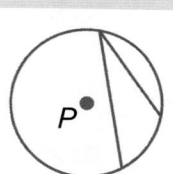	
Center *P* is on the inscribed angle.	Center *P* is inside the inscribed angle.	Center *P* is in the exterior of the inscribed angle.

For each of these cases, the following theorem holds true.

Theorem 5.7: Inscribed Angle Theorem

Words	If an angle is inscribed in a circle, then the measure of the angle equals one half the measure of its intercepted arc.
Example	$m\angle 1 = \frac{1}{2} m\overset{\frown}{AB}$ and $m\overset{\frown}{AB} = 2m\angle 1$ 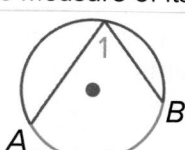

Theorem 5.8

Words	If two inscribed angles of a circle intercept the same arc or congruent arcs, then the angles are congruent.
Example	∠B and ∠C both intercept $\overset{\frown}{AD}$. So, ∠B ≅ ∠C. 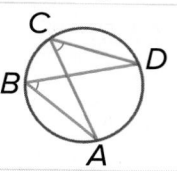

You will prove Theorem 5.8 in Exercise 32.

Today's Goals
- Describe relationships between inscribed angles, and use those relationships to solve problems.
- Identify relationships in inscribed polygons, and use those relationships to solve problems.

Today's Vocabulary
inscribed angle
intercepted arc
inscribed polygon

⊗ **Go Online**
A proof of Theorem 5.7 is available.

⊘ **Talk About It!**
What is the relationship between an inscribed angle and chords of a circle?

Example 1 Use Inscribed Angles to Find Measures

Find each measure.

a. $m\overset{\frown}{CF}$

$$m\overset{\frown}{CF} = 2 \cdot m\angle D \qquad \text{Inscribed Angle Theorem}$$

$$= 2 \cdot 40 \text{ or } 80° \qquad \text{Substitute and simplify.}$$

b. $m\angle C$

$$m\angle C = \frac{1}{2}m\overset{\frown}{DE} \qquad \text{Inscribed Angle Theorem}$$

$$= \frac{1}{2}(98) \text{ or } 49° \qquad \text{Substitute and simplify.}$$

🤔 Think About It!

If an inscribed angle and a central angle in the same circle intercept the same arc, how are they related?

Example 2 Find Measures of Congruent and Inscribed Angles

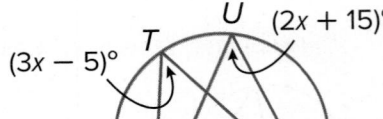

Find $m\angle A$.

$\angle A \cong \angle D$	$\angle A$ and $\angle D$ both intercept $\overset{\frown}{BC}$.
$m\angle A = m\angle D$	Definition of congruent angles
$5x - 12 = 3x + 2$	Substitution
$x = 7$	Solve for x.

So, $m\angle A = 5(7) - 12$ or $23°$

Check

Find $m\angle T$.

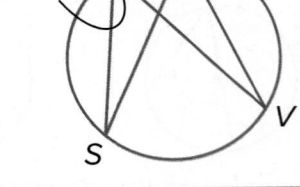

A. 20°   **B.** 25°   **C.** 45°   **D.** 55°

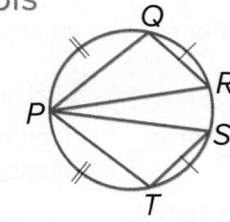

Example 3 Use Inscribed Angles in Proofs

Write a two-column proof.

Given: $\overset{\frown}{QR} \cong \overset{\frown}{ST}$; $\overset{\frown}{PQ} \cong \overset{\frown}{PT}$

Prove: $\triangle PQR \cong \triangle PTS$

Proof:

Statements	Reasons
1. $\overset{\frown}{QR} \cong \overset{\frown}{ST}, \overset{\frown}{PQ} \cong \overset{\frown}{PT}$	1. Given
2. $\angle QPR \cong \angle TPS, \angle QRP \cong \angle TSP$	2. If 2 inscribed $\angle$s of a circle intercept $\cong$ arcs, then the $\angle$s are $\cong$. (Theorem 5.8)
3. $\overline{QR} \cong \overline{ST}$	3. $\cong$ arcs have $\cong$ chords.
4. $\triangle PQR \cong \triangle PTS$	4. AAS

🔎 **Go Online** You can complete an Extra Example online.

Learn Inscribed Polygons

In an **inscribed polygon**, all of the vertices of the polygon lie on a circle. Inscribed triangles and quadrilaterals have special properties.

Theorem 5.9	
Words	An inscribed angle of a triangle intercepts a diameter or semicircle if and only if the angle is a right angle.
Example	If $\overarc{FJH}$ is a semicircle, then $m\angle G = 90°$. If $m\angle G = 90°$, then $\overarc{FJH}$ is a semicircle and $\overline{FH}$ is a diameter.

While many different types of triangles, including right triangles, can be inscribed in a circle, only certain quadrilaterals can be inscribed in a circle.

Theorem 5.10	
Words	If a quadrilateral is inscribed in a circle, then its opposite angles are supplementary.
Example	If quadrilateral $KLMN$ is inscribed in $\odot A$, then $\angle L$ and $\angle N$ are supplementary and $\angle K$ and $\angle M$ are supplementary.

You will prove Theorems 5.9 and 5.10 in Exercises 34 and 33, respectively.

Example 4 Find Angle Measures in Inscribed Triangles

Find $m\angle K$.

$\triangle KLM$ is a right triangle because $\angle L$ intercepts a semicircle.

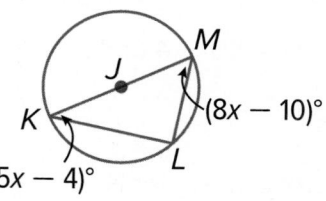

$$m\angle K + m\angle M = 90 \quad \text{Acute } \angle\text{s of a right } \triangle \text{ are complementary.}$$
$$(5x - 4)° + (8x - 10)° = 90° \quad \text{Substitution}$$
$$13x° - 14° = 90° \quad \text{Simplify.}$$
$$13x° = 104° \quad \text{Add 14° to each side.}$$
$$x = 8 \quad \text{Divide each side by 13°.}$$

So, $m\angle K = [5(8) - 4]°$ or $36°$.

Check

Find $m\angle V$.

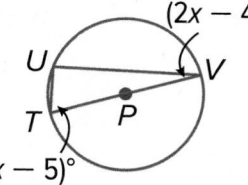

$m\angle V =$ _____?_____

Go Online You can complete an Extra Example online.

Think About It!

What special quadrilaterals can be inscribed in a circle? Justify your argument.

Go Online

You can watch a video to see how to solve problems involving inscribed triangles.

Think About It!

A 45°-45°-90° triangle is inscribed in a circle. What can you say about the arcs into which the circle is divided by the vertices of the triangles?

🌐 Example 5 Find Angle Measures

STAINED GLASS Luca is creating a collection of stained glass ornaments. The ornament shown uses a quadrilateral inscribed in a circle. Find $m\angle F$ and $m\angle G$.

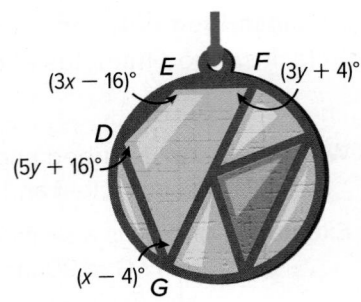

Because *DEFG* is inscribed in a circle, its opposite angles are supplementary.

Step 1 Find $m\angle F$.

$m\angle D + m\angle F = 180$	Definition of supplementary
$5y + 16 + 3y + 4 = 180$	Substitution
$8y + 20 = 180$	Simplify.
$8y = 160$	Subtract 20 from each side.
$y = 20$	Divide each side by 8.

So, $m\angle F = 3(20) + 4$ or $64°$.

Step 2 Find $m\angle G$.

$m\angle G + m\angle E = 180$	Definition of supplementary
$x - 4 + 3x - 16 = 180$	Substitution
$4x - 20 = 180$	Simplify.
$4x = 200$	Add 20 to each side.
$x = 50$	Divide each side by 4.

So, $m\angle G = 50 - 4$ or $46°$.

Check

JEWELRY A designer is making a new line of jewelry with geometric patterns. The ring shown had a quadrilateral gemstone inscribed in a circular piece of metal.

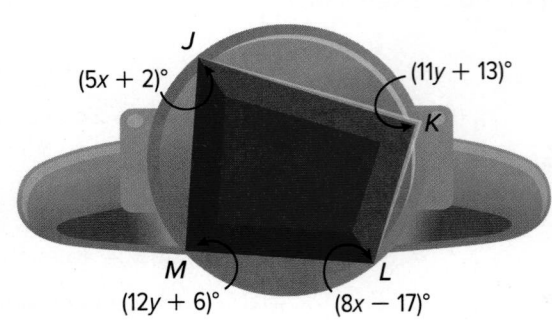

Part A

Find $m\angle J$.

A. 15°   **B.** 42°   **C.** 77°   **D.** 103°

Part B

Find $m\angle K$.

A. 7°   **B.** 46°   **C.** 64°   **D.** 90°

 Go Online You can complete an Extra Example online.

Practice

Go Online You can complete your homework online.

Example 1
Find each measure.

1. $m\widehat{AC}$

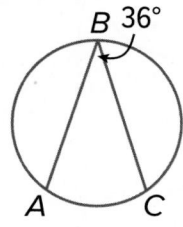

2. $m\angle N$

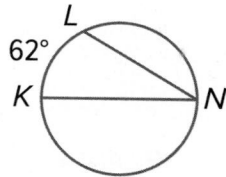

3. $m\widehat{QSR}$

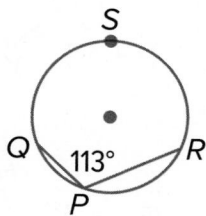

4. $m\widehat{XY}$

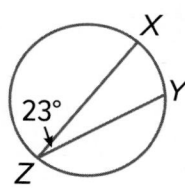

5. $m\angle E$

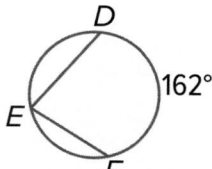

6. $m\angle R$

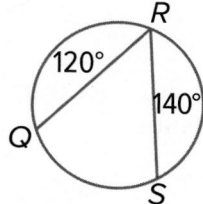

Example 2
Find each measure.

7. $m\angle N$

8. $m\angle L$

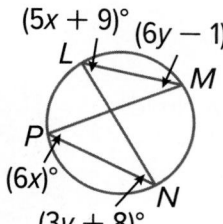

9. $m\angle C$

10. $m\angle A$

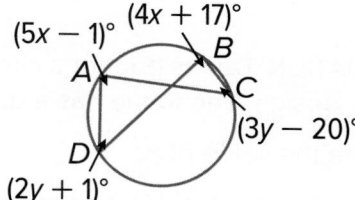

Example 3
PROOF Write the specified type of proof.

11. paragraph proof

Given: $m\angle T = \frac{1}{2} m\angle S$
Prove: $m\widehat{TUR} = 2m\widehat{URS}$

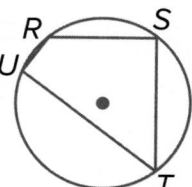

12. two-column proof

Given: $\odot C$
Prove: $\triangle KML \sim \triangle JMH$

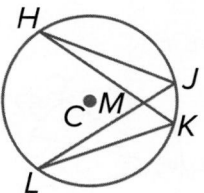

Example 4

Find each value.

13. x

14. m∠W

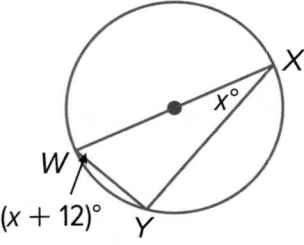

15. x

16. m∠T

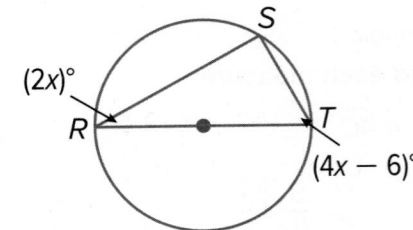

17. m∠J

18. m∠K

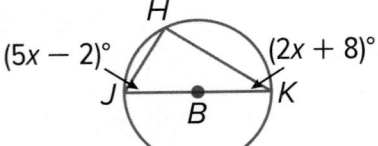

19. m∠A

20. m∠C

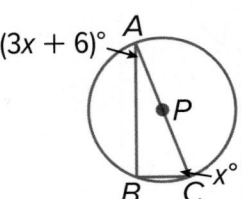

Example 5

Find each measure.

21. m∠R

22. m∠S

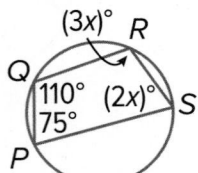

23. m∠W

24. m∠X

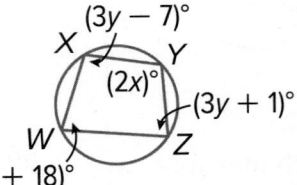

25. USE ESTIMATION Darius bought a circular picture frame with a geometric design. The frame has a quadrilateral inscribed in a circle.

a. Estimate the value of x.

b. Find the exact value of x and m∠J.

c. Is your answer reasonable? Justify your argument.

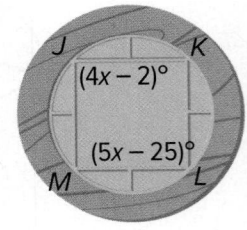

Mixed Exercises

Find each measure.

26. $m\widehat{CE}$

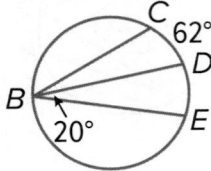

27. $m\widehat{JM}$

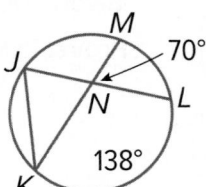

28. m∠QPR

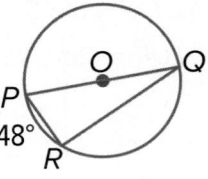

29. $m\widehat{UW}$

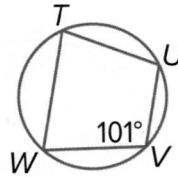

30. m∠AFE

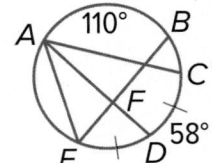

31. m∠KJH

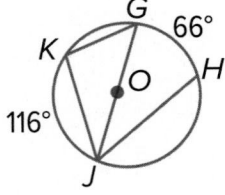

PROOF Write the specified type of proof to prove each theorem.

32. two-column proof Theorem 10.8

Given: $\overset{\frown}{JL} \cong \overset{\frown}{RP}$; $\angle JKL$ and $\angle RQP$ are inscribed.

Prove: $\angle JKL \cong \angle RQP$

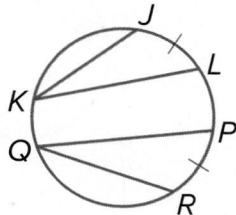

33. paragraph proof Theorem 10.10

Given: Quadrilateral *DEFG* is inscribed in ⊙C.

Prove: $\angle D$ and $\angle F$ are supplementary.
$\angle E$ and $\angle G$ are supplementary.

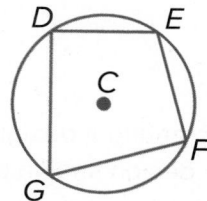

34. PROOF Write a paragraph proof to prove each part of Theorem 10.9.

a. Given: $\overset{\frown}{XWZ}$ is a semicircle.

Prove: $\angle XYZ$ is a right angle.

b. Given: $\angle XYZ$ is a right angle.

Prove: $\overset{\frown}{XWZ}$ is a semicircle.

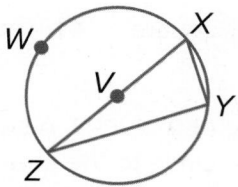

35. PROOF Write a paragraph proof to prove that if *PQRS* is an inscribed quadrilateral and $\angle Q \cong \angle S$, then $\overline{PR}$ is a diameter of the circle.

36. REASONING A landscaping crew is installing a circular garden with three paths that form a triangle. The figure shows the plan for the garden and the paths. The leader of the crew wants to determine the radius they should use when they install the circular garden.

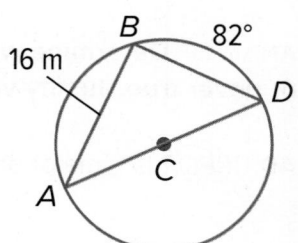

a. Find the measures of the angles in △*ABD*.

b. Use trigonometry to find the radius of the circular garden. Round to the nearest tenth meter, if necessary.

c. Given that the path will go along the outside of the garden, as well as along the three interior paths, find the total length of the path that will need to be installed.

37. **ARENA** A concert arena is lit by five lights equally spaced around the perimeter. What is $m\angle 1$?

38. **JEWELRY** Alyssa makes earrings by bending wire into various shapes. She often bends the wire to form a circle with an inscribed quadrilateral as shown. She would like to know how she can find $m\widehat{ADC}$ if she knows $m\angle ADC$. Write a formula for finding $m\widehat{ADC}$ given that $m\angle ADC = x°$.

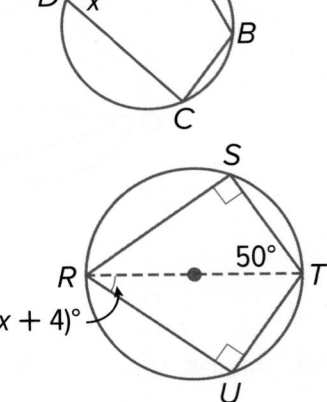

39. **TAPESTRY** Helga is creating a design for a tapestry she is making for her art class. The design shown uses a kite inscribed in a circle. Find $m\angle SRU$ and the value of x.

🧠 **Higher-Order Thinking Skills**

ANALYZE Determine whether the quadrilateral can *always, sometimes,* or *never* be inscribed in a circle. Justify your argument.

40. square   41. rectangle   42. parallelogram   43. rhombus   44. kite

45. **PERSEVERE** A square is inscribed in a circle. What is the ratio of the area of the circle to the area of the square?

46. **WRITE** A 45°-45°-90° right triangle is inscribed in a circle. If the radius of the circle is given, explain how to find the lengths of the legs of the triangle.

47. **CREATE** Draw an inscribed polygon. Then find the measure of each intercepted arc.

48. **WRITE** Compare and contrast inscribed angles and central angles of a circle. If they intercept the same arc, how are they related?

ANALYZE Determine whether each statement is *always, sometimes,* or *never* true. Justify your argument.

49. If $\widehat{PQR}$ is a major arc of a circle, then $\angle PQR$ is obtuse.

50. If $\overline{AB}$ is a diameter of circle O, and X is any point on circle O other than A or B, then $\triangle AXB$ is a right triangle.

51. When an equilateral triangle is inscribed in a circle it partitions the circle into three minor arcs that each measure 120°.

Tangents

Learn Tangents

A **tangent to a circle** is a line or segment in the plane of a circle that intersects the circle in exactly one point and does not contain any points in the interior of the circle. For a line that intersects a circle in one point, the **point of tangency** is the point at which they intersect.

A **common tangent** is a line or segment that is tangent to two circles in the same plane. More than one line can be tangent to the same circle, and the shortest distance from a tangent to the center of a circle is the radius drawn to the point of tangency.

Theorem 5.11

In a plane, a line is tangent to a circle if and only if it is perpendicular to a radius drawn to the point of tangency.

Theorem 5.12: Tangent to a Circle Theorem

If two segments from the same exterior point are tangent to a circle, then they are congruent.

You will prove Theorem 5.12 in Exercise 28.

Go Online A proof of Theorem 5.11 is available.

Example 1 Identify Common Tangents

Identify the number of common tangents that exist between each pair of circles. If no common tangent exists, state *no common tangent*.

a.
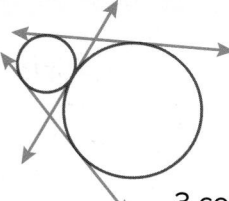
3 common tangents

b.
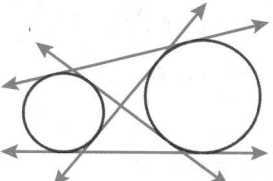
4 common tangents

Example 2 Identify a Tangent

$\overline{AB}$ is a radius of $\odot A$. Determine whether $\overline{BC}$ is tangent to $\odot A$. Justify your answer.

Test to see if $\triangle ABC$ is a right triangle.

$9^2 + 12^2 \overset{?}{=} (9+6)^2$ — Pythagorean Theorem

$81 + 144 \overset{?}{=} 15^2$ — Multiply and add.

$225 = 225$ ✓ — Simplify.

$\triangle ABC$ is a right triangle with right $\angle ABC$. So, segment BC is perpendicular to radius $\overline{AB}$ at point B. Therefore, by Theorem 5.11, $\overline{BC}$ is tangent to $\odot A$.

Go Online You can complete an Extra Example online.

Today's Goals
- Describe relationships between radii and tangents, and use those relationships to solve problems.
- Describe relationships between central and circumscribed angles, and use those relationships to solve problems.

Today's Vocabulary
tangent to a circle
point of tangency
common tangent
circumscribed angle
circumscribed polygon

Go Online You may want to complete the Concept Check to check your understanding.

Talk About It! What is the relationship between a circle and its tangent? When will two circles not have a common tangent? Justify your argument.

Example 3 Use a Tangent to Find Missing Values

$\overline{QS}$ is tangent to $\odot R$ at Q. Find the value of x.

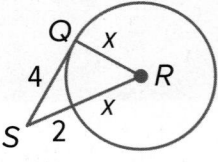

By Theorem 10.11, $\overline{RQ} \perp \overline{QS}$. So, $\triangle RQS$ is a right triangle.

$$RQ^2 + QS^2 = RS^2 \qquad \text{Pythagorean Theorem}$$
$$x^2 + 4^2 = (x + 2)^2 \qquad RQ = x, QS = 4, \text{ and } RS = x + 2$$
$$x^2 + 16 = x^2 + 4x + 4 \qquad \text{Multiply.}$$
$$12 = 4x \qquad \text{Simplify.}$$
$$3 = x \qquad \text{Divide each side by 4.}$$

Check

$\overline{BC}$ is tangent to $\odot A$ at C. Find the value of x to the nearest hundredth.

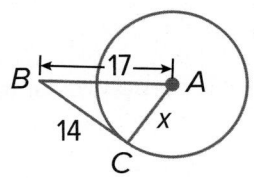

$x =$ ___?___

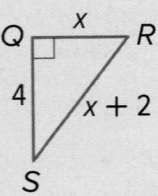

🌐 **Example 4** Use Congruent Tangents to Find Measures

PHOTOGRAPHY A photographer wants to take a picture of a local fountain. She positions herself at point A so that the fountain will be centered in the picture. $\overline{AB}$ and $\overline{AC}$ are tangent to the fountain as shown. If the lengths of the tangents are given in feet, find AB.

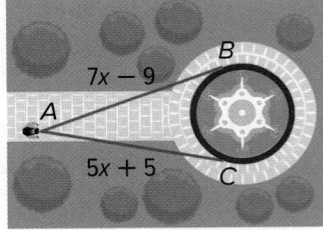

Because tangents $\overline{AB}$ and $\overline{AC}$ are from the same exterior point A, $\overline{AB} \cong \overline{AC}$.

$$AB = AC \qquad \text{Definition of congruent segments}$$
$$7x - 9 = 5x + 5 \qquad \text{Substitution}$$
$$x = 7 \qquad \text{Solve.}$$

So, $AB = 7(7) - 9$ or 40 feet.

Check

LANDSCAPING A landscape designer is creating a tiled patio with a circular design pattern. A corner of the patio is shown. $\overline{DE}$ and $\overline{FE}$ are tangent to $\odot G$, and the lengths of the tangents are given in feet. Find DE.

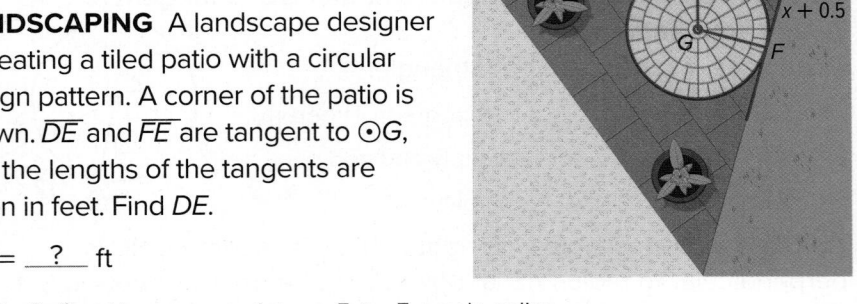

$DE =$ ___?___ ft

🔴 **Go Online** You can complete an Extra Example online.

Problem-Solving Tip

Solve a Simpler Problem
You can *solve a simpler problem* by sketching and labeling the right triangle in the example without the circle. A drawing of the triangle is shown below.

Explore Tangents and Circumscribed Angles

🔘 **Online Activity** Use dynamic geometry software to complete the Explore.

> ⊘ **INQUIRY** What is the relationship between circumscribed angles, tangents, and the radii of a circle? ×

Learn Circumscribed Angles

A **circumscribed angle** is an angle with sides that are tangent to a circle.

Theorem 10.13	
Words	If two segments or lines are tangent to a circle, then the circumscribed angle and the central angle that intercept the arc formed by the points of tangency are supplementary.
Example	If $\overline{QS}$ and $\overline{RS}$ are tangent to ⊙P, then $m\angle P + m\angle S = 180°$. 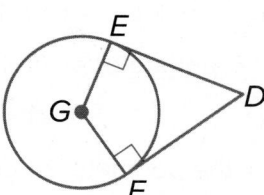

You will prove Theorem 10.13 in Exercise 29.

A **circumscribed polygon** has vertices outside the circle and sides that are tangent to the circle.

Circumscribed Polygons	Polygons Not Circumscribed

Example 5 Use Circumscribed Angles to Find Measures

If $m\angle EGF = (19x + 9)°$ and $m\angle D = (10x - 3)°$, find $m\angle D$.

Because $\overline{ED}$ and $\overline{FD}$ are tangent to circle G, $\angle EGF$ and $\angle D$ are supplementary.

$m\angle EGF + m\angle D = 180$	Definition of supplementary angles
$19x + 9 + 10x - 3 = 180$	Substitution
$29x + 6 = 180$	Simplify.
$29x = 174$	Subtract 6 from each side.
$x = 6$	Divide each side by 29.

So, $m\angle D = 10(6) - 3$ or $57°$.

🔘 **Go Online** You can complete an Extra Example online.

Watch Out!

Identifying Circumscribed Polygons Just because a circle is tangent to one or more of the sides of a polygon does not mean that the polygon is circumscribed about the circle, as shown in the second set of figures.

💭 **Think About It!**

If you do not remember Theorem 10.13, how can you use logic to determine the relationship between a central angle and a circumscribed angle that intercept the same arc?

Check

If $m\angle XWZ = 7x + 10$ and $m\angle Y = 4x + 5$, find $m\angle Y$.

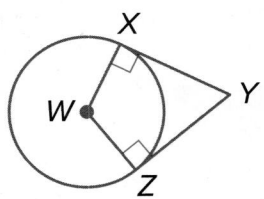

$m\angle Y =$ ___?___

😊 **Think About It!**

How can you check your answer?

Example 6 Find Measures in Circumscribed Polygons

△JKL is circumscribed about ⊙Q.
Find the perimeter of △JKL.

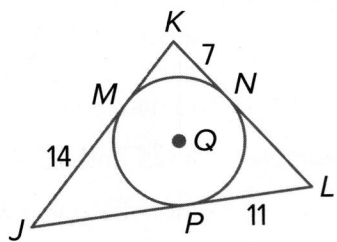

Step 1 Find the missing measures.

Because △JKL is circumscribed about ⊙Q, $\overline{JM}$ and $\overline{JP}$ are tangent to ⊙Q, as are $\overline{KM}$, $\overline{KN}$, $\overline{LN}$, and $\overline{LP}$. Therefore, $\overline{JM} \cong \overline{JP}$, $\overline{KM} \cong \overline{KN}$, and $\overline{LN} \cong \overline{LP}$.

So, $JM = JP = 14$ feet, $KM = KN = 7$ feet, and $LN = LP = 11$ feet.

🎥 **Go Online**
You can watch a video to see how to solve problems involving circumscribed triangles.

Step 2 Find the perimeter of △JKL.

By the Segment Addition Postulate, $JK = JM + KM = 14 + 7$ or 21 units, $KL = KN + LN = 7 + 11$ or 18 units, and $JL = JP + LP = 14 + 11$ or 25 units.

$$\text{perimeter} = JK + KL + JL$$
$$= 21 + 18 + 25 \text{ or } 64 \text{ units}$$

So, the perimeter of △JKL is 64 units.

Check

Quadrilateral *RSTU* is circumscribed about ⊙J. If the perimeter is 18 units, find the value of *x*.

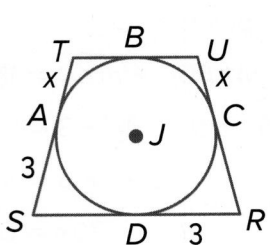

🎥 **Go Online**
You may want to complete the construction activities for this lesson.

A. 1.5 units   **B.** 2 units   **C.** 3 units   **D.** 6 units

🎥 **Go Online** You can complete an Extra Example online.

Practice

⬤ **Go Online** You can complete your homework online.

Example 1

Identify the number of common tangents that exist between each pair of circles. If no common tangent exists, state *no common tangent*.

1.

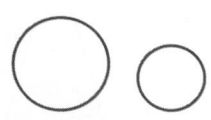

2.

3.

4.

Example 2

Determine whether each segment is tangent to the given circle. Justify your answer.

5. $\overline{HI}$

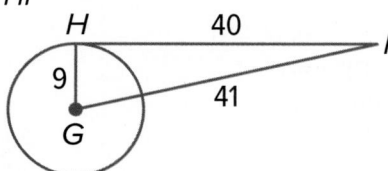

6. $\overline{AB}$

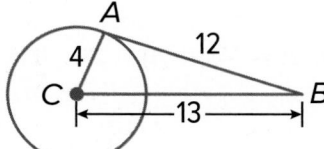

7. $\overline{MP}$

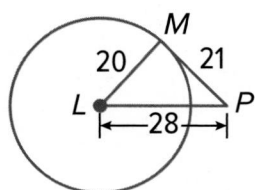

8. $\overline{QR}$

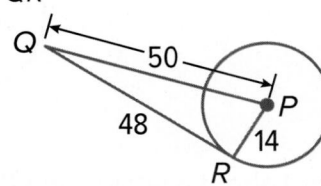

Example 3

Find the value of *x*. Assume that segments that appear to be tangent are tangent. Round your answer to the nearest hundredth, if needed.

9.

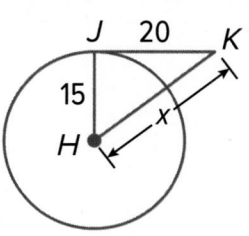

10.

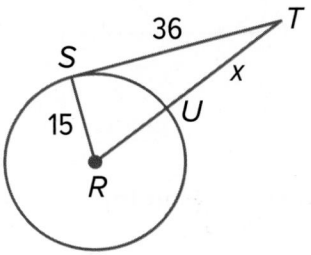

11.

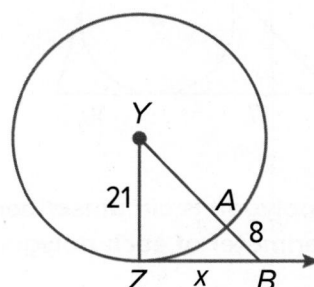

12.

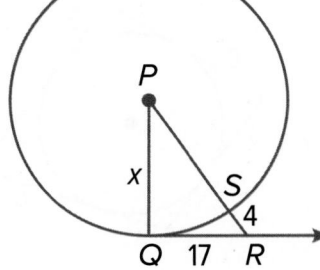

13.

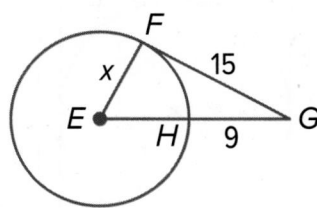

14.

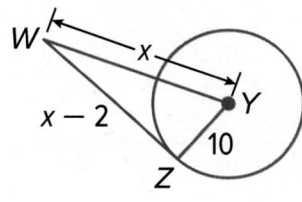

Example 4

Find the value of x. Assume that segments that appear to be tangent are tangent.

15.

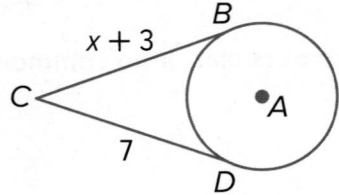

$x + 3$
7

16.

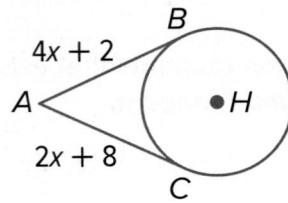

$4x + 2$
$2x + 8$

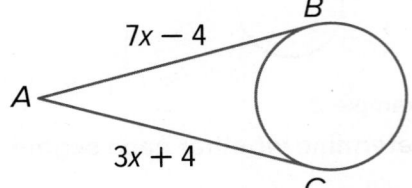

$4x + 4$ $6x - 12$

17. DECORATIONS Federico wants to hang a spherical terrarium using rope. $\overline{JK}$ and $\overline{LK}$ are tangent to the terrarium. If the lengths of the tangents are given in inches, how much rope is needed to hang the terrarium?

18. CARNIVAL GAMES Phoebe is playing a carnival game that involves tossing softballs into a wooden basket. She positions herself at point *A* so $\overline{AB}$ and $\overline{AC}$ are tangent to the basket. If the lengths of the tangents are given in feet, find *AB*.

$7x - 4$
$3x + 4$

Example 5

19. If $m\angle BDC = 12x°$ and $m\angle A = (4x + 4)°$, find $m\angle A$.

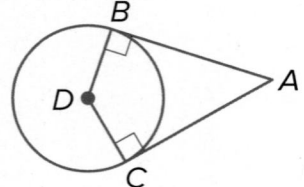

20. If $m\angle QPS = (15x + 8)°$ and $m\angle R = (10x - 3)°$, find $m\angle R$.

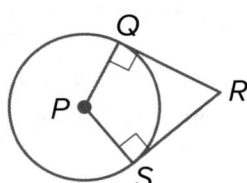

Example 6

Each polygon is circumscribed about a circle. Find the perimeter of each polygon.

21.

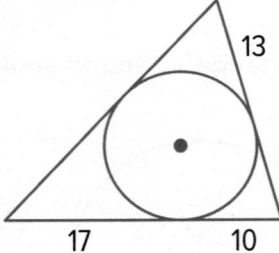

13
17 10

22.

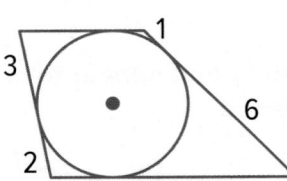

3 1
2 6
2

23.

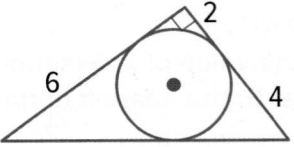

2
6 4

Each polygon is circumscribed about a circle. Find the value of x. Then find the perimeter of each polygon.

24.

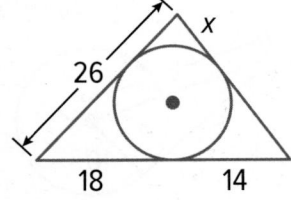

x
26
18 14

25.

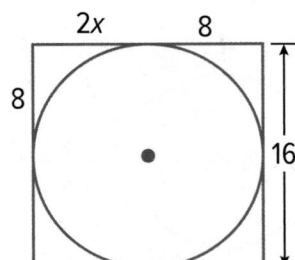

$2x$ 8
8
16

26.

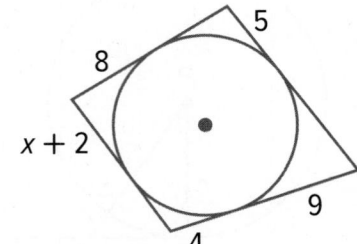

5
8
$x + 2$
4 9

Mixed Exercises

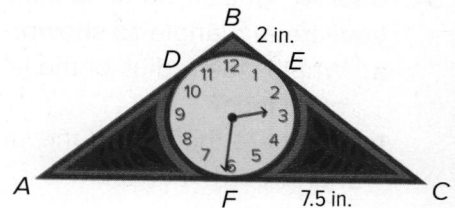

27. REASONING The design shown in the figure is that of a circular clock face inscribed in a triangular base. *AF* and *FC* are equal.

   a. Find *AB*.

   b. Find the perimeter of the clock.

PROOF Write a two-column proof to prove each theorem.

28. Theorem 5.12

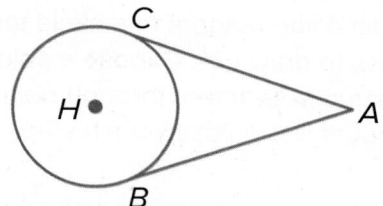

   Given: $\overline{AC}$ is tangent to $\odot H$ at *C*.
             $\overline{AB}$ is tangent to $\odot H$ at *B*.

   Prove: $\overline{AC} \cong \overline{AB}$

29. Theorem 10.13

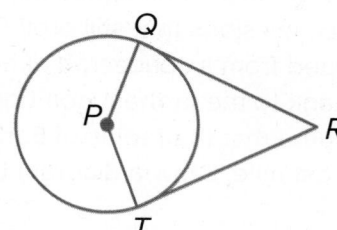

   Given: $\overline{RQ}$ is tangent to $\odot P$ at *Q*.
             $\overline{RT}$ is tangent to $\odot P$ at *T*.

   Prove: $m\angle P + m\angle R = 180°$

30. PROOF Write a two-column proof.

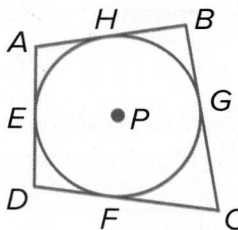

   Given: Quadrilateral *ABCD* is circumscribed about $\odot P$.

   Prove: $AB + CD = AD + BC$

31. JEWELRY Joan is designing a pendant with a circular gem inscribed in a triangle.

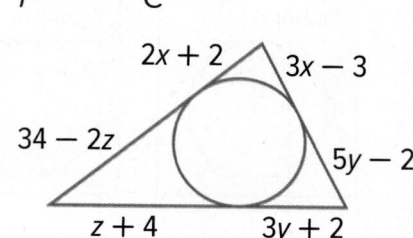

   a. Find the values of *x*, *y*, and *z*.

   b. Find the perimeter of the triangle.

PRECISION Find the value of *x* to the nearest hundredth. Assume that segments that appear to be tangent are tangent.

32.

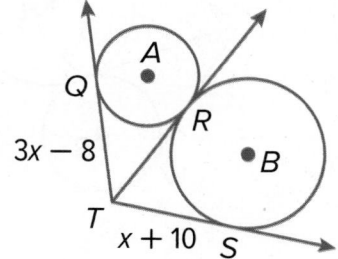

33.

34. DESIGN Ignacio wants to make a design of circles inside an equilateral triangle as shown.

   a. What is the radius of the large circle to the nearest hundredth of an inch?

   b. What are the radii of the smaller circles to the nearest hundredth of an inch?

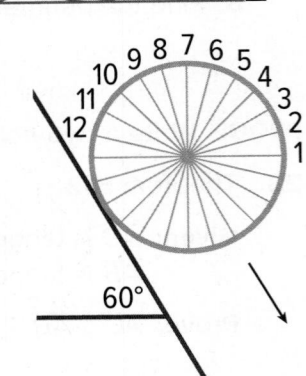

10 in.

35. PERSEVERE A wheel is rolling down an incline as shown in the figure at the right. Twelve evenly spaced radii form spokes of the wheel. When spoke 2 is vertical, which spoke will be perpendicular to the incline?

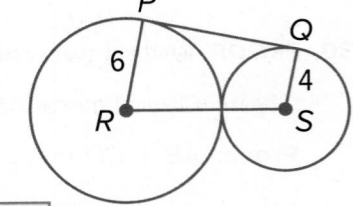

36. USE TOOLS Construct a line tangent to a circle through a point on the circle. Use a compass to draw ⊙A. Choose a point P on the circle and draw $\overleftrightarrow{AP}$. Then construct a segment through point P perpendicular to $\overleftrightarrow{AP}$. Label the tangent line t. Justify each step.

37. USE A MODEL NASA has procedures for limiting orbital debris to mitigate the risk to human life and space missions. *Orbital debris* refers to materials from space missions that still orbit Earth. Suppose an ammonia tank is accidentally discarded from a spacecraft at an altitude of 435 miles. What is the distance from the tank to the farthest point on Earth's surface from which the tank is visible? Assume that the radius of Earth is 4000 miles. Round your answer to the nearest mile. Draw a diagram to represent this situation.

38. PERSEVERE $\overline{PQ}$ is tangent to circles R and S. Find PQ. Explain your reasoning.

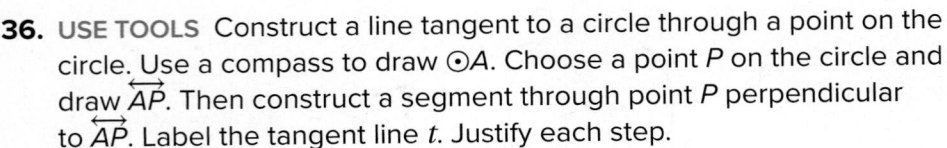

39. WHICH ONE DOESN'T BELONG? Which of the polygons shown below is not circumscribed? Justify your conclusion.

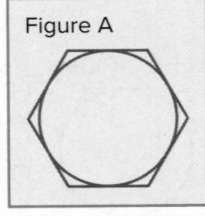

Figure A

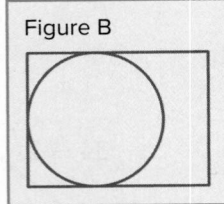

Figure B

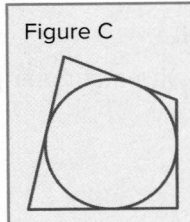

Figure C

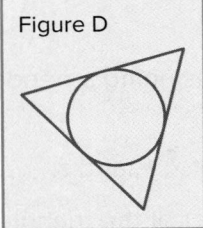

Figure D

40. CREATE Draw a circumscribed triangle and an inscribed triangle.

41. ANALYZE In the figure, $\overline{XY}$ and $\overline{XZ}$ are tangent to ⊙A. $\overline{XZ}$ and $\overline{XW}$ are tangent to ⊙B. Explain how segments $\overline{XY}$, $\overline{XZ}$, and $\overline{XW}$ can all be congruent if the circles have different radii.

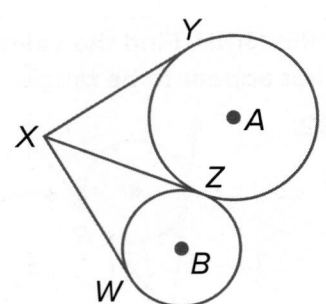

42. WRITE Is it possible to draw a tangent from a point that is located anywhere outside, on, or inside a circle? Explain.

Tangents, Secants, and Angle Measures

Explore Relationships Between Tangents and Secants

🔵 **Online Activity** Use dynamic geometry software to complete the Explore.

> ✕
>
> ⓠ **INQUIRY** How can you calculate the measure of an angle formed when two secants intersect inside a circle or a tangent and a secant intersect on a circle?

Learn Tangents, Secants, and Angle Measures

A **secant** is any line or ray that intersects a circle in exactly two points. Lines j and k are secants of ⊙C.

When two secants intersect inside a circle, the angles formed are related to the arcs they intercept.

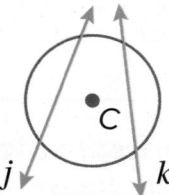

Theorems	
5.14	If two secants or chords intersect in the interior of a circle, then the measure of each angle formed is one half the sum of the measures of the arcs intercepted by the angle and its vertical angle.
5.15	If a secant and a tangent intersect at the point of tangency, then the measure of each angle formed is one half the measure of its intercepted arc.
5.16	If two secants, a secant and a tangent, or two tangents intersect in the exterior of a circle, then the measure of the angle formed is one half the difference of the measures of the intercepted arcs.

You will prove Theorem 5.15 in Exercise 18.

Example 1 Intersecting Chords or Secants

Find the value of x.
Step 1 Find m∠MKP.

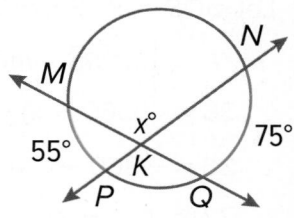

$m\angle MKP = \frac{1}{2}(m\widehat{MP} + m\widehat{NQ})$ Theorem 10.14

$\qquad = \frac{1}{2}(55 + 75)$ Substitution

$\qquad = \frac{1}{2}(130)$ or 65°   Simplify.

Step 2 Find the value of x, the measure of ∠MKN.

∠MKP and ∠MKN are supplementary.

So, $x = 180 - 65$ or 115.

🔵 **Go Online** You can complete an Extra Example online.

Today's Goals
• Use relationships between tangents and secants to solve problems.

Today's Vocabulary
secant

🔵 **Go Online** You may want to complete the Concept Check to check your understanding.

Study Tip

Absolute Value In Theorem 5.16, the measure of each ∠A can also be expressed as half the absolute value of the difference of the arc measures. In this way, the order of the arc measures does not affect the outcome of the calculation.

🔵 **Go Online** Proofs are available for Theorems 5.14 and 5.16.

Check

Find the value of x.

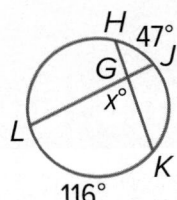

$x = \underline{\quad ? \quad}$

Example 2 Secants and Tangents Intersecting on a Circle

Find $m\widehat{JLK}$.

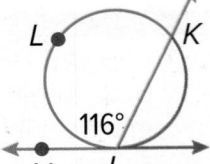

$m\angle HJK = \frac{1}{2} m\widehat{JLK}$ Theorem 10.15

$116 = \frac{1}{2} m\widehat{JLK}$ Substitution

$232° = m\widehat{JLK}$ Solve.

Check

Find $m\widehat{DEF}$ if $m\angle FDC = 64°$.

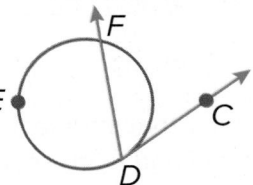

$m\widehat{DEF} = \underline{\quad ? \quad}$

Talk About It!

What is the difference between a tangent and a secant?

Go Online to practice what you've learned about circles in the Put It All Together over Lessons 5-1 through 5-6.

🌐 Example 3 Tangents and Secants Intersecting Outside a Circle

MEMORIALS A photographer is taking a photo of the Thomas Jefferson Memorial in Washington, D.C., from a boat in the Tidal Basin. The photographer's lines of sight are tangent to the memorial at points Q and S. If the camera's viewing angle measures 36°, what portion of the memorial will be visible in the photo?

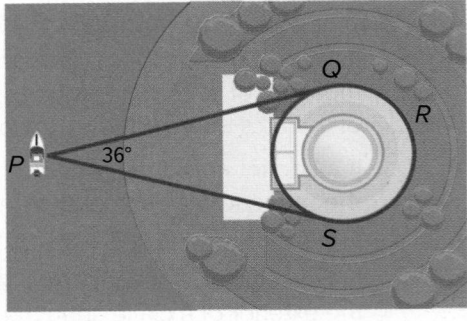

Because the Thomas Jefferson Memorial can be modeled by a circle, the arc measure of the memorial is 360°. So, the portion of the memorial that will be visible in the photo is equal to $\frac{m\widehat{QS}}{360}$.

Let $m\widehat{QS} = x°$. So, $m\widehat{QRS} = 360 - x$.

$m\angle P = \frac{1}{2}m\widehat{QRS} - m\widehat{QS}$ Theorem 5.16

$36 = \frac{1}{2}[(360 - x) - x]$ Substitution

$36 = \frac{1}{2}(360 - 2x)$ Simplify.

$72 = 360 - 2x$ Multiply each side by 2.

$-288 = -2x$ Subtract 360 from each side.

$144 = x$ Divide each side by -2.

So, $m\widehat{QS} = 144°$, and the portion of the memorial that will be visible in the photo is $\frac{144}{360}$ or 40%.

🌐 **Go Online** You can complete an Extra Example online.

Practice

Example 1

Find each measure.

1. $m\angle 2$

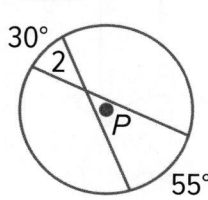
30°
2
P
55°

2. $m\angle 1$

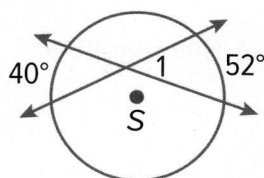
40°
1
52°
S

3. $m\widehat{GH}$

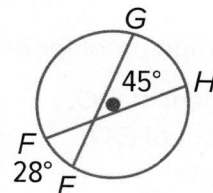
G
45°
H
F
28°
E

4. $m\angle 5$

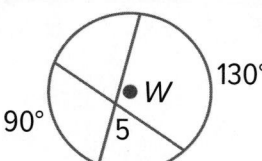
130°
W
90°
5

5. $m\angle 1$

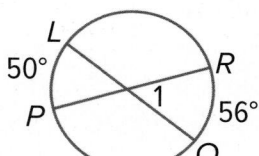
L
50°
R
P
1
56°
Q

6. $m\angle 2$

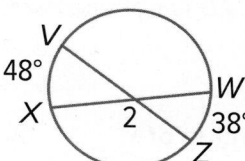
V
48°
W
X
2
38°
Z

Example 2

Find each measure. Assume that segments that appear to be tangent are tangent.

7. $m\angle 1$

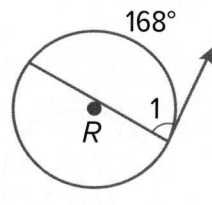
168°
R
1

8. $m\angle 3$

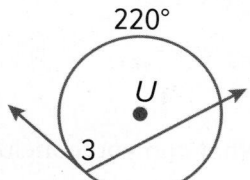
220°
U
3

9. $m\widehat{RT}$

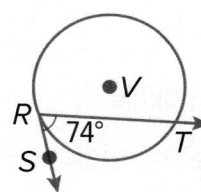
V
R
74°
T
S

10. $m\angle 6$

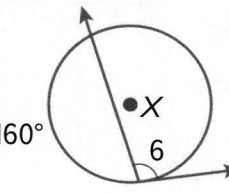
X
160°
6

11. $m\angle 3$

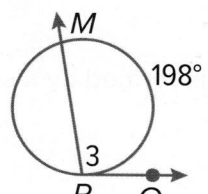
M
198°
3
P   Q

12. $m\angle 4$

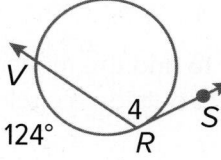
V
124°
4
R
S

Example 3

Find each measure. Assume that segments that appear to be tangent are tangent.

13. $m\angle R$

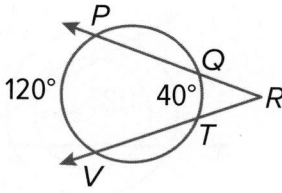
P
120°
Q
40°
R
T
V

14. $m\angle U$

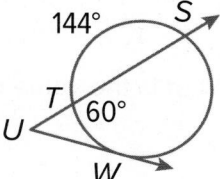
144°
S
T
60°
U
W

15. $m\widehat{DPA}$

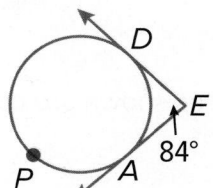
D
E
84°
P
A

16. **FLIGHT** When a plane is flying at an altitude of 5 miles, the lines of sight to the horizon looking north and south make about a 173.7° angle. What portion of the longitude line is visible from 5 miles high?

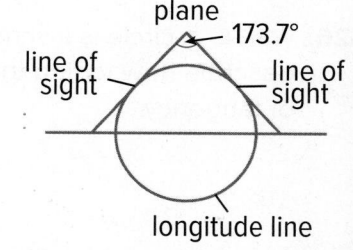
plane
173.7°
line of sight
line of sight
longitude line

Mixed Exercises

17. **REASONING** Salvador places a circular canvas on his A-frame easel and carefully centers it. The top of the easel forms a 30° angle and $m\widehat{BC} = 22°$. What is $m\widehat{AB}$?

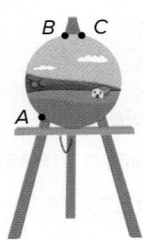

18. **PROOF** Write a paragraph proof for each part of Theorem 10.15.

    a. **Given:** $\overleftrightarrow{AB}$ is a tangent of $\odot O$.
       $\overrightarrow{AC}$ is a secant of $\odot O$.
       $\angle CAE$ is acute.
       Prove: $m\angle CAE = \frac{1}{2}m\widehat{CA}$

    b. Prove that if $\angle CAB$ is obtuse, $m\angle CAB = \frac{1}{2}m\widehat{CDA}$.

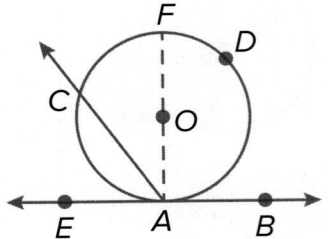

REASONING Find the value of x.

19.

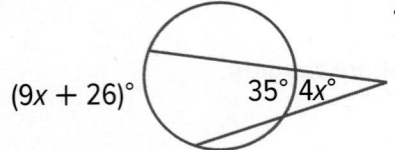

$(9x + 26)°$ $35°$ $4x°$

20. $(5x - 6)°$ $3°$

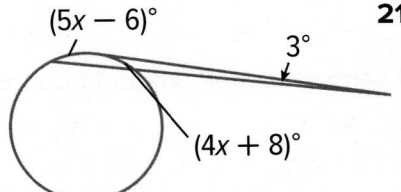

$(4x + 8)°$

21.

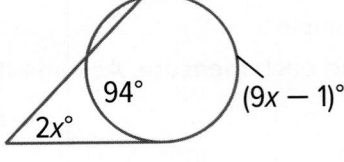

$94°$ $(9x - 1)°$ $2x°$

Higher-Order Thinking Skills

22. **ANALYZE** Isosceles $\triangle ABC$ is inscribed in $\odot D$. What can you conclude about $m\widehat{AB}$ and $m\widehat{BC}$? Explain.

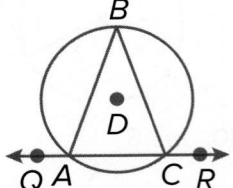

23. **WRITE** Explain how to find the measure of an angle formed by a secant and a tangent that intersect outside a circle.

24. **CREATE** Draw a circle and two tangents that intersect outside the circle. Use a protractor to measure the angle that is formed. Find the measures of the minor and major arcs formed. Explain your reasoning.

25. **PERSEVERE** The circles shown are concentric. What is the value of $\angle A$?

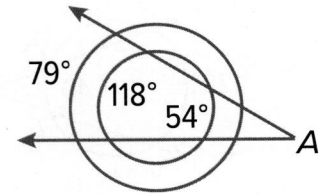

$79°$ $118°$ $54°$ A

26. **WRITE** A circle is inscribed within $\triangle PQR$. If $m\angle P = 50°$ and $m\angle Q = 60°$, describe how to find the measures of the three minor arcs formed by the points of tangency.

Equations of Circles

Explore Deriving Equations of Circles

Online Activity Use the guiding exercises to complete the Explore.

> **INQUIRY** How can you determine whether a point lies on a circle?

Explore Exploring Equations of Circles

Online Activity Use the interactive tool to complete the Explore.

> **INQUIRY** How can you determine the center and radius of a circle from its equation?

Learn Equations of Circles

Key Concept • Equation of a Circle in Standard Form

The standard form of the equation of a circle with center at (h, k) and radius r is

$(x - h)^2 + (y - k)^2 = r^2$.

The standard form of the equation of a circle is also called the *center-radius* form.

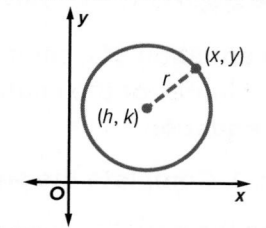

Example 1 Write an Equation by Using the Center and Radius

$\overline{AB}$ is a diameter of the circle. Write the equation of the circle.

Because $\overline{AB}$ is a diameter of the circle, the center of the circle is the midpoint of $\overline{AB}$.

$$M = \left(\frac{x_1 + x_2}{2}, \frac{y_1 + y_2}{2}\right) \qquad \text{Midpoint Formula}$$

$$= \left(\frac{0 + 8}{2}, \frac{-1 + (-1)}{2}\right) \qquad \text{Substitution}$$

$$= (4, -1) \qquad \text{Simplify.}$$

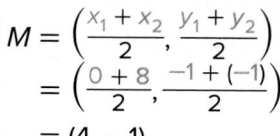

So, the center is at $(4, -1)$ and the radius is 4.

$$(x - h)^2 + (y - k)^2 = r^2 \qquad \text{Equation of a circle}$$

$$(x - 4)^2 + [y - (-1)]^2 = 4^2 \qquad \text{Substitution}$$

$$(x - 4)^2 + (y + 1)^2 = 16 \qquad \text{Simplify.}$$

Check

$\overline{LM}$ is a diameter of the circle. Write the equation of the circle in standard form.

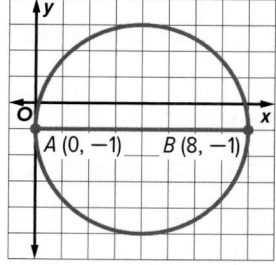

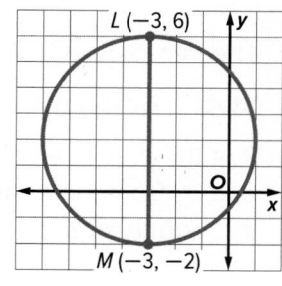

Today's Goals
- Derive the equation of a circle using the Pythagorean Theorem and complete the square to find the center and radius of a circle.

Go Online You can watch a video to see how to write the equation of a circle.

Talk About It!

Leonardo says you can also derive the standard form of the equation of a circle by using the Distance Formula. Do you agree or disagree? Justify your reasoning.

Example 2 Write an Equation by Using the Center and a Point

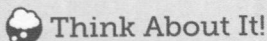 **Think About It!**

Does the graph of a circle represent a function? Justify your reasoning.

Write the equation of the circle with center at (−3, −5) that passes through (0, 0).

Step 1 Find the length of the radius.

Find the distance between the points to determine the radius.

$$r = \sqrt{(x_2 - x_1)^2 + (y_2 - y_1)^2}$$ Distance Formula

$$= \sqrt{[0 - (-3)]^2 + [0 - (-5)]^2}$$ $(x_1, y_1) = (-3, -5)$ and $(x_2, y_2) = (0, 0)$

$$= \sqrt{34}$$ Simplify.

Step 2 Write the equation of the circle.

Write the equation using $h = -3$, $k = -5$, and $r = \sqrt{34}$.

$$(x - h)^2 + (y - k)^2 = r^2$$ Equation of a circle

$$[x - (-3)]^2 + [y - (-5)]^2 = (\sqrt{34})^2$$ Substitution

$$(x + 3)^2 + (y + 5)^2 = 34$$ Simplify.

Example 3 Graph a Circle

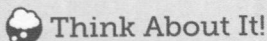 **Think About It!**

A circle has the equation $(x - 5)^2 + (y + 7)^2 = 16$. If the center of the circle is shifted 3 units right and 9 units up, what would be the equation of the new circle? Explain your reasoning.

The equation of a circle is $x^2 + y^2 + 8x - 14y + 40 = 0$. State the coordinates of the center and the measure of the radius. Then graph the equation.

Step 1 Complete the squares.

Write the equation in standard form by completing the squares.

$$x^2 + y^2 + 8x - 14y + 40 = 0$$ Original equation

$$x^2 + y^2 + 8x - 14y = -40$$ Subtract 40 from each side.

$$x^2 + 8x + y^2 - 14y = -40$$ Group terms.

$$x^2 + 8x + 16 + y^2 - 14y + 49 = -40 + 16 + 49$$
Complete the squares.

$$(x + 4)^2 + (y - 7)^2 = 25$$ Factor and simplify.

Step 2 Identify h, k, and r.

$(x + 4)^2 + (y - 7)^2 = 25$ can also be written as $[x - (-4)]^2 + (y - 7)^2 = 5^2$. Now you can identify h, k, and r.

$h = -4$, $k = 7$, and $r = 5$

The center is at $(-4, 7)$, and the radius is 5.

Step 3 Graph the circle.

Plot the center and four points that are 5 units from the center. Sketch the circle through these four points.

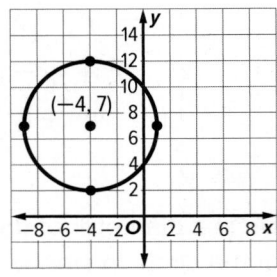

🌐 Apply Example 4 Use a Diameter to Write an Equation

TRANSPORTATION **The school board is determining the new boundary for Riverdale High School's bus transportation. The high school is at point *H* and is the center of the circle that represents the new boundary. The students that live on or within the circle will have to walk to school. Students that live at points *J*(−5, 2), *K*(5, 6), and *L*(5, 2) lie on the boundary, and $\overline{JK}$ is a diameter of ⊙*H*. Write the equation of ⊙*H* in standard form.**

1 What is the task?

Describe the task in your own words. Then list any questions that you may have. How can you find answers to your questions?

Sample answer: I need to find the equation of the circle that describes the boundary for bus transportation. Where is the center of the circle located? What is the length of the radius of the circle?

2 How will you approach the task? What have you learned that you can use to help you complete the task?

Sample answer: Because I know the endpoints of the diameter of the circle, I can use the Midpoint Formula to find the location of the center of the circle. Then, I can use the Distance Formula to find the length of the radius of the circle. Using this information, I can write an equation of the circle in standard form.

3 What is your solution?

Use your strategy to solve the problem.

The coordinates of the center of the circle are (0, 4).

What is the length of the radius of the circle?

$r = \sqrt{29}$ units

Write the equation of the circle in standard form.

$(x - 0)^2 + (y - 4)^2 = 29$

4 How can you know that your solution is reasonable?

✏️ **Write About It!** Write an argument that can be used to defend your solution.

Sample answer: If you graph the equation, the center of the circle is at (0, 4), and the circle passes through *J*(−5, 2), *K*(5, 6), and *L*(5, 2). So, the answer is reasonable.

Check

NAVIGATION A mariner is using a GPS, or global positioning system, to create a nautical chart that documents underwater objects and hazards. The mariner has found three hazards positioned in a circular pattern at points $D(7, 1)$, $E(14, -4)$, and $F(2, -6)$. If $\overline{EF}$ is a diameter of the circle, what is the equation of the circle that contains the three hazards? Write your answer in standard form.

Example 5 Intersections with Circles

Find the point(s) of intersection between $x^2 + y^2 = 9$ and $y = x - 2$.

Step 1 Graph the equations on the same coordinate plane.

The points of intersection are solutions of both equations, and can be estimated to be at about $(-0.9, -2.9)$ and $(2.9, 0.9)$.

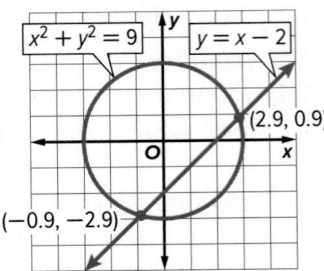

Step 2 Use substitution to find the points of intersection algebraically.

$x^2 + y^2 = 9$	Equation of circle
$x^2 + (x - 2)^2 = 9$	Substitute
$x^2 + x^2 - 4x + 4 = 9$	Multiply.
$2x^2 - 4x - 5 = 0$	Subtract 9 from each side.

Step 3 Use the Quadratic Formula.

Use $\dfrac{-b \pm \sqrt{b^2 - 4ac}}{2a}$ to solve $2x^2 - 4x - 5 = 0$, with $a = 2$, $b = -4$, and $c = -5$.

The solutions are $x = 1 + \dfrac{\sqrt{14}}{2}$ or $x = 1 - \dfrac{\sqrt{14}}{2}$.

Step 4 Find the points of intersection.

Use the equation $y = x - 2$ to find the corresponding y-values.

The points of intersection are $\left(1 + \dfrac{\sqrt{14}}{2}, -1 + \dfrac{\sqrt{14}}{2}\right)$ and $\left(1 - \dfrac{\sqrt{14}}{2}, -1 - \dfrac{\sqrt{14}}{2}\right)$ or at about $(2.87, 0.87)$ and $(-0.87, -2.87)$.

Check

Find the point(s) of intersection between $x^2 + y^2 = 8$ and $y = -x$. If there are no points of intersection, select *no intersection points*.

A. $(2, -2)$ and $(-2, 2)$

B. $(2\sqrt{2}, -2\sqrt{2})$ and $(-2\sqrt{2}, 2\sqrt{2})$

C. $(2, 2)$ and $(-2, -2)$

D. $(2, -2)$

E. no intersection points

🔾 **Go Online** You can complete an Extra Example online.

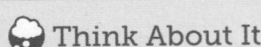

Study Tip

Solving Quadratic Equations In addition to using the Quadratic Formula, you can also solve quadratic equations by taking square roots, completing the square, and factoring.

💭 **Think About It!**

Will a line always intersect a circle in two points? Justify your reasoning.

Practice

 Go Online You can complete your homework online.

Examples 1 and 2

Write the equation of each circle.

1. center at (0, 0), radius 8

2. center at (−2, 6), diameter 8

3.

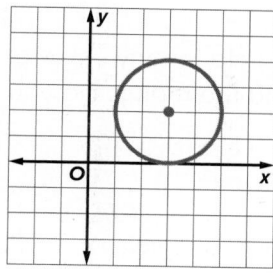

4.

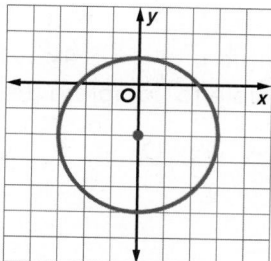

5. center at (3, −4), passes through (−1, −4)

6. center at (0, 3), passes through (2, 0)

7. center at (−4, −1), passes through (−2, 3)

8. center at (5, −2), passes through (4, 0)

Example 3

State the coordinates of the center and the measure of the radius of the circle with the given equation. Then graph the equation.

9. $x^2 + y^2 = 16$

10. $(x − 1)^2 + (y − 4)^2 = 9$

11. $x^2 + y^2 − 4 = 0$

12. $x^2 + y^2 + 6x − 6y + 9 = 0$

Example 4

13. PETS The Villani family is placing a stake in the ground that is connected to a dog leash. The proposed location of the stake is at point Q and is the center of the circle that represents the circular boundary in which their dog will be able to roam. The points $R(−4, 1)$ and $S(8, 7)$ lie on the boundary, and $\overline{RS}$ is a diameter of $\odot Q$. Write the equation of $\odot Q$ in standard form.

14. DELIVERY A new pizza restaurant is determining the boundary for its delivery service. The pizza restaurant is at point P and is the center of the circle that represents the boundary. Customers who are on or within the circle will be eligible for delivery. Customers who live or work at points $A(−2, 2)$, $B(2, −2)$, and $C(6, 2)$ lie on the boundary, and $\overline{AC}$ is a diameter of $\odot P$. Write the equation of $\odot P$ in standard form.

Example 5

Find the point(s) of intersection, if any, between each circle and line with the equations given.

15. $x^2 + y^2 = 9$; $y = 2x + 3$

16. $(x + 4)^2 + (y − 3)^2 = 25$; $y = x + 2$

17. $(x − 5)^2 + (y − 2)^2 = 100$; $y = x − 1$

18. $x^2 + y^2 = 25$; $y = x$

Mixed Exercises

For Exercises 19–20, write the equation of each circle.

19. a circle with a diameter having endpoints at (0, 4) and (6, −4)

20. a circle with $d = 22$ and a center translated 13 units left and 6 units up from the origin

21. STRUCTURE Adam says that $x^2 + y^2 + 4x − 10y = k$ is the equation of a circle for any value of k because it is always possible to complete the square to find the center and the radius. Do you agree? Explain.

22. STRUCTURE The design of a piece of wallpaper consists of circles that can be modeled by $(x − a)^2 + (y − b)^2 = 4$, for all even integers b. Sketch part of the wallpaper on a grid.

23. PRECISION What is the equation of the circle that is inscribed in the square shown?

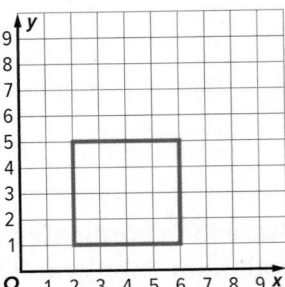

24. REASONING The design for a park is drawn on a coordinate graph. The perimeter of the park is modeled by the equation $(x − 3)^2 + (x − 7)^2 = 225$. Each unit on the graph represents 10 feet. What is the radius of the actual park?

25. FIRE SAFETY A fire station responds to emergencies within a circular area. On a map of the city, the boundary for the area is given by the equation $(x − 8)^2 + (y + 2)^2 = 324$. Each unit on the map represents 1 mile. What is the radius of the boundary?

Write an equation of a circle that contains each set of points with the given diameter. Then graph the circle.

26. $A(−2, 3)$, $B(1, 0)$, $C(4, 3)$; $\overline{AC}$

27. $F(3, 0)$, $G(5, −2)$, $H(1, −2)$; $\overline{GH}$

Higher-Order Thinking Skills

28. FIND THE ERROR The points $P(−1, 2)$, $Q(5, 2)$, and $R(7, −2)$ lie on a circle. Rosalina says that the circle will not intersect $y = 3$. Is Rosalina correct? Explain your reasoning.

29. WRITE Describe how the equation for a circle changes if the circle is translated a units to the right and b units down.

30. CREATE Graph three noncollinear points and connect them to form a triangle. Then construct the circle that circumscribes it. What is the equation of the circle?

31. PERSEVERE The center of a circle is at (2, 3). The point (2, 1) lies on the circle. Find three other points with integer coordinates that lie on the circle. Explain how to do this without finding the equation of the circle.

Equations of Parabolas

Explore Focus and Directrix of a Parabola

▶ **Online Activity** Use dynamic geometry software to complete the Explore.

> ❓ **INQUIRY** How do the focus and directrix affect the shape of a parabola?

Today's Goals
- Derive the equation of a parabola when given a focus and directrix and find the intersection points of linear and quadratic functions.

Today's Vocabulary
parabola
focus
directrix

Learn Equations of Parabolas

A **parabola** is the graph of a quadratic function, such as $y = x^2$. Geometrically, a parabola is the set of all points in a plane equidistant from a fixed point, called the **focus**, and a fixed line, called the directrix. A **directrix** is an exterior line perpendicular to the line containing the foci of a curve.

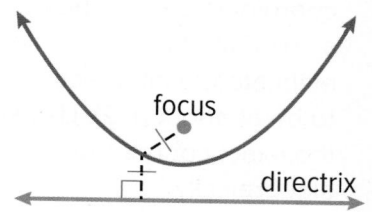

Example 1 Write an Equation for a Parabola

Write an equation for the parabola with the focus at (0, −2) and the directrix $y = 2$.

Step 1 Sketch the parabola.

Graph $F(0, -2)$ and $y = 2$. Sketch a U-shaped curve for the parabola between point F and the directrix as shown. Label a point $P(x, y)$ on the curve.

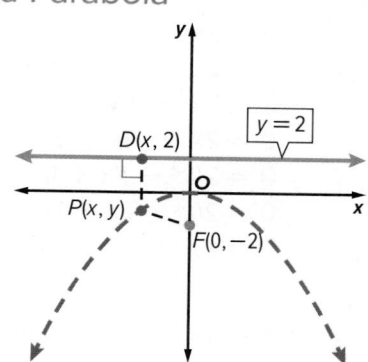

Step 2 Label point D.

Label a point D on $y = 2$ such that $\overline{PD}$ is perpendicular to the line $y = 2$. The coordinates of this point must therefore be $D(x, 2)$.

Step 3 Find PD and PF.

Use the Distance Formula to find PD and PF.

$$PD = \sqrt{(x - x)^2 + (y - 2)^2} \qquad P(x, y) \text{ and } D(x, 2)$$
$$= \sqrt{(y - 2)^2} \qquad \text{Simplify.}$$
$$PF = \sqrt{(x - 0)^2 + [y - (-2)]^2} \qquad P(x, y) \text{ and } F(0, -2)$$
$$= \sqrt{x^2 + (y + 2)^2} \qquad \text{Simplify.}$$

Step 4 Write an equation for the parabola.

Because every point on a parabola is equidistant from the focus and directrix, you can set PD equal to PF. Then solve for y to write an equation for the parabola.

$$\sqrt{(y - 2)^2} = \sqrt{x^2 + (y + 2)^2} \qquad PD = PF$$
$$(y - 2)^2 = x^2 + (y + 2)^2 \qquad \text{Square each side.}$$
$$y^2 - 4y + 4 = x^2 + y^2 + 4y + 4 \qquad \text{Expand } (y - 2)^2 \text{ and } (y + 2)^2.$$
$$y = -\tfrac{1}{8}x^2 \qquad \text{Simplify.}$$

An equation for the parabola with the focus at $(0, -2)$ and the directrix $y = 2$ is $y = -\frac{1}{8}x^2$.

▶ **Go Online** You can complete an Extra Example online.

💬 **Talk About It!**

How can you find the distance between a line and a point not on the line?

▶ **Go Online**

You can watch a video to see how to write the equation of a parabola.

Check

Select the equation of the parabola with the focus at $\left(0, \frac{1}{2}\right)$ and the directrix $y = -\frac{1}{2}$.

A. $y = -\frac{1}{2}x^2$ B. $y = 2x^2$ C. $y = -\frac{1}{2}x^2 - \frac{1}{4}$ D. $y = \frac{1}{2}x^2$

Go Online to see Example 2.

Example 3 Find Points of Intersection

Find the point(s) of intersection, if any, between $y = 2x^2$ and $y = 4x - 2$.

Graph these equations on the same coordinate plane. The point of intersection is a solution of both equations. You can estimate the intersection point on the graph to be at about (1, 2). Use substitution to find the exact coordinates of the point algebraically.

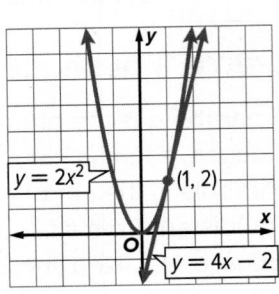

Find the x-coordinate of the intersection point.

$y = 2x^2$	Quadratic Equation
$4x - 2 = 2x^2$	Substitute $4x - 2$ for y.
$-2 = 2x^2 - 4x$	Subtract $4x$ from each side.
$0 = 2x^2 - 4x + 2$	Add 2 to each side.
$0 = 2(x^2 - 2x + 1)$	Factor out the GCF of $2x^2$, $4x$ and 2.
$0 = 2(x - 1)(x - 1)$	Factor $x^2 - 2x + 1$.
$0 = x - 1$	Set the repeated factor equal to zero.
$1 = x$	Add 1 to each side.

Because there is one solution of $4x - 2 = 2x^2$, the parabola and the line intersect in one point.

You can use $y = 4x - 2$ to find the corresponding y-value for $x = 1$.

$y = 4x - 2$	Equation of line
$y = 4(1) - 2$	Substitute.
$y = 2$	Simplify.

So, $y = 2x^2$ and $y = 4x - 2$ intersect at (1, 2).

Check

Find the point(s) of intersection, if any, between $y = -3x^2$ and $y = 6x$. If there are no points of intersection, select *no intersection points*.

A. $(-2, -12)$

B. $(0, 0)$ and $(-2, -12)$

C. $(2, 12)$

D. no intersection points

Go Online You can complete an Extra Example online.

Practice

Go Online You can complete your homework online.

Example 1

Write an equation of the parabola with the given focus and directrix.

1. focus $(0, 3)$, directrix $y = -3$

2. focus $(0, 8)$, directrix $y = -8$

3. focus $(0, -5)$, directrix $y = 5$

4. focus $(0, -9)$, directrix $y = 9$

5. focus $(0, 4)$, directrix $y = -4$

6. focus $(0, 2)$, directrix $y = -2$

Example 2

Write an equation of the parabola with the given focus and directrix.

7. focus $(1, 7)$, directrix $y = -9$

8. focus $(8, 0)$, directrix $y = 4$

9. USE A MODEL A highway is level except for a section that passes through a valley. That section of the highway can be modeled by a parabola with the focus at $(100, 55)$ and the directrix at $y = 5$.

    a. Write an equation for the parabola.

    b. If the level section of the highway is modeled by $y = 70$ and each unit represents a kilometer, what is the maximum width of the valley? Round your answer to the nearest hundredth.

10. USE A MODEL An engineer is using a coordinate plane to design a tunnel in the shape of a parabola. The focus of the tunnel will be at $(4, -2.5)$, and the top of the wall that will contain the tunnel can be modeled by the directrix $y = 2.5$. The base of the tunnel is at ground level and represented by $y = -10$.

    a. Write an equation of the parabola.

    b. Each unit of the coordinate plane represents one foot. Prove or disprove that the width of the tunnel at a height of 5 feet above the ground is exactly 14 feet.

Example 3

Find the point(s) of intersection, if any, between each parabola and line with the given equations.

11. $y = x^2$, $y = x + 2$

12. $y = 2x^2$, $y = 4x - 2$

13. $y = -3x^2$, $y = 6x$

14. $y = -(x + 1)^2$, $y = -x$

Mixed Exercises

15. **HEADLIGHTS** The mirrored parabolic reflector plate of a car headlight has its bulb located at the focus of the parabola as shown. The focus of the parabola is at (2.5, 0) and the directrix is $x = -2.5$. Write an equation that represents the parabolic reflector plate.

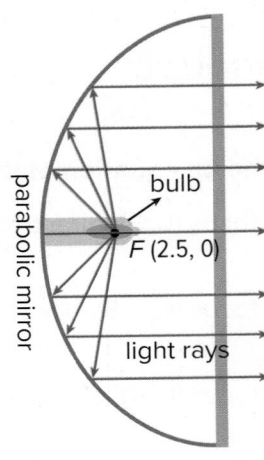

16. **CONSTRUCT ARGUMENTS** Prove or disprove that the point $(\sqrt{3}, -4)$ lies on the parabola with the focus at (0, −3) and the directrix $y = 3$. Justify your argument.

17. **FLASHLIGHTS** The parabolic reflector plate of a flashlight has its bulb located at the focus of the parabola. The focus of the parabola is at (−0.9, 0) and the directrix is $x = 0.9$. Write an equation that represents the reflector plate.

18. **CREATE** Identify the focus and directrix of a parabola if the focus does not lie on the x- or y- axis. Then write the equation of the parabola.

19. **WRITE** The focus of a parabola is at $\left(-\frac{3}{4}, 0\right)$ and the directrix is $x = \frac{3}{4}$. Explain how to determine whether the parabola opens upward, downward, left, or right.

20. **PERSEVERE** The equation of parabola A is $y = \frac{1}{24}x^2$. Parabola B has the same vertex as parabola A and opens in the same direction as parabola A. However, the focus and directrix for parabola B are twice as far apart as they are for parabola A. Write the equation for parabola B. Explain your steps.

21. **ANALYZE** Determine whether the following statement is *sometimes*, *always*, or *never* true. The vertex of a parabola is closer to the focus than it is to the directrix.

Essential Question

How can circles and parts of circles be used to model situations in the real world?

Anything round can be modeled by a circle. Engineers and architects use circles to model moving parts of models they build. Banquet managers use circles to model tables so they know how much room they have to set up an event.

Module Summary

Lessons 5-1 and 5-2

Circles, Arcs, and Angles

- The circumference, or distance around a circle, C equals diameter times pi ($C = \pi d$).

- All circles are similar. Circles with equal radii are congruent.

- To convert a degree measure to radians, multiply by $\frac{\pi \text{ radians}}{180°}$. To convert a radian measure to degrees, multiply by $\frac{180°}{\pi \text{ radians}}$.

- In the same circle or in congruent circles, two minor arcs are congruent if and only if their corresponding chords are congruent.

- If a diameter (or radius) of a circle is perpendicular to a chord, then it bisects the chord and its arc.

- In the same circle or in congruent circles, chords are congruent if and only if they are equidistant from the center.

Lessons 5-3 and 5-4

Inscribed Angles, Tangents, and Secants

- If an angle is inscribed in a circle, then the measure of the angle equals one half the measure of its intercepted arc.

- If two inscribed angles of a circle intercept the same arc or congruent arcs, then the angles are congruent.

- An inscribed angle of a triangle intercepts a diameter or semicircle if and only if the angle is a right angle.

- If a quadrilateral is inscribed in a circle, then its opposite angles are supplementary.

Lessons 5-5 and 5-6

Tangents, Secants, and Angle Measures

- If two segments from the same exterior point are tangent to a circle, then they are congruent.

- If two secants or chords intersect in the interior of a circle, then the measure of each angle formed is one half the sum of the measures of the arcs intercepted by the angle and its vertical angle.

- If a secant and a tangent intersect at the point of tangency, then the measure of each angle formed is one half the measure of its intercepted arc.

- If two secants, a secant and a tangent, or two tangents intersect in the exterior of a circle, then the measure of the angle formed is one half the difference of the measures of the intercepted arcs.

Lessons 5-7 and 5-8

Equations of Circles and Parabolas

- The equation of a circle with center at (h, k) and radius r is $(x - h)^2 + (y - k)^2 = r^2$.

- A parabola is the set of all points in a plane equidistant from a fixed point, called the focus, and a fixed line, called the directrix.

Study Organizer

 **Foldables**

Use your Foldable to review this module. Working with a partner can be helpful. Ask for clarification of concepts as needed.

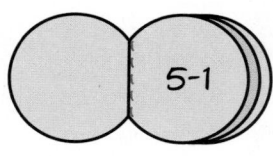

Test Practice

1. **MULTIPLE CHOICE** Kina is building a fence around the circular field used for barrel racing at the rodeo. If the radius of the field is 45 feet, about how many feet of fencing does she need? (Lesson 5-1)

   A. 71 ft

   B. 90 ft

   C. 142 ft

   D. 283 ft

2. **OPEN RESPONSE** The diameter of circle W is 48 centimeters, the diameter of circle Z is 72 centimeters, and YZ is 30 centimeters. What is the length of $\overline{WX}$ in centimeters? (Lesson 5-1)

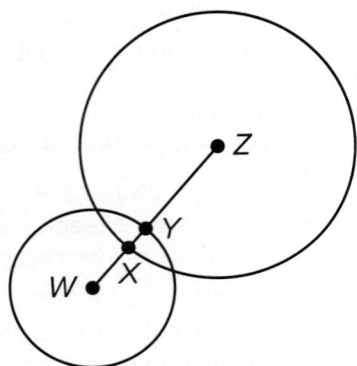

3. **OPEN RESPONSE** What is 240° in radians? Explain your solution process. (Lesson 5-2)

4. **OPEN RESPONSE** $\overline{JK}$ is a diameter of $\odot C$. What is the measure of arc $\overarc{BK}$? (Lesson 5-2)

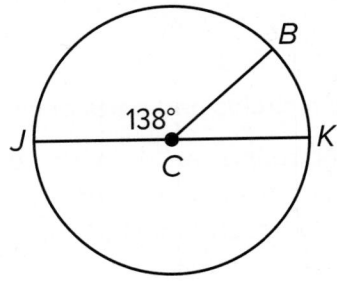

5. **OPEN RESPONSE** Describe the steps to construct a regular hexagon inscribed in a circle. (Lesson 5-1)

6. **MULTIPLE CHOICE** If the $m\overarc{WXY} = 90°$, then what is the $m\overarc{WX}$? (Lesson 5-3)

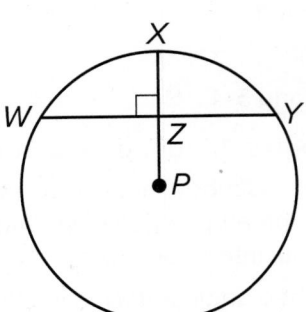

   A. 45°

   B. 90°

   C. 135°

   D. 180°

7. MULTIPLE CHOICE Devon is placing mirrors inside a kaleidoscope as shown.

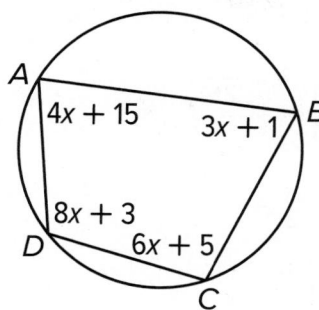

If the four mirrors are placed at the points given, what is the measure of the angle formed at *A*? (Lesson 5-4)

A. 16°

B. 23°

C. 79°

D. 101°

8. MULTIPLE CHOICE Malini and Sedna are visiting a garden. The garden contains a circular area with a fountain at the center and two separate congruent walkways, $\overline{AC}$ and $\overline{DG}$.

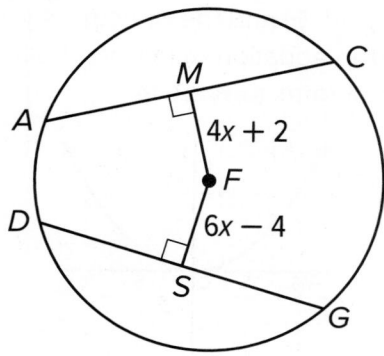

Malini is at point *M* and Sedna is at point *S*. How far is Malini from the fountain?
(Lesson 5-4)

A. 0.2 unit

B. 3 units

C. 6 units

D. 14 units

9. OPEN RESPONSE Use the figure.

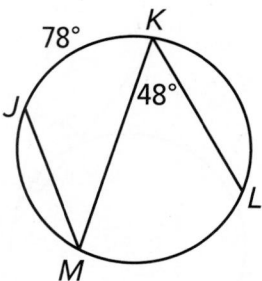

What are $m\widehat{LM}$ and $m\angle JMK$ in degrees?
(Lesson 5-4)

10. MULTI-SELECT Select all the figures where $\overline{HJ}$ is tangent to the circle. (Lesson 5-5)

A.

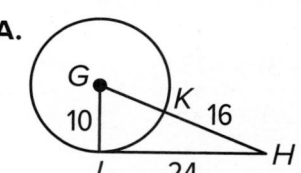

B.

C.

D.

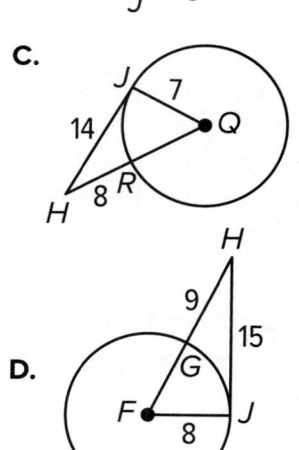

11. MULTIPLE CHOICE What is the measure of $\overset{\frown}{JMK}$? (Lesson 5-6)

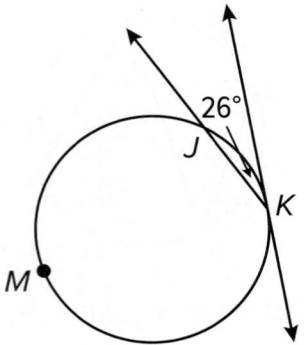

A. 128°

B. 308°

C. 334°

D. 347°

12. MULTI-SELECT A circle has a diameter with endpoints at $(-2, 5)$ and $(4, 1)$. What is the equation of the circle? Select all that apply. (Lesson 5-7)

A. $(x - 1)^2 + (y - 3)^2 = 13$

B. $(x - 3)^2 + (y - 3)^2 = 13$

C. $(x - 1)^2 + (y - 3)^2 = 52$

D. $x^2 + y^2 - 2x - 6y = 8$

E. $x^2 + y^2 - 6x - 6y = 34$

F. $x^2 + y^2 - 2x - 6y = 3$

13. OPEN RESPONSE A meteorologist is tracking a storm and has mapped it to a coordinate plane to make a forecast. At 3 PM, the eye of the storm will be located at $(1, 7)$ and will reach as far as a town located at $(10, -5)$.

What is the equation that could be used to describe this circle? Is the point $(16, 7)$ on the circle? (Lesson 5-7)

14. MULTIPLE CHOICE Which equation of a parabola has the focus at $\left(0, \frac{1}{5}\right)$ and the directrix $y = -\frac{1}{5}$? (Lesson 5-8)

A. $y = \frac{5}{4}x^2 - \frac{8}{25}$

B. $y = \frac{4}{5}x^2$

C. $y = -\frac{5}{4}x^2$

D. $y = \frac{5}{4}x^2$

15. OPEN RESPONSE What is the equation of a parabola with focus $(4, 5)$ and directrix $y = -1$? (Lesson 5-8)

16. OPEN RESPONSE A curve in a highway can be modeled using a parabola where the focus is $(0, 5)$ and the directrix is $y = -5$. Write the equation of the parabola in simplest form. (Lesson 5-8)

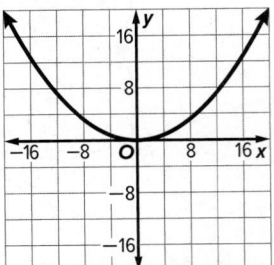

Measurement

How are measurements of two- and three-dimensional figures useful for modeling situations in the real world?

What Will You Learn?

How much do you already know about each topic **before** starting this module?

KEY

— I don't know.   — I've heard of it.   — I know it!

	Before			After		
find areas of quadrilaterals using formulas						
find areas of regular polygons using formulas						
find areas of circles and sectors using formulas						
find surface areas of three-dimensional figures using formulas						
identify cross sections of three-dimensional solids						
identify three-dimensional objects generated by rotations of two-dimensional objects						
find volumes of three-dimensional figures using formulas						
find measures of similar two- and three-dimensional figures						
solve real-world problems involving density using area and volume						

Foldables Make this Foldable to help you organize your notes about area and volume. Begin with one sheet of notebook paper.

1. **Fold** a sheet of paper in half.

2. **Fold** the paper again, two inches from the top.

3. **Unfold** the paper.

4. **Label** as shown.

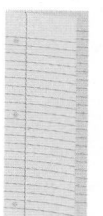 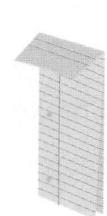 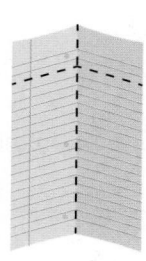 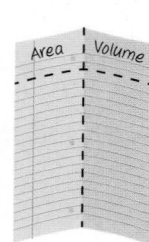

What Vocabulary Will You Learn?

- altitude of a parallelogram
- altitude of a prism or cylinder
- altitude of a pyramid or cone
- apothem
- axis of a cone
- axis of a cylinder
- axis symmetry
- base edge
- base of a parallelogram
- center of a regular polygon
- central angle of a regular polygon
- chord of a sphere
- composite figure
- composite solid
- congruent solids
- conic sections
- cross section
- decomposition
- density
- diameter of a sphere
- height of a parallelogram
- height of a solid
- height of a trapezoid
- lateral area
- lateral edges
- lateral faces
- lateral surface of a cone
- lateral surface of a cylinder
- plane symmetry
- radius of a regular polygon
- radius of a sphere
- regular pyramid
- sector
- similar solids
- slant height of a pyramid or right cone
- solid of revolution
- tangent to a sphere

Are You Ready?

Complete the Quick Review to see if you are ready to start this module. Then complete the Quick Check.

Quick Review

Example 1

Find the value of x. Round to the nearest tenth, if necessary.

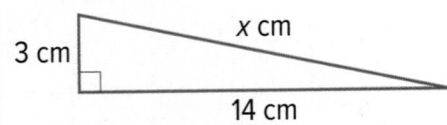

$a^2 + b^2 = c^2$ Pythagorean Theorem

$3^2 + 14^2 = x^2$ Substitute the side lengths.

$9 + 196 = x^2$ Evaluate exponents.

$205 = x^2$ Add.

$14.3 \approx x$ Take the positive square root of each side.

Example 2

Find the value of h.

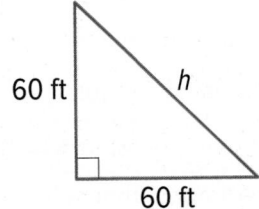

In a 45°-45°-90° triangle, the hypotenuse is $\sqrt{2}$ times the length of a leg.

$h = 60\sqrt{2}$, or about 84.85

So, h is approximately 84.85 feet.

Quick Check

Find the value of x. Round to the nearest tenth, if necessary.

1.

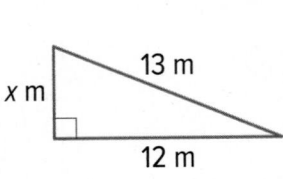

2.

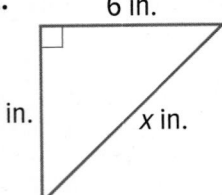

Find the distance between each pair of points. Round to the nearest tenth, if necessary.

3. (2, 1) and (4, 10)

4. (−3, 7) and (5, 4)

5. (−4, −2) and (−1, −2)

6. (0, −6) and (3, 1)

How did you do?

Which exercises did you answer correctly in the Quick Check?

Areas of Quadrilaterals

Explore Deriving the Area Formulas

⬡ **Online Activity** Use dynamic geometry software to complete the Explore.

> @ **INQUIRY** How can you use your knowledge of triangles and rectangles to find areas of parallelograms?

Learn Areas of Parallelograms

A parallelogram is a quadrilateral with both pairs of opposite sides parallel. The **base of a parallelogram** is any side of the parallelogram.

An **altitude of a parallelogram** is defined as a perpendicular segment between any two parallel bases.

The **height of a parallelogram** is the length of an altitude of the parallelogram.

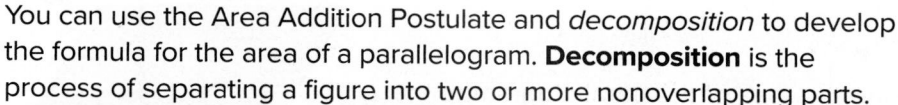

You can use the Area Addition Postulate and *decomposition* to develop the formula for the area of a parallelogram. **Decomposition** is the process of separating a figure into two or more nonoverlapping parts.

Postulate 6.1: Area Addition Postulate

The area of a region is the sum of the areas of its nonoverlapping parts.

To find the area of a parallelogram, imagine cutting off a right triangle from one side of a parallelogram and translating the triangle to the other side to form a rectangle with the same base and height.

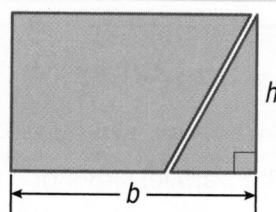

Key Concept • Area of a Parallelogram

The area A of a parallelogram is the product of the base b and its corresponding height h.

Example 1 Area of a Parallelogram

Find the area of the parallelogram.

$h^2 + 7^2 = 25^2$ Pythagorean Theorem

$h = 24$ Solve.

$\overline{DC}$ is one base of $\square ABCD$, and $DC = 23$ feet.

$A = bh$ Area of a parallelogram

$\quad = (23)(24)$ or 552 ft^2 Solve.

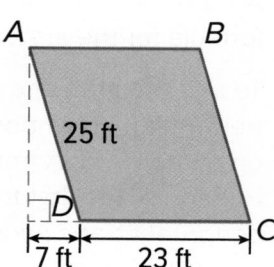

⬡ **Go Online** You can complete an Extra Example online.

Today's Goals

- Find areas of parallelograms.
- Find areas of trapezoids.
- Find the areas of kites and rhombi.

Today's Vocabulary

base of a parallelogram
altitude of a parallelogram
height of a parallelogram
decomposition
height of a trapezoid

💬 **Talk About It!**

What is the relationship between the area of a parallelogram and the area of a rectangle? Justify your answer.

Study Tip

Heights of Figures
The height of a figure can be measured by extending a base. The height of $\square ABCD$ that corresponds to base $\overline{DC}$ can be measured by extending $\overline{DC}$.

Check

Find the area of the parallelogram.

A. 315 cm²

B. 357 cm²

C. 493 cm²

D. 1260 cm²

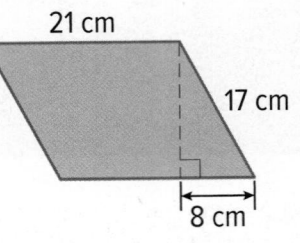
21 cm
17 cm
8 cm

You may need to use trigonometry to find the area of a parallelogram.

Example 2 Use Trigonometry to Find the Area of a Parallelogram

Find the area of the parallelogram.

Step 1 Use special right triangles.

Use a 45°-45°-90° triangle to find the height h of the parallelogram. Recall that if the measure of the leg opposite the 45° angle is h, then the measure of the leg adjacent the 45° angle is also h.

$h = 9$ yd

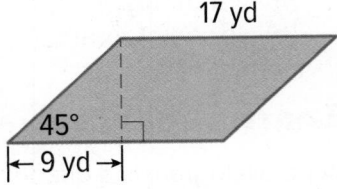
17 yd
45°
9 yd

Step 2 Find the area.

$A = bh$ Area of a parallelogram

$= (17)(9)$ or 153 yd²   Solve.

Check

Find the area of the parallelogram to the nearest tenth.

_____?_____ m²

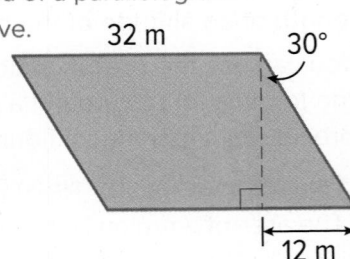
32 m
30°
12 m

Learn Areas of Trapezoids

A trapezoid is a quadrilateral with exactly one pair of parallel sides.

The parallel sides of a trapezoid are called bases.

The **height of a trapezoid** is the perpendicular distance between the bases of a trapezoid.

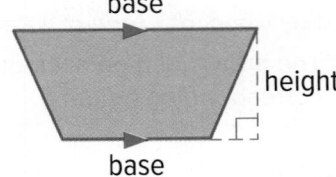
base
height
base

You can use rigid motions and the Area Addition Postulate to develop the formula for the area of a trapezoid.

To find the area of a trapezoid, imagine performing a composition of transformations on a trapezoid. A translation followed by a rotation of the first trapezoid results in two congruent trapezoids that fit together to form a parallelogram.

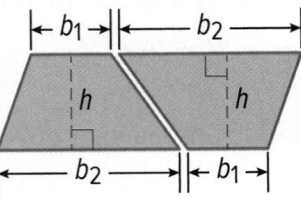
b_1 b_2
h h
b_2 b_1

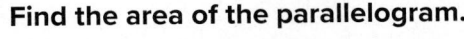

📡 **Go Online** You can complete an Extra Example online.

(continued on the next page)

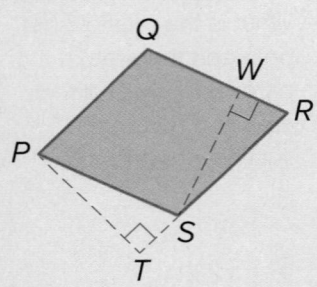

Watch Out!

Units Remember that area is measured in square units such as square feet, square millimeters, and square yards.

🫐 Think About It!

Describe two different ways you could use measurement to find the area of ▱PQRS.

Q
W
R
P
S
T

The area of the parallelogram is the product of the height h and the sum of the two bases, b_1 and b_2. The area of one trapezoid is one half the area of the parallelogram.

Key Concept • Area of a Trapezoid

The area A of a trapezoid is one half the product of the height h and the sum of its bases, b_1 and b_2.

$$A = \frac{1}{2}h(b_1 + b_2)$$

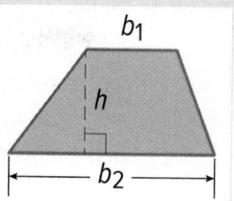

Think About It!

What is the relationship between the area of a parallelogram and the area of a trapezoid? Justify your answer.

Example 3 Area of a Trapezoid

GARDENS Andrea needs enough mulch to cover the garden she planted in a raised bed constructed in the shape of a trapezoid. If one bag of mulch covers 12 square feet at the desired depth, how many bags of mulch does she need to buy?

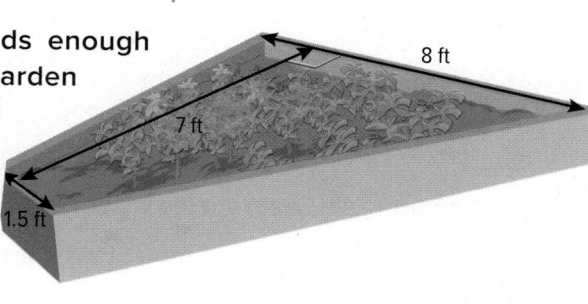

Step 1 Find the area of the garden.

$$A = \frac{1}{2}h(b_1 + b_2)$$ Area of a trapezoid

$$= \frac{1}{2}(7)(8 + 1.5)$$ $h = 7$, $b_1 = 8$, and $b_2 = 1.5$

$$= 33.25 \text{ ft}^2$$ Solve.

The area of the garden is 33.25 square feet.

Step 2 Calculate the number of bags needed.

Use unit analysis to determine how many bags of mulch Andrea should buy.

$$33.25 \text{ ft}^2 \cdot \frac{1 \text{ bag}}{12 \text{ ft}^2} = 2.77 \text{ bags}$$

Round the number of bags up so there is enough mulch. Andrea needs to buy 3 bags of mulch to cover her garden.

Check

ART Miguel wants to cover the top of his desk with butcher paper before working on a project for his art class. If one sheet of butcher paper covers a square meter of work space, how many sheets will Miguel need?

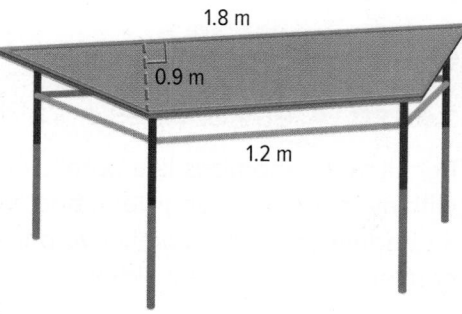

The top of the table has an area of __?__ square meters.
Miguel will need __?__ sheets of butcher paper.

Go Online You can complete an Extra Example online.

Example 4 Use Right Triangles to Find the Area of a Trapezoid

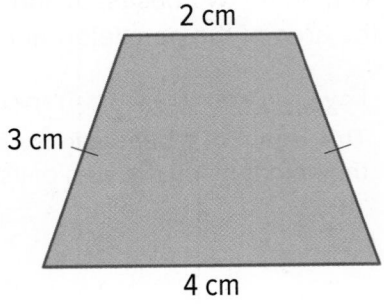

2 cm

3 cm

4 cm

Find the area of the trapezoid.

Step 1 Draw right triangles.

To calculate the height of the trapezoid, draw vertical lines to separate the figure into two right triangles and a rectangle.

Study Tip

Separating Figures
To solve some area problems, you may need to draw in parallel and/or perpendicular lines to find information not provided.

Step 2 Find the height.

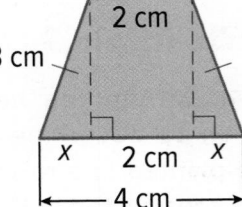

2 cm

3 cm

x 2 cm x

4 cm

$$x + x + 2 = 4 \qquad \text{Length of } b_2$$
$$x = 1 \qquad \text{Solve.}$$

The length of the base of each right triangle is 1 centimeter.

Use the Pythagorean Theorem to calculate the height of the trapezoid.

$$a^2 + b^2 = c^2 \qquad \text{Pythagorean Theorem}$$
$$1^2 + h^2 = 3^2 \qquad \text{Substitute.}$$
$$h = \sqrt{3^2 - 1^2} \text{ or } \sqrt{8} \qquad \text{Solve.}$$

Step 3 Find the area.

$$A = \tfrac{1}{2}h(b_1 + b_2) \qquad \text{Area of a trapezoid}$$
$$= \tfrac{1}{2}(\sqrt{8})(4 + 2) \qquad h = \sqrt{8},\, b_1 = 4, \text{ and } b_2 = 2$$
$$\approx 8.5 \text{ cm}^2 \qquad \text{Solve.}$$

Go Online An alternate method is available for this example.

Check

Find the area of the trapezoid.

A. 4.353 ft²

B. 4.375 ft²

C. 8.706 ft²

D. 8.75 ft²

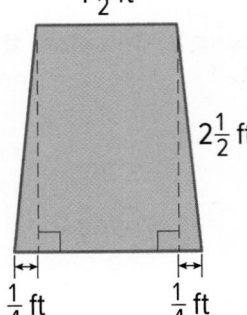

$1\tfrac{1}{2}$ ft

$2\tfrac{1}{2}$ ft

$\tfrac{1}{4}$ ft $\tfrac{1}{4}$ ft

Learn Areas of Kites and Rhombi

Recall that a rhombus is a parallelogram with all four sides congruent, and a kite is a quadrilateral with exactly two pairs of consecutive congruent sides.

The areas of rhombi and kites are related to the lengths of their diagonals.

rhombus   kite

Study Tip

Review Vocabulary
A *diagonal* is a segment that connects any two nonconsecutive vertices in a polygon.

 Go Online You can complete an Extra Example online.

The area A of a rhombus or kite is one half the product of the lengths of its diagonals, d_1 and d_2.

$A = \frac{1}{2} d_1 d_2$

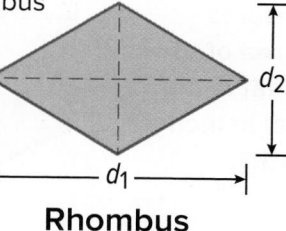

Rhombus

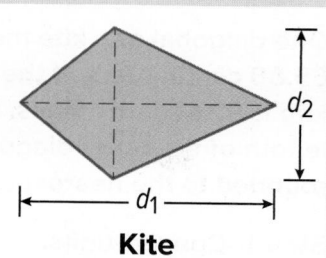

Kite

Example 5 Area of a Rhombus

Find the area of the rhombus.

Step 1 Find the length of each diagonal.
Because the diagonals of a rhombus bisect each other, the lengths of the diagonals are $6 + 6$ or 12 mm and $7 + 7$ or 14 mm.

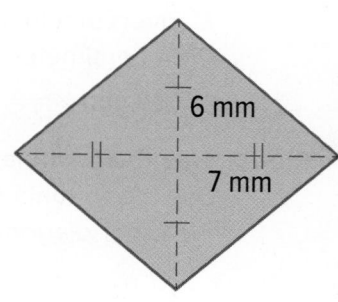

6 mm

7 mm

Step 2 Find the area of the rhombus.

$A = \frac{1}{2} d_1 d_2$ Area of a rhombus

$= \frac{1}{2}(12)(14)$ $d_1 = 12$ and $d_2 = 14$

$= 84$ mm^2 Solve.

Check

Find the area of the rhombus.

A. 45 m^2 **B.** 90 m^2

C. 180 m^2 **D.** 720 m^2

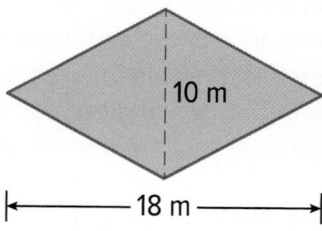

10 m

18 m

Example 6 Area of a Kite

Find the area of the kite.

$A = \frac{1}{2} d_1 d_2$ Area of a kite

$= \frac{1}{2}(16)(9)$ Substitute.

$= 72$ in^2 Solve.

16 in.

9 in.

Check

Find the area of the kite.

$A = \underline{\quad ? \quad}$ m^2

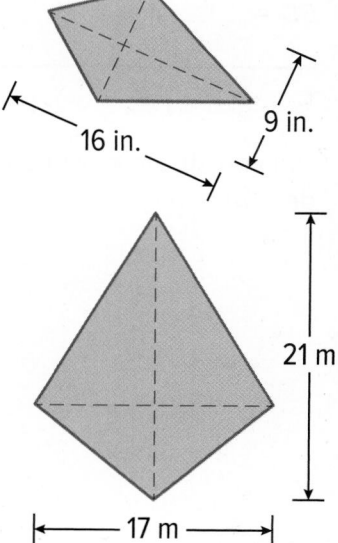

21 m

17 m

Go Online You can complete an Extra Example online.

Example 7 Use Area to Find Missing Measures

One diagonal of a kite measures 55.88 centimeters. If the area of the kite is 92 square inches, what is the length of the other diagonal in inches rounded to the nearest tenth?

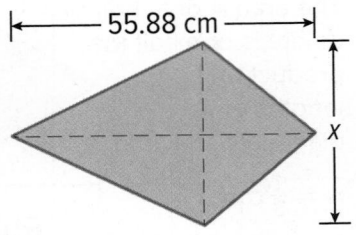

Watch Out!

Converting Units The known length of one of the diagonals is measured in centimeters, which is a metric unit. You are asked to find the measure of the missing diagonal in inches, which is a standard unit. You can use the Internet or other resources to find how to convert from centimeters to inches.

Step 1 Convert units.

The length of one of the diagonals is given in centimeters. Convert that measure to inches. Recall that 1 inch equals 2.54 centimeters.

$$(55.88 \text{ cm}) \times \frac{1 \text{ in.}}{2.54 \text{ cm}} = 22 \text{ in.}$$

Step 2 Use the formula for the area of a kite.

Use the formula for the area of a kite to find the measure of the other diagonal in inches.

$$\frac{1}{2}d_1 d_2 = A \qquad\qquad \text{Area of a kite}$$

$$\frac{1}{2}(22)(x) = 92 \qquad\qquad A = 92, d_1 = 22, \text{ and } d_2 = x$$

$$x \approx 8.4 \qquad\qquad\qquad \text{Solve.}$$

The length of the diagonal is about 8.4 inches.

Check

In rhombus *ABCD*, *AE* = 7.3 inches. If the area of the rhombus is 96 square inches, find the value of *x* and the length of each diagonal. Round to the nearest tenth, if necessary.

$$x \approx \underline{\quad ? \quad} \text{ in.}$$

$$AC = \underline{\quad ? \quad} \text{ in.}$$

$$BD \approx \underline{\quad ? \quad} \text{ in.}$$

Pause and Reflect

Did you struggle with anything in this lesson? If so, how did you deal with it?

Go Online You can complete an Extra Example online.

Practice

Go Online You can complete your homework online.

Examples 1, 2, 4–6

Find the area of each parallelogram, trapezoid, rhombus, or kite. Round to the nearest tenth, if necessary.

1.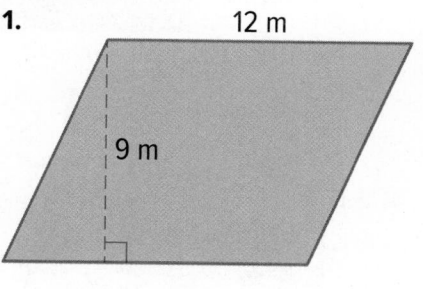
12 m
9 m

2.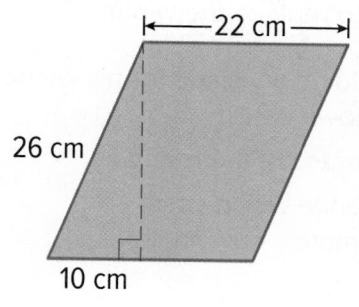
22 cm
26 cm
10 cm

3.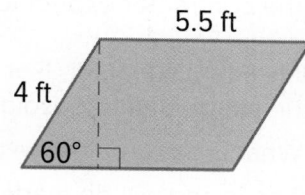
5.5 ft
4 ft
60°

4.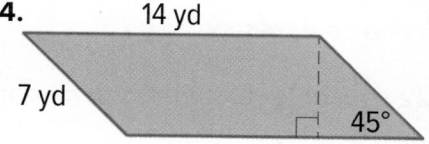
14 yd
7 yd
45°

5.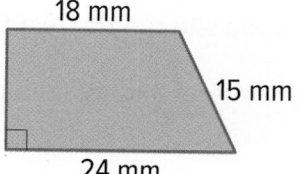
18 mm
15 mm
24 mm

6.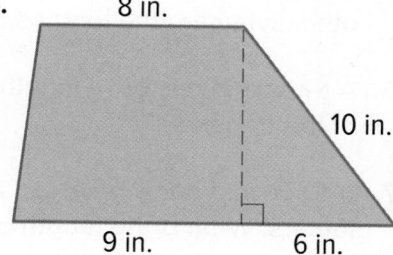
8 in.
10 in.
9 in.
6 in.

7.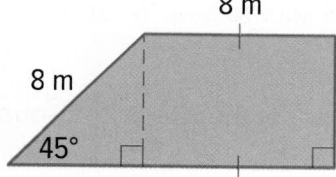
8 m
8 m
45°

8.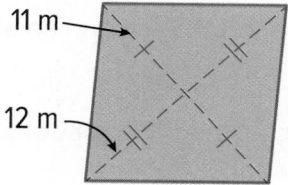
11 m
12 m

9.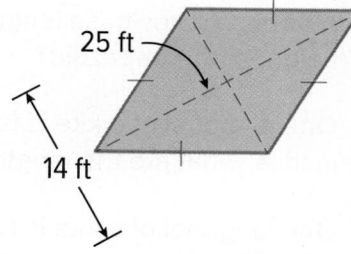
25 ft
14 ft

10.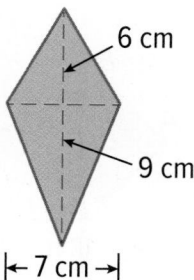
6 cm
9 cm
7 cm

11.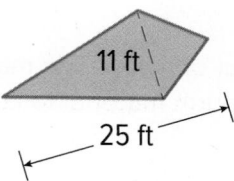
11 ft
25 ft

12.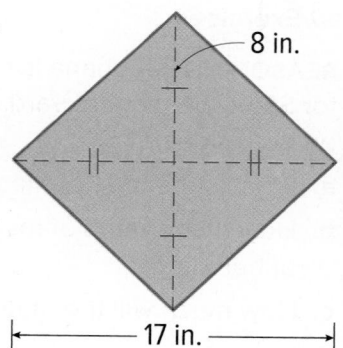
8 in.
17 in.

Example 3

13. **CONSTRUCTION** A contractor is replacing a trapezoidal window on the front of a house.

    a. Find the area of the window.

    b. If the glass for the window costs $7 per square foot, about how much should the contractor expect to pay to replace the window?

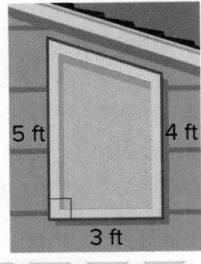

5 ft
4 ft
3 ft

14. **HOME IMPROVEMENT** Iker is building an island for his kitchen in the shape of a trapezoid as shown.

    a. What is the area of the top surface of the island?

    b. The island will be made of granite which costs $65 per square foot. Approximately how much will the granite cost?

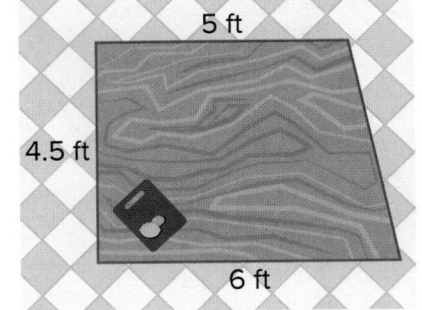

5 ft
4.5 ft
6 ft

Example 7

15. The area of a rhombus is 168 square centimeters. If one diagonal is three times as long as the other, what are the lengths of the diagonals to the nearest tenth of a centimeter?

16. A trapezoid has base lengths of 12 and 14 feet and an area of 322 square feet. What is the height of the trapezoid?

17. A trapezoid has a height of 8 meters, a base length of 12 meters, and an area of 64 square meters. What is the length of the other base?

18. The height of a parallelogram is 10 feet more than the length of the base. If the area of the parallelogram is 1200 square feet, find the length of the base and the height.

19. A trapezoid has base lengths of 4 feet and 19 feet, and an area of 115 square feet. What is the height of the trapezoid?

20. One diagonal of a kite is twice as long as the other diagonal. If the area of the kite is 240 square inches, what are the lengths of the diagonals?

21. One diagonal of a kite is four times as long as the other diagonal. If the area of the kite is 72 square meters, what are the lengths of the diagonals?

Mixed Exercises

22. **REASONING** Meghana is making a kite out of nylon. The material is sold for $5.99 per square yard. She needs enough material to cover her kite on both sides.

    a. What is the area of Meghana's kite?

    b. How many yards of material will Meghana need to cover both sides of her kite?

    c. How much will the material cost before taxes?

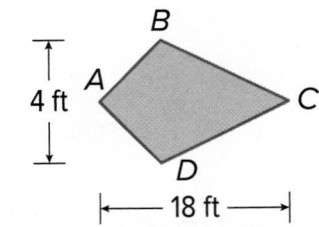

B
A
4 ft
C
D
18 ft

23. **DESIGN** An architect is planning to build an office with a glass façade in the shape of a parallelogram using the given measurements. How much glass will be needed to cover this structure?

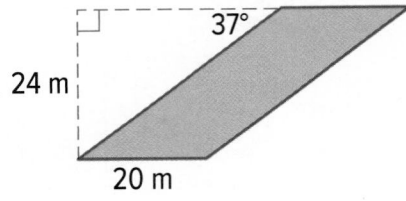

37°
24 m
20 m

24. The diagonal of a square is $6\sqrt{2}$ inches. What is the area of the square?

25. SHADOWS A rectangular billboard casts a shadow on the ground in the shape of a parallelogram. What is the area of the ground covered by the shadow? Round your answer to the nearest tenth.

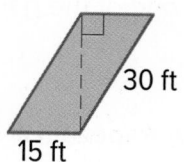

30 ft

15 ft

26. PATHS The concrete path shown is made by joining several parallelograms. What is the total area of the path?

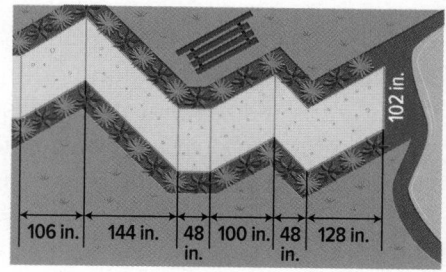

102 in.

106 in.   144 in.   48 in. | 100 in. | 48 in.   128 in.

Find the area of each quadrilateral with the given vertices.

27. $A(-8, 6)$, $B(-5, 8)$, $C(-2, 6)$, and $D(-5, 0)$

28. $W(3, 0)$, $X(0, 3)$, $Y(-3, 0)$, and $Z(0, -3)$

29. REGULARITY Given the area of a geometric figure, describe the method you use to solve for a missing dimension.

30. PRECISION Refer to parallelogram *PQRS*.

    a. Explain how a 30°-60°-90° triangle can be used to find the area of parallelogram *PQRS*.

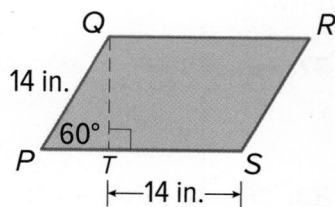

Q   R

14 in.

60°

P   T   S

|—14 in.—|

    b. Find the area of parallelogram *PQRS* to the nearest tenth.

31. STRUCTURE For Bruno's birthday, he got a cake shaped like a kite. He cuts the cake along the diagonals to form 4 pieces. The diagonals are 6 inches and 10 inches long. Which piece(s) is the largest? What is the area of the top of the cake?

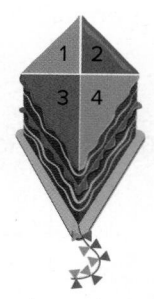

1 2
3 4

32. GARDENS A square landscape plan is composed of three indoor gardens and one walkway that are all congruent. The gardens are centered around a square lounging area. If each side of the lounging area is 15 feet long, what is the area of one of the gardens?

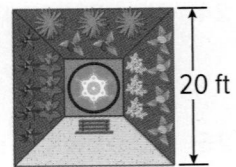

33. STRUCTURE A trapezoid is cut from a rectangle that is 6 inches long and 2 inches wide. The length of one base of the trapezoid is 6 inches. What is the area of the trapezoid?

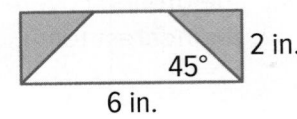

34. REASONING Tile making often requires an artist to find clever ways of dividing a shape into several smaller, congruent shapes. Consider the isosceles trapezoid shown at the right. The trapezoid can be divided into 3 congruent triangles. What is the area of each triangle?

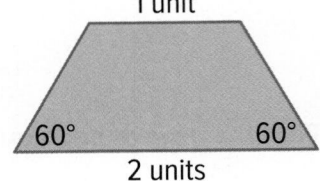

🧠 **Higher-Order Thinking Skills**

35. ANALYZE Will the perimeter of a nonrectangular parallelogram *sometimes, always,* or *never* be greater than the perimeter of a rectangle with the same area and the same height? Justify your argument.

36. FIND THE ERROR Armando and Niran want to draw a trapezoid that has a height of 4 units and an area of 18 square units. Armando says that only one trapezoid will meet the criteria. Niran disagrees and thinks that she can draw several different trapezoids with a height of 4 units and an area of 18 square units. Who is correct? Explain your reasoning.

37. PERSEVERE Find the value of x in parallelogram *ABCD*.

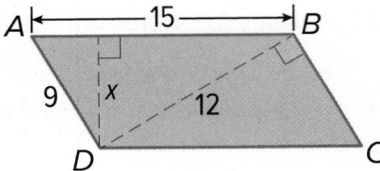

38. CREATE Draw a kite and a rhombus with an area of 6 square inches. Label and justify your drawings.

39. ANALYZE If the areas of two rhombi are equal, are the perimeters *sometimes, always,* or *never* equal? Justify your argument.

40. WRITE How can you use trigonometry to find the area of a figure?

Areas of Regular Polygons

Explore Regular Polygons

🔘 **Online Activity** Use dynamic geometry software to complete the Explore.

> ⊚ **INQUIRY** How can the formula for the area of a regular polygon be derived from the formula for the area of a triangle? ✕

Learn Areas of Regular Polygons

In the figure, a regular pentagon is inscribed in ⊙*P*, and ⊙*P* is circumscribed about the pentagon. The **center of a regular polygon** is the center of the circle circumscribed about the polygon. The **radius of a regular polygon** is the radius of the circle circumscribed about the polygon. The **apothem** of a regular polygon is a perpendicular segment between the center of the polygon and a side of the polygon.

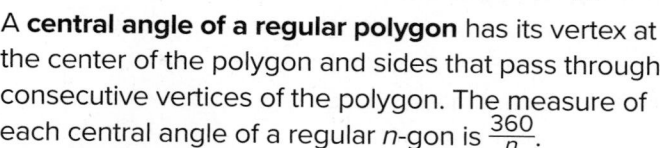

A **central angle of a regular polygon** has its vertex at the center of the polygon and sides that pass through consecutive vertices of the polygon. The measure of each central angle of a regular *n*-gon is $\frac{360}{n}$.

You can find the area of any regular *n*-gon by dividing the polygon into congruent isosceles triangles. This strategy is sometimes called *decomposing the polygon into triangles*. To find the total area of the regular polygon, calculate the area of one triangle and multiply that area by the total number of triangles.

The area of a triangle is $\frac{1}{2}bh$. The base of the triangle in the regular polygon is the length *s* of one side of the polygon and the height of the triangle is the length of an apothem *a*. You can calculate the perimeter *P* of the regular polygon by multiplying the side length *s* by the number of sides *n*. So, you can replace *n* • *s* in the equation with *P*.

Key Concept • Area of a Regular Polygon

The area *A* of a regular *n*-gon with side length *s* is one half the product of the apothem *a* and the perimeter *P*.

$$A = \frac{1}{2}a(ns) \text{ or } A = \frac{1}{2}aP$$

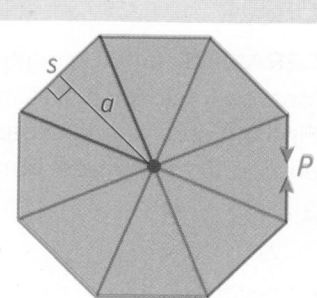

Today's Goals
• Find areas of regular polygons.
• Find areas of composite figures.

Today's Vocabulary
center of a regular polygon
radius of a regular polygon
apothem
central angle of a regular polygon
composite figure

Example 1 Identify Segments and Angles in Regular Polygons

In the figure, regular hexagon *JKLMNP* is inscribed in ⊙*R*. Identify the center, a radius, an apothem, and a central angle of the polygon. Then find the measure of a central angle.

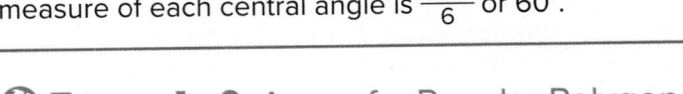

The center of the regular hexagon is point *R*.
A radius of the regular hexagon is $\overline{RK}$.
An apothem of the regular hexagon is *RS*. A central angle of the regular hexagon is ∠*KRL*. A regular hexagon has 6 sides. Thus, the measure of each central angle is $\frac{360°}{6}$ or 60°.

🌐 Example 2 Area of a Regular Polygon

PATCHES Lindsay created a patch for the robotics club at her school. The patch is a regular octagon with a side length of 4 centimeters. Find the area covered by the patch. Round to the nearest tenth.

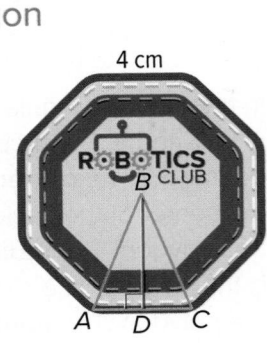

4 cm

Step 1 Find the measure of a central angle.

A regular octagon has 8 congruent central angles, so $m\angle ABC = \frac{360°}{8}$ or 45°.

Step 2 Find the length of the apothem.
Apothem $\overline{BD}$ is the height of isosceles △*ABC*.
It bisects ∠*ABC*, so $m\angle DBC = 45° \div 2$ or 22.5°.
It also bisects $\overline{AC}$, so *DC* = 2 centimeters. Use trigonometric ratios to find the length of the apothem.

$$\tan 22.5° = \frac{2}{a} \qquad\qquad \tan \theta = \frac{\text{opp}}{\text{adj}}$$

$$a = \frac{2}{\tan 22.5°} \qquad \text{Solve for } a.$$

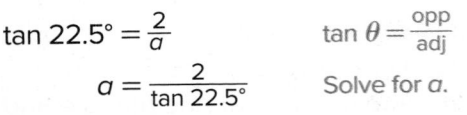

22.5°

2 cm

Step 3 Use the formula for the area of a regular polygon.

$$A = \frac{1}{2}aP \qquad\qquad \text{Area of a regular polygon}$$

$$= \frac{1}{2}\left(\frac{2}{\tan 22.5°}\right)(32) \qquad a = \frac{2}{\tan 22.5°} \text{ and } P = 32$$

$$\approx 77.25 \qquad\qquad \text{Solve.}$$

The patch covers an area of about 77.25 square centimeters.

Check

CERAMICS Imani is crafting coasters for a local craft fair. Each side measures 4 inches. Find the area of each coaster. Round your answer to the nearest tenth, if necessary.

$$A \approx \underline{\ ?\ } \text{ in}^2$$

🔵 **Go Online** You can complete an Extra Example online.

🫧 Think About It!

James calculated the area of the trampoline. His calculations are shown.

$$A = \frac{1}{2}aP$$

$$\approx \frac{1}{2}(4)(27.3)$$

$$\approx 54.6 \text{ ft}$$

Do you agree? Justify your argument.

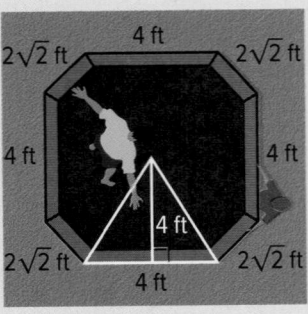

2√2 ft   4 ft   2√2 ft
4 ft   4 ft
4 ft
2√2 ft   4 ft   2√2 ft

Learn Areas of Composite Figures

A **composite figure** is a figure that can be separated into regions that are basic figures, such as triangles, rectangles, trapezoids, and circles. To find the area of a composite figure, find the area of each basic figure and then use the Area Addition Postulate.

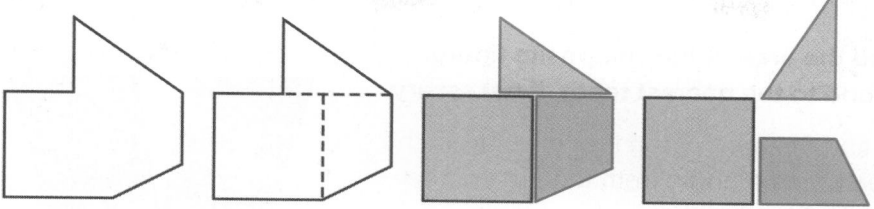

Go Online

You can watch a video to see how to find the area of a composite figure on the coordinate plane.

Example 3 Find the Area of a Composite Figure by Adding

Find the area of the composite figure.

Step 1 Separate the composite figure.

The figure shown is composed of a square, regular hexagon, and trapezoid.

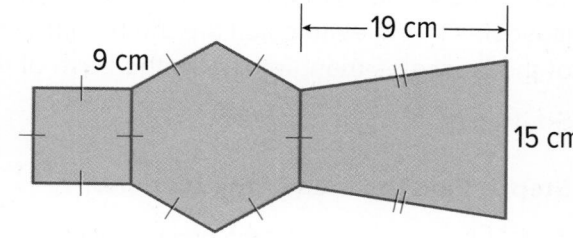

So, area of figure = area of square + area of hexagon + area of trapezoid.

Step 2 Calculate the area of each basic figure.

$A = s^2$	Area of a square
$= 9^2$	$s = 9$
$= 81 \text{ cm}^2$	Solve.
$A = \frac{1}{2}aP$	Area of a regular polygon
$= \frac{1}{2}(4.5\sqrt{3})(54)$	$a = 4.5\sqrt{3}$ and $P = 54$
$\approx 210.4 \text{ cm}^2$	Simplify.
$A = \frac{1}{2}h(b_1 + b_2)$	Area of a trapezoid
$= \frac{1}{2}(19)(15 + 9)$	$h = 19$, $b_1 = 15$, and $b_2 = 9$
$= 228 \text{ cm}^2$	Solve.

Step 3 Calculate the total area of the composite figure.

$$\text{area of figure} = \text{area of square} + \text{area of hexagon} + \text{area of trapezoid}$$
$$\approx 81 + 210.4 + 228$$
$$\approx 519.4 \text{ cm}^2$$

Check

Find the area of the composite figure. Round to the nearest tenth, if necessary.

$A = \underline{\quad ? \quad} \text{ yd}^2$

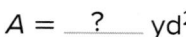

Go Online You can complete an Extra Example online.

The areas of some figures can be found by subtracting the areas of basic figures.

Example 4 Find the Area of a Composite Figure by Subtracting

Find the area of the composite figure. Round to the nearest tenth, if necessary.

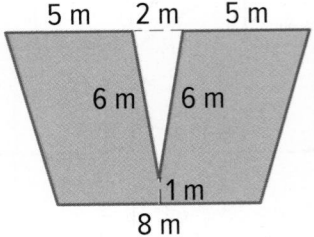

To find the area of the figure, subtract the area of the triangle from the area of the trapezoid.

Step 1 Find the height of the triangle.

You can use the Pythagorean Theorem and what you know about isosceles triangles to calculate the height of the triangle. The altitude of the isosceles triangle bisects the base of the triangle.

So, $1^2 + h^2 = 6^2$ or $h = \sqrt{35}$.

Step 2 Find the area of the triangle.

$A = \frac{1}{2}bh$ Area of a triangle

$= \frac{1}{2}(2)(\sqrt{35})$ $b = 2$ and $h = \sqrt{35}$

$= \sqrt{35}$ m^2 Solve.

Step 3 Find the area of the trapezoid.

$A = \frac{1}{2}h(b_1 + b_2)$ Area of a trapezoid

$= \frac{1}{2}(\sqrt{35} + 1)(12 + 8)$ $b_1 = 12$, $b_2 = 8$, and $h = \sqrt{35} + 1$

$= 10\sqrt{35} + 10$ m^2 Solve.

Step 4 Find the area of the figure.

area of figure = area of trapezoid − area of triangle

$= 10\sqrt{35} + 10 - \sqrt{35}$

≈ 63.2 m^2

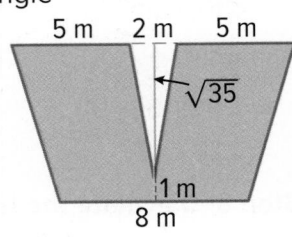

Check

Select the area of the composite figure.

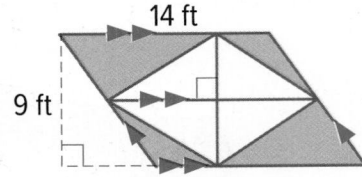

A. 31.5 ft^2 B. 63 ft^2 C. 94.5 ft^2 D. 126 ft^2

🔵 **Go Online** You can complete an Extra Example online.

Drawing Figures
To solve some area problems by subtracting, you may need to draw figures to represent the basic shapes that are being removed from the composite figure. You can use the figures you draw to help you visualize the situation and calculate missing measures.

Practice

Go Online You can complete your homework online.

Example 1

In each figure, a regular polygon is inscribed in a circle. Identify the center, a radius, an apothem, and a central angle of each polygon. Then find the measure of a central angle.

1.

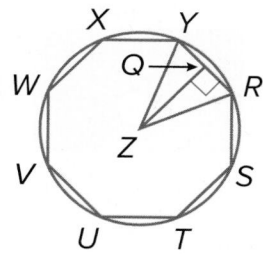

2.

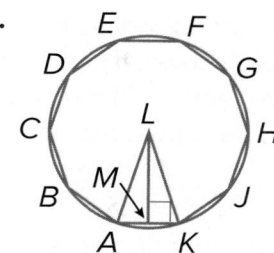

Example 2

Find the area of each regular polygon. Round to the nearest tenth.

3.

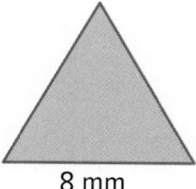

8 mm

4.

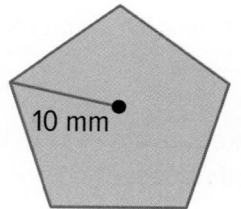

10 mm

5.

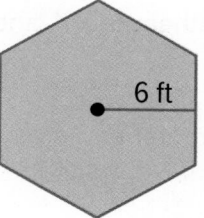

6 ft

6. COINS The Susan B. Anthony dollar coin has a hendecagon (11-gon) inscribed in a circle in its design. Each edge of the hendecagon is approximately 7.46 millimeters. What is the area of this regular polygon? Round to the nearest hundredth.

7. RECREATION A regular octagonal trampoline has a diameter of 16 feet. What is the area of the surface of the trampoline? Round to the nearest tenth.

Examples 3 and 4

Find the area of each figure. Round to the nearest tenth, if necessary.

8.

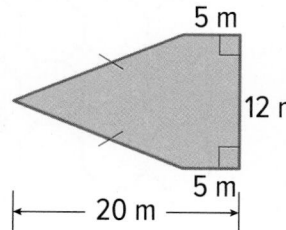

5 m
12 m
5 m
20 m

9.

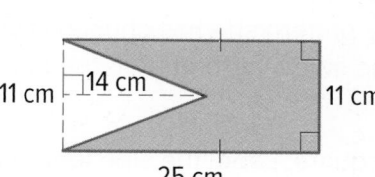

11 cm
14 cm
11 cm
25 cm

10.

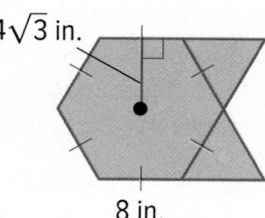

$4\sqrt{3}$ in.
8 in.

11.

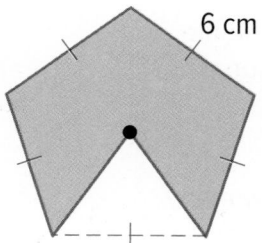

6 cm

12.

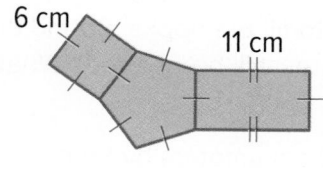

6 cm
11 cm

13.

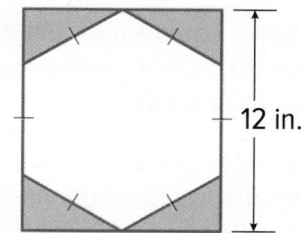

12 in.

Mixed Exercises

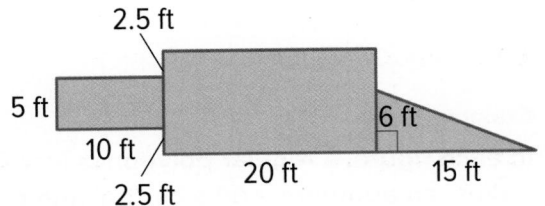

14. LAWN Lalita has to buy grass seed for her lawn. Her lawn is in the shape of the composite figure shown. What is the area of the lawn?

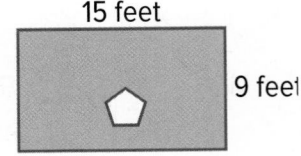

15. PATIO Chenoa is building a patio to surround his fire pit. The patio is in the shape of the composite figure shown. If each side of the fire pit is 2 feet long, how many square feet of patio pavers will Chenoa need to buy to complete the patio? Round to the nearest tenth.

16. Find the area of a regular hexagon with a perimeter of 72 inches. Round to the nearest square inch.

17. USE TOOLS Draw a regular pentagon inscribed in a circle with center X, a radius of $\overline{XV}$, an apothem of $\overline{XY}$, and a central angle of $\angle VXT$ that measures 72°.

18. Find the perimeter and area of a regular hexagon with a side length of 12 centimeters. Round to the nearest tenth, if necessary.

19. Find the total area of the shaded regions. Assume each figure is a regular polygon. Round to the nearest tenth, if necessary.

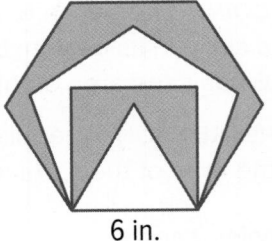

20. HOME IMPROVEMENT Pilar is putting a backsplash in her kitchen made up of octagon and square tiles. The pattern she has chosen is sold in sheets of 4 × 3 octagons as shown. The side length of both the squares and the octagons is 3 centimeters.

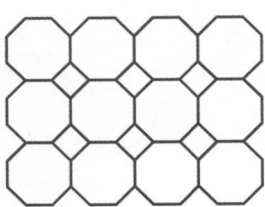

a. Draw one octagon and one square. Label the side length and area of each and the apothem of the octagon. Find the area of one tile sheet to the nearest whole number.

b. Small square tiles will be placed to fill the nooks around the edges of the tile sheet. How many complete squares will be needed? What is the additional area of these tiles?

c. If Pilar's backsplash measures 42 centimeters by 84 centimeters, approximately how many sheets of tile will Pilar need to purchase? How many extra square tiles will need to be purchased?

21. **REASONING** Miguel is planning to renovate his living room. Refer to the given floor plan.

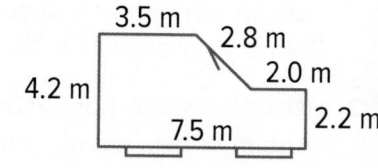

a. How much varnish will he need to stain the hardwood floor? Assume 1 liter of finish covers 4.5 square meters and round to the nearest tenth.

b. The height of the room is 2.6 meters. Approximate how much paint is needed for the walls. Assume that 1 liter of paint covers 7.5 square meters and round to the nearest tenth.

c. Why might Miguel adjust your estimates when he purchases materials?

22. **STATE YOUR ASSUMPTION** Chilam is going to build a shelf for his 15 homerun baseballs. He wants the shelf to be in the shape of a regular triangle. Find the area of the shelf he must create. State any assumptions you make. Round your answer to the nearest tenth.

23. **USE A SOURCE** Research the shape and dimensions of the Pentagon building in Arlington, Virginia.

a. Draw and label a diagram of the Pentagon including the courtyard.

b. Use your diagram in part **a** to find the area in square feet of the Pentagon. Round your answer to the nearest tenth.

24. **REGULARITY** Explain how using the formula for the area of a regular polygon is related to finding the area of a triangle.

25. **SIGNS** A stop sign has side lengths of approximately 12.5 inches. What is the area of a regular octagonal stop sign? Round your answer to the nearest tenth.

26. **DIAMONDS** Mr. Figueroa has bought his wife an anniversary ring with a regular heptagon diamond as shown. If each side of the diamond is 6 millimeters long, what is the area of the face of the diamond?

27. **ARCHITECTURE** Fort Jefferson in the Florida Keys is the largest brick masonry building in the Americas. The Fort was built in the shape of a hexagon with side lengths of 477 feet. What is the area of the hexagon formed by the exterior walls of the fort? Round your answer to the nearest tenth.

28. **FRAME** Darren is hanging a picture frame that is shaped like a regular pentagon with outside lengths of 9 inches. What is the area this picture frame will take up on Darren's wall? Round your answer to the nearest tenth.

29. **FLOWER** Delfina photographed a triangular flower for a photo contest. The flower was approximately regular and had side lengths measuring about 4 centimeters. What is the area of the flower? Round your answer to the nearest tenth.

30. **STAINED GLASS** Kenia makes regular octagonal-shaped stained glass windows. If the apothem of the window is 14 inches, what is the area of the window? Round your answer to the nearest tenth.

31. **GAMING** A 12-sided gaming die is made up of 12 regular pentagons, each with a side length of 1.5 centimeters. What is the area of one face of the gaming die? Round your answer to the nearest hundredth.

32. **HONEYCOMB** A honeycomb is a structure bees make of hexagonal wax cells to contain honey, larvae, and pollen. The height of one hexagon in a honeycomb is approximately 5 millimeters. What is the area of one hexagon? Round your answer to the nearest tenth.

Higher-Order Thinking Skills

33. **FIND THE ERROR** Chenglei and Flavio want to find the area of the hexagon shown. Who is correct? Explain your reasoning.

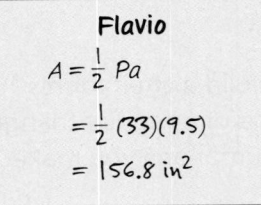
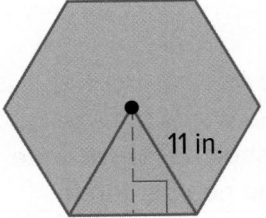

34. **ANALYZE** Is the measure of an apothem of a regular polygon *sometimes*, *always*, or *never* $\frac{\sqrt{3}}{2}s$ where s is a side length of the polygon? Justify your argument.

35. **CREATE** Draw a pair of composite figures that have the same area. Make one composite figure out of a rectangle and a trapezoid, and make the other composite figure out of a triangle and a rectangle. Show the area of each basic figure.

36. **PERSEVERE** Consider the sequence of area diagrams shown. What algebraic theorem do the diagrams prove? Explain your reasoning.

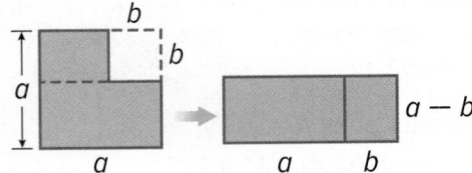

37. **WRITE** How can you find the area of any figure?

38. **WHICH ONE DOESN'T BELONG?** Alden drew the following diagrams to find the area of 4 regular polygons. Which drawing is incorrect. Justify your conclusion.

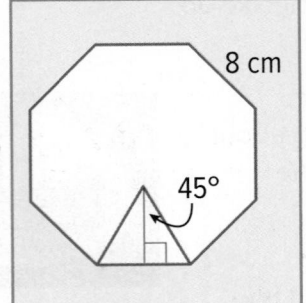
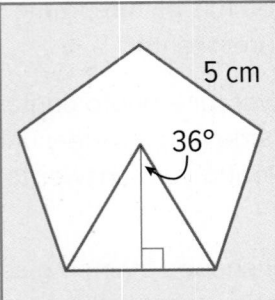
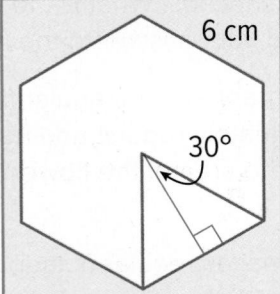
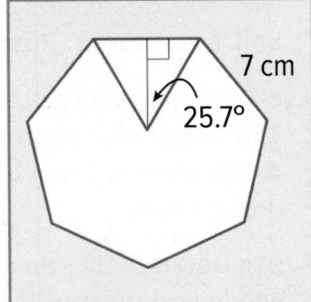

Areas of Circles and Sectors

Explore Areas of Circles

Online Activity Use dynamic geometry software to complete the Explore.

> **INQUIRY** How is the formula for the area of a circle related to the formula for the area of a regular polygon?　×

Learn Areas of Circles

The formula for the circumference C of a circle with radius r is given by $C = 2\pi r$. You can use this formula to develop the formula for the area of a circle.

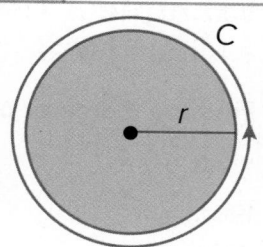

Key Concept • Area of a Circle	
Words	The area A of a circle is equal to π times the square of the radius r.
Symbols	$A = \pi r^2$
Example	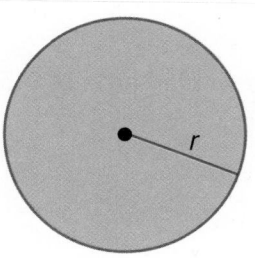

🌎 Example 1 Area of a Circle

PATIO Keon is building a circular patio in his backyard.

Part A What is the area of the patio to the nearest square foot?

The diameter of the circle is 24 feet, so the radius is 12 feet.

$$A = \pi r^2 \qquad \text{Area of a circle}$$
$$= \pi(12)^2 \qquad r = 12$$
$$= 144\pi \qquad \text{Simplify.}$$

So, the area is 144π, or about 452 square feet.

🔘 **Go Online** You can complete an Extra Example online.

Today's Goals
- Find areas of circles.
- Find areas of sectors of circles.

Today's Vocabulary
sector

🔘 **Go Online**
A derivation of the formula for the area of a circle is available.

🔘 **Go Online**
You can watch a video to see how to find the area and circumference of a circle.

Study Tip

Units of Area
Remember that when you convert between units of area that you have to multiply or divide by the square of the conversion for the units of length. For example, 1 yd = 3 ft, so 1 yd^2 = 1 yd • 1 yd = 3 ft • 3 ft or 9 ft^2.

Part B Keon can purchase paving stones for $135 per square yard. About how much will he spend on paving stones for his patio?

Keon measured the area of the patio in square feet, but the price for paving stones is given in square yards, so Keon will have to convert from square feet to square yards to find how much he will spend.

$(144\pi \text{ ft}^2)\left(\dfrac{\text{yd}^2}{9\text{ ft}^2}\right) \approx 16\pi \text{ yd}^2$ $1\text{ yd}^2 = 9\text{ ft}^2$

$(16\pi \text{ yd}^2)\left(\dfrac{\$135}{1\text{ yd}^2}\right) \approx \$6786$ Solve.

So, Keon could expect to spend about $6786 on paving stones.

Example 2 Use the Area of a Circle to Find a Missing Measure

Find the diameter of a circle with an area of 196π square yards.

$A = \pi r^2$	Area of a circle
$196\pi = \pi r^2$	$A = 196\pi$
$\dfrac{196\pi}{\pi} = r^2$	Divide each side by π.
$14 = r$	Solve. Take the positive square root.

The radius is half the diameter, so the diameter of the circle is 28 yards.

Check

Find the diameter of a circle with an area of 915 square feet. Round to the nearest tenth, if necessary.

_____?_____ ft

Pause and Reflect

Did you struggle with anything in this lesson? If so, how did you deal with it?

Go Online You can complete an Extra Example online.

Learn Areas of Sectors

A **sector** is a region of a circle bounded by a central angle and its intercepted arc. The formula for the area of a sector is similar to the formula for arc length.

Key Concept • Area of a Sector

The ratio of the area A of a sector to the area of the whole circle πr^2, is equal to the ratio of the degree measure of the intercepted arc x to 360°.

Proportion: $\frac{A}{\pi r^2} = \frac{x°}{360°}$

Equation: $A = \frac{x°}{360°} \cdot \pi r^2$

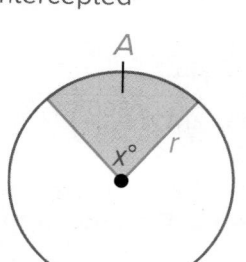

Go Online
A derivation of the formula for the area of a sector is available.

Go Online
You can watch a video to see how to find the area of a sector of a circle.

Think About It!
What assumptions did you make?

Example 3 Area of a Sector

GAMES **Malaya is playing a game where she must track her progress using sectors of a circle and a circular game piece. The game piece has a diameter of 3.5 centimeters and is divided into 6 congruent sectors. What is the area of one sector to the nearest hundredth?**

Step 1 Find the arc measure.

Because the game piece is equally divided into 6 pieces, each piece will have an arc measure of 360° ÷ 6 = 60°.

Step 2 Find the area.

$A = \frac{x°}{360°} \cdot \pi r^2$ Area of a sector

$= \frac{60°}{360°} \cdot \pi (1.75)^2$ $x = 60$ and $r = 1.75$

≈ 1.60 Solve.

So, the area of one sector of this game piece is about 1.60 square centimeters.

Check

ART Jorrie is crafting a wall clock using paint and a set of battery-operated clock hands. The clock will have a diameter of 3 feet. What is the area of each sector to the nearest hundredth?

$A \approx$ _____?_____ ft²

Go Online You can complete an Extra Example online.

McGraw-Hill Education

Example 4 Use the Area of a Sector to Find the Area of a Circle

One sector of a circle has an area of 42 square feet and an arc measure of 45°. Find the area of the circle to the nearest square foot.

Step 1 Find the radius.

Use the area of the sector to find the radius of the circle.

$A = \frac{x^\circ}{360^\circ} \cdot \pi r^2$ Area of a sector

$42 = \frac{45^\circ}{360^\circ} \cdot \pi r^2$ $A = 42$ and $x = 45$

$r = \sqrt{42\left(\frac{360}{45}\right)\frac{1}{\pi}}$ Solve for r.

Step 2 Find the area.

Use the radius to calculate the area of the circle.

$A = \pi r^2$ Area of a circle

$= \pi\left(\sqrt{42\left(\frac{360}{45}\right)\frac{1}{\pi}}\right)^2$ Substitute.

$= 336$ Solve.

The area of the circle is 336 square feet.

Check

One sector of a circle has an area of 860 square inches and an arc measure of 60°. Zari found the area of the circle. Provide the justifications for her calculations.

$A = \frac{x^\circ}{360^\circ} \cdot \pi r^2$

$860 = \frac{60^\circ}{360^\circ} \cdot \pi r^2$

$r = \sqrt{860\left(\frac{360}{60}\right)\frac{1}{\pi}}$

$A = \pi r^2$

$= \pi\left(\sqrt{860\left(\frac{360}{60}\right)\frac{1}{\pi}}\right)^2$

$= 5160 \text{ in}^2$

Go Online
to practice what you've learned about measures of two-dimensional figures in the Put It All Together over Lessons 6-1 through 6-3.

Go Online You can complete an Extra Example online.

Practice

Go Online You can complete your homework online.

Example 1

Find the area of each circle. Round to the nearest tenth.

1.
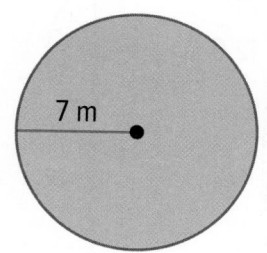
7 m

2.
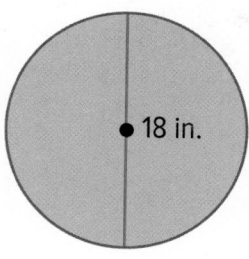
18 in.

3.
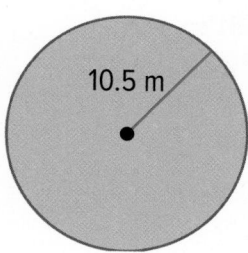
10.5 m

4. **DINING** Maricela is making a tablecloth for a circular table that has a diameter of 8 feet.

   a. Find the area of the tabletop. Round your answer to the nearest tenth.

   b. If a square yard of fabric costs $13.99, what is the minimum Maricela will need to spend to make the tablecloth?

5. **GAMES** Kiyoshi is making circular tiles to display houses for his role-playing game. Each tile has a radius of 2 inches and is being made out of balsa wood that costs $1.99 per square foot.

   a. Find the area of a single tile. Round your answer to the nearest tenth.

   b. How much will it cost Kiyoshi to make 30 tiles?

6. **PORTHOLES** A circular window on a ship is designed with a radius of 8 inches. What is the area of glass needed for the window? Round your answer to the nearest hundredth.

Example 2

Find the indicated measure. Round to the nearest tenth.

7. Find the diameter of a circle with an area of 94 square millimeters.

8. The area of a circle is 132.7 square centimeters. Find the diameter of the circle.

9. The area of a circle is 112 square inches. Find the radius of the circle.

10. Find the diameter of a circle with an area of 1134.1 square millimeters.

11. The area of a circle is 706.9 square inches. Find the radius of the circle.

12. Find the radius of a circle with an area of 2827.4 square feet.

Example 3

Find the area of each shaded sector. Round to the nearest tenth.

13.

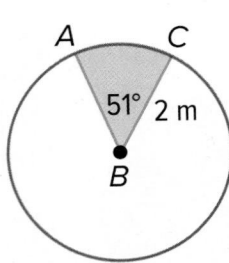

14.

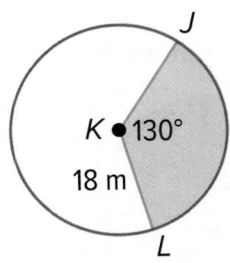

15.

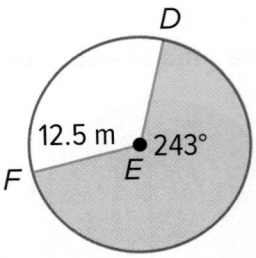

16. SPINNERS Jeremy wants to make a spinner for a new board game he invented. The spinner is a circle divided into 8 congruent pieces. What is the area of each piece to the nearest tenth?

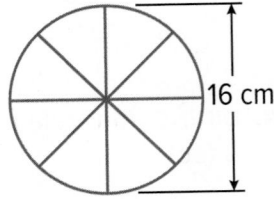

Example 4

Find the area of the circle that contains each sector. Round to the nearest tenth, if necessary.

17. One sector has an area of 210 square centimeters and an arc measure of 30°.

18. One sector has an area of 65 square feet and an arc measure of 270°.

19. One sector has an area of 325 square millimeters and an arc measure of 72°.

20. One sector has an area of 167 square inches and an arc measure of 110°.

21. One sector has an area of 98 square meters and an arc measure of 40°.

22. One sector has an area of 412 square inches and an arc measure of 82°.

Mixed Exercises

23. LOBBY The lobby of a bank features a large marble circular table for displaying brochures.

   a. The diameter of the table is 15 feet. What is the area of the circular table? Round your answer to the nearest tenth.

   b. If the bank manager adds a circular floral arrangement with a diameter of 2 feet, how much space remains for brochure displays? Round your answer to the nearest tenth

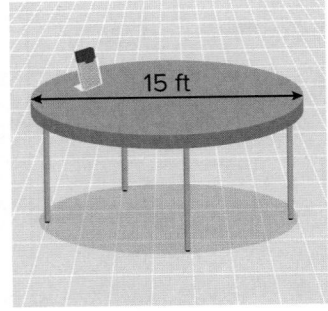

24. STRUCTURE A stained-glass artist is making a circle separated into 3 equal sectors with the bottom sector divided equally in two. Suppose the circle has radius r. What is the area of each of the larger equal sectors?

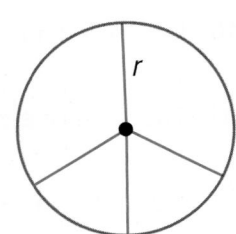

25. SOUP CAN Jaclynn needs to cover the top and bottom of a can of soup with construction paper to include in her art project. Each circle has a diameter of 7.5 centimeters. What is the total area of the can that Jaclynn must cover? Round your answer to the nearest tenth.

26. POOL A circular pool is surrounded by a circular sidewalk. The circular sidewalk is 3 feet wide. The diameter of the sidewalk and the pool is 26 feet.

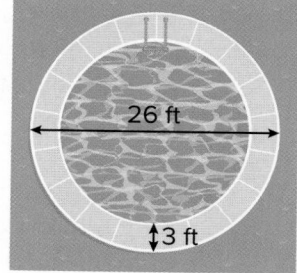

 a. What is the diameter of the pool?

 b. What is the area of the sidewalk and the pool?

 c. What is the area of the pool?

27. REASONING Explain how to find the area of the shaded region if the hexagon is regular. Then find the area of the shaded region. Round to the nearest tenth, if necessary.

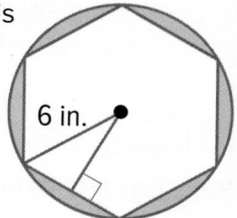

28. REGULARITY A sector of a circle has an area A and a central angle that measures $x°$. Explain how you can find the area of the whole circle.

29. PIE One sector of an apple pie has an area of 8 square inches and an arc measure of 45°. Find the area of the pie.

30. USE ESTIMATION A lawn sprinkler sprays water in an arc as shown in the picture. The area of the sector covered by the spray is approximately 235.6 square feet.

 a. Estimate the area of the yard that can be covered by the sprinkler if the sprinkler is set to water a complete circle.

 b. Find the exact area covered by the sprinkler.

 c. Is your answer reasonable? Justify your argument.

31. PRECISION Luciano wants to use sectors of different circles as part of a mural he is painting. The specifications for each sector are shown.

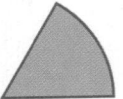

Red:
radius: 14 ft
central angle: 60°

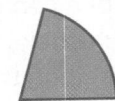

Purple:
radius: 12 ft
central angle: 75°

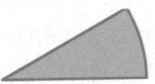

Green:
radius: 18 ft
central angle: 30°

a. Find the area of each sector to the nearest tenth.

b. Was the sector with the largest area the one with the longest radius? Was it the one with the largest central angle? What can you conclude from this?

c. Luciano plans to paint 235 stars on each of the purple sectors and 153 stars on each of the green sectors. To the nearest tenth, how many stars are there per square foot for sectors of each color?

d. Luciano also plans to paint stars on the red sectors. He wants there to be twice as many stars per square foot in the red sectors as there are in the green sectors. How many stars should he paint on each red sector? Round to the nearest whole star.

🌐 Higher-Order Thinking Skills

32. FIND THE ERROR Ketria and Colton want to find the area of a shaded region in the circle shown. Who is correct? Explain your reasoning.

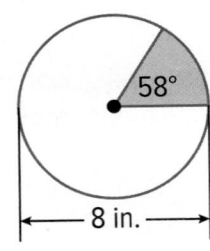

58°

8 in.

Ketria	Colton
$A = \dfrac{x}{360} \cdot \pi r^2$	$A = \dfrac{x}{360} \cdot \pi r^2$
$= \dfrac{58}{360} \cdot \pi(8)^2$	$= \dfrac{58}{360} \cdot \pi(4)^2$
$= 32.4 \text{ in}^2$	$= 8.1 \text{ in}^2$

33. PERSEVERE Find the area of the shaded region. Round to the nearest tenth.

0.5 cm

160°

35 cm

34. ANALYZE A **segment of a circle** is the region bounded by an arc and a chord. Is the area of a sector of a circle *sometimes*, *always*, or *never* greater than the area of its corresponding segment? Justify your argument.

35. WRITE Describe two methods you could use to find the area of the shaded region of the circle. Which method do you think is more efficient? Explain your reasoning.

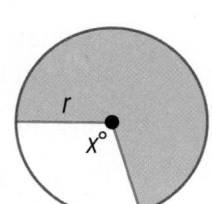

r

x°

36. PERSEVERE Derive the formula for the area of a sector of a circle using the formula for arc length.

37. CREATE Draw a circle with a shaded sector. Label the length of the radius and the measure of the arc intercepted by the sector. Then find the area of the shaded sector.

Surface Area

Learn Surface Areas of Prisms and Cylinders

The **lateral area** L of a prism is the sum of the areas of the **lateral faces**, which are the faces that join the bases of a solid. The **altitude of a prism or cylinder** is the segment perpendicular to the bases that joins the planes of the bases. The height of each solid is the length of the altitude.

Key Concept • Lateral Area of a Right Prism

Words	The lateral area L of a right prism is $L = Ph$, where h is the height of the prism and P is the perimeter of a base.
Symbols	$L = Ph$
Example	

Key Concept • Lateral Area of a Right Cylinder

Words	The lateral area L of a right cylinder is $L = 2\pi rh$, where r is the radius of a base and h is the height of the cylinder.
Symbols	$L = 2\pi rh$
Example	

Key Concept • Surface Area of a Right Prism

Words	The surface area S of a right prism is $S = L + 2B$, where L is the lateral area and B is the area of a base. By substituting $L = Ph$ into the equation, $S = Ph + 2B$, where P is the perimeter of a base and h is the height of the prism.
Symbols	$S = L + 2B$ or $S = Ph + 2B$

Key Concept • Surface Area of a Right Cylinder

Words	The surface area S of a right cylinder is $S = L + 2B$, where L is the lateral area and B is the area of a base. By substituting $L = 2\pi rh$ and $B = \pi r^2$ into the equation, $S = 2\pi rh + 2\pi r^2$, where r is the radius of a base and h is the height of the cylinder.
Symbols	$S = L + 2B$ or $S = 2\pi rh + 2\pi r^2$

Today's Goals
- Find surface areas of prisms and cylinders.
- Find surface areas of pyramids and cones.
- Find surface areas of spheres.

Today's Vocabulary
lateral area
lateral faces
altitude of a prism or cylinder
regular pyramid
altitude of a pyramid or cone
composite solid

Study Tip

Right Solids and Oblique Solids
A solid is a *right solid* if the segment connecting the centers of the bases (for prisms and cylinders) or the center of the base to the vertex (for pyramids and cones) is perpendicular to the base(s). A solid is an *oblique solid* if it is not a right solid.

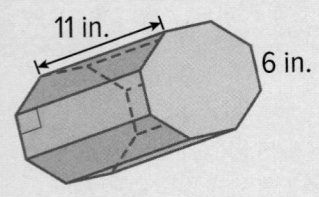

11 in.

6 in.

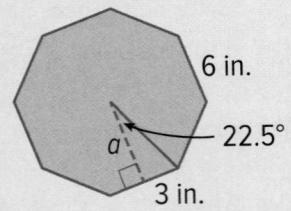

6 in.

22.5°

a

3 in.

Problem-Solving Tip

Use Your Skills To find the lateral area or surface area of a solid, you may have to use other skills to find missing measures. Remember that you can use trigonometric ratios or the Pythagorean Theorem to solve for missing measures in right triangles.

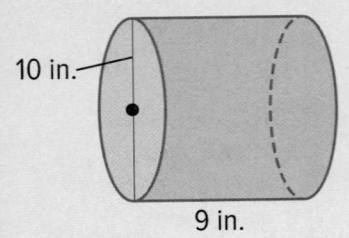

10 in.

9 in.

Go Online

You can complete an Extra Example online.

Example 1 Lateral Area and Surface Area of a Prism

Find the lateral area and surface area of the prism. Round to the nearest tenth, if necessary.

Part A Find the lateral area.

$L = Ph$	Lateral area of a prism
$= (8 \times 6)11$	The base has 8 sides.
$= 528$	Solve.

So, the lateral area of the prism is 528 square inches.

Part B Find the surface area.
Step 1 Find the area of the base.
A central angle of the octagon is $\frac{360°}{8}$ or 45°, so the angle formed in the triangle is 22.5°.

$\tan 22.5° = \frac{3}{a}$	Write a trigonometric ratio to find a.
$a = \frac{3}{\tan 22.5°}$	Solve for a.

$A = \frac{1}{2}Pa$	Area of a regular polygon
$= \frac{1}{2}(48)\left(\frac{3}{\tan 22.5\pi°}\right)$	$P = 48$ and $a \approx = \frac{3}{\tan 22.5\pi°}$
≈ 173.8	Multiply.

So, the area of the base B is approximately 173.8 square inches.

Step 2 Find the surface area of the prism.

$S = L + 2B$	Surface area of a right prism
$\approx 528 + 2(173.8)$	$L = 528$ and $B \approx 173.8$
≈ 875.6	Simplify.

The surface area of the prism is about 875.6 square inches.

Example 2 Lateral Area and Surface Area of a Cylinder

Find the lateral area and surface area of the cylinder. Round to the nearest tenth.

Part A Find the lateral area.

$L = 2\pi rh$	Lateral area of a cylinder
$= 2\pi(5)(9)$	Substitution
≈ 282.7	Solve.

So, the lateral area of the cylinder is 282.7 square inches.

Part B Find the surface area.

$S = 2\pi rh + 2\pi r^2$	Surface area of a cylinder
$\approx 282.7 + 2\pi(5)^2$	Substitute.
≈ 439.8	Solve.

The surface area of the cylinder is about 439.8 square inches.

Go Online to see Example 3.

Explore Cone Patterns

Online Activity Use paper folding to complete the Explore.

> **INQUIRY** How are the formulas for the circumference and area of a circle related to the formula for the surface area of a right cone? ×

Learn Surface Areas of Pyramids and Cones

A **regular pyramid** is a pyramid with a base that is a regular polygon. The height of a pyramid or cone is the length of the **altitude of the pyramid or cone**, which is the segment perpendicular to the base that has the vertex as one endpoint and a point in the plane of the base as the other endpoint.

Talk About It!

Describe the similarities and differences between finding the lateral area of a pyramid and the lateral area of a prism.

Key Concept • Lateral Area of a Regular Pyramid

Words	The lateral area L of a regular pyramid is $L = \frac{1}{2}P\ell$, where ℓ is the slant height and P is the perimeter of the base.	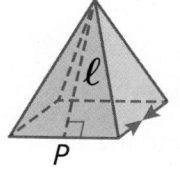
Symbols	$L = \frac{1}{2}P\ell$	

Key Concept • Lateral Area of a Right Cone

Words	The lateral area L of a right circular cone is $L = \pi r \ell$, where r is the radius of the base and ℓ is the slant height.	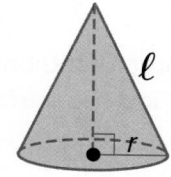
Symbols	$L = \pi r \ell$	

Key Concept • Surface Area of a Regular Pyramid

Words	The surface area S of a regular pyramid is $S = L + B$, where L is the lateral area and B is the area of the base. By substituting $L = \frac{1}{2}P\ell$, into the equation, $S = \frac{1}{2}P\ell + B$, where P is the perimeter of the base and ℓ is the slant height.	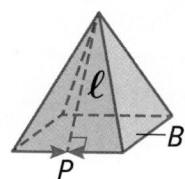
Symbols	$S = L + B$ or $S = \frac{1}{2}P\ell + B$	

Key Concept • Surface Area of a Right Cone

Words	The surface area S of a right circular cone is $S = L + B$, where L is the lateral area and B is the area of the base. By substituting $L = \pi r \ell$ and $B = \pi r^2$ into the equation, $S = \pi r \ell + \pi r^2$, where r is the radius of the base and ℓ is the slant height.	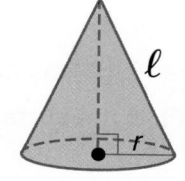
Symbols	$S = L + B$ or $S = \pi r \ell + \pi r^2$	

Example 4 Lateral Area and Surface Area of a Regular Pyramid

Find the lateral area and surface area of the regular pyramid. Round to the nearest tenth, if necessary.

Part A Find the lateral area.

Step 1 Find the perimeter of the base.
The perimeter of the base is 6×3 or 18 feet.

Step 2 Find the lateral area of the pyramid.

$$L = \tfrac{1}{2}P\ell \qquad \text{Lateral area of a regular pyramid}$$
$$= \tfrac{1}{2}(18)(8) \qquad P = 18 \text{ and } \ell = 8$$
$$= 72 \text{ ft}^2 \qquad \text{Solve.}$$

Part B Find the surface area.

Step 1 Find the length of the height of the triangular base and the area of the base.

$$a^2 + b^2 = c^2 \qquad \text{Pythagorean Theorem}$$
$$a^2 = 6^2 - 3^2 \qquad c = 6 \text{ and } b = 3$$
$$a = \sqrt{27} \qquad \text{Solve.}$$

Calculate the area of the base B.

$$B = \tfrac{1}{2}bh \qquad \text{Area of a triangle}$$
$$= \tfrac{1}{2}(6)\sqrt{27} \qquad \text{Substitute.}$$
$$\approx 15.6 \qquad \text{Solve.}$$

So, the area of the base B is approximately 15.6 square feet.

Step 2 Find the surface area of the pyramid.

$$S = L + B \qquad \text{Surface area of a regular pyramid}$$
$$\approx 72 + 15.6 \qquad L = 72 \text{ and } B \approx 15.6$$
$$\approx 87.6 \qquad \text{Simplify.}$$

The surface area of the pyramid is about 87.6 square feet.

Example 5 Lateral Area and Surface Area of a Right Cone

Find the lateral area and surface area of the cone rounded to the nearest tenth.

Part A Find the lateral area.

$$L = \pi r\ell \qquad \text{Lateral area of a cone}$$
$$= \pi(0.8)(2.2) \qquad \text{Substitution}$$
$$\approx 5.5 \qquad \text{Solve.}$$

So, the lateral area of the cone is 5.5 square millimeters.

Part B Find the surface area.

$$S = \pi r\ell + \pi r^2 \qquad \text{Surface area of a cone}$$
$$\approx 5.5 + \pi(0.8)^2 \qquad \text{Substitution}$$
$$\approx 7.5 \qquad \text{Solve.}$$

The surface area of the cone is about 7.5 square millimeters.

 Go Online You can complete an Extra Example online.

🌐 Example 6 Approximate Surface Areas of Pyramids and Cones

AGRICULTURE Specialty watermelons are carefully cultivated in containers so that the fruits form different geometric shapes. Curt wants to build molds to grow watermelons shaped like triangular pyramids.

Part A Find the surface area of the mold in terms of x.

The watermelon can be modeled by using a regular triangular pyramid. The perimeter of the base of the pyramid is $6x$. The slant height of the pyramid is $4x - 4$.

Calculate the area of the triangular base of the pyramid, $A = \frac{1}{2}bh$.

The height of the triangular base is $h = x\sqrt{3}$ and the length is $b = 2x$. Substitute these values into the formula for the area of a triangle to find the area of the base of the pyramid. $B = \frac{1}{2}(2x)(\sqrt{3}x)$ or $\sqrt{3}x^2$

Substitute the values for P, ℓ, and B into the formula for the surface area of a pyramid.

$$S = \frac{1}{2}P\ell + B$$
$$= (12 + \sqrt{3})x^2 - 12x$$

Part B If x = 6, approximate the surface area of the watermelon.

$S = (12 + \sqrt{3})x^2 - 12x$ Surface area from Part A

$\quad = (12 + \sqrt{3})6^2 - 12(6)$ $x = 6$

$\quad \approx 422.4$ Solve.

The surface area of the watermelon is about 422.4 square inches.

Learn Surface Areas of Spheres

A sphere is the set of all points in space that are a given distance from a given point called the center of the sphere.

Key Concept • Surface Area of a Sphere	
Words	The surface area S of a sphere is $S = 4\pi r^2$, where r is the radius.
Symbols	$S = 4\pi r^2$

Example 7 Surface Area of a Sphere

Find the surface area of the sphere to the nearest tenth.

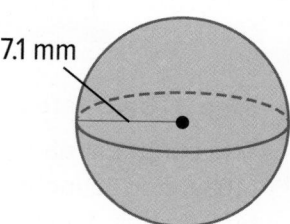

7.1 mm

$S = 4\pi r^2$ Surface area of a sphere

$\quad = 4\pi(7.1)^2$ Substitution

$\quad \approx 633.5$ Solve.

The surface area of the sphere is about 633.5 square millimeters.

Check

Find the surface area of the sphere to the nearest tenth.

$S = \underline{\quad?\quad}$ ft²

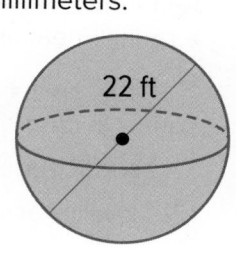

22 ft

🔖 **Go Online** You can complete an Extra Example online.

🔖 **Go Online**
You can watch a video to see how to derive the formula for the surface area of a sphere.

Example 8 Use Formulas to Find the Surface Area of a Sphere

SPORTS If the circumference of the baseball is $C = 3\pi x^2$, find the surface area in terms of x and π.

$C = 2\pi r$	Circumference of a sphere
$3\pi x^2 = 2\pi r$	Substitute.
$\frac{3}{2}x^2 = r$	Solve for r.

Use r to find the surface area of the baseball.

$S = 4\pi r^2$	Surface area of a sphere
$= 4\pi\left(\frac{3}{2}x^2\right)^2$	Substitute.
$= 9\pi x^4$	Solve.

A **composite solid** is a three-dimensional solid that is composed of simpler solids. You can use the formulas you know for calculating the surface areas of prisms, pyramids, cylinders, cones, and spheres to calculate the surface areas of composite solids.

Example 9 Surface Area of a Composite Solid

MANUFACTURING A manufacturer wants to know the approximate area of the metal used to create a basic mailbox. Calculate the surface area of the mailbox.

Step 1 Choose a model.

The mailbox can be modeled by a square prism without a top face and one half of a cylinder.

Step 2 Draw a net of the composite solid.

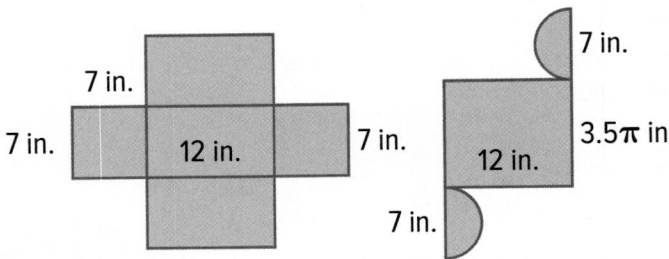

Step 3 Calculate the surface area.

The surface area of the square prism is made up of three rectangles measuring 12 inches by 7 inches and two squares measuring 7 inches by 7 inches. The surface area of the half cylinder is made up of a rectangle measuring 12 inches by 3.5π inches and two half circles with a radius measuring 3.5 inches.

$S = 3(12 \times 7) + 2(7 \times 7) + (12 \times 3.5\pi) + \pi(3.5)^2$ or about 520.4 square inches.

The surface area of the mailbox is about 520.4 square inches.

🔎 **Go Online** You can complete an Extra Example online.

Practice

Go Online You can complete your homework online.

Examples 1, 2, 4, and 5

Find the lateral area and surface area of each solid. Round to the nearest tenth, if necessary.

1.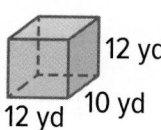
12 yd
12 yd   10 yd

2.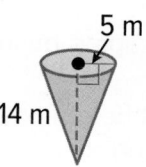
5 m
14 m

3.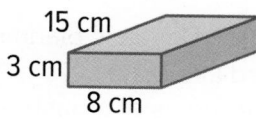
15 cm
3 cm
8 cm

4.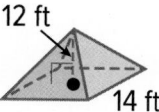
12 ft
14 ft

5.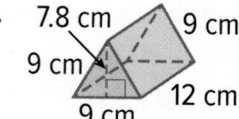
7.8 cm   9 cm
9 cm
12 cm
9 cm

6.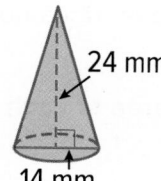
24 mm
14 mm

7.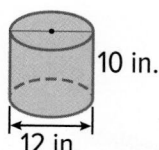
10 in.
12 in.

8.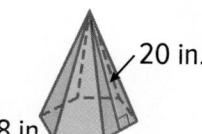
20 in.
8 in.

9.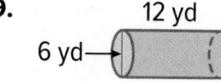
12 yd
6 yd

10.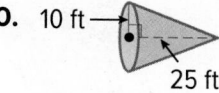
10 ft
25 ft

11.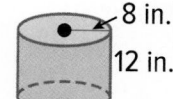
8 in.
12 in.

12.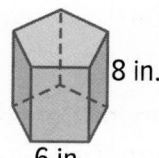
8 in.
6 in.

Examples 3 and 6

13. PAINTING Greg is painting the four walls of his bedroom and the ceiling.

   a. If the height of the walls is x and the edge length of the square ceiling is $2x$, approximate the surface area Greg will be painting in terms of x.

   b. Approximate the surface area that will be painted to the nearest tenth if $x = 8$ feet.

14. MANUFACTURING A food distribution manufacturer is developing a new cylindrical package with a cardboard bottom and sides and a plastic lid. They are evaluating the cost of manufacturing based on the amount of cardboard used.

   a. If the radius of the package is x and the height is $x + 4$, approximate the surface area of the package that will be cardboard in terms of x and π.

   b. Approximate the surface area of the package that will be cardboard to the nearest tenth if $x = 6$ centimeters.

15. CAMPING A company that manufactures camping gear is designing a new tent shaped like a square pyramid with sidewalls made of a waterproof material.

　　a. If the base of the tent is x units long and the slant height of the walls is $1.5x$ units, approximate the surface area of the sidewalls in terms of x.

　　b. Approximate the amount of material needed to manufacture the sidewalls if $x = 9$ feet.

16. TOPIARY Davea is planning to prune her landscaping bushes into topiaries shaped like cones.

　　a. The radius of a bush is $\frac{1}{2}x$ units and the slant height is $4x$ units. Approximate the lateral area of one topiary in terms of x and π.

　　b. A frost is expected, and Davea is making plastic slipcovers to protect her new topiaries. Approximate the surface area of one slipcover to the nearest tenth if the slipcover does not cover the base of the topiary and $x = 0.75$ meter.

Example 7

Find the surface area of each sphere to the nearest tenth.

17.

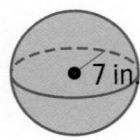
7 in.

18. 32 m

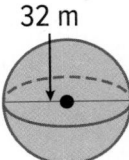

19.

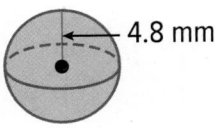
4.8 mm

20. MOONS OF SATURN The planet Saturn has several moons, and they can be modeled accurately by spheres. Saturn's largest moon, Titan, has a radius of about 2575 kilometers. What is the approximate surface area of Titan? Round your answer to the nearest tenth.

Example 8

21. AMUSEMENT PARK Spaceship Earth at Disney's Epcot Center is a sphere. If the circumference of Spaceship Earth is $C = 7\pi x^2$, find the surface area in terms of x and π.

22. BILLIARDS If the circumference of an eight-ball is $C = 4\pi x$, find the surface area in terms of x and π.

Example 9

Find the surface area of each figure to the nearest tenth.

23.

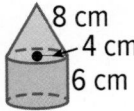
8 cm
4 cm
6 cm

24.

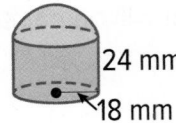
24 mm
18 mm

25. THEATER Carlos is building a prop for the school play, and he needs to calculate the amount of lumber needed to make it. The prop is shaped like a tower and the radius of the base of the tower is 2.5 feet. If the tower is hollow and has no base, what is the approximate surface area of the tower? Round your answer to the nearest tenth.

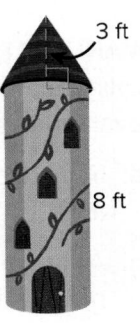
3 ft
8 ft

Mixed Exercises

26. STATE YOUR ASSUMPTION Maddie is painting the shed in her backyard. Approximate the surface area of the shed that will be painted. Explain any assumptions that you make.

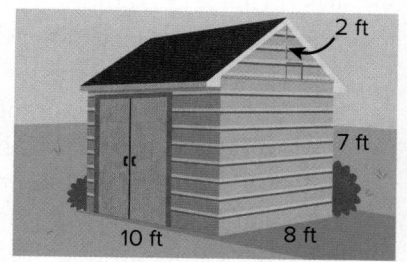

27. STRUCTURE A cylinder has a lateral area of 120π square meters and a height of 7 meters. Find the radius of the cylinder. Round to the nearest tenth.

For Exercises 28–32, find the lateral area and surface area of each solid. Round to the nearest tenth, if necessary.

28. a triangular prism with height of 6 inches, right triangular base with legs of 9 inches and 12 inches

29. a square pyramid with an altitude of 12 inches and a slant height of 18 inches

30. a regular hexagonal pyramid with a base edge of 6 millimeters and a slant height of 9 millimeters

31. a cone with a diameter of 3.4 centimeters and a slant height of 6.5 centimeters

32. a cone with an altitude of 5 feet and a slant height of $9\frac{1}{2}$ feet

33. A *great circle* of a sphere lies on a plane that passes through the center of the sphere. The diameter of a great circle of a sphere is the *diameter of the sphere*.

   a. Find the surface area of a sphere with a great circle that has a circumference of 2π centimeters. Round to the nearest tenth, if necessary.

   b. Find the surface area of a sphere with a great circle that has an area of about 32 square feet. Round to the nearest tenth, if necessary.

34. GREENHOUSE Reina's greenhouse is shaped like a square pyramid with four congruent equilateral triangles for its sides. All of the edges are 6 feet long. What is the total surface area of the greenhouse including the floor? Round your answer to the nearest hundredth.

35. PAPER MODELS Prevan is making a paper model of a castle. Part of the model involves cutting out the net shown and folding it into a pyramid. The pyramid has a square base. What is the surface area of the resulting pyramid?

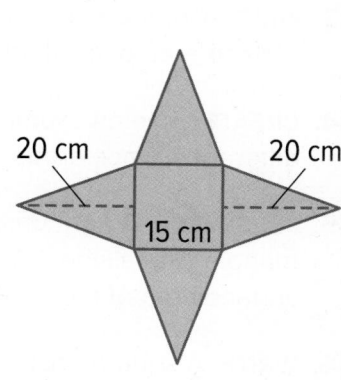

36. CAKES A cake is a rectangular prism with a height of 4 inches and a base that is 12 inches by 15 inches. Wallace wants to apply frosting to the sides and the top of the cake. What is the surface area of the part of the cake that will have frosting?

37. CONSTRUCTION A metal pipe is shaped like a cylinder with a height of 50 inches and a radius of 6 inches. What is the exterior surface area of the pipe? Round your answer to the nearest tenth.

38. INSTRUMENTS A mute for a brass instrument is formed by taking a solid cone with a radius of 10 centimeters and an altitude of 20 centimeters and cutting off the top. The cut is made along a plane that is perpendicular to the altitude of the cone and intersects the altitude 6 centimeters from the vertex. Round your answers to the nearest hundredth.

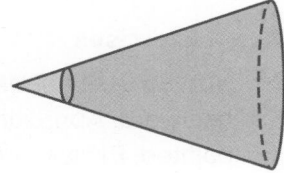

    a. What is the surface area of the original cone?
    b. What is the surface area of the cone that is removed?
    c. What is the surface area of the mute?

39. USE A MODEL The model shows the dimensions of a sofa.

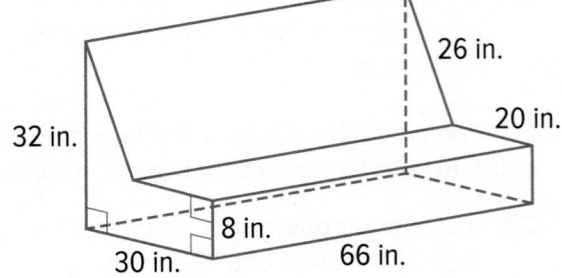

    a. Draw a diagram to show how to calculate the total surface area of the sofa that would be covered by a fitted cover. Explain your technique.
    b. How much fabric is needed for a fitted sofa cover?

40. REASONING Jaylen builds a sphere inside of a cube. The sphere fits snugly inside the cube so that the sphere touches the cube at one point on each face. The length of each edge of the cube is 2 inches.

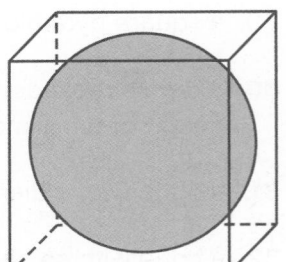

    a. What is the surface area of the cube?
    b. What is the surface area of the sphere? Round your answers to the nearest hundredth.
    c. What is the ratio of the surface area of the cube to the surface area of the sphere? Round your answer to the nearest hundredth.

41. CONSTRUCT ARGUMENTS A cone and a square pyramid have the same surface area. If the areas of their bases are also equal, do they have the same slant height as well? Justify your argument.

42. ANALYZE Classify the following statement as *sometimes*, *always*, or *never* true. Justify your argument.

    The surface area of a cone of radius r and height h is less than the surface area of a cylinder of radius r and height h.

43. WRITE Compare and contrast finding the surface area of a prism and finding the surface area of a cylinder.

44. CREATE Give an example of two cylinders that have the same lateral area and different surface areas. Find the lateral area and surface areas of each cylinder.

45. PERSEVERE A right prism has a height of h units and a base that is an equilateral triangle with a side length of ℓ units. Find the general formula for the total surface area of the prism. Explain your reasoning.

46. WRITE A square prism and a triangular prism are the same height. The base of the triangular prism is an equilateral triangle, with an altitude equal in length to the side of the square. Compare the lateral areas of the prisms.

Cross Sections and Solids of Revolution

Explore Cross Sections

 Online Activity Use concrete models to complete the Explore.

> **INQUIRY** How would you have to cut a polyhedron with *n* faces so that the cross section formed has *n* sides? ×

Learn Cross Sections

A **cross section** is the intersection of a solid figure and a plane. The shape of the cross section formed by the intersection of the plane and the figure depends on the angle of the plane.

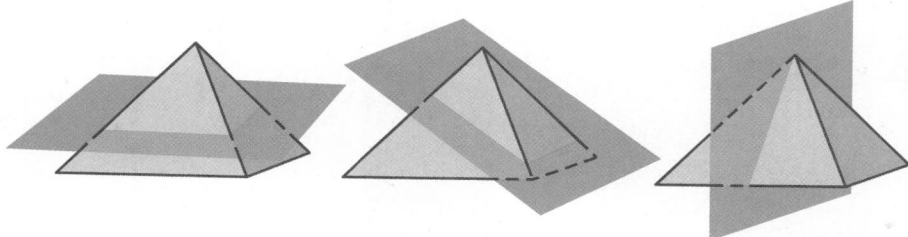

When a plane intersects the face of a solid, one edge of the cross section is formed. When visualizing the shape of the cross section, you can determine the number of sides by counting the number of faces that the plane intersects.

While you must consider a range of possible intersections when determining the cross sections of a solid, one way to start is to determine whether the solid has three-dimensional symmetry.

Three-Dimensional Symmetry: Plane Symmetry	
Words	A three-dimensional figure has **plane symmetry** if the figure can be mapped onto itself by a reflection in a plane.
Model	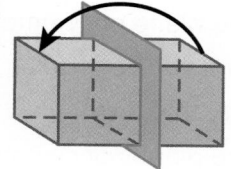

Today's Goals
• Identify the shapes of all cross sections formed by cuts to a solid.
• Identify three-dimensional objects generated by rotations of two-dimensional objects about an axis.

Today's Vocabulary
cross section

plane symmetry

conic sections

conics

solid of revolution

axis symmetry

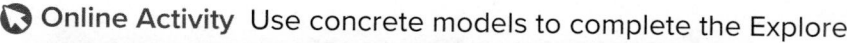 **Go Online**
You can watch a video to see how to identify the shape of a cross section of a three-dimensional figure.

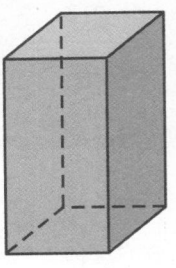

Example 1 Plane Symmetry

Describe each plane of symmetry for the rectangular prism.

The rectangular prism has 3 plane(s) of symmetry.

Choose the correct representation(s) for the plane(s) of symmetry. Select all that apply.

Talk About It!

How could you intersect the right prism to form a pentagonal cross section?

A.

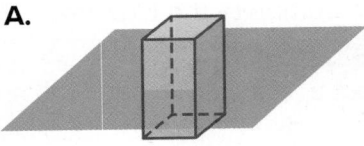

B.

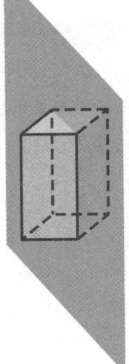

C.

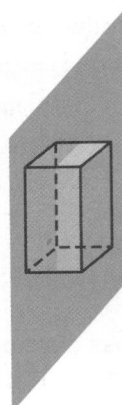

D.

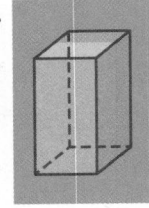

E.

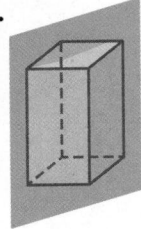

Check

Describe each plane of symmetry for the regular pentagonal prism.

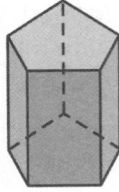

Part A The regular pentagonal prism has ___?___ planes of symmetry.

Part B Complete each sentence using the correct term.

The regular pentagonal prism has one plane of symmetry that is _____?_____ to the base of the prism. The plane of symmetry passes through the _____?_____ of the prism. The regular pentagonal prism has five planes of symmetry that are _____?_____ to the base of the prism. The planes of symmetry each pass through the _____?_____ at one vertex and through the _____?_____ of the opposite _____?_____.

 Go Online You can complete an Extra Example online.

Cross sections of a right circular cone are called **conic sections** or **conics**.

Example 2 Identify Cross Sections

Identify the shape of each cross section of the cone.

Go Online
You may want to complete the Concept Check to check your understanding.

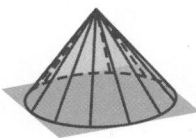

a. intersection of the cone and a plane parallel to the base of the cone

When a plane parallel to the base of the cone intersects the cone, the cross section is a circle. This will always be the case unless the plane intersects the vertex.

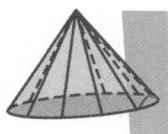

b. intersection of the cone and a plane perpendicular to the base of the cone

When a plane perpendicular to the base of the cone intersects the cone, it does so along its curved lateral surface and through its base. This shape has a curved surface and one flat edge. The curve formed has a special name called a hyperbola. Any cross section formed in this manner, excluding the intersection of a plane and the vertex, will have a flat edge and one hyperbolic curve. The cross section formed by the intersection of a perpendicular plane and the vertex of the cone is a triangle.

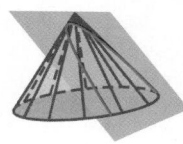

c. intersection of the cone and a plane with the same slope as the slant height of the cone

When the plane has the same slope as the slant height of the cone, the intersection is along its curved lateral surface and through its base. This shape has a curved surface and one flat edge. The curve formed has a special name called a parabola. As the plane moves through the cone at this angle, each cross section has one flat edge and one parabolic curve.

Go Online
You can watch a video to see how to generate a three-dimensional figure by rotating a two-dimensional figure about a line.

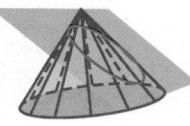

d. intersection of the cone and a plane with slope different than the slant height of the cone

When the plane has a slope different than the slant height of the cone, it can intersect the cone in an oval, or ellipse.

Go Online You can complete an Extra Example online.

Math History
Minute

Russian **Sofia
Kovalevskaya** (1850–
1891) was a great
mathematician, writer,
and passionate
advocate for women's
rights. In 1888, she
entered a paper, "On
the Rotation of a Solid
Body about a Fixed
Point," in a competition
for the Prix Bordin by
the French Academy of
Science, and she won.
Sofia considered it her
greatest personal
triumph.

Study Tip

Use Tools You can use
many different tools to
help you visualize the
solids of revolution.
Straws or dowel rods
could be used to
represent the axis of
rotation. Card stock or
heavy construction
paper can be attached
to the straws or dowel
rods to represent the
two-dimensional figures.
You could also sketch
the figure using dynamic
geometry software or
graph paper.

Learn Solids of Revolution

A **solid of revolution** is a solid figure obtained by rotating a plane
figure or curve around an axis. The shape of the solid of revolution
depends on the location of the axis and the shape of the plane figure
or curve being rotated.

You can determine whether a three-dimensional figure could have
been created by rotation by identifying whether the solid figure has
axis symmetry.

Three-Dimensional Symmetry: Axis Symmetry	
Words	A three-dimensional figure has **axis symmetry** if the figure can be mapped onto itself by a rotation between 0° and 360° in a line.
Model	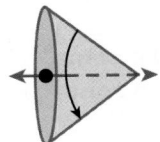

Example 3 Identify Solids of Revolution

**Identify the solid formed by rotating the two-
dimensional shape about line ℓ.**

Imagine rotating the two-dimensional figure about line ℓ.

The solid of revolution is a cylinder.

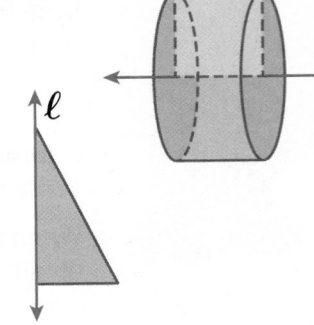

Check

Identify the solid formed by rotating the
two-dimensional shape about line ℓ.

The solid of revolution is a ___?___.

🌐 Example 4 Axis Symmetry

Circle each item with axis symmetry.

Check

Select all of the images that demonstrate axis symmetry.

A.   **B.**   **C.**   **D.**   **E.**

🔄 **Go Online** You can complete an Extra Example online.

Practice

Example 1

Describe each plane of symmetry for each solid.

1.

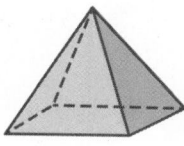

2.

3.

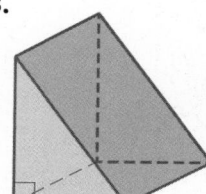

Example 2

Identify the shape of each cross section.

4.

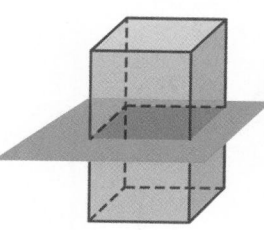

5.

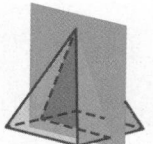

6.

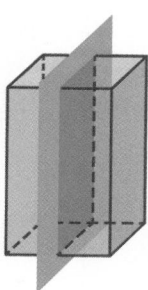

7.

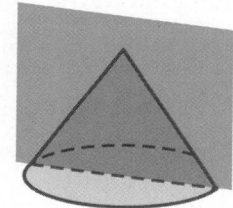

Example 3

Identify the solid formed by rotating each two-dimensional shape about each line.

8.

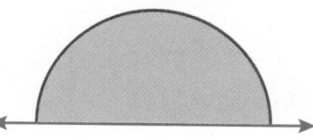

9.

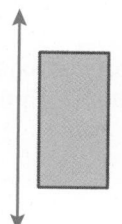

10.

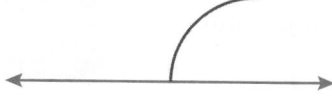

11.

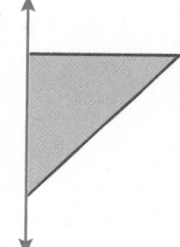

Example 4

Determine whether each item has axis symmetry. Write *yes* or *no*.

12. traffic cone with a base

13. one pound weight

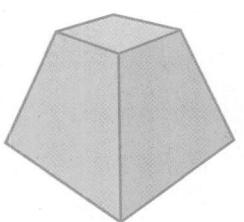

14. flower petal and stem

15. golf ball

16. tree

17. lamp

Mixed Exercises

For Exercises 18 and 19, refer to the solid at the right.

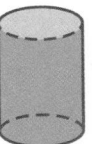

18. Name the shape of the cross section cut parallel to the base.

19. Name the shape of the cross section cut perpendicular to the base.

20. Sketch the cross section from a vertical slice of the solid shown at the right.

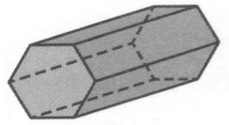

21. Reshan rotated a semicircle about a vertical axis to create a sphere. Describe another way Reshan could have rotated a semicircle about an axis to generate a sphere.

22. WASHER Consider the washer shown. Describe the two-dimensional shape and axis that could be used to generate the washer by rotating the two-dimensional shape around the axis.

23. TRIANGLES Sallie is going to rotate the right triangle around the axis shown. Describe a real-world object that could be generated.

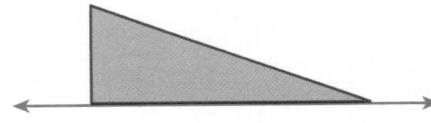

Koya979/Shutterstock; Peredniankina/Shutterstock; McGraw-Hill Education/Mark Steinmetz; Tomaz Kunst//Stock/ Getty Images; Victoroancea/123RF; Richard Hutchings/ Digital Light/McGraw-Hill Education

24. REASONING The rectangle shown will be rotated about the *x*-axis.

    a. Describe the three-dimensional shape that is generated.

    b. What is the length of the radius and height of the figure generated?

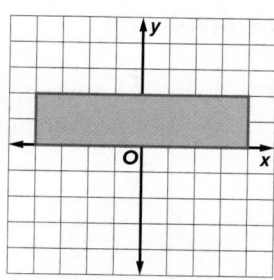

25. USE TOOLS Determine whether each cross section can be made from a cube. If so, sketch the cube and its cross section.

    a. triangle   **b.** square   **c.** rectangle

    d. pentagon   **e.** hexagon   **f.** octagon

26. USE A SOURCE Research the shape and dimensions of the Washington Monument.

    a. Describe each plane of symmetry of the Washington Monument.

    b. Describe the cross section of the Washington Monument from a vertical slice perpendicular to the base through the vertex.

    c. Determine if the Washington Monument has axis symmetry. Write *yes* or *no*.

For Exercises 27 and 28, identify and sketch the cross section of an object made by each cut described.

27. square pyramid cut perpendicular to base but not through the vertex

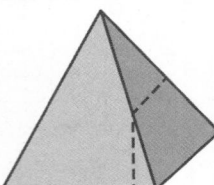

28. rectangular prism cut diagonally from a top edge to a bottom edge on the opposite side

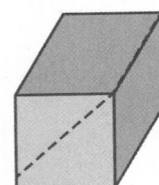

29. You want to cut each geometric object so that the cross section is a circle. Give the name for each object. Then describe the cut that results in a circle.

a.
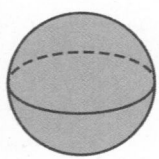

b.

c.
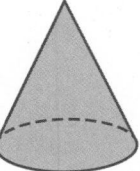

30. **USE TOOLS** Sketch and describe the object that is created by rotating each shape around the indicated axis of rotation.

a.
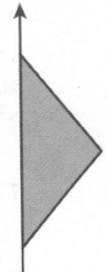

b.
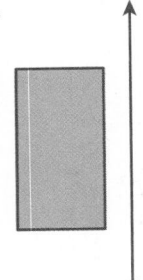

c.
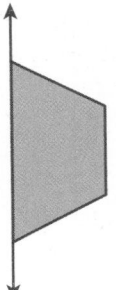

31. **CONSTRUCT ARGUMENTS** A regular polyhedron has axis symmetry of order 3, but does not have plane symmetry. What is the figure? Justify your argument.

32. **PERSEVERE** The figure at the right is a cross section of a geometric solid. Describe a solid and how the cross section was made.

33. **WRITE** A hexagonal pyramid is sliced through the vertex and the base so that the prism is separated into two congruent parts. Describe the cross section. Is there more than one way to separate the figure into two congruent parts? Will the shape of the cross section change? Explain.

34. **CREATE** Sketch a real-world object that has plane symmetry, but not axis symmetry.

35. **ANALYZE** Determine whether the statement below is *true* or *false*. Justify your argument. The only two shapes formed by the cross sections of a square pyramid are a triangle and a square.

Volumes of Prisms and Pyramids

Learn Volume of Prisms

Recall that the volume of a solid is the measure of the amount of space that the solid encloses. Volume is measured in cubic units.

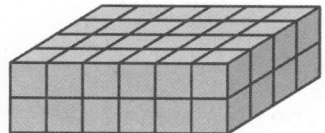

This rectangular prism has 6 · 4 or 24 cubic units in the bottom layer.

Because there are two layers, the total volume is 24 · 2 or 48 cubic units.

Key Concept • Volume of a Prism

Words	The volume V of a prism is $V = Bh$, where B is the area of a base and h is the height of the prism.
Symbols	$V = Bh$
Example	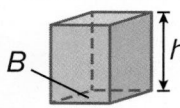

Key Concept • Cavalieri's Principle

Words	If two solids have the same height h and the same cross-sectional area B at every level, then they have the same volume.
Models	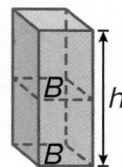 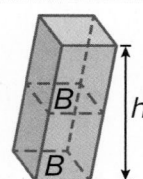 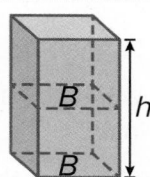

Example 1 Volume of a Prism

Find the volume of the prism.

Step 1 Find the area of the base B.

$B = \frac{1}{2}aP$ Area of a regular polygon

   $= \frac{1}{2}(3.4)(25)$ or 42.5 $a = 3.4$ and $P = 25$

So, the area of the base of the prism is 42.5 square millimeters.

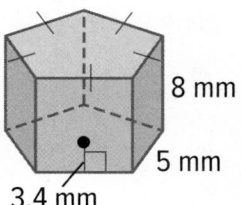

8 mm

5 mm

3.4 mm

Step 2 Find the volume of the prism.

$V = Bh$ Volume of a prism

   $= 42.5(8)$ or 340 $B = 42.5$ and $h = 8$

The volume of the prism is 340 cubic millimeters.

Check

Find the volume of the prism.

$V = \underline{}$ ft^3

10 ft   15 ft   9 ft

 Go Online You can complete an Extra Example online.

Example 2 Volume of an Oblique Prism

Find the volume of the oblique prism.

Part A Find the area of the base.

The base of the prism is a trapezoid.

$A = \frac{1}{2}h(b_1 + b_2)$ Area of a trapezoid

$= \frac{1}{2}(5)(4 + 2)$ or 15 $h = 5, b_1 = 4,$ and $b_2 = 2$

The area of the base is 15 square centimeters.

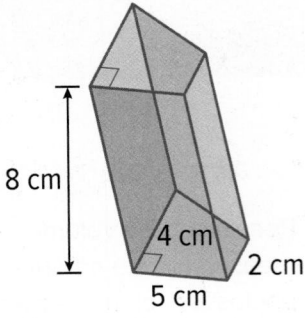

8 cm

4 cm

2 cm

5 cm

Part B Find the volume.

$V = Bh$ Volume of a prism

$= (15)(8)$ $B = 15$ and $h = 8$

$= 120$ cm^3

The volume of the oblique solid is 120 cubic centimeters.

Check

Find the volume of the oblique rectangular prism.

$V = $ ____?____ m^3

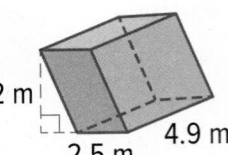

2.2 m

2.5 m

4.9 m

Watch Out!

Area and Volume

Area is two-dimensional, so it is measured in square units. Volume is three-dimensional, so it is measured in cubic units.

🌐 Example 3 Volume of a Prism Using Algebraic Expressions

SNACKS Gustavo wants to calculate the volume of juice in his juice box in cubic inches.

Part A Find the volume of the juice box in terms of x.

Step 1 Find the area of the base.

The base of the juice box is a rectangle.

$A = \ell w$ Area of a rectangle

$= (4x)x$ Substitution

$= 4x^2$ Solve.

PURE

Juice

$4x + 3$

x $4x$

Step 2 Find the volume.

$V = Bh$ Volume of a prism

$= (4x^2)(4x + 3)$ Substitute.

$= 16x^3 + 12x^2$ Simplify.

Part B Find the volume of the juice box if $x = 2$.

$V = 16x^3 + 12x^2$ Volume of the juice box

$= 16(2)^3 + 12(2)^2$ Substitute.

$= 176$ in^3 Simplify.

▶ **Go Online** You can complete an Extra Example online.

Explore Volumes of Square Pyramids

Online Activity Use the guiding exercises to complete the Explore.

> **INQUIRY** How does the formula for the volume of a square pyramid relate to the formula for the volume of a step pyramid with an infinite number of steps?

Learn Volumes of Pyramids

Key Concept • Volume of a Pyramid	
Words	The volume of a pyramid is $V = \frac{1}{3}Bh$, where B is the area of the base and h is the height of the pyramid.
Symbols	$V = \frac{1}{3}Bh$
Model	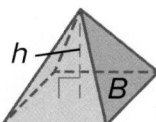 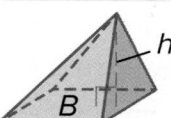

Go Online
You can watch a video to see how you can derive the formula for the volume of a pyramid.

Example 4 Volume of a Pyramid

Find the volume of the pyramid.

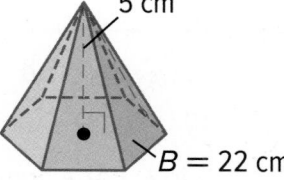

$V = \frac{1}{3}Bh$ Volume of a pyramid

$\approx 36.7 \text{ cm}^3$ Substitute and simplify.

The volume of the pyramid is about 36.7 cubic centimeters.

🌐 Example 5 Volume of a Pyramid Using Algebraic Expressions

SOUVENIRS Martín bought a bank shaped like a square pyramid, and he wants to calculate its volume.

Part A Find the volume of the pyramid in terms of x.

Find the area B of the square base.

$B = (8x)^2$ or $64x^2$

Find the height of the pyramid by using the Pythagorean Theorem.

$a^2 + b^2 = c^2$ Pythagorean Theorem

$(4x)^2 + h^2 = (5x)^2$ $a = 4x$, $b = h$, and $c = 5x$

$h = 3x$ Solve for h.

Find the volume of the pyramid in terms of x.

$V = \frac{1}{3}Bh$ Volume of a pyramid

$= \frac{1}{3}(64x^2)(3x)$ Substitute.

$= 64x^3$ Simplify.

Part B Find the volume in cubic inches if $x = 4$.

Use the formula in **Part A** to find the volume when $x = 4$.

$V = 4096 \text{ in}^3$

Go Online
You can complete an Extra Example online.

Example 6 Volume of a Composite Solid

Find the volume of the composite solid.

The composite solid is a combination of a prism and a square pyramid.

Part A Find the volume of the prism.

$V = Bh$ Volume of a prism

$= 36(3)$ $B = 36$ and $h = 3$

$= 108$ Solve.

So, the volume of the prism is 108 cubic inches.

Part B Find the volume of the pyramid.

$V = \frac{1}{3}Bh$ Volume of a pyramid

$= \frac{1}{3}(36)(4)$ $B = 36$ and $h = 4$

$= 48$ Simplify.

The volume of the pyramid is 48 cubic inches.

Part C Find the volume of the composite solid.

The volume of the composite solid is 156 cubic inches.

🌐 Example 7 Approximate the Volume of a Composite Solid

CHOCOLATE For a competition, a chocolatier created a replica of the Washington Monument made entirely of white chocolate. Approximate the volume of chocolate used to create the sculpture.

The sculpture can be approximated by a composite solid made up of a square prism and a square pyramid.

Step 1 Find the volume of the square prism.

$V = Bh$ Volume of a prism

$= 42.25(36)$ $B = 42.25$ and $h = 36$

$= 1521$ Simplify.

The volume of the square prism is 1521 cubic inches.

Step 2 Find the volume of the square pyramid. Round to the nearest tenth, if necessary.

$V = \frac{1}{3}Bh$ Volume of a pyramid

$= \frac{1}{3}(42.25)(6.5)$ $B = 42.25$ and $h = 6.5$

≈ 91.5 Simplify.

Step 3 Find the volume of the sculpture.

The volume of the sculpture is 1612.5 cubic inches.

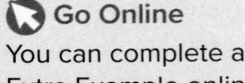
Go Online
You can complete an Extra Example online.

Practice

Go Online You can complete your homework online.

Examples 1, 2, and 4

Find the volume of each prism or pyramid. Round your answer to the nearest tenth, if necessary.

1.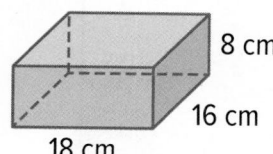
8 cm
16 cm
18 cm

2.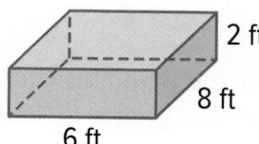
2 ft
8 ft
6 ft

3.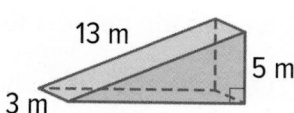
13 m
5 m
3 m

4.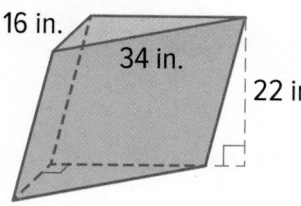
16 in.
34 in.
22 in.

5.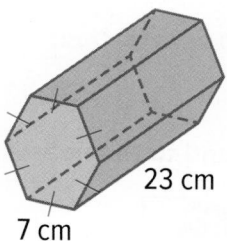
23 cm
7 cm

6.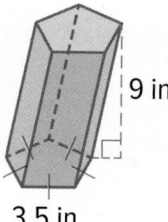
9 in.
3.5 in.

7.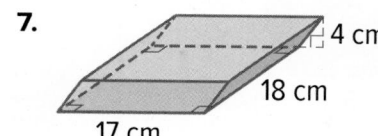
4 cm
18 cm
17 cm

8.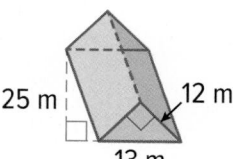
25 m
12 m
13 m

9.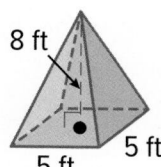
8 ft
5 ft
5 ft

10.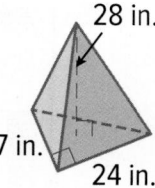
28 in.
7 in.
24 in.

11.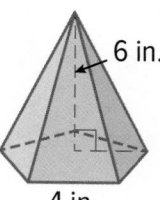
6 in.
4 in.

12.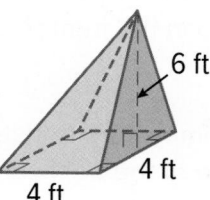
6 ft
4 ft
4 ft

Example 3

13. CHOCOLATE Leah wants to calculate the volume of chocolate in her candy bar.

a. Find the volume of the candy bar in terms of x.

b. Find the volume of the candy bar if $x = 2$ inches.

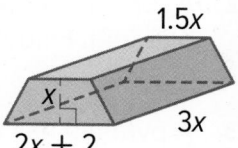
1.5x
x
3x
2x + 2

14. CANDLE Benton wants to calculate the volume of wax needed to make a candle.

a. Find the volume of the candle in terms of x.

b. Find the volume of the candle if $x = 3$ centimeters.

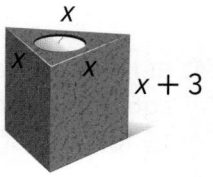
x
x x
x + 3

Example 5

15. SAND ART Noelle wants to calculate the volume of sand needed to make the art decoration within the regular hexagonal pyramid.

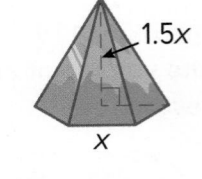

    a. Find the volume of the sand art in terms of x.

    b. Find the volume of the sand art if $x = 5$ inches. Round to the nearest tenth.

16. PUZZLE Roger bought a puzzle shaped like a pyramid. The base of the puzzle is a square and the height of the puzzle is $5x$ units. In the figure, the shorter leg of the triangle shown connects the center of the base to a vertex of the base.

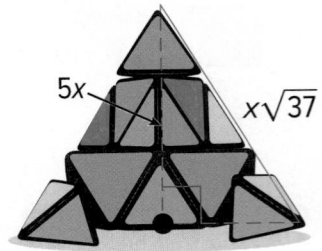

    a. Find the volume of the pyramid in terms of x.

    b. Find the volume of the pyramid if $x = 10$ centimeters.

Example 6

Find the volume of each composite solid. Round to the nearest tenth, if necessary.

17.

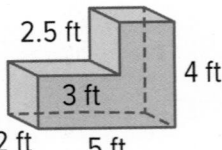

18.

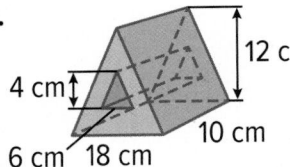

19.

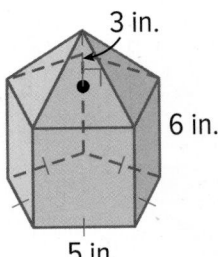

Example 7

20. FRAMES Margaret makes a square frame out of four pieces of wood. Each piece of wood is a rectangular prism with a length of 40 centimeters, a height of 4 centimeters, and a depth of 6 centimeters. What is the total volume of the wood used in the frame?

21. PARKS Grimby Park is installing new animal-proof trashcans. Approximate the volume of the trashcan.

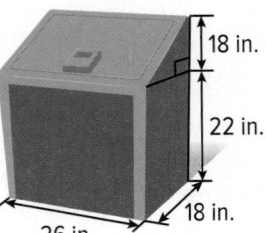

Mixed Exercises

22. A rectangular prism has a length of 16 feet, a width of 9 feet, and a height of 8 feet. Find the volume of the prism.

23. A pyramid has a height of 18 centimeters and a base with an area of 26 square centimeters. Find the volume.

24. BENCH Inside a lobby, there is a bench shaped like a simple block with a square base that measures 6 feet on each side. The height of the bench is $1\frac{3}{5}$ feet. What is the volume of the bench?

25. TUNNELS Construction workers are digging a tunnel through a mountain. The space inside the tunnel is going to be shaped like a rectangular prism. The mouth of the tunnel will be a rectangle 20 feet high and 50 feet wide, and the length of the tunnel will be 900 feet. What volume of rock must be removed to make the tunnel?

26. GREENHOUSES A greenhouse has the shape of a square pyramid. The base has a side length of 30 yards. The height of the greenhouse is 18 yards. What is the volume of the greenhouse?

27. STAGES A solid wooden stage is made out of oak, which has a weight of about 45 pounds per cubic foot. The stage has the form of a square pyramid with the top sliced off along a plane parallel to the base. The side length of the top square is 12 feet, and the side length of the bottom square is 16 feet. The height of the stage is 3 feet.

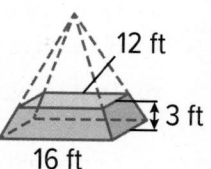

   a. What is the volume of the entire square pyramid from which the stage is made?

   b. What is the volume of the top of the pyramid that is removed to create the stage?

   c. What is the volume of the stage?

   d. What is the weight of the stage?

28. PRECISION Refer to the prisms.

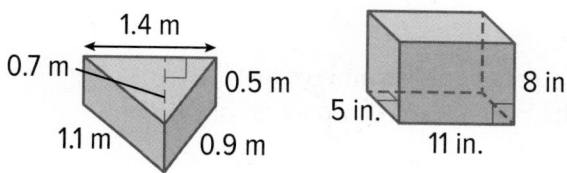

   a. What is the volume of the triangular prism?

   b. What is the volume of the rectangular prism?

   c. Discuss how the formulas that you used to find the volume of each prism are similar.

29. USE A MODEL Benjamin finds that baking soda has a mass of 2.2 grams per cubic centimeter, and corn flakes have a mass of 0.12 gram per cubic centimeter. A box of baking soda is 8 centimeters long by 4 centimeters wide by 12 centimeters high. The dimensions for a box of corn flakes are 30 centimeters long by 6 centimeters wide by 35 centimeters high. Benjamin wants to find the mass of the contents of each box if each box is filled to within 2 centimeters of the top.

   a. How can Benjamin determine the mass of the contents of each box?

   b. Find the mass of the contents of each box.

30. STRUCTURE A model pyramid has a volume of 270 cubic feet and a base area of 90 square feet. What is the height if the pyramid is a right pyramid? What is the height if the pyramid is an oblique pyramid? Explain your reasoning.

31. REASONING Tristan makes and sells sugar-free candies. He packages them in square pyramid-shaped boxes with a height of 3 inches and a base that has a side length of 2 inches. He sells each box for $2.00.

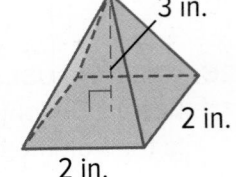

3 in.

2 in.

2 in.

    a. What is the volume of the sugar-free candies in each box? What is the price per cubic inch?

    b. Tristan wants to make a bigger package by doubling the lengths of the sides of the square base. How can he figure out how much to charge if he wants to keep the price per cubic inch the same?

    c. Tristan wants to design a box in the shape of a square pyramid that holds between 7 and 8 cubic inches of sugar-free candies. He wants the height to be within $1\frac{1}{4}$ inches of the length of each side of the base. What is one possible set of dimensions that he can use?

32. STRUCTURE Anisa is building a box in the shape of a right triangular prism for her magic act. She is making a secret compartment inside the box. The compartment will be a pyramid with base $\triangle ABE$ and vertex at point C. After the secret compartment has been made, how is the volume of the space remaining inside the box related to the volume of the secret compartment? Explain your reasoning.

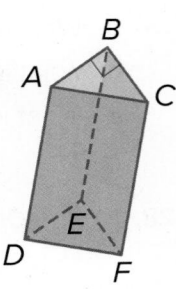

B

A   C

E

D   F

🌑 **Higher-Order Thinking Skills**

33. FIND THE ERROR Francisco and Valerie each calculated the volume of an equilateral triangular prism with a height of 5 units and a base that has an apothem of 4 units. Who is correct? Explain your reasoning.

Francisco	Valerie
$V = Bh$	$V = Bh$
$= \frac{1}{2}aP \cdot h$	$= \frac{\sqrt{3}}{2}s^2 \cdot h$
$= \frac{1}{2}(4)(24\sqrt{3}) \cdot 5$	$= \frac{\sqrt{3}}{2}(4\sqrt{3})^2 \cdot 5$
$= 240\sqrt{3}$ cubic units	$= 120\sqrt{3}$ cubic units

34. CREATE Give an example of a pyramid and a prism that have the same base and the same volume. Explain your reasoning.

35. ANALYZE Make a conjecture about how many pentagonal pyramids will fit inside a pentagonal prism of the same height. Justify your answer.

36. PERSEVERE Write an equation to find the dimensions of a composite solid composed of a cube with a square pyramid on top of equal height. If the volume is equal to 36 cubic inches, what are the dimensions of the solid?

Volumes of Cylinders, Cones, and Spheres

Learn Volumes of Cylinders

Like a prism, the volume of a cylinder can be thought of as consisting of layers. For a cylinder, these layers are congruent circular discs.

Key Concept • Volume of a Cylinder	
Words	The volume V of a cylinder is $V = Bh$ or $V = \pi r^2 h$, where B is the area of the base, h is the height of the cylinder, and r is the radius of the base.
Symbols	$V = Bh$ or $V = \pi r^2 h$
Example	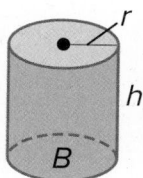

🌎 Example 1 Approximate the Volume of a Cylinder

POSTAL SERVICE **Andrew wants to mail his brother a collection of antique marbles. He needs to calculate the volume of the mail tube before he buys it to ensure that it will hold all of the marbles. Find the volume of the mail tube to the nearest tenth.**

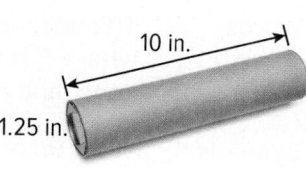

$V = Bh$ Volume of a cylinder

$\quad = \pi(1.25)^2 \cdot 10$ $B = \pi r^2$, $r = 1.25$, and $h = 10$

$\quad \approx 49.1$ Simplify.

The volume of the mail tube is about 49.1 cubic inches.

Example 2 Volume of a Cylinder

Find the volume of a cylinder with a radius of $x - 5$ centimeters and a height of x centimeters.

Part A Find the volume of the cylinder in terms of x and π.

$V = Bh$ Volume of a cylinder

$\quad = \pi(x - 5)^2 x$ $B = \pi r^2$, $r = x - 5$, and $h = x$

So, the volume of the cylinder in terms of x and π is $V = \pi(x - 5)^2 x$ cubic centimeters.

(continued on the next page)

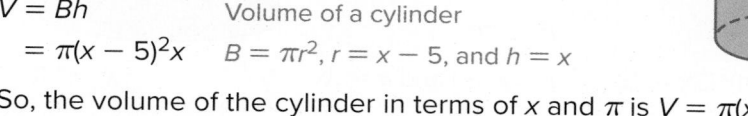 **Go Online** You can complete an Extra Example online.

Today's Goals
- Find volumes of cylinders.
- Find volumes of cones.
- Find volumes of spheres.

💬 Talk About It!
How are the volume formulas for prisms and cylinders similar?

Watch Out!
Multiple Expressions
You could also express the volume of the cylinder in expanded form as $V = \pi x^3 - 10\pi x^2 + 25\pi x$.

Olga Kovalenko/123RF

Part B Find the volume.

Find the volume of the cylinder rounded to the nearest tenth if $x = 8$.

Substitute the value of x into the expression you found in **Part A**.

$V = \pi(x - 5)^2 x$ Expression from Part A

$ = \pi(8 - 5)^2 \cdot 8$ Substitute.

$ \approx 226.2$ Simplify.

The volume of the cylinder is about 226.2 cubic centimeters.

Learn Volumes of Cones

The pyramid and prism shown have the same base area B and height h as the cylinder and cone. You can use Cavalieri's Principle and similar triangles to show that the volume of the cone is equal to the volume of the pyramid.

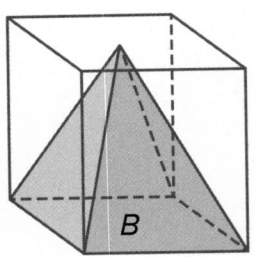 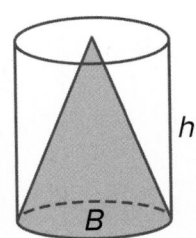

Key Concept • Volume of a Cone	
Words	The volume of a circular cone is $V = \frac{1}{3}Bh$, or $V = \frac{1}{3}\pi r^2 h$, where B is the area of the base, h is the height of the cone, and r is the radius of the base.
Symbols	$V = \frac{1}{3}Bh$ or $V = \frac{1}{3}\pi r^2 h$
Model	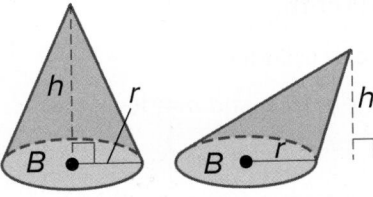

Example 3 Volume of a Cone

Examine the cone.

Part A Find the volume of a cone in terms of x and π.

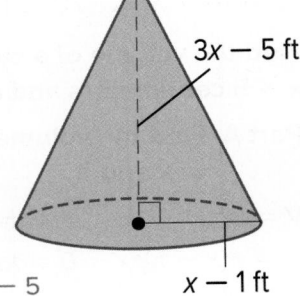

$V = \frac{1}{3}Bh$ Volume of a cone

$ = \frac{1}{3}\pi r^2 h$ $B = \pi r^2$

$ = \frac{1}{3}\pi(x - 1)^2(3x - 5)$ $r = x - 1$ and $h = 3x - 5$

So, the volume of the cone in terms of x and π is

$V = \frac{1}{3}\pi(x - 1)^2(3x - 5)$ cubic feet.

(continued on the next page)

 Go Online You can complete an Extra Example online.

Part B Find the volume of the cone to the nearest tenth if $x = 4$.

Substitute the value of x into the expression you found in **Part A**.

$V = \frac{1}{3}\pi(x-1)^2(3x-5)$ Expression from Part A

$\quad = \frac{1}{3}\pi(4-1)^2[3(4)-5]$ Substitute.

$\quad \approx 66.0$ Simplify.

The volume of the cone is about 66.0 cubic feet.

🌐 Example 4   Approximate the Volume of a Cone

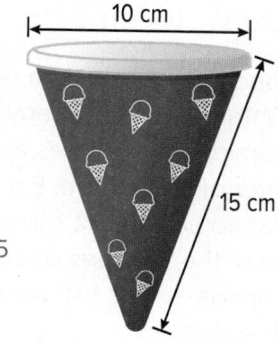

TREATS **Alyssa serves ice cream in paper cones with plastic lids. What is the volume of the paper cone?**

Step 1 Find the height of the paper cone.

$a^2 + b^2 = c^2$ Pythagorean Theorem

$5^2 + h^2 = 15^2$ $a = 5, b = h,$ and $c = 15$

$\quad\quad h = \sqrt{200}$ Solve for h.

$\quad\quad\quad = \sqrt{200}$

Step 2 Find the volume of the paper cone to the nearest tenth.

$V = \frac{1}{3}Bh$ Volume of a cone

$\quad = \frac{1}{3}\pi(5)^2(\sqrt{200})$ $r = 5$ and $h = \sqrt{200}$

$\quad \approx 370.2$ Simplify.

The volume of the paper cone is about 370.2 cubic centimeters.

Check

DÉCOR A soap dispenser can be modeled by a cone with a height of 15 centimeters and a radius of 4 centimeters. What is the approximate volume of soap that the dispenser can hold? Round your answer to the nearest tenth.

$V \approx \underline{\ ?\ }$ cm^3

🔷 Go Online You can complete an Extra Example online.

> ### Watch Out!
> **Rounding** When you find the height of the ice cream container, you may be tempted to round that measure to the nearest tenth. However, rounding at that stage in the calculation will change the final calculation of the volume. Whenever possible, avoid rounding at intermediate steps. Wait to round until you have calculated the final answer.

Explore Volumes of Spheres

Online Activity Use dynamic geometry software to complete the Explore.

> **INQUIRY** A cone with radius r and height r is cut out of a cylinder with radius r and height r. What is the relationship between the volume of the remaining solid and the volume of a sphere with radius r?

Go Online
A derivation of the formula for the volume of a sphere is available.

Learn Volumes of Spheres

Suppose a sphere with radius r contains infinitely many pyramids with vertices at the center of the sphere. Each pyramid has height r. The sum of the volumes of all the pyramids equals the volume of the sphere.

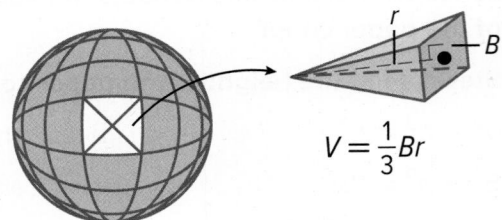

$$V = \frac{1}{3}Br$$

Key Concept • Volume of a Cone	
Words	The volume V of a sphere is $V = \frac{4}{3}\pi r^3$, where r is the radius of the sphere.
Symbols	$V = \frac{4}{3}\pi r^3$
Model	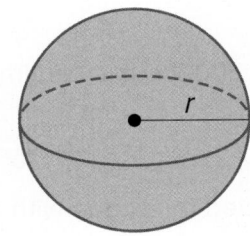

⊕ Example 5 Approximate the Volume of a Sphere

CANDY A chocolate company wants to create bite-sized individually wrapped solid chocolate spheres. If the diameter of the sphere is 1.5 inches, find the volume of chocolate used to make the spheres to the nearest hundredth.

$$V = \frac{4}{3}\pi r^3 \qquad \text{Volume of a sphere}$$

$$= \frac{4}{3}\pi(0.75)^3 \qquad \text{Substitute.}$$

$$\approx 1.77 \qquad \text{Simplify.}$$

The volume of chocolate used is about 1.77 cubic inches.

Go Online You can complete an Extra Example online.

lynx/Iconotec/Glowimages

Example 6 Volume of a Sphere

Examine the sphere.

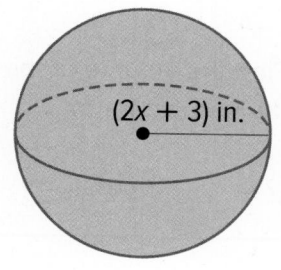
(2x + 3) in.

Part A Find the volume of the sphere in terms of x and π.

$V = \frac{4}{3}\pi r^3$ Volume of a sphere

$= \frac{4}{3}\pi(2x + 3)^3$ $r = 2x + 3$

So, the volume of the sphere in terms of x and π is $V = \frac{4}{3}\pi(2x + 3)^3$ cubic inches.

Part B Find the volume of the sphere if x = 2.2.

Substitute the value of x into the expression you found in **Part A**.

$V = \frac{4}{3}\pi(2x + 3)^3$ Expression from Part A

$= \frac{4}{3}\pi[2(2.2) + 3]^3$ Substitute.

≈ 1697.4 Simplify.

The volume of the sphere is about 1697.4 cubic inches.

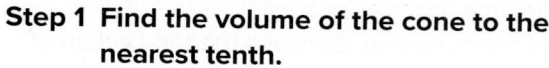

Example 7 Volume of a Composite Solid

Find the volume of the composite solid.

The composite solid is a combination of a cone and a hemisphere.

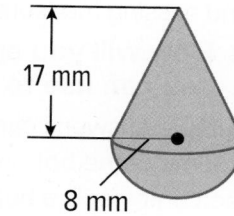
17 mm

8 mm

Step 1 Find the volume of the cone to the nearest tenth.

$V = \frac{1}{3}Bh$ Volume of a cone

$= \frac{1}{3}\pi(r)^2 h$ $B = \pi r^2$

$= \frac{1}{3}\pi(8)^2(17)$ $r = 8$ and $h = 17$

≈ 1139.4 Simplify.

The volume of the cone is about 1139.4 mm³.

Step 2 Find the volume of the hemisphere to the nearest tenth.

$V = \frac{1}{2}\left(\frac{4}{3}\pi r^3\right)$ Volume of a hemisphere

$= \frac{2}{3}\pi(8)^3$ $r = 8$

≈ 1072.3 Simplify.

The volume of the hemisphere is about 1072.3 mm³.

Step 3 Find the volume of the composite solid.

The volume of the composite solid is the sum of the volumes of the cone and the hemisphere.

$V \approx 2211.7$ mm³

🔵 **Go Online** You can complete an Extra Example online.

Watch Out!

Multiple Expressions
You could also express the volume of the sphere in expanded form as

$V = \frac{32\pi x^3 + 144\pi x^2 + 216\pi x + 108\pi}{3}$

cubic inches.

Study Tip

Hemispheres A plane can intersect a sphere in a point or in a circle. If the circle contains the center of the sphere, the intersection is called a *great circle*. A great circle separates a sphere into two congruent halves, called *hemispheres*.

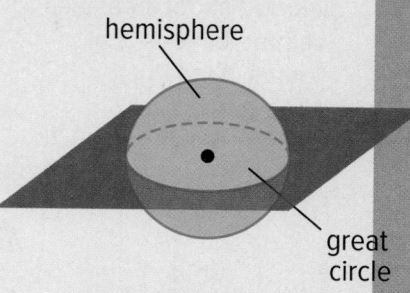
hemisphere

great circle

Check

Find the volume of the composite solid to the nearest tenth.

$V \approx$ _____?_____ cm^3

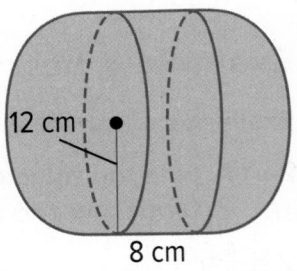

12 cm

8 cm

🌐 Apply Example 8 Approximate the Volume of a Composite Solid

HOME DECOR Mia purchased a trash container for her study area. Find the volume of the trash container rounded to the nearest tenth.

1 What is the task?

Describe the task in your own words. Then list any questions that you may have. How can you find answers to your questions?

Sample answer: I need to find the volume of the trash container. What solids are used to make the container? What is the height of the top portion of the container? I can identify the approximate solids used to make the container and then use the properties of those solids to find missing measures.

4 ft

2 ft

2 How will you approach the task? What have you learned that you can use to help you complete the task?

Sample answer: I can use the volume formula for a cylinder to find the volume of the bottom portion of the container. Because the top portion of the container is a hemisphere, I can use the given diameter and the volume formula for a sphere to find the volume of the top portion of the container. Using this information, I can calculate the total volume of the trash container.

3 What is your solution?

Use your strategy to solve the problem.

What is the volume of the bottom portion of the container in terms of π? 4π

What is the volume of the top portion of the container in terms of π? $\frac{2}{3}\pi$

What is the volume of the trash container to the nearest tenth? 14.7 ft^3

4 How can you know that your solution is reasonable?

✏️ **Write About It!** Write an argument that can be used to defend your solution.

If you approximated the bottom portion of the trash container to be a rectangular prism and the top portion to be a cube, the volume of the trash container would be 24 cubic feet. Because approximating the trash container as a cylinder and hemisphere is more accurate, you would expect the actual volume to be less than 24 cubic feet. So, my answer is reasonable.

🖱 Go Online You can complete an Extra Example online.

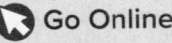 **Go Online**
to practice what you've learned about measures of three-dimensional figures in the Put It All Together over Lessons 6-4 through 6-7.

Practice

🅝 **Go Online** You can complete your homework online.

Example 1

1. **DISPOSAL** The Meyer family uses a kitchen trash can shaped like a cylinder. It has a height of 18 inches and a base diameter of 12 inches. What is the approximate volume of the trash can? Round your answer to the nearest tenth of a cubic inch.

2. **COFFEE** A roasting company sells their coffee in canisters shaped like a cylinder. The radius of the canister is 1.5 inches and the height is 7.5 inches. What is the approximate volume of a coffee canister? Round your answer to the nearest cubic inch.

Example 2

3. Find the volume of a cylinder with a radius of $2x$ millimeters and a height of $x - 2$ millimeters.

   a. Find the volume of the cylinder in terms of x and π.

   b. Find the volume of the cylinder if $x = 10$. Round your answer to the nearest tenth.

4. Find the volume of a cylinder with a diameter that is 6 centimeters shorter than the height x.

   a. Find the volume of the cylinder in terms of x and π.

   b. Find the volume of the cylinder if the height is 14 centimeters. Round your answer to the nearest tenth.

5. Find the volume of a cylinder with a radius of x feet and a height of $3x + 4$ feet.

   a. Find the volume of the cylinder in terms of x and π.

   b. Find the volume of the cylinder if $x = 3$. Round your answer to the nearest tenth.

Examples 3 and 4

6. Examine the cone.

   a. Find the volume of the cone in terms of x and π.

   b. Find the volume of the cone if $x = 4$ feet. Round your answer to the nearest tenth.

7. Examine the cone.

   a. Find the volume of the cone in terms of x and π.

   b. Find the volume of the cone if $x = 6$ meters. Round your answer to the nearest tenth.

8. Examine the cone.

   a. Find the volume of the cone in terms of x and π.

   b. Find the volume of the cone if $x = 3$ inches. Round your answer to the nearest tenth.

9. **DINING** The part of a dish designed for ice cream is shaped like a cone. The base of the cone has a radius of 2 inches and the height is 1.2 inches. What is the volume of the cone? Round your answer to the nearest hundredth.

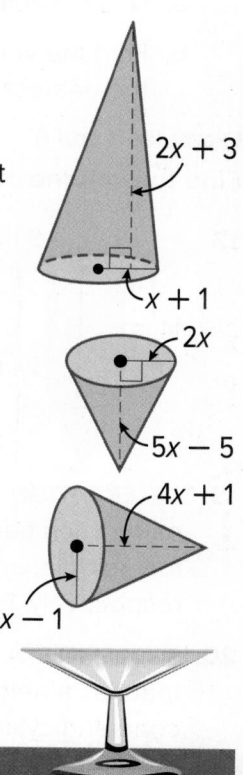

2x + 3

x + 1

2x

5x − 5

4x + 1

3x − 1

10. **AUTOMOBILE** Don uses a funnel to pour oil into his car engine. The funnel has a radius of 6 centimeters and a slant height of 10 centimeters. How much oil, to the nearest cubic centimeter, will the funnel hold?

Examples 5 and 6

11. **ORANGES** Moesha cuts a spherical orange in half along a great circle. The radius of the orange is 2 inches. Find the approximate volume of the orange to the nearest tenth of a cubic inch.

12. **DESIGN** A scale model of a spherical fountain has a radius of 1.5 feet. Find the volume of the actual fountain. Round your answer to the nearest tenth of a cubic foot.

13. Examine the sphere.

    a. Find the volume of the sphere in terms of x and π.

    b. Find the volume of the sphere if $x = 1$ centimeter. Round your answer to the nearest tenth.

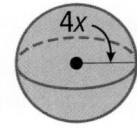

14. Examine the sphere.

    a. Find the volume of the sphere in terms of x and π.

    b. Find the volume of the sphere if $x = 4.3$ inches. Round your answer to the nearest tenth.

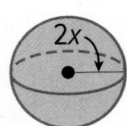

15. Examine the sphere.

    a. Find the volume of the sphere in terms x and π.

    b. Find the volume of the sphere if $x = 0.25$ yards. Round your answer to the nearest tenth.

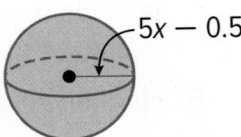

16. Examine the sphere.

    a. Find the volume of the sphere in terms of x and π.

    b. Find the volume of the sphere if $x = 5$ centimeters. Round your answer to the nearest tenth.

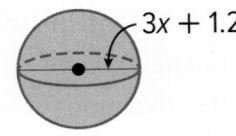

Examples 7 and 8

Find the volume of each composite solid. Round to the nearest tenth, if necessary.

17.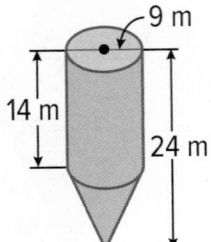
9 m
14 m
24 m

18.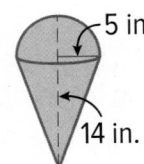
5 in.
14 in.

19. **CELEBRATION** Emilia is having a three-tiered cake made for her quinceañera party. Each tier of the cake will have a height of 4 inches. The diameters of the top, middle, and bottom tiers will measure 6 inches, 10 inches, and 14 inches, respectively. Find the total volume of the cake to the nearest cubic inch.

20. **SCULPTING** A sculptor wants to remove stone from a cylindrical block that has a height of 3 feet to create a cone. The diameter of the base of the cone and cylinder is 2 feet. What is the volume of the stone that the sculptor must remove? Round your answer to the nearest hundredth.

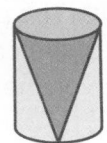

Mixed Exercises

For Exercises 21 and 22, refer to the composite solid shown at the right.

21. STRUCTURE Write a formula for the volume of this solid in terms of the radius r.

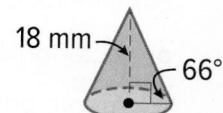

22. PRECISION Explain how you wrote a formula for the volume of this solid.

23. Find the volume of the cone. Round to the nearest tenth.

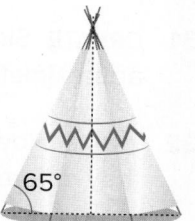

24. REASONING A hemisphere has a base with an area that is 25π square centimeters. Find the volume of the hemisphere. Round to the nearest tenth.

25. TEEPEE Cathy made a teepee for a class project. Her teepee had a diameter of 6 feet. The angle that the side of the teepee made with the ground was 65°. What was the volume of the teepee? Round your answer to the nearest hundredth.

26. SCHOOL SUPPLIES A pencil grip is shaped like a triangular prism with a cylinder removed from the middle. The base of the prism is a right isosceles triangle with leg lengths of 2 centimeters. The diameter of the base of the removed cylinder is 1 centimeter. The heights of the prism and the cylinder are the same and equal to 4 centimeters. What is the exact volume of the pencil grip?

27. CONSTRUCT ARGUMENTS A wooden sphere is carved from a solid cube of wood so that the least amount of wood is carved away.

a. If the block of wood had a volume of 729 cubic inches, what is the volume of the sphere? Explain.

b. Devon says that he can multiply the volume of any cube by $\frac{\pi}{6}$ to find the volume of the sphere that shares the same diameter as the cube's side. Is he correct? Justify your argument.

28. REASONING Reginald is creating a scale model of a building using a scale of 4 feet = 3 inches. The building is in the shape of a cube topped with a hemisphere so that the circular base of the hemisphere is inscribed in the square base of the cube. At its highest point, the building has a height of 30 feet. Find the volume of his scale model to the nearest cubic inch. Explain.

29. REGULARITY A container company manufactures cylindrical containers with a radius of 3 inches and a height of 10 inches. The company decided to produce a different cylindrical container that has a height of 8 inches and the same volume as the original container. What steps would you use to find the radius of the new container? What is the radius of the new container to the nearest tenth?

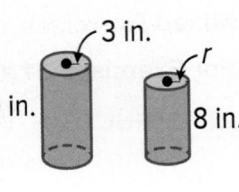

30. PERSEVERE The cylindrical can shown is used to fill a container with liquid. It takes three full cans to fill the container. Describe possible dimensions of the container if it is each of the following shapes.

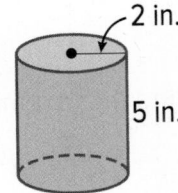

a. rectangular prism

b. square prism

c. triangular prism with a right triangle as the base

31. CREATE Sketch a composite solid made of a cylinder and cone that has an approximate volume of 7698.5 cubic centimeters.

32. WRITE How are the volume formulas for prisms and cylinders similar? How are they different?

33. ANALYZE Determine whether the following statement is *sometimes, always*, or *never* true. Justify your argument.

The volume of a cone with radius r and height h equals the volume of a prism with height h.

34. FIND THE ERROR Alexandra and Cornelio are calculating the volume of the cone at the right. Who is correct? Explain your reasoning.

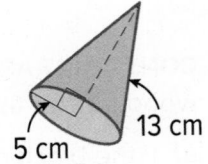

Alexandra	Cornelio
$V = \frac{1}{3} Bh$	$5^2 + 12^2 = 13^2$
$= \frac{1}{3} \pi (5)^2 (13)$	$V = \frac{1}{3} Bh$
$= 340.3 \ cm^3$	$= \frac{1}{3} \pi (5)^2 (12)$
	$= 314.2 \ cm^3$

35. ANALYZE Determine whether the following statement is *true* or *false*. If true, explain your reasoning. If false, provide a counterexample.

If a sphere has radius r, there exists a cone with radius r having the same volume.

Applying Similarity to Solid Figures

Learn Similar Two-Dimensional Figures

Recall that if two polygons are similar, then their perimeters are proportional to the scale factor between them. The areas of two similar polygons share a different relationship.

Theorem 6.1: Areas of Similar Polygons	
Words	If two polygons are similar, then their areas are proportional to the square of the scale factor between them.
Example	If $ABCD \sim FGHJ$, then $\frac{\text{area of } FGHJ}{\text{area of } ABCD} = \left(\frac{FG}{AB}\right)^2$.
Model	

You will prove Theorem 6.1 in Exercise 22.

Today's Goals
- Find measures of similar figures by using scale factors.
- Find measures of similar solids by using scale factors.

Today's Vocabulary
similar solids
congruent solids

Example 1 Use Similar Figures to Find Area

$\square ABCD$ and $\square JKLM$ are similar rectangles. Find the area of $\square JKLM$.

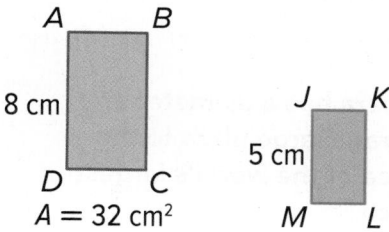

$A = 32$ cm²

Step 1 Find the scale factor from $\square ABCD$ to $\square JKLM$.

The scale factor from $\square ABCD$ to $\square JKLM$ is $\frac{5}{8}$.

Step 2 Find the ratio of the areas.

If two polygons are similar, then their areas are proportional to the square of the scale factor between them. So, the ratio of their areas is $\left(\frac{5}{8}\right)^2$ or $\frac{25}{64}$.

Step 3 Find the area.

$$\frac{\text{area of } \square JKLM}{\text{area of } \square ABCD} = \frac{25}{64} \qquad \text{Write a proportion.}$$

$$\frac{\text{area of } \square JKLM}{32} = \frac{25}{64} \qquad \text{Area of } \square ABCD = 32$$

$$\text{area of } \square JKLM = 12.5 \qquad \text{Multiply and simplify.}$$

So, the area of $\square JKLM$ is 12.5 square centimeters.

Study Tip

Ratios Ratios can be written in different ways. For example, x to y, $x : y$, and $\frac{x}{y}$ are all representations of the ratio of x and y.

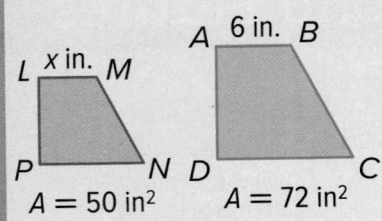

A 6 in. B

L x in. M

P ___ N D ___ C

$A = 50 \text{ in}^2$ $A = 72 \text{ in}^2$

Example 2 Use Areas of Similar Figures

Trapezoids *LMNP* and *ABCD* are similar. Find the scale factor of trapezoid *ABCD* to trapezoid *LMNP* and the value of *x*.

Part A Find the scale factor.

Let *k* be the scale factor from trapezoid *ABCD* to trapezoid *LMNP*.

$\dfrac{\text{area of } LMNP}{\text{area of } ABCD} = k^2$	Theorem 11.1
$\dfrac{50}{72} = k^2$	Substitution
$\dfrac{25}{36} = k^2$	Simplify.
$\dfrac{5}{6} = k$	Take the positive square root.

So, the scale factor from *ABCD* to *LMNP* is $\dfrac{5}{6}$.

Part B Find the value of *x*.

Use the scale factor to find *x*.

$\dfrac{LM}{AB} = k$	The ratio of corresponding lengths of similar polygons is equal to the scale factor between the polygons.
$\dfrac{x}{6} = \dfrac{5}{6}$	Substitution
$x = \dfrac{5}{6} \cdot 6 \text{ or } 5$	Multiply each side by 6.

The value of *x* is 5.

🌐 Example 3 Use Similar Figures to Solve Problems

WORLD RECORDS An average large pizza has a diameter of 14 inches. The scale factor from an average large pizza to the world's largest pizza is $\dfrac{786}{7}$. Find the area of the world's largest pizza rounded to the nearest square inch.

Step 1 Find the area of an average large pizza.

$A = \pi r^2$	Area of a circle
$= \pi(7)^2$	Substitute.

The area of an average large pizza is about 49π or 153.9 square inches.

Step 2 Find the area of the world's largest pizza.

Use the scale factor to find the area of the world's largest pizza rounded to the nearest square inch.

$$\frac{\text{area of the world's largest pizza}}{\text{area of an average large pizza}} = \frac{786^2}{7^2}$$

$$\frac{\text{area of the world's largest pizza}}{49\pi} = \frac{617{,}796}{49}$$

$$\text{area of the world's largest pizza} = \frac{617{,}796(49\pi)}{49}$$

The area of the world's largest pizza is about 1,940,863 in².

Watch Out!

Writing Ratios When finding the ratio of the area of Figure *A* to the area of Figure *B*, be sure to write your ratio as $\dfrac{\text{area of Figure } B}{\text{area of Figure } A}$.

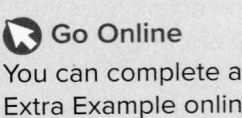

Go Online
You can complete an Extra Example online.

Online Activity Use dynamic geometry software to complete the Explore.

> ⊗ INQUIRY How are the surface areas and volumes of similar solids related?

Learn Similar Three-Dimensional Solids

Similar solids have exactly the same shape but not necessarily the same size. All spheres are similar, and all cubes are similar.

In similar solids, the corresponding linear measures, such as height and radius, are proportional to the scale factor between them.

Similar Solids

Two solids are similar if and only if they have the same shape and the ratios of their corresponding linear measures are equal.

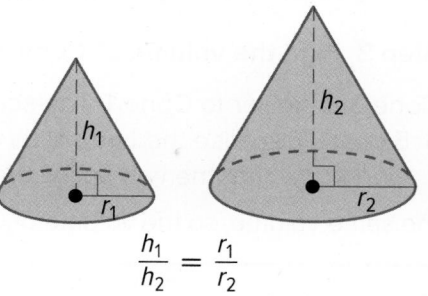

$$\frac{h_1}{h_2} = \frac{r_1}{r_2}$$

Theorem 6.2

Words	If two similar solids have a scale factor of $a : b$, then the surface areas have a ratio of $a^2 : b^2$, and the volumes have a ratio of $a^3 : b^3$.

Example	scale factor	2:3	
	ratio of surface area	4:9	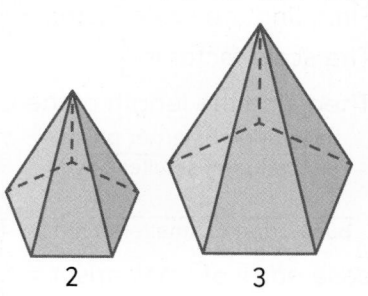
	ratio of volumes	8:27	

2   3

You will prove Theorem 6.2 in Exercise 23.

Key Concept • Characteristics of Congruent Solids

Two solids are congruent if and only if they have the following characteristics.
- Corresponding angles are congruent.
- Corresponding edges are congruent.
- Corresponding faces are congruent.
- Volumes are equal.

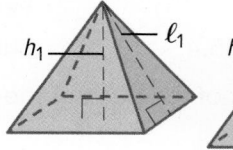 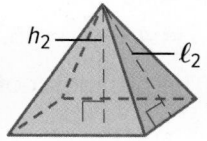

$$\frac{h_1}{h_2} = \frac{\ell_1}{\ell_2} = 1$$

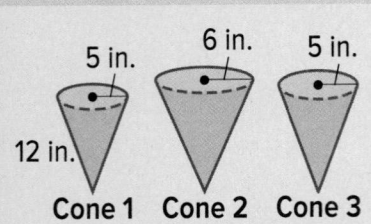

5 in. 6 in. 5 in.

12 in.

Cone 1   Cone 2   Cone 3

Example 4 Use Similar Solids to Find Volume

The three cones are similar. Find the volume of each cone.

Step 1 Find the volume of Cone 1

$$V = \frac{1}{3}Bh \qquad \text{Volume of a cone}$$

$$= \frac{1}{3}\pi(5)^2 \cdot \sqrt{119} \qquad B = \pi r^2, r = 5, \text{ and } h = \sqrt{12^2 - 5^2}$$

$$= \frac{25\pi\sqrt{119}}{3} \qquad \text{Simplify.}$$

Step 2 Find the volume of Cone 2.

Cone 2 is similar to Cone 1. The scale factor from Cone 2 to Cone 1 is 5 : 6. Find the volume of Cone 2.

$$\frac{\text{volume of Cone 1}}{\text{volume of Cone 2}} = \left(\frac{5}{6}\right)^3 \qquad \text{Theorem 11.2}$$

$$\text{volume of Cone 2} = \left(\frac{6}{5}\right)^3 \times \frac{25\pi\sqrt{119}}{3} \qquad \begin{array}{l}\text{Substitute, then solve for} \\ \text{the volume of Cone 2.}\end{array}$$

$$= \frac{72\pi\sqrt{119}}{5} \qquad \text{Simplify.}$$

Step 3 Find the volume of Cone 3.

Cone 3 is similar to Cone 1. The scale factor from Cone 3 to Cone 1 is 5 : 5 or 1 : 1. Because the two solids are similar and have a scale factor of 1 : 1, we know that the two solids are congruent. Congruent solids have the same volume, so the volume of Cone 3 is $\frac{25\pi\sqrt{119}}{3}$ cubic inches.

Example 5 Use Similar Solids to Solve Problems

Two similar rectangular prisms with square bases have surface areas of 98 square centimeters and 18 square centimeters. If one base edge of the larger rectangular prism measures 9 centimeters, what is the perimeter of one base of the smaller prism?

First, find the scale factor. $\frac{\text{surface area of larger prism}}{\text{surface area of smaller prism}} = \frac{98}{18} = \frac{49}{9} = \left(\frac{7}{3}\right)^2$

The scale factor is $\frac{7}{3}$.

Then, find the length of the base edge of the small prism.

$$\frac{\text{base edge of larger prism}}{\text{base edge of smaller prism}} = \frac{7}{3} \qquad \text{The scale factor is } \frac{7}{3}.$$

$$\frac{9}{\text{base edge of smaller prism}} = \frac{7}{3} \qquad \text{Substitute.}$$

$$\text{base edge of small prism} = \frac{27}{7} \qquad \text{Use a proportion}$$

The base edge of the smaller prism is $\frac{27}{7}$ centimeters.

Find the perimeter of the base of the smaller prism to the nearest tenth.

$$P = 4s \qquad \text{Perimeter of square}$$

$$= 4 \cdot \frac{27}{7} \approx 15.4 \qquad \text{Substitute and simplify.}$$

The perimeter of the base of the smaller prism is about 15.4 centimeters.

Check

Two similar cylinders have volumes of 270π and 640π cubic inches, respectively. If the height of the larger cylinder is 10 inches, what is the area of the base of the smaller cylinder?

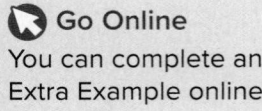

Go Online
You can complete an Extra Example online.

Practice

Example 1

For each pair of similar figures, find the missing area. Round your answer to the nearest tenth, if necessary.

1.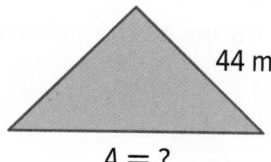
11 m
$A = 20$ m²

44 m
$A = ?$

2.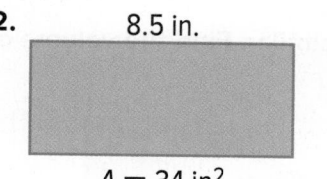
8.5 in.
2 in.
$A = ?$
$A = 34$ in²

3.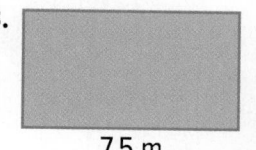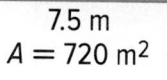
7.5 m
$A = 720$ m²

12 m
$A = ?$

4.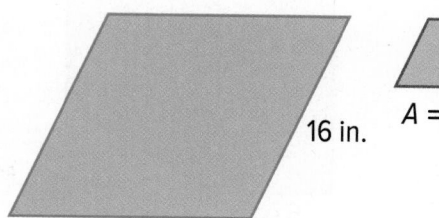
3 in.
$A = ?$
16 in.
$A = 72$ in²

Example 2

For each pair of similar figures, use the given areas to find the scale factor from the figure on the left to the figure on the right. Then find the value of x to the nearest tenth.

5.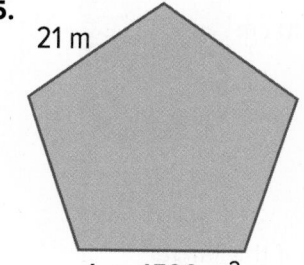
21 m
x
$A = 510$ m²
$A = 4590$ m²

6.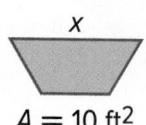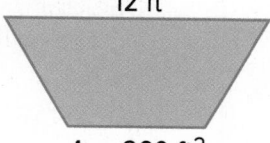
x
12 ft
$A = 10$ ft²
$A = 360$ ft²

7.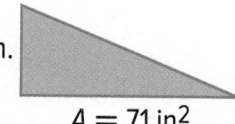
x
9.5 in.
$A = 16$ in²
$A = 71$ in²

8.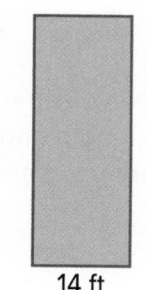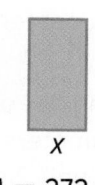
x
$A = 272$ ft²
14 ft
$A = 588$ ft²

Example 3

9. ASSIGNMENTS Matt has two similar posters for his science project. Each poster is a rectangle. The length of the larger poster is 11 inches. The length of the smaller poster is 6 inches. What is the area of the smaller poster if the larger poster is 93.5 square inches?

10. ACCESSORIES Carla has a shirt with decorative pins in the shape of equilateral triangles. The pins come in two sizes. The larger pin has a side length that is three times longer than the smaller pin. If the area of the smaller pin is 6.9 square centimeters, what is the approximate area of the larger pin?

11. QUILT A quilt design has one large rectangle surrounded by four congruent rectangles that are similar to the large rectangle. If the large rectangle has an area of 45 square inches, what is the area of each small rectangle?

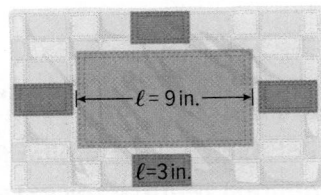

$\ell = 9\,\text{in.}$

$\ell = 3\,\text{in.}$

Example 4

Each pair of solids is similar. Find the volume of each solid. Round to the nearest tenth, if necessary.

12.

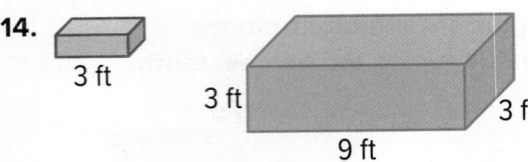

3 cm
4 cm
2 cm
6 cm

13.

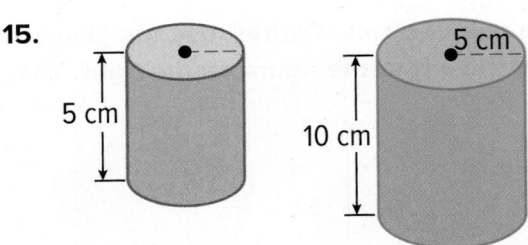

9 cm
6 cm

12 cm

14.

3 ft
3 ft
3 ft
9 ft

15.

5 cm
10 cm
5 cm

Example 5

16. Two similar cylinders have radii of 3 inches and 12 inches. If the height of the larger cylinder is 24 inches, what is the surface area of the smaller cylinder? Round your answer to the nearest tenth.

17. Two similar rectangular prisms have surface areas of 112 square centimeters and 1008 square centimeters. If the length and width of the base of the smaller prism are 4 centimeters and 2 centimeters, respectively, what is the perimeter of one base of the larger prism?

Mixed Exercises

18. Two similar prisms have heights of 12 feet and 20 feet. What is the ratio of the volume of the small prism to the volume of the large prism?

19. Two cubes have surface areas of 81 square inches and 144 square inches. What is the ratio of the volume of the small cube to the volume of the large cube?

20. REGULARITY A polygon has an area of 225 square meters. If the area is tripled, how does each side length change?

21. REASONING Smith's Bakery is baking several cakes for a community festival. The cakes consist of two geometrically similar shapes as shown. If 50 pieces of cake can be cut from the small cake, how many pieces of the same size can be cut from the large cake? Round to the nearest piece of cake.

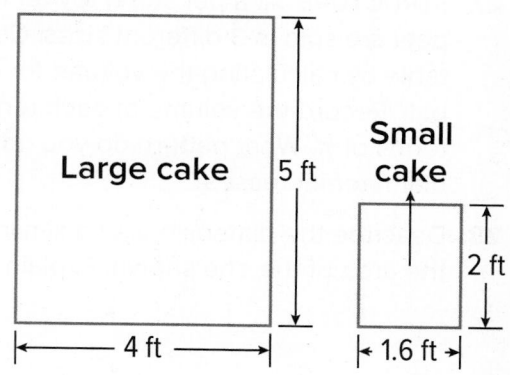

PROOF Write a paragraph proof to prove each theorem.

22. Theorem 11.1

Given: $\triangle DEF \sim \triangle PQR$

Prove: $\dfrac{\text{area of } \triangle DEF}{\text{area of } \triangle PQR} = \left(\dfrac{d}{p}\right)^2$

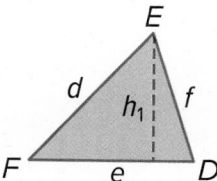

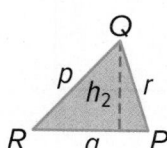

23. Theorem 11.2

Given: rectangular prisms with scale factor $a : b$

Prove: The surface areas have a ratio of $a^2 : b^2$, and the volumes have a ratio of $a^3 : b^3$.

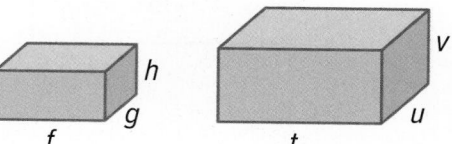

24. ATMOSPHERE About 99% of Earth's atmosphere is contained in a 31-kilometer thick layer that surrounds the planet. The Earth itself is approximately a sphere with a radius of 6378 kilometers. What is the ratio of the volume of the atmosphere to the volume of Earth? Round your answer to the nearest thousandth.

25. SCULPTURE An artist creates metal sculptures in the shapes of regular octagons. The length of each side of the larger sculpture is 7 inches, and the area of the base of the smaller sculpture is 19.28 square inches.

a. What is the length of each side of the smaller sculpture?

b. The artist is going to pack the sculptures in a circular box to take them to an art show. Will the larger sculpture fit in a circular box with a 15-inch diameter? Explain your reasoning.

26. SPORTS Major League Baseball, or MLB, rules state that baseballs must have a circumference of 9 inches. The National Softball Association, or NSA, rules state that softballs must have a circumference not exceeding 12 inches.

a. Find the ratio of the circumference of an MLB baseball to the circumference of a 12-inch NSA softball.

b. Find the ratio of the volume of the MLB baseball to the volume of the NSA softball. Round your answer to the nearest tenth.

27. STRUCTURE At a pet store, toy tennis balls for pets are sold in 3 different sizes. Complete the table by calculating the volume for each size ball. Record the volume of each tennis ball in terms of π. What pattern do you notice as the diameter increases?

Size	Diameter (cm)	Volume (cm³)
Small	3	
Medium	4.5	
Large	6.75	

28. Describe the dimensions of a similar trapezoid that has an area four times the area of the one shown. Explain how you found your answer.

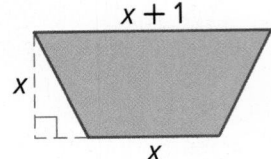

29. FIND THE ERROR Violeta and Gerald are trying to come up with a formula that can be used to find the area of a circle with a radius *r* after it has been enlarged by a scale factor *k*. Is either of them correct? Explain your reasoning.

Violeta	Gerald
$A = k\pi r^2$	$A = \pi(r^2)^k$

30. PERSEVERE If you want the area of a polygon to be $x\%$ of its original area, by what scale factor should you multiply each side length?

31. CREATE Draw a pair of similar figures with areas that have a ratio of 4:1. Explain.

32. WRITE Explain how to find the area of an enlarged polygon if you know the area of the original polygon and the scale factor of the enlargement.

33. PERSEVERE The ratio of the volume of Cylinder A to the volume of Cylinder B is 5:1. Cylinder A is similar to Cylinder C with a scale factor of 2:1, and Cylinder B is similar to Cylinder D with a scale factor of 3:1. What is the ratio of the volume of Cylinder C to the volume of Cylinder D? Explain your reasoning.

Density

Today's Goals
- Solve real-world problems involving density by using area.
- Solve real-world problems involving density by using volume.

Today's Vocabulary
density

Explore Strategies Based on Density

⬤ **Online Activity** Use the video to complete the Explore.

> ② **INQUIRY** How can a knowledge of density help you make decisions in video games and in real-world situations? ×

Learn Density Based on Area

Density is a measure of the quantity of some physical property per unit of length, area, or volume. One example of density is population density, which is the measurement of population per unit of area. Population density is calculated for states, major cities, or other areas, based on data collected from the U.S. Census.

Density Based on Area	
Words	Density is the ratio of objects to area.
Symbols	$\text{density} = \dfrac{\text{number of objects}}{\text{area}}$

🌐 Example 1 Find the Density of an Area

POPULATION **The World Bank reports that the population of Greenland in 2015 was 56,114. If the total land area is about 836,000 square miles, what is the population density of Greenland?**

Calculate population density by adapting the density formula, $\text{population density} = \dfrac{\text{population}}{\text{land area}}$.

The population density of Greenland is $\dfrac{56{,}114}{836{,}000}$ or about 0.067 people per square mile.

Check

VOLCANOES The country with the largest number of active volcanoes is Indonesia with 147. The area of Indonesia is 735,400 square miles. What is the density of active volcanoes in Indonesia?

A. $\approx 0.0002 \frac{\text{volcanoes}}{\text{mi}^2}$

B. $\approx 0.0004 \frac{\text{volcanoes}}{\text{mi}^2}$

C. $\approx 2501.4 \frac{\text{volcanoes}}{\text{mi}^2}$

D. $\approx 5002.7 \frac{\text{volcanoes}}{\text{mi}^2}$

⬤ **Go Online** You can complete an Extra Example online.

Study Tip

Dimensional Analysis In some cases, it may be necessary to use dimensional analysis to convert between units of measurement. To convert from one unit to another, use a numerical quantity known as a conversion factor. For example, if you are trying to find the number of kilograms in a 16-pound box, you can multiply by the conversion factor $\frac{1\text{ kg}}{2.2\text{ lbs}}$, because there are about 2.2 pounds in each kilogram.

$16\text{ lb} \cdot \dfrac{1\text{ kg}}{2.2\text{ lb}} \approx 7.3\text{ kg}$

🌐 Example 2 Use the Density of an Area

GREENHOUSE GASES Masha has a farm with 220 milking cows that produce 286 pounds per acre for a total of 1,412,840 pounds of milk. Due to recent regulations, Masha must pay a fee if she has more than 30 cows per square mile on her farm. Determine whether Masha will have to pay the fee. (*Hint:* There are 640 acres in 1 square mile.)

😮 **Think About It!**

Why do we multiply 4940 by $\frac{1}{640}$ in Step 2?

Step 1 Find the area of the farm.

$$\frac{1{,}412{,}840 \text{ lb}}{286 \text{ lb/acre}} = 4940 \text{ acres on Masha's farm}$$

Step 2 Convert to square miles.

Use a conversion factor.

$$4940 \text{ acres} \cdot \frac{1 \text{ mi}^2}{640 \text{ acres}} \quad \frac{247}{32} \cong 7.7 \text{ mi}^2$$

There are about 7.7 square miles on Masha's farm.

Step 3 Find the density of cows per acre.

Calculate population density by adapting the density formula.

$$\text{population density} = \frac{\text{population}}{\text{land area}} \qquad \text{Population Density Formula}$$

$$= \frac{220}{7.7} \qquad\qquad 220 \text{ cows on 7.7 square miles}$$

$$= 28.57 \qquad\qquad \text{Simplify.}$$

Because 28.57 is fewer than 30, Masha will not have to pay the fee.

Check

DUCK POND For a school carnival game, Adalynn is planning to fill a pool with water and float a layer of numbered rubber ducks on top. She knows that it takes 25 rubber ducks to fill 1 square foot of area. The pool used for the carnival has an area of about 7 square feet. How many rubber ducks should she buy to fill the pool?

Adalynn should buy ___?___ rubber ducks.

POPULATION DENSITY The city of Manila, Philippines, is one of the most densely populated cities on Earth. Its 1,650,000 residents share a space that can be approximated by a rectangle 5.1 miles long by 2.9 miles wide. To the nearest person, what is the approximate density of Manila?

___?___ people per square mile

🌀 **Go Online** You can complete an Extra Example online.

Learn Density Based on Volume

Density is the measure of the quantity of some physical property per unit of length, area, or volume. If two objects have the same volume but different masses, the object with the greater mass will be denser.

Density Based on Volume	
Words	Density is the ratio of mass (or weight) to volume.
Symbols	$\text{density} = \dfrac{\text{mass (or weight)}}{\text{volume}}$

🌐 Example 3 Find the Density of a Solid

ART **Antonio opens a new brick of clay that weighs 25 pounds.**

a. What is the density of the brick of clay?

First find the volume of the clay, which can be approximated by using the formula for the volume of a rectangular prism.

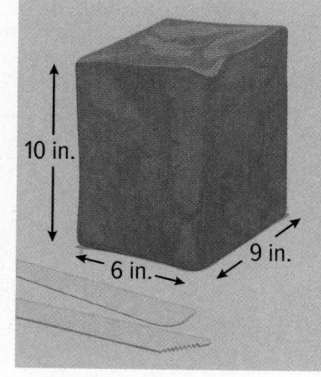

10 in.

6 in.

9 in.

$V = \ell wh$ Volume of a rectangular prism

$V = 6(9)(10)$ $\ell = 6$, $w = 9$, and $h = 10$

$\quad = 540 \text{ in}^3$ Simplify.

Next, use the density formula to calculate the density.

$\text{density} = \dfrac{\text{weight}}{\text{volume}}$ Density Formula

$d = \dfrac{25}{540}$ weight = 25 lb and volume = 540 in^3

The density of the clay is about 0.046 pounds per cubic inch.

b. Antonio uses the same clay to make a foundational cube for a sculpture. If the cube weighs 1.3 pounds, what are the dimensions of the cube?

Use the density formula to find the volume of the clay Antonio is using given the weight and the density of the clay.

$\text{density} = \dfrac{\text{weight}}{\text{volume}}$ Density Formula

$\dfrac{25}{540} = \dfrac{1.3}{V}$ density = $\frac{25}{540}$ and weight = 1.3

$V = 28.08$ Simplify.

Because the foundation is a cube, each edge s must be the same length. Therefore, $s^3 = 28.08$ and $s \approx 3.04$.

Each side of Antonio's cube will be about 3.04 inches long.

🔵 **Go Online** You can complete an Extra Example online.

💭 **Think About It!**

If Antonio decides to change the size of the cube so that its weight is greater than 1.3 pounds, how will this change affect the density? Explain.

Check

GARDENING When Kimani filled her planter with soil, the weight of the planter increased by 90 pounds.

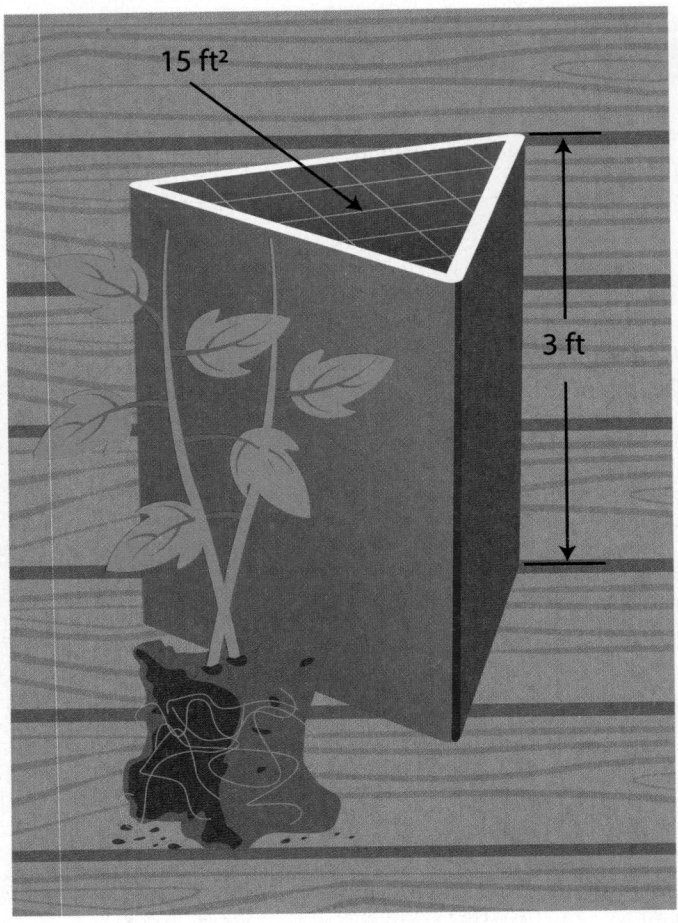

The density of the soil is _____?_____ pounds per cubic foot.

Kimani uses the same soil to fill another planter, and the weight increased by 154 pounds. The volume of the other planter is _____?_____ cubic feet.

Pause and Reflect

Did you struggle with anything in this lesson? If so, how did you deal with it?

Practice

⟶ **Go Online** You can complete your homework online.

Example 1

POPULATION Use the data in the table to find the population density of each city.

1. London, England

2. Paris, France

3. Madrid, Spain

4. Sydney, Australia

City	Population	Area (mi²)
London	8,674,000	607
Paris	2,224,000	40.7
Madrid	3,165,000	234
Sydney	4,293,000	4775

Example 2

5. **WILDLIFE** A town is installing bat boxes to reduce the number of mosquitos in the area. Each bat box will house 150 bats. The town officials approximate that 300 bats per square mile can control the mosquito population. If the area of the town is 12 square miles, how many bat boxes are needed?

6. **OCCUPANCY** Jamero is planning to open a new restaurant. Safety regulations recommend that a restaurant allows for 15 square feet of floor space per person. If Jamero wants his restaurant to be able to accommodate 50 people, what is the smallest area of floor space he will need?

Example 3

7. **FESTIVALS** Every year the Ohio State Fair features sculptures made of butter. While the sculptures vary from year to year, every year a new rendition of a cow and calf are created. In 2017, the sculptures of the cow and calf weighed 2000 pounds.

   5 in. 1.25 in. 1.25 in.

   a. If a stick of butter weighs 4 ounces, what is its approximate density?

   b. What was the total volume of the cow and calf sculptures that were made in 2017? (*Hint:* 1 pound = 16 ounces)

   c. Approximately how many sticks of butter were used to create the sculptures?

8. **ENERGY** A British Thermal Unit (BTU) is equal to the amount of energy used to raise the temperature of one pound of water 1°F. A house is shaped like a rectangular prism with a length, width, and height of 80 feet, 25 feet, and 8 feet, respectively. Approximately 60,000 BTUs are required to heat the house properly.

   a. What is the density of the house in BTUs?

   b. A nearby house requires approximately 52,000 BTUs for heating. If the house is 31 feet long and 25 feet wide, what is the height of the house? Round your answer to the nearest foot.

Mixed Exercises

9. **REGULATIONS** A rectangular national sight-seeing park with a length of 2 miles and a width of 3 miles has a maximum capacity of 250 people. Find the population density of people to the nearest hundredth.

10. **REASONING** A city is divided by a river. The portion of the city east of the river covers 25% of the city and has a population density of 28 people/km². The portion on the west side of the river has a population density of 17 people/km². Find the population density of the city as a whole.

11. **CONSTRUCT ARGUMENTS** The cargo of a semi-trailer can weigh no more than 34,000 pounds. The interior dimensions of a semi-trailer are shown. Suppose a freight company wants to haul a shipment that will completely fill the entire interior of the trailer, and the freight has a known density of 0.006 pounds/in³. Will this proposed load meet the weight restrictions? Justify your argument.

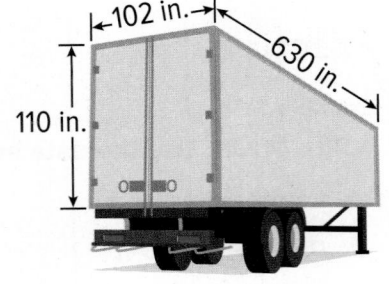

12. **PAPER WEIGHTS** The cylindrical paper weight with dimensions shown has a mass of 606.7 grams.

    a. Find the density of the paperweight.

    b. Use the table to determine which, if any, of the materials listed may have been used to make the paperweight.

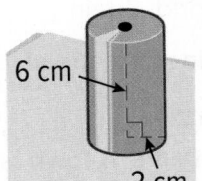

Material	Density
Silver	10.5 g/cm³
Copper	8.96 g/cm³
Steel	8.05 g/cm³

🍩 Higher-Order Thinking Skills

13. **WHICH ONE DOESN'T BELONG?** Jonathan is building his own terrarium for 18 plants, and he wants 120 cubic inches of space per plant. Which of the following proposed structures will not allow Jonathan to meet his requirement? Justify your conclusion.

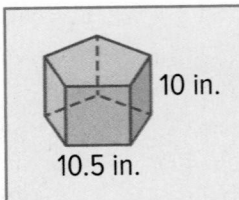
10 in.
10.5 in.

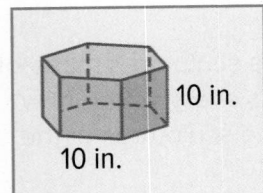
10 in.
10 in.

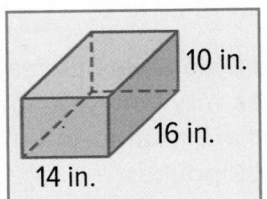
10 in.
16 in.
14 in.

14. **WRITE** Explain why a cubic foot of gas and a cubic foot of gold do not have the same density.

15. **ANALYZE** Determine whether the following statement is *true* or *false*. Justify your argument. Block A has a greater density than block B.

Block A
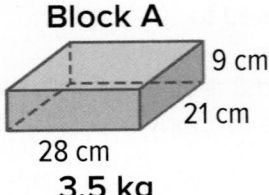
9 cm
21 cm
28 cm
3.5 kg

Block B
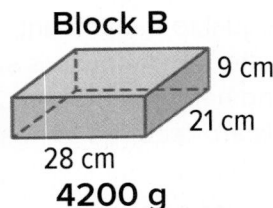
9 cm
21 cm
28 cm
4200 g

16. **PERSEVERE** An engineer is designing a marble fountain for the lobby of a museum. Before the engineer can finalize the design, she must choose a type of marble to use for the fountain. The marble samples that are available for viewing are rectangular prisms. The width, length, and height of each sample is 4 inches, 6 inches, and 1 inch, respectively. The mass of each sample is 1066.32 grams.

    a. What is the density of each marble sample?

    b. The main feature of the fountain will be a marble sphere that has a radius of 2 feet. What will be the mass of the sphere? Round your answer to the nearest kilogram.

Essential Question

How are measurements of two- and three-dimensional figures useful for modeling situations in the real world?

Two- and three-dimensional figures can allow you to visualize or estimate measurements of real-world objects.

Module Summary

Lessons 6-1 through 6-3

Two-Dimensional Areas

- parallelogram: $A = bh$
- trapezoid: $A = \frac{1}{2}h(b_1 + b_2)$
- rhombus or kite: $A = \frac{1}{2}d_1d_2$
- regular n-gon: $A = \frac{1}{2}aP$
- circle: $A = \pi r^2$
- sector: $A = \frac{x°}{360°} \cdot \pi r^2$

Lessons 6-4 and 6-7

Surface Area

- right prism: $S = Ph + 2B$, where P is the perimeter of a base, h is the height, and B is the area of a base
- right cylinder: $S = 2\pi rh + 2\pi r^2$, where r is the radius of a base and h is the height
- regular pyramid: $S = \frac{1}{2}P\ell + B$, where P is perimeter of the base, ℓ is the slant height, and B is the area of the base
- right circular cone: $S = \pi r\ell + \pi r^2$, where r is the radius of the base and ℓ is the slant height
- sphere: $S = 4\pi r^2$, where r is the radius of the sphere

Lessons 6-6 and 6-7

Volume

- prism: $V = Bh$, where B is the area of a base and h is the height of the prism
- pyramid: $V = \frac{1}{3}Bh$, where B is the area of the base and h is the height of the pyramid
- cylinder: $V = Bh$ or $V = \pi r^2h$, where B is the area of the base, h is the height of the cylinder, and r is the radius of the base
- circular cone: $V = \frac{1}{3}Bh$ or $V = \frac{1}{3}\pi r^2h$, where B is the area of the base, h is the height of the cone, and r is the radius of the base.
- sphere: $V = \frac{4}{3}\pi r^3$, where r is the radius of the sphere

Lessons 6-5, 6-8, and 6-9

Other Measurement Topics

- A cross section is the intersection of a solid figure and a plane.
- A solid of revolution is obtained by rotating a plane figure or curve around an axis.
- If two similar solids have a scale factor of $a : b$, then the surface areas have a ratio of $a^2 : b^2$, and the volumes have a ratio of $a^3 : b^3$.
- Density is the ratio of objects to area
- Density is the ratio of mass (or weight) to volume.

Study Organizer

Foldables

Use your Foldable to review this module. Working with a partner can be helpful. Ask for clarification of concepts as needed.

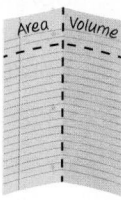

Test Practice

1. MULTIPLE CHOICE The diagram below is a blueprint for the head of a shovel. If the shovel head is cut from a square piece of aluminum that has a side length of 8 inches. What is the area of the wasted aluminum? (Lesson 6-1)

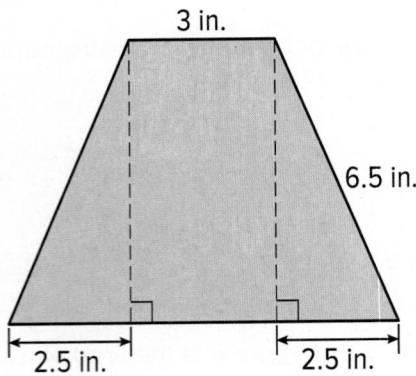

A. 12 in²   B. 16 in²   C. 31 in²   D. 40 in²

2. OPEN RESPONSE A company is using the figure below as the background for its logo. A designer at the company wants to know the area the logo will cover on a document. Find the area, in square centimeters, of the composite figure. (Lesson 6-1)

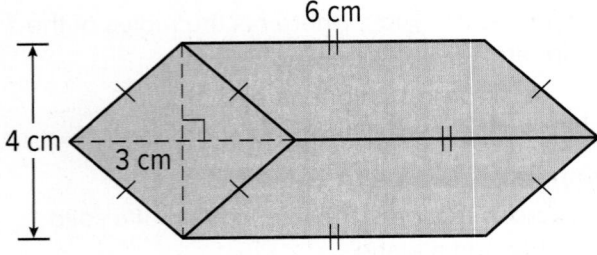

3. MULTIPLE CHOICE A sign-making company wants to know the minimum amount of metal needed to make a stop sign. A stop sign is shaped like a regular octagon. The distance between opposite sides of a stop sign is 30 inches. One side of the stop sign measures approximately 12.4 inches. What is the approximate area of the stop sign to the nearest square inch? (Lesson 6-2)

A. 372 in²   B. 588 in²

C. 744 in²   D. 1488 in²

4. OPEN RESPONSE Describe how you can derive the formula for the area of a sector by using the area of a circle. (Lesson 6-3)

5. MULTI-SELECT Given a circle with radius r and circumference, $C = 2\pi r$, which of the following steps would be included in an informal argument for the formula for the area of a circle? Select all that apply. (Lesson 6-3)

A. As the number of congruent pieces increases, the figure approaches a circle.

B. Divide the circle into equal wedge-shaped pieces.

C. Divide the circle into 4 pieces.

D. As the number of congruent pieces increases, the figure approaches a sector.

E. The wedge-shaped pieces form a regular polygon.

F. The area formula for a regular polygon is $A = \frac{1}{2}aP$.

G. Use the area of each of the 4 sectors to derive the formula for the area of a circle.

6. OPEN RESPONSE A pharmaceutical company is developing a new medicine and needs a capsule in the shape of a sphere to hold the medicine. The company wants to find the amount of material needed to make the capsule. What is the surface area of the capsule if the diameter is 10 millimeters? (Lesson 6-4)

7. MULTIPLE CHOICE A right square pyramid is intersected by a plane perpendicular to the base that passes through the vertex of the pyramid. What shape is the resulting cross section? (Lesson 6-5)

A. isosceles triangle

B. right triangle

C. square

D. trapezoid

8. MULTI-SELECT Which shape could be the cross section of a cube? Select all that apply. (Lesson 6-5)

A. hexagon

B. octagon

C. rectangle

D. triangle

9. MULTI-SELECT Given a triangular prism, which of the following steps would be included in an informal argument for the formula for the volume of a pyramid? Select all that apply. (Lesson 6-6)

A. Separate the prism into three triangular pyramids with congruent bases.

B. By Cavalieri's Principle, two of the pyramids have the same volume because at each level the two pyramids have the same cross-sectional area.

C. Each pyramid has one half the volume of the prism.

D. Separate the prism into three triangular pyramids.

E. Separate a rectangular face of the prism along a diagonal to create two congruent bases for two of the pyramids.

F. The bases of the original triangular prism are the congruent faces of two of the pyramids.

10. MULTIPLE CHOICE The image below represents a toy block. Find the minimum volume, in cubic centimeters, of wood necessary to make 30 blocks. (Lesson 6-6)

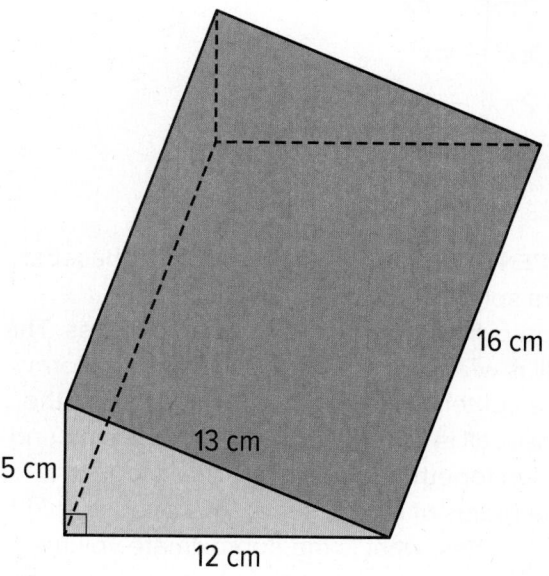

16 cm

13 cm

5 cm

12 cm

A. 8100 cm³

B. 14,400 cm³

C. 16,200 cm³

D. 28,800 cm³

11. OPEN RESPONSE A square pyramid is constructed so that the height has length $x + 2$ and the sides of the base have length x. Write an expression for the volume of the pyramid in terms of x. (Lesson 6-6)

12. MULTIPLE CHOICE The height of a cylinder is 1 inch less than the diameter of the cylinder. Which expression represents the volume of the cylinder in terms of its radius, x? (Lesson 6-7)

A. $2\pi x^3 - \pi x^2$

B. $\pi x^3 - \pi x^2$

C. $2\pi x^3 - 1$

D. $2\pi x^3 + \pi x^2$

13. OPEN RESPONSE At the center of a baseball is a sphere called the pill that has an approximate volume of 1.32 cubic inches. The pill is wrapped with 3 types of string to form the center of the baseball. The center of the baseball is covered with a leather casing and sewn together to make the final product. If the radius of the center of the baseball is 2.9 inches, what is the approximate volume of string, to the nearest cubic inch, that is used to wrap the pill? (Lesson 6-7)

14. MULTIPLE CHOICE An artist has made a scale model of a sculpture shaped like a cone. The ratio of the scale model to the final sculpture is 2 : 5. If the volume of the scale model is approximately 75.4 cubic inches, what is the volume of the final sculpture? (Lesson 6-8)

A. 188.5 in^3

B. 226.2 in^3

C. 471.3 in^3

D. 1178.1 in^3

15. OPEN RESPONSE A candle maker sells sets of candles in the shape of square pyramids. The volume of a smaller candle is 125 cubic centimeters. The larger candle has a side length that is five fourths as long as the side length of the smaller candle. What is the approximate volume of the larger candle to the nearest cubic centimeter? (Lesson 6-8)

16. MULTIPLE CHOICE These cylinders are similar.

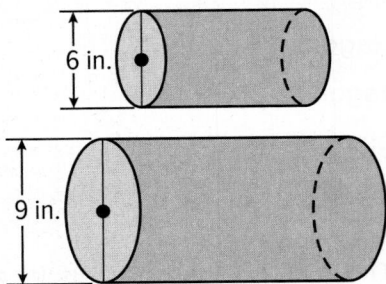

Find the volume of the smaller cylinder given that the larger cylinder has a volume of 190.85 cubic inches. (Lesson 6-8)

A. 28.3 in^3 B. 56.5 in^3

B. 84.8 in^3 D. 91.1 in^3

17. MULTIPLE CHOICE A pile of sand forms a cone with a diameter of 2 meters and a height of 0.7 meter. The mass of the pile is 1170 kilograms. What is the approximate density of the sand in kilograms per cubic meter? (Lesson 6-9)

A. 399 kg/m^3 B. 532 kg/m^3

C. 1140 kg/m^3 D. 1596 kg/m^3

18. OPEN RESPONSE Jacksonville, Florida, has a land area of 875 square miles and a population of 880,619. What is the approximate population density of Jacksonville to the nearest whole number? (Lesson 6-9)

Probability

e Essential Question
How can you use measurements to find probabilities?

What Will You Learn?

How much do you already know about each topic **before** starting this module?

KEY	Before			After		
👎 — I don't know. 👉 — I've heard of it. 👍 — I know it!	👎	👉	👍	👎	👉	👍
describe events using subsets						
solve problems involving using the rule for the probability of complementary events						
find the probability of an event by using lengths of segments and areas						
solve problems involving probabilities of compound events using permutations and combinations						
solve problems involving probability of independent events using the Multiplication Rule						
solve problems involving conditional probability						
solve problems involving mutually exclusive events using the Addition Rule						
solve problems involving events that are not mutually exclusive using the Addition Rule						
solve problems involving conditional probability using the Multiplication Rule						
decide if events are independent and approximate conditional probabilities using two-way frequency tables						

📖 **Foldables** Make this Foldable to help you organize your notes about probability. Begin with one sheet of notebook paper.

1. **Fold** a sheet of paper lengthwise.

2. **Fold** in half two more times.

3. **Cut** along each fold on the left column.

4. **Label** each section with a topic.

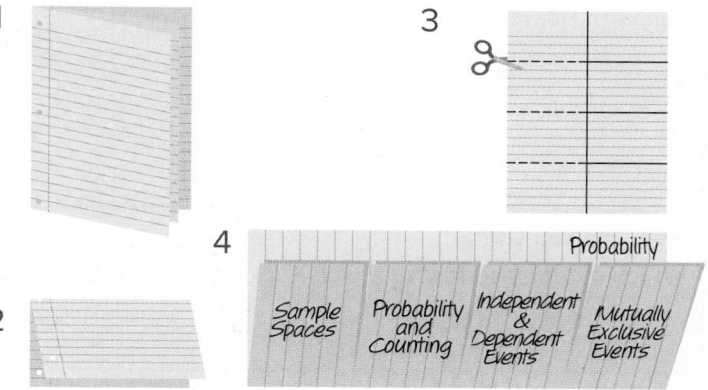

What Vocabulary Will You Learn?

- combination
- complement of A
- compound event
- conditional probability
- dependent events
- event
- experiment
- factorial of n

- finite sample space
- geometric probability
- independent events
- infinite sample space
- intersection of A and B
- joint frequencies
- marginal frequencies
- mutually exclusive

- outcome
- permutation
- relative frequency
- sample space
- two-way frequency table
- union of A and B

Are You Ready?

Complete the Quick Review to see if you are ready to start this module.
Then complete the Quick Check.

Quick Review

Example 1

Suppose a die is rolled. What is the probability of rolling less than 5?

$$P(\text{less than } 5) = \frac{\text{number of favorable outcomes}}{\text{number of possible outcomes}}$$

$$= \frac{4}{6} \text{ or } \frac{2}{3}$$

The probability of rolling less than 5 is $\frac{2}{3}$ or 67%.

Example 2

A spinner numbered 1–6 was spun. Find the experimental probability of landing on a 5.

Outcome	Tally	Frequency
1	IIII	4
2	ЖII	7
3	ЖIII	8
4	IIII	4
5	II	2
6	Ж	5

$$P(5) = \frac{\text{number of times a 5 is spun}}{\text{total number of outcomes}} = \frac{2}{30} \text{ or } \frac{1}{15}$$

The experimental probability of landing on a 5 is $\frac{1}{15}$ or 7%.

Quick Check

A die is rolled. Find the probability of each outcome

1. $P(\text{greater than } 1)$

2. $P(\text{odd})$

3. $P(\text{less than } 2)$

4. $P(1 \text{ or } 6)$

The table shows the results of an experiment in which a spinner numbered 1–4 was spun.

Outcome	Tally	Frequency
1	III	3
2	Ж	5
3	ЖIII	8
4	IIII	4

5. What is the experimental probability that the spinner will land on a 4?

6. What is the experimental probability that the spinner will land on an odd number?

7. What is the experimental probability that the spinner will land on an even number?

How Did You Do?

Which exercises did you answer correctly in the Quick Check?

Sample Spaces

Learn Sample Spaces

An **experiment** is a situation involving chance. An **outcome** is the result of a single performance or trial of an experiment.

The set of all possible outcomes make up the **sample space** of an experiment, which may be finite or infinite. A sample space that contains a countable number of outcomes is a **finite sample space**. A sample space with outcomes that cannot be counted is an **infinite sample space**. An **event** is a subset of the sample space.

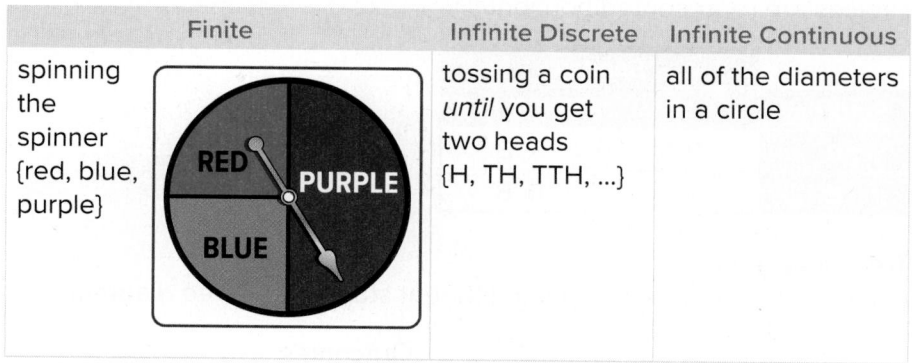

	Finite	Infinite Discrete	Infinite Continuous
spinning the spinner {red, blue, purple}	RED · PURPLE · BLUE	tossing a coin *until* you get two heads {H, TH, TTH, ...}	all of the diameters in a circle

Infinite discrete sample spaces have outcomes that can be arranged in a sequence or counted but go on indefinitely. Infinite continuous sample spaces cannot be counted or defined because there are infinite ways to fill the sample space.

You can represent a sample space by using an organized list, a table, or a tree diagram.

Example 1 Define a Sample Space

A fair die is tossed once.

a. What is the sample space of the experiment?

The sample space S includes all possible outcomes of rolling a die.

$S = \{1, 2, 3, 4, 5, 6\}$

b. What is the sample space for the event of rolling a prime number? Write the outcomes to complete the sample space.

The sample space S (prime number on a die) includes all prime numbers less than 6.

$S(\text{prime number on a die}) = \{2, 3, 5\}$

Go Online You can complete an Extra Example online.

Today's Goals
- Define sample spaces and describe subsets of sample spaces.
- Apply the Fundamental Counting Principle to define sample spaces.

Today's Vocabulary
experiment
outcome
sample space
finite sample space
infinite sample space
event

Study Tip

Empty Set An empty set is a set containing no elements. In probability, if there are no possible outcomes that satisfy an event, then the sample space is described using the empty set ø.

🌐 Example 2 Represent a Sample Space

CLOTHING **Kembe has a black hat and a red hat. He chooses one hat for each day, Saturday and Sunday. Represent the sample space for this experiment by making an organized list, a table, and a tree diagram.**

For each day, Saturday and Sunday, there are two possibilities: the red hat (R) or the black hat (B).

Organized List

Pair each possible outcome from the Saturday's hat choice with the possible outcomes from Sunday's hat choice using coordinates.

$S = \{(R, R), (R, B), (B, B), (B, R)\}$

Table

Saturday's hat choices are represented vertically, and Sunday's hat choices are represented horizontally.

Saturday	Sunday	
	R	B
R	R, R	R, B
B	B, R	B, B

Tree Diagram

Each event is represented by a different stage of the tree diagram.

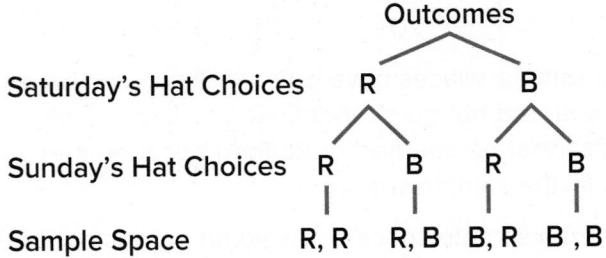

Check

GROUP WORK A geometry teacher always breaks her class up into the red, yellow, and blue groups for class projects. Represent the sample space for the next two class projects by making an organized list.

Enter the outcomes to complete the organized list.

$S = \{(R, R), (R, Y), \text{_____}, (Y, R), \text{_____}, \text{_____}, (B, R), \text{_____}, \text{_____}.$

🔎 **Go Online** You can complete an Extra Example online.

Study Tip

Tree Diagram Notation
Choose notation for outcomes in your tree diagrams that will eliminate confusion. In the example, R stands for red hat and B stands for black hat.

Example 3 Finite and Infinite Sample Spaces

Classify each sample space as *finite* or *infinite*. If it is finite, write the sample space. If it is infinite, classify whether it is *discrete* or *continuous*.

a. A marble is drawn from a bag that contains 3 orange marbles, 5 green marbles, and 4 blue marbles.

There are only three possible outcomes of this experiment: selecting an orange, green, or blue marble. The sample space S is finite, $S = \{$orange, green, blue$\}$.

b. A spinner with four equal parts of green, blue, red, and yellow is spun until it lands on yellow.

There are a(n) infinite number of possible outcomes of this experiment, so its sample space is infinite. Because the experiment ends after a certain number of spins when the spinner lands on yellow, the sample space is discrete.

c. A ball is thrown into the air and its height is recorded in inches.

There are a(n) infinite number of possible outcomes of this experiment, so its sample space is infinite. You can continue to record heights of the thrown ball indefinitely, so the sample space is continuous.

Learn Fundamental Counting Principle

For some large or complicated experiments, listing the entire sample space may not be practical or necessary. To find the *number* of possible outcomes, you can use the Fundamental Counting Principle.

Key Concept • Fundamental Counting Principle	
Words	The number of possible outcomes in a sample space can be found by multiplying the number of possible outcomes from each stage or event.
Symbols	In a k-stage experiment, let $n_1 =$ the number of possible outcomes for the first stage. $n_2 =$ the number of possible outcomes for the second stage after the first stage has occurred. $n_k =$ the number of possible outcomes for the kth stage after the first $k-1$ stages have occurred. Then the total possible outcomes of this k-stage experiment is $n_1 \cdot n_2 \cdot n_3 \cdot \ldots \cdot n_k$.

Think About It!

If each stage of a two-stage experiment has three outcomes, what is the number of total possible outcomes?

Example 4 Use the Fundamental Counting Principle

COLLEGE Santiago lists the number of sections available for the courses he will take in his first semester at college. How many different schedules could Santiago create for this semester?

Course	Sections Offered
Art History	6
French	5
Mathematics	9
Art	4
English	6

You can estimate the total number of different schedules he can make. There are about 10 sections of the mathematics course offered. For each of the other four courses, there are about 5 sections offered. Multiply to estimate that Santiago can create about 6250 schedules.

Find the number of possible outcomes by using the Fundamental Counting Principle to complete the equation.

Art History		French		Mathematics		Art		English		Possible Outcomes
6	×	5	×	9	×	4	×	6	=	6480

Santiago could create 6480 different schedules. Because 6480 is close to the estimate of 6250, the answer is reasonable.

Check

CLOTHING A sneaker company lets you customize your own sneaker on their Web site. Using their most popular sneaker as the base, you have the option to customize the color of each part of the sneaker.

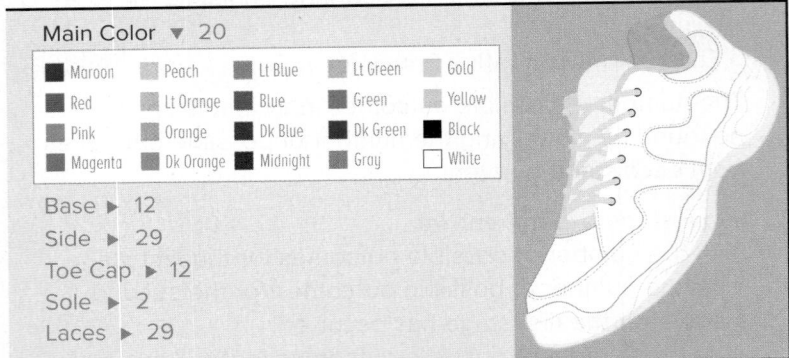

• **Customize Your Shoes**

Main Color ▼ 20

■ Maroon	■ Peach	■ Lt Blue	■ Lt Green	■ Gold
■ Red	■ Lt Orange	■ Blue	■ Green	■ Yellow
■ Pink	■ Orange	■ Dk Blue	■ Dk Green	■ Black
■ Magenta	■ Dk Orange	■ Midnight	■ Gray	□ White

Base ▶ 12
Side ▶ 29
Toe Cap ▶ 12
Sole ▶ 2
Laces ▶ 29

Part A Which is the best estimate for the number of possible customizations?

A. 100 **B.** 3,600,000

C. 5,062,500 **D.** 36,000,000

Part B How many different customizations can be created?

🕹 **Go Online** You can complete an Extra Example online.

Practice

Go Online You can complete your homework online.

Example 1

1. Define the sample space, *S*, of a fair coin being tossed once.

2. A numbered spinner with six equal parts is spun once.
   a. What is the sample space of the experiment?

   b. What is the sample space for the event of landing on a prime number?

3. DODECAGON A regular, 12-sided dodecagon is rolled once.
   a. What is the sample space of the experiment?

   b. What is the sample space for the event of rolling an even number?

4. SPINNERS A lettered spinner with five equal parts is spun once.
   a. What is the sample space of the experiment?

   b. What is the sample space for landing on a vowel?

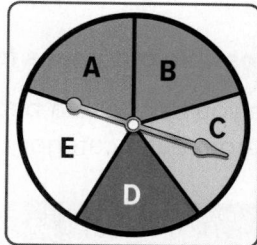

Example 2

5. UNIFORMS For away games, the baseball team can wear blue or white shirts with blue or white pants. Represent the sample space for each experiment by making an organized list, a table, and a tree digram.

6. CHILDCARE Khalid's baby sister can drink either apple juice or milk from a bottle or a toddler cup. Represent the sample space for each experiment by making an organized list, a table, and a tree diagram.

Example 3

Classify each sample space as *finite* or *infinite*. If it is finite, write the sample space. If it is infinite, classify whether it is *discrete* or *continuous*.

7. A color tile is drawn from a cup that contains 1 yellow, 2 blue, 3 green, and 4 red color tiles.

8. A numbered spinner with eight equal parts is spun until it lands on 2.

9. An angler casts a fishing line into a body of water and its distance is recorded in centimeters.

10. A letter is randomly chosen from the alphabet.

Example 4

Find the number of possible outcomes for each situation.

11. A video game lets you decorate a bedroom using one choice from each category.

Bedroom Décor	Number of Choices
Paint color	8
Comforter set	6
Sheet set	8
Throw rug	5
Lamp	3
Wall hanging	5

12. A cafeteria meal at Angela's work includes one choice from each category.

Cafeteria Meal	Number of Choices
Main dish	3
Side dish	4
Vegetable	2
Salad	2
Salad Dressing	3
Dessert	2
Drink	3

13. **SHOPPING** On a website showcasing outdoor patio plans, there are 4 types of stone, 3 types of edging, 5 dining sets, and 6 grills. Kamar plans to order one item from each category. How many different patio sets can Kamar order?

14. **AUDITIONS** The drama club held tryouts for 6 roles in a one-act play. Five people auditioned for lead female, 3 for lead male, 8 for the best friend, 4 for the mother, 2 for the father, and 3 for the humorous aunt. How many different casts can be created from those who auditioned?

Mixed Exercises

15. **BOARD GAMES** The spinner shown is used in a board game. If the spinner is spun 4 times, how many different possible outcomes are there?

16. BASKETBALL In a city basketball league there must be a minimum of 14 players on a team's roster. One 14-player team has three centers, four power forwards, two small forwards, three shooting guards, and the rest of the players are point guards. How many different 5-player teams are possible if one player is selected from each position?

17. VACATION RENTAL Angelica is comparing vacation prices in Boulder, Colorado, and Sarasota, Florida. In Boulder, she can choose a 1- or 2-week stay in a 1- or 2-bedroom suite. In Sarasota, she can choose a 1-, 2-, or 3-week stay in a 2- or 3-bedroom suite, on the beach or not.

   a. How many outcomes are available in Boulder?

   b. How many outcomes are available in Sarasota?

   c. How many total outcomes are available?

18. TRAVEL Maurice packs suits, shirts, and ties that can be mixed and matched. Use his packing list to draw a tree diagram to represent the sample space for possible suit combinations using one article from each category.

> **Maurice's Packing List**
> 1. Suits: Gray, black, khaki
> 2. Shirts: White, light blue
> 3. Ties: Striped (But optional)

Find the number of possible outcomes for each situation.

19. SCHOOL Tala wears a school uniform that consists of a skirt or pants, a white shirt, a blue jacket or sweater, white socks, and black shoes. She has 3 pairs of pants, 3 skirts, 6 white shirts, 2 jackets, 2 sweaters, 6 pairs of white socks, and 3 pairs of black shoes.

20. FOOD A sandwich shop provides its customers with a number of choices for bread, meats, and cheeses. Provided one item from each category is selected, how many different sandwiches can be made?

Bread	Meats	Cheeses
White	Turkey	American
Wheat	Ham	Swiss
Whole Grain	Roast Beef	Provolone
	Chicken	Colby-Jack
		Muenster

21. List six different expressions that could be used to evaluate the area of the composite figure.

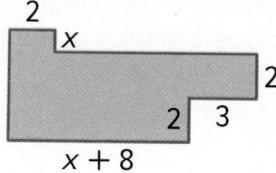

22. LICENSE PLATES One state requires license plates to consist of three letters followed by three numbers. The letter "O" and the number "0" may not be used, but any other combination of letters or numbers is allowed. How many different license plates can be created?

23. COLLEGE Jack has been offered a number of internships that could occur in 3 different months, in 4 different departments, and for 3 different companies. Jack is only available to complete his internship in July. How many different outcomes are there for his internship?

24. **BIKING** Talula got a new bicycle lock that has a four-number combination. Each number in the combination is from 0 to 9.

   a. How many combinations are possible if there are no restriction on the number of times Talula can use each number?

   b. How many combinations are possible if Talula can use each number only once? Explain.

25. **BOARD GAMES** Hugo and Monette are playing a board game in which the player rolls two fair dice per turn.

   a. In one turn, how many outcomes result in a sum of 8?

   b. How many outcomes in one turn result in an odd sum?

26. **WRITING** Explain when it is necessary to show all the possible outcomes of an experiment by using a tree diagram and when using the Fundamental Counting Principle is sufficient.

27. **REASONING** A multistage experiment has n possible outcomes at each stage. If the experiment is performed with k stages, write an equation for the total number of possible outcomes P. Explain.

🐣 **Higher-Order Thinking Skills**

28. **PERSEVERE** A box contains n different objects. If you remove three objects from the box, one at a time, without putting the previous object back, how many possible outcomes exist? Explain your reasoning.

29. **CREATE** Sometimes a tree diagram for an experiment is not symmetrical. Describe a two-stage experiment where the tree diagram is asymmetrical. Include a sketch of the tree diagram. Explain.

30. **WRITE** Explain why it is not possible to represent the sample space for a multi-stage experiment by using a table.

31. **ANALYZE** Determine if the following statement is *sometimes*, *always*, or *never* true. Justify your argument.

   When an outcome falls outside the sample space, it is a failure.

Probability and Counting

Learn Intersections and Unions

When two events *A* and *B* occur, the **intersection of *A* and *B*** is the set of all outcomes in the sample space of event *A* that are also in the sample space of event *B*. In the Venn diagram, the shaded portion represents the intersection.

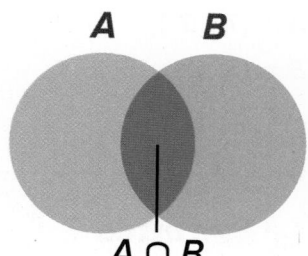

$A \cap B$

To determine the probability of an outcome from the intersection of two or more events, find the ratio of the number of outcomes in both events to the total number of possible outcomes.

> **Key Concept • Probability Rule for Intersections**
>
> The probability of the intersection of two events *A* and *B* occurring is the ratio of the number of outcomes in both *A* and *B* to the total number of possible outcomes.
>
> $$P(A \cap B) = \frac{\text{number of outcomes in } A \text{ and } B}{\text{total number of possible outcomes}}$$

When two events *A* and *B* occur, the **union of *A* and *B*** is the set of all outcomes in the sample space of event *A* combined with all outcomes in the sample space of event *B*. In the Venn diagram, the shaded portion represents the union.

> **Key Concept • Union of Two Events**
>
> The number of elements in the union of two events *A* and *B* is the number of outcomes in both event *A* and *B* minus the number of outcomes in their intersection.
>
> $n(A \cup B) = n(A) + n(B) - n(A \cap B)$

Example 1 Find Intersections

A fair die is rolled once. Let *A* be the event of rolling an odd number, and let *B* be the event of rolling a number greater than 3. Find *A* ∩ *B*.

The possible outcomes for event *A* are all the numbers on a die that are odd, or {1, 3, 5}.

The possible outcomes for event *B* are all the numbers on a die that are greater than 3, or {4, 5, 6}.

A ∩ *B* contains all of the outcomes that are in both sample space *A* and *B*.

$A \cap B = \{5\}$

🅝 **Go Online** You can complete an Extra Example online.

Today's Goals
- Describe events as subsets of sample spaces by using intersections and unions.
- Describe events as subsets of sample spaces by using complements.

Today's Vocabulary
intersection of *A* and *B*

union of *A* and *B*

complement of *A*

Study Tip

Intersection The symbol for intersection is ∩, and it is associated with the word *and*.

$P(A \cap B)$ is read as *the probability of A and B*.

Study Tip

Union The symbol for union is ∪, and it is associated with the word *or*.

$n(A \cup B)$ is read as, *the number of elements in A or B*.

💬 Think About It!

Is the sample space for *A* ∩ *B* finite or infinite? Justify your reasoning.

Go Online
You may want to complete the Concept Check to check your understanding.

Check

Let *A* be the event of the spinner landing on a blue section, and let *B* be the event of the spinner landing on a section with a number divisible by 3. What are the possible outcomes of each event?

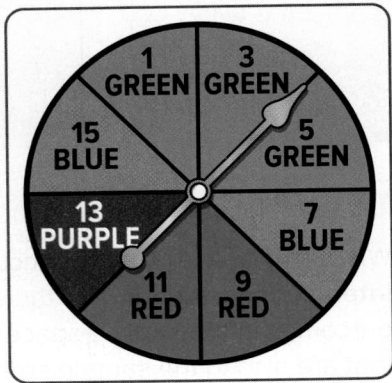

$A = \{7, \underline{}\}$

$B = \{3, \underline{}, 15\}$

$A \cap B = \{\underline{}\}$

🌐 Example 2 Find Probability of Intersections

PLAYING CARDS A card is selected from a standard deck of cards. What is the probability that the card is a queen and is red?

Let *A* be the event of choosing a queen, and let *B* be the event of choosing a red card. The total number of outcomes is the total number of cards in a deck, or 52.

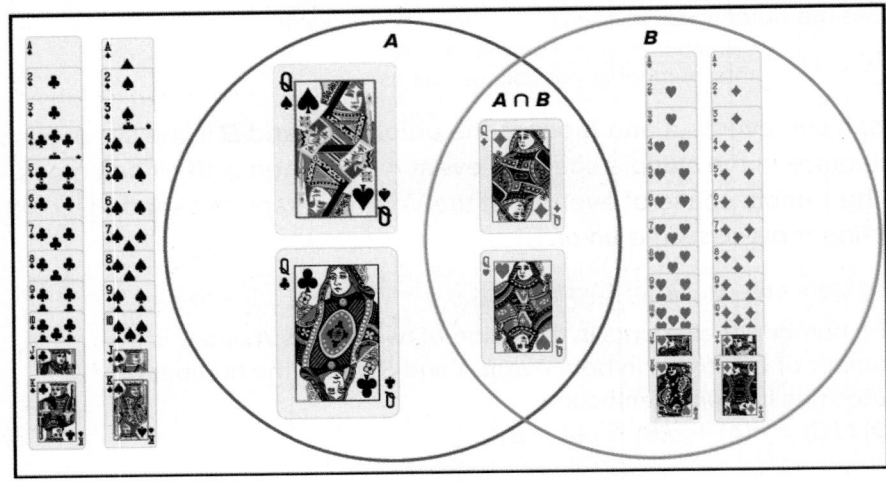

From the diagram, there are only 2 red cards that are also queens.

$$P(A \cap B) = \frac{\text{number of outcomes in } A \text{ and } B}{\text{total number of possible outcomes}} \quad \text{Probability Rule for Intersections}$$

$$= \frac{2}{52} \qquad \text{Substitution}$$

$$= \frac{1}{26} \qquad \text{Simplify.}$$

The probability that the card is both a queen and is red is $\frac{1}{26}$, or about 3.8%.

🌐 **Go Online** You can complete an Extra Example online.

Example 3 Find Unions

A fair die is rolled once. Let _A_ be the event of rolling a number less than 5, and let _B_ be the event of rolling a multiple of 2. Find _A_ ∪ _B_.

The possible outcomes for event _A_ are all the numbers on a die that are less than 5, or {1, 2, 3, 4}.

The possible outcomes for event _B_ are all the numbers on a die that are multiples of 2, or {2, 4, 6}.

_A_ ∪ _B_ contains all of the outcomes that are in either sample space(s) _A_ or _B_.

_A_ ∪ _B_ = {1, 2, 3, 4, 6}

💭 **Think About It!**

Why are 2 and 4 only listed once in _A_ ∪ _B_? Explain.

Check

Let _A_ be the event of the spinner landing on a blue section, and let _B_ be the event of the spinner landing on a section with a number divisible by 3. What are the possible outcomes of each event?

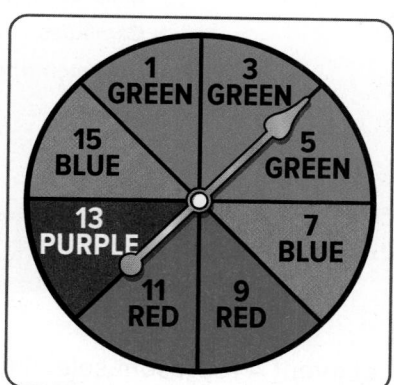

_A_ = {7, __?__}

_B_ = {3, __?__, 15}

_A_ ∪ _B_ = {3, 7, 9, __?__}

Learn Complements

The **complement of _A_** consists of all the outcomes in the sample space that are not included as outcomes of event _A_. The complement of event _A_ can be noted as _A'_, as shown in the Venn diagram.

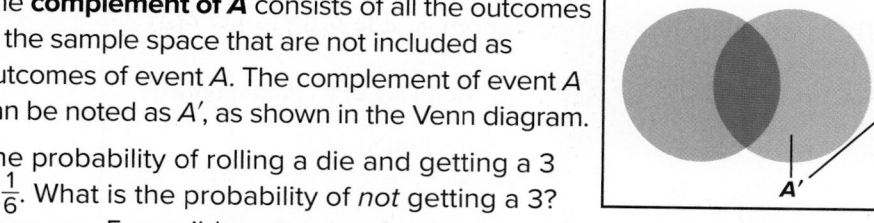

The probability of rolling a die and getting a 3 is $\frac{1}{6}$. What is the probability of _not_ getting a 3? There are 5 possible outcomes for this event: 1, 2, 4, 5, or 6. So, $P(\text{not }3) = \frac{5}{6}$. Notice that this probability is also $1 - \frac{1}{6}$ or $1 - P(3)$.

Key Concept • Probability of the Complement of an Event	
Words	The probability that an event will not occur is equal to 1 minus the probability that the event will occur.
Symbols	For an event _A_, $P(A') = 1 - P(A)$.

🌐 Example 4 Complementary Events

DIGITAL MEDIA **Panju subscribes to a movie streaming service. For movie night, he is going to let the program randomly pick a movie from his list of favorites. What is the probability that a comedy movie will not be chosen?**

My Movie Queue	
GENRES	
▸ Action	44
▸ Anime	109
▸ Children's	8
▸ Comedies	112
▸ Documentaries	13
▸ Dramas	30
▸ Foreign	5
▸ Horror	29

HOME   GENRE   SEARCH   MY LIST

Let event A represent selecting a comedy movie from Panju's favorites. Then find the probability of the complement of A.

There are 112 comedy movies in Panju's favorites list.

There are 350 total movies in Panju's favorites list.

The probability of the complement of A is $P(A') = 1 - P(A)$.

$P(A') = 1 - P(A)$ — Probability of a complement

$\quad = 1 - \dfrac{112}{350}$ — Substitution

$\quad = \dfrac{238}{350} \text{ or } \dfrac{17}{25}$ — Subtract and simplify.

The probability that a comedy movie will not be chosen is $\dfrac{17}{25}$ or 68%.

💬 Talk About It!

Why do you think the probability of the complement of an event is found by subtracting from 1?

Check

RAFFLE The Harvest Fair sold 967 raffle tickets for a chance to win a new TV. Copy and complete the table to find each probability of not winning the TV with the given number of tickets.

Number of Tickets	Probability of Not Winning
20	
200	
100	
1	

🔎 **Go Online** You can complete an Extra Example online.

Practice

🔄 **Go Online** You can complete your homework online.

Example 1

1. A fair die is rolled once. Let A be the event of rolling an even number, and let B be the event of rolling a number greater than 4. Find $A \cap B$.

2. A fair die is rolled once. Let A be the event of rolling an even number, and let B be the event of rolling an odd number. Find $A \cap B$.

Use the spinner.

3. Let A be the event of the spinner landing on 4 or 10, and let B be the event of the spinner landing on a section with a number divisible by 4. What are the possible outcomes of each event?

   a. $A = \{\underline{\ \ ?\ \ }\}$

   b. $B = \{\underline{\ \ ?\ \ }\}$

   c. $A \cap B = \{\underline{\ \ ?\ \ }\}$

4. Let P be the event of the spinner landing on a section with a prime number, and let Q be the event of the spinner landing on a section with a number that is a multiple of 3. What are the possible outcomes of each event?

   a. $P = \{\underline{\ \ ?\ \ }\}$

   b. $Q = \{\underline{\ \ ?\ \ }\}$

   c. $P \cap Q = \{\underline{\ \ ?\ \ }\}$

Example 2

5. A card is selected from a standard deck of cards. What is the probability that the card is a diamond and is a seven?

6. A card is selected from a standard deck of cards. What is the probability that the card has a number on it that is divisible by 2 and is black?

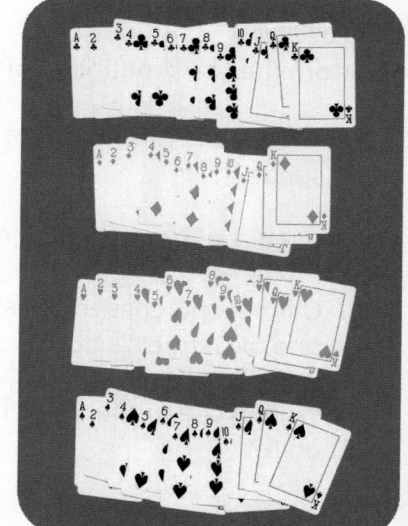

Example 3

Use the spinner.

7. Let *A* be the event that the spinner lands on a vowel. Let *B* be the event that it lands on the letter J. What are the possible outcomes of each event?

   a. $A = \{\underline{\ ?\ }\}$

   b. $B = \{\underline{\ ?\ }\}$

   c. $A \cup B = \{\underline{\ ?\ }\}$

8. Let *X* be the event that the spinner lands on a consonant. Let *Y* be the event that it lands on the letter K. What are the possible outcomes of each event?

   a. $X = \{\underline{\ ?\ }\}$

   b. $Y = \{\underline{\ ?\ }\}$

   c. $X \cup Y = \{\underline{\ ?\ }\}$

9. A random number generator is used to generate one integer between 1 and 20. Let *C* be the event of generating a multiple of 5, and let *D* be the event of generating a number less than 12. What are the possible outcomes of each event?

   a. $C = \{\underline{\ ?\ }\}$

   b. $D = \{\underline{\ ?\ }\}$

   c. $C \cup D = \{\underline{\ ?\ }\}$

10. A random number generator is used to generate one integer between 1 and 100. Let *A* be the event of generating a multiple of 10, and let *B* be the event of generating a factor of 30. What are the possible outcomes of each event?

   a. $A = \{\underline{\ ?\ }\}$

   b. $B = \{\underline{\ ?\ }\}$

   c. $A \cup B = \{\underline{\ ?\ }\}$

Example 4

Determine the probability of each event. Round to the nearest hundredth, if necessary.

11. What is the probability of drawing a card from a standard deck and not getting a spade?

12. What is the probability of flipping a coin and not landing on tails?

13. Carmela purchased 10 raffle tickets. If 250 were sold, what is the probability that one of Carmela's tickets will not be drawn?

14. What is the probability of spinning a spinner numbered 1 to 6 and not landing on 5?

Mixed Exercises

15. **STATISTICS** A survey found that about 90% of the junior class is right-handed. If 1 junior is chosen at random out of 100 juniors, what is the probability that he or she is left-handed?

16. **RAFFLE** Raul bought 24 raffle tickets out of 1545 tickets sold. What is the probability that Raul will not win the grand prize of the raffle?

17. **MASCOT** At Riverview High School, 120 students were asked whether they prefer a lion or a timber wolf as the new school mascot. What is the probability that a randomly-selected student will have voted for a lion as the new school mascot?

	Votes
Lion	78
Timber Wolf	42
Total	120

18. **COLLEGE** In Evan's senior class of 240 students, 85% are planning to attend college after graduation. What is the probability that a senior chosen at random is not planning to attend college after graduation?

19. **DRAMA CLUB** The Venn diagram shows the cast members who are in Acts I and II of a school play. One of the students will be chosen at random to attend a statewide performing arts conference. Let *A* be the event that a cast member is in Act I of the play and let *B* be the event that a cast member is in Act II of the play.

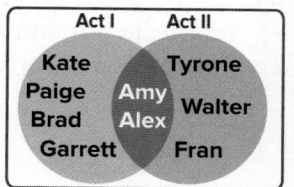

a. Find $A \cap B$.

b. What is the probability that the student who is chosen to attend the conference is a cast member in only one of the two Acts of the play.

20. **GAMES** LaRae is playing a game that uses a spinner. What is the probability that the spinner will land on a prime number on her next spin?

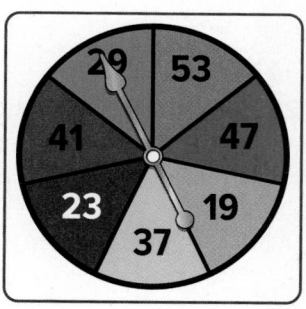

21. SHOPPING Raya asks 40 people outside the mall whether or not they visited for shopping or dining. She records the results in a Venn diagram. One person will be chosen at random to be interviewed on the local evening news. Find the probability that the person chosen will be someone who visited the mall for shopping and dining.

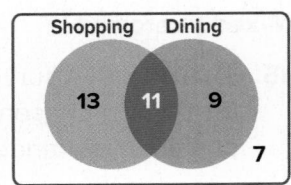

Higher-Order Thinking Skills

22. PERSEVERE Let *A* be the possible integer side measures of the rectangle with perimeter *P* = 52. Let *B* represent the possible integer measures of $\overline{XY}$ in $\triangle XYZ$.

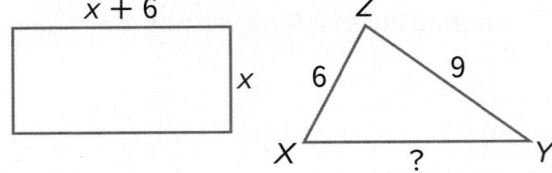

   a. Find $A \cap B$.

   b. Find $A \cup B$.

23. CREATE Let *A* be the months of the year with 31 days and let *B* be the months of the year that begin with the letter J. Create a Venn diagram to display this data.

24. WRITE Suppose you need to explain the concept of *intersections* and *unions* to someone with no knowledge of the topic. Write a brief description of your explanation.

25. ANALYZE Determine if the following statement is *sometimes, always,* or *never* true. Justify your argument.

      The union of two sets has more elements than the intersection of two sets.

26. FIND THE ERROR Let *A* be the event that the spinner lands on a vowel. Let *B* be the event that it lands on the letter J. Truc says $A \cup B$ is {A, E, O, U, J}, and Alan says $A \cup B$ is ∅. Who is correct? Explain.

Geometric Probability

Explore Probability Using Lengths of Segments

 Online Activity Use dynamic geometry software to complete the Explore.

> @ **INQUIRY** How can lengths of segments be used to determine probability? ×

Learn Probability with Length

Probability that involves a geometric measure such as length or area is called **geometric probability**.

Key Concept · Length Probability Ratio	
Words	If a line segment (1) contains another segment (2) and a point on segment (1) is chosen at random, then the probability that the point is on segment (2) is $\dfrac{\text{length of segment (2)}}{\text{length of segment (1)}}$.
Example	If a point E on $\overline{AD}$ is chosen at random, then $P(E \text{ is on } \overline{BC}) = \dfrac{BC}{AD}$. 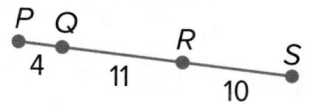

When determining geometric probabilities, we assume
- that the object lands within the target area, and
- it is equally likely that the object will land anywhere in the region.

Example 1 Use Length to Find Geometric Probability

Point X is chosen at random on $\overline{PS}$. Find the probability that X is on $\overline{RS}$.

What is the length of $\overline{PS}$? $PS = 25$

What is the length of $\overline{RS}$? $RS = 10$

$P(X \text{ is on } \overline{RS}) = \dfrac{RS}{PS}$ Length probability ratio

$= \dfrac{10}{25}$ $RS = 10$; $PS = 25$

$= \dfrac{2}{5}$ or 0.4   Simplify.

The probability that X is on $\overline{RS}$ is 40%.

 Go Online You can complete an Extra Example online.

Today's Goals
- Find the probability of an event by using lengths of segments.
- Find the probability of an event by using areas.

Today's Vocabulary
geometric probability

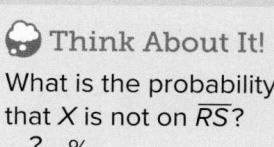 **Think About It!**

What is the probability that X is not on $\overline{RS}$?
___?___ %

Check

Point A is chosen at random on $\overline{WZ}$. Find the probability to the nearest percent that A is not on $\overline{YZ}$.

What is the length of $\overline{WZ}$? $WZ = $ __?__

What is the length of $\overline{YZ}$? $YZ = $ __?__

To the nearest percent, $P(A$ is not on $\overline{YZ})$ is about __?__ %.

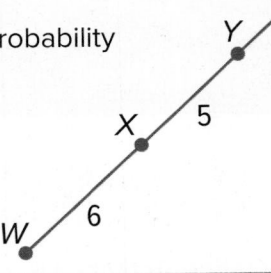

🌐 Example 2 Model Real-World Probabilities

IMAGE SHARING A Web site that hosts an image sharing gallery updates its front page for new content every 6 minutes. Assuming that you open the gallery at a random time, what is the probability that you will have to wait 5 or more minutes for the content to refresh?

You can use a number line to model the situation. Since the page is updated every 6 minutes, the next update will be in 6 minutes or less.

```
      A                         B   D
  ←---+---+---+---+---+---●───●---→
      0   1   2   3   4   5   6
```

The event of waiting 5 or more minutes is modeled by $\overline{BD}$ on the number line.

Find $P(\text{waiting 5 minutes or more}) = \dfrac{BD}{AD}$ Length probability ratio

$$= \frac{1}{6} \qquad BD = 1 \text{ and } AD = 6$$

So, the probability of waiting 5 or more minutes for the gallery to refresh is $\frac{1}{6}$ or about 17%.

Explore Probability and Decision Making

🧭 **Online Activity** Use the guiding exercises to complete the Explore.

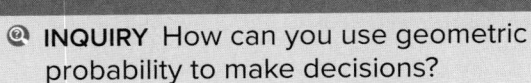

@ **INQUIRY** How can you use geometric probability to make decisions?

Learn Probability with Area

Geometric probability can also involve area.

Key Concept • Area Probability Ratio	
Words	If region A contains a region B and a point E in region A is chosen at random, then the probability that point E is in region B is $\dfrac{\text{area of region } B}{\text{area of region } A}$.
Example	If point E is chosen at random in rectangle A, then $P(\text{point } E \text{ is in circle } B) = \dfrac{\text{area of region } B}{\text{area of region } A}$.

🌐 Example 3 Use Area to Find Geometric Probability

LAWN GAMES Haruko Games is designing a new lawn game where each player will attempt to hit a circular target by tossing a beanbag onto a larger, circular board.

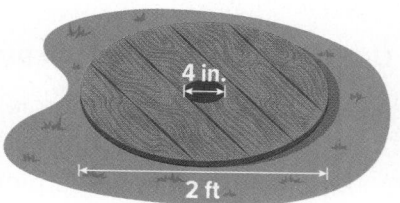

Study Tip

Assumptions When determining the probability that a toss will land on the target, we are assuming that the toss must land on the game board, and not on the ground surrounding the board.

Part A What is the probability that a toss will land on the target?

Find the ratio of the area of the circular target to the area of the board.

$P(\text{toss lands on target}) = \dfrac{\text{area of target}}{\text{area of game board}}$ Area probability ratio

$\qquad\qquad = \dfrac{\pi(2)^2}{\pi(12)^2}$ Area $= \pi r^2$

$\qquad\qquad = \dfrac{4}{144}$ or $\dfrac{1}{36}$ Simplify.

The probability that the toss lands on the target is $\dfrac{1}{36}$ or about 3%.

Study Tip

Units of Measure Notice that the diameter of the target is given in inches, while the diameter of the game board is given in feet. When finding geometric probabilities, be sure to check that all measurements are in the same unit, in order to avoid miscalculations.

Part B To make the game more enjoyable, the company wants to increase the probability of hitting the target. What diameter should they use for the circle so that the probability of a toss landing on the target is 10%? Round to the nearest hundredth of a inch.

Find the diameter of a circle so that the geometric probability of a toss landing in the circle is 10% or 0.1.

$P(\text{toss lands on target}) = \dfrac{\text{area of target}}{\text{area of game board}}$ Area probability ratio

$0.1 = \dfrac{\pi\left(\frac{d}{2}\right)^2}{\pi(12)^2}$ Substitution

$0.1 = \dfrac{d^2}{576}$ Simplify.

$576(0.1) = d^2$ Multiply each side by 576.

$57.6 = d^2$ Simplify.

$7.59 \approx d$ Take the square root of each side.

To increase the probability of a toss landing on the target, the diameter of the target should be about 7.59 inches.

🔾 **Go Online** You can complete an Extra Example online.

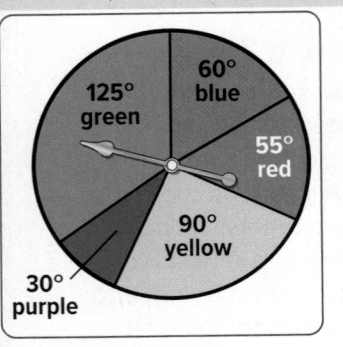

Example 4 Use Angle Measures to Find Geometric Probability

Use the spinner to find the probability of landing in each section.

a. *P*(pointer landing on purple): The angle measure of the purple region is 30°.

$$P(\text{pointer landing on purple}) = \frac{30}{360} \text{ or } 8.3\%.$$

b. *P*(pointer landing on green):
The angle measure of the green region is 125°.

$$P(\text{pointer landing on green}) = \frac{125}{360} \text{ or } 34.7\%.$$

c. *P*(pointer landing on neither yellow nor red):
Combine the angle measures of the yellow and red regions: 90 + 55.

$$P(\text{pointer landing on neither yellow nor red}) = \frac{360 - 145}{360} \text{ or } \frac{215}{360}$$

or 59.7%

🌐 Example 5 Use Probability to Make Decisions

DECISION MAKING **Jayla is visiting the museum and wants to take a guided tour. A friend suggested that she do the tour with Demarcus, rather than Cody, because Demarcus' tour was more informative. Tours with Demarcus depart every 45 minutes, while tours with Cody depart every 30 minutes.**

a. **What is the probability that Jayla will have to wait 20 minutes or less for each tour guide? Explain your reasoning.**

The region below $y = 45$ represents the possible wait time for Demarcus' tour. The region to the left of $x = 30$ represents the possible wait time for Cody's tour. The area formed by the intersection is 45 · 30 or 1350 units².

The region to the left of $x = 20$ and below $y = 20$ represents the possible waiting times of 20 minutes or less for both tour guides. The area of the square is 400 units².

The geometric probability is $\frac{400}{1350}$ or about 30%.

b. **What is the probability that Jayla will have to wait 20 minutes or less for one tour guide? Explain your reasoning.**

The region representing the possible wait times for Demarcus' and Cody's tour is the same as in **part a.**

The region bounded by the lines $x = 30$ and $y = 20$ represents the possibility of waiting 20 minutes or less for Cody's tour. The area of this rectangle is 600 units².

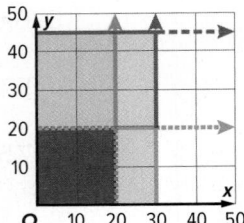

The region bounded by the lines $x = 20$ and $y = 45$ represents the possibility of waiting 20 minutes or less for Demarcus' tour. The area of this rectangle is 900 units².

Because the rectangles that describe Cody and Demarcus' tour times overlap, the waiting time of 20 minutes or less is counted twice. So, the geometric probability is $\frac{600}{1350} + \frac{900}{1350} - \frac{400}{1350} = \frac{1100}{1350}$, or about 81%.

🧭 **Go Online**
You can complete an Extra Example online.

☁️ **Think About It!**

Jayla can wait no more than 20 minutes without risking her favorite exhibit closing. If Cody's tour should depart first, should she wait for Demarcus' tour or go on a tour with Cody? Explain your reasoning.

🧭 **Go Online**
to learn how to use probability tools to make fair decisions in Expand 7-2.

Practice

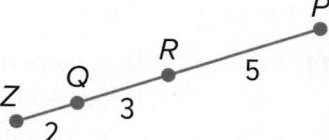 **Go Online** You can complete your homework online.

Example 1

Point _M_ is chosen at random on $\overline{ZP}$. Find the probability of each event.

1. $P(M$ is on $\overline{ZQ})$

2. $P(M$ is on $\overline{QR})$

3. $P(M$ is on $\overline{RP})$

4. $P(M$ is on $\overline{QP})$

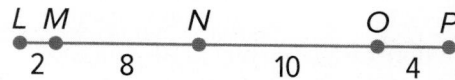

Point _X_ is chosen at random on $\overline{LP}$. Find the probability of each event.

5. $P(X$ is on $\overline{LN})$

6. $P(X$ is on $\overline{MO})$

Example 2

7. WILDLIFE Three frogs are sitting on a 15-foot log. The first two frogs are spaced 5 feet apart and the third frog is 10 feet away from the second one. What is the probability that when a fourth frog hops onto the log, it lands between the first two?

8. LIVESTOCK Four pigs are lined up at a feeding trough. What is the probability that when a fifth pig comes to eat, it lines up between the second and third pig?

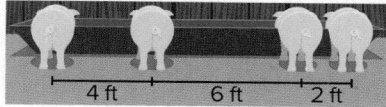

9. DRIVING In a 5-minute traffic cycle, a traffic light is green for 2 minutes 27 seconds, yellow for 6 seconds, and red for 2 minutes 27 seconds. What is the probability that when you get to the light it is green?

10. CARS Once a particular electric car is plugged into a charger, it takes two hours for the battery to have a full charge. If you check the battery level randomly during a charge, what is the probability that the battery will be between $\frac{1}{4}$ and $\frac{1}{2}$ charged?

11. MOVIES A certain store plays a two-hour movie on repeat during store hours. In the movie, there is a song that lasts for 6 minutes and 31 seconds. What is the probability that when a customer randomly enters the store, the song will be playing in the movie?

12. RADIO A radio station is running a contest in which listeners call in when they hear a certain song. The song is 2 minutes 40 seconds long. The radio station claims that they will play the song sometime between noon and 4 P.M. If you randomly turn on that radio station between noon and 4 P.M., what is the probability that the song will be playing?

Example 3

13. **GAMES** One carnival game tasks players with launching a fish charm onto a circular landing pad. The largest prize is awarded if the charm lands in the center circle with a 4-foot diameter. What is the probability the charm lands in the center circle?

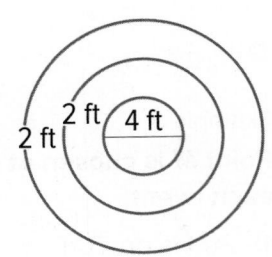

14. **RECREATION** A parachutist is aiming to land in a circular target with a 10-yard radius. The target is in a rectangular field that is 120 yards long and 30 yards wide. Given that the parachutist will land in the field, what is the probability he will land in the target?

Example 4

Use the spinner to find each probability. If the spinner lands on a line it is spun again.

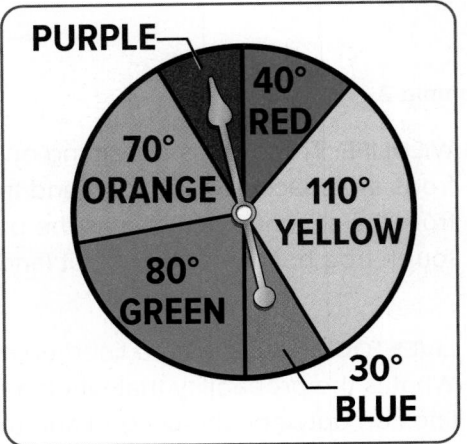

15. *P*(pointer landing on red)

16. *P*(pointer landing on blue)

17. *P*(pointer landing on green)

18. *P*(pointer landing on either green or blue)

19. *P*(pointer landing on neither red nor yellow)

Example 5

20. **STATE YOUR ASSUMPTION** Deangelo is planning a bus ride to the airport. He can choose between the Crimson bus, which arrives at the bus stop every 8 minutes, or the Gold bus, which arrives at the bus stop every 15 minutes. He would prefer to take the Gold bus because the bus makes fewer stops. In order to get to the airport by a certain time, Deangelo cannot wait more than 5 minutes without risking being late for his flight.

a. What assumption do you have to make to solve this problem?

b. To the nearest percent, what is the probability that Deangelo will have to wait 5 minutes or less to see both buses?

c. To the nearest percent, what is the probability that Deangelo will have to wait 5 minutes or less to see only one of the buses?

Mixed Exercises

Describe an event with a 33% probability for each model.

21.

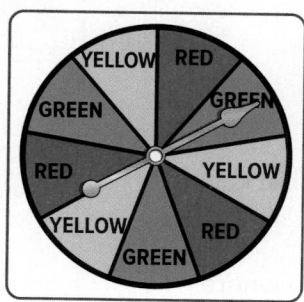

22.

23.

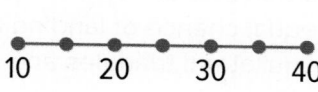

Find the probability that a point chosen at random lies in the shaded region.

24.

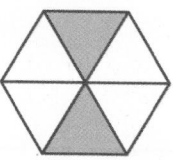

25.

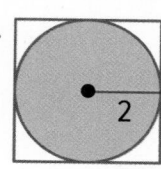

26.

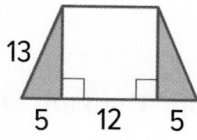

27. DARTS Each sector of a dart board has the same central angle. If a thrown dart has equal probability of hitting any point in play on the dart board, what is the probability that the dart will land in a shaded sector?

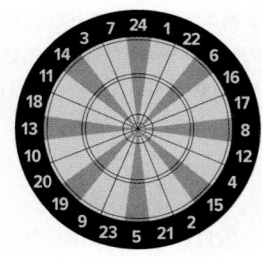

28. GAMES Washers is a popular outdoor game where players attempt to throw washers into a 17-inch by 17-inch square pit. Generally, when a player throws a washer into the square box, he or she scores 1 point; however, if the washer lands within the circular region in the center of the board with a 4-inch diameter, then the player scores 5 points.

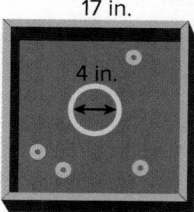

  a. What is the probability a washer is tossed into the circular region? Round to the nearest percent. Use $\pi \approx 3.14$.

  b. What assumption do you have to make to solve this problem?

29. **ENTERTAINMENT** A rectangular dance stage is lit by two lights that light up circular regions of the stage. The circles have radii of the same length and each circle passes through the center of the other. The stage perfectly circumscribes the two circles. A spectator throws a bouquet of flowers onto the stage. Assume the bouquet has an equal chance of landing anywhere on the stage. (*Hint:* Use inscribed equilateral triangles and segments of circles.)

    a. What is the probability that the flowers land on a part of the stage that is illuminated?

    b. What is the probability that the flowers land on the part of the stage where the spotlights overlap?

30. **CONSTRUCT ARGUMENTS** Prove that the probability that a randomly chosen point in the circle will lie in the shaded region is equal to $\frac{x}{360}$.

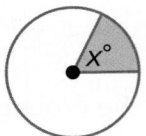

🧠 Higher-Order Thinking Skills

31. **PERSEVERE** Find the probability that a point chosen at random would lie in the shaded area of the figure. Round to the nearest tenth of a percent.

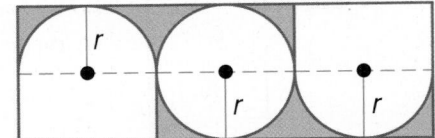

32. **ANALYZE** An isosceles triangle has a perimeter of 32 centimeters. If the lengths of the sides of the triangle are integers, what is the probability that the area of the triangle is exactly 48 square centimeters? Justify your reasoning.

33. **WRITE** Can athletic events be considered random events? Explain.

34. **CREATE** Represent a probability of 20% using three different geometric figures.

35. **WRITE** Explain why the probability of a randomly chosen point falling in the shaded region of either of the squares shown is the same.

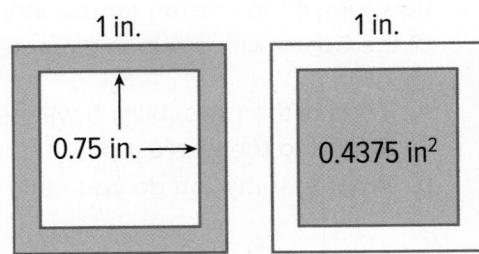

Probability with Permutations and Combinations

Explore Permutations and Combinations

Online Activity Use dynamic geometry software to complete the Explore.

> ×
>
> @ **INQUIRY** How does the order in which objects are arranged affect the sample space of events?

Learn Probability Using Permutations

A **permutation** is an arrangement of objects in which order is important. One permutation for shuffling the songs on a 7-track playlist would be Track 5, Track 7, Track 1, Track 2, Track 6, Track 4, and Track 3. Using the Fundamental Counting Principle, there are $7 \cdot 6 \cdot 5 \cdot 4 \cdot 3 \cdot 2 \cdot 1$ or 5040 possible playlists.

The expression $7 \cdot 6 \cdot 5 \cdot 4 \cdot 3 \cdot 2 \cdot 1$ used to calculate the number of permutations of the 7 tracks can be written as 7!, which is read *7 factorial*.

Factorial	
Words	The **factorial** of a positive integer n is the product of the positive integers less than or equal to n, and is written as $n!$.
Symbols	$n! = n \cdot (n - 1) \cdot (n - 2) \cdot \ldots \cdot 2 \cdot 1$, where $0! = 1$

Suppose you will listen to only four of the seven shuffled tracks. Using the Fundamental Counting Principle, the number of permutations of 4 tracks played from a 7-track playlist is $7 \cdot 6 \cdot 5 \cdot 4$ or 840.

Another way of describing this situation is that this is the number of permutations of 7 tracks taken 4 at a time, denoted $_7P_4$. This number can also be computed by using factorials.

Key Concept • Permutations	
Words	The number of permutations of n distinct objects taken r at a time is denoted by $_nP_r$ and given by $_nP_r = \dfrac{n!}{(n - r)!}$
Symbols	The number of permutations of 7 objects taken 4 at a time is $_7P_4 = \dfrac{7!}{(7 - 4)!} = \dfrac{7 \cdot 6 \cdot 5 \cdot 4 \cdot 3 \cdot 2 \cdot 1}{3 \cdot 2 \cdot 1}$ or 840.

(continued on the next page)

Today's Goals
- Use permutations to compute probabilities.
- Use combinations to compute probabilities.

Today's Vocabulary
permutation
factorial
combination

Go Online
You can watch a video to see how to find a probability using permutations.

Recall that the probability of an event is the ratio of the number of favorable outcomes to the number of total outcomes. To find probabilities with permutations, the number of favorable outcomes and the number of total outcomes can be written as a permutation.

Key Concept • Permutations with Repetition

The number of distinguishable permutations of n objects in which one object is repeated r_1 times, another is repeated r_2 times, and so on, is $\dfrac{n!}{r_1! \cdot r_2! \cdot \ldots \cdot r_k!}$.

🌐 **Example 1** Probability and Permutations of n Objects

PERFORMING ARTS *Tyesha and Liam sign up for an open mic night with 32 available slots that are filled at random. What is the probability that Tyesha will perform first and Liam will perform second?*

Step 1 Find the number of possible outcomes.

The number of possible outcomes in the sample space is the number of permutations of the 32 performers' order, or 32!.

Step 2 Find the number of favorable outcomes.

The number of favorable outcomes is the number of permutations of the other performers' order given that Tyesha performs first and Liam performs second: $(32 - 2)!$ or 30!.

😮 **Think About It!**

How would the probability differ, if at all, of Liam performing first and Tyesha performing second?

Step 3 Calculate the probability.

$P(\text{Tyesha 1, Liam 2}) = \dfrac{30!}{32!}$ ←number of favorable outcomes
←number of possible outcomes

$= \dfrac{\overset{1}{30!}}{32 \cdot 31 \cdot \underset{1}{30!}}$ Expand 32! and divide out common factors.

$= \dfrac{1}{992}$ Simplify.

The probability that Tyesha will perform first and Liam will perform second is $\dfrac{1}{992}$, or about 0.1.

Check

Five geometry students are asked to randomly choose a polygon and describe its properties. What is the probability that the first three students choose the hexagon, the pentagon, and the triangle, in that order?

🔾 **Go Online** You can complete an Extra Example online.

🌐 Example 2 Probability and Permutations with No Repetition

SLIDESHOW **For a project, Rami selects 12 family photographs that will randomly play in a slideshow. The slideshow will not show repeat photos until all 12 photos have been shown. Three photos are of Rami's entire family, two photos are of his brother, three photos are just of him, and four photos are of his sister. What is the probability that the first four pictures in the slideshow will be of Rami's sister?**

Step 1 Find the number of possible outcomes.

Because the photos will not repeat, order in this situation is important. The number of possible outcomes in the sample space is the number of permutations of 12 photos taken 4 at a time, $_{12}P_4$.

$$_{12}P_4 = \frac{(12!)}{(12-4)!} = \frac{12 \cdot 11 \cdot 10 \cdot 9 \cdot 8!}{8!} \text{ or } 11{,}880$$

Step 2 Find the number of favorable outcomes.

The number of favorable outcomes is the number of permutations of the 4 photos of Rami's sister, or 4!.

Step 3 Calculate the probability.

The probability of the four photos of Rami's sister appearing as the first four in the slideshow is $\frac{4!}{11,880}$, or $\frac{1}{495}$.

🌐 Example 3 Probability and Permutations with Repetition

GAME SHOW **One game show has contestants arrange five number tiles to build their guess at the price of a prize. Mishka is given five tiles with the numbers 2, 2, 5, 5, and 8. If Mishka arranges the tiles randomly, what is the probability that she arranges them as the correct price of $25,852?**

Step 1 There is a total of five numbers. Of these numbers, 2 occurs 2 times, 5 occurs 2 times, and 8 occurs 1 time. So, the number of distinguishable permutations of these numbers is

$$\frac{5!}{(2! \cdot 2!)} = \frac{120}{4}, \text{ or } 30 \qquad \text{Simplify.}$$

Step 2 There is only 1 favorable arrangement, the actual price, 25,852.

Step 3 The probability that a permutation of these numbers selected at random results in the correct price of 25,852, is $\frac{1}{30}$.

Check

GAMES The physics team is holding a game night fundraiser. To win a grand prize in a particular game, you must spin the spinner four times and land on blue, red, green, and yellow, in that order. What is the probability that you will spin the winning sequence?

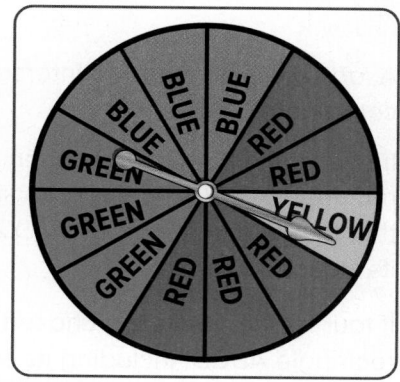

🅡 **Go Online** You can complete an Extra Example online.

> 💭 **Think About It!**
> If the price of the prize were changed to $85,225, would the probability that Mishka arranges the tiles as the correct price increase, decrease, or stay the same? Justify your reasoning.

Study Tip

Randomness When outcomes are decided at random, they are equally likely to occur and their probabilities can be calculated using permutations and combinations.

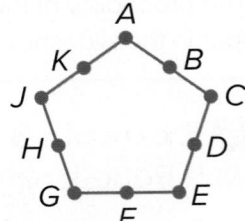

Learn Probability Using Combinations

A **combination** is a selection of objects in which order is *not* important.

A combination of n objects taken r at a time, or $_nC_r$, is calculated by dividing the number of permutations $_nP_r$ by the number of arrangements containing the same elements, $r!$.

Key Concept • Combinations	
Symbols	The number of combinations of n distinct objects taken r at a time is denoted by $_nC_r$ and is given by $_nC_r = \frac{n!}{(n-r)!r!}$.
Example	The number of combinations of 7 objects taken 3 at a time is $_7C_3 = \frac{7!}{(7-3)!3!} = \frac{7!}{4!3!} = \frac{7 \cdot 6 \cdot 5 \cdot 4!}{4! \cdot 6}$ or 35.

Recall that the probability of an event is the ratio of the number of favorable outcomes to the number of total outcomes. To find probabilities with combinations, the number of favorable outcomes and the number of total outcomes can each be written as a combination.

Example 4 Probability and Combinations

If three points are randomly chosen from those named on pentagon *ACEGJ*, what is the probability that they all lie on the same line segment?

Step 1 Find the number of possible outcomes.

Because the order in which the points are chosen does not matter, the number of possible outcomes in the sample space is the number of combinations of 10 points taken 3 at a time, $_{10}C_3$.

$$_{10}C_3 = \frac{10!}{(10-3)!3!} = \frac{10 \cdot 9 \cdot 8 \cdot 7!}{7! \cdot 3!} \text{ or } 120$$

Step 2 Find the number of favorable outcomes.

There are 5 favorable outcomes. The points could lie on $\overline{AC}$, $\overline{CE}$, $\overline{EG}$, $\overline{GJ}$, or $\overline{JA}$.

Step 3 Calculate the probability.

The probability of three randomly chosen points lying on the same segment is $\frac{5}{120}$, or $\frac{1}{24}$.

Check

A *lattice* is a point at the intersection of two or more grid lines in a coordinate plane.

If two lattice points are chosen randomly in rectangle *ABCD*, including its sides, the probability that they are in rectangle *WXYZ*, including its sides, is ____?____ .

If four lattice points are chosen randomly in rectangle *ABCD*, including its sides, the probability that they are *W*, *X*, *Y*, and *Z* is ____?____ .

Practice

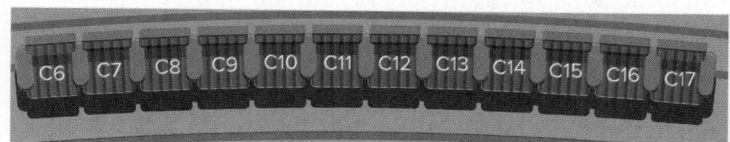 **Go Online** You can complete your homework online.

Example 1

1. **CHEERLEADING** The cheerleading squad is made up of 12 girls. A captain and a co-captain are selected at random. What is the probability that Chantel and Clover are chosen as leaders?

2. **BOOKS** You have a textbook for each of the following subjects: Spanish, English, Chemistry, Geometry, History, and Psychology. If you choose 4 of these books at random to arrange on a shelf, what is the probability that the Geometry textbook will be first from the left and the Chemistry textbook will be second from the left?

3. **RAFFLE** Alfonso and Cordell each bought one raffle ticket at the state fair. If 50 tickets were randomly sold, what is the probability that Alfonso got ticket 14 and Cordell got ticket 23?

4. **CONCERT** Nia and Ciro are going to a concert with their high school's key club. If they choose a seat in the row below at random, what is the probability that Ciro will be in seat C11 and Nia will be in C12?

C6 C7 C8 C9 C10 C11 C12 C13 C14 C15 C16 C17

Examples 2 and 3

5. **PHONE NUMBERS** What is the probability that a 7-digit telephone number generated using the digits 2, 3, 2, 5, 2, 7, and 3 is the number 222-3357?

6. **IDENTIFICATION** A store randomly assigns their employees work identification numbers to track productivity. Each number consists of 5 digits ranging from 1–9. If the digits cannot repeat, find the probability that a randomly generated number is 25938.

7. **STUDENT COUNCIL** The table shows the finalists for class president. The order in which they will give their speeches will be chosen randomly.

   a. What is the probability that Denny, Kelli, and Chaminade are the first 3 speakers, in any order?

   b. What is the probability that Denny is first, Kelli is second, and Chaminade is third?

Class President Finalists
Alan Shepherd
Chaminade Hudson
Denny Murano
Kelli Baker
Tanika Johnson
Jerome Murdock
Marlene Lindeman

Example 4

8. **TROPHIES** Taryn has 15 soccer trophies but she only has room to display 9 of them on a shelf. If she chooses them at random, what is the probability that each of the trophies from the school invitational from the 1st through 9th grades will be chosen?

9. **FROZEN YOGURT** Kali has a choice of 20 flavors for her triple scoop cone. If she chooses the flavors at random, what is the probability that the 3 flavors she chooses will be vanilla, chocolate, and strawberry?

10. **BUSINESS** Kaja has a dog walking business that serves 9 dogs. If she chooses 4 of the dogs at random to take an extra trip to the dog park, what is the probability that Cherish, Taffy, Haunter, and Maverick are chosen?

11. **FOOD TRUCKS** A restaurant critic has 10 new food trucks to try. If she tries half of them this week, what is the probability that she will choose Nacho Best Tacos, Creme Bruleezin, Fre Sha Vaca Do's, You Can't Get Naan, and Grillarious?

12. **DONATIONS** Emily has 20 collectible dolls from different countries that she will donate. If she selects 10 of them at random, what is the probability that she chooses the dolls from Ecuador, Paraguay, Chile, France, Spain, Sweden, Switzerland, Germany, Greece, and Italy?

13. **AMUSEMENT PARK** An amusement park has 12 major attractions: four roller coasters, two carousels, two drop towers, two gravity rides, and two dark rides. The park's app will randomly select attractions for you to visit in order. What is the probability that the four roller coasters are the first four suggested attractions?

Mixed Exercises

14. **BUSINESS TRAVEL** A department manager is selecting team members at random to attend one of four conferences in Los Angeles, Atlanta, Chicago, and New York. If there are 20 team members, what is the probability that Jariah, Sherry, Emilio, and Lavon are chosen for the conferences?

15. **DINING** You are handed 5 pieces of silverware for the formal place setting shown. If you guess their placement at random, what is the probability that the knife and spoon are placed correctly?

16. **CARDS** What is the probability in a line of these five cards that the ace would be first from the left and the 10 would be second from the left?

17. Points A, B, C, D, and E are coplanar, but no 3 are collinear.

   a. What is the total number of lines that can be determined by these points?

   b. What is the probability that $\overleftrightarrow{AB}$ would be chosen at random from all of the possible lines formed?

18. CRAFTING Jaclyn bought some decorative letters for a scrapbook project. If she randomly selected a permutation of the letters shown, what is the probability that they would form the word "photography"?

19. BUSINESS Andres sent emails to 20 of his contacts advertising his new lawn services. If 6 contacts responded to the email, what is the probability that the Michaelsons, the Rodriquezes, the Farooqis, the Salahis, the Kryceks, and the Waltons responded?

20. GAME SHOW The people on the list at the right will be considered to participate in a game show. What is the probability that Wyatt, Gabe, and Isaac will be chosen as the first three contestants?

DAY 1 STANDINGS
MCAFEE, DAVID
FORD, GABE
STANDISH, TRISTAN
NOCHOLS, WYATT
PURCELL, JACK
ANDERSON, BILL
WRIGHT, ISAAC
FILBERT, MITCH

21. SALES The owner of a hair salon advertises that on the first day of each month, the first 6 customers will receive one of the coupons shown at the right for a discount off their total bill. Each coupon is given at random to a different customer.

   a. What is the probability that the first customer on May 1 gets the 10% discount and the second customer gets the 25% discount? Explain your answer using favorable and possible outcomes.

   b. How many different groups of two coupons can the first two customers on August 1 receive regardless of order?

22. DONATIONS As part of a school beautification project, 12 alumni each donated a tree to be planted on the school grounds. The types of trees are shown in the table. There will be a sign next to each tree with the donor's name.

Donated Trees	
Type	Number of Trees
Cherry	5
Dogwood	4
Crabapple	2
Redbud	1

   a. If the trees are planted in a row at random, what is the probability that they will be in alphabetical order by donor name? Explain.

   b. If 4 trees are randomly selected and planted near the school entrance, what is the probability that they will all be dogwood trees? Explain.

23. **PARKING STICKERS** Parking stickers contain randomly generated numbers with 5 digits ranging from 1 to 9. No digits are repeated. What is the probability that a randomly generated number is 54321?

24. **CONSTRUCT ARGUMENTS** Prove that $_nC_{n-r} = {_nC_r}$.

Higher-Order Thinking Skills

25. **PERSEVERE** Fifteen boys and fifteen girls entered a drawing for four free movie tickets. What is the probability that all four tickets were won by girls?

26. **ANALYZE** Is the following statement *sometimes, always,* or *never* true? Justify your reasoning.

$$_nP_r = {_nC_r}$$

27. **WRITE** Compare and contrast permutations and combinations.

28. **CREATE** Describe a situation in which the probability is given by $\frac{1}{_7C_3}$.

29. **PERSEVERE** A student claimed that permutations and combinations were related by $r! \cdot {_nC_r} = {_nP_r}$. Use algebra to show that this is true. Then explain why $_nC_r$ and $_nP_r$ differ by the factor $r!$.

30. **FIND THE ERROR** Charlie claims that the number of ways n objects can be arranged if order matters is equal to the number of permutations of n objects taken $n-1$ at a time. Do you agree with Charlie? Explain your reasoning.

Probability and the Multiplication Rule

Explore Independent and Dependent Events

Online Activity Use the video to complete the Explore.

INQUIRY How can one event affect the probability of a second event? ✕

Learn Independent Events

A **compound event** or *composite event* consists of two or more simple events. **Independent events** are two or more events in which the outcome of one event does not affect the outcome of the other events.

Suppose a coin is tossed and the spinner shown is spun. The sample space for this experiment is {(H, 1), (H, 2), (H, 3), (T, 1), (T, 2), (T, 3)}.

Using the sample space, the probability of the coin landing on heads and the spinner on 2 is $P(\text{H and 2}) = \frac{1}{6}$.

Notice that the same probability can be found by multiplying the probabilities of each simple event.

$$P(H) = \frac{1}{2} \qquad P(2) = \frac{1}{3} \qquad P(\text{H and 2}) = \frac{1}{2} \cdot \frac{1}{3} \text{ or } \frac{1}{6}$$

This example illustrates the first of two Multiplication Rules for Probability.

Key Concept • Probability of Two Independent Events	
Words	The probability that two independent events both occur is the product of the probabilities of each individual event.
Symbols	If two events A and B are independent, then $P(A \text{ and } B) = P(A) \cdot P(B)$.

This rule can be extended to any number of events.

Consider choosing objects from a group of objects. If you replace the object each time, choosing additional objects are independent events.

Today's Goals
- Apply the multiplication rule to situations involving independent events.
- Apply the multiplication rule to situations involving dependent events.

Today's Vocabulary
compound event
independent events
dependent events

Study Tip

and The word *and* often illustrates compound events. For example, if you roll a die, finding the probability of getting an odd number *and* getting a number greater than 5 indicates that the probabilities of the individual events should be multiplied.

🌐 Example 1 Probability of Independent Events

GAMING Ana is a member of a gaming Web site that randomly pairs users together to solve puzzles. Of the 50 other players currently online, Ana is friends with 10 of them. Suppose Ana is paired with a player for a game. Not liking the outcome, she disconnects and is paired with another player.

a. What is the probability that neither player that Ana is paired with is a friend of hers?

These events are independent because the set of possible matches is reset to 50 once Ana disconnects. Let F represent a player who is Ana's friend and NF represent a player who is not Ana's friend.

Complete the equation to determine the probability of independent events.

$$\text{User 1} \quad \text{User 2}$$

$$P(NF \text{ and } NF) = P(NF) \cdot P(NF)$$

$$= \frac{40}{50} \cdot \frac{40}{50} \qquad P(NF) = \frac{40}{50}$$

$$= \frac{1600}{2500} \text{ or } \frac{16}{25} \qquad \text{Simplify.}$$

So, the probability that neither of the two players is Ana's friend is $\frac{16}{25}$ or 64%.

b. What assumption do you have to make in order to solve this problem?

We assume that the same 50 players remain in the set for both selections. If the number of available players changes, or the number of available players who are friends with Ana changes, the probability will change.

Alternate Method

You can also use an area model to calculate the probability that neither player is a friend of Ana's.

The probability that a player is not a friend of Ana's is $\frac{40}{50}$ or $\frac{4}{5}$.

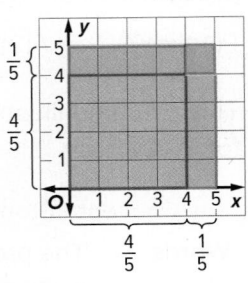

The blue region represents the probability of two sequential players not being friends with Ana. The area of the blue region is $\frac{16}{25}$ of the entire shaded region.

The orange region represents the probability of two sequential players being friends with Ana. The area of the orange region is $\frac{1}{25}$ of the entire shaded region.

Check

WEATHER Paola's weather app tells her that there is a 20% chance of rain on Tuesday and a 50% chance of rain on Wednesday. What is the probability that it will rain on both Tuesday and Wednesday? __?__ or __?__

🌑 **Go Online** You can complete an Extra Example online.

Learn Probability of Dependent Events

Dependent events are two or more events in which the outcome of one event affects the outcome of the other events.

Suppose a marble is chosen from the bag and placed on the table. Then, another marble is chosen from the bag. The sample space for this experiment is {(R, B), (R, R), (B, R), (B, B)}.

When choosing the first marble, there are 7 possible outcomes. The probability of choosing a red marble on the first draw is $\frac{5}{7}$. Because that marble is not returned to the bag, there are only 6 possible outcomes for the second drawing. If a red marble has already been chosen, the probability of choosing a red marble on the second draw is $\frac{4}{6}$.

The second of the Multiplication Rules of Probability addresses the probability of two dependent events.

Key Concept • Probability of Two Dependent Events	
Words	The probability that two dependent events both occur is the product of the probability that the first event occurs and the probability that the second event occurs *after* the first event has already occurred.
Symbols	If two events A and B are dependent, then $P(A \text{ and } B) = P(A) \cdot P(B\|A)$.

This rule can be extended to any number of events.

Example 2 Independent and Dependent Events

Determine whether the events are *independent* or *dependent*. Explain your reasoning.

a. **One spinner is spun twice.**

The outcome of the first spin in no way changes the probability of the outcome of the second spin. Therefore, these two events are independent.

b. **In a raffle, one ticket is drawn for the first place prize, and then another ticket is drawn for the second place prize.**

After the first place prize ticket is drawn, the ticket is removed and cannot be chosen again. This affects the probability of the second place prize winning ticket, because the sample space is reduced by one ticket. Therefore, these two events are dependent.

c. **A random number generator generates two numbers.**

The number for the first generation has no bearing on the number for the second generation. Therefore, these two events are independent.

Go Online You can complete an Extra Example online.

Go Online

Study Tip

Conditional Notation
The notation $P(B\|A)$ is read *the probability that event B occurs given that event A has already occurred*. The "|" symbol should not be interpreted as a division symbol.

Go Online
You can watch a video to see how to use the Multiplication Rule to find the probability of two independent events.

Think About It!
Write an example of a series of three dependent events.

Lesson 7-5 • Probability and the Multiplication Rule **399**

Check

Copy and complete the table. Determine whether the events are *independent* or *dependent*.

	Independent	Dependent
Of the $100 that Rei has to spend, she wants to spend $59 on a blouse and $44 on some jeans.		
Rei asks each of three store associates which handbag they prefer.		
Rei purchases a handbag and a belt.		

🌐 Example 3 Probability of Dependent Events

FOOD The pizza that José and Tessa are eating has 10 slices and is half cheese, half mushroom. Tessa spins the pizza around and randomly selects a slice of mushroom pizza. If José spins the pizza and selects a slice after that, what is the probability that both he and Tessa select a slice of mushroom pizza?

These events are dependent because Tessa does not replace the slice she selected. Let *M* represent a slice of mushroom pizza and *C* represent a slice of cheese pizza.

$P(M \text{ and } M) = P(M) \cdot P(M|M)$ Probability of dependent events

$= \frac{5}{10} \cdot \frac{4}{9} \text{ or } \frac{2}{9}$ After the first slice of mushroom pizza is selected, 9 total pieces remain, and 4 of those slices have mushrooms.

So, the probability that both friends randomly select slices with mushrooms is $\frac{2}{9}$ or about 22%.

Check

SCHOOL On a math test, 5 out of 20 students got all the questions correct. If three students are chosen at random without replacement, what is the probability that all three got all the questions correct on the test?

Pause and Reflect

Did you struggle with anything in this lesson? If so, how did you deal with it?

🔾 **Go Online** You can complete an Extra Example online.

Practice

Go Online You can complete your homework online.

Example 1

1. CLOTHING Omari has two pairs of red socks and two pairs of white socks in a drawer. He has a drawer with 2 red T-shirts and 1 white T-shirt. If he randomly chooses a pair of socks from the sock drawer and a T-shirt from the T-shirt drawer, what is the probability that he gets a pair of red socks and a white T-shirt?

2. Phyllis drops a penny in a pond, and then she drops a nickel in the pond. What is the probability that both coins land with tails showing?

3. A die is rolled and a penny is flipped. Find the probability of rolling a two and landing on a tail.

4. A bag contains 3 red marbles, 2 green marbles, and 4 blue marbles. A marble is drawn randomly from the bag and replaced before a second marble is chosen. Find the probability that both marbles are blue.

5. The forecast predicts a 40% chance of rain on Tuesday and a 60% chance on Wednesday. If these probabilities are independent, what is the chance that it will rain on both days?

Example 2

Determine whether the events are *independent* or *dependent*. Explain your reasoning.

6. You roll an even number on a fair die, and then spin a spinner numbered 1 through 5 and it lands on an odd number.

7. An ace is drawn from a standard deck of 52 cards, and is not replaced. Then, a second ace is drawn.

8. In a bag of 3 green and 4 blue marbles, a blue marble is drawn and not replaced. Then, a second blue marble is drawn.

9. You roll two fair dice and roll a 5 on each.

Example 3

10. LOTTERY Mr. Hanes places the names of four of his students, Joe, Sofia, Hayden, and Bonita, on slips of paper. From these, he intends to randomly select two students to represent his class at the robotics convention. He draws the name of the first student, sets it aside, then draws the name of the second student. What is the probability he draws Sofia, then Joe?

11. **CARDS** A card is drawn from a standard deck of playing cards and is not replaced. Then a second card is drawn. Find the probability the first card is a jack of spades and the second card is black.

12. **INTRAMURAL SPORTS** The table shows the color and number of jerseys available for the intramural volleyball tournament. If each jersey is given away randomly, what is the probability that the first and second jerseys given away are both red?

Jersey Color	Amount
blue	20
white	15
red	25
black	10

Mixed Exercises

13. **SPORTS** The format used to determine an overall champion varies by sport. One type of format used is the *best-of-seven series,* where up to seven games are played and four wins determine a champion. If you assume that each team has an equal chance of winning each game, what is the probability of a team winning the first four games in a best-of-seven series?

14. **BUSINESS** A sales management team consists of three directors and three assistant directors. To ensure that each team member has an equal chance to be chosen to represent the team at a national conference, all 6 names are placed into a hat and four names are drawn at random. What is the probability that those chosen will consist of 3 directors and 1 assistant director?

15. **MAGIC** Iris performs a magic trick in which she holds a standard deck of cards and has each of three people randomly choose a card from the deck. Each person keeps his or her card as the next person draws. What is the probability that all three people will draw a heart?

16. **SCHOOL** The probability that a student takes geometry and French at Saul's school is 0.064. The probability that a student takes French is 0.45. If taking geometry and taking French are dependent events, what is the probability that a student takes geometry if the student takes French?

17. **EXTRACURRICULAR ACTIVITIES** At Bell High School, 43% of the students are in an after-school club and 28% play sports. What is the probability that a student is in an after-school club if he or she also plays a sport if being in a club and playing sports are independent events?

Sunita and Derek work for a company that produces microchips. Part of their job is to estimate the number of defective chips given the total number produced. The table shows the estimated contents of 2 boxes of microchips. Use this information for Exercises 18 and 19.

18. Derek randomly selects one chip from Box B, does not put it back, and then randomly selects another chip from Box B. What is the probability that both chips are defective? Explain.

	Number of Chips	Defective
Box A	100	4%
Box B	150	2%

19. Sunita randomly selects one chip from Box A, and then she randomly selects another chip from Box A. The probability that both chips are defective is $\frac{1}{625}$. Did Sunita replace the first chip before selecting the second one? Explain.

20. TRAVEL A travel agency conducts a survey to determine whether people drive (D) or fly (F) to their vacation destinations. The results indicated that $P(D) = 0.6$, $P(D \cap F) = 0.2$, and the probability that a family did not vacation is 0.1.

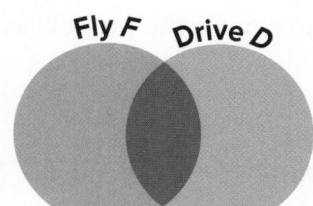

a. What is the probability that a family reached their vacation destination by flying?

b. What is the probability that a family that drives will also fly?

21. BUSINESS TRENDS You are trying to decide whether you should expand a business. If you do not expand and the economy remains good, you expect $2 million in revenue. If the economy is bad, you expect $0.5 million. The cost to expand is $1 million, but the expected revenue after the expansion is $4 million in a good economy and $1 million in a bad economy. You assume that the chances of a good and a bad economy are 30% and 70%, respectively. Create a tree diagram to represent the situation.

22. REASONING If Fred spins the spinner twice, determine the probability that he lands on sections labeled "orange" and "green." Show that $P(\text{orange}) \cdot P(\text{green} \mid \text{orange}) = P(\text{green}) \cdot P(\text{orange} \mid \text{green})$ in this situation and explain why this is true in terms of the model. Justify your answer.

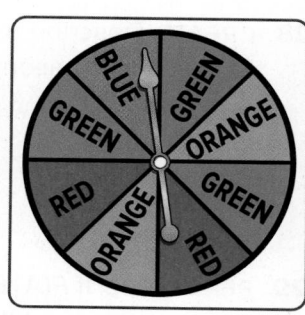

23. TRICKS Sam is doing a trick with a standard deck of 52 playing cards where she begins each trick with a fresh deck of cards. Her friend Tracy randomly selects a card, looks at it, and puts it back in the deck. Then Sam randomly selects the same card. Are the events independent? What is the probability that they both pick the queen of spades? Explain.

24. CONSTRUCT ARGUMENTS Use the formula for the probability of two dependent events $P(A \text{ and } B)$ to derive the conditional probability formula for $P(B|A)$.

25. USE A MODEL A double fault in tennis is when the serving player fails to land their serve "in" without stepping on or over the service line in two chances. Kelly's first serve percentage is 40%, while her second serve percentage is 70%.

  a. Draw a probability tree that shows each outcome.

  b. What is the probability that Kelly will double fault?

🢒 Higher-Order Thinking Skills

26. ANALYZE There are n different objects in a bag. The probability of drawing object A and then object B without replacement is about 2.4%. What is the value of n? Explain.

27. WRITE An article states the chance that a person is left-handed given that his or her parent is left-handed. Explain how you could determine the likelihood that a person being left-handed and their parent being left-handed are independent events.

28. CREATE Describe a pair of independent events and a pair of dependent events. Explain your reasoning.

29. PERSEVERE If $P(A|B)$ is the same as $P(A)$, and $P(B|A)$ is the same as $P(B)$, what can be said about the relationship between events A and B?

Probability and the Addition Rule

Explore Mutually Exclusive Events

Online Activity Use the guiding exercises to complete the Explore.

? INQUIRY How can you find the probability of two or more events that may not share common outcomes?

Learn Probability of Mutually Exclusive Events

To find the probability that one event occurs or another event occurs, you must know how the two events are related. If the two events cannot happen at the same time, they are said to be **mutually exclusive**. That is, the two events have no outcomes in common.

One way of finding the probability of two mutually exclusive events occurring is to examine their sample space.

When a die is rolled, what is the probability of getting a 3 or a 4? From the Venn diagram, you can see that there are two outcomes that satisfy this condition, 3 and 4. So,

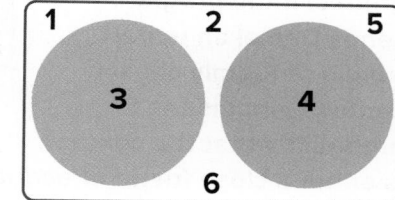

$P(3 \text{ or } 4) = \frac{2}{6}$ or $\frac{1}{3}$.

Notice that this same probability can be found by adding the probabilities of each simple event.

$P(3) = \frac{1}{6}$ $\qquad$ $P(4) = \frac{1}{6}$ $\qquad$ $P(3 \text{ or } 4) = \frac{1}{6} + \frac{1}{6} = \frac{2}{6}$ or $\frac{1}{3}$

This example illustrates the first of two Addition Rules for Probability.

Key Concept • Probability of Mutually Exclusive Events	
Words	If two events A and B are mutually exclusive, then the probability that A or B occurs is the sum of the probabilities of each individual event.
Symbols	If two events A and B are mutually exclusive, then $P(A \text{ or } B) = P(A) + P(B)$.

This rule can be extended to any number of events.

Example 1 Identify Mutually Exclusive Events

A card is drawn from a standard deck of 52 cards. Determine whether the events are *mutually exclusive* or not *mutually exclusive*. Explain your reasoning.

a. drawing a 3 or a 2

There are no common outcomes — a card cannot be both a 2 and a 3. These events are mutually exclusive.

b. drawing a 7 or a red card

The 7 of diamonds is an outcome that both events have in common. These events are not mutually exclusive.

c. drawing a queen or a spade

Because the queen of spades represents both events, they are not mutually exclusive.

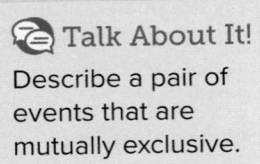
Talk About It!

Describe a pair of events that are mutually exclusive.

🌐 Example 2 Probability of Mutually Exclusive Events

SOCIAL MEDIA Daniel organizes all of his social media contacts into three groups. If the program sends Daniel an update from a randomly chosen contact, what is the probability that the contact is either a close friend or acquaintance?

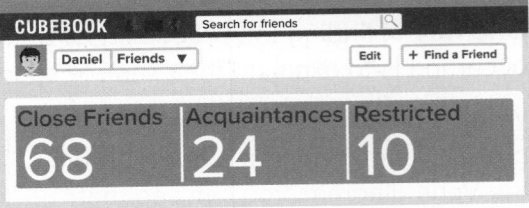

These are mutually exclusive events, because the contacts selected cannot be a close friend and an acquaintance.

Let event F represent selecting a close friend. Let event A represent selecting an acquaintance. There are a total of $10 + 68 + 24$ or 102 contacts.

Because the events are mutually exclusive, you know that

$P(F \text{ or } A) = P(F) + P(A)$.

$$P(F \text{ or } A) = P(F) + P(A) \qquad \text{Probability of mutually exclusive events}$$

$$= \frac{68}{102} + \frac{24}{102} \qquad P(F) = \frac{68}{102} \text{ and } P(A) = \frac{24}{102}$$

$$= \frac{92}{102} \text{ or } \frac{46}{51} \qquad \text{Add.}$$

So the probability that the update is from a close friend or acquaintance is $\frac{46}{51}$ or about 90%.

🔵 **Go Online** You can complete an Extra Example online.

💭 **Think About It!**

What assumption did you make in order to solve this problem?

Check

BIODIVERSITY Of the more than 79,800 species on a Red List of Threatened Species, seabirds are of particular interest because they are indicators of broader marine health issues. The circle graph shows the proportion of seabird species in each Red List category. What is the probability that a randomly selected species of seabird is on the critically endangered list or the endangered list?

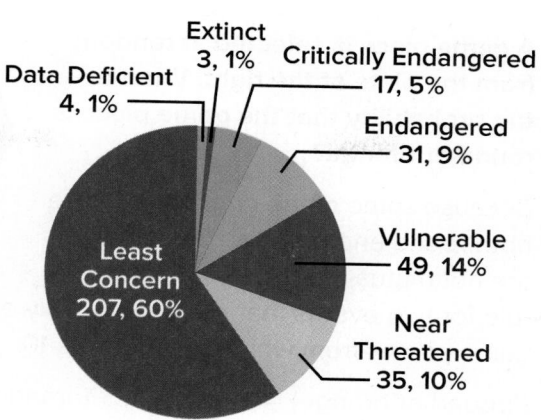

Extinct
3, 1%

Data Deficient
4, 1%

Critically Endangered
17, 5%

Endangered
31, 9%

Vulnerable
49, 14%

Least Concern
207, 60%

Near Threatened
35, 10%

Learn Probability of Events that are Not Mutually Exclusive

If two events can happen at the same time, they are not mutually exclusive.

When a die is rolled, what is the probability of getting a number greater than 2 or an even number? From the Venn diagram, you can see that there are 5 numbers on a die that are either greater than 2 or are an even number: 2, 3, 4, 5, and 6. So, $P(\text{greater than 2 or even}) = \frac{5}{6}$.

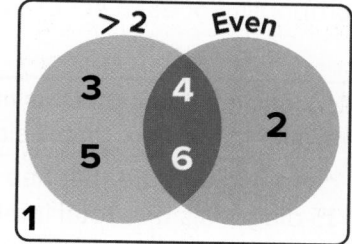

> 2   Even

3 4 2

5 6

1

Because it is possible to roll a number that is greater than 2 and an even number, these events are not mutually exclusive. Consider the probability of each individual event.

$$P(\text{greater than 2}) = \frac{4}{6} \qquad P(\text{even}) = \frac{3}{6}$$

Adding these probabilities results in a number greater than 1 because two of the outcomes, 4 and 6, are in the intersection of the sample spaces—they are both greater than 2 and even. To account for the intersection, subtract the probability of the common outcomes.

$P(\text{greater than 2 or even}) = P(\text{greater than 2}) + P(\text{even}) - P(\text{greater than 2 and even}) = \frac{4}{6} + \frac{3}{6} - \frac{2}{6}$ or $\frac{5}{6}$

This leads to the second of the Addition Rules for Probability.

Key Concept • Probability of Events That Are Not Mutually Exclusive	
Words	If two events A and B are not mutually exclusive, then the probability that A or B occurs is the sum of their individual probabilities minus the probability that both A and B occur.
Symbols	If two events A and B are not mutually exclusive, then $P(A \text{ or } B) = P(A) + P(B) - P(A \text{ and } B)$.

This rule can be extended to any number of events.

 Go Online You can complete an Extra Example online.

Go Online
You can watch a video to see how to use the Addition Rule to find the probability of two events that are not mutually exclusive.

Example 3 Events That Are Not Mutually Exclusive

A game piece is selected at random from the plate at the right. What is the probability that the game piece is round or orange?

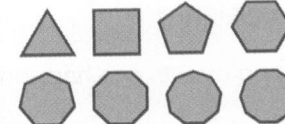

🫧 **Think About It!**
Why is $\frac{2}{10}$ subtracted from the sum of the probabilities P(round) and P(orange)?

Because some of the game pieces are both round and orange, these events are not mutually exclusive. Use the rule for two events that are not mutually exclusive. The total number of game pieces from which to choose is 10.

P(round or orange) = P(round) + P(orange) − P(round and orange)

$$= \frac{5}{10} + \frac{4}{10} - \frac{2}{10}$$

$$= \frac{7}{10}$$

The probability that a game piece will be round or orange is $\frac{7}{10}$ or 70%.

Check

A polygon is chosen at random. Find the probability of each set of events.

choosing a figure that has more than 4 lines of symmetry or more than 7 sides	⟷	?
choosing a figure that has more than 15 diagonals or a total interior angle measure greater than 900°	⟷	?
choosing a figure that has more than 2 pairs of parallel sides or at least 1 diagonal	⟷	?

Pause and Reflect

Did you struggle with anything in this lesson? If so, how did you deal with it?

🔵 **Go Online** You can complete an Extra Example online.

Practice

Go Online You can complete your homework online.

Example 1

Determine whether the events are *mutually exclusive* or *not mutually exclusive*. Explain your reasoning.

1. A die is rolled while a game is being played. The result of the next roll is a 6 or an even number.

2. **SALES** A street vendor is selling T-shirts outside of a concert arena. The colors and sizes of the available T-shirts are shown in the table. The vendor selects a T-shirt that is blue or large.

	Red	Blue	White
Small	1	2	2
Medium	3	2	4
Large	4	5	6
Extra Large	7	6	3

Examples 2 and 3

3. **AWARDS** The student of the month gets to choose one award from 9 gift certificates to area restaurants, 8 T-shirts, 6 water bottles, or 5 gift cards to the mall. What is the probability that the student of the month chooses a T-shirt or a water bottle?

4. **SALES PROMOTIONS** At a grand opening event, a store allows customers to choose an envelope from a bag. Ten of the envelopes contain store coupons, 8 envelopes contain gift cards, and 2 envelopes contain $100. What is the probability that a customer selects an envelope with a gift card or an envelope with $100?

5. **TRAFFIC** If the chance of making a green light at a certain intersection is 35%, what is the probability of arriving when the light is yellow or red?

6. **STUDENTS** In a group of graduate students, 4 out of the 5 females are international students, and 2 out of the 3 men are international students. What is the probability of selecting a graduate student from this group that is a male or an international student?

Mixed Exercises

CARDS **Suppose you pull a card from a standard 52-card deck. Find the probability of each event.**

7. The card is a 4.

8. The card is red.

9. The card is a face card.

10. The card is not a face card.

11. *P*(queen or heart)

12. *P*(jack or spade)

13. *P*(five or prime number)

14. *P*(ace or black)

15. A drawing will take place where one ticket is to be drawn from a set of 80 tickets numbered 1 to 80. If a ticket is drawn at random, what is the probability that the number drawn is a multiple of 4 or a factor of 12?

16. **SCHOOL** The Venn diagram shows the extracurricular activities enjoyed by the senior class at Valley View High School.

   a. How many students are in the senior class?

   b. How many students participate in athletics?

   c. If a student is randomly chosen, what is the probability that the student participates in athletics or drama?

   d. If a student is randomly chosen, what is the probability that the student participates in only drama and band?

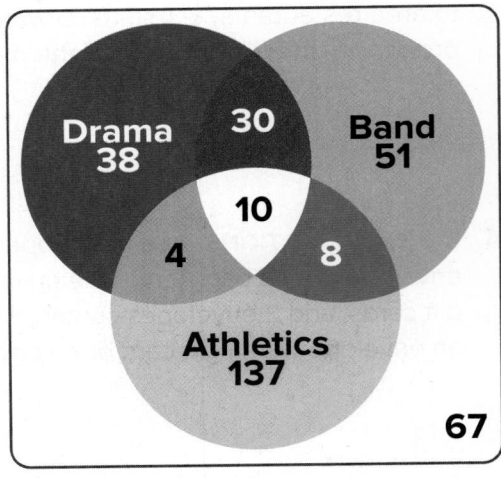

17. **BOWLING** Cindy's bowling records indicate that for any frame, the probability that she will bowl a strike is 30%, a spare 45%, and neither 25%. What is the probability that she will bowl either a spare or a strike for any given frame?

18. **SPORTS CARDS** Dario owns 145 baseball cards, 102 football cards, and 48 basketball cards. What is the probability that he randomly selects a baseball or a football card?

19. **SCHOLARSHIPS** A review committee read 3000 application essays for one $5000 college scholarship. Of the applications reviewed, 2865 essays were the required length, 2577 of the applicants had the minimum required grade-point average, and 2486 had the required length and minimum grade-point average. What is the probability that an application essay selected at random will have the required length or the required gradepoint average?

20. **PETS** Ruby's cat had 8 kittens. The litter included 2 orange females, 3 mixed-color females, 1 orange male, and 2 mixed-color males. Ruby wants to keep one kitten. What is the probability that she randomly chooses a kitten that is female or orange?

21. **SPORTS** The table shows the age and number of participants in each sport at a sporting complex. What is the probability that a player is 14 or plays basketball?

Mason Sports Complex			
Age	Soccer	Volleyball	Basketball
14	28	36	42
15	30	26	33
16	35	41	29

22. **USE A MODEL** Vicente and Kelly are designing a board game. They decide that the game will use a pair of dice and the players will have to find the sum of the numbers rolled. Vicente and Kelly created the table shown to help determine probabilities. Each player will roll the pair of dice twice during that player's turn.

   a. What is the probability of rolling a pair or two numbers that have a sum of seven?

   b. What is the probability of rolling two numbers whose sum is an even number or not rolling a 2? Round to the nearest thousandth.

1, 1	1, 2	1, 3	1, 4	1, 5	1, 6
2, 1	2, 2	2, 3	2, 4	2, 5	2,6
3, 1	3, 2	3, 3	3, 4	3, 5	3, 6
4, 1	4, 2	4, 3	4, 4	4, 5	4, 6
5, 1	5, 2	5, 3	5, 4	5, 5	5, 6
6, 1	6, 2	6, 3	6, 4	6, 5	6, 6

23. **PARKS** The table shows Parks and Recreation Department classes and the number of participants ages 7–9. What is the probability that a participant chosen at random is in drama or is an 8-year-old?

Age	Swimming	Drama	Art
7	40	35	25
8	30	21	14
9	20	44	11

24. **FLOWER GARDEN** Erin is planning her summer garden. The table shows the number of bulbs she has according to type and color of flower. If Erin randomly selects one of the bulbs, what is the probability that she selects a bulb for a yellow flower or a dahlia?

Flower	Orange	Yellow	White
Dahlia	5	4	3
Lily	3	1	2
Gladiolus	2	5	6
Iris	0	1	4

25. PERSEVERE You roll 3 dice. What is the probability that the outcome of at least two of the dice will be less than or equal to 4? Explain your reasoning.

26. FIND THE ERROR Teo and Mason want to determine the probability that a red marble will be chosen out of a bag of 4 red, 7 blue, 5 green, and 2 purple marbles. Is either of them correct? Explain your reasoning.

Teo	Mason
$P(R) = \frac{4}{17}$	$P(R) = 1 - \frac{4}{18}$

ANALYZE **Determine whether the following are mutually exclusive. Explain.**

27. choosing a quadrilateral that is a square and a quadrilateral that is a rectangle

28. choosing a triangle that is equilateral and a triangle that is equiangular

29. choosing a complex number and choosing a natural number

30. WRITE Explain why the sum of the probabilities of two mutually exclusive events is not always 1.

Conditional Probability

Explore Conditional Probability

Online Activity Use the guiding exercises to complete the Explore.

> **INQUIRY** How can you find the probability of an event given that another event has already occurred?

Learn Conditional Probabilities

The **conditional probability** of an event B is the probability that the event will occur given that an event A has already occurred. In addition to finding the probability of two or more dependent events, conditional probability can be used when additional information is known about an event.

Suppose two dice are rolled and it is known that one of the die shows a 5. What is the probability that the sum of the numbers rolled is 7? Because one event, rolling a 5, has already occurred, the sample space for the other event is reduced from 36 to 11 outcomes. This example leads to the following formula.

Key Concept • Conditional Probability

The conditional probability of B given A is $P(B|A) = \frac{P(A \text{ and } B)}{P(A)}$, where $P(A) \neq 0$.

The Venn diagram shows the sample space for both events. The probability of rolling a sum of 7 on two dice given that one die shows a 5, is represented by the probability of the intersection of the two events divided by the probability of the given event.

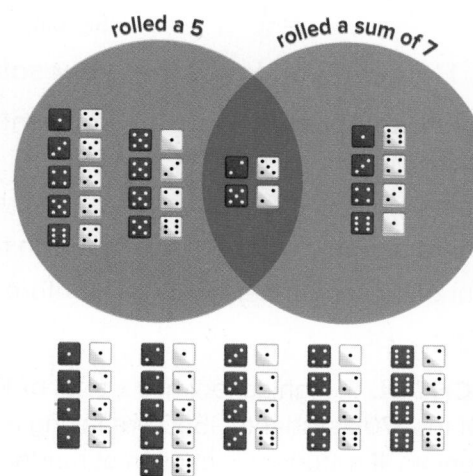

In this case, if event A is rolling a 5 and event B is rolling a sum of 7,

$$P(B|A) = \frac{P(A \text{ and } B)}{P(A)} = \frac{\frac{2}{36}}{\frac{11}{36}} = \frac{2}{11}.$$

🌐 Apply Example 1 Conditional Probability

GROCERY SHOPPING **There are currently 16 customers in line at the deli counter, each holding a numbered ticket from 179 to 194. Naveen will help customers holding tickets with even numbers, and Ellie will help customers holding tickets with odd numbers. If a customer is helped by Naveen, what is the probability that the customer is holding ticket 190?**

1 What is the task?

Describe the task in your own words. Then list any questions that you may have. How can you find answers to your questions?

Sample answer: I need to determine the probability of a customer holding ticket 190 given that the customer is helped by Naveen. What two events are occurring and how do they relate to the formula for conditional probability? I can read through the problem again to identify whether event B will occur given that event A has already occurred.

2 How will you approach the task? What have you learned that you can use to help you complete the task?

Sample answer: I will find $P(A)$, the probability that the number on the ticket is even, then I will find $P(A \text{ and } B)$, the probability that the ticket is both 190 and even. I will use the formula for conditional probability to find the solution.

3 What is your solution?

There are 16 available tickets.

The sample space from event A contains 8 outcomes. From least to greatest, these outcomes are: {180, 182, 184, 186, 188, 190, 192, 194}.

So, $P(A) = \frac{8}{16}$ or $\frac{1}{2}$.

The sample space for $P(A \text{ and } B)$ contains 1 outcome: {190}.

So, $P(A \text{ and } B) = \frac{1}{16}$.

$$P(B|A) = \frac{P(A \text{ and } B)}{P(A)} \qquad \text{Formula for conditional probability}$$

$$= \frac{1}{8} \qquad \text{Simplify.}$$

4 How can you know that your solution is reasonable?

⚡ **Write About It!** Write an argument that can be used to defend your solution.

Sample answer: This situation can be represented with a Venn diagram. There are only eight even numbers in the sample space, and only one out of these numbers is 190. Therefore, the $P(B|A) = \frac{1}{8}$.

Check

SCHOOL A high school has a total of 1700 students, with 450 seniors. Of the 1700 students, 1550 are taking a math class, 280 of which are seniors. If a student is chosen at random, what is the probability that he or she is taking a math class, given that the student is a senior? Write your answer as a fraction or as a percent expressed to the nearest tenth.

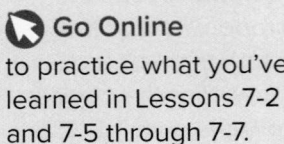
Go Online
to practice what you've learned in Lessons 7-2 and 7-5 through 7-7.

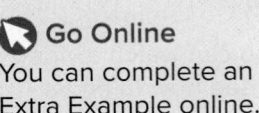
Go Online
You can complete an Extra Example online.

Practice

Go Online You can complete your homework online.

Example 1

1. **CLUBS** The Spanish Club is having a potluck lunch where each student brings in a cultural dish. The 10 students randomly draw cards numbered with consecutive integers from 1 to 10. Students who draw odd numbers will bring main dishes. Students who draw even numbers will bring desserts. If Cynthia is bringing a dessert, what is the probability that she drew the number 10?

2. A card is randomly drawn from a standard deck of 52 cards. What is the probability that the card is a king of diamonds, given that the card drawn is a king?

3. **GAME** In a game, a spinner with the 7 colors of the rainbow is spun. Find the probability that the color spun is blue, given the color is one of the three primary colors: red, yellow, or blue.

4. Fifteen cards numbered 1–15 are placed in a hat. What is the probability that the card has a multiple of 3 on it, given that the card picked is an odd number?

Mixed Exercises

5. A blue marble is selected at random from a bag of 3 red and 9 blue marbles and not replaced. What is the probability that a second marble selected will be blue?

6. A die is rolled. If the number rolled is less than 5, what is the probability that it is the number 2?

7. If two dice are rolled, what is the probability that the sum of the faces is 4, given that the first die rolled is odd?

8. A spinner numbered 1 through 12 is spun. Find the probability that the number spun is an 11 given that the number spun was an odd number.

9. If two dice are rolled, what is the probability that the sum of the faces is 8, given that the first die rolled is even?

10. **PICNIC** A school picnic offers students hamburgers, hot dogs, chips, and a drink.

    a. At the picnic, 60% of the students order a hamburger and 48% of the students order a hamburger and chips. What is the conditional probability that a student who orders a hamburger also orders chips?

    b. If 50% of the students ordered chips, are the events of ordering a hamburger and ordering chips independent? Explain.

    c. If 80% of the students who ordered a hot dog also ordered a drink and 35% of all the students ordered a hot dog, find the probability that a student at the picnic orders a hot dog and drink. Explain.

11. The Venn diagram shows students' favorite places to study, the library (*L*) or home (*H*).

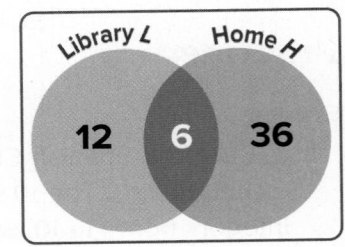

    a. A total of 60 students responded to the survey. Determine the number of students who replied that they study neither at the library nor at home.

    b. What is the probability that if a student selected the library, he or she selected the library and at home? Explain.

    c. A student says that selecting the library and selecting at home are independent events. Do you agree? Explain.

🫧 **Higher-Order Thinking Skills**

12. WRITE Let *A* represent the event of owning a house and let *B* represent the event of owning a car. Are these events independent or dependent? How do you think $P(A|B)$ compares to $P(B|A)$? Explain your reasoning.

13. PERSEVERE Of all the students at North High School, 25% are enrolled in Algebra and 20% are enrolled in Algebra and Health.

    a. If a student is enrolled in Algebra, find the probability that the student is enrolled in Health as well.

    b. If 50% of the students are enrolled in Health, are being enrolled in Algebra and being enrolled in Health independent events? Explain.

    c. Of all the students, 20% are enrolled in Accounting and 5% are enrolled in Accounting and Spanish. If being enrolled in Accounting and being enrolled in Spanish are independent events, what percent of students are enrolled in Spanish? Explain.

14. ANALYZE In a standard deck of playing cards, the face-value cards are the cards numbered 2–10. Two cards are to be randomly drawn without replacing the first card. Find the probability of drawing two face-value cards and the conditional probability that exactly one of those cards is a 4. Explain.

Two-Way Frequency Tables

Explore Two-Way Frequency Tables

Online Activity Use the tables to complete the Explore.

> **INQUIRY** How can you use data in a two-way frequency table to determine whether two events are independent?

Today's Goals
• Construct and interpret two-way frequency tables and use them to determine whether events are independent.

Today's Vocabulary
two-way frequency table
marginal frequencies
joint frequencies
relative frequency

Learn Independent Events in Frequency Tables

A **two-way frequency table**, or **contingency table**, is used to show the frequencies of data from a survey or experiment classified according to two variables, with the rows indicating one variable and the columns indicating the other.

The two-way frequency table shows the results of a survey of 220 men and women about the time of day that they shower. The frequencies reported in the *Totals* row and *Totals* column

Two-Way Frequency Table

	Shower in the Morning	Shower in the Evening	Totals
Women	98	37	135
Men	23	62	85
Totals	121	99	220

are called **marginal frequencies**, with the bottom rightmost cell reporting the total number of observations. Marginal frequencies allow you to analyze with respect to one variable. For example, the marginal frequencies in the right column separate the data by gender.

The frequencies reported in the interior of the table are called **joint frequencies**. These show the frequencies of all possible combinations of the categories for the first variable with the categories for the second variable.

A **relative frequency** is the ratio of the number of observations in a category to the total number of observations.

When survey results are classified according to variables, you may want to decide whether these

Two-Way Relative Frequency Table

	Shower in the Morning	Shower in the Evening	Totals
Women	$\frac{98}{220}$	$\frac{37}{220}$	$\frac{135}{220}$
Men	$\frac{23}{220}$	$\frac{62}{220}$	$\frac{85}{220}$
Totals	$\frac{121}{220}$	$\frac{99}{220}$	$\frac{220}{220}$

variables are independent of each other. Variable A is considered independent of variable B if $P(A \text{ and } B) = P(A) \cdot P(B)$.

🌐 Example 1 Frequency and Relative Frequency Tables

BREAKFAST Francesca asks a random sample of **140** upperclassmen at her high school whether they prefer eating breakfast at home or at school. She finds that **55** juniors and **23** seniors prefer eating breakfast at home before school, while **12** juniors and **50** seniors prefer eating breakfast at school.

Part A Organize the responses in a two-way frequency table.

Identify the variables. The students surveyed can be classified according to *class* and *preference*. Because the survey included only upperclassmen, the variable *class* has two categories: senior or junior. The variable *preference* also has two categories: *prefers eating breakfast at home* and *prefers eating breakfast at school*.

Create a two-way frequency table. Let the rows of the table represent *class* and the columns represent *preference*. Then fill in the cells of the table with the information given.

Add a Totals row and a Totals column to your table and fill in these cells with the correct sums.

	Breakfast at Home	Breakfast at School	Totals
Senior	23	50	73
Junior	55	12	67
Totals	78	62	140

Part B Construct a relative frequency table.

To complete a relative frequency table for these data, start by dividing the frequency reported in each cell by the total number of respondents, 140. Then, write each fraction as a percent rounded to the nearest tenth.

	Breakfast at Home	Breakfast at School	Totals
Senior	$\frac{23}{140} = 16.4\%$	$\frac{50}{140} = 35.7\%$	$\frac{73}{140} = 52.1\%$
Junior	$\frac{55}{140} = 39.3\%$	$\frac{12}{140} = 8.6\%$	$\frac{67}{140} = 47.9\%$
Totals	$\frac{78}{140} = 55.7\%$	$\frac{62}{140} = 44.3\%$	$\frac{140}{140} = 100\%$

💭 Think About It!

What is the probability that a surveyed student is a junior who prefers eating breakfast at home?

 Go Online You can complete an Extra Example online.

Example 2 Independence and Relative Frequency

MUSIC **Anaud polls 240 of his friends on social media about what grade they are in and whether they prefer electronic dance music (EDM) or hip-hop. He posts the results of the survey in the table.**

	EDM	Hip-Hop	Totals
College Freshman	100	107	207
High School Senior	16	17	33
Totals	116	124	240

Use the table to determine whether a respondent's musical preference is independent of his or her grade level.

In a two-way frequency table, you can test for the independence of two variables by comparing the joint relative frequencies with the products of the corresponding marginal relative frequencies.

Divide each reported frequency by 240 to convert the frequency table to a relative frequency table. Enter each fraction as a percent rounded to the nearest tenth. Complete the table.

	EDM	Hip-Hop	Totals
College Freshman	$\frac{100}{240} = 41.7\%$	$\frac{107}{240} = 44.6\%$	$\frac{207}{240} = 86.3\%$
High School Senior	$\frac{16}{240} = 6.7\%$	$\frac{17}{240} = 7.1\%$	$\frac{33}{240} = 13.8\%$
Totals	$\frac{116}{240} = 48.3\%$	$\frac{124}{240} = 51.7\%$	$\frac{240}{240} = 100\%$

Calculate the expected joint relative frequencies if the two variables were independent. Then compare them to the actual relative frequencies.

For example, if 86.3% of respondents were college freshmen and 48.3% of respondents prefer EDM, then one would expect that 86.3% · 48.3% or about 41.7% of respondents are college freshmen who prefer EDM. The table below shows the expected joint relative frequencies.

	EDM	Hip-Hop	Totals
College Freshman	41.7%	44.6%	86.3%
High School Senior	6.7%	7.1%	13.8%
Totals	48.3%	51.7%	100%

Comparing the two tables, the expected and actual joint relative frequencies are the same. Therefore, the musical preferences for these respondents are independent of grade level.

Go Online You can complete an Extra Example online.

Check

SCHOOL Immediately after a physics test, the entire class sits together at lunch and discusses how long each of them studied and how many questions they guessed on. The table shows the responses from the classmates.

	Guessed on < 5 Problems	Guessed on > 5 Problems	Totals
Studied 4 Hours or Less	9	3	12
Studied More Than 4 Hours	12	4	16
Totals	21	7	28

True or False: For these classmates, guessing on more than 5 problems on the physics test is independent of studying 4 hours or less.

🌐 Example 3 Conditional Probability with Two-Way Frequency Tables

You can use joint and marginal relative frequencies to approximate conditional probabilities.

MEMES Abu posts a question to an online forum about the originality of posts to the site. Of the 55 respondents who have posted viral memes, 27 photos and 15 videos were not original content, while 3 photos and 10 videos were original content.

Part A Construct a relative frequency table of the data. Round each percent to the nearest tenth.

	Not Original Content	Original Content	Totals
Video	27.3%	18.2%	45.5%
Photo	49.1%	5.5%	54.5%
Totals	76.4%	23.6%	100%

Part B Find the probability that a viral meme on the forum is not original content given that is it a photo.

The probability that a meme is not original content given that it is a photo is the conditional probability P(not original content|photo).

$$P\text{(not original content|photo)} = \frac{P\text{(not original content and photo)}}{P\text{(photo)}}$$

$$\approx \frac{0.491}{0.545} \text{ or } 90.1\%$$

So the probability that a viral meme on the forum is not original content given that it is a photo is 90.1%.

🧭 **Go Online** You can complete an Extra Example online.

💭 **Think About It!**

Why do we divide by 0.545 when finding the conditional probability in the example?

Practice

Go Online You can complete your homework online.

Example 1

1. **VEHICLES** One hundred people are surveyed about the type of vehicle they drive. The survey finds that 15 males and 40 females drive SUVs, while 35 males and 10 females drive trucks.

   a. Organize the responses in a two-way frequency table.

   b. Construct a relative frequency table.

2. **SOCIAL MEDIA** One hundred students are asked whether or not they have social media accounts. The survey finds that 25 males and 35 females have social media accounts, while 25 males and 15 females do not have social media accounts.

   a. Organize the responses in a two-way frequency table.

   b. Construct a relative frequency table.

Example 2

3. COLLEGE A ride-sharing company surveys 2000 of its customers who are college students. The survey asks the following two questions about the previous academic year:

- Are you attending college in or out of state?
- Did you visit home more than four times this school year?

The survey finds that of the 1260 students who attend an in-state college, 928 visited home more than four times and 332 visited home four or less times. Of the 740 students who attend an out-of-state college, 118 visited home more than four times and 622 visited home four or less times.

a. Organize the responses in a two-way frequency table.

b. Construct a relative frequency table. Enter each fraction as a percent rounded to the nearest tenth, if necessary.

c. Suppose you let event A represent whether the students attend an in-state or an out-of-state college, and event B represent whether the students visit home more than four times or visit home four or fewer times. Use the table to determine whether the number of visits home is independent or dependent on whether the student is attending college at an in-state or out-of-state institution. Justify your response.

Example 3

4. TICKETS A movie theater is keeping track of the last 800 tickets it sold to two different movies. Of the 578 adult tickets sold, 136 of them were for the animated film and 442 were for the documentary film. Of the 222 student tickets sold, 181 of them were for the animated film and 41 were for the documentary film.

a. Copy and complete the two-way frequency table shown.

	Adult	Student	Totals
Animated		181	
Documentary			
Totals	578		800

b. Construct a relative frequency table of your completed two-way frequency table. Round each percent to the nearest tenth, as necessary.

c. Find the probability that a ticket sold is an adult ticket given that it is a documentary ticket. Show your work by writing the formula that you used to perform the calculation.

Mixed Exercises

5. **SCHOOL** The two-way frequency table compares data about students in a class who completed or did not complete homework and those who passed or did not pass an exam. How many students completed their homework and passed the exam? Identify whether marginal or joint frequencies are used.

	Completed Homework	Did Not Complete Homework	Totals
Passed Exam	18	2	20
Did Not Pass Exam	4	2	6
Totals	22	4	26

6. **MOVIES** Raquel surveys 160 people to determine if they prefer drama or comedy movies. The relative frequency table shows the data collected from the survey. Determine whether gender is independent of movie type preference. Explain your reasoning.

	Drama	Comedy	Totals
Male	12.5%	25%	37.5%
Female	46.9%	15.6%	62.5%
Totals	59.4%	40.6%	100%

7. **TECHNOLOGY** For a business report on technology use, Darnell asks a random sample of 72 shoppers whether they own a smart phone and whether they own a tablet computer. His survey shows that out of 51 shoppers who own smart phones, 9 of them also own a tablet, while out of 21 shoppers who do not own smart phones, 15 of them do not own tablets either. Find the conditional probability that a shopper has a tablet, given that he or she has a smart phone. Justify your reasoning.

8. **CONSTRUCT ARGUMENTS** Paz asks a random sample of seniors at her high school whether they own a car and whether they have a job. The results of the survey are shown in the two-way relative frequency table. Paz says that the conditional probability that a student has a job given that he or she has a car is 46.7%. Do you agree? Justify your argument.

	Has a Job	Does Not Have a Job	Totals
Has a Car	21.9%	12.5%	34.4%
Does Not Have a Car	25%	40.6%	65.6%
Totals	46.9%	53.1%	100%

Higher-Order Thinking Skills

9. PERSEVERE Suppose an exit poll held outside a voting area on the day of an election produced these results.

Age and Gender	Votes for Candidate A	Votes for Candidate B
18–30 Male	19	32
18–30 Female	31	18
31–45 Male	51	12
31–45 Female	43	20
46–60 Male	42	35
46–60 Female	20	42
60+ Male	45	21
60+ Female	27	18

a. Which events are mutually exclusive?

b. Find the probability that a male between the ages of 46 and 60 would vote for Candidate A.

c. Find the probability that a female would vote for Candidate A.

d. Find the probability that someone who voted for Candidate B was a female and age 18–30.

e. According to the data, on which demographic(s) does Candidate A need to focus campaign efforts?

f. According to the data, on which demographic(s) does Candidate B need to focus campaign efforts?

10. ANALYZE A market research firm asks a random sample of 240 adults and students at a movie theater whether they would rather see a new summer blockbuster in 2-D or 3-D. The survey shows that 64 adults and 108 students prefer 3-D, while 42 adults and 26 students prefer 2-D.

a. Organize the responses into a two-way frequency table.

b. Convert the table from part **a** into a two-way relative frequency table. Round to the nearest tenth of a percent. Out of every 10 people surveyed, about how many would prefer to see the movie in 3-D? Explain.

c. Find the probability that a person surveyed prefers seeing the movie in 3-D, given that he or she is an adult. Write the formula that you used to perform the calculation.

d. An analyst at the firm claims that the probability that a person surveyed is a student given that he or she does not prefer to see the movie in 3-D is 10.8%. Do you agree? Justify your answer.

e. Is a preference for 2-D or 3-D movies independent of age? Explain your reasoning.

Essential Question

How can you use measurements to find probabilities?

You can find the number of favorable outcomes for an experiment and also find the total number of possible outcomes for an experiment and then find probabilities using a ratio. That ratio can be used to predict how many times a certain event may occur.

Module Summary

Lessons 7-1 through 7-3

Probability of Simple Events

- The number of possible outcomes in a sample space can be found by multiplying the number of possible outcomes from each stage or event.

- For the probability of the intersection of two or more events, find the ratio of the number of outcomes in both events to the total number of possible outcomes.

- When two events A and B occur, the union of A and B is the set of all outcomes in the sample space of event A combined with all outcomes in the sample space of event B.

- If region A contains a region B and a point E in region A is chosen at random, then the probability that point E is in region B is $\dfrac{\text{area of region } B}{\text{area of region } A}$.

Lesson 7-4

Permutations and Combinations

- The number of distinguishable permutations of n objects in which one object is repeated r_1 times, another is repeated r_2 times, and so on, is $\dfrac{n!}{r_1!r_2! \bullet \ldots \bullet r_k!}$.

- The number of permutations of n distinct objects taken r at a time is denoted by $_nP_r$ and given by $_nP_r = \dfrac{n!}{(n-r)!}$.

- The number of combinations of n distinct objects taken r at a time is denoted by $_nC_r$ and is given by $_nC_r = \dfrac{n!}{(n-r)!r!}$.

Lessons 7-5 through 7-7

Probability of Compound Events

- If two events A and B are independent, the $P(A \text{ and } B) = P(A) \bullet P(B)$.

- If two events A and B are dependent, then $P(A \text{ and } B) = P(A) \bullet P(B \mid A)$.

- If two events A or B are mutually exclusive, then $P(A \text{ or } B) = P(A) + P(B)$.

- If two events A or B are not mutually exclusive, then $P(A \text{ or } B) = P(A) + P(B) - P(A \text{ and } B)$.

- The conditional probability of B given A is $P(B \mid A) = \dfrac{P(A \text{ and } B)}{P(A)}$, where $P(A) \neq 0$.

Lesson 7-8

Frequency Tables

- The frequencies in the Totals row and Totals column are marginal frequencies.

- The frequencies in the interior of the table are joint frequencies.

- A relative frequency is the ratio of the number of observations in a category to the total number of observations.

Study Organizer

Foldables

Use your Foldable to review this module. Working with a partner can be helpful. Ask for clarification of concepts as needed.

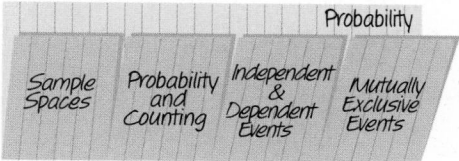

Test Practice

1. **MULTIPLE CHOICE** Two dice are tossed. Which is the sample space of the event that the sum of the outcomes is 5? (Lesson 7-1)

   A. {(5, 5)}

   B. {(1, 4), (2, 3)}

   C. {(1, 4), (2, 3), (3, 2), (4, 1)}

   D. {(1, 5), (2, 5), (3, 5), (4, 5), (5, 5)}

2. **OPEN RESPONSE** A restaurant has a special deal where you can build your own meal from certain selections in the menu. The number of selections available in each category is shown in the table.

Item	Number of Choices
Drink	12
Appetizer	7
Main Entrée	8
Side Dishes	14
Dessert	9

   If a person selects one of each item, how many different meals can be ordered? (Lesson 7-1)

3. **MULTIPLE CHOICE** An integer between 1 and 12 is generated using a random number generator. Let A be the event of generating a multiple of 4, and let B be the event of generating a factor of 12. Which of the following represents $A \cap B$? (Lesson 7-2)

   A. {4, 12}

   B. {4, 8, 12}

   C. {1, 2, 3, 4, 6, 12}

   D. {1, 2, 3, 4, 6, 8, 12}

4. **OPEN RESPONSE** Jenell's birthday is in May. Let W be the event that his birthday lands on a weekend. Let P be the event that his birthday is a prime number.

   What is $W \cap P$? (Lesson 7-2)

May						
Sun	Mon	Tues	Wed	Thurs	Fri	Sat
	1	2	3	4	5	6
7	8	9	10	11	12	13
14	15	16	17	18	19	20
21	22	23	24	25	26	27
28	29	30	31			

5. **MULTIPLE CHOICE** Point J will be placed randomly on $\overline{AD}$.

   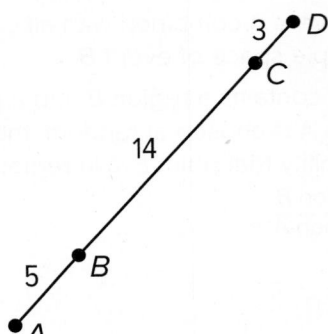

   What is the probability that point J is on $\overline{AC}$ to the nearest percent? (Lesson 7-3)

   A. 14%

   B. 77%

   C. 64%

   D. 86%

6. MULTIPLE CHOICE Josefina is at a carnival trying to win a prize. She must toss a bean bag in the hole to win.

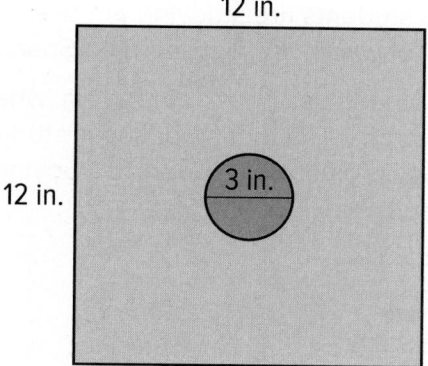

12 in.

12 in.

3 in.

What is the probability that when tossed randomly, the bean bag lands in the hole? Assume that the bean bag lands on the board. (Lesson 7-3)

A. 4.9%

B. 7.065%

C. 19.625%

D. 28.26%

7. MULTIPLE CHOICE If three points are randomly chosen from those named on hexagon *ABCDEF*, what is the probability that they all lie on the same line segment? (Lesson 7-4)

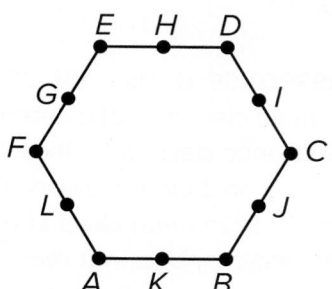

A. $\frac{1}{1320}$

B. $\frac{1}{220}$

C. $\frac{1}{216}$

D. $\frac{3}{110}$

8. OPEN RESPONSE The numbers 0–39 are used to create a locker combination. Show how to determine the probability that the combination is 20-21-22. (Lesson 7-4)

9. MULTI-SELECT Select all the situations that are dependent events. (Lesson 7-5)

A. Two dice are tossed.

B. Two marbles are selected from a bag.

C. Two students are chosen as the captains of a team.

D. A coin is tossed and a card is chosen.

E. Three books are selected from the library.

F. One student from each class is chosen to collect papers.

10. OPEN RESPONSE The table shows the books of several different genres available to read on Imelda's bookshelf.

Genre	Number of Books
Action	8
Mystery	5
Romance	2
Science fiction	12
Horror	3

Imelda selects two different books. What is the probability, as a fraction, that Imelda selects two mysteries? (Lesson 7-5)

11. MULTI-SELECT Suppose a die is tossed once. Which of these events are mutually exclusive? Select all that apply. (Lesson 7-6)

A. Landing on a 4 or a 5

B. Landing on a 2 or an even

C. Landing on a 2 or a prime

D. Landing on 4 or an odd

E. Landing on an odd or a prime

12. MULTIPLE CHOICE A group of college students was surveyed about their browser use. The results are shown on the circle graph.

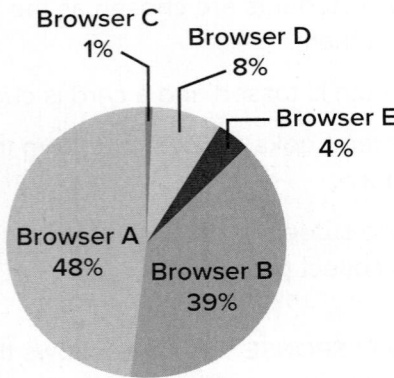

Browser C 1%
Browser D 8%
Browser E 4%
Browser A 48%
Browser B 39%

What is the probability that a student selected randomly will use either Browser A or Browser D? (Lesson 7-6)

A. $\frac{47}{100}$

B. $\frac{12}{25}$

C. $\frac{14}{25}$

D. $\frac{87}{100}$

13. OPEN RESPONSE Two dice have been tossed. What is the probability of tossing a sum of 8 given that at least one die landed on an odd number? (Lesson 7-7)

14. MULTIPLE CHOICE In a recent survey, 1650 students were asked what they are studying. Of the 1650 students, 948 students are learning Spanish. 426 students are studying physics, 378 of whom are also learning Spanish.

If a student is chosen at random, what is the probability that he or she is studying physics, given the student is studying Spanish? (Lesson 7-7)

A. 57.5%

B. 52.9%

C. 39.9%

D. 25.8%

15. MULTIPLE CHOICE The table shows the results of a survey asking whether the respondent preferred to use the Internet on a phone or laptop.

	Phone	Laptop	Total
12–29 years old	85	21	106
30+ years old	124	87	211
Total	209	108	317

What is the probability that a participant is 30+ years old given that they prefer a laptop to use the Internet? (Lesson 7-8)

A. 27.4%

B. 41.2%

C. 59.3%

D. 80.6%

16. OPEN RESPONSE Booker asks 29 students from his math class and 21 different students from his science class when they typically start working on their homework. Of the students, 18 from math class and 8 from science class respond that they do their homework as soon as they return home from school, and 11 from math class and 13 from science class respond that they start working on homework after dinner. What is the probability, as a percent to the nearest tenth, that a student waits to start homework until after dinner, given that they are in Booker's math class? (Lesson 7-7)

Relations and Functions

What Will You Learn?

How much do you already know about each topic **before** starting this module?

	Before			After		
KEY 👎 — I don't know. 👍 — I've heard of it. 👍 — I know it!	👎	👍	👍	👎	👍	👍
identify one-to-one and onto functions						
identify discrete and continuous functions						
identify intercepts of graphs of functions						
identify linear and nonlinear functions						
identify extrema of graphs and functions						
identify end behavior of graphs of functions						
identify graphs that display line or point symmetry						
sketch and compare graphs of functions						
graph linear functions						
graph linear inequalities in two variables						
graph piecewise, step, and absolute value functions						
translate, dilate & reflect the graphs of functions						

Foldables Make this Foldable to help you organize your notes about relations and functions. Begin with four sheets of notebook paper.

1. **Fold** each sheet of paper in half from top to bottom.

2. **Cut** along the fold. Staple the eight half-sheets together to form a booklet.

3. **Cut** tabs into the margin. The top tab is 2 lines deep, the next tab is 6 lines deep, and so on.

4. **Label** each of the tabs with a lesson number, and the final tab *vocabulary*.

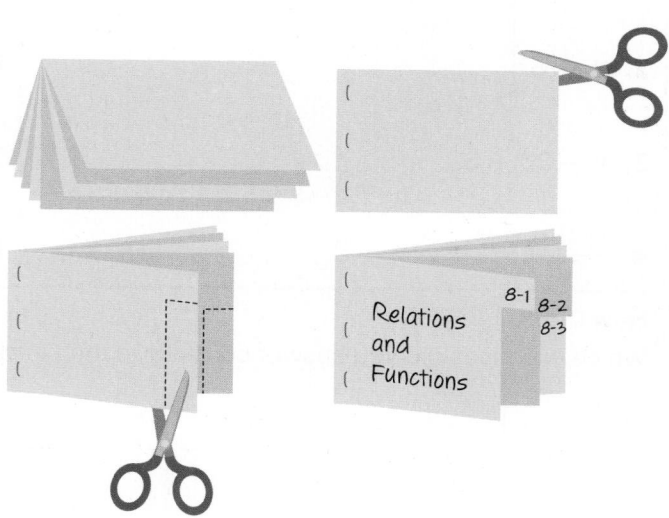

Relations and Functions

8-1 8-2 8-3

What Vocabulary Will You Learn?

- absolute value function
- algebraic notation
- boundary
- closed half-plane
- codomain
- constant function
- constraint
- continuous function
- dilation
- discontinuous function
- discrete function
- domain

- end behavior
- even functions
- extrema
- family of graphs
- greatest integer function
- identity function
- intercept
- interval notation
- line of reflection
- line of symmetry
- line symmetry
- linear equation

- linear function
- linear inequality
- maximum
- minimum
- nonlinear function
- odd functions
- one-to-one function
- onto function
- open half-plane
- parabola
- parent function
- piecewise-defined function

- point of symmetry
- point symmetry
- range
- reflection
- relative maximum
- relative minimum
- set-builder notation
- step function
- symmetry
- translation
- transformation
- x-intercept
- y-intercept

Are You Ready?

Complete the Quick Review to see if you are ready to start this module.
Then complete the Quick Check.

Quick Review

Example 1

Evaluate $3a^2 - 2ab + b^2$ if $a = 4$ and $b = -3$.

$$
\begin{aligned}
3a^2 - 2ab + b^2 &= 3(4^2) - 2(4)(-3) + (-3)^2 \\
&= 3(16) - 2(4)(-3) + 9 \\
&= 48 - (-24) + 9 \\
&= 48 + 24 + 9 \\
&= 81
\end{aligned}
$$

Example 2

Solve $3x + 6y = 24$ for y.

$3x + 6y = 24$	Original equation
$3x + 6y - 3x = 24 - 3x$	Subtract $3x$ from each side.
$6y = 24 - 3x$	Simplify.
$\frac{6y}{6} = \frac{24}{6} - \frac{3x}{6}$	Divide each side by 6.
$y = 4 - \frac{1}{2}x$	Simplify.

Quick Check

Evaluate each expression if $a = -3$, $b = 4$, and $c = -2$.

1. $4a - 3$

2. $2b - 5c$

3. $b^2 - 3b + 6$

4. $\dfrac{2a + 4b}{c}$

Solve each equation for the given variable.

5. $a = 3b + 9$ for b

6. $15w - 10 = 5v$ for v

7. $3x - 4y = 8$ for x

8. $\dfrac{d}{6} + \dfrac{f}{3} = 4$ for d

How Did You Do?

Which exercises did you answer correctly in the Quick Check?

Functions and Continuity

Explore Analyzing Functions Graphically

🔾 **Online Activity** Use graphing technology to complete the Explore.

> ⊚ **INQUIRY** How can you use a graph to analyze the relationship between the domain and range of a function? ✕

Explore Defining and Analyzing Variables

🔾 **Online Activity** Use a real-world situation to complete the Explore.

> ⊚ **INQUIRY** How can you define variables to effectively model a situation? ✕

Learn Functions

A function describes a relationship between input and output values. The **domain** is the set of *x*-values to be evaluated by a function. The **codomain** is the set of all the *y*-values that could possibly result from the evaluation of the function. The codomain of a function is assumed to be all real numbers unless otherwise stated. The **range** is the set of *y*-values that actually result from the evaluation of the function. The range is contained within the codomain.

If each element of a function's range is paired with exactly one element of the domain, then the function is a **one-to-one function**. If a function's codomain is the same as its range, then the function is an **onto function**.

Example 1 Domains, Codomains, and Ranges

Part A Identify the domain, range, and codomain of the graph.

Domain	Range	Codomain
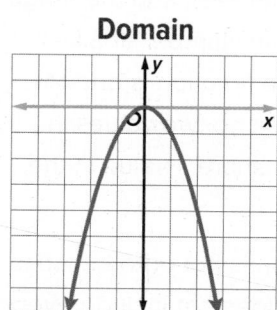	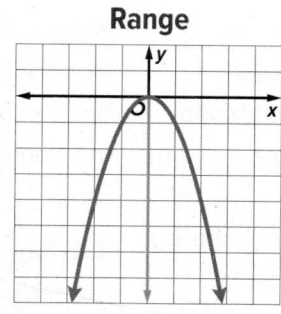	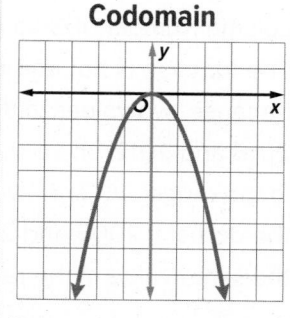
Because there are no restrictions on the *x*-values, the domain is all real numbers.	Because the maximum *y*-value is 0, the range is $y \le 0$.	Because it is not stated otherwise, the codomain is all real numbers.

(continued on the next page)

Today's Goals
• Determine whether functions are one-to-one and/or onto.
• Determine the continuity, domain, and range of functions.
• Write the domain and range of functions by using set-builder and interval notations.

Today's Vocabulary
domain
codomain
range
one-to-one function
onto function
continuous function
discontinuous function
discrete function
algebraic notation
set-builder notation
interval notation

Study Tip

Horizontal Line Test
Place a pencil at the top of the graph and slowly move it down.

• If there are places where the pencil intersects the graph at more than one point, then more than one element of the range is paired with an element of the domain and the function is not one-to-one.

• If there are places where the pencil does not intersect the graph at all, then there are real numbers that are not paired with an element of the domain and the function is not onto.

Part B **Use these values to determine whether the function is onto.**

The range is not the same as the codomain because it does not include the positive real numbers. Therefore, the function is not onto.

Check

For what codomain is $f(x)$ an onto function?

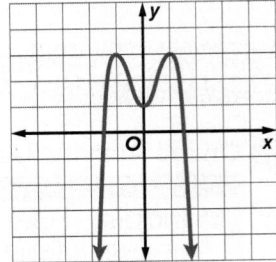

A. $y \leq 3$ **B.** $y \geq 3$

C. all real numbers **D.** $x \leq 3$

🌐 **Example 2** Identify One-to-One and Onto Functions from Tables

OLYMPICS **The table shows the number of medals the United States won at five Summer Olympic Games.**

Year	Number of Gold Medals	Number of Silver Medals	Number of Bronze Medals
2016	46	37	38
2012	46	29	29
2008	36	38	36
2004	36	39	26
2000	37	24	32

Analyze the functions that give the number of gold and silver medals won in a particular year. Define the domain and range of each function and state whether it is *one-to-one, onto, both* or *neither*.

Gold Medals	Silver Medals
Let $f(x)$ be the function that gives the number of gold medals won in a particular year. The domain is in the column Year, and the range is in the column Number of Gold Medals. The function is not one-to-one because two values in the domain, 2016 and 2012, share the same value in the range, 46, and two values in the domain, 2008 and 2004, share the same value in the range, 36. The function is not onto because the range does not include every whole number.	Let $g(x)$ be the function that gives the number of silver medals won in a particular year. The domain is the column Year, and the range is the column Number of Silver Medals. The function is one-to-one because no two values in the domain share a value in the range. The function is not onto because the range does not include every whole number.

🔵 **Go Online** You can complete an Extra Example online.

Use a Source

Choose another country and research the number of medals they won in the Summer Olympic Games from 2000-2016. Are the functions that give the number of each type of medal won in a particular year *one-to-one, onto, both*, or *neither*?

Example 3 Identify One-to-One and Onto Functions from Graphs

Determine whether each function is *one-to-one*, *onto*, *both*, or *neither* for the given codomain.

$f(x)$, where the codomain is all real numbers 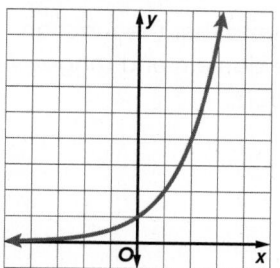	The graph indicates that the domain is all real numbers, and the range is all positive real numbers. Every x-value is paired with exactly one unique y-value, so the function is one-to-one. If the codomain is all real numbers, then the range is not equal to the codomain. So, the function is not onto.
$g(x)$, where the codomain is $\{y \mid y \leq 4\}$ 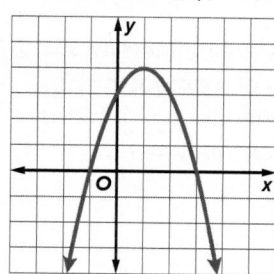	The graph indicates that the domain is all real numbers, and the range is $y \leq 4$. Each x-value is not paired with a unique y-value; for example, both $x = 0$ and $x = 2$ are paired with $y = 3$. So the function is not one-to one. The codomain and range are equal, so the function is onto.
$h(x)$, where the codomain is all real numbers 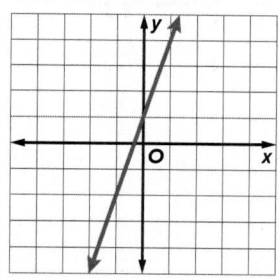	The graph indicates that the domain and range are both all real numbers. Every x-value is paired with exactly one unique y-value, so the function is one-to-one. The codomain and range are equal, so the function is onto.

Learn Discrete and Continuous Functions

Functions can be discrete, continuous, or neither. Real-world situations where only some numbers are reasonable are modeled by discrete functions. Situations where all real numbers are reasonable are modeled by continuous functions.

A **continuous function** is graphed with a line or an unbroken curve. A function that is not continuous is a **discontinuous function**. A **discrete function** is a discontinuous function in which the points are not connected. A function that is neither discrete nor continuous may have a graph in which some points are connected, but it is not continuous everywhere.

🔵 **Go Online** You can complete an Extra Example online.

Study Tip

Intervals An interval is the set of all real numbers between two given numbers. For example, the interval $-2 < x < 5$ includes all values of x greater than -2 but less than 5. Intervals can also continue on infinitely in a direction. For example, the interval $y \geq 1$ includes all values of y greater than or equal to 1. You can use intervals to describe the values of x or y for which a function exists.

Example 4 Determine Continuity from Graphs

Examine the functions. Determine whether each function is *discrete*, *continuous*, or *neither* discrete nor continuous. Then state the domain and range of each function.

a. *f(x)*

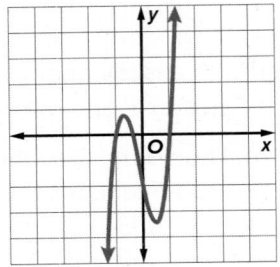

The function is continuous because it is a curve with no breaks or discontinuities.

Because you can assume that the function continues forever, the domain and range are both all real numbers.

b. *g(x)*

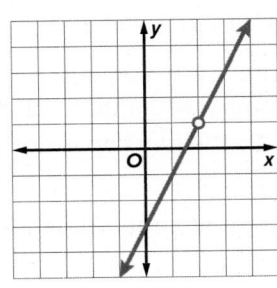

The function is neither because there are continuous sections, but there is a break at (2, 1).

Because the function is not defined for $x = 2$, the domain is all values of x except $x = 2$. The function is not defined for $y = 1$, so the range is all values of y except $y = 1$.

c. *h(x)*

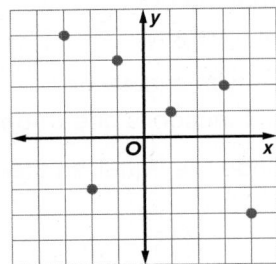

The function is discrete because it is made up of distinct points that are not connected.

The domain is $\{-3, -2, -1, 1, 3, 4\}$ and the range is $\{-3, -2, 1, 2, 3, 4\}$.

🌐 Example 5 Determine Continuity

BUSINESS Determine whether the function that models the cost of coffee beans is *discrete, continuous,* or *neither* discrete nor continuous. Then state the domain and range of the function.

Because customers can purchase any amount of coffee up to 2 pounds, the function is continuous over the interval $0 \leq x \leq 2$.

COFFEE

Weight	Price
Up to 2 lbs	$8/lb
2.5 lbs	$20
3 lbs	$22
5 lbs	$35

🔵 **Go Online** You can complete an Extra Example online.

For larger quantities, the coffee is sold by distinct amounts. This part of the function is discrete.

Since the domain and range are made up of neither a single interval nor individual points, the function is neither discrete nor continuous.

The domain of the function is $0 \leq x \leq 2$ or $x = 2.5, 3, 5$. This represents the possible weights of coffee beans that customers could purchase. The range of the function is $0 \leq y \leq 16$ or $y = 20, 22, 35$. This represents the possible costs of coffee beans.

Think About It!

Why does the range include values from 0 to 16 instead of 0 to 8?

Learn Set-Builder and Interval Notation

Sets of numbers like the domain and range of a function can be described by using various notations. Set-builder notation, interval notation, and algebraic notation are all concise ways of writing a set of values. Consider the set of values represented by the graph.

- In **algebraic notation**, sets are described using algebraic expressions. Example: $x < 2$

- **Set-builder notation** is similar to algebraic notation. Braces indicate the set. The symbol | is read as *such that*. The symbol ∈ is read *is an element of*. Example: $\{x \mid x < 2\}$

- In **interval notation** sets are described using endpoints with parentheses or brackets. A parenthesis, (or), indicates that an endpoint *is not* included in the interval. A bracket, [or], indicates that an endpoint is included in the interval. Example: $(-\infty, 2)$

Example 6 Set-Builder and Interval Notation for Continuous Intervals

Write the domain and range of the graph in set-builder and interval notation.

Domain

The graph will extend to include all *x*-values.

The domain is all real numbers.

$\{x \mid x \in \mathbb{R}\}$

$(-\infty, \infty)$

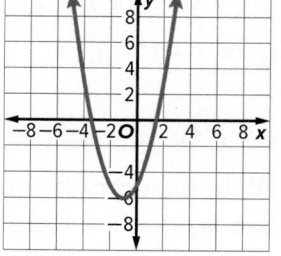

Range

The least *y*-value for this function is −6.

The range is all real numbers greater than or equal to −6.

$\{y \mid y \geq -6\}$

$[-6, \infty)$

Study Tip

Using Symbols You can use the symbol $\mathbb{R}$ to represent all real numbers in set-builder notation. In interval notation, the symbol ∪ indicates the union of two sets. Parentheses are always used with ∞ and −∞ because they do not include endpoints.

Go Online You can complete an Extra Example online.

Check

State the domain and range of each graph in set-builder and interval notation.

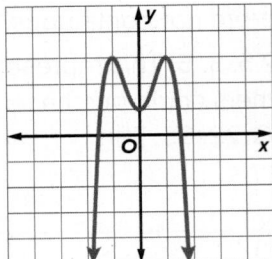

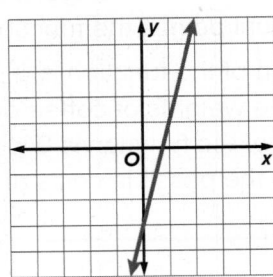

Example 7 Set-Builder and Interval Notation for Discontinuous Intervals

Write the domain and range of the graph in set-builder and interval notation.

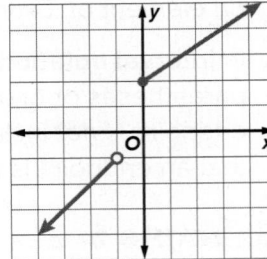

Domain

The domain is all real numbers less than −1 or greater than or equal to 0.

$\{x \mid x < -1 \text{ or } x \geq 0\}$

$(-\infty, -1) \cup [0, \infty)$

Range

The range is all real numbers less than −1 or greater than or equal to 2.

$\{y \mid y < -1 \text{ or } y \geq 2\}$

$(-\infty, -1) \cup [2, \infty)$

Check

State the domain and range of the graph in set-builder and interval notation.

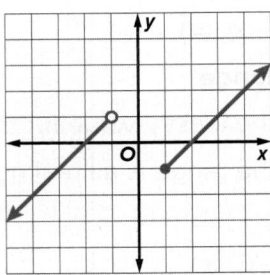

Go Online You can complete an Extra Example online.

Practice

Go Online You can complete your homework online.

Example 1

Identify the domain, range, and codomain in each graph. Then use the codomain and range to determine whether the function is onto.

1.
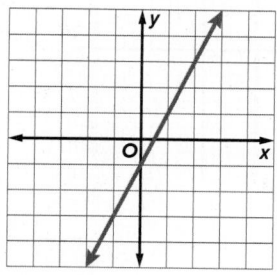

2.
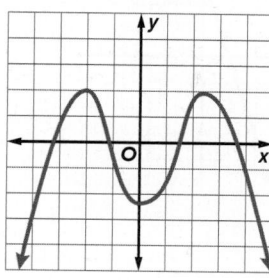

3.
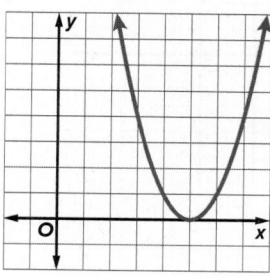

Example 2

4. **SALES** Cool Athletics introduced the new Power Sneaker in one of their stores. The table shows the sales for the first 6 weeks. Define the domain and range of the function and state whether it is *one-to-one, onto, both* or *neither*.

Week	1	2	3	4	5	6
Pairs Sold	8	10	15	22	15	44

5. **TEMPERATURES** The table shows the low temperatures in degrees Fahrenheit for the past week in Sioux Falls, Idaho. Define the domain and range of the function and state whether it is *one-to-one, onto, both,* or *neither*.

Day	1	2	3	4	5	6	7
Low Temp.	56	52	44	41	43	46	53

6. **PLANETS** The table shows the orbital period of the eight major planets in our Solar System given their mean distance from the Sun. Define the domain and range of the function and state whether it is *one-to-one, onto, both* or *neither*.

Planet	Mean Distance from Sun (AU)	Orbital Period (years)
Mercury	0.4	0.241
Venus	0.7	0.615
Earth	1.0	1.0
Mars	1.5	1.881
Jupiter	5.2	11.75
Saturn	9.5	29.5
Uranus	19.2	84
Neptune	30	164.8

Example 3

Determine whether each function is *one-to-one, onto, both,* or *neither*.

7.
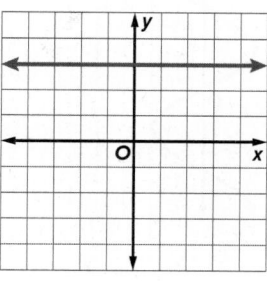

8.
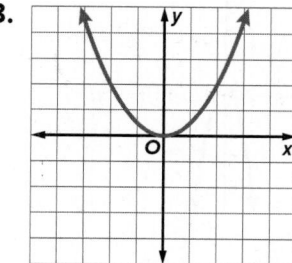

9.
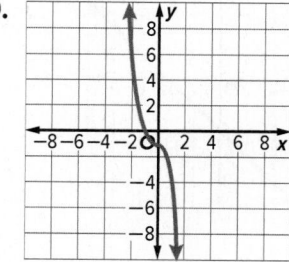

Example 4

Examine the graphs. Determine whether each function is *discrete, continuous,* or *neither* discrete nor continuous. Then state the domain and range of each function.

10.

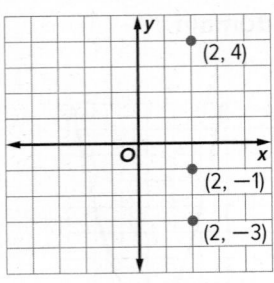

11.

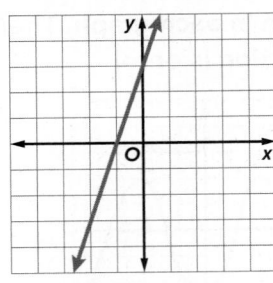

12.

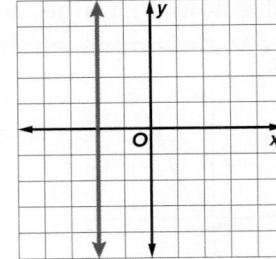

Example 5

13. PROBABILITY The table shows the outcome of rolling a number cube. Determine whether the function that models the outcome of each roll is *discrete, continuous,* or *neither* discrete nor continuous. Then state the domain and range of the function.

Roll	Outcome
1	4
2	3
3	6
4	3
5	5
6	4

14. AMUSEMENT PARK The table shows the price of tickets to an amusement park based on the number of people in the group. Determine whether the function that models the price of tickets is *discrete, continuous,* or *neither* discrete nor continuous. Then state the domain and range of the function.

Group Size	Price
up to 15 people	$45
16-50 people	$38
51-100 people	$30
101 or more people	$26

15. GROCERIES A local grocery store sells grapes for $1.99 per pound. Determine whether the function that models the cost of grapes is *discrete, continuous,* or *neither* discrete nor continuous. Then state the domain and range of the function.

Examples 6 and 7

Write the domain and range of the graph in set-builder and interval notation.

16.

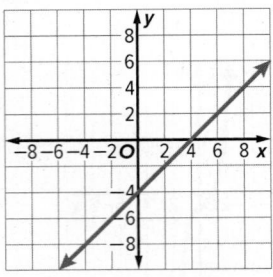

17.

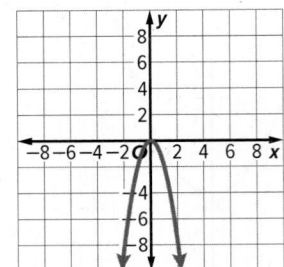

18.

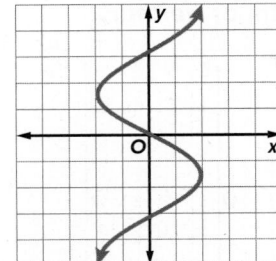

Write the domain and range of the graph in set-builder and interval notation.

19.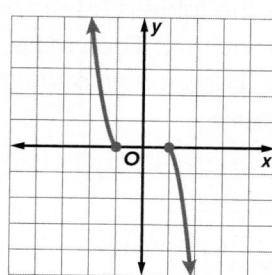

20.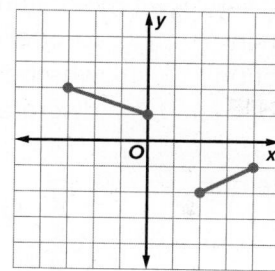

21.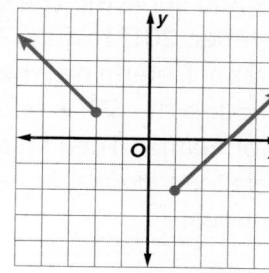

Mixed Exercises

STRUCTURE **Write the domain and range of each function in set-builder and interval notation. Determine whether each function is *one-to-one, onto, both,* or *neither*. Then state whether it is *discrete, continuous,* or *neither* discrete nor continuous.**

22.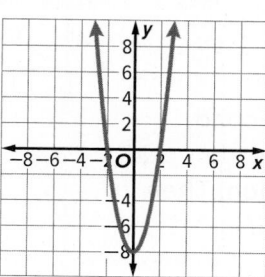

23.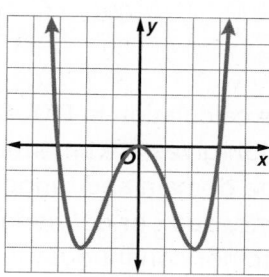

24.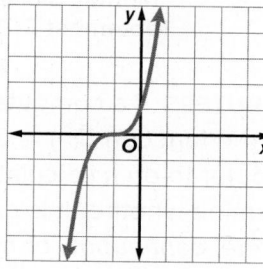

25.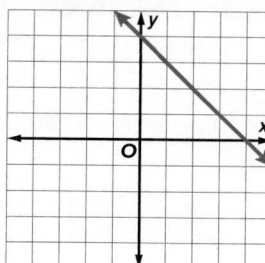

26.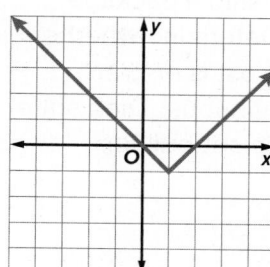

27.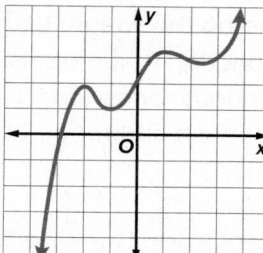

28. USE A SOURCE Research the total number of games won by a professional baseball team each season for five consecutive years. Determine the domain, range, and continuity of the function that models the number of wins.

29. SPRINGS When a weight up to 15 pounds is attached to a 4-inch spring, the length L, in inches, that the spring stretches is represented by the function $L(w) = \frac{1}{2}w + 4$, where w is the weight, in pounds, of the object. State the domain and range of the function. Then determine whether it is *one-to-one, onto, both,* or *neither* and whether it is *discrete, continuous,* or *neither* discrete nor continuous.

30. CASHEWS An airport snack stand sells whole cashews for $12.79 per pound. Determine whether the function that models the cost of cashews is *discrete, continuous,* or *neither* discrete nor continuous. Then state the domain and range of the function in set-builder and interval notation.

31. PRICES The Consumer Price Index (CPI) gives the relative price for a fixed set of goods and services. The CPI from September, 2017 to July, 2018 is shown in the graph. Determine whether the function that models the CPI is *one-to-one, onto, both,* or *neither.* Then state whether it is *discrete, continuous,* or *neither* discrete nor continuous.

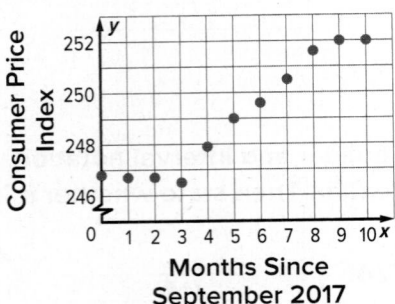

Months Since September 2017

32. LABOR A town's annual jobless rate is shown in the graph. Determine whether the function that models the jobless rate is *one-to-one, onto, both,* or *neither.* Then state whether it is *discrete, continuous,* or *neither* discrete nor continuous.

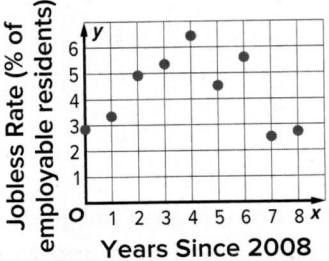

Years Since 2008

33. COMPUTERS If a computer can do one calculation in 0.0000000015 second, then the function $T(n) = 0.0000000015n$ gives the time required for the computer to do n calculations. State the domain and range of the function. Then determine whether it is *one-to-one, onto, both,* or *neither* and whether it is *discrete, continuous,* or *neither* discrete nor continuous.

34. SHIPPING The table shows the cost to ship a package based on the weight of the package. Determine whether the function that models the shipping cost is *discrete, continuous,* or *neither* discrete nor continuous. Then state the domain and range of the function in set-builder notation.

Package Weight (lbs)	Cost
up to 5 pounds	$4
5-10 pounds	$6
exceeds 10 pounds	$0.65/lb

😀 Higher-Order Thinking Skills

35. CREATE Sketch the graph of a function that is onto, but not one-to-one, if the codomain is restricted to values greater than or equal to −3.

36. ANALYZE Determine whether the following statement is *true* or *false*. Explain your reasoning.

If a function is onto, then it must be one-to-one as well.

37. PERSEVERE Consider $f(x) = \frac{1}{x}$. State the domain and the range of the function. Determine whether the function is *one-to-one, onto, both,* or *neither.* Determine whether the function is *discrete, continuous,* or *neither* discrete nor continuous.

38. PERSEVERE Use the domain {−4, −2, 0, 2, 4}, the codomain {−4, −2, 0, 2, 4}, and the range {0, 2, 4} to create a function that is neither one-to-one nor onto.

39. WRITE Compare and contrast the vertical and horizontal line tests.

Linearity, Intercepts, and Symmetry

Explore Symmetry and Functions

🔵 **Online Activity** Use graphing technology to complete the Explore.

> ⓠ **INQUIRY** How can you tell whether the graph of a function is symmetric? ×

Learn Linear and Nonlinear Functions

In a **linear function**, no variable is raised to a power other than 1. Any linear function can be written in the form $f(x) = mx + b$, where m and b are real numbers. Linear functions can be represented by **linear equations**, which can be written in the form $Ax + By = C$. The graph of a linear equation is a straight line.

A function that is not linear is called a **nonlinear function**. The graph of a nonlinear function includes a set of points that cannot all lie on the same line. A nonlinear function cannot be written in the form $f(x) = mx + b$. A **parabola** is the graph of a quadratic function, which is a type of nonlinear function.

Example 1 Identify Linear Functions from Equations

Determine whether each function is a linear function. Justify your answer.

a. $f(x) = \dfrac{6x - 5}{3}$

$f(x) = \dfrac{6x - 5}{3}$ Original equation

$f(x) = \dfrac{6}{3}x - \dfrac{5}{3}$ Distribute the denominator of 3.

$f(x) = 2x - \dfrac{5}{3}$ Simplify.

The function can be written in the form $f(x) = mx + b$, so it is a linear function.

b. $5y = 4 + 3x^3$

$5y = 4 + 3x^3$ Original equation

$5y = 3x^3 + 4$ Commutative Property

The function cannot be written in the form $f(x) = mx + b$ because the independent variable x is raised to a whole number power greater than 1. So, it is a nonlinear function.

🔵 **Go Online** You can complete an Extra Example online.

Today's Goals
- Identify linear and nonlinear functions.
- Identify and interpret the intercepts of functions.
- Identify whether graphs of functions possess line or point symmetry and determine whether functions are even, odd, or neither.

Today's Vocabulary
linear function
linear equation
nonlinear function
parabola
intercept
x-intercept
y-intercept
symmetry
line symmetry
line of symmetry
point symmetry
point of symmetry
even functions
odd functions

Study Tip

Linear Functions To write any linear equation in function form, solve the equation for y and replace the variable y with $f(x)$.

Example 2 Identify Linear Functions from Graphs

Determine whether each graph represents a *linear* or *nonlinear* function.

Think About It!

Why is $f(x) = \sqrt{2x} + 3$ not a linear function?

a.

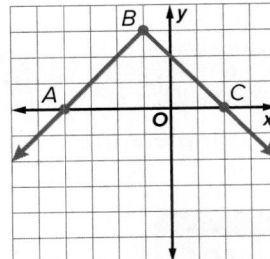

There is no straight line that will contain the chosen points *A*, *B*, and *C*, so this graph represents a nonlinear function.

b.

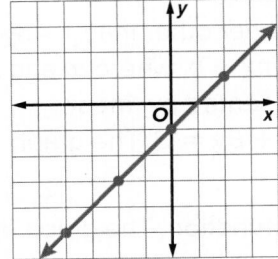

The points on this graph all lie on the same line, so this graph represents a linear function.

🌐 Example 3 Identify Linear Functions from Tables

EARNINGS **Makayla has started working part-time at the local hardware store. Her time at work steadily increases for the first five weeks. The table shows her total earnings each of those weeks. Are her weekly earnings modeled by a *linear* or *nonlinear* function?**

Think About It!

Are negative *x*- or *y*-values possible in the context of the situation?

Week	1	2	3	4	5
Earnings ($)	85	119	153	187	221

Graph the points that represent the week and total earnings and try to draw a line that contains all the points.

Since there is a line that contains all the points, Makayla's earning can be modeled by a linear function.

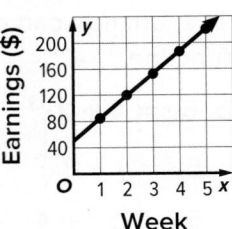

🧭 **Go Online** You can complete an Extra Example online.

Learn Intercepts of Graphs of Functions

A point at which the graph of a function intersects an axis is called an **intercept**. An *x-intercept* is the x-coordinate of a point where the graph crosses the x-axis, and a *y-intercept* is the y-coordinate of a point where the graph crosses the y-axis.

A linear function has at most one x-intercept while a nonlinear function may have more than one x-intercept.

Example 4 Find Intercepts of a Linear Function

Use the graph to estimate the *x*- and *y*-intercepts.

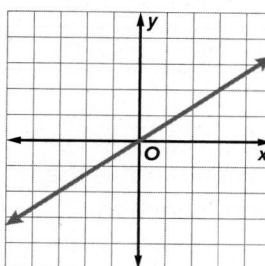

The graph intersects the x-axis at (0, 0), so the x-intercept is 0.

The graph intersects the y-axis at (0, 0), so the y-intercept is 0.

Example 5 Find Intercepts of a Nonlinear Function

Use the graph to estimate the *x*- and *y*-intercepts.

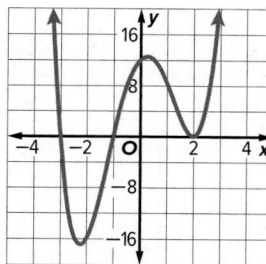

The graph appears to intersect the x-axis at (−3, 0), (−1, 0), and (2, 0), so the function has x-intercepts of −3, −1, and 2.

The graph appears to intersect the y-axis at (0, 12), so the function has a y-intercept of 12.

Check

Estimate the *x*- and *y*-intercepts of each graph.

a.

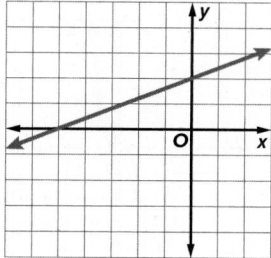

x-intercept(s): ___?___

y-intercept(s): ___?___

b.

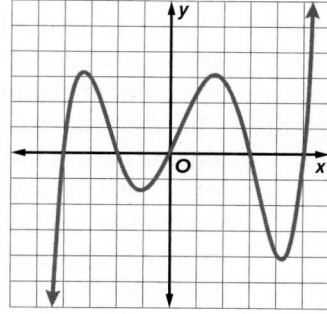

x-intercept(s): _____?_____

y-intercept(s): ___?___

 Go Online You can complete an Extra Example online.

Study Tip

Point or Coordinate
Intercept may refer to the point or one of its coordinates. The context of the situation will often dictate which form to use.

 Think About It!

Describe a line that does not have two distinct intercepts.

 Think About It!

The graph of the nonlinear function has three x-intercepts. Can the graph have more than one y-intercept? Explain your reasoning.

🌐 **Example 6** Interpret the Meaning of Intercepts

MODEL ROCKETS **Ricardo launches a rocket from a balcony. The table shows the height of the rocket after each second of its flight.**

Time (s)	Height (ft)
0	15
1	60
2	130
3	180
4	210
5	170
6	110
7	55
8	0

Part A Identify the x- and y-intercepts of the function that models the flight of the rocket.

In the table, the x-coordinate when $y = 0$ is 8. Thus, the x-intercept is 8.

In the table, the y-coordinate when $x = 0$ is 15. Thus, the y-intercept is 15.

Part B What is the meaning of the intercepts in the context of the rocket's flight?

The x-intercept is the number of seconds after the rocket is launched that it returns to the ground. The y-intercept is the height of the balcony from which the rocket is launched.

Learn Symmetry of Graphs of Functions

A figure has **symmetry** if there exists a rigid motion—reflection, translation, rotation, or glide reflection—that maps the figure onto itself.

Key Concept • Symmetry

Type of Symmetry	Description	Example
A graph has **line symmetry** if it can be reflected in a vertical line so that each half of the graph maps exactly to the other half.	The line dividing the graph into matching halves is called the **line of symmetry**. Each point on one side is reflected in the line to a point equidistant from the line on the opposite side.	
A graph has **point symmetry** when a figure is rotated 180° about a point and maps onto itself.	The point about which the graph is rotated is called the **point of symmetry**. The image of each point on one side of the point of symmetry can be found on the line through the point and the point of symmetry, equidistant from the point of symmetry.	

Sidebar:

🌑 **Think About It!**

Describe the domain of the function that models the rocket's height over time.

Watch Out!

Switching Coordinates A common mistake is to switch the coordinates for the intercepts. Remember that for the x-intercept, the y-coordinate is 0, and for the y-intercept, the x-coordinate is 0.

💬 **Talk About It**

Can the graph of a function be symmetric in a horizontal line? Justify your answer.

Key Concept • Even and Odd Functions

Type of Function	Algebraic Test	Example
Functions that have line symmetry with respect to the y-axis are called **even functions**.	For every x in the domain of f, $f(-x) = f(x)$.	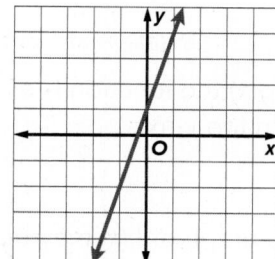
Functions that have point symmetry with respect to the origin are called **odd functions**.	For every x in the domain of f, $f(-x) = -f(x)$.	

Example 7 Identify Types of Symmetry

Identify the type of symmetry in the graph of each function. Explain.

a. $f(x) = 3x + 1$

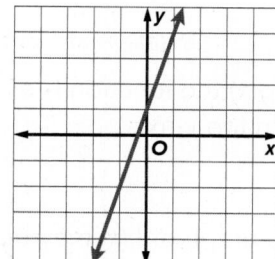

point symmetry: A 180° rotation about any point on the graph is the original graph.

b. $g(x) = -x^2 - 4x - 2$

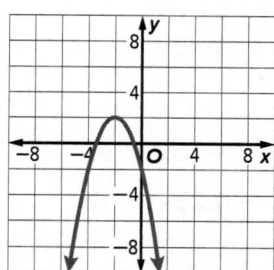

line symmetry: The reflection in the line $x = -2$ coincides with the original graph.

c. $h(x) = 3x^4 + 4x^3 - 12x^2 + 13$

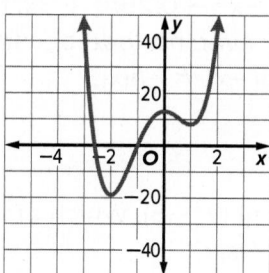

no symmetry: There is no line or point of symmetry.

d. $j(x) = x^3 - 2$

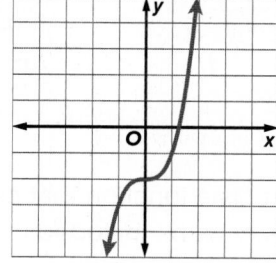

point symmetry: A 180° rotation about the point $(0, -2)$ is the original graph.

Go Online You can complete an Extra Example online.

Think About It!

How would knowing the type of symmetry help you graph a function?

Example 8 Identify Even and Odd Functions

Determine whether each function is *even*, *odd*, or *neither*. Confirm algebraically. If the function is odd or even, describe the symmetry.

a. $f(x) = x^3 - 4x$

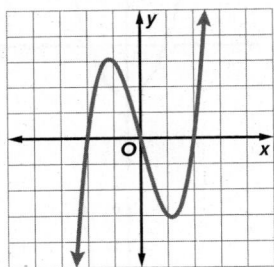

It appears that the graph of $f(x)$ is symmetric about the origin. Substitute $-x$ for x to test this algebraically.

$$f(-x) = (-x)^3 - 4(-x)$$
$$= -x^3 + 4x \qquad \text{Simplify.}$$
$$= -(x^3 - 4x) \qquad \text{Distribute.}$$
$$= -f(x) \qquad f(x) = x^3 - 4x$$

Because $f(-x) = -f(x)$ the function is odd and is symmetric about the origin.

b. $g(x) = 2x^4 - 6x^2$

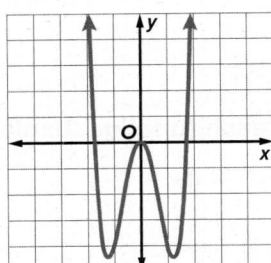

It appears that the graph of $g(x)$ is symmetric about the y-axis. Substitute $-x$ for x to test this algebraically.

$$g(-x) = 2(-x)^4 - 6(-x)^2$$
$$= 2x^4 - 6x^2 \qquad \text{Simplify.}$$
$$= g(x) \qquad g(x) = 2x^4 - 6x^2$$

Because $g(-x) = g(x)$ the function is even and is symmetric in the y-axis.

c. $h(x) = x^3 + 0.25x^2 - 3x$

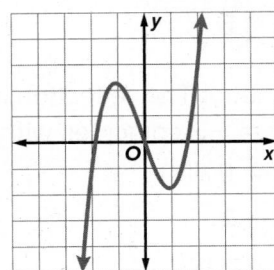

It appears that the graph of $h(x)$ may be symmetric about the origin. Substitute $-x$ for x to test this algebraically.

$$h(-x) = (-x)^3 + 0.25(-x)^2 - 3(-x)$$
$$= -x^3 - 0.25x^2 + 3x$$

Because $-h(x) = -x^3 - 0.25x^2 + 3x$, the function is neither even nor odd because $h(-x) \neq h(x)$ and $h(-x) \neq -h(x)$.

> **Watch Out!**
>
> **Even and Odd Functions** Always confirm symmetry algebraically. Graphs that appear to be symmetric may not actually be.

Check

Assume that f is a function that contains the point $(2, -5)$. Which of the given points must be included in the function if f is:

even? _____?_____   odd? _____?_____

$(-2, -5)$ $(-2, 5)$ $(2, 5)$ $(-5, -2)$ $(-5, 2)$

⊗ **Go Online** You can complete an Extra Example online.

Practice

Go Online You can complete your homework online.

Example 1

Determine whether each function is a linear function. Justify your answer.

1. $y = 3x$

2. $y = -2 + 5x$

3. $2x + y = 10$

4. $y = 4x^2$

Example 2

Determine whether each graph represents a *linear* or *nonlinear* function.

5.

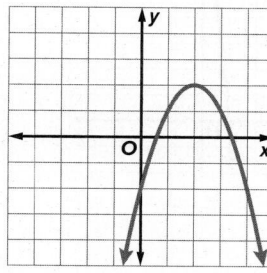

6.

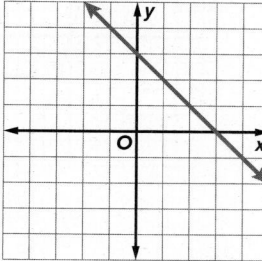

7.

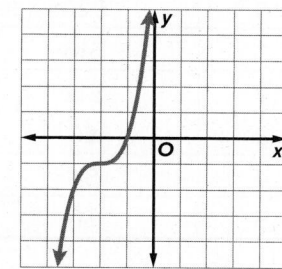

8.

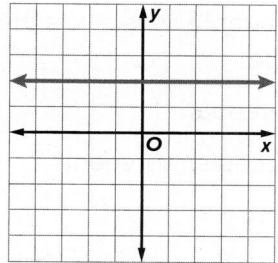

Example 3

9. **MEASUREMENT** The table shows a function modeling the number of inches and feet. Can the relationship be modeled by a *linear* or *nonlinear* function? Explain.

Inches	0	1	2	3	4
Feet	0	12	24	36	48

10. **ASTRONOMY** The table shows the velocity of *Cassini 2* space probe as it passes Saturn. Is the velocity modeled by a *linear* or *nonlinear* function? Explain.

Cassini 2 Velocity					
Time (s)	5	10	15	20	25
Velocity (mph)	50,000	60,000	70,000	60,000	50,000

Examples 4 and 5

Use the graph to estimate the *x*- and *y*-intercepts.

11.
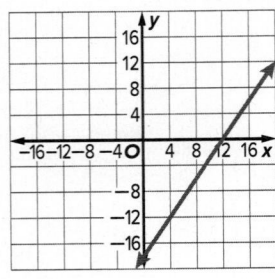

12.
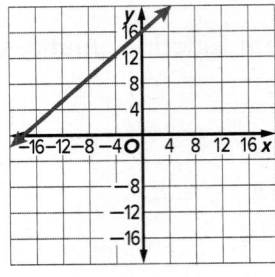

13.
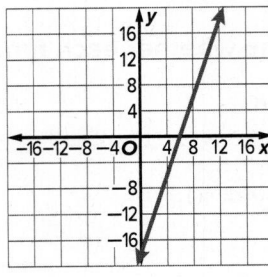

14.
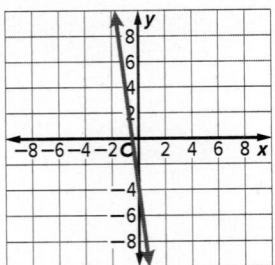

15.
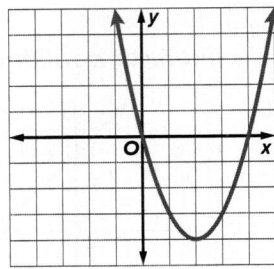

16.
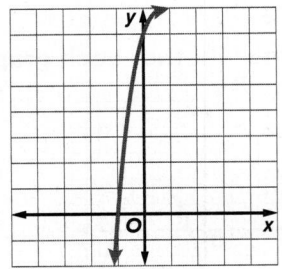

Example 6

17. **MONEY** At the beginning of the week, Aksa's parents deposited $20 into Aksa's lunch account. The amount of money Aksa had left after each day is shown in the table, where *x* is the number of days and *y* is the remaining balance.

   a. What are the *x*- and *y*-intercepts?

   b. What do the *x*- and *y*-intercepts represent?

Days	Account Balance
0	$20
1	$16
2	$12
3	$8
4	$4
5	$0

18. **GOLF** In golf, the first shot on every hole can be hit off a tee. The table shows the height *y* of the golf ball *x* seconds after it has been hit off the tee.

Time (s)	0	1	3	5	7
Height (in.)	3	20	36	28	0

   a. What are the *x*- and *y*-intercepts?

   b. What do the *x*- and *y*-intercepts represent in the context of the situation?

Example 7

Identify the type of symmetry for the graph of each function.

19.
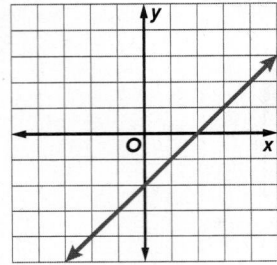

20.
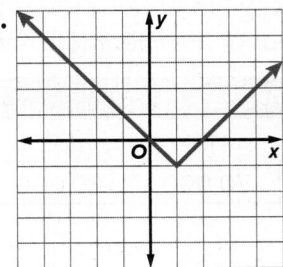

21.
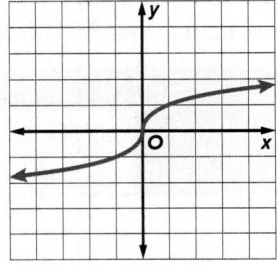

Example 8

Determine whether each function is *even*, *odd*, or *neither*. Confirm algebraically. If the function is odd or even, describe the symmetry.

22. $f(x) = 2x^3 - 8x$

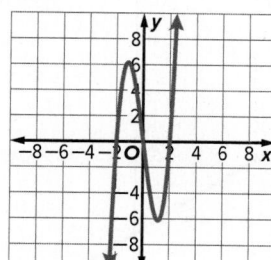

23. $f(x) = x^3 + x^2$

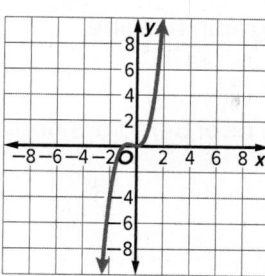

24. $f(x) = x^2 + 2$

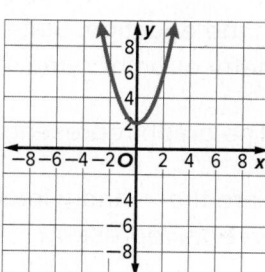

Mixed Exercises

Determine whether each equation represents a linear function. Justify your answer. Algebraically determine whether each equation is *even*, *odd*, or *neither*.

25. $-\frac{3}{x} + y = 15$

26. $x = y + 8$

27. $y = 8$

28. $y = \sqrt{x} + 3$

29. $y = 3x^2 - 1$

30. $y = 2x^3 + x + 1$

Determine whether each graph represents a *linear* or *nonlinear* function. Use the graph to estimate the x- and y-intercepts. Identify the type of symmetry in each graph.

31.

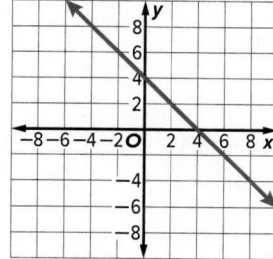

32.

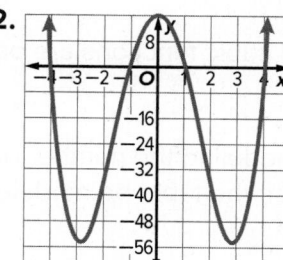

33.

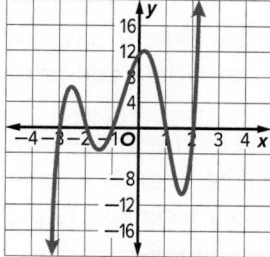

34. GAMES Pedro is creating an online racquetball game. In one play, the motion of the ball across the screen is partially modeled by the graph shown. State whether the graph has line symmetry or point symmetry, and identify any lines or points of symmetry.

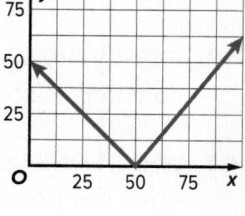

35. BASKETBALL Tiana tossed a basketball. The graph shows the height of the basketball as a function of time. If the graph of the function is extended to include the domain of all real numbers, would it have line or point symmetry? If so, identify the line or point of symmetry.

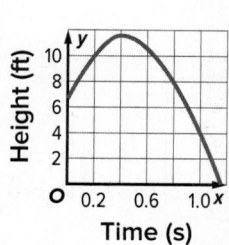

36. PROFIT Stefon charges people $25 to test the air quality in their homes. The device he uses to test air quality cost him $500. The function $y = 25x - 500$ describes Stefon's net profit, y, as a function of the number of clients he gets, x.

a. State whether the function is a linear function. Write *yes* or *no*. Explain.

b. What do the x- and y-intercepts of the function represent in terms of the situation?

37. PLAYGROUNDS A playground is shaped as shown. The total perimeter is 500 feet.

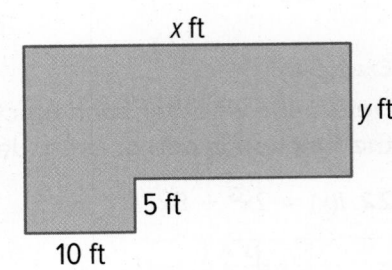

a. Write an equation that relates x and y.

b. Is the equation that relates x and y linear? Explain.

c. Graph the equation. State whether the graph has line symmetry or point symmetry.

38. POOLS The graph represents a 720-gallon pool being drained.

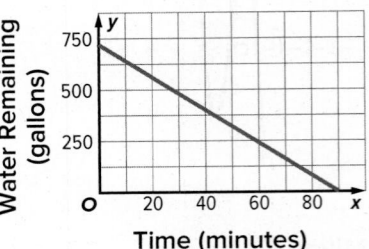

a. What are the x- and y-intercepts? What do the x- and y-intercepts represent?

b. Does the graph display line symmetry? Explain why or why not in terms of the situation.

39. VOLUME The function, $f(r) = \frac{4}{3}\pi r^3$ describes the relationship between the volume $f(r)$ and radius r of a sphere. Determine whether the function is *odd*, *even*, or *neither*. Explain your reasoning.

40. USE A SOURCE Research online to find an equation that models a car's braking distance in relation to its speed. Then identify and interpret the y-intercept of the equation.

🧠 Higher-Order Thinking Skills

41. FIND THE ERROR Javier claimed that all cubic functions are odd. Is he correct? If not, provide a counterexample.

42. ANALYZE The table shows a function modeling the number of gifts y Cornell can wrap if he spends x hours wrapping. Can the table be modeled by a *linear* or *nonlinear* function? Explain.

Hours	0	1	2	3	4
Gifts	0	14	28	42	56

43. PERSEVERE Determine whether an equation of the form $x = a$, where a is a constant, is *sometimes, always,* or *never* a linear function. Explain your reasoning.

44. WHICH ONE DOESN'T BELONG? Of the four functions shown, identify the one that does not belong. Explain your reasoning.

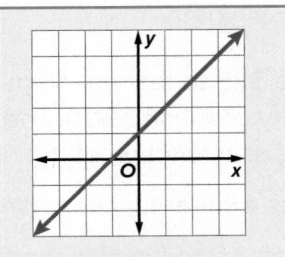

| $y = 2x + 3$ |

x	y
0	4
1	2
2	0
3	−2

| $y = 2xy$ |

Extrema and End Behavior

Learn Extrema of Functions

Graphs of functions can have high and low points where they reach a maximum or minimum value. The maximum and minimum values of a function are called **extrema.**

The **maximum** is at the highest point on the graph of a function. The **minimum** is at the lowest point on the graph of a function.

A **relative maximum** is located at a point on the graph of a function where no other nearby points have a greater y-coordinate. A **relative minimum** is located at a point on the graph of a function where no other nearby points have a lesser y-coordinate.

Example 1 Find Extrema from Graphs

Identify and estimate the x- and y-values of the extrema. Round to the nearest tenth if necessary.

f(x)
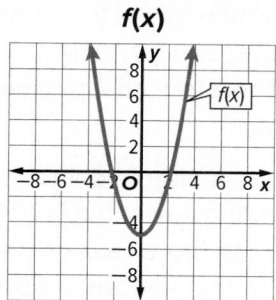

g(x)
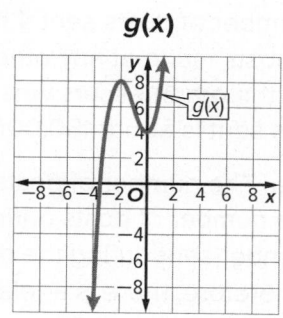

$f(x)$: The function $f(x)$ is decreasing as it approaches $x = 0$ from the left and increasing as it moves away from $x = 0$. Further, $(0, -5)$ is the lowest point on the graph, so $(0, -5)$ is a minimum.

$g(x)$: The function $g(x)$ is increasing as it approaches $x = -2$ from the left and decreasing as it moves away from $x = -2$. Further, there are no greater y-coordinates surrounding $(-2, 8)$. However, $(-2, 8)$ is not the highest point on the graph, so $(-2, 8)$ is a relative maximum.

The function $g(x)$ is decreasing as it approaches $x = 0$ from the left and increasing as it moves away from $x = 0$. Further, there are no lesser y-coordinates surrounding $(0, 4)$. However, $(0, 4)$ is not the lowest point on the graph, so $(0, 4)$ is a relative minimum.

⭐ **Go Online** You can complete an Extra Example online.

Today's Goals
- Identify extrema of functions.
- Identify end behavior of graphs.

Today's Vocabulary
extrema

maximum

minimum

relative maximum

relative minimum

end behavior

Watch Out!

No Extrema Some functions, like $f(x) = x^3$, have no extrema.

Study Tip

Reading in Math In this context, *extrema* is the plural form of *extreme point*. The plural of *maximum* and *minimum* are *maxima* and *minima*, respectively.

💭 Think About It!

Why are the extrema identified on the graph of $g(x)$ relative maxima and minima instead of maxima and minima?

Example 2 Find and Interpret Extrema

SOCIAL MEDIA Use the table and graph to estimate the extrema of the function that relates the number hours since midnight *x* to the number of posts being uploaded *y*. Describe the meaning of the extrema in the context of the situation.

x	y
0	2.8
4	1.8
8	3.1
12	11.5
14	9.1
16	10.2
20	5.8
24	2.8

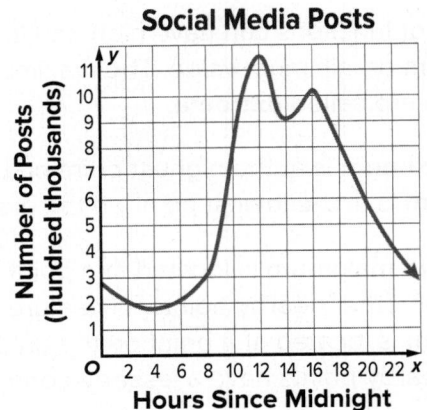

Social Media Posts

Number of Posts (hundred thousands) vs *Hours Since Midnight*

maxima The number of posts sent 12 hours after midnight is greater than the number of posts made at any other time during the day. The highest point on the graph occurs when $x = 12$. Therefore, the maximum number of posts sent is about 1,150,000 at noon.

minima The number of posts sent 4 hours after midnight is less than the number of posts made at any other time during the day. The lowest point on the graph occurs when $x = 4$. Therefore, the minimum number of posts sent is about 180,000 at 4:00 A.M.

relative maxima The number of posts sent 16 hours after midnight is greater than the number of posts during surrounding times, but is not the greatest number sent during the day. The graph has a relative peak when $x = 16$. Therefore, there is a relative peak in the number of posts sent, or relative maximum, at 4:00 P.M. of about 1,020,000 posts.

relative minima The number of posts sent 14 hours after midnight is less than the number of posts during surrounding times, but is not the least number sent during the day. The graph dips when $x = 14$. Therefore, there is a relative low in the number of posts sent, or relative minimum, at 2:00 P.M. of about 910,000 posts.

Explore End Behavior of Linear and Quadratic Functions

Online Activity Use graphing technology to complete the Explore.

INQUIRY Given the behavior of a linear or quadratic function as *x* increases towards infinity, how can you find the behavior as *x* decreases toward negative infinity or vice versa?

Go Online
You can complete an Extra Example online.

Learn End Behavior of Graphs of Functions

End behavior is the behavior of a graph as x approaches positive or negative infinity. As you move right along the graph, the values of x are increasing toward infinity. This is denoted as $x \to \infty$. At the left end, the values of x are decreasing toward negative infinity, denoted as $x \to -\infty$. When a function $f(x)$ increases without bound, it is denoted as $f(x) \to \infty$. When a function $f(x)$ decreases without bound, it is denoted as $f(x) \to -\infty$.

Example 3 End Behavior of Linear Functions

Describe the end behavior of each linear function.

a. $f(x)$

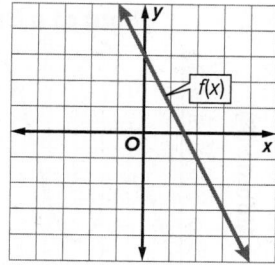

b. $g(x)$

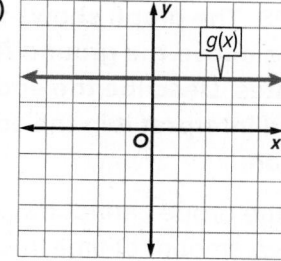

As x decreases, $f(x)$ increases, and as x increases $f(x)$ decreases. Thus, as $x \to -\infty$, $f(x) \to \infty$ and as $x \to \infty$, $f(x) \to -\infty$.

As x decreases or increases, $g(x) = 2$. Thus, as $x \to -\infty$, $g(x) = 2$, and as $x \to \infty$, $g(x) = 2$.

Check

Use the graph to describe the end behavior of the function.

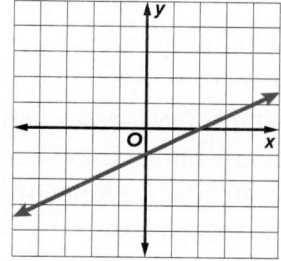

Example 4 End Behavior of Nonlinear Functions

Describe the end behavior of each nonlinear function.

a. $f(x)$

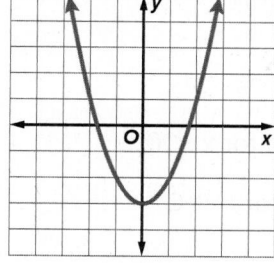

b. $g(x)$

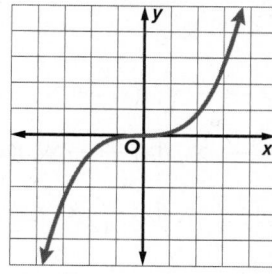

As you move left or right on the graph, $f(x)$ increases. Thus as $x \to -\infty$, $f(x) \to \infty$, and as $x \to \infty$, $f(x) \to \infty$.

As $x \to -\infty$, $g(x) \to -\infty$, and as $x \to \infty$, $g(x) \to \infty$.

Go Online You can complete an Extra Example online.

Think About It!

For $f(x) = a$, where a is a real number, describe the end behavior of $f(x)$ as $x \to \infty$ and as $x \to -\infty$.

Talk About It!

In part **a**, the function's end behavior as $x \to -\infty$ is the opposite of the end behavior as $x \to \infty$. Do you think this is true for all linear functions where $m \neq 0$? Explain your reasoning.

Math History Minute
Júlio César de Mello e Souza (1895–1974) was a Brazilian mathematician who is known for his books on recreational mathematics. His most famous book, *The Man Who Counted*, includes problems, puzzles, and curiosities about math. The State Legislature of Rio de Janeiro declared that his birthday, May 6, be Mathematician's Day.

Think About It!

If the graph of a function is symmetric about a vertical line, what do you think is true about the end behavior of $f(x)$ as $x \rightarrow -\infty$ and as $x \rightarrow \infty$?

Check

Use the graph to describe the end behavior of the function.

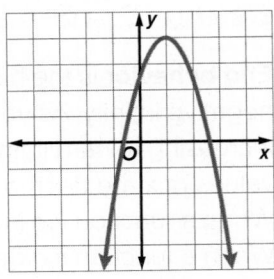

⊕ Example 5 Determine and Interpret End Behavior

DRONES **The graph shows the altitude of a drone above the ground $f(x)$ after x minutes. Describe the end behavior of $f(x)$ and interpret it in the context of the situation.**

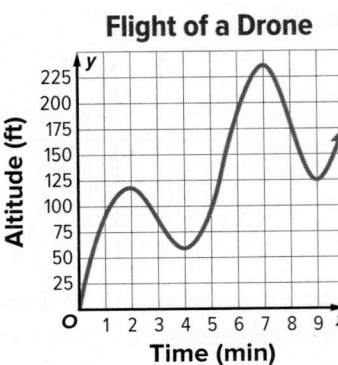

Since the drone cannot travel for a negative amount of time, the function is not defined for $x < 0$. So, there is no end behavior as $x \rightarrow -\infty$.

As $x \rightarrow \infty$, $f(x) \rightarrow \infty$. The drone is expected to continue to fly higher.

Study Tip

Assumptions Assuming that the drone can continue to fly for an infinite amount of time and to an infinite altitude lets us analyze the end behavior as $x \rightarrow \infty$. While there are maximum legal altitudes that a drone can fly as well as limited battery life, assuming that the time and altitude will continue to increase allows us to describe the end behavior.

Check

RIDESHARING Mika and her friends are using a ride-sharing service to take them to a concert. The function models the cost of the ride $f(x)$ after x miles. Describe the end behavior of $f(x)$ and interpret it in the context of the situation.

Part A

What is the end behavior of the function?

A. as $x \rightarrow -\infty$, $f(x) \rightarrow -\infty$; as $x \rightarrow \infty$, $f(x) \rightarrow -\infty$

B. as $x \rightarrow -\infty$, $f(x) \rightarrow \infty$; as $x \rightarrow \infty$, $f(x) \rightarrow \infty$

C. as $x \rightarrow \infty$, $f(x) \rightarrow -\infty$; $f(x)$ is not defined for $x < 0$

D. as $x \rightarrow \infty$, $f(x) \rightarrow \infty$; $f(x)$ is not defined for $x < 0$

Part B

What does the end behavior represents in the context of the situation?

Go Online

to practice what you've learned about analyzing graphs in the Put It All Together over Lessons 8-1 through 8-3.

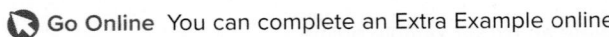

🐦 **Go Online** You can complete an Extra Example online.

Practice

Go Online You can complete your homework online.

Examples 1 and 2

Identify and estimate the *x*- and *y*-values of the extrema. Round to the nearest tenth if necessary.

1.
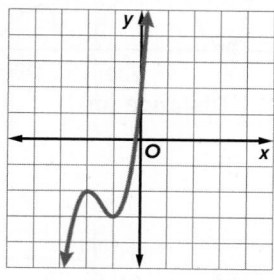

2.
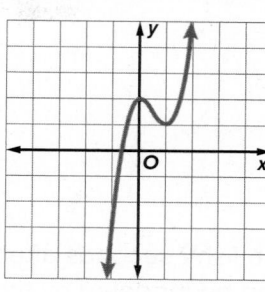

3.
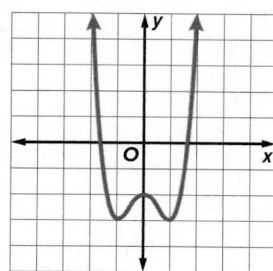

4.
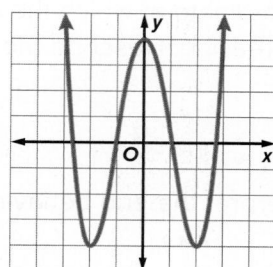

5. **LANDSCAPES** Jalen uses a graph of a function to model the shape of two hills in the background of a videogame that he is writing. Estimate the *x*-coordinates at which the relative maxima and relative minima occur. Describe the meaning of the extrema in the context of the situation.

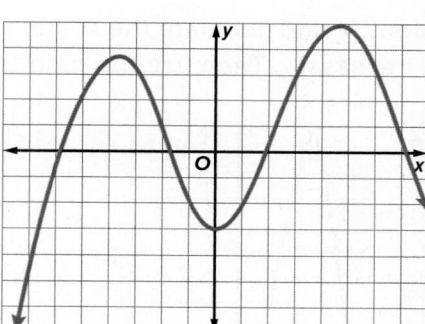

Examples 3-5

Describe the end behavior of each function.

6.
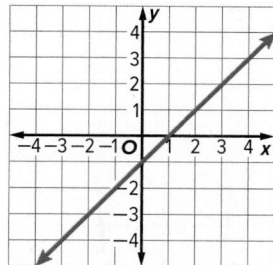

7.
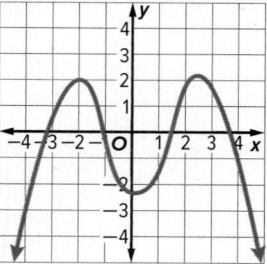

8.
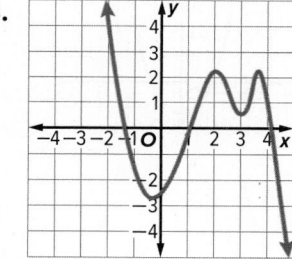

9.
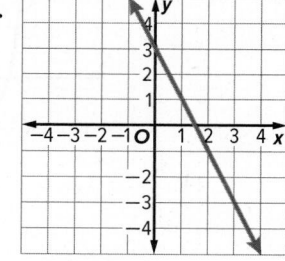

Lesson 8-3 • Extrema and End Behavior **455**

10. ROLLER COASTER The graph shows the height of a roller coaster in terms of its distance away from the starting point. Describe and interpret the end behavior in the context of the situation.

Roller Coaster Height

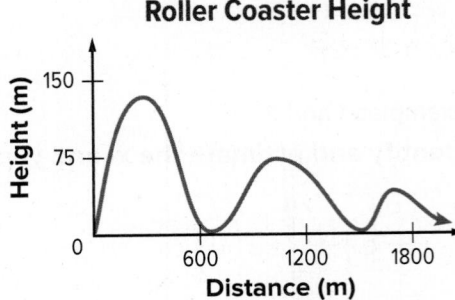

Mixed Exercises

11. MODEL The height of a fish t seconds after it is thrown to a dolphin from a 64-foot-tall platform can be modeled by the equation $h(t) = -16t^2 + 48t + 64$, where $h(t)$ is the height of the fish in feet. The graph of the function is shown.

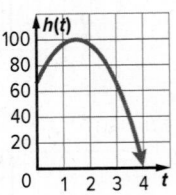

a. Estimate the t-coordinate at which the height of the fish changes from increasing to decreasing. Describe the meaning in terms of the context of the situation.

b. Describe and interpret the end behavior of $h(t)$ in the context of the situation.

Identify and estimate the x- and y-values of the extrema. Round to the nearest tenth if necessary. Then use the graphs to describe the end behavior of each function.

12.

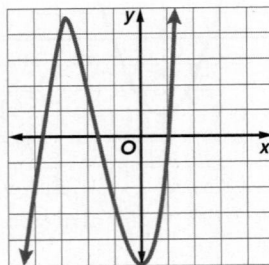

13.

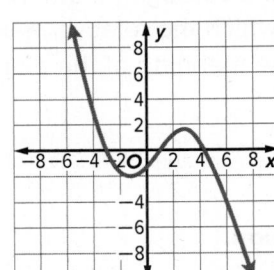

14.

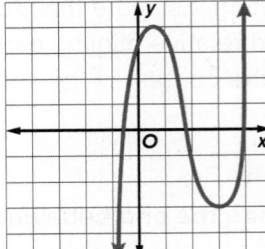

15.

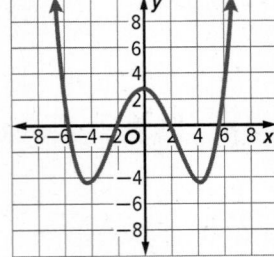

16.

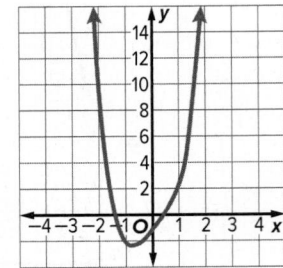

17.

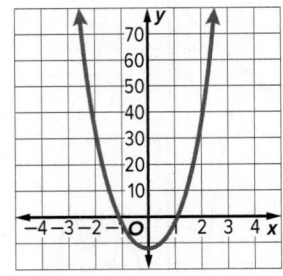

18. BUBBLES The volume of a soap bubble can be estimated by the formula $V = 4\pi r^2$, where r is its radius. The graph shows the function of the bubble's volume. Describe the end behavior of the graph.

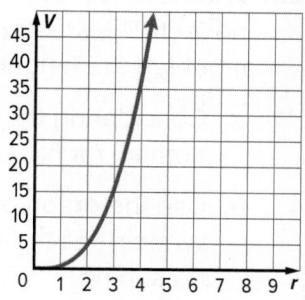

19. SCIENCE The table shows the density of water at its saturation pressure for various temperatures. Interpret the end behavior of the graph of the function as temperature increases.

Temperature (°C)	0	50	100	150	200	250	300	350
Density (g/cm³)	1.000	0.988	0.958	0.917	0.865	0.799	0.713	0.573

Identify and estimate the x- and y-values of the extrema. Round to the nearest tenth if necessary. Then describe the end behavior of each function.

20.

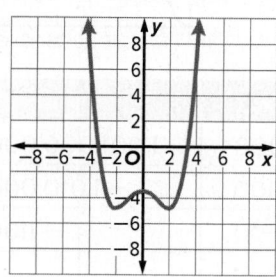

21.

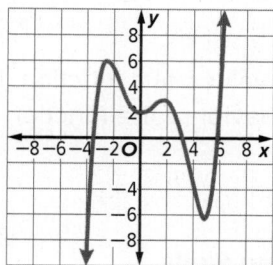

USE ESTIMATION Use a graphing calculator to estimate the x-coordinates at which any extrema occur for each function. Round to the nearest hundredth.

22. $f(x) = x^3 + 3x^2 - 6x - 6$

23. $f(x) = -2x^3 + 8$

24. $f(x) = -2x^4 + 5x^3 - 4x^2 + 3x - 7$

25. $f(x) = x^5 - 4x^3 + 3x^2 - 8x - 6$

26. CONSTRUCT ARGUMENTS Sheena says that in the graph of $f(x)$ shown below, the graph has relative maxima at B and G, and a relative minimum at A. Is she correct? Explain.

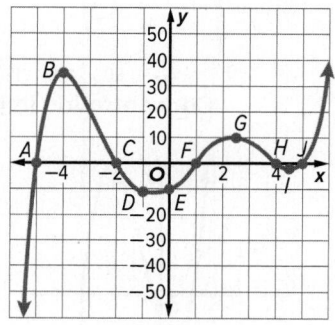

27. CHEMISTRY Dynamic pressure is generated by the velocity of a moving fluid and is given by $q(v) = \frac{1}{2}pv^2$, where p is the density of the fluid and v is its velocity. Water has a density of 1 g/cm³. What happens to the dynamic pressure of water when the velocity continuously increases?

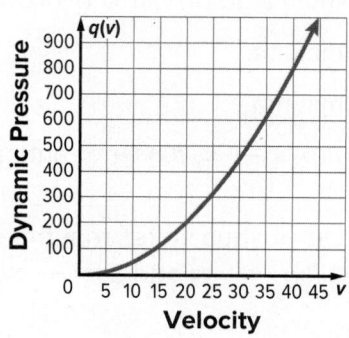

28. ENGINEERING Several engineering students built a catapult for a class project. They tested the catapult by launching a watermelon and modeled the height h of the watermelon in feet over time t in seconds.

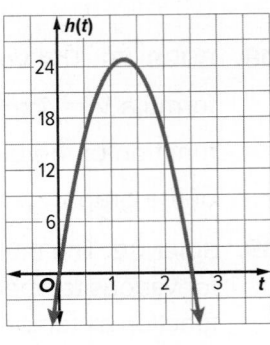

   a. Considering the context of the problem, what is an appropriate domain for $h(t)$? Explain your reasoning.

   b. Use the graph of $h(t)$ to find the maximum height of the watermelon. When does the watermelon reach the maximum height? Explain your reasoning.

29. DRILLING The volume of a drill bit can be estimated by the formula for a cone, $V = \frac{1}{3}\pi h r^2$, where h is the height of the bit and r is its radius. Substituting $\frac{\sqrt{3}}{3}r$ for h, the volume of the drill bit is estimated as $\frac{\sqrt{3}}{9}\pi r^3$. The graph shows the function of drill bit volume. Describe the end behavior.

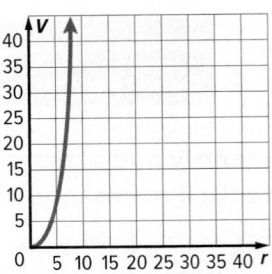

30. The table shows the values of a function. Use the table to describe the end behavior of the function.

x	y
−1000	−1,001,000,000
−100	−1,010,000
−10	−1100
−1	−2
1	0
10	900
100	990,000
1000	999,000,000

31. WRITE Describe what the end behavior of a graph is and how it is determined.

32. CREATE Sketch a graph of a linear function and a nonlinear function with the following end behavior: as $x \to -\infty$, $f(x) \to \infty$ and as $x \to \infty$, $f(x) \to -\infty$.

33. ANALYZE A catalyst is used to increase the rate of a chemical reaction. The reaction rate, or the speed at which the reaction is occurring, is given by $R(x) = \frac{0.5x}{x + 10}$, where x is the concentration of the catalyst solution in milligrams of solute per liter. What does the end behavior of the graph mean in the context of this experiment?

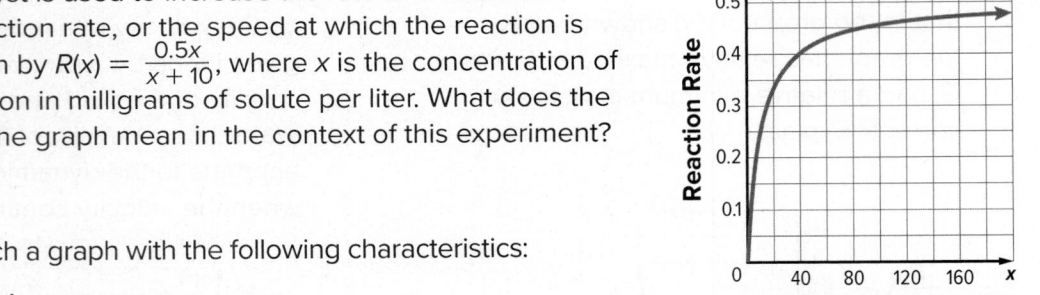
Concentration (mg/L)

34. PERSEVERE Sketch a graph with the following characteristics:

- 2 relative maxima
- 2 relative minima
- end behavior: $x \to \infty$, $f(x) \to \infty$ and as $x \to -\infty$, $f(x) \to -\infty$

35. FIND THE ERROR Joshua states that the end behavior of the graph is: as $x \to -\infty$, $f(x) \to -\infty$ and as $x \to \infty$, $f(x) \to \infty$. What error did he make?

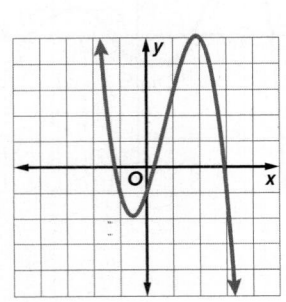

Sketching Graphs and Comparing Functions

Explore Using Technology to Examine Key Features of Graphs

🔖 **Online Activity** Use graphing technology to complete the Explore.

> @ **INQUIRY** What can key features of a function tell you about its graph? ✕

Learn Sketching Graphs of Functions

You can use key features of a function to sketch its graph.

Key Concept • Using Key Features

Key Feature	What it tells you about the graph of f(x)
Domain	the values of x for which $f(x)$ is defined
Range	the values that $f(x)$ take as x varies
Intercepts	where the graph crosses the x- or y-axis
Symmetry	where one side of the graph is a reflection or rotation of the other side
End Behavior	what the graph is doing at the right and left sides as x approaches infinity or negative infinity
Extrema	high or low points where the graph changes from increasing to decreasing or vice versa
Increasing/ Decreasing	where the graph is going up or down as x increases
Positive/Negative	where the graph is above or below the x-axis

Example 1 Sketch a Linear Function

Use the key features of the function to sketch its graph.

y-intercept: $(0, -70)$

Linearity: linear

Positive: for values of x such that $x < -30$

Decreasing: for all values of x

End Behavior: As $x \to \infty$, $f(x) \to -\infty$.
As $x \to -\infty$, $f(x) \to \infty$.

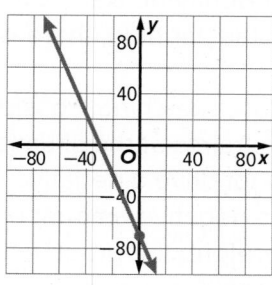

🔖 **Go Online** You can complete an Extra Example online.

Study Tip

Scales and Axes
Before you sketch a function, consider the scales or axes that best fit the situation. You want to capture as much information as possible, so you want the scales to be big enough to easily see the extrema and x- and y intercepts, but not so big that you cannot determine the values.

💬 **Talk About It!**

Given the y-intercept and for what values of x the function is positive, what other information do you need to sketch a linear function? Explain your reasoning.

Example 2 Sketch a Nonlinear Function

Use the key features of the function to sketch its graph.

y-intercept: (0, 3)

Linearity: nonlinear

Continuity: continuous

Positive: for all values of *x*

Decreasing: for all values of *x*
such that $x < 0$

Extrema: minimum at (0, 3)

End Behavior: As $x \to \infty$, $f(x) \to \infty$.
As $x \to -\infty$, $f(x) \to \infty$.

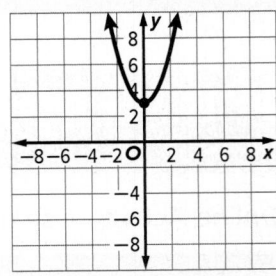

 Think About It!

Explain why the end behavior is not defined in the context of this situation.

Study Tip

Assumptions When sketching the function using the given key features, assumptions must be made. As in this example, the same key features could describe many different graphs. The key features could also be represented by a parabola, a curve that is narrower or wider, or an absolute value function.

🌐 Example 3 Sketch a Real-World Function

TEST DRIVE **Hae is test driving a car she is thinking of buying. She decides to accelerate to 60 miles per hour and then decelerate to a stop to test its acceleration and brakes. It takes her 15 seconds to reach her maximum speed and 15 additional seconds to come to a stop. Use the key features to sketch a graph that shows the speed *y* as a function of time *x*.**

y-intercept: Hae starts her test drive at a speed of 0 miles per hour.

Linear or Nonlinear: The function that models the situation is nonlinear.

Extrema: Hae's maximum speed is 60 miles per hour, which she reaches 15 seconds into her test drive.

Increasing: Hae increases the speed at a uniform rate for the first 15 seconds.

Decreasing: Hae decreases the speed at a uniform rate for the next 15 seconds until she reaches a stop.

End Behavior: Because Hae starts at 0 miles per hour and ends at 0 miles per hour, there is no end behavior.

 Think About It!

Based on the graph, the speed of the car at 10 seconds is 40 miles per hour. Is it appropriate to assume that the car is traveling that exact speed at a specific time? Explain.

Before sketching, consider the constraints of the situation. Hae cannot drive a negative speed or for a negative amount of time. Therefore, the graph only exists for positive *x*- and *y*-values.

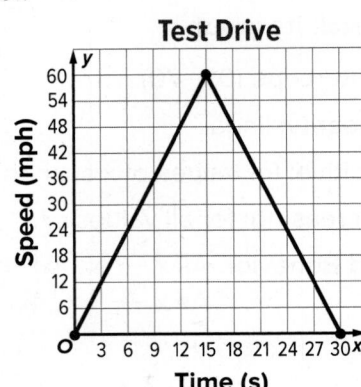

🅑 **Go Online** You can complete an Extra Example online.

Example 4 Compare Properties of Linear Functions

Use the table and graph to compare the two functions.

😀 **Think About It!**

How would a function that passes through (1, 0) with a slope of −4 compare to f(x) and g(x)?

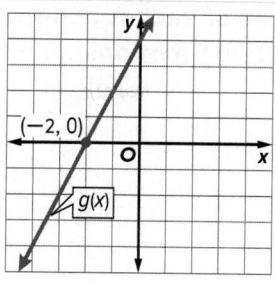

x	f(x)
−6	−3
−3	−2
0	−1
3	0
6	1

x-intercept of f(x): 3
x-intercept of g(x): −2.

So, f(x) intersects the x-axis at a point farther to the right than g(x).

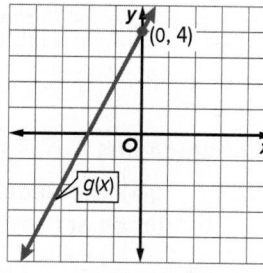

x	f(x)
−6	−3
−3	−2
0	−1
3	0
6	1

y-intercept of f(x): −1,
y-intercept of g(x): 4.

So, g(x) intersects the y-axis at a higher point than f(x).

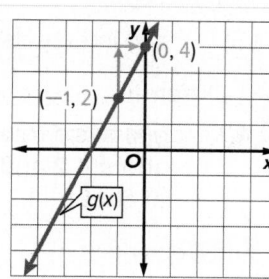

x	f(x)
−6	−3
−3	−2
0	−1
3	0
6	1

slope of f(x): $\frac{1}{3}$
slope of g(x): 2
Each function is increasing, but the slope of g(x) is greater than the slope of f(x).

So, g(x) increases faster than f(x).

Check

Use the table and graph to compare the two functions.

x	f(x)
−2	−6
−1	−4
0	−2
1	0
2	2

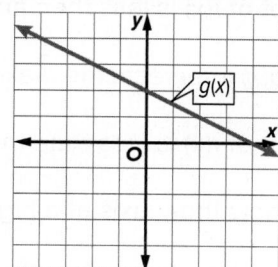

Which statements about f(x) and g(x) are true? Select all that apply.

A. g(x) has a faster rate of change than f(x).

B. The x-intercept of g(x) is greater than the x-intercept of f(x).

C. The y-intercept of f(x) is greater than the x-intercept of g(x).

D. Both functions are decreasing.

E. f(x) increases while g(x) decreases.

F. f(x) has a faster rate of change than g(x).

🔾 **Go Online** You can complete an Extra Example online.

Example 5 Compare Properties of Nonlinear Functions

Examine the categories to see how to use the description and the graph to identify key features of each function. Then complete the statements to compare the two functions.

$f(x)$	$g(x)$
x-intercept: (−3.4, 0) y-intercept: (0, 1.5) relative maximum: (−2.3, 4.7) relative minimum: (−0.4, 1.1) end behavior: as $x \rightarrow -\infty$, $f(x) \rightarrow -\infty$ and as $x \rightarrow \infty$, $f(x) \rightarrow \infty$	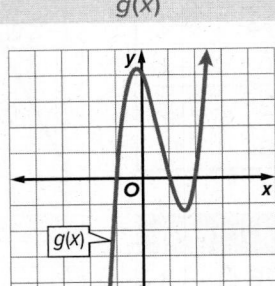

x-intercept(s)	
f(x) intersects the x-axis once at (−3.4, 0).	g(x) intersects the x-axis three times at (−1, 0), (1, 0), and (2, 0).

y-intercept	
f(x) intersects the y-axis at (0, 1.5).	g(x) intersects the y-axis at (0, 4).

Extrema	
f(x) has a relative maximum of 4.7 and a relative minimum of 1.1.	g(x) has a relative maximum of about 4.2 and a relative minimum of about −1.2.

End Behavior	
As $x \rightarrow -\infty$, $f(x) \rightarrow -\infty$, and as $x \rightarrow \infty$, $f(x) \rightarrow \infty$.	As $x \rightarrow -\infty$, $g(x) \rightarrow -\infty$, and as $x \rightarrow \infty$, $g(x) \rightarrow \infty$.

- The x-intercept of f(x) is less than any of the x-intercepts of g(x).

- The graph of g(x) intersects the x-axis more times than f(x).

- The y-intercept of f(x) is less than the y-intercept of g(x).

- So, g(x) intersects the y-axis at a higher point than f(x).

- The relative maximum of f(x) is greater than the relative maximum of g(x). The relative minimum of f(x) is greater than the relative minimum of g(x).

- The two functions have the same end behavior.

Go Online You can complete an Extra Example online.

Practice

Go Online You can complete your homework online.

Examples 1 and 2

Use the key features of each function to sketch its graph.

1. **x-intercept:** (2, 0)
   y-intercept: (0, −6)
   Linearity: linear
   Continuity: continuous
   Positive: for values $x > 2$
   Increasing: for all values of x
   End Behavior: As $x \rightarrow \infty$, $f(x) \rightarrow \infty$
   and as $x \rightarrow -\infty$, $f(x) \rightarrow -\infty$.

2. **x-intercept:** (0, 0)
   y-intercept: (0, 0)
   Linearity: linear
   Continuity: continuous
   Positive: for values $x < 0$
   Negative: for values of $x > 0$
   Decreasing: for all values of x
   End Behavior: As $x \rightarrow \infty$, $f(x) \rightarrow -\infty$
   and as $x \rightarrow -\infty$, $f(x) \rightarrow \infty$.

3. **x-intercept:** (5, 0)
   y-intercept: (0, 5)
   Linearity: linear
   Continuity: continuous
   Positive: for values $x < 5$
   Decreasing: for all values of x
   End Behavior: As $x \rightarrow \infty$, $f(x) \rightarrow -\infty$
   and as $x \rightarrow -\infty$, $f(x) \rightarrow \infty$.

4. **x-intercept:** (5, 0)
   y-intercept: (0, 2)
   Linearity: linear
   Continuity: continuous
   Positive: for values $x < 5$
   Decreasing: for all values of x
   End Behavior: As $x \rightarrow \infty$, $f(x) \rightarrow -\infty$
   and as $x \rightarrow -\infty$, $f(x) \rightarrow \infty$.

5. **x-intercept:** (−1, 0) and (1, 0)
   y-intercept: (0, 1)
   Linearity: nonlinear
   Continuity: continuous
   Symmetry: symmetric about the line $x = 0$
   Positive: for values $-1 < x < 1$
   Negative: for values of $x < -1$ and $x > 1$
   Increasing: for all values of $x < 0$
   Decreasing: for all values of $x > 0$
   Extrema: maximum at (0, 1)
   End Behavior: As $x \rightarrow \infty$, $f(x) \rightarrow -\infty$
   and as $x \rightarrow -\infty$, $f(x) \rightarrow -\infty$.

6. **x-intercept:** (−3, 0) and (2, 0)
   y-intercept: (0, −4)
   Linearity: nonlinear
   Continuity: continuous
   Positive: for values $x < -3$ and $x > 2$
   Negative: for values of $-3 < x < 2$
   Increasing: for all values of $x > 0$
   Decreasing: for all values of $x < 0$
   Extrema: minimum at (0, −4)
   End Behavior: As $x \rightarrow \infty$, $f(x) \rightarrow \infty$
   and as $x \rightarrow -\infty$, $f(x) \rightarrow \infty$.

7. **x-intercept:** (1, 0)
   y-intercept: (0, −1)
   Linearity: linear
   Continuity: continuous
   Positive: for values $x > 1$
   Increasing: for all values of x
   End Behavior: As $x \rightarrow -\infty$, $f(x) \rightarrow -\infty$
   and as $x \rightarrow \infty$, $f(x) \rightarrow \infty$.

8. **x-intercept:** (−2, 0) and (2, 0)
   y-intercept: (0, −1)
   Linearity: nonlinear
   Continuity: continuous
   Symmetry: symmetric about the
   line $x = 0$
   Positive: for values $x < -2$ and $x > 2$
   Negative: for values of $-2 < x < 2$
   Increasing: for all values of $x > 0$
   Decreasing: for all values of $x < 0$
   Extrema: minimum at (0, −1)
   End Behavior: As $x \rightarrow -\infty$, $f(x) \rightarrow \infty$
   and as $x \rightarrow \infty$, $f(x) \rightarrow \infty$.

Example 3

9. PELICANS A pelican descends to the ground. The pelican starts at a height of 6 feet. The pelican reaches the ground, at a height of 0 feet, after 3 seconds. The function that models the situation is linear. Use the key features to sketch a graph.

10. SCOOTERS Greg rides his motorized scooter for 20 minutes. Greg starts riding at 0 mph. Greg's maximum speed is 35 mph, which he reaches 5 minutes after he starts riding. Greg's speed increases steadily for 5 minutes. At the 10-minute mark, Greg decreases his speed for 2.5 minutes, then he stays at 20 mph for 5 minutes. At the 17.5-minute mark, he again decreases his speed for 2.5 minutes until he stops. Use the key features to sketch a graph.

Examples 4 and 5

11. Compare the key features of the functions represented with a graph and a table.

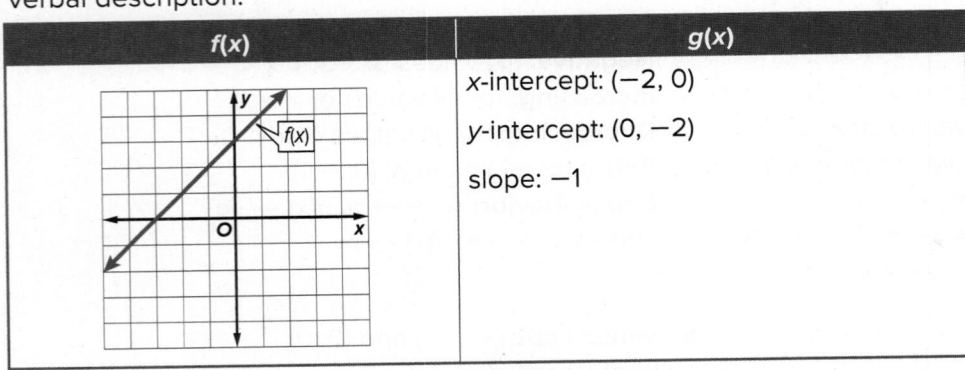

f(x)	g(x)	
	x	**g(x)**
	−2	−4
	−1	−1
	0	2
	1	5
	2	8

12. Compare the key features of the functions represented with a graph and a verbal description.

f(x)	g(x)
	x-intercept: (−2, 0)
	y-intercept: (0, −2)
	slope: −1

13. Compare the key features of the functions represented with a table and a verbal description.

f(x)		g(x)
x	**f(x)**	*x*-intercept: (1, 0)
−4	0	*y*-intercept: (0, −7)
−3	−3	relative maximum: none
−2	−4	relative minimum: none
−1	−3	end behavior:
0	0	as $x \to -\infty$, $g(x) \to -\infty$ and as $x \to \infty$, $g(x) \to \infty$

14. Compare the key features of the functions represented with a graph and a verbal description.

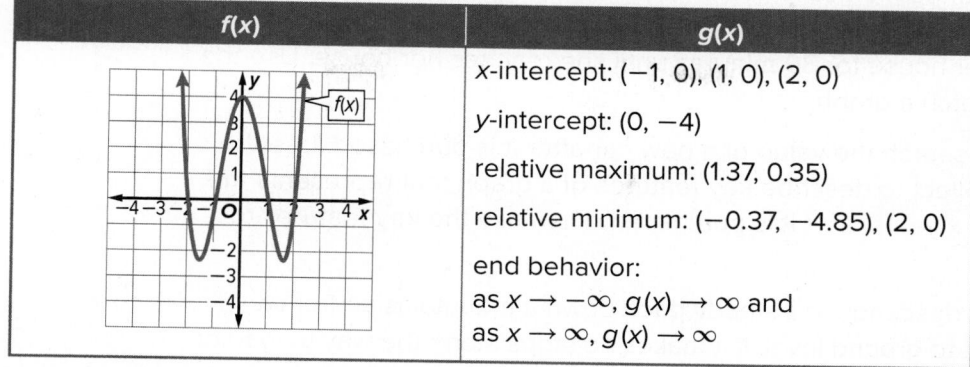

f(x)	g(x)
	x-intercept: (−1, 0), (1, 0), (2, 0)
	y-intercept: (0, −4)
	relative maximum: (1.37, 0.35)
	relative minimum: (−0.37, −4.85), (2, 0)
	end behavior: as $x \to -\infty$, $g(x) \to \infty$ and as $x \to \infty$, $g(x) \to \infty$

15. Compare the key features of the functions represented with a table and a verbal description.

f(x)		g(x)
x	**f(x)**	*x*-intercept: $\left(\frac{3}{8}, 0\right)$
$-\frac{2}{3}$	1	*y*-intercept: $\left(0, \frac{1}{2}\right)$
$-\frac{1}{3}$	$\frac{3}{4}$	slope: $-\frac{4}{3}$
0	$\frac{1}{2}$	
$\frac{1}{3}$	$\frac{1}{4}$	
$\frac{2}{3}$	0	

16. Compare the key features of the functions represented with a graph and a table.

f(x)	g(x)	
	x	**g(x)**
	−1	7
	−0.56	0
	0	−3
	1.89	0
	2	1

Mixed Exercises

17. **USE A MODEL** Sketch the graph of a linear function with the following key features. The *x*-intercept is 2. The *y*-intercept is 2. The function is decreasing for all values of *x*. The function is positive for *x* < 2. As $x \to -\infty$, $f(x) \to \infty$ and as $x \to \infty$, $f(x) \to -\infty$.

18. **WATER** Sia filled a pitcher with water. The pitcher started with 0 ounces of water. After 8 seconds the pitcher contains 64 ounces of water. The function that models the situation is linear.

a. Use the key features to sketch a graph.

b. What is the end behavior of the graph? Explain.

19. **USE TOOLS** Monica walks for 60 minutes. She starts walking from her house. The maximum distance Monica is from her house is 2 miles, which she reaches 30 minutes after she starts walking. At the 30-minute mark, Monica starts walking back to her house for 30 minutes until she reaches her house. Use the key features to sketch a graph.

20. **USE A SOURCE** Research the value of a new car after it is purchased. Use the information you collect to describe key features of a graph that represents the value of a new car x years after it is purchased. Then use the key features to sketch a graph.

21. **SKI LIFTS** A ski lift descends at a steady pace down a mountainside from a height of 1800 feet to ground level. If it makes no stops along the way to load or unload passengers, then the time it takes to complete its descension is 4 minutes.

    a. Is the graph that relates the lift's height as a function of time linear or nonlinear? Explain.

    b. Use the key features to sketch a graph.

22. **CONSTRUCT ARGUMENTS** Keisha babysits for her aunt for an hourly rate of $9. The graph shows Keisha's earnings y as a function of hours spent babysitting x. Explain why the graph only exists for positive x- and y-values.

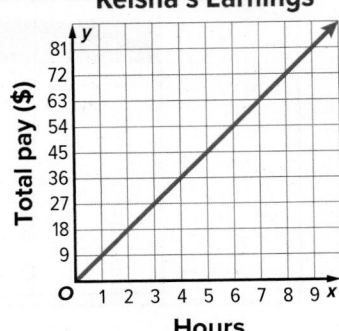

Keisha's Earnings

🌎 **Higher-Order Thinking Skills**

23. **CREATE** Choose a function and create a list of key features to describe the function. Then sketch the function.

24. **WRITE** Describe the relationship between the slope of a linear function and when the function is increasing or decreasing.

25. **ANALYZE** Determine whether the statement is *always*, *sometimes*, or *never* true.

    A graph that has more than one x-intercept is represented by a nonlinear function.

26. **PERSEVERE** Deborah filled an empty tub with water for 30 minutes. The maximum amount of water in the tub is 50 gallons, which is reached 10 minutes after Deborah starts filling the tub. The amount of water in the tub increases steadily for 10 minutes. At the 10-minute mark, the amount of water in the tub starts decreasing for 20 minutes until there is no water left in the tub.

    a. Use the key features to sketch a graph.

    b. Describe an event that could have occurred at the 10-minute mark if Deborah continues filling the tub at the same rate from the 10-minute mark to the 30-minute mark as the rate from the 0-minute mark to the 10-minute mark.

27. **FIND THE ERROR** Linda and Rubio sketched a graph with the following key features. The x-intercept is 2. The y-intercept is −9. The function is positive for $x > 2$. As $x \rightarrow -\infty$, $f(x) \rightarrow -\infty$ and as $x \rightarrow \infty$, $f(x) \rightarrow \infty$. Is either graph correct based on the key features? Explain your reasoning.

Linda's Graph   **Rubio's Graph**

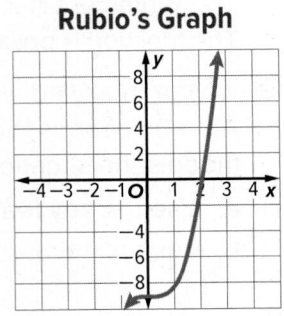

Graphing Linear Functions and Inequalities

Learn Graphing Linear Functions

The graph of a function represents all ordered pairs that are true for the function. You can use various methods to graph a linear function.

Example 1 Graph by Using a Table

Graph $x + 3y - 6 = 0$ by using a table.

Solve the equation for y. Then make a table and complete the graph.

$x + 3y - 6 = 0$ Original function

    $y = -\frac{1}{3}x + 2$ Solve.

x	$-\frac{1}{3}x + 2$	y
-6	$-\frac{1}{3}(-6) + 2$	4
-3	$-\frac{1}{3}(-3) + 2$	3
0	$-\frac{1}{3}(0) + 2$	2
3	$-\frac{1}{3}(3) + 2$	1
6	$-\frac{1}{3}(6) + 2$	0

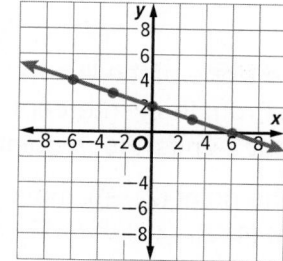

Check

Graph $6x + 2y = 10$ by using a table.

x	y
-1	?
0	?
1	?
3	?

Example 2 Graph by Using Intercepts

Graph $3x - 2y = -12$ by using the x- and y-intercepts.

To find the x-intercept, let $y = 0$.   To find the y-intercept, let $x = 0$.

$3x - 2y = -12$ Original function $3x - 2y = -12$

$3x - 2(0) = -12$ Replace with 0. $3(0) - 2y = -12$

    $x = -4$ Simplify. $y = 6$

(continued on the next page)

Go Online You can complete an Extra Example online.

Today's Goals
- Graph linear functions.
- Graph linear inequalities in two variables.

Today's Vocabulary
linear inequality

boundary

closed half-plane

open half-plane

constraint

Go Online
You can watch a video to see how to graph linear functions.

Study Tip
Slope Recall that **slope** is the ratio of the change in the y-coordinates (rise) to the corresponding change in the x-coordinates (run) as you move from one point to another along a line.

Think About It!
Explain why -6, -3, 0, 3, and 6 were selected for the x-values in the table.

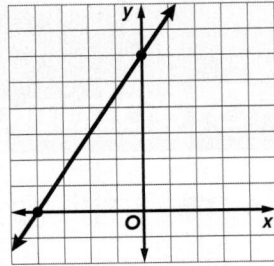

The x-intercept is −4, and the y-intercept is 6. This means that the graph passes through (−4, 0) and (0, 6).

Plot the two intercepts.

Draw a line through the points.

Example 3 Graph by Using the Slope and y-intercept

Graph $y = \frac{3}{2}x - 4$ by using m and b.

Follow these steps.

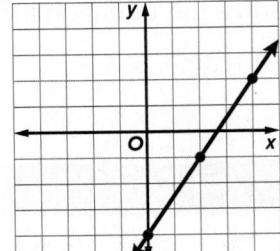

- Begin by identifying the slope m and y-intercept b of the function.

  $m = \frac{3}{2}$ $b = -4$

- Use the value of b to plot the y-intercept (0, −4).

- Use the slope of the line $m = \frac{3}{2}$ to plot more points. From the y-intercept, move up 3 units and right 2 units. Plot a point at (2, −1).

- From the point (2, −1), move up 3 units and right 2 units. Plot a point at (4, 2).

- Draw a line through the points.

Explore Shading Graphs of Linear Inequalities

Online Activity Use graphing technology to complete the Explore.

> **INQUIRY** How can you use a point to test the graph of an inequality?

Learn Graphing Linear Inequalities in Two Variables

The graph of a **linear inequality** is a half-plane with a boundary that is a straight line. The half-plane is shaded to indicate that all points contained in the region are solutions of the inequality. A **boundary** is a line or curve that separates the coordinate plane into two half-planes.

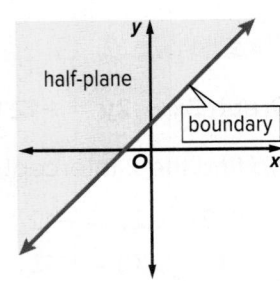

Go Online You can complete an Extra Example online.

The boundary is solid when the inequality contains ≤ or ≥ to indicate that the points on the boundary are included in the solution, creating a **closed half-plane**. The boundary is dashed when the inequality contains < or > because the points on the boundary do not satisfy the inequality. This results in an **open half-plane**.

A **constraint** is a condition that a solution must satisfy. Each solution of the inequality represents a viable, or possible, option that satisfies the constraint.

Example 4 Graph an Inequality with an Open Half-Plane

Graph $12 - 4y > x$.

Step 1 Graph the boundary.

$12 - 4y > x$ Original inequality

$-4y > x - 12$ Subtract 12 from each side.

$y < -\frac{1}{4}x + 3$ Divide each side by -4, and reverse the inequality symbol.

The boundary of the graph is $y = -\frac{1}{4}x + 3$. Because the inequality symbol is >, the boundary is dashed.

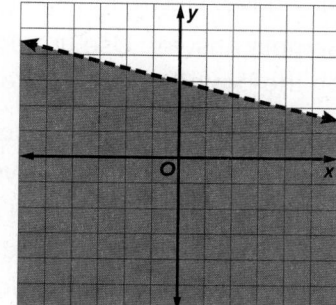

Step 2 Use a test point and shade.

Test (0, 0).

$12 - 4y > x$ Original inequality

$12 - 4(0) \overset{?}{>} 0$ Substitute.

$12 > 0$ True

Because (0, 0) is a solution of the inequality, shade the half-plane that contains the test point.

Check: Check by selecting another point in the shaded region to test.

Example 5 Graph an Inequality with a Closed Half-Plane

Graph $9 + 3y \le 8x$.

Step 1 Graph the boundary.

Solve for y in terms of x and graph the related function.

$9 + 3y \le 8x$ Original inequality

$3y \le 8x - 9$ Subtract 9 from each side.

$y \le \frac{8}{3}x - 3$ Divide each side by 3.

The related equation of $y \le \frac{8}{3}x - 3$ is $y = \frac{8}{3}x - 3$, and the boundary is solid.

(continued on the next page)

Study Tip

Above or Below
Usually the shaded half-plane of a linear inequality is said to be *above* or *below* the line of the related equation. However, if the equation of the boundary is $x = c$ for some constant c, then the function is a vertical line. In this case, the shading is considered to be *to the left* or *to the right* of the boundary.

Talk About It!
Can a linear inequality ever be a function? Explain your reasoning.

Think About It!
Why should you not test a point that is on the boundary?

Think About It!

Is (3, 5) a solution of the inequality? Explain.

Go Online

You can watch a video to see how to graph an inequality using a graphing calculator.

Step 2 Use a test point and shade.

Select a test point, such as (0, 0).

$9 + 3y \leq 8x$ Original inequality

$9 + 3(0) \overset{?}{\leq} 8(0)$ $(x, y) = (0, 0)$.

$9 \nleq 0$ False

Shade the side of the graph that does not contain the test point.

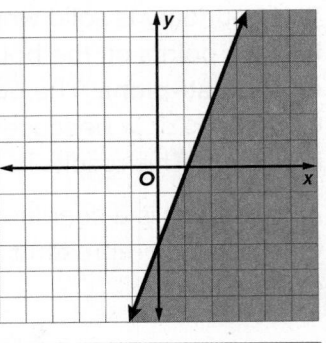

🌐 Apply Example 6 Linear Inequalities

GRADES Malik's algebra teacher determines semester grades by finding the sum of 70% of a student's test grade average and 30% of a student's homework grade average. If Malik wants a semester grade of 90% or better, write and graph the inequality that represents the constraints for Malik's test grade _x_ and homework grade _y_.

1 What is the task?

Describe the task in your own words. Then list any questions that you may have. How can you find answers to your questions?

Sample answer: Use the description to write the inequality. Find points on the boundary and use a test point to create the graph.

2 How will you approach the task? What have you learned that you can use to help you complete the task?

Sample answer: Write and graph an inequality to represent the constraints on Malik's grades. How do the test and homework grades relate to the semester grade?

3 What is your solution?

Use your strategy to solve the problem. What inequality represents the constraints for Malik's test and homework grades? Use the grid to graph the inequality.

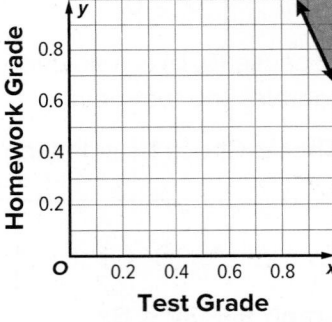

$0.7x + 0.3y \geq 0.9$

Which of these are viable solutions for Malik's test and homework grades?

- ■ 88% test, 100% homework
- ■ 90% test, 90% homework
- ☐ 90% test, 80% homework
- ☐ 95% test, 70% homework
- ■ 95% test, 80% homework
- ■ 100% test, 70% homework

4 How can you know that your solution is reasonable?

✏️ **Write About It!** Write an argument that can be used to defend your solution.

Sample answer: I can select a point in the shaded region, such as (0.95, 0.8) and test it in the inequality.

Go Online

You can complete an Extra Example online.

Practice

Go Online You can complete your homework online.

Example 1

Graph each equation by using a table.

1. $4x - 1 = y$

2. $-3 = 5x - y$

3. $y - 4 = -2x$

4. $y + x = 1$

5. $y + 3x = 1$

6. $y + 4x - 1 = 4x + 2$

Example 2

Graph each equation by using the x- and y-intercepts.

7. $3y - x = 6$

8. $2x - 3y = 6$

9. $y - x = -3$

10. $-2x + y = 4$

11. $y - 2x = -3$

12. $\frac{1}{2}x + y = 2$

Example 3

Graph each equation by using m and b.

13. $y = -\frac{5}{3}x + 12$

14. $y = \frac{2}{3}x + 6$

15. $y = 4x - 15$

16. $y - 2x = -1$

17. $y - x = -4$

18. $4 = 3x - y$

Examples 4 and 5

Graph each inequality.

19. $y > 1$

20. $y \leq x + 2$

21. $x + y \leq 4$

22. $x + 3 < y$

23. $2 - y < x$

24. $y \geq -x$

25. $x - y > -2$

26. $9x + 3y - 6 \leq 0$

27. $y + 1 \geq 2x$

28. $y - 7 \leq -9$

29. $x > -5$

30. $y + x > 1$

Example 6

31. CRAFT FAIR Kylie is going to try to sell two of her oil paintings at the local craft fair. She is hoping to earn at least $400.

    a. Write the inequality that represents the constraint of the situation, where x is the price of the first oil painting, and y is the price of the second.

    b. Graph the inequality that represents the constraint on the sale.

32. **BUILDING CODES** A city has a building code that limits the height of buildings around the central park. The code says that the height of a building must be less than 0.1x, where x is the distance in hundreds of feet of the building from the center of the park. Assume that the park center is located at $x = 0$. Graph the inequality that represents the building code.

33. **WEIGHT** A delivery crew is going to load a truck with tables and chairs. The trucks weight limitations are represented by the inequality $200t + 60c < 1200$, where t is the number of tables and c is the number of chairs. Graph this inequality.

34. **ART** An artist can sell each drawing for $100 and each watercolor for $400. He hopes to make at least $2000 every month.

    a. Write an inequality that expresses how many drawings and/or watercolors the artist needs to sell each month to reach his goal.

    b. Graph the inequality.

    c. If the artist sells three watercolors one month, how many drawings would he have to sell in the same month to reach $2000?

Mixed Exercises

Graph each equation or inequality.

35. $x + y = 1$

36. $y \geq -3x - 2$

37. $x + 2y > 6$

38. $y + 2 = 3x + 3$

39. $y + 3 = 0$

40. $4x - 3y > 12$

41. $x + y = 3$

42. $2y - x = 2$

43. $4x + 3y = 12$

44. $\frac{1}{2}x + \frac{1}{4}y = 8$

45. $-\frac{1}{2}x + y = -2$

46. $y \geq \frac{3}{4}x + 6$

47. $y = -2x + 3$

48. $2x - y = 1$

49. $2y + 3 \leq 11$

50. $y + 2 = -x + 1$

51. $6x + 4y \leq -24$

52. $-2x + 5y = 2$

53. **REASONING** Name the x- and y-intercept for the linear equation given by $6x - 2y = 12$. Use the intercepts to graph the equation and describe the graph as *increasing, decreasing,* or *constant.*

54. **ANIMALS** During the winter, a horse requires about 36 liters of water per day and a sheep requires about 3.6 liters per day. A farmer is able to supply his horses and sheep with a total of 300 liters of water each day.

    a. Write an equation or inequality that represents the possible number of each type of animal that the farmer can keep.

    b. Graph the equation or inequality.

55. COMPUTERS A school system is buying new computers. They will buy desktop computers costing $1000 per unit, and notebook computers costing $1200 per unit. The total cost of the computers cannot exceed $80,000.

    a. Write an inequality that describes this situation.

    b. Graph the inequality.

    c. If the school wants to buy 50 desktop computers and 25 notebook computers, will they have enough money? Explain.

56. BAKED GOODS Mary sells giant chocolate chip and peanut butter cookies for $1.25 and $1.00, respectively, at a local bake shop. She wants to make at least $25 a day.

    a. Write and graph an inequality that represents the number of cookies Mary needs to sell each day.

    b. If Mary decides to charge $1.50 for chocolate chip cookies rather than $1.25, what impact will this have on the graph of the solution set? Give an (x, y) pair that is not in the original solution set, but is in the solution set of the new revised scenario.

    c. How does the graph of the inequality change if Mary wants to make at least $50 a day? How does the graph of the inequality change if Mary wants to make no more than $25 a day?

57. FUNDRAISING The school drama club is putting on a play to raise money. Suppose it will cost $400 to put on the play and that 300 students and 150 adults will attend.

    a. Write an equation to represent revenue from ticket sales if the club wants to raise $1400 after expenses.

    b. Graph your equation. Then determine four possible prices that could be charged for student and adult tickets to earn $1400 in profit.

58. CONSTRUCTION You want to make a rectangular sandbox area in your backyard. You plan to use no more than 20 linear feet of lumber to make the sides of the sandbox.

    a. Write and graph a linear inequality to describe this situation.

    b. What are two possible sizes for the sandbox?

    c. Can you make a sandbox that is 7 feet by 6 feet? Justify your answer.

    d. What can you conclude about the intercepts of your graph?

59. SPIRITWEAR A company makes long-sleeved and short-sleeved shirts. The profit on a long-sleeved shirt is $7 and the profit on a short-sleeved shirt is $4. How many shirts must the company sell to make a profit of at least $280?

    a. Write and graph a linear inequality to describe this situation.

    b. Write two possible solutions to the problem.

    c. Which values are reasonable for the domain and for the range? Explain.

    d. The point $(-10, 90)$ is in the shaded region. Is it a solution of the problem? Explain your reasoning.

60. MONEY Gemma buys candles and soaps online. The scented candles cost $9, and the hand soaps cost $4. To qualify for free shipping, Gemma needs to spend at least $50.

    a. Write an inequality that represents the constraints on the number of candles x and the number of soaps y that Gemma must buy in order to qualify for free shipping.

    b. Graph the inequality.

    c. Suppose Gemma decides not to buy any soaps. Determine the number of candles she needs to buy in order to qualify for free shipping. Explain.

    d. If Gemma decides not to buy any candles how many soaps will she need to buy in order to qualify for free shipping? Explain.

    e. Will Gemma qualify for free shipping if she buys 2 candles and 8 soaps? Explain how you can be sure.

🌐 Higher-Order Thinking Skills

61. FIND THE ERROR Paulo and Janette are graphing $x - y \geq 2$. Is either of them correct? Explain your reasoning.

Paulo

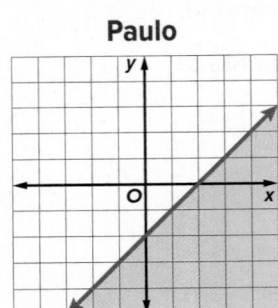

Janette

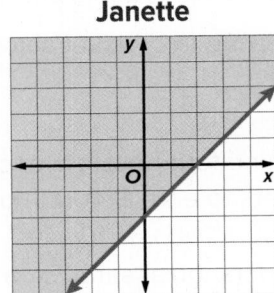

62. CREATE Write an inequality that has a graph with a dashed boundary line. Then graph the inequality.

63. WRITE You can graph a line by making a table, using the x- and y-intercepts, or by using m and b. Which method do you prefer? Explain your reasoning.

64. ANALYZE Write a counterexample to show that the following statement is false. *Every point in the first quadrant is a solution for $3y > -x + 6$.*

65. PERSEVERE Write an equation of the line that has the same slope as $2x - 8y = 7$ and the same y-intercept as $4x + 3y = 15$.

Special Functions

Today's Goals
- Write and graph piecewise-defined functions.
- Write and graph step functions.
- Graph and analyze absolute value functions.

Today's Vocabulary
piecewise-defined function
step function
greatest integer function
absolute value function
parent function

Explore Using Tables to Graph Piecewise Functions

🔾 **Online Activity** Use a table and graphing technology to complete the Explore.

> ✕
> @ **INQUIRY** How can you write a piecewise function when given a table of values?

Learn Graphing Piecewise-Defined Functions

A function that is written using two or more expressions is called a **piecewise-defined function**. Each of the expressions is defined for a distinct interval of the domain.

A dot is used if a point is included in the graph. A circle is used for a point that is not included in the graph.

Example 1 Graph a Piecewise-Defined Function

Graph $f(x) = \begin{cases} x - 3 \text{ if } x \leq 1 \\ 2x \text{ if } x > 1 \end{cases}$. **Then, analyze the key features.**

Step 1 Graph $f(x) = x - 3$ **for** $x \leq 1$.

Find $f(x)$ for the endpoint of the domain interval, $x = 1$.

$f(x) = x - 3$

$\quad = 1 - 3 \text{ or } -2$

Since 1 satisfies the inequality, place a dot at $(1, -2)$. Because $x - 3$ is defined for values of x less than or equal to 1, graph the linear function with a slope of 1 to the left of the dot.

Step 2 Graph $f(x) = 2x$ **for** $x > 1$.

Find $f(x)$ for the endpoint of the domain interval, $x = 1$.

$f(x) = 2x$

$\quad = 2(1) \text{ or } 2$

Since 1 is not included in the inequality, place a circle at $(1, 2)$. Because $2x$ is defined for values greater than 1, graph the linear function with a slope of 2 to the right of the circle.

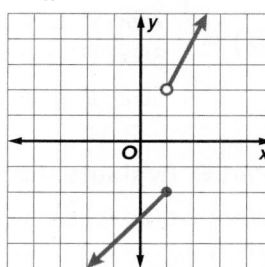

(continued on the next page)

🔾 **Go Online** You can complete an Extra Example online.

💬 Talk About It!

Can the piecewise-defined function in the example have $x \geq 1$ instead of $x > 1$ after the second expression? Explain.

Step 3 Analyze key features.

The function is defined for all values of x, so the domain is all real numbers.

The range is all real numbers less than or equal to -2 and all real numbers greater than 2. This can be represented symbolically as

$\{f(x) \mid f(x) \leq -2 \text{ or } f(x) > 2\}$.

The y-intercept is -3, and there is no x-intercept.

The function is increasing for all values of x.

Check

Graph $f(x) = \begin{cases} \frac{2}{3}x \text{ if } x \leq 0 \\ 3 \text{ if } 1 \leq x \leq 3 \\ -2x + 5 \text{ if } x \geq 4 \end{cases}$. Then, analyze the key features.

The domain is _____?_____.

The range is _____?_____.

The x- intercept is __?__.

The y-intercept is __?__.

For $\{x \mid x \leq 0\}$, the function is ____?____.

For $\{x \mid 1 \leq x \leq 3\}$, the function is ____?____.

For $\{x \mid x \geq 4\}$, the function is ____?____.

🌐 **Think About It!**

What do the domain and range represent in the context of this situation?

🌐 **Example 2** Model by Using a Piecewise-Defined Function

UNIFORMS The football coach is ordering new jerseys for the new season. The manufacturer charges $88 for each jersey when five or fewer are ordered, $75 each for an order of six to 11 jerseys, $65 each for an order of 12 to 29 jerseys, and $56 each when thirty or more jerseys are ordered.

Part A Write a piecewise-defined function describing the cost of the jerseys.

$f(x) = \begin{cases} 88x \text{ if } 0 < x \leq 5 \\ 75x \text{ if } 5 < x \leq 11 \\ 65x \text{ if } 11 < x \leq 29 \\ 56x \text{ if } x > 29 \end{cases}$

Part B Evaluate the function.

What would it cost to purchase 11 jerseys? $825

What would it cost to purchase 25 jerseys? $1625

Evaluate $f(29)$. $f(29) = \$1885$

Watch Out!

Evaluating Endpoints of Intervals When evaluating a piecewise-defined function for a value of x that is an endpoint for two consecutive intervals, be careful to evaluate the function that contains that point.

🧭 **Go Online** You can complete an Extra Example online.

Learn Graphing Step Functions

A common type of piecewise function is a step function. A **step function** has a graph that is a series of horizontal line segments that may resemble a staircase.

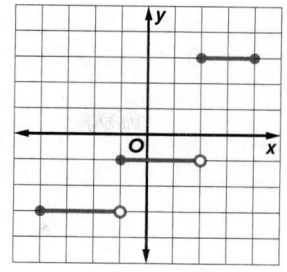

A step function is defined by a set of constant functions. The domain of a step function is an interval of real numbers. The range of a step function is a discrete set of real numbers. The graph of a step function is discontinuous because it cannot be drawn without lifting your pencil.

The **greatest integer function**, written $f(x) = [\![x]\!]$, is one kind of step function in which $f(x)$ is the greatest integer less than or equal to x. For example, $[\![10.7]\!] = 10$, $[\![-6.35]\!] = -7$, and $[\![5]\!] = 5$.

😮 **Think About It!**

Why is the range of the function not expressed as $\{y|\ -3 \le y \le 3\}$?

😮 **Think About It!**

Explain why the value of $[\![4.3]\!]$ is 4, but the value of $[\![-4.3]\!]$ is −5.

🌐 Example 3 Graph a Step Function

POSTAL RATES **The cost of mailing a first-class letter is determined by rates adopted by the U.S. Postal Service. The rates adopted in 2016 charge $0.47 for letters not over 1 ounce, $0.68 if not over 2 ounces, $0.89 if not over 3 ounces, and $1.10 if not over 3.54 ounces. Complete the table and draw a graph that represents the charges. State the domain and range.**

Step 1 Make a table.

Let x be the weight of a first-class letter and $C(x)$ represent the cost for mailing it. Use the given rates to make a table.

x	$C(x)$
$0 < x \le 1$	$0.47
$1 < x \le 2$	$0.68
$2 < x \le 3$	$0.89
$3 < x \le 3.54$	$1.10

Step 2 Make a graph.

Graph the first step of the function. Place a circle at (0, 0.47) since there is no charge for not mailing a letter. Place a dot at (1, 0.47) since a letter weighing one ounce will cost $0.47 to mail. Draw a segment that connects the points.

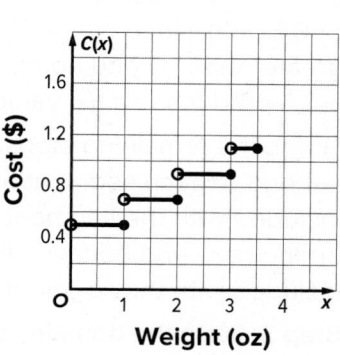

Graph the remaining steps.

Place a circle on the left end of each segment as that domain value is included with the segment below it.

(continued on the next page)

Step 3 State the domain and range.

The constraints for the weight of a first-class letter are more than 0 ounces up to and including 3.54 ounces. Therefore, the domain is $\{x \mid 0 < x \le 3.54\}$.

Because the only viable solutions for the cost of mailing a first-class letter are $0.47, $0.68, $0.89, and $1.10, the range is $\{C(x) = 0.47, 0.68, 0.89, 1.10\}$.

Check

FIGURINES Chris and Joaquin design figurines for board game and toy companies. The rate they charge $R(x)$ depends on the number of hours x they spend creating the figurines. Draw a graph that represents the charges. State the domain and range.

x	R(x)
$0 < x \le 5$	500
$5 < x \le 15$	1400
$15 < x \le 30$	2500
$30 < x \le 50$	4000

Think About It!

Would $C(x)$ still be a function if the open points at (0, 0.47), (1, 0.68), (2, 0.89), and (3, 1.10) were closed points? Justify your argument.

Think About It!

Will the range of a greatest integer function always be all integers? If not, provide a counterexample.

Example 4 Graph a Greatest Integer Function

Complete the table and graph $f(x) = [\![2x - 1]\!]$. State the domain and range.

Step 1 Make a table.

Make a table of the intervals of x and associated values of $f(x)$.

Step 2 Make a graph.

Graph the first step. Place a dot at $(-1, -3)$, because $f(-1) = [\![2(-1) - 1]\!] = [\![-3]\!] = -3$. Place a circle at $(-0.5, -3)$, since every decimal value greater than -1 and up to but not including -0.5 produces an $f(x)$ value of -3.

Graph the remaining steps. Place a dot on the left end of each segment as that point is included with the segment, and place a circle on the right end because that domain value is included with the segment above it.

x	f(x)
$-1 \le x < -0.5$	-3
$-0.5 \le x < 0$	-2
$0 \le x < 0.5$	-1
$0.5 \le x < 1$	0
$1 \le x < 1.5$	1
$1.5 \le x < 2$	2

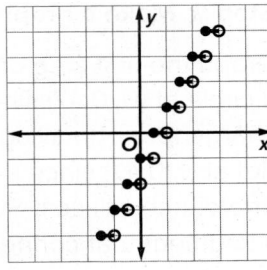

Step 3 State the domain and range.

The domain of $f(x) = [\![2x - 1]\!]$ is all real numbers. The range is all integers.

Go Online

You can watch a video to see how to graph step functions.

Go Online You can complete an Extra Example online.

Learn Graphing Absolute Value Functions

An **absolute value function** is a function that contains an algebraic expression within absolute value symbols. It can be defined and graphed as a piecewise function.

For an absolute value function, $f(x) = |x|$ is the **parent function**, which is the simplest of the functions in a family.

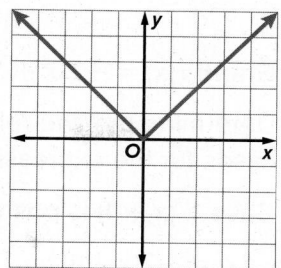

Key Concept • Parent Function of Absolute Value Functions	
parent function	$f(x) = |x|$ or $f(x) = \begin{cases} x \text{ if } x \geq 0 \\ -x \text{ if } x < 0 \end{cases}$
domain	all real numbers
range	all nonnegative real numbers
intercepts	x-intercept: $x = 0$, y-intercept: $y = 0$

Think About It!

Describe the line of symmetry of any absolute value function and compare it with the line of symmetry of the parent function.

Example 5 Graph an Absolute Value Function, Positive Coefficient

Graph $f(x) = \left|\frac{3}{4}x\right| + 3$. State the domain and range.

Create a table of values. Plot the points and connect them with two rays.

x	$f(x) = \left|\frac{3}{4}x\right| + 3$
-4	$\left|\frac{3}{4}(-4)\right| + 3 = 6$
-2	$\left|\frac{3}{4}(-2)\right| + 3 = 4\frac{1}{2}$
0	$\left|\frac{3}{4}(0)\right| + 3 = 3$
2	$\left|\frac{3}{4}(2)\right| + 3 = 4\frac{1}{2}$
4	$\left|\frac{3}{4}(4)\right| + 3 = 6$

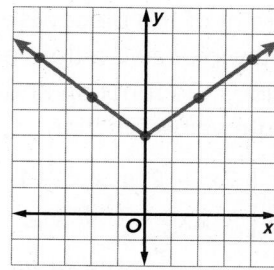

The function is defined for all values of x, so the domain is all real numbers. The function is defined only for values of $f(x)$ such that $f(x) \geq 3$, so the range is $\{f(x) \mid f(x) \geq 3\}$.

Check

Graph $f(x) = |x - 1| + 3$. State the domain and range.

Go Online

An alternate method is available for this example.

Go Online You can complete an Extra Example online.

Example 6 Graph an Absolute Value Function, Negative Coefficient

Graph $f(x) = -2|x + 1|$. State the domain and range.

Determine the two related linear equations using the two possible cases for the expression inside of the absolute value.

Case 1: $(x + 1)$ is positive.

$f(x) = -2(x + 1)$	$x + 1$ is positive, so $|x + 1| = x + 1$.
$\quad = -2x - 2$	Simplify.

Case 2: $(x + 1)$ is negative.

$f(x) = -2[-(x + 1)]$	$x + 1$ is negative, so $|x + 1| = -(x + 1)$.
$\quad = -2(-x - 1)$	Distributive Property
$\quad = 2x + 2$	Simplify.

The x-coordinate of the vertex is the value of x where the two cases of the absolute value are equal.

$-2x - 2 = 2x + 2$	Set Case 1 equal to Case 2.
$-2x = 2x + 4$	Add 2 to each side.
$-4x = 4$	Subtract $2x$ from each side.
$x = -1$	Divide each side by -4.

The x-coordinate of the vertex represents the constraint of the piecewise-defined function. Write the piecewise-defined function that describes the function and use it to graph the absolute value function.

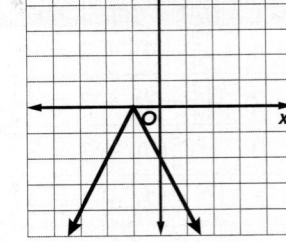

$$f(x) = \begin{cases} 2x + 2 & \text{if } x < -1 \\ -2x - 2 & \text{if } x \geq -1 \end{cases}$$

The function is defined for all values of x, so the domain is all real numbers. The function is defined only for values of $f(x)$ such that $f(x) \leq 0$, so the range is $\{f(x) \mid f(x) \leq 0\}$.

Check

Graph $f(x) = 0.25|8x| - 3$. State the domain and range.

D = _____?_____

R = _____?_____

Go Online You can complete an Extra Example online.

Think About It!

How does multiplying the absolute value by a negative number affect the shape of the graph? the range?

Practice

Go Online You can complete your homework online.

Examples 1 and 2

Graph each function. Then, analyze the key features.

1. $f(x) = \begin{cases} -1 & \text{if } x \leq 0 \\ 2x & \text{if } 0 < x \leq 3 \\ 6 & \text{if } x > 3 \end{cases}$

2. $f(x) = \begin{cases} -x & \text{if } x < -1 \\ 0 & \text{if } -1 \leq x \leq 1 \\ x & \text{if } x > 1 \end{cases}$

3. $f(x) = \begin{cases} x & \text{if } x < 0 \\ 2 & \text{if } x \geq 0 \end{cases}$

4. $h(x) = \begin{cases} 3 & \text{if } x < -1 \\ x + 1 & \text{if } x > 1 \end{cases}$

5. TILE Mark is purchasing new tile for his bathrooms. The home improvement store charges $48 for each box of tiles when three or fewer boxes are purchased, $45 for each box when 4 to 8 boxes are purchased, $42 for each box when 9 to 19 boxes are purchased, and $38 for each box when more than nineteen boxes are purchased.

   a. Write a piecewise-defined function describing the cost of the boxes of tile.

   b. What is the cost of purchasing 5 boxes of tile? What is the cost of purchasing 19 boxes of tile?

6. BOOKLETS A digital media company is ordering booklets to promote their business. The manufacturer charges $0.50 for each booklet when 50 or fewer are ordered, $0.45 for each booklet when 51 to 100 booklets are ordered, $0.40 for each booklet when 101 to 250 booklets are ordered, and $0.35 for each booklet when 251 or more booklets are ordered. Each order consists of a $10 shipping charge, no matter the size of the order.

   a. Write a piecewise-defined function describing the cost of ordering booklets.

   b. What is the cost of purchasing 132 booklets? What is the cost of purchasing 518 booklets?

Examples 3 and 4

Graph each function. State the domain and range.

7. $f(x) = [\![x]\!] - 6$

8. $h(x) = [\![3x]\!] - 8$

9. $f(x) = [\![x + 1]\!]$

10. $f(x) = [\![x - 3]\!]$

11. PARKING The rates at a short-term parking garage are $5.00 for 2 hours or less, $10.00 for 4 hours or less, $15.00 for 6 hours or less, and $20.00 for 8 hours or less. Draw a graph that represents the charges. State the domain and range.

12. BOWLING The bowling alley offers special team rates. They charge $30 for one hour or less of team bowling, $45 for 2 hours or less, and $60 for unlimited bowling after 2 hours of play. Draw a graph that represents the charges. State the domain and range.

Examples 5 and 6

Graph each function. State the domain and range.

13. $f(x) = |x - 5|$

14. $g(x) = |x + 2|$

15. $h(x) = |2x| - 8$

16. $k(x) = |-3x| + 3$

17. $f(x) = 2|x - 4| + 6$

18. $h(x) = -3|0.5x + 1| - 2$

19. $g(x) = 2|x|$

20. $f(x) = |x| + 1$

Mixed Exercises

Graph each function. State the domain and range.

21. $f(x) = \begin{cases} -3x & \text{if } x \le -4 \\ x & \text{if } 0 < x \le 3 \\ 8 & \text{if } x > 3 \end{cases}$

22. $f(x) = \begin{cases} 2x & \text{if } x \le -6 \\ 5 & \text{if } -6 < x \le 2 \\ -2x + 1 & \text{if } x > 4 \end{cases}$

23. $g(x) = \begin{cases} 2x + 2 & \text{if } x < -6 \\ x & \text{if } -6 \le x \le 2 \\ -3 & \text{if } x > 2 \end{cases}$

24. $g(x) = \begin{cases} -2 & \text{if } x < -4 \\ x - 3 & \text{if } -1 \le x \le 5 \\ 2x - 15 & \text{if } x > 7 \end{cases}$

25. $f(x) = \begin{cases} -0.5x + 1.5 & \text{if } x \le 1 \\ x - 4 & \text{if } x > 1 \end{cases}$

26. $f(x) = |x - 2|$

27. $f(x) = [\![x + 2]\!]$

28. $g(x) = 2[\![0.5x + 4]\!]$

29. $f(x) = [\![|0.5x|]\!]$

30. $g(x) = |[\![2x]\!]|$

31. $g(x) = \begin{cases} [\![x]\!] & \text{if } x < -4 \\ x + 1 & \text{if } -4 \le x \le 3 \\ -x & \text{if } x > 3 \end{cases}$

32. $h(x) = \begin{cases} -|x| & \text{if } x < -6 \\ |x| & \text{if } -6 \le x \le 2 \\ |-x| & \text{if } x > 2 \end{cases}$

33. Identify the domain and range of $h(x) = |x + 4| + 2$.

34. **FINANCE** For every transaction, a certain financial advisor gets a 5% commission, regardless of whether the transaction is a deposit or withdrawal. Write a formula using the absolute value function for the advisor's commission C. Let D represent the value of one transaction.

35. **GAMING** The graph shows the monthly fee that an an online gaming site charges based on the average number of hours spent online per day. Write the function represented by the graph.

Monthly Fee ($) vs Gaming Time (h)

36. **ROUNDING** A science teacher instructs students to round their measurements as follows: If the decimal portion of a measurement is less than 0.5 mm, round down to the nearest whole millimeter. If the decimal portion of a measurement is exactly 0.5 or greater, round up to the next whole millimeter. Write a formula to represent the rounded measurements.

37. **REUNIONS** The cost to reserve a banquet hall is $500. The catering cost per guest is $17.50 for the first 40 guests and $14.75 for each additional guest.

  a. Write a piecewise-defined function describing the cost C of the reunion.

  b. Use a graphing calculator to graph the function.

  c. If the Cramers can spend up to $900 on an event, what is the greatest number of guests that can attend? Explain.

38. SAVINGS Nathan puts $200 into a checking account when he gets his paycheck each month. The value of his checking account is modeled by $v = 200 \llbracket m \rrbracket$, where m is the number of months that Nathan has been working. After 105 days, how much money is in the account?

39. POLITICS The approval rating $R(t)$, measured as a percent, of a class officer during her 9-month term starting in September is described by the graph, where t is her time in office.

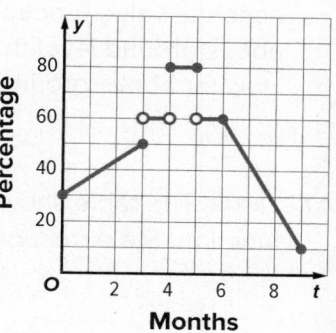

a. Formulate a piecewise-defined function $R(t)$ describing the approval rating of this class officer. Then, identify the range.

b. During which months is the approval rating increasing?

40. STRUCTURE Consider the functions $f(x) = 3\llbracket x \rrbracket$ and $g(x) = \llbracket 3x \rrbracket$ for $0 \le x \le 2$.

a. Graph each function.

b. What effect does this 3 appear to have on the graphs?

c. Consider the functions $f(x) = 4\llbracket x \rrbracket$ and $g(x) = \llbracket 4x \rrbracket$ for $0 \le x \le 2$. Graph each function.

d. What effect does this 4 appear to have on the graphs?

e. Generalize your findings from **parts a through d** to explain the differences between $f(x) = n\llbracket x \rrbracket$ and $g(x) = \llbracket nx \rrbracket$ for $0 \le x \le 2$, where n is any positive integer greater than or equal to 2.

41. USE TOOLS Use a graphing calculator to graph the absolute value of the greatest integer of x, or $f(x) = |\llbracket x \rrbracket|$. Is the graph what you expected? Explain.

Write a piecewise-defined function for each graph.

42.

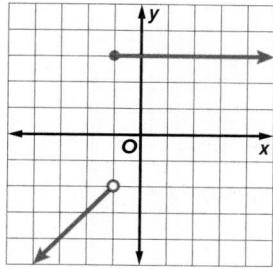

43.

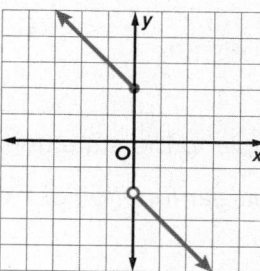

44.

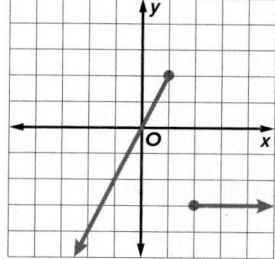

45.

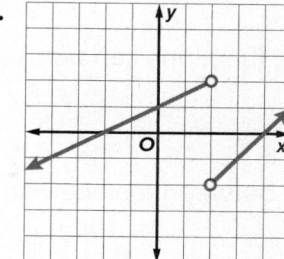

46. SKYSCRAPERS To clean windows of skyscrapers, some companies use a carriage. A carriage is mounted on a railing on the roof of a skyscraper and moves up and down using cables. A crew plans to start at the 12th floor and move the carriage down as they clean the windows on the west side of a building. If the crew members clean windows at a constant rate of 0.75 floor per hour, the absolute value function $f(t) = |0.75t - 12|$ represents the the number of floors above ground level that the carriage is after t hours. Graph the function. How far above ground level is the carriage after 4 hours?

m	$C(m)$
$0 < m \le 1$	$2.00
$1 < m \le 2$	$4.00
$2 < m \le 3$	$6.00
$3 < m \le 4$	$8.00
$4 < m \le 5$	$10.00
$5 < m \le 6$	$12.00

47. TAXIS The table shows the cost C of a taxi ride of m miles. Graph the function. State the domain and range.

48. WALKING Jackson left his house and walked at a constant rate. After 20 minutes, he was 2 miles from his house. Jackson then walked back towards his house at a constant rate. After another 30 minutes he arrived at his house.

 a. Jackson wants to write a function to model the distance from his house d as after t minutes. Should Jackson write an absolute value function or a piecewise-defined function? Explain your reasoning.

 b. Write an appropriate function to model the distance from his house d as after t minutes.

 c. Graph the function.

 d. State the domain and range.

🌐 **Higher-Order Thinking Skills**

49. CREATE Write an absolute value relation in which the domain is all nonnegative numbers and the range is all real numbers.

50. PERSEVERE Graph $|y| = 2|x + 3| - 5$.

51. ANALYZE Find a counterexample to the statement and explain your reasoning.

 In order to find the greatest integer function of x when x is not an integer, round x to the nearest integer.

52. CREATE Write an absolute value function in which $f(5) = -3$.

53. WRITE Explain how piecewise functions can be used to accurately represent real-world problems.

54. WRITE Explain the difference between a piecewise function and step function.

Transformations of Functions

Learn Translations of Functions

A **family of graphs** includes graphs and equations of graphs that have at least one characteristic in common. The parent graph is transformed to create other members in a family of graphs.

Key Concept • Parent Functions

Constant Function	Identity Function
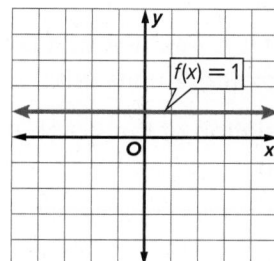	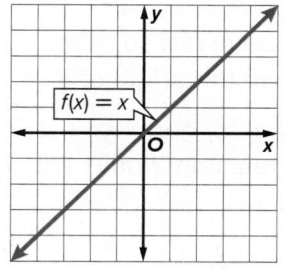
The general equation of a **constant function** is $f(x) = a$, where a is any number. Domain: all real numbers Range: $\{f(x) \mid f(x) = a\}$	The **identity function** $f(x) = x$ includes all points with coordinates (a, a). It is the parent function of most linear functions. Domain: all real numbers Range: all real numbers

Absolute Value Function	Quadratic Function
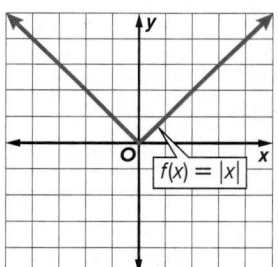	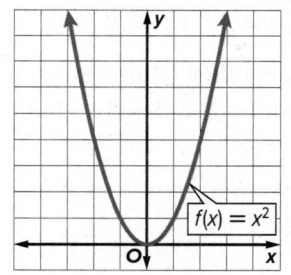
The parent function of absolute value functions is $f(x) = |x|$. Domain: all real numbers Range: $\{f(x) \mid f(x) \geq 0\}$	The parent function of quadratic functions is $f(x) = x^2$. Domain: all real numbers Range: $\{f(x) \mid f(x) \geq 0\}$

A **translation** is a transformation in which a figure is slid from one position to another without being turned.

Key Concept • Translations

Translation	Change to Parent Graph
$f(x) + k$; $k > 0$	The graph is translated k units up.
$f(x) + k$; $k < 0$	The graph is translated $|k|$ units down.
$f(x - h)$; $h > 0$	The graph is translated h units right.
$f(x - h)$; $h < 0$	The graph is translated $|h|$ units left.

Today's Goals
• Apply translations to the graphs of functions.
• Apply dilations to the graphs of functions.
• Apply compositions of transformations to the graphs of functions and use transformations to write equations from graphs.

Today's Vocabulary
family of graphs
constant function
identity function
transformations
translation
dilation
reflection
line of reflection

▶ **Go Online**
You may want to complete the Concept Check to check your understanding.

▶ **Go Online**
You can watch a video to see how to describe translations of functions.

💬 **Think About It!**
Describe the vertex and axis of symmetry of a translated quadratic function in terms of h and k.

Example 1 Translations

Describe the translation in $g(x) = (x + 2)^2 - 4$ as it relates to the graph of the parent function.

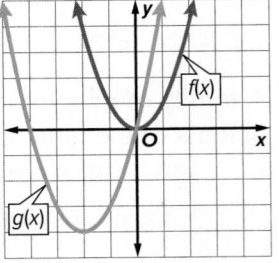

Since $g(x)$ is quadratic, the parent function is $f(x) = x^2$.

Since $f(x) = x^2$, $g(x) = f(x - h) + k$, where $h = -2$ and $k = -4$.

The constant k is added to the function after it has been evaluated, so k affects the output, or y-values. The value of k is less than 0, so the graph of $f(x) = x^2$ is translated down 4 units.

The value of h is subtracted from x before it is evaluated and is less than 0, so the graph of $f(x) = x^2$ is also translated 2 units left.

The graph of $g(x) = (x + 2)^2 - 4$ is the translation of the graph of the parent function 2 units left and 4 units down.

Example 2 Identify Translated Functions from Graphs

Use the graph of the function to write its equation.

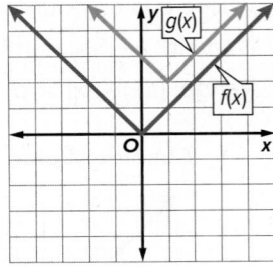

The graph is an absolute value function with a parent function of $f(x) = |x|$. Notice that the vertex of the function has been shifted both vertically and horizontally from the parent function.

To write the equation of the graph, determine the values of h and k in $g(x) = |x - h| + k$.

The translated graph has been shifted 2 units up and 1 unit right. So, $h = 1$ and $k = 2$. Thus, $g(x) = |x - 1| + 2$.

Learn Dilations and Reflections of Functions

A **dilation** is a transformation that stretches or compresses the graph of a function. Multiplying a function by a constant dilates the graph with respect to the x- or y-axis.

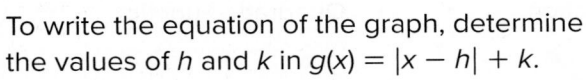

Dilation	Change to Parent Graph
$af(x)$, $\|a\| > 1$	The graph is stretched vertically.
$af(x)$, $0 < \|a\| < 1$	The graph is compressed vertically.
$f(ax)$, $\|a\| > 1$	The graph is compressed horizontally.
$f(ax)$, $0 < \|a\| < 1$	The graph is stretched horizontally.

🗗 **Go Online** You can complete an Extra Example online.

A **reflection** is a transformation where a figure, line, or curve, is flipped in a **line of reflection**. Often the reflection is in the x- or y-axis.

When a parent function $f(x)$ is multiplied by -1, the result $-f(x)$ is a reflection of the graph in the x-axis. When only the variable is multiplied by -1, the result $f(-x)$ is a reflection of the graph in the y-axis.

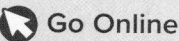
Go Online
You can watch videos to see how to describe dilations or reflections of functions.

| Key Concept • Reflections | |
Reflection	Change to Parent Graph
$-f(x)$	reflection in the x-axis
$f(-x)$	reflection in the y-axis

Example 3 Vertical Dilations

Describe the dilation and reflection in $g(x) = -\frac{2}{5}x$ as it relates to the parent function.

Since $g(x)$ is a linear function, the parent function is $f(x) = x$.

Since $f(x) = x$, $g(x) = -1 \cdot a \cdot f(x)$ where $a = \frac{2}{5}$.

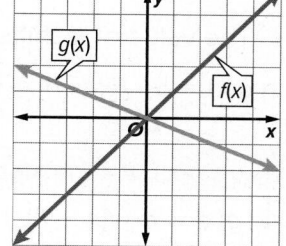

The function is multiplied by -1 and the constant a after it has been evaluated. $0 < |a| < 1$, so the graph is compressed vertically and reflected in the x-axis.

The graph of $g(x) = -\frac{2}{5}x$ is the graph of the parent function compressed vertically and reflected in the x-axis.

Think About It!
Describe the effect of multiplying the same value of a, $-\frac{2}{5}$, by a different parent function such as $f(x) = |x|$.

Example 4 Horizontal Dilations

Describe the dilation and reflection in $g(x) = (-2.5x)^2$ as it relates to the parent function.

Since $g(x)$ is a quadratic function, the parent function is $f(x) = x^2$. Since $f(x) = x^2$, $g(x) = f(-1 \cdot a \cdot x)$, where $a = 2.5$.

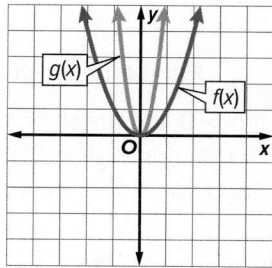

x is multiplied by -1 and the constant a before the function is performed and $|a|$ is greater than 1, so the graph of $f(x) = x^2$ is compressed horizontally and reflected in the y-axis.

Think About It!
Why does the graph of $g(x) = (-2.5x)^2$ appear the same as $j(x) = (2.5x)^2$?

Go Online You can complete an Extra Example online.

Explore Using Technology to Transform Functions

Online Activity Use graphing technology to complete the Explore.

> ×
> @ **INQUIRY** How does performing an operation on a function change its graph?

Think About It!

Do the values of a, h, and k affect various parent functions in different ways? Explain.

Go Online

You can watch a video to see how to graph transformations of functions using a graphing calculator.

Learn Transformations of Functions

The general form of a function is $g(x) = a \cdot f(x - h) + k$, where $f(x)$ is the parent function. Each constant in the equation affects the parent graph.

- The value of $|a|$ stretches or compresses (dilates) the parent graph.
- When the value of a is negative, the graph is reflected across the x-axis.
- The value of h shifts (translates) the parent graph left or right.
- The value of k shifts (translates) the parent graph up or down.

In $g(x) = a \cdot f(x - h) + k$, each constant affects the graph of $f(x) = x^2$.

Key Concept • Transformations of Functions

In $g(x) = a \cdot f(x - h) + k$, each constant affects the graph of $f(x) = x^2$.

Dilation, a	**Reflection, a**
If $|a| > 1$, the graph of $f(x)$ is stretched vertically. If $0 < |a| < 1$, the graph of $f(x)$ is compressed vertically.	If $a > 0$, the graph of $f(x)$ opens up. If $a < 0$, the graph of $f(x)$ opens down.

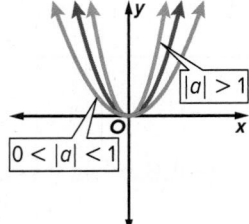

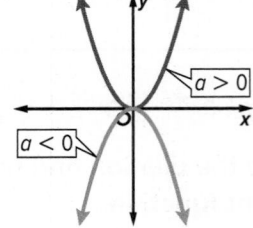

Horizontal Translation, h	**Vertical Translation, k**
If $h > 0$, the graph of $f(x)$ is translated h units right. If $h < 0$, the graph of $f(x)$ is translated $|h|$ units left.	If $k > 0$, the graph of $f(x)$ is translated k units up. If $k < 0$, the graph of $f(x)$ is translated $|k|$ units down.

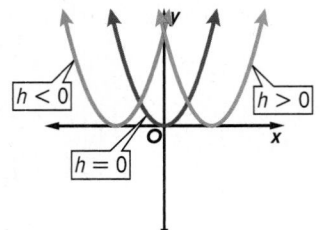

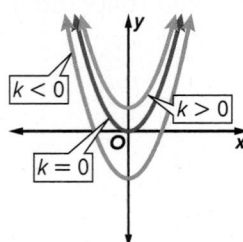

Example 5 Multiple Transformations of Functions

Describe how the graph of $g(x) = -\frac{2}{3}|x + 3| + 1$ is related to the graph of the parent function.

The parent function is $f(x) = |x|$.

Since $f(x) = |x|$, $g(x) = af(x - h) + k$ where
$a = -\frac{2}{3}$, $h = -3$ and $k = 1$.

The graph of $g(x) = -\frac{2}{3}|x + 3| + 1$ is the graph
of the parent function compressed vertically,
reflected in the x-axis, and translated 3 units
left and 1 unit up.

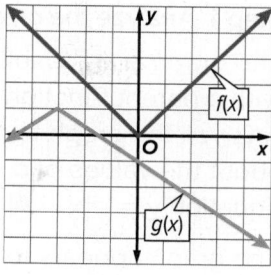

Check

Describe how $g(x) = -(0.4x + 2)$ is related to the graph of the parent function.

🌐 Example 6 Apply Transformations of Functions

DOLPHINS Suppose the path of a dolphin during a jump is modeled by $g(x) = -0.125(x - 12)^2 + 18$, where x is the horizontal distance traveled by the dolphin and $g(x)$ is its height above the surface of the water. Describe how $g(x)$ is related to its parent function and interpret the function in the context of the situation.

Because $f(x) = x^2$ is the parent function, $g(x) = af(x - h) + k$, where
$a = -0.125$, $h = 12$, and $k = 18$.

Translations

$12 > 0$, so the graph of $f(x) = x^2$ is translated 12 units right.

$18 > 0$, so the graph of $f(x) = x^2$ is translated 18 units up.

Dilation and Reflection

$0 < |-0.125| < 1$, so the graph of $f(x) = x^2$ is compressed vertically.

$a < 0$, so the graph of $f(x) = x^2$ is a reflection in the x-axis.

Interpret the Function

Because a is negative, the path of the dolphin is modeled by a
parabola that opens down. This means that the vertex of the parabola
(h, k) represents the maximum height of the dolphin, 18 feet, at 12 feet
from the starting point of the jump.

🅡 **Go Online** You can complete an Extra Example online.

Go Online
You can watch a video
to see how to use
transformations to
graph an absolute
value function.

💭 Think About It!
Write an equation for a
quadratic function that
opens down, has been
stretched vertically by
a factor of 4, and is
translated 2 units right
and 5 units down.

Study Tip
Interpretations When
interpreting
transformations, analyze
how each value
influences the function
and alters the graph.
Then determine what
you think each value
might mean in the
context of the situation.

Example 7 Identify an Equation from a Graph

Write an equation for the function.

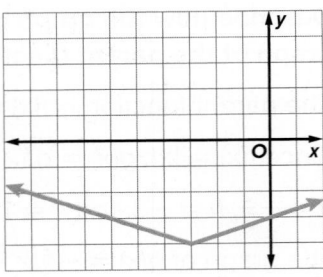

Step 1 Analyze the graph.

The graph is an absolute value function with a parent function of $f(x) = |x|$. Analyze the graph to make a prediction about the values of a, h, and k in the equation $y = a|x - h| + k$.

The graph appears to be wider than the parent function, implying a vertical compression, and is not reflected. So, a is positive and $0 < |a| < 1$.

The graph has also been shifted left and down from the parent graph. So, $h < 0$ and $k < 0$.

Step 2 Identify the translation(s).

Identify the horizontal and vertical translations to find the values of h and k.

The vertex is shifted 3 units left, so $h = -3$.

It is also shifted 4 units down, so $k = -4$.

$y = a|x - h| + k$	General form of the equation
$y = a|x - (-3)| + (-4)$	$h = -3$ and $k = -4$
$y = a|x + 3| - 4$	Simplify.

Step 3 Identify the dilation and/or reflection.

Use the equation from Step 2 and a point on the graph to find the value of a.

The point $(0, -3)$ lies on the graph. Substitute the coordinates in for x and y to solve for a.

$y = a|x + 3| - 4$	Original equation
$-3 = a|0 + 3| - 4$	$(0, -3) = (x, y)$
$-3 = a|3| - 4$	Add.
$-3 = 3a - 4$	Evaluate the absolute value.
$\frac{1}{3} = a$	Solve.

Step 4 Write an equation for the function.

Since $a = \frac{1}{3}$, $h = -3$ and $k = -4$, the equation is $g(x) = \frac{1}{3}|x + 3| - 4$.

Check

Use the graph to write an equation for $g(x)$.

 **Go Online** You can complete an Extra Example online.

Watch Out!

Choosing a Point
When substituting for x and y in the equation, use a point other than the vertex.

Think About It!

How does the equation you found compare to the prediction you made in step 1?

Practice

Go Online You can complete your homework online.

Example 1

Describe each translation as it relates to the graph of the parent function.

1. $y = x^2 + 4$

2. $y = |x| - 3$

3. $y = x - 1$

4. $y = x + 2$

5. $y = (x - 5)^2$

6. $y = |x + 6|$

Example 2

Use the graph of each translated parent function to write its equation.

7.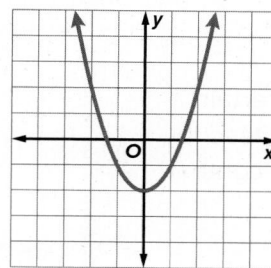

8.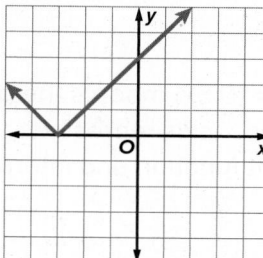

9.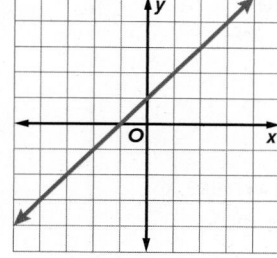

10.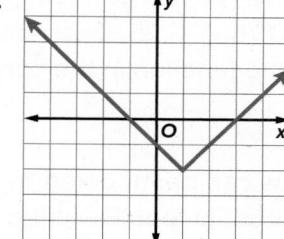

11.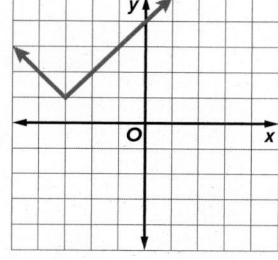

12.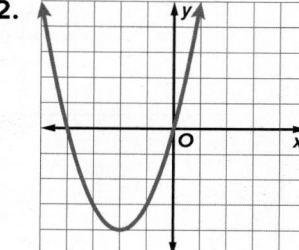

Examples 3 and 4

Describe each dilation and reflection as it relates to the parent function.

13. $y = (-3x)^2$

14. $y = -6x$

15. $y = -4|x|$

16. $y = |-2x|$

17. $y = -\frac{2}{3}x$

18. $y = -\frac{1}{2}x^2$

19. $y = \left|-\frac{1}{3}x\right|$

20. $y = \left(-\frac{3}{4}x\right)^2$

Example 5

Describe each transformation as it relates to the graph of the parent function.

21. $y = -6|x| - 4$

22. $y = 3x + 11$

23. $y = \frac{1}{3}x^2 - 2$

24. $y = \frac{1}{2}|x - 1| + 14$

25. $y = -0.8(x + 3)$

26. $y = (1.5x)^2 + 22$

Example 6

27. BILLIARDS The function $g(x) = |0.5x|$ models the path of a cue ball in a certain shot on a pool table, where the x-axis represents the edge of the table. Describe how $g(x)$ is related to its parent function and interpret the function in the context of the situation.

28. SALAD The cost for a salad depends on its weight, x, in ounces, and is described by $c(x) = 4.5 + 0.32x$. Describe how $c(x)$ is related to its parent function and interpret the function in the context of the situation.

29. TRAVEL The cost to travel x miles east or west on a train is the same. The function for the cost is $c(x) = 0.75|x| + 25$. Describe how $c(x)$ is related to its parent function and interpret the function in the context of the situation.

30. ARCHERY The path of an arrow can be modeled by $h(x) = -0.03x^2 + 6$, where x is distance and $h(x)$ is height, both in feet. Describe how $h(x)$ is related to its parent function and interpret the function in the context of the situation.

Example 7

Write an equation for each function.

31.

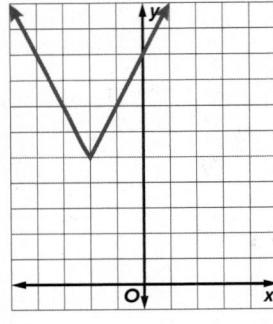

32.

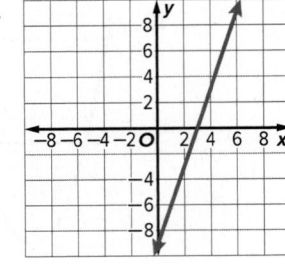

Write an equation for each function.

33.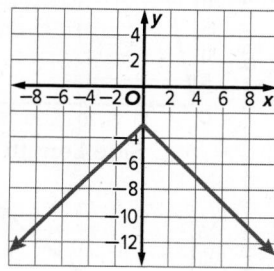

34.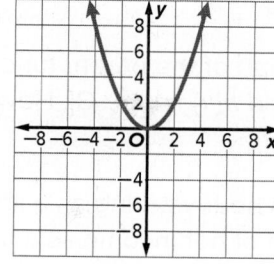

35.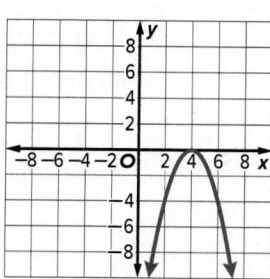

36.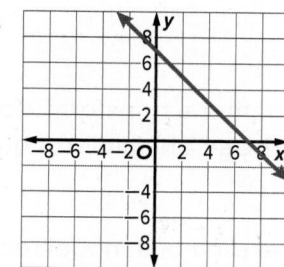

Mixed Exercises

Describe each transformation as it relates to the graph of the parent function. Then graph the function.

37. $y = |x| - 2$

38. $y = (x + 1)^2$

39. $y = -x$

40. $y = -|x|$

41. $y = 3x$

42. $y = 2x^2$

43. Describe the translation in $y = x^2 - 4$ as it relates to the parent function.

44. Describe the reflection in $y = -x^3$ as it relates to the parent function.

45. Describe the type of transformation in the function $f(x) = (5x)^2$.

46. ARCHITECTURE The cross-section of a roof is shown in the figure. Write an absolute value function that models the shape of the roof.

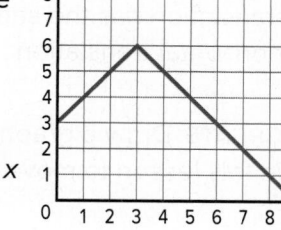

47. SPEED The speedometer in Henry's car is broken. The function $y = |x - 8|$ represents the difference y between the car's actual speed x and the displayed speed.
  a. Describe the translation. Then graph the function.
  b. Interpret the function and the translation in terms of the context of the situation.

48. GRAPHIC DESIGN Kassie sketches the function $f(x) = -1.25(x - 1)^2 + 18.75$ as part of a new logo design. Describe the transformations she applied to the parent function in creating her function.

49. GEOMETRY Chen made a graph to show how the perimeter of a square changes as the length of sides increase. How is this graph related to the parent function $y = x$?

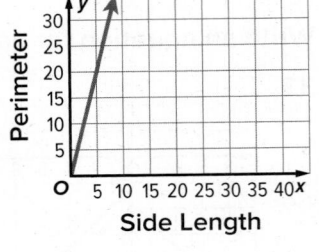

50. REASONING Compare the graph of the parent function $f(x) = |x|$ with the graphs of $g(x) = |x + 2|$ and $h(x) = |x - 3|$. How are the graphs similar? How are they different?

51. BUSINESS Maria earns $10 an hour working as a lifeguard. She drew the graph to show the relation of her income as a function of the hours she works. How did she modify the function $y = x$ to create her graph?

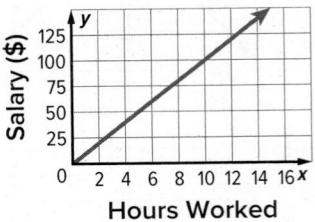

52. ANALYZE What determines whether a transformation will affect the graph vertically or horizontally? Use the family of quadratic functions as an example.

53. PERSEVERE Laura sketches the path of a model rocket that she launches.

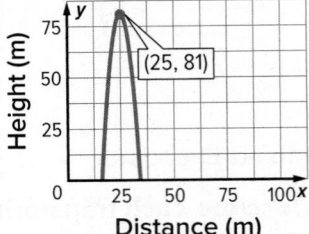

   a. What type of function does the graph show?
   b. In which axis has the parent function been reflected?
   c. How has the graph been translated? Assume that the function has not been dilated.
   d. What is the equation for the curve shown on the graph?

54. ANALYZE Graph $g(x) = -3|x + 5| - 1$. Describe the transformations of the parent function $f(x) = |x|$ that produce the graph of $g(x)$. What are the domain and range?

55. ANALYZE Consider $f(x) = |2x|$, $g(x) = x + 2$, $h(x) = 2x^2$, and $k(x) = 2x^3$.

   a. Graph each function and its reflection in the y-axis.
   b. Analyze the functions and the graphs. Determine whether each function is *odd*, *even*, or *neither*.
   c. Recall that if for all values of x, $f(-x) = f(x)$ the function $f(x)$ is an even function. If for all values of x, $f(-x) = -f(x)$ the function $f(x)$ is an odd function. Explain why this is true.

56. PERSEVERE Explain why performing a horizontal translation followed by a vertical translation has the same results as performing a vertical translation followed by a horizontal translation.

57. CREATE Draw a graph in Quadrant II. Use any of the transformations you learned in this lesson to move your figure to Quadrant IV. Describe your transformation.

58. ANALYZE Study the parent graphs at the beginning of this lesson. Select a parent graph with positive y-values when $x \rightarrow -\infty$ and positive y-values when $x \rightarrow \infty$.

59. WRITE Explain why the graph of $g(x) = (-x)^2$ appears the same as the graph of $f(x) = x^2$. Is this true for all reflections of quadratic functions? If not, describe a case when it is false.

Essential Question

How can analyzing a function help you understand the situation it models?

You can analyze a function by observing the domain and range, the intercepts, end behavior, and so on. These observations can help you understand the situation it is modeling by seeing what values make sense in the situation, how the quantities will change as the domain values change, and so on.

Module Summary

Lesson 8-1 through 8-3

Function Behavior

- The graph of a continuous function is a line or curve. The domain of a continuous function is a single interval of all real numbers.

- A linear function is a function in which no independent variable is raised to a whole number power greater than 1.

- If a vertical line intersects the graph of a relation more than once, then the relation is not a function.

- An x-intercept occurs when the graph intersects the x-axis, and a y-intercept occurs when the graph intersects the y-axis.

- A graph has line symmetry if each half of the graph on either side of a line matches the other side exactly.

- A graph has point symmetry when a figure is rotated 180° about a point and maps exactly onto the other part.

- A point is a relative maximum if there are no other nearby points with a greater y-coordinate. A point is a relative minimum if there are no other nearby points with a lesser y-coordinate.

- End behavior is the behavior of the graph at its ends. At the right end, the values of x are increasing toward infinity. This is denoted as $x \rightarrow \infty$. At the left end, the values of x are decreasing toward negative infinity, denoted as $x \rightarrow -\infty$.

Lessons 8-4 through 8-7

Graphs of Functions

- You can use key features of a function to sketch its graph. Features such as intercepts, symmetry, end behavior, extrema, and intervals where the function is increasing, decreasing, positive, or negative provide information for sketching the graph.

- A function that is written using two or more expressions is a piecewise-defined function.

- A step function has a graph that is a series of horizontal line segments.

- An absolute value function is a function that contains an algebraic expression within absolute value symbols.

- A translation moves a figure up, down, left, or right.

- A dilation shrinks or enlarges a figure proportionally. Multiplying a function by a constant dilates the graph with respect to the x- or y-axis.

- A reflection is a transformation that flips a figure in a line of reflection.

Study Organizer

Foldables

Use your Foldable to review this module. Working with a partner can be helpful. Ask for clarification of concepts as needed.

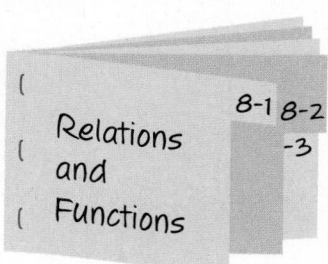

Test Practice

1. MULTIPLE CHOICE What is the domain of the function shown? (Lesson 8-1)

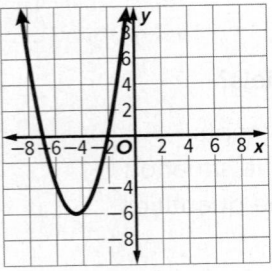

A. (−8, 0)

B. [−6, ∞)

C. (−∞, ∞)

D. (−∞, 0)

2. MULTIPLE CHOICE Salvatore is a plumber. He charges $100 for all work that is completed in less than 2 hours. He charges $250 for work that requires 2 to 5 hours, and he charges $400 for work that takes between 5 and 8 hours.

Which best describes the domain of the function that models Salvatore's price scale? (Lesson 8-1)

A. The domain is {100, 150, 400}.

B. The domain is $\{x|\ 0 < x \leq 8\}$.

C. The domain is $\{x|\ 100 \leq x \leq 400\}$.

D. The domain is $\{x|\ x = 2, 5, 8\}$.

3. OPEN RESPONSE The table shows the amount of money Tia owed her friend over time after borrowing the money to go to a theme park. (Lesson 8-2)

Week	0	1	2	3	4	5	6
Amount	$80	$68	$52	$39	$21	$10	$0

What are the coordinates of the x-intercept and the coordinates of the y-intercept? What is the meaning of the intercepts in context of the situation?

4. MULTIPLE CHOICE What type of symmetry is shown? (Lesson 8-2)

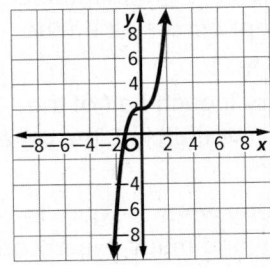

A. line symmetry

B. point symmetry

C. both line and point symmetry

D. no symmetry

5. OPEN RESPONSE Determine whether each of the x-values $(-5, -4, -2, 0, 1, 5)$ is a *relative maximum*, *relative minimum* or *neither*. (Lesson 8-3)

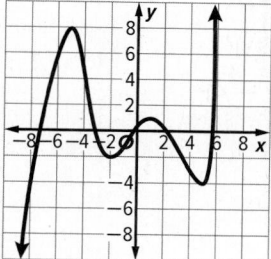

6. OPEN RESPONSE The graph shows the height of a ball after being thrown from a height of 26 feet.

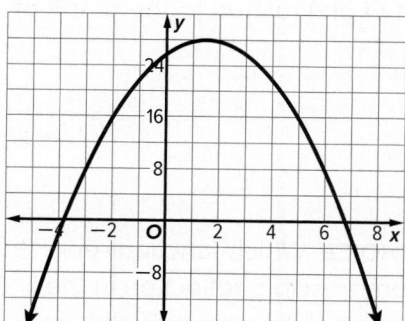

Explain why the end behavior does or does not make sense in this context. (Lesson 8-3)

7. OPEN RESPONSE Compare the key features of $f(x)$ and $g(x)$. (Lesson 8-4)

$f(x)$	$g(x)$
$f(x) = -2x + 4$	x-intercept: $(1, 0)$
	y-intercept: $(0, -5)$
	slope: 5

8. MULTIPLE CHOICE Sofia is sketching the graph of a function. She knows that as $x \to \infty, y \to -\infty$ and that the function has a y-intercept at $(0, 8)$. Which other feature fits the sketch of the graph? (Lesson 8-4)

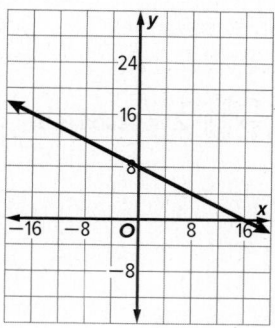

A. as $x \to -\infty, y \to -\infty$

B. x-intercept at $(-14, 0)$

C. increases for y in the interval $8 < x < 16$

D. decreases for y in the interval $-16 < x < 16$

9. MULTIPLE CHOICE A guidance counselor tells Rebekah that she needs a combined score of at least 1210 on the two portions of her college entrance exam to be eligible for the college of her choice. Suppose x represents Rebekah's score on the verbal portion and y represents her score on the math portion. Choose the equation or inequality that represents the scores Rebekah needs for eligibility. (Lesson 8-5)

A. $x + y \geq 1210$

B. $x + y = 1210$

C. $x + y > 1210$

D. $x + y \leq 1210$

10. OPEN RESPONSE The value of the Kim's portfolio is found by finding the sum of 60% of Stock A's value and 40% of Stock B's value. Kim wants her portfolio to be more than $500,000. Write the inequality that represents the constraints for Stock A's value x and Stock B's value y. (Lesson 8-5)

11. GRAPH Graph $f(x) = \left[\!\left[\frac{1}{3}x \right]\!\right] + 2$. (Lesson 8-6)

12. OPEN RESPONSE Determine if each transformation is a *translation*, *reflection*, or *dilation* from the parent function $f(x) = x^2$. (Lesson 8-7)

$$g(x) = -x^2$$
$$h(x) = (3x)^2$$
$$j(x) = (x - 4)^2$$

13. OPEN RESPONSE Describe the transformation(s) from the parent function to $f(x) = 3(x - 2)^2 + 9$. (Lesson 8-7)

14. OPEN RESPONSE If $g(x) = f(0.75x)$ then how is the graph of $g(x)$ related to the graph of $f(x)$? (Lesson 8-7)

15. MULTIPLE CHOICE Which function represents a vertical compression, reflection in the x-axis, and a translation down 3 units in relation to the parent function? (Lesson 8-7)

A. $y = -\frac{2}{3}(x - 3)^2$

B. $y = -\left(\frac{2}{3}x\right)^2 - 3$

C. $y = -\frac{2}{3}x^2 - 3$

D. $y = \frac{2}{3}(-x + 3)^2$

Linear Equations, Inequalities, and Systems

e Essential Question

How are equations, inequalities, and systems of equations or inequalities best used to model to real-world situations?

What Will You Learn?

How much do you already know about each topic **before** starting this module?

KEY	Before			After		
👎 — I don't know. 👍 — I've heard of it. 👍 — I know it!	👎	👍	👍	👎	👍	👍
solve linear equations						
solve linear inequalities						
solve absolute value equations and inequalities						
write equations of linear functions in standard, slope-intercept, and point-slope form						
solve systems of equations by graphing, by substitution, and by elimination						
solve systems of inequalities in two variables						
use linear programming to find maximum and minimum values of a function						
solve systems of equations in three variables						

📖 **Foldables** Make this Foldable to help you organize your notes about equations and inequalities. Begin with one sheet of paper.

1. **Fold** 2-inch tabs on each of the short sides.

2. **Fold** in half in both directions.

3. **Open** and cut as shown.

4. **Refold** along the width. Staple each pocket. Label pockets as *Solving Equations and Inequalities, Writing Equations for Functions, Systems of Equations and Inequalities,* and *Solve and Graph Inequalities.* Place index cards for notes in each pocket.

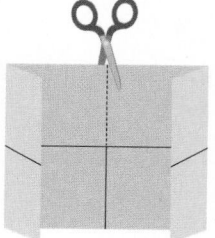
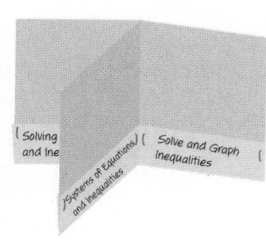

What Vocabulary Will You Learn?

- absolute value
- bounded
- consistent
- dependent
- elimination
- empty set
- equation
- extraneous solution
- feasible region
- inconsistent
- independent
- inequality
- linear programming
- optimization
- ordered triple
- root
- solution
- substitution
- system of equations
- system of inequalities
- unbounded
- zero

Are You Ready?

Complete the Quick Review to see if you are ready to start this module.
Then complete the Quick Check.

Quick Review

Example 1

Graph $2y + 5x = -10$.

Find the x- and y-intercepts.

$$2(0) + 5x = -10 \qquad 2y + 5(0) = -10$$
$$5x = -10 \qquad\qquad 2y = -10$$
$$x = -2 \qquad\qquad\quad y = -5$$

The graph crosses
the x-axis at $(-2, 0)$
and the y-axis at
$(0, -5)$. Use these
ordered pairs to
graph the equation.

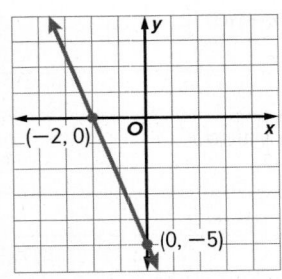

Example 2

Graph $y \geq 3x - 2$.

The boundary is the graph of $y = 3x - 2$. Since the inequality symbol is $\geq$, the boundary will be solid.

Test the point $(0, 0)$.

$$0 \geq 3(0) - 2 \quad (x, y) = (0, 0)$$

$$0 \geq -2$$

Shade the region
that includes $(0, 0)$.

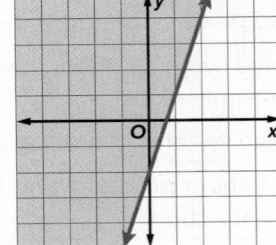

Quick Check

Graph each equation.

1. $x + 2y = 4$

2. $y = -x + 6$

3. $3x + 5y = 15$

4. $3y - 2x = -12$

Graph each inequality.

5. $y < 3$

6. $x + y \geq 1$

7. $3x - y > 6$

8. $x + 2y \leq 5$

How Did You Do?

Which exercises did you answer correctly in the Quick Check?

Solving Linear Equations and Inequalities

Learn Solving Linear Equations

An **equation** is a mathematical sentence stating that two mathematical expressions are equal. The **solution** of an equation is a value that makes the equation true. To solve equations, use the properties of equality to isolate the variable on one side.

Key Concept • Properties of Equality

Property of Equality	Symbols
Addition Property of Equality	For any real numbers a, b, and c, if $a = b$, then $a + c = b + c$.
Subtraction Property of Equality	For any real numbers a, b, and c, if $a = b$, then $a - c = b - c$.
Multiplication Property of Equality	For any real numbers a, b, and c, $c \neq 0$, if $a = b$, then $ac = bc$.
Division Property of Equality	For any real numbers a, b, and c, $c \neq 0$, if $a = b$, then $\frac{a}{c} = \frac{b}{c}$.

Example 1 Solve a Linear Equation

Solve $\frac{1}{3}(2x - 57) + \frac{1}{3}(6 - x) = -4$.

$\frac{1}{3}(2x - 57) + \frac{1}{3}(6 - x) = -4$	Original equation
$\frac{2}{3}x - 19 + 2 - \frac{1}{3}x = -4$	Distributive Property
$\frac{1}{3}x - 17 = -4$	Combine like terms.
$\frac{1}{3}x = 13$	Add 17 to each side and simplify.
$x = 39$	Multiply each side by 3 and simplify.

🌐 Example 2 Write and Solve an Equation

SPACE **The diameter of Earth is 828 kilometers less than twice the diameter of Mars. If Earth has a diameter of 12,756 kilometers, what is the diameter of Mars?**

Part A Write an equation that represents the situation.

Words	The diameter of Earth	is	828 less than twice the diameter of Mars.
Variable	Let m = the diameter of Mars.		
Equation	12,756	=	$2m - 828$

🌐 Go Online You can complete an Extra Example online.

(continued on the next page)

Today's Goals
- Solve linear equations.
- Solve linear equations by examining graphs of the related functions.
- Solve linear inequalities.

Today's Vocabulary
equation
solution
root
zero
inequality

Study Tip

Justifications The properties of equality can be used as justifications. However, in most future solutions, the justifications for steps will read as "Subtract c from each side," or "Divide each side by c."

🌐 **Go Online**
You can watch a video to see how to solve equations in one variable using a graphing calculator.

Part B Solve the equation.

$12{,}756 = 2m - 828$	Original equation
$12{,}756 + 828 = 2m - 828 + 828$	Add 828 to each side.
$13{,}584 = 2m$	Simplify.
$\dfrac{13{,}584}{2} = \dfrac{2m}{2}$	Divide each side by 2.
$6792 = m$	Simplify.

The diameter of Mars is 6792 kilometers. This is a reasonable solution because 12,756 is a little less than $6792 \cdot 2 = 13{,}584$, as indicated in the problem.

Check

BASKETBALL In 1962, Wilt Chamberlain set the record for the most points scored in a single NBA game. He scored 28 points from free throws and made x field goals, worth two points each. If Wilt Chamberlain scored 100 points, how many field goals did he make? Which equation represents the number of field goals that Chamberlain scored?

A. $100 = 28 + 2x$ **B.** $100 = 28x + 2$ **C.** $28 = 2x$ **D.** $100 = 2x$

How many field goals did Chamberlain score? __?__ field goals

---

Example 3 Solve for a Variable

GEOMETRY The formula for the perimeter of a parallelogram is $P = 2a + 2b$ where a and b represent the measures of the bases. Solve the equation for b.

$P = 2a + 2b$	Original equation
$P - 2a = 2a + 2b - 2a$	Subtract $2a$ from each side.
$P - 2a = 2b$	Simplify.
$\dfrac{P}{2} - \dfrac{2a}{2} = \dfrac{2b}{2}$	Divide each side by 2.
$\dfrac{P}{2} - a = b$	Simplify.

Check

GEOMETRY The formula for the area A of a trapezoid is solved for h. Fill in the missing justification.

$A = \frac{1}{2}h(a + b)$	Original equation
$2A = 2 \cdot \frac{1}{2}h(a + b)$	Multiplication Property of Equality
$2A = h(a + b)$	Simplify.
$\dfrac{2A}{(a + b)} = \dfrac{h(a + b)}{(a + b)}$	__?__
$\dfrac{2A}{(a + b)} = h$	Simplify.

Go Online You can complete an Extra Example online.

Learn Solving Linear Equations by Graphing

The solution of an equation is called a **root**. You can find the root of an equation by examining the graph of its related function $f(x)$. A related function is found by solving the equation for 0 and then replacing 0 with $f(x)$. A related function for $2x + 13 = 9$ is $f(x) = 2x + 4$ or $y = 2x + 4$.

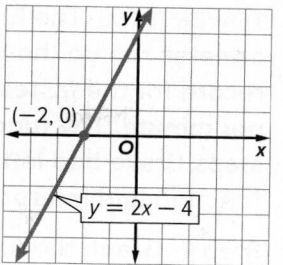

Values of x for which $f(x) = 0$ are called **zeros** of the function f. The zero of a function is the x-intercept of its graph. The solution and root of a linear equation are the same as the zero and x-intercept of its related function.

Example 4 Solve a Linear Equation by Graphing

Solve $\frac{1}{2}x - 11 = -8$ by graphing.

Step 1 Find a related function.

Rewrite the equation with 0 on the right side.

$$\frac{1}{2}x - 11 = -8 \qquad \text{Original equation.}$$

$$\frac{1}{2}x - 11 + 8 = -8 + 8 \qquad \text{Add 8 to each side.}$$

$$\frac{1}{2}x - 3 = 0 \qquad \text{Simplify.}$$

Replacing 0 with $f(x)$ gives the related function, $f(x) = \frac{1}{2}x - 3$.

Step 2 Graph the related function.

Since the graph of $f(x) = \frac{1}{2}x - 3$ intersects the x-axis at 6, the solution of the equation is 6.

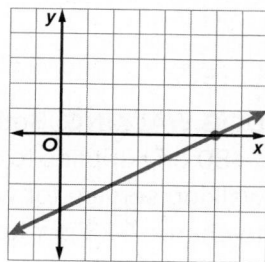

Check

Graph the related function of $2x - 5 = -11$.
Use the graph to solve the equation.

The related function is _____?_____.

The solution is __?__.

Go Online You can complete an Extra Example online.

🗨 Talk About It!

Because there is typically more than one way to solve an equation for 0, there may be more than one related function for an equation. What is another possible related function of $2x + 13 = 9$? How does the zero of this function compare to the zero of $f(x) = 2x + 4$?

🧠 Think About It!

Explain why −3 is *not* a zero of the function.

Watch Out!

Intercepts Be careful not to mistake y-intercepts for zeros of functions. The y-intercept on a graph occurs when $x = 0$. The x-intercepts are the zeros of a function because they are where $f(x) = 0$.

Example 5 Estimate Solutions by Graphing

TOWER RACE The Empire State Building Run-Up is a race in which athletes run up the building's 1576 stairs. In 2003, Paul Crake set the record for the fastest time, running up an average of about 165 stairs per minute. The function $c = 1576 - 165m$ represents the number of steps Crake had left to climb c after m minutes. Find the zero of the function and interpret its meaning in the context of the situation.

Step 1 Graph the function.

Step 2 Estimate the zero.

The graph appears to intersect the x-axis at about 9.5. This means that Paul Crake finished the race in about 9.5 minutes, or 9 minutes and 30 seconds.

Tower Race

(graph: y-axis "Steps Remaining" from 0 to 1800; x-axis "Time (min)" from 0 to 9; line decreasing from about 1576 to the x-axis near 9.5)

Step 3 Solve algebraically.

Use the equation. Substitute 0 for c, and solve algebraically to check your solution.

$c = 1576 - 165m$	Original equation
$0 = 1576 - 165m$	Replace c with 0.
$165m = 1576$	Add $165m$ to each side.
$m \approx 9.55$	Divide each side by 165.

The solution is about 9.55. So, Paul Crake completed the Empire State Building Run-Up in about 9.55 minutes, or 9 minutes and 33 seconds. This is close to the estimated time of 9.5 minutes.

Check

DOG WALKING Bethany spends $480 on supplies to start a dog walking service for which she plans to charge $23 per hour. The function $y = 23x - 480$ represents Bethany's profit after x hours of dog walking.

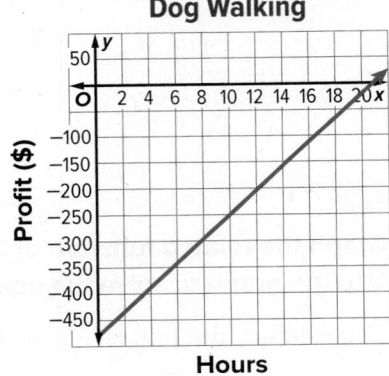

Dog Walking

Part A The graph appears to intersect the x-axis at about ___?___.

Part B Solve algebraically to verify your estimate. Round to the nearest hundredth.

Go Online You can complete an Extra Example online.

Explore Comparing Linear Equations and Inequalities

⏻ **Online Activity** Use a comparison to complete the Explore.

> ×
>
> @ **INQUIRY** How do the solution methods and
> the solutions of linear equations and
> inequalities in one variable compare?

Learn Solving Linear Inequalities

An **inequality** is a mathematical sentence that contains the symbol
$<, \leq, >, \geq$, or $\neq$. Properties of inequalities allow you to perform
operations on each side of an inequality without changing the truth of
the inequality.

Key Concept • Addition and Subtraction Properties of Inequality	
Symbols	For any real numbers, a, b, and c:
	If $a > b$, then $a + c > b + c$. If $a > b$, then $a - c > b - c$.
	If $a < b$, then $a + c < b + c$. If $a < b$, then $a - c < b - c$.
Examples	$2 > 0$ $9 > 6$
	$2 + 1 > 0 + 1$ $9 - 4 > 6 - 4$
	$3 > 1$ $5 > 2$

Key Concept • Multiplication and Division Properties of Inequality	
Symbols	For any real numbers, a, b, and c:
	where c is positive:
	If $a > b$, then $ac > bc$. If $a > b$, then $\frac{a}{c} > \frac{b}{c}$.
	If $a < b$, then $ac < bc$. If $a < b$, then $\frac{a}{c} < \frac{b}{c}$.
	where c is negative:
	If $a > b$, then $ac < bc$. If $a > b$, then $\frac{a}{c} < \frac{b}{c}$.
	If $a < b$, then $ac > bc$. If $a < b$, then $\frac{a}{c} > \frac{b}{c}$.
Examples	$8 > -2$ $-4 < -2$
	$8(3) > -2(3)$ $-4(-8) > -2(-8)$
	$24 > -6$ $32 > 16$

The solution sets of inequalities can be expressed by using set-builder
notation. For example, $\{x \mid x > 1\}$ represents the set of all numbers x
such that x is greater than 1. The solution sets can also be graphed on
number lines. Circles and dots are used to indicate whether an
endpoint is included in the solution. The circle at 1 means that this
point is *not* included in the solution set.

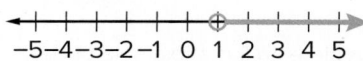

$$-5\ -4\ -3\ -2\ -1\ \ 0\ \ 1\ \ 2\ \ 3\ \ 4\ \ 5$$

Go Online
You may want to
complete the Concept
Check to check your
understanding.

🗨 **Think About It!**
What does the arrow
on the graph of a
solution set represent?

Think About It!

What does a dot on the graph of a solution set indicate?

Example 6 Solve a Linear Inequality

Solve $-5.6n + 12.9 \geq -71.1$. Graph the solution set on a number line.

$-5.6n + 12.9 \geq -71.1$	Original inequality.
$-5.6n + 12.9 - 12.9 \geq -71.1 - 12.9$	Subtract 12.9 from each side.
$-5.6n \geq -84$	Simplify.
$\dfrac{-5.6n}{-5.6} \leq \dfrac{-84}{-5.6}$	Divide each side by -5.6, reversing the inequality symbol.
$n \leq 15$	Simplify.

The solution set is $\{n \mid n \leq 15\}$. Graph the solution set.

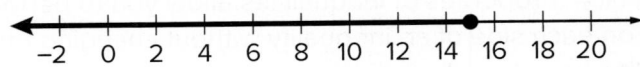

Check

What is the solution of $-p - 3 \geq -4(p + 6)$? Graph the solution set.

Study Tip

Reversing the Inequality Symbol

Adding the same number to, or subtracting the same number from, each side of an inequality does not change the truth of the inequality. Multiplying or dividing each side of an inequality by a positive number does not change the truth of the inequality. However, multiplying or dividing each side of an inequality by a negative number requires that the order of the inequality be reversed. In Example 6, $\geq$ was replaced with $\leq$.

🌐 Example 7 Write and Solve an Inequality

NUTRITION The recommended daily intake of calcium for teens is 1300 mg. Jake gets 237 mg of calcium from a multivitamin he takes each morning and 302 mg from each glass of skim milk that he drinks. How many glasses of milk would Jake need to drink to meet the recommendation?

Step 1 Write an inequality to represent the situation.

Let $g =$ the number of glasses of milk Jake needs.

$237 + 302g \geq 1300$

Step 2 Solve the inequality.

$237 + 302g \geq 1300$	Original inequality
$302g \geq 1063$	Subtract 237 from each side.
$g \geq 3.52$	Divide each side by 302.

Step 3 Interpret the solution in the context of the situation.

Jake will need to drink slightly more than 3.5 glasses of milk to intake at least the recommended daily amount of calcium. This is a viable solution because Jake can pour part of a full glass of milk.

Watch Out!

Reading Math Be sure to always read problems carefully. The term *at least* is used here, and can be confusing since it actually means *greater than or equal to*, and is represented by $\geq$. In this instance, Jake should intake at least 1300 mg, which means he must intake an amount greater than or equal to 1300 mg.

🔁 **Go Online** You can complete an Extra Example online.

Practice

📡 **Go Online** You can complete your homework online.

Example 1

Solve each equation. Check your solution.

1. $6x - 5 = 7 - 9x$

2. $-1.6r + 5 = -7.8$

3. $\frac{3}{4} - \frac{1}{2}n = \frac{5}{8}$

4. $\frac{5}{6}c + \frac{3}{4} = \frac{11}{12}$

5. $2.2n + 0.8n + 5 = 4n$

6. $6y - 5 = -3(2y + 1)$

7. $-6(2x + 4) + \frac{1}{2}(8 + 3x) = -20$

8. $7(-1 + 4x) - 12x = 5$

9. $-4(10 + 3x) - (x + 8) = -9$

Example 2

Solve each problem.

10. REASONING The length of a rectangle is twice the width. Find the width if the perimeter is 60 centimeters. Define a variable, write an equation, and solve the problem.

11. GOLF Sergio and three friends went golfing. Two of the friends rented clubs for $6 each. The total cost of the rented clubs and the green fees was $76. What was the cost of the green fees for each person? Define a variable, write an equation, and solve the problem.

Example 3

Solve each equation or formula for the specified variable.

12. BANKING The formula for simple interest I is $I = Prt$, where P is the principal, r is the interest rate, and t is time. Solve for P.

13. MEAN The mean A of two numbers, x and y, is given by $A = \frac{x + y}{2}$. Solve for y.

14. SLOPE The slope m between two points (x_1, y_1) and (x_2, y_2) is $m = \frac{y_2 - y_1}{x_2 - x_1}$. Solve for y_2.

15. CYLINDERS The surface area of a cylinder A is given by $A = 2\pi r^2 + 2\pi rh$, where r is radius and h is height. Solve for h.

16. PHYSICS The height h of a falling object is given by $h = vt - gt^2$, where v is the initial velocity of the object, t is time, and g is the gravitational constant. Solve for v.

Example 4

Find the related function for each equation. Then graph the related function. Use the graph to solve the equation.

17. $2x + 5 = -7$

18. $-x + 3 = 6$

19. $\frac{1}{2}x + 4 = 10$

20. $\frac{1}{3}x + 1 = -5$

21. $-3x - 1 = 1$

22. $-\frac{1}{4}x + 1 = -2$

Example 5

Solve each problem.

23. LUNCH ACCOUNT At the beginning of the quarter, Nahla deposited $100 into her lunch account. If Nahla spends an average of $18 per week on lunches, then $m = 100 - 18w$ represents the amount of money Nahla has left in her lunch account m after w weeks of school.

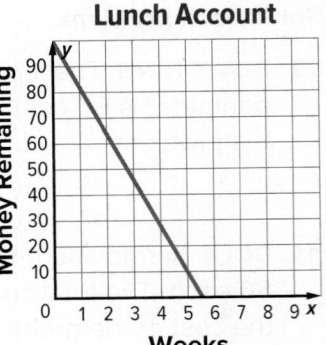

a. Use the graph to estimate to the nearest tenth the number of weeks that Nahla can purchase school lunches with the money she deposited at the beginning of the quarter.

b. Solve algebraically to verify your estimate. Round to the nearest hundredth.

24. READING Mario has a 500-page novel which he is required to read over summer break for his upcoming language arts class.

a. If Mario reads 24 pages each day, write and graph a function to represent the number of pages that Mario has left to read p after d days.

b. Estimate the zero of the function by using the graph. Justify your response.

c. Find the zero algebraically. Interpret its meaning in the context of the situation.

Example 6

Solve each inequality. Graph the solution set on a number line.

25. $\frac{z}{-4} \geq 2$

26. $3a + 7 \leq 16$

27. $20 - 3n > 7n$

28. $7f - 9 > 3f - 1$

29. $0.7m + 0.3m \geq 2m - 4$

30. $4(5x + 7) \leq 13$

Example 7

Solve each problem.

31. **INCOME** Manuel takes a job translating English instruction manuals to Spanish. He will receive $15 per page plus $100 per month. Manuel plans to work for 3 months during the summer and wants to make at least $1500. Write and solve an inequality to find the minimum number of pages Manuel must translate in order to reach his goal. Then, interpret the solution in the context of the situation.

32. **STRUCTURE** On a conveyor belt, there can only be two boxes moving at a time. The total weight of the boxes cannot be more than 300 pounds. Let x and y represent the weights of two boxes on the conveyor belt.

    a. Write an inequality that describes the weight limitation in terms of x and y.

    b. Write an inequality that describes the limit on the average weight a of the two boxes.

    c. Two boxes are to be placed on the conveyor belt. The first box weighs 175 pounds. What is the maximum weight of the second box?

Mixed Exercises

Solve each equation. Check your solution by graphing the related function.

33. $-3b + 7 = -15 + 2b$ 34. $a - \frac{2a}{5} = 3$ 35. $2.2n + 0.8n + 5 = 4n$

Solve each inequality. Graph the solution set on a number line.

36. $\frac{4x - 3}{2} \geq -3.5$ 37. $1 + 5(x - 8) \leq 2 - (x + 5)$ 38. $-36 - 2(w + 77) > -4(2w + 52)$

39. **REASONING** An ice rink offers open skating several times a week. An annual membership to the skating rink costs $60. The table shows the cost of one session for members and non-members.

Open Ice Skating Sessions	
members	$6
non-members	$10

Kaliska plans to spend no more than $90 on skating this year. Define a variable then write and solve inequalities to find the number of sessions she can attend with and without buying a membership. Should Kaliska buy a membership?

40. PRECISION The formula to convert temperature in degrees Fahrenheit to degrees Celsius is $\frac{5}{9}(F - 32) = C$.

a. Solve the equation for F.

b. Use your result from part a to determine the temperature in degrees Fahrenheit when the Celsius temperature is 30.

c. At a certain temperature, a Fahrenheit thermometer and a Celsius thermometer will read the same temperature. Write and solve an equation to find the temperature.

Higher-Order Thinking Skills

41. FIND THE ERROR Steven and Jade are solving $A = \frac{1}{2}h(b_1 + b_2)$ for b_2. Is either of them correct? Explain your reasoning.

Steven	Jade
$A = \frac{1}{2}h(b_1 + b_2)$	$A = \frac{1}{2}h(b_1 + b_2)$
$\frac{2A}{h} = (b_1 + b_2)$	$\frac{2A}{h} = (b_1 + b_2)$
$\frac{2A - b_1}{h} = b_2$	$\frac{2A}{h} - b_1 = b_2$

42. CREATE Write an equation involving the Distributive Property that has no solution and another example that has infinitely many solutions.

43. PERSEVERE Solve $d = \sqrt{(x_2 - x_1)^2 + (y_2 - y_1)^2}$ for y_1.

44. ANALYZE Vivek's teacher made the statement, "Four times a number is less than three times a number." Vivek quickly responded that the answer is *no solution*. Do you agree with Vivek? Write and solve an inequality to justify your argument.

45. WRITE Why does the inequality symbol need to be reversed when multiplying or dividing by a negative number?

46. PERSEVERE Given $\triangle ABC$ with sides $AB = 3x + 4$, $BC = 2x + 5$, and $AC = 4x$, determine the values of x such that $\triangle ABC$ exists.

Solving Absolute Value Equations and Inequalities

Learn Solving Absolute Value Equations Algebraically

The **absolute value** of a number is its distance from zero on the number line. The definition of absolute value can be used to solve equations that contain absolute value expressions by constructing two cases. For any real numbers a and b, if $|a| = b$ and $b \geq 0$, then $a = b$ or $a = -b$.

Step 1 Isolate the absolute value expression on one side of the equation.

Step 2 Write the two cases.

Step 3 Use the properties of equality to solve each case.

Step 4 Check your solutions.

Absolute value equations may have one, two, or no solutions.

- An absolute value equation has one solution if one of the answers does not meet the constraints of the problem. Such an answer is called an **extraneous solution**.

- An absolute value equation has no solution if there is no answer that meets the constraints of the problem. The solution set of this type of equation is called the **empty set**, symbolized by {} or Ø.

Example 1 Solve an Absolute Value Equation

Solve $2|5x + 1| - 9 = 4x + 17$. Check your solutions. Then graph the solution set.

$2|5x + 1| - 9 = 4x + 17$	Original equation
$2|5x + 1| = 4x + 26$	Add 9 to each side.
$|5x + 1| = 2x + 13$	Divide each side by 2.

Case 1	Case 2
$5x + 1 = 2x + 13$	$5x + 1 = -(2x + 13)$
$3x + 1 = 13$	$7x + 1 = -13$
$x = 4$	$x = -2$

CHECK Substitute each value in the original equation.

$2|5(4) + 1| - 9 \stackrel{?}{=} 4(4) + 17$ $2|5(-2) + 1| - 9 \stackrel{?}{=} 4(-2) + 17$

$33 = 33$ True $9 = 9$ True

Both solutions make the equation true. Thus, the solution set is {4, −2}. The solution set can be graphed by graphing each solution on a number line.

$-5\ -4\ -3\ -2\ -1\ \ 0\ \ 1\ \ 2\ \ 3\ \ 4\ \ 5$

Go Online You can complete an Extra Example online.

Today's Goals
- Write and solve absolute value equations, and graph the solutions on a number line.
- Write and solve absolute value inequalities, and graph the solutions on a number line.

Today's Vocabulary
absolute value
extraneous solution
empty set

Watch Out!
Distribute the Negative For Case 2, remember to use the Distributive Property to multiply the entire expression on the right side of the equation by −1.

Check

Graph the solution set of $|5x - 3| - 6 = -2x + 12$.

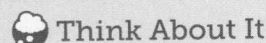 **Think About It!**

What would the graph of the solution set of an absolute value equation with only one solution look like?

Example 2 Extraneous Solution

Solve $2|x + 1| - x = 3x - 4$. Check your solutions.

$2\|x + 1\| - x = 3x - 4$	Original equation
$2\|x + 1\| = 4x - 4$	Add x to each side.
$\|x + 1\| = 2x - 2$	Divide each side by 2.

Case 1	Case 2
$x + 1 = 2x - 2$	$x + 1 = -(2x - 2)$
$1 = x - 2$	$x + 1 = -2x + 2$
$3 = x$	$3x + 1 = 2$
	$x = \frac{1}{3}$

There appear to be two solutions, 3 and $\frac{1}{3}$.

CHECK Substitute each value in the original equation.

$2|3 + 1| - 3 \overset{?}{=} 3(3) - 4 \qquad 2\left|\frac{1}{3} + 1\right| - \left(\frac{1}{3}\right) \overset{?}{=} 3\left(\frac{1}{3}\right) - 4$

$\qquad\qquad 5 = 5 \text{ True} \qquad\qquad\qquad \frac{7}{3} \neq -3 \text{ False}$

Because $\frac{7}{3} \neq -3$, the only solution is 3. Thus, the solution set is {3}.

Example 3 The Empty Set

Solve $|4x - 7| + 10 = 2$.

$\|4x - 7\| + 10 = 2$	Original equation
$\|4x - 7\| = -8$	Subtract 10 from each side.

Because the absolute value of a number is always positive or zero, this sentence is never true. The solution is $\varnothing$.

Talk About It!

Is the following statement *always*, *sometimes*, or *never* true? Justify your argument. For real numbers a, b, and c, $|ax + b| = -c$ has no solution.

Check

Solve each absolute value equation.

a. $|x + 10| = 4x - 8$

b. $3|4x - 11| + 1 = 9x + 13$

c. $|2x + 5| - 18 = -3$

d. $-5|7x - 2| + 3x = 3x + 10$

Go Online You can complete an Extra Example online.

🌐 Apply Example 4 Write and Solve an Absolute Value Equation

FOOTBALL **The NFL regulates the inflation, or air pressure, of footballs used during games. It requires that footballs have an air pressure of 13 pounds per square inch (PSI), plus or minus 0.5 PSI. What is the greatest and least acceptable air pressure of a regulation NFL football?**

1 What is the task?

Describe the task in your own words. Then list any questions that you may have. How can you find answers to your questions?

Sample answer: I need to find the greatest and least acceptable air pressure for an NFL football. How can I write an absolute value equation to find the solution? Will there be one solution or two that make sense in this problem? I can find the answers to my questions by referencing other examples in the lesson and by checking my solutions.

2 How will you approach the task? What have you learned that you can use to help you complete the task?

Sample answer: I will write an equation to represent the situation. I have learned how to write and solve an equation involving absolute value.

3 What is your solution?

Use your strategy to solve the problem.

What absolute value equation represents the greatest and least acceptable air pressure?

$|x - 13| = 0.5$

How are the solutions of the equation represented on a graph?

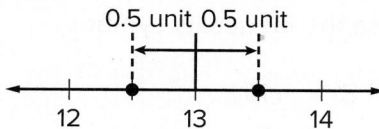

Interpret your solution. What are the greatest and least acceptable air pressures for an NFL football?

The greatest air pressure an NFL football can have is 13.5 PSI and the least is 12.5 PSI.

4 How can you know that your solution is reasonable?

🔵 **Write About It!** Write an argument that can be used to defend your solution. Sample answer: Because the distance between 13 and each solution is 0.5, both solutions satisfy the constraints of the equation. I can substitute each solution back into the original equation and check that the value makes the equation true. The pressures of 12.5 PSI and 13.5 PSI are within 0.5 PSI of 13 PSI.

🧭 **Go Online** You can complete an Extra Example online.

Learn Solving Absolute Value Inequalities Algebraically

When solving absolute value inequalities, there are two cases to consider. These two cases can be rewritten as a compound inequality.

Key Concept • Absolute Value Inequalities

For all real numbers a, b, c and x, $c > 0$, the following statements are true.

Absolute Value Inequality	Case 1	Case 2	Compound Inequality
$\lvert ax + b \rvert < c$	$ax + b < c$	$-(ax + b) < c$ $\dfrac{-(ax + b)}{-1} > \dfrac{c}{-1}$ $ax + b > -c$	$ax + b < c$ and $ax + b > -c$ which is $-c < ax + b < c$
$\lvert ax + b \rvert > c$	$ax + b > c$	$-(ax + b) > c$ $\dfrac{-(ax + b)}{-1} < \dfrac{c}{-1}$ $ax + b < -c$	$ax + b > c$ or $ax + b < -c$

These statements are also true for $\leq$ and $\geq$, respectively.

Think About It!

Describe a shortcut you could use to write case 2.

Go Online

An alternate method is available for this example.

Study Tip

Check Your Solutions
Remember to check your solutions by substituting values in each interval in the original inequality.

Example 5 Solve an Absolute Value Inequality ($<$ or $\leq$)

Solve $\lvert 4x - 8 \rvert - 5 < 11$. Then graph the solution set.

$$\lvert 4x - 8 \rvert - 5 < 11 \qquad \text{Original inequality}$$
$$\lvert 4x - 8 \rvert < 16 \qquad \text{Add 5 to each side.}$$

Because the inequality uses $<$, rewrite it as a compound inequality joined by the word *and*. For the case where the expression inside the absolute value symbols is negative, reverse the inequality symbol.

$4x - 8 < 16$	and	$4x - 8 > -16$
$4x < 24$		$4x > -8$
$x < 6$		$x > -2$

So, $x < 6$ and $x > -2$. The solution set is $\{x \mid -2 < x < 6\}$. All values of x between -2 and 6 satisfy the original inequality.

The solution set represents the interval between two numbers. Because the $<$ symbols indicate that -2 and 6 are not solutions, graph the endpoints of the interval on a number line using circles. Then, shade the interval from -2 to 6.

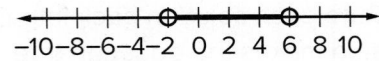

Go Online You can complete an Extra Example online.

Example 6 Solve an Absolute Value Inequality (> or ≥)

Solve $\dfrac{|6x + 3|}{2} + 5 \geq 14$. Then graph the solution set.

$\dfrac{|6x + 3|}{2} + 5 \geq 14$	Original inequality
$\dfrac{|6x + 3|}{2} \geq 9$	Subtract 5 from each side.
$|6x + 3| \geq 18$	Multiply each side by 2.

Since the inequality uses ≥, rewrite it as a compound inequality joined by the word *or*. For the case where the expression inside the absolute value symbols is negative, reverse the inequality symbol.

$$6x + 3 \geq 18 \qquad \text{or} \qquad 6x + 3 \leq -18$$
$$6x \geq 15 \qquad\qquad\qquad 6x \leq -21$$
$$x \geq \frac{5}{2} \qquad\qquad\qquad x \leq -\frac{7}{2}$$

So, $x \geq \dfrac{5}{2}$ or $x \leq -\dfrac{7}{2}$. The solution set is $\left\{ x \mid x \leq -\dfrac{7}{2} \text{ or } x \geq \dfrac{5}{2} \right\}$.

All values of x less than or equal to $-\dfrac{7}{2}$ as well as values of x greater than $\dfrac{5}{2}$ satisfy the constraints of the original inequality.

The solution set represents the union of two intervals. Since the ≤ and ≥ symbols indicate that $-\dfrac{7}{2}$ and $\dfrac{5}{2}$ are solutions, graph the endpoints of the interval on a number line using dots. Then, shade all points less than $-\dfrac{7}{2}$ and all points greater than $\dfrac{5}{2}$.

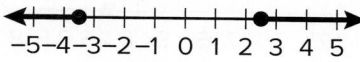

Check

Match each solution set with the appropriate absolute value inequality.

$-8|x + 14| + 7 \geq -17$

$\dfrac{|2x - 8|}{3} - 10 < 6$

$5|2x + 28| + 6 \geq -24$

$\dfrac{|3x - 12|}{4} - 13 > 5$

A. $\{x \mid x \text{ is a real number.}\}$ B. $\{x \mid 28 < x < -20\}$

C. $\{x \mid x < -20 \text{ or } x > 28\}$ D. $\{x \mid x \leq -11 \text{ or } x \geq -17\}$

E. $\{x \mid -20 < x < 28\}$ F. $\{x \mid -17 \leq x \leq -11\}$

Go Online You can complete an Extra Example online.

> **Watch Out!**
>
> **Isolate the Expression** Remember to isolate the absolute value expression on one side of the inequality symbol before determining whether to rewrite an absolute value inequality using *and* or *or*. When transforming the inequality, you might divide or multiply by a negative number, causing the inequality symbol to be reversed.

🌐 Example 7 Write and Solve an Absolute Value Inequality

SLEEP You can find how much sleep you need by going to sleep without turning on an alarm. Once your sleep pattern has stabilized, record the amount of time you spend sleeping each night. The amount of time you sleep plus or minus 15 minutes is your sleep need. Suppose you sleep 8.5 hours per night. Write and solve an inequality to represent your sleep need, and graph the solution on a number line.

Part A Write an absolute value inequality to represent the situation.

The difference between your actual sleep need and the amount of time you sleep is less than or equal to 15 minutes. So, 8.5 hours is the central value and 15 minutes, or 0.25 hour, is the acceptable range.

The difference between your actual sleep need and 8.5 hours is 0.25 hour. Let n = your actual sleep need.

$$|n - 8.5| \leq 0.25$$

Part B Solve the inequality and graph the solution set.

Rewrite $|n - 8.5| \leq 0.25$ as a compound inequality.

$$n - 8.5 \leq 0.25 \qquad \text{and} \qquad n - 8.5 \geq -0.25$$

$$n \leq 8.75 \qquad\qquad\qquad n \geq 8.25$$

The solution set represents the interval between two numbers. Since the $\leq$ and $\geq$ symbols indicate that 8.25 and 8.75 are solutions, graph the endpoints of the interval on a number line using dots. Then, shade the interval from 8.25 to 8.75.

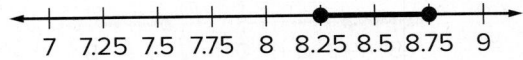

This means that you need between 8.25 and 8.75 hours of sleep per night, inclusive.

Check

FOOD A survey found that 58% of American adults eat at a restaurant at least once a week. The margin of error was within 3 percentage points.

Part A Write an absolute value inequality to represent the range of the percent of American adults who eat at a restaurant once a week, where x is the actual percent.

Part B Use your inequality from Part **A** to find the range of the percent of American adults who eat at a restaurant once a week.

The actual percent of American adults who eat out at least once a week is _____?_____.

🔘 **Go Online** You can complete an Extra Example online.

Practice

Examples 1–3

Solve each equation. Check your solutions.

1. $|8 + p| = 2p - 3$

2. $|4w - 1| = 5w + 37$

3. $4|2y - 7| + 13 = 9$

4. $-2|7 - 3y| - 6 = -14$

5. $2|4 - n| = -3n$

6. $5 - 3|2 + 2w| = -7$

7. $5|2r + 3| - 5 = 0$

8. $3 - 5|2d - 3| = 4$

Example 4

Solve each problem.

9. **WEATHER** The packaging of a thermometer claims that the thermometer is accurate within 1.5 degrees of the actual temperature in degrees Fahrenheit. Write and solve an absolute value equation to find the least and greatest possible temperature if the thermometer reads 87.4° F.

10. **OPINION POLLS** Public opinion polls reported in newspapers are usually given with a margin of error. A poll for a local election determined that Candidate Morrison will receive 51% of the votes. The stated margin of error is ±3%. Write and solve an absolute value equation to find the minimum and maximum percent of the vote that Candidate Morrison can expect to receive.

Examples 5 and 6

Solve each inequality. Graph the solution set on a number line.

11. $|2x + 2| - 7 \leq -5$

12. $\left|\frac{x}{2} - 5\right| + 2 > 10$

13. $|3b + 5| \leq -2$

14. $|x| > x - 1$

15. $|4 - 5x| < 13$

16. $|3n - 2| - 2 < 1$

17. $|3x + 1| > 2$

18. $|2x - 1| < 5 + 0.5x$

Example 7

Solve each problem.

19. **RAINFALL** For 90% of the last 30 years, the rainfall at Shell Beach has varied no more than 6.5 inches from its mean value of 24 inches. Write and solve an absolute value inequality to describe the rainfall in the other 10% of the last 30 years, and graph the solution on a number line.

20. **MANUFACTURING** A food manufacturer's guidelines state that each can of soup produced cannot vary from its stated volume of 14.5 fluid ounces by more than 0.08 fluid ounce. Write and solve an absolute value inequality to describe acceptable volumes, and graph the solution on a number line.

Mixed Exercises

Solve. Check your solutions.

21. $8x = 2|6x - 2|$

22. $-6y + 4 = |4y + 12|$

23. $8z + 20 > -|2z + 4|$

24. $-3y - 2 \leq |6y + 25|$

REASONING **Write an absolute value equation to represent each situation. Then solve the equation and discuss the reasonableness of your solution given the constraints of the absolute value equation.**

25. The absolute value of the sum of 4 times a number and 7 is the sum of 2 times a number and 3.

26. The sum of 7 and the absolute value of the difference of a number and 8 is −2 times a number plus 4.

27. **MODELING** A carpenter cuts lumber to the length of 36 inches. For her project, the lumber must be accurate within 0.125 inch.

    a. Write an inequality to represent the acceptable length of the lumber. Explain your reasoning.

    b. Solve the inequality. Then state the maximum and minimum length for the lumber.

28. **SAND** A home improvement store sells bags of sand, which are labeled as weighing 35 pounds. The equipment used to package the sand produces bags with a weight that is within 8 ounces of the labeled weight.

    a. Write an absolute value equation to represent the maximum and minimum weight for the bags of sand.

    b. Solve the equation and interpret the result.

29. CONSTRUCT ARGUMENTS Megan and Yuki are solving the equation $|x - 9| = |5x + 6|$. Megan says that there are 4 cases to consider because there are two possible values for each absolute value expression. Yuki says only 2 cases need to be considered. With which person, do you agree? Will they both get the same solution(s)?

Solve each inequality. Graph the solution set on a number line.

30. $3|2z - 4| - 6 > 12$

31. $6|4p + 2| - 8 < 34$

32. $\dfrac{|5f - 2|}{6} > 4$

33. $\dfrac{|2w + 8|}{5} \geq 3$

34. $-\dfrac{3x|6x + 1|}{5} < 12x$

35. $-\dfrac{7}{8}|2x + 5| > 14$

36. TIRES The recommended inflation of a car tire is no more than 35 pounds per square inch. Depending on weather conditions, the actual reading of the tire pressure could fluctuate up to 3.4 psi. Write and solve an absolute value equation to find the maximum and minimum tire pressure.

37. PROJECTILE An object is launched into the air and then falls to the ground. Its velocity is modeled by the equation $v = 200 - 32t$, where the velocity v is measured in feet per second and time t is measured in seconds. The object's speed is the absolute value of its velocity. Write and solve a compound inequality to determine the time intervals in which the speed of the object will be between 40 and 88 feet per second. Interpret your solution in the context of the situation.

38. USE A SOURCE Research to find a poll with a margin of error. Describe the poll then write an absolute value inequality to represent the actual results.

39. CONSTRUCT ARGUMENTS Roberto claims that the solution to $|3c - 4| > -4.5$ is the same as the solution to $|3c - 4| \geq 0$, because an absolute value is always greater than or equal to zero. Is he correct? Explain your reasoning.

40. WRITE Summarize the difference between *and* compound inequalities and *or* compound inequalities.

41. WHICH ONE DOESN'T BELONG? Identify the compound inequality that does not share the same characteristics as the other three. Justify your conclusion.

$-3 < x < 5$	$x > 2$ and $x < 3$	$x > 5$ and $x < 1$	$x > -4$ and $x > -2$

42. FIND THE ERROR Ana and Ling are solving $|3x + 14| = -6x$. Is either of them correct? Explain your reasoning.

Ana	Ling
$|3x + 14| = -6x$	$|3x + 14| = -6x$
$3x + 14 = -6x$ or $3x + 14 = 6x$	$3x + 14 = -6x$ or $3x + 14 = 6x$
$9x = -14 \qquad 14 = 3x$	$9x = -14 \qquad 14 = 3x$
$x = -\dfrac{14}{9} \qquad x = \dfrac{14}{3}$	$x = -\dfrac{14}{9} \qquad x = \dfrac{14}{3}$

43. PERSEVERE Solve $|2x - 1| + 3 = |5 - x|$. List all cases and resulting equations.

ANALYZE **If *a*, *x*, and *y* are real numbers, determine whether each statement is *sometimes*, *always*, or *never* true. Justify your argument.**

44. If $|a| > 7$, then $|a + 3| > 10$.

45. If $|x| < 3$, then $x + 3 > 0$.

46. If y is between 1 and 5, then $|y - 3| \leq 2$.

47. CREATE Write an absolute value inequality with a solution of $a \leq x \leq b$.

Equations of Linear Functions

Explore Arithmetic Sequences

Online Activity Use a real-world situation to complete the Explore.

> ✕
>
> **@ INQUIRY** How can you write formulas that relate to the numbers in an arithmetic sequence?

Learn Linear Equations in Standard Form

Any linear equation can be written in standard form, $Ax + By = C$, where $A \geq 0$, A and B are not both 0, and A, B, and C are integers with a greatest common factor of 1.

Example 1 Write Linear Equations in Standard Form

Write $y = \frac{2}{5}x + 14$ in standard form. Identify A, B, and C.

$y = \frac{2}{5}x + 14$	Original equation
$-\frac{2}{5}x + y = 14$	Subtract $\frac{2}{5}x$ from each side.
$2x - 5y = -70$	Multiply each side by -5.

$A = 2 \qquad B = -5 \qquad C = -70$

Check

Write $2y = 10x - 16$ in standard form. Identify A, B, and C.

equation in standard form: _____?_____

$A = \underline{\ ?\ } \qquad B = \underline{\ ?\ } \qquad C = \underline{\ ?\ }$

Learn Linear Equations in Slope-Intercept Form

The equation of a linear function can be written in slope-intercept form, $y = mx + b$, where m is the slope and b is the y-intercept.

The slope is $\frac{\text{rise}}{\text{run}} = \frac{2}{3}$. This value can be substituted for m in the slope-intercept form.

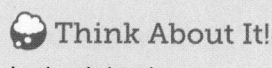

The line intersects the y-axis at 1. This value can be substituted for b in the slope-intercept form.

Go Online You can complete an Extra Example online.

Today's Goals
- Write linear equations in standard form and identify values of A, B, and C.
- Create linear equations in slope-intercept form and by using the coordinates of two points.
- Create linear equations in point-slope form by using two points on the line or the slope and a point on the line.

> 💭 **Think About It!**
>
> Is $-2x + 2y = 2$ written in standard form? Why or why not?

> 💭 **Think About It!**
>
> Is the b in slope-intercept form equivalent to the B in standard form, $Ax + By = C$? If yes, explain your reasoning. If no, provide a counterexample.

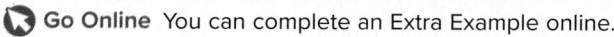

Example 2 Write Linear Equations in Slope-Intercept Form

Write $12x - 4y = 24$ in slope-intercept form. Identify the slope m and y-intercept b.

$$12x - 4y = 24 \qquad \text{Original equation}$$
$$-4y = -12x + 24 \qquad \text{Subtract } 12x \text{ from each side.}$$
$$y = 3x - 6 \qquad \text{Divide each side by } -4.$$
$$m = 3 \qquad b = -6$$

Check

Write $4x = -2y + 22$ in slope-intercept form. $y = -\underline{\ ?\ }x + \underline{\ ?\ }$

🌐 Example 3 Interpret an Equation in Slope-Intercept Form

SHOES The equation $3246x - 2y = -152{,}722$ can be used to estimate shoe sales in Europe from 2010 to 2015, where x is the number of years after 2010 and y is the revenue in millions of dollars.

Part A Write the equation in slope-intercept form.

$$3246x - 2y = -152{,}722 \qquad \text{Original equation}$$
$$-2y = -3246x - 152{,}722 \qquad \text{Subtract } 3246x \text{ from each side.}$$
$$y = 1623x + 76{,}361 \qquad \text{Divide each side by } -2.$$

Part B Interpret the parameters in the context of the situation.

1623 represents that sales increased by $1623 million each year.

76,361 represents that in year 0, or in 2010, sales were $76,361 million.

🌐 Example 4 Use a Linear Equation in Slope-Intercept Form

SMARTPHONES In 2013, there were 1.31 billion smartphone users worldwide. By 2017, there were 2.38 billion smartphone users. Write and use an equation to estimate the number of users in 2025.

Step 1 Define the variables. Because you want to estimate the number of users in 2025, write an equation that represents the number of smartphone users y after x years. Let x be the number of years after 2013 and let y be the number of billions of smartphone users.

Step 2 Find the slope. Since x is the years after 2013, (0, 1.31) and (4, 2.38) represent the number of smartphone users in 2013 and 2017, respectively. Round to the nearest hundredth.

$$m = \frac{2.38 - 1.31}{4 - 0} = 0.27$$

So, the number of users is increasing at a rate of 0.27 billion per year.

🔇 **Go Online** You can complete an Extra Example online.

Study Tip

Assumptions
Assuming that the rate at which the number of smartphone users increases is constant allows us to represent the situation using a linear equation. While the rate at which the number of smartphone users increases may vary each year, using a constant rate allows for a reasonable equation that can be used to estimate future data.

💭 **Think About It!**
When using the equation to estimate the number of smartphone users in the future, what constraint does the world's population place on the possible number of users?

Step 3 Find the *y*-intercept. The *y*-intercept represents the number of smartphone users when $x = 0$, or in 2013. So, $b = 1.31$.

Step 4 Write an equation. Use $m = 0.27$ and $b = 1.31$ to write the equation.

$$y = 0.27x + 1.31 \qquad m = 0.27, b = 1.31$$

Step 5 Estimate. Since 2025 is 12 years after 2013, substitute 12 for *x*.

$$y = 0.27(12) + 1.31; y = 4.55$$

If the trend continues, there will be about 4.55 billion users in 2025.

Suppose the data spanned 2 years instead of 4 years. That is, there were 1.31 billion smartphone users in 2013 and 2.38 billions users in 2015. How would this affect the rate of change abnd your estimate in **Step 5**?

Learn Linear Equations in Point-Slope Form

The equation of a linear function can be written in point-slope form, $y - y_1 = m(x - x_1)$, where *m* is the slope and (x_1, y_1) are the coordinates of a point on the line.

Example 5 Point-Slope Form Given Slope and One Point

Write the equation of a line that passes through (3, −5) and has a slope of 11 in point-slope form.

$$y - y_1 = m(x - x_1) \qquad \text{Point-slope form}$$
$$y - (-5) = 11(x - 3) \qquad m = 11; (x_1, y_1) = (3, -5)$$
$$y + 5 = 11(x - 3) \qquad \text{Simplify.}$$

Check

Write the equation of a line that passes through (13, −5) and has a slope of 4.5 in point-slope form.

Example 6 Point-Slope Form Given Two Points

Write an equation of a line that passes through (1, 1) and (7, 13) in point-slope form.

Step 1 Find the slope.

$$m = \frac{y_2 - y_1}{x_2 - x_1} \qquad \text{Slope formula}$$
$$= \frac{13 - 1}{7 - 1} \qquad (x_1, y_1) = (1, 1); (x_2, y_2) = (7, 13)$$
$$= \frac{12}{6} \qquad \text{Simplify.}$$
$$= 2 \qquad \text{Simplify.}$$

(continued on the next page)

Go Online You can complete an Extra Example online.

Talk About It!

What other values would you need to write the equation of this line in slope-intercept form? Could you determine those values from the given information?

Step 2 Write an equation.

Substitute the slope for m and the coordinates of either of the given points for (x_1, y_1) in the point-slope form.

$$y - y_1 = m(x - x_1)$$ Point-slope form

$$y - 1 = 2(x - 1)$$ $m = 2; (x_1, y_1) = (1, 1)$

Check

Select all the equations for the line that passes through $(-1, 1)$ and $(-2, 13)$.

A. $x - 1 = -12(y + 1)$ **B.** $y - 1 = -12(x + 1)$ **C.** $x + 1 = -12(y - 1)$

D. $y + 1 = -12(x - 1)$ **E.** $y - 2 = -12(x + 13)$ **F.** $x - 2 = -12(y + 13)$

G. $x + 2 = -12(y - 13)$ **H.** $y - 13 = -12(x + 2)$

🌐 Example 7 Write and Interpret a Linear Equation in Point-Slope Form

ARCHITECTURE The Tower of Pisa began tilting during its construction in 1178 and continued to move until a restoration effort reduced the lean and stabilized the structure. The Tower of Pisa leaned 5.4 meters in 1993 compared to a lean of just 1.4 meters in 1350. Write an equation in point-slope form that represents the lean y in meters of the Tower of Pisa x years after its construction in 1178.

Step 1 Find the slope. Round to the nearest hundredth.

The tower was leaning 1.4 meters in 1350, 172 years after 1178.

The tower was leaning 5.4 meters in 1993, 815 years after 1178.

$m = \frac{5.4 - 1.4}{815 - 172} = 0.006$ The lean of the Tower of Pisa increased at a rate of 0.006 meter per year.

Think About It!

Could this equation be used to estimate the lean of the Tower of Pisa for any year? Explain your reasoning.

Step 2 Write an equation.

Substitute the slope for m and the coordinates of either of the given points for (x_1, y_1) in the point-slope form.

$$y - y_1 = m(x - x_1)$$ Point-slope form

$$y - 1.4 = 0.006(x - 172)$$ $m = 0.006; (x_1, y_1) = (172, 1.4)$

Check

SOCIAL MEDIA In 2011, the Miami Marlins had about 11,000 followers on a social media site. In 2016, they had about 240,000 followers. Which equation represents the number of followers y the Miami Marlin's had x years after they joined the site in 2009?

A. $y - 11{,}000 = 45{,}800(x - 2)$ **B.** $y - 45{,}800 = 11{,}000(x - 2)$

C. $y - 11{,}000 = 45{,}800(x - 2011)$ **D.** $y - 2 = 45{,}800(x - 11{,}000)$

🔖 **Go Online** You can complete an Extra Example online.

Practice

Go Online You can complete your homework online.

Example 1

Write each equation in standard form. Identify A, B, and C.

1. $-7x - 5y = 35$

2. $8x + 3y + 6 = 0$

3. $10y - 3x + 6 = 11$

4. $\frac{2}{3}y - \frac{3}{4}x + \frac{1}{6} = 0$

5. $\frac{4}{5}y + \frac{1}{8}x = 4$

6. $-0.08x = 1.24y - 3.12$

Example 2

Write each equation in slope-intercept form. Identify the slope m and the y-intercept b.

7. $6x + 3y = 12$

8. $14x - 7y = 21$

9. $\frac{2}{3}x + \frac{1}{6}y = 2$

10. $5x + 10y = 20$

11. $6x + 9y = 15$

12. $\frac{1}{5}x + \frac{1}{2}y = 4$

Example 3

13. **CHARITY** The linear equation $y - 20x = 83$ relates the number of shirts collected during a charity clothing drive, where x is the number of hours since noon and y is the total number of shirts collected. Write the equation in slope-intercept form and interpret the parameters of the equation in the context of the situation.

14. **GROWTH** Suppose the body length y in inches of a baby snake is given by $4x - 2y = -3$, where x is the age of the snake in months until it becomes 1 year old. Write the equation in slope-intercept form and interpret the parameters of the equation in the context of the situation.

Example 4

15. **PLUMBER** Two neighbors, Camila and Conner, hire the same plumber for household repairs. The plumber worked at Camila's house for 3 hours and charged her $191. The plumber worked at Conner's house for 1 hour and charged him $107.

    a. Define variables to represent the situation.

    b. Find the slope and y-intercept. Then, write an equation.

    c. How much would it cost to hire the plumber for 5 hours of work?

16. **HIKING** Tim began a mountain hike near Big Bear Lake, California at 9:00 A.M. By 10:30 A.M., his elevation is 7200 feet above sea level. At 11:15 A.M., he is at an elevation of 7425 feet above sea level.

    a. Define variables to represent the situation.

    b. Write an equation in slope-intercept form that represent Tim's elevation since he began hiking.

    c. If Tim's altitude continues to increase at the same rate, estimate his altitude at 12:30 P.M.

Example 5

Write an equation in point-slope form for the line that satisfies each set of conditions.

17. slope of -5, passes through $(-3, -8)$ **18.** slope of $\frac{4}{5}$, passes through $(10, -3)$

19. slope of $-\frac{2}{3}$, passes through $(6, -8)$ **20.** slope of 0, passes through $(0, -10)$

Example 6

Write an equation in point-slope form for a line that passes through each set of points.

21. $(2, -3)$ and $(1, 5)$ **22.** $(3, 5)$ and $(-6, -4)$

23. $(-1, -2)$ and $(-3, 1)$ **24.** $(-2, -4)$ and $(1, 8)$

Example 7

Solve each problem.

25. **SALES** Light truck is a vehicle classification for trucks weighing up to 8500 pounds. In 2011, 5.919 million light trucks were sold in the U.S. In 2017, 11.055 million light trucks were sold. Write an equation in point-slope form that represents the number of light trucks y sold x years after 2010.

26. **RESTAURANTS** In 2012, a popular pizza franchise had 2483 restaurants. In 2017, there were 2606 franchised restaurants. Write an equation in point-slope form that represents the number of restaurants y that are franchised x years after 2010.

Mixed Exercises

REGULARITY **Write linear equations in standard form, slope-intercept form, and point-slope form that satisfy each set of conditions.**

27. slope of -2, passes through $(6, -7)$

28. x-intercept: 3, y-intercept: 5

29. passes through $(4, -1)$ and $(8, 3)$

30. slope of 0.6, passes through $(1, 1)$

31. **STATE YOUR ASSUMPTION** The surface of Grand Lake is at an elevation of 648 feet. During a drought, the water level drops at a rate of 3 inches per day. Write an equation in slope-intercept form that gives the elevation in feet y of the surface of Grand Lake after x days. Explain any assumptions you made to write the equation.

32. **USE A SOURCE** Go online to find the population of your city in 2010 and 2015. Write an equation in slope-intercept form to represent the population y of the city x years after 2010. State assumptions you made to write the equation.

33. **PRECISION** Use the graph shown to write the equation of the line in standard form that passes through the two points.

34. **USE ESTIMATION** In May, Jacalyn opens a savings account with an initial balance of $200 and deposits about the same amount each month. The table shows her account balance at the beginning of each month.

Month	June	July	Aug.	Sept.	Oct.
Balance ($)	533.95	871.86	1204.59	1541.55	1882.74

a. Write a linear equation in slope-intercept form that relates the balance of her savings account y in months x since she opened her account.

b. Estimate the balance of Jacalyn's account at the beginning of December.

35. **CONSTRUCT ARGUMENTS** Consider the line that passes through $(3, 1)$ and $(0, 7)$.

a. Given the two points, find a third point on the line. Justify your argument.

b. Explain why both $y - 1 = -2(x - 3)$ and $y - 7 = -2(x - 0)$ can be used to represent the line.

36. USE A MODEL Joe and Alisha are reading novels for book reports. Joe records the number of pages he has remaining to read after each day in the table below. Alisha records the number of pages she has remaining each day on the graph at the right.

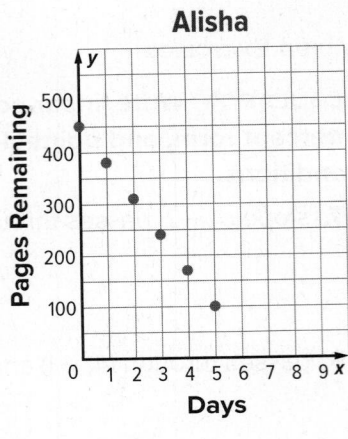

Alisha

Joe

Days	0	1	2	3	4	5
Pages Remaining	585	520	455	390	325	260

a. Describe the function that models the number of pages remaining for each student.

b. What is the y-intercept for each function? Interpret its meaning in the context of the problem.

c. Write a linear equation in slope-intercept form for the function that can be used to model the pages remaining for each student. Then write each equation in standard form.

d. After how many days will each finish with their reading? What feature of the function represents this event? Explain your answer.

e. Who is the faster reader and by how many pages per day? Support your answer.

🧠 **Higher-Order Thinking Skills**

37. CONSTRUCT ARGUMENTS Determine whether an equation in the form $x = a$, where a is a constant is *sometimes, always,* or *never* a function. Explain your reasoning.

38. PERSEVERE Write an equation in point-slope form of a line that passes through $(a, 0)$ and $(0, b)$.

39. CREATE Write an equation in point-slope form of a line with an x-intercept of 3.

40. WRITE Consider the relationship between hours worked and earnings. When would this situation represent a linear relationship? When would this situation represent a nonlinear relationship? Explain your reasoning.

41. FIND THE ERROR Dan claims that since $y = x + 1$ and $y = 3x + 2$ are both linear functions, the function $y = (x + 1)(3x + 2)$ must also be linear. Is he correct? Explain your reasoning.

42. PERSEVERE Write $y = ax + b$ in point-slope form.

43. WRITE Why is it important to be able to represent linear equations in more than one form?

Solving Systems of Equations Graphically

Explore Solutions of Systems of Equations

Online Activity Use graphing technology to complete the Explore.

> **@ INQUIRY** How is the solution of a system of equations represented on a graph? ×

Learn Solving Systems of Equations in Two Variables by Graphing

A **system of equations** is a set of two or more equations with the same variables. One method for solving a system of equations is to graph the related function for each equation on the same coordinate plane. The point of intersection of the two graphs represents the solution.

Today's Goal
• Solve systems of linear equations by graphing.

Today's Vocabulary
system of equations
consistent
inconsistent
independent
dependent

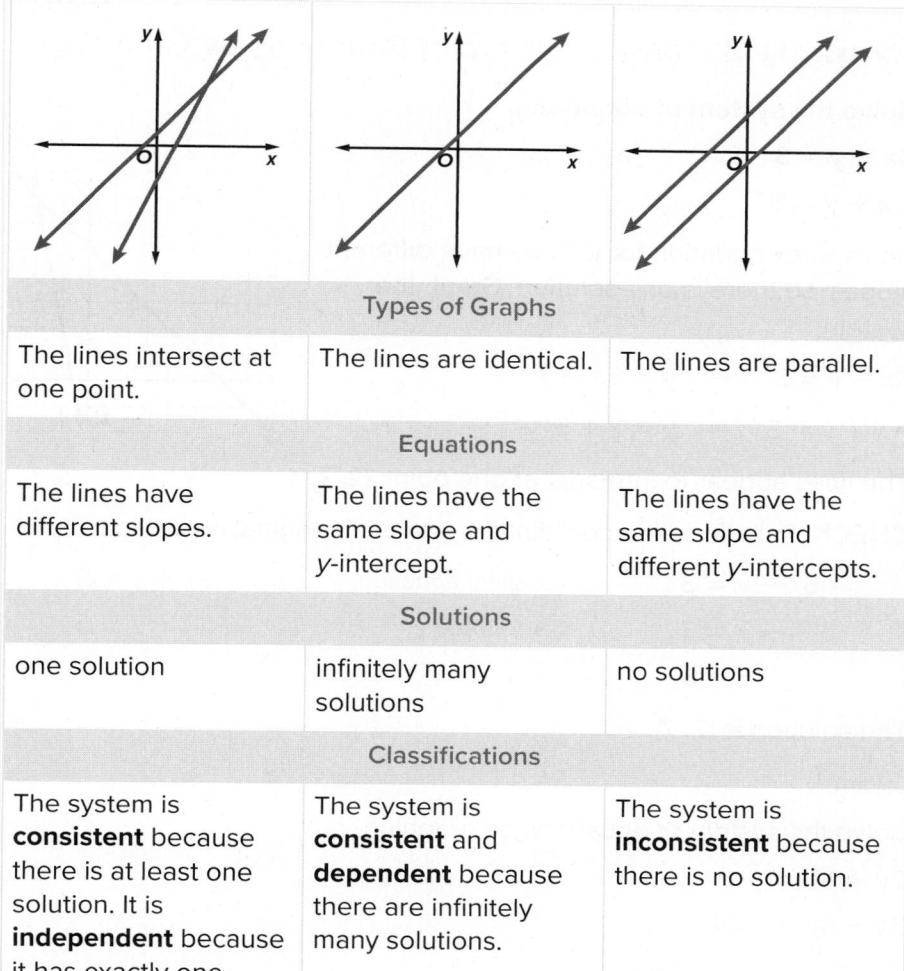

Types of Graphs		
The lines intersect at one point.	The lines are identical.	The lines are parallel.
Equations		
The lines have different slopes.	The lines have the same slope and y-intercept.	The lines have the same slope and different y-intercepts.
Solutions		
one solution	infinitely many solutions	no solutions
Classifications		
The system is **consistent** because there is at least one solution. It is **independent** because it has exactly one solution.	The system is **consistent** and **dependent** because there are infinitely many solutions.	The system is **inconsistent** because there is no solution.

> **Talk About It!**
> Explain why the intersection of the two graphs is the solution of the system of equations.

Example 1 Classify Systems of Equations

Determine the number of solutions the system has. Then state whether the system of equations is *consistent* or *inconsistent* and whether it is *independent* or *dependent*.

$2y = 6x - 14$

$3x - y = 7$

Solve each equation for y.

$2y = 6x - 14 \quad \rightarrow \quad y = 3x - 7$

$3x - y = 7 \quad\quad \rightarrow \quad y = 3x - 7$

The equations have the same slope and y-intercept. Thus, both equations represent the same line and the system has infinitely many solutions. The system is consistent and dependent.

Check

Determine the number of solutions and classify the system of equations.

$3x - 2y = -7$

$4y = 9 - 6x$

Example 2 Solve a System of Equations by Graphing

Solve the system of equations.

$5x - y = 3$

$-x + y = 5$

Solve each equation for y. They have different slopes, so there is one solution. Graph the system.

$5x - y = 3 \quad \rightarrow \quad y = 5x - 3$

$-x + y = 5 \quad \rightarrow \quad y = x + 5$

The lines appear to intersect at one point, (2, 7).

CHECK Substitute the coordinates into each original equation.

	Original equation	
$5x - y = 3$		$-x + y = 5$
$5(2) - 7 \overset{?}{=} 3$	$x = 2$ and $y = 7$	$-(2) + 7 \overset{?}{=} 5$
$3 = 3$	True	$5 = 5$

The solution is (2, 7).

Check

Solve the system of equations by graphing.

$2y + 14x = -6$

$8x - 4y = -24$

The solution is (___?___).

Study Tip

Number of Solutions
By first determining the number of solutions a system has, you can make decisions about whether further steps need to be taken to solve the system. If a system has one solution, you can graph to find it. If a system has infinitely many solutions or no solution, no further steps are necessary. However, you can graph the system to confirm.

Example 3 Solve a System of Equations

Solve the system of equations.

7x + 2y = 16

−21x − 6y = 24

Solve each equation for y to determine the number of solutions the system has.

$7x + 2y = 16 \rightarrow y = -3.5x + 8$

$-21x - 6y = 24 \rightarrow y = -3.5x + -4$

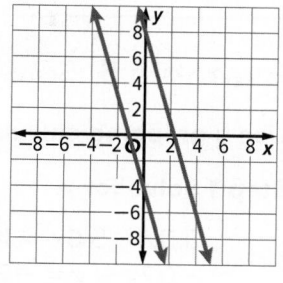

The equations have the same slope and different y-intercepts. So, these equations represent parallel lines, and there is no solution. You can graph each equation on the same grid to confirm that they do not intersect.

Study Tip

Parallel Lines Graphs of lines with the same slope and different intercepts are, by definition, parallel.

🧁 **Think About It!**

What would the graph of a system of linear equations with infinitely many solutions look like? Explain your reasoning.

🌐 Example 4 Write and Solve a System of Equations by Graphing

CARS **Suppose an electric car costs $29,000 to purchase and $0.036 per mile to drive, and a gasoline-powered car costs $19,000 to purchase and $0.08 per mile to drive. Estimate after how many miles of driving the total cost of each car will be the same.**

Part A Write equations for the total cost of owning each type of car.

Let y = the total cost of owning the car and x = the number of miles driven.

So, the equation is y = 0.036x + 29,000 for the electric car and y = 0.08x + 19,000 for the gasoline car.

Part B Examine the graph to estimate the number of miles you would have to drive before the cost of owning each type of car would be same.

The graphs appear to intersect at approximately (225,000, 37,500). This means that after driving about 225,000 miles, the cost of owning each car will be the same.

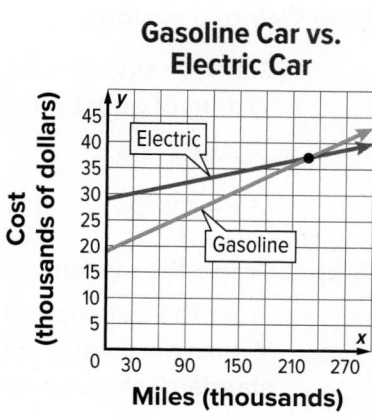

Gasoline Car vs. Electric Car

🧁 **Think About It!**

Explain what the two equations represent in the context of the situation.

(continued on the next page)

CHECK Substitute the coordinates into each original equation.

$$0.036x + 29,000 = y \qquad\qquad 0.08x + 19,000 = y$$

$$0.036(225,000) + 29,000 \overset{?}{=} 37,500 \quad\Big|\quad 0.08(225,000) + 19,000 \overset{?}{=} 37,500$$

$$37,100 \approx 37,500 \qquad\qquad 37,000 \approx 37,500$$

The estimated number of miles makes both equations approximately true. So, our estimate is reasonable.

Example 5 Solve a System by Using Technology

Use a graphing calculator to solve the system of equations.

Step 1 Solve for y.

$$3.5y - 5.6x = 18.2 \quad\rightarrow\quad y = 1.6x + 5.2$$
$$-0.7x - y = -2.4 \quad\rightarrow\quad y = -0.7x + 2.4$$

Step 2 Graph the system.

Enter the equations in the **Y=** list and graph in the standard viewing window.

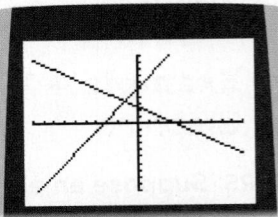

[−10, 10] scl: 1 by [−10, 10] scl: 1

Step 3 Find the intersection.

Use the **intersect** feature from the **CALC** menu to find the coordinates of the point of intersection. When prompted, select each line. Press $\boxed{\textbf{enter}}$ to see the intersection. The solution is approximately $(-1.22, 3.25)$.

Example 6 Solve a Linear Equation by Using a System

Use a graphing calculator to solve $4.5x - 3.9 = 6.5 - 2x$ by using a system of equations.

Step 1 Write a system.

Set each side of $4.5x - 3.9 = 6.5 - 2x$ equal to y to create a system of equations.

$$y = 4.5x - 3.9$$

$$y = 6.5 - 2x$$

Step 2 Graph the system.

Enter the equations in the **Y=** list and graph in the standard viewing window.

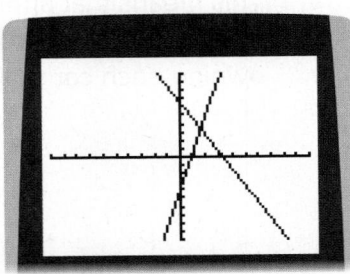

Step 3 Find the intersection.

The solution is the x-coordinate of the intersection, which is 1.6.

[−10, 10] scl: 1 by [−10, 10] scl: 1

 Go Online You can complete an Extra Example online.

Watch Out!

Solving by Graphing Solving a system of equations by graphing does not usually give an exact solution. Remember to substitute the solution into both of the original equations to verify the solution or use an algebraic method to find the exact solution.

Go Online

to see how to use a graphing calculator with Examples 5 and 6.

Study Tip

Window Dimensions If the point of intersection is not visible in the standard viewing window, zoom out or adjust the window settings manually until it is visible. If the lines appear to be parallel, zoom out to verify that they do not intersect.

Practice

Go Online You can complete your homework online.

Example 1

Determine the number of solutions for each system. Then state whether the system of equations is *consistent* or *inconsistent* and whether it is *independent* or *dependent*.

1. $y = 3x$
$y = -3x + 2$

2. $y = x - 5$
$-2x + 2y = -10$

3. $2x - 5y = 10$
$3x + y = 15$

4. $3x + y = -2$
$6x + 2y = 10$

5. $x + 2y = 5$
$3x - 15 = -6y$

6. $3x - y = 2$
$x + y = 6$

Examples 2 and 3

Solve the system of equations by graphing.

7. $x - 2y = 0$
$y = 2x - 3$

8. $-4x + 6y = -2$
$2x - 3y = 1$

9. $2x + y = 3$
$y = \frac{1}{2}x - \frac{9}{2}$

10. $y - x = 3$
$y = 1$

11. $2x - 3y = 0$
$4x - 6y = 3$

12. $5x - y = 4$
$-2x + 6y = 4$

Example 4

Solve each problem.

13. USE ESTIMATION Mr. Lycan is considering buying clay from two art supply companies. Company A sells 50-pound containers of clay for $24, plus $42 to ship the total order. Company B sells the same clay for $28, plus $25 to ship the total order.

   a. Write equations for the total cost of ordering clay from each company.

   b. Graph the equations on the same coordinate plane. Examine the graph to estimate how much Mr. Lycan would have to order for the cost of ordering clay from each company to be the same.

   c. Check your estimate by substituting into each original equation. How reasonable is your estimation? Justify your reasoning.

14. USE ESTIMATION Two moving truck companies offer the same vehicle at different rates. At Haul-n-Save, the truck can be rented for $30, plus $0.79 per mile. At Rent It Trucks, the truck can be rented for $75, plus $0.55 per mile.

   a. Write equations for the total cost of renting a truck from each company.

   b. Graph the equations on the same coordinate plane. Examine the graph to estimate after how many miles of driving the total rental cost will be the same from each company.

   c. Check your estimate. How reasonable is your estimation? Justify your reasoning.

Example 5

USE TOOLS **Use a graphing calculator to solve each system of equations. Round the coordinates to the nearest hundredth, if necessary.**

15. $12y = 5x - 15$
$4.2y + 6.1x = 11$

16. $-3.8x + 2.9y = 19$
$6.6x - 5.4y = -23$

17. $5.8x - 6.3y = 18$
$-4.3x + 8.8y = 32$

Example 6

USE TOOLS **Use a graphing calculator to solve each equation by using a system of equations. Round to the nearest hundredth, if necessary.**

18. $-4.7x + 16 = 16.79x - 80.2$

19. $0.0019x + 3.55 = 0.27x + 2.81$

20. $471 - 63x = -50.5x + 509$

21. $-47.83x - 9 = 33x + 71.019$

Mixed Exercises

Solve each system of equations by graphing.

22. $x - 3y = 6$
$2x - y = -3$

23. $2x - y = 3$
$x + 2y = 4$

24. $4x + y = -2$
$2x + \frac{y}{2} = -1$

25. LASERS A machinist programs a laser cutting machine to focus two laser beams at the same point. One beam is programmed to follow the path $y = 0.5x + -3.15$ and the other is programmed to follow $10x + 5y = 63$. Graph both equations and find the point at which the lasers are focused.

26. REASONING A high school band was selling ride tickets for the school fair. On the first day, 250 children's tickets and 150 adult tickets were sold for a total of $550. On the second day, 180 children's tickets and 120 adult tickets were sold for a total of $420. What is the price for each child ticket and each adult ticket?

   a. Write a system of equations to represent this situation.

   b. Graph the system of equations.

   c. Find the intersection of the graphs. What does the point of intersection represent?

🧠 Higher-Order Thinking Skills

27. ANALYZE For linear functions a, b, and c, if a is consistent and dependent with b, b is inconsistent with c, and c is consistent and independent with d, then a will *sometimes*, *always*, or *never* be consistent and independent with d. Explain your reasoning.

28. WRITE Explain how to find the solution to a system of linear equations by graphing.

29. ANALYZE Determine if the following statement is *sometimes*, *always*, or *never* true. Explain your reasoning.

   A system of linear equations in two variables can have exactly two solutions.

30. CREATE Write a system of equations that has no solution.

Solving Systems of Equations Algebraically

Learn Solving Systems of Equations in Two Variables by Substitution

One algebraic method to solve a system of equations is a process called **substitution,** in which one equation is solved for one variable in terms of the other.

> **Key Concept • Substitution Method**
>
> **Step 1** When necessary, solve at least one equation for one of the variables.
>
> **Step 2** Substitute the resulting expression from Step 1 into the other equation to replace the variable. Then solve the equation.
>
> **Step 3** Substitute the value from Step 2 into either equation, and solve for the other variable. Write the solution as an ordered pair.

Example 1 Substitution When There Is One Solution

Use substitution to solve the system of equations.

$8x - 3y = -1$ Equation 1
$x + 2y = -12$ Equation 2

Step 1 Solve one equation for one of the variables.

Because the coefficient of x in Equation 2 is 1, solve for x in that equation.

$x + 2y = -12$ Equation 2
$x = -2y - 12$ Subtract $2y$ from each side.

Step 2 Substitute the expression. Substitute for x. Then solve for y.

$8x - 3y = -1$ Equation 1
$8(-2y - 12) - 3y = -1$ $x = -2y - 12$
$-16y - 96 - 3y = -1$ Distributive Property
$-19y - 96 = -1$ Simplify.
$-19y = 95$ Add 96 to each side.
$y = -5$ Divide each side by -19.

Step 3 Substitute to solve. Use one of the original equations to solve for x.

$x + 2y = -12$ Equation 2
$x + 2(-5) = -12$ $y = -5$
$x = -2$ Simplify.

The solution is $(-2, -5)$. Substitute into the original equations to check.

 Go Online You can complete an Extra Example online.

Today's Goals
- Solve systems of equations by using the substitution method.
- Solve systems of equations by using the elimination method.

Today's Vocabulary
substitution
elimination

 Go Online
You can watch a video to see how to use algebra tiles to solve a system of equations by using substitution.

Talk About It!
Describe the benefit of solving a system of equations by substitution instead of graphing when the coefficients are not integers.

Check

Use substitution to solve the system of equations.

$-5x + y = -3$

$3x - 8y = 24$

Think About It!

What can you conclude about the slopes and y-intercepts of the equations when a system of equations has no solution? when a system of equations has infinitely many solutions?

Example 2 Substitution When There Is Not Exactly One Solution

Use substitution to solve the system of equations.

$$-5x + 2.5y = -15 \qquad \text{Equation 1}$$
$$y = 2x - 11 \qquad \text{Equation 2}$$

Equation 2 is already solved for y, so substitute $2x - 11$ for y in Equation 1.

$-5x + 2.5y = -15$	Equation 1
$-5x + 2.5(2x - 11) = -15$	$y = 2x - 11$
$-5x + 5x - 27.5 = -15$	Distributive Property
$-27.5 = -15$	False

This system has no solution because $-27.5 = -15$ is not true.

🌐 **Example 3** Apply the Substitution Method

CHEMISTRY **Ms. Washington is preparing a hydrochloric acid (HCl) solution. She will need 300 milliliters of a 5% HCl solution for her class to use during a lab. If she has a 3.5% HCl solution and a 7% HCl solution, how much of each solution should she use in order to make the solution needed?**

Step 1 Write two equations in two variables.

Let x be the amount of 3.5% solution and y be the amount of 7% solution.

$x + y = 300$	Equation 1
$0.035x + 0.07y = 0.05(300)$	Equation 2

Step 2 Solve one equation for one of the variables.

$x + y = 300$	Equation 1
$x = -y + 300$	Subtract y from each side.

Step 3 Substitute the resulting expression and solve.

$0.035x + 0.07y = 15$	Equation 2
$0.035(-y + 300) + 0.07y = 15$	$x = -y + 300$
$-0.035y + 10.5 + 0.07y = 15$	Distributive Property
$0.035y = 4.5$	Simplify.
$y \approx 128.57$	Divide each side by 0.035.

Think About It!

Explain what approximations were made while solving this problem and how they affect the solution.

 **Go Online** You can complete an Extra Example online.

(continued on the next page)

Step 4 Substitute to solve for the other variable.

$$x + y = 300 \qquad \text{Equation 1}$$
$$x + 128.57 \approx 300 \qquad y \approx 128.57$$
$$x \approx 171.43 \qquad \text{Simplify.}$$

The solution of the system is (171.43, 128.57). Ms. Washington should use 171.43 mL of the 3.5% solution and 128.57 mL of the 7% solution.

Learn Solving Systems of Equations in Two Variables by Elimination

Systems of equations may be solved algebraically using **elimination,** which is the process of using addition or subtraction to eliminate one variable.

Key Concept • Elimination Method

Step 1 Multiply one or both of the equations by a number to result in two equations that contain opposite or equal terms.

Step 2 Add or subtract the equations, eliminating one variable. Then solve the equation.

Step 3 Substitute the value from Step 2 into either equation, and solve for the other variable. Write the solution as an ordered pair.

Example 4 Elimination When There Is One Solution

Use elimination to solve the system of equations.

$$-2x - 9y = -25 \qquad \text{Equation 1}$$
$$-4x - 9y = -23 \qquad \text{Equation 2}$$

Step 1 Multiply the equation.

Multiply Equation 2 by −1 to get opposite terms −9y and 9y.

$$-4x - 9y = -23 \quad \boxed{\text{Multiply by } -1.} \longrightarrow \quad 4x + 9y = 23$$

Step 2 Add the equations.

Add the equations to eliminate the y-term and solve for x.

$$-2x - 9y = -25 \qquad \text{Equation 1}$$
$$\underline{(+)\ 4x + 9y = 23} \qquad \text{Equation 2} \times (-1)$$
$$2x \quad\ = -2 \qquad \text{Add the equations.}$$
$$x \quad\ = -1 \qquad \text{Divide each side by 2.}$$

Step 3 Substitute and solve.

$$-4x - 9y = -23 \qquad \text{Substitute } -1 \text{ for } x \text{ in Equation 2.}$$
$$-4(-1) - 9y = -23 \qquad x = -1$$
$$-9y = -27 \qquad \text{Simplify.}$$
$$y = 3 \qquad \text{Divide each side by } -9.$$

The solution of the system is (−1, 3).

Think About It!

When using elimination, when should you add the equations, and when should you subtract the equations?

Think About It!

Describe the benefit of using elimination instead of substitution to solve this problem.

Go Online

You can complete an Extra Example online.

Example 5 Multiply Both Equations Before Using Elimination

Use elimination to solve the system of equations.

$2x + 5y = 1$ Equation 1
$3x - 4y = -10$ Equation 2

Step 1 Multiply one or both equations.

Multiply Equation 1 by 3 and Equation 2 by 2.

$2x + 5y = 1$	Original equations	$3x - 4y = -10$
$3(2x + 5y) = 3(1)$	Multiply.	$2(3x - 4y) = 2(-10)$
$6x + 15y = 3$	Distribute.	$6x - 8y = -20$

Step 2 Eliminate one variable and solve.

In order to eliminate the *x*-terms, subtract the equations. Then, solve for *y*.

$$6x + 15y = 3 \qquad \text{Equation 1} \times 3$$
$$\underline{(-)\ 6x - 8y = -20} \qquad \text{Equation 2} \times 2$$
$$23y = 23 \qquad \text{Subtract the equations.}$$
$$y = 1 \qquad \text{Divide each side by 23.}$$

Step 3 Substitute and solve.

Substitute $y = 1$ in either of the original equations and solve for *x*.

$$2x + 5y = 1 \qquad \text{Equation 1}$$
$$2x + 5(1) = 1 \qquad y = 1$$
$$2x + 5 = 1 \qquad \text{Multiply.}$$
$$x = -2 \qquad \text{Solve for } x.$$

The solution of the system is $(-2, 1)$.

Example 6 Elimination Where There is Not Exactly One Solution

Use elimination to solve the system of equations.

$18x + 21y = 14$ Equation 1
$6x + 7y = 2$ Equation 2

Steps 1 and 2 Multiply one or both equations and add them.

Multiply Equation 2 by -3. Then add the equations.

$18x + 21y = 14$ **Multiply by -3.** $18x + 21y = 14$
$6x + 7y = 2$ $\underline{(+)\ -18x - 21y = -6}$
                                            $0 \neq 8$

Because $0 \neq 8$, this system has no solution.

🖱 **Go Online** You can complete an Extra Example online.

💭 **Think About It!**

Describe the graph of this system of equations.

🖱 **Go Online**

to practice what you've learned about solving systems of equations in the Put It All Together over Lessons 9-4 and 9-5.

Practice

Go Online You can complete your homework online.

Examples 1 and 2

Use substitution to solve each system of equations.

1. $2x - y = 9$
  $x + 3y = -6$

2. $2x - y = 7$
  $6x - 3y = 14$

3. $2x + y = 5$
  $3x - 3y = 3$

4. $3x + y = 7$
  $4x + 2y = 16$

5. $4x - y = 6$
  $2x - \frac{y}{2} = 4$

6. $2x + y = 8$
  $3x + \frac{3}{2}y = 12$

Example 3

Solve each problem.

7. BAKE SALE Cassandra and Alberto are selling pies for a fundraiser. Cassandra sold 3 small pies and 14 large pies for a total of $203. Alberto sold 11 small pies and 11 large pies for a total of $220. Determine the cost of each pie.

   a. Write a system of equations and solve by using substitution.

   b. What does the solution represent in terms of this situation?

   c. How can you verify that the solution is correct?

8. STOCKS Ms. Patel invested a total of $825 in two stocks. At the time of her investment, one share of Stock A was valued at $12.41 and a share of Stock B was valued at $8.62. She purchased a total of 79 shares.

   a. Write a system of equations and solve by substitution.

   b. How many shares of each stock did Ms. Patel buy? How much did she invest in each of the two stocks?

Examples 4-6

Use elimination to solve each system of equations.

9. $3x - 2y = 4$
  $5x + 3y = -25$

10. $5x + 2y = 12$
  $-6x - 2y = -14$

11. $7x + 2y = -1$
  $21x + 6y = -9$

12. $3x - 5y = -9$
  $-7x + 3y = 8$

13. $x - 3y = -12$
  $2x + y = 11$

14. $6w - 8z = 16$
  $3w - 4z = 8$

Mixed Exercises

Use substitution or elimination to solve each system of equations.

15. $0.5x + 2y = 5$
$x - 2y = -8$

16. $h - z = 3$
$-3h + 3z = 6$

17. $-r + t = 5$
$-2r + t = 4$

18. $3r - 2t = 1$
$2r - 3t = 9$

19. $5g + 4k = 10$
$-3g - 5k = 7$

20. $4m - 2p = 0$
$-3m + 9p = 5$

21. The sum of two numbers is 12. The difference of the same two numbers is -4. Find the two numbers.

22. Twice a number minus a second number is -1. Twice the second number added to three times the first number is 9. Find the two numbers.

23. REASONING Mr. Janson paid for admission to the high school football game for his family. He purchased 3 adult tickets and 2 student tickets for a total of $22. Ms. Pham purchased 5 adult tickets and 3 student tickets for a total of $35.75. What is the cost of each adult ticket and each student ticket?

24. USE A MODEL The Newton City Park has 11 basketball courts, which are all in use. There are 54 people playing basketball. Some are playing one-on-one, and some are playing in teams. A one-on-one game requires 2 players, and a team game requires 10 players.

    a. Write a system of equations that represents the number of one-on-one and team games being played.

    b. Solve the system of equations and interpret your results.

25. FIND THE ERROR Gloria and Syreeta are solving the system $6x - 4y = 26$ and $-3x + 4y = -17$. Is either of them correct? Explain your reasoning.

Gloria	
$6x - 4y = 26$	$-3(3) + 4y = -17$
$-3x + 4y = -17$	$-9 + 4y = -17$
$\overline{3x = 9}$	$4y = -8$
$x = 3$	$y = -2$
The solution is $(3, -2)$.	

Syreeta	
$6x - 4y = 26$	$6(-3) - 4y = 26$
$-3x + 4y = -17$	$-18 - 4y = 26$
$\overline{3x = -9}$	$-4y = 44$
$x = -3$	$y = -11$
The solution is $(-3, -11)$.	

26. CREATE Write a system of equations in which one equation should be multiplied by 3 and the other should be multiplied by 4 in order to solve the system with elimination. Then solve your system.

27. WRITE Why is substitution sometimes more helpful than elimination?

Solving Systems of Inequalities

Explore Solutions of Systems of Inequalities

Online Activity Use a graph to complete the Explore.

> **INQUIRY** How is a graph used to determine viable solutions of a system of inequalities? ✕

Learn Solving Systems of Inequalities in Two Variables

A **system of inequalities** is a set of two or more inequalities with the same variables. The **feasible region** is the intersection of the graphs. Ordered pairs within the feasible region are viable solutions. The feasible region may be **bounded**, if the graph of the system is a polygonal region, or **unbounded** if it forms a region that is open.

Key Concept • Solving Systems of Inequalities

Step 1 Graph each inequality by graphing the related equation and shading the correct region.

Step 2 Identify the feasible region that is shaded for all of the inequalities. This represents the solution set of the system.

Example 1 Unbounded Region

Solve the system of inequalities.

$y \leq 4x - 3$ Inequality 1

$-2y > x$ Inequality 2

Use a solid line to graph the first boundary $y = 4x - 3$. The appropriate half-plane is shaded yellow. Use a dashed line to graph the second boundary $y = -0.5x$. The appropriate half-plane is shaded blue.

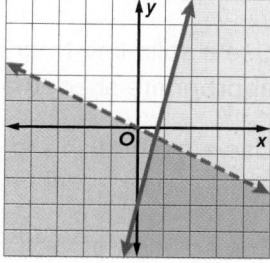

The solution of the system is the set of ordered pairs in the intersection of the graphs shaded in green. The feasible region is unbounded.

(continued on the next page)

Today's Goal
• Solve systems of linear inequalities in two variables.

Today's Vocabulary
system of inequalities
feasible region
bounded
unbounded

Study Tip

Related Equation A related equation of the inequality $y \leq mx + b$ is $y = mx + b$. The inequalities $y < mx + b$, $y \geq mx + b$, and $y > mx + b$ all share this same related equation.

Go Online
You can watch a video to see how to graph a system of linear inequalities.

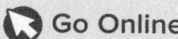

Study Tip

Boundaries The boundaries of inequalities with symbols < and > are graphed using dashed lines, indicating that the ordered pairs on the boundary are not included in the feasible region.

CHECK

Test the solution by substituting the coordinate of a point in the unbounded region, such as (2, −3), into the system of inequalities. If the point is viable for both inequalities, it is a solution of the system.

$y \leq 4x - 3$	Original inequality	$-2(y) > x$
$-3 \overset{?}{\leq} 4(2) - 3$	$x = 2$ and $y = -3$	$-2(-3) \overset{?}{>} 2$
$-3 \leq 5$	True	$6 > 2$

Check

Graph the solution of the system of inequalities.

$y \leq \frac{1}{3}x + 2$

$y > x$

$y \leq 1$

🦫 **Think About It!**

How can you find the coordinates of the vertices of a polygon formed by the system of inequalities?

Example 2 Bounded Region

Solve the system of inequalities.

$y < -\frac{4}{3}x + 5$ Inequality 1
$y \geq x - 2$ Inequality 2
$x > 1$ Inequality 3

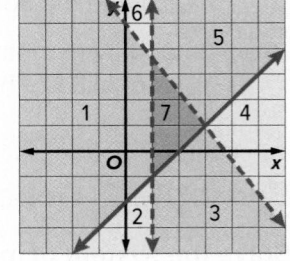

Use a dashed line to graph the first boundary $y = -\frac{4}{3}x + 5$.

The appropriate shaded area contains regions 1, 2, 3, and 7.

Use a solid line to graph the second boundary $y = x - 2$. The appropriate shaded area contains regions 1, 5, 6, and 7.

Use a dashed line to graph the third boundary $x = 1$. The appropriate shaded area contains regions 3, 4, 5 and 7.

The solution of the system is the set of ordered pairs in the intersection of the graphs, represented by region 7. The feasible region is bounded.

🧭 **Go Online** You can complete an Extra Example online.

Check

Graph the solution of the system of inequalities.

$$y \geq \frac{4}{5}x - 3$$

$$y < -\frac{2}{3}x + 2$$

$$x \geq 0$$

🌐 Example 3 Use Systems of Inequalities

TOURS A Niagara Falls boat tour company charges $19.50 for adult tickets and $11 for children's tickets. Each boat has a capacity of 600 passengers, including 8 crew members. Suppose the company's operating cost for one boat tour is $2750. Write and graph a system of inequalities to represent the situation so the company will make a profit on each tour. Then, identify some viable solutions.

Part A Write the system of inequalities.

Let a represent the number of adult tickets and c represent the number of children's tickets.

Inequality 1: $a + c + 8 \leq 600$ Inequality 2: $19.5a + 11c > 2750$

Inequality 3: $a \geq 0$ Inequality 4: $c \geq 0$

Part B Graph the system of inequalities.

Graph the inequalities.

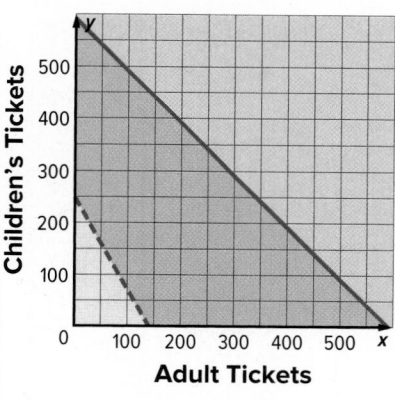

Identify feasible region.

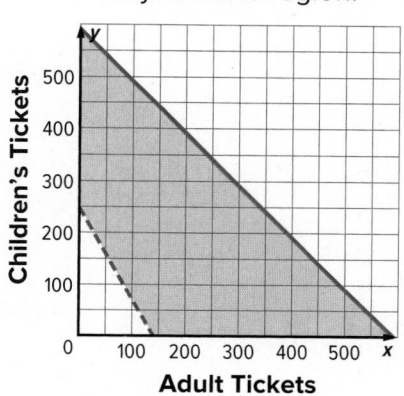

Part C Identify some viable solutions.

Passengers	Viable	Nonviable
60 adults, 100 children	☐	☒
210 adults, 350 children	☒	☐
415 adults, 200 children	☐	☒
390 adults, 240 children	☐	☒
550 adults, 0 children	☒	☐

🧭 **Go Online** You can complete an Extra Example online.

🗨 **Talk About It!**

Why is it important to label the axes given the context of this problem? Explain.

Study Tip

Consider the Context
While the feasible region represents the viable solutions, the solution may be limited to only integers or only positive numbers. In this case, the touring company cannot sell a fraction of a ticket. So the solution must be given as whole numbers.

Check

FUNDRAISER The international club raised $1200 to buy livestock for a community in a different part of the world. The club can buy an alpaca for $160 and a sheep for $120. If the club wants to donate at least 8 animals, determine the system of inequalities to represent the situation.

Part A

Graph of the system of inequalities that represents the possible combinations of animals the club can donate.

Part B

Select all of the viable solutions given the constraints of the club's funds.

A. 0 alpacas, 10 sheep

B. 1 alpaca, 8 sheep

C. 3 alpacas, 6 sheep

D. 6 alpacas, 3 sheep

E. 8 alpacas, 0 sheep

Pause and Reflect

Did you struggle with anything in this lesson? If so, how did you deal with it?

Go Online You can complete an Extra Example online.

Practice

Go Online You can complete your homework online.

Example 1
Solve each system of inequalities.

1. $x - y \leq 2$

   $x + 2y \geq 1$

2. $3x - 2y \leq -1$

   $x + 4y \geq -12$

3. $y \geq \frac{x}{2} - 3$

   $y < 2x$

4. $y < \frac{x}{3} + 2$

   $y < -2x + 1$

5. $x + y \geq 4$

   $2x - y > 2$

6. $x + 3y < 3$

   $x - 2y \geq 4$

Example 2
Solve each system of inequalities.

7. $y \geq -3x + 7$

   $y > \frac{1}{2}x$

   $y < 2$

8. $x > -3$

   $y < -\frac{1}{3}x + 3$

   $y > x - 1$

9. $y < -\frac{1}{2}x + 3$

   $y > \frac{1}{2}x + 1$

   $y < 3x + 10$

10. $y \leq 0$

    $x \leq 0$

    $y \geq -x - 1$

11. $y \leq 3 - x$

    $y \geq 3$

    $x \geq -5$

12. $x \geq -2$

    $y \geq x - 2$

    $x + y \leq 2$

Example 3

13. **TICKETS** The high school auditorium has 800 seats. Suppose that the drama club has a goal of making at least $3400 each night of their spring play. Student tickets are $4 and adult tickets are $7.

    a. Write a system of inequalities to represent the situation.

    b. Graph the system of inequalities. In which quadrant(s) is the solution?

    c. Could the club meet its goal by selling 200 adult and 475 student tickets? Explain.

14. **CONSTRUCT ARGUMENTS** Anthony charges $15 an hour for tutoring and $10 an hour for babysitting. He can work no more than 14 hours a week. How many hours should Anthony spend on each job if he wants to earn at least $125 each week?

    a. Write a system of inequalities to represent this situation.

    b. Graph the system of inequalities and highlight the solution.

    c. Determine whether (4, 5), (7, 6), and (5, 10) are viable solutions given the constraints of the situation. Explain.

Mixed Exercises

Solve each system of inequalities.

15. $y \geq |2x + 4| - 2$
$3y + x \leq 15$

16. $y \geq |6 - x|$
$y \leq 4$

17. $y > -3x + 1$
$4y \leq x - 8$
$3x - 5y < 20$

18. $|x| > y$
$y \leq 6$
$y \geq -2$

19. FINANCE Sheila plans to invest $2000 or less in two different accounts. The low risk account pays 3% annual simple interest, and the high risk account pays 12% annual simple interest. Sheila wants to make at least $150 in interest this year.

  a. Define the variables, then write and graph a system of inequalities to show how Sheila can split her investment between the accounts.

  b. Explain why your graph for this situation is restricted to Quadrant I.

  c. Give three viable solutions to meet the constraints of Sheila's investments.

20. STRUCTURE Write a system of inequalities for the graph shown.

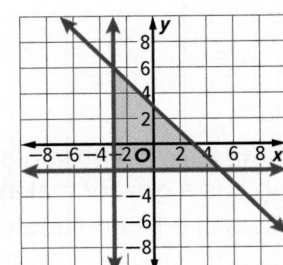

🧠 **Higher Order Thinking Skills**

21. PERSEVERE Find the area of the region defined by the following inequalities.

  $y \geq -4x - 16$
  $4y \leq 26 - x$
  $3y + 6x \leq 30$
  $4y - 2x \geq -10$

22. CREATE Write systems of two inequalities in which the solution:

  a. lies only in the third quadrant.

  b. does not exist.

  c. lies only on a line.

23. ANALYZE Determine whether the statement is *true* or *false*. Justify your argument.

  A system of two linear inequalities has either no points or infinitely many points in its solution.

24. WRITE Explain how you would determine whether $(-4, 6)$ is a solution of a system of inequalities.

Optimization with Linear Programming

Learn Finding Maximum and Minimum Values

Linear programming is the process of finding the maximum or minimum values of a function for a region defined by a system of linear inequalities.

Key Concept • Linear Programming

Step 1 Graph the inequalities.

Step 2 Determine the coordinates of the vertices.

Step 3 Evaluate the function at each vertex.

Step 4 For a bounded region, determine the maximum and minimum. For an unbounded region, test other points within the feasible region to determine which vertex represents the maximum or minimum.

Example 1 Maximum and Minimum Values for a Bounded Region

Graph the system of inequalities. Name the coordinates of the vertices of the feasible region. Find the maximum and minimum values of the function for this region.

$-2 \leq x \leq 4$

$y \leq x + 2$

$y \geq -0.5x - 3$

$f(x, y) = -2x + 6y$

Steps 1 and 2 Graph the inequalities and determine the vertices.

The vertices of the feasible region are $(-2, -2)$, $(-2, 0)$, $(4, -5)$ and $(4, 6)$.

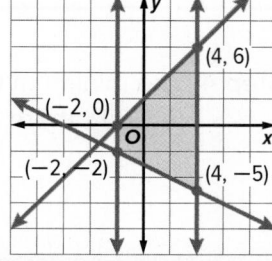

Step 3 Evaluate the function at each vertex.

(x, y)	$-2x + 6y$	$f(x, y)$
$(-2, -2)$	$-2(-2) + 6(-2)$	-8
$(-2, 0)$	$-2(-2) + 6(0)$	4
$(4, -5)$	$-2(4) + 6(-5)$	-38
$(4, 6)$	$-2(4) + 6(6)$	28

Step 4 Determine the maximum and minimum.

The maximum value is 28 at $(4, 6)$. The minimum value is -38 at $(4, -5)$.

Today's Goals
- Find maximum and minimum values of a function over a region.
- Solve real-world optimization problems by graphing systems of inequalities maximizing or minimizing constraints.

Today's Vocabulary
linear programming

optimization

Study Tip

Unbounded Regions An unbounded feasible region does not necessarily contain a maximum or minimum.

Study Tip

Feasible Region To determine the feasible region, you can shade the solution set of each inequality individually, and then find where they all overlap. Shading each inequality using a different color or shading style can help you easily determine the feasible region.

Example 2 Maximum and Minimum Values for an Unbounded Region

Graph the system of inequalities. Name the coordinates of the vertices of the feasible region. Find the maximum and minimum values of the function for this region.

$1 \leq y \leq 3$

$y \leq -x$

$y \geq 0.5x + 3$

$f(x, y) = -x + y$

Steps 1 and 2 Graph the inequalities and determine the vertices.

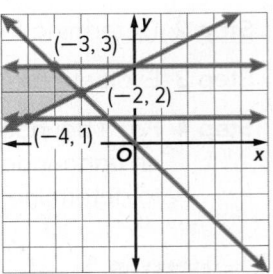

The vertices of the feasible region are

$(-3, 3)$, $(-2, 2)$, and $(-4, 1)$.

Notice that the region is unbounded. This may indicate that there is no minimum or maximum value.

Step 3 Evaluate the function.

Evaluate at each vertex and a point in the feasible region.

(x, y)	$-x + y$	$f(x, y)$
$(-3, 3)$	$-(-3) + 3$	6
$(-2, 2)$	$-(-2) + 2$	4
$(-4, 1)$	$-(-4) + 1$	5
$(-10, 2)$	$-(-10) + 2$	12

Step 4 The minimum value is 4 at $(-2, 2)$. As shown by the test point $(-10, 2)$, there is no maximum value.

Explore Using Technology with Linear Programming

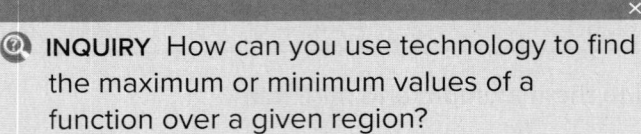

🅡 **Online Activity** Use graphing technology to complete the Explore.

> ❓ **INQUIRY** How can you use technology to find the maximum or minimum values of a function over a given region?

🅡 **Go Online** You can complete an Extra Example online.

Study Tip

Feasible Region To determine whether an unbounded region has a maximum or minimum for the function $f(x, y)$, you need to test several points in the feasible region to see if any values of $f(x, y)$ are greater than or less than the values of $f(x, y)$ for the vertices.

 **Talk About It!**

The function in the Example has a minimum but no maximum on the unbounded region. Is there a function that has a maximum, but no minimum on the region? Does the function $f(x) = y$ have a maximum and/or a minimum on the region? Justify your reasoning.

Learn Linear Programming

Optimization is the process of seeking the optimal value of a function subject to given constraints.

> **Key Concept • Optimization with Linear Programming**
>
> **Step 1** Define the variables.
>
> **Step 2** Write a system of inequalities.
>
> **Step 3** Graph the system of inequalities.
>
> **Step 4** Find the coordinates of the vertices of the feasible region.
>
> **Step 5** Write a linear function to be maximized or minimized.
>
> **Step 6** Evaluate the function at each vertex by substituting the coordinates into the function.
>
> **Step 7** Interpret the results.

🌐 Example 3 Optimizing with Linear Programming

GARDENING Avoree has a 30-square-foot plot in the school greenhouse and wants to plant lettuce and cucumbers while minimizing the amount of water she uses for them. Each cucumber requires 2.25 square feet of space and uses 25 gallons of water over the lifetime of the plant. Each lettuce plant requires 1.5 square feet of space and uses 17 gallons of water. She wants to grow at least 4 of each type of plant and at least 16 plants in total. Determine how many of each plant Avoree should plot in order to minimize her water usage.

Step 1 Define the variables.

Because the number of plants of different types determine the water usage, the independent variables should be the numbers of plants. The dependent variable in the function to be minimized should be total water used. Let c represent the number of cucumber plants and t represent the number of lettuce plants. Let $f(c, t)$ represent the water used for c cucumber plants and t lettuce plants.

Step 2 Write a system of inequalities.

Avoree wants to have at least 4 of each type of plant, so 4 must be included as minimums for both c and t in the inequalities. The total number of plants must be at least 16. Each cucumber requires 2.25 square feet of space and each lettuce plant requires 1.5 square feet of space. The total planting area of the plants must be less than or equal to 30 ft^2.

$c \geq 4$

$t \geq 4$

$c + t \geq 16$

$2.25c + 1.5t \leq 30$

🔖 **Go Online** You can complete an Extra Example online.

Math History Minute

In the 1960s, **Christine Darden (1942—)** became one of the "human computers" who crunched numbers for engineers at NASA's Langley Research Center. After earning a doctorate degree in mechanical engineering, Darden became one of few female aerospace engineers at NASA Langley. For most of her career, her focus was sonic boom minimization.

Step 3 Graph the system of inequalities.

Step 4 Find the coordinates of the vertices of the feasible region.

The vertices of the feasible region are (4, 12), (4, 14), and (8, 8).

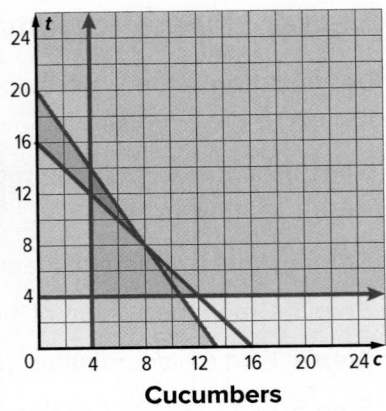

Cucumbers

Step 5 Write a linear function to be minimized.

Because Avoree wants to minimize her water usage, the linear function will be the sum of the water usage for each plant.

$f(c, t) = 25c + 17t$

Step 6 Evaluate the function at each vertex.

(c, t)	25c + 17t	f(c, t)
(4, 12)	25(4) + 17(12)	304
(4, 14)	25(4) + 17(14)	338
(8, 8)	25(8) + 17(8)	336

Step 7 Interpret the results.

Avoree should plant 4 cucumber plants and 12 lettuce plants to minimize her water usage.

Check

SOCCER A new soccer team is being created for a professional soccer league, and they need to hire at least ten new players. They need five to eight defenders and seven to ten forwards, and they want to minimize the amount they spend on these players' salaries so they have enough money remaining to hire goalkeepers and midfielders. Determine the number of forwards f and defenders d they should hire to minimize the cost.

Position	Minimum	Maximum	Salary per Player ($)
forward f	5	8	120,000
defender d	7	10	100,000

The least amount of money that the team can spend is $ ___?___

by hiring __?__ forwards and __?__ defenders.

Think About It!

Does this solution seem reasonable? Explain.

 Go Online You can complete an Extra Example online.

Practice

🜂 **Go Online** You can complete your homework online.

Example 1

Graph the system of inequalities. Name the coordinates of the vertices of the feasible region. Find the maximum and minimum values of the function for this region.

1. $y \geq 2$
$1 \leq x \leq 5$
$y \leq x + 3$
$f(x, y) = 3x - 2y$

2. $y \geq -2$
$y \geq 2x - 4$
$x - 2y \geq -1$
$f(x, y) = 4x - y$

3. $x + y \geq 2$
$4y \leq x + 8$
$y \geq 2x - 5$
$f(x, y) = 4x + 3y$

4. $x \geq 2$
$x \leq 5$
$y \geq 1$
$y \leq 4$
$f(x, y) = x + y$

5. $x \geq 1$
$y \leq 6$
$y \geq x - 2$
$f(x, y) = x - y$

6. $x \geq 0$
$y \geq 0$
$y \leq 7 - x$
$f(x, y) = 3x + y$

Example 2

Graph the system of inequalities. Name the coordinates of the vertices of the feasible region. Find the maximum and minimum values of the function for this region.

7. $x \geq -1$
$x + y \leq 6$
$f(x, y) = x + 2y$

8. $y \leq 2x$
$y \geq 6 - x$
$y \leq 6$
$f(x, y) = 4x + 3y$

9. $y \leq 3x + 6$
$4y + 3x \leq 3$
$x \geq -2$
$f(x, y) = -x + 3y$

Example 3

10. PAINTING A painter has exactly 32 units of yellow dye and 54 units of green dye. He plans to mix as many gallons as possible of color A and color B. Each gallon of color A requires 4 units of yellow dye and 1 unit of green dye. Each gallon of color B requires 1 unit of yellow dye and 6 units of green dye. Find the maximum number of gallons he can mix.
   a. Define the variables and write a system of inequalities.
   b. Graph the system of inequalities and find the coordinates of the vertices of the feasible region.
   c. Find the maximum number of gallons the painter can make.

11. REASONING A jewelry company makes and sells necklaces. For one type of necklace, the company uses clay beads and glass beads. Each necklace has no more than 10 clay beads and at least 4 glass beads. For every necklace, four times the number of glass beads is less than or equal to 8 more than twice the number of clay beads. Each clay bead costs $0.20 and each glass bead costs $0.40. The company wants to find the minimum cost to make a necklace with clay and glass beads and find the combination of clay and glass beads in a necklace that costs the least to make.
   a. Define the variables and write a system of inequalities. Then write an equation for the cost C.
   b. Graph the system of inequalities and find the coordinates of the vertices of the feasible region.
   c. Find the number of clay beads and glass beads in a necklace that costs the least to make.

Mixed Exercises

12. **REASONING** Juan has 8 days to make pots and plates to sell at a local craft fair. Each pot weighs 2 pounds and each plate weighs 1 pound. Juan cannot carry more than 50 pounds to the fair. Each day, he can make at most 5 plates and 3 pots. He will make $12 profit for every plate and $25 profit for every pot that he sells.

   a. Write linear inequalities to represent the number of pots p and plates a Juan can bring to the fair.
   b. List the coordinates of the vertices of the feasible region.
   c. How many pots and plates should Juan make to maximize his potential profit?

13. **USE A MODEL** A trapezoidal park is built on a slight incline. The ground elevation above sea level is given by $f(x, y) = x - 3y + 20$ feet. What are the coordinates of the highest point in the park and what is the elevation at that point?

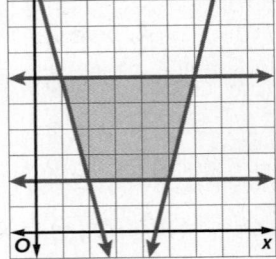

14. **FOOD** A zoo is mixing two types of food for the animals. Each serving is required to have at least 60 grams of protein and 30 grams of fat. Custom Foods has 15 grams of protein and 10 grams of fat and costs 80 cents per unit. Zookeeper's Best contains 20 grams of protein and 5 grams of fat and costs 50 cents per unit.

   a. The zoo wants to minimize their costs. Define the variables and write the inequalities that represent the constraints of the situation.
   b. Graph the inequalities. What does the unbound region represent? Determine how much of each type of food should be used to minimize costs.

15. **ANALYZE** Determine whether the following statement is *sometimes*, *always*, or *never* true. Explain your reasoning.

   An unbounded region will not have both a maximum and minimum value.

16. **WHICH ONE DOESN'T BELONG?** Identify the system of inequalities that is not the same as the other three. Explain your reasoning.

   a. 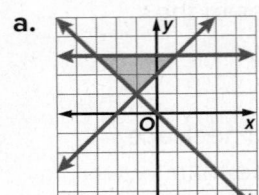   b. 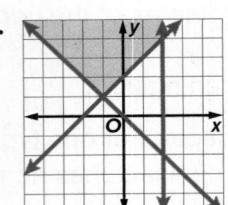   c. 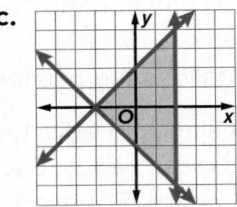   d.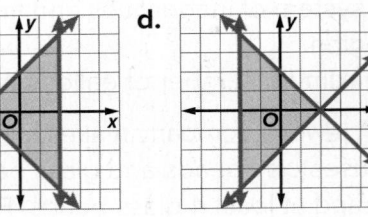

17. **WRITE** Upon determining a bounded feasible region, Kelvin noticed that vertices $A(-3, 4)$ and $B(5, 2)$ yielded the same maximum value for $f(x, y) = 16y + 4x$. Kelvin confirmed that his constraints were graphed correctly and his vertices were correct. Then he said that those two points were not the only maximum values in the feasible region. Explain how this could have happened.

18. **CREATE** Create a set of inequalities that forms a bounded region with an area of 20 units^2 and lies only in the fourth quadrant.

Systems of Equations in Three Variables

Explore Systems of Equations Represented as Lines and Planes

Online Activity Use concrete models to complete the Explore.

> **❓ INQUIRY** How does the way that lines or planes intersect affect the solution of a system of equations? ×

Learn Solving Systems of Equations in Three Variables

The graph of an equation in three variables is a plane. The graph of a system of equations in three variables is an intersection of planes.

- If the three individual planes intersect at a specific point, then there is one solution given as an ordered triple (x, y, z). An **ordered triple** is three numbers given in a specific order to locate a point in space.
- If the planes intersect in a line, every point on the line represents a solution of the system. If they intersect in the same plane, every equation is equivalent and every coordinate in the plane represents a solution of the system.
- If there are no points in common with all three planes, then there is no solution.

Example 1 Solve a System with One Solution

Solve the system of equations.

$$4x + y + 6z = 12 \qquad \text{Equation 1}$$
$$2x - 10y - 2z = 12 \qquad \text{Equation 2}$$
$$3x + 8y + 19z = 38 \qquad \text{Equation 3}$$

Step 1 Eliminate one variable.

Select two of the equations and eliminate one of the variables.

$4x + y + 6z = 12$ Equation 1 $2x - 10y - 2z = 12$ Equation 2

Multiply Equation 1 by 10 to eliminate y.

$4x + y + 6z = 12$ **Multiply by 10.** $40x + 10y + 60z = 120$

Add the equations to eliminate y.

$$
\begin{array}{ll}
\quad 40x + 10y + 60z = 120 & \text{Equation 1} \times 10 \\
(+)\quad 2x - 10y - 2z = 12 & \text{Equation 2} \\
\hline
\quad 42x + 58z = 132 & \text{Add the equations.}
\end{array}
$$

(continued on the next page)

Today's Goal
- Solve systems of linear equations in three variables.

Today's Vocabulary
ordered triple

🗨 Talk About It!

Is it possible for a system of equations in three variables to have exactly three solutions? Justify your argument.

Use a different combination of the original equations to create another equation in two variables. Eliminate y again.

$4x + y + 6z = 12$ Equation 1

$3x + 8y + 19z = 38$ Equation 3

Multiply Equation 1 by -8 and add the equations to eliminate y.

$4x + y + 6z = 12$ **Multiply by -8.** ➡ $-32x - 8y - 48z = -96$

Add the equations to eliminate y.

$-32x - 8y - 48z = -96$	Equation 1 × (−8).
(+) $3x + 8y + 19z = 38$	Equation 3
$-29x + {} - 29z = -58$	Add the equations.

Step 2 Solve the system of two equations.

Multiply the second equation by 2 and add the equations to eliminate z.

$42x + 58z = 132$ $42x + 58z = 132$

$-29x - 29z = -58$ **Multiply by 2.** ➡   (+) $-58x - 58z = -116$

$$-16x = 16$$
$$x = -1$$

Use substitution to solve for z.

$42x + 58z = 132$	Equation in two variables
$42(-1) + 58z = 132$	$x = -1$
$-42 + 58z = 132$	Multiply.
$58z = 174$	Add 42 to each side.
$z = 3$	Divide each side by 58.

The result is $x = -1$ and $z = 3$.

Step 3 Solve for y.

Substitute the two values into one of the original equations to find y.

$4x + y + 6z = 12$	Equation 1
$4(-1) + y + 6(3) = 12$	$x = -1, z = 3$
$-4 + y + 18 = 12$	Multiply.
$y = -2$	Subtract 14 from each side.

The ordered triple is $(-1, -2, 3)$.

◗ **Go Online** You can complete an Extra Example online.

Example 2 Solve a System with Infinitely Many Solutions

Solve the system of equations.

$-x + 5y - 4z = -2$ Equation 1

$4x - 20y + 16z = 8$ Equation 2

$-x - y - z = -1$ Equation 3

Step 1 Eliminate one variable.

Select two of the equations and eliminate one of the variables.

Multiply Equation 1 by 4 and add the equations to eliminate y.

$-x + 5y - 4z = -2$ **Multiply by 4.** ➡ $-4x + 20y - 16z = -8$

Add the equations to eliminate x.

$$-4x + 20y - 16z = -8 \qquad \text{Equation 1} \times 4.$$
$$\underline{(+) \quad 4x - 20y + 16z = 8} \qquad \text{Equation 2}$$
$$0 = 0 \qquad \text{Add the equations.}$$

The equation $0 = 0$ is always true. This indicates that the first two equations represent the same plane.

Step 2 Check the third plane.

Multiply Equation 1 by -1 and add the equations to eliminate x.

$-x + 5y - 4z = -2$ **Multiply by -1.** ➡ $x - 5y + 4z = 2$

Add the equations to eliminate x.

$$x - 5y + 4z = 2 \qquad \text{Equation 1} \times (-1).$$
$$\underline{(+) \quad -x - y - z = -1} \qquad \text{Equation 3}$$
$$-6y + 3z = 1 \qquad \text{Add the equations.}$$

The planes intersect in a line, because the resultant equation is in two variables. So, there are an infinite number of solutions.

Check

Solve each system of equations.

$3x - 18y + 6z = 7$ $3x - y - z = 2$
$7x - 2y + z = 1$ $-4x - 2y + 3z = 19$
$-x + 6y - 2z = -2$ $5x + 3y + z = 8$

Go Online to see Example 3.

😎 **Think About It!**

Is it possible to have two planes that coincide and yet the system of three equations has no solution? Explain.

🐦 **Go Online**

You can complete an Extra Example online.

🌐 Example 4 Write and Solve a System of Equations

MUSEUM MEMBERSHIPS In 2016, Dali Museum in St. Petersburg, Florida offered individual, dual, and family memberships, which cost $60, $80, and $100, respectively. Suppose in one month the museum sells a total of 81 new memberships, for a total of $6420. The number of dual memberships purchased is twice that of individual memberships. Write and solve a system of equations to determine the number of new individual memberships x, dual memberships y, and family memberships z.

Step 1 Write the system of equations.

a total of 81 new memberships: $x + y + z = 81$

The number of dual memberships purchased is twice that of individual memberships: $y = 2x$

individual, dual, and family memberships, which cost $60, $80, and $100, respectively, for a total of $6420: $60x + 80y + 100z = 6420$

Step 2 Eliminate one variable.

Substitute $y = 2x$ into Equation 1 and Equation 3 to eliminate y.

$x + y + z = 81$	Equation 1
$x + 2x + z = 81$	$y = 2x$
$3x + z = 81$	Add.

$60x + 80y + 100z = 6420$	Equation 3
$60x + 80(2x) + 100z = 6420$	$y = 2x$
$220x + 100z = 6420$	Simplify.

Step 3 Solve the resulting system of two equations.

$-300x - 100z = -8100$	Multiply new Equation 1 by -100.
$(+)\ 220x + 100z = 6420$	
$-80x = -1680$	Add to eliminate z.
$x = 21$	Solve for x.

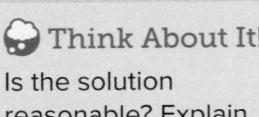
Think About It!

Is the solution reasonable? Explain.

Step 4 Substitute to find z.

$3x + z = 81$	Remaining equation in two variables
$3(21) + z = 81$	$x = 21$
$z = 18$	Simplify.

Step 5 Substitute to find y.

$y = 2x$	Equation 2
$y = 2(21)$	$x = 21$
$y = 42$	Multiply.

The solution is (21, 42, 18). So, the museum sold 21 individual memberships, 42 dual memberships, and 18 family memberships.

🕐 **Go Online** You can complete an Extra Example online.

Practice

Go Online You can complete your homework online.

Examples 1–3

Solve each system of equations.

1. $2x + 3y - z = 0$

$x - 2y - 4z = 14$

$3x + y - 8z = 17$

2. $2p - q + 4r = 11$

$p + 2q - 6r = -11$

$3p - 2q - 10r = 11$

3. $a - 2b + c = 8$

$2a + b - c = 0$

$3a - 6b + 3c = 24$

4. $3s - t - u = 5$

$3s + 2t - u = 11$

$6s - 3t + 2u = -12$

5. $2x - 4y - z = 10$

$4x - 8y - 2z = 16$

$3x + y + z = 12$

6. $p - 6q + 4r = 2$

$2p + 4q - 8r = 16$

$p - 2q = 5$

7. $2a + c = -10$

$b - c = 15$

$a - 2b + c = -5$

8. $x + y + z = 3$

$13x + 2z = 2$

$-x - 5z = -5$

9. $2m + 5n + 2p = 6$

$5m - 7n = -29$

$p = 1$

10. $f + 4g - h = 1$

$3f - g + 8h = 0$

$f + 4g - h = 10$

11. $-2c = -6$

$2a + 3b - c = -2$

$a + 2b + 3c = 9$

12. $3x - 2y + 2z = -2$

$x + 6y - 2z = -2$

$x + 2y = 0$

Example 4

13. ANIMAL NUTRITION A veterinarian wants to make a food mix for guinea pigs that contains 23 grams of protein, 6.2 grams of fat, and 16 grams of moisture. The composition of three available mixtures are shown in the table. How many grams of each mix should be used to make the desired new mix?

	Protein (g)	Fat (g)	Moisture (g)
Mix A	0.2	0.02	0.15
Mix B	0.1	0.06	0.10
Mix C	0.15	0.05	0.05

14. ENTERTAINMENT At the arcade, Marcos, Sara, and Darius played video racing games, pinball, and air hockey. Marcos spent $6 for 6 racing games, 2 pinball games, and 1 game of air hockey. Sara spent $12 for 3 racing games, 4 pinball games, and 5 games of air hockey. Darius spent $12.25 for 2 racing games, 7 pinball games, and 4 games of air hockey. How much did each of the games cost?

15. FOOD A natural food store makes its own brand of trail mix from dried apples, raisins, and peanuts. A one-pound bag of the trail mix costs $3.18. It contains twice as much peanuts by weight as apples. If a pound of dried apples costs $4.48, a pound of raisins is $2.40, and a pound of peanuts is $3.44, how many ounces of each ingredient are contained in 1 pound of the trail mix?

Lesson 9-8 · Systems of Equations in Three Variables **557**

Mixed Exercises

Solve each system of equations.

16. $-x - 5z = -5$

$y - 3x = 0$

$13x + 2z = 2$

17. $-3x + 2z = 1$

$4x + y - 2z = -6$

$x + y + 4z = 3$

18. $x - y + 3z = 3$

$-2x + 2y - 6z = 6$

$y - 5z = -3$

19. REASONING A newspaper company has three printing presses that together can produce 3500 newspapers each hour. The fastest printer can print 100 more than twice the number of papers as the slowest press. The two slower presses combined produce 100 more papers than the fastest press. How many newspapers can each printing press produce in 1 hour?

20. USE A SOURCE A shop is having a sale on pool accessories. The table shows the orders of three customers and their total price before tax. Research the sales tax in your area to determine whether a customer who has $200 could buy 1 chlorine filter, 1 raft, and 1 large lounge chair after sales tax is applied. Justify your response.

Combo	Price Before Tax
1 Raft and 2 Chlorine Filters	$220
1 Chlorine Filter and 2 Large Lounge Chairs	$245
1 Raft and 4 Large Lounge Chairs	$315

21. TICKETS Three kinds of tickets are available for a concert: orchestra seating, mezzanine seating, and balcony seating. The orchestra tickets cost $2 more than the mezzanine tickets, while the mezzanine tickets cost $1 more than the balcony tickets. Twice the cost of an orchestra ticket is $1 less than 3 times the cost of a balcony ticket. Determine the price of each kind of ticket.

22. CONSTRUCT ARGUMENTS Consider the following system. Prove that if $b = c = -a$, then $ty = a$.

$$rx + ty + vz = a$$
$$rx - ty + vz = b$$
$$rx + ty - vz = c$$

23. PERSEVERE The general form of an equation for a parabola is $y = ax^2 + bx + c$, where (x, y) is a point on the parabola. If three points on a parabola are $(2, -10)$, $(-5, -101)$, and $(6, -90)$, determine the values of a, b, and c and write the equation of the parabola in general form.

24. WRITE Use your knowledge of solving a system of three linear equations with three variables to explain how to solve a system of four equations with four variables.

25. CREATE Write a system of three linear equations that has a solution of $(-5, -2, 6)$. Show that the ordered triple satisfies all three equations.

Solving Absolute Value Equations and Inequalities by Graphing

Today's Goals
• Solve absolute value equations.
• Solve absolute value inequalities.

Learn Solving Absolute Value Equations by Graphing

The graph of a related function can be used to solve an absolute value equation. The graph of an absolute value function may intersect the x-axis once or twice, or it may not intersect it at all. The number of times the graph intersects the x-axis corresponds to the number of solutions of the equation.

two solutions	one solution	no solution
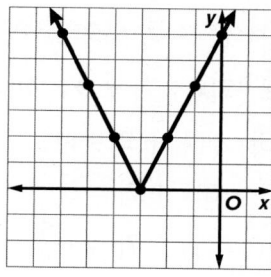		

Example 1 Solve an Absolute Value Equation by Graphing

Solve $5 + |2x + 6| = 5$ by graphing.

Step 1 Find the related function.

Rewrite the equation with 0 on the right side.

$$5 + |2x + 6| = 5 \qquad \text{Original equation}$$

$$|2x + 6| = 0 \qquad \text{Subtract 5 from each side.}$$

Replacing 0 with $f(x)$ gives the related function $f(x) = |2x + 6|$.

Step 2 Graph the related function.

Make a table of values for $f(x) = |2x + 6|$. Then graph the ordered pairs and connect them.

x	f(x)
−6	6
−5	4
−4	2
−3	0
−2	2
−1	4
0	6

Since the graph of $f(x) = |2x + 6|$ only intersects the x-axis at −3, the equation has one solution. The solution set of the equation is {−3}.

⯈ **Go Online** You can complete an Extra Example online.

😮 Think About It!
How could you use the table of values to solve the equation?

Check

Solve $|x + 1| + 9 = 13$ by graphing.

Part A Graph the related function.

Part B What is the solution set of the equation?

Go Online
to see how to use a
graphing calculator with
Examples 2 and 3.

Example 2 Solve an Absolute Value Equation by Using Technology

Use a graphing calculator to solve $\frac{4}{5}|x - 1| + 8 = 11$.

Rewrite the equation results in the related function as the system $f(x) = \frac{4}{5}|x - 1| + 8$ and $g(x) = 11$.

Enter the functions in the **Y =** list and graph. To enter the absolute value symbols, press [math] and select **abs(** from the **NUM** menu.

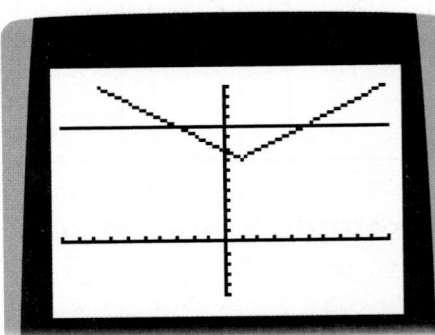

Use the **intersect** feature from the **CALC** menu to find the x-coordinates of the points of intersection. When prompted, use the arrow keys to move the cursor close to each point of intersection and press [enter] three times.

[−10, 10] scl: 1 by [−5, 15] scl: 1

Go Online
An alternate method is
available for this
example.

The graphs intersect where $x = -2.75$ and $x = 4.75$. So, the solution set of the equation is $\{-2.75, 4.75\}$.

Check

Use a graphing calculator to solve $-\left|\frac{2}{3}x + 5\right| - 16 = -10$.

Example 3 Confirm Solutions by Using Technology

Solve $-3|x + 7| + 9 = 14$. Check your solutions graphically.

$-3|x + 7| + 9 = 14$	Original equation
$-3|x + 7| = 5$	Subtract 9 from each side.
$|x + 7| = -\frac{5}{3}$	Divide each side by −3.

Because the absolute value of a number is always positive or zero, this sentence is *never* true. The solution is ∅.

(continued on the next page)

Go Online You can complete an Extra Example online.

Use a graphing calculator to confirm that there is no solution.

Step 1 Find and graph the related function.

Rewriting the equation results in the related function $f(x) = -3|x + 7| - 5$. Enter the related function in the **Y=** list and graph.

Step 2 Find the zeros.

The graph does not appear to intersect the x-axis. Use the **ZOOM** feature or adjust the window manually to see this more clearly. Since the related function never intersects the x-axis, there are no real zeros. This confirms that the equation has no solution.

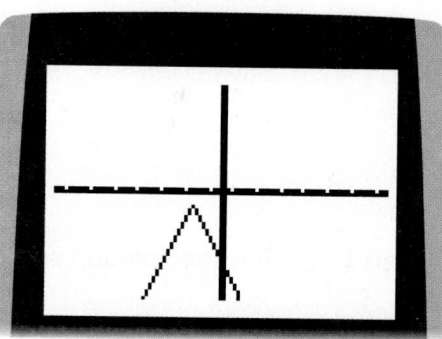

$[-40, 40]$ scl: 1 by $[-40, 40]$ scl: 1

💭 **Think About It!**

How could you use a calculator to confirm the solutions of an equation with one or two real solutions?

Learn Solving Absolute Value Inequalities by Graphing

The related functions of absolute value inequalities are found by solving the inequality for 0, replacing the inequality symbol with an equals sign, and replacing 0 with $f(x)$.

For $<$ and $\leq$, identify the x-values for which the graph of the related function lies below the x-axis. For $\leq$, include the x-intercepts in the solution. For $>$ and $\geq$, identify the x-values for which the graph of the related function lies *above* the x-axis. For $\geq$, include the x-intercepts in the solution.

Example 4 Solve an Absolute Value Inequality by Graphing

Solve $|3x - 9| - 6 \leq 0$ by graphing.

The solution set consists of x-values for which the graph of the related function lies *below* the x-axis, including the x-intercepts. The related function is $f(x) = |3x - 9| - 6$. Graph $f(x)$ by making a table.

x	f(x)
0	3
1	0
2	−3
3	−6
4	−3
5	0
6	3

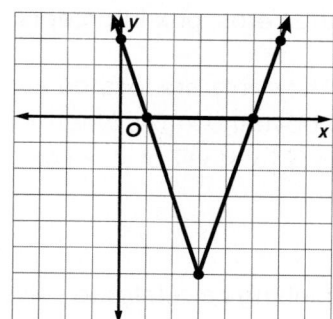

💬 **Talk About It!**

Would the solution change if the inequality symbol was changed from $\leq$ to $\geq$? Explain your reasoning.

The graph lies below the x-axis between $x = 1$ and $x = 5$. Thus, the solution set is $\{x \mid 1 \leq x \leq 5\}$ or $[1, 5]$. All values of x between 1 and 5 satisfy the constraints of the original inequality.

▶ **Go Online** You can complete an Extra Example online.

Check

Solve $|x - 2| - 3 \leq 0$ by graphing.

Part A Graph a related function.

Part B What is the solution set of $|x - 2| - 3 \leq 0$?

💭 Think About It!

The inequality in the example is solved for 0. What additional step(s) would you need to take if the given inequality had a nonzero term on the right side of the inequality symbol?

🔗 Go Online

to see how to use a graphing calculator with this example.

Example 5 Solve an Absolute Value Equation by Using Technology

Use a graphing calculator to solve $\left|\frac{5}{7}x + 2\right| - 3 > 0$.

Step 1 Graph the related function.

Rewriting the inequality results in the related function $f(x) = \left|\frac{5}{7}x + 2\right| - 3$.

Step 2 Find the zeros.

The > symbol indicates that the solution set consists of x-values for which the graph of the related function lies *above* the x-axis, not including the x-intercepts.

Use the **zero** feature from the **CALC** menu to find the zeros, or x-intercepts.

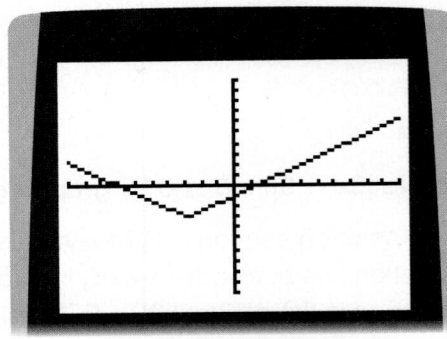

$[-10, 10]$ scl: 1 by $[-10, 10]$ scl: 1

The zeros are located at $x = -7$ and $x = 1.4$. The graph lies above the x-axis when $x < -7$ and $x > 1.4$. So the solution set is $\{x \mid x < -7 \text{ or } x > 1.4\}$.

Check

Use a graphing calculator to solve $\frac{1}{2}|4x + 1| - 5 > 0$.

🔗 **Go Online** You can complete an Extra Example online.

Practice

Go Online You can complete your homework online.

Example 1

Solve each equation by graphing.

1. $|x - 4| = 5$

2. $|2x - 3| = 17$

3. $3 + |2x + 1| = 3$

4. $|x - 1| + 6 = 4$

5. $7 + |3x - 1| = 7$

6. $|x + 2| + 5 = 13$

Example 2

USE TOOLS Use a graphing calculator to solve each equation.

7. $\frac{1}{2}|x - 1| + 5 = 9$

8. $\frac{3}{4}|x + 1| + 1 = 7$

9. $\frac{2}{3}|x - 2| - 4 = 4$

10. $2|x + 2| = 10$

11. $\frac{1}{5}|x + 6| - 1 = 9$

12. $3|x + 5| - 1 = 11$

Example 3

USE TOOLS Solve each equation algebraically. Use a graphing calculator to check your solutions.

13. $|3x - 6| = 42$

14. $7|x + 3| = 42$

15. $-3|4x - 9| = 24$

16. $-6|5 - 2x| = -9$

17. $5|2x + 3| - 5 = 0$

18. $|15 - 2x| = 45$

Example 4

Solve each inequality by graphing.

19. $|2x - 6| - 4 \leq 0$

20. $|x - 1| - 3 \leq 0$

21. $|2x - 1| \geq 4$

22. $|3x + 2| \geq 6$

23. $2|x + 2| < 8$

24. $3|x - 1| < 12$

Example 5

USE TOOLS Use a graphing calculator to solve each inequality.

25. $\left|\frac{1}{4}x + 4\right| - 1 > 0$

26. $\frac{2}{5}|x - 5| + 1 > 0$

27. $|3x - 1| < 2$

28. $|4x + 1| \leq 1$

29. $\frac{1}{6}|x - 1| + 1 \leq 0$

30. $\frac{1}{4}|x + 5| - 1 \leq 1$

Mixed Exercises

Solve by graphing.

31. $0.4|x - 1| = 0.2$

32. $0.16|x + 1| = 4.8$

33. $0.78|2x + 0.1| + 2.3 = 0$

34. $\left|\frac{1}{3}x + 3\right| + 1 = 0$

35. $\frac{1}{2}|6 - 2x| \leq 1$

36. $|3x - 2| < \frac{1}{2}$

USE TOOLS **Solve each equation or inequality algebraically. Use a graphing calculator to check your solutions.**

37. $\left|\frac{5}{9}x + 1\right| - 5 > 0$

38. $\frac{2}{7}\left|\frac{1}{2}x - 1\right| < 1$

39. $0.28|0.4x - 2| = 10.08$

40. REASONING A pet store sells bags of dog food that are labeled as weighing 50 pounds. The equipment used to package the dog food produces bags with a weight that is within 0.75 lbs of the advertised weight. Write an absolute value equation to determine the acceptable maximum and minimum weight for the bags of dog food. Then, use a graph to find the minimum and maximum weights.

41. SPACE The mean distance from Mars to the Sun is 1.524 astronomical units (au). The distance varies during the orbit of Mars by 0.147 au. Write an absolute value inequality to represent the distance of Mars from the Sun as it completes an orbit. Then, use a graph to solve the inequality.

42. CONSTRUCT ARGUMENTS Ms. Uba asked her students to write an absolute value equation with the solutions and related function shown in the graph. Sawyer wrote $|x + 2| = 6$ and Kaleigh wrote $2|x + 2| = 12$. Is either student correct? Justify your argument.

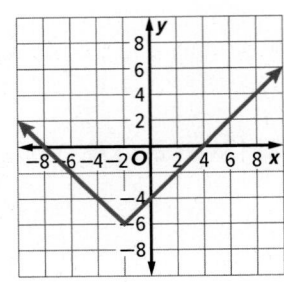

43. WRITE Compare and contrast the solution sets of absolute value equations and absolute value inequalities.

44. ANALYZE How can you tell that an absolute value equation has no solutions without graphing or completely solving it algebraically?

45. CREATE Create an absolute value equation for which the solution set is {9, 11}.

Essential Question

How are equations, inequalities, and systems of equations or inequalities best used to model to real-world situations?

Equations allow you to find quantities that fit constraints, such as costs for given numbers of items. Inequalities represent situations involving intervals of numbers, such as quantities for which cost is below a given value. Systems allow you to have multiple constraints for the same situation.

Module Summary

Lessons 9-1 and 9-2

Linear and Absolute Value Equations and Inequalities

- The solution of an equation or an inequality is any value that, when substituted into the equation, results in a true statement.

- An absolute value equation or inequality is solved by writing it as two cases. An absolute value equation may have 0, 1, or 2 solutions.

Lesson 9-3

Equations of Linear Functions

- Standard form: $Ax + By = C$, where A, B, and C are integers with a greatest common factor of 1, $A \geq 0$, and A and B are not both 0

- Slope-intercept form: $y = mx + b$, where m is the slope and b is the y-intercept

- Point-slope form: $y - y_1 = m(x - x_1)$, where m is the slope and (x_1, y_1) are the coordinates of a point on the line

Lessons 9-4 through 9-8

Systems of Equations and Inequalities

- The point of intersection of the two graphs a system of equations represents the solution.

- In the substitution method, one equation is solved for a variable and substituted to find the value of another variable.

- In the elimination method, one variable is eliminated by adding or subtracting the equations.

- To solve a system of inequalities, graph each inequality. Viable solutions to the system of inequalities are in the intersection of all the graphs.

- Linear programming is a method for finding maximum or minimum values of a function over a given system of inequalities with each inequality representing a constraint.

- Systems of equations in three variables can have infinitely many solutions, no solution, or one solution which is written as an ordered triple (x, y, z).

- Systems of equations in three variables can be solved by using elimination and substitution.

Lesson 9-9

Solving Absolute Value Equations and Inequalities by Graphing

- The graph of the related absolute value function can be used to solve an equation or inequality. The x-intercept(s) of the function give the solution(s) of the equation.

Study Organizer

 Foldables

Use your Foldable to review this module. Working with a partner can be helpful. Ask for clarification of concepts as needed.

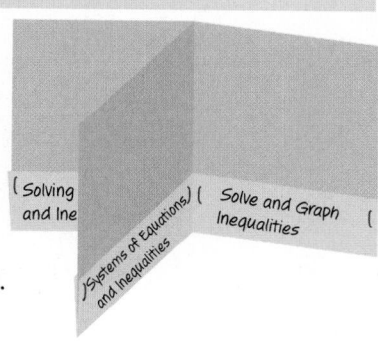

Test Practice

1. **OPEN RESPONSE** Explain each step in solving the equation $3(2x + 9) + \frac{1}{2}(4x - 8) = 55$. (Lesson 9-1)

2. **MULTIPLE CHOICE** In a single football game, a team scored 14 points from touchdowns and made x field goals, which are worth three points each. The team scored 23 points. How many field goals did the team make? (Lesson 9-1)

   A. 2

   B. 3

   C. 4

   D. 5

3. **OPEN RESPONSE** A temperature in degrees Celsius, C, is equal to five-ninths times the difference of the temperature in degrees Fahrenheit, F, and 32. Write an equation to relate the temperature in degrees Celsius to the temperature in degrees Fahrenheit. Then determine the Fahrenheit temperature that is equivalent to 25°C. (Lesson 9-1)

4. **OPEN RESPONSE** The temperature of an oven varies by as much as 7.5 degrees from the temperature shown on its display. If the oven is set to 425°, write an equation to find the minimum and maximum actual temperature of the oven. (Lesson 9-2)

5. **OPEN RESPONSE** The equation $2x + y = 10$ can be used to find the number of miles Allie has left to jog this week to reach her goal, where x is the number of days and y is the number of miles left to reach the goal. Write the equation in slope-intercept form. Then interpret the parameters in the context of the situation. (Lesson 9-3)

6. **MULTIPLE CHOICE** Thomas is driving his truck at a constant speed. The table below gives the distance remaining to his destination y in miles x minutes after he starts driving.

x	15	30	45
y	186.5	173	159.5

   Which equation models the distance remaining after any number of minutes? (Lesson 9-3)

   A. $y = \frac{9}{10}x - 200$

   B. $y = -\frac{10}{9}x + 200$

   C. $y = \frac{10}{9}x - 200$

   D. $y = -\frac{9}{10}x + 200$

7. **MULTIPLE CHOICE** Describe the system of equations shown in the graph. (Lesson 9-4)

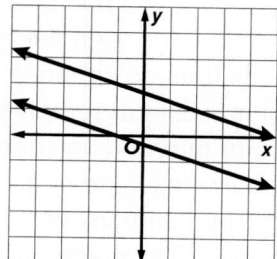

A. It has no solution, and is consistent and dependent.

B. It has no solution, and is inconsistent.

C. It has infinitely many solutions, and is consistent and dependent.

D. It has infinitely many solutions, and is inconsistent.

8. **MULTIPLE CHOICE** Last season, the volleyball team paid $5 for each pair of socks and $17 for each pair of shorts. They spent a total of $315. This season, the price of socks is $6 per pair and shorts are $18. The team spent $342 for the same number of socks and shorts as the previous season. How many pairs of socks and shorts did the team buy each season? (Lesson 9-5)

A. 12 pairs of socks; 15 shorts

B. 15 pairs of socks; 12 shorts

C. 24 pairs of socks; 11 shorts

D. 29 pairs of socks; 10 shorts

9. **OPEN RESPONSE** Solve this system of equations.

$5x - 2y = 1$
$2x + 8y = 7$

Round to the nearest hundredth, if necessary.
(Lesson 9-5)

10. **MULTIPLE CHOICE** A choir concert will be held in a venue with 500 seats. The goal of the choir director is to make at least $3250. Adult tickets cost $8 and student tickets cost $5. Select the system of inequalities that represents the situation, where x is the number of adult tickets sold and y is the number of student tickets sold. (Lesson 9-6)

A. $x + y \geq 3250$
   $8x + 5y \leq 500$

B. $x + y \leq 3250$
   $8x + 5y \geq 500$

C. $x + y \geq 500$
   $8x + 5y \leq 3250$

D. $x + y \leq 500$
   $8x + 5y \geq 3250$

11. **OPEN RESPONSE** Jazmine earns $20 per hour for landscaping and $15 per hour for painting. She can work no more than 12 hours per week. If Jazmine wants to earn at least $225 in a week, could she landscape for 6 hours and paint for 6 hours to meet her goal? Justify your response. (Lesson 9-6)

12. MULTIPLE CHOICE Which system of inequalities represents the graph shown? (Lesson 9-6)

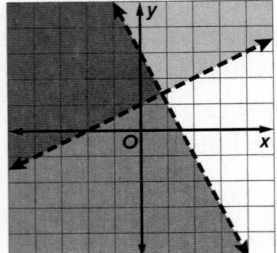

A. $x - 2y < -2$
   $2x + y < 3$

B. $x - 2y < -2$
   $2x + y > 3$

C. $x - 2y > -2$
   $2x + y < 3$

D. $x - 2y > -2$
   $2x + y > 3$

13. MULTI-SELECT The shaded region represents the feasible region for a linear programming problem.

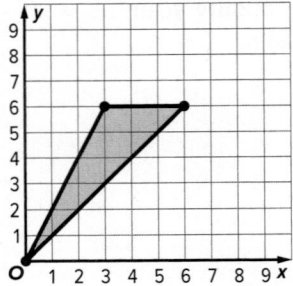

Select all the points at which a minimum or maximum value may occur. (Lesson 9-7)

A. (0, 0)

B. (1, 5)

C. (2, 6)

D. (3, 6)

E. (4, 5)

F. (5, 3)

G. (6, 6)

14. MULTI-SELECT A shoe manufacturer makes two types of athletic shoes—a cross-training shoe and a running shoe. Each pair of shoes is assembled by machine and then finished by hand. For the cross-training shoes, it takes 15 minutes for machine assembly and 6 minutes by hand. For the running shoes, it takes 9 minutes on the machine and 12 minutes by hand. The company can allocate no more than 900 machine hours and 500 hand hours each day. The profit is $10 for each pair of cross-training shoes and $15 for each pair of running shoes. Let x represent the number of cross-training shoes and y represent the number of running shoes manufactured each day.

Select the inequalities that form the boundary of the feasible region. (Lesson 9-7)

A. $x \geq 0$

B. $y \geq 0$

C. $0.1x + 0.2y \leq 500$

D. $0.25x + 0.15y \leq 900$

E. $6x + 12y \leq 500$

F. $15x + 9y \leq 900$

15. OPEN RESPONSE Last month, Jeremy took 2 piano lessons, 3 guitar lessons, and 1 drum lesson and spent a total of $285. D'Asia took 1 piano lesson, 5 guitar lessons, and 2 drum lessons and spent a total of $400. Raj took 3 piano lessons, 2 guitar lessons, and 4 drum lessons and spent a total of $440. Write a system of equations that could be used to find the prices of each piano lesson, x, guitar lesson, y, and drum lesson, z. (Lesson 9-8)

16. OPEN RESPONSE Use a graphing calculator to solve the equation $2|x - 3| - 1 = 5$. (Lesson 9-9)

Exponents and Roots

e Essential Question

How do you perform operations and represent real-world situations with exponents?

What will you learn?

How much do you already know about each topic **before** starting this module?

KEY

👎 — I don't know.   👊 — I've heard of it.   👍 — I know it!

	Before			After		
	👎	👊	👍	👎	👊	👍
use the Product of Powers Property						
use the Power of a Power Property						
use the Power of a Product Property						
use the Quotient of Powers Property						
use the Power of a Quotient Property						
simplify expressions with zero exponents						
simplify expressions with negative exponents						
use rational exponents to solve problems						
simplify radical expressions						
solve exponential equations						

📖 **Foldables** Make this Foldable to help you organize your notes about exponents and roots. Begin with seven sheets of notebook paper.

1. **Arrange** the paper into a stack.

2. **Staple** along the left side. Starting with the second sheet of paper, **cut** along the right side to form tabs.

3. **Label** the cover sheet "Exponents and Roots" and label each tab with a lesson number.

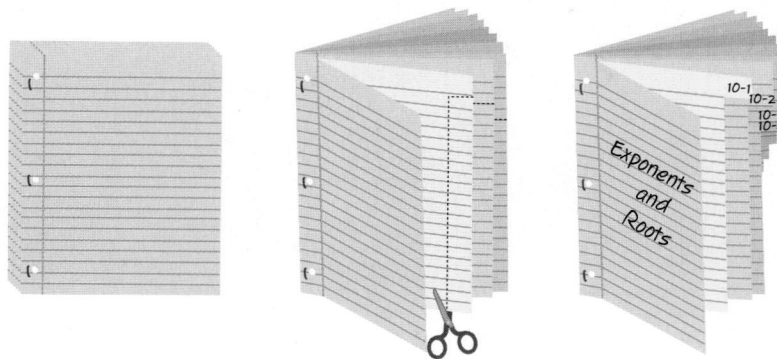

What Vocabulary Will You Learn?

- cube root
- exponential equation
- index
- monomial
- negative exponent

- nth root
- perfect cube
- perfect square
- principal square root
- radical expression

- radicand
- rational exponent
- square root

Are You Ready?

Complete the Quick Review to see if you are ready to start this module.
Then complete the Quick Check.

Quick Review	
Example 1	**Example 2**
Write $5 \cdot 5 \cdot 5 \cdot 5 + y \cdot y$ using exponents.	Evaluate $\left(\frac{5}{7}\right)^2$.
There are 4 factors of 5.	$\left(\frac{5}{7}\right)^2 = \frac{5}{7} \cdot \frac{5}{7}$ Expand the expression.
There are 2 factors of y.	$= \frac{5 \cdot 5}{7 \cdot 7}$ Multiply the numerators and multiply the denominators.
$5 \cdot 5 \cdot 5 \cdot 5 = 5^4$	
$y \cdot y = y^2$	$= \frac{25}{49}$ Simplify.
$5 \cdot 5 \cdot 5 \cdot 5 + y \cdot y = 5^4 + y^2$	

Quick Check	
Write each expression using exponents.	**Evaluate each expression.**
1. $4 \cdot 4 \cdot 4 \cdot 4 \cdot 4$	**5.** 2^5
2. $x \cdot x \cdot x$	**6.** $(-5)^2$
3. $m \cdot m \cdot m \cdot p \cdot p \cdot p \cdot p \cdot p$	**7.** $\left(\frac{1}{2}\right)^4$
4. $\left(\frac{1}{5}\right) \cdot \left(\frac{1}{5}\right) \cdot \left(\frac{1}{5}\right) \cdot \left(\frac{1}{5}\right) \cdot \left(\frac{1}{5}\right) \cdot \left(\frac{1}{5}\right)$	**8.** $(-4)^3$

How Did You Do?

Which exercises did you answer correctly in the Quick Check?

Multiplication Properties of Exponents

Explore Products of Powers

Online Activity Use an interactive tool to complete the Explore.

> ×
>
> **INQUIRY** How can you determine the product of two powers a^m and a^p?

Learn Product of Powers

A **monomial** is a number, a variable, or a product of a number and one or more variables. It has only one term. The term a^4 is a monomial. An expression that involves division by a variable is not a monomial. The term $\frac{ab}{c}$ is not a monomial.

Key Concept • Product of Powers	
Words	To multiply two powers that have the same base, add their exponents.
Symbols	For any real number a and any integers m and p, $a^m \cdot a^p = a^{m+p}$.
Examples	$b^2 \cdot b^4 = b^{2+4}$ or b^6; $d^3 \cdot d^7 = d^{3+7}$ or d^{10}

Example 1 Product of Powers

Simplify each expression.

a. $(3n^4)(4n^7)$

$$(3n^4)(4n^7) = (3 \cdot 4)(n^4 \cdot n^7) \qquad \text{Group coefficients and variables.}$$

$$= (3 \cdot 4)(n^{4+7}) \qquad \text{Product of Powers}$$

$$= 12n^{11} \qquad \text{Simplify.}$$

b. $(7xy^2)(2x^4y^3)$

$$(7xy^2)(2x^4y^3) = (7 \cdot 2)(x \cdot x^4)(y^2 \cdot y^3) \qquad \text{Group coefficients and variables.}$$

$$= (7 \cdot 2)(x^{1+4})(y^{2+3}) \qquad \text{Product of Powers}$$

$$= 14x^5y^5 \qquad \text{Simplify.}$$

Check

Simplify $(7n^7)(-7n)$.   Simplify $(11x^6y^6)(xy^9)$.

Today's Goals
- Find products of monomials.
- Find the power of a power.
- Find the power of a product.

Today's Vocabulary
monomial

🗨 Think About It!
How does the process for simplifying $b^2 \cdot b^4$ differ from simplifying $b^2 + b^4$?

Study Tip
Terminology Recall that for the expression b^2, b is called the *base* and the 2 is called the *exponent* or *power*.

🗨 Think About It!
What properties allow you to group the coefficients and the variables in the first step of each solution?

STARS **The fastest recorded star in the Milky Way galaxy is US 708, which travels 2,700,000 miles per hour. How far does US 708 travel in 1 year? Write your answer in scientific notation and round to the nearest tenth. (Hint: 1 year = 8760 hours)**

Step 1 Convert the speed of US 708 and the number of hours in a year to scientific notation.

Speed of US 708

$$2{,}700{,}000 = 2.7 \times 1{,}000{,}000 \qquad\qquad 1 \leq 2.7 < 10$$
$$= 2.7 \times 10^6 \qquad\qquad 1{,}000{,}000 = 10^6$$

Hours in a Year

$$8760 = 8.76 \times 1000 \qquad\qquad 1 \leq 8.76 < 10$$
$$= 8.76 \times 10^3 \qquad\qquad 1000 = 10^3$$

Step 2 Use the formula $d = rt$ to find approximately how far US 708 travels in a year.

$d = rt$	distance = rate • time
$d = (2.7 \times 10^6) \times (8.76 \times 10^3)$	$r = 2.7 \times 10^6$ mph, $t = 8.76 \times 10^3$ hrs
$= (2.7 \times 8.76) \times (10^6 \times 10^3)$	Comm. and Assoc. Properties
$= 23.652 \times (10^6 \times 10^3)$	Multiply.
$= 23.652 \times 10^{6+3}$	Product of Powers
$= 23.652 \times 10^9$	Simplify.
$= 2.3652 \times 10^1 \times 10^9$	$23.652 = 2.3652 \times 10^1$
$= 2.3652 \times 10^{1+9}$	Product of Powers
$= 2.3652 \times 10^{10}$	Simplify.
$\approx 2.4 \times 10^{10}$	Round to the nearest tenth.

Check

COMPUTERS Typically, a 4K computer monitor describes a screen with a resolution of 3840 × 2160. This means that the display's width is 3840 pixels and the height is 2160 pixels. How many pixels are there on a typical 4K computer monitor?

A. 2.2×10^3 pixels

B. 3.8×10^3 pixels

C. 8.3×10^6 pixels

D. 8.3×10^7 pixels

🌐 **Go Online** You can complete an Extra Example online.

Learn Power of a Power

Key Concept • Power of a Power	
Words	To find a power of a power, multiply the exponents.
Symbols	For any real number a and any integers m and p, $(a^m)^p = a^{mp}$.
Examples	$(b^2)^4 = b^{2 \cdot 4}$ or b^8; $(d^3)^7 = d^{3 \cdot 7}$ or d^{21}

Example 3 Power of a Power

Simplify each expression.

a. $(n^5)^3$

$$(n^5)^3 = n^{5 \cdot 3} \qquad \text{Power of a Power}$$

$$= n^{15} \qquad \text{Simplify.}$$

b. $(x^4)^2$

$$(x^4)^2 = x^{4 \cdot 2} \qquad \text{Power of a Power}$$

$$= x^8 \qquad \text{Simplify.}$$

Check

Simplify $(n^4)^{11}$.

Simplify $(x^{-7})^{-3}$.

Think About It!

Is $(x^4)^2$ equivalent to $(x^2)^4$? Explain your reasoning.

Learn Power of a Product

Key Concept • Power of a Product	
Words	To find a power of a product, find the power of each factor and multiply.
Symbols	For any real numbers a and b and any integer m, $(ab)^m = a^m b^m$.
Examples	$(-5x^2y)^3 = (-5)^3 x^6 y^3$ or $-125x^6 y^3$

Think About It!

Is $2(xy)^3$ equivalent to $(2xy)^3$? Explain your reasoning.

Example 4 Power of a Product

Simplify each expression.

a. $(3x^5y^2)^5$

$$(3x^5y^2)^5 = 3^5 x^{5 \cdot 5} y^{2 \cdot 5} \qquad \text{Power of a Product}$$

$$= 243 x^{25} y^{10} \qquad \text{Simplify.}$$

(continued on the next page)

Go Online You can complete an Extra Example online.

b. $(-5ab^4)^2$

$$(-5ab^4)^2 = (-5)^2 a^{1\cdot 2} b^{4\cdot 2} \qquad \text{Power of a Product}$$

$$= 25a^2 b^8 \qquad \text{Simplify.}$$

Check

Simplify $(4x^3 y^{-2})^3$.

Simplify $(-3a^{-3}b)^6$.

🌐 Example 5 Power of a Product and Area

If the side of each smaller square is x inches, and the side of the whole canvas is s inches, then what is the area of the painting in terms of x?

x in.   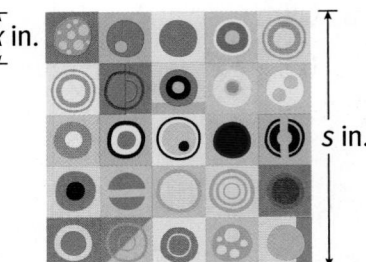 s in.

The side length of the canvas, s, can also be described as $5x$.

$$A = s^2 \qquad\qquad \text{Area of a square}$$

$$= (5x)^2 \qquad\quad s = 5x$$

$$= 5^2 x^{1\cdot 2} \qquad\quad \text{Power of a Product}$$

$$= 25x^2 \qquad\qquad \text{Simplify.}$$

Check

GAME DESIGN Jeanine is designing an early version of a game and wants to use a cube to stand in for the character, which she will design later. She bases the dimensions of the cube around the height of the character, which she defines as $\frac{1}{8}x$ pixels, where x is the height of the total game screen. What is the volume of the cube in terms of x?

A. $\frac{1}{64}x^2$ pixels

B. $\frac{1}{512}x^3$ pixels

C. $\frac{1}{8}x^3$ pixels

D. $512x^3$ pixels

🔴 **Go Online** You can complete an Extra Example online.

Practice

Go Online You can complete your homework online.

Examples 1, 3, and 4

Simplify each expression.

1. $(q^2)(2q^4)$

2. $(-2u^2)(6u^6)$

3. $(9w^2x^8)(w^6x^4)$

4. $(y^6z^9)(6y^4z^2)$

5. $(b^8c^6d^5)(7b^6c^2d)$

6. $(14fg^2h^2)(3f^4g^2h^2)$

7. $(j^5k^7)^4$

8. $(n^3p)^4$

9. $[(2^2)^2]^2$

10. $[(3^2)^2]^4$

11. $[(4r^2t)^3]^2$

12. $[(-2xy^2)^3]^2$

13. $(y^2z)(yz^2)$

14. $(\ell^2k^2)(\ell^3k)$

15. $(-5m^3)(3m^8)$

16. $(-2c^4d)(-4cd)$

17. $(3pr^2)^2$

18. $(2b^3c^4)^2$

Example 2

19. **COMMUNITY SERVICE** During the school year, each student planted 1.2×10^2 flowers as part of a community service project. If there are 1.5×10^3 students in the school, how many flowers did they plant in total?

20. **CHERRIES** A farmer has 350 cherry trees on his farm. If each cherry tree yields 6200 cherries per year, how many cherries can the farmer harvest per year? Write your answer in scientific notation.

21. **PANDAS** When born, a panda cub weighs only 2.2×10^{-1} pounds. As they grow older, adult pandas, on average, weigh 1.1×10^3 times more than a newborn panda. How much does an average adult panda weigh?

22. **SALES** An automobile company sold 2.3 million new cars in a year. If the average price per car was $21,000, how much money did the company make that year? Write your answer in scientific notation.

23. **APPLE JUICE** An apple juice company produced 8.3 billion individual-size bottles of apple juice last year. If each bottle contains 355 milliliters of juice, how many milliliters of apple juice did the company produce?

Example 5

24. **BLOCKS** Building blocks are in the shape of a cube. The dimensions of the building blocks are based on the length of the packaging box, which is defined as $\frac{1}{12}x^2$ centimeters, where x is the length of the packaging box. What is the volume of a building block in terms of x?

25. FIELDS A field is in the shape of a square. The length of the field is $5x^3y^5$. What is the area of the field in terms of x and y?

26. PICTURE FRAME Michelle is purchasing a new picture frame to hang in her bedroom. The picture frame is square shaped. What is the area of the interior of the picture frame in terms of c and d?

$\leftarrow 5c^3d^2 \rightarrow$

27. SHIPPING A shipping box is in the shape of a cube. What is the volume of the shipping box in terms of m and n?

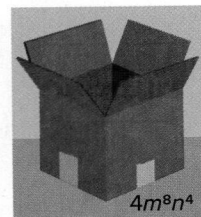

$4m^8n^4$

Mixed Exercises

Simplify each expression.

28. $(2a^3)^4(a^3)^3$

29. $(c^3)^2(-3c^5)^2$

30. $(2gh^4)^3[(-2g^4h)^3]^2$

31. $(5k^2m)^3[(4km^4)^2]^2$

32. $(p^5r^2)^4(-7p^3r^4)(6pr^3)$

33. $(5x^2y)^2(2xy^3z)^3(4xyz)$

34. $(5a^2b^3c^4)^4(6a^3b^4c^2)$

35. $(10xy^5z^3)(3x^4y^6z^3)$

36. $(0.5x^3)^2$

37. $(0.4h^5)^3$

38. $\left(-\dfrac{3}{4c}\right)^3$

39. $\left(\dfrac{4}{5}a^2\right)^2$

40. $(8y^3)(-3x^2y^2)\left(\dfrac{3}{8}xy^4\right)$

41. $\left(\dfrac{4}{7}m\right)^2(49m)(17p)\left(\dfrac{1}{34}p^5\right)$

42. $(-3r^3w^4)^3(2rw)^2(-3r^2)^3(4rw^2)^3(2r^2w^3)$ **43.** $(3ab^2c)^2(-2a^2b^4)^2(a^4c^2)^3(a^2b^4c^5)^2(2a^3b^2c^4)^3$

STRUCTURE Determine whether each pair of expressions is equivalent. Write *yes* or *no*.

44. $(a^7)(a^7)(a^7)(a^7)$ and a^{2401}

45. $(-j^9)(-j^9)(-j^9)(-j^9)$ and $-j^{27}$

46. $(5p^3q)(4p^5q^9)$ and $20p^8q^{10}$

47. $(6w^5)^2$ and $36w^{25}$

48. $(x^{10})^3$ and $(x^3)^{10}$

49. $[(2n^2)^3]^2$ and $(4n^4)^3$

50. GRAVITY An object that has been falling for x seconds has dropped at an average speed of $16x$ feet per second. If the object is dropped from a great height, its total distance traveled is the product of the average rate times the time. Write a simplified expression to show the distance the object has traveled after x seconds.

51. ELECTRICITY An electrician uses the formula $W = I^2R$, where W is the power in watts, I is the current in amperes, and R is the resistance in ohms.

    a. Find the power in a household circuit that has $2x^2$ amperes of current and $5x^3$ ohms of resistance.

    b. If the current is reduced by one-half, what happens to the power?

52. MOLECULES A glass of water contains 0.25 liter of water. If 1 milliliter of water contains 3.3×10^{22} water molecules, how many water molecules are there in the glass of water?

53. CIVIL ENGINEERING A developer is planning a sidewalk for a new development. The sidewalk can be installed in rectangular sections that have a fixed width of 3 feet and a length that can vary. Assuming that each section is the same length, express the area of a 4-section sidewalk as a monomial.

54. USE A SOURCE Suppose each student in your school is required to complete 60 hours of community service before they graduate. Find the number of students in the senior class at your school. Based on this number, calculate total minimum number of community service hours completed by the senior class by the time they graduate. Write your answer in scientific notation.

55. REGULARITY Recall that both multiplication and addition are commutative and associative. Multiplication also distributes over addition.

    a. What would it mean for the operation of raising one number to an exponent to be commutative? Decide and explain whether this operation is commutative.

    b. What would it mean for the operation of raising one number to an exponent to be associative? Decide and explain whether this operation is associative.

    c. What would it mean for the process of raising one exponent to another to distribute over addition? Decide and explain whether this is true.

    d. What would it mean for the process of raising one exponent to another to distribute over multiplication? Decide and explain whether this is true.

56. STATE YOUR ASSUMPTION Consider the expressions $(m^p)(m^q)$ and $(m^p)^q$.

    a. What must be true about p and q for $(m^p)(m^q) = (m^p)^q$? Give an example.

    b. Assuming m, p, and q are all positive integers, can $(m^p)(m^q) > (m^p)^q$ ever be true? Explain.

🍩 **Higher-Order Thinking Skills**

57. PERSEVERE For any nonzero real numbers a and b and any integers m and t, simplify the expression $\left(\dfrac{-a^m}{b^t}\right)^{2t}$. Explain.

58. ANALYZE Consider the equations in the table.

    a. For each equation, copy and complete the table to write the related expression and record the power of x.

Equation	Related Expression	Power of x	Linear or Nonlinear
$y = x$			
$y = x^2$			
$y = x^3$			

    b. Graph each equation using a graphing calculator.

    c. Classify each graph as *linear* or *nonlinear*.

    d. Explain how to determine whether an equation, or its related expression, is linear or nonlinear without graphing.

59. CREATE Write three different expressions that can be simplified to x^6.

60. WRITE Write a product of powers that is positive for all nonzero values of the variable. Explain your reasoning.

61. FIND THE ERROR Jade and Sal are each writing an equivalent form of g^5gh^6. Is either correct? Explain your reasoning.

Jade	Sal
$g^5gh^6 =$ $(g \cdot g \cdot g \cdot g \cdot g) \cdot (g) \cdot (h \cdot h \cdot h \cdot h \cdot h \cdot h)$	$g^5gh^6 =$ $(g \cdot g \cdot g \cdot g \cdot g) \cdot (gh \cdot gh \cdot gh \cdot gh \cdot gh \cdot gh)$

Division Properties of Exponents

Today's Goals
- Find quotients of monomials.
- Find powers of quotients.

Explore Quotients of Powers

Online Activity Use an interactive tool to complete the Explore.

×

INQUIRY How can you determine the quotient of two powers a^m and a^p?

Learn Quotient of Powers

You can use repeated multiplication and the principles for reducing fractions to simplify the quotients of monomials with the same base, like $\frac{2^8}{2^3}$. First, expand the numerator and the denominator. Then, divide the common factors.

$$\frac{2^8}{2^3} = \frac{\overbrace{2 \cdot 2 \cdot 2 \cdot 2 \cdot 2 \cdot 2 \cdot 2 \cdot 2}^{8 \text{ factors}}}{\underbrace{2 \cdot 2 \cdot 2}_{3 \text{ factors}}}$$

$$= 2 \cdot 2 \cdot 2 \cdot 2 \cdot 2$$

$$= 2^5$$

$$\frac{r^5}{r^4} = \frac{\overbrace{r \cdot r \cdot r \cdot r \cdot r}^{5 \text{ factors}}}{\underbrace{r \cdot r \cdot r \cdot r}_{4 \text{ factors}}}$$

$$= r$$

These examples demonstrate the Quotient of Powers Property.

Key Concept • Quotient of Powers	
Words	To divide two powers with the same base, subtract the exponents.
Symbols	For any nonzero number a, and any integers m and p, $\frac{a^m}{a^p} = a^{m-p}$.
Examples	$\frac{b^{12}}{b^9} = b^{12-9} = b^3$; $\frac{w^6}{w^2} = w^{6-2} = w^4$

Think About It!

What steps would you take to simplify $\frac{10^{12}}{10^9}$?

Go Online You can complete an Extra Example online.

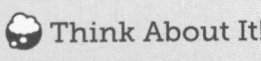 Think About It!

In the expression $\frac{b^5c^7}{b^2c}$, why can you not subtract $5 - 7$ in the numerator and $2 - 1$ in the denominator and simplify?

Example 1 Quotient of Powers

Simplify $\frac{b^5c^7}{b^2c}$. Assume that the denominator does not equal zero.

Step 1 Group powers with the same base.

$$\frac{b^5c^7}{b^2c} = \left(\frac{b^5}{b^2}\right)\left(\frac{c^7}{c}\right)$$

Step 2 Use the Quotient of Powers Property.

$$\left(\frac{b^5}{b^2}\right)\left(\frac{c^7}{c}\right) = (b^{5-2})(c^{7-1})$$ Subtract the exponents in each group.

$$= b^3c^6$$ Simplify.

Check

Simplify $\frac{a^2b^9\,cd^4}{b^8cd}$. Assume that the denominator does not equal zero.

A. abd^3

B. a^2bcd^3

C. a^2bcd^4

D. a^2bd^3

Think About It!

Why must you assume that the denominator does not equal 0?

🌐 Example 2 Apply Division of Monomials

CHEMISTRY At sea level, there are about 10^{25} molecules in a cubic liter of air. In the stratosphere, about 30 kilometers above the Earth's surface, the same cubic liter of air has about 10^{23} molecules. Approximately how many times as many molecules are there in a cubic liter of air at sea level as there are in the stratosphere?

$$\frac{a^m}{a^p} = a^{m-p} = 10^{25-23} = 10^2$$

There are about 100 times as many molecules in a cubic liter of air at sea level as there are in the stratosphere.

Check

SHOPPING Canada's West Edmonton Mall claims to have the largest parking lot in the world, with about 3^9 parking spaces. California's Glendale Galleria has about 3^8 parking spaces.

The West Edmonton Mall has ____?____ times as many parking spaces as the Glendale Galleria.

🧭 Go Online You can complete an Extra Example online.

Learn Power of a Quotient

You can use the Product of Powers Property to find the powers of quotients for monomials.

$$\left(\frac{2}{5}\right)^3 = \overbrace{\left(\frac{2}{5}\right)\left(\frac{2}{5}\right)\left(\frac{2}{5}\right)}^{3 \text{ factors}} = \underbrace{\frac{2 \cdot 2 \cdot 2}{5 \cdot 5 \cdot 5}}_{3 \text{ factors}} = \frac{2^3}{5^3}$$

$$\left(\frac{p}{q}\right)^2 = \overbrace{\left(\frac{p}{q}\right)\left(\frac{p}{q}\right)}^{2 \text{ factors} \; 2 \text{ factors}} = \underbrace{\frac{p \cdot p}{q \cdot q}}_{2 \text{ factors}} = \frac{p^2}{q^2}$$

These examples demonstrate the Powers of a Quotient Property.

Key Concept • Power of a Quotient	
Words	To find the power of a quotient, find the power of the numerator and the power of the denominator.
Symbols	For any real numbers a and $b \neq 0$, and any integer m, $\left(\frac{a}{b}\right)^m = \frac{a^m}{b^m}$.
Examples	$\left(\frac{1}{4}\right)^5 = \frac{1^5}{4^5}$; $\left(\frac{c}{d}\right)^6 = \frac{c^6}{d^6}$

Example 3 Power of a Quotient

Simplify $\left(\frac{5a^2}{6}\right)^3$.

$$\left(\frac{5a^2}{6}\right)^3 = \frac{(5a^2)^3}{6^3} \qquad \text{Power of a Quotient}$$

$$= \frac{5^3(a^2)^3}{6^3} \qquad \text{Power of a Product}$$

$$= \frac{125a^6}{216} \qquad \text{Power of a Power}$$

Check

Simplify $\left(\frac{5j^3l^2}{7k^5}\right)^4$. Assume that the denominator does not equal zero.

A. $\dfrac{5j^{12}l^8}{7k^{20}}$

B. $\dfrac{625j^{12}l^8}{2401k^{20}}$

C. $\dfrac{625j^7l^6}{2401k^9}$

D. $\dfrac{625j^{12}l^8}{7k^5}$

🔾 **Go Online** You can complete an Extra Example online.

😃 **Think About It!**

How would you simplify $\left(\frac{1}{2}\right)^2$?

Study Tip

Power Rules with Variables The power rules apply to variables and numbers. For example, $\left(\frac{4m}{2n}\right)^4 = \frac{(4m)^4}{(2n)^4} = \frac{4^4m^4}{2^4n^4} = \frac{256m^4}{16n^4}$ or $\frac{16m^4}{n^4}$.

Example 4 Power of a Quotient with Variables

Simplify $\left(\dfrac{x^4y}{xyz}\right)^2$. Assume that the denominator does not equal zero.

Write the appropriate justification next to each step.

$$\left(\frac{x^4y}{xyz}\right)^2 = \frac{(x^4y)^2}{(xyz)^2} \qquad \text{Power of a Quotient}$$

$$= \frac{(x^4)^2y^2}{x^2y^2z^2} \qquad \text{Power of a Product}$$

$$= \frac{x^8y^2}{x^2y^2z^2} \qquad \text{Power of a Power}$$

$$= \frac{x^6}{z^2} \qquad \text{Quotient of Powers}$$

Check

Simplify $\left(\dfrac{4m^2n^2p^2}{3mp}\right)^4$. Assume that the denominator does not equal zero.

A. $\dfrac{256mn^2p}{81}$

B. $\dfrac{256m^4n^6p^4}{81}$

C. $\dfrac{256m^4n^8p^4}{81}$

D. $\dfrac{256m^4n^8p^4}{81mp}$

Pause and Reflect

Did you struggle with anything in this lesson? If so, how did you deal with it?

🔊 **Go Online** You can complete an Extra Example online.

Practice

🔘 **Go Online** You can complete your homework online.

Examples 1, 3, and 4

Simplify each expression. Assume that no denominator equals zero.

1. $\dfrac{m^4p^2}{m^2p}$

2. $\dfrac{p^{12}t^3r}{p^2tr}$

3. $\dfrac{c^4d^4f^3}{c^2d^4f^3}$

4. $\left(\dfrac{3xy^4}{5z^2}\right)^2$

5. $\left(\dfrac{p^2t^7}{10}\right)^3$

6. $\dfrac{a^7b^8c^8}{a^5bc^7}$

7. $\left(\dfrac{3np^3}{7q^2}\right)^2$

8. $\left(\dfrac{2r^3t^6}{5u^9}\right)^4$

9. $\left(\dfrac{3m^5r^3}{4p^8}\right)^4$

10. $\dfrac{p^{12}t^7r^2}{p^2t^7r}$

11. $\dfrac{k^4m^3p^2}{k^2m^2}$

12. $\dfrac{m^7p^2}{m^3p^2}$

13. $\dfrac{32x^3y^2z^5}{-8xyz^2}$

14. $\left(\dfrac{4p^7}{7r^2}\right)^2$

15. $\dfrac{9d^7}{3d^6}$

16. $\dfrac{12n^5}{36n}$

17. $\dfrac{w^4x^3}{w^4x}$

18. $\dfrac{a^3b^5}{ab^2}$

Example 2

19. **SPACE** The Moon is approximately 25^4 kilometers away from Earth on average. The Olympus Mons volcano on Mars stands 25 kilometers high. How many Olympus Mons volcanoes, stacked on top of one another, would fit between the surface of Earth and the Moon?

20. **GEOMETRY** Write the ratio of the area of a circle with radius r to the circumference of the same circle.

21. **COMBINATIONS** The number of four-letter combinations that can be formed with the English alphabet is 26^4. The number of six-letter combinations that can be formed is 26^6. How many times more six-letter combinations can be formed than four-letter combinations?

22. **BLOOD COUNT** A lab technician draws a sample of blood. A cubic millimeter of the blood contains 22^5 red blood cells and 22^3 white blood cells. How many times more red blood cells are there than white blood cells?

Mixed Exercises

Simplify each expression. Assume that no denominator equals zero.

23. $\dfrac{-4w^{12}}{12w^3}$

24. $\dfrac{13r^7}{39r^4}$

25. $\left(\dfrac{2a^4c^3}{5b^2d^2}\right)^2$

26. $\dfrac{m^6n^3}{m^2n^6}$

27. $\left(\dfrac{24a^{11}b^{16}c^6}{18a^6b^6c^6}\right)^3$

28. $\left(\dfrac{q^2r^3}{qr^2}\right)^5$

29. $\dfrac{-16x^7y^2}{-6xy^5}$

30. $\dfrac{21d^{18}e^5}{7d^{11}e^6}$

31. $\left(\dfrac{2x^3y^2z^5}{3xyz}\right)^3$

32. $\dfrac{8c^4d^2f^9}{4cd^2f^3}$

33. $\left(\dfrac{7c^4}{14d^2}\right)^6$

34. $\left(\dfrac{6j^5}{7m^6n^3}\right)^2$

35. SOUND Decibels are used to measure sound. The softest sound that can be heard is rated at 0 decibels, or a relative loudness of 1. Ordinary conversation is rated at about 60 decibels, or a relative loudness of 10^6. A stock car race is rated at about 130 decibels, or a relative loudness of 10^{13}. How many times greater is the relative loudness of a stock car race than the relative loudness of ordinary conversation?

36. COMPUTERS The byte is the fundamental unit of computer processing. Almost all aspects of a computer's performance and specifications are measured in bytes or multiples of bytes. The byte is based on powers of 2, as shown in the table. How many times greater is a megabyte than a kilobyte?

Memory Term	Number of Bytes
byte	2^0 or 1
kilobyte	2^{10}
megabyte	2^{20}
gigabyte	2^{30}

37. AREA The area of the triangle shown is $6x^5y^3$. Find the base of the triangle.

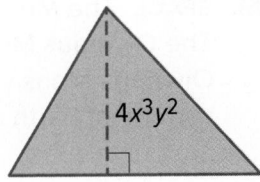

$4x^3y^2$

38. AREA The area of the rectangle in the figure is $32xy^3$ square units. Find the width of the rectangle.

$8xy$

39. USE A MODEL An investment is expected to increase in value by 4% every year.

    a. Write an expression that represents the value of the investment after t years if the initial value was n dollars.

    b. By what percent does the value of the investment change between the end of year 2 and the end of year 8? Round your answer to the nearest tenth of a percent, and show your work.

40. INVESTMENTS A poor investment is expected to decrease in value by 5% every year.

    a. If the initial value of the investment was $100, what does it mean for it to decrease in value by 5% in the first year?

    b. Erik claims that rather than multiplying by 0.05 and subtracting, we can simply multiply the investment by 0.95. Is he correct? Explain.

    c. Write an expression that represents the value of the investment after t years if the initial value was n dollars.

    d. By what percent does the value of the investment change between the end of year 2 and the end of year 10? Round your answer to the nearest tenth of a percent, and show your work.

41. PAPER FOLDING If you fold a sheet of paper in half, you have a thickness of 2 sheets. Folding again, you have a thickness of 4 sheets. Fold the paper in half one more time. How many times thicker is a sheet that has been folded 3 times than a sheet that has not been folded?

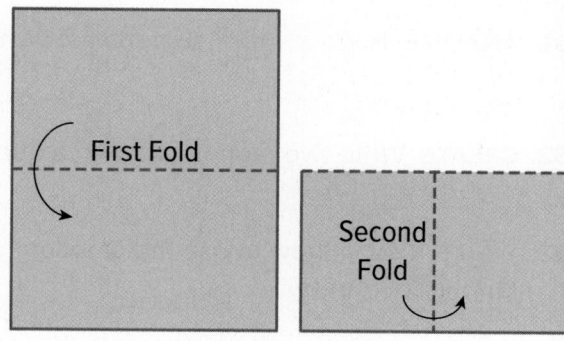
First Fold

Second Fold

42. USE TOOLS Create a table in a computer spreadsheet program to show the powers of 2^0 through 2^{10}. Use the formula functions in the program to show the Quotient of Powers Property is true for five different division problems using the powers you already entered.

Higher-Order Thinking Skills

43. CONSTRUCT ARGUMENTS Is $\left(\dfrac{x}{y^2}\right)^3$ the same as $\left(\dfrac{x}{y^3}\right)^2$? Justify your argument.

44. STRUCTURE Find the value of x that makes $\dfrac{8^{5x}}{8^{4x+1}} = 8^9$ true. Explain.

45. REASONING Which expression does not have the same answer as the others when simplified using exponent rules?

    A. $\left(\dfrac{3x^4y}{5}\right)^2$
    B. $\dfrac{-18x^{14}y^{10}}{-50x^6y^8}$
    C. $\dfrac{-3^2x^9y^5}{(-5)^2xy^3}$

46. PRECISION What error did the student make in simplifying the expression $\dfrac{5^4}{5} = 1^4 = 1$? What is the correct value of the expression?

47. REGULARITY Simplify the expression $\dfrac{a^{x+y}}{a^y}$.

48. CREATE Write three quotients of powers expressions that are equivalent to 2^5.

49. REGULARITY Consider the equation $\dfrac{3^h}{3^k} = 3^2$.

    a. Find two numbers h and k that satisfy the equation.

    b. Are there any other pairs of numbers that satisfy the equation? Explain.

50. FIND THE ERROR Kathryn and Salvador used different methods to simplify $\left(\dfrac{p^9}{p^5}\right)^2$. Is either correct? Explain your reasoning.

Kathryn	Salvador
$\left(\dfrac{p^9}{p^5}\right)^2 = \dfrac{p^{18}}{p^{10}} = p^8$	$\left(\dfrac{p^9}{p^5}\right)^2 = (p^4)^2 = p^8$

51. ANALYZE Is $x^y \cdot x^z = x^{yz}$ sometimes, always, or never true? Justify your argument.

52. CREATE Write two monomials with a quotient of $24a^2b^3$.

53. WRITE Explain how to use the Quotient of Powers Property and the Power of a Quotient Property.

54. PERSEVERE Is the expression $\left(\dfrac{p^7}{p^3}\right)^5$ positive or negative for all nonzero values of p? Explain.

55. WHICH ONE DOESN'T BELONG? Which quotient does *not* belong with the other three? Justify your conclusion.

$\dfrac{6^7}{6^2}$	$\dfrac{(-7)^3}{(-7)^2}$	$\dfrac{6^5}{6^3}$	$\dfrac{(-3)^8}{(-2)^4}$

56. FIND THE ERROR Andrew and Mateo are trying to simplify the expression $\dfrac{9^5}{9^3}$. Is either correct? Explain your reasoning.

Andrew	Mateo
$\dfrac{9^5}{9^3} = 9^2$	$\dfrac{9^5}{9^3} = \dfrac{1}{9^2}$

Negative Exponents

Learn Zero Exponent

Key Concept • Zero Exponent Property	
Words	Any nonzero number raised to the zero power is equal to 1.
Symbols	For any nonzero number a, $a^0 = 1$.
Examples	$30^0 = 1; \left(\frac{x}{y}\right)^0 = 1; \left(\frac{8}{3}\right)^0 = 1$

Example 1 Zero Exponent

Simplify each expression. Assume that no denominator equals zero.

a. $\left(-\dfrac{8m^2np^8}{9k^2mn^4}\right)^0$

$\left(-\dfrac{8m^2np^8}{9k^2mn^4}\right)^0 = 1$ $a^0 = 1$

b. $\dfrac{a^4b^0}{a^2}$

$\dfrac{a^4b^0}{a^2} = \dfrac{a^4(1)}{a^2}$ $b^0 = 1$

$= a^2$ Quotient of Powers

Check

Select the simplified form of $\dfrac{56g^2hj^{11}}{8g^0h^0j^0}$. Assume that the denominator does not equal zero.

A. $7g^2hj^{11}$

B. $7gj^{10}$

C. $7ghj^{10}$

D. $7g^2hj^{10}$

Explore Simplifying Expressions with Negative Exponents

🔎 **Online Activity** Use an interactive tool to complete the Explore.

❓ INQUIRY How can you simplify expressions with negative exponents?

🔎 **Go Online** You can complete an Extra Example online.

Today's Goals
- Simplify expressions containing zero and negative exponents.
- Simplify expressions containing negative exponents.

Today's Vocabulary
negative exponent

💬 **Talk About It!**
Why is the answer to part **b** not 1?

🔎 **Go Online** An alternate method is available for this example.

Learn Negative Exponents

Any nonzero real number can be raised to a negative power. That power is called a **negative exponent**.

Key Concept • Negative Exponent Property	
Words	For any nonzero number a and any integer n, a^{-n} is the reciprocal of a^n. Also, the reciprocal of a^{-n} is a^n.
Symbols	For any nonzero number a and any integer n, $a^{-n} = \frac{1}{a^n}$.
Examples	$3^{-5} = \frac{1}{3^5} = \frac{1}{243}; \frac{1}{m^{-2}} = m^2$

An expression is considered simplified when:

- it contains only positive exponents.

- each base appears exactly once.

- there are no powers of powers.

- all fractions are in simplest form.

😮 **Think About It!**

Describe how to simplify an expression of the form a^{-n}.

Example 2 Negative Exponents

Simplify $\frac{a^3 b^{-4}}{c^{-2}}$. Assume that the denominator does not equal zero.

$$\frac{a^3 b^{-4}}{c^{-2}} = \left(\frac{a^3}{1}\right)\left(\frac{b^{-4}}{1}\right)\left(\frac{1}{c^{-2}}\right) \qquad \text{Write as a product of fractions.}$$

$$= \left(\frac{a^3}{1}\right)\left(\frac{1}{b^4}\right)\left(\frac{c^2}{1}\right) \qquad a^{-n} \text{ is } \frac{1}{a^n}; \frac{1}{a^{-n}} = a^n$$

$$= \frac{a^3 c^2}{b^4} \qquad \text{Multiply.}$$

Check

Select the simplified form of $\frac{42 r^{-2} t^{-6} u^3}{14 r^2 u^{-3}}$. Assume that the denominator does not equal zero.

A. $\frac{3}{r^4 t^6 u^6}$

B. $\frac{3u^6}{r^4 t^6}$

C. $\frac{3 t^6 u^6}{r^4}$

D. $3 r^4 t^6 u^6$

Math History Minute

In the 15th century, French physician **Nicolas Chuquet (c. 1445–1488)** wrote a book called *Triparty en la science des nombres* or *A Three-Part Book on the Science of Numbers*, which contained early notation for zero and negative exponents.

▶ **Go Online** You can complete an Extra Example online.

Example 3 Simplify an Expression with Negative Exponents

Simplify $\dfrac{3g^{-3}h^2}{-36g^3hj^{-4}}$. Assume that the denominator does not equal zero.

$$\frac{3g^{-3}h^2}{-36g^3hj^{-4}} = \left(-\frac{3}{36}\right)\left(\frac{g^{-3}}{g^3}\right)\left(\frac{h^2}{h}\right)\left(\frac{1}{j^{-4}}\right)$$

Group powers with the same base.

$$= \left(-\frac{1}{12}\right)(g^{-3-3})(h^{2-1})(j^4)$$

Quotient of Powers and Negative Exponent Properties

$$= \left(-\frac{1}{12}\right)g^{-6}hj^4$$

Simplify.

$$= \left(-\frac{1}{12}\right)\left(\frac{1}{g^6}\right)hj^4$$

Negative Exponent Property

$$= -\frac{hj^4}{12g^6}$$

Multiply.

Think About It!

Why do you not leave the answer as $\left(-\frac{1}{12}\right)g^{-6}hj^4$? Justify your argument.

Check

Select the simplified form of $\dfrac{35h^0j^{-5}k^{-2}}{14h^2m^5}$. Assume that the denominator does not equal zero.

A. $\dfrac{5}{2hj^5m^5k^2}$

B. $\dfrac{5h^2}{2hj^5m^5k^2}$

C. $\dfrac{5}{2h^2j^5m^5k^2}$

D. $\dfrac{5h^2m^5}{2j^5k^2}$

Order of magnitude is used to compare measures and to estimate and perform rough calculations. The **order of magnitude** of a quantity is the number rounded to the nearest power of 10. For example, the power of 10 closest to 105,000,000 is 10^8, or 100,000,000. So the order of magnitude of 105,000,000 is 10^8.

🌐 Apply Example 4 Apply Properties of Exponents

SPEED **The maximum speed of a peregrine falcon is about 90 meters per second. The maximum speed of a garden snail is about 0.01 meter per second. How many orders of magnitude as fast as a peregrine falcon is a garden snail?**

1. What is the task?
Describe the task in your own words. Then list any questions that you may have. How can you find answers to your questions?

Sample answer: The garden snail is how many orders of magnitude as fast as the falcon? What are the orders of magnitude of each animal? And then what is the ratio of the two orders of magnitude?

(continued on the next page)

🧭 **Go Online** You can complete an Extra Example online.

2. **How will you approach the task? What have you learned that you can use to help you complete the task?**

Sample answer: First, I will compute the orders of magnitude for each animal. Then I will find the ratio of the orders of magnitude of the snail to the falcon. I have learned how to find the order of magnitude.

3. **What is your solution?**
Use your strategy to solve the problem.

The maximum speed of a snail is 0.01 m/s. So, the order of magnitude of the speed of the snail is 10^{-2}.

The maximum speed of a falcon is close to 100 m/s. So, the order of magnitude of the speed of the falcon is 10^2.

What is the ratio of the order of magnitude of the snail to the order of magnitude of the falcon?

$$\frac{\text{order of magnitude of snail}}{\text{order of magnitude of falcon}} = \frac{10^{-2}}{10^2}$$

A snail is approximately 0.0001 times as fast as a falcon, or a snail is -4 orders of magnitude as fast as a falcon.

4. **How can you know that your solution is reasonable?**

✏️ **Write About It!** Write an argument that can be used to defend your solution.

Sample answer: The ratio of the snail's speed to the falcon's speed is $\frac{0.01}{90} \approx 0.0001$ or 10^{-4}. Because approaching the problem in a different way yields the same result, the answer is reasonable.

Check

LENGTH The radius of Earth is about 10,000,000 meters. The radius of a virus is about 0.0000001 meter. Approximately how many orders of magnitude as long as the radius of a virus is the radius of Earth?

Part A Using estimation, how many times greater is the radius of Earth than the radius of a virus?

    A. 0.0001

    B. 10

    C. 1,000,000

    D. 100,000,000,000,000

Part B The radius of Earth is about ___?___ orders of magnitude greater than the radius of a virus.

🧭 **Go Online** You can complete an Extra Example online.

Practice

Go Online You can complete your homework online.

Examples 1–3

Simplify each expression. Assume that no denominator equals zero.

1. $\dfrac{r^6 n^{-7}}{r^4 n^2}$

2. $\dfrac{h^3}{h^{-6}}$

3. $\dfrac{f^{-7}}{f^4}$

4. $\left(\dfrac{16 p^5 w^2}{2 p^3 w^3}\right)^0$

5. $\dfrac{f^{-5} g^4}{h^{-2}}$

6. $\dfrac{15 x^6 y^{-9}}{5 x y^{-11}}$

7. $\dfrac{-15 t^0 u^{-1}}{5 u^3}$

8. $\dfrac{(z^2 w^{-1})^3}{(z^3 w^2)^2}$

9. $\dfrac{-10 m^{-1} y^0 r}{-14 m^{-7} y^{-3} r^{-4}}$

10. $\dfrac{51 x^{-1} y^3}{17 x^2 y}$

11. $\dfrac{3 m^{-3} r^4 p^2}{12 t^4}$

12. $\left(\dfrac{3 t^6 u^2 v^5}{9 t u v^{21}}\right)^0$

13. $\dfrac{x^{-4} y^9}{z^{-2}}$

14. $\left(-\dfrac{5 f^9 g^4 h^2}{f g^2 h^3}\right)^0$

15. $\dfrac{p^4 t^{-3}}{r^{-2}}$

16. $-\dfrac{5 c^2 d^5}{8 c d^5 f^0}$

17. $\dfrac{-2 f^3 g^2 h^0}{8 f^2 g^2}$

18. $\dfrac{g^0 h^7 j^{-2}}{g^{-5} h^0 j^{-2}}$

Example 4

19. **METRIC MEASUREMENT** Consider a dust mite that measures 10^{-3} millimeters in length and a gecko that measures 10 centimeters long. How many orders of magnitude as long as the mite is the gecko?

20. **CHEMISTRY** The nucleus of a certain atom is 10^{-13} centimeters across. If the nucleus of a different atom is 10^{-11} centimeters across, how many orders of magnitude as great is the second nucleus?

21. **WEIGHT** A paper clip weighs about 10^{-3} kilograms. A draft horse weighs about 10^3 kilograms. How many orders of magnitude as heavy is a draft horse than a paper clip?

22. **GDP** Gross Domestic Product (GDP) is a measure of a country's wealth. Indonesia had an estimated GDP in 2017 of 1,020,515,000,000. Madagascar had an estimated GDP of 10,372,000,000 in 2017. About how many orders of magnitude as great is Indonesia's GDP?

23. **GROWTH** An old oak tree is 1012 inches tall. A younger oak tree near it is 98 inches tall. About how many orders of magnitude as tall is the old oak tree?

Mixed Exercises

Simplify each expression. Assume that no denominator equals zero.

24. $\dfrac{3wy^{-2}}{(w^{-1}y)^3}$

25. $\dfrac{(4k^3m^2)^3}{(5k^2m^{-3})^{-2}}$

26. $\dfrac{-12c^3d^0f^{-2}}{6c^5d^{-3}f^4}$

27. $\dfrac{20qr^{-2}t^{-5}}{4q^0r^4t^{-2}}$

28. $\dfrac{(5pr^{-2})^{-2}}{(3p^{-1}r)^3}$

29. $\dfrac{(2g^3h^{-2})^2}{(g^2h^0)^{-3}}$

30. $\left(\dfrac{2a^{-2}b^4c^2}{-4a^{-2}b^{-5}c^{-7}}\right)^{-1}$

31. $\left(\dfrac{-3x^{-6}y^{-1}z^{-2}}{6x^{-2}yz^{-5}}\right)^{-2}$

32. $\left(\dfrac{4^0c^2d^3f}{2c^{-4}d^{-5}}\right)^{-3}$

33. $\dfrac{(16x^2y^{-1})^0}{(4x^0y^{-4}z)^{-2}}$

34. RATIOS Yvonne is comparing the weights of a semi-truck trailer tire and a mobile home. A semi-truck trailer tire weighs about 10^2 pounds, and a mobile home weighs about 10^4 pounds. What is the ratio of the weight of a semi-truck trailer tire compared to the weight of a mobile home? Write your answer as a monomial.

35. MARKETING Jana's marketing plan has a goal that each person who hears of their new product tells 10 people about it, each of those 10 people tells another 10, and so on. The level of spread is the number of times this repeats, as shown in the table. How many orders of magnitude greater is the level 4 spread compared to the level 1 spread?

Level	0	1	2	3	4
Spread	1	10	100	1000	10,000
Powers	10^0	10^1	10^2	10^3	10^4

36. ASTRONOMY The diagram shows the distance from Earth to the Sun and the distance from Saturn to the Sun. Approximately what portion of the distance from Saturn to the Sun is the distance from Earth to the Sun?

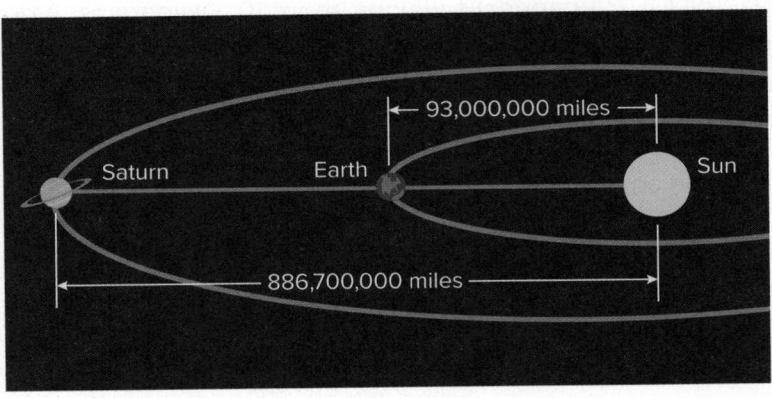

37. COMPUTERS In 1995, standard capacity for a personal computer hard drive was 40 megabytes (MB). In 2010, a standard hard drive capacity was 500 gigabytes (GB). Refer to the table.

Memory Capacity Approximate Conversions
8 bits = 1 byte
103 bytes = 1 kilobyte
103 kilobytes = 1 megabyte
103 megabytes = 1 gigabyte
103 gigabytes = 1 terabyte
103 terabytes = 1 petabyte

a. The newer hard drives have about how many times the capacity of the 1995 drives?

b. Predict the hard drive capacity in the year 2025 if the capacity continues to grow by the same factor you found in part **a**.

c. One kilobyte of memory is what fraction of one terabyte?

38. MICROSCOPES Cheveyo is looking at a slide of red blood cells under a microscope. One of the red blood cells has an actual radius of about one millionth of a meter, or 10^{-6} meters. Through the microscope, the radius of that red blood cell appears to be 10^{-2} meters. How many times larger does the microscope make the blood cells appear?

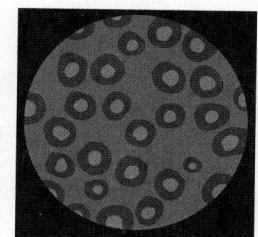

39. STRUCTURE Are $4n^5$ and $4n^{-5}$ reciprocals? Explain.

40. REGULARITY Explain why $3^0 = 1$.

41. STRUCTURE Use the Power of a Power Property to rewrite $\frac{1}{p^{-3}}$ with only positive exponents.

42. REASONING Angelo believes that $y^6 \geq y^{-6}$ for all nonzero values of y. Do you agree? Explain.

43. BACTERIA The bacterial population in a petri dish was measured at 1500 per cubic centimeter Monday morning. Each day the population doubled from the previous day. The model $P = 1500(2^d)$ where d is the day and P is the population. If the population was first measured when $d = 0$, find the population of the bacteria the day before, when $d = -1$.

44. VEHICLE DEPRECIATION Jay buys a used car for $12,000. If the car loses about 10% of its value every year, the value of the car can be modeled by the equation $V = 12{,}000(0.90^y)$, where y is the number of years and V is the value of the car.

a. Use the equation to determine the value of Jay's car 2 years before he purchased it.

b. How much money did Jay save by waiting to purchase the car?

45. Consider the expression $\left(\frac{4a^3}{2a^{-2}}\right)^4$.

   a. Simplify $\left(\frac{4a^3}{2a^{-2}}\right)^4$ by using the Quotient of Powers Property first. Then use the Power of a Power Property.

   b. Simplify $\left(\frac{4a^3}{2a^{-2}}\right)^4$ by using the Power of a Quotient Property first. Then use the Quotient of Powers Property.

   c. Write a statement that generalizes the results of part **a** and part **b**.

46. REASONING Dalip evaluated -5^0 on his calculator.

   a. What answer did the calculator give Dalip when he evaluated -5^0?

   b. Is this a counterexample that contradicts the Zero Exponent Property? Explain.

🧠 **Higher-Order Thinking Skills**

47. FIND THE ERROR Colleen and Tyler are using different steps to simplify $\left(\frac{x^2}{x^9}\right)^3$. Their work is shown in the boxes.

   a. Which student is correct? Explain your reasoning.

   b. Which student's answer is in simplest form?

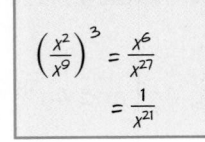

Colleen	Tyler
$\left(\frac{x^2}{x^9}\right)^3 = \frac{x^6}{x^{27}}$	$\left(\frac{x^2}{x^9}\right)^3 = \left(\frac{1}{x^7}\right)^3$
$= \frac{1}{x^{21}}$	$= x^{-21}$

48. PERSEVERE Use the Quotient of Powers Property to explain why $x^{-n} = \frac{1}{x^n}$.

49. FIND THE ERROR Lola claims that we never need to use the Division Property of Exponents. She says that exponents in the denominator can be written as negative exponents, and then the Multiplication Property of Exponents can be used instead. Is Lola correct? If so, give an example. If not, explain why.

50. WRITE Explain how to evaluate the expression 0^0.

51. CREATE Measure something large and something very small in the same units. Write and solve a problem comparing the order of magnitude of the two measurements.

52. ANALYZE Consider the expression, $\frac{a^b c^{-d}}{w^x y^{-z}}$, where all values are nonnegative. Make a conjecture about why only c and w cannot be zero.

53. ANALYZE Dale wrote the value of several powers of 2 on a piece of paper. Describe the pattern shown. Then explain how to use the pattern to find 2^0.

$$2^4 = 2 \times 2 \times 2 \times 2 = 16$$
$$2^3 = 2 \times 2 \times 2 = 8$$
$$2^2 = 2 \times 2 = 4$$
$$2^1 = 2 = 2$$

Rational Exponents

Today's Goals
• Rewrite expressions involving *n*th roots and rational exponents.

• Rewrite expressions involving powers of *n*th roots and rational exponents.

Today's Vocabulary
rational exponent
*n*th root
index
radicand

Explore Expressions with Rational Exponents

🔘 **Online Activity** Use an interactive tool to complete the Explore.

×

@ **INQUIRY** How can you simplify expressions with rational exponents?

Learn *n*th Roots

Not all exponents are integers. Exponents that are expressed as a fraction are called **rational exponents**. One of the most commonly used rational exponents is $\frac{1}{2}$. Expressions with an exponent of $\frac{1}{2}$ can also be represented as a square root.

Key Concept • Powers of One Half	
Words	For any nonnegative real number b, $b^{\frac{1}{2}} = \sqrt{b}$.
Example	$64^{\frac{1}{2}} = \sqrt{64}$ or 8

You are likely familiar with the square root of a number a. In the same way, you can find other roots of numbers. For example, if $a^3 = b$, then a is the cube root of b, and if $a^n = b$ for a positive integer n, then a is the **nth root** of b.

For example, $\sqrt[5]{18}$ is read as *the fifth root of 18*. In this example, 5 is the **index** and 18 is the **radicand**, which is the expression inside the radical symbol.

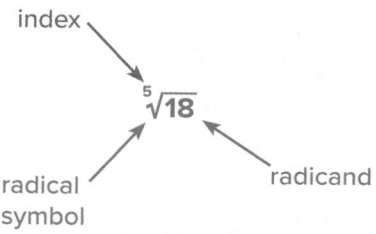

index

$\sqrt[5]{18}$

radical symbol

radicand

Key Concept • *n*th Roots	
Words	For any real numbers a and b and any positive integer n, if $a^n = b$, then a is the nth root of b.
Symbols	If $a^n = b$, then $a = \sqrt[n]{b}$.
Example	Because $3^4 = 81$, 3 is the fourth root of 81; $\sqrt[4]{81} = 3$

🔘 **Go Online** You may want to complete the Concept Check to check your understanding.

🔘 **Go Online** You can complete an Extra Example online.

Since $3^2 = 9$ and $(-3)^2 = 9$, both 3 and -3 are square roots of 9. Similarly, since $2^4 = 16$ and $(-2)^4 = 16$, both 2 and -2 are fourth roots of 16. The positive roots are called *principal roots*. Radical symbols indicate principal roots, so $\sqrt[4]{16} = 2$.

Key Concept • Rational Exponents	
Words	For any nonnegative real number b and any positive integer n, $b^{-n} = \sqrt[n]{b}$.
Example	$32^{\frac{1}{5}} = \sqrt[5]{32} = \sqrt[5]{2 \cdot 2 \cdot 2 \cdot 2 \cdot 2}$ or 2

Example 1 Radical and Exponential Forms

Write each expression in radical or exponential form.

a. $10^{\frac{1}{2}}$

By the definition of $b^{\frac{1}{2}}$, $10^{\frac{1}{2}} = \sqrt{10}$.

b. $\sqrt{7k^2}$

Because $7k^2$ is all under the radical symbol, both 7 and k^2 must be raised to the $\frac{1}{2}$ power. So, $\sqrt{7k^2} = 7^{\frac{1}{2}} K$

Check

Write each expression in radical or exponential form.

a. $\sqrt{16p}$ = ___?___

b. $34^{\frac{1}{2}}$ = ___?___

c. $8\sqrt{p}$ = ___?___

d. $(3p)^{\frac{1}{2}}$ = ___?___

Example 2 Evaluate *n*th Roots

Evaluate each expression.

a. $\sqrt[5]{243}$

$\sqrt[5]{243} = \sqrt[5]{3 \cdot 3 \cdot 3 \cdot 3 \cdot 3}$ $243 = 3 \cdot 3 \cdot 3 \cdot 3 \cdot 3$

$= 3$ Simplify.

b. $\sqrt[4]{625}$

$\sqrt[4]{625} = \sqrt[4]{5 \cdot 5 \cdot 5 \cdot 5}$ $625 = 5 \cdot 5 \cdot 5 \cdot 5$

$= 5$ Simplify.

Check

Evaluate each expression.

a. $\sqrt[4]{4096}$

b. $\sqrt[5]{16,807}$

Go Online You can complete an Extra Example online.

Think About It!

How do $(6x)^{\frac{1}{2}}$ and $6x^{\frac{1}{2}}$ differ?

Study Tip

Calculators You can use a calculator to find *n*th roots. On many calculators, enter the value of *n*, press MATH, and choose $\sqrt[x]{\ }$. Then enter the radicand and compute.

Example 3 Evaluate Exponential Expressions with Rational Exponents

Evaluate $4096^{\frac{1}{6}}$.

$$4096^{\frac{1}{6}} = \sqrt[6]{4096}$$

$$\sqrt[6]{4096} = \sqrt[6]{4 \cdot 4 \cdot 4 \cdot 4 \cdot 4 \cdot 4}$$

$$= 4$$

$b^{\frac{1}{n}} = \sqrt[n]{b}$

$2096 = 4 \cdot 4 \cdot 4 \cdot 4 \cdot 4 \cdot 4$

Simplify.

Talk About It!

Malik says that $\sqrt[3]{64} = \sqrt[3]{8 \cdot 8} = 8$. Is he correct? If not, correct his mistake.

Check

Evaluate each expression.

a. $\left(\frac{81}{256}\right)^{\frac{1}{4}}$

b. $100{,}000^{\frac{1}{5}}$

Learn Powers of nth Roots

Key Concept • Powers of nth Roots	
Words	For any real number b and any integers m and $n > 1$, $b^{\frac{m}{n}} = \left(\sqrt[n]{b}\right)^m$ or $\sqrt[n]{b^m}$.
Examples	$216^{\frac{4}{3}} = \left(\sqrt[3]{216}\right)^4 = 6^4$ or 1296 $32^{\frac{2}{5}} = \sqrt[5]{32^2} = \sqrt[5]{1024} = \sqrt[5]{4 \cdot 4 \cdot 4 \cdot 4 \cdot 4}$ or 4

Sometimes, the exponent can be simplified before it is evaluated. For example,

$$4^{\frac{15}{5}} = 4^3$$

$$= 4 \cdot 4 \cdot 4$$

$$= 64$$

$\frac{15}{5} = 3$

$4^3 = 4 \cdot 4 \cdot 4$

Simplify.

Think About It!

Determine the value of $b^{\frac{m}{n}}$ if $m = n$. Explain your reasoning.

Example 4 Evaluate Expressions of Powers of nth Roots

Evaluate each expression.

a. $729^{\frac{5}{6}}$

$$729^{\frac{5}{6}} = \left(\sqrt[6]{729}\right)^5$$

$$= \left(\sqrt[6]{3 \cdot 3 \cdot 3 \cdot 3 \cdot 3 \cdot 3}\right)^5$$

$$= \left(\sqrt[6]{3^6}\right)^5$$

$$= 3^5 \text{ or } 243$$

$b^{\frac{m}{n}} = \left(\sqrt[n]{b}\right)^m$

$729 = 3 \cdot 3 \cdot 3 \cdot 3 \cdot 3 \cdot 3$

$3 \cdot 3 \cdot 3 \cdot 3 \cdot 3 \cdot 3 = 3^6$

$\sqrt[6]{3^6} = 3$

(continued on the next page)

Go Online You can watch a video to learn how to evaluate expressions involving rational exponents.

b. $\left(\dfrac{36}{49}\right)^{\frac{3}{2}}$

$$\left(\dfrac{36}{49}\right)^{\frac{3}{2}} = \dfrac{36^{\frac{3}{2}}}{49^{\frac{3}{2}}}$$ Power of a Quotient

$$= \dfrac{(\sqrt{36})^3}{(\sqrt{49})^3}$$ $b^{\frac{m}{n}} = \left(\sqrt[n]{b}\right)^m$

$$= \dfrac{6^3}{7^3}$$ $\sqrt{36} = 6$ and $\sqrt{49} = 7$

$$= \dfrac{216}{343}$$ Simplify.

Check

Evaluate each expression.

a. $49^{\frac{3}{2}}$

b. $243^{\frac{3}{5}}$

🌐 Example 5 Apply Rational Exponents

BIOLOGY For insects, the resting metabolic rate can be determined by $r = 4.14m^{\frac{2}{3}}$, where r is the resting metabolic rate in cubic milliliters of oxygen per hour and m is the body mass of the insect in milligrams. Determine the resting metabolic rate of a 125-mg ebony jewelwing damselfly.

$$r = 4.14m^{\frac{2}{3}}$$ Original equation

$$= 4.14(125)^{\frac{2}{3}}$$ $m = 125$

$$= 4.14\left(\sqrt[3]{125}\right)^2$$ $125^{\frac{2}{3}} = \left(\sqrt[3]{125}\right)^2$

$$= 4.14(5)^2$$ $\sqrt[3]{125} = 5$

$$= 4.14 \cdot 25$$ Simplify.

$$= 103.5$$ Simplify.

The resting metabolic rate of a ebony jewelwing damselfly is 103.5 cubic milliliters of oxygen per hour.

Use a Source

Find the weight of another insect, in milligrams, and determine its resting metabolism. Provide the name, weight, and resting metabolism of the insect, and compare it to the ebony jewelwing damselfly.

Check

PROFITS The profit p, in thousands of dollars, of a company is modeled by the equation $p = 3.7c^{\frac{6}{5}}$, where c is the number of customers in thousands. Determine the profits of the company if they have 32,000 customers.

A. $142,080

B. $236,800

C. $307,625.29

D. $942,717,779.90

🌐 **Go Online** You can complete an Extra Example online.

Practice

Example 1

Write each expression in radical or exponential form.

1. $15^{\frac{1}{2}}$

2. $24^{\frac{1}{2}}$

3. $4k^{\frac{1}{2}}$

4. $(12y)^{\frac{1}{2}}$

5. $\sqrt{26}$

6. $\sqrt{44}$

7. $2\sqrt{ab}$

8. $\sqrt{3xyz}$

Examples 2–4

Evaluate each expression.

9. $\left(\frac{1}{16}\right)^{\frac{1}{4}}$

10. $\sqrt[5]{3125}$

11. $729^{\frac{1}{3}}$

12. $\left(\frac{1}{32}\right)^{\frac{1}{5}}$

13. $\sqrt[6]{4096}$

14. $1024^{\frac{1}{5}}$

15. $\left(\frac{16}{625}\right)^{\frac{1}{4}}$

16. $\sqrt[6]{15,625}$

17. $117,649^{\frac{1}{6}}$

18. $\sqrt[4]{\frac{16}{81}}$

19. $\left(\frac{1}{81}\right)^{\frac{1}{4}}$

20. $\left(\frac{3125}{32}\right)^{\frac{1}{5}}$

21. $729^{\frac{5}{6}}$

22. $256^{\frac{3}{8}}$

23. $125^{\frac{4}{3}}$

24. $49^{\frac{5}{2}}$

25. $\left(\frac{9}{100}\right)^{\frac{3}{2}}$

26. $\left(\frac{8}{125}\right)^{\frac{4}{3}}$

Example 5

27. VELOCITY The velocity v in feet per second of a freely falling object that has fallen h feet can be represented by $v = 8h^{\frac{1}{2}}$. Find the velocity of an object if it has fallen a distance of 144 feet.

28. GEOMETRY The surface area S of a cube in square inches can be determined by $S = 6V^{\frac{2}{3}}$, where V is the volume of the cube in cubic inches. Find the surface area of a cube that has a volume of 4096 cubic inches.

29. PLANETS The average distance d in astronomical units that a planet is from the Sun can be modeled by $d = t^{\frac{2}{3}}$, where t is the number of Earth years that it takes for the planet to orbit the Sun. Find the average distance a planet is from the Sun if the planet has an orbit of 27 Earth years.

30. BIOLOGY The relationship between the mass m in kilograms of an organism and its metabolism P in Calories per day can be represented by $P = 73.3\sqrt[4]{m^3}$. Find the metabolism of an organism that has a mass of 16 kilograms.

31. PROFIT The profit P of a company, in thousands of dollars, can be modeled by $P = 12.75\sqrt[5]{c^2}$, where c is the number of customers in hundreds. What is the profit of the company if the company has 3200 customers?

32. TIRE MARKS When a driver applies the brakes, the tires lock but the car will continue to slide, leaving skid marks on the road. You can approximate the speed at which a car was traveling on a dry road based on the length of a skid mark left by the car using the formula Speed $= (30 \cdot \text{length} \cdot 0.75)^{\frac{1}{2}}$, where speed is measured in miles per hour and length is measured in feet. At approximately what speed was a car traveling if it left a 50-foot long skid mark? Round to the nearest tenth.

Mixed Exercises

Write each expression in radical form, or write each radical in exponential form.

33. $17^{\frac{1}{3}}$

34. $q^{\frac{1}{4}}$

35. $7b^{\frac{1}{3}}$

36. $m^{\frac{2}{3}}$

37. $\sqrt[3]{29}$

38. $\sqrt[5]{h}$

39. $2\sqrt[3]{a}$

40. $\sqrt[3]{xy^2}$

Simplify.

41. $\sqrt[3]{0.027}$

42. $\sqrt[4]{\dfrac{n^4}{16}}$

43. $a^{\frac{1}{3}} \cdot a^{\frac{2}{3}}$

44. $c^{\frac{1}{2}} \cdot c^{\frac{3}{2}}$

45. $(8^2)^{\frac{2}{3}}$

46. $\left(y^{\frac{3}{4}}\right)^{\frac{1}{2}}$

47. $9^{-\frac{1}{2}}$

48. $16^{-\frac{3}{2}}$

49. $(3^2)^{-\frac{3}{2}}$

50. $\left(81^{\frac{1}{4}}\right)^{-2}$

51. $k^{-\frac{1}{2}}$

52. $\left(d^{\frac{4}{3}}\right)^0$

53. USE A MODEL In economics, the Cobb-Douglas production function is commonly used to relate input to output. The general form of the function is $P = bL^{ax}K^y$. The values in the table are given for the U.S. economy for the years 1900 and 1920. For both parts, assume a is 1.

		1900	1920
L	Labor input	105	194
K	Capital output	107	407
b	Total factor productivity	1.01	1.01
x	Output elasticity of labor	$\frac{3}{4}$	$\frac{3}{4}$
y	Output elasticity of capital	$\frac{1}{4}$	$\frac{1}{4}$

a. Find the total production P for 1900, to the nearest whole number.
b. Find the total production P for 1920, to the nearest whole number.

54. BASKETBALL The formula $S = 4\pi\left(\frac{3V}{4\pi}\right)^{\frac{2}{3}}$ can be used to find the surface area of a sphere, where V represents its volume. A regulation basketball has a volume of about 456 cubic inches. How much leather is needed (surface area) to make a regulation basketball? Round your answer to the nearest tenth.

55. BIOLOGY The function $h(x) = 0.4x^{\frac{2}{3}}$ can be used to find the height h in meters of a female giraffe with mass x kilograms. What is the height of a female giraffe with a mass of 32.8 kilograms? Round your answer to the nearest tenth of a meter.

56. GRAPH The graph of $y = x^{\frac{2}{5}}$ is shown over the interval $0 < x < 10$. Copy and complete the table to find the values. Round to the nearest tenth if necessary.

x	0		5	
y		1		2.3

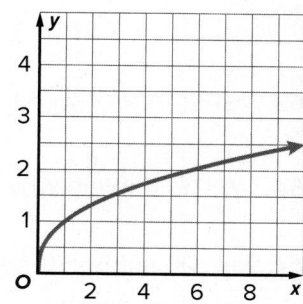

57. CONSTRUCT ARGUMENTS Use the properties of exponents to show which of the expressions are equivalent. Explain your reasoning.

a. $p^{\frac{f}{k}}$ b. $p^{\frac{f}{5}}$ c. $\sqrt[k]{p^f}$ d. $\left(\sqrt[k]{p}\right)^f$

58. USE TOOLS The amount of material needed to make a party hat in the shape of a cone can be found by calculating the hat's lateral area. The formula $LA = \pi r(h^2 + r^2)^{\frac{1}{2}}$ represents the lateral area of a cone for a given height h and radius r. Use your calculator to find the lateral area of a party hat that has a radius of 2 inches and a height of 7 inches. Round the answer to the nearest tenth of a square inch.

iStockphoto/Getty Images

59. GEOMETRY The surface area S of a cylinder in square centimeters can be determined by $S = 2\left(\frac{V}{h}\right) + 2\pi\left(\frac{V}{\pi h}\right)^{\frac{1}{2}}h$, where V is the volume of the cylinder in cubic centimeters and h is the height of the cylinder is centimeters. Find the surface area of a cylinder that has a volume of 160π cubic centimeters and a height of 10 centimeters. Write the surface area in terms of π.

60. PENDULUMS A pendulum is a weight hanging from a point of suspension so that it can swing freely. The formula $T = 2\pi\left(\frac{L}{32}\right)^{\frac{1}{2}}$ gives the time T in seconds it takes for a pendulum of length L in feet to complete one full cycle. How long will it take the pendulum shown to complete one full cycle? Write the length of time in terms of π.

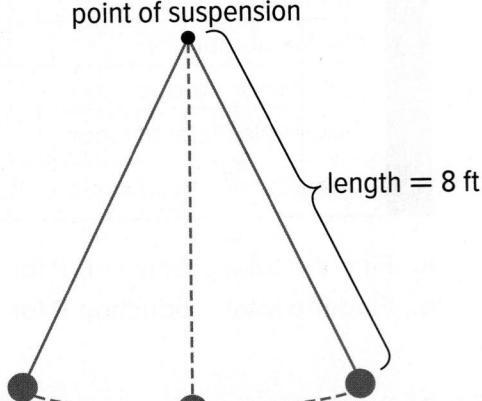

point of suspension

length = 8 ft

🧠 **Higher-Order Thinking Skills**

61. CREATE Write two different expressions with rational exponents equal to $\sqrt{2}$.

62. ANALYZE Determine whether each statement is *sometimes*, *always*, or *never* true. Assume that $x > 0$. Justify your argument.

a. $x^2 = x^{\frac{1}{2}}$

b. $x^{-2} = x^{\frac{1}{2}}$

c. $x^{\frac{1}{3}} = x^{\frac{1}{2}}$

d. $\sqrt{x} = x^{\frac{1}{2}}$

e. $\left(x^{\frac{1}{2}}\right)^2 = x$

f. $x^{\frac{1}{2}} \cdot x^2 = x$

63. PERSEVERE For what values of x is $x = x^{\frac{1}{3}}$?

64. WRITE Explain why $16^{\frac{1}{4}}$ will be less than 3.

65. FIND THE ERROR Sachi and Makayla are evaluating $27^{\frac{2}{3}}$. Is either correct? Explain your reasoning.

Sachi
$27^{\frac{2}{3}} = \sqrt[3]{27^2}$
$= \sqrt[3]{729}$
$= 9$

Makayla
$27^{\frac{2}{3}} = \sqrt[3]{27^2}$
$= 3^2$
$= 9$

66. WHICH ONE DOESN'T BELONG? Consider the four expressions shown. Which one does not belong? Explain your reasoning.

$\left(\sqrt[4]{w}\right)^5$

$\sqrt[4]{w^5}$

$w^{\frac{4}{5}}$

$w^{\frac{5}{4}}$

Simplifying Radical Expressions

Today's Goals
- Simplify square roots.
- Simplify cube roots.

Today's Vocabulary
radical expression
square root
perfect square
principal square root
cube root
perfect cube

Explore Square Roots and Negative Numbers

Online Activity Use an interactive tool to complete the Explore.

> ⊘ **INQUIRY** How can you simplify algebraic expressions with square roots?

Learn Simplifying Square Root Expressions

A **radical expression** contains a radical such as a square root. A **square root** of a number is a value that, when multiplied by itself, gives the number. So, because $7 \cdot 7 = 49$, a square root of 49 is 7. A number like 49 is sometimes called a perfect square. A **perfect square** is a rational number with a square root that is a rational number. Recall that the square root of a positive number has two possible solutions. For $\sqrt{a^2}$ where $a > 0$, $\sqrt{a^2} = \pm a$, because $a^2 = (-a)^2$. The positive solution a is called the **principal square root**.

Key Concept • Product Property of Square Roots	
Words	For any nonnegative real numbers a and b, the square root of ab is the square root of a times the square root of b.
Symbols	$\sqrt{ab} = \sqrt{a} \cdot \sqrt{b}$ if $a \geq 0$ and $b \geq 0$
Example	$\sqrt{4 \cdot 16} = \sqrt{4} \cdot \sqrt{16} = 2 \cdot 4$ or 8

Key Concept • Quotient Property of Square Roots	
Words	For any real numbers a and b, where $a \geq 0$ and $b > 0$, the square root of $\frac{a}{b}$ is equal to the square root of a divided by the square root of b.
Symbols	$\sqrt{\frac{a}{b}} = \frac{\sqrt{a}}{\sqrt{b}}$ if $a \geq 0$ and $b > 0$
Example	$\sqrt{\frac{100}{4}} = \sqrt{25} = 5$ or $\sqrt{\frac{100}{4}} = \frac{\sqrt{100}}{\sqrt{4}} = \frac{10}{2} = 5$

A square root is in simplest form if these three conditions are true.

- The square root contains no perfect square factors other than 1.

- The square root contains no fractions.

- There are no square roots in the denominator of a fraction.

Go Online You can complete an Extra Example online.

Study Tip

Prime Factorization
If you are trying to determine the prime factorization of a number and the number is even, then you know that the prime factorization contains at least one 2. Divide the number by 2 and repeat until you have an odd number, and then see if the odd number can be factored any further as a product of primes.

Example 1 Simplify Square Roots

Simplify $\sqrt{72}$.

$$\sqrt{72} = \sqrt{2 \cdot 2 \cdot 2 \cdot 3 \cdot 3}$$ Prime factorization of 72

$$= \sqrt{2} \cdot \sqrt{2} \cdot \sqrt{2} \cdot \sqrt{3} \cdot \sqrt{3}$$ Product Property of Square Roots

$$= \sqrt{2^2} \cdot \sqrt{2} \cdot \sqrt{3^2}$$ Product Property of Square Roots

$$= 2 \cdot \sqrt{2} \cdot 3 = 6\sqrt{2}$$ Simplify.

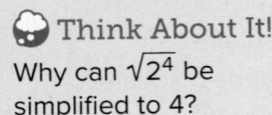 **Think About It!**

Why can $\sqrt{2^4}$ be simplified to 4?

Example 2 Multiply Square Roots

Simplify $\sqrt{5} \cdot \sqrt{48}$.

$$\sqrt{5} \cdot \sqrt{48} = \sqrt{5} \cdot (\sqrt{2} \cdot \sqrt{2} \cdot \sqrt{2} \cdot \sqrt{2} \cdot \sqrt{3})$$ Prime factorizations of 5 and 48

$$= \sqrt{5} \cdot (\sqrt{2^2} \cdot \sqrt{2^2} \cdot \sqrt{3})$$ Product Property of Square Roots

$$= 4\sqrt{5} \cdot \sqrt{3} = 4\sqrt{15}$$ Simplify.

Check

Simplify $\sqrt{8} \cdot \sqrt{14}$.

Example 3 Divide Square Roots

Simplify $\sqrt{\dfrac{24}{27}}$.

$$\sqrt{\frac{24}{27}} = \sqrt{\frac{8}{9}}$$ Reduce the radicand.

$$= \frac{\sqrt{8}}{\sqrt{9}}$$ Quotient Property of Square Roots

$$= \frac{2\sqrt{2}}{3}$$ Product Property of Square Roots

Example 4 Simplify Square Roots with Variables

Simplify $\sqrt{525x^4y^5z^5}$.

$$\sqrt{525x^4y^5z^5} = \sqrt{3 \cdot 5^2 \cdot 7 \cdot x^4 \cdot y^5 \cdot z^5}$$

$$= \sqrt{3} \cdot \sqrt{5^2} \cdot \sqrt{7} \cdot \sqrt{x^4} \cdot \sqrt{y^5} \cdot \sqrt{z^5}$$

$$= \sqrt{3} \cdot 5 \cdot \sqrt{7} \cdot \sqrt{x^4} \cdot \sqrt{y^4} \cdot \sqrt{y} \cdot \sqrt{z^4} \cdot \sqrt{z}$$

$$= \sqrt{3} \cdot 5 \cdot \sqrt{7} \cdot x^2 \cdot y^2 \cdot \sqrt{y} \cdot z^2 \cdot \sqrt{z}$$

$$= 5x^2y^2z^2\sqrt{21yz}$$

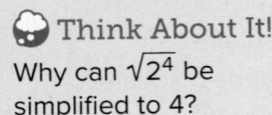 **Think About It!**

Why does the final answer include $\sqrt{21}$ when $\sqrt{21}$ does not appear in the previous step?

Check

Simplify $\sqrt{147x^3y^4z^5}$.

🔵 **Go Online** You can complete an Extra Example online.

⊕ Example 5 Write and Solve a Radical Equation

FINANCE **Sarah uses an online calculator to determine how much interest she would accrue in a savings account after two years. She plans to invest $750, and the calculator determines that she could have $780 after two years. She can find the interest rate r of the savings account by using the formula $r = \sqrt{\frac{a}{p}} - 1$, where a is the amount after two years and p is the initial investment. What is the interest rate?**

$$r = \sqrt{\frac{a}{p}} - 1 \qquad \text{Original formula}$$

$$= \sqrt{\frac{780}{750}} - 1 \qquad a = 780; \ p = 750$$

$$= \sqrt{\frac{26}{25}} - 1 \qquad \text{Simplify the radicand.}$$

$$= \frac{\sqrt{26}}{\sqrt{25}} - 1 \qquad \text{Quotient Property of Square Roots}$$

$$= \frac{\sqrt{26}}{5} - 1 \qquad \text{Simplify.}$$

$$\approx 0.02 \qquad \text{Simplify.}$$

The interest rate is approximately 2%.

Learn Simplifying Cube Root Expressions

The **cube root** of a number is the value that, when multiplied by itself twice, gives the number. A **perfect cube** is a rational number with a cube root that is an integer. So, because $6 \cdot 6 \cdot 6 = 216$, the cube root of 216 is 6 and 216 is a perfect cube.

Key Concept • Product Property of Radicals	
Words	For any nonnegative real numbers a and b, the nth root of ab is equal to the nth root of a times the nth root of b.
Symbols	$\sqrt[n]{ab} = \sqrt[n]{a} \cdot \sqrt[n]{b}$ if $a \geq 0$ and $b \geq 0$
Examples	$\sqrt[3]{8 \cdot 64} = \sqrt[3]{512}$ or 8; $\sqrt[3]{8 \cdot 64} = \sqrt[3]{8} \cdot \sqrt[3]{64} = 2 \cdot 4$ or 8

Key Concept • Quotient Property of Radicals	
Words	For any real numbers a and b, where $a \geq 0$ and $b > 0$, the nth root of $\frac{a}{b}$ is equal to the nth root of a divided by the nth root of b.
Symbols	$\sqrt[n]{\frac{a}{b}} = \frac{\sqrt[n]{a}}{\sqrt[n]{b}}$ if $a \geq 0$ and $b > 0$
Examples	$\sqrt[3]{\frac{216}{8}} = \sqrt[3]{27}$ or 3; $\sqrt[3]{\frac{216}{8}} = \frac{\sqrt[3]{216}}{\sqrt[3]{8}} = \frac{6}{2}$ or 3

Example 6 Find Cube Roots

Simplify $\sqrt[3]{343}$.

$$\sqrt[3]{343} = \sqrt[3]{7 \cdot 7 \cdot 7} \qquad \text{Prime factorization of 343}$$

$$= 7 \qquad \text{Simplify.}$$

⊙ **Go Online** You can complete an Extra Example online.

⊜ Talk About It!

Could this equation be solved if $\sqrt{\frac{a}{p}}$ could not be simplified?

⊜ Talk About It!

What must be true for a cube root expression to be in simplest form?

Watch Out!

Cube Roots Remember, $\sqrt[3]{64}$ is not the same as $3\sqrt{64}$. The former means *the cube root of 64* or 4, and the latter means *3 times the square root of 64* or 24.

Think About It!

Explain why $\sqrt[3]{3} \cdot \sqrt[3]{3} \cdot \sqrt[3]{3}$ simplifies to 3.

Example 7 Simplify Cube Roots

Simplify $\sqrt[3]{135}$.

$$\sqrt[3]{135} = \sqrt[3]{3 \cdot 3 \cdot 3 \cdot 5} \qquad \text{Prime factorization of 135}$$

$$= \sqrt[3]{3} \cdot \sqrt[3]{3} \cdot \sqrt[3]{3} \cdot \sqrt[3]{5} \qquad \text{Product Property of Radicals}$$

$$= 3\sqrt[3]{5} \qquad \text{Simplify.}$$

Think About It!

Why are you able to simplify the radicand before finding the cube root?

Example 8 Multiply Cube Roots

Simplify $\sqrt[3]{2} \cdot \sqrt[3]{36}$.

$$\sqrt[3]{2} \cdot \sqrt[3]{36} = \sqrt[3]{2} \cdot \left(\sqrt[3]{2 \cdot 2 \cdot 3 \cdot 3}\right) \qquad \text{Prime factorizations of 2 and 36}$$

$$= \sqrt[3]{2} \cdot \sqrt[3]{2} \cdot \sqrt[3]{2} \cdot \sqrt[3]{3} \cdot \sqrt[3]{3} \qquad \text{Product Property of Radicals}$$

$$= \sqrt[3]{8} \cdot \sqrt[3]{9} \qquad \text{Product Property of Radicals}$$

$$= 2\sqrt[3]{9} \qquad \text{Simplify.}$$

Think About It!

How would your answer change if you simplified $\sqrt[3]{\dfrac{15x^2y^3z^4}{81x^4}}$ instead of $\sqrt[3]{\dfrac{15x^2y^3z^4}{81x^5}}$? Is this answer in simplest form? Why or why not?

Example 9 Divide Cube Roots

Simplify $\sqrt[3]{\dfrac{168}{375}}$.

$$\sqrt[3]{\frac{168}{375}} = \sqrt[3]{\frac{2 \cdot 2 \cdot 2 \cdot 3 \cdot 7}{3 \cdot 5 \cdot 5 \cdot 5}} \qquad \text{Prime factorizations of 168 and 375}$$

$$= \sqrt[3]{\frac{2 \cdot 2 \cdot 2 \cdot 7}{5 \cdot 5 \cdot 5}} \qquad \text{Simplify the radicand.}$$

$$= \frac{\sqrt[3]{2 \cdot 2 \cdot 2 \cdot 7}}{\sqrt[3]{5 \cdot 5 \cdot 5}} \qquad \text{Quotient Property of Radicals}$$

$$= \frac{\sqrt[3]{2} \cdot \sqrt[3]{2} \cdot \sqrt[3]{2} \cdot \sqrt[3]{7}}{\sqrt[3]{5} \cdot \sqrt[3]{5} \cdot \sqrt[3]{5}} \qquad \text{Product Property of Radicals}$$

$$= \frac{2\sqrt[3]{7}}{5} \qquad \text{Simplify.}$$

Example 10 Simplify Cube Roots with Variables

Simplify $\sqrt[3]{\dfrac{15x^2y^3z^4}{81x^5}}$.

$$\sqrt[3]{\frac{15x^2y^3z^4}{81x^5}} = \sqrt[3]{\frac{3 \cdot 5 \cdot x \cdot x \cdot y \cdot y \cdot y \cdot z \cdot z \cdot z \cdot z}{3 \cdot 3 \cdot 3 \cdot 3 \cdot x \cdot x \cdot x \cdot x \cdot x}}$$

$$= \sqrt[3]{\frac{5 \cdot y \cdot y \cdot y \cdot z \cdot z \cdot z \cdot z}{3 \cdot 3 \cdot 3 \cdot x \cdot x \cdot x}}$$

$$= \sqrt[3]{\frac{5 \cdot y \cdot y \cdot y \cdot z \cdot z \cdot z \cdot z}{3 \cdot 3 \cdot 3 \cdot x \cdot x \cdot x}}$$

$$= \frac{\sqrt[3]{5 \cdot y \cdot y \cdot y \cdot z \cdot z \cdot z \cdot z}}{\sqrt[3]{3 \cdot 3 \cdot 3 \cdot x \cdot x \cdot x}}$$

$$= \frac{\sqrt[3]{5} \cdot \sqrt[3]{y} \cdot \sqrt[3]{y} \cdot \sqrt[3]{y} \cdot \sqrt[3]{z} \cdot \sqrt[3]{z} \cdot \sqrt[3]{z} \cdot \sqrt[3]{z}}{\sqrt[3]{3} \cdot \sqrt[3]{3} \cdot \sqrt[3]{3} \cdot \sqrt[3]{x} \cdot \sqrt[3]{x} \cdot \sqrt[3]{x}}$$

$$= \frac{yz\sqrt[3]{5z}}{3x}$$

Go Online You can complete an Extra Example online.

Watch Out!

Simplifying Cube Roots Be sure to simplify the fraction before finding the cube root.

Practice

🔄 **Go Online** You can complete your homework online.

Examples 1–4

Simplify each expression.

1. $\sqrt{52}$

2. $\sqrt{56}$

3. $\sqrt{72}$

4. $\sqrt{162}$

5. $\sqrt{243}$

6. $\sqrt{245}$

7. $\sqrt{5} \cdot \sqrt{10}$

8. $\sqrt{10} \cdot \sqrt{20}$

9. $\sqrt{2} \cdot \sqrt{10}$

10. $\sqrt{5} \cdot \sqrt{60}$

11. $\sqrt{8} \cdot \sqrt{12}$

12. $\sqrt{11} \cdot \sqrt{22}$

13. $\sqrt{\frac{75}{49}}$

14. $\sqrt{\frac{8}{81}}$

15. $\sqrt{\frac{192}{216}}$

16. $\sqrt{\frac{88}{18}}$

17. $\sqrt{\frac{84}{121}}$

18. $\sqrt{\frac{75}{20}}$

19. $\sqrt{16b^4}$

20. $\sqrt{40x^4y^8}$

21. $\sqrt{81a^{12}d^4}$

22. $\sqrt{75qr^3}$

23. $\sqrt{28a^4b^3}$

24. $\sqrt{w^{13}x^5z^8}$

Example 5

25. NATURE In 2010, an earthquake below the ocean floor initiated a devastating tsunami in Sumatra. Scientists can approximate the velocity V in feet per second of a tsunami in water of depth d feet with the formula $V = \sqrt{16} \cdot \sqrt{d}$. Determine the velocity of a tsunami in 300 feet of water. Write your answer in simplified radical form.

26. PHYSICAL SCIENCE The average velocity V of gas molecules is represented by the formula $V = \sqrt{\frac{3kt}{m}}$. The velocity is measured in meters per second, t is the temperature in Kelvin, and m is the molar mass of the gas in kilograms per mole. The variable k represents a value called the molar gas constant, which is about 8.3. Write a simplified radical expression that represents the average velocity of gas molecules that have a temperature of 300 Kelvin and a mass of 0.045 kilogram per mole. Round to the nearest meter per second.

27. CLOCKS A grandfather clock has a pendulum that swings back and forth. The time t in seconds it takes the pendulum to complete one full cycle is given by the formula $t = 0.2\sqrt{p}$, where p is the length of the pendulum in centimeters. How long is a full cycle if the pendulum is 22 cm long? Round your answer to the nearest tenth of a second.

28. HORIZON The distance, d, in miles that a person can see to the horizon can be modeled by the formula $d = \sqrt{\dfrac{3h}{2}}$ where h is the person's height above sea level in feet. How far, in miles, to the horizon can a person see if they are 60 feet above sea level? Write your answer in simplified radical form.

Examples 6–10

Simplify each expression.

29. $\sqrt[3]{8}$

30. $\sqrt[3]{125}$

31. $\sqrt[3]{216}$

32. $\sqrt[3]{64}$

33. $\sqrt[3]{27}$

34. $\sqrt[3]{1000}$

35. $\sqrt[3]{243}$

36. $\sqrt[3]{4000}$

37. $\sqrt[3]{162}$

38. $\sqrt[3]{875}$

39. $\sqrt[3]{80}$

40. $\sqrt[3]{384}$

41. $\sqrt[3]{3} \cdot \sqrt[3]{32}$

42. $\sqrt[3]{24} \cdot \sqrt[3]{18}$

43. $\sqrt[3]{12} \cdot \sqrt[3]{6}$

44. $\sqrt[3]{16} \cdot \sqrt[3]{20}$

45. $\sqrt[3]{9} \cdot \sqrt[3]{3}$

46. $\sqrt[3]{2} \cdot \sqrt[3]{6}$

47. $\sqrt[3]{\dfrac{162}{375}}$

48. $\sqrt[3]{\dfrac{648}{750}}$

49. $\sqrt[3]{\dfrac{18}{128}}$

50. $\sqrt[3]{\dfrac{162}{3}}$

51. $\sqrt[3]{\dfrac{192}{3}}$

52. $\sqrt[3]{\dfrac{500}{2}}$

53. $\sqrt[3]{54b^8}$

54. $\sqrt[3]{10yz^5} \cdot \sqrt[3]{4y^3}$

55. $\sqrt[3]{\dfrac{192x^7}{1029x^{10}}}$

56. $\sqrt[3]{\dfrac{640a^3b^8}{5ab^4}}$

57. $\sqrt[3]{54x^2y^8} \cdot \sqrt[3]{5x^5y^4}$

58. $\sqrt[3]{25p^2r^5} \cdot \sqrt[3]{30q^3r}$

Mixed Exercises

Simplify each expression.

59. $\sqrt{245m^9n^5}$

60. $2\sqrt{5} \cdot 7\sqrt{10}$

61. $\sqrt{\dfrac{96d^4e^2f^8}{75de^6}}$

62. $-11\sqrt[3]{250}$

63. $\sqrt[3]{320x^{14}y^{17}z^{20}}$

64. $\sqrt[3]{45k^4m^{10}} \cdot \sqrt[3]{32k^7m^3}$

65. $\sqrt[3]{\dfrac{264w^6xy^7}{3w^6x^4y^3}}$

66. $\sqrt[3]{\dfrac{48a^7}{125b^9}}$

67. **AREA** The base of a triangle measures $6\sqrt{2}$ meters and the height measures $3\sqrt{6}$ meters. What is the area?

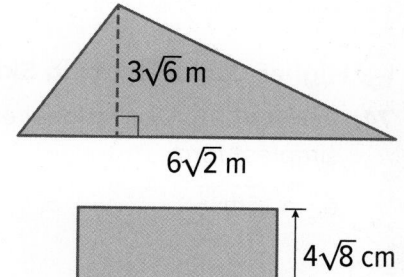

68. **AREA** The length of a rectangle measures $8\sqrt{12}$ centimeters and the width measures $4\sqrt{8}$ centimeters. What is the area of the rectangle?

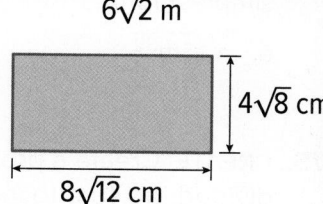

69. **MEAN** The geometric mean of two numbers h and k can be found by evaluating $\sqrt{h \cdot k}$. Find the geometric mean of 32 and 14 in simplified radical form.

70. **THEATRE CREW** Fernanda oversees painting props for a theatre production. She needs to create a cube that will be used as a prop by the actors. She has enough paint to cover 5 square yards. The formula $s = \sqrt{\dfrac{A}{6}}$ gives the longest side length s in yards of a cubic prop Fernanda can make, where A is the surface area to be covered with paint. What is the longest side length of a cube Fernanda could make and not have to purchase any more paint? Round to the nearest tenth of a yard.

71. **SCALE MODEL** While on vacation, Deon decided to purchase a scale model of the Empire State Building. Before making the purchase, Deon wants to determine the maximum height of a model that will fit inside his suitcase. Deon's suitcase measures 22 inches long by 14 inches wide by 9 inches tall. The diagonal length D is given by $D = \sqrt{L^2 + W^2 + H^2}$, where L is the length in inches, W is the width in inches, and H is the height in inches. Find the length of the tallest scale model that will fit inside Deon's suitcase. Round your answer to the nearest tenth of an inch.

72. **BALLOONS** The radius of a sphere r is given in terms of its volume V by the formula $r = \sqrt[3]{\dfrac{0.75V}{\pi}}$. By how many inches has the radius of a spherical balloon increased when the amount of air in the balloon is increased from 4.5 cubic feet to 4.7 cubic feet? Round your answer to the nearest hundredth.

73. REGULARITY The Product Property of Square Roots and the Quotient Property of Square Roots can be written in symbols as $\sqrt{ab} = \sqrt{a} \cdot \sqrt{b}$ and $\sqrt{\frac{a}{b}} = \frac{\sqrt{a}}{\sqrt{b}}$, respectively.

a. Explain the Product Property of Square Roots and discuss any limitations of a and b for this property.

b. Explain the Quotient Property of Square Roots and discuss any limitations of a and b for this property.

c. Discuss any similarities of the two properties.

Higher-Order Thinking Skills

74. PERSEVERE Use rational exponents to find an equivalent radical expression in simplest form.

a. $\sqrt[4]{16m^{32}}$

b. $(\sqrt{x})(\sqrt[3]{x})$

c. $\sqrt[3]{\sqrt{b}}$

75. CREATE Create a problem where two square roots are being either multiplied or divided. Be sure to include at least one variable in your problem. Solve your problem.

76. PERSEVERE Margarita takes a number, subtracts 4, multiplies by 4, takes the square root, and takes the reciprocal to get $\frac{1}{2}$. What number did she start with? Write a formula to describe the process.

77. ANALYZE Find a counterexample to show that the following statement is false. *If you take the square root of a number, the result will always be less than the original number.*

78. ANALYZE Order the expressions from least to greatest. $\sqrt{47}, 9, \sqrt[3]{421}, \sqrt{85}$

79. PERSEVERE If the area of a rectangle is $144\sqrt{5}$ square inches, what are possible dimensions of the rectangle? Explain your reasoning.

80. WRITE Describe the required conditions for a radical expression to be in simplest form.

Operations With Radical Expressions

Learn Adding and Subtracting Radical Expressions

To add or subtract radical expressions, the radicands must be alike in the same way that monomial terms must be alike to add or subtract.

Monomials

$$5a + 3a = (5 + 3)a$$
$$= 8a$$
$$7b - 2b = (7 - 2)b$$
$$= 5b$$

Radical Expressions

$$5\sqrt{3} + 3\sqrt{3} = (5 + 3)\sqrt{3}$$
$$= 8\sqrt{3}$$
$$7\sqrt{5} - 2\sqrt{5} = (7 - 2)\sqrt{5}$$
$$= 5\sqrt{5}$$

Notice that when adding and subtracting radical expressions, the radicand does not change. This is the same as when adding or subtracting monomials.

Example 1 Add and Subtract Expressions with Like Radicands

Simplify $4\sqrt{5} + 3\sqrt{7} - 2\sqrt{5} + 7\sqrt{7}$.

$$4\sqrt{5} + 3\sqrt{7} - 2\sqrt{5} + 7\sqrt{7} = (4 - 2)\sqrt{5} + (3 + 7)\sqrt{7}$$
$$= 2\sqrt{5} + 10\sqrt{7}$$

Check

Simplify $17\sqrt{19} - 14\sqrt{6} + 11\sqrt{6} - 3\sqrt{19}$.

A. 0

B. $14\sqrt{19} - 3\sqrt{6}$

C. $11\sqrt{25}$

D. $28\sqrt{19} - 17\sqrt{6}$

Go Online You can complete an Extra Example online.

Today's Goals
- Add and subtract radical expressions.
- Multiply radical expressions.

Think About It!
Can you simplify $6\sqrt{3} + 7\sqrt{5}$? If so, simplify it, and if not, explain why.

Watch Out!
Radicands Before applying the Distributive Property, be certain that you group expressions with like radicands.

Think About It!
Jaylen tries to simplify $3\sqrt{5} - 3\sqrt{5}$ and determines that the simplest form is $0\sqrt{5}$. Is he correct? Why or why not?

Example 2 Add and Subtract Expressions with Unlike Radicands

Simplify $3\sqrt{75} - 6\sqrt{48} - \sqrt{27}$.

$3\sqrt{75} - 6\sqrt{48} - \sqrt{27}$

$$= 3(\sqrt{5^2}) \cdot \sqrt{3}) - 6(\sqrt{4^2} \cdot \sqrt{3}) - (\sqrt{3^2}) \cdot \sqrt{3}) \quad \text{Product Property of Square Roots}$$

$$= 3(5\sqrt{3}) - 6(4\sqrt{3}) - 3\sqrt{3} \quad \text{Simplify.}$$

$$= 15\sqrt{3} - 24\sqrt{3} - 3\sqrt{3} \quad \text{Multiply.}$$

$$= (15 - 24 - 3)\sqrt{3} \quad \text{Distributive Property}$$

$$= -12\sqrt{3} \quad \text{Simplify.}$$

Check

Simplify $-11\sqrt{50} + 2\sqrt{32} - 18\sqrt{8}$.

A. $99\sqrt{2}$

B. $-83\sqrt{2}$

C. $-27\sqrt{2}$

D. $-11\sqrt{50} - 14\sqrt{8}$

🌐 Example 3 Use Radical Expressions

DECORATIONS **Kate is decorating a rectangular pavilion for a party and wants to string lights along the edge of the roof. Two of the sides have a length of $13\sqrt{5}$ feet, and the other two sides have a length of $6\sqrt{45}$ feet. How many feet of lights will Kate need to decorate the entire pavilion?**

Because there are two sides $13\sqrt{5}$ feet long and two sides $6\sqrt{45}$ feet long, the expression $13\sqrt{5} + 13\sqrt{5} + 6\sqrt{45} + 6\sqrt{45}$ represents the total perimeter of the roof.

$$13\sqrt{5} + 13\sqrt{5} + 6\sqrt{45} + 6\sqrt{45} = (13 + 13)\sqrt{5} + (6 + 6)\sqrt{45}$$

$$= 26\sqrt{5} + 12\sqrt{45}$$

$$= 26\sqrt{5} + 12(\sqrt{5} \cdot \sqrt{3^2})$$

$$= 26\sqrt{5} + 36\sqrt{5}$$

$$= 62\sqrt{5}$$

Kate will need $62\sqrt{5}$ feet of lights to decorate the pavilion.

🅝 **Go Online** You can complete an Extra Example online.

Talk About It!

If a radical expression has addition or subtraction with unlike radicands, what must you check before determining whether the terms can be added or subtracted? Explain your reasoning.

Check

CONSTRUCTION A business is adding wheelchair ramps to their building. The three ramps will require $10\sqrt{28}$ ft³, $13\sqrt{7}$ ft³, and $4\sqrt{112}$ ft³ of concrete. How much concrete will be needed to build the three ramps?

A. 27 ft³

B. $27\sqrt{147}$ ft³

C. $49\sqrt{7}$ ft³

D. $117\sqrt{7}$ ft³

Learn Multiplying Radical Expressions

Multiplying radical expressions is similar to multiplying monomials. Let $x \geq 0$.

Monomials

$$5x(3x) = 5 \cdot 3 \cdot x \cdot x$$

$$= 15x^2$$

Radical Expressions

$$5\sqrt{x}(3\sqrt{x}) = 5 \cdot 3 \cdot \sqrt{x} \cdot \sqrt{x}$$

$$= 15x$$

You can apply the Distributive Property to radical expressions. You can also multiply radical expressions with more than one term in each factor. This is similar to multiplying two binomials.

Binomials

$$(2x + 3x)(4x + 5x) = 2x(4x) + 2x(5x) + 3x(4x) + 3x(5x)$$

$$= 8x^2 + 10x^2 + 12x^2 + 15x^2$$

$$= 45x^2$$

Radical Expressions

$$(2\sqrt{2} + 3\sqrt{2})(4\sqrt{2} + 5\sqrt{2})$$

$$= 2\sqrt{2}(4\sqrt{2}) + 2\sqrt{2}(5\sqrt{2}) + 3\sqrt{2}(4\sqrt{2}) + 3\sqrt{2}(5\sqrt{2})$$

$$= 8\sqrt{2^2} + 10\sqrt{2^2} + 12\sqrt{2^2} + 15\sqrt{2^2}$$

$$= 45\sqrt{2^2}$$

$$= 90$$

Notice that when multiplying radical expressions, you multiply the radicands.

 Go Online You can complete an Extra Example online.

Example 4 Multiply Radical Expressions

Simplify $5\sqrt{3} \cdot 4\sqrt{6}$.

$$5\sqrt{3} \cdot 4\sqrt{6} = (5 \cdot 4)(\sqrt{3} \cdot \sqrt{6})$$
$$= 20\sqrt{18}$$
$$= 20(3\sqrt{2})$$
$$= 60\sqrt{2}$$

Check
Simplify $3\sqrt{10} \cdot (-9\sqrt{6})$.

Example 5 Multiply Radical Expressions by Using the Distributive Property

Simplify $4\sqrt{7}(2\sqrt{8} + 3\sqrt{7})$.

$$4\sqrt{7}(2\sqrt{8} + 3\sqrt{7}) = (4\sqrt{7} \cdot 2\sqrt{8}) + (4\sqrt{7} \cdot 3\sqrt{7})$$
$$= [(4 \cdot 2)(\sqrt{7} \cdot \sqrt{8})] + [(4 \cdot 3)(\sqrt{7} \cdot \sqrt{7})]$$
$$= 8\sqrt{56} + 12\sqrt{49}$$
$$= 16\sqrt{14} + 12(7)$$
$$= 16\sqrt{14} + 84$$

Check
Simplify $-4\sqrt{6}(7\sqrt{12} - 4\sqrt{8})$.

Go Online
to learn how to find the sums and products of rational and irrational numbers in Expand 10-6.

Pause and Reflect

Did you struggle with anything in this lesson? If so, how did you deal with it?

Go Online You can complete an Extra Example online.

Practice

📡 **Go Online** You can complete your homework online.

Examples 1 and 2

Simplify.

1. $7\sqrt{7} - 2\sqrt{7}$

2. $3\sqrt{13} + 7\sqrt{13}$

3. $7\sqrt{5} + 4\sqrt{5}$

4. $2\sqrt{6} + 9\sqrt{6}$

5. $12\sqrt{r} - 9\sqrt{r}$

6. $9\sqrt{6a} - 11\sqrt{6a} + 4\sqrt{6a}$

7. $3\sqrt{5} - 2\sqrt{20}$

8. $3\sqrt{50} - 3\sqrt{32}$

9. $2\sqrt{13} + 4\sqrt{2} - 5\sqrt{13} + \sqrt{2}$

10. $5\sqrt{8} + 2\sqrt{20} - \sqrt{8}$

11. $7\sqrt{3} - 2\sqrt{2} + 3\sqrt{2} + 5\sqrt{3}$

12. $8\sqrt{12} - \sqrt{5} - 4\sqrt{3}$

Example 3

13. ARCHITECTURE The Pentagon is the building that houses the U.S. Department of Defense. If the building is a regular pentagon with each side measuring $23\sqrt{149}$ meters, find the perimeter. Leave your answer as a radical expression.

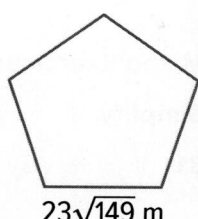

$23\sqrt{149}$ m

14. BIKING Iker rode his bike on three trails this week. The first was $2\sqrt{3}$ kilometers, the second was $4\sqrt{3}$ kilometers, and the third was $3\sqrt{3}$ kilometers. How long did Iker ride this week? Give your answer as a radical expression.

15. RAMP A ramp at a dog park is made of three sections. The incline and decline pieces are the same length, $13\sqrt{37}$ inches. The center is $10\sqrt{41}$ inches long. What is the total length of the ramp? Give your answer as a radical expression.

$10\sqrt{41}$ in.

$13\sqrt{37}$ in. $13\sqrt{37}$ in.

16. DISTANCE The diagonal length of a football field, including both end zones, is $40\sqrt{97}$ feet. The diagonal length of a practice field is $29\sqrt{97}$ feet. How much longer is the diagonal length of a football field than the diagonal length of the practice field? Give your answer as a radical expression.

17. CIRCLES A large circle has an area of 400 cm^2 and a small circle has an area of 200 cm^2. The radius of the large circle is $\sqrt{\frac{400}{\pi}}$ cm. The radius of the small circle is $\sqrt{\frac{200}{\pi}}$ cm. Find the difference of the radius of the large circle and the radius of the small circle.

18. GARDEN Camila put a frame around a triangular garden in her yard. The two legs of the triangle are 8 feet long each, and the third side is equal to the length of a leg times $\sqrt{2}$. What is the perimeter of the triangle? Give your answer as a simplified radical expression.

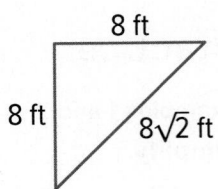

8 ft

8 ft

$8\sqrt{2}$ ft

Examples 4 and 5

Simplify.

19. $2\sqrt{3} \cdot 3\sqrt{15}$

20. $5\sqrt{3} \cdot 2\sqrt{21}$

21. $6\sqrt{7} \cdot 2\sqrt{8}$

22. $7\sqrt{10} \cdot 4\sqrt{10}$

23. $11\sqrt{6} \cdot 3\sqrt{12}$

24. $10\sqrt{5} \cdot 5\sqrt{11}$

25. $\sqrt{2}(\sqrt{8} + \sqrt{6})$

26. $\sqrt{5}(\sqrt{10} - \sqrt{3})$

27. $\sqrt{5}(\sqrt{2} + 4\sqrt{2})$

28. $\sqrt{6}(2\sqrt{10} + 3\sqrt{2})$

29. $4\sqrt{5}(3\sqrt{5} + 8\sqrt{2})$

30. $5\sqrt{3}(6\sqrt{10} - 6\sqrt{3})$

Mixed Exercises

Simplify.

31. $\sqrt{\frac{1}{25}} - \sqrt{5}$

32. $\sqrt{\frac{2}{9}} + \sqrt{6}$

33. $2 + 2\sqrt{2} - \sqrt{8}$

34. $8\sqrt{\frac{5}{4}} + 3\sqrt{20} - \sqrt{45}$

35. $\sqrt{24} - 5\sqrt{12} + 4\sqrt{2} - 3\sqrt{3}$

36. $\frac{1}{2}\sqrt{8} - \sqrt{\frac{2}{9}} + 9\sqrt{2}$

37. $5\sqrt{3} \cdot 3\sqrt{5}$

38. $8\sqrt{7} \cdot \frac{2}{3}\sqrt{7}$

39. $\sqrt{\frac{3}{4}}(\sqrt{6} + 2\sqrt{20})$

40. $6\sqrt{6}(9\sqrt{2} - 5\sqrt{12})$

41. $(a\sqrt{3} + b\sqrt{5})(c\sqrt{8} + d\sqrt{5})$

42. $(6\sqrt{2} + 2\sqrt{3})(\sqrt{10} - 4\sqrt{7})$

43. GEOMETRY The area of a trapezoid is found by multiplying its height by the average length of its bases. Find the area of the deck attached to Mr. Wilson's house. Give your answer as a simplified radical expression.

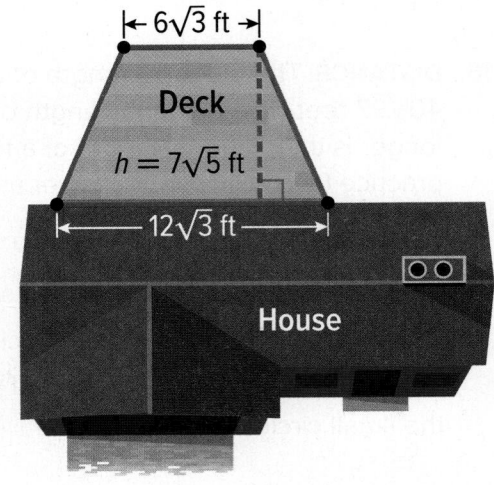

$6\sqrt{3}$ ft

Deck

$h = 7\sqrt{5}$ ft

$12\sqrt{3}$ ft

House

44. TRAVEL Lucia used the straight-line distance between Lincoln, Nebraska, and Houston, Texas, and the straight-line distance between Houston and Tallahassee, Florida, to estimate the straight-line distance between Lincoln and Tallahassee, as shown on the map. Lucia traveled from Lincoln to Tallahassee for vacation. A week after returning home from vacation, Lucia traveled from Lincoln to Houston to visit a relative. About how far did Lucia travel after returning home from Houston? Give your answer as a radical expression.

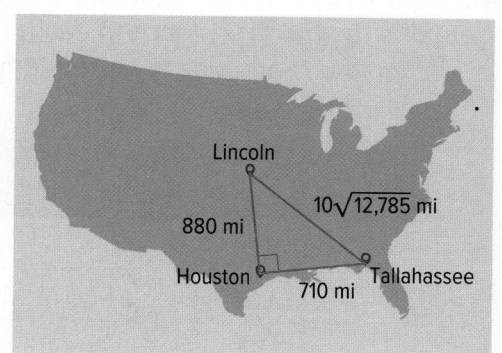

45. FREE FALL A ball is dropped from a building window 800 feet above the ground. Another ball is dropped from a lower window 288 feet high. Both balls are released at the same time. Assume air resistance is not a factor, and use the formula $t = \frac{1}{4}\sqrt{h}$ to find how many seconds t it will take a ball to fall h feet.

a. How much longer after the first ball hits the ground does the second ball land?

b. Find a decimal approximation of the answer for part **a**. Round your answer to the nearest tenth.

46. REASONING Deja is putting a fence around her yard. The long side of the yard is $12\sqrt{2}$ meters, the two shorter sides are each $5\sqrt{2}$ meters, and there is a $\sqrt{2}$-meter section that runs up to the house. When Deja buys the fencing, she approximates the length of fencing she needs. Write and solve an equation to find the total length of fencing, f, needed for the yard. Give your answer as a radical expression. Then approximate the amount of fencing Deja should buy to the nearest meter.

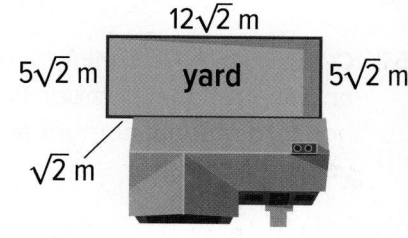

47. SHADOWS The distance from the top of Gino's head to the end of his head's shadow is the hypotenuse of a 45°-45°-90° triangle, or Gino's height times $\sqrt{2}$. Gino is 4.5 feet tall. The distance from a nearby tree to the end of its shadow is 2.5 times as long as the distance from the top of Gino's head to the end of his shadow. How far is it from the top of the tree to the end of its shadow? Give your answer as a radical expression.

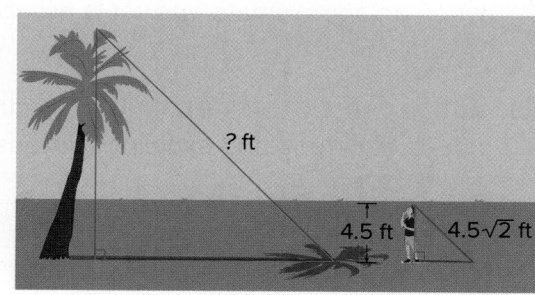

48. **STRUCTURE** The table shows several radical expressions. Copy and complete the table by writing *yes* or *no* in each row to indicate whether the expression in column A and the expression in column B have the same radicand when simplified.

Expression A	Expression B	Simplify to have same radicand? (Write Yes or No)
$4\sqrt{24}$	$6\sqrt{28}$	
$3\sqrt{18}$	$2\sqrt{32}$	
$5\sqrt{15}$	$12\sqrt{30}$	
$10\sqrt{20}$	$7\sqrt{45}$	

49. **SCALING** Austin has a model car that uses a 1:24 scale. The real car's hood is $20\sqrt{5}$ inches long. How long is the model car's hood? Give your answer as a radical expression.

50. **FIND THE ERROR** Jhumpa said $3\sqrt{32}$ cannot be subtracted from $8\sqrt{8}$ because they have different radicands. Kenyon said they can be subtracted. Is either correct? Explain your reasoning.

51. **PERSEVERE** Determine whether the following statement is *true* or *false*. Provide a proof or a counterexample to support your answer.

$(x + y)^2 > \left(\sqrt{x^2 + y^2}\right)^2$ when $x > 0$ and $y > 0$

52. **CONSTRUCT ARGUMENTS** Make a conjecture about the sum of a rational number and an irrational number. Is the sum *rational* or *irrational*? Is the product of a nonzero rational number and an irrational number *rational* or *irrational*? Explain your reasoning.

53. **CREATE** Write an equation that shows a sum of two radicals with different radicands. Explain how you could combine these terms.

54. **WRITE** A radical expression is an exact value. Rounding a radical gives an approximate value. Give examples of when an exact answer is preferred and when an approximate answer is acceptable.

55. **WHICH ONE DOESN'T BELONG?** Consider the expressions. Identify which one does not belong and explain your reasoning.

$10\sqrt{5}$	$2\sqrt{20}$	$6\sqrt{12}$	$8\sqrt{45}$

Exponential Equations

Today's Goals
• Solve exponential equations.

Today's Vocabulary
exponential equation

Explore Solving Exponential Equations

Online Activity Use an interactive tool to complete the Explore.

×

@ INQUIRY How can you solve an equation where the variable is in the exponent?

Learn Exponential Equations

In an **exponential equation**, the independent variable is an exponent. The Power Property of Equality and the other properties of exponents can be used to solve exponential equations.

Go Online You may want to complete the Concept Check to check your understanding.

Key Concept • Power Property of Equality

Words	For any real number $b > 0$ and $b \neq 1$, $b^x = b^y$ if and only if $x = y$.
Examples	If $4^x = 4^5$, then $x = 5$. If $n = \frac{2}{5}$, $5^n = 5^{\frac{2}{5}}$. If $x^{\frac{1}{2}} = 3^{\frac{1}{2}}$, then $x = 3$.

Example 1 Solve a One-Step Exponential Equation

Solve $5^x = 625$.

$$5^x = 625$$ Original equation

$$5^x = 5^4$$ Rewrite 625 as 5^4.

$$x = 4$$ Simplify.

Check

Solve $2^x = 512$.

$x = \underline{\ ?\ }$

Go Online You can complete an Extra Example online.

Think About It!

Can you use the Power
Property of Equality if
the two bases are not
equal? Justify your
argument.

Example 2 Solve a Multi-Step Exponential Equation

Solve $27^{x-1} = 3$.

$27^{x-1} = 3$	Original equation
$(3^3)^{x-1} = 3$	Rewrite 27 as 3^3.
$3^{3x-3} = 3$	Power of a Power Property, Distributive Property
$3^{3x-3} = 3^1$	Rewrite 3 as 3^1.
$3x - 3 = 1$	Power Property of Equality
$3x = 4$	Add 3 to each side.
$x = \frac{4}{3}$	Divide each side by 3.

Example 3 Solve a Real-World Exponential Equation

AEROSPACE In the 1950s, scientists proposed a space station that could house a crew of approximately 80 people. The station could produce artificial gravity by rotating at a speed of $w = \sqrt{gr}$, where g is 32 feet per second squared, and r is the radius of the station. If the station design required a rotating speed of approximately 64 feet per second to simulate gravity on Earth, what would the radius need to be?

$w = \sqrt{gr}$	Original equation
$64 = \sqrt{32r}$	$w = 64, g = 32$
$64 = (32r)^{\frac{1}{2}}$	$\sqrt[n]{b} = b^{\frac{1}{n}}$
$8^2 = (32r)^{\frac{1}{2}}$	$64 = 8^2$
$(8^4)^{\frac{1}{2}} = (32r)^{\frac{1}{2}}$	$8^2 = (8^4)^{\frac{1}{2}}$
$8^4 = 32r$	Power Property of Equality
$4096 = 32r$	$8^4 = 4096$
$128 = r$	Divide each side by 32.

The radius of the space station would need to be approximately 128 feet.

Need Help?

In order to apply the
Power Property of
Equality in this
example, the two
bases must be raised
to the same exponent.
Thus, you rewrite 8^2
as $(8^4)^{\frac{1}{2}}$ so that you can
set the bases equal to
each other.

Check

SUNSCREEN The degree to which sunscreen protects your skin is measured by its sun protection factor (SPF). For a sunscreen with SPF f, the percentage of UV-B rays absorbed p is $p = 50f^{\frac{1}{5}}$. What SPF absorbs 100% of UV-B rays?

$f = $ ___?___

 Go Online You can complete an Extra Example online.

Practice

Go Online You can complete your homework online.

Examples 1 and 2
Solve each equation.

1. $2^x = 512$

2. $3^x = 243$

3. $6^x = 46{,}656$

4. $5^x = 125$

5. $3^x = 6561$

6. $16^x = 4$

7. $3^{x-3} = 243$

8. $4^{x-1} = 1024$

9. $6^{x-1} = 1296$

10. $4^{2x+1} = 1024$

11. $2^{4x+3} = 2048$

12. $3^{3x+3} = 6561$

Example 3

13. ELECTRICITY The relationship of the current, power, and resistance of an appliance can be modeled by $I\sqrt{R} = \sqrt{P}$, where I is the current in amperes, P is the power in watts, and R is the resistance in ohms. Find the resistance that an appliance is using if the current is 2.5 amps and the power is 100 watts.

14. VIDEO Felipe uploaded a funny video of his dog. The relationship between the elapsed time in days, d, since the video was first uploaded and the total number of views, v, that the video received is modeled by $v = 4^{1.25d}$. Find the number of days it took Felipe's video to get 1024 views.

15. CONSTRUCTION A large plot of land has been purchased by developers. They roll out a schedule of construction. The relationship between the area of the undeveloped land in hectares, A, and the elapsed time in months, t, since the construction began is modeled by the function $A = 6250 \cdot 10^{-0.1t}$. How many months of construction will there be before the area of the undeveloped land decreases to 62.5 hectares?

Mixed Exercises

Solve each equation.

16. $2^{5x} = 8^{2x-4}$

17. $81^{2x-3} = 9^{x+3}$

18. $2^{4x} = 32^{x+1}$

19. $16^x = \frac{1}{2}$

20. $25^x = \frac{1}{125}$

21. $6^{8-x} = \frac{1}{216}$

22. $5^x = 125$

23. $2^{5x-4} = 64$

24. $4^{x+1} = 256$

25. $3^{4x-2} = 729$

26. USE A MODEL Without advertising, a Web site had 96 total visits. Today, the owners of the site are starting a new promotion, which is expected to double the total number of visits to their Web site every 5 days.

    a. Write an equation that relates the total number of visits, v, to the number of days the promotion has been running, d.

    b. Use your equation from part **a** to find how many days the promotion should be run in order to increase the traffic to the Web site to 12,288 total visits.

27. PHYSICS The velocity v of an object dropped from a tall building is given by the formula $v = \sqrt{64d}$, where d is the distance the object has dropped. What distance was the object dropped from if it has a velocity of 49 feet per second? Round your answer to the nearest hundredth.

28. FENCING Representatives from the neighborhood have requested that the city install a fence around a newly-built playground. The equation $f = 4\sqrt{A}$ represents the amount of fence f needed based on the area A of the playground. If the playground has 324 feet of fencing, find the area of the playground.

29. CREATE Write an equation equivalent to $8^{(2 + x)} = 16^{(2.5 - 0.5x)}$ with a base of 4. Then solve for x. Justify your answer.

30. FIND THE ERROR Zari and Jenell are solving $128^x = 4$. Is either correct? Explain your reasoning.

Zari	Jenell
$128^x = 4$	$128^x = 4$
$(2^7)^x = 2^2$	$(2^7)^x = 4$
$2^{7x} = 2^2$	$2^{7x} = 4^1$
$7^x = 2$	$7^x = 1$
$x = \frac{2}{7}$	$x = \frac{1}{7}$

31. CONSTRUCT ARGUMENTS Make a conjecture about why the equation $2^b = 5^{b-1}$ cannot be solved by the methods used in this section.

32. CREATE Write and solve an exponential equation where both bases need to be changed.

33. PERSEVERE Find the solutions to the equation $32^{(x^2 + 4x)} = 16^{(x^2 + 4x + 3)}$.

34. CREATE Write an exponential equation that has a solution of $x = -1$.

35. WHICH ONE DOESN'T BELONG? Which exponential equation does not belong? Justify your conclusion.

$3^{2x + 1} = 243$	$4^{2x - 1} = 1024$	$2^{3x + 2} = 2048$	$5^{x + 2} = 3125$

36. WRITE Explain how to solve the exponential equation $9^{4n - 3} = 3^6$. Include the solution in your explanation.

Review

@ Essential Question

How do you perform operations and represent real-world situations with exponents?

Exponents are a way of representing repeated multiplication. There are rules which allow exponential expressions to be combined to result in a single expression.

Module Summary

Lessons 10-1 and 10-2

Properties of Exponents

- To multiply two powers with the same base, add their exponents.
- To find the power of a power, multiply the exponents.
- To find the power of a product, find the power of each factor and multiply.
- To divide two powers with the same base, subtract their exponents.
- To find the power of a quotient, find the power of the numerator and the power of the denominator.

Lessons 10-3 and 10-4

Negative and Rational Exponents

- Any nonzero number raised to the zero power is 1.
- A negative exponent is the reciprocal of a number raised to the positive exponent.
- A rational exponent is an exponent that is a rational number.
- If $a^2 = b$, then a is the square root of b.
- If $a^3 = b$, then a is the cube root of b.
- If $a^n = b$ for any positive integer n, then a is an nth root of b.
- The nth root of b can be written as $b^{\frac{1}{n}}$ or $\sqrt[n]{b}$.
- For any positive real number b and any integers m and $n > 1$, $b^{\frac{m}{n}} = \left(\sqrt[n]{b}\right)^m$ or $\sqrt[n]{b^m}$.

Lessons 10-5 and 10-6

Radical Expressions

- For any nonnegative real numbers a and b, $\sqrt{ab} = \sqrt{a} \cdot \sqrt{b}$.
- For any nonnegative real numbers a and b, $\sqrt{\frac{a}{b}} = \frac{\sqrt{a}}{\sqrt{b}}$, if $b \neq 0$.
- A square root is in simplest form if it contains no perfect square factors other than 1, it contains no fractions, and there are no square roots in the denominator of a fraction.

Lesson 10-7

Exponential Equations

- In an exponential equation, the independent variable is an exponent.
- The Power Property of Equality and the other properties of exponents can be used to solve exponential equations.
- If you square both sides of an equation, the resulting equation is still true.

Study Organizer

⬜ Foldables

Use your Foldable to review this module. Working with a partner can be helpful. Ask for clarification of concepts as needed.

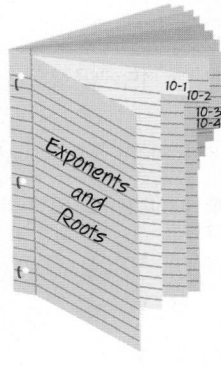

Test Practice

1. MULTI-SELECT Select all expressions equivalent to $(5xy^3)(3xy^3)$. (Lesson 10-1)

A. $(5 \cdot 3)(xy^3 \cdot xy^3)$

B. $15x^2y^6$

C. $15xy^6$

D. $(5 + 3)(xy^3 + xy^3)$

E. $15xy^9$

2. OPEN RESPONSE Earth's mass is about 5,973,600,000,000,000,000,000,000 kg. Write this mass in scientific notation.
(Lesson 10-1)

3. MULTIPLE CHOICE The volume of a cube is $V = s^3$, where s is the length of one side. Which expression represents the volume of the cube? (Lesson 10-1)

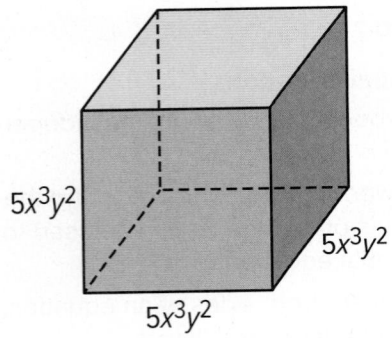

A. $5x^9y^6$

B. $125x^9y^6$

C. $25x^9y^6$

D. $15x^3y^2$

4. MULTIPLE CHOICE Simplify $(w^3)^5$. (Lesson 10-1)

A. w^8

B. w^{15}

C. w^{27}

D. w^{243}

5. OPEN RESPONSE A manufacturer sells square trivets in various sizes. In the corner of each trivet sold, they engrave the company's square logo. The diagram shows the ratio of the size of each trivet to the size of the company's logo.

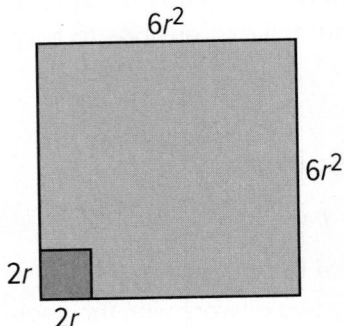

What is the ratio of the area of the trivet to the area of the logo in simplified form?
(Lesson 10-2)

Ratio of area of trivet to area of logo:

__?__ r^2 to 1.

6. OPEN RESPONSE Simplify $\frac{k^6r^3t^5}{kr^2t^4}$. (Lesson 10-2)

7. MULTIPLE CHOICE Simplify $\left(\frac{2x^2y^3}{3x^5}\right)^0$.

(Lesson 10-3)

A. 0

B. $\frac{2}{3}$

C. 1

D. $3x^5$

8. MULTIPLE CHOICE Simplify $\left(\dfrac{2m^2j^3}{3p^5}\right)^4$.

(Lesson 10-2)

A. $\dfrac{8m^8j^{12}}{12p^{20}}$

B. $\dfrac{16m^8j^{12}}{81p^{20}}$

C. $\dfrac{16m^8j^{12}}{3p^5}$

D. $\dfrac{16m^6j^7}{81p^9}$

9. OPEN RESPONSE The maximum speed of an object made of material A when pushed by wind is about 0.001 foot per minute, so the order of magnitude is about 10^{-3}.

The maximum speed of an object made of material B when pushed by the same wind is about 1,000 feet per minute, so the order of the magnitude is about 10^3 feet per minute.

What is the ratio of the maximum speed of the object made of material A to the maximum speed of the object made of material B? Write your answer in decimal form. (Lesson 10-3)

10. OPEN RESPONSE Explain whether the following equation is true or false. If false, explain how to make it true. (Lesson 10-3)

$$\dfrac{49x^{-3}y^4z^{-1}}{4x^2y^{-2}} = \dfrac{7y^6}{2x^5yz}$$

11. OPEN RESPONSE Explain how to simplify the following expression in your own words.

(Lesson 10-4)

$729^{\frac{1}{6}}$

12. MULTIPLE CHOICE Evaluate $\sqrt[4]{81}$. (Lesson 10-4)

A. 4

B. $\dfrac{1}{4}$

C. 81

D. 3

13. MULTIPLE CHOICE Evaluate $\sqrt[6]{4096}$.

(Lesson 10-4)

A. 2

B. 4

C. 16

D. 64

14. MULTIPLE CHOICE Simplify $25^{-\frac{3}{2}}$. (Lesson 10-4)

A. -125

B. -37.5

C. $\dfrac{1}{125}$

D. $\dfrac{1}{15}$

15. OPEN RESPONSE Evaluate $81^{\frac{3}{4}}$. (Lesson 10-4)

16. OPEN RESPONSE Simplify $\sqrt{98}$. (Lesson 10-5)

17. MULTIPLE CHOICE Simplify $\sqrt{121a^5b^3c}$. (Lesson 10-5)

A. $11a^2b\sqrt{abc}$

B. $11a^2b^4$

C. $11ab\sqrt{ab^3c}$

D. $11b\sqrt{a^3b^3c}$

18. MULTIPLE CHOICE Simplify $\sqrt[3]{\dfrac{125a^6}{27b^3}}$. (Lesson 10-5)

A. $\dfrac{a^2}{b}$

B. $\dfrac{5a^6}{3b^3}$

C. $\dfrac{\sqrt[3]{a^2}}{\sqrt[3]{b}}$

D. $\dfrac{5a^2}{3b}$

19. OPEN RESPONSE Simplify

$8\sqrt{5} + 2\sqrt{7} - 6\sqrt{5} + 4\sqrt{7}$. (Lesson 10-6)

20. OPEN RESPONSE Simplify $4\sqrt{3}(5\sqrt{8} + 2\sqrt{3})$. (Lesson 10-6)

21. MULTIPLE CHOICE Solve $7 = 343^{x-2}$. (Lesson 10-7)

A. 5

B. $\dfrac{7}{3}$

C. 2

D. 1

22. MULTIPLE CHOICE Solve $4 = 16^{x+2}$. (Lesson 10-7)

A. -4

B. $-\dfrac{3}{2}$

C. $\dfrac{2}{3}$

D. 4

23. OPEN RESPONSE The volume of a pyramid is $V = \frac{1}{3}Bh$, where B is the area of the base and h is the height. What is the value of x if the volume of the pyramid, in simplest form, is equivalent to $6^{\frac{x}{3}}$ cm³? (Lesson 10-7)

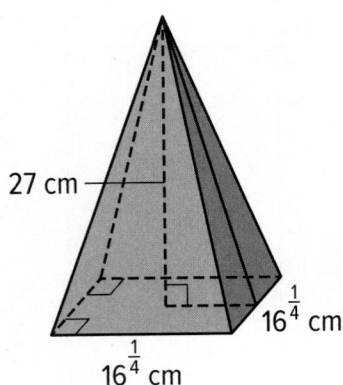

27 cm

$16^{\frac{1}{4}}$ cm

$16^{\frac{1}{4}}$ cm

Polynomials

℮ Essential Question

How can you perform operations on polynomials and use them to represent real-world situations?

What Will You Learn?

How much do you already know about each topic **before** starting this module?

KEY

👎 — I don't know.   ✊ — I've heard of it.   👍 — I know it!

	Before			After		
	👎	✊	👍	👎	✊	👍
write polynomials in standard form						
add polynomials						
subtract polynomials						
multiply polynomials by a monomial						
solve equations with polynomial expressions						
multiply binomials						
multiply polynomials						
factor polynomials using the Distributive Property						
factor quadratic trinomials by grouping						
factor polynomials that are the result of special products						

📖 **Foldables** Make this Foldable to help you organize your notes about polynomials. Begin with four sheets of grid paper.

1. **Fold** in half along the width. On the first two sheets, cut 5 centimeters along the fold at the ends. On the second two sheets cut in the center, stopping 5 centimeters from the ends.

2. **Insert** the first sheets through the second sheets and align the folds. Label the front Module 11, Polynomials. Label the pages with lesson numbers and the last page vocabulary.

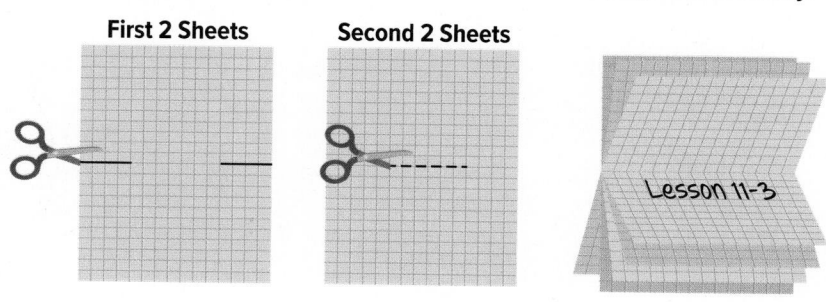

First 2 Sheets   Second 2 Sheets

Lesson 11-3

What Vocabulary Will You Learn?

- binomial
- degree of a monomial
- degree of a polynomial
- difference of two squares
- factoring

- factoring by grouping
- leading coefficient
- perfect square trinomials
- polynomial
- prime polynomial

- quadratic expression
- standard form of a polynomial
- trinomial

Are You Ready?

Complete the Quick Review to see if you are ready to start this module.
Then complete the Quick Check.

Quick Review

Example 1

Rewrite $6x(-3x - 5x - 5x^2 + x^3)$ using the Distributive Property. Then simplify.

$6x(-3x - 5x - 5x^2 + x^3)$

$= 6x(-3x) + 6x(-5x) + 6x(-5x^2) + 6x(x^3)$

$= -18x^2 + (-30x^2) - 30x^3 + 6x^4$

$= -48x^2 - 30x^3 + 6x^4$

Example 2

Simplify $8c + 6 - 4c + 2c^2$.

$8c + 6 - 4c + 2c^2$	Original expression
$= 2c^2 + 8c - 4c + 6$	Rewrite in descending order.
$= 2c^2 + (8 - 4)c + 6$	Use the Distributive Property.
$= 2c^2 + 4c + 6$	Combine like terms.

Quick Check

Rewrite each expression using the Distributive Property. Then simplify.

1. $a(a + 5)$

2. $2(3 + x)$

3. $n(n - 3n^2 + 2)$

4. $-6(x^2 - 5x + 4)$

Simplify each expression. If not possible, write *simplified*.

5. $3x + 10x$

6. $4w^2 + w + 15w^2$

7. $6m^2 - 8m$

8. $2x^2 + 5 + 11x^2 + 7$

How did you do?

Which exercises did you answer correctly in the Quick Check?

Adding and Subtracting Polynomials

Learn Types of Polynomials

A **polynomial** is a monomial or the sum of two or more monomials. Some polynomials have special names.

- A monomial is a number, a variable, or a product of a number and one or more variables.
- A **binomial** is the sum of *two* monomials.
- A **trinomial** is the sum of *three* monomials.

The **degree of a monomial** is the sum of the exponents of all its variables. A nonzero constant term has degree 0, and zero has no degree.

The **degree of a polynomial** is the greatest degree of any term in the polynomial. You can find the degree of a polynomial by finding the degree of each term. Polynomials are named by their degree.

Degree	Name
0	constant
1	linear
2	quadratic
3	cubic
4	quartic
5	quintic
6 or more	6th degree, 7th degree, . . .

Addition is commutative, and therefore the terms of a polynomial can be written in any order. However, the **standard form of a polynomial** has the terms written in order from greatest degree to least degree. When a polynomial is in standard form, the coefficient of the first term is called the **leading coefficient**.

Example 1 Identify Polynomials

Determine whether each expression is a polynomial. If it is a polynomial, find the degree and determine whether it is a monomial, binomial, or trinomial.

a. **$8ab - 2c$**

$8ab - 2c$ is the sum of two monomials, $8ab$ and $-2c$, so this is a polynomial. Degree: 2; binomial.

Today's Goals
- Identify and write polynomials by using the standard form.
- Add polynomials.
- Subtract polynomials.

Today's Vocabulary
polynomial
binomial
trinomial
degree of a monomial
degree of a polynomial
standard form of a polynomial
leading coefficient

 Think About It!

Is $4x - 2x^2 + 9$ written in standard form? Justify your argument.

 Talk About It!

Explain why $8ab - 2c$ is a 2nd degree polynomial and not a 1st degree polynomial.

b. −11.25

−11.25 is a real number, so this is a polynomial. Degree: 0; monomial.

c. $2x^{-2} + 3xy$

$2x^{-2} = \frac{2}{x^2}$, which is not a monomial, so this is not a polynomial.

d. $9x^3 - 8x + 5x - 27$

The simplified form is $9x^3 - 3x - 27$, which is the sum of three monomials, so this is a polynomial. Degree: 3; trinomial.

e. $2m^2 + 2mn - n^2$

$2m^2 + 2mn - n^2$ is the sum of three monomials, so this is a polynomial. Degree: 2; trinomial.

Check

Copy and complete the table. Determine whether each expression is a polynomial. If it is a polynomial, find the degree and determine whether it is a *monomial, binomial,* or *trinomial.*

Expression	Is it a polynomial ?	Degree	Classification
a. $3z^{-2}$			
b. $2x^3 + x - 12$			
c. $9b$			
d. $9x - 2$			

Example 2 Standard Form of a Polynomial

Write $4x + 12 + 2x^3 - 3x^2$ in standard form. Identify the leading coefficient.

To write the polynomial in standard form, rewrite the terms in order from greatest degree, 3, to least degree, 0. The polynomial can be rewritten as $2x^3 - 3x^2 + 4x + 12$ with a leading coefficient of 2.

Check

Part A Write $5b - 10b^2 + 35 - b^3$ in standard form.

Part B Identify the leading coefficient in $5b - 10b^2 + 35 - b^3$.

The leading coefficient of the polynomial is _____?_____.

Go Online You can complete an Extra Example online.

Watch Out!

Degree of a Polynomial
Remember that the degree of a monomial is the sum of the exponents of all its variables, and the degree of the polynomial in which it is a term is the greatest degree of any term in it. Be careful not to identify the degree of a polynomial by looking only at the greatest exponent.

Think About It!

In standard form, why is the constant term at the end of the polynomial rather than the beginning?

📡 **Online Activity** Use algebra tiles to complete the Explore.

⊘ **INQUIRY** How are the processes for adding and subtracting polynomials similar?

Learn Adding Polynomials

Adding polynomials involves adding like terms. When adding polynomials, you can group like terms by using a horizontal or vertical format.

Method 1 Horizontal Method

Group and combine like terms.

$$(3x^2 + 9x + 27) + (2x^2 + 4x - 12)$$

$$= [3x^2 + 2x^2] + [9x + 4x] + [27 + (-12)] \qquad \text{Group like terms.}$$

$$= 5x^2 + 13x + 15 \qquad \text{Combine like terms.}$$

Method 2 Vertical Method

Align like terms in columns and combine.

$$\begin{array}{r} 3x^2 + 9x + 27 \\ (+) \quad 2x^2 + 4x - 12 \\ \hline 5x^2 + 13x + 15 \end{array}$$

Align like terms.

Combine like terms.

> 🫐 **Think About It!**
>
> How can writing polynomials in standard form be helpful when adding?

Example 3 Add Polynomials

Find each sum.

a. **$(3x^2 - 4) + (x^2 - 9)$**

$$(3x^2 - 4) + (x^2 - 9) = [3x^2 + x^2] + [-4 + (-9)] \qquad \text{Group like terms.}$$

$$= 4x^2 - 13 \qquad \text{Combine like terms.}$$

b. **$(8 - x^2) + (4x + 2x^2 - 9)$**

$$
\begin{array}{lll}
-x^2 + 8 & \rightarrow & -x^2 + 0x + 8 \\
4x + 2x^2 - 9 & \rightarrow & (+) \ 2x^2 + 4x - 9 \\
& & \overline{ x^2 + 4x - 1}
\end{array}
$$

Insert a placeholder to align the terms.

Align and combine like terms.

> **Study Tip**
>
> **Placeholders** When adding polynomials, it may be necessary to insert a placeholder to help align the terms. For example, if one of the polynomials does not have an x^2 term, add $0x^2$ to keep the terms aligned.

📡 **Go Online** You can complete an Extra Example online.

Check

Find each sum. Write your answer in standard form.

$(12y + 20y^2 - 2) + (-13y^2 + y - 10)$

$(-4b - b^2 + 2) + 2(b^2 + 2b - 1)$

$(-f + 5f^2 + 5) + (3f^3 - f + f^2)$

Learn Subtracting Polynomials

You can subtract a polynomial by adding its additive inverse. To find the additive inverse of a polynomial, write the opposite of each term.

Select a method to find $(11x - 13 - 7x^3 - 8x^2) - (2x + 8x^2 + 20)$.

Method 1 Horizontal method

Subtract $2x + 8x^2 + 20$ by adding its additive inverse.

$(11x - 13 - 7x^3 - 8x^2) - (2x + 8x^2 + 20)$ The additive inverse of
$2x + 8x^2 + 20$ is $-2x - 8x^2 - 20$.

$$= (11x - 13 - 7x^3 - 8x^2) + (-2x - 8x^2 - 20)$$
$$= -7x^3 + [-8x^2 + (-8x^2)] + [11x + (-2x)] + [-13 + (-20)]$$
$$= -7x^3 - 16x^2 + 9x - 33$$

Method 2 Vertical method

Align like terms in columns and subtract by adding the additive inverse.

$$\begin{array}{r} -7x^3 - 8x^2 + 11x - 13 \\ (-)\,0x^3 + 8x^2 + 2x + 20 \\ \hline \end{array} \quad\rightarrow\quad \begin{array}{r} -7x^3 - 8x^2 + 11x - 13 \\ (+)\,-0x^3 - 8x^2 - 2x - 20 \\ \hline -7x^3 - 16x^2 + 9x - 33 \end{array}$$

Adding or subtracting integers results in an integer, so the set of integers is closed under addition and subtraction. Similarly, when you add or subtract polynomials, you are combining like terms. This results in a polynomial with the same variables and exponents as the original polynomials, but possibly different coefficients. Thus, the sum or difference of two polynomials is always a polynomial, and the set of polynomials is closed under addition and subtraction.

Example 4 Subtract Polynomials Horizontally

Find $(6x - 11) - (2x - 19)$.

Subtract $(2x - 19)$ by adding its additive inverse.

$(6x - 11) - (2x - 19) = (6x - 11) + (-2x + 19)$

$\qquad\qquad\qquad = [6x + (-2x)] + [-11 + 19]$ Group like terms.

$\qquad\qquad\qquad = 4x + 8$ Combine like terms.

🔵 **Go Online** You can complete an Extra Example online.

> 💭 **Think About It!**
>
> In the example, why is the term $0x^3$ introduced when subtracting the polynomials?

Example 5 Subtract Polynomials Vertically

Find $(x + 2) - (7x - 3x^2 + 14)$.

Align like terms in columns and subtract by adding the additive inverse.

$$0x^2 + x + 2 \qquad \rightarrow \qquad 0x^2 + x + 2$$
$$\underline{(-) \; -3x^2 + 7x + 14} \qquad \quad \underline{(+) \; 3x^2 - 7x - 14}$$
$$\qquad \qquad \qquad \qquad \qquad \qquad 3x^2 - 6x - 12$$

Check

Find $(z^2 + 2z - 5) - (9z - 3z^2)$. Write your answer in standard form.

Find $(8r - 14 + 7r^2) - (-16r^2 - 7r - 3)$. Write your answer in standard form.

Find $(h - 2h - h^2) - (5h^2 - 2 + 8h)$. Write your answer in standard form.

🌐 Example 6 Add and Subtract Polynomials

ALBUM SALES Today's recording artists can sell hard copies H and digital copies D of their albums. The equations $H = 9w + 53$ and $D = 13w + 126$ represent the number of albums (in thousands) one artist sold in w weeks. Write an equation that shows how many more digital albums were sold than hard copies S. Then predict how many more digital albums are sold than hard copies in 52 weeks.

To write an equation that represents how many more digital albums were sold than hard copies S, subtract the equation for the number of hard copies H sold from the equation for the number of digital albums D sold.

$$S = (13w + 126) - (9w + 53)$$

$$= 4w + 73$$

Substitute 52 for w to predict how many more digital albums are sold than hard copies in 52 weeks.

There will be 281,000 more digital albums sold than hard copies in 52 weeks.

🧭 **Go Online** You can complete an Extra Example online.

😮 **Think About It!**

What is the first step for finding the difference of polynomials?

Study Tip

Units Pay attention to the language in the question. The equations represent the number of hard copy and digital albums sold in the thousands, so your answer should represent album sales in the thousands.

😮 **Think About It!**

What assumption did you make about the trend of sales over the 52 weeks? Can you determine the sales of hard copy and digital albums for a specific week? Explain.

Check

COLLEGE LIVING The total number of students T who attend a college consists of two groups: students who live in dorm rooms on campus D and students who live in apartments off campus A. The number (in hundreds) of students who live in dorm rooms and the total number of students enrolled in the college can be modeled by the following equations, where n is the number of years since 2001.

$$T = 17n + 23$$
$$D = 11n + 8$$

Part A Write an equation that models the number of students who live in apartments.

Part B Predict the number of students who will live in apartments in 2020.

Part C What do you need to assume in order to predict the number of students who will live on campus in 2020?

A. The total number of students does not include students who commute.

B. Students do not share dorm rooms.

C. The number of students enrolled in the college remains the same.

D. Many students live at home during the summer.

Pause and Reflect

Did you struggle with anything in this lesson? If so, how did you deal with it?

Go Online You can complete an Extra Example online.

Practice

Go Online You can complete your homework online.

Example 1

Determine whether each expression is a polynomial. If it is a polynomial, find the degree and determine whether it is a *monomial*, *binomial*, or *trinomial*.

1. $\frac{5y^3}{x^2} + 4x$

2. 21

3. $c^4 - 2c^2 + 1$

4. $d + 3d^c$

5. $a - a^2$

6. $5n^3 + nq^3$

Example 2

Write each polynomial in standard form. Identify the leading coefficient.

7. $5x^2 - 2 + 3x$

8. $8y + 7y^3$

9. $4 - 3c - 5c^2$

10. $-y^3 + 3y - 3y^2 + 2$

11. $11t + 2t^2 - 3 + t^5$

12. $2 + r - r^3$

13. $\frac{1}{2}x - 3x^4 + 7$

14. $-9b^2 + 10b - b^6$

Examples 3–5

Find each sum or difference.

15. $(2x + 3y) + (4x + 9y)$

16. $(6s + 5t) + (4t + 8s)$

17. $(5a + 9b) - (2a + 4b)$

18. $(11m - 7n) - (2m + 6n)$

19. $(m^2 - m) + (2m + m^2)$

20. $(x^2 - 3x) - (2x^2 + 5x)$

21. $(d^2 - d + 5) - (2d + 5)$

22. $(2h^2 - 5h) + (7h - 3h^2)$

23. $(5f + g - 2) + (-2f + 3)$

24. $(6k^2 + 2k + 9) + (4k^2 - 5k)$

25. $(2c^2 + 6c + 4) + (5c^2 - 7)$

26. $(2x + 3x^2) - (7 - 8x^2)$

Find each sum or difference.

27. $(3c^3 - c + 11) - (c^2 + 2c + 8)$

28. $(z^2 + z) + (z^2 - 11)$

29. $(2x - 2y + 1) - (3y + 4x)$

30. $(4a - 5b^2 + 3) + (6 - 2a + 3b^2)$

31. $(x^2y - 3x^2 + y) + (3y - 2x^2y)$

32. $(-8xy + 3x^2 - 5y) + (4x^2 - 2y + 6xy)$

33. $(5n - 2p^2 + 2np) - (4p^2 + 4n)$

34. $(4rxt - 8r^2x + x^2) - (6rx^2 + 5rxt - 2x^2)$

Example 6

35. PROFIT Company A and Company B both started their businesses in the same year. The profit P, in millions, of Company A is given by the equation $P = 3.2x + 12$, where x is the number of years in business. The profit P, in millions, of Company B is given by the equation $P = 2.7x + 10$, where x is the number of years in business.

   a. Write a polynomial equation to give the difference in profit D after x years.

   b. Predict the difference in profit after 10 years.

36. ENVELOPES An office supply company produces yellow document envelopes. The envelopes come in a variety of sizes, but the length is always 4 centimeters more than double the width, x.

   a. Write a polynomial equation to give the perimeter P of any of the envelopes.

   b. Predict the perimeter of an envelope with a width of 6 centimeters.

Mixed Exercises

Classify each polynomial according to its degree and number of terms.

37. $4x - 3x^2 + 5$

38. $11z^3$

39. $9 + y^4$

40. $3x^3 - 7x$

41. $-2x^5 - x^2 + 5x - 8$

42. $10t - 4t^2 + 6t^3$

Find each sum or difference.

43. $(4x + 2y - 6z) + (5y - 2z + 7x) + (-9z - 2x - 3y)$

44. $(5a^2 - 4) + (a^2 - 2a + 12) + (4a^2 - 6a + 8)$

45. $(3c^2 - 7) + (4c + 7) - (c^2 + 5c - 8)$

46. ROCKETS Two toy rockets are launched straight up into the air. The height, in feet, of each rocket at t seconds after launch is given by the polynomial equations shown. Write an equation to find the difference in height of Rocket A and Rocket B. Predict the difference in height after 5 seconds.

Rocket A: $D_1 = -16t^2 + 122t$

Rocket B: $D_2 = -16t^2 + 84t$

47. INDUSTRY Two identical right cylindrical steel drums containing oil need to be covered with a fire−resistant sealant. In order to determine how much sealant to purchase, George must find the surface area of the two drums. The surface area, including the top and bottom bases, is given by the formula $S = 2\pi rh + 2\pi r^2$.

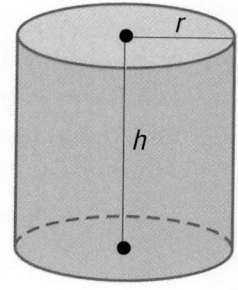

a. Write a polynomial to represent the total surface area of the two drums.

b. Find the total surface area, to the nearest tenth of a square meter, if the height of each drum is 2 meters and the radius of each is 0.5 meter.

48. MANUFACTURING A company delivers their product in cubic boxes that have volume x^3. When the company begins to manufacture a second product, manufacturing designs a new shipping box that is 3 inches longer in one dimension and 1 inch shorter in another dimension. The volume of the new box is $x^3 + 2x^2 - 3x$.

a. Write an expression to represent the total volume of 4 of each kind of box.

b. Find an expression that shows the difference in volume between the two boxes.

49. FIND THE ERROR Claudio says that polynomials are not closed under addition and gives this counterexample, which he says is not a polynomial: $(x^2 - 2x) + (-x^2 + 2x) = 0$. Is Claudio correct? Explain your reasoning.

50. VOLUME The volume of a sphere with radius x is $\frac{4}{3}\pi x^3$ units^3, and the volume of a cube with side length x is x^3 units^3.

a. Find the total combined volume of the sphere and cube.

b. How much more volume does the sphere contain?

51. **STRUCTURE** Compute the following differences.

    a. $(5x^2 - 3x + 7) - (18x^2 - 2x - 3)$

    b. $(18x^2 - 2x - 3) - (5x^2 - 3x + 7)$

    c. What do you notice about **part a** compared to **part b**? How does this relate to the structure of the integers?

52. **REGULARITY** In the set of integers, every integer has an additive inverse. In other words, for every integer n, there is another integer $-n$ such that $n + (-n) = 0$. Is this true in the set of polynomials? Does every polynomial have an additive inverse? Demonstrate with an example.

🫐 **Higher-Order Thinking Skills**

53. **FIND THE ERROR** Cheyenne and Nicolas are finding $(2x^2 - x) - (3x + 3x^2 - 2)$. Is either correct? Explain your reasoning.

Cheyenne	Nicolas
$(2x^2 - x) - (3x + 3x^2 - 2)$	$(2x^2 - x) - (3x + 3x^2 - 2)$
$= (2x^2 - x) + (-3x + 3x^2 - 2)$	$= (2x^2 - x) + (-3x - 3x^2 - 2)$
$= 5x^2 - 4x - 2$	$= -x^2 - 4x - 2$

54. **ANALYZE** Determine whether each of the following statements is *true* or *false*. Justify your argument.

    a. A binomial can have a degree of zero.

    b. The order in which polynomials are subtracted does not matter.

55. **PERSEVERE** Write a polynomial that represents the sum of $2n + 1$ and the next two consecutive odd integers.

56. **WRITE** Why would you add or subtract equations that represent real-world situations? Explain.

57. **WRITE** Describe how to add and subtract polynomials using both the vertical and horizontal methods.

58. **CREATE** Write two polynomials that can be added to have a sum of $-11x^3 - x^2 + 5x + 6$.

Multiplying Polynomials by Monomials

Explore Using Algebra Tiles to Find Products of Polynomials and Monomials

 Online Activity Use algebra tiles to complete the Explore.

> **INQUIRY** How can you use the Distributive Property to find the product of a polynomial and a monomial? ×

Learn Multiplying a Polynomial by a Monomial

To find the product of a monomial and a binomial, you can use the Distributive Property. You can also use the Distributive Property to find the product of a monomial and a longer polynomial. When polynomials are multiplied, the product is also a polynomial. Therefore, the set of polynomials is closed under multiplication. This is similar to the system of integers, which is also closed under multiplication.

Example 1 Multiply a Polynomial by a Monomial

Simplify $-2x(4x^2 + 3x - 5)$.

$-2x(4x^2 + 3x - 5)$	Original expression
$= -2x(4x^2) + (-2x)(3x) - (-2x)(5)$	Distributive Property
$= -8x^3 + (-6x^2) - (-10x)$	Multiply.
$= -8x^3 - 6x^2 + 10x$	Simplify.

Go Online
An alternate method is available for this example.

Example 2 Simplify Expressions

Simplify $3n(6n^3 - 4n) - 2(9n^4 - 11)$.

$3n(6n^3 - 4n) - 2(9n^4 - 11)$	
$= 3n(6n^3) - 3n(4n) + (-2)(9n^4) + (-2)(-11)$	Distributive Property
$= 18n^4 - 12n^2 - 18n^4 + 22$	Multiply.
$= (18n^4 - 18n^4) - 12n^2 + 22$	Commutative and Associative properties
$= -12n^2 + 22$	Combine like terms.

Watch Out!

Negatives If the monomial has a negative coefficient, remember to distribute the negative sign to each term in the polynomial.

 Go Online You can complete an Extra Example online.

Example 3 Write and Evaluate a Polynomial Expression

Use a Source

Find the height of the basket building and use it to determine the area of one side.

ARCHITECTURE The world's largest basket is a building. Each face of the building is in the shape of a trapezoid, with the largest face having a height of h and two base lengths, $h + 90$ and $2h + 84$. Write and simplify an expression to represent the area of one side of the building.

Let $h =$ the height of a trapezoid, $a = h + 90$ and $b = 2h + 84$.

$$A = \tfrac{1}{2}h(a + b) \qquad \text{Area of a trapezoid}$$
$$= \tfrac{1}{2}h[(h + 90) + (2h + 84)] \qquad a = h + 90 \text{ and } b = 2h + 84$$
$$= \tfrac{1}{2}h(3h + 174) \qquad \text{Add and simplify.}$$
$$= \tfrac{3}{2}h^2 + 87h \qquad \text{Distributive Property}$$

The area of one side of the building is $\tfrac{3}{2}h^2 + 87h$.

Example 4 Solve Equations with Polynomial Expressions

Solve each equation.

a. $-16p = -2(p + 3) + 3(6p - 30)$

Study Tip

Check To ensure that your answer is correct, substitute it into the original equation and verify that the simplified expressions are equal.

$$-16p = -2(p + 3) + 3(6p - 30) \qquad \text{Original equation}$$
$$-16p = -2p - 6 + 18p - 90 \qquad \text{Distributive Property}$$
$$-16p = 16p - 96 \qquad \text{Combine like terms.}$$
$$-32p = -96 \qquad \text{Subtract } 16p \text{ from each side.}$$
$$p = 3 \qquad \text{Divide each side by } -32.$$

b. $-2q(4q - 9) = q(-8q + 15) - 3(-4q - 6)$

$$-2q(4q - 9) = q(-8q + 15) - 3(-4q - 6) \qquad \text{Original equation}$$
$$-8q^2 + 18q = -8q^2 + 15q + 12q + 18 \qquad \text{Distributive Property}$$
$$-8q^2 + 18q = -8q^2 + 27q + 18 \qquad \text{Combine like terms.}$$
$$18q = 27q + 18 \qquad \text{Add } 8q^2 \text{ to each side.}$$
$$-9q = 18 \qquad \text{Subt. } 27q \text{ from each side.}$$
$$q = -2 \qquad \text{Divide each side by } -9.$$

Check

Solve each equation.

$2p = 3(4p - 10)$

$p = \underline{\quad ? \quad}$

$-3q(2q + 5) = 2(-3q^2 + 15) - 5(10q + 6)$

$q = \underline{\quad ? \quad}$

Go Online You can complete an Extra Example online.

Practice

Go Online You can complete your homework online.

Example 1

Simplify each expression.

1. $b(b^2 - 12b + 1)$

2. $f(f^2 + 2f + 25)$

3. $-3m^3(2m^3 - 12m^2 + 2m + 25)$

4. $2j^2(5j^3 - 15j^2 + 2j + 2)$

5. $2pr^2(2pr + 5p^2r - 15p)$

6. $4t^3u(2t^2u^2 - 10tu^4 + 2)$

Example 2

Simplify each expression.

7. $-3(5x^2 + 2x + 9) + x(2x - 3)$

8. $a(-8a^2 + 2a + 4) + 3(6a^2 - 4)$

9. $-4d(5d^2 - 12) + 7(d + 5)$

10. $-9g(-2g + g^2) + 3(g^3 + 4)$

11. $2j(7j^2k^2 + jk^2 + 5k) - 9k(-2j^2k^2 + 2k^2 + 3j)$

12. $4n(2n^3p^2 - 3np^2 + 5n) + 4p(6n^2p - 2np^2 + 3p)$

Example 3

13. NUMBER THEORY The sum of the first n whole numbers is given by the expression $\frac{1}{2}(n^2 + n)$. Expand the equation by multiplying, then find the sum of the first 12 whole numbers.

14. COLLEGE Troy's grandfather gave him \$700 to start his college savings account. Troy's grandfather also gives him \$40 each month to add to the account. Troy's mother gives him \$50 each month, but has been doing so for 4 fewer months than Troy's grandfather. Write a simplified expression for the amount of money Troy has received from his grandfather and mother after m months.

15. MARKET Sophia went to the farmers' market to purchase some vegetables. She bought peppers and potatoes. The peppers were \$0.39 each and the potatoes were \$0.29 each. She spent \$3.88 on vegetables, and bought 4 more potatoes than peppers. If $x =$ the number of peppers, write and solve an equation to find out how many of each vegetable Sophia bought.

16. GEOMETRY The volume of a pyramid can be found by multiplying the area of its base B by one-third of its height. The area of the rectangular base of a pyramid is given by the polynomial equation $B = x^2 - 4x - 12$.

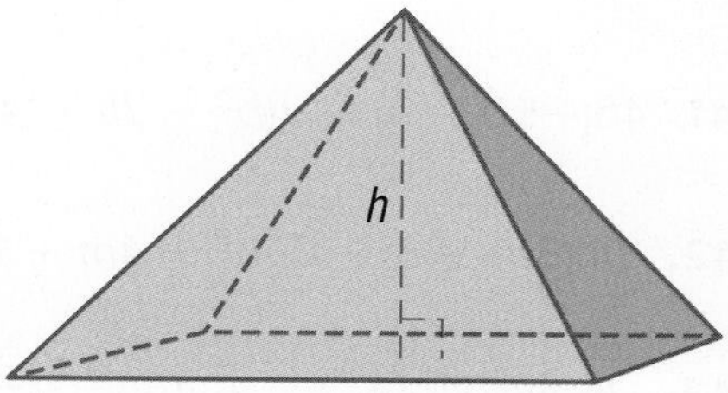

a. Write a polynomial equation to represent the volume of the pyramid V if its height is 10 meters.

b. Find the volume of the pyramid if $x = 12$ m.

Example 4

Solve each equation.

17. $7(t^2 + 5t - 9) + t = t(7t - 2) + 13$

18. $w(4w + 6) + 2w = 2(2w^2 + 7w - 3)$

19. $5(4z + 6) - 2(z - 4) = 7z(z + 4) - z(7z - 2) - 48$

20. $9c(c - 11) + 10(5c - 3) = 3c(c + 5) + c(6c - 3) - 30$

21. $2f(5f - 2) - 10(f^2 - 3f + 6) = -8f(f + 4) + 4(2f^2 - 7f)$

22. $2k(-3k + 4) + 6(k^2 + 10) = k(4k + 8) - 2k(2k + 5)$

Mixed Exercises

Simplify each expression.

23. $a(4a + 3)$

24. $-c(11c + 4)$

25. $x(2x - 5)$

26. $2y(y - 4)$

27. $-3n(n^2 + 2n)$

28. $4h(3h - 5)$

29. $3x(5x^2 - x + 4)$

30. $7c(5 - 2c^2 + c^3)$

31. $-4b(1 - 9b - 2b^2)$

32. $6y(-5 - y + 4y^2)$

33. $2m^2(2m^2 + 3m - 5)$

34. $-3n^2(-2n^2 + 3n + 4)$

Simplify each expression.

35. $w(3w + 2) + 5w$

36. $f(5f - 3) - 2f$

37. $-p(2p - 8) - 5p$

38. $y^2(-4y + 5) - 6y^2$

39. $2x(3x^2 + 4) - 3x^3$

40. $4a(5a^2 - 4) + 9a$

41. $4b(-5b - 3) - 2(b^2 - 7b - 4)$

42. $3m(3m + 6) - 3(m^2 + 4m + 1)$

43. $-5q^2w^3(4q + 7w) + 4qw^2(7q^2w + 2q) - 3qw(3q^2w^2 + 9)$

44. $-x^2z(2z^2 + 4xz^3) + xz^2(xz + 5x^3z) + x^2z^3(3x^2z + 4xz)$

Solve each equation.

45. $3(a + 2) + 5 = 2a + 4$

46. $2(4x + 2) - 8 = 4(x + 3)$

47. $5(y + 1) + 2 = 4(y + 2) - 6$

48. $4(b + 6) = 2(b + 5) + 2$

49. $6(m - 2) + 14 = 3(m + 2) - 10$

50. $3(c + 5) - 2 = 2(c + 6) + 2$

51. LANDSCAPING The courtyard on a college campus has a sculpture surrounded by a circle of 50 flags. The university plans to install a new sidewalk 12 feet wide around the perimeter of the outside of the circle of flags. If the outside circumference of the sidewalk is 1.10 times the circumference of the circle of flags, write an equation for the outside circumference of the sidewalk. Solve the equation for the radius of the circle of flags. Recall that the circumference of a circle is $2\pi r$.

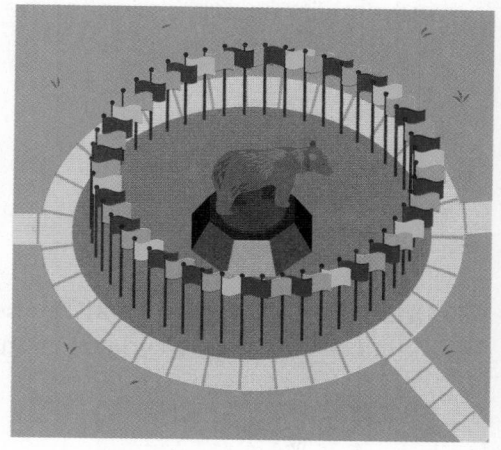

52. STRUCTURE The base lengths of the trapezoid shown are given by polynomial expressions in terms of the trapezoid's height h.

a. Write and simplify an expression for the area of the trapezoid.

b. If the height of the trapezoid is 4 units, what is the area of the trapezoid?

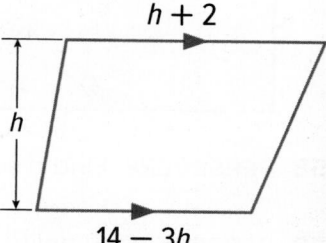

53. FIND THE ERROR Andres simplified the expression $12y^2(3y - 2y^2) - 3y^3(4 - 2y)$. Is he correct? Explain your reasoning.

$$12y^2(3y - 2y^2) - 3y^3(4 - 2y)$$
$$36y^3 - 24y^2 - 12y^3 - 6y^4$$
$$24y^3 - 24y^2 - 6y^4$$

54. STRUCTURE The diagram shows the dimensions of a right rectangular prism.

a. Write and simplify an expression for the volume of the prism.

b. If the height of the rectangular prism is 6 units, what is the volume of the rectangular prism?

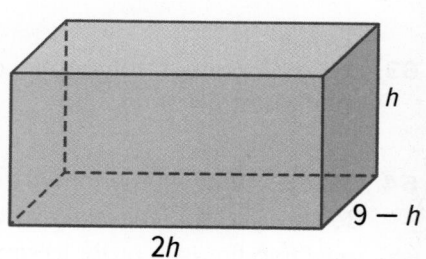

55. USE TOOLS Through market research, a company finds that it can expect to sell $45 - 5x$ products if each is priced at $1.25x$ dollars.

    a. Write and simplify an expression for the expected revenue.

    b. Determine the price of each product, to the nearest cent, when $x = 4.5$.

    c. Determine the revenue when the expected number of products are sold, to the nearest cent, when $x = 4.5$.

56. STRUCTURE An area is enclosed by the fence shown at the right, represented by the solid line segments. Write and simplify an expression for the area of the enclosure.

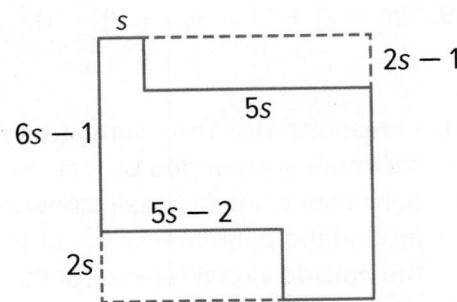

57. FIND THE ERROR Pearl and Ted both simplified $2x^2(3x^2 + 4x + 2)$. Is either of them correct? Explain your reasoning.

Pearl	Ted
$2x^2(3x^2 + 4x + 2)$	$2x^2(3x^2 + 4x + 2)$
$6x^4 + 8x^2 + 4x^2$	$6x^4 + 8x^3 + 4x^2$
$6x^4 + 12x^2$	

58. PERSEVERE Find p such that $3x^p(4x^{2p + 3} + 2x^{3p - 2}) = 12x^{12} + 6x^{10}$.

59. PERSEVERE Simplify $4x^{-3}y^2(2x^5y^{-4} + 6x^{-7}y^6 - 4x^0y^{-2})$.

60. ANALYZE Is there a value of x that makes the statement $(x + 2)^2 = x^2 + 2^2$ true? If so, find a value for x. Justify your argument.

61. CREATE Write a monomial and a polynomial using n as the variable. Find their product.

62. WRITE Describe the steps to multiply a polynomial by a monomial.

63. CREATE Write a polynomial equation, with variables on both sides, that has a solution of $t = 9$.

64. CREATE Write a polynomial expression that can be simplified to $-3c^3 + 74c^2 - 4c$. Be sure your polynomial expression requires the use of the Distributive Property at least two times in order to simplify.

Multiplying Polynomials

Today's Goal
- Multiply binomials by using the Distributive Property and the FOIL Method.

Today's Vocabulary
quadratic expression

Explore Using Algebra Tiles to Find Products of Two Binomials

⬤ **Online Activity** Use algebra tiles to complete the Explore.

> ⓧ
> ② **INQUIRY** How can you use the Distributive Property to find the product of two binomials?

Learn Multiplying Binomials

Binomials can be multiplied horizontally or vertically. Multipy $(x - 2)(x + 6)$.

Method 1 Vertical Method

$$\begin{array}{r} x - 2 \\ (\times)\ \ x + 6 \\ \hline 6x - 12 \\ (+)\ \ x^2 - 2x \\ \hline x^2 + 4x - 12 \end{array}$$

Multiply by 6.
Multiply by x.
Combine like terms.

Method 2 Horizontal Method

$(x - 2)(x + 6) = x(x + 6) - 2(x + 6)$ Rewrite as the sum of two products.

$= x^2 + 6x - 2x - 12$ Distributive Property

$= x^2 + 4x - 12$ Combine like terms.

You can also use a shortcut version of the Distributive Property, called the FOIL method, to multiply binomials.

Key Concept • FOIL Method

To multiply two binomials, find the sum of the products of **F** the *First* terms, **O** the *Outer terms*, **I** the *Inner terms*, and **L** the *Last terms*.

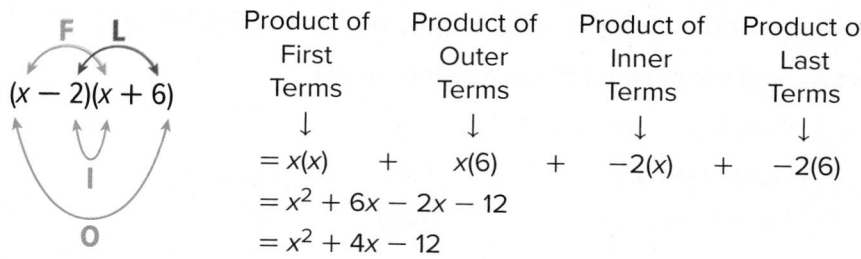

Product of First Terms	Product of Outer Terms	Product of Inner Terms	Product of Last Terms
↓	↓	↓	↓
$= x(x)$ +	$x(6)$ +	$-2(x)$ +	$-2(6)$

$= x^2 + 6x - 2x - 12$

$= x^2 + 4x - 12$

An expression in one variable with a degree of 2 is called a **quadratic expression**.

⬤ **Go Online** You can complete an Extra Example online.

Think About It!
How does the horizontal method using the Distributive Property differ from the FOIL method?

Study Tip
FOIL Method
The FOIL method can be used only when multiplying two binomials. The FOIL method cannot be used when multiplying, for example, two trinomials since six products are needed.

Example 1 Multiply Binomials by Using the Vertical Method

Find $(x - 1)(x + 7)$ by using the vertical method.

$$
\begin{array}{r}
x \quad - \quad 1 \\
(\times) \quad x \quad + \quad 7 \\
\hline
7x \quad - \quad 7 \\
x^2 \quad - \quad x \qquad\quad \\
\hline
x^2 \quad + \quad 6x \quad - \quad 7
\end{array}
$$

Multiply by 7.

Multiply by x.

Combine like terms.

Example 2 Multiply Binomials by Using the Horizontal Method

Find $(3x - 4)(4x - 10)$ by using the horizontal method.

$(3x - 4)(4x - 10)$

$= 3x(4x - 10) + -4(4x - 10)$ Rewrite as sum of two products.

$= 12x^2 - 30x - 16x + 40$ Distributive Property

$= 12x^2 - 46x + 40$ Combine like terms.

Study Tip

Signs Notice that in the first step of the solution, you added the product of -4 and $4x$, and in the second step, that turned into a subtraction of $16x$. Remember that $-4(4x) = -16x$.

Example 3 Multiply Binomials by Using the FOIL Method

Find $(2a - 12)(5a + 3)$ by using the FOIL method.

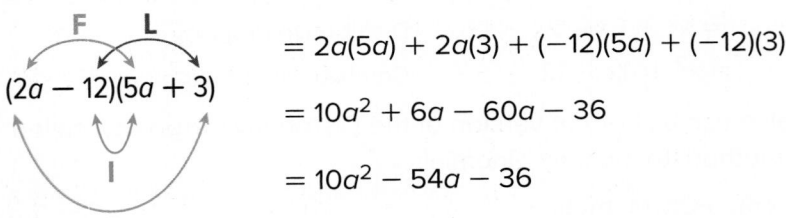

$= 2a(5a) + 2a(3) + (-12)(5a) + (-12)(3)$

$= 10a^2 + 6a - 60a - 36$

$= 10a^2 - 54a - 36$

Study Tip

FOIL Method
The FOIL method is a memory device that can help you remember to find all four products when multiplying two binomials. The order in which the terms are multiplied is not important.

Talk About It!

Is your answer complete after you use the FOIL method to multiply? Explain.

Check

Find the product of $(3p - 9)(2p + 6)$ by using the FOIL Method.

Part A Find the product of each pair of terms.

First Terms: $(3p)(\underset{\text{?}}{\rule{2cm}{0.4pt}})$

Outer Terms: $(\underset{\text{?}}{\rule{2cm}{0.4pt}}) (6)$

Inner Terms: $(\underset{\text{?}}{\rule{2cm}{0.4pt}}) (2p)$

Last Terms: $(-9) (\underset{\text{?}}{\rule{2cm}{0.4pt}})$

Part B What is the product of $(3p - 9)(2p + 6)$ written in standard form?

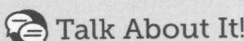 **Go Online** You can complete an Extra Example online.

Apply Example 4 Use the FOIL Method

SNOW REMOVAL **A town worker is clearing the snow from a parking lot and the surrounding sidewalk. The sidewalk extends x feet on every side of the parking lot. Write an expression for the total area of the parking lot and sidewalk.**

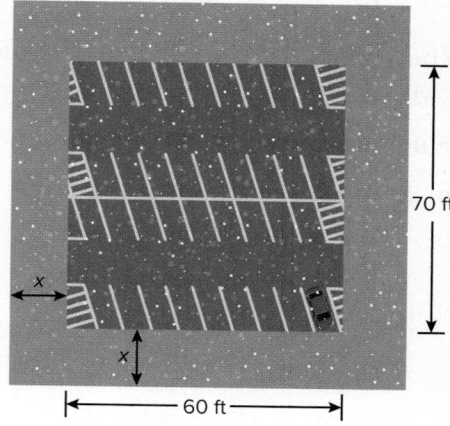

70 ft

x

x

60 ft

1 What is the task?

Describe the task in your own words. Then list any questions that you may have. How can you find answers to your questions?

Sample answer: the worker needs to shovel the snow from the parking lot and the sidewalk. We need to find the total area that needs to be shoveled.

2 How will you approach the task? What have you learned that you can use to help you complete the task?

Sample answer: first I'll find the total length and the total width. Then I'll multiply the length and the width to find the area. I have learned how to write and multiply binomials.

3 What is your solution?

Use your strategy to solve the problem.

Length $= 2x + 60$ Width $= 2x + 70$

The total area of the parking lot and sidewalk is $4x^2 + 260x + 4200$.

4 How can you know that your solution is reasonable?

✏️ **Write About It!** Write an argument that can be used to defend your solution.

Sample answer: Let $x = 4$, substituting 4 for x in the equation results in $(2(4) + 60)(2(4) + 70) = 4(4^2) + 260(4) + 4200$. This simplifies to $5304 = 5304$.

Check

FRAME Jacinta is framing her newest painting. The dimensions of the frame with the painting are represented by a width of $5y - 5$ and a length of $2y + 6$. Write an expression in standard form that represents the area of the frame with the painting.

🔎 **Go Online** You can complete an Extra Example online.

Example 5 Multiply Polynomials by Using the Distributive Property

The Distributive Property can also be used to multiply any two polynomials.

Find each product.

a. $(x + 3)(x^2 - x - 9)$

$\quad (x + 3)(x^2 - x - 9)$

$\quad\quad = x(x^2 - x - 9) + 3(x^2 - x - 9)$ Distributive Property

$\quad\quad = x^3 - x^2 - 9x + 3x^2 - 3x - 27$ Multiply.

$\quad\quad = x^3 + 2x^2 - 12x - 27$ Combine like terms.

b. $(3t^2 - 5t + 9)(4t^2 - 4t + 14)$

$\quad (3t^2 - 5t + 9)(4t^2 - 4t + 14)$

$\quad\quad = 3t^2(4t^2 - 4t + 14) - 5t(4t^2 - 4t + 14) + 9(4t^2 - 4t + 14)$

$\quad\quad = 12t^4 - 12t^3 + 42t^2 - 20t^3 + 20t^2 - 70t + 36t^2 - 36t + 126$

$\quad\quad = 12t^4 - 32t^3 + 98t^2 - 106t + 126$

Think About It!

When multiplying two trinomials, do you have to apply the Distributive Property in any certain order?

Check

Find each product.

a. $(m - 3)(m^2 + m - 5)$

     A. $m^3 - 2m^2 - 8m + 15$

     B. $m^3 - 4m^2 - 8m + 15$

     C. $m^3 - 2m^2 + 8m + 15$

     D. $m^3 - 2m^2 - 8m - 15$

b. $(6v^2 + 4v - 3)(v^2 - v + 6)$

     A. $6v^4 + 10v^3 + 29v^2 + 27v - 18$

     B. $6v^4 - 2v^3 + 32v^2 + 27v - 18$

     C. $6v^4 - 2v^3 + 29v^2 + 27v - 18$

     D. $6v^4 - 2v^3 + 29v^2 + 24v - 18$

Go Online You can complete an Extra Example online.

Practice

⊙ **Go Online** You can complete your homework online.

Examples 1–3

Find each product.

1. $(3c - 5)(c + 3)$

2. $(g + 10)(2g - 5)$

3. $(6a + 5)(5a + 3)$

4. $(4x + 1)(6x + 3)$

5. $(5y - 4)(3y - 1)$

6. $(6d - 5)(4d - 7)$

7. $(3m + 5)(2m + 3)$

8. $(7n - 6)(7n - 6)$

9. $(12t - 5)(12t + 5)$

10. $(5r + 7)(5r - 7)$

11. $(8w + 4x)(5w - 6x)$

12. $(11z - 5y)(3z + 2y)$

Example 4

13. PLAYGROUND The dimensions of a playground are represented by a width of $9x + 1$ feet and a length of $5x - 2$ feet. Write an expression that represents the area of the playground.

14. THEATER The Loft Theater has a center seating section with $3c + 8$ rows and $4c - 1$ seats in each row. Write an expression for the total number of seats in the center section.

15. CRAFTS Suppose a rectangular quilt made up of squares has a length-to-width ratio of 5 to 4. The length of the quilt is $5x$ inches. The quilt can be made slightly larger by adding a border of 1-inch squares all the way around the perimeter of the quilt. Write a polynomial expression for the area of the larger quilt.

16. FLAG CASE A United States flag is sometimes folded into a triangle shape and displayed in a triangular display case. If a display case has dimensions shown in inches, write a polynomial expression that represents the area of wall space covered by the display case.

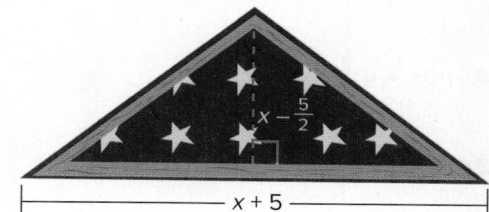

17. NUMBER THEORY Think of a whole number. Subtract 2. Write down this number. Take the original number and add 2. Write down this number. Find the product of the numbers you wrote down. Subtract the square of the original number. The result is always -4. Use polynomials to show how this number trick works.

Example 5

Find each product.

18. $(2y - 11)(y^2 - 3y + 2)$

19. $(4a + 7)(9a^2 + 2a - 7)$

20. $(m^2 - 5m + 4)(m^2 + 7m - 3)$

21. $(x^2 + 5x - 1)(5x^2 - 6x + 1)$

22. $(3b^3 - 4b - 7)(2b^2 - b - 9)$

23. $(6z^2 - 5z - 2)(3z^3 - 2z - 4)$

Mixed Exercises

Find each product.

24. $(m + 4)(m + 1)$

25. $(x + 2)(x + 2)$

26. $(b + 3)(b + 4)$

27. $(t + 4)(t - 3)$

28. $(r + 1)(r - 2)$

29. $(n - 5)(n + 1)$

30. $(3c + 1)(c - 2)$

31. $(2x - 6)(x + 3)$

32. $(d - 1)(5d - 4)$

33. $(2\ell + 5)(\ell - 4)$

34. $(3n - 7)(n + 3)$

35. $(q + 5)(5q - 1)$

36. $(3b + 3)(3b - 2)$

37. $(2m + 2)(3m - 3)$

38. $(4c + 1)(2c + 1)$

39. $(5a - 2)(2a - 3)$

40. $(4h - 2)(4h - 1)$

41. $(x - y)(2x - y)$

42. $(w + 4)(w^2 + 3w - 6)$

43. $(t + 1)(t^2 + 2t + 4)$

44. $(k - 4)(k^2 + 5k - 2)$

45. $(m + 3)(m^2 + 3m + 5)$

46. $(2x + 1)(x^2 - 3x - 4)$

47. $(3b + 4)(2b^2 - b + 4)$

Simplify.

48. $(m + 2)[(m^2 + 3m - 6) + (m^2 - 2m + 4)]$

49. $[(t^2 + 3t - 8) - (t^2 - 2t + 6)](t - 4)$

Find each product.

50. $(a - 2b)^2$

51. $(3c + 4d)^2$

52. $(x - 5y)^2$

53. $(2r - 3t)^3$

54. $(5g + 2h)^3$

55. $(4y + 3z)(4y - 3z)^2$

56. PRECISION Write each expression as a simplified polynomial.

    a. $(3c - 2)(4c^2 - c^3 + 3)$

    b. $(5x - y)(3x^2 - 2xy) + (2x + y)(y^2 - 4x^2)$

    c. $-2x(3 - x^2)(2x + 4)$

    d. $(z - 1)(2 - z)(z + 1)$

57. ART The museum where Julia works plans to have a large wall mural painted in its lobby. First, Julia wants to paint a large frame around where the mural will be. She only has enough paint for the frame to cover 100 square feet of wall surface. The mural's length will be 5 feet longer than its width, and the frame will be 2 feet wide on all sides.

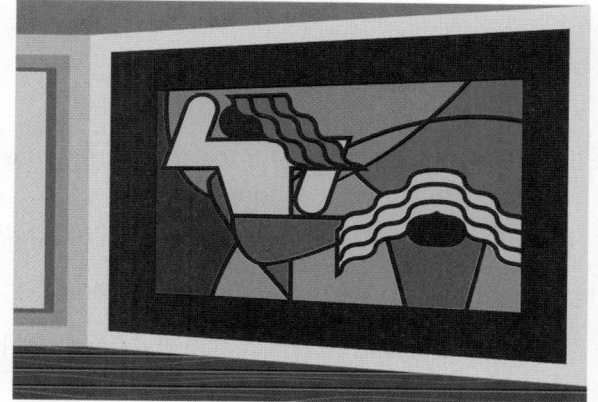

    a. Write an expression for the area of the mural.

    b. Write an expression for the area of the frame.

    c. Write and solve an equation to find how large the mural can be.

58. STRUCTURE The dimensions of the composite figure shown are given in terms of the triangle's height, h.

    a. Write and simplify a quadratic expression for the area of the figure.

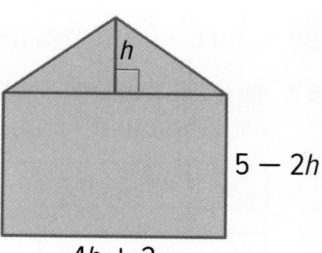

    b. If $h = 1.42$ units, what is the area of the figure? Round to the nearest hundredth, if necessary.

59. STRUCTURE Consider the expression $x^{4p + 1}(x^{1 - 2p})^{2p + 3}$.

    a. Use the laws of exponents to simplify the expression.

    b. Find any integer values of p that make this expression equal to 1 for all values of x.

60. USE A MODEL The relationship between monthly profit P, monthly sales n, and unit price p is $P = n(p - U) - F$, where U is the unit cost per sale and F is a fixed cost that does not depend on the number of sales. For an online business advice service, the unit cost is $30 per hour-long session and the monthly fixed cost is $3000.

 a. Given a model for monthly sales of $n = 5000 - 40p$ for a given price p per session, write and simplify a quadratic expression for P in terms of p.

 b. If the unit price, p, is $77.50, what is the monthly profit, P?

61. STRUCTURE Find and simplify an expression for the volume of the rectangular prism shown.

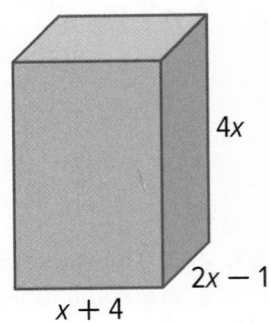

$4x$

$2x - 1$

$x + 4$

62. ANALYZE Determine if the following statement is *sometimes*, *always*, or *never* true. Justify your argument.

  The FOIL method can be used to multiply a binomial and a trinomial.

63. PERSEVERE Find $(x^m + x^p)(x^{m-1} - x^{1-p} + x^p)$.

64. CREATE Write a binomial and a trinomial involving a single variable. Then find their product.

65. WRITE Compare and contrast the procedure used to multiply a trinomial by a binomial using the vertical method with the procedure used to multiply a three-digit number by a two-digit number.

66. WRITE Summarize the methods that can be used to multiply polynomials.

67. WHICH ONE DOESN'T BELONG? Which polynomial expression does not belong with the other expressions? Explain your reasoning.

$(2x + 11)(3x - 7)$	$(6y + 1)(y - 4)$
$(3y - 4)(2y + 5)$	$(2x - 1)(x^2 + x - 1)$

68. FIND THE ERROR Jariah and Malia are multiplying the expression $(2x + 1)(x - 4)$. Is either of them correct? Explain your reasoning.

Jariah	Malia
$(2x + 1)(x - 4) = 2x^2 + 8x + 1x - 4$	$(2x + 1)(x - 4) = 2x^2 - 8x + 1x - 4$
$= 2x^2 + 9x - 4$	$= 2x^2 - 7x - 4$

Special Products

Today's Goals
- Multiply binomials by applying the pattern formed by squares of sums.
- Multiply binomials by applying the pattern formed by squares of differences.

Explore Using Algebra Tiles to Find the Squares of Sums

🌐 **Online Activity** Use algebra tiles to complete the Explore.

ⓠ **INQUIRY** How can you write the square of a sum?

Learn Square of a Sum

Key Concept • Square of a Sum	
Words	The square of $a + b$ is the square of a plus twice the product of a and b plus the square of b.
Symbols	$(a + b)^2 = (a + b)(a + b)$ $= a^2 + 2ab + b^2$
Example	$(x + 3)^2 = (x + 3)(x + 3)$ $= x^2 + 6x + 9$

Example 1 Square of a Sum

Find each product.

a. $(x + 6)^2$

$(a + b)^2 = a^2 + 2ab + b^2$ Square of a sum

$(x + 6)^2 = (x)^2 + 2(x)(6) + 6^2$ $a = x, b = 6$

$ = x^2 + 12x + 36$ Simplify.

b. $(3g + 10h)^2$

$(a + b)^2 = a^2 + 2ab + b^2$ Square of a sum

$(3g + 10h)^2 = (3g)^2 + 2(3g)(10h) + (10h)^2$ $a = 3g, b = 10h$

$ = 9g^2 + 60gh + 100h^2$ Simplify.

💭 **Think About It!**

How can you check that using the square of a sum pattern gives the correct product?

Check

Find $(2x + 9)^2$. Select the correct product.

A. $2x^2 + 18x + 18$

B. $2x^2 + 36x + 81$

C. $4x^2 + 18x + 81$

D. $4x^2 + 36x + 81$

Find $(6m + 11n)^2$. Select the correct product.

A. $6m^2 + 132mn + 11n^2$

B. $12m^2 + 66mn + 121n^2$

C. $36m^2 + 66mn + 121n^2$

D. $36m^2 + 132mn + 121n^2$

🌐 Example 2 Use Squares of Sums

GENETICS A Punnett square is used to predict the probability of offspring inheriting certain genetic characteristics. In Doberman Pinschers, black fur _B_ is dominant over the recessive gene for brown fur _b_. If two parents have both a dominant and a recessive gene, use the square of a sum to determine the possible combinations of their offspring.

	_B_	_b_
_B_	_BB_	_Bb_
_b_	_bB_	_bb_

Both parents can be represented by $(B + b)$, and the combinations of the offspring are the product of $(B + b)^2$.

$(a + b)^2 = a^2 + 2ab + b^2$ Square of a sum

$(B + b)^2 = B^2 + 2Bb + b^2$ $a = B, b = b$

Check

SURFACE AREA The surface area of a cube is given by $A = 6s^2$, where s is the length of one side. Select the expression that represents the surface area of a cube with side length $3n + 6$.

A. $6(3n^2 + 18n + 36)$

B. $6(9n^2 + 36)$

C. $6(9n^2 + 36n + 36)$

D. $6(9n^2 + 81n + 36)$

🌐 **Go Online** You can complete an Extra Example online.

Explore Using Algebra Tiles to Find the Squares of Difference

Online Activity Use algebra tiles to complete the Explore.

INQUIRY How can you write the square of a difference?

Learn Square of a Difference

Key Concept • Square of a Difference

Words	The square of $a - b$ is the square of a minus twice the product of a and b plus the square of b.
Symbols	$(a - b)^2 = (a - b)(a - b)$
	$= a^2 - 2ab + b^2$
Example	$(x - 8)^2 = (x - 8)(x - 8)$
	$= x^2 - 2 \cdot x \cdot 8 + 8^2$
	$= x^2 - 16x + 64$

Example 3 Square of a Difference

Find $(7d - 2f)^2$.

$(a - b)^2 = a^2 - 2ab + b^2$ Square of a difference

$(7d - 2f)^2 = (7d)^2 - 2(7d)(2f) + (2f)^2$ $a = 7d, b = 2f$

$\quad\quad = 49d^2 + 28df + 4f^2$ Simplify.

Check

Find $(4k - 1)^2$.

A. $4k^2 - 8k - 1$

B. $16k^2 - 1$

C. $16k^2 - 8k + 1$

D. $16k^2 - 16k + 1$

Study Tip

Watching Signs Since the square of a sum and the square of a difference vary only by the sign of the middle term, pay close attention to the sign being used within the square of the trinomial.

Think About It!

Demarco says that the Square of a Difference pattern is a special case of the Square of a Sum pattern. Is he correct? Explain.

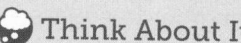

Go Online You can complete an Extra Example online.

Explore Using Algebra Tiles to Find Products of Sums and Differences

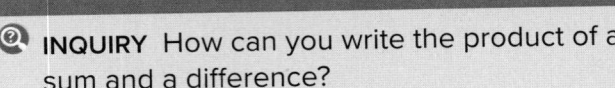 **Online Activity** Use algebra tiles to complete the Explore.

> **⊘ INQUIRY** How can you write the product of a sum and a difference?

Learn Product of a Sum and a Difference

Key Concept • Product of a Sum and a Difference

Words	The product of $a + b$ and $a - b$ is the square of a minus the square of b.
Symbols	$(a + b)(a - b) = a^2 - ab + ab - b^2$ $\qquad\qquad\qquad = a^2 - b^2$
Example	$(x + 6)(x - 6) = x^2 - 6x + 6x - 6^2$ $\qquad\qquad\qquad = x^2 - 36$

Example 4 Product of a Sum and a Difference

Find each product.

a. $(z + 5)(z - 5)$

$(a + b)(a - b) = a^2 - b^2$ Product of a sum and a difference.

$(z + 5)(z - 5) = z^2 - 5^2$ $a = z, b = 5$

$\qquad\qquad\quad = z^2 - 25$ Simplify.

b. $(3y^3 + 4)(3y^3 - 4)$

$(a + b)(a - b) = a^2 - b^2$ Product of a sum and a difference.

$(3y^3 + 4)(3y^3 - 4) = (3y^3)^2 - (4)^2$ $a = 3y^2, b = 4$

$\qquad\qquad\qquad\quad = 9y^6 - 16$ Simplify.

Check

Find each product.

a. $(x^2 + 3y)(x^2 - 3y)$

b. $(2x + 9y^2)(2x - 9y^2)$

c. $(x + 9y)(x - 9y)$

d. $(x + 9)(x - 9)$

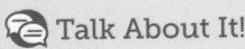 **Go Online** You can complete an Extra Example online.

Study Tip

Identical Values Note that the values of a and b must be identical within each quantity to use the pattern for the product of a sum and a difference. The only difference between the quantities is the operation between a and b.

Watch Out!

Power of a Power Remember to square all parts of a and b, including powers. When a power is raised to another power, multiply the exponents.

ⓔ Talk About It!

Explain how you would tell someone to find the product of a sum and a difference.

Practice

Go Online You can complete your homework online.

Examples 1 and 3

Find each product.

1. $(a + 10)(a + 10)$
2. $(b - 6)(b - 6)$
3. $(h + 7)^2$
4. $(x + 6)^2$
5. $(8 - m)^2$
6. $(9 - 2y)^2$
7. $(2b + 3)^2$
8. $(5t - 2)^2$
9. $(8h - 4n)^2$
10. $(4m - 5n)^2$

Example 2

11. **ROUNDABOUTS** A city planner is proposing a roundabout to improve traffic flow at a busy intersection. Write a polynomial equation for the area A of the traffic circle if the radius of the outer circle is r and the width of the road is 18 feet.

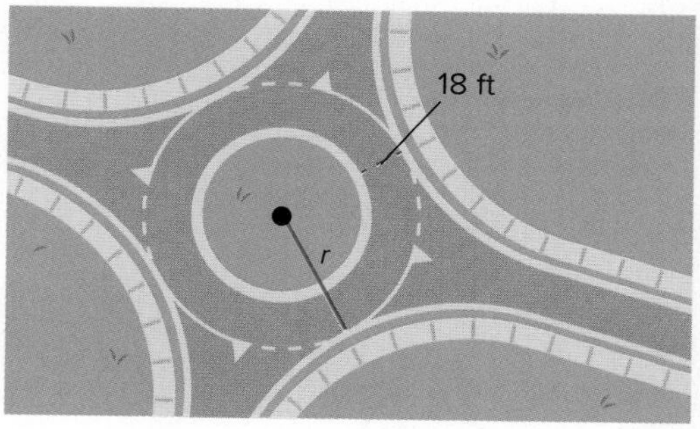

18 ft
r

12. **NUMBER CUBES** Kivon has two number cubes. Each edge of number cube A is 3 millimeters less than each edge of number cube B. Each edge of number cube B is x millimeters. Write an equation that models the surface area of number cube A.

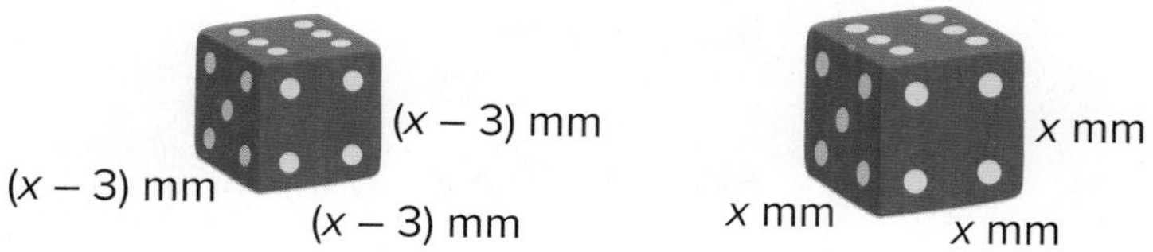

Number Cube A
(x – 3) mm
(x – 3) mm
(x – 3) mm
Number Cube B
x mm
x mm
x mm

13. **PROBABILITY** The spinner has two equal sections, blue (B) and red (R). Use the square of a sum to determine the possible combinations of spinning the spinner two times.

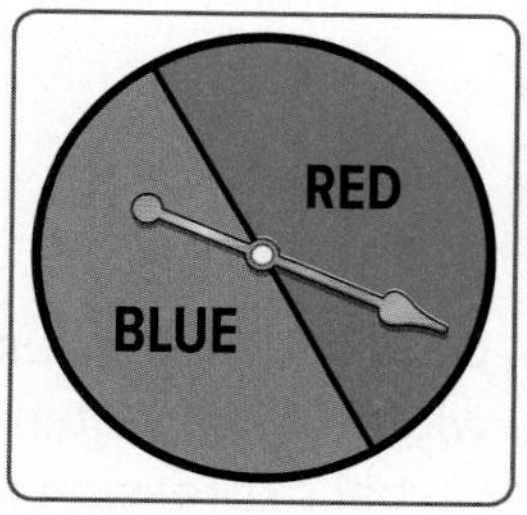

RED
BLUE

14. **BUSINESS** The Combo Lock Company finds that its profit data from 2015 to the present can be modeled by the function $y = (2n + 11)^2$, where y is the profit n years since 2015. Which special product does this polynomial demonstrate? Simplify the polynomial.

Example 4

Find each product.

15. $(u + 3)(u - 3)$

16. $(b + 7)(b - 7)$

17. $(2 + x)(2 - x)$

18. $(4 - x)(4 + x)$

19. $(2q + 5r)(2q - 5r)$

20. $(3a^2 + 7b)(3a^2 - 7b)$

Mixed Exercises

Find each product.

21. $(n + 3)^2$

22. $(x + 4)(x + 4)$

23. $(y - 7)^2$

24. $(t - 3)(t - 3)$

25. $(b + 1)(b - 1)$

26. $(a - 5)(a + 5)$

27. $(p - 4)^2$

28. $(z + 3)(z - 3)$

29. $(\ell + 2)(\ell + 2)$

30. $(r - 1)(r - 1)$

31. $(3g + 2)(3g - 2)$

32. $(2m - 3)(2m + 3)$

33. $(6 + u)^2$

34. $(r + t)^2$

35. $(3q + 1)(3q - 1)$

36. $(c - d)^2$

37. $(2k - 2)^2$

38. $(w + 3h)^2$

39. $(3p - 4)(3p + 4)$

40. $(t + 2u)^2$

41. $(x - 4y)^2$

42. $(3b + 7)(3b - 7)$

43. $(3y - 3g)(3y + 3g)$

44. $(n^2 + r^2)^2$

45. $(2k + m^2)^2$

46. $(3t^2 - n)^2$

47. GEOMETRY The length of a rectangle is the sum of two whole numbers. The width of the rectangle is the difference of the same two whole numbers. Write a verbal expression for the area of the rectangle.

48. Find the product of $(10 - 4t)$ and $(10 + 4t)$. What type of special product does this represent?

49. STORAGE A cylindrical tank is placed along a wall. A cylindrical PVC pipe will be hidden in the corner behind the tank. See the side-view diagram shown. The radius of the tank is r inches, and the radius of the PVC pipe is s inches.

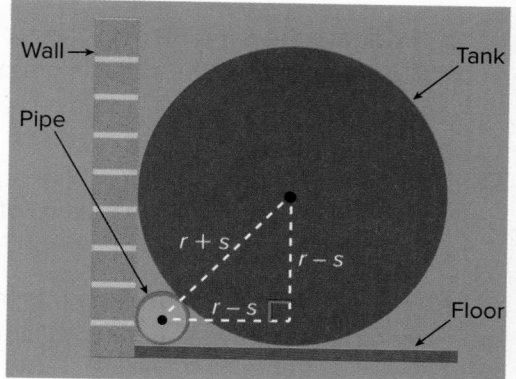

a. Use the Pythagorean Theorem to write an equation for the relationship between the two radii. Simplify your equation so that there is a zero on one side of the equal sign.

b. Write a polynomial equation you could solve to find the radius s of the PVC pipe if the radius of the tank is 20 inches.

Find each product.

50. $(2a - 3b)^2$

51. $(5y + 7)^2$

52. $(8 - 10a)^2$

53. $(10x - 2)(10x + 2)$

54. $(3t + 12)(3t - 12)$

55. $(a + 4b)^2$

56. $(3q - 5r)^2$

57. $(2c - 9d)^2$

58. $(g + 5h)^2$

59. $(6y - 13)(6y + 13)$

60. $(3a^4 - b)(3a^4 + b)$

61. $(5x^2 - y^2)^2$

62. $(8a^2 - 9b^3)(8a^2 + 9b^3)$

63. $\left(\frac{3}{4}k + 8\right)^2$

64. $\left(\frac{2}{5}y - 4\right)^2$

65. $(7z^2 + 5y^2)(7z^2 - 5y^2)$

66. $(2m + 3)(2m - 3)(m + 4)$

67. $(r + 2)(r - 5)(r - 2)(r + 5)$

Find each product.

68. $(c + d)(c + d)(c + d)$

69. $(2a - b)^3$

70. $(f + g)(f - g)(f + g)$

71. $(k - m)(k + m)(k - m)$

72. $(n - p)^2(n + p)$

73. $(q - r)^2(q - r)$

74. Consider the product $(a + b)(a - b)(a + b)(a - b)$.

    a. Show the steps required to determine the product.

    b. Evaluate the original expression for $a = 5$ and $b = 2$.

    c. Evaluate the simplified expression you wrote in **part a** for $a = 5$ and $b = 2$. Compare this to the result from **part b**.

75. STRUCTURE Tanisha is investigating growth patterns for the area of a square. She begins with a square of side length s and looks at the effects of enlarging the side length by one unit at a time.

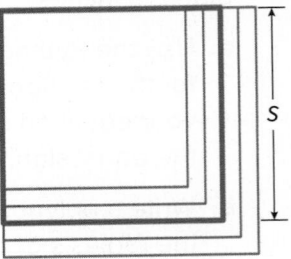

    a. How much more area does the square with side length $s + 1$ have compared to the square with side length s?

    b. How much more area does the square with side length $s + 2$ have compared to the square with side length $s + 1$?

    c. How much more area does the square with side length $s + 3$ have compared to the square with side length $s + 2$?

🧠 Higher-Order Thinking Skills

76. WHICH ONE DOESN'T BELONG? Which expression does not belong? Explain your reasoning.

$(2c - d)(2c - d)$	$(2c + d)(2c - d)$	$(2c + d)(2c + d)$	$(c + d)(c + d)$

77. PERSEVERE Does a pattern exist for the cube of a sum $(a + b)^3$?

    a. Investigate this question by finding the product $(a + b)(a + b)(a + b)$.

    b. Use the pattern you discovered in **part a** to find $(x + 2)^3$.

    c. Draw a diagram of a geometric model for $(a + b)^3$.

    d. What is the pattern for the cube of a difference $(a - b)^3$?

78. ANALYZE The square of a sum is called a *perfect square trinomial*. Find c that makes $25x^2 - 90x + c$ a perfect square trinomial.

79. CREATE Write two binomials with a product that is a binomial. Then write two binomials with a product that is not a binomial.

80. WRITE Describe how to square the sum of two quantities, how to square the difference of two quantities, and how to find the product of a sum of two quantities and a difference of two quantities.

Using the Distributive Property

Today's Goals
- Factor polynomials by using the Distributive Property
- Factor polynomials by using the Distributive Property and grouping

Today's Vocabulary
factoring

factoring by grouping

Explore Using Algebra Tiles to Factor Polynomials

Online Activity Use algebra tiles to complete the Explore.

> ×
>
> **INQUIRY** How is factoring polynomials related to multiplying polynomials?

Learn Factoring by Using the Distributive Property

You can use the Distributive Property to multiply a polynomial by a monomial.

$$3y(2y + 5) = 3y(2y) + 3y(5)$$
$$= 6y^2 + 15y$$

You can also use the Distributive Property to factor a polynomial. **Factoring** is the process of expressing a polynomial as the product of monomials and polynomials.

$$6y^2 + 15y = 3y(2y) + 3y(5)$$
$$= 3y(2y + 5)$$

So, $3y(2y + 5)$ is the factored form of $6y^2 + 15y$. When factoring a polynomial, it must be factored completely. If you are using the Distributive Property to factor, one factor will be the greatest common factor (GCF) for all terms of the polynomial.

Study Tip

Greatest Common Factor To find the GCF of a polynomial, write all factors of each term as prime numbers or variables to the first degree.

Example 1 Use the Distributive Property

Use the Distributive Property to factor each polynomial.

a. $12a^2 + 16a$

Step 1 Factor each term.

$$12a^2 = 2 \cdot 2 \cdot 3 \cdot a \cdot a$$
$$16a = 2 \cdot 2 \cdot 2 \cdot 2 \cdot a$$

Step 2 Underline the common terms.

$$12a^2 = \underline{2} \cdot \underline{2} \cdot 3 \cdot \underline{a} \cdot a$$
$$16a = \underline{2} \cdot \underline{2} \cdot 2 \cdot 2 \cdot \underline{a}$$

Step 3 Find the GCF.

$$GCF = 2 \cdot 2 \cdot a \text{ or } 4a$$

(continued on the next page)

Step 4 Write each term as the product of the GCF and the remaining factors. Use the Distributive Property.

$$12a^2 + 16a = 4a(3a) + 4a(4) \qquad \text{Rewrite each term using the GCF.}$$
$$= 4a(3a + 4) \qquad \text{Distributive Property.}$$

b. $20x^2y^2 - 45x^2y - 35x^2$

$$20x^2y^2 = 2 \cdot 2 \cdot 5 \cdot x \cdot x \cdot y \cdot y$$
$$- 45x^2y = -1 \cdot 3 \cdot 3 \cdot 5 \cdot x \cdot x \cdot y$$
$$- 35x^2 = -1 \cdot 5 \cdot 7 \cdot x \cdot x$$
$$GCF = 5x^2$$

Write each term as the product of the GCF and the remaining factors. Use the Distributive Property.

$$20x^2y^2 - 45x^2y - 35x^2 = 5x^2(4y^2) + 5x^2(-9y) + 5x^2(-7)$$
$$= 5x^2(4y^2 - 9y - 7)$$

Factor each polynomial.

a. $33n^3 - 121n^2$

b. $14a^2b^2c - 6ac^2 + 10ac$

🌐 **Example 2** Use Factoring

VOLCANOS The 1980 eruption of Washington's Mt. St. Helens had an initial lateral blast with a velocity of 440 feet per second. The expression $440t - 16t^2$ models the height of a rock erupted from the volcano after t seconds. Factor the expression.

$$440t = 2 \cdot 2 \cdot 2 \cdot 5 \cdot 11 \cdot t$$
$$-16t^2 = -1 \cdot 2 \cdot 2 \cdot 2 \cdot 2 \cdot t \cdot t$$
$$GCF = 8t$$
$$440t - 16t^2 = 8t(55) + 8t\,(-2t)$$
$$= 8t(55 - 2t)$$

Check

GOLF A golfer hits a golf ball with a velocity of 112 feet per second. The expression $112t - 16t^2$ represents the height of the golf ball after t seconds. Factor the expression.

🔵 **Go Online** You can complete an Extra Example online.

662 Module 11 · Polynomials

Watch Out!

Factoring Completely
Once you have factored the polynomial, check the remaining polynomial for other common factors you may have missed. *Factoring* means to factor completely.

Learn Factor by Grouping

When a polynomial has four or more terms, you can sometimes use a method called **factoring by grouping**. Similar terms are grouped, and the Distributive Property is applied to a common binomial.

Key Concept • Factor by Grouping	
Words	A polynomial can be factored by grouping only if all of the following conditions exist. • There are four or more terms. • Terms have common factors that can be grouped together. • There are at least two common factors that are identical or additive inverses of each other.
Symbols	$ax + bx + ay + by = (ax + bx) + (ay + by)$ $= x(a + b) + y(a + b)$ $= (x + y)(a + b)$

Example 3 Factor by Grouping

Factor $2uv + 6u + 5v + 15$.

$2uv + 6u + 5v + 15 = (2uv + 6u) + (5v + 15)$ Group terms with common factors

$= 2u(v + 3) + 5(v + 3)$ Factor the GCF from each group.

Notice that $(v + 3)$ is common to both groups, so it becomes the GCF.

$= (2u + 5)(v + 3)$ Distributive Property.

Check

Factor $tw + 10t - 2w - 20$.

Go Online You can complete an Extra Example online.

Talk About It!

Explain why you cannot use factoring by grouping on a polynomial with three terms.

Study Tip

Creating Groups If you are unable to get identical or additive inverse binomials after factoring out the GCF, try grouping the terms in a different way.

Think About It!

Which property is used to simplify $3m[(-1)(p-7)] + 4(p-7)$ to $-3m(p-7) + 4(p-7)$?

Example 4 Factor by Grouping with Additive Inverses

It is helpful to be able to recognize when binomials are additive inverses of each other. For example, $5 - x = -1(x - 5)$.

Factor $21m - 3mp + 4p - 28$.

$21m - 3mp + 4p - 28$	Original expression
$= (21m - 3mp) + (4p - 28)$	Group terms with common factors.
$= 3m(7 - p) + 4(p - 7)$	Factor the GCF from each group.
$= 3m[(-1)(p - 7)] + 4(p - 7)$	$7 - p = -1(p - 7)$
$= -3m(p - 7) + 4(p - 7)$	Associative Property
$= (-3m + 4)(p - 7)$	Distributive Property

Alternate Method

$21m - 3mp + 4p - 28$	Original expression
$= 4p - 3mp + 21m - 28$	Rearrange the expression.
$= (4p - 3mp) + (21m - 28)$	Group terms with common factors.
$= p(4 - 3m) + 7(3m - 4)$	Factor the GCF from each group.
$= p(-3m + 4) + 7[-1(-3m + 4)]$	$4 - 3m = -1(3m - 4)$
$= p(-3m + 4) - 7(-3m + 4)$	Associative Property
$= (p - 7)(-3m + 4)$	Distributive Property
$= (-3m + 4)(p - 7)$	Symmetric Property

Check

Factor $-3x^3 + 33x^2 + 4x - 44$.

Pause and Reflect

Did you struggle with anything in this lesson? If so, how did you deal with it?

Go Online to learn how to prove the Elimination Method in Expand 11-5.

Go Online You can complete an Extra Example online.

Practice

Go Online You can complete your homework online.

Example 1
Use the Distributive Property to factor each polynomial.

1. $16t - 40y$

2. $30v + 50x$

3. $2k^2 + 4k$

4. $5z^2 + 10z$

5. $4a^2b^2 + 2a^2b - 10ab^2$

6. $5c^2v - 15c^2v^2 + 5c^2v^3$

Example 2

7. PHYSICS The distance d an object falls after t seconds is given by $d = 16t^2$ (ignoring air resistance). To find the height of an object launched upward from ground level at a rate of 32 feet per second, use the expression $32t - 16t^2$, where t is the time in seconds. Factor the expression.

8. SWIMMING POOL The area of a rectangular swimming pool is given by the expression $12w - w^2$, where w is the width of one side. Factor the expression.

9. VERTICAL JUMP Your vertical jump height is measured by subtracting your standing reach height from the height of the highest point you can reach by jumping without taking a running start. Typically, NBA players have vertical jump heights of up to 34 inches. If an NBA player jumps this high, his height in inches above his standing reach height after t seconds can be modeled by the expression $162t - 192t^2$. Factor the expression.

10. PETS Conner is playing with his dog. He tosses a treat upward with an initial velocity of 13.7 meters per second. His hand starts at the same height as the dog's mouth, so the height of the treat above the dog's mouth in meters after t seconds is given by the expression $13.7t - 4.9t^2$. Factor the expression.

Examples 3 and 4
Factor each polynomial.

11. $fg - 5g + 4f - 20$

12. $a^2 - 4a - 24 + 6a$

13. $hj - 2h + 5j - 10$

14. $xy - 2x - 2 + y$

15. $45pq - 27q - 50p + 30$

16. $24ty - 18t + 4y - 3$

17. $3dt - 21d + 35 - 5t$

18. $8r^2 + 12r$

19. $21th - 3t - 35h + 5$

20. $vp + 12v + 8p + 96$

21. $5br - 25b + 2r - 10$

22. $2nu - 8u + 3n - 12$

23. $b^2 - 2b + 3b - 6$

24. $2j^2 + 2j + 3j + 3$

25. $2a^2 - 4a + a - 2$

Mixed Exercises

Factor each polynomial.

26. $7x + 49$

27. $8m - 6$

28. $5a^2 - 15$

29. $10q - 25q^2$

30. $a^2b^2 + a$

31. $x + x^2y + x^3y^2$

32. $3p^2r^2 + 6pr + p$

33. $4a^2b^2 + 16ab + 12a$

34. $10h^3n^3 - 2hn^2 + 14hn$

35. $48a^2b^2 - 12ab$

36. $6x^2y - 21y^2w + 24xw$

37. $x^2 + 3x + x + 3$

38. $2x^2 - 5x + 6x - 15$

39. $3n^2 + 6np - np - 2p^2$

40. $4x^2 - 1.2x + 0.5x - 0.15$

41. $9x^2 - 3xy + 6x - 2y$

42. $3x^2 + 24x - 1.5x - 12$

43. $2x^2 - 0.6x + 3x - 0.9$

44. ARCHERY The height, in feet, of an arrow can be modeled by the expression $80t - 16t^2$, where t is the time in seconds. Factor the expression.

Higher-Order Thinking Skills

45. REGULARITY You have factored some polynomials using the Distributive Property, and others were factored by grouping. What are the similarities between the two methods?

46. CREATE Write a four-term polynomial that can be factored by grouping. Then factor the polynomial.

47. FIND THE ERROR Given the polynomial expression $3x^2 + 7x - 18x - 42$, Theresa says you need to factor by grouping using the binomials $(3x^2 - 18x)$ and $(7x - 42)$. Akash says you need to use the binomials $(3x^2 + 7x)$ and $(-18x - 42)$. Is either of them correct? Justify your answer.

48. PRECISION Choose a value, q, so that $2x^2 + qx + 7x - 21$ can be factored by grouping. Then factor the expression.

49. WRITE Explain how to factor the polynomial $12a^2b^2 - 16a^2b^3$.

Factoring Quadratic Trinomials

Explore Using Algebra Tiles to Factor Trinomials

🔘 **Online Activity** Use algebra tiles to complete the Explore.

> ⊘ **INQUIRY** How can you use the constant and coefficients in a polynomial to find its factors?

Learn Factoring Trinomials with a Leading Coefficient of 1

Key Concept • Factoring Trinomials with a Leading Coefficient of 1	
Words	To factor trinomials in the form $x^2 + bx + c$, find two integers, m and p, with a sum of b and a product of c. Then write $x^2 + bx + c$ as $(x + m)(x + p)$.
Symbols	$x^2 + bx + c = (x + m)(x + p)$ when $m + p = b$ and $mp = c$.
Example	$x^2 + 8x + 15 = (x + 3)(x + 5)$ because $3 + 5 = 8$ and $3 \cdot 5 = 15$.

A polynomial that cannot be written as a product of two polynomials with integer coefficients is called a **prime polynomial**.

Example 1 c Is Positive

Factor $x^2 - 9x + 18$.

In this trinomial, $b = -9$ and $c = 18$. Because c is positive and b is negative, you need to find two negative factors with a sum of -9 and a product of 18.

Complete the table to make an organized list of the factors of 18 and look for the pair of factors with a sum of -9.

Factors of 18	Sum of Factors
−1, −18	−19
−2, −9	−11
−3, −6	−9

The correct factors are −3 and −6.

$$x^2 - 9x + 18 = (x + m)(x + p) \qquad \text{Write the pattern.}$$
$$= [x + (-3)][x + (-6)] \qquad m = -3 \text{ and } p = -6$$
$$= (x - 3)(x - 6) \qquad \text{Simplify.}$$

🔘 **Go Online** You can complete an Extra Example online.

Today's Goals
- Determine the factors of trinomials with a leading coefficient of 1.
- Determine the factors of trinomials with a leading coefficient not equal to 1.

Today's Vocabulary
prime polynomial

🔘 **Go Online**
You may want to complete the Concept Check to check your understanding.

🟣 **Think About It!**
If c is negative, then what is the relationship between the signs of the factors? Explain your reasoning.

Problem-Solving Tip
Guess and Check
When factoring a trinomial, make an educated guess, check for reasonableness, and then adjust the guess until you find the correct answer.

Example 2 *c* Is Negative and *b* Is Positive

Factor $x^2 + 5x - 14$.

In this trinomial, $b = 5$ and $c = -14$. Since c is negative, the factors m and p have opposite signs. So, either m or p is negative, but not both. Since b is positive, the factor with the greater absolute value is also positive.

Complete the table to make a list of the factors of -14, where one factor of each pair is negative and the factor with the greater absolute value is positive. Look for the pair of factors with a sum of 5.

The correct factors are -2 and 7.

Factors of -14	Sum of Factors
$-1, 14$	13
$-2, 7$	5

$$x^2 + 5x - 14 = (x + m)(x + p)$$
$$= (x - 2)(x + 7)$$

Example 3 *c* Is Negative and *b* Is Negative

Factor $x^2 - 3x - 4$.

In this trinomial, $b = -3$ and $c = -4$. Either m or p is negative, but not both. Since b is negative, the factor with the greater absolute value is also negative.

Complete the table to make a list of the factors of -4, where one factor of each pair is negative and the factor with the greater absolute value is negative. Look for the pair of factors with a sum of -3.

The correct factors are 1 and -4.

Factors of -4	Sum of Factors
$1, -4$	-3
$2, -2$	0

$$x^2 - 3x - 4 = (x + m)(x + p)$$
$$= (x + 1)(x - 4)$$

Example 4 Factor a Polynomial

Factor $x^2 - 4x + 8$, if possible. If the polynomial cannot be factored using integers, write *prime*.

In this trinomial, $b = -4$ and $c = 8$. Since b is negative, $m + p$ is negative. Since c is positive, mp is positive. So, m and p are both negative.

Next, list the factors of 8. Look for the pair with a sum of -4.

Factors of 8	Sum of Factors
$-1, -8$	-9
$-2, -4$	-6

There are no factors with a sum of -4. So, the trinomial cannot be factored using integers. Therefore, $x^2 - 4x + 8$ is prime.

Check

Write the factored form of each polynomial. If the polynomial cannot be factored using integers, write *prime*.

a. $x^2 + 7x + 6$ b. $x^2 - 8x + 12$ c. $x^2 + 3x - 40$

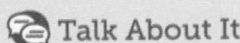

Think About It!

Is $(x + 1)(x - 4)$ equal to $(x - 4)(x + 1)$? to $(x - 1)(x + 4)$? Explain your reasoning.

Talk About It

How does the process of factoring $ax^2 + bx + c$ compare to the process of factoring $x^2 + bx + c$?

🌐 Example 5 Solve a Problem by Factoring

FLAG DESIGN Switzerland's flag has a very unique shape; it is a square. However, the flag used by the country's naval vessels is rectangular, as shown. If the area of the square flag is $x^2 - 6x + 9$ square feet, and the length is increased by 4 feet, then what is the area of the naval flag in terms of x?

Step 1 Factor $x^2 - 6x + 9$. In this trinomial, $b = -6$ and $c = 9$. Because c is positive and b is negative, you need to find two negative factors with a sum of -6 and a product of 9.

Factors of 9	Sum of Factors
$-1, -9$	-10
$-3, -3$	-6

The correct factors are -3 and -3.

$$x^2 - 6x + 9 = (x + m)(x + p) \qquad \text{Write the pattern.}$$
$$= (x - 3)(x - 3) \qquad m = -3 \text{ and } p = -3$$

Step 2 Increase length and multiply. The length is increased by 4 feet, so the factor representing the length must be increased by 4.

$$(x - 3 + 4)(x - 3) = (x + 1)(x - 3) \qquad \text{Add 4 to the length.}$$
$$= x^2 - 3x + x - 3 \qquad \text{FOIL}$$
$$= x^2 - 2x - 3 \qquad \text{Simplify.}$$

The new area is $x^2 - 2x - 3$ square feet.

Robert_Ford/Getty Images.

Learn Factoring Trinomials

Key Concept • Factoring Trinomials with a Leading Coefficient ≠ 1

To factor trinomials in the form $ax^2 + bx + c$, find two integers, m and p, with a sum of b and a product of ac. Then write $ax^2 + bx + c$ as $ax^2 + mx + px + c$, and factor by grouping.

Example 6 *c* Is Negative

Factor $4x^2 + 18x - 10$.

In this trinomial, a GCF of 2 can be factored out.

$$4x^2 + 18x - 10 = 2(2x^2 + 9x - 5)$$

Then in the trinomial $2x^2 + 9x - 5$, $a = 2$, $b = 9$, and $c = -5$. You need to find two numbers with a sum of 9 and a product of $2(-5)$ or -10.

(continued on the next page)

Study Tip

Assumptions Because the polynomial represents the area of a flag, you can assume that it can be factored. The area of a rectangle can always be written as the product of two sides.

💭 Think About It!

Which property is applied to go from $2[2x(x + 5) - (x + 5)]$ to $2(2x - 1)(x + 5)$?

Complete the table to make a list of the factors of −10.

Factors of −10	Sum of Factors
1, −10	−9
2, −5	−3
5, −2	3
10, −1	9

Look for a pair of factors with a sum of 9. The correct factors are 10 and −1.

$4x^2 + 18x − 10 = 2(2x^2 + mx + px − 5)$	Write the pattern.
$= 2(2x^2 + 10x + (−1)x − 5)$	$m = 10$ and $p = −1$
$= 2[(2x^2 + 10x) + (−x − 5)]$	Group terms with common factors.
$= 2[2x(x + 5) − (x + 5)]$	Factor the GCFs.
$= 2(2x − 1)(x + 5)$	$x + 5$ is the common factor.

Example 7 *c* Is Positive

Factor $2x^2 − 17x + 21$.

In this trinomial, $a = 2$, $b = −17$, and $c = 21$. Since b is negative, $m + p$ will be negative. Since c is positive, mp will be positive. To determine m and p, list the negative factors of ac. The sum of m and p should be equal to b.

Complete the table to make a list of the negative factors of 42, and look for a pair of factors with a sum of −17.

Factors of 42	Sum of Factors
−1, −42	−43
−2, −21	−23
−3, −14	−17
−6, −7	−13

The correct factors are −3 and −14.

$2x^2 − 17x + 21 = 2x^2 + mx + px + 21$	Write the pattern.
$= 2x^2 + (−3)x + (−14)x + 21$	$m = −3$ and $p = −14$
$= (2x^2 − 14x) + (−3x + 21)$	Group terms with common factors.
$= 2x(x − 7) + (−3)(x − 7)$	Factor the GCFs.
$= (2x − 3)(x − 7)$	$x − 7$ is the common factor.

🔵 **Go Online** An alternate method is available for this example.

🔵 **Go Online** You can complete an Extra Example online.

Practice

Examples 1–4

Factor each polynomial, if possible. If the polynomial cannot be factored using integers, write *prime*.

1. $x^2 + 17x + 42$

2. $y^2 - 17y + 72$

3. $a^2 + 8a - 48$

4. $n^2 - 2n - 35$

5. $44 + 15h + h^2$

6. $40 - 22x + x^2$

7. $-24 - 5x + x^2$

8. $-42 - m + m^2$

9. $t^2 + 8t + 12$

10. $d^2 + 5d - 13$

11. $y^2 - 6y + 17$

12. $n^2 + 7n + 12$

13. $b^2 - 12b - 101$

14. $p^2 + 9p + 20$

15. $h^2 + 9h + 18$

16. $c^2 + c + 21$

Example 5

17. COSMETICS CASE The top of a cosmetics case is a rectangle in which the width is 2 centimeters greater than the length. The expression $x^2 + 26x - 168$ represents the area of the top of the case. Factor the expression.

18. CARPENTRY Miko wants to build a crate to hold record albums. The expression $2x^2 - 6x - 80$ represents the volume of the crate. Factor the expression.

19. BRIDGE ENGINEERING A suspension bridge is a bridge in which the deck is supported by cables with towers spaced throughout the span of the bridge. The height of a cable n inches above the deck measured at distance d in yards from the first tower is given by $d^2 - 36d + 324$. Factor the expression.

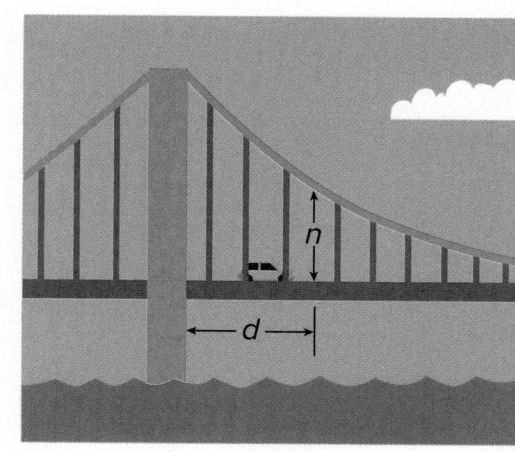

20. FINANCE The break-even point for a business occurs when the revenues equal the cost. A local children's museum studied their costs and revenues from paid admission. They found that their break-even point is given by the expression $2h^2 - 2h - 24$, where h is the number of hours the museum is open per day. Factor the expression.

Factor each polynomial, if possible. If the polynomial cannot be factored using integers, write _prime_.

21. $5x^2 + 34x + 24$

22. $2x^2 + 19x + 24$

23. $4x^2 + 22x + 10$

24. $4x^2 + 38x + 70$

25. $2x^2 - 3x - 9$

26. $4x^2 - 13x + 10$

27. $2x^2 + 3x + 6$

28. $5x^2 + 3x + 4$

29. $12x^2 + 69x + 45$

30. $4x^2 - 5x + 7$

31. $3x^2 - 8x + 15$

32. $5x^2 + 23x + 24$

33. $2x^2 + 3x - 6$

34. $2t^2 + 9t - 5$

35. $2y^2 + y - 1$

36. $4h^2 + 8h - 5$

Mixed Exercises

Factor each polynomial, if possible. If the polynomial cannot be factored using integers, write _prime_.

37. $n^2 + 3n - 18$

38. $x^2 + 2x - 8$

39. $r^2 + 4r - 12$

40. $x^2 - x - 12$

41. $w^2 - w - 6$

42. $y^2 - 6y + 8$

43. $t^2 - 15t + 56$

44. $-4 - 3m + m^2$

45. $2x^2 + 5x + 2$

46. $3n^2 + 5n + 2$

47. $3g^2 - 7g + 2$

48. $2t^2 - 11t + 15$

49. $4x^2 - 3x - 3$

50. $4b^2 + 15b - 4$

Factor each polynomial, if possible. If the polynomial cannot be factored using integers, write *prime*.

51. $9p^2 + 6p - 8$

52. $6q^2 - 13q + 6$

53. $a^2 - 10a + 21$

54. $x^2 + 2x - 15$

55. $2x^2 + 7x + 3$

56. $6x^2 + x + 2$

57. $x^2 + x - 20$

58. $x^2 - 6x - 7$

59. $p^2 - 10p + 21$

60. $5x^2 - 6x + 1$

61. $q^2 + 11qr + 18r^2$

62. $x^2 - 14xy - 51y^2$

63. $x^2 - 6xy + 5y^2$

64. $a^2 + 10ab - 39b^2$

65. $-6x^2 - 23x - 20$

66. $-4x^2 - 15x - 14$

67. $-5x^2 + 18x + 8$

68. $-6x^2 + 31x - 35$

69. $-4x^2 + 5x - 12$

70. $-12x^2 + x + 20$

71. **MONUMENTS** Susan is designing a pyramidal stone monument for a local park. The design specifications tell her that the height needs to be 9 feet and the width of the base must be 5 feet less than the length. The expression $3x^2 - 15x - 150$ represents the volume of the pyramidal stone monument.

   a. Factor the expression that represents the volume of the pyramidal stone monument that Susan is designing for a local park.

   b. What does each factored expression represent?

72. **PROJECTILES** The height of a projectile in feet is given by $-16t^2 + vt + h_0$, where t is the time in seconds, v is the initial upward velocity in feet per second, and h_0 is the initial height in feet. A T-shirt is propelled from 32 feet above ground level into the air at an initial velocity of 16 feet per second.

   a. Write an expression to represent how much time the T-shirt is in the air.

   b. Factor the expression that represents the amount of time the T-shirt is in the air.

 **Higher-Order Thinking Skills**

73. **WHICH ONE DOESN'T BELONG?** Which expression does not belong with the others? Explain your reasoning.

$$x^2 + 2x - 24$$ $$x^2 + 11x + 24$$ $$x^2 - 10x - 24$$ $$x^2 + 12x + 24$$

74. **FIND THE ERROR** Jamaall and Charles have factored $x^2 + 6x - 16$. Is either of them correct? Explain your reasoning.

Jamaall	Charles
$x^2 + 6x - 16 = (x + 2)(x - 8)$	$x^2 + 6x - 16 = (x - 2)(x + 8)$

ANALYZE Find all values of k so that each polynomial can be factored using integers.

75. $x^2 + kx - 19$

76. $x^2 + kx + 14$

77. $x^2 - 8x + k, k > 0$

78. $x^2 - 5x + k, k > 0$

79. **ANALYZE** For any factorable trinomial, $x^2 + bx + c$, will the absolute value of b *sometimes*, *always*, or *never* be less than the absolute value of c? Justify your argument.

80. **CREATE** Give an example of a trinomial that can be factored using the factoring techniques presented in this lesson. Then factor the trinomial.

81. **PERSEVERE** Factor $(4y - 5)^2 + 3(4y - 5) - 70$.

82. **WRITE** Explain how to factor trinomials of the form $x^2 + bx + c$ and how to determine the signs of the factors of c.

83. **ANALYZE** A square has an area of $9x^2 + 30xy + 25y^2$ square inches. The dimensions are binomials with positive integers coefficients. What is the perimeter of the square? Explain.

84. **PERSEVERE** Find all values of k so that $2x^2 + kx + 12$ can be factored as two binomials using integers.

85. **WRITE** Explain how to determine which values should be chosen for m and p when factoring a polynomial of the form $ax^2 + bx + c$.

Factoring Special Products

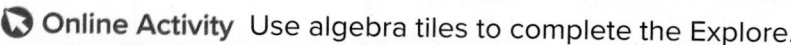

Explore Using Algebra Tiles to Factor Differences of Squares

Online Activity Use algebra tiles to complete the Explore.

> ✕
>
> **@ INQUIRY** How is factoring a difference of squares related to the product of a sum and a difference?

Learn Factoring Differences of Squares

The product of the sum and difference of two quantities results in a **difference of two squares.** So, the factored form of a difference of squares is the product of the sum and difference of two quantities.

Key Concept · Factoring Differences of Squares	
Symbols	$a^2 - b^2 = (a + b)(a - b)$ or $(a - b)(a + b)$
Examples	$x^2 - 9 = (x + 3)(x - 3)$ or $(x - 3)(x + 3)$
	$4u^2 - 1 = (2u + 1)(2u - 1)$ or $(2u - 1)(2u + 1)$

Example 1 Factor Differences of Squares

Factor each polynomial.

a. $81v^2 - 64w^2$

$$81v^2 - 64w^2 = (9v)^2 - (8w)^2 \qquad \text{Write in the form } a^2 - b^2.$$
$$= (9v + 8w)(9v - 8w) \qquad \text{Factor the difference of squares.}$$

b. $1 - 144q^2$

$$1 - 144q^2 = (1)^2 - (12q)^2 \qquad \text{Write in the form } a^2 - b^2.$$
$$= (1 + 12q)(1 - 12q) \qquad \text{Factor the difference of squares.}$$

Check

Factor each polynomial.

a. $25x^2 - 64y^2$

b. $196 - g^4$

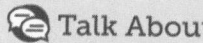 **Go Online** You can complete an Extra Example online.

Today's Goals
- Factor binomials that are differences of squares.
- Factor trinomials that are perfect squares.

Today's Vocabulary
difference of two squares

perfect square trinomials

⊘ Talk About It!

How can you check that the factors of a polynomial are correct?

Example 2 Factor More Than Once

Factor $x^4 - 256$.

$$x^4 - 256 = (x^2)^2 - (16)^2 \qquad \text{Write in the form } a^2 - b^2.$$

$$= (x^2 + 16)(x^2 - 16) \qquad \text{Factor the difference of squares.}$$

$$= (x^2 + 16)(x^2 - 4^2) \qquad x^2 - 16 \text{ is also a difference of squares.}$$

$$= (x^2 + 16)(x + 4)(x - 4) \qquad \text{Factor the difference of squares.}$$

Check

Factor each polynomial.

$81n^4 - 1$

$16x^8 - y^4$

🌐 Example 3 Use Factors to Find Area

AREA **Rosita's family is buying carpet for their living room, which is 6 meters wide and 6 meters long. They plan to carpet the whole area except for a square area near the door, which will be tiled. Find the factors representing the area of the living room that will be carpeted.**

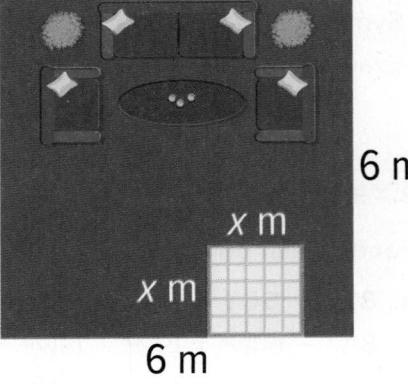

6 m

x m

x m

6 m

The area of the living room is $6 \cdot 6$ or 36 square meters, and the tiled area is $x \cdot x$ or x^2 square meters. So, the carpet will cover an area of $36 - x^2$.

$$36 - x^2 = (6)^2 - (x)^2 \qquad \text{Write in the form } a^2 - b^2.$$

$$= (6 + x)(6 - x) \qquad \text{Factor the difference of squares.}$$

Check

VOLUME A rectangular solid has a volume of $x^4 - 16$. Factor the polynomial to determine the dimensions of the rectangular solid.

A. $x^2(x + 4)(x - 4)$

B. $(x^2 + 4)(x + 2)(x - 2)$

C. $(x^2 - 4)(x + 2)(x + 2)$

D. $(x^3 + 4)(x - 2)(x - 2)$

🡲 **Go Online** You can complete an Extra Example online.

Watch Out!

Sum of Squares The sum of squares cannot be factored; that is, $a^2 + b^2 \neq (a + b)(a + b)$. The sum of squares is a prime polynomial.

🗯 **Think About It!**

What would the factors be if Rosita's family knew that they needed to tile a 1-meter square area near the door, but did not know the dimensions of their square living room?

Learn Factoring Perfect Squares

Squares of binomials, such as $(a + b)^2$ and $(a - b)^2$, have special products called **perfect square trinomials.**

$$(a + b)^2 = (a + b)(a + b) \qquad\qquad (a - b)^2 = (a - b)(a - b)$$

$$= a^2 + ab + ab + b^2 \qquad\qquad\qquad = a^2 - ab - ab + b^2$$

$$= a^2 + 2ab + b^2 \qquad\qquad\qquad\quad = a^2 - 2ab + b^2$$

For a trinomial to be factorable as a perfect square, the following must be true:

- The first term is a perfect square.

- The last term is a perfect square.

- The middle term is two times the product of the square roots of the first and last terms.

Key Concept • Factoring Perfect Square Trinomials	
Symbols	$a^2 + 2ab + b^2 = (a + b)(a + b) = (a + b)^2$
	$a^2 - 2ab + b^2 = (a - b)(a - b) = (a - b)^2$
Examples	$4x^2 - 28x + 49 = (2x - 7)(2x - 7) = (2x - 7)^2$
	$x^2 + 10x + 25 = (x + 5)(x + 5) = (x + 5)^2$

Example 4 Identify a Perfect Square Trinomial

Determine whether $4j^2 + 8j + 16$ is a perfect square trinomial. If so, factor it.

Is the first term a perfect square?

Is the second term equal to $2(2j)(4)$?   No, $8j \neq 2(2j)(4)$

Is the last term a perfect square?

Since this trinomial does not satisfy all conditions, it is not a perfect square trinomial.

🔵 **Go Online** You can complete an Extra Example online.

💭 Think About It!

To make $?j^2 + 8j + 16$ a perfect square trinomial, what would the coefficient of the first term need to be? __?__

To make $4j^2 + ?j + 16$ a perfect square trinomial, what would the coefficient of the middle term need to be? __?__

To make $4j^2 + 8j + ?$ a perfect square trinomial, what would the constant term need to be? __?__

Go Online to practice what you've learned about factoring quadratic expressions in the Put It All Together over Lessons 11-6 and 11-7.

Example 5 Recognize and Factor a Perfect Square Trinomial

Determine whether $36h^2 - 12h + 1$ is a perfect square trinomial. If so, factor it.

Is $36h^2$ a perfect square? yes

Is $-12h$ equal to $-2(6h)(1)$? yes

Is 1 a perfect square? yes

This trinomial satisfies all the conditions for a perfect square trinomial.

$$36h^2 - 12h + 1 = (6h)^2 - 2(6h)(1) + (1)^2 \qquad \text{Write as } a^2 - 2ab + b^2.$$
$$= (6h - 1)^2 \qquad \text{Factor using the pattern.}$$

Check

If the trinomial is a perfect square trinomial, factor it. If not, write *not a perfect square trinomial.*

$36x^2 - 36x + 9$

$36x^2 - 18x + 9$

$36x^2 + 36x + 9$

Pause and Reflect

Did you struggle with anything in this lesson? If so, how did you deal with it?

Go Online You can complete an Extra Example online.

Practice

Go Online You can complete your homework online.

Examples 1 and 2

Factor each polynomial.

1. $q^2 - 121$

2. $r^4 - k^4$

3. $w^4 - 625$

4. $r^2 - 9t^2$

5. $h^4 - 256$

6. $2x^3 - x^2 - 162x + 81$

7. $x^2 - 4y^2$

8. $3c^3 + 2c^2 - 147c - 98$

9. $f^3 + 2f^2 - 64f - 128$

10. $r^3 - 5r^2 - 100r + 500$

11. $3t^3 - 7t^2 - 3t + 7$

12. $a^2 - 49$

13. $4m^3 + 9m^2 - 36m - 81$

14. $3x^3 + x^2 - 75x - 25$

Example 3

15. TICKETING A ticketing company for sporting events analyzes the ticket purchasing patterns. The expression $9a^2 - 4b^2$ is developed to help officials calculate the likely number of people who will buy tickets for a certain sporting event. Factor the expression.

16. BASKETBALL COURT A half-court basketball court is a square of pavement with an area represented by $x^2 - 25$. Factor the expression.

17. DECORATING Marvin saw a rug in a store that he would like to purchase. It has an area represented by the expression shown on the rug. He cannot remember the length and width, but he remembers that the length and the width were the same.

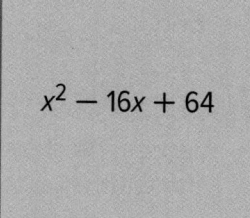

$x^2 - 16x + 64$

   a. Factor the expression that represents the area of the rug.

   b. What do the factors in the factored expression represent?

Examples 4 and 5

Determine whether each trinomial is a perfect square trinomial. Write *yes* or *no*. If so, factor it.

18. $4x^2 - 42x + 110$

19. $16x^2 - 56x + 49$

20. $81x^2 - 90x + 25$

21. $x^2 + 26x + 168$

Mixed Exercises

Factor each polynomial, if possible. If the polynomial cannot be factored using integers, write *prime*.

22. $36t^2 - 24t + 4$

23. $4h^2 - 56$

24. $17a^2 - 24ab$

25. $q^2 - 14q + 36$

26. $y^2 + 24y + 144$

27. $6d^2 - 96$

28. $1 - 49d^2$

29. $-16 + p^2$

30. $k^2 + 25$

31. $36 - 100w^2$

32. $64m^2 - 9y^2$

33. $4h^2 - 25g^2$

34. $x^3 + 3x^2 - 4x - 12$

35. $8x^2 - 72p^2$

36. $20q^2 - 5r^2$

37. $32a^2 - 50b^2$

38. $16b^2 - 100$

39. $49x^2 - 64y^2$

40. $3n^4 - 42n^3 + 147n^2$

41. $8m^3 - 24m^2 + 18m$

42. GARDEN DESIGN Marren is planning to build a raised garden bed. The area of the rectangular plot can be represented by $x^2 - 49$. Factor the expression to determine the possible length and width of the garden bed.

43. PARKING LOT The area of a rectangular parking lot is represented by the expression $a^2 - 25$, where the length is longer than the width. Factor the expression to determine the possible dimensions of the length and width of the parking lot. If the length of the parking lot is 105 yards, what is the width of the parking lot?

44. USE A SOURCE Research the dimensions of the outside diameter and inside diameter of metal washers. Write an expression for the surface area of the top of a metal washer with outside diameter D and inside diameter d. Factor your expression. Then use your expression and the dimensions you researched to find the surface area of the top of a metal washer.

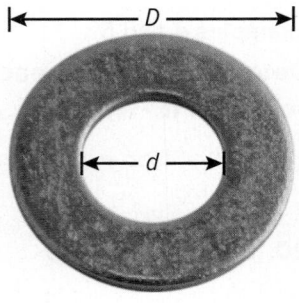

Determine whether each trinomial is a perfect square trinomial. Write *yes* or *no*. If so, factor it.

45. $m^2 - 6m + 9$

46. $r^2 + 4r + 4$

47. $g^2 - 14g + 49$

48. $2w^2 - 4w + 9$

49. $4d^2 - 4d + 1$

50. $9n^2 + 30n + 25$

51. $9z^2 - 6z + 1$

52. $36x^2 - 60x + 25$

53. $49r^2 + 14r + 4$

54. $a^2 + 14a + 49$

55. $t^2 - 18t + 81$

56. $4c^2 + 2cd + d^2$

57. ARCHITECTURE The drawing shows a triangular roof truss with a base measuring the same as its height. The expression $\frac{1}{2}x^2 - 98$ represents the area of the triangular roof truss, where x is the length of the base and the height. Factor the expression.

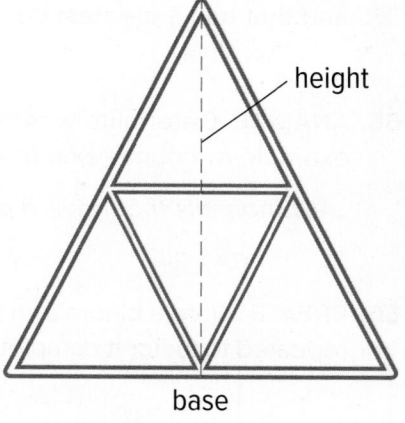

height

base

58. A company that manufactures cardboard boxes sells three sizes of boxes. The volume of the small box is represented by $n^3 + 4n^2 - 16n - 64$ in³. The volume of the medium box is represented by $n^3 + 6n^2 - 36n - 216$ in³. The volume of the large box is represented by $n^3 + 8n^2 - 64n - 512$ in³. The company wants to start selling an extra-large box. Predict the dimensions and the volume of the extra-large box.

    a. Find the dimensions of the small, medium, and large boxes. Explain the method you used.

    b. Look for a pattern in the areas. What pattern is represented by the areas of the small, medium, and large boxes? Use this pattern to predict the dimensions of the extra-large box.

    c. Use your prediction in **part b** to find the volume of the extra-large box.

59. STATE YOUR ASSUMPTION Paul suggests that to factor $x^4 - 81$, you can use difference of squares twice. He says that the result will have the same factors as $x^2 - 9$. Prove or disprove his statement.

60. REASONING Debony claims the expression $9x^2 + 50x + 25$ is a perfect square trinomial. Is she correct? If she is incorrect, show how the expression can be changed so that it is a perfect square.

61. STRUCTURE Find a value for m that will make the expression $4x^4 - 44x^2 + m$ a perfect square. Then use the value to factor the expression completely.

Higher-Order Thinking Skills

62. FIND THE ERROR Elizabeth and Lorenzo are factoring an expression. Is either of them correct? Explain your reasoning.

Elizabeth	Lorenzo
$16x^4 - 25y^2 =$	$16x^4 - 25y^2 =$
$(4x - 5y)(4x + 5y)$	$(4x^2 - 5y)(4x^2 + 5y)$

63. PERSEVERE Factor and simplify $9 - (k + 3)^2$, a difference of squares.

64. ANALYZE Write and factor a binomial that is the difference of two perfect squares and that has a greatest common factor of $5mk$.

65. ANALYZE Determine whether the following statement is *true* or *false*. Give an example or counterexample to justify your answer.

All binomials that have a perfect square in each of the two terms can be factored.

66. CREATE Write a binomial in which the difference of squares pattern must be repeated to factor it completely. Then factor the binomial.

67. WRITE Describe why the difference of squares has no middle term.

68. WHICH ONE DOESN'T BELONG? Identify the trinomial that does not belong. Explain.

$4x^2 - 36x + 81$	$25x^2 + 10x + 1$	$4x^2 + 10x + 4$	$9x^2 - 24x + 16$

69. WRITE Explain how to determine whether a trinomial is a perfect square trinomial.

70. PERSEVERE Use the difference of squares to factor and simplify the expression $121x^2y^6z^4 - 16y^2z^2$.

Essential Question

How can you perform operations on polynomials and use them to represent real-world situations?

Much like combining like terms, polynomials can be added or subtracted. The Distributive Property is used to multiply and factor polynomials. Polynomials can represent areas and volumes of three-dimensional solids.

Module Summary

Lesson 11-1

Adding and Subtracting Polynomials

- The degree of a monomial is the sum of the exponents of all its variables.
- The degree of a polynomial is the greatest degree of any term in the polynomial.
- When adding polynomials, you can group like terms by using a horizontal or vertical format.
- You can subtract a polynomial by adding its additive inverse.
- Adding or subtracting polynomials results in a polynomial, so the set of polynomials is closed under addition and subtraction.

Lessons 11-2 through 11-4

Multiplying Polynomials

- To find the product of a monomial and a binomial, use the Distributive Property.
- The FOIL method helps to multiply binomials: To multiply two binomials, find the sum of the products of F the *First Terms*, O the *Outer terms*, I the *Inner terms*, and L the *Last terms*.
- Square of a sum: $(a + b)^2 = (a + b)(a + b)$
  $= a^2 + 2ab + b^2$
- Square of a difference: $(a - b)^2 =$
  $(a - b)(a - b) = a^2 - 2ab + b^2$
- The product of $a + b$ and $a - b$ is the square of a minus the square of b: $(a + b)(a - b) =$
  $a^2 - ab + ab - b^2 = a^2 - b^2$

Lessons 11-5 through 11-7

Factoring Polynomials

- By using the Distributive Property in reverse, you can factor a polynomial as the product of a monomial and a polynomial.
- To factor trinomials in the form $x^2 + bx + c$, find two integers, m and p, with a sum of b and a product of c. Then write $x^2 + bx + c$ as $(x + m)(x + p)$.
- To factor trinomials in the form $ax^2 + bx + c$, find two integers, m and p, with a sum of b and a product of ac. Then write $ax^2 + bx + c$ as $ax^2 + mx + px + c$, and factor by grouping.
- The factored form of a difference of squares is the product of the sum and difference of two quantities.
  $$a^2 - b^2 = (a + b)(a - b) \text{ or } (a - b)(a + b)$$
- Squares of binomials, such as $(a + b)^2$ and $(a - b)^2$, have special products called perfect square trinomials.
  $$a^2 + 2ab + b^2 = (a + b)(a + b) = (a + b)^2$$
  $$a^2 - 2ab + b^2 = (a - b)(a - b) = (a - b)^2$$

Study Organizer

📖 **Foldables**

Use your Foldable to review this module. Working with a partner can be helpful. Ask for clarification of concepts as needed.

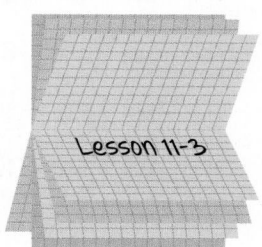

Lesson 11-3

Test Practice

1. **MULTI-SELECT** Select all of the expressions that are polynomials. (Lesson 11-1)

   A. $4xy - 3x$

   B. $4ab + \frac{5}{x^2}$

   C. $45 + 3d^2 + d - w^3$

   D. 12

   E. $\frac{x^{-3} y^2}{5}$

2. **OPEN RESPONSE** The perimeter of this triangular plot of land can be represented by $3x^2 + 10x + 20$. (Lesson 11-1)

   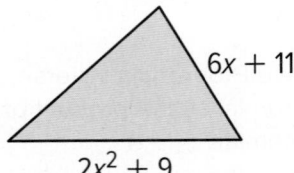

   Write a polynomial that represents the measure of the third side of the plot of land.

3. **MULTIPLE CHOICE** Find the difference. $(5d - 8) - (-2d + 3)$ (Lesson 11-1)

   A. $3d - 5$

   B. $7d - 11$

   C. $7d - 5$

   D. $7d + 11$

4. **MULTIPLE CHOICE** Find the sum. $(x^2 + 3x - 9) + (4x^2 - 6x + 1)$ (Lesson 11-1)

   A. $4x^2 - 3x - 8$

   B. $5x^2 - 3x - 8$

   C. $5x^2 + 9x - 10$

   D. $4x^2 + 9x - 10$

5. **OPEN RESPONSE** Mr. Soto is installing a new trapezoid shaped window. He plans to cover the window with a protective coating that prevents UV rays from entering the house. The height h of the window is 24 inches.

   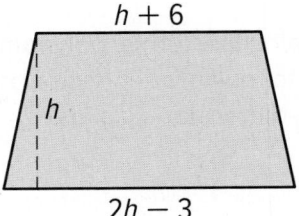

   The formula for the area of a trapezoid is $A = \frac{1}{2} h (b_1 + b_2)$ where h is the height and b_1 and b_2 are the bases of the trapezoid. How many square inches of protective coating will Mr. Soto need? (Lesson 11-2)

6. **OPEN RESPONSE** Use the Distributive Property to simplify $-3(g^2 - 4g + 1)$. (Lesson 11-2)

7. **MULTIPLE CHOICE** Find the product of $(2x + 7)$ and $(x - 5)$. (Lesson 11-3)

   A. $3x + 2$

   B. $2x^2 - 3x - 35$

   C. $2x^2 - 10x - 35$

   D. $2x^2 - 35$

8. **OPEN RESPONSE** Saurabh is designing a rectangular vegetable garden with a stone path around it. The total area of the path is 292 square feet. Use the diagram to determine the value of x. (Lesson 11-3)

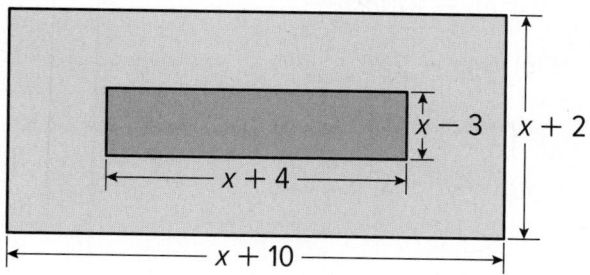

9. **OPEN RESPONSE** Find the product.
$(2r - 3)(r^2 - 5r + 1)$ (Lesson 11-3)

10. **OPEN RESPONSE** *True* or *false*:
$(8k^2 + 3k - 1)(k^2 - k + 1)$ is equivalent to $8k^4 - 5k^3 + 4k^2 + 4k - 1$. (Lesson 11-3)

11. **MULTIPLE CHOICE** Find $(8a + 3b)^2$.
(Lesson 11-4)

A. $73a^2b^2 + 48ab$

B. $64a^2 + 48ab + 9b^2$

C. $64a^2 + 9b^2$

D. $32a^2 + 22ab + 12b^2$

12. **OPEN RESPONSE** Find the product.
$(6h + 7)(6h - 7)$ (Lesson 11-4)

13. **OPEN RESPONSE** Describe a general rule for finding the square of a sum. (Lesson 11-4)

14. **MULTIPLE CHOICE** Kala is making a tile design for her kitchen floor. Each tile has sides that are 3 inches less than twice the side length of the smaller square inside the design. (Lesson 11-4)

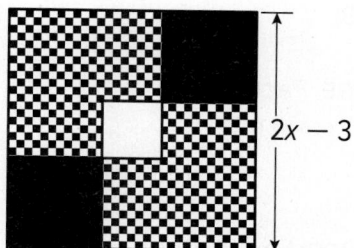

Select the polynomial that represents the area of the tile.

A. $2x^2 - 3x$

B. $4x^2 - 12x + 9$

C. $4x^2 + 12x + 9$

D. $4x^2 - 9$

15. **OPEN RESPONSE** Factor
$-10xy - 15x + 12y + 18$. (Lesson 11-5)

16. **OPEN RESPONSE** Factor $36x^3 - 24x^2$.
(Lesson 11-5)

17. MULTI-SELECT A golf ball is hit with a velocity of 128 feet per second. The expression $128t - 16t^2$ represents the height of the ball after t seconds. Select all that are equivalent to the expression. (Lesson 11-5)

A. $16t(8 - t)$

B. $-16t(t - 8)$

C. $t(t - 8)(8 - t)$

D. $8t(7 - 2t)$

E. $8(-2t + 16)$

18. OPEN RESPONSE Factor $x^2 + 8x + 15$. (Lesson 11-6)

19. MULTIPLE CHOICE Mrs. Torres wants to add x feet onto the length and width of an existing rectangular patio. The new patio would have an area of $x^2 + 14x + 48$ square feet. (Lesson 11-6)

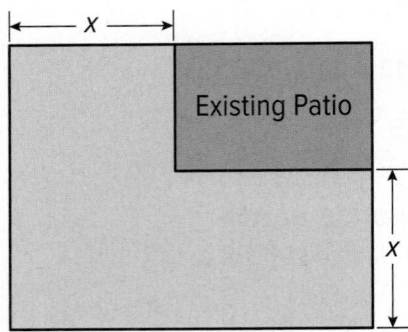

What are the dimensions of the existing patio?

A. 14 ft by 48 ft

B. 6 ft by 8 ft

C. 7 ft by 24 ft

D. 4 ft by 12 ft

20. OPEN RESPONSE Factor $9x^2 + 9x + 2$. (Lesson 11-6)

21. OPEN RESPONSE Identify if each polynomial is a difference of squares. Write *yes* or *no*. (Lesson 11-7)

a) $16y^2 - 25b^2$

b) $m^2 - 100n^2$

c) $64r^2 - 225$

d) $49z^2 - 36a^3$

e) $-4y^2 - k^2$

22. MULTI-SELECT Select all of the perfect square trinomials. (Lesson 11-7)

A. $49x^2 + 112x + 64$

B. $16x^2 - 24x + 9$

C. $49x^2 + 30x + 64$

D. $9x^2 - 6x + 16$

E. $x^2y^2 - 10xy^2 + 25y^2$

23. MULTIPLE CHOICE A square piece of cloth has an area of $4y^2 - 28y + 49$ square meters. Find the length of each side. (Lesson 11-7)

A. $(2y + 7)$ m

B. $(2y - 7)$ m

C. $(4y - 49)$ m

D. $(2y - 14)$ m

Quadratic Functions

e Essential Question

Why is it helpful to have different methods to analyze quadratic functions and solve quadratic equations?

What Will You Learn?

How much do you already know about each topic **before** starting this module?

KEY

👎 — I don't know.   👊 — I've heard of it.   👍 — I know it!

	Before			After		
	👎	👊	👍	👎	👊	👍
graph quadratic equations using a table						
graph quadratic equations using key features						
transform quadratic functions						
solve systems of linear and quadratic equations						
solve quadratic equations by factoring						
solve quadratic equations by completing the square						
find maximum and minimum values						
solve quadratic equations using the quadratic formula						
fit a quadratic function to data						
combine functions						

📖 **Foldables** Make this Foldable to help you organize your notes about quadratic functions. Begin with a sheet of notebook paper.

1. **Fold** the sheet of paper along the length so that the edge of the paper aligns with the margin rule of the paper.

2. **Fold** the sheet twice widthwise to form four sections.

3. **Unfold** the sheet and cut along the folds on the front flap only.

4. **Label** each section as shown.

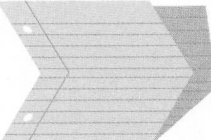

What Vocabulary Will You Learn?

- axis of symmetry
- coefficient of determination
- completing the square
- curve fitting
- discriminant
- double root
- maximum
- minimum
- parabola
- quadratic equation
- quadratic function
- standard form of a quadratic function
- vertex
- vertex form

Are You Ready?

Complete the Quick Review to see if you are ready to start this module.
Then complete the Quick Check.

Quick Review

Example 1

Describe the translation in $g(x) = 5^x + 3$ as it relates to the graph of the parent function $f(x) = 5^x$.

A constant term is being added to the function, so it is a translation of that many units up.

The graph of $g(x) = 5^x + 3$ is the translation of the graph of the parent function $f(x) = 5^x$ 3 units up.

Determine whether $x^2 - 10x + 25$ is a perfect square trinomial. Write *yes* or *no*. If so, factor it.

Is the first term, x^2, a perfect square? yes

Is the last term, 25, a perfect square? yes

Is the middle term equal to the opposite of twice the product of the first and last terms?

yes; $2(5)(x) = 10x$ and the opposite of $10x$ is $-10x$

$x^2 - 10x + 25 = (x - 5)^2$

Quick Check

Describe the translation in $g(x)$ as it relates to the graph of the parent function $f(x) = 2^x$.

1. $g(x) = 2^{x-4}$
2. $g(x) = -2^x$
3. $g(x) = 2^x - 1$
4. $g(x) = 0.25 \cdot 2^x$

Determine whether each trinomial is a perfect square trinomial. Write *yes* or *no*. If so, factor it.

5. $a^2 + 12a + 36$

6. $w^2 + 5w + 25$

7. $m^2 - 22m + 121$

8. $5t^2 + 12t + 100$

How Did You Do?

Which exercises did you answer correctly in the Quick Check?

Graphing Quadratic Functions

Learn Analyzing Graphs of Quadratic Functions

A second-degree polynomial, for example $x^2 + 2x - 8$, is a quadratic polynomial, which has a related quadratic function $f(x) = x^2 + 2x - 8$. The graph of a quadratic function is called a parabola.

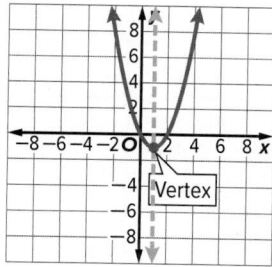

Parabolas are symmetric about a central line called the **axis of symmetry**. Every point on the parabola to the left of the axis of symmetry has a corresponding point on the right half.

The axis of symmetry intersects a parabola at the vertex. The vertex is either the lowest point or the highest point on a parabola.

When $a > 0$, the graph of $y = ax^2 + bx + c$ opens up. In this case, the graph has a **minimum** at the lowest point. When $a < 0$, the graph opens down. In this case, the graph has a **maximum** at the highest point.

The leading coefficient determines the end behavior of a quadratic function.

Example 1 Identify Characteristics: Graph with *x*-intercept

Identify the axis of symmetry, the vertex, and the *y*-intercept of the graph. Then describe the end behavior.

The equation of the axis of symmetry is $x = -2$.

The vertex is located at the maximum point, $(-2, 8)$.

The parabola crosses the *y*-axis at $(0, 4)$, so the *y*-intercept is 4.

As *x* increases or decreases, *y* decreases.

Today's Goals
- Analyze graphs of quadratic functions.
- Graph quadratic functions by using key features and tables.
- Use graphing calculators to analyze key features of quadratic functions.

Today's Vocabulary
quadratic function

parabola

axis of symmetry

vertex

minimum

maximum

end behavior

standard form of a quadratic function

🍎 **Think About It!**

What do you notice about the equation of the axis of symmetry and the *x*-coordinate of the vertex?

Check

Identify the axis of symmetry, the vertex, and the y-intercept of the graph. Then describe the end behavior.

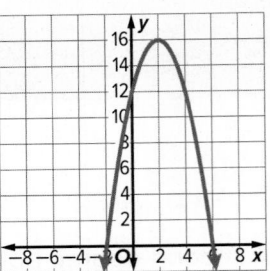

axis of symmetry: $x =$ __?__

vertex: (__?__, __?__).

y-intercept:

end behavior:

As x increases, y __?__.

As x decreases, y __?__.

Example 2 Identify Characteristics: Graph with No x-intercept

Identify the axis of symmetry, the vertex, and the y-intercept of the graph. Then describe the end behavior of the function.

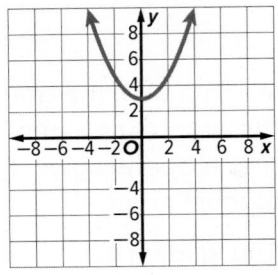

The equation of the axis of symmetry is $x = 0$.

The vertex is located at the minimum point, (0, 3).

The parabola crosses the y-axis at (0, 3), so the y-intercept is 3.

The parabola opens up.

As x increases, y increases.

As x decreases, y increases.

Check

Identify the axis of symmetry, the vertex, and the y-intercept of the graph. Then describe the end behavior of the function.

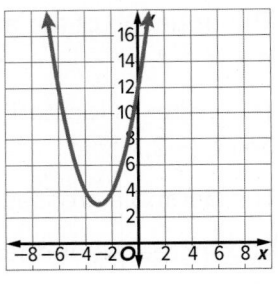

axis of symmetry: $x =$ __?__

vertex: (__?__, __?__).

y-intercept:

end behavior:

As x increases, y __?__.

As x decreases, y __?__.

Go Online You can complete an Extra Example online.

Explore Graphing Parabolas

Online Activity Use graphing technology to complete the Explore.

> ×
> **INQUIRY** How can you use the equation of a quadratic function to visualize its graph?

Learn Graphing Quadratic Functions

Key Concept • Standard Form of a Quadratic Function

Words

- The **standard form of a quadratic function** is $f(x) = ax^2 + bx + c$, where a, b, and c are integers and $a \neq 0$.

Examples

- In $f(x) = 3x^2 - 2x + 9$, $a = 3$, $b = -2$, and $c = 9$.
- In $f(x) = -16x^2$, $a = -16$, $b = 0$, and $c = 0$.

Key Concept • Graphing Quadratic Functions

Step 1 Find the equation of the axis of symmetry.

Step 2 Find the vertex, and determine whether it is a maximum or minimum.

Step 3 Find the y-intercept.

Step 4 Use symmetry to find additional points on the graph, if necessary.

Step 5 Connect the points with a smooth curve.

Example 3 Graph a Quadratic Function by Using Key Features

Graph $f(x) = x^2 + 2x - 6$.

Step 1 Find the axis of symmetry.

Use the formula to find the equation of the axis of symmetry.

$x = -\dfrac{b}{2a}$ Equation of the axis of symmetry

$x = -\dfrac{2}{2(1)} = -1$ $a = 1$ and $b = 2$

Step 2 Find the vertex.

Use the value for the axis of symmetry as the x-coordinate of the vertex. Find the y-coordinate using the original equation.

$f(x) = x^2 + 2x - 6$ Original Equation

$= (-1)^2 + 2(-1) - 6$ $x = -1$

$= -7$ Simplify.

The vertex lies at $(-1, -7)$. Because a is positive, the graph opens up. So the vertex is a minimum.

(continued on the next page)

Study Tip

Open and Closed Intervals The interval at which a function increases or decreases is always an open interval along the x-axis and is expressed using parentheses. A closed interval is expressed using brackets.

Talk About It!

Compare and contrast the graphs of exponential and quadratic functions.

Watch Out!

Minimum and Maximum Values Don't forget to find both coordinates of the vertex (x, y). The minimum or maximum value is the y-coordinate.

Step 3 Find the y-intercept.

$$f(x) = x^2 + 2x - 6 \qquad \text{Original equation}$$
$$= (0)^2 + 2(0) - 6 \qquad x = 0$$
$$= -6 \qquad \text{Simplify.}$$

The y-intercept is −6.

Step 4 Find additional points.

Let $x = 2$.

$$f(x) = x^2 + 2x - 6 \qquad \text{Original equation}$$
$$= (2)^2 + 2(2) - 6 \qquad x = 2$$
$$= 2 \qquad \text{Simplify.}$$

Step 5 Connect the points.

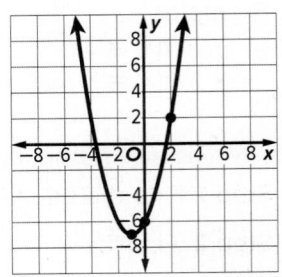

Go Online
You can watch a video to see how to graph quadratic functions by using a table.

Example 4 Graph a Quadratic Function by Using a Table

Use a table of values to graph $y = 2x^2 - 8x + 2$.

First find the x-coordinate of the vertex by using the equation for the axis of symmetry.

$$x = -\frac{b}{2a} \qquad \text{Equation of the axis of symmetry}$$
$$x = -\frac{-8}{2(2)} = 2 \qquad a = 2, b = -8$$

Complete the table.

x	0	1	2	3	4
y	2	−4	−6	−4	2

Think About It!
Why do you think that the range is $\{y \mid y \geq -6\}$?

y-intercept (0, 2)

The vertex of this function is a minimum.

Graph the function.

The parabola extends to infinity, so the domain is all real numbers. The range is $\{y \mid y \geq -6\}$.

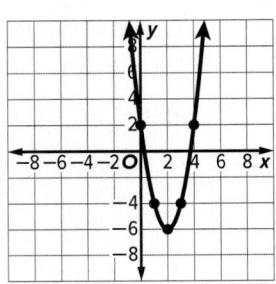

 Go Online You can complete an Extra Example online.

Example 5 Use the Graph of a Quadratic Function

CATAPULT In 1304, Edward I built Warwolf, one of the largest catapults ever used in battle. It had the capacity to launch a 300-pound boulder over 900 feet. The height of a projectile launched from Warwolf can be modeled by the function $h(x) = -16x^2 + 96x + 40$, where $h(x)$ represents the height in feet of the projectile after x seconds. Graph the function. Interpret the key features of the graph in terms of the quantities.

Step 1 Find the axis of symmetry and vertex.

$$x = -\frac{b}{2a}$$ Equation of the axis of symmetry

$$x = -\frac{96}{2(-16)} = 3$$ $a = -16$ and $b = 96$

$$h(x) = -16x^2 + 96x + 40$$ Original equation

$$= -16(3)^2 + 96(3) + 40$$ $x = 3$

$$= 184$$ Simplify.

The vertex is at (3, 184).

Step 2 Complete the table by substituting each x-value.

x	0	1	2	3	4	5
$h(x)$	40	120	168	184	168	120

Step 3 Graph the function.

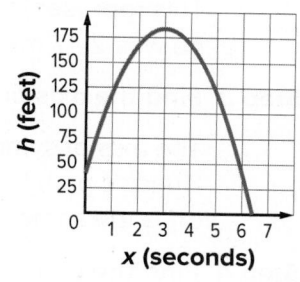

Step 4 Interpret the key features.

intercepts: $x = 6.4$, so the projectile landed about 6.4 seconds after it was launched.

vertex: (3, 184), so the projectile reached its maximum height of 184 feet 3 seconds after launch.

increasing: The function is increasing for $x < 3$, so it is gaining height up to 3 seconds after launch.

decreasing: The function is decreasing for $x > 3$, so it is falling 3 seconds after launch.

positive: intervals where the function represents the flight

end behavior: represents the projectile starting and returning to the ground

domain: all nonnegative numbers less than 6.4

range: {$h(x) \mid 0 \le h(x) \le 184$}

Go Online You can watch a video to see how to graph and analyze a quadratic function on a graphing calculator.

Go Online to see how to use a graphing calculator with this example.

Math History Minute

During the 19th century, several designers of the Brooklyn Bridge were unable to complete it. But **Emily Warren Roebling** (1843–1903) supervised the bridge's construction until its completion. The bridge was the first major suspension bridge constructed in the U.S. Its cables form a catenary, which approximates a parabola.

Learn Analyzing Key Features of Quadratic Functions

You can use a graphing calculator to analyze key features of quadratic functions by graphing or by using the table feature.

Example 6 Interpret the Graph of a Quadratic Function

HORSES **In 1949, Huaso, ridden by Alberto Larraguibel Morales, set the world record for the highest jump by a horse when he cleared a 2.47-meter-high obstacle. His path can be approximately modeled by $f(x) = -0.418x^2 + 2.033x$, where $f(x)$ is the height above the ground in meters and x is the distance from the point of take-off in meters. Find and interpret the key features of the path of Huaso's jump.**

Step 1 Graph the function.

The graph only exists in Quadrant I in the context of the situation.

Step 2 Find the vertex.

Use the maximum feature from the CALC menu to find the vertex.

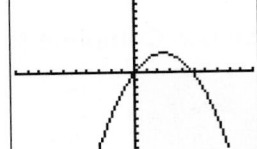

The vertex is located at about (2.43, 2.47). This represents Huaso's maximum height of 2.47 meters when he was 2.43 meters from the point of take-off.

Step 3 Find the axis of symmetry.

The axis of symmetry is located at $x = 2.43$. This means that Huaso's path before he is 2.43 meters from the point of take-off is the same as after he reaches 2.43 meters.

Step 4 Find the y-intercept.

Use the value feature from the CALC menu to find the y-intercept. Enter 0 for x. The y-intercept is located at (0, 0), so the y-intercept is 0. This means that when Huaso took off, he was 0 meters high and 0 meters from the point of take-off.

Step 5 Find the zeros.

Use the zero feature from the CALC menu to find the zeros, or x-intercepts.

The zeros are located at x equals 0 and x equals 4.86. These represent the points of take-off and landing, when Huaso's height above the ground was 0 meters.

Step 6 Examine the end behavior.

As x increases or decreases from the maximum, the value of y decreases until it reaches 0. This represents Huaso taking off and returning to the ground.

Practice

⬧ **Go Online** You can complete your homework online.

Examples 1 and 2

Identify the axis of symmetry, the vertex, and the *y*-intercept of each graph. Then describe the end behavior.

1.

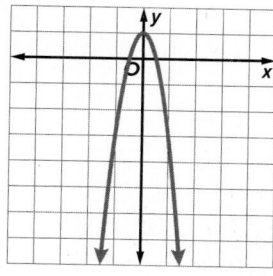

2.

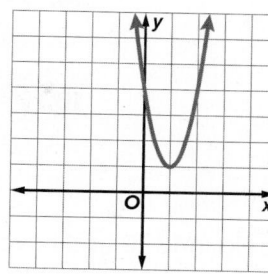

3.

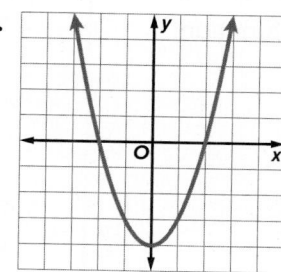

4.

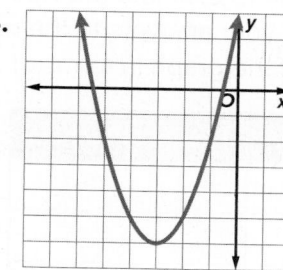

Example 3

Graph each function.

5. $y = -3x^2 + 6x - 4$

6. $y = -2x^2 - 4x - 3$

7. $y = -2x^2 - 8x + 2$

8. $y = x^2 + 6x - 6$

9. $y = x^2 - 2x + 2$

10. $y = 3x^2 - 12x + 5$

Example 4

Use a table of values to graph each function. State the domain and range.

11. $y = x^2 + 4x + 6$

12. $y = 2x^2 + 4x + 7$

13. $y = 2x^2 - 8x - 5$

14. $y = 3x^2 + 12x + 5$

15. $y = 3x^2 - 6x - 2$

16. $y = x^2 - 2x - 1$

Examples 5 and 6

17. OLYMPICS Olympics were held in 1896 and have been held every four years except 1916, 1940, and 1944. The winning height y in men's pole vault at any number Olympiad x can be approximated by the equation $y = 0.37x^2 + 4.3x + 126$. Copy and complete the table to estimate the winning pole vault heights in each of the Olympic Games. Round your answers to the nearest tenth. Graph the function. Interpret the key features of the graph in terms of the quantities.

Year	Olympiad (x)	Height (y inches)
1896	1	
1900	2	
1924	7	
1936	10	
1964	15	
2008	26	
2012	27	
2016	28	

18. PHYSICS Mrs. Capwell's physics class investigates what happens when a ball is given an initial push, rolls up, and then rolls back down an inclined plane. The class finds that $y = -x^2 + 6x$ accurately predicts the ball's position y after rolling x seconds. Graph the function. Interpret the key features of the graph in terms of the quantities.

19. ARCHITECTURE A hotel's main entrance is in the shape of a parabolic arch. The equation $y = -x^2 + 10x$ models the arch height y, in feet, for any distance x, in feet, from one side of the arch. Graph the function. Interpret the key features of the graph in terms of the quantities.

20. SOFTBALL Olympic softball gold medalist Michele Smith pitches a curveball with a speed of 64 feet per second. If she throws the ball straight upward at this speed, the ball's height h in feet after t seconds is given by $h = -16t^2 + 64t$. Graph the function. Interpret the key features of the graph in terms of the quantities.

Mixed Exercises

21. CONSTRUCTION Teddy is building the rectangular deck shown.

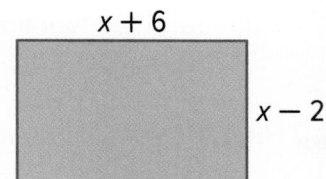

$x + 6$

$x - 2$

   a. Write an equation representing the area of the deck y.

   b. What is the equation of the axis of symmetry?

   c. Graph the equation and label its vertex.

22. USE TOOLS Write a quadratic function whose graph opens up. Write an exponential function. Graph both functions on the same coordinate plane. Compare and contrast the domains and ranges of the two functions and any symmetry of the two graphs.

23. STRUCTURE Consider the quadratic function $y = -x^2 - 2x + 2$.

   a. Find the equation for the axis of symmetry.

   b. Find the coordinates of the vertex and determine if it is a maximum or minimum.

   c. Graph the function.

Identify the axis of symmetry, the vertex, and the y-intercept of each graph. Then describe the end behavior.

24. $y = 2x^2 - 8x + 6$ **25.** $y = x^2 + 4x + 6$ **26.** $y = -3x^2 - 12x + 3$

27. STRUCTURE DeMarcus is sitting in a lifeguard's chair at the beach. He tosses a beanbag into the air and lets it land on the beach below him. The function $h(t) = -16t^2 + 8t + 12$ models the beanbag's height in feet, t seconds after DeMarcus tosses it into the air.

   a. Graph the function on a coordinate plane. Be sure to label the x- and y-axes and provide a scale for each axis.

   b. Find the intercepts of the graph. Describe what they represent in the context of the situation.

   c. What is the maximum value of the function, and what does it represent? At what time is the maximum reached?

   d. What is the value of $h(0.5)$? Describe its meaning in the context of the situation.

   e. State a reasonable domain for this function. What does it represent in the context of the situation?

28. REGULARITY Write the equation for a quadratic function that has a *y*-intercept of 0 and a minimum value at *x* = 2. Explain the steps you used to write the equation, and graph the function on a coordinate plane.

 **Higher-Order Thinking Skills**

29. CREATE Write a quadratic function for which the graph has an axis of symmetry of $x = -\frac{3}{8}$. Summarize your steps.

30. FIND THE ERROR Noelia thinks that the parabolas represented by the graph and the description have the same axis of symmetry. Chase disagrees. Is either correct? Explain your reasoning.

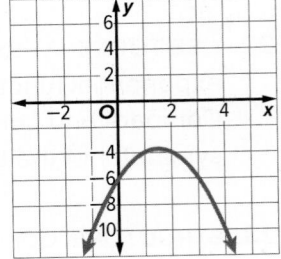

> a parabola that opens downward, passing through (0, 6) and having a vertex at (2, 2)

31. PERSEVERE Using the axis of symmetry, the *y*-intercept, and one *x*-intercept, write an equation for the graph shown.

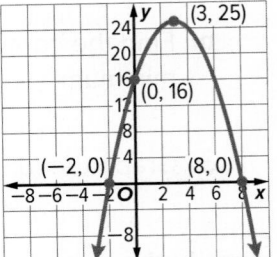

32. ANALYZE The graph of a quadratic function has a vertex (2, 0). One point on the graph is (5, 9). Find another point on the graph. Explain how you found it.

33. CREATE Describe a real-world situation that involves a quadratic equation. Explain what the vertex represents.

34. ANALYZE Provide a counterexample that shows that the following statement is false. Justify your argument.

> *The vertex of a parabola is always the minimum of the graph.*

35. WRITE Use tables and graphs to compare and contrast an exponential function $f(x) = ab^x + c$, where $a \neq 0$, $b > 0$, and $b \neq 1$, a quadratic function $g(x) = ax^2 + c$, and a linear function $h(x) = ax + c$. Include intercepts, portions of the graph where the functions are increasing, decreasing, positive, negative, relative maxima, relative minima, symmetries, and end behavior. Which function eventually exceeds the others?

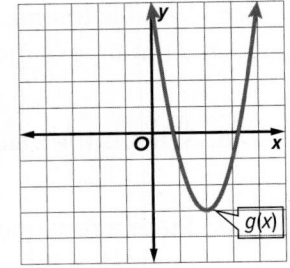

36. ANALYZE Consider $g(x)$ shown in the graph and $f(x) = -x^2 + 3x - 2$. Determine which function has the lesser minimum.

Transformations of Quadratic Functions

Explore Transforming Quadratic Functions

Online Activity Use graphing technology to complete the Explore.

INQUIRY How does performing an operation on a quadratic function change its graph?

Learn Translations of Quadratic Functions

Key Concept • Vertical Translations of Quadratic Functions

The graph of $g(x) = x^2 + k$ is the graph of $f(x) = x^2$ translated vertically.

If $k > 0$, the graph of $f(x)$ is translated k units up.

If $k < 0$, the graph of $f(x)$ is translated $|k|$ units down.

Key Concept • Horizontal Translations of Quadratic Functions

The graph of $g(x) = (x - h)^2 + k$ is the graph of $f(x) = x^2$ translated horizontally.

If $h > 0$, the graph of $f(x)$ is translated h units right.

If $h < 0$, the graph of $f(x)$ is translated $|h|$ units left.

Example 1 Vertical Translations of Quadratic Functions

Describe the translation in $g(x) = x^2 - 4$ as it relates to the graph of the parent function.

Graph the parent function, $f(x) = x^2$, for quadratic functions.

Since $f(x) = x^2$, $g(x) = f(x) + k$ where $k = -4$.

The constant k is added to the function after it has been evaluated, so k affects the output, or y-values. The value of k is less than 0, so the graph of $f(x) = x^2$ is translated down 4 units.

$g(x) = x^2 - 4$ is the translation of the graph of the parent function 4 units down.

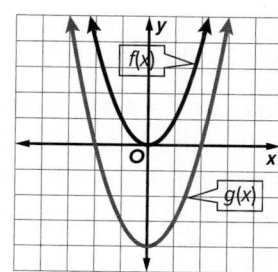

Today's Goals
- Apply translations to quadratic functions.
- Apply dilations to quadratic functions.
- Use transformations to identify quadratic functions from graphs and write equations of quadratic functions.

Today's Vocabulary
vertex form

Go Online
You can watch a video to see how to translate functions.

Think About It!
Is the following statement *sometimes*, *always*, or *never* true? Explain.
The graph of $g(x) = x^2 + k$ has its vertex at the origin.

Check

Describe the transformation of $g(x) = x^2 - 8$ as it relates to the graph of the parent function.

The graph of $g(x) = x^2 - 8$ is the translation of the graph of the parent function ___?___ units ___?___.

Example 2 Horizontal Translations of Quadratic Functions

Describe the translation in $g(x) = (x + 2)^2$ as it relates to the graph of the parent function.

Graph the parent function, $f(x) = x^2$, for quadratic functions.

Since $f(x) = x^2$, $g(x) = f(x - h)$ where $h = -2$.

The constant h is subtracted from x before the function is performed, so h affects the input, or x-values. The value of h is less than 0, so the graph of $f(x) = x^2$ is translated $|-2|$ units left.

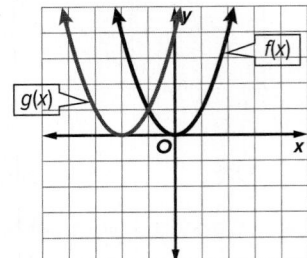

$g(x) = (x + 2)^2$ is the translation of the graph of the parent function 2 units left.

Check

Describe the transformation of $g(x) = (x - 6)^2$ as it relates to the graph of the parent function.

The graph of $g(x) = (x - 6)^2$ is the translation of the graph of the parent function ___?___ units ___?___.

Example 3 Multiple Translations of Quadratic Functions

Describe the translation in $g(x) = (x - 3)^2 + 1$ as it relates to the graph of the parent function.

Graph the parent function, $f(x) = x^2$, for quadratic functions.

Since $f(x) = x^2$, $g(x) = f(x - h) + k$ where $h = 3$ and $k = 1$.

The constant h is subtracted from x before the function is performed and is less than 0, so the graph of $f(x) = x^2$ is translated 3 units right.

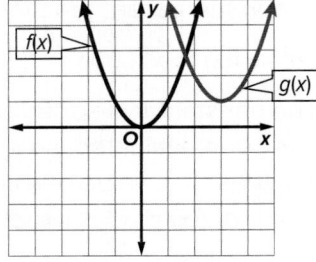

The constant k is added to the function after it has been evaluated and is greater than 0, so the graph of $f(x) = x^2$ is also translated 1 unit up.

🔵 **Go Online** You can complete an Extra Example online.

> **Think About It!**
>
> What do you notice about the vertex of horizontally translated quadratic functions compared to the vertex of the parent function?

Check

Describe the translation in $g(x) = (x + 3)^2 - 5$ as it relates to the graph of the parent function.

The graph of $g(x) = (x + 3)^2 - 5$ is the translation of the graph of the parent function 3 units __?__ and 5 units __?__.

Graph $g(x) = (x + 3)^2 - 5$.

Learn Dilations of Quadratic Functions

Key Concept • Vertical Dilations of Quadratic Functions

The graph of $g(x) = ax^2$ is the graph of $f(x) = x^2$ stretched or compressed vertically by a factor of $|a|$.

If $|a| > 1$, the graph of $f(x)$ is stretched vertically away from the x-axis.

If $0 < |a| < 1$, the graph of $f(x)$ is compressed vertically toward the x-axis.

Key Concept • Horizontal Dilations of Quadratic Functions

The graph of $g(x) = (ax)^2$ is the graph of $f(x) = x^2$ stretched or compressed vertically by a factor of $\frac{1}{|a|}$.

If $|a| > 1$, the graph of $f(x)$ is compressed horizontally toward the y-axis.

If $0 < |a| < 1$, the graph of $f(x)$ is stretched horizontally away from the y-axis.

> **Go Online**
> You can watch a video to see how to describe dilations of functions.

Example 4 Vertical Dilations of Quadratic Functions

Describe the dilation in $g(x) = 3x^2$ as it relates to the graph of the parent function.

Graph the parent function, $f(x) = x^2$, for quadratic functions.

Since $f(x) = x^2$, $g(x) = a \cdot f(x)$ where $a = 3$.

The function is multiplied by the positive constant a after it has been evaluated and $|a|$ is greater than 1, so the graph of $f(x) = x^2$ is stretched vertically by a factor of $|a|$, or 3.

$g(x) = 3x^2$ is a vertical stretch of the graph of the parent function.

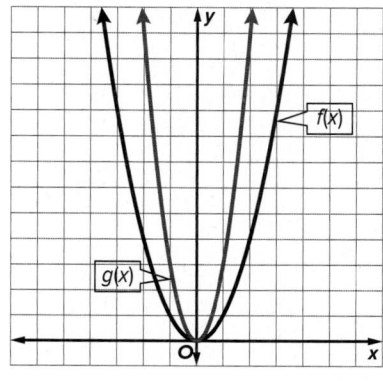

> **Study Tip**
> **Vertical Dilations** For a vertical dilation, if you multiply each y-coordinate of the function $f(x)$ by a, you'll get the corresponding y-coordinate of the function $g(x)$. For example, for the function above, the point (2, 4) on $f(x)$ corresponds to the point (2, 12) on $g(x)$. The y-coordinate of $f(x)$, 4, is multiplied by a, which is 3.

Check

Graph $g(x) = \frac{1}{4}x^2$.

Example 5 Horizontal Dilations of Quadratic Functions

Describe the dilation in $g(x) = \left(\frac{1}{2}x\right)^2$ as it relates to the graph of the parent function.

Graph the parent function, $f(x) = x^2$, for quadratic functions.

Since $f(x) = x^2$, $g(x) = f(a \cdot x)$ where $a = \frac{1}{2}$.

x is multiplied by the positive constant a before the function is performed and $|a|$ is between 0 and 1, so the graph of $f(x) = x^2$ is stretched horizontally by a factor of $\frac{1}{|a|}$, or 2.

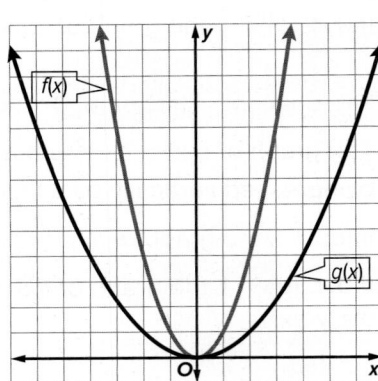

$g(x) = \left(\frac{1}{2}x\right)^2$ is a horizontal stretch of the graph of the parent function.

Check

Graph $g(x) = (2x)^2$.

Go Online You can complete an Extra Example online.

Talk About It!

Could the graph of $g(x) = \left(\frac{1}{2}x\right)^2$ also be described by a vertical dilation? Justify your argument.

Study Tip

Horizontal Dilations For a horizontal dilation, if you multiply each x-coordinate of the function $f(x)$ by $\frac{1}{|a|}$, you'll get the corresponding x-coordinate of the function $g(x)$. For example, for the function above, the point $(-1, 1)$ on $f(x)$ corresponds to the point $(-2, 1)$ on $g(x)$. The x-coordinate of $f(x)$, -1, is multiplied by $\frac{1}{|a|}$.

Learn Reflections of Quadratic Functions

Key Concept • Reflections of Quadratic Functions Across the *x*-axis

The graph of $-f(x)$ is the reflection of the graph of $f(x) = x^2$ across the *x*-axis.

Key Concept • Reflections of Quadratic Functions Across the *y*-axis

The graph of $f(-x)$ is the reflection of the graph of $f(x) = x^2$ across the *y*-axis.

Example 6 Vertical Reflections of Quadratic Functions

Describe how the graph of $g(x) = -\frac{2}{3}x^2$ is related to the graph of the parent function.

Graph the parent function, $f(x) = x^2$, for quadratic functions.

Because $f(x) = x^2$, $g(x) = -1 \cdot a \cdot f(x)$ where $a = \frac{2}{3}$.

The function is multiplied by -1 and the constant a after it has been evaluated and $|a|$ is between 0 and 1, so the graph of $f(x)$ is compressed vertically and reflected across the *x*-axis.

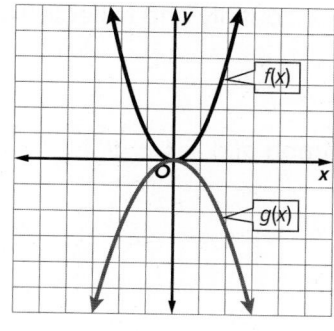

$g(x) = -\frac{2}{3}x^2$ is the graph of the parent function compressed vertically and reflected across the *x*-axis.

Check

Graph $g(x) = -\frac{3}{5}x^2$.

Go Online You can complete an Extra Example online.

> **Think About It!**
>
> How does a reflection of the parent function $f(x) = x^2$ across the *x*-axis affect the end behavior of the graph?

Example 7 Horizontal Reflections of Quadratic Functions

Describe how the graph of $g(x) = (-3x)^2$ is related to to the graph of the parent function.

Graph the parent function, $f(x) = x^2$, for quadratic functions.

Because $f(x) = x^2$, $g(x) = f(-1 \cdot a \cdot x)^2$ where $a = 3$.

x is multiplied by -1 and the constant a before the function is performed and $|a|$ is greater than 1, so the graph of $f(x)$ is compressed horizontally and reflected across the y-axis.

$g(x) = (-3x)^2$ is the graph of the parent function compressed horizontally and reflected across the y-axis.

Check

Graph $g(x) = \left(-\dfrac{7}{2}x\right)^2$.

Learn Transformations of Quadratic Functions

A quadratic function in the form $f(x) = a(x - h)^2 + k$ is written in **vertex form**. Each constant in the equation affects the parent graph.

- The value of $|a|$ stretches or compresses (dilates) the parent graph.

- When the value of a is negative, the graph is reflected across the x-axis.

- The value of k shifts (translates) the parent graph up or down.

- The value of h shifts (translates) the parent graph left or right.

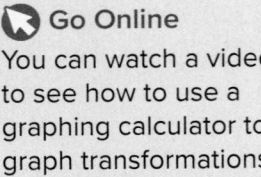

Go Online
You can watch a video to see how to use a graphing calculator to graph transformations of quadratic functions.

Example 8 Multiple Transformations of Quadratic Functions

Describe how the graph of $g(x) = -3(x - 1)^2 - 2$ is related to the graph of the parent function.

Graph the parent function, $f(x) = x^2$.

Since $f(x) = x^2$, $g(x) = a\, f(x - h) + k$ where $a = -3$.

$a < 0$ and $|a| > 1$, so the graph of $f(x) = x^2$ is compressed vertically and reflected across the x-axis and stretched vertically by a factor of $|a|$, or 3.

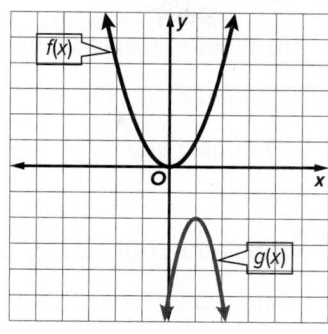

$h > 0$, so the graph is then translated h units right, or 1 unit right.

$k < 0$, so the graph is then translated $|k|$ units down, or 2 units down.

$g(x) = -3(x - 1)^2 - 2$ is the graph of the parent function stretched vertically, reflected across the x-axis, and translated 1 unit right and 2 units down.

Check

Graph $g(x) = -\frac{1}{3}(x + 4)^2 - 1$.

☁ Think About It!

Write a quadratic function that opens down and is translated 6 units left and 1 unit down.

☀ Example 9 Apply Transformations of Quadratic Functions

FOOTBALL Although they may appear flat, properly designed football fields arc to allow water to drain. Fields rise from each sideline to the center of the field, known as the crown, which should be between 1 and 1.5 feet in height. A cross section of a football field that is 160 feet wide and has a 1.5 foot crown can be modeled by $g(x) = -0.000234(x - 80)^2 + 1.5$, where $g(x)$ is the height of the field and x is the distance from the sideline in feet. Describe how $g(x)$ is related to the graph of $f(x) = x^2$.

$a < 0$, so the graph of $f(x) = x^2$ is a reflection across the x-axis.

$0 < |-0.000234| < 1$, so the graph of $f(x) = x^2$ is compressed vertically.

$80 > 0$, so the graph of $f(x) = x^2$ is translated 80 units right.

$1.5 > 0$, so the graph of $f(x) = x^2$ is translated 1.5 units up.

ℝ Go Online You can complete an Extra Example online.

☁ Think About It!

What do the values of h and k represent in the context of the situation?

Check

BRIDGES The lower arch of the Sydney Harbor Bridge can be modeled by $g(x) = -0.0018(x - 251.5)^2 + 118$. Select all of the transformations that occur in $g(x)$ as it relates to the graph of $f(x) = x^2$.

A. vertical compression

B. translation down 251.5 units

C. translation up 118 units

D. reflection across the x-axis

E. vertical stretch

F. translation right 251.5 units

G. reflection across the y-axis

Watch Out!

Choosing a Point
When substituting for x and y in the equation, use a point other than the vertex.

🗨 **Think About It!**

How does the equation you found compare to the prediction you made in Step 1?

Example 10 Identify a Quadratic Equation from a Graph

Use the graph of the function to write its equation.

Step 1 Analyze the graph.

The graph appears to be narrower than the parent function, implying a vertical stretch, and is reflected across the x-axis. So, $a < 0$ and $|a| > 1$. The graph has also been shifted left and up from the parent graph. So, $h < 0$ and $k > 0$.

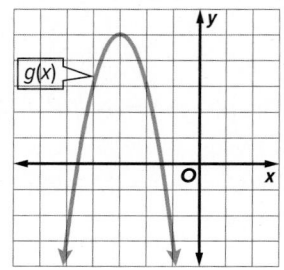

Step 2 Identify the translations.

The vertex is shifted 3 units left, so $h = -3$.

It is also shifted 5 units up, so $k = 5$.

Step 3 Identify the dilation and/or reflection.

The point $(-2, 3)$ lies on the graph. Substitute the coordinates in for x and y to solve for a.

$y = a(x + 3)^2 + 5$	Vertex form of the graph
$3 = a(-2 + 3)^2 + 5$	$(-2, 3) = (x, y)$
$3 = a(1)^2 + 5$	Add.
$3 = a + 5$	Evaluate 1^2.
$-2 = a$	Subtract 5 from each side.

So, the equation is $g(x) = -2(x + 3)^2 + 5$.

🌐 **Go Online** You can complete an Extra Example online.

Practice

Examples 1–3

Describe the translation in each function as it relates to the graph of the parent function.

1. $g(x) = -10 + x^2$

2. $g(x) = x^2 + 2$

3. $g(x) = (x - 1)^2$

4. $g(x) = x^2 - 8$

5. $g(x) = (x + 3)^2$

6. $g(x) = x^2 + 7$

7. $g(x) = (x + 2.5)^2$

8. $g(x) = (x + 5)^2 - 2$

9. $g(x) = 6 + x^2$

10. $g(x) = -4 + (x - 3)^2$

11. $g(x) = (x - 9.5)^2$

12. $g(x) = (x - 1.5)^2 + 3.5$

Examples 4 and 5

Describe the dilation in each function as it relates to the graph of the parent function.

13. $g(x) = 7x^2$

14. $g(x) = \frac{1}{5}x^2$

15. $g(x) = (4x)^2$

16. $g(x) = \left(\frac{1}{2}x\right)^2$

17. $g(x) = \left(\frac{5}{3}x\right)^2$

18. $g(x) = 5x^2$

19. $g(x) = \frac{3}{4}x^2$

20. $g(x) = \left(\frac{7}{8}x\right)^2$

Examples 6 and 7

Describe how the graph of each function is related to the graph of the parent function.

21. $g(x) = -6x^2$

22. $g(x) = (-9x)^2$

23. $g(x) = -\frac{1}{3}x^2$

24. $g(x) = \left(-\frac{2}{3}x\right)^2$

25. $g(x) = -2x^2$

26. $g(x) = \left(-\frac{6}{5}x\right)^2$

Example 8

Describe how the graph of each function is related to the graph of the parent function.

27. $h(x) = -7 - x^2$

28. $g(x) = 2(x - 3)^2 + 8$

29. $h(x) = 6 + \frac{2}{3}x^2$

30. $g(x) = -5 - \frac{4}{3}x^2$

31. $h(x) = 3 + \frac{5}{2}x^2$

32. $g(x) = -x^2 + 3$

Example 9

33. SPRINGS The potential energy stored in a spring is given by $U_s = \frac{1}{2}kx^2$, where k is a constant known as the spring constant, and x is the distance the spring is stretched or compressed from its initial position. How is the graph of the function for a spring where $k = 10$ newtons/meter related to the graph of the function for a spring where $k = 2$ newtons/meter?

34. PHYSICS A ball is dropped from a height of 20 feet. The function $h = -16t^2 + 20$ models the height of the ball in feet after t seconds. Compare this graph to the graph of its parent function.

35. ACCELERATION The distance d in feet a car accelerating at 6 ft/s^2 travels from the start of a race after t seconds is modeled by the function $d = 3t^2$. Suppose a second car begins the race at the same time 100 feet ahead of the first car and accelerating at 4 ft/s^2. The distance the second car is from the starting line after t seconds is modeled by the function $d = 2t^2 + 100$.

    a. Explain how each graph is related to the graph of $d = t^2$.

    b. After how many seconds will the first car pass the second car?

Example 10

Use the graph of each function to write its equation.

36.

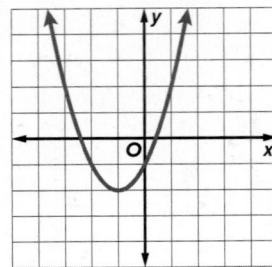

37.

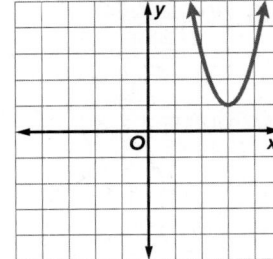

38.

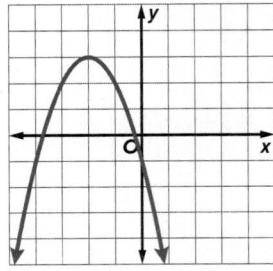

Mixed Exercises

Describe how the graph of each function is related to the graph of the parent function.

39. $g(x) = 5 - \frac{1}{5}x^2$

40. $g(x) = 4(x - 1)^2$

41. $g(x) = 0.25x^2 - 1.1$

42. $h(x) = 1.35(x + 1)^2 + 2.6$

43. $g(x) = \frac{3}{4}x^2 + \frac{5}{6}$

44. $h(x) = 1.01x^2 - 6.5$

STRUCTURE Use transformations to graph each quadratic function. Then describe the transformation.

45. $h(x) = -2(x + 2)^2 + 2$ **46.** $g(x) = \left(\frac{1}{2}x\right)^2 - 3$ **47.** $h(x) = -\frac{1}{4}(x - 1)^2 + 4$

Match each function to its graph.

A.
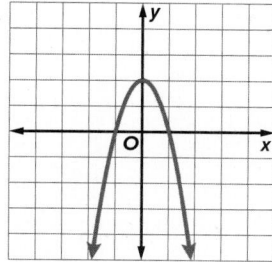

B.
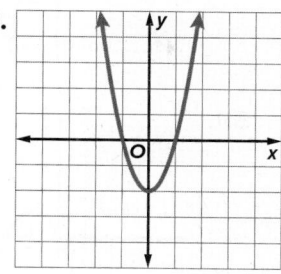

C.
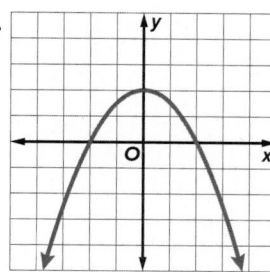

D.
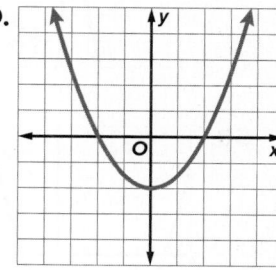

48. $f(x) = 2x^2 - 2$

49. $g(x) = \frac{1}{2}x^2 - 2$

50. $h(x) = -\frac{1}{2}x^2 + 2$

51. $j(x) = -2x^2 + 2$

52. PRECISION Write a set of instructions that a classmate could use to transform the graph of $f(x) = x^2$ and end up with the graph of $g(x) = -10(x - 17)^2$.

53. CONSTRUCT ARGUMENTS A function is an even function if $f(-x) = f(x)$ for all x in the domain of the function. A function is an odd function if $f(-x) = -f(x)$ for all x in the domain of the function. Is the parent function of a quadratic function an even function or an odd function? Justify your answer.

54. USE A MODEL An animator is using a coordinate plane to design a scene in a movie. In the scene, a comet enters the screen in Quadrant II, moves around the screen in a parabolic path, and leaves the screen in Quadrant I.

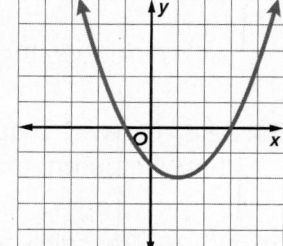

a. The figure shows the path of the comet. What equation represents the path?

b. The animator decides to translate the path of the comet 8 units up and 7 units left. What equation represents the new path?

Use the graph of each function to write its equation.

55.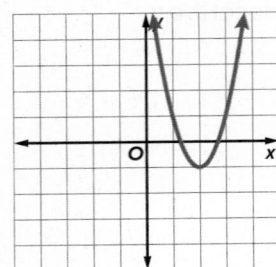

56.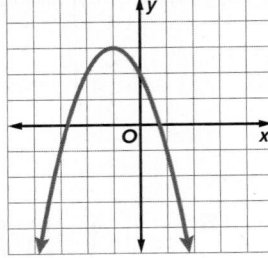

57.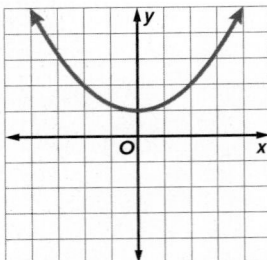

58.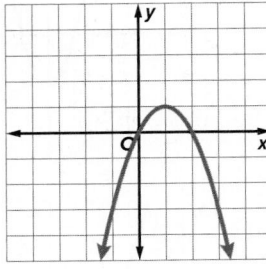

59.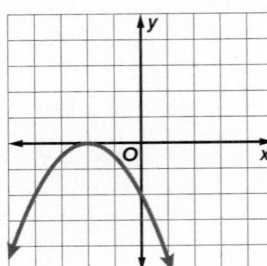

60.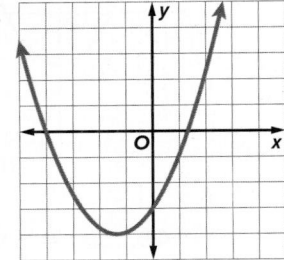

🌐 **Higher-Order Thinking Skills**

61. ANALYZE Are the following statements *sometimes*, *always*, or *never* true? Justify your argument.

a. The graph of $y = x^2 + k$ has its vertex at the origin.

b. The graphs of $y = ax^2$ and its reflection across the x-axis are the same width.

c. The graph of $f(x) = x^2 + k$, where $k \geq 0$, and the graph of a quadratic function $g(x)$ with vertex at $(0, -3)$ have the same maximum or minimum.

62. PERSEVERE Write a function of the form $y = ax^2 + k$ with a graph that passes through the points $(-2, 3)$ and $(4, 15)$.

63. ANALYZE Determine whether all quadratic functions that are reflected across the y-axis produce the same graph. Justify your argument.

64. CREATE Write a quadratic function with a graph that opens down and is wider than the parent graph.

65. WRITE Describe how the values of a and k affect the graphical and tabular representations of the functions $y = ax^2$, $y = x^2 + k$, and $y = ax^2 + k$.

Solving Quadratic Equations by Graphing

Explore Roots and Zeros of Quadratics

Online Activity Use graphing technology to complete the Explore.

> @ **INQUIRY** How can you use the *x*-intercepts of a quadratic function to identify the solutions of its related equation? ✕

Learn Solving Quadratic Equations by Graphing

A **quadratic equation** can be written in the standard form $ax^2 + bx + c = 0$, where $a \neq 0$. Because the graphs of these quadratic functions have zero, one, or two zeros, the corresponding quadratic equations also have zero, one, or two solutions.

Key Concept • Solutions of Quadratic Equations

two unique real solutions	*one* unique real solution	*no* real solutions
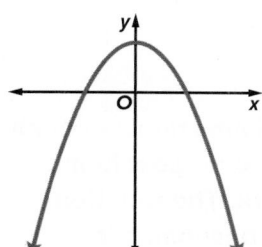	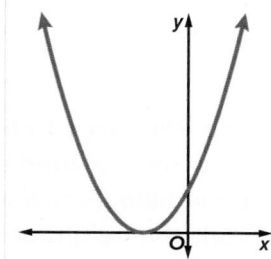	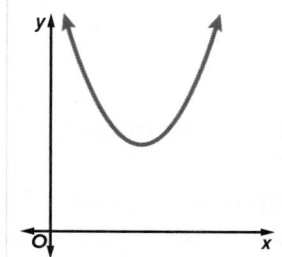

When the vertex of the related quadratic function lies on the *x*-axis, the solution is a **double root**, which means that there are two roots of a quadratic equation that are the same number.

Example 1 Solve a Quadratic Equation with Two Roots

Solve $-x^2 + 4x + 5 = 0$ by graphing.

Graph the related function $f(x) = -x^2 + 4x + 5$. The *x*-intercepts of the graph appear to be at −1 and 5, so the solutions are −1 and 5.

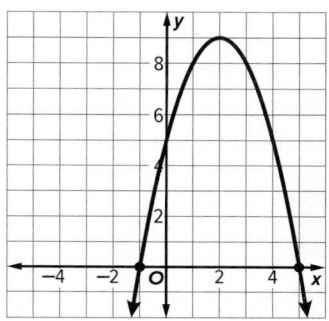

Check

Use the graph of the related function to solve $x^2 + 5x + 6 = 0$ by graphing. The solutions are __?__ and __?__.

Today's Goal
• Solve quadratic equations by graphing.

Today's Vocabulary
quadratic equation
double root

💭 **Think About It!**
How can the solutions of a quadratic equation help you graph the related function?

💭 **Think About It!**
How do you know that the solutions are correct?

Example 2 Solve a Quadratic Equation with a Double Root

Solve $x^2 - 8x = -16$ by graphing.

Rewrite the equation in standard form and then graph the related function $f(x) = x^2 - 8x + 16$.

Notice that the vertex of the parabola is the only x-intercept. Therefore, there is only one solution, 4.

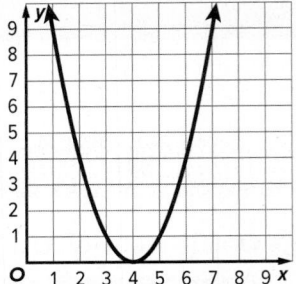

Talk About It!

The function $f(x) = x^2 - 8x + 16$ is a perfect square trinomial. Explain why there is a double root for this type of function.

Example 3 Solve a Quadratic Equation with No Real Roots

Solve $-2x^2 - x - 2 = 0$ by graphing.

Graph the related function $f(x) = -2x^2 - x - 2$.

This graph has no x-intercepts. Therefore, this equation has no real number solutions.

Go Online An alternate method is available for this example.

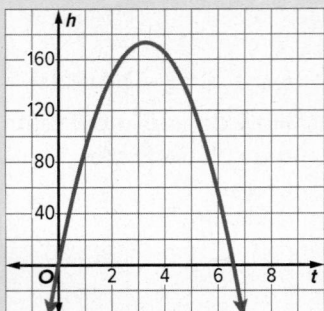

⊕ Example 4 Approximate Roots of Quadratic Functions

BASEBALL In 1941, the Boston Red Sox's Ted Williams hit a baseball 172 meters. The hit would later be regarded as the longest home run ever hit at the team's home field, Fenway Park. The function $h = -16t^2 + 105t + 1.5$ models the height of the baseball h in meters after t seconds. If it falls to the ground, approximately how long would the ball be in the air?

You need to find the roots of the equation $-16t^2 + 105t + 1.5 = 0$. Graph the related function $h = -16t^2 + 105t + 1.5$, and use estimation to approximate the zeros.

The t-intercept that shows when the ball would hit the ground is located between 6 and 7 seconds. To find a more exact time when the ball would hit the ground, complete the table using an increment of 0.1 for the t-values between 6 and 7.

t	6.3	6.4	6.5	6.6	6.7	6.8	6.9
h	27.96	18.14	8	−2.46	−13.24	−24.34	−35.76

The function value that is closest to zero when the sign changes is −2.46. Thus, the ball would be in the air for approximately 6.6 seconds before hitting the ground.

Go Online You can complete an Extra Example online.

Practice

Go Online You can complete your homework online.

Examples 1–3
Solve each equation by graphing.

1. $x^2 + 7x + 14 = 0$
2. $x^2 + 2x - 24 = 0$
3. $x^2 + 16x + 64 = 0$
4. $x^2 - 5x + 12 = 0$
5. $x^2 + 14x = -49$
6. $x^2 = 2x - 1$
7. $x^2 - 10x = -16$
8. $-2x^2 - 8x = 13$
9. $2x^2 - 16x = -30$
10. $2x^2 = -24x - 72$
11. $-3x^2 + 2x = 15$
12. $x^2 = -2x + 80$

Example 4

13. **SOCCER** Claudia kicked a soccer ball off of a platform. The equation $y = -x^2 + 3x + 12$ models the height of the ball y in feet after x seconds. Approximately how long is the ball in the air?

14. **TRAMPOLINE** A gymnast jumped on a trampoline. The equation $y = -16x^2 + 58x$ models the height of the gymnast y in feet after x seconds for one of the jumps. Approximately how long is the gymnast in the air?

Mixed Exercises

Estimate the solution(s) to each equation by graphing the related function. Round to the nearest tenth.

15. $p^2 + 4p + 2 = 0$
16. $x^2 + x - 3 = 0$
17. $d^2 + 6d = -3$
18. $h^2 + 1 = 4h$
19. $3x^2 - 5x = -1$
20. $x^2 + 1 = 5x$

21. **FARMING** In order for Mr. Moore to decide how much fertilizer to apply to his corn crop this year, he reviews records from previous years. His crop yield y depends on the amount of fertilizer he applies to his fields x according to the equation $y = -x^2 + 4x + 12$. Graph the function, and find the point at which Mr. Moore gets the highest yield possible. Then find and interpret the zero(s) of the function.

22. **FRAMING** A rectangular photograph is 7 inches long and 6 inches wide. The photograph is framed using a material that is x inches wide. If the area of the frame and photograph combined is 156 square inches, what is the width of the framing material?

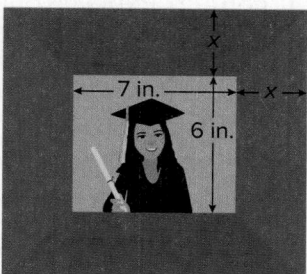

23. **WRAPPING PAPER** Can a rectangular piece of wrapping paper with an area of 81 square inches have a perimeter of 60 inches? (*Hint*: Let length = 30 − width.) Explain.

24. **ENGINEERING** The shape of a satellite dish is often parabolic because of the reflective qualities of parabolas. Suppose a particular satellite dish is modeled by the function $0.5x^2 = 2 + y$.
    a. Approximate the zeros of this function by graphing.
    b. On the coordinate plane, translate the parabola so that there is only one zero. Label this curve A.
    c. Translate the parabola so that there are no real zeros. Label this curve B.

25. REASONING The three equations below are shown on the graph.

$y = x^2 + 12x + m; y = 2x^2 - nx + 72; y = -x^2 + 3x - 10$

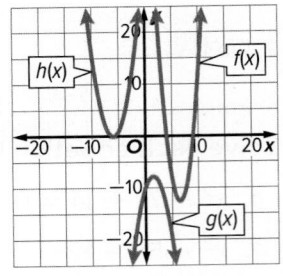

a. Describe the solutions for each function. Then, match each graph to its related equation. Explain your reasoning.

b. How could you choose an appropriate value for m?

c. How could you choose an appropriate value for n?

26. ROCKETS The height h of a model rocket launched from ground level into the air after t seconds can be modeled by the equation $h = -16t^2 + 160t$. The equation is graphed on the coordinate grid.

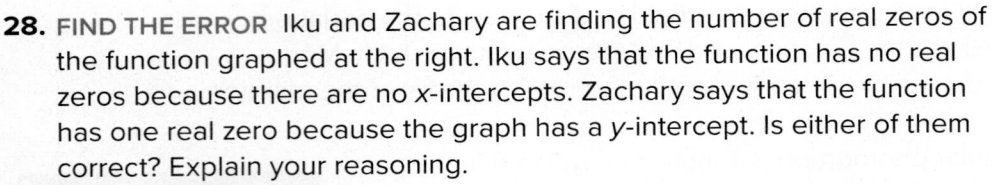

a. Where does the graph intersect the x-axis? What do these points represent?

b. How long does it take the rocket to reach its maximum height? Explain your reasoning.

27. USE TOOLS Through market research, a company finds that its profit in dollars can be modeled by the function $P(x) = -50,000x^2 + 300,000x - 250,000$, where x is the price in dollars at which they sell their product.

a. Describe how to use graphing technology to graph the function, and then provide a sketch of the graph.

b. Find the x-intercepts. What do these represent in the context of the situation?

c. Write an equation for the price at which the company should sell their product to make a profit of $150,000. Sketch a graph of the function relating to this equation and use it to solve the equation graphically.

d. Is there a price at which the company can sell their product to make a profit of $300,000? Explain your reasoning, and use a graph to support your answer.

28. FIND THE ERROR Iku and Zachary are finding the number of real zeros of the function graphed at the right. Iku says that the function has no real zeros because there are no x-intercepts. Zachary says that the function has one real zero because the graph has a y-intercept. Is either of them correct? Explain your reasoning.

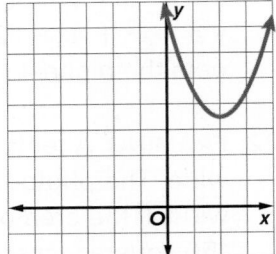

29. CREATE Describe a real-world situation in which a thrown object travels in the air. Write an equation that models the height of the object with respect to time, and determine how long the object travels in the air.

30. ANALYZE The graph shown is a *quadratic inequality*. Analyze the graph, and find 3 solutions with y-values greater than 2.

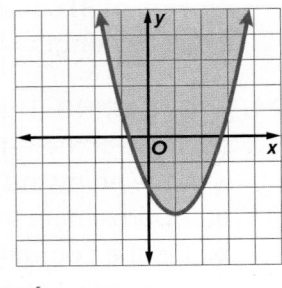

31. PERSEVERE Write a quadratic equation that has the roots described.

a. one double root

b. no real solutions

c. two unique real solutions

32. WRITE Explain how to approximate the roots of a quadratic equation when the roots are not integers.

Solving Quadratic Equations by Factoring

Learn Solving Quadratic Equations by Using the Square Root Property

Key Concept • Square Root Property

Words: To solve a quadratic equation in the form $x^2 = n$, take the square root of each side.

Symbols: For any number $n \geq 0$, if $x^2 = n$, then $x = \pm\sqrt{n}$.

Example:

$x^2 = 49$

$x = \pm\sqrt{49}$ or $x = \pm 7$

In the equation $x^2 = n$, if n is a perfect square, you will have an exact answer. If n is not a perfect square, you will need to approximate the square root. You can use a calculator or estimation to find an approximation.

Example 1 Use the Square Root Property

Solve $(y + 5)^2 = 21$.

$(y + 5)^2 = 21$	Original equation
$y + 5 = \pm\sqrt{21}$	Square Root Property
$y = -5 \pm \sqrt{21}$	Subtract 5 from each side.
$y = -5 + \sqrt{21}$ or $y = -5 - \sqrt{21}$	Separate into two equations.

The solutions are $-5 + \sqrt{21}$ or $= -5 - \sqrt{21}$.
Using a calculator $-5 + \sqrt{21} \approx -0.42$ and $-5 - \sqrt{21} \approx -9.58$.

Check

Select all the solutions of $(x - 2)^2 = 16$.

A. −6   **B.** −4

C. −2   **D.** 2

E. 4   **F.** 6

Select all the solutions of $(y + 18)^2 = 77$.

A. ≈ -26.77 **B.** ≈ -9.23

C. ≈ -7.68 **D.** ≈ 7.68

E. ≈ 9.23 **F.** ≈ 26.77

 Go Online You can complete an Extra Example online.

Today's Goals
- Solve quadratic equations by using the Square Root Property.
- Solve quadratic equations by factoring and sketch graphs of quadratic functions by using the zeros.
- Write equations of quadratic functions given their graphs or roots.

🌧 Think About It!
Without using a calculator, describe how can you approximate $\sqrt{30}$.

Study Tip
Reading Math $\pm\sqrt{30}$ is read as "plus or minus the square root of 30."

Watch Out!
Square Root Be careful not to confuse the *roots* of an equation with taking the *square root* of an expression. The *roots* of an equation are any values of the equation that make it true. The *square root* of an expression is one of two equal factors.

Think About It!

Why is $t \approx -1.37$ not a solution?

EGG DROP During the Egg Drop Competition at Anika's school, students must create a container for an egg that prevents the egg from breaking when dropped from a height of 30 feet. The formula $h = -16t^2 + h_0$ can be used to approximate the number of seconds t it takes for the container to reach height h from an initial height of h_0 in feet. Find the time it takes the container to reach the ground.

At ground level, $h = 0$ and the initial height is 30, so $h_0 = 30$.

$h = -16t^2 + h_0$	Original equation
$0 = -16t^2 + 30$	$h = 0$ and $h_0 = 30$
$-30 = -16t^2$	Subtract 30 from each side.
$1.875 = t^2$	Divide each side by -16.
$1.37 \approx t$	Use the Square Root Property.

It takes ≈ 1.37 seconds for the container to reach the ground.

Study Tip

Modeling The function $h = -16t^2 + h_0$ does not model the path of the container after it is dropped. This function is used to model the height of the container as a function of time.

Explore Using Factors to Solve Quadratic Equations

🅝 **Online Activity** Use graphing technology to complete the Explore.

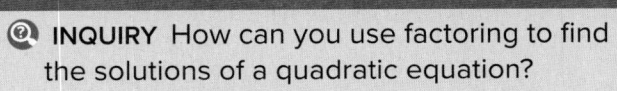

    ⓐ **INQUIRY** How can you use factoring to find the solutions of a quadratic equation?

Learn Solving Quadratic Equations by Factoring

Key Concept • Zero Product Property

Words: If the product of two factors is 0, then at least one of the factors must be 0.

Symbols: For any real numbers a and b, if $ab = 0$, then $a = 0$, $b = 0$, or both a and b equal zero.

Factor Using the Distributive Property

$ax + bx + zy + by$
$= x(a + b) + y(a + b)$
$= (a + b)(x + y)$

Factor Quadratic Trinomials

$ax^2 + bx + c = ax^2 + mx + px + c$ when $m + p = b$ and $mp = ac$

Factor Differences of Squares

$a^2 - b^2 = (a + b)(a - b)$

Factor Perfect Squares

$a^2 + 2ab + b^2 = (a + b)^2$

Think About It!

Why do you think that there are so many different methods for factoring an equation?

🅝 **Go Online** You can complete an Extra Example online.

Example 3 Solve a Quadratic Equation by Using the Distributive Property

Solve $4x^2 - 12x = 0$. Check your solution.

$4x^2 - 12x = 0$	Original equation
$4x(x - 3) = 0$	Factor by using the Distributive Property.
$4x = 0$ and $x - 3 = 0$	Zero Product Property
$x = 0 \qquad x = 3$	Simplify.

Check the roots by substituting them into the original equation.

$4(0)^2 - 12(0) = 0 \checkmark \quad 4(3)^2 - 12(3) = 0 \checkmark$

Example 4 Solve a Quadratic Equation by Factoring a Trinomial

Part A Solve $x^2 - 2x - 8 = 0$. Check your solution.

$x^2 - 2x - 8 = 0$	Original equation
$(x + 2)(x - 4) = 0$	Factor the trinomial.
$x + 2 = 0$ and $x - 4 = 0$	Zero Product Property
$x = -2 \qquad x = 4$	Simplify.

Check the roots by substituting them into the original equation.

Part B Use the roots of $x^2 - 2x - 8 = 0$ to sketch the graph of the related function.

Graph $(-2, 0)$ and $(4, 0)$. The vertex is $(1, -9)$. Because a is positive, the graph opens up.

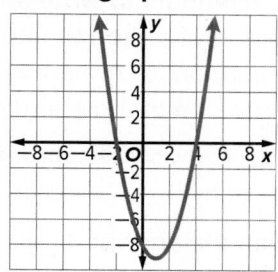

Check

Solve $x^2 - 10x + 24 = 0$.

Problem-Solving Tip

Make a Chart Making a chart to organize the factors of c can help you identify the pair of factors that have a sum of b.

Example 5 Solve a Quadratic Equation by Factoring a Difference of Squares

Solve $49 - x^2 = 0$. Check your solution.

$49 - x^2 = 0$	Original equation
$7^2 - x^2 = 0$	Write in the form $a^2 - b^2$.
$(7 + x)(7 - x) = 0$	Difference of squares
$7 + x = 0$ and $7 - x = 0$	Zero Product Property
$x = -7$ and $x = 7$	Simplify.

Check the roots by substituting them into the original equation.

Example 6 Solve a Quadratic Equation by Factoring a Perfect Square Trinomial

Solve $25y^2 - 60y + 81 = 45$. Check your solution.

$25y^2 - 60y + 81 = 45$	Original equation
$25y^2 - 60y + 36 = 0$	Subtract 45 from each side.
$(5y)^2 - 2(5y)(6) + 6^2 = 0$	Factor the perfect square trinomial.
$(5y - 6)^2 = 0$	Factor the trinomial.
$\sqrt{(5y - 6)^2} = \sqrt{0}$	Square Root Property
$y = \frac{6}{5}$	Simplify.

Check the root by substituting it into the original equation.

Think About It!

Why does $25y^2 - 60y + 81 = 45$ have one solution instead of two?

🌐 Apply Example 7 Factor a Trinomial to Solve a Problem

The area of an isosceles triangle is 108 square feet. Find the perimeter of the triangle if the length of each leg is 15 feet, the length of the base is $4x + 2$ feet, and the height is $3x$ feet.

1. **What is the task?**

Describe the task in your own words. Then list any questions that you may have. How can you find answers to your questions?

Sample answer: I need to use the given information about the area of the triangle to write and solve an equation for x. Then I need to use this information to find the perimeter of the triangle. How can I write an equation for the area of the triangle? I can review the formula for the area of a triangle and how to solve equations.

2. **How will you approach the task? What have you learned that you can use to help you complete the task?**

Sample answer: I will approach this task by first drawing a diagram. I will use what I have learned about factoring to help me solve the equation.

Think About It!

Can $x = -\frac{9}{2}$ ever be a solution of $108 = \frac{1}{2}(4x + 2)(3x)$?

3. **What is your solution?**

Use your strategy to solve the problem. Label the drawing and write an equation for the area of the triangle.

$108 = \frac{1}{2}(4x + 2)(3x)$

Solve your equation by factoring.

$x = -\frac{9}{2}$ or $x = 4$

The perimeter of the triangle is 48 ft.

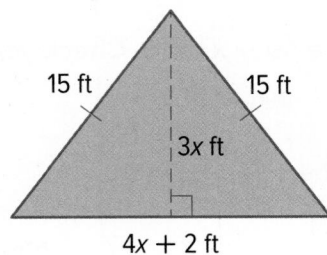

4. **How can you know that your solution is reasonable?**

✏️ **Write About It!** Write an argument that can be used to defend your solution.

Sample answer: I can substitute $x = 4$ into the formula for the area to check my answer. $108 = \frac{1}{2}[4(4) + 2][3(4)]$

🔗 Go Online

You can complete an Extra Example online.

Check

The surface area of a prism is 264 square centimeters. Find the volume of the prism if its height is 15 centimeters, its length is $x + 4$ centimeters, and its width is x centimeters. (Hint: $S = Ph + 2B$ and $V = Bh$)

A. 2 cm³

B. 26 cm³

C. 180 cm³

D. 264 cm³

Learn Writing Quadratic Functions Given the Zeros

You can write the equation of a quadratic function if you know its zeros.

Key Concept • Writing Equations of Quadratic Functions

Step 1 Find the factors of a related expression.

Step 2 Determine whether the value of a is positive or negative.

Step 3 Use another point on the graph to determine the value of a.

Talk About It!

How many points on the graph of a quadratic function must be identified in order to write its equation? Explain.

Example 8 Write a Quadratic Function Given a Graph

Write a quadratic function for the given graph.

Step 1 Find the factors of a related expression.

The two zeros on the graph are -2 and 6, so $(x + 2)$ and $(x - 6)$ are factors of the related expression.

The function $f(x) = a(x + 2)(x - 6)$ represents the graph.

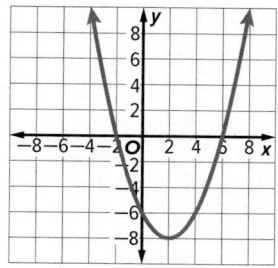

Step 2 Determine whether the value of a is positive or negative.
The graph opens upward, so a must be positive.

Step 3 Use another point on the graph to determine the value of a.
The point $(2, -8)$ is on the graph.

$$f(x) = a(x + 2)(x - 6)$$ Quadratic function with zeros of -2 and 6

$$-8 = a(2 + 2)(2 - 6)$$ $[x, f(x)] = (2, -8)$

$$-8 = -16a$$ Simplify.

$$\frac{1}{2} = a$$ Divide each side by -16.

One equation for the function is $f(x) = \frac{1}{2}(x + 2)(x - 6)$, or $f(x) = \frac{1}{2}x^2 - 2x - 6$.

 Go Online You can complete an Extra Example online.

Think About It!

How would this process have been different if the graph opened downward?

Check

Which quadratic function represents the given graph?

A. $f(x) = -\frac{1}{4}x^2 - \frac{3}{4}x + 1$

B. $f(x) = \frac{1}{4}x^2 - \frac{3}{4}x + 1$

C. $f(x) = -\frac{1}{4}x^2 + \frac{3}{4}x + 1$

D. $f(x) = \frac{1}{4}x^2 + \frac{3}{4}x + 1$

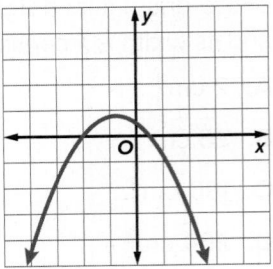

Example 9 Write a Quadratic Function Given Points

Write a quadratic function for the graph that contains (−7, 0), (−3, −32), and (1, 0).

We know that the zeros of a function occur at the *x*-intercepts, or when $y = 0$. Therefore, −7 and 1 are the zeros of this function, and $x + 7$ and $x - 1$ are factors of the related function. To find the value of a in the related function $f(x) = a(x + 7)(x - 1)$, use the third point (−3, −32).

$f(x) = a(x + 7)(x - 1)$	Quadratic function with zeros of −7 and 1
$-32 = a(-3 + 7)(-3 - 1)$	$[x, f(x)] = (-3, -32)$
$-32 = -16a$	Simplify.
$2 = a$	Divide each side by −16.

One function that passes through the given points is
$f(x) = 2(x + 7)(x - 1)$ or $f(x) = 2x^2 + 12x - 14$.

Check

Which quadratic function has a graph that contains $\left(-\frac{5}{2}, 0\right)$, (1, 35), and $\left(\frac{7}{2}, 0\right)$?

A. $f(x) = x^2 + x - \frac{35}{4}$

B. $f(x) = -4x^2 + 4x + 35$

C. $f(x) = 4x^2 + 4x - 35$

D. No quadratic function exists that passes through these points.

🔵 **Go Online** You can complete an Extra Example online.

Practice

Go Online You can complete your homework online.

Examples 1, 3, 5, 6

Solve each equation. Check your solutions.

1. $x^2 = 36$

2. $x^2 = 81$

3. $(k + 1)^2 = 9$

4. $81 - 4b^2 = 0$

5. $3b(9b - 27) = 0$

6. $(7x + 3)(2x - 6) = 0$

7. $b^2 = -3b$

8. $a^2 = 4a$

9. $x^2 - 6x = 27$

10. $a^2 + 11a = -18$

11. $n^2 - 120 = 7n$

12. $d^2 + 56 = -18d$

13. $y^2 - 90 = 13y$

14. $h^2 + 48 = 16h$

15. $x^2 + 9x + 18 = 0$

16. $4x^2 + 28x + 49 = 0$

17. $-9x^2 = 30x + 25$

18. $-x^2 - 12 = -8x$

19. $36w^2 = 121$

20. $100 = 25w^2$

21. $64x^2 - 1 = 0$

22. $4y^2 - \frac{9}{16} = 0$

23. $\frac{1}{4}b^2 = 16$

24. $81 - \frac{1}{25}x^2 = 0$

Example 4

25. Consider the equation $c^2 + 10c + 9 = 0$.
   a. Solve the equation by factoring.
   b. Use the roots to sketch the related function.

26. Consider the equation $x^2 - 18x = -32$.
   a. Solve the equation by factoring.
   b. Use the roots to sketch the related function.

Examples 2 and 7

27. NUMBER THEORY The product of the two consecutive positive integers is 11 more than their sum. What are the numbers?

28. LADDERS A ladder is resting against a wall. The top of the ladder touches the wall at a height of 15 feet, and the length of the ladder is one foot more than twice its distance from the wall. Find the distance from the wall to the bottom of the ladder. (*Hint*: Use the Pythagorean Theorem to solve the problem.)

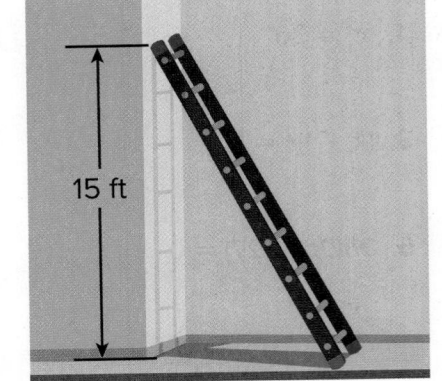

15 ft

29. FREE FALL The function $f(t) = -16t^2 + 576$ represents the height of a freely falling ballast bag that was accidentally dropped from a hot-air balloon 576 feet above the ground. After how many seconds t does the ballast bag hit the ground?

30. VOLUME Catalina can make an open-topped box out of a square piece of cardboard by cutting 3-inch squares from the corners and folding up the sides to meet. The volume of the resulting box is $V = 3x^2 - 36x + 108$, where x is the original length and width of the cardboard.

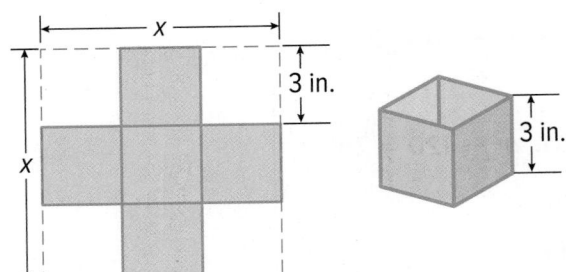

a. Factor the polynomial expression from the volume equation.

b. What is the volume of the box if the original length of each side of the cardboard was 9 inches?

Example 8

Write a quadratic function for the given graph.

31.

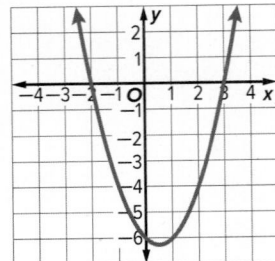

32.

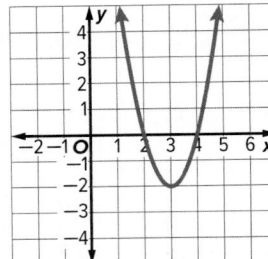

33.

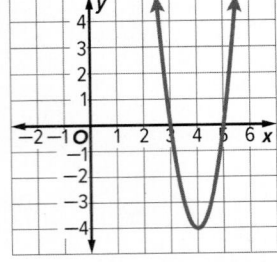

34.

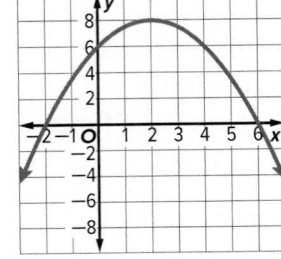

Example 9

Write a quadratic function that contains the given points.

35. $(-9, 0), (-3, -30), (2, 0)$ **36.** $(-5, 0), (-4, 0), (3, 56)$

37. $(-4, 0), (2, -9), (8, 0)$ **38.** $(1, 0), (3, -18), (6, 0)$

Mixed Exercises

39. CONSTRUCT ARGUMENTS Kayla says the zeros of the equation $2m^2 - 12m = 0$ are 0 and -12. Is she correct? Explain your reasoning.

40. STRUCTURE Write an expression for the perimeter of a rectangle that has an area $A = x^2 + 20x + 96$. Explain how you solved the problem.

Solve each equation. Check the solutions.

41. $25p^2 - 16 = 0$ **42.** $4p^2 + 4p + 1 = 0$

43. $n^2 - \frac{9}{25} = 0$ **44.** $9y^2 + 18y - 12 = 6y$

45. $6h^2 + 8h + 2 = 0$ **46.** $16d^2 = 4$

47. $\frac{1}{16} y^2 = 81$ **48.** $8p^2 - 16p = 10$

49. $x^2 + 30x + 150 = -75$ **50.** $10b^2 - 15b = 8b - 12$

51. $y^2 - 16y + 64 = 81$ **52.** $(m - 4)^2 = 9$

53. $4x^2 = 80x - 400$ **54.** $9 - 54x = -81x^2$

55. $4c^2 + 4c + 1 = 9$ **56.** $x^2 - 16x + 64 = 4$

57. STRUCTURE A triangle's height is 10 feet more than its base. If the area of the triangle is 100 square feet, use factoring to find its dimensions.

58. REGULARITY Explain how to use the Zero Product Property to solve a quadratic equation.

59. REGULARITY Consider a quadratic equation written in standard form, $ax^2 + bx + c = 0$, where $a \neq 1$. Explain how solving an equation by factoring a trinomial where $a \neq 1$ differs from solving a quadratic equation by factoring a trinomial where $a = 1$.

60. STRUCTURE A square has an area of $4x^2 + 16xy + 16y^2$ square inches. The dimensions are binomials with positive integer coefficients. Find the perimeter of the square. Is the perimeter a multiple of $(x + 2y)$? Explain.

61. STRUCTURE The equation $x^2 + qx - 12 = 6x$ can be solved by factoring. How can you use number relationships to predict values for q?

62. USE A MODEL The rectangle at the right has a perimeter of $14x + 30$ centimeters and an area of 225 square centimeters. Find the dimensions of the rectangle. Explain how you solved the problem.

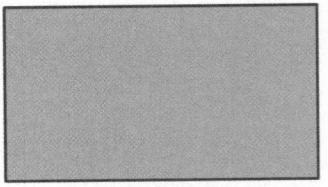

63. AREA A triangle has an area of 64 square feet. If the height of the triangle is 8 feet more than its base, x, what are its height and base?

$6x + 15$ cm

64. CONSTRUCT ARGUMENTS Jarrod claims that the quadratic equation $x^2 - 7x + 5 = 0$ has no solution because the left side does not factor. Do you agree or disagree? Use a graph to justify your argument.

Higher-Order Thinking Skills

65. PERSEVERE Given the equation $(ax + b)(ax - b) = 0$, solve for x. What do you know about the values of a and b?

66. WRITE Explain how to solve the quadratic equation $x^2 + 16x = -64$.

67. FIND THE ERROR Ignatio and Samantha are solving $6x^2 - x = 12$. Is either of them correct? Explain your reasoning.

Ignatio	Samantha
$6x^2 - x = 12$	$6x^2 - x = 12$
$x(6x - 1) = 12$	$6x^2 - x - 12 = 0$
$x = 12$ or $6x - 1 = 12$	$(2x - 3)(3x + 4) = 0$
$6x = 13$	$2x - 3 = 0$ or $3x + 4 = 0$
$x = \frac{13}{6}$	$x = \frac{3}{2}$ $x = -\frac{4}{3}$

68. ANALYZE What should you consider when solving a quadratic equation that models a real-world situation?

69. CREATE Write a perfect square trinomial equation in which the coefficient of the middle term is negative and the last term is a fraction. Solve the equation.

70. PERSEVERE Factor the polynomial $x^3 + x^2 - 6x$. Make a table. Then sketch the graph of the related function.

Solving Quadratic Equations by Completing the Square

Explore Using Algebra Tiles to Complete the Square

 Online Activity Use algebra tiles to complete the Explore.

> ×
>
> @ **INQUIRY** How does forming a square to create a perfect square trinomial help you solve quadratic equations?

Learn Solving Quadratic Equations by Completing the Square

Key Concept • Completing the Square

To **complete the square** for any quadratic expression of the form $x^2 + bx$, follow the steps below.

Step 1 Find one-half of b, the coefficient of x.

Step 2 Square the result from Step 1.

Step 3 Add the result of Step 2 to $x^2 + bx$.

The pattern to complete the square is represented by $x^2 + bx + \left(\frac{b}{2}\right)^2 = \left(x + \frac{b}{2}\right)^2$.

Example 1 Complete the Square

Find the value of c that makes $x^2 + 8x + c$ a perfect square trinomial. Use the completing-the-square algorithm.

Step 1 Find half of 8. $\qquad\qquad\qquad \frac{8}{2} = 4$

Step 2 Square the result of Step 1. $\qquad\quad 4^2 = 16$

Step 3 Add the result of Step 2 to $x^2 + 8x$. $\quad x^2 + 8x + 16$

Thus, $c = 16$. Notice that $x^2 + 8x + 16 = (x + 4)^2$.

Check

Find the value that makes the expression a perfect square trinomial.

$x^2 + 40x +$ _?_

 Go Online You can complete an Extra Example online.

Today's Goals
- Solve quadratic equations by completing the square.
- Identify key features of quadratic functions by writing quadratic equations in vertex form.

Today's Vocabulary
completing the square

🗨 **Think About It!**

To complete the square, what constant would you add to $x^2 + 5x$?

🗨 **Think About It!**

How is the process of completing the square different when b is even and when b is odd?

Example 2 Solve an Equation by Completing the Square

Solve $x^2 - 10x + 14 = 5$ by completing the square.

In order to solve a quadratic equation by completing the square, first isolate $x^2 - bx$ on one side.

$x^2 - 10x + 14 = 5$ Original equation

$x^2 - 10x = -9$ Subtract 14 from each side.

$x^2 - 10x + 25 = -9 + 25$ Because $\left(\frac{-10}{2}\right)^2 = 25$, add 25 to each side.

$(x - 5)^2 = 16$ Factor $x^2 - 10x + 25$

$x - 5 = \pm 4$ Take the square root of each side.

$x = 5 \pm 4$ Add 5 to each side.

$x = 9$ or 1 Separate the solutions.

Example 3 Solve an Equation with a Not Equal to 1

Solve $-4x^2 + 32x - 72 = 0$ by completing the square.

To solve a quadratic equation when the leading coefficient is not 1, divide or multiply each term to eliminate the coefficient. Then, isolate $x^2 - bx$ and complete the square.

$-4x^2 + 32x - 72 = 0$ Original equation

$x^2 - 8x + 18 = 0$ Divide each side by -4.

$x^2 - 8x = -18$ Subtract 18 from each side.

$x^2 - 8x + 16 = -18 + 16$ Add $\left(\frac{8}{2}\right)^2$ to each side.

$(x - 4)^2 = -2$ Factor $x^2 - 8x + 16$

No real number has a negative square. So, this equation has no real solutions.

Learn Finding the Maximum or Minimum Value

Key Concept • Use Vertex Form to Graph

Step 1 Complete the square to write the function in vertex form.

Step 2 Identify the axis of symmetry and extrema based on the function in vertex form. When the leading coefficient is positive, the parabola will open up and the vertex will be a minimum. When the leading coefficient is negative, the parabola will open down and the vertex will be a maximum.

Step 3 Solve for x to find the zeros. The zeros are the x-intercepts of the graph.

Step 4 Use the key features to graph the function.

Think About It!

What do the zeros of a quadratic function tell you about its graph?

Study Tip

Multiplicative Inverse When the leading coefficient is a fraction, remember that you can either divide all of the terms by the fraction or multiply each term by the multiplicative inverse.

Go Online You can complete an Extra Example online.

Example 4 Find a Minimum

Write $y = x^2 + 2x - 5$ in vertex form. Identify the axis of symmetry, extrema, and zeros. Then, use the key features to graph the function.

Step 1 Complete the square to write the function in vertex form.

$$y = x^2 + 2x - 5$$ Original function

$$y + 5 = x^2 + 2x$$ Add 5 to each side.

$$y + 5 + 1 = x^2 + 2x + 1$$ Add $\left(\frac{b}{2}\right)^2$ to each side.

$$y + 6 = (x + 1)^2$$ Factor.

$$y = (x + 1)^2 - 6$$ Subtract 6 from each side to write in vertex form.

Step 2 Identify the axis of symmetry and extrema.

In vertex form the vertex of the parabola (h, k) or $(-1, -6)$. Since the x^2-term is positive, the vertex is a minimum. The axis of symmetry is $x = h$ or $x = -1$.

Step 3 Solve for x to find the zeros.

$$(x + 1)^2 - 6 = 0$$

$$(x + 1)^2 = 6$$

$$x + 1 = \pm\sqrt{6}$$

$$x \approx -3.45 \text{ or } 1.45$$

Step 4 Use the key features to graph the function.

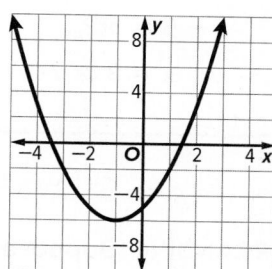

💬 **Talk About It**

Explain how you can determine whether a parabola will open up or open down.

Example 5 Find a Maximum

Write $y = -2x^2 + 20x - 42$ in vertex form. Identify the axis of symmetry, extrema, and zeros. Then, use the key features to graph the function.

Step 1 Complete the square to write the function in vertex form.

$$y = -2x^2 + 20x - 42$$ Original function

$$y + 42 = -2x^2 + 20x$$ Add 42 to each side.

$$y + 42 = -2(x^2 - 10x)$$ Factor out -2.

$$y + 42 - 50 = -2(x^2 - 10x + 25)$$ Since $-2\left[\left(\frac{10}{2}\right)^2\right] = -50$, add -50 to each side.

$$y - 8 = -2(x - 5)^2$$ Factor $x^2 - 10x + 25$.

$$y = -2(x - 5)^2 + 8$$ Add 8 to each side to write in vertex form.

Watch Out!

Leading Coefficients
When completing the square of a quadratic equation where $a \neq 1$, be sure to multiply the constant added by the coefficient before adding it to both sides.

(continued on the next page)

Step 2 Identify the axis of symmetry and extrema. In vertex form the vertex of the parabola is at (h, k) or $(5, 8)$. Since the x^2-term is negative, the vertex is a maximum. The axis of symmetry is $x = h$ or $x = 5$.

Step 3 Solve for x to find the zeros.

$$0 = -2(x - 5)^2 + 8$$
$$2(x - 5)^2 = 8$$
$$(x - 5)^2 = 4$$
$$x - 5 = \pm 2$$
$$x = 3 \text{ or } 7$$

Step 4 Use the key features to graph the function.

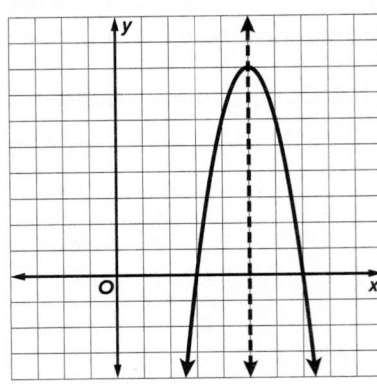

🌎 Example 6 Use Extrema and Key Features

LUNAR LANDING **Suppose the path of a golf ball can be represented by $y = -0.001x^2 + 0.248x$, where y is the height of the ball in meters and x is its horizontal distance in meters. Determine the maximum height of the golf ball and the horizontal distance it traveled.**

Complete the square to write the equation of the function in vertex form.

$y = -0.001x^2 + 0.248x$	Original equation
$y = -0.001(x^2 - 248x)$	Factor out -0.001.
$y - 15.376 = -0.001(x^2 - 248x + 15{,}376)$	Add $-15{,}376$ to each side.
$y - 15.376 = -0.001(x - 124)^2$	Factor $x^2 - 248x + 15{,}376$.
$y = -0.001(x - 124)^2 + 15.376$	Add 15.376 to each side.

The vertex of the parabola is at $(124, 15.376)$. Since the vertex represents the maximum, the ball reached a height of 15.376 meters.

To find the horizontal distance the ball traveled, find the zeros.

$$0 = -0.001(x - 124)^2 + 15.376$$
$$0.001(x - 124)^2 = 15.376$$
$$(x - 124)^2 = 15{,}376$$
$$x - 124 = \pm 124$$
$$x = 0 \text{ or } 248$$

Since $x = 0$ represents the golf ball's initial point, the golf ball traveled 248 meters before hitting the surface of the Moon.

🔵 **Go Online** You can complete an Extra Example online.

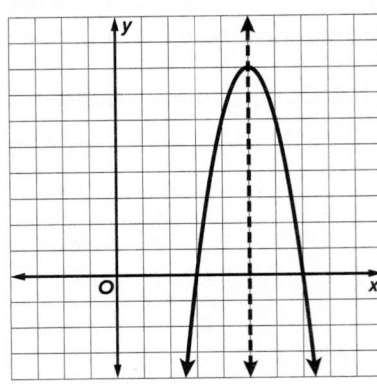

☁️ **Think About It!**

Suppose Alan Shepard had hit a golf ball standing on a platform 3 meters above the lunar surface. How would this affect the vertex form of the equation?

Practice

Go Online You can complete your homework online.

Example 1

Find the value of c that makes each trinomial a perfect square.

1. $x^2 + 26x + c$

2. $x^2 - 24x + c$

3. $x^2 - 19x + c$

4. $x^2 + 17x + c$

5. $x^2 + 5x + c$

6. $x^2 - 13x + c$

7. $x^2 - 22x + c$

8. $x^2 - 15x + c$

9. $x^2 + 24x + c$

Examples 2 and 3

Solve each equation by completing the square. Round to the nearest tenth, if necessary.

10. $x^2 + 6x - 16 = 0$

11. $x^2 - 2x - 14 = 0$

12. $x^2 - 8x - 1 = 8$

13. $x^2 + 3x + 21 = 22$

14. $x^2 - 11x + 3 = 5$

15. $5x^2 - 10x = 23$

16. $2x^2 - 2x + 7 = 5$

17. $3x^2 + 12x + 81 = 15$

18. $4x^2 + 6x = 12$

19. $4x^2 + 6 = 10x$

20. $-2x^2 + 10x = -14$

21. $-3x^2 - 12 = 14x$

Examples 4 and 5

Write each equation in vertex form. Identify the axis of symmetry, extrema, and zeros. Then, use the key features to graph the function.

22. $y = x^2 + 8x + 7$

23. $y = x^2 - 12x + 16$

24. $y = -x^2 - 4x + 5$

25. $y = x^2 - 8x + 10$

Example 6

26. MARS On Mars, the gravity acting on an object is less than that on Earth. On Earth, a golf ball hit with an initial upward velocity of 26 meters per second will hit the ground in about 5.4 seconds. The height h of an object on Mars that leaves the ground with an initial velocity of 26 meters per second is given by the equation $h = -1.9t^2 + 26t$. How much longer will it take for the golf ball hit on Mars to reach the ground? What is the maximum height of the golf ball on Mars? Round your answer to the nearest tenth.

27. FROGS A frog sitting on a stump 3 feet high hops off and lands on the ground. During its leap, its height h in feet is given by $h = -0.5d^2 + 2d + 3$, where d is the distance from the base of the stump. How far is the frog from the base of the stump when it lands on the ground? What is the maximum height of the frog during its leap?

28. FALLING OBJECTS Keisha throws a rock down an abandoned well. The distance d in feet the rock falls after t seconds can be represented by $d = 16t^2 + 64t$. If the water in the well is 80 feet below ground, how many seconds will it take for the rock to hit the water?

Mixed Exercises

29. **GARDENING** Peggy is planning a rectangular vegetable garden using 200 feet of fencing material. She only needs to fence three sides of the garden since one side borders an existing fence. Let x = the width of the rectangle. For what widths would the area of Peggy's garden equal 4800 square feet if she uses all the fencing material?

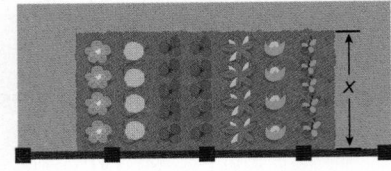

30. **REASONING** Find the value of q that makes $0.5x^2 + 0.5qx + 72$ a perfect square trinomial. Show that the same value of q makes $4x^2 + 24x + 1.5q$ a perfect square trinomial.

31. **AREA** The area of the rectangle shown is 352 square inches. What are the dimensions of the rectangle?

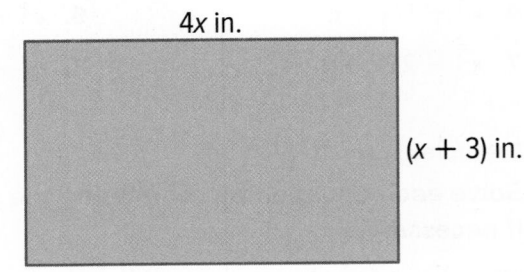

4x in.

$(x + 3)$ in.

32. **CONSTRUCT ARGUMENTS** Use completing the square to show that no two consecutive positive even integers can have a product of 27.

33. **STRUCTURE** Lawrence writes a trinomial that can be solved by completing the square and cannot be solved by factoring. One of the solutions of his equation is between 4 and 5.

    a. Write a possible equation.

    b. What are the solutions to the equation?

    c. Explain why your equation meets Lawrence's criteria.

34. **STRUCTURE** The quadratic function $f(t) = -5000t^2 + 70,000t + 5000$ is used to model the number of products a company sells in years t since the product was released. Complete the square to find the zeros of $f(t)$ and interpret each in the context of the situation. Then find the maximum sales for the product.

35. **PERSEVERE** Given $y = ax^2 + bx + c$ with $a \neq 0$, derive the equation for the axis of symmetry by completing the square, and rewrite the equation in the form $y = a(x - h)^2 + k$.

36. **ANALYZE** Determine the number of solutions $x^2 + bx = c$ has if $c < - \left(\frac{b}{2}\right)^2$. Justify your argument.

37. **WHICH ONE DOESN'T BELONG?** Identify the expression that does not belong with the other three. Explain your reasoning.

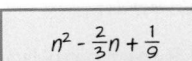

| $n^2 - n + \frac{1}{4}$ | $n^2 + n + \frac{1}{4}$ | $n^2 - \frac{2}{3}n + \frac{1}{9}$ | $n^2 + \frac{1}{3}n + \frac{1}{9}$ |

38. **CREATE** Write a quadratic equation for which the only solution is 4.

39. **WRITE** Compare and contrast the following strategies for solving $x^2 - 5x - 7 = 0$: completing the square, graphing, and factoring.

Solving Quadratic Equations by Using the Quadratic Formula

Learn Solving Quadratic Equations by Using the Quadratic Formula

Completing the square of the quadratic equation produces a formula that allows you to find the solutions of *any* quadratic equation. This formula is called the Quadratic Formula, $x = \dfrac{-b \pm \sqrt{b^2 - 4ac}}{2a}$.

Example 1 Use the Quadratic Formula

Solve $x^2 - 12x = -11$ by using the Quadratic Formula.

Step 1 Rewrite the equation in standard form.

$$x^2 - 12x = -11 \qquad \text{Original equation}$$

$$x^2 - 12x + 11 = 0 \qquad \text{Add 11 to each side.}$$

Step 2 Apply the Quadratic Formula.

$$x = \dfrac{-b \pm \sqrt{b^2 - 4ac}}{2a} \qquad \text{Quadratic Formula}$$

$$= \dfrac{-(-12) \pm \sqrt{(-12)^2 - 4(1)(11)}}{2(1)} \qquad a = 1, b = -12, c = 11$$

$$= \dfrac{12 \pm \sqrt{144 - 44}}{2} \qquad \text{Multiply.}$$

$$= \dfrac{12 \pm \sqrt{100}}{2} \qquad \text{Subtract.}$$

$$= \dfrac{12 \pm 10}{2} \qquad \text{Take the square root.}$$

$$x = \dfrac{12 + 10}{2} \text{ or } x = \dfrac{12 - 10}{2} \qquad \text{Separate the solutions.}$$

$$= 11 \qquad\qquad = 1 \qquad \text{Simplify.}$$

The solutions are 1 and 11.

Check

Solve the equation by using the Quadratic Formula.

$$x^2 - 19x = -70$$

$$x = \underline{\;\;?\;\;}, \underline{\;\;?\;\;}$$

Today's Goals
- Solve quadratic equations by using the Quadratic Formula.
- Use the discriminant to determine the number of solutions of a quadratic equation.

Today's Vocabulary
discriminant

 Talk About It!

What does the $\pm$ symbol indicate? How does that affect the number of factors determined by using the Quadratic Formula? Explain.

Think About It!

If the expression under the square root simplified to 0, then what would be the relation between the two factors?

Example 2 Use the Quadratic Formula When a Is Not Equal to 1

Watch Out!

Squaring a Negative
For the expression b^2, the result will be positive regardless of whether b is negative or positive.

Solve $2x^2 + 5x = 12$ by using the Quadratic Formula.

Step 1 Rewrite the equation in standard form.

$$2x^2 + 5x - 12 = 0$$

Step 2 Apply the Quadratic Formula.

$$x = \frac{-b \pm \sqrt{b^2 - 4ac}}{2a}$$ Quadratic Formula

$$= \frac{-(5) \pm \sqrt{(5)^2 - 4(2)(-12)}}{2(2)}$$ $a = 2, b = 5, c = -12$

$$= \frac{-5 \pm \sqrt{25 + 96}}{4}$$ Multiply.

$$= \frac{-5 \pm \sqrt{121}}{4}$$ Subtract.

$$= \frac{-5 \pm 11}{4}$$ Take the square root.

$$x = \frac{-5 - 11}{4} \text{ or } x = \frac{-5 + 11}{4}$$ Separate the solutions.

$$= -4 \qquad\qquad = \frac{3}{2}$$ Simplify.

The solutions are -4 and $\frac{3}{2}$.

🌐 Example 3 Solve a Quadratic Equation with Irrational Roots

HEALTH The normal blood pressure of an adult woman P in millimeters of mercury given her age t in years can be modeled by the equation $P = 0.01t^2 + 0.5t + 107$. If Seiko's blood pressure is 120, how old might she be?

Step 1 Rewrite the equation in standard form.

$$0.01t^2 + 0.5t + 107 = 120$$ Original equation

$$0.01t^2 + 0.5t - 13 = 0$$ Subtract 120 from each side.

Step 2 Apply the Quadratic Formula.

$$x = \frac{-b \pm \sqrt{b^2 - 4ac}}{2a}$$ Quadratic Formula

$$= \frac{-0.5 \pm \sqrt{(0.5)^2 - 4(0.01)(-13)}}{2(0.01)}$$ $a = 0.01, b = 0.5, c = -13$

$$= \frac{-0.5 \pm \sqrt{0.25 + 0.52}}{0.02}$$ Multiply.

$$= \frac{-0.5 \pm \sqrt{0.77}}{0.02}$$ Simplify.

$$x = \frac{-0.5 + \sqrt{0.77}}{0.02} \text{ or } x = \frac{-0.5 - \sqrt{0.77}}{0.02}$$ Separate the solutions.

$$\approx 18.9 \qquad\qquad\qquad \approx -68.9$$ Simplify.

The solutions are 18.9 and -68.9.

🔵 **Go Online** You can complete an Extra Example online.

Step 3 Eliminate unreasonable solutions.

Because t measures age in years, it is reasonable to assume that Seiko is 18.9 years old.

Because t measures age in years, which cannot be negative, -68.9 is an unreasonable solution.

Check

FOOTBALL The quarterbacks on a high school football team want to see who can get the most hang-time when throwing the ball. Connor knows that using his initial velocity and height, he can model the trajectory of his throw with the equation $h = -16t^2 + 95t + 5.2$, where h is the height of the ball and t is the time after throwing. Solve the equation by using the Quadratic Formula, eliminating any extraneous solutions. Round your answer(s) to the nearest tenth.

$t = \underline{\ ?\ }$

Think About It!

If $P = 0$, then there are no real solutions to the equation. Does this make sense in the context of the situation?

Explore Deriving the Quadratic Formula Algebraically

Online Activity Use an interactive tool to complete the Explore.

> **INQUIRY** How does having a formula make it possible to solve quadratic equations where the other methods are not easy to apply?

Explore Deriving the Quadratic Formula Visually

Online Activity Use an interactive tool to complete the Explore.

> **INQUIRY** Why can you use the Quadratic Formula to solve any quadratic equation?

Go Online You can complete an Extra Example online.

Learn The Discriminant

In the Quadratic Formula, $x = \frac{-b \pm \sqrt{b^2 - 4ac}}{2a}$, the expression under the radical sign, $b^2 - 4ac$, is called the **discriminant**. You can use the discriminant to determine the number of real solutions of a quadratic equation.

Key Concept • Using the Discriminant

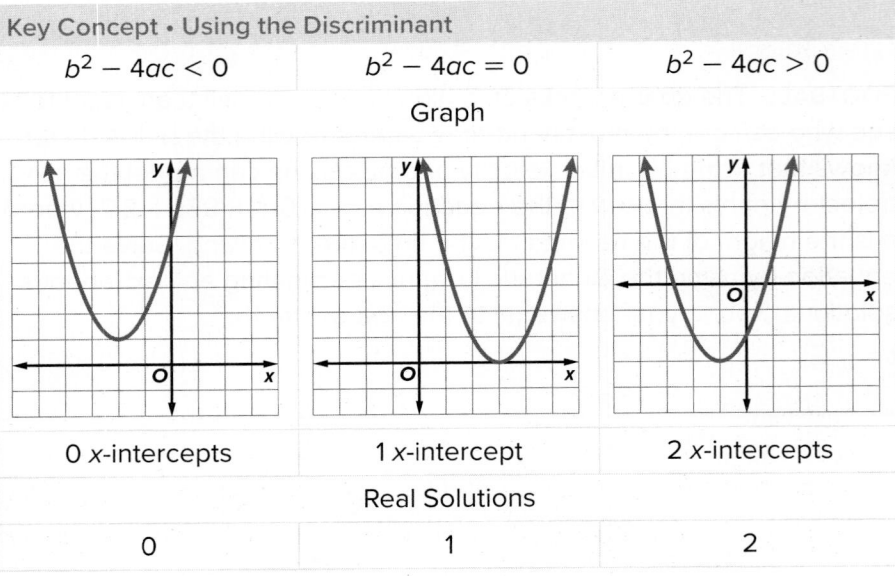

$b^2 - 4ac < 0$	$b^2 - 4ac = 0$	$b^2 - 4ac > 0$
Graph		
0 x-intercepts	1 x-intercept	2 x-intercepts
Real Solutions		
0	1	2

Example 4 Use the Discriminant

State the value of the discriminant of $4x^2 - 3x = -1$. Then determine the number of real solutions of the equation.

Step 1 Rewrite the equation in standard form.

$$4x^2 - 3x = -1 \qquad \text{Original equation}$$

$$4x^2 - 3x + 1 = 0 \qquad \text{Add 1 to each side.}$$

Step 2 Find the discriminant.

$$b^2 - 4ac = (-3)^2 - 4(4)(1) \qquad a = 4, b = -3, c = 1$$

$$= -7 \qquad \text{Simplify.}$$

Because the discriminant is negative, the equation has no real solution.

Check

State the value of the discriminant of $8x^2 - 15x = -9$.

The discriminant is ___?___.

Determine the number of real solutions of the equation.

The equation has ___?___ solution(s).

Go Online to practice what you've learned about solving quadratic equations in the Put It All Together over Lessons 12-3 through 12-6.

Go Online You can complete an Extra Example online.

Examples 1 and 2

Solve each equation by using the Quadratic Formula. Round to the nearest tenth, if necessary.

1. $x^2 - 49 = 0$

2. $x^2 - x - 20 = 0$

3. $x^2 - 5x - 36 = 0$

4. $4x^2 + 5x - 6 = 0$

5. $x^2 + 16 = 0$

6. $6x^2 - 12x + 1 = 0$

7. $5x^2 - 8x = 6$

8. $2x^2 - 5x = -7$

9. $5x^2 + 21x = -18$

10. $81x^2 = 9$

11. $8x^2 + 12x = 8$

12. $4x^2 = -16x - 16$

13. $10x^2 = -7x + 6$

14. $-3x^2 = 8x - 12$

15. $2x^2 = 12x - 18$

Example 3

16. BUSINESS Tanya runs a catering business. Based on her records, her weekly profit can be approximated by the function $f(x) = x^2 + 2x - 37$, where x is the number of meals she caters. If $f(x)$ is negative, it means that the business has lost money. What is the least number of meals that Tanya needs to cater in order to make a profit?

17. AERONAUTICS At liftoff, the space shuttle *Discovery* has a constant acceleration of 16.4 feet per second squared and an initial velocity of 1341 feet per second due to the rotation of Earth. If the distance *Discovery* has traveled t seconds after liftoff is given by the equation $d(t) = 1341t + 8.2t^2$, how long after liftoff has *Discovery* traveled 40,000 feet? Round your answer to the nearest tenth.

18. ARCHITECTURE The Golden Ratio appears in the design of the Greek Parthenon because the width and height of the façade are related by the equation $\frac{W + H}{W} = \frac{W}{H}$. If the height of a model of the Parthenon is 16 inches, what is its width? Round your answer to the nearest tenth.

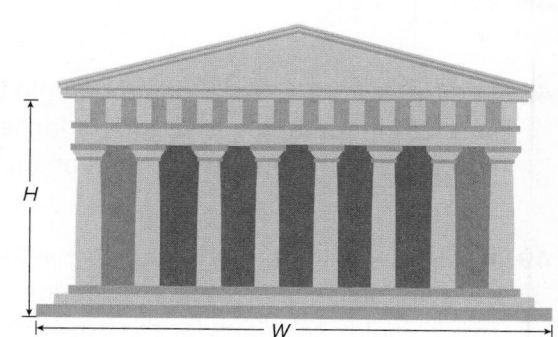

19. CRAFTS Ariadna cut a 60-inch chenille stem into two unequal pieces, and then she used each piece to make a square. The sum of the areas of the squares was 117 square inches. Let x be the length of one piece. Write and solve an equation to represent the situation and find the lengths of the two original pieces.

Example 4

State the value of the discriminant for each equation. Then determine the number of real solutions of the equation.

20. $0.2x^2 - 1.5x + 2.9 = 0$

21. $2x^2 - 5x + 20 = 0$

22. $x^2 - \frac{4}{5}x = 3$

23. $0.5x^2 - 2x = -2$

24. $2.25x^2 - 3x = -1$

25. $2x^2 = \frac{5}{2}x + \frac{3}{2}$

26. $x^2 + 2x + 1 = 0$

27. $x^2 - 4x + 10 = 0$

28. $x^2 - 6x + 7 = 0$

29. $x^2 - 2x - 7 = 0$

30. $x^2 - 10x + 25 = 0$

31. $2x^2 + 5x - 8 = 0$

32. $2x^2 + 6x + 12 = 0$

33. $2x^2 - 4x + 10 = 0$

34. $3x^2 + 7x + 3 = 0$

Mixed Exercises

Solve each equation by using the Quadratic Formula. Round to the nearest tenth, if necessary.

35. $x^2 + 4x = -1$

36. $x^2 - 9x + 22 = 0$

37. $x^2 + 6x + 3 = 0$

38. $2x^2 + 5x - 7 = 0$

39. $2x^2 - 3x = -1$

40. $2x^2 + 5x + 4 = 0$

41. $2x^2 + 7x = 9$

42. $3x^2 + 2x - 3 = 0$

43. $3x^2 - 7x - 6 = 0$

Without graphing, determine the number of x-intercepts of the graph of the related function for each equation.

44. $x^2 + 4x + 3 = 0$

45. $4.25x + 3 = -3x^2$

46. $x^2 + \frac{2}{25} = \frac{3}{5}x$

47. $0.25x^2 + x = -1$

48. RECTANGLES The base of a rectangle is 4 inches greater than the height. The area of the rectangle is 15 square inches. What are the dimensions of the rectangle to the nearest tenth of an inch?

49. REASONING For the equation $3x^2 + 2x + q = 0$, find all values of q so that there are two real solutions to the equation. Then find all values of q so that there are two complex solutions for the equation. Explain.

50. SITE DESIGN The town of Smallport plans to build a new water treatment plant on a rectangular piece of land 75 yards wide and 200 yards long. The buildings and facilities need to cover an area of 10,000 square yards. The town's zoning board wants the site designer to allow as much room as possible between each edge of the site and the buildings and facilities. Let x represent the width of the border.

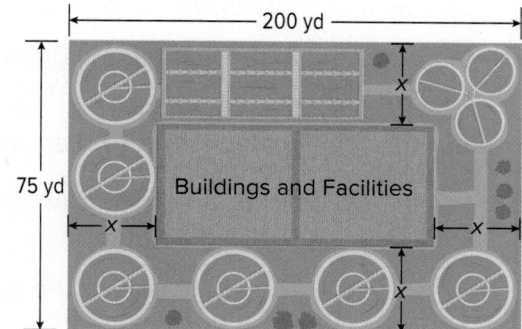

a. Write an equation to represent the area covered by the building and facilities.

b. Write the standard form of the quadratic equation.

c. Find the width of the border. Round your answer to the nearest tenth.

51. USE A MODEL New City's public works is putting together a fireworks presentation to celebrate the Fourth of July. They will be launching fireworks at different heights. The path of one of the fireworks is defined by the equation $h_1 = -16t^2 + 90t + 120$, with t in seconds and h_1 in feet.

a. The equation provides the height the firework is above the ground at any time between when it is launched and when it hits the ground. What does 120 represent in the context of the situation?

b. Write the equation that models when the firework will hit the ground. Solve the equation. Round to the nearest hundredth.

c. How high is the firework when it is at its highest point? How long did it take to get to this point? Round to the nearest hundredth. Explain.

d. Without graphing the equation, use the information from parts **a, b,** and **c** to describe what the graph would look like. Explain.

e. Write and solve an equation for the times at which the firework reaches a height of 200 feet. Round to the nearest tenth.

52. STRUCTURE A cylinder is filled completely with water. Its base has a radius of x cm, and its height is 8 cm. After Ella removes $40x$ cm^3 of water, the volume of the remaining water is 250 cm^3. What is the radius of the cylinder? Round to the nearest whole number, if necessary.

53. REGULARITY Solve the equation $x^2 + bx + c = 0$ for x in terms of b and c. Then determine whether each statement is *sometimes, always,* or *never* true.

a. If both b and c are negative, there will be at least one real solution.

b. If both b and c are positive, there will be complex nonreal solutions.

54. REASONING Braden wrote three quadratic equations: $-2x^2 - 3x - 8 = 0$, $-x^2 + 6x - 9 = 0$, and $-2x^2 + 4x + 3 = 0$. He graphs one of the equations as shown. He shows Ava the graph and the three equations and challenges her to match the correct equation with the graph. Does Ava need to solve all the equations to match one with the graph? Explain your reasoning.

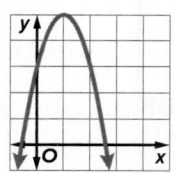

55. STRUCTURE A rectangular area is to be enclosed with its length 20 feet less than its width, as shown.

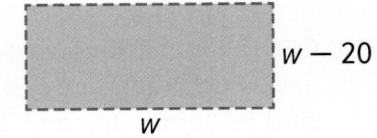

$w - 20$

w

   a. What should the width be in order for the enclosure to have an area of 100 ft²?

   b. Find the width for which the area enclosed is R ft² for $R > 0$.

56. STRUCTURE A frame is made with width w inches and length $15 - w$ inches, as shown in the figure.

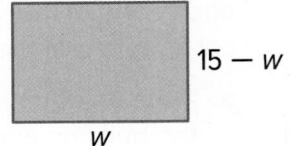

$15 - w$

w

   a. Is there any width for which the frame will have an area of 150 in²? Explain your reasoning.

   b. Find the maximum area of the frame and the dimensions that produce the maximum area.

57. USE A SOURCE The St. Louis Arch can be approximated by a parabola. Research its dimensions and use the dimensions to write a quadratic equation representing the St. Louis Arch. Then set the equation equal to 0 and use the Quadratic Formula to solve for x. Interpret the solutions.

Determine whether there are *two*, *one*, or *no* real solutions of each equation.

58. The graph of the related quadratic function does not have an x-intercept.

59. The graph of the related quadratic function touches, but does not intersect the x-axis.

60. The graph of the related quadratic function intersects the x-axis twice.

61. Both a and b are greater than 0 and c is less than 0 in a quadratic equation.

62. WRITE Why can the discriminant be used to confirm the number of real solutions of a quadratic equation?

63. ANALYZE Use factoring techniques to determine the number of real zeros of $f(x) = x^2 - 8x + 16$. Compare this method to using the discriminant.

64. PERSEVERE Find all values of k such that $2x^2 - 3x + 5k = 0$ has two solutions.

65. WRITE Describe the advantages and disadvantages of each method of solving quadratic equations. Why are the methods equivalent? Which method do you prefer, and why?

66. CREATE Write a quadratic equation that has no real roots, and find its discriminant. Explain how the discriminant shows that a quadratic equation has no real roots.

67. CREATE Write a quadratic equation that has one real root, and find its discriminant. Explain why the Quadratic Formula yields only one solution when the discriminant of a quadratic equation is equal to zero.

68. CREATE Write a quadratic equation that has two real roots. Determine whether completing the square, graphing, or factoring would be the best method to use to solve your quadratic equation.

Solving Systems of Linear and Quadratic Equations

Today's Goals
- Solve systems of linear and quadratic equations by graphing.
- Solve systems of linear and quadratic equations by using algebraic methods.

Explore Using Algebra Tiles to Solve Systems of Linear and Quadratic Equations

🎯 **Online Activity** Use algebra tiles to complete the Explore.

> ⊘ **INQUIRY** How can you use algebra tiles to solve systems of linear and quadratic equations?

Think About It!

Can you put the steps in any other order? Explain your reasoning.

Learn Solving Systems of Linear and Quadratic Equations by Graphing

To solve a system of linear and quadratic equations by graphing, you can follow a set of steps.

Step 1 Graph the quadratic function.

Step 2 Graph the linear function.

Step 3 Find the point(s) of intersection.

Check Substitute the values in the original equation.

Example 1 Solve a System of Linear and Quadratic Equations Graphically

Solve the system of equations by graphing.

$y = x^2 + 4x - 1$

$y = 2x + 2$

Step 1 Graph $y = x^2 + 4x - 1$.

Step 2 Graph $y = 2x + 2$.

Step 3 Find the points of intersection.

The graphs appear to intersect at $(-3, -4)$ and $(1, 4)$

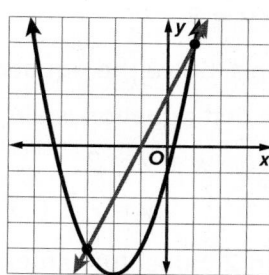

Talk About It!

Cordell says that the graphs of a linear function and a quadratic function always have 2 points of intersection. Do you agree or disagree? Justify your reasoning or provide a counterexample.

🎯 **Go Online** You can complete an Extra Example online.

Check

The graphs of $f(x) = -2x + 3$ and $g(x) = x^2 - 4x - 5$ are shown.

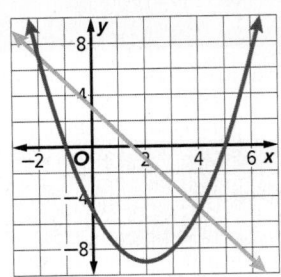

Complete the ordered pair(s) to represent each situation.

$f(x) = 0$ (? , ?)

$g(x) = 0$ (? , ?) (? , ?)

$f(x) = g(x)$ (? , ?) (? , ?)

Think About It!

What does it mean if there is no solution when you substitute one expression for y in the other equation and solve?

Learn Solving Systems of Linear and Quadratic Equations Algebraically

To solve a system of linear and quadratic equations algebraically, you can follow a set of steps.

Step 1 Solve the equations for y.

Step 2 Substitute one expression for y in the other equation.

Step 3 Solve for x.

Step 4 Substitute the x-value(s) in either equation.

Step 5 Solve for y.

Check Graph the equations.

Example 2 Solve a System of Linear and Quadratic Equations Algebraically

Solve the system of equations algebraically.

$$y = x^2 - 2x - 3$$

$$x + y = 3$$

Step 1 Solve the equations for y.

The first equation is already solved for y.

$x + y = 3$	Original second equation
$y = -x + 3$	Subtract x from each side.

Step 2 Substitute an expression for y.

$y = -x + 3$	Second equation solved for y
$x^2 - 2x - 3 = -x + 3$	Substitution

Go Online You can complete an Extra Example online.

Step 3 Solve for x.

$x^2 - 2x - 3 = -x + 3$	Original equation
$x^2 - x - 3 = +3$	Add x to each side.
$x^2 - x - 6 = 0$	Subtract 3 from each side.
$(x - 3)(x + 2) = 0$	Factor.
$x - 3 = 0 \quad x + 2 = 0$	Zero Product Property
$x = 3 \qquad x = -2$	Simplify.

Step 4 Substitute the x-value(s) in either equation.

$$y = -x + 3 \qquad\qquad y = -x + 3$$
$$y = -3 + 3 \qquad\qquad y = -(-2) + 3$$

Step 5 Solve for y.

$$y = -3 + 3 \qquad\qquad y = -(-2) + 3$$
$$y = 0 \qquad\qquad\qquad y = 5$$

Check Graph the equations.

The graphs of the functions intersect at $(3, 0)$ and $(-2, 5)$, so $(3, 0)$ and $(-2, 5)$ are solutions of this system.

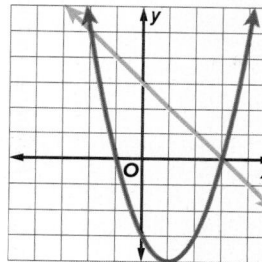

Check

Solve the system of equations algebraically.

$$y = x^2 - 8x + 19 \qquad x - 0.5y = 3$$

$(\underline{\ ?\ }, \underline{\ ?\ })$

🌐 **Example 3** Use a System of Linear and Quadratic Equations

SALES Bathing suit sales at a clothing store can be modeled by the function $y = -x^2 + 12x + 25$, and gift card sales can be modeled by the function $y = 5x + 7$, where x represents the number of months past January and y represents the total revenue in thousands of dollars. Solve a system of equations algebraically to find the month when the revenue from bathing suit sales is equal to the revenue from gift card sales.

(continued on the next page)

▶ **Go Online** You can complete an Extra Example online.

Set the expressions equal to each other and solve for x.

$-x^2 + 12x + 25 = 5x + 7$	Substitute.
$12x + 25 = x^2 + 5x + 7$	Add $-x^2$ to each side.
$25 = x^2 - 7x + 7$	Subtract $12x$ from each side.
$0 = x^2 - 7x - 18$	Subtract 25 from each side.
$0 = (x - 9)(x + 2)$	Factor.
$x - 9 = 0 \qquad x + 2 = 0$	Zero Product Property
$x = 9 \qquad\quad x = -2$	Simplify.

Because the expressions are equivalent when $x = 9$, the revenue of bathing suit sales and gift card sales is equal 9 months past January, or in October.

Graph the equations to check your solution.

Sales

Check

SALES Revenue from single-day ticket sales to a local amusement park can be modeled by the function $y = -x^2 + 11x + 42$, and revenue from season pass sales can be modeled by the function $y = -5x + 81$, where x represents the number of months past January and y represents the total revenue in tens of thousands of dollars. Solve a system of equations algebraically to find the first month when the revenue from single-day tickets is equal to the revenue from season pass sales.

A. March

B. April

C. May

D. June

 Go Online You can complete an Extra Example online.

Think About It!

Find the revenue from bathing suit sales in the month when the revenue from bathing suit sales is equal to the revenue from gift card sales.

Watch Out!

Labeling Axes The x-axis represents the months past January, so January is represented by $x = 0$, not $x = 1$.

Practice

Go Online You can complete your homework online.

Example 1

Solve each system of equations by graphing.

1. $y = x^2 - 4$
$y = -3$

2. $y = x^2 + x - 2$
$y = -x + 1$

3. $y = 2x^2 + 1$
$y = 1$

4. $y = x^2 + 3x + 1$
$y = x + 1$

Example 2

Solve each system of equations algebraically.

5. $y = x^2 - 2x - 5$
$y = 3$

6. $y = x^2 + 4x - 1$
$y = 3x + 1$

7. $y = x^2 - 6x + 5$
$x + y = -1$

8. $y = x^2 + x + 1$
$y - 1 = x$

9. $y + 3x = x^2 - 3$
$y = -2x + 3$

10. $y - 1 = 2x^2 - x$
$-2x + y = 3$

Example 3

11. **GYMNASTICS** A gymnast throws a baton up into the air during her floor routine. The height of the baton can be modeled by $y = -16x^2 + 24x + 4$, where x is the time in seconds and y is the height of the baton in feet since it was released. The ceiling of the gym can be represented by the function $y = 42$.

   a. Solve the system of equations algebraically.

   b. Will the baton reach the ceiling of the gym? Explain your reasoning.

12. **DISC GOLF** The height y in feet of a disc x seconds after it was thrown can be modeled by $y = -\frac{1}{30}x^2 + \frac{1}{2}x + 5\frac{2}{15}$. The height of Rodrigo's hands as he runs to catch the disc can be represented by the function $y = \frac{1}{2}x + 3$.

   a. Solve the system of equations algebraically.

   b. At what height will Rodrigo catch the disc?

13. **ZOO** Revenue from single-day ticket sales at a local zoo can be modeled by the function $y = -x^2 + 25x + 80$, and revenue from season pass sales can be modeled by the function $y = -4x + 200$. In both functions, x represents the number of months past January, and y represents the total revenue in thousands of dollars. Solve the system of equations algebraically to find the first month when the revenue from single-day ticket sales is equal to the revenue from season pass sales.

Mixed Exercises

Solve each system of equations.

14. $y = x^2$
$y = 2x$

15. $y = -2x^2 + 7x - 2$
$y = 3 - 4x$

16. $y = -x^2 + 4$
$y = \frac{1}{5}x + 5$

17. $y = -x^2 + 4x - 4$
$y = 2x - 3$

18. $y = -x^2 - 5x - 6$
$y = -3x - 1$

19. $y = 2x^2 - 4$
$y = 2x$

20. $y = x^2 + 7x + 12$
$y = 2x + 8$

21. $y = x^2 - x - 20$
$y = 3x + 12$

22. $y = 3x^2 - x - 2$
$y = -2x + 2$

23. $y = x^2 - x - 18$
$y = x - 3$

24. $y = x^2 - 3x + 1$
$y = x + 1$

25. $y = x^2 - 4x + 6$
$y = 2x - 3$

26. $y = -x^2 + 2x + 3$
$y = x + 2$

27. $y = x^2 + 4x - 1$
$y = 3x$

Solve each equation by using a system of equations.

28. $x^2 - 4x + 3 = x - 1$

29. $x^2 + 2x - 1 = x - 2$

30. $x^2 + 4x + 4 = 4$

31. $x^2 = -2x - 1$

32. $\frac{1}{2}x^2 - 4 = 3x + 4$

33. $x^2 + 5x + 5 = -x - 8$

34. WINGSUIT A wingsuit flyer jumps off a tall cliff. He falls freely for a few seconds
before deploying the wingsuit and slowing his descent. His height during the
freefall can be modeled by the function $y = -4.9x^2 + 420$, where y is the height
above the ground in meters and x is the time in seconds. After deploying the
wingsuit, the flyer's height is given by the function $y = -3x + 200$.

   a. To the nearest whole number, how long after jumping does the flyer deploy the
   wingsuit?

   b. To the nearest whole number, what was the flyer's height when he deployed
   the wingsuit?

35. REASONING The graph shows a quadratic function and a linear function $y = k$.

a. How many solutions are there to the system?

b. If the linear function were changed to $y = k + 5$, how many solutions would the system have?

c. If the linear function were changed to $y = k - 2$, how many solutions would the system have?

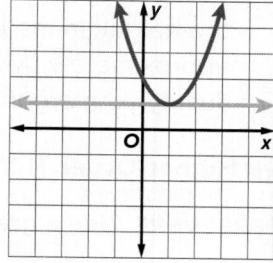

36. USE A MODEL The function $R(x) = -x^2 + 5x + 14$ represents the revenue earned by a manufacturing company, where R is the revenue in millions of dollars and x is the number of items produced in millions. It cost the company $4 million to produce 3 million items.

a. Write a function to represent the cost C in millions of dollars of producing x million items.

b. Solve the system involving the revenue function and cost function. What does the solution represent?

37. USE TOOLS In a video game, players fling virtual rubber bands at a small target that moves around the screen. The rubber bands are all launched from the point $(-4, 0)$ and follow a path along the line $y = \frac{1}{2}x + 2$. The target moves around the screen following a parabolic path represented by the equation $y = -x^2 - 2x + 4$.

a. Graph the path of the rubber bands and the path of the target.

b. If a player hits the target, at approximately what point or points on the plane will this take place? Explain.

c. Use your calculator to find the coordinates of the point or points at which the rubber bands may hit the target. Round to the nearest tenth. How do your results compare to your answer in part **b**?

38. ROCKETS Jamie is using binoculars to watch a friend set off a model rocket from a distance. Jamie's line of sight through the binoculars is a straight line from 50 meters above the launch point, decreasing $\frac{1}{2}$ a meter in height for every 1 meter of lateral distance to the point where Jamie is standing. The rocket's path follows a projectile motion formula, $-4.9t^2 + vt + h$, where v is the initial velocity in meters per second and h is the launch height in meters. The rocket is launched at a velocity of 40 meters per second from the ground. At what height(s), in meters, will the rocket be in Jamie's line of sight?

a. Write a system of equations to represent the situation.

b. Find the solution to the system of equations graphically.

c. Find the solution to the system of equations algebraically.

d. Explain what the solution(s) mean in the context of the situation.

39. REGULARITY Explain a method for determining the number of solutions that exist for a system of equations from a graph of the related functions.

40. STATE YOUR ASSUMPTION Kasia found two solutions to a system of equations she used to determine whether her slinky would travel far enough to get to the next stair. One solution has a negative x-value, and one solution has a positive x-value. Which solution should Kasia use? What assumption did you make to decide?

41. CONSTRUCT ARGUMENTS Rafaela kicked a rock off the top of a cliff to a deserted river below. The rock's path is modeled by $h(x) = -0.1x^2 + x + 600$, where $h(x)$ is the height, in meters, of the rock after x seconds. The cliff face is modeled approximately by $h(x) = 600 - 9x$. Will the rock land on the cliff or in the river? Justify your argument.

Higher-Order Thinking Skills

PERSEVERE Use a graphing calculator to solve other types of systems.

42. $y = x^2 + 3x - 5$
$y = -x^2$

43. $y = \frac{3}{4}x$
$x^2 + y^2 = 1$ (*Hint*: Enter as two functions, $y = \sqrt{1 - x^2}$ and $y = -\sqrt{1 - x^2}$.)

44. ANALYZE For what value of k does the system $y = x^2 + x$ and $y = -2x + k$ have exactly one solution? Explain your reasoning.

45. ANALYZE For what value of k does the system $y = x^2 + 2$ and $y = 3x + k$ have exactly one solution? Explain your reasoning.

46. CREATE Below are a graph and a table of values that describe two equations, y_1 and y_2, one linear and one quadratic. Write a problem that could be solved using the graph and table of values, then solve your problem.

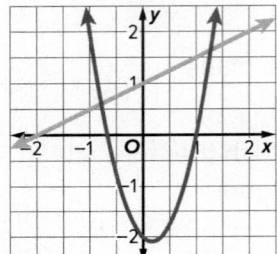

x	y_1	y_2
-2	0	12
0	1	-2
1	1.5	0

47. WRITE What is the maximum number of solutions that a system of one linear and one quadratic equation can have? What is the maximum number of solutions to a system of two linear functions? Write a paragraph to explain your reasoning.

48. FIND THE ERROR Xavier found the solution to a system of equations using the steps shown. In which step did Xavier make an error? Correct his error.

$y = x^2 - 4x - 6$
$y = 9x + 8$

Step 1: $x^2 - 4x - 6 = 9x + 8$
Step 2: $x^2 - 4x - 9x - 6 - 8 = 0$
Step 3: $x^2 - 13x - 14 = 0$
Step 4: $(x - 14)(x + 1) = 0$

Step 5: $y = 14, y = -1$
Step 6: $14 = 9x + 8; x = \frac{2}{3}$
Step 7: $-1 = 9x + 8; x = -1$

49. WHICH ONE DOESN'T BELONG? Analyze the three systems of equations given, and identify which system does not belong. Explain your reasoning.

$y = x^2 - 3x - 4$
$y = 5x + 1$

$y = -x^2 - x + 11$
$y = 2x - 10$

$y = 4x^2 + 4x - 2$
$y = x - 7$

Modeling and Curve Fitting

Explore Using Differences and Ratios to Model Data

Online Activity Use an interactive tool to complete the Explore.

> **⊘ INQUIRY** How can differences and ratios of successive y-values be used to write a model?

Learn Modeling Real-World Situations

Different types of functions can be used to model data. You can use the equation or graph of a data set to determine the type of function that represents the data.

Concept Summary • Linear and Nonlinear Functions

Linear	Quadratic	Exponential
Function		
$y = mx + b$	$y = ax^2 + bx + c$	$y = ab^x$, where $b > 0$
Graph		
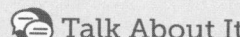		
Description		
y increases or decreases at a constant rate.	y increases or decreases by a square of x.	y increases or decreases by a power of x.
Differences/Ratios		
First differences are equal.	Second differences are equal.	Successive ratios are equal.
Uses of Difference/Ratio		
The first difference is the slope m.	Half of the second difference is a.	The common ratio is the base b of the function.

Go Online You can complete an Extra Example online.

Example 1 Determine a Model by Using First Differences

Think About It!

How would the process of finding the function that represents the data below differ from this example?

x	9.75	10	10.25	10.5
y	−9	−5	−1	3

Look for a pattern in the data table to determine the kind of model that best describes the data. Then, write the function.

x	5	6	7	8	9
y	−1	−4	−7	−10	−13

Part A Determine the model.

Notice that the x-values are increasing by 1, so the model can be determined by examining successive differences or ratios.

$$-1 \quad -4 \quad -7 \quad -10 \quad -13$$
$$-3 \quad -3 \quad -3 \quad -3$$

The first differences are equal. This means that the data can be represented by a linear function.

Part B Write the function.

Substitute for m.

$y = mx + b$	Slope-intercept form of a linear function
$y = -3x + b$	Since the first difference is −3, the slope is −3.

Find the y-intercept.

$-1 = -3(5) + b$	$(x_1, y_1) = (5, -1)$
$-1 = -15 + b$	Simplify.
$14 = b$	Add 15 to each side.

Write the equation in slope-intercept form.

$y = mx + b$	Slope-intercept form
$y = -3x + 14$	Replace m with −3 and b with 14.

The data are modeled by $y = -3x + 14$.

Go Online You can complete an Extra Example online.

Example 2 Determine a Model by Using Second Differences

Look for a pattern in the data table to determine the kind of model that best describes the data. Then, write the function.

x	−2	−1	0	1	2
y	−1	0	4	11	21

Part A Determine the model.

The x-values are increasing by 1, so the model can be determined by examining successive differences or ratios.

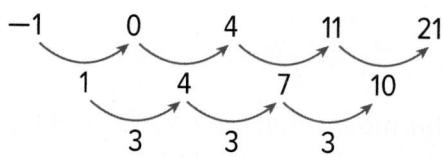

The second differences are equal. This means that the data can be modeled by a quadratic function.

Part B Write the function.

$y = ax^2 + bx + c$	Standard form of a quadratic function
$y = 1.5x^2 + bx + 4$	a is one-half of the second difference, and c is the y-intercept from the given data point.

Find b.

$-1 = 1.5(-2)^2 + b(-2) + 4$	Use data points $(-2, -1)$ for (x, y).
$-1 = 6 - 2b + 4$	Simplify.
$-1 = 10 - 2b$	Simplify.
$-11 = -2b$	Subtract 10 from each side.
$5.5 = b$	Divide each side by −2.

Write the function in standard form.

$y = ax^2 + bx + c$	Standard form of a quadratic function
$y = 1.5x^2 + 5.5x + 4$	$a = 1.5, b = 5.5, c = 4$

The data are modeled by $y = 1.5x^2 + 5.5x + 4$.

 Go Online You can complete an Extra Example online.

Example 3 Determine a Model by Using Ratios

Look for a pattern in the data table to determine the kind of model that best describes the data. Then, write the function.

x	3	4	5	6	7
y	400	100	25	6.25	1.5625

Part A Determine the model.

Since the *x*-values are increasing by 1, the model can be determined by examining successive differences or ratios.

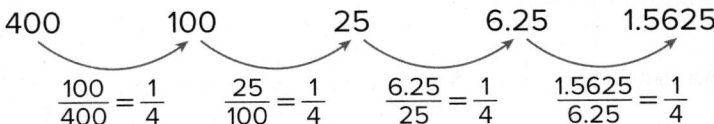

400   100   25   6.25   1.5625

−300   −75   −18.75   −4.6875

225   56.25   14.0625

The first differences are not equal. This means that the data cannot be modeled by a linear function. The second differences are not equal. This means that the data cannot be modeled by a quadratic function.

400   100   25   6.25   1.5625

$$\frac{100}{400} = \frac{1}{4} \qquad \frac{25}{100} = \frac{1}{4} \qquad \frac{6.25}{25} = \frac{1}{4} \qquad \frac{1.5625}{6.25} = \frac{1}{4}$$

The ratios of successive *y*-values are equal. Therefore, the data can be modeled by an exponential function.

Part B Write the function.

$$y = a(b)^x \qquad\qquad \text{Exponential function}$$

$$y = a\left(\frac{1}{4}\right)^x \qquad\qquad \text{Since the ratio is } \frac{1}{4}, b = \frac{1}{4}.$$

Solve for *a*.

$$400 = a\left(\frac{1}{4}\right)^3 \qquad\qquad (x_1, y_1) = (3, 400)$$

$$400 = \left(\frac{1}{64}\right)a \qquad\qquad \text{Simplify.}$$

$$25{,}600 = a \qquad\qquad \text{Multiply each side by 64.}$$

Write the equation.

$$y = a(b)^x \qquad\qquad \text{Exponential function}$$

$$y = 25{,}600\left(\frac{1}{4}\right)^x \qquad\qquad \text{Replace } a \text{ with 25,600 and } b \text{ with } \frac{1}{4}.$$

The data are modeled by $y = 25{,}600\left(\frac{1}{4}\right)^x$.

Check

Determine the kind of model that best describes the data set. Then write the function.

x	0	1	2	3
y	−3	−6	−12	−24

Go Online You can complete an Extra Example online.

Watch Out!

Constant Difference
Before checking differences and ratios, check that the *x*-values are increasing or decreasing by a constant value. In order to use the differences or ratios in the model function, *x* must be increasing by 1.

Learn Curve Fitting

You can use a graphing calculator to find a regression equation for a set of data that is approximated by a function. This process is called **curve fitting.**

The **coefficient of determination,** R^2, indicates how well the function fits the data. The closer R^2 is to 1, the better the model.

The table shows a scatter plot and graphs of linear, exponential, and quadratic functions. Which type of function is the best fit for the data?

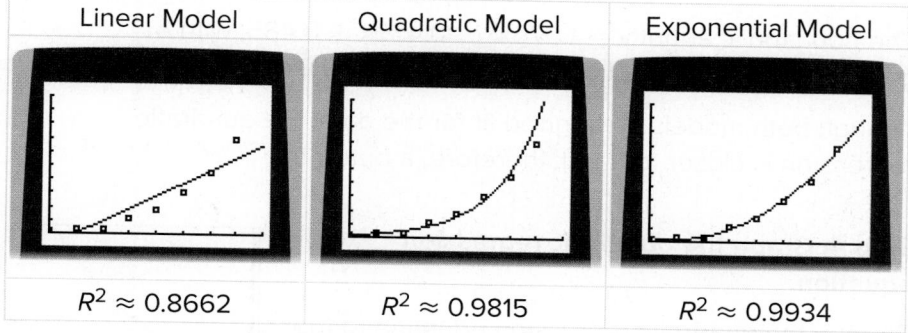

Linear Model	Quadratic Model	Exponential Model
$R^2 \approx 0.8662$	$R^2 \approx 0.9815$	$R^2 \approx 0.9934$

Based on the coefficients of determination, the data are best modeled by a quadratic function.

🌐 Example 4 Find the Best Model

VIDEO STREAMING **The table shows the number of households worldwide that subscribe to a video streaming service. Write a model that fits the data.**

Year	2007	2008	2009	2010	2011	2012	2013	2014	2015
Households (millions)	8	9	12	19	23	33	44	57	75

Step 1 Enter the data.

Enter the data by pressing STAT and selecting the Edit option.

Let the year 2007 be represented by 0.

Enter the years since 2007 into List 1 (L1).

Enter the number of households in List 2 (L2).

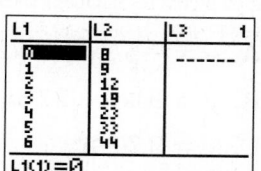

Step 2 Make a scatter plot.

Graph the scatter plot. Turn on Plot 1 under the STAT PLOT menu and choose the scatter plot feature.

Change the viewing window so that all data are visible by pressing ZOOM and then selecting ZoomStat.

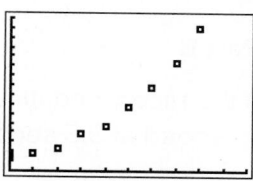

(continued on the next page)

🡢 **Go Online**

You can watch a video to see how to fit a curve to a set of data.

🡢 **Go Online**

to see how to use a graphing calculator with this example.

Use a Source

Use an outside source to find the number of subscribers or members of other services, such as ride-sharing companies, Internet service, or social media networks, over time. Then determine whether the data can be modeled by a linear, quadratic, exponential function, or none of these.

Step 3 Find the regression equation.

Exponential Regression

Select ExpReg and press ENTER.

The equation is about $y = 7.306(1.342)^x$, with a coefficient of determination of approximately 0.994.

Quadratic Regression

Select Quadreg and press ENTER.

The equation is about $y = 1.076x^2 - 0.439x + 8.485$, with a coefficient of determination of approximately 0.998.

Though both models are a good fit for the data, the quadratic regression is closer to 1 and, therefore, a better fit.

Step 4 Graph the quadratic regression equation.

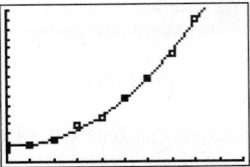

To copy the quadratic regression equation to the Y= list, press **Y=**, **VARS**, and choose **Statistics**.

From the **EQ** menu, choose **RegEQ**.

Press GRAPH.

Check

RACING The table shows the speed in kilometers per hour of a racecar after x seconds.

Time x	0.5	1	1.5	2	2.5
Speed y	12	32	58	88	122

Part A

Select the model that best fits the data.

A. $y = -5.3x^2 + 0.2$

B. $y = 9.1x^2 + 27.8x - 4.4$

C. $y = 8.7(3.1)^x$

D. $y = 0.6(1.0)^x$

Part B

If the racecar continues to accelerate following the same trend, predict its speed at 3.5 seconds. Round to the nearest kilometer.

____?____ kph

📡 Go Online You can complete an Extra Example online.

📡 **Go Online**
to learn about exponential growth patterns in Expand 12-8.

Practice

Go Online You can complete your homework online.

Examples 1–3

Look for a pattern in each table of values to determine which kind of model best describes the data. Then write an equation for the function that models the data.

1.

x	−3	−2	−1	0
y	−8.8	−8.6	−8.4	−8.2

2.

x	−2	−1	0	1	2
y	10	2.5	0	2.5	10

3.

x	−1	0	1	2	3
y	0.75	3	12	48	192

4.

x	−2	−1	0	1	2
y	0.008	0.04	0.2	1	5

5.

x	0	1	2	3	4
y	0	4.2	16.8	37.8	67.2

6.

x	−3	−2	−1	0	1
y	14.75	9.75	4.75	−0.25	−5.25

7.

x	−3	−2	−1	0	1	2
y	32	16	8	4	2	1

8.

x	−1	0	1	2	3
y	7	3	−1	−5	−9

9.

x	−3	−2	−1	0	1
y	−27	−12	−3	0	−3

10.

x	−2	−1	0	1	2
y	−8	−4	0	4	8

11.

x	0	1	2	3	4
y	0.5	1.5	4.5	13.5	40.5

12.

x	−1	0	1	2	3
y	27	9	3	1	$\frac{1}{3}$

13.

x	2	3	4	5	6
y	12	27	48	75	108

14.

x	0	1	2	3	4
y	80	73	66	59	52

Example 4

15. DRONES The table shows the time after a drone was launched (in seconds) x and the height of the drone, in feet, above the ground y.

x	y
1	30
2	40
3	50
4	55
5	50
6	40

   a. Make a scatter plot of the data.

   b. Which regression equation has an R^2 value closest to 1?

   c. Find an appropriate regression equation, and state the coefficient of determination. Based on the regression equation, what are the relevant domain and range?

   d. Predict the height of the drone 7 seconds after it was launched. Round to the nearest foot.

16. BAKING Alyssa baked a cake and is waiting for it to cool so she can ice it. The table shows the temperature of the cake every 5 minutes after Alyssa took it out of the oven.

Time (min)	Temperature (°F)
0	350
5	244
10	178
15	137
20	112
25	96
30	89

a. Make a scatter plot of the data.

b. Which regression equation has an R^2 value closest to 1? Is this the equation that best fits the context of the problem? Explain your reasoning.

c. Find an appropriate regression equation, and state the coefficient of determination. Based on the regression equation, what are the relevant domain and range?

d. Alyssa will ice the cake when it reaches room temperature (70°F). Use the regression equation to predict when she can ice her cake.

Mixed Exercises

USE TOOLS Use a graphing calculator to determine whether to use a *linear*, *quadratic*, or *exponential* regression equation. State the coefficient of determination.

17.

x	y
0	1.1
2	3.3
4	2.9
6	5.6
8	11.9
10	19.8

18.

x	y
1	1.67
5	2.59
9	4.37
13	6.12
17	5.48
21	3.12

19.

x	y
−2	0.2
−1	0.5
0	1
1	2
2	4
3	7.5

20. WEATHER The San Mateo weather station records the amount of rainfall since the beginning of a thunderstorm. Data for a storm is recorded as a series of ordered pairs (2, 0.3), (4, 0.6), (6, 0.9), (8, 1.2), (10, 1.5), where the x-value is the time in minutes since the start of the storm, and the y-value is the amount of rain in inches that has fallen since the start of the storm. Determine which kind of model best describes the data.

21. INVESTING The value of a certain parcel of land has been increasing in value ever since it was purchased. The table shows the value of the land parcel over time.

Year Since Purchasing	0	1	2	3	4
Land Value (thousands $)	$1.05	$2.10	$4.20	$8.40	$16.80

Look for a pattern in the table of values to determine which model best describes the data. Then write an equation for the function that models the data.

22. BOATS The value of a boat typically depreciates over time. The table shows the value of a boat over a period of time.

Years	0	1	2	3	4
Boat Value ($)	8250	6930	5821.20	4889.81	4107.44

Write an equation for the function that models the data. Then use the equation to determine how much the boat is worth after 9 years.

23. NUCLEAR WASTE Radioactive material slowly decays over time. The amount of time needed for an amount of radioactive material to decay to half its initial quantity is known as its half-life. Consider a 20-gram sample of a radioactive isotope.

Half-Lives Elapsed	0	1	2	3	4
Amount of Isotope Remaining (grams)	20	10	5	2.5	1.25

a. Is radioactive decay a *linear* decay, a *quadratic* decay, or an *exponential* decay?

b. Write an equation to determine how many grams y of the radioactive isotope will be remaining after x half-lives.

c. How many grams of the isotope will remain after 11 half-lives?

d. Plutonium-238 is one of the most dangerous waste products of nuclear power plants. If the half-life of plutonium-238 is 87.7 years, how long would it take for a 20-gram sample of plutonium-238 to decay to 0.078 gram?

24. USE A MODEL The table shows the populations of two towns from 2015 to 2018.

a. For each town, write a function that models the town's population x years after 2015.

b. Based on the function types you used in **part a**, what can you conclude about the populations of the towns as time goes on? Explain.

Year	Population	
	Dixon	Midville
2015	80,000	96,000
2016	84,000	96,070
2017	88,200	96,280
2018	92,610	96,630

25. **STRUCTURE** The table shows the height of an elevator above ground level at various times.

Time (s), x	0	1	2	3	4
Height (ft), y	142	124	106	88	70

a. What type of function best models the data in the table? Why?

b. The average rate of change is the change in the value of the dependent variable divided by the change in the value of the independent variable. What is the average rate of change of the function over the interval $x = 0$ to $x = 4$? What does this tell you about the motion of the elevator?

c. Write a function that models the height of the elevator as a function of time. How is the coefficient of x related to your answers to **parts a** and **b**?

26. **SOCCER** The table shows the total number of games a soccer team played and the number of total goals scored in a season.

Games Played	Goals Scored
1	1
2	3
3	6
4	10
5	14

a. Make a scatter plot of the data.

b. Which regression equation has an R^2 value closest to 1? What is the value?

c. Find the appropriate regression equation. What are the domain and range?

d. The team's goal is to score 25 goals for the season. Use the regression equation to predict in which game they will score their 25th goal.

27. **USE A SOURCE** Look up hourly temperature data for a starting time of 6:00 A.M. and every hour after that for 8 hours. Plot the points, with x representing the time in hours after 6:00 A.M. and y representing the temperature. Then, identify what type of model best describes the data.

28. **PRECISION** Jase used a graphing calculator to find the predicted dollar value of a used vehicle after 7 years. The calculator reported the value as 5127.45928113. How should Jase decide how many digits to report?

29. **CONSTRUCT ARGUMENTS** Diego claims that no function can have equal non-zero first differences and equal non-zero second differences. Is Diego correct? Use examples or counterexamples to justify your argument.

30. **PERSEVERE** Write a function that has constant second differences, first differences that are not constant, a y-intercept of -5, and contains the point (2, 3).

31. **ANALYZE** What type of function will have constant third differences, but not constant second differences? Justify your argument.

32. **CREATE** Write a linear function that has a constant first difference of 4.

33. **PERSEVERE** Explain why linear functions grow by equal differences over equal intervals, and exponential functions grow by equal factors over equal intervals. (*Hint*: Let $y = ax$ represent a linear function, and let $y = a^x$ represent an exponential function.)

34. **WRITE** How can you determine whether a given set of data should be modeled by a *linear* function, a *quadratic* function, or an *exponential* function?

Combining Functions

Today's Goals
- Combine standard function types by using addition and subtraction.
- Combine standard function types by using multiplication.

Explore Using Graphs to Combine Functions

 Online Activity Use graphing technology to complete the Explore.

> ×
> @ **INQUIRY** How can you use the graphs of functions to determine their sum, difference, or product?

Learn Adding and Subtracting Functions

Some situations are best modeled by the sum or difference of functions.

Consider $f(x) = x^2 - 2x + 4$ and $g(x) = -5x + 1$. To find the value of the sum of the functions when $x = -3$ or $(f + g)(-3)$, you can find $f(-3)$ and $g(-3)$ and add them.

$f(x) = x^2 - 2x + 4$	Original function	$g(x) = -5x + 1$
$f(-3) = (-3)^2 - 2(-3) + 4$	Substitute -3 for x.	$g(-3) = -5(-3) + 1$
$f(-3) = 19$	Simplify.	$g(-3) = 16$

The sum is given by $f(-3) + g(-3) = 19 + 16$ or 35.

To find a function that gives the sum for all values of the domain, add the two functions and combine the like terms.

$$(f + g)(x) = f(x) + g(x) \qquad \text{Addition of functions}$$
$$= (x^2 - 2x + 4) + (-5x + 1) \qquad \text{Substitution}$$
$$= x^2 - 2x - 5x + 4 + 1 \qquad \text{Combine like terms.}$$
$$= x^2 - 7x + 5 \qquad \text{Simplify.}$$

Solve for $(f + g)(-3)$ to check that this method results in the same solution as $f(-3) + g(-3)$.

$$(f + g)(-3) = (-3)^2 - 7(-3) + 5 \qquad \text{Substitute } -3 \text{ for } x.$$
$$= 9 + 21 + 5 \qquad \text{Simplify.}$$
$$= 35 \qquad \text{Simplify.}$$

So, $f(x) + g(x) = (f + g)(x)$.

Example 1 Add Functions

Given $f(x) = 3^x + 1$ and $g(x) = 6x^2 + x - 2$, find $(f + g)(x)$.

$$(f + g)(x) = f(x) + g(x) \qquad \text{Addition of functions}$$
$$= (3^x + 1) + (6x^2 + x - 2) \qquad \text{Substitution}$$
$$= 3^x + 6x^2 + x - 1 \qquad \text{Simplify.}$$

Think About It!

What similarities do you notice about $(f - g)(x)$ and $(g - f)(x)$? Will those similarities exist between any two functions? Justify your answer.

Example 2 Subtract Functions

Given $f(x) = -x^2 + 4x - 5$ and $g(x) = x - 7$, find each function.

a. $(f - g)(x)$

$(f - g) = f(x) - g(x)$

$= (-x^2 + 4x - 5) - (x - 7)$

$= -x^2 + 4x - 5 - x + 7$

$= -x^2 + 3x + 2$

b. $(g - f)(x)$

$(g - f) = g(x) - f(x)$

$= (x - 7) - (-x^2 + 4x - 5)$

$= x - 7 + x^2 - 4x + 5$

$= x^2 - 3x - 2$

Check

Given $f(x) = 5 \cdot 2^x - 1$, $g(x) = -3x^2 + 2x - 8$, and $h(x) = 4x - 5$, find each function. Write each answer in standard form.

a. $(f + h)(x)$

b. $(h - g)(x)$

Think About It!

How could you use the degrees of polynomials to determine the degree of their combined function?

Learn Multiplying Functions

Consider $f(x) = x^2 + x + 3$ and $g(x) = 4x$. To find $(f \cdot g)(2)$ you can find the product of $f(2)$ and $g(2)$.

$f(x) = x^2 + x + 3$	Original function	$g(x) = 4x$
$f(2) = (2)^2 + (2) + 3$	Substitute 2 for x.	$g(2) = 4(2)$
$f(2) = 9$	Simplify.	$g(2) = 8$

$f(2) \cdot g(2) = 9 \cdot 8$ or 72.

Multiply the two functions and combine like terms.

$(f \cdot g)(x) = f(x) \cdot g(x)$	Multiplication of functions
$= (x^2 + x + 3) \cdot (4x)$	Substitution
$= x^2(4x) + x(4x) + 3(4x)$	Distributive property.
$= 4x^3 + 4x^2 + 12x$	Simplify.

Solve for $(f \cdot g)(2)$.

$(f \cdot g)(2) = 4(2)^3 + 4(2)^2 + 12(2)$	Substitute 2 for x.
$= 32 + 16 + 24$ or 72	Simplify.

So, $f(x) \cdot g(x) = (f \cdot g)(x)$.

Go Online You can complete an Extra Example online.

Example 3 Multiply Linear and Quadratic Functions

Given $f(x) = x^2 + 2x - 7$ and $g(x) = 3x - 10$, find $(f \cdot g)(x)$.

$(f \cdot g)(x) = f(x) \cdot g(x)$

$\qquad = (x^2 + 2x - 7) \cdot (3x - 10)$

$\qquad = x^2(3x) + x^2(-10) + 2x(3x) + 2x(-10) - 7(3x) - 7(-10)$

$\qquad = 3x^3 - 10x^2 + 6x^2 - 20x - 21x + 70$

$\qquad = 3x^3 - 4x^2 - 41x + 70$

Think About It!

Consider the degree of the polynomials $f(x)$ and $g(x)$. Does the degree of the product meet your expectations? Explain.

Example 4 Multiply Linear and Exponential Functions

Given $g(x) = 3x - 10$ and $h(x) = \left(\frac{1}{3}\right)^x + x$, find $(h \cdot g)(x)$.

$(h \cdot g)(x) = h(x) \cdot h(x)$

$\qquad = \left[\left(\frac{1}{3}\right)^x + x\right] \cdot (3x - 10)$

$\qquad = \left(\frac{1}{3}\right)^x (3x) + \left(\frac{1}{3}\right)^x (-10) + x(3x) + x(-10)$

$\qquad = 3x\left(\frac{1}{3}\right)^x - 10\left(\frac{1}{3}\right)^x + 3x^2 - 10x$

🌐 Example 5 Combine Functions

FINANCE **According to the Wall Street Journal, the average student loan debt was more than \$35,000 in 2015. Hugo graduated from college with \$28,500 in student loan debt. He decides to defer his payments while he is in graduate school. However, he still accrues interest on his student loans at an annual rate of 5.65%.**

Part A **Write an exponential function $r(t)$ to express the amount of money Hugo owes on his student loans after time t, where t is the number of years after interest began accruing.**

$r(t) = a(1 + r)^t$ Equation for exponential growth

$\qquad = 28{,}500(1 + 0.0565)^t$ $a = 28{,}500$ and $r = 5.65\%$ or 0.0565

$\qquad = 28{,}500(1.0565)^t$ Simplify.

Watch Out!

Percentages $5.65\% = 0.0565$

Part B **While he is in graduate school, Hugo's parents also lend him \$400 a month for rent. His parents decide not to charge him interest on this loan. Write a function $p(t)$ to represent this loan, where t is the time in years that Hugo borrows money from his parents.**

Since t is the time in years, first find the amount of money Hugo borrows from his parents each year.

$12(400) = 4800$

So, $p(t) = 4800t$ represents the loan from his parents as a function of time.

📲 **Go Online**
You can complete an Extra Example online.

(continued on the next page)

Talk About It!
What assumptions did you make in **Part A**?

Part C Find $C(t) = r(t) + p(t)$. What does this new function represent?

$C(t) = 28{,}500(1 + 0.0565)^t + 4800t$

$C(t)$ represents the total amount of money Hugo has to repay after t years.

Part D If Hugo spends 3 years in graduate school, find the total amount of money he will have to repay.

Because Hugo spends 3 years in graduate school, $t = 3$.

$$C(t) = 28{,}500(1 + 0.0565)^t + 4800t$$

$$C(3) = 28{,}500(1 + 0.0565)^3 + 4800(3)$$

$$= 33{,}608.83 + 14{,}400$$

$$= \$48{,}008.83$$

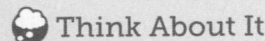 **Example 6** Combine Two Functions

SANDWICHES **A sandwich shop charges \$7 for a large sub and sells an average of 360 large subs per day. The shop predicts that they will sell 20 fewer subs for every \$0.25 increase in the price.**

Part A Let x represent each \$0.25 price increase. Write a function $P(x)$ to represent the price of a large sub.

The price of a large sub is \$7 plus \$0.25 times each price increase.

$P(x) = 7 + 0.25x$

Part B Write a function $T(x)$ to represent the number of subs sold.

The number of subs sold is 360 minus 20 times the number of price increases.

$T(x) = 360 - 20x$

Part C Write a function $R(x)$ that can be used to maximize the revenue from sales of large subs.

The revenue from sales of large subs will be equal to the price times the number of subs sold.

$$R(x) = P(x) \cdot T(x)$$

$$= (7 + 0.25x)(360 - 20x)$$

$$= -5x^2 - 50x + 2520$$

Part D If the sandwich shop charges \$8.50 for a large sub, find the revenue from sales of large subs. x represents each \$0.25 price increase. So, $x = 6$.

$$R(x) = -5x^2 - 50x + 2520$$

$$R(6) = -5(6)^2 - 50(6) + 2520$$

$$= -180 - 300 + 2520 \text{ or } \$2040$$

Think About It
How could you determine the amount of money the sub shop should charge for a large sub?

 Go Online You can complete an Extra Example online.

29. STRUCTURE Tyree makes a pattern, as shown, using shaded tiles and white tiles.

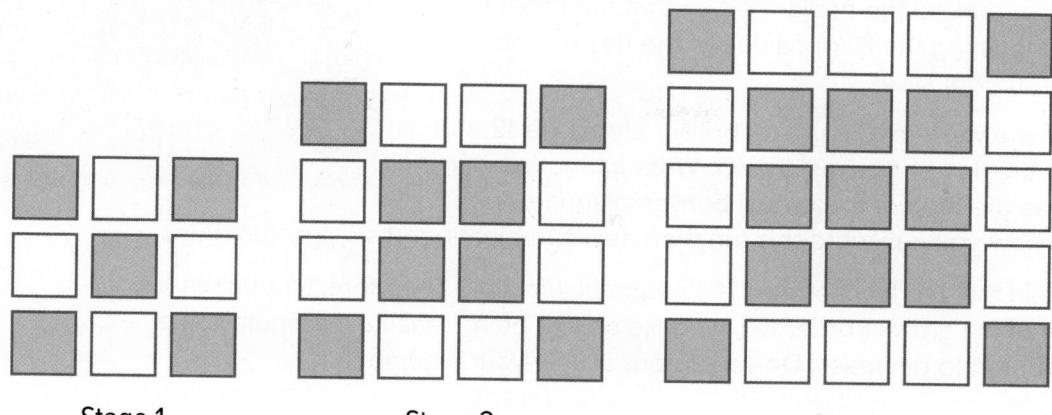

Stage 1   Stage 2   Stage 3

a. Write a function $f(n)$ that gives the number of shaded tiles needed to make stage n of the pattern and a function $g(n)$ that gives the number of white tiles needed to make stage n of the pattern.

b. Explain how to write a function $h(n)$ that gives the total number of tiles needed to make stage n of the pattern. Use the function to find the total number of tiles needed to make stage 10.

c. Explain how you know your answer to **part b** is correct.

30. USE TOOLS A chemist works in a lab that maintains a constant temperature of 22°C. She heats a saline solution and then lets the solution cool. As the solution cools, she records its temperature every 3 minutes. Her data is shown in the table.

Time (minutes)	0	3	6	9	12	15
Temperature of Solution (°C)	82.0	79.1	76.3	73.6	70.9	68.4
Temperature Above the Lab Temperature	60.0	57.1	54.3	51.6	48.9	46.4

a. Write a function $P(x)$ that models the temperature of the solution above the lab temperature in degrees Celsius x minutes after the chemist starts recording the data. Write a second function $S(x)$ that models the temperature of the solution.

b. Explain how the function you wrote for $S(x)$ in **part a** is a combination of two functions.

c. The chemist wants to know approximately how long it will take until the solution cools to a temperature of 45°C. Explain how to use your calculator to estimate this time to the nearest minute.

31. CONSTRUCT ARGUMENTS A rectangular flower bed in a garden is 12 feet long and 8 feet wide. Kazuo plans to add a gravel border around the flower bed so that the border is twice as wide along the 8-foot sides of the flower bed as it is along the 12-foot sides.

a. Let x be the width of the gravel border along the 12-foot sides of the flower bed, as shown. Write a function $A(x)$ that gives the area of the gravel border in square feet. Explain how you can write this function as a combination of simpler functions.

b. Kazuo said that the function $A(x)$ is a quadratic function. Therefore, as x increases, the area of the gravel border will increase up to a point, reach a maximum value, and then start to decrease. Do you agree? Justify your argument.

32. REGULARITY Describe the difference between the method for adding a linear and a quadratic function and the method for multiplying a linear and a quadratic function.

33. STRUCTURE Adding the linear function $j(x) = 5$ to the exponential function $k(x) = 3^x - 7$, gives a new exponential function $m(x)$, which is a translation of $k(x)$. What is the equation of the translated function, $m(x)$?

34. FRAME An 8×10 picture frame has a glass center that is 8 inches wide and 10 inches long. The frame that surrounds the glass can be a variety of different sizes. How much larger is the area of a framed picture than its perimeter if the frame extends x inches outside the glass in each direction?

🌐 Higher-Order Thinking Skills

35. CREATE Define $f(x)$ and $g(x)$ if $(f - g)(x) = x^2 + 14x - 12$, where $f(x)$ is a quadratic function and $g(x)$ is a linear function.

36. CREATE Define $f(x)$ and $g(x)$ if $(f \cdot g)(x) = x^3 - x^2 - 4x + 4$, where $f(x)$ is a quadratic function and $g(x)$ is a linear function.

37. ANALYZE Determine whether the following statement is *sometimes*, *always*, or *never* true. Justify your argument.

Multiplying a linear equation with two terms by a quadratic equation with three terms will result in an equation that has three terms.

38. WRITE You learned that when a linear function and a quadratic function are multiplied, the product is a third-degree polynomial. What do you think would be the degree of the product of two quadratic functions? What about the degree of the sum of two quadratic functions? Explain your reasoning.

39. FIND THE ERROR Delfina multiplied the linear function $f(x) = 6x - 8$ by the quadratic function $g(x) = 4x^2 + 7x - 19$. She found that $(f \cdot g)(x) = 24x^3 + 42x^2 + 152$. Is Delfina's solution correct? Explain your reasoning.

40. WHICH ONE DOESN'T BELONG? Consider the functions $a(x)$, $b(x)$, and $c(x)$, below, and their products, $(a \cdot b)(x)$, $(a \cdot c)(x)$, and $(b \cdot c)(x)$. Determine which of the products doesn't belong. Justify your conclusion.

$a(x) = 4x + 8$	$b(x) = -x^2 + 3x - 5$	$c(x) = 6x - 7$

Inverses of Quadratic Functions

Learn Inverse Relations and Functions

Recall that the inverse of a function can be found by exchanging the domain and range. The inverse of $f(x)$ is written as $f^{-1}(x)$.

Not all functions have an inverse function. The inverse of a function will also be a function if it passes the horizontal line test.

> **Key Concept • Horizontal Line Test**
>
> If a horizontal line intersects the graph of a function more than once, the inverse of the function is not a function.

If a function fails the horizontal line test, you can restrict the domain of the function to make the inverse a function. Choose a portion of the domain on which the function is one-to-one. There may be more than one possible domain.

Example 1 Inverses with Restricted Domains

Consider $f(x) = x^2 + 4x - 5$.

Part A Find the inverse of $f(x)$.

$f(x) = x^2 + 4x - 5$	Original function
$y = x^2 + 4x - 5$	Replace $f(x)$ with y.
$x = y^2 + 4y - 5$	Exchange x and y.
$x + 5 = y^2 + 4y$	Add 5 to each side.
$x + 5 + 4 = y^2 + 4y + 4$	Complete the square.
$x + 9 = (y + 2)^2$	Simplify.
$\pm\sqrt{x + 9} = y + 2$	Take the square root of each side.
$-2 \pm \sqrt{x + 9} = y$	Subtract 2 from each side.
$f^{-1}(x) = -2 \pm \sqrt{x + 9}$	Replace y with $f^{-1}(x)$.

Part B If necessary, restrict the domain so that the inverse is a function.

Graph $f(x)$.

Because $f(x)$ fails the horizontal line test, $f^{-1}(x)$ is not a function. Restrict the domain of $f(x)$.

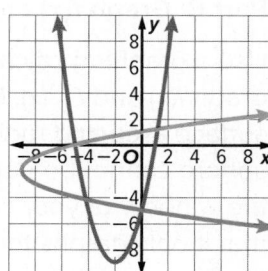

Find the restricted domain of $f(x)$ so that $f^{-1}(x)$ will be a function. Look for a portion of the graph that is one-to-one.

If the domain of $f(x)$ is restricted to $x \le -2$, then the inverse is $f^{-1}(x) = -2 - \sqrt{x + 9}$.

If the domain of $f(x)$ is restricted to $x \ge -2$, then the inverse is $f^{-1}(x) = -2 + \sqrt{x + 9}$.

 Think About It!

Write a function that does not pass the horizontal line test.

Watch Out!

Inverse Functions f^{-1} is read *f inverse* or *the inverse of f*. Note that -1 is not an exponent.

Check

Examine $f(x) = x^2 - 6x + 2$.

Part A Select the inverse of $f(x) = x^2 - 6x + 2$. ___

A. $f^{-1}(x) = \pm\sqrt{x + 10}$

B. $f^{-1}(x) = 1 \pm\sqrt{x + 4}$

C. $f^{-1}(x) = 2 \pm\sqrt{x - 3}$

D. $f^{-1}(x) = 3 \pm\sqrt{x + 7}$

Part B Restrict the domain so that the inverse is a function.

If the domain of $f(x)$ is restricted to _____, then the inverse is $f^{-1}(x) = 3 - \sqrt{x + 7}$.

If the domain of $f(x)$ is restricted to _____, then the inverse is $f^{-1}(x) = 3 + \sqrt{x + 7}$.

🌐 Example 2 Interpret Inverse Functions

SURFACE AREA **The world's largest rubber band ball is made up of 700,000 rubber bands. It is in the shape of a sphere. The formula $f(x) = 4\pi x^2$ can be used to find the surface area of the sphere for a given radius x.**

Part A Find the inverse of $f(x)$, and describe its meaning.

$f(x) = 4\pi x^2$	Original function
$y = 4\pi x^2$	Replace $f(x)$ with y.
$x = 4\pi y^2$	Exchange x and y.
$\frac{x}{4\pi} = y^2$	Divide each side by 4π.
$\pm\sqrt{\frac{x}{4\pi}} = y$	Take the square root of each side.
$f^{-1}(x) = \pm\sqrt{\frac{x}{4\pi}}$	Replace y with $f^{-1}(x)$.

$f^{-1}(x)$ can be used to find the radius of the sphere, given the surface area. Because negative values of x and $f^{-1}(x)$ do not make sense in the context of the situation, restrict the domain to $x \geq 0$.

💬 Talk About It!

Find the domain of $f(x)$ and its inverse. Explain your reasoning.

Part B Graph $f(x)$ and $f^{-1}(x)$.

Use a graphing calculator to graph $f(x)$ and $f^{-1}(x)$. Because the surface area and radius of a sphere cannot be negative, adjust the window of your calculator so that it only graphs the functions in the first quadrant.

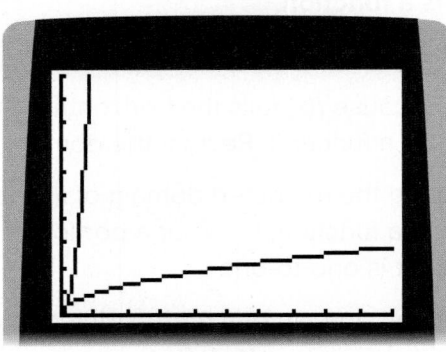

[0, 10] scl: 1 by [0, 10] scl: 1

Check

ENERGY The formula $f(x) = \frac{1}{2}kx^2$ can be used to find elastic potential energy, where k is the spring constant and x is the length an object is stretched or compressed.

Part A

Find the inverse of $f(x)$. _____

A. $f^{-1}(x) = \pm\sqrt{\frac{k}{2x}}$

B. $f^{-1}(x) = \pm\sqrt{\frac{2k}{x}}$

C. $f^{-1}(x) = \pm\sqrt{\frac{2x}{k}}$

D. $f^{-1}(x) = \pm\sqrt{\frac{x}{2k}}$

Part B

Describe the meaning of the inverse in the context of the situation.

$f^{-1}(x)$ can be used to find the _____
given the _____ .

Learn Verifying Inverses

You can determine whether two functions are inverses by finding both of their compositions. If both compositions equal the identity function $f(x) = x$, then the functions are inverse functions.

Key Concept • Inverse Functions	
Words	Two functions f and g are inverse functions if and only if both of their compositions are the identity function.
Symbols	$f(x)$ and $g(x)$ are inverses if and only if $[f \circ g](x) = x$ and $[g \circ f](x) = x$

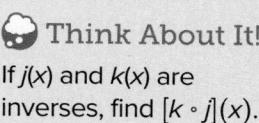

Think About It!

If $j(x)$ and $k(x)$ are inverses, find $[k \circ j](x)$.

Example 3 Composition of Functions

Determine whether the functions are inverse functions.

a. $f(x) = 3x + 8$ and $g(x) = \frac{x-8}{3}$

To determine whether the functions are inverses, verify that the compositions of $f(x)$ and $g(x)$ result in the identity function.

Find $[f \circ g](x)$.

$$[f \circ g](x) = f[g(x)]$$
$$= f\left(\frac{x-8}{3}\right)$$
$$= 3\left(\frac{x-8}{3}\right) + 8$$
$$= x - 8 + 8$$
$$= x$$

(continued on the next page)

Study Tip

Inverse Functions
Recall that the identity function is $f(x) = x$. So, if two functions are inverses, both of their compositions will be equal to x.

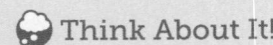
Find $[g \circ f](x)$.

$$[g \circ f](x) = g[f(x)]$$
$$= g(3x + 8)$$
$$= \frac{3x + 8 - 8}{3}$$
$$= \frac{3x}{3}$$
$$= x$$

Because $[f \circ g](x)$ and $[g \circ f](x)$ are both equal to x, $f(x)$ and $g(x)$ are inverses.

b. $h(x) = \sqrt{x + 17}$ and $k(x) = (x - 17)^2$

Find $[h \circ k](x)$.

$$[h \circ k](x) = h[k(x)]$$
$$= h[(x - 17)^2]$$
$$= \sqrt{(x - 17)^2 + 17}$$
$$= \sqrt{x^2 - 34x + 289 + 17}$$
$$= \sqrt{x^2 - 34x + 306}$$

Because $[h \circ k](x)$ is not the identity function, $h(x)$ and $k(x)$ are not inverses.

Check

Determine whether $f(x) = \frac{x}{4} + \frac{9}{2}$ and $g(x) = 4x + 18$ are inverses.

Explain your reasoning. _____

A. Yes; $[f \circ g](x) = x$ and $[g \circ f](x) = x$.

B. Yes; $[f \circ g](x) = x + 9$ and $[g \circ f](x) = x + 36$.

C. No; $[f \circ g](x) = x$ and $[g \circ f](x) = x$.

D. No; $[f \circ g](x) = x + 9$ and $[g \circ f](x) = x + 36$.

Determine whether $f(x) = \sqrt{x} - 64$ and $g(x) = (x + 64)^2$ are inverses. Explain your reasoning.

$f(x)$ and $g(x)$ _____ inverses because $[f \circ g](x) = $ __ and $[g \circ f](x) = $ __.

Practice

Go Online You can complete your homework online.

Example 1

Find the inverse of $f(x)$. If necessary, restrict the domain so that the inverse is a function.

1. $f(x) = x^2 + 4$

2. $f(x) = x^2 - 8$

3. $f(x) = 2x^2 - 1$

4. $f(x) = 4x^2 + 3$

5. $f(x) = (x - 5)^2$

6. $f(x) = (x + 6)^2$

7. $f(x) = x^2 + 8x + 1$

8. $f(x) = x^2 - 4x + 1$

9. $f(x) = x^2 - 4x - 2$

10. $f(x) = x^2 + 8x - 2$

11. $f(x) = -x^2 + 2x + 5$

12. $f(x) = -x^2 + 6x + 3$

13. $f(x) = 2x^2 + 4x + 3$

14. $f(x) = 3x^2 - 12x + 1$

Example 2

15. **SURFACE AREA** The formula $S = 6x^2$ can be used to find the surface area of a cube with sides x units long.

    a. Find the inverse of the function, and describe its meaning.

    b. Identify the domain of $f(x)$ that makes sense in the context of the situation. Explain your reasoning.

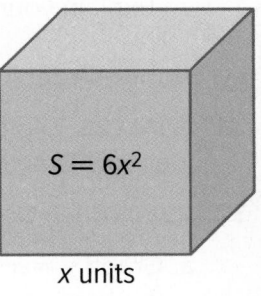

$S = 6x^2$

x units

16. **MEDICINE** You can use the formula $c = -0.05t^2 + 2t + 2$ to calculate the concentration in parts per million of a certain drug in the bloodstream after t hours.

    a. Find the inverse of the function, and describe its meaning.

    b. Identify the domain of the function that makes sense in the context of the situation. Explain your reasoning.

Example 3

Determine whether the functions are inverse functions.

17. $f(x) = x^2 - 9; g(x) = \pm\sqrt{x + 9}$

18. $f(x) = \frac{1}{2}x + 2; g(x) = x^2 - 2$

19. $f(x) = (x - 9)^2; g(x) = 9 \pm \sqrt{x}$

20. $f(x) = (x - 4)^2; g(x) = 4 \pm \sqrt{x}$

21. $f(x) = (x - 16)^2; g(x) = 4 \pm \sqrt{x}$

22. $f(x) = (x - 4)^2; g(x) = -2 \pm \sqrt{x + 3}$

23. $f(x) = \frac{1}{2}x^2 + 4; g(x) = 2x^2 - 2$

24. $f(x) = \frac{1}{2}x^2 + 4; g(x) = \sqrt{x - 4}$

25. $f(x) = (x - 12)^2; g(x) = 12 \pm \sqrt{x}$

26. $f(x) = (4x - 1)^2; g(x) = 4 \pm 2\sqrt{x}$

Mixed Exercises

Sketch the graph of the inverse of each relation. Then, if the inverse is not a function, identify a domain of $f(x)$ for which it is a function.

27.
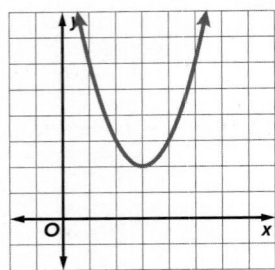

28.
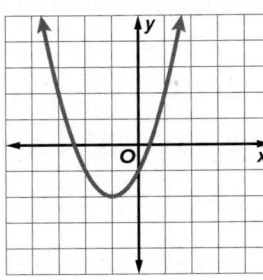

29.
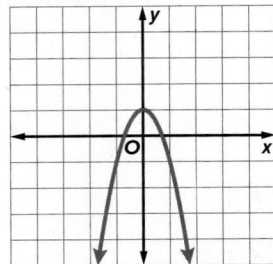

30. **WORLD RECORDS** In 2011, David Marvin, Jr., set a world record by being fired from a cannon a distance of 59.05 meters. His initial upward velocity was approximately 33.32 meters per second. The height $h(t)$ can be represented by the function $h(t) = 33.32t - 9.8t^2$, where t represents the number of seconds that have passed.

   a. Find the inverse of $h(t)$, and describe its meaning.

   b. Identify the domain of $h(t)$ that makes sense in the context of the situation. Explain your reasoning.

🧠 Higher-Order Thinking Skills

31. **ANALYZE** Explain the difference between an inverse relation and an inverse function. Is an inverse function also an inverse relation? Justify your argument.

32. **USE TOOLS** Consider the functions $y = x^n$ for $n = 0, 1, 2, \ldots$.

   a. Graph $y = x^n$ for $n = 0, 1, 2, 3,$ and 4. Determine whether the inverse of each function is a function. Complete the table at the right.

   b. Make a conjecture about the values of n for which the inverse of $f(x) = x^n$ is a function. Assume that n is a whole number.

Function	Inverse a function?
$y = x^0$ or $y = 1$	
$y = x^1$ or $y = x$	
$y = x^2$	
$y = x^3$	
$y = x^4$	

33. **CREATE** If possible, create a quadratic function that is its own inverse. If it is not possible, explain why it is not.

34. **FIND THE ERROR** Doug and Tandi are finding the inverse of $f(x) = 2x^2 - 8x + 4$. Is either of them correct? Explain your reasoning.

Doug

$f(x) = 2x^2 - 8x + 4$

$y = 2x^2 - 8x + 4$

$x = 2y^2 - 8y + 4$

$x - 4 = 2y^2 - 8y$

$x - 4 + 16 = 2y^2 - 8y + 16$

$x + 12 = 2(y - 4)^2$

$\frac{1}{2}x + 6 = (y - 4)^2$

$\pm\sqrt{\frac{1}{2}x + 6} = y - 4$

$4 \pm \sqrt{\frac{1}{2}x + 6} = y$

$f^{-1}(x) = 4 \pm \sqrt{\frac{1}{2}x + 6}$

Tandi

$f(x) = 2x^2 - 8x + 4$

$y = 2x^2 - 8x + 4$

$x = 2y^2 - 8y + 4$

$x - 4 = 2y^2 - 8y$

$x - 4 + 2(4) = 2(y^2 - 4y + 4)$

$x + 4 = 2(y - 2)^2$

$\frac{1}{2}x + 2 = (y - 2)^2$

$\pm\sqrt{\frac{1}{2}x + 2} = y - 2$

$2 \pm \sqrt{\frac{1}{2}x + 2} = y$

$f^{-1}(x) = 2 \pm \sqrt{\frac{1}{2}x + 2}$

Essential Question

Why is it helpful to have different methods to analyze quadratic functions and solve quadratic equations?

Depending on the information given, one method may be easier to use than another. It also depends on whether you need an approximate or exact answer. For example, you can approximate the answer using a graph or mental math, and you can find an exact answer using algebraic techniques.

Module Summary

Lessons 12-1 and 12-2

Graphing Quadratic Functions

- The graph of a quadratic function is a parabola. The axis of symmetry intersects a parabola at the vertex. The vertex is either the lowest point or the highest point on a parabola.

- The standard form of a quadratic function is $f(x) = ax^2 + bx + c$, where a, b, and c are integers and $a \neq 0$.

- Quadratic functions can be translated like other functions.

- A quadratic function in the form $f(x) = a(x - h)^2 + k$ is in vertex form.

Lessons 12-3 through 12-6

Solving Quadratic Equations

- The solutions or roots of an equation can be identified by finding the x-intercepts of the graph of the related function.

- If the product of two factors is 0, then at least one of the factors must be 0.

- To complete the square for any quadratic expression of the form $x^2 + bx$, find one-half of b, the coefficient of x. Square the result. Then add the result to $x^2 + bx$.

- Quadratic Formula: $x = \frac{-b \pm \sqrt{b^2 - 4ac}}{2a}$.

Lesson 12-7

Solving Systems of Linear and Quadratic Equations

- The solution to a system of linear and quadratic equations is at the point of intersections of the graphs of each equation.

- You can use the Substitution Method or the Elimination Method to solve a system of linear and quadratic equations.

Lesson 12-8

Modeling and Curve Fitting

- The coefficient of determination, R^2, indicates how well the function fits the data.

Lesson 12-9

Combining Functions

- You can add two functions and combine the like terms.

- You can multiply two functions.

Study Organizer

 Foldables

Use your Foldable to review this module. Working with a partner can be helpful. Ask for clarification of concepts as needed.

Key Features

Transformations

Solve by Factoring

Completing the Square

Test Practice

1. **GRAPH** Graph $f(x) = x^2 + 4x - 2$. (Lesson 12-1)

2. **MULTIPLE CHOICE** The height of a baseball is modeled by the function $h(x) = -16x^2 + 50x + 5$, where $h(x)$ represents the height, in feet, of the ball x seconds after being hit by a bat. Which of the following describes the most appropriate domain for this function? (Lesson 12-1)

   A. all integers

   B. positive integers

   C. all real numbers

   D. positive real numbers

3. **MULTIPLE CHOICE** Find the vertex of the graph of $f(x) = 4x^2 - 24x - 2$. (Lesson 12-1)

   A. $(6, -2)$

   B. $(3, -38)$

   C. $(3, -62)$

   D. $(-3, 106)$

4. **MULTIPLE CHOICE** Find the axis of symmetry of the graph of $f(x) = 4x^2 - 24x - 2$. (Lesson 12-1)

   A. $x = 6$

   B. $x = 3$

   C. $x = \frac{1}{3}$

   D. $x = -3$

5. **OPEN RESPONSE** Find the y-intercept of $f(x) = 4x^2 - 24x - 2$. (Lesson 12-1)

6. **MULTIPLE CHOICE** If $f(x) = -4x^2$, write a function $k(x)$ representing a reflection of $f(x)$ across the x-axis. (Lesson 12-2)

   A. $k(x) = -\frac{1}{4}x^2$

   B. $k(x) = \frac{1}{4}x^2$

   C. $k(x) = (-4x)^2$

   D. $k(x) = 4x^2$

7. **GRAPH** Given the parent function, $f(x) = x^2$, graph the translation 3 units to the right, $g(x)$. (Lesson 12-2)

8. **MULTI-SELECT** Use the graph of the related function to solve $x^2 + x = 20$. Select all the solutions that apply. (Lesson 12-3)

   A. -20

   B. -5

   C. 1

   D. 4

   E. 9

9. **MULTIPLE CHOICE** Which quadratic function models the graph? (Lesson 12-3)

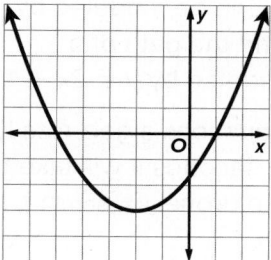

A. $y = \frac{1}{3}x^2 - \frac{4}{3}x + \frac{5}{3}$

B. $y = \frac{1}{3}x^2 - \frac{4}{3}x - \frac{5}{3}$

C. $y = \frac{1}{3}x^2 + \frac{4}{3}x - \frac{5}{3}$

D. $y = \frac{1}{3}x^2 + \frac{4}{3}x + \frac{5}{3}$

10. **OPEN RESPONSE** During a thunderstorm, a branch fell from a tree. Chantel estimates the branch fell from 25 feet above the ground.

The formula $h = -16t^2 + h_0$ can be used to approximate the number of seconds t it takes for the branch to reach height h from an initial height of h_0 in feet. Find the time it takes the branch to reach the ground. Round to the nearest hundredth, if necessary. (Lesson 12-4)

11. **MULTIPLE CHOICE** The area of the rectangle shown is 120 square inches.

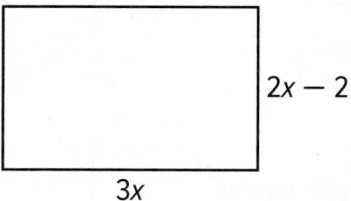

What is the perimeter of the rectangle? (Lesson 12-4)

A. 5 inches

B. 23 inches

C. 46 inches

D. 120 inches

12. **MULTIPLE CHOICE** Solve $x^2 + 6x = 7$ by completing the square. (Lesson 12-5)

A. 3, 9

B. 6, 7

C. 1, −7

D. 3, 4

13. **OPEN RESPONSE** Write $y = 2x^2 + 4x + 6$ in vertex form. Identify the extrema, and explain whether it is a minimum or maximum. (Lesson 12-5)

14. **OPEN RESPONSE** Solve $x^2 - 20x + 27 = 8$ by completing the square. (Lesson 12-5)

15. **OPEN RESPONSE** A cat that is sitting on a boulder 5 feet high jumps off and lands on the ground. During its jump, its height h in feet is given by $h = -0.5d^2 + 2d + 5$, where d is the distance from the base of the boulder. (Lesson 12-5)

How far is the cat from the base of the boulder when it lands on the ground? Round to the nearest tenth.

What is the maximum height of the cat during its jump?

16. MULTIPLE CHOICE A local company manufactures brake drums. Based on their records, their daily profit can be approximated by the function $f(x) = x^2 + 3x - 19$, where x is the number of brake drums they manufacture. If $f(x)$ is negative, it means the company has lost money. What is the least number of brake drums that the company needs to manufacture each day in order to make a profit? (Lesson 12-6)

A. 3

B. 4

C. 7

D. 19

17. OPEN RESPONSE How many real solutions does the equation $2x^2 + 5x + 7 = 0$ have? (Lesson 12-6)

18. MULTIPLE CHOICE Find the solution(s) of the system shown.

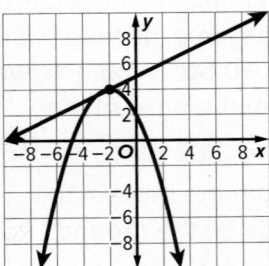

(Lesson 12-7)

A. $(-2, 4)$

B. $(-2, 4)$ and $(-4, 2)$

C. $(0, 2)$

D. $(0, 5)$

19. OPEN RESPONSE Let x represent the time in seconds and y represent the height in feet. The path of a drone can be modeled by $y = -x^2 + 19x + 1$ and the path of a projectile can be modeled by $y = 2x + 1$.

Solve a system of equations algebraically to calculate how many seconds it will take the projectile to meet the drone. (Lesson 12-7)

20. MULTIPLE CHOICE The table shows the profit earned y as it relates to the number of magazines sold x.

x	0	1	2	3	4
y	0.2	1	5	25	125

Which model best describes the data in the table? (Lesson 12-8)

A. Linear

B. Exponential

C. Quadratic

D. Square

21. MULTIPLE CHOICE Which coefficient of determination shows the best fit of the data? (Lesson 12-8)

A. $R^2 = 0.589$

B. $R^2 = 0.880$

C. $R^2 = 0.989$

D. $R^2 = 1.54$

22. OPEN RESPONSE Given $f(x) = 7x^2 + 22x - 6$ and $g(x) = 3x - 9$, find $(g - f)(x)$. (Lesson 12-9)

Trigonometric Identities and Equations

e Essential Question
How are trigonometric identities similar to and different from other equations?

What Will You Learn?

How much do you already know about each topic **before** starting this module?

KEY	Before			After		
👎 — I don't know. 👍 — I've heard of it. 👍 — I know it!	👎	👍	👍	👎	👍	👍
find trigonometric values using trigonometric identities						
simplify trigonometric expressions using trigonometric identities						
verify trigonometric identities by transforming equations						
find trigonometric values using sum and difference identities						
verify identities using sum and difference identities						
find values of sine and cosine using double-angle and half-angle identities						
solve equations and determine extraneous solutions using trigonometric identities						

Foldables Make this Foldable to help you organize your notes about trigonometric identities and equations. Begin with one sheet of 11" × 17" paper and four sheets of grid paper.

1. **Fold** the short sides of the 11" × 17" paper to meet in the middle.
2. **Cut** each tab in half as shown.
3. **Cut** four sheets of grid paper in half and fold the half-sheets in half.
4. **Insert** two folded half-sheets under each of the four tabs and staple along the fold.
5. **Label** each tab as shown.

 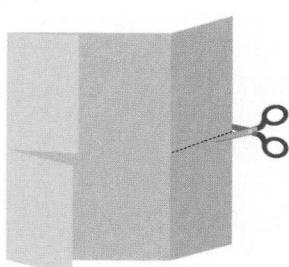 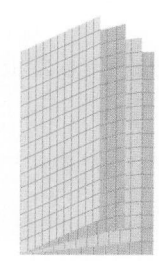 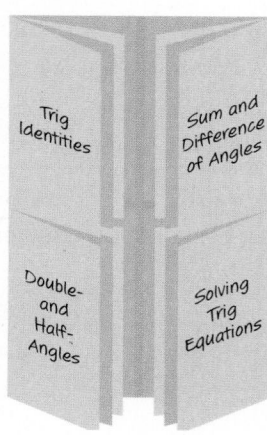

What Vocabulary Will You Learn?

- cofunction identities
- Pythagorean identities
- trigonometric equation
- trigonometric identity

Are You Ready?

Complete the Quick Review to see if you are ready to start this module.

Then complete the Quick Check.

<table>
<tr><td colspan="2">Quick Review</td></tr>
<tr>
<td>

Example 1

Factor $x^3 + 2x^2 - 24x$ completely.

$x^3 + 2x^2 - 24x = x(x^2 + 2x - 24)$

The product of the coefficients of the x-terms in factors of $x^2 + 2x - 24$ must be -24, and their sum must be 2. The product of 6 and -4 is -24, and their sum is 2.

$x(x^2 + 2x - 24) = x(x + 6)(x - 4)$

</td>
<td>

Example 2

Find the exact value of cos 135°.

The reference angle is $180° - 135°$, or $45°$.

$\cos 45°$ is $\frac{\sqrt{2}}{2}$. Since $135°$ is in the second quadrant, $\cos 135° = -\frac{\sqrt{2}}{2}$.

</td>
</tr>
<tr><td colspan="2">Quick Check</td></tr>
<tr>
<td>

Factor. Assume that no variable equals zero.

1. $-16a^2 + 4a$

2. $5x^2 - 20$

3. $x^3 + 9$

4. $2y^2 - y - 15$

</td>
<td>

Find the exact value of each trigonometric function.

5. $\sin 45°$

6. $\cos 225°$

7. $\tan 150°$

8. $\sin 120°$

</td>
</tr>
<tr>
<td colspan="2">

How Did You Do?

Which exercises did you answer correctly in the Quick Check?

</td>
</tr>
</table>

Trigonometric Identities

Explore Pythagorean Identity

🔗 **Online Activity** Use graphing technology to complete the Explore.

> ❔ **INQUIRY** How can the unit circle be used to justify trigonometric identities?   ✕

Learn Using Trigonometric Identities to Find Values

A **trigonometric identity** is an equation involving trigonometric functions that is true for all values for which every expression in the equation is defined. A *counterexample* can be used to show that an equation is false and, therefore, is not an identity.

Key Concept • Quotient and Reciprocal Identities

Quotient Identities	
$\tan \theta = \frac{\sin \theta}{\cos \theta}$; $\cos \theta \neq 0$	$\cot \theta = \frac{\cos \theta}{\sin \theta}$; $\sin \theta \neq 0$

Reciprocal Identities	
$\csc \theta = \frac{1}{\sin \theta}$; $\sin \theta \neq 0$	$\sin \theta = \frac{1}{\csc \theta}$; $\csc \theta \neq 0$
$\sec \theta = \frac{1}{\cos \theta}$; $\cos \theta \neq 0$	$\cos \theta = \frac{1}{\sec \theta}$; $\sec \theta \neq 0$
$\cot \theta = \frac{1}{\tan \theta}$; $\tan \theta \neq 0$	$\tan \theta = \frac{1}{\cot \theta}$; $\cot \theta \neq 0$

The identity $\tan \theta = \frac{\sin \theta}{\cos \theta}$ is true except for angle measures 90°, 270°, ... , 90° + k180°, where k is an integer. The cosine of each of these angle measures is 0, so $\tan \theta$ is not defined when $\cos \theta = 0$. Similarly, $\cot \theta = \frac{\cos \theta}{\sin \theta}$ is undefined when $\sin \theta = 0$ for angle measures 0°, 180°, ... , k180°, where k is an integer.

The **Pythagorean identities** express the Pythagorean Theorem in terms of the trigonometric functions.

Key Concept • Pythagorean Identities

$\cos^2 \theta + \sin^2 \theta = 1$	$\tan^2 \theta + 1 = \sec^2 \theta$	$\cot^2 \theta + 1 = \csc^2 \theta$

Today's Goals
- Find trigonometric values by using trigonometric identities.
- Simplify trigonometric expressions by using trigonometric identities.

Today's Vocabulary
trigonometric identity

Pythagorean identities

cofunction identities

Study Tip

Reading Trigonometric Functions $\sin^2 \theta$ is read as *sine squared theta*. It has the same value and meaning as the square of the quantity $\sin \theta$, or $(\sin \theta)^2$.

🔗 **Go Online**
to see a proof of the Pythagorean identity $\cos^2 \theta + \sin^2 \theta = 1$.

A trigonometric function f is a cofunction of another trigonometric function g if $f(A) = g(B)$ when A and B are complementary angles. The **cofunction identities** show the relationships between sine and cosine, tangent and cotangent, and secant and cosecant.

All six of the trigonometric functions are either odd or even. Recall that a function is even if $f(-x) = f(x)$ is true for every value in the domain. A function is odd if $f(-x) = -f(x)$ is true for every value in the domain. These relationships are given in the negative-angle identities, which are also sometimes called odd-even identities.

Key Concept • Cofunction Identities and Negative-Angle Identities		
Cofunction Identities		
$\sin\left(\frac{\pi}{2} - \theta\right) = \cos\theta$	$\cos\left(\frac{\pi}{2} - \theta\right) = \sin\theta$	$\tan\left(\frac{\pi}{2} - \theta\right) = \cot\theta$
Negative-Angle Identities		
$\sin(-\theta) = -\sin\theta$	$\cos(-\theta) = \cos\theta$	$\tan(-\theta) = -\tan\theta$

Example 1 Use the Pythagorean Identities

Find the exact value of $\cos\theta$ if $\sin\theta = \frac{2}{9}$ and $90° < \theta < 180°$.

You can use the unit circle to estimate the value of θ and $\cos\theta$.

Because $90° < \theta < 180°$, θ is in Quadrant II. $\sin\theta$ is $\frac{2}{9}$, which is between 0 and $\frac{1}{2}$, so θ will be between 150° and 180°.

Therefore, $\cos\theta$ is between -1 and $-\frac{\sqrt{3}}{2}$.

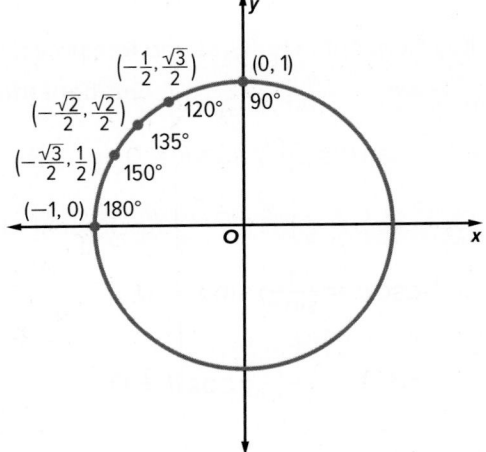

To find the exact value of $\cos\theta$, use a Pythagorean identity.

$\sin^2\theta + \cos^2\theta = 1$ Pythagorean Identity

$\quad \cos^2\theta = 1 - \sin^2\theta$ Subtract $\sin^2\theta$ from each side.

$\quad \cos^2\theta = 1 - \left(\frac{2}{9}\right)^2$ Substitute $\frac{2}{9}$ for $\sin\theta$.

$\quad \cos^2\theta = 1 - \frac{4}{81}$ Square $\frac{2}{9}$.

$\quad \cos^2\theta = \frac{77}{81}$ Subtract.

$\quad\quad \cos\theta = \pm\frac{\sqrt{77}}{9}$ Take the square root of each side.

Because θ is in Quadrant II, $\cos\theta$ is negative. Therefore, $\cos\theta = -\frac{\sqrt{77}}{9}$.

Using the unit circle, the estimated value of $\cos\theta$ was between -1 and $-\frac{\sqrt{3}}{2}$. The solution lies within this interval.

🅦 **Go Online** You can complete an Extra Example online.

Math History Minute
Swiss clockmaker **Jost Bürgi (1552–1632)** was able to calculate sines using several algorithms. He explained them in his work *Fundamentum Astronomiae* (1592). Bürgi is also known for constructing a table of logarithms around 1600. Few copies of this work were saved, and it went virtually unnoticed by the mathematics community.

You can use a graphing calculator to approximate values and check the solution. Find $\cos\left[\sin^{-1}\left(\frac{2}{9}\right)\right]$ using a calculator and compare it to $-\frac{\sqrt{77}}{9}$. Because the two values are approximately equal, our solution is correct.

Check

Find the exact value of $\sin\theta$ if $\cot\theta = 2\sqrt{6}$ and $0 < \theta < 90°$.

$\sin\theta = \underline{\quad?\quad}$

Example 2 Use the Cofunction and Negative-Angle Identities

Find the exact value of $\cot\left(\theta - \frac{\pi}{2}\right)$ if $\tan\theta = 1.79$.

$$\cot\left(\theta - \frac{\pi}{2}\right) = \frac{\cos\left(\theta - \frac{\pi}{2}\right)}{\sin\left(\theta - \frac{\pi}{2}\right)} \qquad \text{Quotient Identity}$$

$$= \frac{\cos\left[-\left(\frac{\pi}{2} - \theta\right)\right]}{\sin\left[-\left(\frac{\pi}{2} - \theta\right)\right]} \qquad \begin{array}{l}\text{Factor } -1 \text{ from each angle}\\ \text{measure expression.}\end{array}$$

$$= \frac{\cos\left(\frac{\pi}{2} - \theta\right)}{-\sin\left(\frac{\pi}{2} - \theta\right)} \qquad \text{Negative-Angle Identity}$$

$$= -\frac{\sin\theta}{\cos\theta} \qquad \text{Cofunction Identity}$$

$$= -\tan\theta \qquad \text{Quotient Identity}$$

$$= -1.79 \qquad \tan\theta = 1.79$$

> **Think About It!**
>
> Why did we not need to know the quadrant in which θ was located in order to solve the problem?

Check

Find the exact value of $-\sin\theta$ if $\cos\left(\theta - \frac{\pi}{2}\right) = 1.4$ and $0° < \theta < \frac{\pi}{2}$.

$-\sin\theta = \underline{\quad?\quad}$

Explore Negative-Angle Identity

🔗 **Online Activity** Use graphing technology to complete the Explore.

> ❓ **INQUIRY** How can the graphs of trigonometric functions be used to justify trigonometric identities?

Learn Using Trigonometric Identities to Simplify

You can use the basic trigonometric identities along with algebra to simplify trigonometric expressions. In particular, you should familiarize yourself with the Quotient and Reciprocal Identities and the Pythagorean Identities.

(continued on the next page)

🔗 **Go Online** You can complete an Extra Example online.

To simplify trigonometric expressions, you will want to use these strategies.

- Recognize and use the Pythagorean Identities. Because $\sin^2 x + \cos^2 x = 1$, $\cos^2 x = 1 - \sin^2 x$ is also true.

- Factor. For example, given an expression like $\sin x - \sin x \cos^2 x$, you can factor and then use a Pythagorean Identity to simplify.

- Rewrite. When an expression includes tangent or one of the reciprocal trigonometric functions, it may help to rewrite the expression in terms of sine or cosine.

- Separate or combine. Writing two fractions over a common denominator or splitting one fraction into two fractions, like $\frac{a-b}{c} = \frac{a}{c} - \frac{b}{c}$, may also help.

Simplifying an expression that contains trigonometric functions means that the expression is written as a numerical value or in terms of a single trigonometric function, if possible.

Example 3 Simplify a Trigonometric Expression

Simplify $\frac{\csc\theta\cos\theta}{\sin\theta\cot\theta}$.

$\dfrac{\csc\theta\cos\theta}{\sin\theta\cot\theta} = \dfrac{\frac{1}{\sin\theta}\cdot\cos\theta}{\sin\theta\,\cot\theta}$	Reciprocal Identity, $\csc\theta = \frac{1}{\sin\theta}$
$= \dfrac{\frac{\cos\theta}{\sin\theta}}{\sin\theta\,\cot\theta}$	Simplify the numerator.
$= \dfrac{\frac{\cos\theta}{\sin\theta}}{\sin\theta\cdot\frac{\cos\theta}{\sin\theta}}$	$\cot\theta = \frac{\cos\theta}{\sin\theta}$
$= \dfrac{\frac{\cos\theta}{\sin\theta}}{\cos\theta}$	Simplify.
$= \dfrac{\frac{\cos\theta}{\sin\theta}\cdot\frac{1}{\cos\theta}}{\cos\theta\cdot\frac{1}{\cos\theta}}$	Multiply by $\frac{\frac{1}{\cos\theta}}{\frac{1}{\cos\theta}}$.
$= \dfrac{1}{\sin\theta}$	Simplify.
$= \csc\theta$	Reciprocal Identity, $\frac{1}{\sin\theta} = \csc\theta$

Check

Simplify $\dfrac{\tan^2 x \csc^2 x - 1}{\sec^2 x}$.

Example 4 Simplify a Trigonometric Expression by Factoring

Simplify $-\sec\left(\frac{\pi}{2} - \theta\right) - \cot^2\theta\csc\theta.$

$-\sec\left(\frac{\pi}{2} - \theta\right) - \cot^2\theta\csc\theta$ Original expression

$= -\dfrac{1}{\cos\left(\frac{\pi}{2} - \theta\right)} - \cot^2\pi\csc\theta$ $\sec\theta = \frac{1}{\cos\theta}$

$= -\dfrac{1}{\sin\theta} - \cot^2\theta\csc\theta$ $\cos\left(\frac{\pi}{2} - \theta\right) = \sin\theta$

$= -\csc\theta - \cot^2\theta\csc\theta$ $\dfrac{1}{\sin\theta} = \csc\theta$

$= -\csc\theta\,(1 + \cot^2\theta)$ Factor $-\csc\theta$ from each term.

$= -\csc\theta\,(\csc^2\theta)$ $\cot^2\theta + 1 = \csc^2\theta$

$= -\csc^3\theta$ Simplify.

Check

Simplify $\cos x\tan^2 x - \cos x\sec^2 x.$

Pause and Reflect

Did you struggle with anything in this lesson so far? If so, how did you deal with it?

Talk About It!

Without calculating the answer, would you expect $-\csc^3\frac{8\pi}{7}$ to be positive or negative? Explain.

Go Online You can complete an Extra Example online.

🌍 Example 5 Use a Trigonometric Expression

ART GALLERY An art gallery is rearranging its large canvas murals. One of the gallery's hallways that is 8 feet wide meets another hallway that is 10 feet wide at a right angle. The length L of the longest canvas mural that can be maneuvered around the corner, without tipping the canvas, can be represented by $L = \frac{8 \cot \theta \cos \theta}{1 - \sin^2 \theta} + 10 \csc \left(\frac{\pi}{2} - \theta\right)$, where θ is the angle between the canvas and the wall of the narrower hallway. Rewrite the equation in terms of sine and cosine.

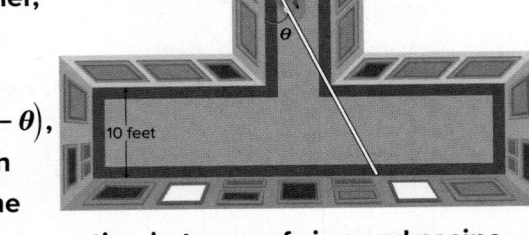

🍩 **Think About It!**

Find L if the angle between the canvas and the wall of the narrower hallway is 30°. Round to the nearest tenth.

$L = \dfrac{8 \cot \theta \cos \theta}{1 - \sin^2 \theta} + 10 \csc\left(\dfrac{\pi}{2} - \theta\right)$ Original equation

$= \dfrac{8 \cot \theta \cos \theta}{1 - \sin^2 \theta} + 10 \cdot \dfrac{1}{\sin\left(\frac{\pi}{2} - \theta\right)}$ $\csc \theta = \dfrac{1}{\sin \theta}$

$= \dfrac{8 \cot \theta \cos \theta}{1 - \sin^2 \theta} + 10 \cdot \dfrac{1}{\cos \theta}$ $\sin\left(\frac{\pi}{2} - \theta\right) = \cos \theta$

$= \dfrac{8 \cot \theta \cos \theta}{1 - \sin^2 \theta} + \dfrac{10}{\cos \theta}$ Multiply.

$= \dfrac{8 \cot \theta \cos \theta}{\cos^2 \theta} + \dfrac{10}{\cos \theta}$ $\cos^2 \theta + \sin^2 \theta = 1$

$= \dfrac{8 \cot \theta}{\cos \theta} + \dfrac{10}{\cos \theta}$ $\dfrac{\cos \theta}{\cos \theta} = 1$

$= \dfrac{8\left(\frac{\cos \theta}{\sin \theta}\right)}{\cos \theta} + \dfrac{10}{\cos \theta}$ $\cot \theta = \dfrac{\cos \theta}{\sin \theta}$

$= \dfrac{8\left(\frac{\cos \theta}{\sin \theta}\right) \cdot \frac{1}{\cos \theta}}{\cos \theta \cdot \frac{1}{\cos \theta}} + \dfrac{10}{\cos \theta}$ Multiply the numerator and denominator by $\frac{1}{\cos \theta}$.

$= \dfrac{8}{\sin \theta} + \dfrac{10}{\cos \theta}$ Simplify.

The length of the longest mural canvas can be expressed as $L = \dfrac{8}{\sin \theta} + \dfrac{10}{\cos \theta}$.

Check

FOOTBALL A football player kicks a field goal at an initial velocity of 55 feet per second. The maximum height h in feet of the football is given by the equation $h = \dfrac{47 \tan^2 x}{\sec^2 x}$, where x is the degree measure of the angle at which the ball is kicked. Write the simplified form of the expression of the equation.

$h = $ _____?_____

🔎 **Go Online** You can complete an Extra Example online.

Problem-Solving Tip

Identify Subgoals
Often there are multiple ways to simplify a trigonometric expression. Before beginning a problem, it may help to identify the subgoals and possible identities that can be used to simplify the expression. This will help you determine the shortest and clearest way to simplify.

Practice

Go Online You can complete your homework online.

Examples 1 and 2

Find the exact value of each expression if $0° < \theta < 90°$.

1. If $\sin \theta = \frac{3}{5}$, find $\cos \theta$.   **2.** If $\sin \theta = \frac{1}{2}$, find $\tan \theta$.   **3.** If $\cos \theta = \frac{3}{5}$, find $\csc \theta$.

Find the exact value of each expression if $180° < \theta < 270°$.

4. If $\sin \theta = -\frac{1}{2}$, find $\cos \theta$. **5.** If $\cos \theta = -\frac{3}{5}$, find $\csc \theta$. **6.** If $\sec \theta = -3$, find $\tan \theta$.

Find the exact value of each expression if $270° < \theta < 360°$.

7. If $\csc \theta = -\frac{5}{3}$, find $\cos \theta$. **8.** If $\cos \theta = \frac{5}{13}$, find $\sin \theta$.   **9.** If $\tan \theta = -1$, find $\sec \theta$.

10. Find the exact value of $\tan \left(\theta - \frac{\pi}{2} \right)$ if $\cot \theta = 1.53$.

11. Find the exact value of $\cos \theta$ if $\sin \left(\theta - \frac{\pi}{2} \right) = 2.5$ and $0 < \theta < \frac{\pi}{2}$.

12. Find the exact value of $\cos \theta$ if $\sin \theta = \frac{2}{3}$ and $90° < \theta < 180°$.

13. Find $\sin \theta$ if $\cos \theta = \frac{3}{4}$ and $0° \leq \theta < 90°$.

14. Find $\cos \theta$ if $\sin \theta = \frac{1}{2}$ and $90° \leq \theta < 180°$.

15. Find $\cot \theta$ if $\tan \theta = 2$ and $180° < \theta < 270°$.

Examples 3 and 4

Simplify each expression.

16. $\sec \theta \tan^2 \theta + \sec \theta$ **17.** $\cos \left(\frac{\pi}{2} - \theta \right) \cot \theta$ **18.** $\sin \theta \sec \theta$

19. $\cot \theta \sec \theta$ **20.** $(\sin \theta)(1 + \cot^2 \theta)$ **21.** $\csc \theta \sin \theta$

22. $\sin \left(\frac{\pi}{2} - \theta \right) \sec \theta$ **23.** $\frac{\cos (-\theta)}{\sin (-\theta)}$ **24.** $\frac{\cos \theta}{\sec \theta}$

25. $4(\tan^2 \theta - \sec^2 \theta)$ **26.** $\frac{1 + \tan^2 \theta}{\csc^2 \theta}$ **27.** $\csc \theta \tan \theta - \tan \theta \sin \theta$

Example 5

28. WAVES The path P of a wave in the ocean is given by the equation

$P = \frac{1 + \sin^2 \theta \sec^2 \theta}{\sec^2 \theta} - \cos^2 \theta$, where θ is the angle between sea level and the wave. Simplify the equation.

29. LIGHT WAVE The distance in feet a light wave d is from its source is given by the equation $d = \sin^2 \theta + \tan^2 \theta + \cos^2 \theta$, where θ is the angle between the source and the light wave. Simplify the equation.

Mixed Exercises

STRUCTURE **Simplify each expression.**

30. $\dfrac{1 - \sin^2 \theta}{\sin \theta + 1}$

31. $\csc \theta + \cot \theta$

32. $\dfrac{1 - \sin^2 \theta}{\sin^2 \theta}$

33. $\tan \theta \csc \theta$

34. $\dfrac{1}{\sin^2 \theta} - \dfrac{\cos^2 \theta}{\sin^2 \theta}$

35. $2(\csc^2 \theta - \cot^2 \theta)$

36. $(1 + \sin \theta)(1 - \sin \theta)$

37. $2 - 2 \sin^2 \theta$

38. $\dfrac{\tan \left(\frac{\pi}{2} - \theta\right) \sec \theta}{1 - \csc^2 \theta}$

39. $\dfrac{\cos \left(\frac{\pi}{2} - \theta\right) - 1}{1 + \sin (-\theta)}$

40. $\dfrac{\sec \theta \sin \theta + \cos \left(\frac{\pi}{2} - \theta\right)}{1 + \sec \theta}$

41. $\dfrac{\cot \theta \cos \theta}{\tan (-\theta) \sin \left(\frac{\pi}{2} - \theta\right)}$

42. PERSEVERE Find a counterexample to show that $1 - \sin x = \cos x$ is not an identity.

43. WRITE Pythagoras is most famous for the Pythagorean Theorem. The identity $\cos^2 \theta + \sin^2 \theta = 1$ is an example of a Pythagorean identity. Why do you think that this identity is classified in this way?

44. PERSEVERE Prove that $\tan (-a) = -\tan a$ by using the quotient and negative angle identities.

45. CREATE Write two expressions that are equivalent to $\tan \theta \sin \theta$.

46. ANALYZE Explain how you can use division to rewrite $\sin^2 \theta + \cos^2 \theta = 1$ as $1 + \cot^2 \theta = \csc^2 \theta$.

47. FIND THE ERROR Jordan and Ebony are simplifying $\dfrac{\sin^2 \theta}{\cos^2 \theta + \sin^2 \theta}$. Is either of them correct? Explain your reasoning.

Jordan	Ebony
$\dfrac{\sin^2 \theta}{\cos^2 \theta + \sin^2 \theta} = \dfrac{\sin^2 \theta}{\cos^2 \theta} \dfrac{\sin^2 \theta}{\sin^2 \theta}$ $= \tan^2 \theta + 1$ $= \sec^2 \theta$	$\dfrac{\sin^2 \theta}{\cos^2 \theta + \sin^2 \theta} = \dfrac{\sin^2 \theta}{1}$ $= \sin^2 \theta$

48. PERSEVERE Prove that $\tan^2 \theta + 1 = \sec^2 \theta$ and $\cot^2 \theta + 1 = \csc^2 \theta$.

Verifying Trigonometric Identities

Learn Verify Trigonometric Identities by Transforming One Side

You can use the basic trigonometric identities along with the definitions of the trigonometric functions to verify identities. By proving that both sides of the equation are equal for all defined values of the variable, you can show that the identity is true.

Key Concept • Strategies for Verifying Trigonometric Identities

• First, try to verify the identity by simplifying one side. It is usually best to start with the more complicated side and simplify it to match the less complicated side.

• Try substituting basic identities. Rewriting everything on the more complicated side in terms of sine and cosine may make the steps easier.

• Use what you have learned from algebra to simplify. Factor, multiply, add, simplify, or combine fractions as necessary.

• When a term includes $1 + \sin\theta$ or $1 + \cos\theta$, think about multiplying the numerator and denominator by the conjugate. Then you can use a Pythagorean Identity.

Think About It!

Why should every step be given a reason, usually another verified trigonometric identity, algebraic operation, or definition, when verifying an identity?

Example 1 Verify a Trigonometric Identity by Transforming One Side

Verify that $\cot\theta\,(\cot\theta + \tan\theta) = \csc^2\theta$ is an identity.

$\cot\theta\,(\cot\theta + \tan\theta) \overset{?}{=} \csc^2\theta$ Original equation

$\cot^2\theta + \cot\theta\tan\theta \overset{?}{=} \csc^2\theta$ Distributive Property

$\cot^2\theta + \dfrac{\sin\theta}{\cos\theta}\cdot\dfrac{\cos\theta}{\sin\theta} \overset{?}{=} \csc^2\theta$ Quotient Identities

$\cot^2\theta + \dfrac{\cos\theta\,\sin\theta}{\sin\theta\,\cos\theta} \overset{?}{=} \csc^2\theta$ Multiply.

$\cot^2\theta + 1 \overset{?}{=} \csc^2\theta$ Simplify.

$\csc^2\theta = \csc^2\theta$ Pythagorean Identity

Because $\csc^2\theta = \csc^2\theta$ is true for all values of θ, $\cot\theta\,(\cot\theta + \tan\theta) = \csc^2\theta$ is an identity.

Think About It!

As you are verifying an identity, what are some ways to quickly check that the simplified identity is equivalent to the original identity?

🖱 **Go Online** You can complete an Extra Example online.

Alternate Method

$$\cot \theta \, (\cot \theta + \tan \theta) \overset{?}{=} \csc^2 \theta \quad \text{Original equation}$$

$$\cot \theta \left(\frac{\cos \theta}{\sin \theta} + \frac{\sin \theta}{\cos \theta} \right) \overset{?}{=} \csc^2 \theta \quad \text{Quotient Identities}$$

$$\cot \theta \left(\frac{\cos \theta}{\sin \theta} \cdot \frac{\cos \theta}{\cos \theta} + \frac{\sin \theta}{\cos \theta} \cdot \frac{\sin \theta}{\sin \theta} \right) \overset{?}{=} \csc^2 \theta \quad \text{Find LCD.}$$

$$\cot \theta \left(\frac{\cos^2 \theta}{\sin \theta \cos \theta} + \frac{\sin^2 \theta}{\sin \theta \cos \theta} \right) \overset{?}{=} \csc^2 \theta \quad \text{Simplify.}$$

$$\cot \theta \left(\frac{\cos^2 \theta + \sin^2 \theta}{\sin \theta \cos \theta} \right) \overset{?}{=} \csc^2 \theta \quad \text{Combine fractions.}$$

$$\cot \theta \left(\frac{1}{\sin \theta \cos \theta} \right) \overset{?}{=} \csc^2 \theta \quad \text{Pythagorean Identity}$$

$$\left(\frac{\cos \theta}{\sin \theta} \right) \left(\frac{1}{\sin \theta \cos \theta} \right) \overset{?}{=} \csc^2 \theta \quad \text{Quotient Identity}$$

$$\frac{1}{\sin^2 \theta} \overset{?}{=} \csc^2 \theta \quad \text{Simplify.}$$

$$\csc^2 \theta = \csc^2 \theta \quad \text{Reciprocal Identity}$$

Examine the Alternate Method. Compare and contrast the methods.

The alternate method simplified the expression within the parentheses instead of distributing the cotangent function. The alternate method required more steps, but both methods proved that the equation is an identity.

Check

Write the explanations to verify that $\csc \theta \cdot \sec \theta = \tan \theta + \cot \theta$ is an identity.

$$\csc \theta \cdot \sec \theta \overset{?}{=} \tan \theta + \cot \theta \qquad \text{Original equation}$$

$$\csc \theta \cdot \sec \theta \overset{?}{=} \frac{\sin \theta}{\cos \theta} + \frac{\cos \theta}{\sin \theta} \qquad \underline{\hspace{3cm} ? \hspace{3cm}}$$

$$\csc \theta \cdot \sec \theta \overset{?}{=} \frac{\sin \theta}{\cos \theta} \cdot \frac{\sin \theta}{\sin \theta} + \frac{\cos \theta}{\sin \theta} \cdot \frac{\cos \theta}{\cos \theta} \qquad \text{Find common denominator.}$$

$$\csc \theta \cdot \sec \theta \overset{?}{=} \frac{\sin^2 \theta}{\cos \theta \sin \theta} + \frac{\cos^2 \theta}{\cos \theta \sin \theta} \qquad \text{Simplify.}$$

$$\csc \theta \cdot \sec \theta \overset{?}{=} \frac{1}{\cos \theta \sin \theta} \qquad \underline{\hspace{3cm} ? \hspace{3cm}}$$

$$\csc \theta \cdot \sec \theta = \csc \theta \cdot \sec \theta \qquad \underline{\hspace{3cm} ? \hspace{3cm}}$$

Example 2 Verify a Trigonometric Identity by Transforming One Side

Verify that $2 \sec^2 x = \dfrac{1}{1 - \sin x} + \dfrac{1}{1 + \sin x}$ **is an identity.**

$2 \sec^2 x \overset{?}{=} \dfrac{1}{1 - \sin x} + \dfrac{1}{1 + \sin x}$ Original equation

$2 \sec^2 x \overset{?}{=} \dfrac{1}{1 - \sin x} \cdot \dfrac{1 + \sin x}{1 + \sin x} + \dfrac{1}{1 + \sin x} \cdot \dfrac{1 - \sin x}{1 - \sin x}$ Find common denominators.

$2 \sec^2 x \overset{?}{=} \dfrac{1 + \sin x}{1 - \sin^2 x} + \dfrac{1 - \sin x}{1 - \sin^2 x}$ Simplify.

$2 \sec^2 x \overset{?}{=} \dfrac{1 + \sin x + 1 - \sin x}{1 - \sin^2 x}$ Combine fractions.

$2 \sec^2 x \overset{?}{=} \dfrac{2}{1 - \sin^2 x}$ Simplify the numerator.

$2 \sec^2 x \overset{?}{=} \dfrac{2}{\cos^2 x}$ Pythagorean Identity

$2 \sec^2 x = 2 \sec^2 x$ Reciprocal Identity

$2 \sec^2 x = \dfrac{1}{1 - \sin x} + \dfrac{1}{1 + \sin x}$ is an identity.

Talk About It!

In the first step, the fractions on the right side of the equation were multiplied by two different fractions, $\dfrac{1 + \sin x}{1 + \sin x}$ and $\dfrac{1 - \sin x}{1 - \sin x}$, while the left side of the equation was not multiplied by anything. Justify how equality was maintained.

Check

Complete each statement or reason to verify that $\dfrac{\sin \theta}{1 - \cos \theta} = \csc \theta + \cot \theta$ is an identity.

$\dfrac{\sin \theta}{1 - \cos \theta} \overset{?}{=} \csc \theta + \cot \theta$ Original equation

$\underline{\hspace{3cm} ? \hspace{3cm}} \overset{?}{=} \csc \theta + \cot \theta$ Multiply the numerator and denominator by $1 + \cos \theta$.

$\dfrac{\sin \theta + \sin \theta \cos \theta}{1 - \cos^2 \theta} \overset{?}{=} \csc \theta + \cot \theta$ Multiply.

$\underline{\hspace{2cm} ? \hspace{2cm}} \overset{?}{=} \csc \theta + \cot \theta$ $\underline{\hspace{3cm} ? \hspace{3cm}}$

$\underline{\hspace{2cm} ? \hspace{2cm}} \overset{?}{=} \csc \theta + \cot \theta$ $\underline{\hspace{3cm} ? \hspace{3cm}}$

$\underline{\hspace{1.5cm} ? \hspace{1.5cm}} \overset{?}{=} \csc \theta + \cot \theta$ Divide by the common factor.

$\csc \theta + \cot \theta = \csc \theta + \cot \theta$ $\underline{\hspace{3cm} ? \hspace{3cm}}$

Study Tip

Transforming When transforming an equation to verify an identity, try to simplify the more complicated side of the equation until both sides are the same.

Study Tip

Additional Steps
Because there are often many alternative methods to verify an identity, you may notice that the same identity can be verified using fewer or more steps. If each step is mathematically sound and the final equation is true, then the verification should be correct, regardless of the number of steps.

Example 3 Verify a Trigonometric Identity by Transforming Each Side

Verify that $\frac{\sin\theta\tan\theta}{1-\cos\theta} = (1+\cos\theta)\sec\theta$ **is an identity.**

$\frac{\sin\theta\tan\theta}{1-\cos\theta} \overset{?}{=} (1+\cos\theta)\sec\theta$	Original equation
$\frac{\sin\theta\cdot\frac{\sin\theta}{\cos\theta}}{1-\cos\theta} \overset{?}{=} (1+\cos\theta)\sec\theta$	Reciprocal Identity
$\frac{\frac{\sin^2\theta}{\cos\theta}}{1-\cos\theta}\cdot\frac{1+\cos\theta}{1+\cos\theta} \overset{?}{=} (1+\cos\theta)\sec\theta$	Multiply by $\frac{1+\cos\theta}{1+\cos\theta}=1$.
$\frac{\frac{\sin^2\theta}{\cos\theta}+\sin^2\theta}{1-\cos^2\theta} \overset{?}{=} (1+\cos\theta)\sec\theta$	Distributive Property
$\frac{\frac{\sin^2\theta}{\cos\theta}+\sin^2\theta}{\sin^2\theta} \overset{?}{=} (1+\cos\theta)\sec\theta$	Pythagorean Identity
$\frac{\sin^2\theta}{\cos\theta}\cdot\frac{1}{\sin^2\theta}+\sin^2\theta\cdot\frac{1}{\sin^2\theta} \overset{?}{=} (1+\cos\theta)\sec\theta$	Simplify the complex fraction.
$\frac{1}{\cos\theta}+1 \overset{?}{=} (1+\cos\theta)\sec\theta$	Simplify.
$\sec\theta+1 \overset{?}{=} (1+\cos\theta)\sec\theta$	Reciprocal Identity

Isolate the right side of the equation, $(1+\cos\theta)\sec\theta$, with the goal of simplifying the expression to $\sec\theta + 1$.

$\frac{\sin\theta\tan\theta}{1-\cos\theta} \overset{?}{=} (1+\cos\theta)\sec\theta$	Original equation
$\sec\theta+1 \overset{?}{=} \sec\theta+\cos\theta\sec\theta$	Distribute.
$\sec\theta+1 \overset{?}{=} \sec\theta+\cos\theta\cdot\frac{1}{\cos\theta}$	Reciprocal Identity
$\sec\theta+1 = \sec\theta+1$	Simplify.

Because both sides of the equation can be simplified to $\sec\theta + 1$, $\frac{\sin\theta\tan\theta}{1-\cos\theta} = (1+\cos\theta)\sec\theta$ is an identity.

Check

Verify that $\sin^2\theta+\cos^2\theta-\sec^2\theta = \frac{1+\tan^2\theta}{\csc^2\theta}$ **is an identity.**

$\sin^2\theta+\cos^2\theta-\sec^2\theta \overset{?}{=} \frac{1+\tan^2\theta}{\csc^2\theta}$	Original equation
$\underline{\quad?\quad} \overset{?}{=} \frac{1+\tan^2\theta}{\csc^2\theta}$	$\underline{\quad?\quad}$
$\tan^2\theta \overset{?}{=} \frac{1+\tan^2\theta}{\csc^2\theta}$	$\underline{\quad?\quad}$
$\tan^2\theta \overset{?}{=} \frac{\sec^2\theta}{\csc^2\theta}$	$\underline{\quad?\quad}$
$\tan^2\theta \overset{?}{=} \frac{\frac{1}{\cos^2\theta}}{\frac{1}{\sin^2\theta}}$	$\underline{\quad?\quad}$
$\tan^2\theta \overset{?}{=} \frac{\sin^2\theta}{\cos^2\theta}$	Simplify the complex fraction.
$\tan^2\theta = \tan^2\theta$	$\underline{\quad?\quad}$

Go Online You can complete an Extra Example online.

Practice

Examples 1 and 2

Verify that each equation is an identity by transforming one side.

1. $\cos^2 \theta + \tan^2 \theta \cos^2 \theta = 1$

2. $\cot \theta (\cot \theta + \tan \theta) = \csc^2 \theta$

3. $1 + \sec^2 \theta \sin^2 \theta = \sec^2 \theta$

4. $\sin \theta \sec \theta \cot \theta = 1$

5. $\dfrac{1 - \cos \theta}{1 + \cos \theta} = (\csc \theta - \cot \theta)^2$

6. $\dfrac{1 - 2\cos^2 \theta}{\sin \theta \cos \theta} = \tan \theta - \cot \theta$

7. $(\sin \theta - 1)(\tan \theta + \sec \theta) = -\cos \theta$

8. $\cos \theta \cos (-\theta) - \sin \theta \sin (-\theta) = 1$

9. $\sec \theta - \tan \theta = \dfrac{1 - \sin \theta}{\cos \theta}$

10. $\dfrac{1 + \tan \theta}{\sin \theta + \cos \theta} = \sec \theta$

Example 3

Verify that each equation is an identity by transforming each side.

11. $\left(\sin \theta + \dfrac{\cot \theta}{\csc \theta}\right)^2 = \dfrac{2 + \sec \theta \csc \theta}{\sec \theta \csc \theta}$

12. $\dfrac{\cos \theta}{1 - \sin \theta} = \dfrac{1 + \sin \theta}{\cos \theta}$

13. $\csc^2 \theta - 1 = \dfrac{\cot^2 \theta}{\csc \theta \sin \theta}$

14. $\cos \theta \cot \theta = \csc \theta - \sin \theta$

15. $\csc^2 \theta = \cot^2 \theta + \sin \theta \csc \theta$

16. $\dfrac{\sec \theta - \csc \theta}{\csc \theta \sec \theta} = \dfrac{\sin \theta - \cos \theta}{\sin^2 \theta + \cos^2 \theta}$

Mixed Exercises

Verify that each equation is an identity.

17. $\tan \theta \cos \theta = \sin \theta$

18. $\cot \theta \tan \theta = 1$

19. $(\tan \theta)(1 - \sin^2 \theta) = \sin \theta \cos \theta$

20. $\dfrac{\csc \theta}{\sec \theta} = \cot \theta$

21. $\dfrac{\sin^2 \theta}{1 - \sin^2 \theta} = \tan^2 \theta$

22. $\dfrac{\cos^2 \theta}{1 - \sin \theta} = 1 + \sin \theta$

23. **OPTICS** The polarizing angle for any substance can be found using Brewster's Law. It states that the relationship between the two indices of fractions n_1 and n_2 and the polarizing angle θ_p is $\tan \theta_p = \dfrac{n_2}{n_1}$. Use the law of refraction $n_1 \sin \theta_p = n_2 \sin \theta_r$ to prove Brewster's Law.

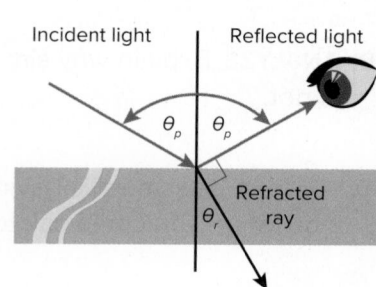

24. **STRUCTURE** The graph of $y = \frac{\cos^2 x}{1 - \sin x}$ is shown below on the left. The graph of $y = \sin x$ is shown below on the right. Use the graphs to write an identity involving $\frac{\cos^2 x}{1 - \sin x}$ and $\sin x$. Then verify the identity.

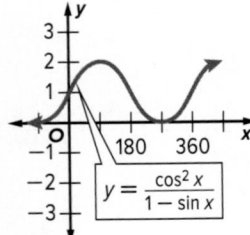

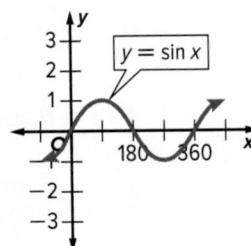

25. **CONSTRUCT ARGUMENTS** Show two different methods of verifying that $\frac{1}{1 - \sin^2 \theta} = \tan^2 \theta + 1$ is a trigonometric identity.

Determine whether each equation is an identity. Justify your argument.

26. $\dfrac{\tan\left(\frac{\pi}{2} - \theta\right) \csc \theta}{\csc^2 \theta} = \cos \theta$

27. $\dfrac{1 + \tan \theta}{1 + \cot \theta} = \tan \theta$

28. $\sin \theta \csc (-\theta) = 1$

29. $\dfrac{\sec^2 \theta - \tan^2 \theta}{\cos^2 \theta + \sin^2 \theta} = 1$

30. $\tan \theta + \cos \theta = \sin \theta$

31. $\cot (-\theta) \cot \left(\frac{\pi}{2} - \theta\right) = 1$

32. $\sec \theta \sin \left(\frac{\pi}{2} - \theta\right) = 1$

33. $\dfrac{1 + \tan^2 \theta}{\csc^2 \theta} = \sin^2 \theta$

34. $\sec^2 (-\theta) - \tan^2 (-\theta) = 1$

35. **REASONING** Diego decides that if $\sin^2 A + \cos^2 B = 1$, and A and B both have measures between 0° and 180°, then $A = B$. Is he correct? Explain your reasoning.

36. **USE TOOLS** How can you use your calculator to show that $\sin^2 \theta + \csc^2 \theta = 1$ is not an identity? Explain your reasoning.

Higher-Order Thinking Skills

37. **WHICH ONE DOESN'T BELONG?** Identify the equation that does not belong with the other three. Justify your conclusion.

$\sin^2\theta + \cos^2\theta = 1$	$1 + \cot^2\theta = \csc^2\theta$
$\sin^2\theta - \cos^2\theta = 2\sin^2\theta$	$\tan^2\theta + 1 = \sec^2\theta$

38. **PERSEVERE** Transform the right side of $\tan^2 \theta = \frac{\sin^2 \theta}{\cos^2 \theta}$ to show that $\tan^2 \theta = \sec^2 \theta - 1$.

39. **ANALYZE** Explain why $\sin^2 \theta + \cos^2 \theta = 1$ is an identity, but $\sin \theta = \sqrt{1 - \cos^2 \theta}$ is not.

40. **WRITE** A classmate is having trouble trying to verify a trigonometric identity. Write a question you could ask to help her work through the problem.

41. **CREATE** Let $x = \frac{1}{2} \tan \theta$, where $-\frac{\pi}{2} < \theta < \frac{\pi}{2}$. Write $f(x) = \dfrac{x}{\sqrt{1 + 4x^2}}$ in terms of a single trigonometric function of θ.

42. **CREATE** A statement such as $\cos \theta = 2$ is a *contradiction*. Write two contradictions involving the sine or tangent function.

Sum and Difference Identities

Learn Use Sum and Difference Identities to Find Trigonometric Values

By writing angle measures as the sums or differences of more familiar angle measures, you can use these sum and difference identities to find exact values of trigonometric functions for angles that are less common.

Key Concept • Sum and Difference Identities

$\sin (A + B) = \sin A \cos B + \cos A \sin B$

$\cos (A + B) = \cos A \cos B - \sin A \sin B$

$\tan (A + B) = \dfrac{\tan A + \tan B}{1 - \tan A \tan B}$, $A, B \neq 90° + 180n°$, where n is any integer

$\sin (A - B) = \sin A \cos B - \cos A \sin B$

$\cos (A - B) = \cos A \cos B + \sin A \sin B$

$\tan (A - B) = \dfrac{\tan A - \tan B}{1 + \tan A \tan B}$, $A, B \neq 90° + 180n°$, where n is any integer

Example 1 Use a Sum Identity

Find the exact value of sin 165°.

$\sin 165° = \sin (45° + 120°)$ $A = 45°, B = 120°$

$\qquad = \sin 45° \cos 120° + \cos 45° \sin 120°$ Sum Identity for Sine

$\qquad = \dfrac{\sqrt{2}}{2} \cdot \left(-\dfrac{1}{2}\right) + \dfrac{\sqrt{2}}{2} \cdot \dfrac{\sqrt{3}}{2}$ Evaluate each expression.

$\qquad = -\dfrac{\sqrt{2}}{4} + \dfrac{\sqrt{6}}{4}$ Multiply.

$\qquad = \dfrac{\sqrt{6} - \sqrt{2}}{4}$ Combine the fractions.

Check

Find the exact value of tan 255°.

A. undefined

B. $\sqrt{3} + 2$

C. $\dfrac{2}{3}$

D. $\sqrt{3} - 2$

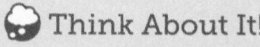 **Go Online** You can complete an Extra Example online.

Today's Goals
- Find values of sine and cosine by using sum and difference identities.
- Verify trigonometric identities by using sum and difference identities.

🗪 Talk About It!
Which of the sum and difference identities are undefined for certain angle measures? Explain your reasoning.

🡢 Go Online
to see a common error to avoid.

💭 Think About It!
How can you check your solution?

Example 2 Use a Difference Identity

Find the exact value of cos $\frac{5\pi}{12}$.

$\cos\frac{5\pi}{12} = \cos\left(\frac{2\pi}{3} - \frac{\pi}{4}\right)$ $A = \frac{2\pi}{3}, B = \frac{\pi}{4}$

$\qquad\quad = \cos\frac{2\pi}{3}\cos\frac{\pi}{4} - \sin\frac{2\pi}{3}\sin\frac{\pi}{4}$ Difference Identity for Cosine

$\qquad\quad = \left(-\frac{1}{2}\cdot\frac{\sqrt{2}}{2}\right) - \left(\frac{\sqrt{3}}{2}\cdot\frac{\sqrt{2}}{2}\right)$ Evaluate each expression.

$\qquad\quad = -\frac{\sqrt{2}}{4} - \frac{\sqrt{6}}{4}$ Multiply.

$\qquad\quad = -\frac{\sqrt{2}+\sqrt{6}}{4}$ Combine the fractions.

Check
Find the exact value of $\sin\left(-\frac{7\pi}{12}\right)$.

🌐 Example 3 Use Sum and Difference Identities

ELECTRICITY For a certain circuit carrying alternating current, the formula $c = 4\sin 25t°$ can be used to find the current c after t seconds. Find the exact current in amperes at $t = 3$ seconds.

$c = 4\sin 25t°$ Original equation

$\quad = 4\sin(25\cdot 3)°$ $t = 3$

$\quad = 4\sin 75°$ Multiply.

$\quad = 4\sin(30° + 45°)$ $30° + 45° = 75°$

$\quad = 4(\sin 30°\cos 45° + \cos 30°\sin 45°)$ Sum Identity for Sine

$\quad = 4\left(\frac{1}{2}\cos 45° + \cos 30°\cdot\frac{\sqrt{2}}{2}\right)$ Evaluate sin 30° and sin 45°.

$\quad = 4\left(\frac{1}{2}\cdot\frac{\sqrt{2}}{2} + \frac{\sqrt{3}}{2}\cdot\frac{\sqrt{2}}{2}\right)$ Evaluate cos 30° and cos 45°.

$\quad = 4\left(\frac{\sqrt{2}}{4} + \frac{\sqrt{6}}{4}\right)$ Multiply.

$\quad = 4\left(\frac{\sqrt{2}+\sqrt{6}}{4}\right)$ Add.

$\quad = \sqrt{2} + \sqrt{6}$ Distribute.

The exact current after 3 seconds is $\sqrt{2} + \sqrt{6}$ amperes.

Check

ART A digital artist uses tessellations of basic shape patterns to create fractal art. To make one shape, he combines two right triangles as shown, with lengths measured in pixels. In order to complete his work, he needs to find the exact value of the sine of angle *BAC*. What is it?

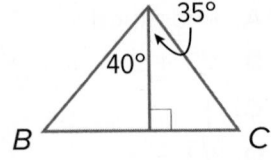

🔘 **Go Online** You can complete an Extra Example online.

Think About It!
How would you find the exact value of csc 105°?

Think About It!
Why is the degree symbol necessary in $c = 4\sin 25t°$ in the context of this problem?

Learn Use Sum and Difference Identities to Verify Trigonometric Identities

Sum and difference identities can be used to rewrite trigonometric expressions in which one of the angles is a multiple of 90°. The resulting identity is called a reduction identity because it reduces the complexity of the expression.

Example 4 Verify a Cofunction Identity

Verify that tan (90° − θ) = cot θ is an identity.

$$\tan (90° - \theta) \overset{?}{=} \cot \theta \qquad \text{Original equation}$$

$$\frac{\sin (90° - \theta)}{\cos (90° - \theta)} \overset{?}{=} \cot \theta \qquad \text{Quotient Identity}$$

$$\frac{\sin 90° \cos \theta - \cos 90° \sin \theta}{\cos 90° \cos \theta + \sin 90° \sin \theta} \overset{?}{=} \cot \theta \qquad \text{Difference Identities}$$

$$\frac{\sin 90° \cos \theta - 0 \cdot \sin \theta}{0 \cdot \cos \theta + \sin 90° \sin \theta} \overset{?}{=} \cot \theta \qquad \cos 90° = 0$$

$$\frac{1 \cdot \cos \theta - 0 \cdot \sin \theta}{0 \cdot \cos \theta + 1 \cdot \sin \theta} \overset{?}{=} \cot \theta \qquad \sin 90° = 1$$

$$\frac{\cos \theta}{\sin \theta} \overset{?}{=} \cot \theta \qquad \text{Simplify.}$$

$$\cot \theta = \cot \theta \qquad \text{Quotient Identity}$$

Check

Complete the statements to verify that cos (90° − θ) = sin θ is an identity.

$$\cos (90° - \theta) \overset{?}{=} \sin \theta \qquad \text{Original equation}$$

$$\underline{\hspace{3cm}?\hspace{3cm}} \overset{?}{=} \sin \theta \qquad \underline{\hspace{2cm}?\hspace{2cm}}$$

$$\underline{\hspace{2.5cm}?\hspace{2.5cm}} \overset{?}{=} \sin \theta \qquad \cos 90° = 0;\ \sin 90° = 1$$

$$\underline{\hspace{1cm}?\hspace{1cm}} = \sin \theta \qquad \text{Simplify.}$$

Go Online You can complete an Extra Example online.

Think About It!

Why is it easier to simplify an expression when one of the angles is a multiple of 90°?

Think About It!

Why can't the tangent difference identity be used to verify tan (90° − θ) = cot θ? Justify your argument.

Example 5 Verify a Reduction Identity

Verify that tan $(\theta - 180°) = $ tan θ.

$\tan(\theta - 180°) \overset{?}{=} \tan\theta$ Original equation

$\dfrac{\tan\theta - \tan 180°}{1 + \tan\theta \tan 180°} \overset{?}{=} \tan\theta$ Difference Identity

$\dfrac{\tan\theta - 0}{1 + \tan\theta \cdot 0} \overset{?}{=} \tan\theta$ tan 180° = 0

$\dfrac{\tan\theta}{1} \overset{?}{=} \tan\theta$ Simplify the numerator and denominator.

$\tan\theta = \tan\theta$ Simplify.

Check

Complete the statements to verify that cot $(\theta - 180°) = $ cot θ.

$\cot(\theta - 180°) \overset{?}{=} \cot\theta$ Original equation

$\dfrac{1}{\tan(\theta - 180°)} \overset{?}{=} \cot\theta$ _____?_____

_____?_____ $\overset{?}{=} \cot\theta$ Difference Identity

_____?_____ $\overset{?}{=} \cot\theta$ Simplify.

$\dfrac{1 + \tan\theta \cdot 0}{\tan\theta - 0} \overset{?}{=} \cot\theta$ tan 180° = 0

$\underline{\quad?\quad} \overset{?}{=} \cot\theta$ Simplify.

$\cot\theta = \cot\theta$ _____?_____

Pause and Reflect

Did you struggle with anything in this lesson? If so, how did you deal with it?

🅝 **Go Online** You can complete an Extra Example online.

Practice

🔵 **Go Online** You can complete your homework online.

Examples 1 and 2

Find the exact value of each expression.

1. $\sin 135°$

2. $\cos 165°$

3. $\cos \frac{7\pi}{12}$

4. $\sin \frac{\pi}{12}$

5. $\tan 195°$

6. $\cos\left(-\frac{\pi}{12}\right)$

Example 3

7. ART As part of a mosaic that an artist is making, she places two right triangular tiles together to make a new triangular piece. One tile has lengths of 3 inches, 4 inches, and 5 inches. The other tile has lengths 4 inches, $4\sqrt{3}$ inches, and 8 inches. The pieces are placed with the sides of 4 inches against each other as shown in the figure.

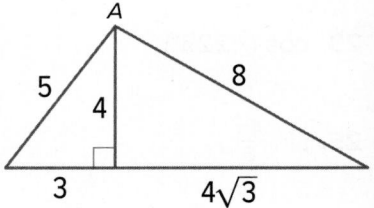

   a. What are the exact values of $\sin A$ and $\cos A$?

   b. What is the measure of angle A?

   c. Is the new triangle formed from the two triangular tiles also a right triangle? Explain.

8. CAMERAS Security cameras are being installed in the Community Center. One camera will be placed on the wall 4 yards above the pool deck. If the pool deck is 5 yards wide, through what angle θ must the camera rotate to view the entire length of the pool? Round to the nearest tenth of a degree.

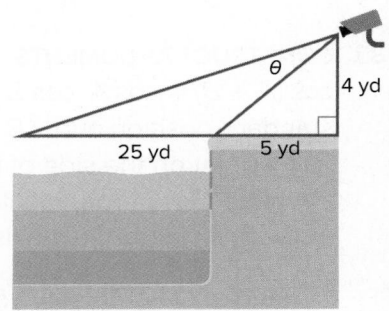

Examples 4 and 5

Verify that each equation is an identity.

9. $\cos\left(\frac{\pi}{2} + \theta\right) = -\sin\theta$

10. $\cos(60° + \theta) = \sin(30° - \theta)$

11. $\cos(180° + \theta) = -\cos\theta$

12. $\tan(\theta + 45°) = \frac{1 + \tan\theta}{1 - \tan\theta}$

Mixed Exercises

Find the exact value of each expression.

13. $\sin 330°$

14. $\cos (-165°)$

15. $\sin (-225°)$

16. $\cos 135°$

17. $\sin (-45)°$

18. $\cos 210°$

19. $\cos (-135°)$

20. $\tan 75°$

21. $\sin (-195°)$

22. $\sin 75°$

23. $\cos (-225°)$

24. $\tan 210°$

25. $\sin \frac{4\pi}{3}$

26. $\sin \frac{23\pi}{12}$

Verify that each equation is an identity.

27. $\sin (90° + \theta) = \cos \theta$

28. $\sin (180° + \theta) = -\sin \theta$

29. $\cos (270° - \theta) = -\sin \theta$

30. $\cos (\theta - 90°) = \sin \theta$

31. $\sin \left(\theta - \frac{\pi}{2}\right) = -\cos \theta$

32. $\cos (\pi + \theta) = -\cos \theta$

33. CONSTRUCT ARGUMENTS You can use the figure to prove that $\cos (A + B) = \cos A \cos B - \sin A \sin B$. Angle A was drawn in standard position and $\angle B$ shares a side with the $\angle A$, as shown. P is a point on the side of the angle with measure $A + B$ so that $OP = 1$. $\overline{PQ}$ is perpendicular to the terminal side of $\angle A$. $\overline{QS}$ is perpendicular to the x-axis. $\overline{QT}$ is perpendicular to $\overline{PR}$.

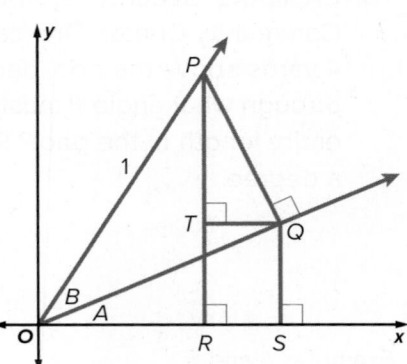

a. Find $m\angle TPQ$. Justify your answer.

b. Explain how to write PQ and OQ in terms of $\sin B$ and/or $\cos B$. (*Hint*: Focus on $\triangle POQ$.)

c. Use the side lengths of $\triangle QOS$ to write a ratio for $\cos A$. Then use this, and the expressions for PQ and OQ from **part b**, to write an expression for OS that involves sines and/or cosines of A and/or B.

d. Use the side lengths of $\triangle TPQ$ to write a ratio for $\sin A$. Then use this, and the expressions for PQ and OQ from **part b**, to write an expression for QT that involves sines and/or cosines of A and/or B.

e. Prove the sum formula for cosines by using your work in the previous steps.

Find the exact value of each expression.

34. tan 165°

35. sec 1275°

36. sin 735°

37. tan $\frac{23\pi}{12}$

38. csc $\frac{5\pi}{12}$

39. cot $\frac{113\pi}{12}$

Verify that each equation is an identity.

40. $\sin (A + B) = \frac{\tan A + \tan B}{\sec A \sec B}$

41. $\cos (A + B) = \frac{1 - \tan A \tan B}{\sec A \sec B}$

42. $\sec (A - B) = \frac{\sec A \sec B}{1 + \tan A \tan B}$

43. $\sin (A + B) \sin (A - B) = \sin^2 A - \sin^2 B$

44. CONSTRUCT ARGUMENTS Explain how to use the sum formula for cosines to prove the difference formula for cosines.

45. CONSTRUCT ARGUMENTS You can use the sum and difference formulas for sine and cosine to prove the sum and difference formulas for tangent.
 a. Prove the sum formula for tangent.

 b. Prove the difference formula for tangent.

46. REASONING The figure shows the graphs of $y = \sin \theta$ and $y = \cos \theta$.

 a. Explain how to use the graphs to find the value of h in the equation $\sin (\theta - h) = \cos \theta$.

 b. Use one or more sum or difference formulas to prove that you found the correct value of h.

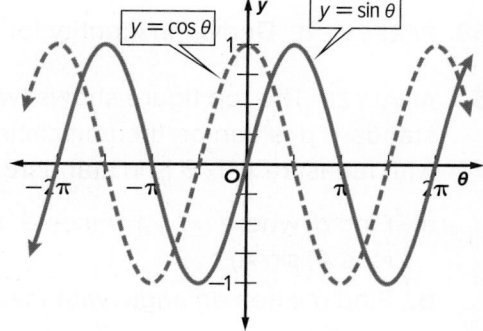

47. STRUCTURE Show how to find the exact value of $\sin \frac{13\pi}{12}$.

48. USE A MODEL Demetri stands 6 feet from the base of a flagpole and sights the top of the pole with an angle of elevation of 75°. His eyes are 5 feet above the ground. Find the exact height of the flagpole. Then find the height to the nearest tenth of a foot.

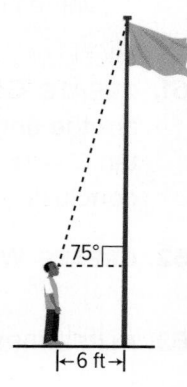

49. STRUCTURE Solve $\cos (\theta + \pi) = \sin (\theta - \pi)$ algebraically, given that $0 \le \theta \le \pi$. Explain your steps.

50. PRECISION Use sum or difference identities to prove that the equation $\cos \left(\frac{\pi}{2} - \theta\right) = \sin \theta$ is an identity.

STRUCTURE Use sum or difference identities to write each expression as a single trigonometric expression. Then find the exact value of each expression.

51. $\sin 19° \cos 11° + \cos 19° \sin 11°$

52. $\dfrac{\tan 144° - \tan 9°}{1 + \tan 144° \tan 9°}$

53. $\cos 111° \cos 21° + \sin 111° \sin 21°$

54. $\sin 108° \cos 18° - \cos 108° \sin 18°$

55. $\dfrac{\tan \frac{\pi}{16} + \tan \frac{3\pi}{16}}{1 - \tan \frac{\pi}{16} \tan \frac{3\pi}{16}}$

56. $\cos \frac{\pi}{3} \cos \frac{2\pi}{3} - \sin \frac{\pi}{3} \sin \frac{2\pi}{3}$

🧠 **Higher-Order Thinking Skills**

57. ANALYZE Simplify the following expression without expanding any of the sums or differences. $\sin \left(\frac{\pi}{3} - \theta\right) \cos \left(\frac{\pi}{3} + \theta\right) - \cos \left(\frac{\pi}{3} - \theta\right) \sin \left(\frac{\pi}{3} + \theta\right)$

58. WRITE You may have experienced a wireless Internet provider temporarily losing the signal. Waves that pass through the same place at the same time cause interference. Interference occurs when two waves combine to have a greater, or smaller, amplitude than either of the component waves. *Constructive interference occurs* when two waves combine to have a greater amplitude than either of the component waves. *Destructive interference* occurs when the component waves combine to have a smaller amplitude. Explain how the sum and difference identities are used to describe wireless Internet interference. Include an explanation of the difference between constructive and destructive interference.

59. PERSEVERE Derive an identity for $\cot (A + B)$ in terms of $\cot A$ and $\cot B$.

60. ANALYZE The top figure shows two angles A and B in standard position on the unit circle. In the bottom figure, an angle with measure $A - B$ is in standard position.

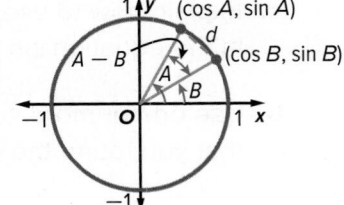

a. Find d, where $(x_1, y_1) = (\cos B, \sin B)$ and $(x_2, y_2) = (\cos A, \sin A)$.

b. Find d when an angle with measure $A - B$ is in standard position.

c. Find and simplify d^2 for each expression. Then equate the values of d^2 to derive a formula for $\cos (A - B)$.

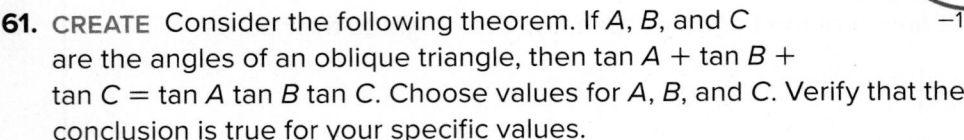

61. CREATE Consider the following theorem. If A, B, and C are the angles of an oblique triangle, then $\tan A + \tan B + \tan C = \tan A \tan B \tan C$. Choose values for A, B, and C. Verify that the conclusion is true for your specific values.

62. CREATE Write an expression that has an exact value of $\dfrac{\sqrt{6} - \sqrt{2}}{4}$.

63. PERSEVERE Verify that $\sin (90° - \theta) = \cos \theta$ is an identity using the Difference Identity.

Double-Angle and Half-Angle Identities

Today's Goals
- Find values of sine and cosine by using double-angle identities.
- Find values of sine and cosine by using half-angle identities.

Explore Proving the Double-Angle Identity for Cosine

 **Online Activity**

> **INQUIRY** How can you use an Angle Sum Identity to find a Double-Angle Identity?

Learn Double-Angle Identities

Double-angle formulas can help you find the value of a function of twice an angle measure. The double-angle formulas are identities, so they are true for all real numbers.

Key Concept • Double-Angle Identities

The following identities hold true for all values of θ.

$\sin 2\theta = 2 \sin \theta \cos \theta$ $\qquad \cos 2\theta = 2 \cos^2 \theta - 1$

$\cos 2\theta = \cos^2 \theta - \sin^2 \theta$ $\qquad \cos 2\theta = 1 - 2 \sin^2 \theta$

$\tan 2\theta = \dfrac{2 \tan \theta}{1 - \tan^2 \theta}$, $\theta \neq 45° + 90n°$, where n is any integer

Study Tip

Multiple Identities Note that three variations of the cosine double-angle identity are provided because all three can be used to derive other identities.

Example 1 Use a Double-Angle Identity to Find an Exact Value

Find the exact value of $\sin 2\theta$ if $\sin \theta = -\dfrac{4}{9}$ and $180° < \theta < 270°$.

You can use the double-angle identity for sine.

Step 1 Find $\cos \theta$.

$\cos^2 \theta = 1 - \sin^2 \theta$ $\qquad$ Pythagorean Identity

$\cos^2 \theta = 1 - \left(-\dfrac{4}{9}\right)^2$ $\qquad$ Substitute.

$\cos^2 \theta = 1 - \dfrac{16}{81}$ $\qquad$ Square $-\dfrac{4}{9}$.

$\cos^2 \theta = \dfrac{65}{81}$ $\qquad$ Subtract.

$\cos \theta = \pm \dfrac{\sqrt{65}}{9}$ $\qquad$ Take the square root.

Step 2 Determine the sign.

Because $180° < \theta < 270°$, $360° < 2\theta < 540°$.

Because 2θ is in Quadrant I or II, $\sin 2\theta$ is positive.

(continued on the next page)

 **Go Online** You can complete an Extra Example online.

Step 3 Find sin 2θ.

$$\sin 2\theta = 2 \sin \theta \cos \theta \qquad \text{Double-Angle Identity}$$

$$= 2\left(-\frac{4}{9}\right) \cdot \left(-\frac{\sqrt{65}}{9}\right) \qquad \sin \theta = -\frac{4}{9} \text{ and } \cos \theta = -\frac{\sqrt{65}}{9}$$

$$= \frac{8\sqrt{65}}{81} \qquad \text{Multiply.}$$

🌐 **Example 2** Use a Double-Angle Identity to Rewrite an Identity

BASEBALL **The distance *d* in meters that a baseball travels from the time it leaves the bat to the time it returns to the batted height is represented by $d = \dfrac{v_0^2 \sin \theta \cos \theta}{0.5g}$, where v_0 is the initial velocity, and *g* is the acceleration due to gravity. Rewrite this formula in terms of *d* and 2θ.**

Step 1 Determine the correct identity.

Note that the numerator has $\sin \theta \cos \theta$ as a factor, which is exactly half of $\sin 2\theta = 2 \sin \theta \cos \theta$. Thus, you can use the identity if you multiply each side of the equation by 2.

Step 2 Rewrite the equation.

$$d = \frac{v_0^2 \sin \theta \cos \theta}{0.5g} \qquad \text{Original equation}$$

$$2d = 2\left(\frac{v_0^2 \sin \theta \cos \theta}{0.5g}\right) \qquad \text{Multiply each side by 2.}$$

$$2d = \frac{v_0^2 (2 \sin \theta \cos \theta)}{0.5g} \qquad \text{Commutative Property}$$

$$2d = \frac{v_0^2 \sin 2\theta}{0.5g} \qquad \sin 2\theta = 2 \sin \theta \cos \theta$$

$$d = \frac{v_0^2 \sin 2\theta}{g} \qquad \text{Multiply each side by } \frac{1}{2}.$$

Thus, $d = \dfrac{v_0^2 \sin 2\theta}{g}$.

Check

ELECTRICITY The power *P* delivered to a resistor in a certain AC circuit two seconds after it has been started is given by the equation $P = R \sin^2 2\theta$, where *R* is the resistance. Select the formula that is rewritten in terms of *P* and θ.

A. $P = \dfrac{R}{\tan^2 \theta}$

B. $P = 4R \sin^2 \theta \cos^2 \theta$

C. $P = R \sin^2 \theta \cos^2 \theta$

D. $P = 2R \sin \theta \cos \theta$

🔎 **Go Online** You can complete an Extra Example online.

Watch Out!

Calculator Check You can check your answer by using a calculator. Find $\sin^{-1}\left(-\frac{4}{9}\right)$ to get $-26.39°$, but θ is in Quadrant III. Using the symmetry of the unit circle, $\sin 206.39°$ also equals $-\frac{4}{9}$, so $\theta = 206.39°$, $2\theta = 412.78°$, and $\sin 2\theta = 0.796$. This is equal to $\frac{8\sqrt{65}}{81}$.

Study Tip

Identities Write down the identities so you can examine the equation you want to rewrite and find which identity is best to use.

🌧 **Think About It!**

What are the units for *g* and v_0 in $d = \dfrac{v_0^2 \sin 2\theta}{g}$?

Learn Half-Angle Identities

Half-angle formulas can help you find the value of a function of half an angle measure. The half-angle formulas are identities, so they are true for all real numbers.

> **Key Concept • Half-Angle Identities**
>
> The following identities hold true for all values of θ.
>
> $$\sin\frac{\theta}{2} = \pm\sqrt{\frac{1-\cos\theta}{2}} \qquad\qquad \cos\frac{\theta}{2} = \pm\sqrt{\frac{1+\cos\theta}{2}}$$
>
> $$\tan\frac{\theta}{2} = \pm\sqrt{\frac{1-\cos\theta}{1+\cos\theta}}, \theta \neq 180° + 360n°, \text{ where } n \text{ is an integer}$$

Example 3 Use a Half-Angle Identity to Find an Exact Value

Find the exact value of $\cos\frac{\theta}{2}$ if $\sin\theta = \frac{12}{13}$ and $90° < \theta < 180°$.

You can use the half-angle identity for sine.

Step 1 Find $\cos\theta$.

$$\cos^2\theta = 1 - \sin^2\theta \qquad\qquad \text{Pythagorean Identity}$$

$$\cos^2\theta = 1 - \left(\frac{12}{13}\right)^2 \qquad\qquad \text{Substitute.}$$

$$\cos^2\theta = 1 - \frac{144}{169} \qquad\qquad \text{Square } \frac{12}{13}.$$

$$\cos^2\theta = \frac{25}{169} \qquad\qquad \text{Subtract.}$$

$$\cos\theta = \pm\frac{5}{13} \qquad\qquad \text{Take the square root.}$$

Step 2 Determine the signs.

Because $90° < \theta < 180°$, $45° < \frac{\theta}{2} < 90°$. Because $\frac{\theta}{2}$ is in Quadrant I and θ is in Quadrant II, $\cos\frac{\theta}{2}$ is positive and $\cos\theta = -\frac{5}{13}$.

Step 3 Find $\cos\frac{\theta}{2}$.

$$\cos\frac{\theta}{2} = \pm\sqrt{\frac{1+\cos\theta}{2}} \qquad\qquad \text{Half-Angle Identity}$$

$$= \pm\sqrt{\frac{1+\left(-\frac{5}{13}\right)}{2}} \qquad\qquad \cos\theta = -\frac{5}{13}$$

$$= \pm\sqrt{\frac{\frac{8}{13}}{2}} \qquad\qquad \text{Add.}$$

$$= \pm\sqrt{\frac{4}{13}} \qquad\qquad \text{Simplify the fraction.}$$

$$= \pm\frac{2}{\sqrt{13}} \qquad\qquad \text{Simplify the numerator.}$$

$$= \pm\frac{2\sqrt{13}}{13} \qquad\qquad \text{Rationalize the denominator.}$$

$$= \frac{2\sqrt{13}}{13} \qquad\qquad \frac{\theta}{2} \text{ is in Quadrant I.}$$

Go Online You can complete an Extra Example online.

Talk About It!

Why is it necessary to know the quadrant in which θ lies? Provide an example where a different quadrant for θ can lead to a different answer.

Study Tip

Signs To determine which sign is appropriate using a half-angle identity, check quadrant $\frac{\theta}{2}$ and not θ.

Check

Find the exact value of $\tan \frac{\theta}{2}$ if $\sin \theta = -\frac{7}{11}$ and $180° < \theta < 270°$.

Example 4 Use a Half-Angle Identity to Evaluate an Expression

Find the exact value of sin 22.5°.

Step 1 Find cos θ.

To use the half-angle identity for sine, the expression must be of the form $\sin \frac{\theta}{2}$. Because $22.5 = \frac{45}{2}$, $\theta = 45$.

$$\sin \frac{\theta}{2} = \pm \sqrt{\frac{1 - \cos \theta}{2}} \qquad \text{Half-Angle Identity}$$

$$\sin \frac{45°}{2} = \pm \sqrt{\frac{1 - \cos 45°}{2}} \qquad \theta = 45$$

Step 2 Determine the signs.

Because $\frac{\theta}{2}$ and θ, or 22.5° and 45°, are in Quadrant I, $\sin \frac{\theta}{2}$ is positive and $\cos \theta$ is positive.

Step 3 Use the half-angle identity.

$$\sin 22.5° = \pm \sqrt{\frac{1 - \cos 45°}{2}} \qquad \begin{array}{l}\text{Half-Angle Identity}\\ \theta \text{ is in Quadrant I, so } \sin \theta \text{ is positive.}\end{array}$$

$$= \sqrt{\frac{1 - \frac{\sqrt{2}}{2}}{2}} \qquad \text{Evaluate cos 45°.}$$

$$= \sqrt{\frac{\frac{2}{2} - \frac{\sqrt{2}}{2}}{2}} \qquad \text{The least common denominator is 2.}$$

$$= \sqrt{\frac{\frac{2 - \sqrt{2}}{2}}{2}} \qquad \text{Combine the fractions.}$$

$$= \sqrt{\frac{2 - \sqrt{2}}{4}} \qquad \text{Divide.}$$

$$= \frac{\sqrt{2 - \sqrt{2}}}{\sqrt{4}} \qquad \text{Quotient Property of Radicals}$$

$$= \frac{\sqrt{2 - \sqrt{2}}}{2} \qquad \text{Simplify.}$$

Check

Use a calculator to check your answer.

$\sin 22.5° = \frac{\sqrt{2 - \sqrt{2}}}{2}$, so the answer is correct.

Check

Find the exact value of cos 22.5°.

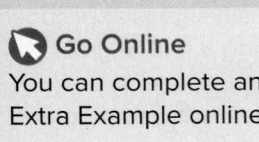

Practice

Go Online You can complete your homework online.

Examples 1 and 3

Find the exact values of sin 2θ, cos 2θ, sin $\frac{\theta}{2}$, and cos $\frac{\theta}{2}$.

1. $\sin \theta = \frac{2}{3}; 90° < \theta < 180°$

2. $\sin \theta = -\frac{15}{17}; \pi < \theta < \frac{3\pi}{2}$

3. $\cos \theta = \frac{3}{5}; \frac{3\pi}{2} < \theta < 2\pi$

4. $\cos \theta = \frac{1}{5}; 270° < \theta < 360°$

5. $\tan \theta = \frac{4}{3}; 180° < \theta < 270°$

6. $\tan \theta = \frac{3}{5}; \frac{\pi}{2} < \theta < \pi$

7. $\cos \theta = \frac{7}{25}, 0° < \theta < 90°$

8. $\sin \theta = -\frac{4}{5}, 180° < \theta < 270°$

9. $\sin \theta = \frac{40}{41}, 90° < \theta < 180°$

10. $\cos \theta = \frac{3}{7}, 270° < \theta < 360°$

Example 2

11. **SOUND WAVES** The sound waves produced by vibrating a tuning fork is represented by $S = 2 \sin 2\theta$. Rewrite this formula in terms of S and 2θ.

12. **MONUMENTS** The World War II Memorial consists of 56 granite pillars arranged around a plaza with two triumphal arches on opposite sides. When the shadow from one of the pillars is 16.1 meters long, the angle of elevation to the Sun is θ. When the shadow is 7.2 meters long, the angle of elevation is 2θ. What is the height of the pillar?

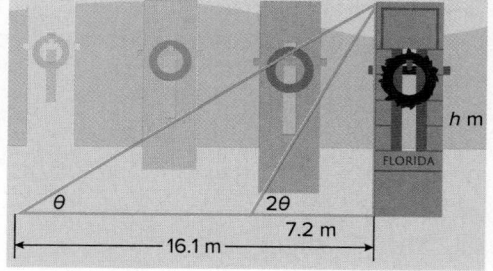

Example 4

Find the exact value of each expression.

13. $\sin 75°$

14. $\sin \frac{3\pi}{8}$

15. $\sin \frac{7\pi}{12}$

16. $\tan 165°$

17. $\tan \frac{5\pi}{12}$

18. $\tan 22.5°$

19. $\cos 22.5°$

20. $\sin 165°$

21. $\cos 105°$

22. $\sin \frac{\pi}{8}$

23. $\sin \frac{15\pi}{8}$

24. $\cos 75°$

Mixed Exercises

Find the exact values of sin 2θ, cos 2θ, and tan 2θ.

25. $\cos \theta = \frac{4}{5}, 0° < \theta < 90°$

26. $\sin \theta = \frac{1}{3}, 0 < \theta < \frac{\pi}{2}$

27. $\tan \theta = -3, 90° < \theta < 180°$

28. $\sec \theta = -\frac{4}{3}, 90° < \theta < 180°$

29. $\csc \theta = -\frac{5}{2}, \frac{3\pi}{2} < \theta < 2\pi$

30. $\cot \theta = \frac{3}{2}, 180° < \theta < 270°$

31. STRUCTURE The large triangle is an isosceles right triangle. The small triangle inside the large triangle was formed by bisecting each of the acute angles of the right triangle.

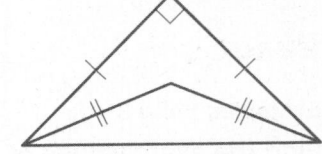

a. What are the exact values of sine and cosine for either of the congruent angles of the small triangle?

b. What are the exact values of sine and cosine for the obtuse angle of the small triangle?

32. RAMPS A ramp for loading goods onto a truck was mistakenly built with the dimensions shown. The degree measure of the angle the ramp makes with the ground should have been twice the degree measure of the angle used.

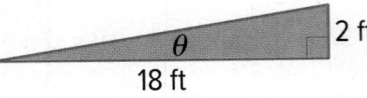

a. Find the exact values of the sine and cosine of the angle the ramp should have made with the ground.

b. If the ramp had been built properly, what would the degree measures of the two acute angles have been?

33. Show how to find the exact value of sin 240° by each method indicated.

a. using a sum of angles formula   b. using a difference of angles formula

c. using a double-angle formula   d. using a half-angle formula

34. FIND THE ERROR Teresa and Armando are calculating the exact value of sin 15°. Is either of them correct? Explain your reasoning.

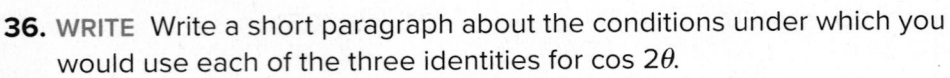

Teresa

$\sin (A - B) = \sin A \cos B - \cos A \sin B$

$\sin (45° - 30°) = \sin 45° \cos 30° - \cos 45° \sin 30°$

$= \frac{\sqrt{2}}{2} \cdot \frac{\sqrt{3}}{2} - \frac{\sqrt{2}}{2} \cdot \frac{1}{2}$

$= \frac{\sqrt{4}}{4}$

Armando

$\sin \frac{A}{2} = \pm \frac{\sqrt{1 - \cos A}}{2}$

$\sin \frac{30°}{2} = \pm \frac{\sqrt{1 - \frac{1}{2}}}{2}$

$= 0.5$

35. PERSEVERE Circle O is a unit circle. Use the figure to prove that $\tan \frac{1}{2}\theta = \frac{\sin \theta}{1 + \cos \theta}$.

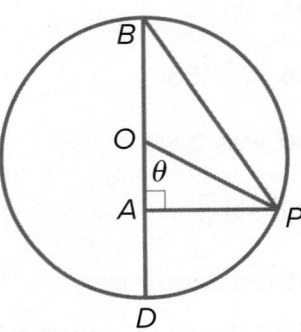

36. WRITE Write a short paragraph about the conditions under which you would use each of the three identities for cos 2θ.

37. PERSEVERE Use the sum identity for sin (A + B) to derive the double-angle identity for sin 2θ. Then use the sum identity for cos (A + B) to derive the double-angle identity for cos 2θ.

38. ANALYZE Derive the half-angle identities from the double-angle identities.

39. PERSEVERE Suppose a golfer consistently hits the ball so that it leaves the tee with an initial velocity of 115 feet per second. If $d = \frac{2v^2 \sin \theta \cos \theta}{g}$, explain why the maximum distance is attained when $\theta = 45°$.

40. CREATE Choose an integer n greater than 2 and write an identity for a trigonometric ratio involving nθ.

Solving Trigonometric Equations

Learn Solving Trigonometric Equations

Trigonometric identities are equations that are true for all values of the variable for which both sides are defined. Typically, **trigonometric equations** are true for only certain values of the variable. Solving these equations resembles solving algebraic equations.

Trigonometric equations are usually solved for values of the variable between 0° and 360° or between 0 and 2π radians. There may be solutions outside that interval. These other solutions differ by integral multiples of the period of the function.

Key Concept • Some Approaches to Solving Trigonometric Equations

- Factor and use the Zero Product Property.
- Use identities to write in terms of one common angle.
- Use identities to write in terms of one common trigonometric function.
- Take the square root of both sides to reduce the power of a function.
- Square both sides to convert to one common trigonometric function.
- Graph a system of equations to approximate the solutions, or check a solution algebraically.

Example 1 Solve an Equation Over a Given Interval

Solve $2 \sin \theta \cos^2 \theta - \cos^2 \theta = 0$ if $0 \le \theta < 360°$.

Step 1 Factor.

$2 \sin \theta \cos^2 \theta - \cos^2 \theta = 0$	Original equation
$\cos^2 \theta (2 \sin \theta - 1) = 0$	Factor.

Step 2 Solve.

By the Zero Product Property, θ is a solution when $\cos^2 \theta = 0$ or $2 \sin \theta - 1 = 0$.
Find the values of θ such that $\cos^2 \theta = 0$.

$\cos^2 \theta = 0$	Zero Product Property
$\cos \theta = 0$	Take the square root of each side.
$\theta = 90°$ or $270°$	$\cos \theta = 0$ for $[0, 360)$

Find the values of θ such that $2 \sin \theta - 1 = 0$.

$2 \sin \theta - 1 = 0$	Zero Product Property
$2 \sin \theta = 1$	Add.
$\sin \theta = \frac{1}{2}$	Divide.
$\theta = 30°$ or $150°$	$\sin \theta = \frac{1}{2}$ for $[0, 360)$

The solutions are 30°, 90°, 150°, and 270°.

Go Online You can complete an Extra Example online.

Today's Goal
- Solve trigonometric equations.

Today's Vocabulary
trigonometric equation

Study Tip

Check Your Answers
You can check your answers algebraically by entering the solutions into the original equation and graphically by graphing each side of the original equation and noting the intersections.

Go Online
to watch a video to see how to solve trigonometric equations by using a graphing calculator.

Think About It!

Why are there infinitely many solutions of $2 \sin \theta \cos^2 \theta - \cos^2 \theta = 0$ when θ is not restricted?

Check

Solve $\tan \theta \sin 2\theta - \sin \theta = 0$ for $0 \leq \theta < 360°$.

$\theta = \underline{\ ?\ }°, \underline{\ ?\ }°, \underline{\ ?\ }°, \underline{\ ?\ }°$

Example 2 Solve an Equation for All Values of θ

Solve $\sin 2\theta + \sin \theta = 0$ for all values of θ with θ measured in radians.

Step 1 Factor.

$$\sin 2\theta + \sin \theta = 0 \qquad \text{Original equation}$$

$$2 \sin \theta \cos \theta + \sin \theta = 0 \qquad \text{Double-Angle Identity}$$

$$\sin \theta (2 \cos \theta + 1) = 0 \qquad \text{Factor.}$$

Step 2 Solve.

By the Zero Product Property, θ is a solution when $\sin \theta = 0$ or $2 \cos \theta + 1 = 0$.

Solve each equation.

The period of $\sin \theta$ is 2π. Find the solutions within $[0, 2\pi]$.

$$\sin \theta = 0 \qquad \text{Zero Product Property}$$

$$\theta = 0, \pi, 2\pi \qquad \sin \theta = 0 \text{ for } [0, 2\pi]$$

The solutions will repeat every period, so the solutions can be written as $0 + 2k\pi$ and $\pi + 2k\pi$, where k is any integer. This can be simplified to $k\pi$, where k is any integer.

The period of $2 \cos \theta + 1$ is 2π. Find the solutions within $[0, 2\pi]$.

$$2 \cos \theta + 1 = 0 \qquad \text{Zero Product Property}$$

$$2 \cos \theta = -1 \qquad \text{Subtract.}$$

$$\cos \theta = -\frac{1}{2} \qquad \text{Divide.}$$

$$\theta = \frac{2\pi}{3} \text{ or } \frac{4\pi}{3} \qquad \cos \theta = -\frac{1}{2} \text{ for } [0, 2\pi]$$

The solutions will repeat every period, so the solutions can be written as $\frac{2\pi}{3} + 2k\pi$ and $\frac{4\pi}{3} + 2k\pi$, where k is any integer.

The solutions of $\sin 2\theta + \sin \theta = 0$ are $k\pi$, $\frac{2\pi}{3} + 2k\pi$ and $\frac{4\pi}{3} + 2k\pi$, where k is any integer.

Check

Solve $\frac{2\sqrt{3}}{3} \sin \theta = 1$ for all values of θ. Assume that k is an integer. Select all that apply.

A. $\theta = 2k\pi$

B. $\theta = \frac{\pi}{3} + 2k\pi$

C. $\theta = \frac{\pi}{2} + 2k\pi$

D. $\theta = \frac{2\pi}{3} + 2k\pi$

E. $\theta = \pi + 2k\pi$

F. $\theta = \frac{3\pi}{2} + 2k\pi$

 Go Online You can complete an Extra Example online.

🌐 Apply Example 3 Use a Trigonometric Equation

JUMP ROPE **During a jump rope activity, the height of the center of the rope can be modeled by** $y = 4 \sin 3\pi\left(t - \frac{1}{6}\right) + 4$, **where t is the time in seconds. How often does the rope hit the ground? How many times will it hit the ground in one minute?**

1 What is the task?

Describe the task in your own words. Then list any questions that you may have. How can you find answers to your questions?

Sample answer: I need to use the equation that models the height of the jump rope to find when the rope hits the ground and the number of times the rope hits the ground in one minute.

What height represents the rope hitting the ground? How many times does it hit in one period? How many periods are in one minute?

2 How will you approach the task? What have you learned that you can use to help you complete the task?

Sample answer: I will find the period of the equation. Then I will find height and times at which the rope hits the ground. I will use that information to find all the times during which the rope hits the ground.

3 What is your solution?

Use your strategy to solve the problem.

How long is one period, or one revolution of the jump rope? $\frac{2}{3}$ second

At what height will the jump rope hit the ground? 0 feet

What equation can be used to find the times the rope hits the ground during one period? $4 \sin 3\pi\left(t - \frac{1}{6}\right) + 4 = 0$

In one minute, how many times will the jump rope hit the ground?
90 times

4 How can you know that your solution is reasonable?

✏️ **Write About It!** Write an argument that can be used to defend your solution.

Sample answer: Ground height is represented by $y = 0$. By substituting $y = 0$ in the given equation and solving, I know that the rope hits the ground at $t = \frac{2}{3}$ in the interval $\left(0, \frac{2}{3}\right]$. That means that the rope hits the ground once each period. Because there are 60 seconds in one minute and a period is $\frac{2}{3}$ second, I found $60 \div \frac{2}{3} = 90$, which is the number of times the rope hits the ground in one minute.

 Think About It!
What assumption did you make while solving this problem?

Problem-Solving Tip

Determine Reasonable Answers When you have solved a problem, check your answer for reasonableness. In this example, the rope hits the ground 90 times per minute, which is a reasonable number of times.

Example 4 Solve a Trigonometric Inequality

Solve 2 sec $\theta \le 4$ for $0° \le \theta < 360°$.

Step 1 Isolate the trigonometric expression.

$2 \sec \theta \le 4$	Original inequality
$\sec \theta \le 2$	Divide each side by 2.

Step 2 Identify the solutions of the related equation.

Because $\sec \theta = \dfrac{1}{\cos \theta}$, $\sec \theta = 2$ when $\cos \theta = \dfrac{1}{2}$.

So, the two solutions within $[0, 360)$ are $60°$ and $300°$.

Step 3 Identify where sec θ is undefined.

For $0° \le \theta < 360°$, sec θ is undefined when $\cos \theta = 0$, or when $\theta = 90°$ or $270°$.

Step 4 Create a sign chart.

Label solutions of the related equation and undefined values.

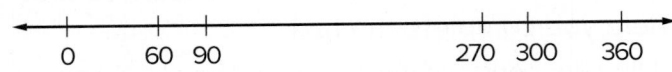

Because secant is continuous on its domain, sec θ will be a constant sign between each interval.

Step 5 Test values within each interval.

Because $\sec \theta \le 2$, the inequality is valid when $\sec \theta - 2$ is 0 or negative.

$\sec 30° - 2 \approx -0.845 \qquad \sec 75° - 2 \approx 1.864 \qquad \sec 180° - 2 = -3$

$\sec 280° - 2 \approx 3.759 \qquad \sec 330° - 2 \approx -0.845$

Step 6 Complete the sign chart and identify the solutions.

$$(-)\quad(+)\qquad\qquad(-)\qquad\qquad(+)\quad(-)$$
$$\begin{array}{ccccc} 0 & 60\ 90 & & 270\ 300 & 360 \end{array}$$

$\sec \theta - 2 \le 0$, so the solution set includes intervals where $\sec \theta - 2$ is negative. The solution set is $[0, 60°]$, $(90°, 270°)$, $[300°, 360°)$.

Step 7 Check the solutions with a graphing calculator.

Graph $y = \sec \theta - 2$ and note where it is negative within $[0, 360°)$.

$[0, 360]$ scl: 30 by $[-5, 5]$ scl:1

Check

Solve $\csc \theta < \sqrt{2}$ for $45° < \theta < 135°$ and $180° < \theta < 360°$. Select all that apply.

A. $(0, 45°]$ **B.** $[45°, 90°)$ **C.** $(90°, 270°)$ **D.** $[135°, 180°)$

E. $(180°, 360°)$ **F.** $[180°, 360°)$ **G.** $(270°, 315°]$

Go Online
You can complete an Extra Example online.

Some trigonometric equations have no solution. Other times, a value may appear to be a solution, but is extraneous. Check all of the solutions within a period to identify extraneous solutions.

Example 5 Determine Whether a Solution Exists Over a Given Interval

Solve $\cos \theta - 1 = \sin \theta$ for $[0, 360°)$.

Step 1 Factor.

$$\cos \theta - 1 = \sin \theta \qquad \text{Original equation}$$
$$(\cos \theta - 1)^2 = \sin^2 \theta \qquad \text{Square each side.}$$
$$\cos^2 \theta - 2 \cos \theta + 1 = \sin^2 \theta \qquad \text{Multiply.}$$
$$\cos^2 \theta - 2 \cos \theta + 1 = 1 - \cos^2 \theta \qquad \sin^2 \theta = 1 - \cos^2 \theta$$
$$2 \cos^2 \theta - 2 \cos \theta + 1 = 1 \qquad \text{Add } \cos^2 \theta \text{ to each side.}$$
$$2 \cos^2 \theta - 2 \cos \theta = 0 \qquad \text{Subtract 1 from each side.}$$
$$2 \cos \theta (\cos \theta - 1) = 0 \qquad \text{Factor.}$$

Step 2 Solve.

By the Zero Product Property, θ is a solution when $2 \cos \theta = 0$ or $\cos \theta - 1 = 0$.

Solve each equation.

Case 1		Case 2
$2 \cos \theta = 0$	Zero Product Property	$\cos \theta - 1 = 0$
$\cos \theta = 0$	Simplify.	$\cos \theta = 1$
$\theta = 90°$ or $270°$	Solve.	$\theta = 0°$

Step 3 Check for extraneous solutions.

Check each solution by substituting it into the original equation.

$\theta = 0°$

$$\cos \theta - 1 = \sin \theta \qquad \text{Original equation}$$
$$\cos 0° - 1 \overset{?}{=} \sin 0° \qquad \theta = 0°$$
$$1 - 1 \overset{?}{=} 0 \qquad \text{Evaluate } \cos 0° \text{ and } \sin 0°.$$
$$0 = 0 \qquad \text{Subtract.}$$

$\theta = 90°$

$$\cos \theta - 1 = \sin \theta \qquad \text{Original equation}$$
$$\cos 90° - 1 \overset{?}{=} \sin 90° \qquad \theta = 90°$$
$$0 - 1 \overset{?}{=} 1 \qquad \text{Evaluate } \cos 90° \text{ and } \sin 90°.$$
$$-1 \neq 1 \qquad \text{Subtract.}$$

$\theta = 270°$

$$\cos \theta - 1 = \sin \theta \qquad \text{Original equation}$$
$$\cos 270° - 1 \overset{?}{=} \sin 270° \qquad \theta = 270°$$
$$0 - 1 \overset{?}{=} -1 \qquad \text{Evaluate } \cos 270° \text{ and } \sin 270°.$$
$$-1 = -1 \qquad \text{Subtract.}$$

The solutions are 0° and 270°. 90° is an extraneous solution.

(continued on the next page)

 Go Online You can complete an Extra Example online.

Talk About It!

How do you think an extraneous solution was introduced during the process of solving this problem? Explain your reasoning.

Check Confirm with a graph.

The graphs of $y = \cos\theta - 1$ and $y = \sin\theta$ intersect at two locations within $[0°, 360°)$. Those intersections, 0° and 270°, are the solutions of $\cos\theta - 1 = \sin\theta$ on $[0°, 360°)$.

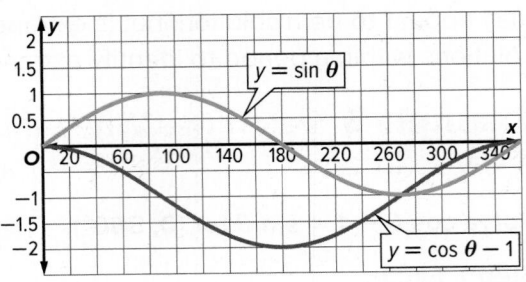

Example 6 Determine Whether a Solution Exists for All Values of θ

Solve $2\sin^4\theta - \cos^2\theta = 0$ for all values of θ with θ measured in degrees.

Step 1 Factor.

$2\sin^4\theta - \cos^2\theta = 0$	Original equation
$2\sin^4\theta - (1 - \sin^2\theta) = 0$	$\cos^2\theta = 1 - \sin^2\theta$
$2\sin^4\theta + \sin^2\theta - 1 = 0$	Distributive Property
$(2\sin^2\theta - 1)(\sin^2\theta + 1) = 0$	Factor.

Step 2 Solve.

By the Zero Product Property, θ is a solution when $2\sin^2\theta - 1 = 0$ or $\sin^2\theta + 1 = 0$. Solve each equation.

$2\sin^2\theta - 1 = 0$	Zero Product Property
$2\sin^2\theta = 1$	Add 1 to each side.
$\sin^2\theta = \frac{1}{2}$	Divide each side by 2.
$\sin\theta = \pm\frac{\sqrt{2}}{2}$	Take the square root of each side.
$\theta = 45°, 135°, 225°, \text{ or } 315°$	$\sin\theta = \pm\frac{\sqrt{2}}{2}$ for $[0, 360°)$
$\sin^2\theta + 1 = 0$	Zero Product Property
$\sin^2\theta = -1$	Subtract 1 from each side.

$\sin^2\theta$ is never negative, so there are no real solutions for $\sin^2\theta + 1 = 0$.

Step 3 Check the solutions.

Because all four values check, 45°, 135°, 225°, and 315° are all solutions. Thus, the solutions are $45° + 90k°$, where k is any integer. You can also confirm the solutions with a graph and check that the zeros are $45° + 90k°$.

Go Online You can complete an Extra Example online.

🫧 Think About It!

Why do $\theta = 135°, 225°,$ and 315° not need to be checked once 45° was confirmed as a solution? Explain your reasoning.

Study Tip

Patterns Look for patterns in your solutions. Look for pairs of solutions that differ by exactly π or 2π and write your solutions with the simplest possible pattern.

Practice

Go Online You can complete your homework online.

Example 1

Solve each equation for the given interval.

1. $\cos^2 \theta = \frac{1}{4}$; $0° \le \theta \le 360°$

2. $2 \sin^2 \theta = 1$; $90° \le \theta \le 270°$

3. $\sin 2\theta - \cos \theta = 0$; $0 \le \theta \le 2\pi$

4. $3 \sin^2 \theta = \cos^2 \theta$; $0 \le \theta \le \frac{\pi}{2}$

5. $2 \sin \theta + \sqrt{3} = 0$; $180° \le \theta \le 360°$

6. $4 \sin^2 \theta - 1 = 0$; $180° \le \theta \le 360°$

Example 2

Solve each equation for all values of θ if θ is measured in radians.

7. $\cos 2\theta + 3 \cos \theta = 1$

8. $2 \sin^2 \theta = \cos \theta + 1$

9. $\cos^2 \theta - \frac{3}{2} = \frac{5}{2} \cos \theta$

10. $3 \cos \theta - \cos \theta = 2$

Solve each equation for all values of θ if θ is measured in degrees.

11. $\sin \theta - \cos \theta = 0$

12. $\tan \theta - \sin \theta = 0$

13. $\sin^2 \theta = 2 \sin \theta + 3$

14. $4 \sin^2 \theta = 4 \sin \theta - 1$

Example 3

15. SANDCASTLES The water level on Sunset Beach can be modeled by the function $y = 7 + 7 \sin \frac{\pi}{6} t$, where y is the distance in feet of the waterline above the low tide mark and t is the number of hours past 6 A.M. At 2 P.M., Nashiko built her sandcastle 10 feet above the low tide mark. At what time will the waterline reach Nashiko's sandcastle?

16. BATTERY The light of a battery indicator pulses while the battery is charging. This can be modeled by the equation $y = 60 + 60 \sin \frac{\pi}{4} t$, where y is the lumens emitted from the bulb and t is the number of seconds since the beginning of a pulse. At what time will the amount of light emitted be equal to 110 lumens?

Example 4

17. Solve $3 \csc \theta \le 2$ for $0° \le \theta \le 360°$.

18. Solve $\sqrt{3} \sec \theta \le 2$ for $0° \le \theta \le 360°$.

Example 5

Solve each equation over the given interval.

19. $\cos \theta + 1 = \sin \theta$; $[0, 2\pi)$

20. $\cos \theta + \sin \theta = 1$; $[0, 2\pi)$

21. $1 - \cos \theta = \sqrt{3} \sin \theta$; $[0, 2\pi)$

22. $\tan \frac{\theta}{2} - 1 = 0$; $[0, 2\pi)$.

Example 6

Solve each equation for θ with θ measured in degrees.

23. $\sin \theta + \sqrt{2} = -\sin \theta$

24. $\tan 3\theta = 1$

Mixed Exercises

Solve each equation for the given interval.

25. $\sin \theta = \frac{\sqrt{2}}{2}$, $0° \leq \theta \leq 360°$

26. $2 \cos \theta = -\sqrt{3}$, $90° < \theta < 180°$

27. $\tan^2 \theta = 1$, $180° < \theta < 360°$

28. $2 \sin \theta = 1$, $0 \leq \theta < \pi$

29. $\sin^2 \theta + \sin \theta = 0$, $\pi \leq \theta < 2\pi$

30. $2 \cos^2 \theta + \cos \theta = 0$, $0 \leq \theta < \pi$

Solve each equation for all values of θ if θ is measured in radians.

31. $2 \cos^2 \theta - \cos \theta = 1$

32. $\sin^2 \theta - 2 \sin \theta + 1 = 0$

33. $\sin \theta + \sin \theta \cos \theta = 0$

34. $\sin^2 \theta = 1$

35. $4 \cos \theta = -1 + 2 \cos \theta$

36. $\tan \theta \cos \theta = \frac{1}{2}$

Solve each equation for all values of θ if θ is measured in degrees.

37. $2 \sin \theta + 1 = 0$

38. $2 \cos \theta + \sqrt{3} = 0$

39. $\sqrt{2} \sin \theta + 1 = 0$

40. $4 \sin^2 \theta = 3$

41. $2 \cos^2 \theta = 1$

42. $\cos 2\theta = -1$

Solve each equation.

43. $3 \cos^2 \theta - \sin^2 \theta = 0$

44. $\sin \theta + \sin 2\theta = 0$

45. $2 \sin^2 \theta = \sin \theta + 1$

46. $\cos \theta + \sec \theta = 2$

47. Find all solutions for $\sin \theta = \cos 2\theta$ if $0° \leq \theta < 360°$.

48. Find all solutions for $4 \cos^2 \theta = 1$ if $0 \leq \theta < 2\pi$.

49. Solve $\cos 2\theta = \cos \theta$ for all values of θ if θ is measured in degrees.

50. Solve $\cos 2\theta = 3 \sin \theta - 1$ for all values of θ if θ is measured in radians.

51. SOUND The sound from stringed instruments is generated by waves that travel along the strings. The wave traveling on a guitar string can be modeled by $D = 0.5 \sin (6.5x)° \sin (2500t)°$, where D is the displacement in millimeters at the position x millimeters from the bridge of the guitar at time t seconds. Find the first positive time when the point 50 centimeters from the bridge has a displacement of 0.01 millimeter.

52. RIDES The original Ferris wheel had a diameter of about 260 feet and would make a complete revolution in 9 minutes. The motion of the wheel can be modeled by $h = 134 - 130 \cos \frac{2\pi t}{9}$, where h is the height of the rider in feet above the ground t minutes after they begin the ride. When is the rider about 175 feet above the ground?

53. USE A MODEL The table shows the number of hours of daylight on various days of the year in Seattle, Washington. The function $f(x) = 3.725 \sin (0.016x - 1.180) + 11.932$ models the number of hours of daylight in Seattle on day x of the calendar year.

Number of Hours of Daylight in Seattle						
Date	Jan 1	Feb 11	May 3	Aug 22	Oct 15	Nov 29
Day of Year	1	42	123	234	288	333
Hours of Daylight	8.5	10.1	14.6	13.9	10.9	8.8

a. What is the maximum value of $f(x)$? What does this tell you?

b. During what period of the year does Seattle get more than 14 hours of daylight per day? Explain.

54. USE A SOURCE Research the outside temperature several times during the day. Make a table of the data, where x is the number of hours since midnight and y is the temperature in degrees Fahrenheit. Use the data and your calculator to write a sine function $T(x)$ that models the temperature. Approximate the high temperature for the day. At approximately what time did the high temperature occur? Justify your reasoning.

55. BUSINESS Carissa has a small business installing and repairing air conditioners. She has kept track of her monthly profit since she started the business. The function $P(x) = 1808.831 \sin(0.543x - 2.455) + 1942.476$ models the monthly profit in dollars, x months since Carissa started the business.

   a. Determine the approximate period of the function. What does this tell you about Carissa's business?

   b. Next month will mark the 6th anniversary of the day that Carissa started the business. What profit should she expect to make next month? Explain.

56. STRUCTURE A garden sprinkler is attached to a spigot by a straight hose that is 3 meters long. The sprinkler makes 6 complete rotations every minute. As it rotates, it sprays a stream of water that hits the wall, as shown. Let d be the length of the stream of water from the sprinkler to the wall, and assume the stream of water hits the spigot when the sprinkler is first turned on. The function $d(x) = 3 \sec \frac{\pi}{5}x$ models the length of the stream to strike the wall in meters as a function of the time x in seconds. After the sprinkler is turned on, what is the first time when the stream of water that strikes the wall is 5 meters long?

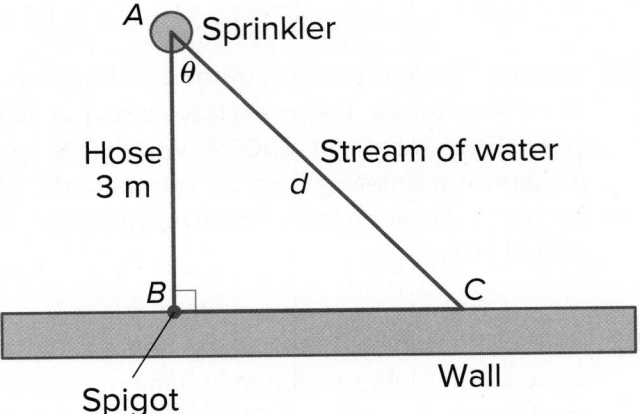

57. PERSEVERE Solve $\sin 2x < \sin x$ for $0 \le x \le 2\pi$ without a calculator.

58. ANALYZE Compare and contrast solving trigonometric equations with solving linear and quadratic equations. What techniques are the same? What techniques are different? How many solutions do you expect?

59. WRITE Why do trigonometric equations often have infinitely many solutions?

60. CREATE Write an example of a trigonometric equation that has exactly two solutions if $0° \le \theta \le 360°$.

61. PERSEVERE How many solutions in the interval $0° \le \theta < 360°$ should you expect for $a \sin(b\theta + c) = d$, if $a \ne 0$ and b is a positive integer?

62. FIND THE ERROR Ms. Rollins divided her students into four groups, asking each to solve the equation $\sin \theta \cot \theta = \cos^2 \theta$. Do any of the groups have the correct solution? Explain your reasoning.

Group A:	$0° + k \cdot 360°$, $90° + k \cdot 360°$, $270° + k \cdot 360°$
Group B:	$0° + k \cdot 360°$, $90° + k \cdot 180°$
Group C:	$90° + k \cdot 180°$
Group D:	$90° + k \cdot 360°$, $270° + k \cdot 360°$

Essential Question

How are trigonometric identities similar to and different from other equations?

Trigonometric identities are similar to equations in that the Properties of Equality hold true. You can perform operations on both sides to maintain a true statement. They are different in that in a trigonometric identity, there is no unknown for which to solve. The identity is true for any value of a variable.

Module Summary

Lessons 13-1 and 13-2

Trigonometric Identities

- A trigonometric identity holds true for all values for which every expression in the equation is defined.

- The Pythagorean identities are $\cos^2 \theta + \sin^2 \theta = 1$, $\tan^2 \theta + 1 = \sec^2 \theta$, and $\cot^2 \theta + 1 = \csc^2 \theta$.

- To verify a trigonometric identity:
  - First, try to verify the identity by simplifying one side.
  - Try substituting basic identities.
  - Use what you have learned from algebra to simplify.
  - When a term includes $1 + \sin \theta$ or $1 + \cos \theta$, think about multiplying the numerator and denominator by the conjugate. Then you can use a Pythagorean Identity.

Lesson 13-3

Sum and Difference Identities

- $\sin (A + B) = \sin A \cos B + \cos A \sin B$
- $\cos (A + B) = \cos A \cos B - \sin A \sin B$
- $\tan (A + B) = \frac{\tan A + \tan B}{1 - \tan A \tan B}$, $A, B \neq 90° + 180n°$, where n is any integer
- $\sin (A - B) = \sin A \cos B - \cos A \sin B$
- $\cos (A - B) = \cos A \cos B + \sin A \sin B$
- $\tan (A - B) = \frac{\tan A - \tan B}{1 + \tan A \tan B}$, $A, B \neq 90° + 180n°$, where n is any integer

Lesson 13-4

Double-Angle and Half-Angle Identities

- $\sin 2\theta = 2 \sin \theta \cos \theta$
- $\cos 2\theta = \cos^2 \theta - \sin^2 \theta = 2 \cos^2 \theta - 1 = 1 - 2 \sin^2 \theta$
- $\tan 2\theta = \frac{2 \tan \theta}{1 - \tan^2 \theta}$, $\theta \neq 45° + 90n°$, where n is any integer
- $\sin \frac{\theta}{2} = \pm\sqrt{\frac{1 - \cos \theta}{2}}$, $\cos \frac{\theta}{2} = \pm\sqrt{\frac{1 + \cos \theta}{2}}$, $\tan \frac{\theta}{2} = \pm\sqrt{\frac{1 - \cos \theta}{1 + \cos \theta}}$, $\theta \neq 180° + 360n°$, where n is any integer

Lesson 13-5

Solving Trigonometric Equations

- To solve a trigonometric equation:
  - Factor and use the Zero Product Property.
  - Use identities to write in terms of one common angle or trig function.
  - Take the square root of both sides to reduce the power of a function.
  - Square both sides to convert to one common trigonometric function.
  - Graph a system of equations to approximate the solutions or check a solution algebraically.

Study Organizer

📖 **Foldables**

Use your Foldable to review this module. Working with a partner can be helpful. Ask for clarification of concepts as needed.

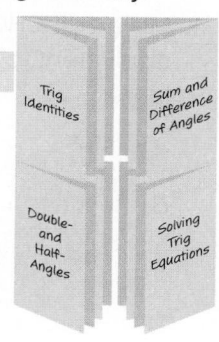

Test Practice

1. **OPEN RESPONSE** What is the exact value of $\cot \theta$ given $\cos\left(\frac{\pi}{2} - \theta\right) = -\frac{8}{9}$ and $180° < \theta < 270°$? (Lesson 13-1)

2. **MULTIPLE CHOICE** Which expression is equivalent to $\sin x\,(\cot x - \csc x)$? (Lesson 13-1)

   A. $\sin x$

   B. $-\sin x$

   C. $\cos x - \tan x$

   D. $\cos x - 1$

3. **MULTIPLE CHOICE** If $\tan \theta = \frac{8}{15}$, then what is the value of $\csc \theta$ when $0° < \theta < 90°$?
   (Lesson 13-1)

   A.   0.471

   B.   0.533

   C.   1.133

   D.   2.125

4. **MULTIPLE CHOICE** Which expression is equivalent to $\left(\tan\left(\frac{\pi}{2} - x\right)\right)(\cos x)(\sin x) - 1$?
   (Lesson 13-1)

   A. $\sin^2 x$

   B. $-\sin^2 x$

   C. $\sec^2 x$

   D. $-\sec^2 x$

5. **MULTI-SELECT** Select the reasons that justify each step. (Lesson 13-2)

   $$\tan^2 \theta \cos^2 \theta \stackrel{?}{=} 1 - \cos^2 \theta \quad \text{Original equation}$$
   $$\frac{\sin^2 \theta}{\cos^2 \theta} \cdot \cos^2 \theta \stackrel{?}{=} 1 - \cos^2 \theta \quad \textit{Step 2 Reason}$$
   $$\sin^2 \theta \stackrel{?}{=} 1 - \cos^2 \theta \quad \textit{Step 3 Reason}$$
   $$1 - \cos^2 \theta = 1 - \cos^2 \theta \quad \text{Pythagorean Identity}$$

   A.   Step 2: Cofunction Identity

   B.   Step 2: Quotient Identity

   C.   Step 2: Reciprocal Identity

   D.   Step 3: Simplify.

   E.   Step 3: Quotient Identity

   F.   Step 3: Pythagorean Identity

6. **OPEN RESPONSE** The verification of $\dfrac{\cos \theta}{1 + \sin \theta} + \tan \theta = \sec \theta$ is shown.

   $$\frac{\cos \theta}{1 + \sin \theta} + \tan \theta \stackrel{?}{=} \sec \theta$$
   $$\textit{Missing Step} \stackrel{?}{=} \sec \theta$$
   $$\frac{\cos^2 \theta}{\cos \theta\,(1 + \sin \theta)} + \frac{\sin \theta(1 + \sin \theta)}{\cos \theta(1 + \sin \theta)} \stackrel{?}{=} \sec \theta$$
   $$\frac{\cos^2 \theta + \sin \theta + \sin^2 \theta}{\cos \theta\,(1 + \sin \theta)} \stackrel{?}{=} \sec \theta$$
   $$\frac{1 + \sin \theta}{\cos \theta\,(1 + \sin \theta)} \stackrel{?}{=} \sec \theta$$
   $$\frac{1}{\cos \theta} \stackrel{?}{=} \sec \theta$$
   $$\sec \theta = \sec \theta$$

   What is the missing step? (Lesson 13-2)

7. OPEN RESPONSE The verification of $\frac{\cot^2\theta}{1+\csc\theta} = \frac{1-\sin\theta}{\sin\theta}$ is shown. Give the reason that justifies each step. (Lesson 13-2)

Step 1: $\quad\quad\quad\quad\frac{\cot^2\theta}{1+\csc\theta} \overset{?}{=} \frac{1-\sin\theta}{\sin\theta}$

Step 2: $\quad\quad\quad\quad\frac{\csc^2\theta-1}{1+\csc\theta} \overset{?}{=} \frac{1-\sin\theta}{\sin\theta}$

Step 3: $\quad\frac{(\csc\theta+1)(\csc\theta-1)}{1+\csc\theta} \overset{?}{=} \frac{1-\sin\theta}{\sin\theta}$

Step 4: $\quad\quad\quad\quad\csc\theta-1 \overset{?}{=} \frac{1-\sin\theta}{\sin\theta}$

Step 5: $\quad\quad\quad\quad\csc\theta-1 \overset{?}{=} \frac{1}{\sin\theta} - \frac{\sin\theta}{\sin\theta}$

Step 6: $\quad\quad\quad\quad\csc\theta-1 \overset{?}{=} \frac{1}{\sin\theta} - 1$

Step 7: $\quad\quad\quad\quad\csc\theta-1 = \csc\theta-1$

8. MULTI-SELECT Which expression(s) could be used to determine cos 75°? Select all that apply. (Lesson 13-3)

A. cos 45° cos 30° + sin 45° sin 30°

B. cos 45° cos 30° − sin 45° sin 30°

C. cos 90° cos 15° + sin 90° sin 15°

D. cos 90° cos 15° − sin 90° sin 15°

E. sin 45° cos 30° + cos 45° sin 30°

F. sin 45° cos 30° − cos 45° sin 30°

G. sin 90° cos 15° + cos 90° sin 15°

H. sin 90° cos 15° − cos 90° sin 15°

9. MULTIPLE CHOICE What is the exact value of $\tan\frac{7\pi}{12}$? (Lesson 13-3)

A. $-2 + \sqrt{3}$

B. $-2 - \sqrt{3}$

C. $1 + \sqrt{3}$

D. $1 - \sqrt{3}$

10. OPEN RESPONSE Verify that $\tan(2\pi - x) = -\tan x$ is an identity. (Lesson 13-3)

11. MULTIPLE CHOICE Find the exact value of $\cos 2\theta$. (Lesson 13-4)

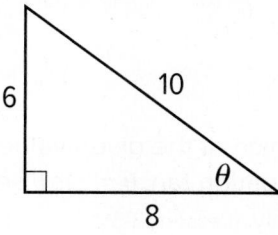

A. 0.28

B. 0.32

C. 0.56

D. 0.64

12. MULTIPLE CHOICE What is the value of $\sin 2\theta$ if $\sin \theta = \frac{8}{17}$ and $0° < \theta < 90°$? (Lesson 13-4)

A. $-\frac{161}{289}$

B. $\frac{161}{289}$

C. $\frac{240}{289}$

D. $\frac{16}{17}$

13. OPEN RESPONSE What is the exact value of $\sin 157.5°$? (Lesson 13-4)

14. MULTIPLE CHOICE Monisha drew the triangle shown. She then drew a line bisecting θ. What is the value of the tangent of the new angle formed? (Lesson 13-4)

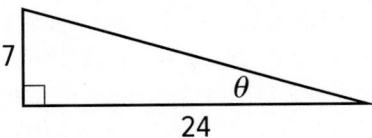

A. $\frac{\sqrt{2}}{10}$

B. $\frac{7\sqrt{2}}{10}$

C. $\frac{1}{7}$

D. $\frac{7}{48}$

15. MULTI-SELECT Which of the given values is a solution to the equation $\tan^2\theta = \sin\theta \sec\theta$? Select all that apply. (Lesson 13-5)

A. 45°

B. 135°

C. 225°

D. 315°

16. MULTIPLE CHOICE The distance a baseball travels after being thrown can be modeled by $d = \frac{v^2}{16} \sin\theta \cos\theta$, where d is is the horizontal distance in feet and v is the initial velocity. If the ball was thrown with an initial velocity of 72 feet per second and traveled 81 feet, at what angle (in degrees) was the ball thrown? (Lesson 13-5)

A. 15°

B. 30°

C. 45°

D. 50°

17. MULTI-SELECT Solve $\tan^2 x + \tan x = 0$ for $0° \le x < 360°$. Select all that apply. (Lesson 13-5)

A. 0°

B. 45°

C. 90°

D. 135°

E. 150°

F. 180°

G. 225°

H. 315°

18. OPEN RESPONSE Solve $\sin \theta \cos \theta = \frac{1}{2}$ for all values of θ, if θ is measured in radians. (Lesson 13-5)

Module 1

Quick Check

1. 9 **3.** 10 ft **5.** $x > 9$ **7.** $x < 41$

Lesson 1-1

1. 2.8 **3.** 4.2 **5.** 7 **7.** 9.8 **9.** (4, 1.8) **11.** 7

13. No; we need to know if the segment bisector is a perpendicular bisector.

15. It is given that $\overleftrightarrow{CD}$ is the perpendicular bisector of $\overline{AB}$. By the definition of bisector, E is the midpoint of $\overline{AB}$. Thus, $\overline{AE} \cong \overline{BE}$ by the Midpoint Theorem. Because two points determine a line, you can draw line segments from A to C and from B to C. $\angle CEA$ and $\angle CEB$ are right angles by the definition of perpendicular. Because all right angles are congruent, $\angle CEA \cong \angle CEB$. By the Reflexive Property of Congruence, $\overline{CE} \cong \overline{CE}$. Thus, $\triangle CEA \cong \triangle CEB$ by SAS. $\overline{CA} \cong \overline{CB}$ by CPCTC, and by the definition of congruence, $CA = CB$. By the definition of equidistant, C is equidistant from A and B.

17. The equation of a line of one of the perpendicular bisectors is $y = 3$. The equation of the line of another perpendicular bisector is $x = 5$. These lines intersect at (5, 3). The circumcenter is located at (5, 3).

19. a plane perpendicular to the plane in which $\overline{CD}$ lies and bisecting $\overline{CD}$ **33.** $\left(\frac{39}{10}, \frac{19}{10}\right)$

21.

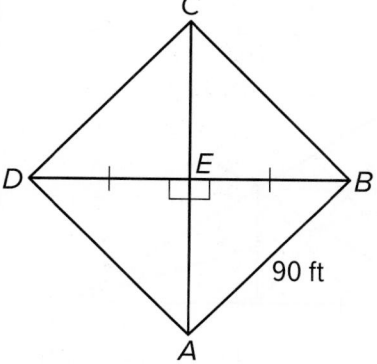

Sample answer: A is equidistant from B and D. Therefore, $AD = AB$. So, $AD = 90$

feet. AD is the distance from home plate to third base. Therefore, it is 90 feet from third base to home plate.

23. Proof:

Statements (Reasons)

1. Plane Y is a perpendicular bisector of $\overline{DC}$. (Given)

2. $\angle DBA$ and $\angle CBA$ are right angles, and $\overline{DB} \cong \overline{CB}$ (Definition of perpendicular bisector)

3. $\angle DBA \cong \angle CBA$ (All right angles are congruent.)

4. $\overline{AB} \cong \overline{AB}$ (Reflexive Property of $\cong$)

5. $\triangle DBA \cong \triangle CBA$ (SAS)

6. $\angle ADB \cong \angle ACB$ (CPCTC)

Lesson 1-2

1. 43° **3.** 19 **5.** 7 **7.** 40° **9.** 28° **11.** 13 **13.** (5.5, 6) **15.** 6

17. Sample answer: Given that $\overline{BD}$ is the angle bisector of $\angle ABC$, $\angle ABD \cong \angle CBD$ by the definition of angle bisector. Because there is a perpendicular line between any line and a point not on the line by the Perpendicular Postulate, let F be on $\overline{BC}$ such that $\overline{DF} \perp \overline{BC}$, and let E be on $\overline{AB}$ such that $\overline{DE} \perp \overline{AB}$. Therefore, $\angle BFD$ and $\angle BED$ are right angles by the definition of perpendicular, and $m\angle BFD = 90°$ and $m\angle BED = 90°$ by the definition of right angle. Further, $m\angle BFD = m\angle BED$ by substitution, and $\angle BFD \cong \angle BED$ by the definition of congruence. Also, $\overline{BD} \cong \overline{BD}$ by the Reflexive Property of Congruence. Because $\angle EBD \cong \angle DBF$, $\angle BFD \cong \angle BED$, and $\overline{BD} \cong \overline{BD}$, $\triangle BED \cong \triangle BFD$ by AAS. Therefore, $\overline{ED} \cong \overline{FD}$ by CPCTC. $ED = FD$ by the definition of congruence, so D is equidistant from $\overline{AB}$ and $\overline{BC}$ by the definition of equidistant.

19. Proof:

Statements (Reasons)

1. $\overline{AD}, \overline{BE},$ and $\overline{CF}$ are all angle bisectors, and $\overline{KP} \perp \overline{AB}, \overline{KQ} \perp \overline{BC},$ and $\overline{KR} \perp \overline{AC}$. (Given)

2. $KP = KQ$, $KQ = KR$, $KP = KR$ (Any point on the angle bisector is equidistant from the sides of the angle.)

3. $KP = KQ = KR$ (Transitive Property of Equality)

21. Sample answer: I constructed the angle bisector of each angle. The intersection point is the incenter. To verify the construction, I could construct segments perpendicular to each side through the incenter and then measure each distance with the ruler to verify they are all equal.

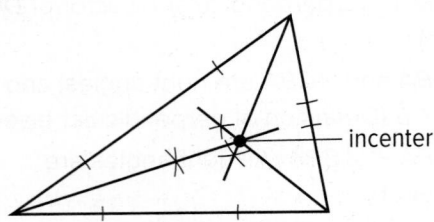
incenter

23. Sometimes; sample answer: If the triangle is equilateral, then this is true, but if the triangle is isosceles or scalene, the statement is false.

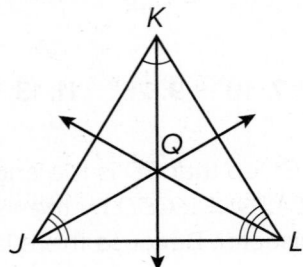

$JQ = KQ = LQ$

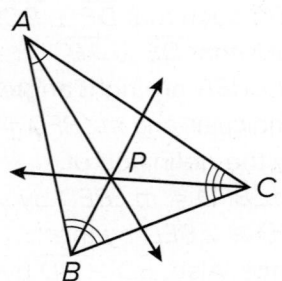

$AP \neq BP \neq CP$

25. Proof: Statements (Reasons)

1. Plane Z is an angle bisector of $\angle KJH$; $\overline{KJ} \cong \overline{HJ}$. (Given)

2. $\angle KJM \cong \angle HJM$ (Definition of angle bisector)

3. $\overline{JM} \cong \overline{JM}$ (Reflexive Property of Congruence)

4. $\triangle KJM \cong \triangle HJM$ (SAS)

5. $\overline{MH} \cong \overline{MK}$ (CPCTC)

27. A point is on the bisector of an angle if and only if it is equidistant from the sides of the angle.

Lesson 1-3

1. 24 **3.** 14 **5.** 12 **7.** 6 **9.** (1, 10) **11.** (1, 4)

13. (1, 0) **15.** (−9, −3) **17.** 3 **19.** $\frac{1}{2}$

21. 6; no; because $m\angle ECA = 92°$

23. altitude **25.** median **27.** 32

29. Proof: Because $\triangle XYZ$ is isosceles, $\overline{XY} \cong \overline{YZ}$. By the definition of angle bisector, $\angle XYW \cong \angle ZYW$. $\overline{YW} \cong \overline{YW}$ by the Reflexive Property. So, by SAS, $\triangle XYW \cong \triangle ZYW$. By CPCTC, $\overline{XW} \cong \overline{ZW}$. By the definition of midpoint, W is the midpoint of $\overline{XZ}$. By the definition of median, $\overline{WY}$ is a median.

31. Sample answer: Kareem is correct. According to the Centroid Theorem, $AP = \frac{2}{3}AD$. The segment lengths are transposed.

33a.

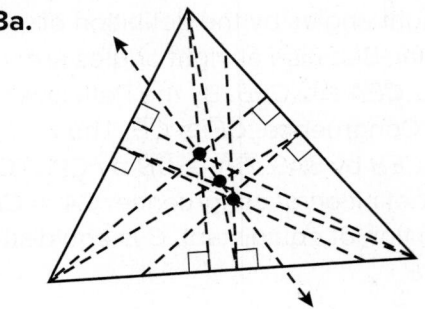

33b.

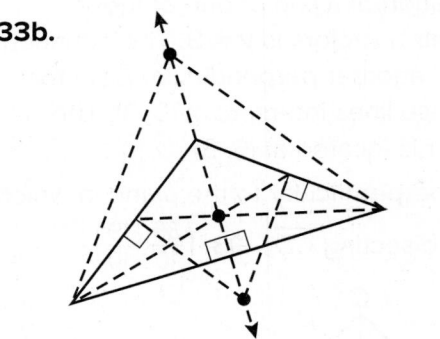

33c.

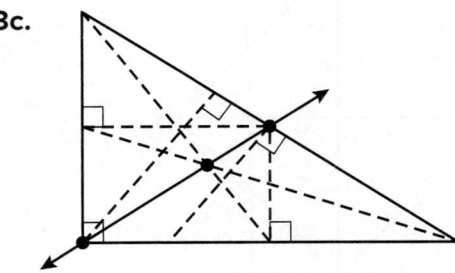

33d. Sample answer: The circumcenter, centroid, and orthocenter are all collinear.

35. perpendicular bisectors/incenter; The point of concurrency for perpendicular bisectors is the circumcenter. The incenter is the point of concurrency for the angle bisectors of a triangle.

Lesson 1-4

1. $\angle 4$, $\angle 7$ **3.** $\angle 7$ **5.** $\angle 4$ **7.** $\angle T$, $\angle R$, $\angle S$, $\overline{RS}$, $\overline{ST}$, $\overline{RT}$ **9.** $\angle E$, $\angle C$, $\angle D$, $\overline{CD}$, $\overline{DE}$, $\overline{CE}$ **11.** $\angle Q$, $\angle P$, $\angle R$, $\overline{PR}$, $\overline{RQ}$, $\overline{QP}$ **13.** They are equal.
15. $\angle X$, $\angle Y$, $\angle Z$, $\overline{YZ}$, $\overline{XZ}$, $\overline{XY}$
17. Sample answer: Because $BC > AC$, $m\angle BAC > m\angle ABC$ by Theorem 6.9. $\overline{AY}$ and $\overline{BX}$ are angle bisectors, so $m\angle ABX = \frac{1}{2} m\angle ABC$ and $m\angle BAY = \frac{1}{2} m\angle BAC$. Then $m\angle BAY < m\angle ABX$.

19. Sample answer: $\overline{AB}$ is the longest. $m\angle BQY > m\angle ABQ$ and $m\angle BQY > m\angle BAQ$ by the Exterior Angle Inequality Theorem. Therefore, if $m\angle AQB > m\angle BQY$, then $m\angle AQB > m\angle ABQ$ and $m\angle ABQ > m\angle BAQ$ by the Transitive Property of Inequality. Then, by Theorem 6.10, $AB > AQ$ and $AB > BQ$. Therefore, $\overline{AB}$ is the longest side of $\triangle ABQ$.

21. 2, 1, 3 **23.** $\angle 2$ **25.** $\angle 3$ **27.** $\angle 8$
29. $m\angle BCF > m\angle CFB$
31. $m\angle DBF < m\angle BFD$ **33.** $RP > MP$

35. $RM > RQ$ **37.** Sample answer: 10; $m\angle C > m\angle B$, so if $AB > AC$, then Theorem 6.10 is satisfied. Because $10 > 6$, $AB > AC$.
39. $m\angle 1$, $m\angle 2 = m\angle 5$, $m\angle 4$, $m\angle 6$, $m\angle 3$; Sample answer: The side opposite $\angle 5$ is the smallest side in that triangle, and $m\angle 2 = m\angle 5$; so, we know that $m\angle 4$ and $m\angle 6$ are both greater than $m\angle 2$ and $m\angle 5$. The side opposite $m\angle 6$ is greater than the side opposite $m\angle 4$. Because the side opposite $\angle 2$ is greater than the side opposite $\angle 1$, we know that $m\angle 1 < m\angle 2$ and $m\angle 5$. Because $m\angle 2 = m\angle 5$, $m\angle 1 + m\angle 3 = m\angle 4 + m\angle 6$. Because $m\angle 1 < m\angle 4$, then $m\angle 3 > m\angle 6$.

Lesson 1-5

1. Proof:

Step 1: Assume that $x > 2$.

Step 2: For all values of $x > 2$, $x^2 + 8 > 12$, so the assumption contradicts the given information for $x > 2$.

Step 3: The assumption of $x > 2$ leads to a contradiction of the given information that $x^2 + 8 \leq 12$. Therefore, the assumption of $x > 2$ must be false, so the original conclusion of $x \leq 2$ must be true.

3. Proof:

Step 1: Assume that $x < -2$ or $x = -2$.

Step 2: When $x = -2$, $2x - 7 = -11$. Because $-11 \not> -11$, the assumption contradicts the given information for $x = -2$. For all values of $x < -2$, $2x - 7 < -11$, so the assumption contradicts the given information for $x < -2$.

Step 3: In both cases, the assumption leads to a contradiction of the given information that $2x - 7 > -11$. Therefore, the assumption of $x \leq -2$ must be false, so the original conclusion of $x > -2$ must be true.

5. Proof:

Step 1: Assume that $x < -1$ or $x = -1$.

Step 2: When $x = -1$, $-3x + 4 = 7$. Because $7 \not< 7$, the assumption contradicts the given information for $x = -1$. For all values of $x < -1$, $-3x + 4 > 7$, so the assumption contradicts the given information for $x < -1$.

Step 3: In both cases, the assumption leads to a contradiction of the given information that $-3x + 4 < 7$. Therefore, the assumption of $x \leq -1$ must be false, so the original conclusion of $x > -1$ must be true.

7. Let x be the cost of one bracelet and the cost of the other bracelet be y.

Step 1: Given: $x + y > 40$; Prove: $x > 20$ or $y > 20$; Indirect Proof: Assume that $x \leq 20$ and $y \leq 20$.

Step 2: If $x \leq 20$ and $y \leq 20$, then $x + y \leq 20 + 20$ or $x + y \leq 40$. This is a contradiction because we know that $x + y > 40$.

Step 3: Because the assumption that $x \leq 20$ and $y \leq 20$ leads to a contradiction of a known fact, the assumption must be false. Therefore, the conclusion that $x > 20$ or $y > 20$ must be true. Thus, at least one of the bracelets had to cost more than $20.

9. Step 1: Assume that x and y are not both odd integers. That is, assume that either x or y is an even integer, or that x and y are both even integers.

Step 2: Case 1: You only need to show that the assumption that x is an even integer leads to a contradiction, because the argument for y is an even integer follows the same reasoning. So, assume that x is an even integer and that y is an odd integer. This means that $x = 2k$ for some integer k and that $y = 2m + 1$ for some integer m.

$xy = (2k)(2m + 1)$ Statement of assumption

$\quad = 4km + 2k$ Distributive Property

$\quad = 2(2km + k)$ Distributive Property

Because k and m are integers, $2km + k$ is also an integer. Let p represent the integer $2km + k$. So, xy can be represented by $2p$. This means that xy is an even integer, which contradicts the given information that xy is an odd integer.

Case 2: Assume that x and y are both even integers. This means that $x = 2k$ and $y = 2m$ for some integers k and m.

$xy = (2k)(2m)$ Statement of assumption

$\quad = 4km$ Simplify.

$\quad = 2(2km)$ Distributive Property

Because k and m are integers, $2km$ is also an integer. Let p represent the integer $2km$. So, xy can be represented by $2p$. This means that xy is an even integer which contradicts the given information that xy is an odd integer.

Step 3: In both cases, the assumption leads to a contradiction of the given information, so the original conclusion that both x and y are odd integers must be true.

11. Step 1: Assume that x is divisible by 4. That is, assume that 4 is a factor of x.

Step 2: Let $x = 4n$ for some integer n. So, $x = 2(2n)$. So, 2 is a factor of x, which means that x is an even number. This contradicts the given information.

Step 3: Because the assumption that x is divisible by 4 leads to a contradiction of the given information, the original conclusion that x is not divisible by 4 must be true.

13.

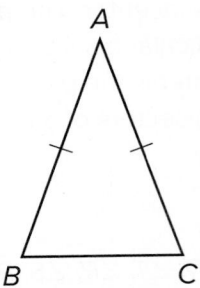

Step 1: Assume that $\angle B$ is a right angle.

Step 2: By the Isosceles Triangle Theorem, $\angle C$ is also a right angle. This contradicts the fact that a triangle can have no more than one right angle.

Step 3: Because the assumption that $\angle B$ is a right angle must be false, the original conclusion that neither of the base angles is a right angle must be true.

15. Step 1: Assume that $\angle A$ is a right angle.

Step 2: Show that this leads to a contradiction. If $\angle A$ is a right angle, then $m\angle A = 90°$ and $m\angle C + m\angle A = 100 + 90 = 190°$. Thus, the sum of the measures of the angles of $\triangle ABC$ is greater than 180°.

Step 3: The conclusion that the sum of the measures of the angles of $\triangle ABC$ is greater than 180° is a contradiction of a known property. The assumption that $\angle A$ is a right angle must be false, which means that the statement $\angle A$ *is not a right angle* must be true.

17. Assume that it is 20°C when Enrique hears the siren; then show that at this temperature it will take more than 5 seconds for the sound of the siren to reach him. Because the assumption is false, it must not be 20°C when Enrique hears the siren.

19. Given: $\triangle ABC$; $m\angle A > m\angle ABC$

Prove: $BC > AC$

Proof: Assume that $BC \ngtr AC$. By the Comparison Property, $BC = AC$ or $BC < AC$.

Case 1: If $BC = AC$, then $\angle ABC \cong \angle A$ by the Isosceles Triangle Theorem. But, $\angle ABC \cong \angle A$ contradicts the given statement that $m\angle A > m\angle ABC$. So, $BC \neq AC$.

Case 2: If $BC < AC$, then there must be a point D between A and C such that $\overline{DC} \cong \overline{BC}$. Draw the auxiliary segment $\overline{BD}$.

Because $DC = BC$, by the Isosceles Triangle Theorem, $\angle BDC \cong \angle DBC$. Now, $\angle BDC$ is an exterior angle of $\triangle BAD$ and by the Exterior Angles Inequality Theorem $m\angle BDC > m\angle A$. By the Angle Addition Postulate, $m\angle ABC = m\angle ABD + m\angle DBC$. Then, by the definition of inequality, $m\angle ABC > m\angle DBC$. By substitution and the Transitive Property of Inequality, $m\angle ABC > m\angle A$. But this contradicts the given statement that $m\angle A > m\angle ABC$. In both cases, a contradiction was found, and hence our assumption must have been false. Therefore, $BC > AC$.

21. Sample answer: First identify the statement you need to prove and assume that this statement is false by assuming that the opposite of the statement is true. Next, reason logically until you reach a contradiction. Finally, conclude that the statement you wanted to prove must be true because the contradiction proves that the assumption you made was false.

23. Step 1: Let x be a nonzero rational number such that $x = \frac{a}{b}$ for some integers a and b, $b \neq 0$. Let y represent an irrational number. Substituting, $xy = \frac{ay}{b}$. Assume that xy is a rational number such that $xy = \frac{c}{d}$ for some integers c and d, $d \neq 0$.

Step 2:

$xy = \frac{ay}{b}$ x is a rational number.

$\frac{c}{d} = \frac{ay}{b}$ Substitution of assumption

$cb = ayd$ Multiply each side by db. This is possible because $d \neq 0$ and $b \neq 0$.

$\frac{cb}{ad} = y$ Divide each side by ad. $a \neq 0$ because $x = \frac{a}{b}$ and x is nonzero.

Because a, b, c, and d are integers, $a \neq 0$, and $d \neq 0$, $\frac{cb}{ad}$ is the quotient of two integers. Therefore, y is a rational number. This contradicts the given statement that y is an irrational number.

Step 3: Because the assumption that xy is a rational number leads to a contradiction of the given, the original conclusion that xy is irrational must be true.

Lesson 1-6

1. yes   **3.** yes   **5.** yes   **7.** no; $0.7 + 1.4 = 2.1$
9. 13 ft $< n <$ 25 ft   **11.** 14 in. $< n <$ 40 in.
13. 90 mi
15. Proof:

Statements (Reasons)

1. $\overline{PL} \parallel \overline{MT}$ (Given)

2. $\angle P \cong \angle T$ (Alternate Interior Angles Theorem)

3. K is the midpoint of $\overline{PT}$ (Given)

4. $PK = KT$ (Definition of midpoint)

5. $\angle PKL \cong \angle MKT$ (Vertical Angles Theorem)

6. $\triangle PKL \cong \triangle TKM$ (ASA)

7. $PK + KL > PL$ (Triangle Inequality Theorem)

8. $KL = KM$ (CPCTC)

9. $PK + KM > PL$ (Substitution)

17. $2 < x < 10$ **19.** yes; $XY + YZ > XZ$, $XY + XZ > YZ$, and $XZ + YZ > XY$ **21.** 3   **23.** $\frac{1}{7}$
25. Sample answer: whether or not the side lengths actually form a triangle, what the smallest and largest angles are, whether the triangle is equilateral, isosceles, or scalene

Lesson 1-7

1. the second plane; Sample answer: The legs are congruent. If $x < 30$, then the measure of the included angle, $(180 - x)°$, is greater for the second plane; so, by the Hinge Theorem, the second plane if farther away from the airstrip.

3. Proof:

Statements (Reasons)

1. $\angle SXT$ and $\angle RXT$ are supplementary. (Def. of linear pair)

2. $m\angle SXT + m\angle RXT = 180°$ (Def. of supplementary)

3. $m\angle SXT = 97°$ (Given)

4. $97° + m\angle RXT = 180°$ (Substitution)

5. $m\angle RXT = 83°$ (Subtraction)

6. $97 > 83$ (Inequality)

7. $m\angle SXT > m\angle RXT$ (Substitution)

8. $RX = XS$ (Given)

9. $TX = TX$ (Reflexive Property of Equality)

10. $ST > RT$ (Hinge Theorem)

5. $x > 12.5$ **7.** Proof:

Statements (Reasons)

1. $\overline{XU} \cong \overline{VW}$, $\overline{XU} \parallel \overline{VW}$ (Given)

2. $\angle UXV \cong \angle XVW$, $\angle XUW \cong \angle UWV$ (Alternate Interior Angles Theorem)

3. $\triangle XZU \cong \triangle VZW$ (ASA)

4. $\overline{XZ} \cong \overline{VZ}$ (CPCTC)

5. $\overline{WZ} \cong \overline{WZ}$ (Reflexive Property)

6. $VW > XW$ (Given)

7. $m\angle VZW > m\angle XZW$ (Converse of Hinge Theorem)

8. $\angle VZW \cong \angle XZU$, $\angle XZW \cong \angle VZU$ (Vertical angles are congruent.)

9. $m\angle VZW = m\angle XZU$, $m\angle XZW = m\angle VZU$ (Definition of congruent angles)

10. $m\angle XZU > m\angle UZV$ (Substitution Property)

9. $MR > RP$ **11.** $m\angle C < m\angle Z$

13. $m\angle BXA < m\angle DXA$ **15.** It is given that $\overline{EF} \cong \overline{GH}$. Also, $\overline{FG} \cong \overline{FG}$ by the Reflexive Property. It is also given that $m\angle F > m\angle G$. Therefore, by the Hinge Theorem, $EG > FH$.

17. 8:00

19. Given: $\overline{RS} \cong \overline{UW}$
$\overline{ST} \cong \overline{WV}$
$\overline{RT} > \overline{UV}$

Prove: $m\angle S > m\angle W$

Indirect Proof

Step 1: Assume that $m\angle S \leq m\angle W$.

Step 2: If $m\angle S \leq m\angle W$, then either $m\angle S < m\angle W$ or $m\angle S = m\angle W$.

Case 1: If $m\angle S < m\angle W$, then $RT < UV$ by SAS Inequality.

Case 2: If $m\angle S = m\angle W$, then $\triangle RST = \triangle UVW$ by SAS, and $\overline{RT} \cong \overline{UV}$ by CPCTC. Thus, $RT = UV$.

Step 3: Both cases contradict the given $RT > UV$. Therefore, the assumption must be false, and the conclusion, $m\angle S < m\angle W$, must be true.

21a. Northwest; sample answer: Submarine B travels farther than submarine A by the Hinge Theorem, and on an overall heading closer to east-west, so it ends up farther west than submarine A's starting position. Also submarine B's southwesterly leg is nearer to the westerly direction, so it does not travel as far south, and finishes to the north of submarine A's starting position. **21b.** Sample answer: Submarine B would have traveled farther than submarine A and would not be to the northeast or southeast of submarine A's starting position.

23. Right or obtuse; sample answer: If $RT = RS$, then the triangle is isosceles, and the median is also perpendicular to $\overline{TS}$. That would mean that both triangles formed by the median, $\triangle RQT$ and $\triangle RQS$, are right. If $RT > RS$, that means that $m\angle RQT > m\angle RQS$. Because they are a linear pair and the sum of the angle measures must be 180°, $m\angle RQT$ must be greater than 90° and $\triangle RQT$ is obtuse.

25. Never; sample answer: From the Converse of the Hinge Theorem, $m\angle ADB < m\angle BDC$. $\angle ADB$ and $\angle BDC$ form a linear pair. So, $m\angle ADB + m\angle BDC = 180°$. Because $m\angle BDC > m\angle ADB$, $m\angle BDC$ must be greater than 90° and $m\angle ADB$ must be less than 90°. So, by the definition of obtuse and acute angles, $\angle BDC$ is always obtuse and $\angle ADB$ is always acute.

Module 1 Review

1. 4.8 **3.** C **5.** C

7.

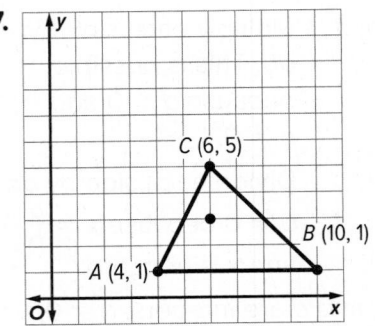

9. $DG = 12$, $GC = 24$

11. Antonia must assume that one of the angles is greater than 90°.

13. D

15. C

Module 2

Quick Check

1. 150   **3.** 1

Lesson 2-1

1. $m\angle Q = 121°$; $m\angle R = 58°$; $m\angle S = 123°$; $m\angle T = 58°$ **3.** $m\angle A = 90°$; $m\angle B = 90°$; $m\angle C = 128°$; $m\angle D = 74°$; $m\angle E = 158°$
5. 135°   **7.** 128.6°   **9.** 10   **11.** 18   **13.** 6
15. 37   **17.** 44   **19.** 72°   **21.** 60°   **23.** 40°
25. 51.4°; 128.6°   **27.** 25.7°; 154.3°   **29.** 30
31. 186°; 137°; 40°; 54°; 123°   **33.** 360°
35. Consider the sum of the measures of the exterior angles N for an n-gon. N = sum of measures of linear pairs − sum of measures of interior angles

$$= 180n° − 180°(n − 2)$$
$$= 180n° − 180n° + 360°$$
$$= 360°$$

So, the sum of the exterior angle measures is 360° for any convex polygon.   **37.** 15   **39.** 360°
41. Liang; sample answer: By the Exterior Angles Sum Theorem, the sum of the exterior angle measures of any convex polygon is 360°.

43. Sample answer:

8; The sum of the interior angles is $(5 − 2) \cdot$ 180° or 540° Twice this sum is 2(540°) or 1080°. A polygon with this interior angles sum is the solution to $(n − 2) \cdot 180° = 1080°$. So, $n = 8$.

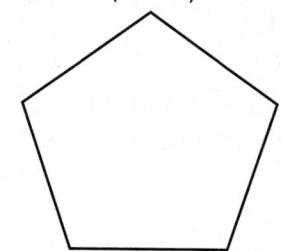

45. Always; sample answer: By the Polygon Exterior Angles Sum Theorem, $m\angle QPR = 60°$ and $m\angle QRP = 60°$. Because the sum of the interior angle measures of a triangle is 180°, the measure of $\angle PQR = 180° − m\angle QPR − m\angle QRP = 180° − 60° − 60° = 60°$. So, $\triangle PQR$ is an equilateral triangle.

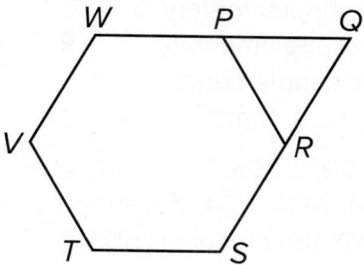

Lesson 2-2

1. 52°   **3.** 5   **5.** $x°$
7. Proof:

Statements (Reasons)

1. $WXTV$ and $YZTV$ are parallelograms. (Given)

2. $\overline{WX} \cong \overline{VT}$, $\overline{VT} \cong \overline{YZ}$ (Opp. sides of a parallelogram are ≅.)

3. $\overline{WX} \cong \overline{YZ}$ (Transitive Property of Congruence)

9. $x = 5$, $y = 17$ **11.** $x = 58$, $y = 63.5$
13. 203.32 m; 2125 m²
15. Proof:

Statements (Reasons)

1. $\square PQRS$ (Given)

2. Draw auxiliary segment PR. (Diagonal of $PQRS$)

3. $\overline{PQ} \parallel \overline{SR}$, $\overline{PS} \parallel \overline{QR}$ (Opp. sides of a parallelogram are ∥.)

4. $\angle QPR \cong \angle SRP$, $\angle SPR \cong \angle QRP$ (Alt. Interior Angles Thm.)

5. $\overline{PR} \cong \overline{PR}$ (Reflexive Property of ≅)

6. $\triangle QPR \cong \triangle SRP$ (ASA)

7. $\overline{PQ} \cong \overline{RS}$, $\overline{QR} \cong \overline{SP}$ (CPCTC)

17. Proof:

Statements (Reasons)

1. □GKLM (Given)

2. $\overline{GK} \parallel \overline{ML}, \overline{GM} \parallel \overline{KL}$ (Opp. sides of a parallelogram are ∥.)

3. ∠G and ∠K are supplementary.
∠K and ∠L are supplementary.
∠L and ∠M are supplementary.
∠M and ∠G are supplementary.
(Consecutive Int. ∠s Thm.)

19. 3 **21.** 131° **23.** 29° **25a.** $JP = \sqrt{13}$, $LP = \sqrt{13}$, $MP = \sqrt{34}$, $KP = \sqrt{34}$; because $JP = LP$ and $MP = KP$, the diagonals bisect each other. **25b.** No; $JP + LP \neq MP + KP$. **25c.** No; sample answer: The slope of $\overline{JK} = 0$, and the slope of $\overline{JM} = 2$. The slopes are not negative reciprocals of each other, so the consecutive sides are not perpendicular. **27.** 7 **29.** Sample answer: In a parallelogram, the opp. sides and ∠s are ≅. Two consecutive ∠s in a □ are supplementary. If one angle of a □ is right, then all the angles are right. The diagonals of a parallelogram bisect each other. **31.** $m\angle 1 = 116°$, $m\angle 10 = 115°$; Sample answer: $m\angle 8 = 64°$ because alternate interior angles are congruent. ∠1 is supplementary to ∠8 because consecutive angles in a parallelogram are supplementary, so $m\angle 1$ is 116°. ∠10 is supplementary to the 65°-angle because consecutive angles in a parallelogram are supplementary, so $m\angle 10 = 180° - 65°$ or 115°.

Lesson 2-3

1. Yes; a pair of opposite sides are parallel and congruent. **3.** No; none of the tests for parallelograms are fulfilled. **5.** Yes; the diagonals bisect each other. **7.** $x = 20$; $y = 45$ **9.** $x = -6$; $y = 23$

11. Yes; sample answer: Use the Slope Formula.

$$m = \frac{y_2 - y_1}{x_2 - x_1}$$

slope of $\overline{AD} = \frac{3 - 0}{-2 - (-3)} = \frac{3}{1} = 3$

slope of $\overline{BC} = \frac{2 - (-1)}{3 - 2} = \frac{3}{1} = 3$

slope of $\overline{AB} = \frac{2 - 3}{3 - (-2)} = \frac{1}{5}$

slope of $\overline{CD} = \frac{-1 - 0}{2 - (-3)} = -\frac{1}{5}$

Because opposite sides have the same slope, $\overline{AB} \parallel \overline{CD}$ and $\overline{AD} \parallel \overline{BC}$. Therefore, $ABCD$ is a parallelogram by definition.

13. Yes; $SR = ZT$ and the slopes of $\overline{SR}$ and $\overline{ZT}$ are equal, so one pair of opposite sides is parallel and congruent.

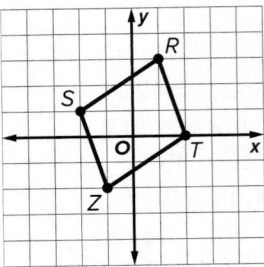

15. No; slope $\overline{XY} = -\frac{3}{5}$ and slope of $\overline{WZ} = -\frac{1}{3}$, so opposite sides are not parallel.

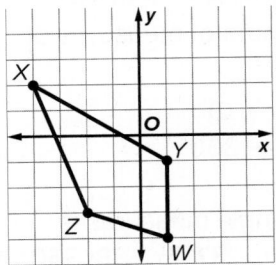

17. Given: $ABCD$ is a parallelogram.
Prove: ∠B, ∠C, and ∠D are right angles.
Proof:

slope of $\overline{AD} = \frac{0 - 0}{a - 0} = 0$

slope of $\overline{BC} = \frac{b - b}{a - 0} = 0$

slope of $\overline{AB}$ is undefined
slope of $\overline{CD}$ is undefined

Therefore, $\overline{AB} \perp \overline{BC}$, $\overline{BC} \perp \overline{CD}$, and $\overline{AB} \parallel \overline{CD}$. So, ∠B, ∠C, and ∠D are right angles.

19. Proof:

Statements (Reasons)

1. $\overline{PR}$ bisects $\overline{TQ}$; $\overline{TQ}$ bisects $\overline{PR}$. (Given)

2. $\overline{PV} \cong \overline{VR}, \overline{TV} \cong \overline{VQ}$ (Def. of bisector)

3. ∠PVT ≅ ∠RVQ, ∠TVR ≅ ∠QVP (Vertical Angles Thm.)

4. △PVT ≅ △RVQ, △TVR ≅ △QVP (SAS)

5. $\overline{PQ} \cong \overline{RT}, \overline{PT} \cong \overline{RQ}$ (CPCTC)

6. PQRT is a parallelogram. (If both pairs of opp. sides are ≅, then quad. is a □.)

21. (4, −1), (0, 3), or (−4, −5) **23.** −4
25. 28 cm **27.** Yes; sample answer: The lengths of the opposite sides are congruent. When the coordinate plane is placed over the map, the street lamps align perfectly with the points on the grid.
29. No; sample answer: Madison and Angela have to be the same distance from the center and Nikia and Shelby have to be the same distance from the center, but Nikia and Shelby's distance from the center does not have to be equal to Madison and Angela's distance.
31. Given: *RSTV* is a quadrilateral. *A, B, C,* and *D* are midpoints of sides $\overline{RS}, \overline{ST}, \overline{TV},$ and $\overline{VR}$, respectively.

Prove: *ABCD* is a parallelogram.

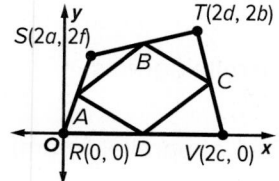

Proof:

Place quadrilateral *RSTV* on the coordinate plane and label the coordinates as shown. By the Midpoint Formula, the coordinates of *A, B, C,* and *D* are:

$A\left(\frac{2a}{2}, \frac{2f}{2}\right) = (a, f);$

$B\left(\frac{2d + 2a}{2}, \frac{2f + 2b}{2}\right) = (d + a, f + b);$

$C\left(\frac{2d + 2c}{2}, \frac{2b}{2}\right) = (d+c, b); \text{and} D\left(\frac{2c}{2}, \frac{0}{2}\right) = (c, 0).$

Find the slopes of $\overline{AB}$ and $\overline{DC}$.

slope of $\overline{AB} = \frac{(f + b) - f}{(d + a) - a} = \frac{b}{d}$

slope of $\overline{DC} = \frac{0 - b}{c - (d + c)} = \frac{-b}{-d} = \frac{b}{d}$

The slopes of $\overline{AB}$ and $\overline{DC}$ are the same, so the segments are parallel. Use the Distance Formula to find *AB* and *DC*.

$AB = \sqrt{(d + a - a)^2 + (f + b - f)^2} = \sqrt{d^2 + b^2}$

$DC = \sqrt{(d + c - c)^2 + (b - 0)^2} = \sqrt{d^2 + b^2}$

Thus, $AB = DC$ and $\overline{AB} \cong \overline{DC}$. Therefore, *ABCD* is a parallelogram because if one pair of opposite sides of a quadrilateral are both parallel and congruent, then the quadrilateral is a parallelogram.

33. Sample answer: The theorems are converses of each other. The hypothesis of Theorem 7.3 is *a figure is a parallelogram*, and the hypothesis of Theorem 7.9 is *both pairs of opposite sides of a quadrilateral are congruent*. The conclusion of Theorem 7.3 is *opposite sides are congruent*, and the conclusion of Theorem 7.9 is *the quadrilateral is a parallelogram*. **35.** (−3, 1) and (−2, −2)

Lesson 2-4

1. 2 ft **3.** 50° **5.** 39 ft **7.** 52° **9.** $x = 7$
11. 77° **13.** 180°
15. Proof:

Statements (Reasons)

1. *ABCD* is a rectangle. (Given)

2. *ABCD* is a parallelogram. (Def. of rectangle)

3. $\overline{AD} \cong \overline{BC}$ (Opp. sides of a parallelogram are ≅.)

4. $\overline{DC} \cong \overline{DC}$ (Reflexive Property of ≅)

5. $\overline{AC} \cong \overline{BD}$ (Diag. of a rectangle are ≅.)

6. $\triangle ADC \cong \triangle BCD$ (SSS)

17. Yes; sample answer: Opposite sides are parallel and consecutive sides are perpendicular.

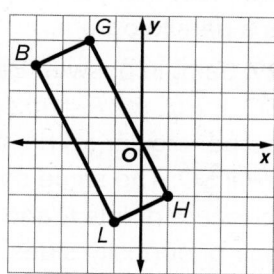

19. No; sample answer: Diagonals are not congruent.

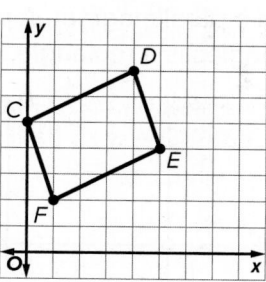

21. Yes; sample answer: Both pairs of opposite sides are congruent and diagonals are congruent.

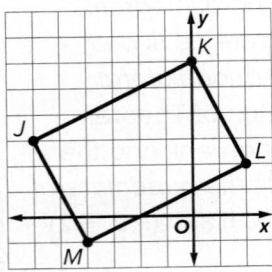

23. Proof:

Statements (Reasons)

1. *ABCD* is a rectangle. (Given)
2. $\overline{AD} \cong \overline{BC}$ (If a quad. is a ▱, its opp. sides are ≅.)
3. $\overline{DC} \cong \overline{DC}$ (Reflexive Property of ≅)
4. ∠*ADC* and ∠*BCD* are rt. angles. (Def. of rectangle)
5. ∠*ADC* ≅ ∠*BCD* (All rt. angles are ≅.)
6. △*ADC* ≅ △*BCD* (SAS)
7. $\overline{AC} \cong \overline{BD}$ (CPCTC)

25. No; sample answer: If you only know that opposite sides are congruent and parallel, then the most that you can conclude is that the plot is a parallelogram.

27. Sample answer: Because $\overline{RP} \perp \overline{PQ}$ and $\overline{SQ} \perp \overline{PQ}$, m∠*P* = m∠*Q* = 90°. Lines that are perpendicular to the same line are parallel, so $\overline{RP} \parallel \overline{SQ}$. The same compass setting was used to locate points *R* and *S*, so $\overline{RP} \cong \overline{SQ}$. If one pair of opposite sides of a quadrilateral is both parallel and congruent, then the quadrilateral is a parallelogram. A parallelogram with right angles is a rectangle. Thus, *PRSQ* is a rectangle.

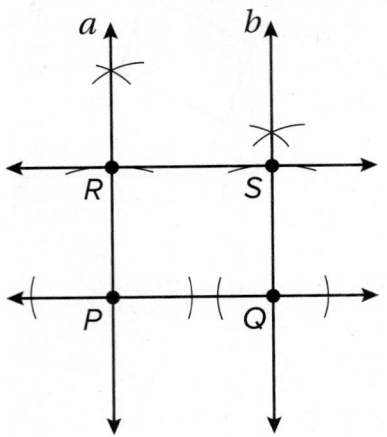

29. 5   **31.** They should be equal.   **33.** (−6, 3)

35. $x = 6$; $y = -10$ **37.** Sample answer: All rectangles are parallelograms because, by definition, both pairs of opposite sides are parallel. Parallelograms with right angles are rectangles, so some parallelograms are rectangles, but others with nonright angles are not.

39. Yes; sample answer: By Theorem 7.6, if a parallelogram has one right angle, then it has four right angles. Therefore, a parallelogram with one right angle must be a rectangle.

Lesson 2-5

1. 60°   **3.** 24   **5.** 32°   **7.** 10   **9.** 21

11. Proof:

Statements (Reasons)

1. *ACDH* and *BCDF* are parallelograms; $\overline{BF} \cong \overline{AB}$. (Given)
2. $\overline{BF} \cong \overline{CD}$, $\overline{CD} \cong \overline{AH}$ (Def. of parallelogram)
3. $\overline{BF} \cong \overline{AH}$ (Transitive Property of ≅)
4. $\overline{BC} \cong \overline{FD}$, $\overline{AC} \cong \overline{HD}$ (Def. of parallelogram)
5. *BC* = *FD*, *AC* = *HD* (Def. of ≅ segments)
6. *AC* = *AB* + *BC*, *HD* = *HF* + *FD* (Seg. Add. Post.)
7. *AB* + *BC* = *HF* + *FD* (Substitution)
8. *AB* + *FD* = *HF* + *FD* (Substitution)
9. *AB* = *HF* (Subtraction Property of Equality)
10. $\overline{AB} \cong \overline{HF}$ (Def. of ≅ segments)
11. $\overline{AH} \cong \overline{BF}$, $\overline{AB} \cong \overline{HF}$ (Substitution)
12. *ABFH* is a rhombus. (Def. of rhombus)

13. Proof:

Statements (Reasons)

1. $\overline{WZ} \parallel \overline{XY}$, $\overline{WX} \parallel \overline{ZY}$, $\overline{WX} \cong \overline{ZY}$ (Given)
2. *WXYZ* is a parallelogram. (Both pairs of opposite sides are parallel.)
3. *WXYZ* is a rhombus. (If one pair of consecutive sides of a parallelogram are congruent, then the parallelogram is a rhombus.)

15. Sample answer: Because consecutive sides are congruent, the garden is a rhombus. Jorge needs to know if the diagonals of the garden are congruent to determine whether it is a square.   **17.** Rectangle, rhombus, square; the four sides are congruent, and consecutive sides are perpendicular.

19. Rhombus; all sides are congruent, and the diagonals are perpendicular, but not congruent.   **21.** Rhombus, rectangle, square; all sides are congruent, and the diagonals are perpendicular and congruent.

23. 55   **25.** 31   **27.** 6   **29.** 90°

31. Proof:

Statements (Reasons)

1. *ABCD* is a rhombus. (Given)

2. *ABCD* is a parallelogram. (Def. of rhombus)

3. $\angle DAB \cong \angle DCB$, $\angle ABC \cong \angle ADC$ (If a quad. is a ▱, its opp. $\angle$s are $\cong$.)

4. $\overline{AB} \cong \overline{BC} \cong \overline{CD} \cong \overline{AD}$ (Def. of rhombus)

5. $\triangle DAB \cong \triangle DCB$, $\triangle ABC \cong \triangle ADC$ (SAS)

6. $\angle 8 \cong \angle 7$, $\angle 3 \cong \angle 4$, $\angle 1 \cong \angle 2$, $\angle 5 \cong \angle 6$ (CPCTC)

7. $\overline{AC}$ bisects $\angle DAB$ and $\angle DCB$. $\overline{BD}$ bisects $\angle ABC$ and $\angle ADC$. (Def. of angle bisector)

33. Proof:

Statements (Reasons)

1. *RSTU* is a parallelogram; $\overline{RS} \cong \overline{ST}$. (Given)

2. $\overline{RS} \cong \overline{UT}$, $\overline{RU} \cong \overline{ST}$ (If a quad. is a ▱, its opp. sides are $\cong$.)

3. $\overline{RS} \cong \overline{RU}$ (Transitive Property of $\cong$)

4. *RSTU* is a rhombus. (Def. of rhombus)

35. The figure consists of 15 congruent rhombi.

37. square

39.

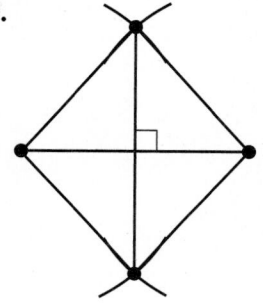

Sample answer: The diagonals bisect each other, so the quadrilateral is a parallelogram. Because the diagonals of the parallelogram are perpendicular to each other, the parallelogram is a rhombus.   **41.** right triangles

43. Parallelogram: Opposite sides of a parallelogram are parallel and congruent. Opposite angles of a parallelogram are congruent. The diagonals of a parallelogram bisect each other and each diagonal separates a parallelogram into two congruent triangles.

Rectangle: A rectangle has all the properties of a parallelogram. A rectangle has four right angles. The diagonals of a rectangle are congruent.

Rhombus: A rhombus has all the properties of a parallelogram. All sides of a rhombus are congruent. The diagonals of a rhombus are perpendicular and bisect the angles of the rhombus.

Square: A square has all of the properties of a parallelogram. A square has all the properties of a rectangle. A square has all of the properties of a rhombus.   **45.** True; sample answer: A rectangle is a quadrilateral with four right angles and a square is both a rectangle and a rhombus, so a square is always a rectangle.

Converse: If a quadrilateral is a rectangle, then it is a square. False; sample answer: A rectangle is a quadrilateral with four right angles. It is not necessarily a rhombus, so it is not necessarily a square.

Inverse: If a quadrilateral is not a square, then it is not a rectangle. False; sample answer: A quadrilateral that has four right angles and two pairs of congruent sides is not a square, but it is a rectangle.

Contrapositive: If a quadrilateral is not a rectangle, then it is not a square. True; sample answer: If a quadrilateral is not a rectangle, it is also not a square by definition.

Lesson 2-6

1a. Proof:

Statements (Reasons)

1. *WXYZ* is an isosceles trapezoid. (Given)

2. $\overline{WZ} \cong \overline{XY}$ (Def. of isosceles trapezoid)

3. $WZ = XY$ (Def. of congruent segments)

4. $4x + 5 = 5x - 3$ (Substitution)

5. $5 = x - 3$ (Subtraction Prop. of Equality)

6. $8 = x$ (Addition Prop. of Equality)

7. $x = 8$ (Symmetric Prop. of Equality)

1b. 74° **1c.** 110 in. **3.** 112°

5a. $\overline{AD} \parallel \overline{BC}$, but $\overline{AB} \nparallel \overline{CD}$. **5b.** yes; $AB = 5$ and $CD = 5$

7. 11 **9.** 13 **11.** $(-3, -0.5)$ and $(2, 2)$

13. 101° **15a.** $\sqrt{65}$ **15b.** 42.2

17. 15 **19.** 100° **21.** 17.5 ft

23. Proof:

Statements (Reasons)

1. $TUVW$ is a trapezoid; $\angle W \cong \angle V$. (Given)

2. Draw auxiliary line $\overline{UX} \parallel \overline{TW}$. (Parallel Postulate)

3. $\angle UXV \cong \angle W$ (Corresponding Angles Thm.)

4. $\angle UXV \cong \angle V$ (Transitive Property of $\cong$)

5. $\overline{UX} \cong \overline{UV}$ (Converse of Isosceles Triangle Thm.)

6. $TUXW$ is a parallelogram. (Def. of parallelogram)

7. $\overline{TW} \cong \overline{UX}$ (If a quad. is a $\square$, its opp. sides are $\cong$.)

8. $\overline{TW} \cong \overline{UV}$ (Transitive Property of $\cong$)

9. Trapezoid $TUVW$ is isosceles. (Def. of isosceles trapezoid)

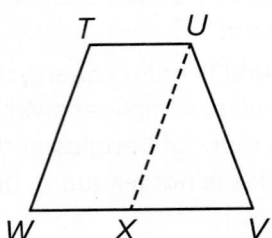

25. Proof: It is given that $LMNP$ is a kite. By the definition of kite, $\overline{LM} \cong \overline{LP}$ and $\overline{MN} \cong \overline{PN}$. By the Reflexive Property of Congruence, $\overline{LN} \cong \overline{LN}$. Therefore, $\triangle LMN \cong \triangle LPN$ by SSS. $\angle M \cong \angle P$ by CPCTC. If both pairs of opposite angles of a quadrilateral are congruent, then the quadrilateral is a parallelogram. So, if $\angle MLP \cong \angle MNP$, then $LMNP$ is a parallelogram. It is given the $LMNP$ is a kite. Therefore, $\angle MLP \ncong \angle MNP$.

27. Sample answer: $y = 3$, $y = 1$, $y = x$, $y = -x + 10$; The intersection points of the four lines are $A(1, 1)$, $B(3, 3)$, $C(7, 3)$, and $D(9, 1)$. The slope of $\overline{AD}$ is 0, and the slope of $\overline{BC}$ is 0. Therefore, $\overline{AD} \parallel \overline{BC}$. The slope of $\overline{AB}$ is 1, and the slope of $\overline{CD}$ is -1. So, $\overline{AB} \nparallel \overline{CD}$. $AB = 2\sqrt{2}$, and $CD = 2\sqrt{2}$. Therefore, by definition, $ABCD$ is an isosceles trapezoid.

29. Belinda; sample answer: Because $ABCD$ is a kite, $m\angle D = m\angle B$. So, $m\angle A + m\angle B + m\angle C + m\angle D = 360°$, or $m\angle A + 100° + 45° + 100° = 360°$. Therefore, $m\angle A = 115°$.

31. Sample answer: A quadrilateral must have exactly one pair of sides parallel to be a trapezoid. If the legs are congruent, then the trapezoid is an isosceles trapezoid. If a quadrilateral has exactly two pairs of consecutive congruent sides with the opposite sides not congruent, the quadrilateral is a kite. A trapezoid and a kite both have four sides. In a trapezoid and isosceles trapezoid, both have exactly one pair of parallel sides.

33. Never; sample answer: Exactly two pairs of adjacent sides are congruent.

Module 2 Review

1. C **3.** C **5.** $m\angle Q = 108°$ and $m\angle R = 72°$

7. A **9.** A **11.** A

13. B, C, D, E **15.** A, B, C, E **17.** B

Module 3

Quick Check

1. $\frac{1}{4}$ **3.** $\frac{3}{5}$ **5.** $\frac{4}{5}$ **7.** 40 in.

Lesson 3-1

1. enlargement; 3   **3.** reduction; $\frac{2}{3}$

5. 50%; The perimeter of the updated blueprint will be 26 units.

7. $S'(0, 0)$, $T'(-5, -5)$, $V'(-10, -10)$

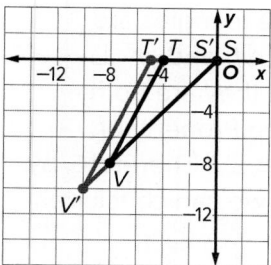

9. $D'(3, 3)$, $F'(0, 0)$, $G'(6, 0)$

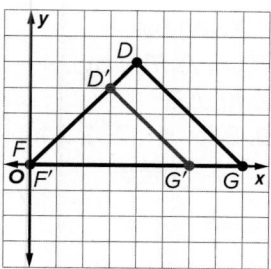

11. 1.5

13.

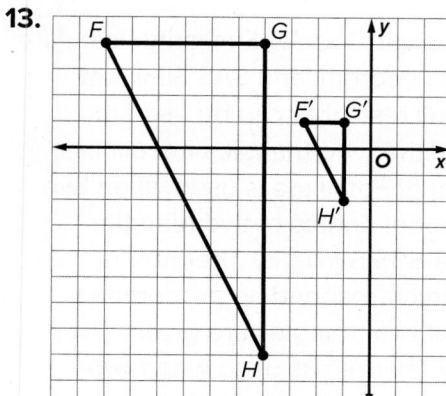

15.

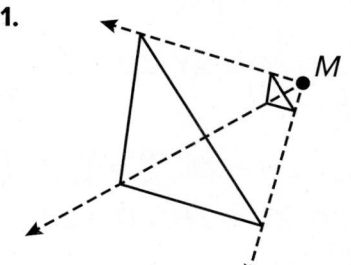

17. $\frac{2}{3}$ **19.** enlargement; 2

21.

23a. $\frac{b}{a}$; Sample answer: The coordinates of P' are (ka, kb).

The slope of $\overleftrightarrow{PP'}$ is $\frac{kb - b}{ka - a} = \frac{b(k - 1)}{a(k - 1)} = \frac{b}{a}$.

23b. Sample answer: If P lies on the y-axis, then its image P' will also lie on the y-axis. In this case, $\overleftrightarrow{PP'}$ will be vertical, and the slope will be undefined.

25. $y = 4x - 3$

27a. Always; sample answer: Points remain invariant under the dilation.

27b. Always; sample answer: Because the rotation is centered at B, point B will always remain invariant under the rotation.

27c. Sometimes; sample answer: If one of the vertices is on the x-axis, then that point will remain invariant under reflection. If two vertices are on the x-axis, then the two vertices located on the x-axis will remain invariant under reflection.

27d. Never; sample answer: When a figure is translated, all points move an equal distance. Therefore, no points can remain invariant under translation.

27e. Sometimes; sample answer: If one of the vertices of the triangle is located at the origin, then that vertex would remain invariant under the dilation. If none of the points of $\triangle XYZ$ are located at the origin, then no points will remain invariant under the dilation.

29. Sample answer: Translations, reflections, and rotations produce congruent figures because the sides and angles of the preimage are congruent to the corresponding sides and angles of the image.

Lesson 3-2

1. $\angle A \cong \angle W$, $\angle B \cong \angle X$, $\angle C \cong \angle Y$, $\angle D \cong \angle Z$, $\frac{AB}{WX} = \frac{BC}{XY} = \frac{CD}{YZ} = \frac{AD}{WZ}$

3. $\angle F \cong \angle J$, $\angle G \cong \angle K$, $\angle H \cong \angle L$, $\frac{FG}{JK} = \frac{GH}{KL} = \frac{FH}{JL}$

5. $\angle A \cong \angle F$, $\angle B \cong \angle G$, $\angle C \cong \angle H$, $\angle D \cong \angle J$, $\frac{AB}{FG} = \frac{BC}{GH} = \frac{CD}{HJ} = \frac{AD}{FJ}$

7. no; $\angle W \not\cong \angle L$ **9.** no; $\frac{MN}{GH} \neq \frac{NP}{HJ}$ **11.** 3

13. 5   **15.** 8.5 km   **17.** 18.9

19. $x = 63$, $y = 32$ **21.** Because $\triangle ABC \sim \triangle DEF$, $\frac{AB}{DE} = \frac{BC}{EF} = \frac{AC}{DF}$. So, $\frac{AB}{DE} = \frac{BC}{EF} = \frac{AC}{DF} = \frac{m}{n}$. By the Multiplication Property of Equality, $AB = DE\left(\frac{m}{n}\right)$, $BC = EF\left(\frac{m}{n}\right)$, and $AC = DF\left(\frac{m}{n}\right)$. Using substitution, the perimeter of $\triangle ABC = DE\left(\frac{m}{n}\right) + EF\left(\frac{m}{n}\right) + DF\left(\frac{m}{n}\right) = \frac{m}{n}(DE + EF + DF)$. The ratio of the perimeters is $\frac{\text{perimeter of } \triangle ABC}{\text{perimeter of } \triangle DEF} = \frac{\left(\frac{m}{n}\right)(DE + EF + DF)}{DE + EF + DF} = \frac{m}{n}$.   **23.** $\frac{3}{2}$

25. Yes; the ratio of the longer dimensions of the rinks is $\frac{20}{17}$, and the ratio of the smaller dimensions of the rinks is $\frac{20}{17}$.   **27.** 4

29. Sample answer: $\frac{4}{3} \neq \frac{4}{10}$

4 cm ▢ 4 cm   3 cm ▭ 10 cm

31. Sample answer: The figures could be described as congruent if they are the same size and shape, similar if their corresponding angles are congruent and their corresponding sides are proportional, and equal if they are the same figure, or none of those.

Lesson 3-3

1. Yes; $\triangle FGH \sim \triangle JHK$ by AA Similarity.

3. No; the triangles would be similar by AA Similarity if $\overline{AB} \parallel \overline{DF}$.   **5.** Yes; the triangles are similar by AA Similarity.   **7.** 135 ft

9. $\triangle ABC \sim \triangle DBE$; 16   **11.** $\triangle DEF \sim \triangle GEH$; 9
13. Sample answer: $m\angle ADB = 108°$, so $m\angle DBA = 36°$ because base angles of an isosceles triangle are congruent. Thus, $m\angle ABC = 72°$. Similarly, $m\angle DCB = 36°$ and $m\angle ACB = 72°$. So, $m\angle BEC = 72°$ and $m\angle BAC = 36°$. Therefore, $\triangle ABC$ and $\triangle BCE$ are similar.

15. Yes; sample answer: The triangles are similar. It is given that $\overline{KM} \perp \overline{JL}$ and $\overline{JK} \perp \overline{KL}$. $\angle JKL \cong \angle JMK$ because they are both right angles. By the Reflexive Property of Congruence, we know $\angle J \cong \angle J$. Therefore, by the AA Similarity Postulate, we can conclude that $\triangle JKL \sim \triangle JMK$.

Lesson 3-4

1. Yes; $\triangle RST \sim \triangle WSX$ (or $\triangle XSW$) by SAS Similarity   **3.** Yes; $\triangle STU \sim \triangle JPM$ by SAS Similarity   **5.** $\triangle HIJ \sim \triangle KLJ$; 5
7. $\triangle RST \sim \triangle UVW$; 11.25   **9.** 5 ft   **11.** 202.2 in.
13. Sample answer: $m\angle TSU = m\angle QSR$ because they are vertical angles. $\frac{ST}{SQ}$ is proportional to $\frac{SU}{SR}$. Therefore, $\triangle STU$ and $\triangle SQR$ are similar by the SAS Similarity Theorem.

15. Sample answer: The AA Similarity Postulate, SSS Similarity Theorem, and SAS Similarity Theorem are all tests that can be used to determine whether two triangles are similar. The AA Similarity Postulate is used when two pairs of congruent angles of two triangles are given. The SSS Similarity Theorem is used when the corresponding proportional side lengths of two triangles are given. The SAS Similarity Theorem is used when two corresponding proportional side lengths and the included angle of two triangles are given.   **17.** 6

Lesson 3-5

1. 6   **3.** yes; $\frac{PN}{NM} = \frac{QR}{RM} = \frac{1}{2}$

5. 50   **7.** 1.35   **9.** 1.12 km   **11.** $x = 6$, $y = 6.5$

13. 15   **15.** 8   **17.** 12 ft

19. Because $\frac{AD}{DB} = \frac{AE}{EC}$, an equivalent proportion

is $\frac{DB}{AD} = \frac{EC}{AE}$. Add 1 to each side of the

proportion as follows: $\frac{DB}{AD} + \frac{AD}{AD} = \frac{EC}{AE} + \frac{AE}{AE}$.

Therefore, $\frac{DB + AD}{AD} = \frac{EC + AE}{AE}$. By the Segment

Addition Postulate, this is equivalent to

$\frac{AB}{AD} = \frac{AC}{AE}$. Because $\frac{AB}{AD} = \frac{AC}{EC}$ and $\angle A \cong \angle A$ by

the Reflexive Property of Congruence, $\triangle ADE$
~ $\triangle ABC$ by the SAS Similarity Theorem.
Therefore, $\angle ADE \cong \angle ABC$ because they are
corresponding angles of similar triangles;
so $\overline{DE} \parallel \overline{BC}$, because if corresponding angles
are congruent, then the lines are parallel.

21. Proof: We are given that $\overleftrightarrow{AE} \parallel \overleftrightarrow{BF} \parallel \overleftrightarrow{CG}$.
Draw $\overline{AG}$ so that $\overline{AG}$ intersects $\overline{BF}$ at D.

In $\triangle ACG$, $\overline{CG} \parallel \overline{BD}$. By the Triangle
Proportionality Theorem, $\frac{AB}{BC} = \frac{AD}{DG}$.
In $\triangle AGE$, $\overline{AE} \parallel \overline{DF}$. By the Triangle
Proportionality Theorem, $\frac{AD}{DG} = \frac{EF}{FG}$. Therefore,
by the Transitive Property, $\frac{AB}{BC} = \frac{EF}{FG}$.

23. 39 cm; Sample answer: Because J, K, and L
are midpoints of their respective sides,

$JK = \frac{1}{2}QR$, $KL = \frac{1}{2}PQ$, and $JL = \frac{1}{2}PR$ by the

Triangle Midsegment Theorem. So $JK + KL +$

$JL = \frac{1}{2}(QR + PQ + PR) = \frac{1}{2}(78) = 39$.

25a. $ST = 8$ $UV = 4$ $WX = 2$

25b. Based on the pattern, the length of the
midsegment of $\triangle WXR = 1$.

27. Always; sample answer: FH is a
midsegment. Let $BC = x$, then $FH = \frac{1}{2}x$.

$FHCB$ is a trapezoid, so $DE = \frac{1}{2}(BC + FH) =$

$\frac{1}{2}\left(x + \frac{1}{2}x\right) = \frac{1}{2}x + \frac{1}{4}x = \frac{3}{4}x$. Therefore,

$DE = \frac{3}{4}BC$.

29. Sample answer: By Corollary 8.1, $\frac{a}{b} = \frac{c}{d}$.

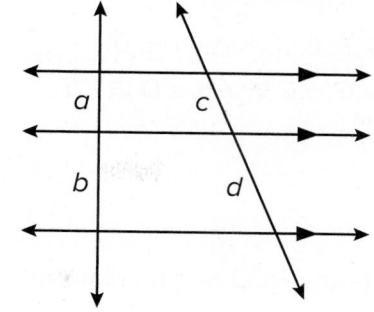

Lesson 3-6

1. 16.5   **3.** 11   **5.** 8.4   **7.** 7 yd

9. 8.4   **11.** 5.3 cm

13. Proof:

Statements (Reasons)

1. $\triangle ABC \sim \triangle RST$, $\overline{AD}$ is a median of $\triangle ABC$
and $\overline{RU}$ is a median of $\triangle RST$. (Given)

2. $CD = DB$, $TU = US$ (Definition of median)

3. $\frac{AB}{RS} = \frac{CB}{TS}$ (Definition of similar triangles)

4. $CB = CD + DB$, $TS = TU + US$ (Segment
Addition Postulate)

5. $\frac{AB}{RS} = \frac{CD + DB}{TU + US}$ (Substitution)

6. $\frac{AB}{RS} = \frac{DB + DB}{US + US} = \frac{2(DB)}{2(US)}$ (Substitution)

7. $\frac{AB}{RS} = \frac{DB}{US}$ (Substitution)

8. $\angle B \cong \angle S$ (Definition of similar triangles)

9. $\triangle ABD \sim \triangle RSU$ (SAS Similarity)

10. $\frac{AD}{RU} = \frac{AB}{RS}$ (Definition of similar triangles)

15. Chun; sample answer: By the Triangle
Angle Bisector Theorem, the correct

proportion is $\frac{5}{8} = \frac{15}{x}$.

17. $PS = 18.4$, $RS = 24$

Module 3 Review

1. D   **3.** $A'(4, 3.2)$, $B'(8, 6.4)$, and $C'(16, 0)$

5. $\angle A \cong \angle F$, $\angle B \cong \angle G$, $\angle C \cong \angle H$, $\angle D \cong \angle E$

$$\frac{AB}{FG} = \frac{BC}{GH} = \frac{CD}{HE} = \frac{DA}{EF}$$

$$\frac{9}{12} = \frac{3}{4} = \frac{6}{8} = \frac{12}{16}$$

$$\frac{3}{4} = \frac{3}{4} = \frac{3}{4} = \frac{3}{4}$$

Therefore, quadrilateral $ABCD \sim$ quadrilateral $FGHE$.

7. C   **9.** C   **11.** B

13. $\overline{AB} = 1.6$, $\overline{BD} = 3.6$, $\overline{BC} = 2.4$

15. 8   **17.** A

Module 4

Quick Check

1. $x = 18$ **3.** 10   **5.** 14   **7.** 17

Lesson 4-1

1. $\sqrt{24}$ or $2\sqrt{6} \approx 4.9$ **3.** 10   **5.** $\sqrt{51} \approx 7.1$

7. $\triangle ACB \sim \triangle CDB \sim \triangle ADC$

9. $\triangle EGF \sim \triangle GHF \sim \triangle EHG$

11. $x = \sqrt{184}$ or $2\sqrt{46} \approx 13.6$; $y = \sqrt{248}$ or $2\sqrt{62} \approx 15.7$; $z = \sqrt{713} \approx 26.7$

13. $x = 4.5$; $y = \sqrt{13} \approx 3.6$; $z = 6.5$

15. 7.2 ft
Sample answer:

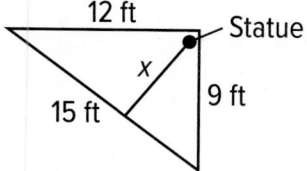

17. 2.88 mi

19. Proof: It is given that $\triangle ADC$ is a right triangle and $\overline{DB}$ is an altitude of $\triangle ADC$. $\angle ADC$ is a right angle by the definition of a right triangle. Therefore, $\triangle ADB \sim \triangle DCB$, because if the altitude is drawn from the vertex of the right angle to the hypotenuse of a right triangle, then the two triangles formed are similar to the given triangle and to each other. So, $\frac{AB}{DB} = \frac{DB}{CB}$ by the definition of similar triangles.

21. $x = 5.2$; $y = 6.8$; $z = 11.1$

23. Sample answer: 9 and 4, 8 and 8; For two whole numbers to result in a whole-number geometric mean, their product must be a perfect square.

25a. Never; sample answer: The geometric mean of two consecutive integers is $\sqrt{x(x + 1)}$, and the average of two consecutive integers is $\frac{x + (x + 1)}{2}$. If you set the two expressions equal to each other, the equation has no solution.

25b. Always; sample answer: Because $\sqrt{ab}$ is equal to $\sqrt{a} \cdot \sqrt{b}$, the geometric mean for two perfect squares will always be the product of two positive integers, which is a positive integer.

25c. Sometimes; sample answer: When the product of two integers is a perfect square, the geometric mean will be a positive integer.

Lesson 4-2

1. $\sqrt{18}$ or $3\sqrt{2} \approx 4.2$ **3.** 60   **5.** $\sqrt{1345} \approx 36.7$

7. 15   **9.** 100   **11.** 6   **13.** about 4.6 feet high

15. Yes; right; sample answer: The measures of the sides would be $XY = \sqrt{8} \approx 2.83$, $YZ = \sqrt{2} \approx 1.41$, and $XZ = \sqrt{10} \approx 3.16$. Because $2.83 + 1.41 > 3.16$, $2.83 + 3.16 > 1.41$, and $1.41 + 3.16 > 2.83$, the side lengths can form a triangle. Because $(\sqrt{8})^2 + (\sqrt{2})^2 = (\sqrt{10})^2$, we know that the triangle is a right triangle.

17. Yes; obtuse; sample answer: The measures of the sides would be $XY = 5$, $YZ = 2$, and $XZ = \sqrt{41} \approx 6.40$. Because $5 + 2 > 6.40$, $5 + 6.40 > 2$, and $2 + 6.40 > 5$, the side lengths can form a triangle. Because $5^2 + 2^2 < (\sqrt{41})^2$, we know that the triangle is an obtuse triangle.

19. 30 ft   **21.** yes, right; $(\sqrt{12})^2 = (\sqrt{8})^2 + 2^2$

23. Proof:

Statements (Reasons)

1. In $\triangle DEF$, $f^2 < d^2 + e^2$ where f is the length of the longest side. In $\triangle LMN$, $\angle M$ is a right angle. (Given)

2. $d^2 + e^2 = x^2$ (Pythagorean Thm.)

3. $f^2 < x^2$ (Substitution)

4. $f < x$ (Take the positive square root.)

5. $m\angle M = 90°$ (Def. of right angle)

6. $m\angle F < m\angle M$ (Conv. of the Hinge Thm.)

7. $m\angle F < 90°$ (Substitution)

8. $\angle F$ is an acute angle. (Def. of acute angle)

9. $m\angle D < m\angle F$, $m\angle E < m\angle F$ (If one side of a $\triangle$ is longer than another side, then the $\angle$ opp. the longer side has a greater measure than the $\angle$ opp. the shorter side.)

10. $m\angle D < 90°$, $m\angle E < 90°$ (Transitive Prop. of Inequality)

11. $\triangle DEF$ is an acute triangle. (Def. of acute $\triangle$)

25. 2129   **27.** P = 36 units; A = 60 units²

29. 15   **31a.** Because the height of the tree is 20 m, $JL = 20 - x$. By the Pythagorean Theorem, $16^2 + x^2 = (20 - x)^2$.

31b. $16^2 + x^2 = (20 - x)^2$, so $256 + x^2 = 400 - 40x + x^2$. Subtracting x^2 from both sides gives $256 = 400 - 40x$. Therefore, $-144 = -40x$. Dividing both sides by -40 gives $3.6 = x$. The stump of the tree is 3.6 meters tall. **33.** 10 **35.** $\frac{1}{2}$

37. 12.04 miles

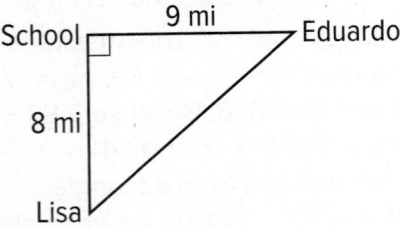

39a. Sample answer: $a^2 + b^2 = (m^2 - n^2)^2 + (2mn)^2 = m^4 - 2m^2n^2 + n^4 + 4m^2n^2 = m^4 + 2m^2n^2 + n^4$ and $c^2 = (m^2 + n^2)^2 = m^4 + 2m^2n^2 + n^4$. This means that $a^2 + b^2 = c^2$, so a, b, and c do form the sides of a right triangle by the Converse of the Pythagorean Theorem.

39b.

m	n	a	b	c
2	1	3	4	5
3	1	8	6	10
3	2	5	12	13
4	1	15	8	17
4	2	12	16	20
4	3	7	24	25
5	1	24	10	26

39c. Sample answer: Take $m = 24$ and $n = 7$ to get $a = 527$, $b = 336$, and $c = 625$.

41.

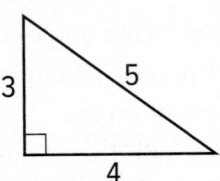

Right; sample answer: If you double or halve the side lengths, all three sides of the new triangles are proportional to the sides of the original triangle. Using the Side-Side-Side Similarity Theorem, you know that both of the new triangles are similar to the original triangle, so they are both right.

43. Sample answer: Incommensurable magnitudes are magnitudes of the same kind that do not have a common unit of measure. Irrational numbers were invented to describe geometric relationships, such as ratios of incommensurable magnitudes that cannot be described using rational numbers. For example, to express the measures of the sides of a square with an area of 2 square units, the irrational number $\sqrt{2}$ is needed.

Lesson 4-3

1.

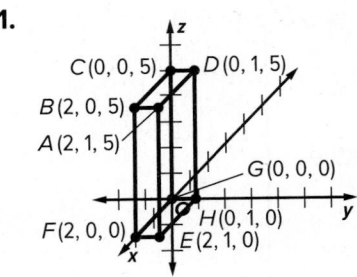

3.

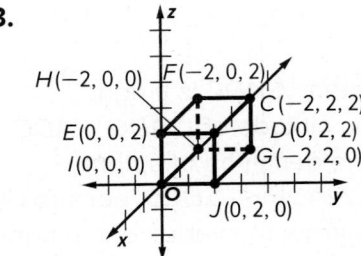

5.

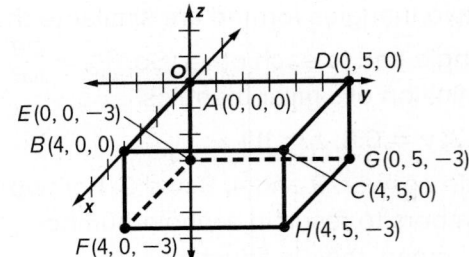

7. $\sqrt{29}$ **9.** $\sqrt{46}$ **11.** 20.2 mi **13.** $(1, -1, -2)$

15. $\left(4, \frac{7}{2}, \frac{17}{2}\right)$ **17.** $\left(5, 4, \frac{19}{2}\right)$

19. $PQ = \sqrt{36}$ or 6; $(-3, -1, 1)$

21. $FG = \sqrt{10}$; $\left(\frac{3}{10}, \frac{3}{2}, \frac{2}{5}\right)$

23. $BC = \sqrt{39}$; $\left(-\frac{\sqrt{3}}{2}, 3, 3\sqrt{2}\right)$

25. Proof: Points $A(x_1, y_1, z_1)$ and $B(x_2, y_2, z_2)$ are given. It is also given that M is the midpoint of $\overline{AB}$. To find the coordinates of point M, you must find the coordinates of a point that is $\frac{1}{2}$ of the distance from point A to point B in three-dimensional space. The difference between the x-, y-, and z-coordinates of points A and B are $x_2 - x_1$, $y_2 - y_1$, and $z_2 - z_1$, respectively. Because point M is halfway between points A and B, the distances between the x-, y-, and z-coordinates of points A and M are $\frac{1}{2}(x_2 - x_1)$, $\frac{1}{2}(y_2 - y_1)$, and $\frac{1}{2}(z_2 - z_1)$, respectively. To find the coordinates of point M, add the fractional distances to the coordinates of point A.

coordinates of point M =

$\left(x_1 + \frac{1}{2}(x_2 - x_1), y_1 + \frac{1}{2}(y_2 - y_1), z_1 + \frac{1}{2}(z_2 - z_1)\right) =$

$\left(\frac{2x_1 + x_2 - x_1}{2}, \frac{2y_1 + y_2 - y_1}{2}, \frac{2z_1 + z_2 - z_1}{2}\right) =$

$\left(\frac{x_1 + x_2}{2}, \frac{y_1 + y_2}{2}, \frac{z_1 + z_2}{2}\right)$

27. Teion; sample answer: Camilla made a mistake when substituting the value for y_1.
29. $2\sqrt{19}$

Lesson 4-4

1. $7\sqrt{2}$ **3.** $6\sqrt{2}$ **5.** 10   **7.** $18\sqrt{2}$ **9.** $25\sqrt{2}$
11. 2   **13.** $5\sqrt{2}$ **15.** $2\sqrt{2}$ **17.** 16   **19.** $x = 8$;
$y = 16$ **21.** $x = \frac{15\sqrt{3}}{2}$; $y = \frac{15}{2}$ **23.** $x = 24\sqrt{3}$;
$y = 48$ **25.** 20 ft   **27.** No; sample answer: The height of the frame is only 10 centimeters, and the height of the certificate is about 10.4 centimeters. So, the certificate will not fit.
29. $6\sqrt{2}$ yd or about 8.49 yd   **31a.** x feet
31b. $\sqrt{3}$ ft   **31c.** 20.5 ft   **33.** $x = \frac{13\sqrt{2}}{2}$; $y = 45$
35. $x = 3$; $y = 1$ **37.** $x = 6\sqrt{3}$; $y = 3$
39. Proof: A 45°-45°-90° triangle with a leg of length ℓ and a hypotenuse of length h is given. Because the measure of each acute angle in the triangle is 45°, by the definition of congruent angles, the angles are congruent. Therefore, by the Converse of the Isosceles Triangle Theorem, the legs of the triangle are congruent. Because the legs are congruent, the measure of each leg is ℓ. By the Pythagorean Theorem, $\ell^2 + \ell^2 = h^2$. This equation simplifies to $2\ell^2 = h^2$. Take the positive square root of each side to get

$\sqrt{2\ell^2} = \sqrt{h^2}$. So, $\ell\sqrt{2} = h$.
By the Symmetric Property of Equality, $h = \ell\sqrt{2}$.
41. (6, 9)
43. $RS \approx 5.464$ ft; $ST \approx 9.464$ ft; $RT \approx 10.928$ ft
45. 59.8   **47.** Carmen; sample answer: Because the three angles of the larger triangle are congruent, it is an equilateral triangle and the right triangles formed by the altitude are 30°-60°-90° triangles. The hypotenuse is 6, so the shorter leg is 3 and the longer leg x is $3\sqrt{3}$.
49. Sample answer:

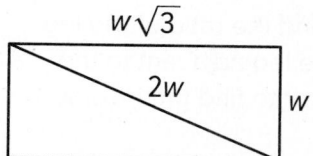

Let ℓ represent the length.
$\ell^2 + w^2 = (2w)^2$; $\ell^2 = 3w^2$; $\ell = w\sqrt{3}$

Lesson 4-5

1. $\sin L = \frac{5}{13} \approx 0.38$; $\cos L = \frac{12}{13} \approx 0.92$;
$\tan L = \frac{5}{12} \approx 0.42$; $\sin M = \frac{12}{13} \approx 0.92$;
$\cos M = \frac{5}{13} \approx 0.38$; $\tan M = \frac{12}{5} = 2.4$
3. $\sin R = \frac{8}{17} \approx 0.47$; $\cos R = \frac{15}{17} \approx 0.88$;
$\tan R = \frac{8}{15} \approx 0.53$; $\sin S = \frac{15}{17} \approx 0.88$;
$\cos S = \frac{8}{17} \approx 0.47$; $\tan S = \frac{15}{8} \approx 1.88$
5. $\frac{1}{2}$; 0.5   **7.** $\frac{1}{2}$; 0.5   **9.** $\frac{\sqrt{3}}{3}$; 0.58   **11.** 22.55
13. 24.15   **15.** 15.5 ft   **17.** 33.6°   **19.** 35.5°
21. 67.0°   **23.** $WX = 15.1$; $XZ = 9.8$; $m\angle W = 33°$
25. $RT = 3.7$; $ST = 5.9$; $m\angle R = 58°$
27. $NQ = 25.5$; $MQ = 18.0$; $m\angle N = 45°$
29. 56.3°   **31.** 28.52 cm; 23.39 cm²
33a. $\cos 26°$ or $\sin 64°$ **33b.** $\cos 64°$ or $\sin 26°$ **33c.** $\tan 64°$ **35.** Sample answer: The sine of an angle equals the cosine of its complement. So, $\sin \beta = \cos \alpha = 0.7660$. Similarly, $\cos \beta = \sin \alpha = 0.6428$.   **37.** $x = 9.2$; $y = 11.7$ **39.** Yes, they are both correct; sample answer: Because $27 + 63 = 90$, the sine of 27° is the same ratio as the cosine of 63°.
41. Yes; sample answer: Because the values of sine and cosine are both calculated by

dividing one of the legs of a right triangle by the hypotenuse, and the hypotenuse is always the longest side of a right triangle, the values will always be less than 1. You will always be dividing the smaller number by the larger number.

43. Sample answer: To find the measure of an acute angle of a right triangle, you can find the ratio of the leg opposite the angle to the hypotenuse and use a calculator to find the inverse sine of the ratio; you can find the ratio of the leg adjacent to the angle to the hypotenuse and use a calculator to find the inverse cosine of the ratio; or you can find the ratio of the leg opposite the angle to the leg adjacent to the angle and use a calculator to find the inverse tangent of the ratio.

Lesson 4-6

1. 64 m   **3.** about 21 ft   **5.** about 35°
7. 14.8°   **9.** 10 feet tall   **11.** 19 ft
13. 62.3 units^2 **15.** 71.5 units^2 **17.** 10.0 ft^2
19. 32.9 cm^2 **21.** 106.4 ft^2 **23a.** Sample answer: Because the horizontal distance between the hiker and the park ranger is known, I can use the 32° angle of depression to estimate the distance that the backpack fell.   **23b.** 71 ft   **25.** 12.2°   **27.** 4.8 cm^2
29. 9.1 ft^2 **31.** 22.4 ft^2 **33.** Sample answer: What is the relationship between the angle of elevation and the angle of depression?
35. 7.8

Lesson 4-7

1. $x \approx 102.1$ **3.** $x \approx 22.9$ **5.** $x \approx 4.1$
7. $x \approx 22.8$ **9.** $x \approx 15.1$ **11.** $x \approx 2.0$
13. 6.2 mi   **15.** two solutions; $m\angle B \approx 64°$, $m\angle C \approx 66°$, $c \approx 40.4$; $m\angle B \approx 116°$, $m\angle C \approx 14°$, $c \approx 11.0$ **17.** one solution; $m\angle B \approx 34°$, $m\angle C \approx 21°$, $c \approx 9.6$ **19.** one solution; $m\angle B = 90°$, $m\angle C = 60°$, $c \approx 3.5$ **21.** one solution; $m\angle B \approx 34°$, $m\angle C \approx 108°$, $c \approx 15.4$
23. one solution; $m\angle B \approx 35°$, $m\angle C \approx 12°$, $c \approx 2.6$ **25.** one solution; $m\angle B \approx 31°$, $m\angle C \approx 40°$, $c \approx 16.4$ **27a.** Definition of sine
27b. Multiplication Property   **27c.** Substitution

27d. Division Property   **29.** 24.3   **31a.** 402 m
31b. 676 m   **33.** 119.4° or 16.6°

35. 2; $b \sin A = 16 \sin 55°$ or about 13.1; Because $\angle A$ is acute, $14 < 16$, and $14 > 13.1$, the measures define 2 triangles.   **37.** 2; $b \sin A = 25 \sin 39°$ or about 15.7; Because $\angle A$ is acute, $22 < 25$, and $22 > 15.7$, the measures define 2 triangles.

39. 1; Because $\angle A$ is acute and $a = b = 10$, the measures define 1 triangle.
41. 0; $b \sin A = 17 \sin 52°$ or about 13.4; Because $\angle A$ is acute and $13 < 13.4$, the measures define 0 triangles.
43. 2; $b \sin A = 15 \sin 33°$ or about 8.2; Because $\angle A$ is acute, $10 < 15$, and $10 > 8.2$, the measures define 2 triangles.
45. Cameron; sample answer: $\angle R$ is acute and $r > t$, so there is one solution.
47a. $m\angle B \approx 70°$, $m\angle C \approx 68°$, $c \approx 20.9$
47b. $m\angle B \approx 110°$, $m\angle C \approx 28°$, $c \approx 10.4$
49a. Sample answer: $a = 22$, $b = 25$, $m\angle A = 70°$ **49b.** Sample answer: $a = 25$, $b = 22$, $m\angle A = 95°$ **49c.** Sample answer: $a = 22$, $b = 30$, $m\angle A = 43°$

Lesson 4-8

1. $x \approx 29.9$ **3.** $x \approx 74$ **5.** $x \approx 20$ **7.** 69°
9. $m\angle A \approx 41°$, $m\angle C \approx 54°$, $b \approx 6.1$ **11.** $c \approx 12.8$, $m\angle A \approx 67°$, $m\angle B \approx 33°$ **13.** $m\angle N = 42°$, $n \approx 35.8$, $m \approx 24.3$ **15.** Law of Sines; $m\angle B = 142°$, $a \approx 21.0$, $b \approx 67.8$ **17.** Law of Sines; $m\angle B \approx 33°$, $m\angle C \approx 110°$, $c \approx 31.2$ **19a.** about 11.3 yd   **19b.** about 40.9° **21.** 5.6   **23.** Sample answer: When solving a right triangle, you can use the Pythagorean Theorem to find missing side lengths and trigonometric ratios to find missing side lengths or angle measures. To solve any triangle, you can use the Law of Sines or the Law of Cosines, depending on what measures you are given.

Module 4 Review

1. A   **3.** A, D   **5.** Multiplication Property of Equality; Distributive Property   **7.** $\frac{9}{2}\sqrt{5}$
9. 8.7 feet   **11.** $\frac{\sqrt{2}}{2}$ **13.** 45 ft   **15.** 5.4   **17.** D

Module 5

Quick Check

1. 8 in.   **3.** -2.4, 1.6   **5.** -1.8, 2.2

Lesson 5-1

1. $\odot O$ **3.** $\overline{AB}$ and $\overline{CD}$ **5.** $\overline{PA}$, $\overline{PB}$, and $\overline{PC}$
7. $\overline{AB}$ **9.** 9 mm   **11.** Yes; all diameters of the same circle are congruent.   **13.** 9.5 m
15. 13 in.; 81.68 in.   **17.** 12.73 in.; 6.37 in.
19. 4.97 m; 2.49 m   **21.** 25.31 yd; 12.65 yd
23. 0.5   **25.** 6   **27.** 4.25 in.; 26.70 in.
29. 199.90 m; 99.95 m   **31.** 11.14x cm; 5.57x cm
33. congruent   **35.** neither   **37.** $9\sqrt{2}\pi$ in.
39. 11π yd   **41.** 2π cm   **43.** Sample answer: The face of the clock in Elizabeth Tower in London, England, has a circumference of 72.26 feet.   **45.** chord   **47a.** 0.796 m
47b. 0.398 m   **49.** 24 units   **51.** Sample answer: First apply the translation $(x, y) \rightarrow (x - 3, y - 4)$ to $\odot D$. The dilation should have a scale factor of 3 and be centered at $(-1, -1)$. The translation and dilation are similarity transformations, so $\odot D$ is similar to $\odot E$.

Lesson 5-2

1. 80   **3.** 138   **5.** minor arc; 50°   **7.** major arc; 210°   **9.** semicircle; 180°   **11.** 108°
13. 180°   **15.** 50°   **17.** 130°   **19.** 320°
21. 2.09 yd   **23.** 12.57 ft   **25.** 13.09 in.
27. $\frac{\pi}{4}$ radians   **29.** $\frac{\pi}{2}$ radians   **31.** $\frac{5\pi}{4}$ radians
33. 270°   **35.** 150°   **37.** 15°   **39.** 2.79 in.
41. 5.24 in.   **43.** 98.4°   **45a.** 62.83 in.
45b. 7.85 in.   **45c.** 57.3°
47. Proof:

   Statements (Reasons)

    1. $\angle BAC \cong \angle DAE$ (Given)

    2. $m\angle BAC \cong m\angle DAE$ (Def. of congruent angles)

    3. $m\widehat{BAC} = m\widehat{BC}$ and $m\widehat{DAE} = m\widehat{DE}$ (Def. of arc measure)

    4. $m\widehat{BC} = m\widehat{DE}$ (Substitution)

    5. $\widehat{BC} \cong \widehat{DE}$ (Def. of congruent arcs)

49. 150°   **51.** 52   **53.** 128
55a. 10π or about 31.4 ft   **55b.** Sample answer: She can find the circumference of the entire circle and then use the arc measure to find the appropriate fraction of the circle. If the radius of the circle is r and the measure of the arc is x, then the arc length is $\frac{x}{360} \cdot 2\pi r$.
57. Never; sample answer: Obtuse angles intersect arcs that measure between 90° and 180°.
59. Selena; sample answer: The circles are not congruent because they do not have congruent radii. So, the arcs are not congruent.
61. Sample answer:

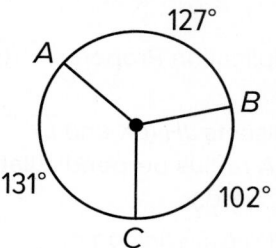

63. $m\widehat{LM} = 150°$; $m\widehat{MN} = 90°$; $m\widehat{NL} = 120°$

Lesson 5-3

1. 148   **3.** 82   **5.** 7   **7.** 2   **9.** 13   **11.** 5
13. 45   **15.** 4.47   **17.** 7   **19.** Sample answer: Draw two chords on the arc and find the perpendicular bisector of each chord. The perpendicular bisectors will intersect at the center of the original circle. Then use a compass and the radius to draw the circle for the original size plate.

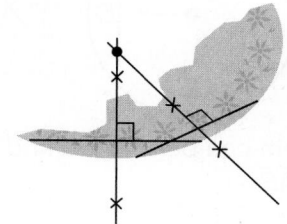

21. Proof:
Statements (Reasons)

1. $\overline{FG}$ and $\overline{JH}$ are equidistant from center L. (Given)

2. $\overline{LX} \cong \overline{LY}$ (Definition of equidistant)

3. $\overline{LG} \cong \overline{LH}$ (All radii of a circle are congruent.)

4. $\overline{LX} \perp \overline{FG}, \overline{LY} \perp \overline{JH}$ (Definition of equidistant)

5. $\angle LXG$ and $\angle LYH$ are right angles. (Definition of perpendicular lines)

6. $\triangle XGL \cong \triangle YHL$ (HL)

7. $\overline{XG} \cong \overline{YH}$ (CPCTC)

8. $XG = YH$ (Definition of congruent segments)

9. $2(XG) = 2(YH)$ (Multiplication Property of Equality)

10. $\overline{LX}$ bisects $\overline{FG}$; $\overline{LY}$ bisects $\overline{JH}$ ($\overline{LX}$ and $\overline{LY}$ are contained in radii. A radius perpendicular to a chord bisects the chord.)

11. $FG = 2(XG)$; $JH = 2(YH)$ (Definition of segment bisector)

12. $FG = JH$ (Substitution)

13. $\overline{FG} \cong \overline{JH}$ (Definition of congruent segments)

23. Sample answer: Neil can draw a second chord perpendicular to the first chord that he drew through the mark that he made. If the midpoint of the first chord is also the midpoint of the second chord, then the first chord that he drew is a diameter.

25. Sample answer:
$BC = 4$ cm and
$AD = 2$ cm;
Because $\overline{AD}$ is perpendicular to $\overline{BC}$, $\overline{AD}$ bisects $\overline{BC}$, and $BD = 2$.

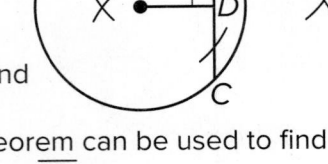

The Pythagorean Theorem can be used to find the length of the radius $\overline{AB}$.
$$AB^2 = AD^2 + BD^2$$
$$= 2^2 + 2^2$$
$$= 8$$
$$AB = 2\sqrt{2} \text{ or about } 2.83 \text{ cm}$$

Lesson 5-4

1. 72°   **3.** 226°   **5.** 81°   **7.** 17°   **9.** 43°

11. Proof: It is given that $m\angle T = \frac{1}{2}m\angle S$. This means that $m\angle S = 2m\angle T$. Because $m\angle S = \frac{1}{2}m\widehat{TUR}$ and $m\angle T = \frac{1}{2}m\widehat{URS}$, the equation becomes $\frac{1}{2}m\widehat{TUR} = 2\left(\frac{1}{2}m\widehat{URS}\right)$. Multiplying each side of the equation by 2 results in $m\widehat{TUR} = 2m\widehat{URS}$.

13. 39   **15.** 16   **17.** 58°   **19.** 69°   **21.** 105°
23. 72°   **25a.** 22.5   **25b.** 23; 90°   **25c.** Yes; sample answer: Because my estimate for the value of x is close to the exact value and $\angle J$ appears to be a right angle, my answer is reasonable.   **27.** 82°   **29.** 158°   **31.** 65°

33. Proof: By arc addition and the definitions of arc measure and the sum of central angles, $m\widehat{EFG} + m\widehat{EDG} = 360°$. By Theorem 10.7, $m\angle D = \frac{1}{2}m\widehat{EFG}$ and $m\angle F = \frac{1}{2}m\widehat{EDG}$. So, $m\angle D + m\angle F = \frac{1}{2}m\widehat{EFG} + \frac{1}{2}m\widehat{EDG}$ or $\frac{1}{2}(m\widehat{EFG} + m\widehat{EDG})$. By substitution, $m\angle D + m\angle F = \frac{1}{2}(360)$ or 180°. By the definition of supplementary angles, $\angle D$ and $\angle F$ are supplementary. Because the sum of the measures of the interior angles of a quadrilateral is 360°, $m\angle D + m\angle F + m\angle E + m\angle G = 360°$. By substitution, $180 + m\angle E + m\angle G = 360°$. By the Subtraction Property of Equality, $m\angle E + m\angle G = 180°$. Therefore, $\angle E$ and $\angle G$ are supplementary by the definition of supplementary angles.

35. Proof: Because $PQRS$ is inscribed in a circle, $\angle Q$ is supplementary to $\angle S$ because they are opposite angles. Because $\angle Q \cong \angle S$, $\angle Q$ and $\angle S$ must be right angles. So, $\angle Q$ intercepts $\widehat{PSR}$, $m\angle Q = 90°$ and $m\widehat{PSR} = 180°$ by the Inscribed Angle Theorem. This means that $\widehat{PSR}$ is a semicircle, so $\overline{PR}$ must be a diameter.   **37.** 72°

39. $m\angle SRU = 80°$; $x = 12$ **41.** Always; sample answer: Rectangles have right angles at each vertex, therefore each pair of opposite angles will be supplementary and inscribed in a circle.

43. Sometimes; sample answer: A rhombus can be inscribed in a circle as long as it is a square. Because the opposite angles of rhombi that are not squares are not supplementary, they cannot be inscribed in a circle.

45. $\frac{\pi}{2}$

47. Sample answer:

$m\widehat{EHG} = m\widehat{HGF} = m\widehat{GFE} = m\widehat{FEH} = 180°$

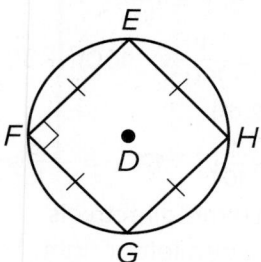

49. Never; sample answer: $\widehat{PR}$ is a minor arc, so $m\widehat{PR} < 180°$. By the Inscribed Angle Theorem, $m\angle PQR < 90°$.   **51.** Always; each angle of the triangle measures 60°, so each intercepted arc measures 120°.

Lesson 5-5

1. 4   **3.** 3   **5.** yes; $9^2 + 40^2 = 41^2$ **7.** no; $20^2 + 21^2 \ne 28^2$ **9.** 25   **11.** 20   **13.** 8   **15.** 4

17. 72 in.   **19.** 48°   **21.** 80   **23.** 24   **25.** 4; 64

27a. 9.5 in.   **27b.** 34 in.

29. Proof:

Statements (Reasons)

1. $\overline{RQ}$ is tangent to $\odot P$ at Q; $\overline{RT}$ is tangent to $\odot P$ at T. (Given)

2. $\overline{PQ} \perp \overline{RQ}$, $\overline{PT} \perp \overline{RT}$ (A tangent is $\perp$ to the radius drawn to the point of tangency.)

3. $\angle Q$ and $\angle T$ are rt. angles. ($\perp$ segments form right angles.)

4. $m\angle Q = 90°$, $m\angle T = 90°$ (Def. of right angle)

5. $PQRT$ is a quadrilateral. (Def. of quadrilateral)

6. $m\angle P + m\angle Q + m\angle R + m\angle T = 360°$ (Polygon Interior Angles Sum Thm.)

7. $m\angle P + 90° + m\angle R + 90° = 360°$ (Substitution)

8. $m\angle P + m\angle R + 180° = 360°$ (Substitution)

9. $m\angle P + m\angle R = 180°$ (Subtraction Property of Equality)

31a. 5; 2; 10   **31b.** 68

33. 8.06   **35.** spoke 10   **37.** 1916 mi;

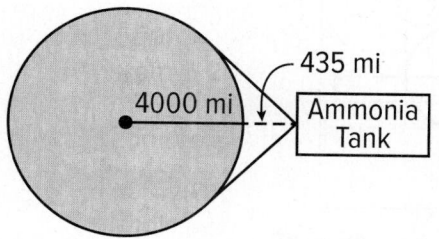

39. Figure B; because one of the sides of the quadrilateral is not tangent to the circle, the quadrilateral is not circumscribed about the circle.

41. By Theorem 10.12, if two segments from the same exterior point are tangent to a circle, then they are congruent. So, $\overline{XY} \cong \overline{XZ}$ and $\overline{XZ} \cong \overline{XW}$. Thus, $\overline{XY} \cong \overline{XZ} \cong \overline{XW}$.

Lesson 5-6

1. 42.5°   **3.** 62°   **5.** 53°   **7.** 84°   **9.** 148°

11. 99°   **13.** 40°   **15.** 264°   **17.** 128°   **19.** 9

21. 19   **23.** Sample answer: Find the difference of the two intercepted arcs and divide by 2.

25. 32

Lesson 5-7

1. $x^2 + y^2 = 64$ **3.** $(x - 3)^2 + (y - 2)^2 = 4$

5. $(x - 3)^2 + (y + 4)^2 = 16$

7. $(x + 4)^2 + (y + 1)^2 = 20$

9. $(0, 0)$; $r = 4$

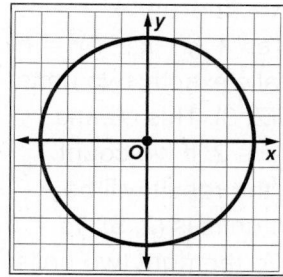

11. $(0, 0)$; $r = 2$

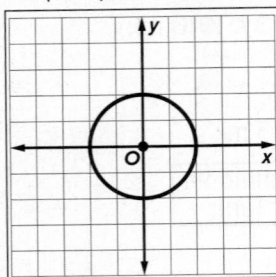

13. $(x - 2)^2 + (y - 4)^2 = 45$
15. $(0, 3)$, $(-2.4, -1.8)$ **17.** $(-3, -4)$, $(11, 10)$
19. $(x - 3)^2 + y^2 = 25$ **21.** Disagree; sample
answer: Completing the square shows that the
equation can be written as $x^2 + 4x + 4 +$
$y^2 - 10y + 25 = k + 29$ or $(x + 2)^2 + (y - 5)^2 =$
$k + 29$. The radius of the circle is $\sqrt{k + 29}$, and
this expression results in a positive radius only
when $k + 29 > 0$. So, the equation represents
a circle only if $k > -29$.
23. $(x - 4)^2 + (y - 3)^2 = 4$
25. 18 mi **27.** $(x - 3)^2 + (y + 2)^2 = 4$

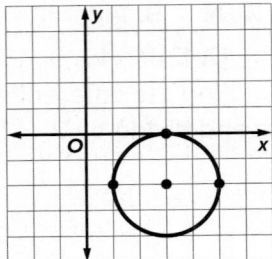

29. Sample answer: The equation for a circle
is $(x - h)^2 + (y - k)^2 = r^2$. When the circle
is translated a units to the right, the new
x-coordinate of the center is $x + a$. When
the circle is translated b units down, the new
y-coordinate of the center is $y - b$. The new
equation for the circle is $(x - (h + a))^2 +$
$(y - (k - b))^2 = r^2$, or $(x - h - a)^2 + (y - k + b)^2 = r^2$.
31. Sample answer: $(2, 1)$ falls exactly two units
below the center point of $(2, 3)$. This means
that the radius of the circle is 2. If we count
two units in any direction, the result will be a
point on the circle. Up two units is $(2, 5)$. To
the left two units is $(0, 3)$. To the right two units
is $(4, 3)$. These three points all have integer
coordinates.

Lesson 5-8

1. $y = \frac{1}{12}x^2$ **3.** $y = -\frac{1}{20}x^2$ **5.** $y = \frac{1}{16}x^2$
7. $y = \frac{1}{32}(x - 1)^2 - 1$ **9a.** $y = \frac{1}{100}(x - 100)^2 + 30$
9b. 126.49 km **11.** $(-1, 1)$, $(2, 4)$
13. $(-2, -12)$, $(0, 0)$ **15.** $x = \frac{1}{10}y^2$ **17.** $x = -\frac{5}{18}y^2$
19. Sample answer: Because the directrix is
vertical, the parabola must open left or right.
The focus is $\left(-\frac{3}{4}, 0\right)$, which is to the left of the
directrix, so the parabola opens to the left.
21. Never; sample answer: Because the
vertex is a point on the parabola, it is always
equidistant from the focus and directrix.

Module 5 Review

1. D **3.** $\frac{4\pi}{3}$; Sample answer: I used the
constant of proportionality $\frac{\pi}{180}$ multiplied by
240 to find the measure in radians.
5. Sample answer: Use a compass to construct
a circle with center C. Place point P on the
circle. Keeping the same compass setting
that was used to construct circle C, place
the point of the compass on P and construct
an arc through the circle. Label the point of
intersection Q. Without changing the compass
setting, place the compass point on Q and
intersect the circle again labeling the new point
R. Repeat this step for points S, T, and U. Use
a straightedge to connect each point to the
next consecutive point. The result is a regular
hexagon. **7.** C
9. $m\widehat{JK} = 39°$, $m\angle LKM = 96°$

11. B **13.** $(x - 1)^2 + (y - 7)^2 = 225$; yes
15. $y = \frac{1}{12}(x - 4)^2 + 2$

Module 6

Quick Check

1. $x = 5$ **3.** 9.2   **5.** 3

Lesson 6-1

1. 108 m²   **3.** 19.1 ft²   **5.** 288.7 mm²   **7.** 61.3 m²
9. 350 ft²   **11.** 137.5 ft²   **13a.** 13.5 ft²
13b. \$94.50   **15.** 10.6 cm, 31.7 cm   **17.** 4 m
19. 10 ft   **21.** 6 m; 24 m   **23.** 480 m²
25. 389.7 ft²   **27.** 24 units²   **29.** Sample
answer: Using the area formula for the given
figure, you can substitute the known area
and all the other known dimensions into the
formula. Use algebraic properties to solve for
the missing dimension.   **31.** Pieces 3
and 4 are the largest. The area of the top of
the cake is 30 square inches.   **33.** 8 in²
35. Always; sample answer: The area of a
parallelogram is given by $A = bh$, where b
is the length of the base and h is the height
of the parallelogram. Because the areas and
the heights of the rectangle and the non-
rectangular parallelogram are equal, the
lengths of the bases must be equal. Consider
the sides of each figure. The sides of the
rectangle are the base and the height of the
rectangle, so the perimeter of the rectangle
is given by $P = 2b + 2h$. The sides of the
non-rectangular parallelogram are the base
of the parallelogram and the hypotenuse of
the right triangle formed by the height of the
parallelogram and a portion of the base of
the parallelogram. Let x equal the length of
the hypotenuse of the right triangle. Because
the hypotenuse is the longest side of a right
triangle, $x > h$. So, the perimeter of the non-
rectangular parallelogram is given by $P = 2b +
2x$. Because $x > h$, the perimeter of the non-
rectangular parallelogram is always greater
than the perimeter of a rectangle given that
the figures have equal areas and heights.
37. 7.2   **39.** Sometimes; sample answer: If the
areas are equal, it means that the products of
the diagonals are equal. The only time that the
perimeters will be equal is when the diagonals
are also equal, or when the two rhombi are
congruent.

Lesson 6-2

1. Sample answer: center: point Z, radius: $\overline{ZY}$,
apothem: $\overline{ZQ}$, central angle: $\angle YZR$, 45°
3. 27.7 mm²   **5.** 124.7 ft²
7. 181.0 ft²   **9.** 198 cm²   **11.** 49.5 cm²
13. 31.2 in²   **15.** 128.1 ft²
17. Sample answer:

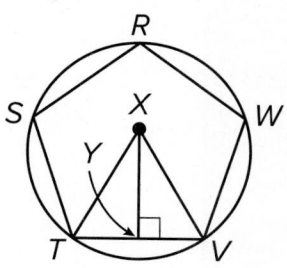

19. 52.0 in²   **21a.** Sample answer: Floor
area $= (7.5)(2.2) + (3.5)(4.2 - 2.2) +$
$\frac{1}{2}(7.5 - 3.5 - 2.0)(4.2 - 2.2) = 25.5$ m²;
$\frac{x}{1.0 \text{ L}} = \frac{25.5 \text{ m}^2}{4.5 \text{ m}^2}$, so $x = \frac{25.5 \text{ m}^2}{4.5 \text{ m}^2}$ (1.0 L) $\approx$ 5.7 L.
21b. Sample answer: Perimeter $= 7.5 + 2.2 +$
$2.0 + 2.8 + 3.5 + 4.2 = 22.2$ m. Area to
be painted (ignoring door and windows) $=$
$(22.2)(2.6) = 57.72$ m². $\frac{x}{1.0}$ L $= \frac{57.72 \text{ m}^2}{7.5 \text{ m}^2}$, so
$x = \frac{57.72 \text{ m}^2}{7.5 \text{ m}^2} \approx 7.7$ L.   **21c.** Sample answer:
Miguel may need to adjust the estimate
because I did not account for the area taken
up by doors and windows. So, the original
estimate is too high.
23a. Sample answer:

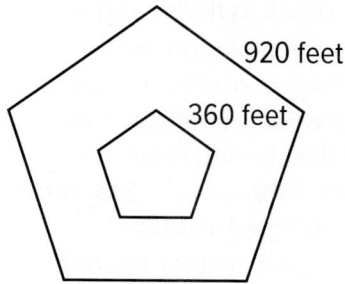

23b. 1,233,238.2 ft²   **25.** 754.4 in²
27. 591,137.7 ft²   **29.** 6.9 cm²   **31.** 3.87 cm²
33. Chenglei; sample answer: The measure
of each angle of a regular hexagon is 120°, so
the segments from the center of each vertex
form 60°-angles. The triangles formed by the
segments from the center to each vertex are
equilateral, so each side of the hexagon is
11 in. The perimeter of the hexagon is 66 in.

Using technology, the length of the apothem is about 9.5 in. Substituting the values into the formula for the area of a regular polygon and simplifying, the area is about 313.5 in².

35. Sample answer:

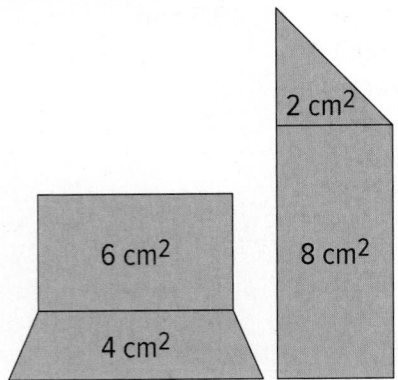

37. Sample answer: You can decompose the figure into shapes of which you know the area formulas. Then, you can add all of the areas to find the total area of the figure.

Lesson 6-3

1. 153.9 m²   **3.** 346.4 m²   **5a.** 12.6 in²
5b. about $5.21   **7.** 10.9 mm   **9.** 6.0 in.
11. 15.0 in.   **13.** 1.8 m²   **15.** 331.3 m²
17. 2520 cm²   **19.** 1625 mm²   **21.** 882 m²
23a. 176.7 ft²   **23b.** 173.6 ft²   **25.** 88.4 cm²
27. Sample answer: The area of the circle is 36π. The area of the hexagon is $A = \frac{1}{2}aP$. Because a regular hexagon can be divided into six equilateral triangles, use the 30°-60°-90° triangle ratios to find a and the length of one side of the hexagon. So, $a = 3\sqrt{3}$ in., each side is 6 in., and $P = 36$ in. By substitution, the area of the hexagon is 93.5 in². The area of the shaded region is $36\pi - 93.5 \approx 19.6$ in².   **29.** 64 in²   **31a.** red: 102.6 ft², purple: 94.2 ft², green: 84.8 ft²
31b. The largest sector was neither the one with the largest radius nor the one with the largest central angle. We can conclude that *both* factors affect the area of the sector.   **31c.** purple: 2.5 stars per square foot; green: 1.8 stars per square foot   **31d.** $2 \times 1.8 = 3.6$; 3.6(area of the red sector) ≈ 369 stars   **33.** 30.1 cm²   **35.** Sample answer: You can find the shaded area of the circle by subtracting x from 360 and using the resulting measure in the formula for

the area of a sector. You could also find the shaded area by finding the area of the entire circle, finding the area of the unshaded sector using the formula for the area of a sector, and subtracting the area of the un-shaded sector from the area of the entire circle. The first method is more efficient. It requires less steps, is faster, and there is a lower probability for error.

37. Sample answer: $A \approx 9.42$ cm²

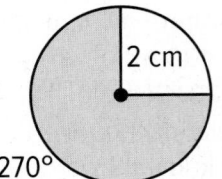

Lesson 6-4

1. $L = 528$ yd²; $S = 768$ yd²   **3.** $L = 138$ cm²; $S = 378$ cm²   **5.** $L = 324$ cm²; $S = 394.2$ cm²
7. $L \approx 377.0$ in²; $S \approx 603.2$ in²   **9.** $L \approx 226.2$ yd²; $S \approx 282.7$ yd²   **11.** $L \approx 603.2$ in²; $S \approx 1005.3$ in²   **13a.** $12x^2$ **13b.** 768 ft²
15a. $3x^2$ **15b.** 243 ft²   **17.** $S \approx 615.8$ in²
19. $S \approx 289.5$ mm²   **21.** $49\pi x^4$
23. $S \approx 301.6$ cm²   **25.** 156.3 ft²
27. 8.6 m   **29.** $L \approx 996.0$ in²; $S \approx 1686.0$ in²   **31.** $L \approx 34.7$ cm²; $S \approx 43.8$ cm²
33a. $S = 12.6$ cm²   **33b.** $S = 128$ ft²
35. 825 cm²   **37.** 2111.15 in²   **39a.** Sample answer: I drew a figure showing all of the surfaces of the sofa, using simple geometric figures with the dimensions labeled. To find the total surface area, I can find and add the areas of the simple figures.

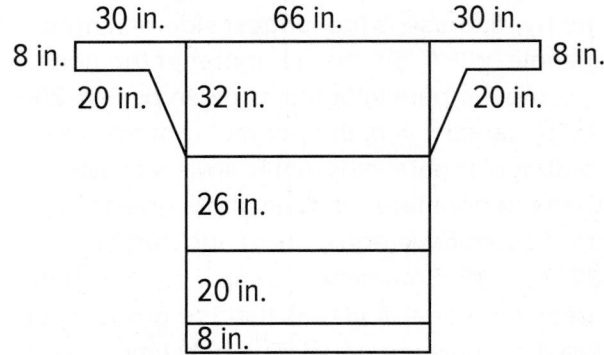

39b. Sample answer: Area = $(66)(8 + 20 + 26 + 32) + 2(30)(8) + 2\left[\frac{1}{2}(30 - 20)(32 - 8)\right]$ = 6396 in²; I did not include the base in my calculation.   **41.** No; sample answer: The slant height of the cone is $\frac{2\sqrt{\pi}}{\pi}$ or about 1.13 times greater than the slant height of the square pyramid.   **43.** Sample answer: To find the surface area of any solid figure, find the area of the base (or bases) and add the area of the lateral faces of the figure. The lateral bases of a rectangular prism are rectangles. Because the bases of a cylinder are circles, the lateral face of a cylinder is a rectangle.

45. $\frac{\sqrt{3}}{2}\ell^2 + 3\ell h$; Sample answer: The area of an equilateral triangle with side length ℓ is $\frac{\sqrt{3}}{4}\ell^2$, and the perimeter of the triangle is 3ℓ. So, the total surface area is $\frac{\sqrt{3}}{2}\ell^2 + 3\ell h$.

Lesson 6-5

1. The square pyramid has 2 planes of symmetry. One plane of symmetry is perpendicular to the base of the pyramid and passes through the vertex. Another plane of symmetry is perpendicular to the base of the pyramid, perpendicular to the first plane of symmetry, and passes through the vertex.
3. The triangular prism has 1 plane of symmetry. The plane of symmetry is parallel to the base of the prism and passes through the center of the prism.   **5.** triangle   **7.** triangle   **9.** cylinder with a cylinder removed from the middle
11. cone   **13.** yes   **15.** yes   **17.** yes
19. rectangle   **21.** Sample answer: Reshan could have rotated the semicircle around a horizontal axis.   **23.** Sample answer: ice cream cone

25a. yes

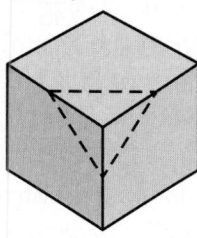 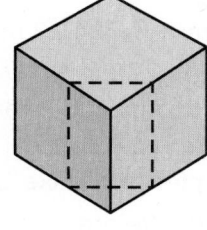

25c. yes

25b. yes

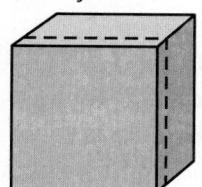

25f. no
27. trapezoid

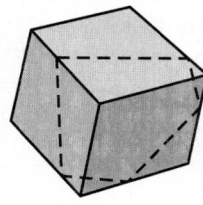

25d. yes

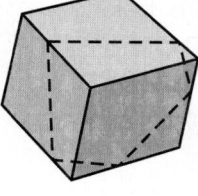

25e. yes

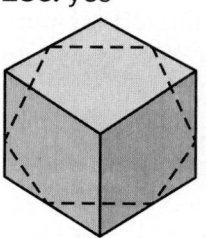

29a. sphere; any cut   **29b.** cylinder; cut parallel to base   **29c.** cone; cut parallel to base
31. Equilateral triangular pyramid; sample answer: Because the figure has axis symmetry of order 3, the base has to be an equilateral triangle. Because it does not have plane symmetry, you know that it is a pyramid instead of a prism. Therefore, the figure must be an equilateral triangular pyramid.   **33.** Sample answer: The cross section is a triangle. There are six different ways to slice the pyramid so that two equal parts are formed because the figure has six planes of symmetry. In each case, the cross section is an isosceles triangle. Only the side lengths of the triangles change.
35. False; sample answer: If a plane has a slope different than the slant height of the pyramid and intersects a lateral face of the pyramid, then the cross section is a trapezoid.

Lesson 6-6

1. 2304 cm³   **3.** 90 m³   **5.** 2928.0 cm³
7. 1224 cm³   **9.** 66.7 ft³   **11.** 55.1 in³
13a. $5.25x^3 + 3x^2$ **13b.** 54 in³

15a. $\frac{3\sqrt{3}}{4}x^3$ **15b.** 162.4 in³   **17.** 25 ft³
19. 301.1 in³   **21.** 14,508 in³   **23.** 156 cm³
25a. 900,000 ft³   **25b.** 883,573 ft³
27a. 1024 ft³   **27b.** 432 ft³   **27c.** 592 ft³
27d. about 26,640 pounds   **29a.** Sample
answer: He can find the volume of the contents
of each box by multiplying the length times
the width times 2 less than the height of the
box and then multiply by the mass per cubic
centimeter of the contents.   **29b.** baking
soda: volume = 8 · 4 · 10 = 320 cm³,
weight = 320 · 2.2 = 704 grams; corn flakes:
volume = 30 · 6 · 33 = 5940 cm³, weight =
5940 · 0.12 = 712.8 grams

31a. $V = \frac{1}{3} \cdot 2^2 \cdot 3 = 4$; the volume is 4 in³,
and the price per cubic inch is $0.50.
31b. Sample answer: Doubling the length and
the width multiplies the volume by 4. He would
need to multiply the price by 4 and charge
$8.00 for the larger box.   **31c.** Sample
answer: base has sides of 3 inches, and the
height is 2.5 inches; $V = \frac{1}{3} \cdot 3^2 \cdot 2.5 = 7.5$;
the volume is 7.5 in³.   **33.** Francisco; sample
answer: Valerie incorrectly used $4\sqrt{3}$ as the
length of one side of the triangular base.
Francisco used a different approach, but his
solution is correct.

35. Sample answer: Because the volume of a
pyramid is one-third the volume of the same
prism, 3 of the pentagonal pyramids would fit in
a prism of the same height.

Lesson 6-7

1. 2035.8 in³   **3a.** $2\pi x^2 - 4\pi x$ mm³
3b. 502.7 mm³   **5a.** $3\pi x^2 + 4\pi x$ ft³
5b. 122.5 ft³   **7a.** $\frac{1}{3}\pi(2x)^2(5x - 5)$ cubic units
7b. 3769.9 m³   **9.** 5.03 in³   **11.** 33.5 in³
13a. $\frac{4}{3}\pi(4x)^3$ cubic units   **13b.** 268.1 cm³
15a. $\frac{4}{3}\pi(5x - 0.5)^3$ cubic units   **15b.** 1.8 yd³
17. 1102.7 m³   **19.** 1043 in³   **21.** $2\pi r^3 + \frac{2\pi r^3}{3} +$
$\frac{\pi r^3}{3}$ or $3\pi r^3$ cubic units   **23.** 1210.6 mm³
25. 60.63 ft³   **27a.** About 382 in³; sample
answer: The cube root of 729 is 9, so the
side length of the cube of wood is 9 inches.

The largest sphere that could be carved has
a radius of 4.5 inches. So, $V = \frac{4}{3}\pi(4.5)^3$ or
about 382 in³.   **27b.** Yes; sample answer: The
radius of the sphere is exactly one half the
length of the side of the cube. If the cube has
a side length s, then the volume of the sphere
is $\frac{4}{3}\pi\left(\frac{s}{2}\right)^3 = \frac{4}{3}\pi\frac{s^3}{8} = \frac{s^3\pi}{6} = s^3\frac{\pi}{6}$. Therefore,
the volume of a sphere that shares a diameter
with the side of a cube is always $\frac{\pi}{6}$ times the
volume of the cube.

29. Sample answer: I would find the volume of
the original container. Then, I would substitute
that volume and the 8-inch height in the
volume formula and solve for r; 3.4 inches.

31. Sample answer:

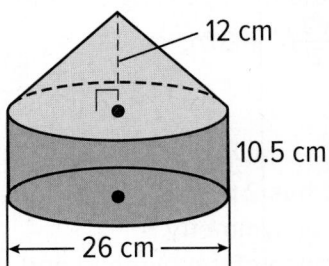

33. Sometimes; sample answer: The statement
is true if the base area of the cone is 3 times
as great as the base area of the prism. For
example, if the base of the prism has an area
of 10 square units, then its volume is 10h cubic
units. So, the cone must have a base area of
30 square units so that its volume is $\frac{1}{3}$(30)h or
10h cubic units.   **35.** True; sample answer: A
cone of radius r and height $4r$ has the same
volume, $\frac{4}{3}\pi r^3$, as a sphere with radius r.

Lesson 6-8

1. 320 m²   **3.** 1843.2 m²   **5.** $\frac{1}{3}$; 7 m   **7.** $\sqrt{\frac{71}{16}}$;
4.5 in.   **9.** 27.8 in²   **11.** 5 in²   **13.** small: 108 cm³;
large: 256 cm³   **15.** small: 98.2 cm³; large:
785.4 cm³   **17.** 36 cm   **19.** 27 : 64
21. 313 pieces of cake

23. Proof: We are given rectangular prisms with
a scale factor of $a : b$. By the definition of similar
solids, $\frac{f}{t} = \frac{g}{u} = \frac{h}{v} = \frac{a}{b}$, so $t = f \cdot \frac{b}{a}$, $u = g \cdot \frac{b}{a}$, and
$v = h \cdot \frac{b}{a}$.

Then, $\dfrac{\text{surface area of smaller prism}}{\text{surface area of larger prism}} =$

$\dfrac{2fg + 2fh + 2gh}{2tu + 2tv + 2uv}.$

By substitution, the right side of the equation becomes

$$\dfrac{2fg + 2fh + 2gh}{2\left(f\cdot\frac{b}{a}\right)\left(g\cdot\frac{b}{a}\right) + 2\left(f\cdot\frac{b}{a}\right)\left(h\cdot\frac{b}{a}\right) + 2\left(g\cdot\frac{b}{a}\right)\left(h\cdot\frac{b}{a}\right)}.$$

Simplify this fraction to get

$$\dfrac{2fg + 2fh + 2gh}{2fg\cdot\frac{b^2}{a^2} + 2fh\cdot\frac{b^2}{a^2} + 2gh\cdot\frac{b^2}{a^2}}.$$

By using the Distributive Property, this becomes

$$\dfrac{2fg + 2fh + 2gh}{(2fg + 2fh + 2gh)\cdot\frac{b^2}{a^2}} = \dfrac{a^2}{b^2}.$$

Also, the ratio of the volume is

$\dfrac{\text{volume of smaller prism}}{\text{volume of larger prism}} = \dfrac{fgh}{tuv}.$ Using the definition

of similar solids, we can rewrite the right side of the equation above as $\dfrac{fgh}{\left(f\cdot\frac{b}{a}\right)\left(g\cdot\frac{b}{a}\right)\left(h\cdot\frac{b}{a}\right)}.$

After multiplying, this becomes

$\dfrac{fgh}{fgh\cdot\frac{b^3}{a^3}}$, which simplifies to $\dfrac{a^3}{b^3}$. Thus, the surface

areas have a ratio of $a^2 : b^2$, and the volumes have a ratio of $a^3 : b^3$.

25a. 2 in.

25b. No; sample answer: The larger sculpture has an apothem of about 8.5 inches. This means that the octagonal shape of the larger sculpture is about 17 inches across. This is greater than the diameter of the box.   **27.** Sample answer: As the diameter increases, the volume increases by the cube of the ratio of the diameters. For example, from size Small to Medium, the ratio of the diameters is $\dfrac{4.5}{3}$ or 1.5. The volume of the Medium is $4.5\pi(1.5)^3$ or 15.1875π.

Size	Diameter (cm)	Volume (cm³)
Small	3	4.5π
Medium	4.5	15.1875π
Large	6.75	51.2578125π

29. Neither; sample answer: To find the area of the enlarged circle, you can multiply the radius by the scale factor and substitute it into the area formula, or you can multiply the area formula by the scale factor squared.

The formula for the area of the enlargement is $A = \pi(kr)^2$ or $A = k^2r^2$.   **31.** Sample answer: Because the ratio of the areas should be 4 : 1, the ratio of the lengths of the sides will be $\sqrt{4} : \sqrt{1}$ or 2 : 1. Thus, a 0.5-inch by 1-inch rectangle and a 1-inch by 2-inch rectangle are similar, and the ratio of their areas is 4 : 1.

$A = 0.5$ in²

0.5 in. ☐

1 in.

1 in. $A = 2$ in²

2 in.

33. 8 : 135; Sample answer: The volume of Cylinder C is 8 times the volume of Cylinder A, and the volume of Cylinder D is 27 times the volume of Cylinder B. If the original ratio of volumes was $1x : 5x$, the new ratio is $8x : 135x$. So, the ratio of volumes is 8 : 135.

Lesson 6-9

1. $\approx 14{,}290.0$ people/mi²
3. $\approx 13{,}525.6$ people/mi²   **5.** 24
7a. 0.512 oz/in³   **7b.** 62,500 in³   **7c.** 8000
9. 41.67 people/mi²   **11.** No; sample answer: Because the load will weigh 42,411.6 pounds, which exceeds the limit of 34,000 pounds, it will not meet the weight restrictions.
13. Pentagonal prism; the pentagonal prism would only allow for 105.4 cubic inches per plant.   **15.** False; Block B has a density of 0.79 g/cm³, and Block A has a density of 0.66 g/cm³. So, Block B has a greater density.

Module 6 Review

1. C   **3.** C   **5.** A, B, E, F   **7.** A   **9.** B, D, E, F
11. $\dfrac{x^2(x + 2)}{3}$ **13.** 101 in³   **15.** 244 cm³   **17.** D

Module 7

Quick Check

1. $\frac{5}{6}$ or 83%   **3.** $\frac{1}{6}$ or 17%   **5.** $\frac{1}{5}$ or 20%

7. $\frac{11}{20}$ or 55%

Lesson 7-1

1. $S = \{H, T\}$ **3a.** $S = \{1, 2, 3, 4, 5, 6, 7, 8, 9, 10, 11, 12\}$ **3b.** $S(\text{even number}) = \{2, 4, 6, 8, 10, 12\}$

5. BB, BW, WB, WW

	Blue Pants	White Pants
Blue Shirts	B, B	B, W
White Shirts	W, B	W, W

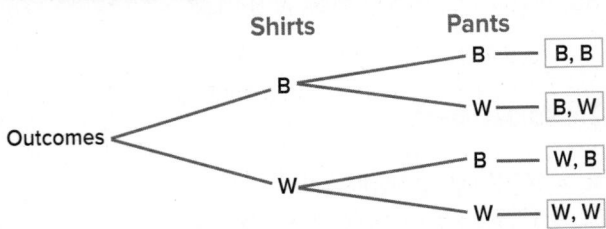

7. finite, $S = \{\text{yellow, blue, green, red}\}$

9. infinite; continuous

11. 28,800   **13.** 360   **15.** 1296   **17a.** 4
17b. 12   **17c.** 16   **19.** 2592

21. Sample answer: 6 different ways

$2(x + 4) + 4(x + 6) + 2(3);$

$2(x + 4) + 2(x + 6) + 2(x + 6) + 2(3);$

$2(x) + 2(4) + 4(x + 6) + 2(3);$

$2(x) + 2(4) + 2(x + 6) + 2(x + 6) + 2(3);$

$2(x) + 2(2) + 2(2) + 4(x + 6) + 2(3);$

$2(x) + 2(2) + 2(2) + 2(x + 6) + 2(x + 6) + 2(3)$

23. 12   **25a.** 5   **25b.** 18

27. $P = n^k$; Sample answer: The total number of possible outcomes is the product of the number of outcomes for each of the stages 1 through k. Because there are k stages, you are multiplying n by itself k times which is n^k.

29. Sample answer: In an experiment, you choose between a blue box and a red box. You then remove a ball from the box that you chose without looking into the box. The blue box contains a red ball, a purple ball, and a green ball. The red box contains a yellow ball and an orange ball.

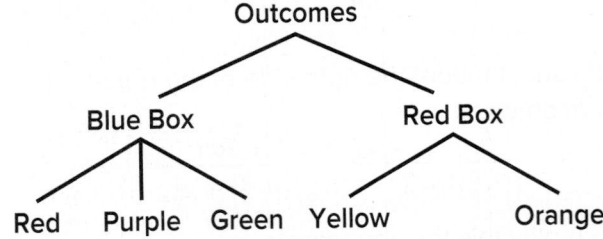

31. Never; sample answer: The sample space is the set of all possible outcomes. An outcome cannot fall outside the sample space. A failure occurs when the outcome is in the sample space, but is not a favorable outcome.

Lesson 7-2

1. $A \cap B = \{6\}$ **3a.** 4, 10   **3b.** 4, 8, 12   **3c.** 4
5. $\frac{1}{52}$ or 1.92%   **7a.** A, E, O, U   **7b.** J
7c. A, E, O, U, J   **9a.** 5, 10, 15, 20
9b. 1, 2, 3, 4, 5, 6, 7, 8, 9, 10, 11
9c. 1, 2, 3, 4, 5, 6, 7, 8, 9, 10, 11, 15, 20

11. $\frac{39}{52}$ or $\frac{3}{4}$ or 0.75   **13.** $\frac{240}{250}$ or 0.96

15. $\frac{10}{100}$ or $\frac{1}{10}$ or 0.10   **17.** $\frac{13}{20}$ or 0.65

19a. {Amy, Alex}   **19b.** $P(\text{one musical}) = 1 - P(\text{both}) = 1 - \frac{2}{9} = \frac{7}{9}$ or about 0.78

21. 0.275

23.

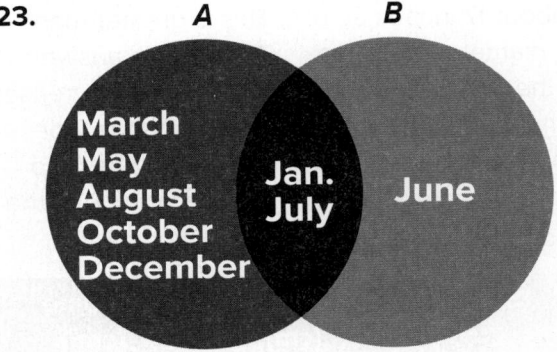

25. Sometimes: sample answer: While it is usually the case that the union of two sets will consist of more elements than their intersection, there are exceptions. For example, let A be the days of the week that end in *day*, and let B be the days of the week that begin with the letters *F, M, S, T,* or *W*. Because the two sets have the same list of items, their intersection and union will also be this same list.

Lesson 7-3

1. $\frac{1}{5}$, 0.2, or 20%   **3.** $\frac{1}{2}$, 0.5, or 50%

5. $\frac{5}{12} = 0.41\overline{6}$ or about 42%   **7.** $\frac{1}{3} = 0.\overline{3}$ or about 33%

9. $\frac{49}{100}$, 0.49, or 49%

11. approximately 0.054 or about 5%

13. $\frac{1}{9} = 0.\overline{1}$ or about 11%

15. $\frac{1}{9} = 0.\overline{1}$ or about 11%

17. $\frac{2}{9} = 0.\overline{2}$ or about 22%

19. $\frac{7}{12} = 0.583\overline{3}$ or about 58%

21. Sample answer: spinner landing on yellow

23. Sample answer: a point between 10 and 20

25. $\frac{\pi}{4} \approx 0.79$ or about 79%

27. $\frac{1}{3}$, $0.3\overline{3}$, or about 33%

29a. 0.842   **29b.** 0.205   **31.** 14.3%

33. No; sample answer: Athletic events should not be considered random because there are other factors involved such as pressure and ability that have an impact on the success of the event.

35. Sample answer: The probability of a randomly chosen point lying in the shaded region of the square on the left is found by subtracting the area of the unshaded square from the area of the larger square and finding the ratio of the difference of the areas to the area of the larger square. The probability is $\frac{1^2 - 0.75^2}{1^2}$ or 43.75%. The probability of a randomly chosen point lying in the shaded region of the square on the right is the ratio of the area of the shaded square to the area of the larger square, which is $\frac{0.4375}{1}$ or 43.75%. Therefore, the probability of a randomly chosen point lying in the shaded area of either square is the same.

Lesson 7-4

1. $\frac{1}{66}$ **3.** $\frac{1}{2450}$ **5.** $\frac{1}{420}$

7a. $\frac{1}{35}$ **7b.** $\frac{1}{210}$ **9.** $\frac{1}{1140}$ **11.** $\frac{1}{252}$

13. $\frac{1}{495}$ **15.** $\frac{1}{20}$ **17a.** 10   **17b.** $\frac{1}{10}$ **19.** $\frac{1}{38,760}$

21a. $\frac{1}{30}$; Sample answer: There are 6 possible coupons for the first customer, 5 possible for the second customer, and so forth, so the total number of possible outcomes is 6! = 720. If the first customer gets the 10% coupon and the second customer gets the 25% coupon, then there are 4! = 24 ways the remaining four customers can get coupons so the total number of favorable outcomes is 24. The probability is $\frac{\text{number of favorable outcomes}}{\text{number of possible outcomes}} = \frac{24}{720} = \frac{1}{30}$.

21b. $_6C_2 = 15$ **23.** $\frac{1}{15,120}$

25. $\frac{13}{261}$

27. Sample answer: Both permutations and combinations are used to find the number of possible arrangements of a group of objects. The order of the objects is important in permutations, but not in combinations.

29. Sample answer:

$$r! \cdot {_nC_r} = r! \cdot \frac{n!}{(n-r)!r!}$$
$$= \frac{n!r!}{(n-r)!r!}$$
$$= \frac{n!}{(n-r)!}$$
$$= {_nP_r}$$

$_nC_r$ and $_nP_r$ differ by the factor $r!$ because there are always $r!$ ways to order the groups that are selected. Therefore, there are $r!$ permutations of each combination.

Lesson 7-5

1. $\frac{1}{6}$ or about 17%   **3.** $\frac{1}{12}$ or about 8%

5. $\frac{6}{25}$ or 24%

7. Dependent; sample answer: Because the first ace drawn was not replaced, the probability of drawing the second card is affected.

9. Independent; sample answer: These two rolls have no bearing on each other.

11. $\frac{25}{2652}$ or about 1%

13. $\frac{1}{16}$ or about 6.25%

15. $\frac{11}{850}$ or about 1.29%   **17.** 0.43

19. Yes; sample answer: The probability of selecting a defective chip from Box A is $\frac{1}{25}$. Because $\frac{1}{625} = \frac{1}{25} \cdot \frac{1}{25}$, $P(A \text{ and } B) = P(A) \cdot P(B)$ and the events are independent.

21.

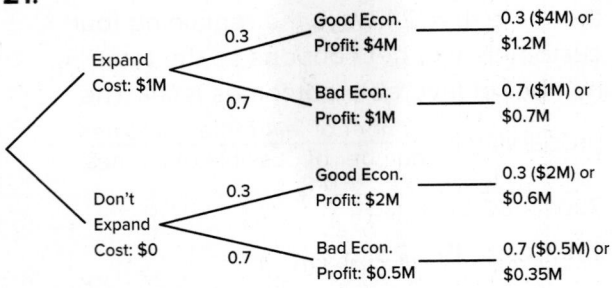

Sample answer: The expected value of choosing to expand is $1.2M + $0.7M or $1.9M, and the expected value of choosing not to expand is $0.95M. When we subtract the costs of expanding and of not expanding, we find the net expected value of expanding is $1.9M − $1M or $0.9M and the net expected value of not expanding is $0.95M − $0 or $0.95M. Because $0.9M < $0.95M, you should not expand the business.

23. $\frac{1}{2704}$; Sample answer: The card is replaced, so the events are independent; the probability of each picking the queen of spades is $\frac{1}{52}$, so the probability of them both picking it is $\frac{1}{52} \cdot \frac{1}{52} = \frac{1}{2704}$.

25a. The first branch of the tree should be the first serve, while the second set of branches should be the second serve. The second serve only occurs when the first serve is a fault.

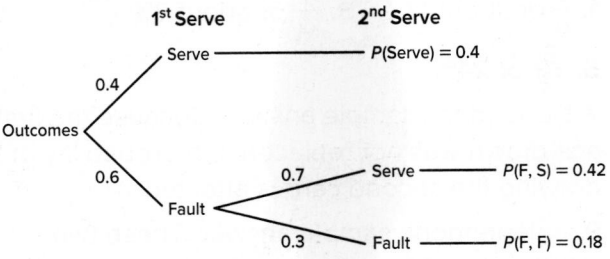

25b. A double fault is back-to-back faults, or $P(F, F)$, which equals 0.18 or 18%.

27. Sample answer: In order for the events to be independent, two things must be true: 1) the chance that a person is left-handed is the same as the chance that a person is left-handed given that the person's parent is left-handed, and 2) the chance that a person's parent is left-handed is the same as the chance that a person's parent is left-handed given that the person is left-handed.

29. A and B are independent events.

Lesson 7-6

1. Not mutually exclusive; sample answer: These events are not mutually exclusive because a 6 is also an even number.

3. $\frac{1}{2}$ or 0.50   **5.** 0.65   **7.** $\frac{1}{13}$ or 7.7%

9. $\frac{3}{13}$ or 23.1%   **11.** $\frac{4}{13}$ or about 31%

13. $\frac{4}{13}$ or about 31%   **15.** 0.30   **17.** $\frac{3}{4}$ or 75%

19. $\frac{739}{750}$ or about 98.5%   **21.** 56%   **23.** 0.6

25. 0.74; Sample answer: There are three outcomes in which the values of two or more of the dice are less than or equal to 4 and one outcome where the values of all three of the dice are less than or equal to 4. You have to find the probability of each of the four scenarios and add them together.

27. Not mutually exclusive; sample answer: Because squares are rectangles, but rectangles are not necessarily squares, a quadrilateral can be both a square and a rectangle, and a quadrilateral can be a rectangle but not a square.

29. Not mutually exclusive; sample answer: A natural number is also a complex number.

Lesson 7-7

1. $\frac{1}{5}$ or 20%   **3.** $\frac{1}{3}$ **5.** $\frac{8}{11}$ or about 73%

7. $\frac{1}{4}$ or 25%   **9.** $\frac{1}{6}$ **11a.** 6 students

11b. $P(H|L) = \frac{1}{3}$; $P(L) = \frac{3}{10}$ and $P(H \text{ and } L) = \frac{1}{10}$.
So, $P(H|L) = \frac{P(H \text{ and } L)}{P(L)} = \frac{\frac{1}{10}}{\frac{3}{10}}$ or $\frac{1}{3}$.

11c. No; sample answer: If $P(H|L)$ is the same as $P(H)$ or $P(L|H)$ is the same as $P(L)$, then H and L are independent. Because $P(H|L)$ is $\frac{1}{3}$ and $P(H)$ is $\frac{7}{10}$, H and L are not independent.

13a. 80%; $P(Algebra) = 0.25$ and $P(Algebra$ and $Health) = 0.2$, so $P(Health|Algebra) =$
$$\frac{P(Algebra \text{ and } Health)}{P(Algebra)} = \frac{0.2}{0.25} = 0.8.$$

13b. No; $P(Health|Algebra) = 0.8$ and $P(Health) = 0.5$. Because the two probabilities are not equal, the two events are dependent.

13c. 25%; $P(Accounting) = 0.2$ and $P(Accounting$ and $Spanish) = 0.05$, so $P(Spanish|Accounting) =$
$$\frac{P(Accounting \text{ and } Spanish)}{P(Accounting)} = \frac{0.05}{0.2} = 0.25.$$
Because the events are independent, $P(Spanish|Accounting) = P(Spanish) = 0.25$.

Lesson 7-8

1a.

	SUV	Truck	Totals
Male	15	35	50
Female	40	10	50
Totals	55	45	100

1b.

	SUV	Truck	Totals
Male	$\frac{15}{100} = 15\%$	$\frac{35}{100} = 35\%$	$\frac{50}{100} = 50\%$
Female	$\frac{40}{100} = 40\%$	$\frac{10}{100} = 10\%$	$\frac{50}{100} = 50\%$
Totals	$\frac{55}{100} = 55\%$	$\frac{45}{100} = 45\%$	$\frac{100}{100} = 100\%$

3a.

	More Than 4	4 or Fewer	Totals
In State	928	332	1260
Out of State	118	622	740
Totals	1046	954	2000

3b.

	More Than 4	4 or Fewer	Totals
In State	$\frac{928}{2000} = 46.4\%$	$\frac{332}{2000} = 16.6\%$	$\frac{1260}{2000} = 63\%$
Out of State	$\frac{118}{2000} = 5.9\%$	$\frac{622}{2000} = 31.1\%$	$\frac{740}{2000} = 37\%$
Totals	$\frac{1046}{2000} = 52.3\%$	$\frac{954}{2000} = 47.7\%$	$\frac{2000}{2000} = 100\%$

3c. Sample answer: 63% of the college students surveyed are attending an in-state college. 52.3% of the college students surveyed visited home more than 4 times. It would be expected that 63% · 52.3%, or about 33% would be in-state students that visited home more than 4 times. Because $P(A$ and $B) \neq P(A) \cdot P(B)$, or 33% ≠ 46.4%, the events are dependent.

5. 18; joint frequency

7. 17.7%; $P(\text{has a tablet} \mid \text{has a smart phone}) =$
$$\frac{P(\text{has a tablet and has a smart phone})}{P(\text{has a smart phone})} \approx \frac{0.125}{0.708} \text{ or }$$
17.7%

9a. voting for candidate A or B

9b. 55% **9c.** 55% **9d.** 9%

9e. Sample answer: males age 18–30 and females age 46–60

9f. Sample answer: females age 18–30, males and females age 31–45, and anyone over 60

Module 7 Review

1. C   **3.** A   **5.** D   **7.** D

9. B, C, E   **11.** A, D   **13.** $\frac{2}{27}$ **15.** D

Module 8

Quick Check

1. −15   **3.** 10   **5.** $b = \frac{a}{3} - 3$ **7.** $x = \frac{4}{3}y + \frac{8}{3}$

Lesson 8-1

1. D = {all real numbers}; R = {all real numbers};
Codomain = {all real numbers}; onto
3. D = {all real numbers}, R = {y | y ≥ 0},
Codomain = {all real numbers}; not onto
5. D = {1, 2, 3, 4, 5, 6, 7}; R = {56, 52, 44, 41, 43, 46, 53}; one-to-one
7. neither   **9.** both
11. continuous; D = {all real numbers},
R = {all real numbers}
13. discrete; D = {1, 2, 3, 4, 5, 6}, R = {3, 4, 5, 6}
15. continuous; D = all positive real numbers,
R = {y | y ≥ 0}
17. D = {x | x ∈ ℝ} or (−∞, ∞);
R = {y | y ≤ 0} or (−∞, 0]
19. D = {x | x ≤ −1 or x ≥ 1} or (−∞, −1] U [1, ∞);
R = {y | y ∈ ℝ} or [−∞, ∞)
21. D = {x | x ≤ −2 or x ≥ 1} or (−∞, −2] U [1, ∞);
R = {y | y ≥ −2} or [−2, ∞)
23. D = {x | x ∈ ℝ} or (−∞, ∞);
R = {y | y ≥ −4} or [−4, ∞); neither one-to-one
nor onto; continuous
25. D = {x | x ∈ ℝ} or (−∞, ∞); R = {y | y ∈ ℝ}
or [−∞, ∞); both; continuous
27. D = {x | x ∈ ℝ} or (−∞, ∞); R = {y | y ∈ ℝ}
or [−∞, ∞); onto; continuous
29. D = {w | 0 ≤ w ≤ 15}; R = {L | 4 ≤ L ≤ 11.5};
one-to-one; continuous
31. neither one-to-one nor onto; discrete
33. D = {n | n ≥ 0}; R = {T(n) | T(n) ≥ 0}; both
(within the restrictions of the domain and
codomain); continuous
35.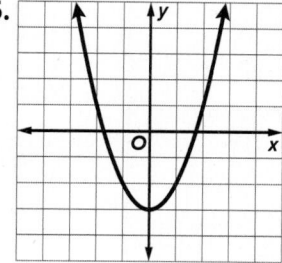

37. D = {x | x ≠ 0} or (−∞, 0) U (0, ∞);
R = {y | y ≠ 0} or (−∞, 0) U (0, ∞); neither;
continuous
39. Sample answer: The vertical line test is
used to determine whether a relation is a
function. If no vertical line intersects a graph
in more than one point, the graph represents
a function. The horizontal line test is used to
determine whether a function is one-to-one.
If no horizontal line intersects the graph more
than once, then the function is one-to one.
The horizontal line test can also be used to
determine whether a function is onto. If every
horizontal line intersects the graph at least
once, then the function is onto.

Lesson 8-2

1. Yes; it can be written in $y = mx + b$ form.
3. Yes; it can be written in $y = mx +$ form.
5. nonlinear   **7.** nonlinear   **9.** The number of
inches and corresponding number of feet is a
linear function because when graphed, a line
contains all of the points shown in the table.
11. x-int: 12; y-int: −18   **13.** x-int: 6; y-int: −18
15. x-int: 0, 4; y-int: 0   **17a.** x-int: 4; y-int: 20
17b. The x-intercept represents the number
of days until Aksa will run out of money. The
y-intercept represents the total amount Aksa
had in her lunch account at the beginning of
the week.   **19.** point symmetry
21. point symmetry   **23.** neither; $f(−x) = (−x)^3 + (−x)^2 = −x^3 + x^2 \neq f(x)$ and $\neq −f(x)$ **25.** No;
x is in a denominator. The equation is neither
even nor odd.   **27.** Yes; it can be written in
$y = mx + b$ form. The equation is even.
29. No; there is an x^2 term. The equation
is even.   **31.** linear; x-int: $\frac{10}{3}$; y-int: $\frac{30}{7}$; point
symmetry   **33.** nonlinear; x-int: −3, −2, −1,
1, 2; y-int: 12; neither point nor line symmetry
35. line symmetry; $x = 0.4$
37a. $2x + 2y + 10 = 500$
37b. Yes; it can be written in $y = mx + b$ form.

37c. point symmetry;

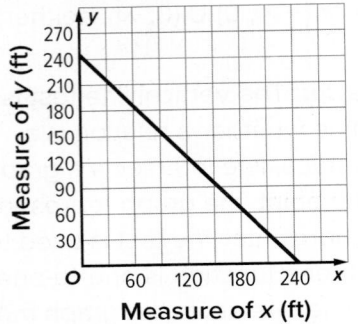

Measure of x (ft)

39. Odd; $f(-r) = -f(r)$
41. No; sample answer: $f(x) = x^3 + 2x - 5$
43. Never; sample answer: the graph of $x = a$ is a vertical line, so it is not a function.

Lesson 8-3

1. rel. max. at $(-2, -2)$ rel. min. at $(-1, -3)$
3. rel. max. at $(0, -2)$, min. at $(-1, -3)$ and $(1, -3)$
5. The relative maxima occur at $x = -3.7$ and $x = 4.5$, and the relative minimum occurs at $x = 0$. The relative maxima at $x = -3.7$ and $x = 4.5$ represents the top of two hills. The relative minimum at $x = 0$ represents a valley between the hills.
7. As $x \to -\infty, y \to -\infty$ and as $x \to \infty, y \to -\infty$.
9. As $x \to -\infty, y \to \infty$ and as $x \to \infty, y \to -\infty$.
11a. $t = 1.5$; The fish reaches its maximum height 1.5 seconds after it is thrown.
11b. As $t \to -\infty, h(t) \to -\infty$ and as $t \to \infty$, $h(t) \to -\infty$; The height cannot be negative because we are considering the path of the fish above the surface of the water, $h = 0$. Time cannot be negative because we're measuring from the initial time the fish was thrown at $t = 0$.
13. rel. max. at $x = 2.7$, rel. min. at $x = -1.2$; As $x \to -\infty, f(x) \to \infty$ and as $x \to \infty, f(x) \to -\infty$.
15. rel. max. at $x = -2$, rel. min. at $x = -1$; As $x \to -\infty, f(x) \to \infty$ and as $x \to \infty, f(x) \to \infty$.
17. no rel. max, min: $x = 0$; As $x \to -\infty, f(x) \to \infty$ and as $x \to \infty, f(x) \to \infty$.

19. As temperature increases, density decreases.

21. rel. max: $(-2.8, 6)$, $(1.8, 3)$; rel. min: $(0, 2)$, $(5, -6)$; As $x \to -\infty, y \to -\infty$ and as $x \to \infty$, $y \to \infty$.
23. no relative max or min

25. rel. max: $x = -1.87$; rel. min: $x = 1.52$
27. The dynamic pressure would approach ∞.
29. as $r \to \infty, V \to \infty$
31. Sample answer: The end behavior of a graph describes the output values the input values approach negative and positive infinity. It can be determined by examining the graph.
33. As the concentration of the catalyst is increased, the reaction rate approaches 0.5.
35. Joshua switched the $f(x)$ values. He read the graph from right to left instead of left to right.

Lesson 8-4

1.

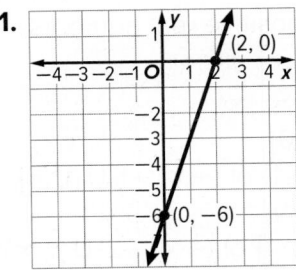

3.

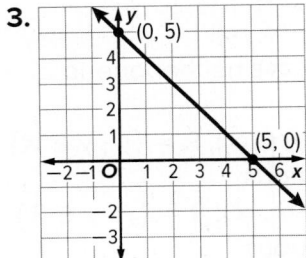

5.

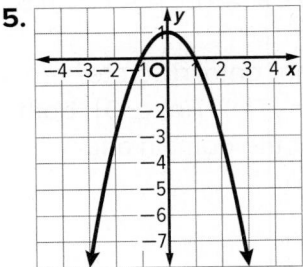

7.

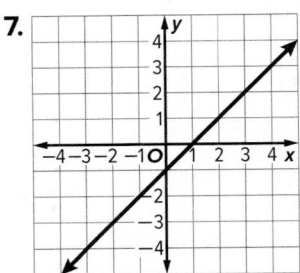

9.

Pelican's Height

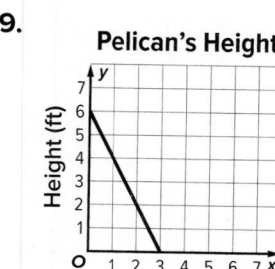

11. The x-intercept of $f(x)$ is 2, and the x-intercept of $g(x)$ is $-\frac{2}{3}$. The x-intercept of $f(x)$ is greater than the x-intercept of $g(x)$. So, $f(x)$ intersects the x-axis at a point farther to the right than $g(x)$. The y-intercept of $f(x)$ is -1, and the y-intercept of $g(x)$ is 2. The y-intercept of $g(x)$ is greater than the y-intercept of $f(x)$. So, $g(x)$ intersects the y-axis at a higher point than $f(x)$. The slope of $f(x)$ is $\frac{1}{2}$ and the slope of $g(x)$ is 3. Each function is increasing, but the slope of $g(x)$ is greater than the slope of $f(x)$. So, $g(x)$ increases faster than $f(x)$. **13.** Both x-intercepts of $f(x)$ are less than the x-intercept of $g(x)$. The graph of $f(x)$ intersects the x-axis more times than $g(x)$. The y-intercept of $g(x)$ is less than the y-intercept of $f(x)$. So, $f(x)$ intersects the y-axis at a higher point than $g(x)$. Neither function has a relative maximum. $f(x)$ has a minimum at $(-2, -4)$. The two functions have the opposite end behaviors as $x \rightarrow -\infty$. The two functions have the same end behavior as $x \rightarrow \infty$.

15. The x-intercept of $f(x)$ is $\frac{2}{3}$, and the x-intercept of $g(x)$ is $\frac{3}{8}$. The x-intercept of $f(x)$ is greater than the x-intercept of $g(x)$. So, $f(x)$ intersects the x-axis at a point farther to the right than $g(x)$. The y-intercept of $f(x)$ and $g(x)$ is $\frac{1}{2}$. So, $f(x)$ and $g(x)$ intersect the y-axis at the same point. The slope of $f(x)$ is $\frac{3}{4}$ and the slope of $g(x)$ is $-\frac{4}{3}$. Each function is decreasing, but the slope of $g(x)$ is less than the slope of $f(x)$. So, $g(x)$ decreases faster than $f(x)$.

17.

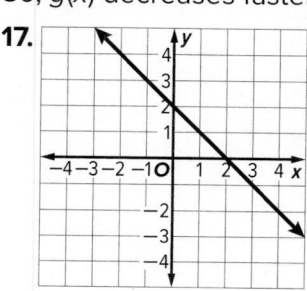

19.

Monica's Walk

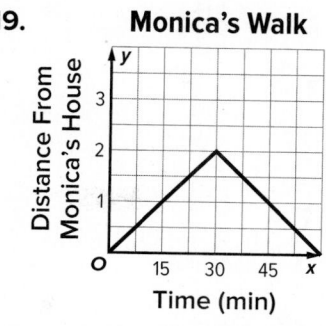

21a. linear; Sample answer: It is linear because it makes no stops along the way, and it descends at a steady pace, which indicates a constant rate of change, or slope.

21b.

Ski Lift Height

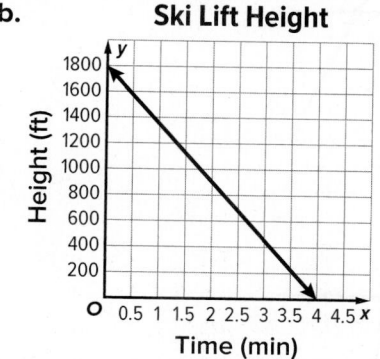

23. Sample answer: The function is continuous. The function has a y-intercept at -3. The function has a maximum at $(-3, 1)$. The function has a minimum at $(1.4, -4)$. As $x \rightarrow -\infty$, $f(x) \rightarrow -\infty$ and as $x \rightarrow \infty$, $f(x) \rightarrow \infty$.

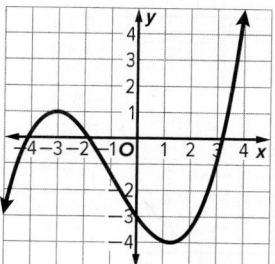

25. Always; Sample answer: A linear function cannot cross the x-axis more than once. So, if a function has more than one x-intercept, it is a nonlinear function.

27. Both Linda and Rubio sketched correct graphs. Both graphs have an x-intercept at 2, a y-intercept at -9, are positive for $x > 2$, and have an end behavior of as $x \rightarrow -\infty$, $f(x) \rightarrow -\infty$ and as $x \rightarrow \infty$, $f(x) \rightarrow \infty$.

Lesson 8-5

1.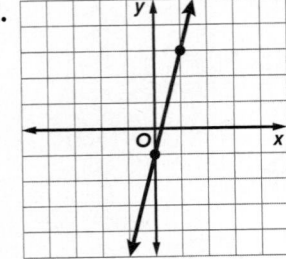

3.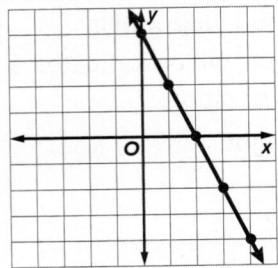

5.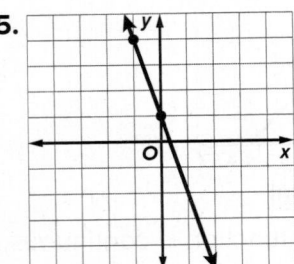

7.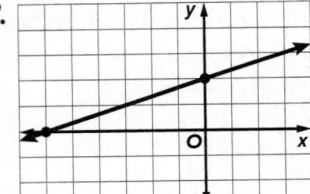

9.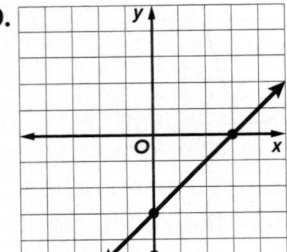

11.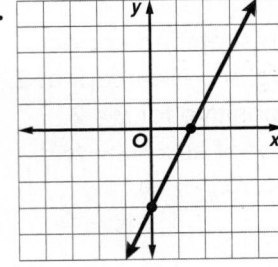

13.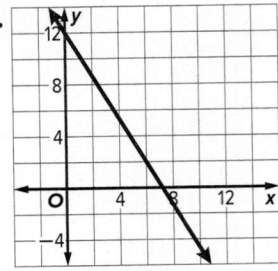

15.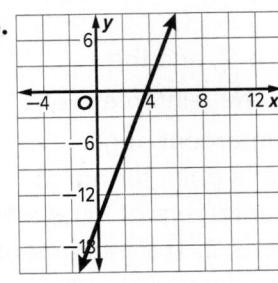

17.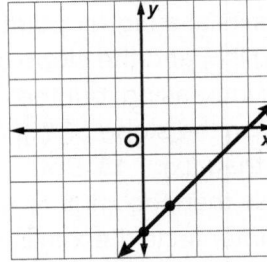

19.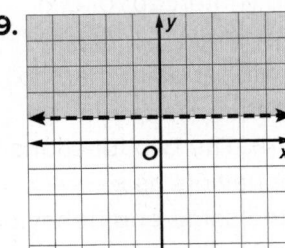

21.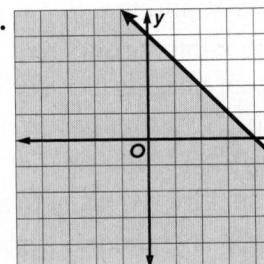

23.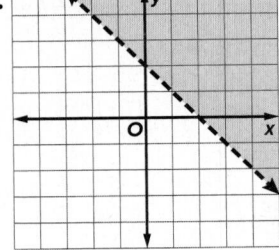

25.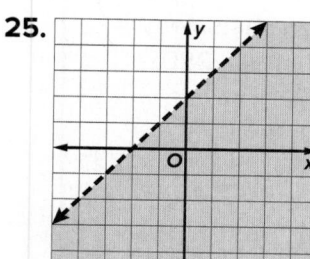

27.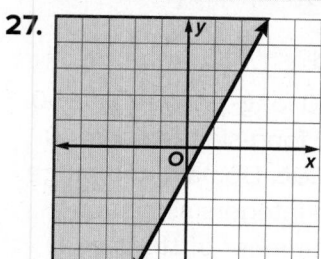

29.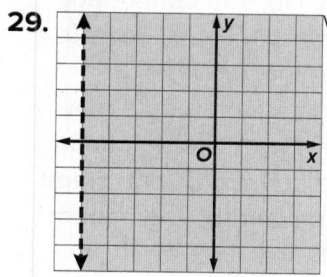

31a. $x + y \geq 400$

31b.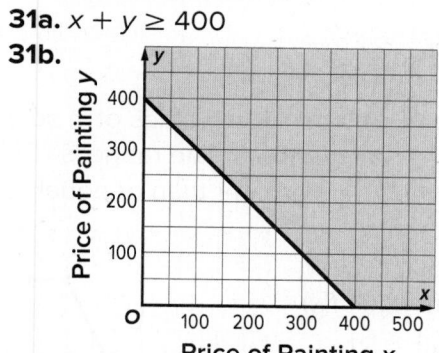

33.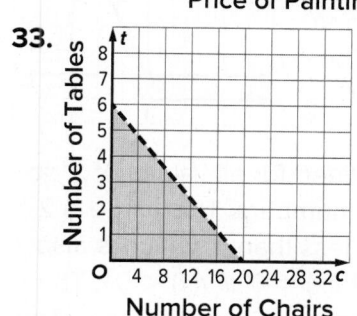

35.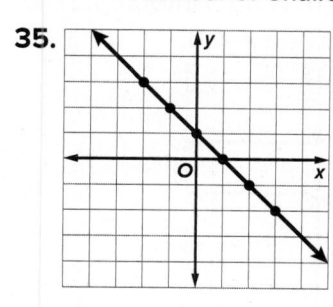

37.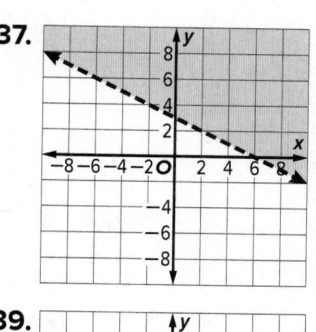

39.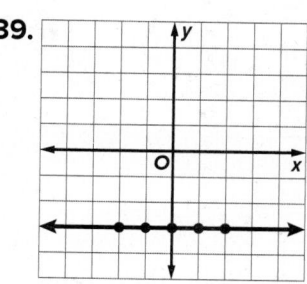

41.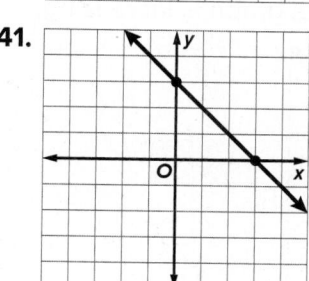

43.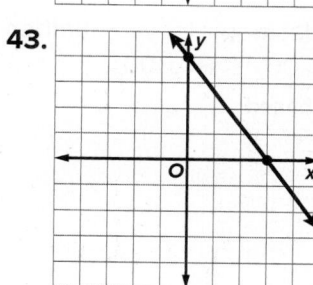

45.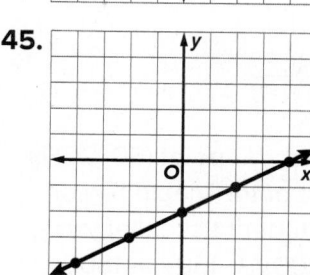

47.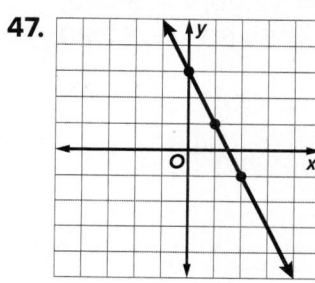

49.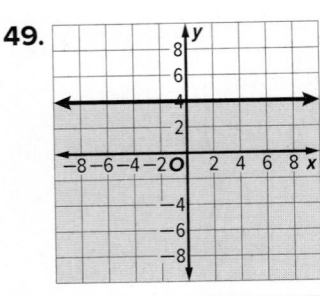

51.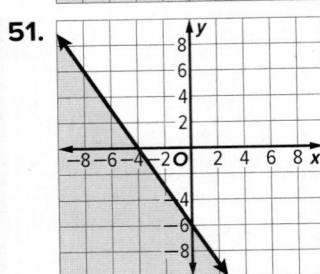

53. x-int = 2, y-int = −6; graph is increasing;

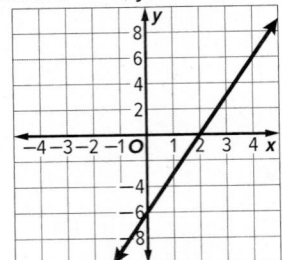

55a. Let x = number of desktops; let y = number of notebooks; $1000x + 1200y \leq 80{,}000$.

55b.

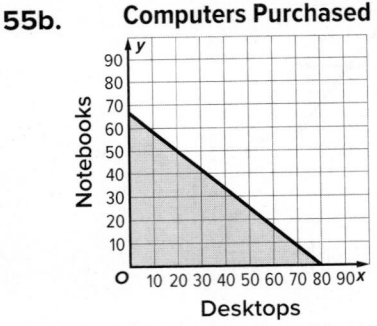

55c. Yes; Sample answer: the point (50, 25) is on the line, which is part of the viable region.

57a. Let x = cost of a student ticket; let y = cost of an adult ticket; $300x + 150y = 1800$.

57b. Sample answer: x = \$2.00, y = \$8.00; x = \$3.00, y = \$6.00; x = \$4.60, y = \$2.80; x = \$6.00, y = \$0.00

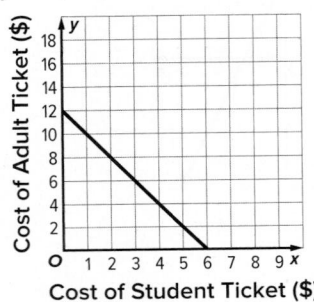

59a. Let x = long-sleeved shirts; let y = short-sleeved shirts; $7x + 4y \geq 280$;

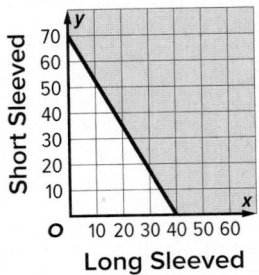

59b. Sample answer: 30 long-sleeved and 50 short-sleeved shirts; 60 long-sleeved and 40 short-sleeved shirts.   **59c.** Domain and range values must be positive integers since you cannot buy a negative number of shirts or a portion of a shirt.   **59d.** No, you cannot buy −10 long-sleeved shirts.   **61.** Paulo; Janette shaded the incorrect region.   **63.** Sample answer: If given the x- and y-intercepts of a linear function, I already know two points on the graph. To graph the equation, I only need to graph those two points and connect them with a straight line.   **65.** $y = \frac{1}{4}x + 5$

Lesson 8-6

1. The function is defined for all values of x, so the domain is all real numbers. The range is −1 and all real numbers greater than or equal to 0 and less than or equal to 6, which is also represented as $\{f(x) \mid f(x) = -1$ or $0 \leq f(x) \leq 6\}$. The y-intercept is 0, and the x-intercept is 0. The function is increasing when $0 \leq x \leq 3$.

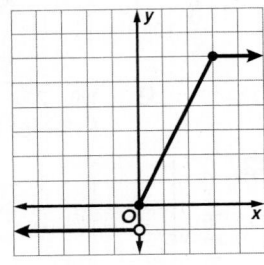

3. The function is defined for all values of x, so the domain is all real numbers. The range is 2 and all real numbers less than 0, which is also represented as $\{f(x) \mid f(x) = 0$ or $f(x) < 0\}$. The y-intercept is 2, and there is no x-intercept. The function is decreasing when $x < 0$.

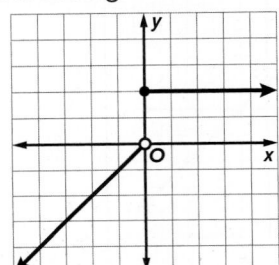

5a. $f(x) = \begin{cases} 48x & \text{if } 0 < x \le 3 \\ 45x & \text{if } 3 < x \le 8 \\ 42x & \text{if } 8 < x \le 19 \\ 38x & \text{if } x > 19 \end{cases}$

5b. $225; $798

7. D = {all real numbers}; R = {all integers}

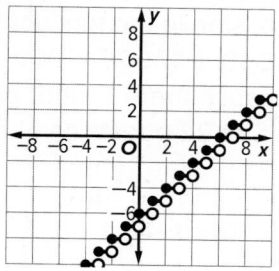

9. D = {all real numbers}; R = {all integers}

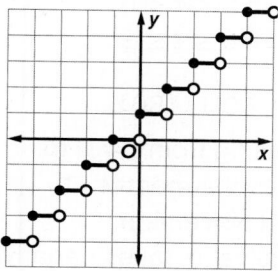

11. D = $\{x \mid 0 < x \le 8\}$;
R = $\{y \mid y = 5.00, 10.00, 15.00, 20.00\}$

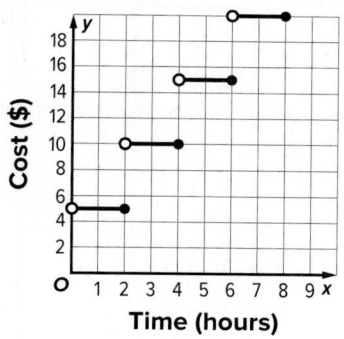

13. D = {all real numbers}; R = $\{f(x) \mid f(x) \ge 0\}$

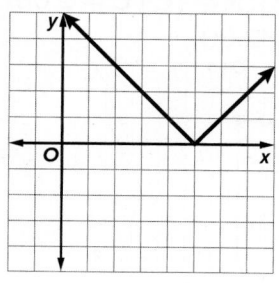

15. D = {all real numbers}; R = $\{h(x) \mid h(x) \ge -8\}$

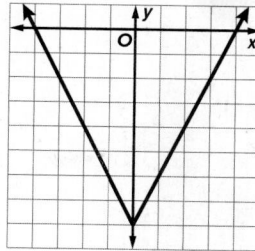

17. D = {all real numbers}; R = $\{f(x) \mid f(x) \ge 6\}$

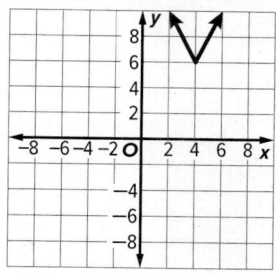

19. D = {all real numbers}; R = $\{g(x) \mid g(x) \ge 0\}$

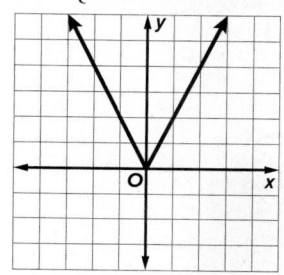

21. D = $\{x \mid x \le -4 \text{ or } 0 < x\}$;
R = $\{f(x) \mid f(x) \ge 12, f(x) = 8, \text{ or } 0 < f(x) \le 3\}$

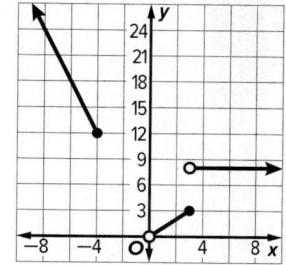

23. D = {all real numbers};
R = $\{g(x) \mid g(x) < -10 \text{ or } -6 \le g(x) \le 2\}$

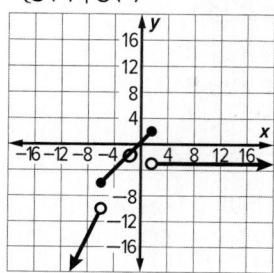

25. D = {all real numbers}; R = {f(x) | f(x) > −3}

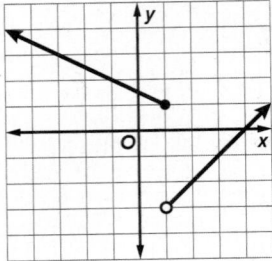

27. D = {all real numbers}; R = {all integers}

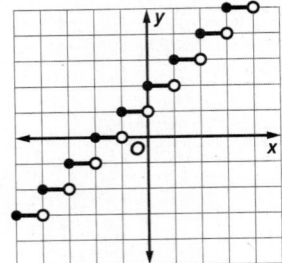

29. D = {all real numbers};
R = {all whole numbers}

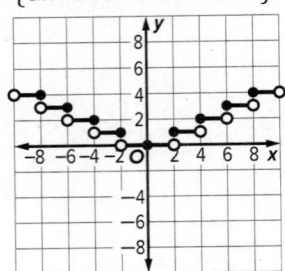

31. D = {all real numbers}; R = {g(x)| g(x) ≤ 4}

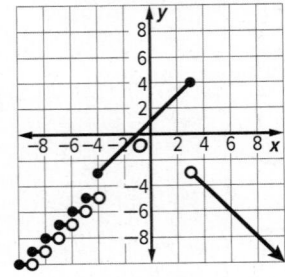

33. D = {all real numbers}; R = {h(x) | h(x) ≥ 2}

35. $C(x) = \begin{cases} 5 & \text{if } 0 \le x \le 1 \\ 7.5 & \text{if } 1 < x \le 2 \\ 10 & \text{if } 2 < x \le 3 \\ 12.5 & \text{if } 3 < x \le 4 \\ 15 & \text{if } 4 < x \le 24 \end{cases}$

37a. $C(x) = \begin{cases} 500 + 17.50x & \text{if } 0 \le x \le 40 \\ 1200 + 14.75(x-40) & \text{if } x \ge 41 \end{cases}$

37b.

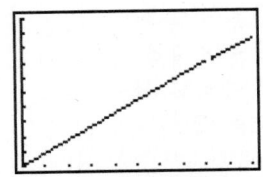

[0, 50] scl: 5 by [500, 1500] scl: 100

37c. Because it costs $1200 for 40 guests to attend, use the first expression in the function C(x). Solve the equation 500 + 17.50x = 900 to obtain about 22.9. Because there cannot be a fraction of a guest, at most 22 guests can be invited to the reunion.

39a. $R(t) = \begin{cases} \frac{20}{3}t + 30 & \text{if } 0 \le t \le 3 \\ 60 & \text{if } 3 < t < 4 \\ 80 & \text{if } 4 \le t \le 5 \\ 60 & \text{if } 5 < t < 6 \\ -\frac{50}{3} + 160 & \text{if } 6 \le t \le 9 \end{cases}$

Range = [10, 60] ∪ {80}

39b. The graph is increasing from t = 0 to t = 3. This corresponds to the months of September, October, and November.

41. Because the absolute value takes negative f(x)-values and makes them positive, the graph retains the step-like nature of the greatest integer function, but it also has the "v" shape of the absolute value.

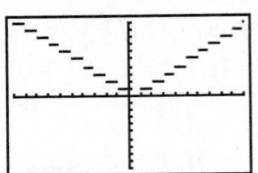

[−10, 10] scl: 1 by [−10, 10] scl: 1

43. $f(x) = \begin{cases} -x + 2 & \text{if } x \le 0 \\ -x - 2 & \text{if } x > 0 \end{cases}$

45. $f(x) = \begin{cases} \frac{1}{2}x + 1 & \text{if } x < 2 \\ x - 4 & \text{if } x > 2 \end{cases}$

47. D = {m | 0 < m ≤ 6}; R = {C | C = 2.00, 4.00, 6.00, 8.00, 10.00, 12.00}

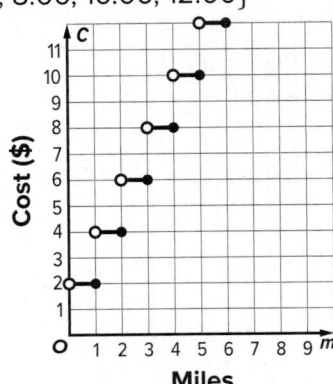

49. Sample answer: $|y| = x$ **51.** Sample answer: 8.6; The greatest integer function asks for the greatest integer less than or equal to the given value; thus 8 is the greatest integer. If we were to round this value to the nearest integer, we would round up to 9.   **53.** Sample answer: Piecewise functions can be used to represent the cost of items when purchased in quantities, such as a dozen eggs.

Lesson 8-7

1. translation of the graph of $y = x^2$ up 4 units
3. translation of the graph of $y = x$ down 1 unit
5. translation of the graph of $y = x^2$ right 5 units
7. $y = x^2 - 2$ **9.** $y = x + 1$ **11.** $y = |x + 3| + 1$
13. compressed horizontally and reflected in the y-axis   **15.** stretched vertically and reflected in the x-axis   **17.** compressed vertically and reflected in the x-axis
19. stretched horizontally and reflected in the y-axis   **21.** reflected in the x-axis, stretched vertically, and translated down 4 units
23. compressed vertically and translated down 2 units   **25.** reflected in the x-axis, compressed vertically, and translated left 3 units   **27.** stretched horizontally by a factor of 0.5; The absolute value function shows the ball bouncing off the edge of the pool table and the stretch shows the wide angle.
29. compressed vertically by a factor of 0.75 and translated up 25; There is a $25 fixed cost, plus $0.75 per mile, regardless of direction.
31. $y = 2|x + 2| + 5$ **33.** $y = -|x| - 3$
35. $y = -(x - 4)^2$
37. translation of $y = |x|$ down 2 units

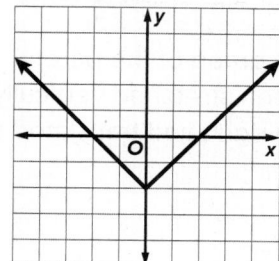

39. reflection of $y = x$ in the x-axis

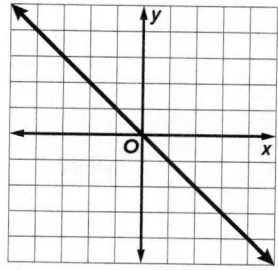

41. vertical stretch of $y = x$

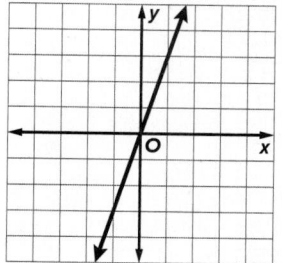

43. translation of $y = x^2$ down 4 units
45. horizontal compression of the graph of $y = x^2$
47a. translation of $y = |x|$ right 8 units

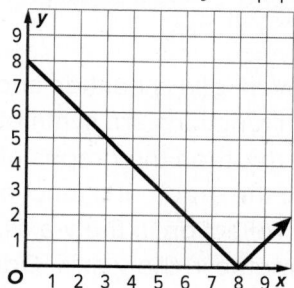

47b. Sample answer: The speedometer is stuck at 8 mph.
49. stretched vertically by a factor of 4
51. Maria stretched the function vertically by a factor of 10.   **53a.** quadratic   **53b.** x-axis
53c. right 25 units and up 81 units
53d. $y = -(x - 25)^2 + 81$
55a.

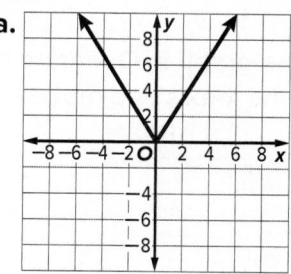

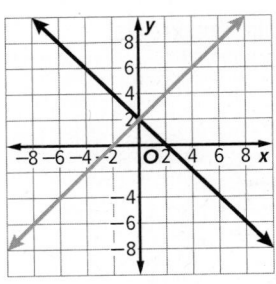

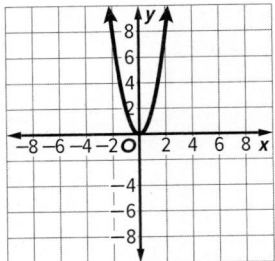

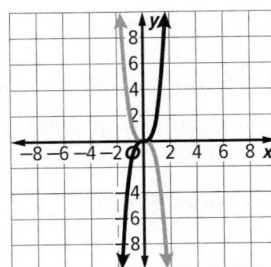

55b. $f(x)$ and $h(x)$ are even, $g(x)$ is neither, and $k(x)$ is odd. **55c.** Even functions are symmetric in the y-axis. If $f(-x) = f(x)$, then the graphs of $f(-x)$ and $f(x)$ coincide. If the graph of a function coincides with its own reflection in the y-axis, then the graph is symmetric in the y-axis. Odd functions are symmetric in the origin, which means that the graph of an odd function coincides with its rotation of 180° about the origin. A rotation of 180° is equivalent to reflection in two perpendicular lines. $f(-x)$ is a reflection in the y-axis and $-f(x)$ is a reflection in the x-axis. Thus if the graphs of $f(-x)$ and $-f(x)$ coincide, $f(x)$ is symmetric about the origin.

57. Sample answer: The graph in Quadrant II has been reflected in the x-axis and moved right 10 units.

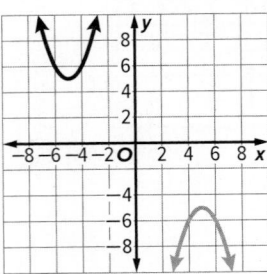

59. Sample answer: Because the graph of $g(x)$ is symmetric about the y-axis, reflecting in the y-axis results in a graph that appears the same. It is not true for all quadratic functions. When the axis of symmetry of the parabola is not along the y-axis, the graph and the graph reflected in the y-axis will be different.

Module 8 Review

1. C   **3.** Sample answer: The x-intercept is (6, 0). This means that after 6 weeks, Tia owes her friend $0. The y-intercept is (0, 80). This means that Tia initially owed her friend $80, or that Tia borrowed $80 from her friend to go to a theme park.

5. $x = -5$ is a relative maximum. $x = -4$ is neither. $x = -2$ is a relative minimum. $x = 0$ is neither. $x = 1$ is a relative maximum. $x = 5$ is a relative minimum.

7. The x-intercept of $f(x)$ is 2, and the x-intercept of $g(x)$ is 1. The x-intercept of $f(x)$ is greater than the x-intercept of $g(x)$. So, $f(x)$ intersects the x-axis at a point farther to the right than $g(x)$. The y-intercept of $f(x)$ is 4, and the y-intercept of $g(x)$ is -5. The y-intercept of $f(x)$ is greater than the y-intercept of $g(x)$. So, $f(x)$ intersects the y-axis at a higher point than $g(x)$. The slope of $f(x)$ is -2 and the slope of $g(x)$ is 5. $f(x)$ is decreasing and $g(x)$ is increasing. The slope of $g(x)$ is greater than the slope of $f(x)$.   **9.** A

11.

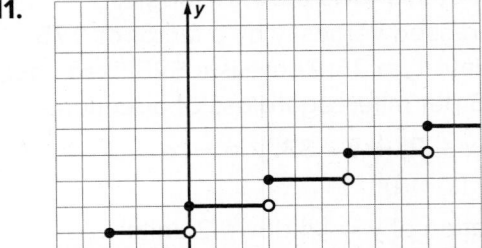

13. Sample answer: The graph of the parent function $f(x) = x^2$ has been stretched vertically, translated 2 units to the right, and translated up 9 units.   **15.** C

Module 9
Quick Check

1.

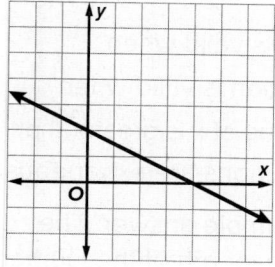

3.

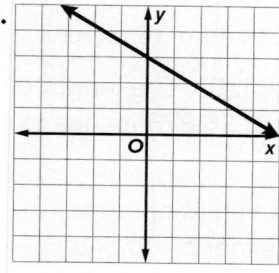

5.

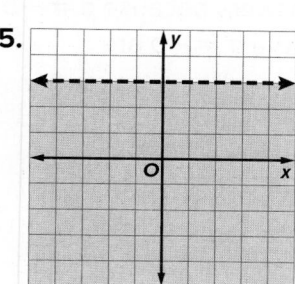

7.

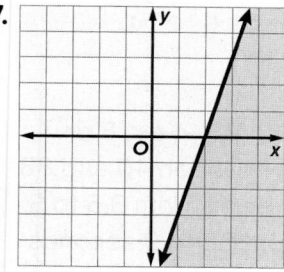

Lesson 9-1

1. $\frac{4}{5}$ **3.** $\frac{1}{4}$ **5.** 5   **7.** 0   **9.** -3

11. g = green fees per person;
$6(2) + 4g = 76$; $16

13. $y = 2A - x$

15. $h = \frac{A - 2\pi r^2}{2\pi r}$

17. $f(x) = 2x + 12$ The solution is -6.

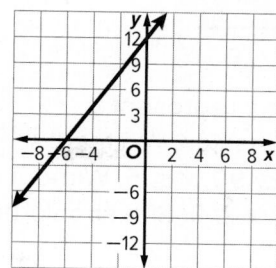

19. $f(x) = \frac{1}{2}x - 6$ The solution is 12.

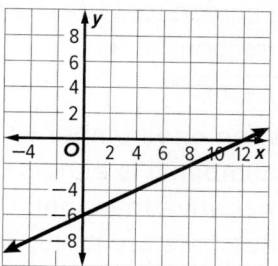

21. $f(x) = -3x - 2$ The solution is $x = -\frac{2}{3}$.

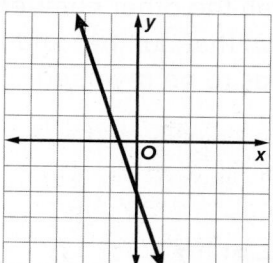

23a. Sample answer: about 5.5 weeks

23b. $w \approx 5.56$

25. $\{z \mid z \leq -8\}$

$-9\ -8\ -7\ -6\ -5\ -4\ -3\ -2\ -1$

27. $\{n \mid n < 2\}$

$-4\ -3\ -2\ -1\ \ 0\ \ 1\ \ 2\ \ 3\ \ 4$

29. $\{m \mid m \leq 4\}$

$-2\ -1\ \ 0\ \ 1\ \ 2\ \ 3\ \ 4\ \ 5\ \ 6$

31. $15P + 300 \geq 1500$; $P \geq 80$; Manuel must translate at least 80 pages.

33. $\frac{22}{5}$ **35.** 5

37. $\{x \mid x \le 6\}$

39. Sample answer: Let x represent the number of skating session. With a membership: $6x + 60 \le 90$; $x \le 5$. Without a membership: $10x \le 90$; $x \le 9$. She should not buy a membership.

41. Jade; Sample answer: in the last step, when Steven subtracted b_1 from each side, he mistakenly put b_1 in the numerator instead of subtracting it from the fraction.

43. $y_1 = y_2 - \sqrt{d^2 - (x_2 - x_1)^2}$

45. Sample answer: When one number is greater than another number, it is either more positive or less negative than that number. When these numbers are multiplied by a negative value, their roles are reversed. That is, the number that was more positive is now more negative than the other number. Thus, it is now *less than* that number and the inequality symbol needs to be reversed.

Lesson 9-2

1. $\{11\}$ **3.** $\{3, 4\}$ **5.** $\{-8\}$ **7.** $\{-2, -1\}$

9. $|x - 87.4| \le 1.5$; $85.9°$ F; $88.9°$ F

11. $\{x \mid -2 \le x \le 0\}$

13. $\varnothing$

15. $\{x \mid -1.8 < x < 3.4\}$

17. $\left\{ x \mid x < -1 \text{ or } x > \frac{1}{3} \right\}$

19. $|r - 24| > 6.5$; $\{r \mid r < 17.5 \text{ or } r > 30.5\}$

21. $\left\{ 1, \frac{1}{5} \right\}$

23. $\left\{ z \mid z > -\frac{8}{3} \right\}$

25. $|4x + 7| = 2x + 3$; $x = -2$, $x = -\frac{5}{3}$; The absolute value equation is valid when $2x + 3 \ge 0$, so the equation is valid when $x \ge -\frac{3}{2}$. Since neither value of x is greater than or equal to $-\frac{3}{2}$, both solutions are extraneous.

27a. $|x - 36| \le 0.125$; Sample answer: The inequality shows that the length of the lumber x could be as much as 0.125 inch greater than 36 inches or 0.125 inch less than 36 inches.

27b. $\{x \mid 35.875 \le x \le 36.125\}$; The length of the lumber can range from 35.875 inches to 36.125 inches.

29. Yuki; Sample answer: Yuki is correct because if $|a| = |b|$, then either $a = b$ or $a = -b$. They will get the same answers because $a = -b$ and $b = -a$ and $a = b$ and $-a = -b$ are equivalent equations.

31. $\left\{ p \mid -\frac{9}{4} < p < \frac{5}{4} \right\}$

33. $\left\{ w \mid w \le -\frac{23}{2} \text{ or } w \ge \frac{7}{2} \right\}$

35. $\varnothing$

37. $40 < |200 - 32t| < 88$; $3.5 < t < 5$ or $7.5 < t < 9$; The speed is between 40 and 88 feet per second in the intervals from 3.5 to 5 seconds going up and from 7.5 to 9 seconds coming down.

39. Roberto is correct. Sample answer: The solution set for each inequality is all real numbers. For any value of c (positive, negative, or zero), each inequality will be true.

41. $x > 5$ and $x < 1$; Sample answer: Each of these has a non-empty solution set except for $x > 5$ and $x < 1$. There are no values of x that are simultaneously greater than 5 and less than 1.

43. The 4 potential solutions are:

1. $(2x - 1) \geq 0$ and $(5 - x) \geq 0$
2. $(2x - 1) \geq 0$ and $(5 - x) < 0$
3. $(2x - 1) < 0$ and $(5 - x) \geq 0$
4. $(2x - 1) < 0$ and $(5 - x) < 0$

The resulting equations corresponding to these cases are:

1. $2x - 1 + 3 = 5 - x : x = 1$
2. $2x - 1 + 3 = x - 5 : x = -7$
3. $1 - 2x + 3 = 5 - x : x = -1$
4. $1 - 2x + 3 = x - 5 : x = 3$

The solutions from case 1 and case 3 work. The others are extraneous. The solution set is $\{-1, 1\}$.

45. Always; if $|x| < 3$, then x is between -3 or 3. Adding 3 to the absolute value of any of the numbers in this set will produce a positive number.

47. Sample answer: $\left| x - \dfrac{a + b}{2} \right| \leq b - \dfrac{a + b}{2}$

Lesson 9-3

1. $7x + 5y = -35; A = 7, B = 5, C = -35$

3. $3x - 10y = -5; A = 3, B = -10, C = -5$

5. $5x + 32y = 160; A = 5, B = 32, C = 160$

7. $y = -2x + 4; m = -2, b = 4$

9. $y = -4x + 12; m = -4, b = 12$

11. $y = -\dfrac{2}{3}x + \dfrac{5}{3}; m = -\dfrac{2}{3}, b = \dfrac{5}{3}$

13. $y = 20x + 83$; There were 83 shirts collected before noon. There were 20 shirts collected each hour after noon.

15a. Let x represent the number of hours the plumber spends working at a job site, and y represent the total cost for the services.

15b. slope: 42; y-intercept: 65; $y = 42x + 65$

15c. $275

17. $y + 8 = -5(x + 3)$

19. $y + 8 = -\dfrac{2}{3}(x - 6)$

21. $y + 3 = -8(x - 2)$ or $y - 5 = -8(x - 1)$

23. $y + 2 = -\dfrac{3}{2}(x + 1)$ or $y - 1 = -\dfrac{3}{2}(x + 3)$

25. $y - 5.919 = 0.856(x - 1)$ or $y - 11.055 = 0.856(x - 7)$

27. $2x + y = 5; y = -2x + 5;$ $y + 7 = -2(x - 6)$

29. $x - y = 5; y = x - 5; y + 1 = 1(x - 4)$ or $y - 3 = 1(x - 8)$

31. $y = -0.25x + 648$; Sample answer: I assumed that the water level continues to drop at a constant rate.

33. $16x - 19y = -41$

35a. Sample answer: $(8, -9)$; I used the given two points to write an equation. Then, I substituted 8 for x and solved for y.

35b. Sample answer: The equations are equivalent when simplified.

37. Never; sample answer: The graph of $x = a$ is a vertical line.

39. Sample answer: $y - 0 = 2(x - 3)$

41. No; Sample answer: You can choose points on the graph and show on a coordinate plane that they do not fall on a single line. For instance, the points (0, 2), (1, 10), (2, 24), and (3, 49) do not lie on a straight line.

43. Sample answer: Depending on what information is given and what the problem is, it might be easier to represent a linear equation in one form over another. For example, if you are given the slope and the y-intercept, you could represent the equation in slope-intercept form. If you are given a point and the slope, you could represent the equation in point-slope form. If you are trying to graph an equation using the x- and y-intercepts, you could represent the equation in standard form.

Lesson 9-4

1. 1; consistent and independent

3. 1; consistent and independent

5. infinitely many; consistent and dependent

7. (2, 1)

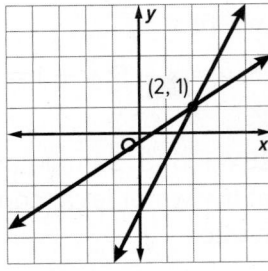

9. (3, −3)

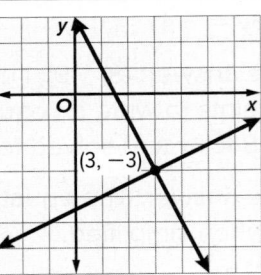

11. no solution

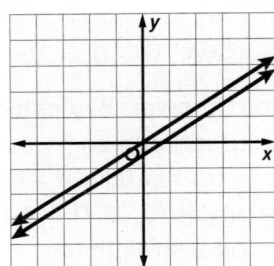

13a. Company A: $y = 24x + 42$;
Company B: $y = 28x + 25$

13b. about 4 containers

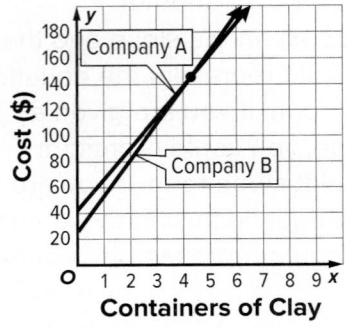

13c. Sample answer: I estimated that the cost would be the same when ordering 4 containers. By substituting $x = 4$ in the equations, the cost at Company A is $138 and the cost at Company B is $137. These values are approximately equal, so the estimate is reasonable.

15. (2.07, −0.39)

17. (15.03, 10.98)

19. 2.76

21. −0.99

23. (2, 1)

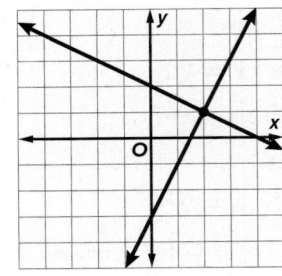

25. (3.78, 5.04)

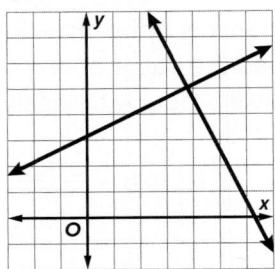

27. Always; Sample answer: a and b are the same line. b is parallel to c, so a is also parallel to c. Since c and d are consistent and independent, then c is not parallel to d and, thus, intersects d. Since a and c are parallel, then a cannot be parallel to d, so, a must intersect d and must be consistent and independent with d.

29. Never; Sample answer: Lines cannot intersect at exactly two distinct points. Lines intersect once (one solution), coincide (infinite solutions), or never intersect (no solution).

Lesson 9-5

1. (3, −3)   **3.** (2, 1)   **5.** no solution
7a. Cassandra: $3x + 14y = 203$;
Alberto: $11x + 11y = 220$; $x = 7, y = 13$
7b. The cost of each small pie is $7. The cost of each large pie is $13.
7c. Sample answer: By substituting the solution into each equation in the system, you can verify that it is correct. $3(7) + 14(13) = 203$, and $11(7) + 11(13) = 220$.

9. $(-2, -5)$ **11.** no solution **13.** $(3, 5)$
15. $(-2, 3)$ **17.** $(1, 6)$ **19.** $(6, -5)$ **21.** 4, 8
23. adult ticket $5.50; student ticket $2.75.
25. Gloria is correct.; Sample answer: Syreeta subtracted 26 from 17 instead of 17 from 26 and got $3x = -9$ instead of $3x = 9$. She proceeded to get a value of -11 for y. She would have found her error if she had substituted the solution into the original equations.
27. Sample answer: It is more helpful to use substitution when one of the variables has a coefficient of 1 or if a coefficient can easily be reduced to 1.

Lesson 9-6

1.

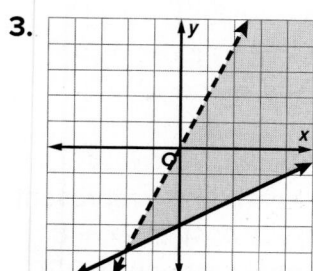

3.

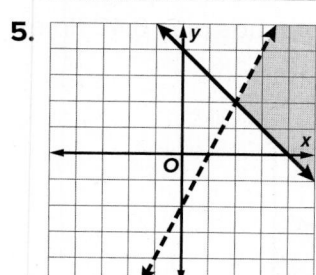

5.

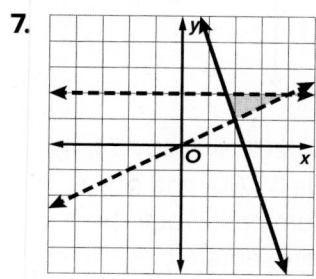

7.

9.

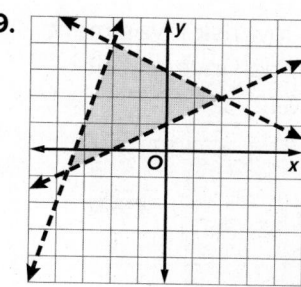

11.

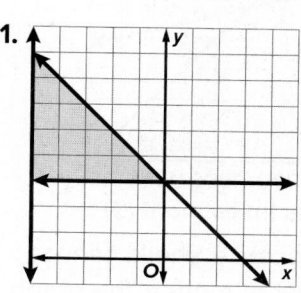

13a. Let x be student tickets and y be adult tickets. $x + y \leq 800$; $4x + 7y \geq 3400$

13b. Quadrant I

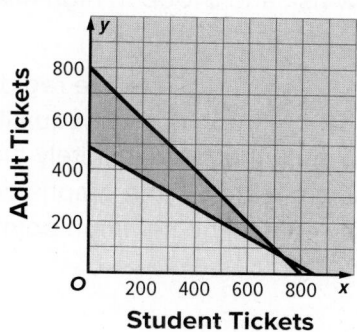

13c. No; they would only make $3300.

15.

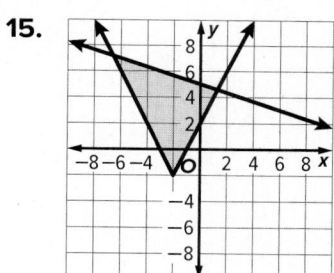

17.

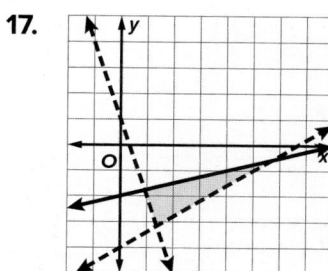

19a. Let x represent the low risk investment and y represent the high risk investment. $x + y \leq 2000$; $0.03x + 0.12y \geq 150$

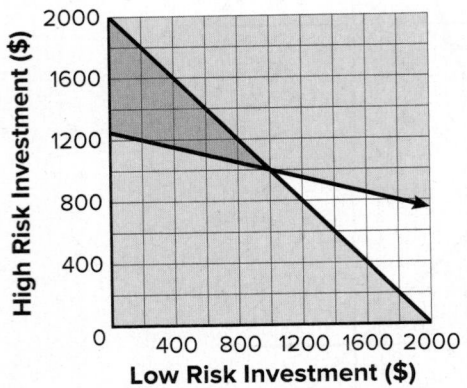

19b. Sample answer: Because Sheila cannot invest a negative amount of money, the graph is limited to positive values of x and y.

19c. Sample answer: $200 in low risk and $1700 in high risk; $400 in low risk and $1600 in high risk; $1000 in low risk and $1000 in high risk

21. 75 units^2

23. True; sample answer: The feasible region is the intersection of the graph of the inequalities. If the graphs intersect, there are infinitely many points in the feasible region. If the graphs do not intersect, they contain no common points and there is no solution.

Lesson 9-7

1. vertices: (1, 2), (1, 4), (5, 8), (5, 2); max: 11; min: −5

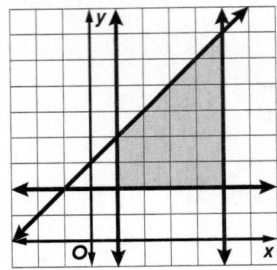

3. vertices: (0, 2), (4, 3), $\left(\frac{7}{3}, -\frac{1}{3}\right)$; max: 25; min: 6

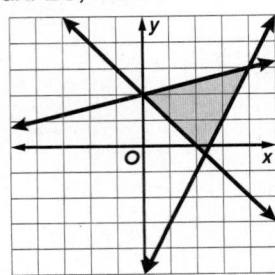

5. vertices: (1, −1), (1, 6), (8, 6); max: 2; min: −5

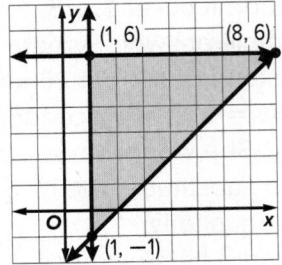

7. vertex: (−1, 7); max: 13; no min.

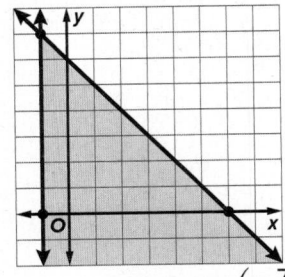

9. vertices: (−2, 0); $\left(-\frac{7}{5}, \frac{9}{5}\right)$;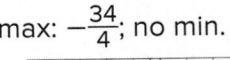

max: $-\frac{34}{4}$; no min.

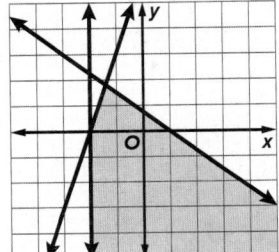

11a. Let x represent clay beads and y represent glass beads.; $0 \leq x \leq 10$; $y \geq 4$; $4y \leq 2x + 8$ $C = 0.20x + 0.40y$; The total cost equals 0.20 times the number of clay beads plus 0.40 times the number of glass beads.

11b. (4, 4), (10, 4), and (10, 7)

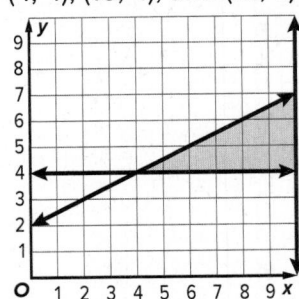

11c. Substitute into $C = 0.20x + 0.40y$: (4, 4) yields $2.40, (10, 4) yields $3.60, and (10, 7) yields $4.80. The minimum cost would be $2.40 at (4, 4), which represents 4 clay beads and 4 glass beads.

13. (5, 2); 19 feet

15. Always; Sample answer: if a point on the unbounded region forms a minimum, then a maximum cannot also be formed because of the unbounded region. There will always be a value in the solution that will produce a higher value than any projected maximum.

17. Sample answer: Even though the region is bounded, multiple maximums occur at A and B and all of the points on the boundary of the feasible region containing both A and B. This happened because that boundary of the region has the same slope as the function.

Lesson 9-8

1. $(4, -3, -1)$ **3.** infinitely many solutions

5. no solution   **7.** $(5, -5, -20)$

9. $(-3, 2, 1)$ **11.** $(2, -1, 3)$

13. 60 grams of Mix A, 50 grams of Mix B, and 40 grams of Mix C

15. 3 oz of apples, 7 oz of raisins, 6 oz of peanuts

17. 9, 6, 3   **19.** fastest press: 1700 papers; slower press: 1000 papers; slowest press: 800 papers

21. orchestra ticket: $10; mezzanine ticket: $8; balcony ticket: $7

23. $a = -3, b = 4, c = -6; y = -3x^2 + 4x - 6$

25. Sample answer:
$$3x + 4y + z = -17$$
$$2x - 5y - 3z = -18$$
$$-x + 3y + 8z = 47$$

$$3x + 4y + z = -17$$
$$3(-5) + 4(-2) + 6 = -17$$
$$-15 + (-8) + 6 = -17$$
$$-17 = -17 \checkmark$$

$$2x - 5y - 3z = -18$$
$$2(-5) - 5(-2) - 3(6) = -18$$
$$-10 + 10 - 18 = -18$$
$$-18 = -18 \checkmark$$

$$-x + 3y + 8z = 47$$
$$-(-5) + 3(-2) + 8(6) = 47$$
$$5 - 6 + 48 = 47$$
$$47 = 47 \checkmark$$

Lesson 9-9

1. $\{-1, 9\}$

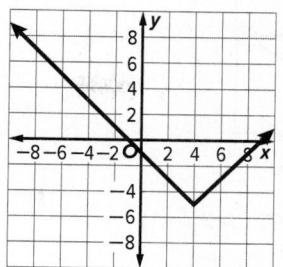

3. $\left\{-\dfrac{1}{2}\right\}$

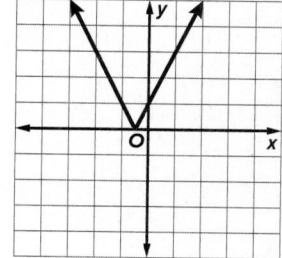

5. $\left\{\dfrac{1}{3}\right\}$

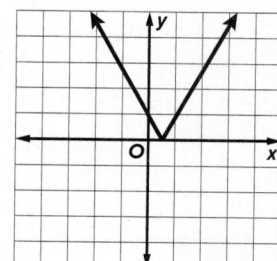

7. $\{-7, 9\}$

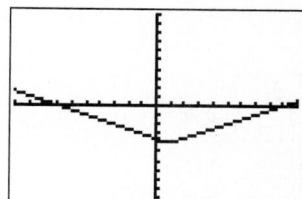

$[-10, 10]$ scl: 1 by $[-10, 10]$ scl: 1

9. $\{-10, 14\}$

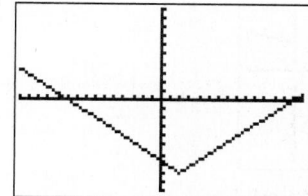

$[-15, 15]$ scl: 1 by $[-10, 10]$ scl: 1

11. $\{-56, 44\}$

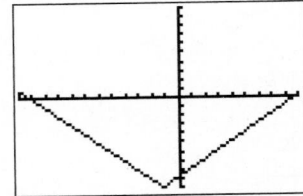

[−60, 45] scl: 5 by [−10, 10] scl: 1

13. $\{-12, 16\}$ **15.** $\varnothing$ **17.** $\{-2, -1\}$

19. $\{x \mid 1 \leq x \leq 5\}$

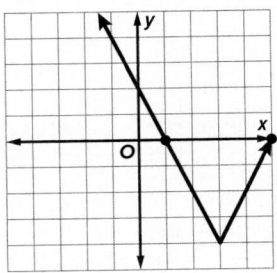

21. $\left\{ x \mid x \leq -\frac{3}{2} \text{ or } x \geq \frac{5}{2} \right\}$

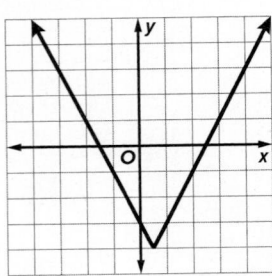

23. $\{x \mid -6 < x < 2\}$

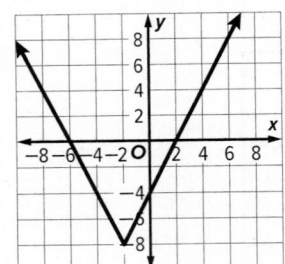

25. $\{x \mid x < -20 \text{ or } x > 12\}$

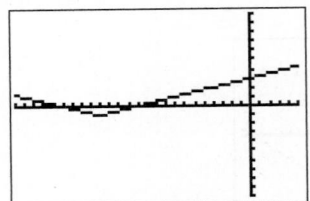

[−25, 5] scl: 2 by [−10, 10] scl: 1

27. $\left\{ x \mid -\frac{1}{3} < x < 1 \right\}$

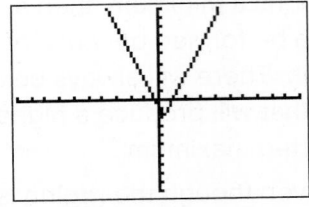

[−10, 10] scl: 1 by [−10, 10] scl: 1

29. $\varnothing$

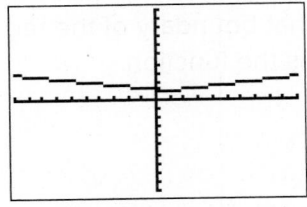

[−10, 10] scl: 1 by [−10, 10] scl: 1

31. $\{0.5, 1.5\}$

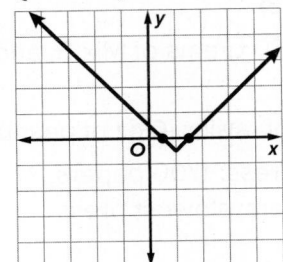

33. $\varnothing$

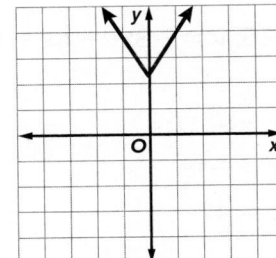

35. $\{x \mid 2 \leq x \leq 4\}$

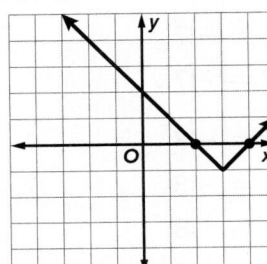

37. $\left\{ x \mid x < -10\frac{4}{5} \text{ or } x > 7\frac{1}{5} \right\}$

39. $\{-85, 95\}$

41. $|x - 1.524| \leq 0.147;\ \{x \mid 1.377 \leq x \leq 1.671\}$

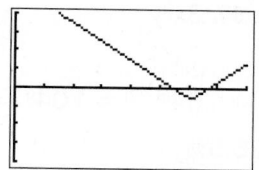

[0, 2] scl: 0.25 by [−1, 1] scl: 0.25

43. Sample answer: The process by which the equation or inequality is set to zero to represent $f(x)$ and making a table of values is the same. However, the graph of an absolute value equation is restricted to having either 0, 1, or 2 solutions; whereas the absolute value inequalities can have infinitely many solutions.

45. Sample answer: $|x - 10| = 1$

Module 9 Review

1. Sample answer: First, use the Distributive Property: $6x + 27 + 2x - 4 = 55$. Combine like terms: $8x + 23 = 55$. Then use the Subtraction Property of Equality: $8x = 32$. Finally, use the Division Property of Equality: $x = 4$.

3. $C = \frac{5}{9}(F - 32)$; 77°F

5. $y = -2x + 10$; -2 represents the miles Allie is running each day. 10 represents that Allie's goal at the beginning of the week is to jog 10 miles.

7. B   **9.** (0.5, 0.75)

11. No; Sample answer: Jazmine would only earn $210.

13. A, D, G

15. $2x + 3y + z = 285$

$x + 5y + 2z = 400$

$3x + 2y + 4z = 440$

Module 10

Quick Check

1. 4^5 **3.** $m^3 p^5$ **5.** 32 **7.** $\frac{1}{16}$

Lesson 10-1

1. $2q^6$ **3.** $9w^8x^{12}$ **5.** $7b^{14}c^8d^6$ **7.** $j^{20}k^{28}$
9. 2^8 or 256 **11.** $4096r^{12}t^6$ **13.** y^3z^3
15. $-15m^{11}$ **17.** $9p^2r^4$ **19.** 1.8×10^5 flowers
21. 242 pounds **23.** 2.9465×10^{12} ml
25. $25x^6y^{10}$ **27.** $64m^{24}n^{12}$ **29.** $9c^{16}$
31. $32{,}000k^{10}m^{19}$ **33.** $800x^8y^{12}z^4$
35. $30x^5y^{11}z^6$ **37.** $0.064h^{15}$ **39.** $\frac{16}{25}a^4$
41. $8m^3p^6$ **43.** $288a^{31}b^{26}c^{30}$ **45.** no **47.** no
49. yes **51a.** $20x^7$ **51b.** The power is one-fourth the previous power: $\frac{1}{4}(20x^7)$, or $5x^7$.
53. $12x$ ft^2 **55a.** If the process of raising a number to an exponent were commutative, then $a^b = b^a$ for all numbers a and b. This is not true, because $2^3 = 8$, whereas $3^2 = 9$.
55b. If the process of raising a number to an exponent were associative, then $(a^b)^c = a^{(b^c)}$ for all numbers a, b, and c. This is not true, because $(4^3)^2 = 64^2 = 4096$, whereas $4^{(3^2)} = 4^9 = 262{,}144$. **55c.** If the process of raising a number to an exponent were to distribute over addition, then $(a + b)^c = a^c + b^c$ for all numbers a, b, and c. This is not true, because $(1 + 2)^3 = 3^3 = 27$, but $1^3 + 2^3 = 1 + 8 = 9$. **55d.** If the process of raising a number to an exponent were to distribute over multiplication, then $(ab)^c = a^cb^c$ for all numbers a, b, and c. This is true. It is a Power of a Product Property.
57. Sample answer: Use the Power of a Power Property to simplify the expression to $\frac{a^{2tm}}{b^{2t^2}}$.
59. Sample answer: $x^4 \cdot x^2$; $x^5 \cdot x$; $(x^3)^2$
61. Jade is correct; Sample answer: Only the h is raised to the power of 6. The exponent of the second g is 1.

Lesson 10-2

1. m^2p **3.** c^2 **5.** $\frac{p^6t^{21}}{1000}$ **7.** $\frac{9n^2p^6}{49q^4}$ **9.** $\frac{81m^{20}r^{12}}{256p^{32}}$
11. k^2mp^2 **13.** $-4x^2yz^3$ **15.** $3d$ **17.** x^2

19. $25^3 = 15{,}625$ **21.** $26^2 = 676$ **23.** $-\frac{w^9}{3}$
25. $\frac{4a^8c^9}{25b^6d^6}$
27. $\frac{64a^{15}b^{30}}{27}$ **29.** $\frac{8x^6}{3y^3}$ **31.** $\frac{8x^6y^3z^{12}}{27}$
33. $\frac{c^{24}}{64d^{12}}$ **35.** 10^7 **37.** $3x^2y$
39a. $n(1.04^t)$
39b. 26.5%; $\frac{n(1.04^8)}{n(1.04^2)} = 1.04^{(8-2)} = 1.04^6 \approx 1.2653$;
$126.5\% - 100\% = 26.5\%$
41. 8
43. No; sample answer: The denominator of both expressions is y^6, but the numerators are different. The first numerator is x^3 and the second is x^2.
45. C
47. a^x
49a. Sample answer: $h = 3$ and $k = 1$
49b. Yes; sample answer: As long as $h - k = 2$, then any numbers would work.
51. Sometimes; sample answer: The equation is true when $x = 0$ or 1, or when $y = 2$ and $z = 2$, but it is false in all other cases.
53. Sample answer: The Quotient of Powers Property is used when dividing two powers with the same base. The exponents are subtracted. The Power of a Quotient Property is used to find the power of a quotient. You find the power of the numerator and the power of the denominator.
55. $\frac{(-3)^8}{(-2)^4}$; The numerator and denominator do not have the same base.

Lesson 10-3

1. $\frac{r^2}{n^9}$ **3.** $\frac{1}{f^{11}}$ **5.** $\frac{g^4h^2}{f^5}$ **7.** $-\frac{3}{u^4}$ **9.** $\frac{5m^6y^3r^5}{7}$
11. $\frac{r^4p^2}{4m^3t^4}$ **13.** $\frac{y^9z^2}{x^4}$ **15.** $\frac{p^4r^2}{t^3}$ **17.** $\frac{-f}{4}$
19. 5 **21.** 6 **23.** 1
25. $1600k^{13}$ **27.** $\frac{5q}{r^6t^3}$ **29.** $\frac{4g^{12}}{h^4}$ **31.** $\frac{4x^8y^4}{z^6}$
33. $\frac{16z^2}{y^8}$ **35.** 3 **37a.** 12,500
37b. 6.25 petabytes
37c. $\frac{1}{10^9}$
39. No; sample answer: The exponent does not affect the coefficient in $4n^{-5}$, so the two expressions simplify to $4n^5$ and $\frac{4}{n^5}$.

41. $\frac{1}{p^{-3}} = \frac{1}{(p^3)^{-1}} = \frac{1}{\left(\frac{1}{p^3}\right)} = 1 \cdot \frac{p^3}{1} = p^3$

43. 750

45a. $\left(\frac{4a^3}{2a^{-2}}\right)^4 = (2a^5)^4 = 16a^{20}$

45b. $\left(\frac{4a^3}{2a^{-2}}\right)^4 = \frac{(4a^3)^4}{(2a^{-2})^4} = \frac{256a^{12}}{16a^{-8}} = 16a^{20}$

45C. Sample answer: When simplifying monomials, the order of applying the Quotient of Powers Property and Power of a Quotient Property does not matter.

47a. Both; sample answer: Colleen applied the Power of a Power Property first, while Tyler applied the Quotient of a Power Property first. Their answers are equivalent.

47b. Sample answer: Colleen's answer is in simplest form. Tyler's answer uses negative exponents, so it is not considered simplest form. **49.** Sample answer: Lola is correct. If we are asked to simplify $\frac{a^7}{a^2}$, instead of using the Division Property of Exponents ($a^{7-2} = a^5$), we could rewrite the fraction as a^7a^{-2}, and then use the Multiplication Property of Exponents to get $a^{7+(-2)} = a^5$. **51.** Sample answer: Students measure the width of their classroom and the width of a pencil, then write and solve a question such as: How many times wider is the classroom than the pencil?

53. Each power is 1 less than the previous power. Each value is half the previous value. So, each time 1 is subtracted from the power, divide the value by 2. Since $2^1 = 2$, then 2^{1-1}, or $2^0 = 2 \div 2$, or 1.

Lesson 10-4

1. $\sqrt{15}$ **3.** $4\sqrt{k}$ **5.** $26^{\frac{1}{2}}$ **7.** $2(ab)^{\frac{1}{2}}$ **9.** $\frac{1}{2}$ **11.** 9
13. 4 **15.** $\frac{2}{5}$ **17.** 7 **19.** $\frac{1}{3}$ **21.** 243 **23.** 625
25. $\frac{27}{1000}$ **27.** 96 ft/s
29. 9 astronomical units **31.** \$51,000 **33.** $\sqrt[3]{17}$
35. $7\sqrt[3]{b}$ **37.** $29^{\frac{1}{3}}$ **39.** $2a^{\frac{1}{3}}$ **41.** 0.3 **43.** a
45. 16 **47.** $\frac{1}{3}$ **49.** $\frac{1}{27}$ **51.** $\frac{1}{\sqrt{k}}$ **53a.** 107
53b. 236 **55.** 4.1 m **57.** Sample answer: The expressions in a, c, and d are equivalent; The expression in b is not equivalent to the others. Let $p = 4$, $f = 3$, and $k = 2$, then $p^{\frac{f}{k}} = 8$, $p^{\frac{f}{5}} \approx 2.30$, $\sqrt[k]{p^f} = 8$, and $(\sqrt[k]{p})^f = 8$.
59. 112π cm² **61.** Sample answer: $2^{\frac{1}{2}}$ and

$4^{\frac{1}{4}}$ **63.** −1, 0, 1 **65.** Sample answer: Both Sachi and Makayla are correct. Sachi addressed the numerator of the rational exponent first and then the denominator. Makayla addressed the denominator of the rational exponent first and then the numerator. Both methods will lead to the correct answer, 9.

Lesson 10-5

1. $2\sqrt{13}$ **3.** $6\sqrt{2}$ **5.** $9\sqrt{3}$ **7.** $5\sqrt{2}$ **9.** $2\sqrt{5}$
11. $4\sqrt{6}$ **13.** $\frac{5\sqrt{3}}{7}$ **15.** $\frac{2\sqrt{2}}{3}$ **17.** $\frac{2\sqrt{21}}{11}$ **19.** $4b^2$
21. $9a^6d^2$ **23.** $2a^2b\sqrt{7b}$ **25.** $40\sqrt{3}$ ft/s
27. 0.9 second **29.** 2 **31.** 6 **33.** 3 **35.** $3\sqrt[3]{9}$
37. $3\sqrt[3]{6}$ **39.** $2\sqrt[3]{10}$ **41.** $2\sqrt[3]{12}$ **43.** $2\sqrt[3]{9}$ **45.** 3
47. $\frac{3\sqrt[3]{2}}{5}$ **49.** $\frac{\sqrt[3]{9}}{4}$ **51.** 4 **53.** $3b^2\sqrt[3]{2b^2}$ **55.** $\frac{4}{7x}$
57. $3x^2y^4\sqrt[4]{10x}$ **59.** $7m^4n^2\sqrt{5mn}$ **61.** $\frac{4df^4\sqrt{2d}}{5e^2}$
63. $4x^4y^5z^6\sqrt[3]{5x^2y^2z^2}$ **65.** $\frac{2y^3\sqrt{11y}}{x}$ **67.** $18\sqrt{3}$ m²
69. $8\sqrt{7}$ **71.** 27.5 in.

73a. The Product Property of Square Roots states that the square root of a product is equal to the product of the square roots of the factors. For this property, $a \geq 0$ and $b \geq 0$.
73b. The Quotient Property of Square Roots states that the square root of a quotient is equal to the quotient of the square roots of the numerator and denominator. For this property, $a \geq 0$ and $b > 0$. **73c.** Both properties state that finding a square root can be done before or after certain other operations.
75. Sample answer: $\sqrt{5b} \cdot \sqrt{15b^4} = 5b^2\sqrt{3b}$
77. Sample answer: 0.25. Any number between 0 and 1 will have a square root larger than itself.
79. Sample answer: Because $4\sqrt{12} \times 6\sqrt{15} = 144\sqrt{5}$, $4\sqrt{12}$ in. is a possible length and $6\sqrt{15}$ in. is a possible width.

Lesson 10-6

1. $5\sqrt{7}$ **3.** $11\sqrt{15}$ **5.** $3\sqrt{r}$ **7.** $-\sqrt{5}$
9. $-3\sqrt{13} + 5\sqrt{2}$ **11.** $12\sqrt{3} + \sqrt{2}$
13. $115\sqrt{149}$ m **15.** $26\sqrt{37} + 10\sqrt{41}$ in.
17. $\frac{20 - 10\sqrt{2}}{\sqrt{\pi}}$ cm or $\frac{20\sqrt{\pi} - 10\sqrt{2\pi}}{\pi}$ cm
19. $18\sqrt{5}$ **21.** $24\sqrt{14}$ **23.** $198\sqrt{2}$
25. $4 + 2\sqrt{3}$ **27.** $5\sqrt{10}$ **29.** $60 + 32\sqrt{10}$

31. $-\dfrac{4\sqrt{5}}{5}$ **33.** $\sqrt{2}$ **35.** $2\sqrt{6} - 13\sqrt{3} + 4\sqrt{2}$
37. $15\sqrt{15}$ **39.** $\dfrac{3\sqrt{2}}{2} + 2\sqrt{15}$
41. $2ac\sqrt{6} + ad\sqrt{15} + 2bc\sqrt{10} + 5bd$
43. $63\sqrt{15}$ ft^2 **45a.** $2\sqrt{2}$ s   **45b.** about 2.8 s
47. $11.25\sqrt{2}$ ft   **49.** $\dfrac{5}{6}\sqrt{5}$ in.
51. True; $(x + y)^2 > \left(\sqrt{x^2 + y^2}\right)^2 \rightarrow x^2 +$
$2xy + y^2 > x^2 + y^2 \rightarrow 2xy > 0$; Because
$x > 0$ and $y > 0$, $2xy > 0$ is always true.
So, $(x + y)^2 > \left(\sqrt{x^2 + y^2}\right)^2$ is always true for all
$x > 0$ and $y > 0$.
53. Sample answer: $\sqrt{12} + \sqrt{27} = 5\sqrt{3}$;
Simplify $\sqrt{12}$ to get $2\sqrt{3}$. Simplify $\sqrt{27}$ to get
$3\sqrt{3}$.
Because $2\sqrt{3}$ and $3\sqrt{3}$ have the same
radicand, they can be combined.
55. $6\sqrt{12}$; Sample answer: $\sqrt{12}$ is the only
radical that does not simplify to $k\sqrt{5}$.

Lesson 10-7

1. 9   **3.** 6   **5.** 8   **7.** 8   **9.** 5   **11.** 2   **13.** 16 ohms
15. 20 months   **17.** 3   **19.** $-\dfrac{1}{4}$ **21.** 11   **23.** 2
25. 2   **27.** 37.52 ft   **29.** Use 4 as the common
base. Replace 8 with $4^{\frac{3}{2}}$ and 16 with 4^2 because
$4^{\frac{3}{2}} = (\sqrt{4})^3$, or 8 and $4^2 = 16$. So, $4^{\frac{3}{2}(2 + x)} =$
$4^{2(2.5 - 0.5x)}$. Distribute and get $4^{(3 + 1.5x)} = 4^{(5 - x)}$.
Set the exponents equal to each other: $3 +$
$1.5x = 5 - x$. Solve for x: $x = 0.8$, or $\dfrac{4}{5}$.
31. Sample answer: There is no way to make a
common base between 2 and 5.   **33.** $x = -6$
and $x = 2$ **35.** $3^{2x + 1} = 243$; This equation has
a solution of $x = 2$. The other three equations
have a solution of $x = 3$.

Module 10 Review

1. A, B   **3.** B   **5.** 9   **7.** C   **9.** 0.000001
11. Sample answer: Because $b^{\frac{1}{n}} = \sqrt[n]{b}$, rewrite
$729^{\frac{1}{6}}$ as $\sqrt[6]{729}$. Because $729 = 3 \cdot 3 \cdot 3 \cdot 3 \cdot 3 \cdot$
3, you can write $\sqrt[6]{729}$ as $\sqrt[6]{3 \cdot 3 \cdot 3 \cdot 3 \cdot 3 \cdot 3}$,
which equals 3.   **13.** B   **15.** 27
17. A   **19.** $2\sqrt{5} + 6\sqrt{7}$ **21.** B   **23.** 6

Module 11

Quick Check

1. $a^2 + 5a$ **3.** $n^2 - 3n^3 + 2n$ **5.** $13x$
7. simplified

Lesson 11-1

1. no   **3.** yes; 4; trinomial   **5.** yes; 2; binomial
7. $5x^2 + 3x - 2$; 5   **9.** $-5c^2 - 3c + 4$; -5
11. $t^5 + 2t^2 + 11t - 3$; 1   **13.** $-3x^4 + \frac{1}{2}x + 7$; -3
15. $6x + 12y$ **17.** $3a + 5b$ **19.** $2m^2 + m$
21. $d^2 - 3d$ **23.** $3f + g + 1$ **25.** $7c^2 + 6c - 3$
27. $3c^3 - c^2 - 3c + 3$ **29.** $-2x - 5y + 1$
31. $-x^2y - 3x^2 + 4y$ **33.** $-6p^2 + 2np + n$
35a. $D = 0.5x + 2$ **35b.** \$7,000,000
37. quadratic trinomial   **39.** quartic binomial
41. quintic polynomial   **43.** $9x + 4y - 17z$
45. $2c^2 - c + 8$ **47a.** $4\pi rh + 4\pi r^2$
47b. 15.7 m²   **49.** Both $0x^2 + 0x$ and 0 are
polynomials.   **51a.** $-13x^2 - x + 10$;
$(5x^2 - 3x + 7) - (18x^2 - 2x - 3) = (5x^2 - 18x^2)$
$+ (-3x + 2x) + (7 + 3)$ **51b.** $13x^2 + x - 10$;
$(18x^2 - 2x - 3) - (5x^2 - 3x + 7) = (18x^2 - 5x^2)$
$+ (-2x + 3x) + (-3 - 7)$ **51c.** The second
result is the negative of the first. Reversing the
order of subtraction with integers yields the
negative of the original difference. For a and b
integers, $(a - b) = -(b - a)$.   **53.** No; neither
of them found the additive inverse correctly.
All terms should have been multiplied by
-1.   **55.** $6n + 9$ **57.** Sample answer: To add
polynomials in a horizontal method, combine
like terms. For the vertical method, write the
polynomials in standard form, align like terms
in columns, and combine like terms. To subtract
polynomials in a horizontal method, find the
additive inverse of the polynomial that is being
subtracted, and then combine like terms. For
the vertical method, write the polynomials in
standard form, align like terms in columns, and
subtract by adding the additive inverse.

Lesson 11-2

1. $b^3 - 12b^2 + b$
3. $-6m^6 + 36m^5 - 6m^4 - 75m^3$

5. $4p^2r^3 + 10p^3r^3 - 30p^2r^2$ **7.** $-13x^2 - 9x - 27$
9. $-20d^3 + 55d + 35$ **11.** $14j^3k^2 + 2j^2k^2 -$
$17jk + 18j^2k^3 - 18k^3$ **13.** $\frac{n^2}{2} + \frac{n}{2}$; 78
15. $3.88 = x(0.39) + (x + 4)(0.29)$; 4 peppers
and 8 potatoes   **17.** 2   **19.** $\frac{43}{6}$ **21.** $\frac{30}{43}$
23. $4a^2 + 3a$ **25.** $2x^2 - 5x$ **27.** $-3n^3 - 6n^2$
29. $15x^3 - 3x^2 + 12x$ **31.** $-4b + 36b^2 + 8b^3$
33. $4m^4 + 6m^3 - 10m^2$ **35.** $3w^2 + 7w$
37. $-2p^2 + 3p$ **39.** $3x^3 + 8x$
41. $-22b^2 + 2b + 8$
43. $-q^3w^3 - 35q^2w^4 + 8q^2w^2 - 27qw$ **45.** -7
47. -5 **49.** -2 **51.** $1.10(2\pi r) = 2\pi(r + 12)$;
$r = 120$ ft   **53.** The Distributive Property has
been misapplied. The second term should be
$12y^2(-2y^2) = -24y^4$, and the fourth term should
be $-3y^3(-2y) = 6y^4$. So, it simplifies to
$24y^3 - 18y^4$.   **55a.** $1.25x(45 - 5x) = 56.25x -$
$6.25x^2$ **55b.** \$5.63   **55c.** \$126.56   **57.** Ted;
Pearl used the Distributive Property incorrectly.
59. $8x^2y^{-2} + 24x^{-10}y^8 - 16x^{-3}$ **61.** Sample
answer: $3n$, $4n + 1$; $12n^2 + 3n$ **63.** Sample
answer: $4t(t - 5) - 2t(3t + 2) + 108 =$
$3t(2t - 5) - t(8t - 3)$

Lesson 11-3

1. $3c^2 + 4c - 15$ **3.** $30a^2 + 43a + 15$
5. $15y^2 - 17y + 4$ **7.** $6m^2 + 19m + 15$
9. $144t^2 - 25$ **11.** $40w^2 - 28wx - 24x^2$
13. $45x^2 - 13x - 2$ **15.** $20x^2 + 18x + 4$ in²
17. $(x - 2)(x + 2) - x^2 = -4$; $x^2 - 2x + 2x -$
$4 - x^2 = -4$; $-4 = -4$
19. $36a^3 + 71a^2 - 14a - 49$
21. $5x^4 + 19x^3 - 34x^2 + 11x - 1$
23. $18z^5 - 15z^4 - 18z^3 - 14z^2 + 24z + 8$
25. $x^2 + 4x + 4$ **27.** $t^2 + t - 12$
29. $n^2 - 4n - 5$ **31.** $2x^2 - 18$
33. $2\ell^2 - 3\ell - 20$ **35.** $5q^2 + 24q - 5$
37. $6m^2 - 6$ **39.** $10a^2 - 19a + 6$
41. $2x^2 - 3xy + y^2$ **43.** $t^3 + 3t^2 + 6t + 4$
45. $m^3 + 6m^2 + 14m + 15$
47. $6b^3 + 5b^2 + 8b + 16$ **49.** $5t^2 - 34t + 56$
51. $9c^2 + 24cd + 16d^2$
53. $8r^3 - 36r^2t + 42rt^2 - 27t^3$
55. $64y^3 - 48y^2z - 36yz^2 + 27z^3$

57a. Let w represent the width of the mural;
$w(w + 5)$ ft²   **57b.** $(w + 9)(w + 4) - (w + 5)w$ ft²
57c. $(w + 9)(w + 4) - (w + 5)w = 100$; 8 ft by 13 ft

59a. $x^{4p+1}(x^{1-2p})^{2p+3} = x^{4p+1+(1-2p)(2p+3)} =$
$x^{4p+1+(2p+3)-2p(2p+3)} = x^{6p+4-2p(2p)-2p(3)} =$
$x^{6p+4-4p^2-6p} = x^{-4p^2+4}$ **59b.** $x^0 = 1$ for all
values of x, so find p such that $-4p^2 + 4 = 0$.
So, $p = \pm 1$. **61.** $8x^3 + 28x^2 - 16x$ units^3;
Volume $= 4x(2x - 1)(x + 4) =$
$4x[2x(x + 4) - 1(x + 4)] = 4x[2x^2 + 8x -$
$x - 4] = 4x(2x^2 + 7x - 4) = 4x(2x^2) + 4x(7x) +$
$4x(-4) = 8x^3 + 28x^2 - 16x$ **63.** $x^{2m-1} -$
$x^{m-p+1} + x^{m+p} + x^{m+p-1} - x + x^{2p}$
65. The three monomials that make up a
trinomial are similar to the three digits that
make up the 3-digit number. The single
monomial is similar to a 1-digit number. With
each procedure, you perform 3 multiplications.
The difference is that polynomial multiplication
involves variables and the resulting product
is often the sum of two or more monomials,
while numerical multiplication results in a single
number. **67.** The expression, $(2x - 1)(x^2 +$
$x - 1)$, does not belong with the other three
polynomial expressions because the product
of this expression has 4 terms, and the products
of the other three polynomial expressions have
3 terms.

Lesson 11-4

1. $a^2 + 10a + 100$ **3.** $h^2 + 14h + 49$
5. $64 - 16m + m^2$ **7.** $4b^2 + 12b + 9$
9. $64h^2 - 64hn + 16n^2$ **11.** $A = \pi r^2 - \pi(r - 18)^2$
13. $B^2 + 2BR + R^2$ **15.** $u^2 - 9$ **17.** $4 - x^2$
19. $4q^2 - 25r^2$ **21.** $n^2 + 6n + 9$
23. $y^2 - 14y + 49$ **25.** $b^2 - 1$ **27.** $p^2 - 8p + 16$
29. $\ell^2 + 4\ell + 4$ **31.** $9g^2 - 4$ **33.** $36 + 12u + u^2$
35. $9q^2 - 1$ **37.** $4k^2 - 8k + 4$ **39.** $9p^2 - 16$
41. $x^2 - 8xy + 16y^2$ **43.** $9y^2 - 9g^2$
45. $4k^2 + 4km^2 + m^4$ **47.** Sample answer:
The area of the rectangle is the square of the
larger number minus the square of the smaller
number. **49a.** $0 = r^2 - 6rs + s^2$
49b. $0 = s^2 - 120s + 400$ **51.** $25y^2 + 70y + 49$
53. $100x^2 - 4$ **55.** $a^2 + 8ab + 16b^2$
57. $4c^2 - 36cd + 81d^2$ **59.** $36y^2 - 169$
61. $25x^4 - 10x^2y^2 + y^4$ **63.** $\frac{9}{16}k^2 + 12k + 64$
65. $49z^4 - 25y^4$ **67.** $r^4 - 29r^2 + 100$
69. $8a^3 - 12a^2b + 6ab^2 - b^3$
71. $k^3 - k^2m - km^2 + m^3$

73. $q^3 - 3q^2r + 3qr^2 - r^3$
75a. $2s + 1$; $(s + 1)^2 - s^2 = (s^2 + 2(s)(1) + 1^2) - s^2$
$= 2s + 1$ **75b.** $2s + 3$; $(s + 2)^2 - (s + 1)^2 =$
$(s^2 + 2(s)(2) + 2^2) - (s^2 + 2s + 1) = s^2 + 4s + 4 -$
$s^2 - 2s - 1 = 2s + 3$ **75c.** $2s + 5$; $(s + 3)^2 -$
$(s + 2)^2 = (s^2 + 2(s)(3) + 3^2) - (s^2 + 4s + 4) =$
$s^2 + 6s + 9 - s^2 - 4s - 4 = 2s + 5$
77a. $a^3 + 3a^2b + 3ab^2 + b^3$
77b. $x^3 + 6x^2 + 12x + 8$
77c. Sample answer:

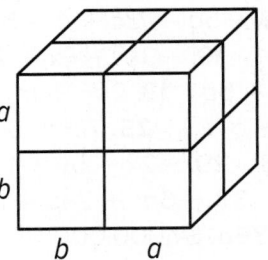

77d. $a^3 - 3a^2b + 3ab^2 - b^3$ **79.** Sample
answer: $(x - 2)(x + 2) = x^2 - 4$ and
$(x - 2)(x - 2) = x^2 - 4x + 4$

Lesson 11-5

1. $8(2t - 5y)$ **3.** $2k(k + 2)$ **5.** $2ab(2ab + a - 5b)$
7. $16t(2 - t)$ **9.** $6t(27 - 32t)$ **11.** $(g + 4)(f - 5)$
13. $(h + 5)(j - 2)$ **15.** $(9q - 10)(5p - 3)$
17. $(3d - 5)(t - 7)$ **19.** $(3t - 5)(7h - 1)$
21. $(r - 5)(5b + 2)$ **23.** $(b + 3)(b - 2)$
25. $(2a + 1)(a - 2)$ **27.** $2(4m - 3)$
29. $5q(2 - 5q)$ **31.** $x(1 + xy + x^2y^2)$
33. $4a(ab^2 + 4b + 3)$ **35.** $12ab(4ab - 1)$
37. $(x + 1)(x + 3)$ **39.** $(3n - p)(n + 2p)$
41. $(3x + 2)(3x - y)$ **43.** $(2x + 3)(x - 0.3)$
45. In both cases you are trying to build
an expression equivalent to the original
expression that has two or more factors.

47. Both are correct. Akash's grouping: $3x^2 +$
$7x - 18x - 42 = (3x^2 + 7x) - (18x + 42) =$
$x(3x + 7) - 6(3x + 7) = (x - 6)(3x + 7)$.
Theresa's grouping: $3x^2 - 18x + 7x - 42 = (3x^2$
$- 18x) + (7x - 42) = 3x(x - 6) + 7(x - 6) =$
$(3x + 7)(x - 6)$. **49.** Find the GCF of $12a^2b^2$
and $-16a^2b^3$, which is $4a^2b^2$. Write each term
as the product of the GCF and the remaining
factors using the Distributive Property:
$4a^2b^2(3) + 4a^2b^2(-4b) = 4a^2b^2(3 - 4b)$.

Lesson 11-6

1. $(x + 3)(x + 14)$ **3.** $(a - 4)(a + 12)$
5. $(h + 4)(h + 11)$ **7.** $(x + 3)(x - 8)$
9. $(t + 2)(t + 6)$ **11.** prime **13.** prime
15. $(h + 6)(h + 3)$ **17.** $(x + 14)(x + 12)$
19. $(d - 18)(d - 18)$ **21.** $(5x + 4)(x + 6)$
23. $2(2x + 1)(x + 5)$ **25.** $(2x + 3)(x - 3)$
27. prime **29.** $3(4x + 3)(x + 5)$
31. prime **33.** prime **35.** $(2y - 1)(y + 1)$
37. $(n + 6)(n - 3)$ **39.** $(r + 6)(r - 2)$
41. $(w - 3)(w + 2)$ **43.** $(t - 8)(t - 7)$
45. $(2x + 1)(x + 2)$ **47.** $(3g - 1)(g - 2)$
49. prime **51.** $(3p - 2)(3p + 4)$
53. $(a - 3)(a - 7)$ **55.** $(2x + 1)(x + 3)$
57. $(x - 4)(x + 5)$ **59.** $(p - 3)(p - 7)$
61. $(q + 2r)(q + 9r)$ **63.** $(x - y)(x - 5y)$
65. $-(2x + 5)(3x + 4)$ **67.** $-(x - 4)(5x + 2)$
69. prime **71a.** $3(x + 5)(x - 10)$ **71b.** possible
lengths of the pyramidal stone monument
73. $x^2 + 12x + 24$; The other 3 expressions
can be factored, but this expression cannot be
factored. **75.** $-18, 18$ **77.** 7, 12, 15, 16
79. Sometimes; sample answer: The trinomial
$x^2 + 10x + 9 = (x + 1)(x + 9)$ and $10 > 9$. The
trinomial $x^2 + 7x + 10 = (x + 2)(x + 5)$ and
$7 < 10$. **81.** $(4y - 5)^2 + 3(4y - 5) - 70 =$
$[(4y - 5) + 10][(4y - 5) - 7] = (4y + 5)(4y - 12) =$
$4(4y + 5)(y - 3)$
83. $(12x + 20y)$ in.; The area of the square
equals $(3x + 5y)(3x + 5y)$ in^2, so the length of
one side is $(3x + 5y)$ in. The perimeter is
$4(3x + 5y)$ or $(12x + 20y)$ in.
85. Sample answer: Find two numbers, m and
p, with a product of ac and a sum of b.

Lesson 11-7

1. $(q + 11)(q - 11)$ **3.** $(w^2 + 25)(w + 5)(w - 5)$
5. $(h^2 + 16)(h + 4)(h - 4)$ **7.** $(x + 2y)(x - 2y)$
9. $(f + 8)(f - 8)(f + 2)$ **11.** $(t + 1)(t - 1)(3t - 7)$
13. $(m + 3)(m - 3)(4m + 9)$
15. $(3a + 2b)(3a - 2b)$ **17a.** $(x - 8)(x - 8)$
17b. the length and width of the rug

19. yes; $(4x - 7)^2$ **21.** no **23.** $4(h^2 - 14)$
25. prime **27.** $6(d + 4)(d - 4)$
29. $(-4 + p)(4 + p)$ **31.** $(6 + 10w)(6 - 10w)$
33. $(2h + 5g)(2h - 5g)$ **35.** $8(x + 3p)(x - 3p)$
37. $2(4a + 5b)(4a - 5b)$ **39.** $(7x + 8y)(7x - 8y)$
41. $2m(2m - 3)^2$ **43.** $(a - 5)(a + 5)$; 95 yards
45. yes; $(m - 3)^2$ **47.** yes; $(g - 7)^2$
49. yes; $(2d - 1)^2$ **51.** yes; $(3z - 1)^2$ **53.** no
55. yes; $(t - 9)^2$ **57.** $\frac{1}{2}(x + 14)(x - 14)$
59. $x^4 - 81$ can be factored by using the
difference of squares twice. First $(x^2 - 9)(x^2 + 9)$, then $(x + 3)(x - 3)(x^2 + 9)$. $x^2 - 9$ can
be factored using the difference of squares:
$(x^2 - 9) = (x + 3)(x - 3)$. The factors of $x^4 - 81$
include the factors of $x^2 - 9$, but there is an
additional factor of $x^2 + 9$.
61. $m = 121$; Let $n^2 = m$. Then $2(2x^2)(n) = -44x^2$,
so $n = -11$. Because $n^2 = m$, $(-11)^2 = 121$, so
$m = 121$. $4x^4 - 44x^2 + 121 = (2x^2 - 11)(2x^2 - 11)$
63. $[3 + (k + 3)][3 - (k + 3)] = (k + 6)(-k) =$
$-k^2 - 6k$ **65.** false; $a^2 + b^2$

67. Sample answer: When the difference
of squares is multiplied together using the
FOIL method, the outer and inner terms are
opposites of each other. When these terms are
added together, the sum is zero.

69. Determine if the first and last terms are
perfect squares. Then determine if the middle
term is equal to ± 2 times the product of the
principal square roots of the first and last terms.
If these three criteria are met, the trinomial is a
perfect square trinomial.

Module 11 Review

1. A, C, D **3.** B **5.** 900 square inches **7.** B
9. $2r^3 - 13r^2 + 17r - 3$ **11.** B **13.** Sample
answer: The square of a sum is the square of
the first term plus two times the product of the
first and second terms plus the square of the
second term.

15. $(-5x + 6)(2y + 3)$ **17.** A, B **19.** B
21. yes, yes, yes, no, no **23.** B

Module 12

Quick Check

1. translation right 4 units

3. translation down 1 unit

5. yes; $(a + 6)^2$

7. yes; $(m - 11)^2$

Lesson 12-1

1. axis of symmetry: $x = 0$; vertex: $(0, 1)$; y-intercept: 1; end behavior: As x increases or decreases, y decreases.

3. axis of symmetry: $x = 0$; vertex: $(0, -4)$; y-intercept: -4; end behavior: As x increases or decreases, y increases.

5.
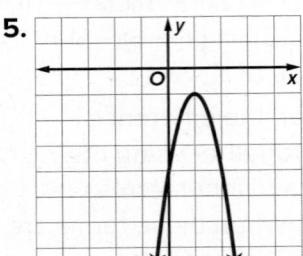

7.
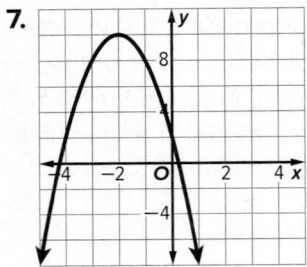

9.
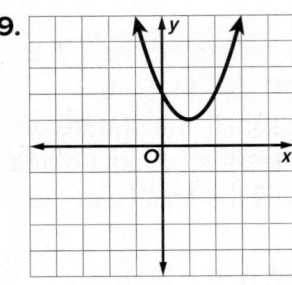

11. D = {all real numbers}; R = $\{y \mid y \geq 2\}$

x	-4	-3	-2	-1	0
y	6	3	2	3	6

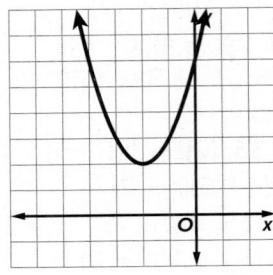

13. D = {all real numbers}; R = $\{y \mid y \geq -13\}$

x	4	3	2	1	0
y	-5	-11	-13	-11	-5

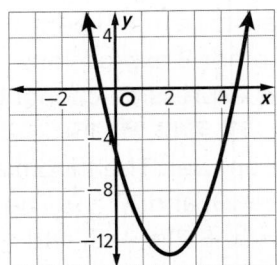

15. D = {all real numbers}; R = $\{y \mid y \geq -5\}$

x	3	2	1	0	-1
y	7	-2	-5	-2	7

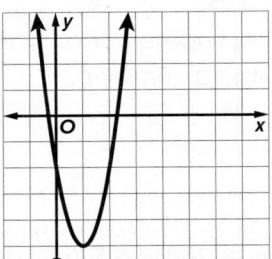

17.

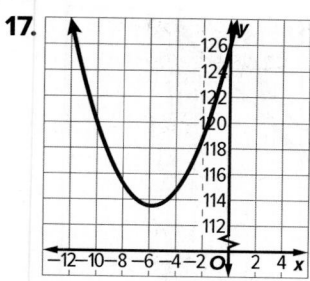

The vertex is located at about $(-5.8, 113.5)$. The vertex does not make sense in this case because there cannot be an Olympiad before the first Olympiad. The axis of symmetry is located at $x = -5.8$. The axis of symmetry does not make sense in this case because there cannot be an Olympiad before the first Olympiad. The y-intercept is located at $(0, 126)$, so the y-intercept is 126. This does not make sense in this case because there cannot be an Olympiad before the first Olympiad. As x increases from the minimum, the value of y increases. This represents that the winning pole vault height increases for each Olympiad. The graph is always positive, which means the winning pole vault height is always greater than 0 inches. The relevant domain for the situation is all nonnegative whole numbers greater than 0. This represents each Olympiad since the Olympic Games started. The relevant range is all numbers greater than or equal to about 130.7. This means that the winning pole vault height is greater than or equal to about 130.7 inches.

19.

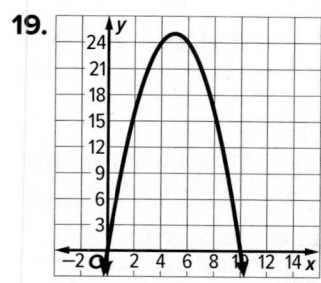

The vertex is located at $(5, 25)$. The vertex represents the arch's height of 25 feet when the distance from one side of the arch is 5 feet. The axis of symmetry is located at $x = 5$. This means that the arch's height from a distance 0 feet to 5 feet away from one side of the arch is the same as the arch's height from a distance

5 feet to 10 feet away from one side of the arch. The y-intercept is located at $(0, 0)$, so the y-intercept is 0. This means that the initial height of the arch is 0 feet when the distance from one side of the arch is 0 feet. As x increases and decreases from the maximum, the value of y decreases. This represents the height of the arch based on the distance from one side of the arch. The zeros of the graph are 0 and 10. This means the bases of the arch are a distance of 0 feet and 10 feet from one side of the arch. The relevant domain for the situation is all numbers greater than or equal to 0 and less than or equal to 10. This represents the distance from one side of the arch to the other side of the arch. The relevant range is all numbers greater than or equal to 0 and less than or equal to 25. This means that the height of the arch is between 0 feet and 25 feet.

21a. $y = (x - 2)(x + 6)$ or $y = x^2 + 4x - 12$

21b. $x = -2$

21c.

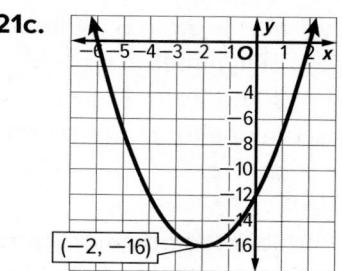

23a. $x = -1$

23b. $(-1, 3)$; maximum

23c.

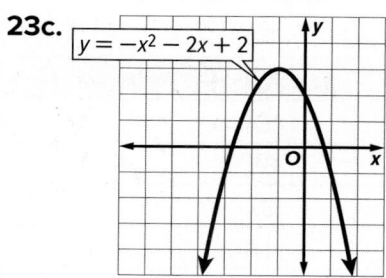

25. axis of symmetry: $x = -2$; vertex: $(-2, 2)$; y-intercept: 6; end behavior: As x increases or decreases, y increases.

27a.

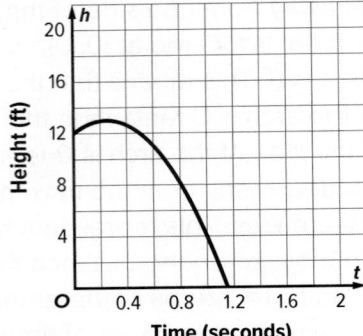

27b. The *y*-intercept is 12. This represents the height, in feet, at which the beanbag is tossed into the air. The *x*-intercept is approximately 1.15. This represents the time, in seconds, at which the beanbag lands on the beach.
27c. The maximum value is 13. The maximum height of the beanbag is 13 feet. It reaches this height 0.25 second after it is tossed into the air.
27d. $h(0.5) = 12$; This means it takes 0.5 second for the beanbag to come back down to the height at which it was tossed. **27e.** The domain is $\{t \mid 0 \leq t \leq 1.15\}$. The domain represents the length of time the beanbag is in the air.
29. Sample answer: $y = 4x^2 + 3x + 5$; Write the equation of the axis of symmetry, $x = -\frac{b}{2a}$. From the equation, $b = 3$ and $2a = 8$, so $a = 4$. Substitute these values for a and b into the equation $y = ax^2 + bx + c$.
31. $y = -x^2 + 6x + 16$
33. Sample answer: A quadratic equation can be used to model the path of a football that is kicked during a game. The vertex represents the maximum height of the ball.
35. Sample answer: Suppose $a = 1$, $b = 2$, and $c = 1$.

x	$f(x) = 2^x + 1$	$g(x) = x^2 + 1$	$h(x) = x + 1$
−10	1.00098	101	−9
−8	1.00391	65	−7
−6	1.01563	37	−5
−4	1.0625	17	−3
−2	1.25	5	−1
0	2	1	1
2	5	5	3
4	17	17	5
6	65	37	7
8	257	65	9
10	1205	101	11

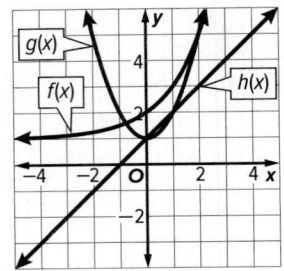

Intercepts: $f(x)$ and $g(x)$ have no *x*-intercepts, $h(x)$ has one at −1 because $c = 1$. $g(x)$ and $h(x)$ have one *y*-intercept at 1 and $f(x)$ has one *y*-intercept at 2. The graphs are all translated up 1 unit from the graphs of the parent functions because $c = 1$.

Increasing/Decreasing: $f(x)$ and $h(x)$ are increasing on the entire domain. $g(x)$ is increasing to the right of the vertex and decreasing to the left.

Positive/Negative: The function values for $f(x)$ and $g(x)$ are all positive. The function values of $h(x)$ are negative for $x < -1$ and positive for $x > 1$.

Maxima/Minima: $f(x)$ and $h(x)$ have no maxima or minima. $g(x)$ has a minimum at (0, 1).

Symmetry: $f(x)$ and $g(x)$ have no symmetry. $g(x)$ is symmetric about the *y*-axis.

End behavior: For $f(x)$ and $h(x)$, as x increases, y increases and as x decreases, y decreases. For $g(x)$ as x increases, y decreases and as x decreases, y increases.

The exponential function $f(x)$ eventually exceeds the others.

Lesson 12-2

1. translated down 10 units **3.** translated right 1 unit **5.** translated left 3 units **7.** translated left 2.5 units **9.** translated up 6 units **11.** translated right 9.5 units **13.** stretched vertically **15.** compressed horizontally **17.** compressed horizontally **19.** compressed vertically **21.** stretched vertically and reflected across the *x*-axis **23.** compressed vertically and reflected across the *x*-axis **25.** stretched vertically and reflected across the *x*-axis **27.** reflected across the *x*-axis and translated down 7 units **29.** compressed vertically and translated up 6 units

31. stretched vertically and translated up 3 units   **33.** The graph of $U_s = 5x^2$ is a vertical stretch of the graph of $U_s = x^2$.   **35a.** $d = 3t^2$ is a vertical stretch of $d = t^2$; $d = 2t^2 + 100$ is a vertical stretch of $d = t^2$ translated up 100 units (feet).   **35b.** after 10 seconds   **37.** $f(x) = 2(x - 3)^2 + 1$

39. compressed vertically, reflected across the x-axis, and translated up 5 units

41. compressed vertically and translated down 1.1 units   **43.** compressed vertically and translated up $\frac{5}{6}$ unit

45. Apply a vertical stretch of 2 to the parent function. Reflect across the x-axis. Translate left 2 units and up 2 units.

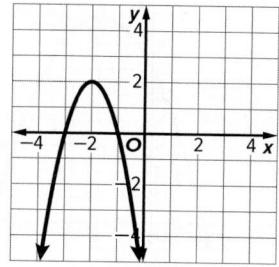

47. Apply a vertical compression of $\frac{1}{4}$ to the parent function. Reflect over the x-axis. Translate right 1 and up 4 units.

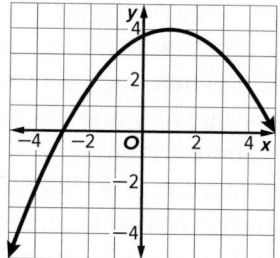

49. D   **51.** A

53. The parent quadratic function is an even function since $f(-x) = (-x)^2 = x^2$, so $f(-x) = f(x)$.
55. $y = 2(x - 2)^2 - 1$ **57.** $y = 0.25x^2 + 1$
59. $y = -0.5(x + 2)^2$ **61a.** Sometimes; sample answer: This occurs only if $k = 0$. For any other value, the graph will be translated up or down.
61b. Always; sample answer: The reflection does not affect the width. Both graphs are dilated by a factor of a.
61c. Never; sample answer: Because $k >= 0$, the graph of $f(x)$ has a minimum point with a y-coordinate greater than or equal to 0. The graph of $g(x)$ has a vertex with a y-coordinate of -3. Even if the graph of $g(x)$ opens up, the graphs can never have the same minimum.   **63.** Sample answer: Not all reflections over the y-axis produce the same graph. If the vertex of the original graph is not

on the y-axis, the graph will not have the y-axis as its axis symmetry, and its reflection across the y-axis will be a different parabola.
65. Sample answer: For $y = ax^2$, the parent graph is stretched vertically if $a > 1$ or compressed vertically if $0 < a < 1$. The y-values in the table will all be multiplied by a factor of a. For $y = x^2 + k$, the parent graph is translated up if k is positive and moved down if k is negative. The y-values in the table will all have the constant k added to them or subtracted from them. For $y = ax^2 + k$, the graph will either be stretched vertically or compressed vertically based upon the value of a and then will be translated up or down depending on the value of k. The y-values in the table will be multiplied by a factor of a and the constant k will be added to them.

Lesson 12-3

1. no solutions

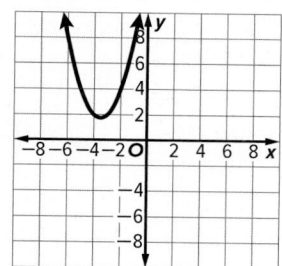

3. -8

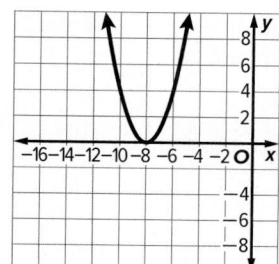

5. -7

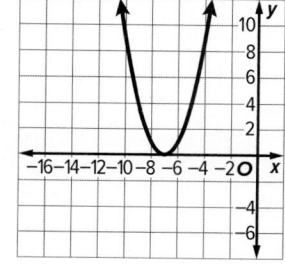

7. 2, 8

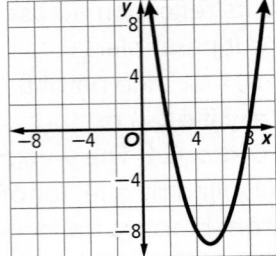

9. 3, 5

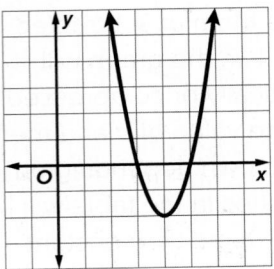

11. no solutions

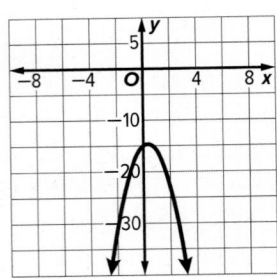

13. 5.3 seconds

15. −3.4, −0.6

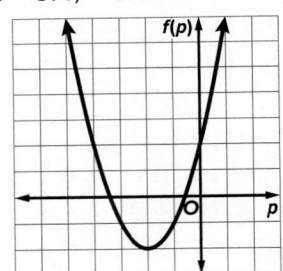

17. −5.4, −0.6

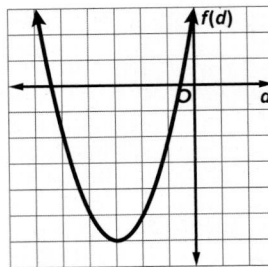

19. 0.2, 1.4

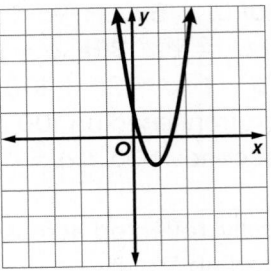

21. point of highest yield: (2, 16); There is a zero at (6, 0), which means that 6 units of fertilizer will yield 0 crops.

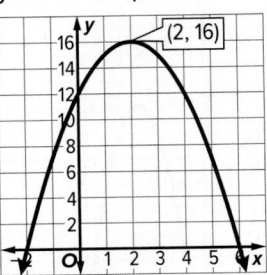

23. Yes; solving the equation $(30 - w)w = 81$ gives $w = 3$, or 27. A 3 in. by 27 in. sheet of paper has an area of 81 in^2 and a perimeter of 60 in. **25a.** Sample answer: $f(x)$ has two real solutions at $x = 4$ and $x = 9$. $g(x)$ has no real solutions. $h(x)$ has one real solution at $x = -6$. Because $h(x)$ has one solution, it could be a perfect square trinomial; $y = x^2 + 12x + m$ is the only equation that could be a perfect square. $g(x)$ has no real solutions, and $y = -x^2 + 3x - 10$ is not factorable, so it is likely $g(x)$. I can double check that by substituting x-values into the equation and seeing if the resulting ordered pair is on the graph. $f(x)$ has two real solutions, and since $y = 2x^2 - nx + 72$ could be made into a factorable equation that is not a perfect square, it is likely $f(x)$. **25b.** For $h(x)$, because the solution is $x = -6$, the first term of the equation has a coefficient of 1, and the middle term is positive, the factored equation could be $y = (x + 6)(x + 6)$, which would mean that $m = 36$. **25c.** For $f(x)$, the solutions are $x = 9$ and $x = 4$, and the first term of the equation has a coefficient of 2. So, the first term of one of the factors must be x and the other must be $2x$. I can multiply 2 by 72 and find suitable factor pairs, and then apply trial and error until a pair results in the solutions of 4 and 9, $n = 26$.

27a. Sample answer: Press Y= and enter the function $-50{,}000x^2 + 300{,}000x - 250{,}000$.

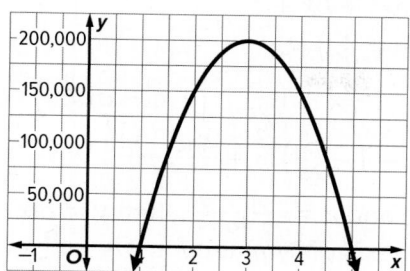

27b. The x-intercepts are $x = 1$ and $x = 5$. Both represent selling prices of the product that result in zero profit.

27c. $-50{,}000x^2 + 300{,}000x - 250{,}000 = 150{,}000$. The function relating to this equation is $f(x) = -50{,}000x^2 + 300{,}000x - 400{,}000$. The x-values where the graph intersects the x-axis are solutions to the equation, $x = 2$ and $x = 4$.

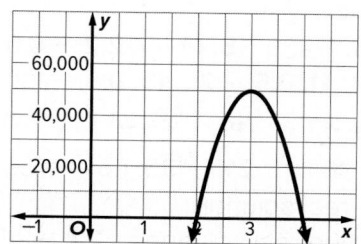

27d. No; sample answer: To make a profit of \$300,000, the product should be sold at a price that is a solution to $-50{,}000x^2 + 300{,}000x - 550{,}000 = 0$, but the function $f(x) = -50{,}000x^2 + 300{,}000x - 550{,}000$ has no real solutions.

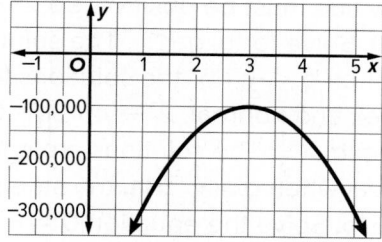

29. Sample answer: An athlete hits a tennis ball into the air. The height of the tennis ball can be modeled by the equation $h = -16t^2 + 25t + 2$. The ball is in the air for about 1.64 seconds.

31a. Sample answer: $x^2 + 8x + 16 = 0$
31b. Sample answer: $x^2 + 25 = 0$
31c. Sample answer: $x^2 - 7x + 12 = 0$

Lesson 12-4

1. ± 6 **3.** $-4, 2$ **5.** $0, 3$ **7.** $0, -3$ **9.** $-3, 9$
11. $-8, 15$ **13.** $-5, 18$ **15.** $-3, -6$ **17.** $-\frac{5}{3}$
19. $\frac{11}{6}, -\frac{11}{6}$ **21.** $\frac{1}{8}, -\frac{1}{8}$ **23.** $-8, 8$ **25a.** $-1, -9$
25b.

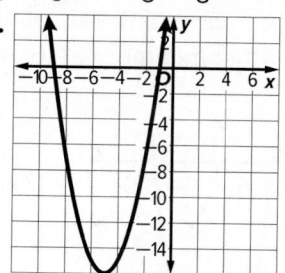

27. 4 and 5   **29.** after 6 seconds
31. $f(x) = x^2 - x - 6$ **33.** $f(x) = 4x^2 - 32x + 60$
35. $f(x) = x^2 + 7x - 18$
37. $f(x) = \frac{1}{4}x^2 - x - 8$ **39.** No; Kayla factored incorrectly. The correct solutions are 0 and 6.   **41.** $\pm\frac{4}{5}$
43. $\pm\frac{3}{5}$ **45.** $-1, -\frac{1}{3}$ **47.** ± 36 **49.** -15
51. $-1, 17$ **53.** 10   **55.** $\frac{-1 \pm \sqrt{15}}{2}$
57. Sample answer: $\frac{1}{2}(x)(x + 10) = 100$; $\frac{1}{2}(x^2 + 10x) = 100$; $x^2 + 10x = 200$; $x^2 + 10x - 200 = 0$; $(x - 10)(x + 20) = 0$. $x = 10$ or $x = -20$. Use $x = 10$. So, the dimensions are 10 feet and 20 feet.   **59.** Sample answer: When $a \neq 1$, the coefficients of x in each factor may not be 1 and must have a product equal to a. The product of m and b must be equal to ac rather than c.
61. Write the equation as $x^2 - 6x + qx - 12 = 0$. Find factor pairs of -12 that have -6 as a factor. The other factor is q since $-12 = (-6)(2)$. So, $q = 2$.   **63.** 8 ft by 16 ft; $\frac{1}{2}(x)(x + 8) = 64$, $\frac{1}{2}(x^2 + 8x) = 64$, $x^2 + 8x = 128$, $x^2 + 8x - 128 = 0$; The factor pair with a sum of 8 is -8 and 16, so $(x - 8)(x + 16) = 0$ and $x = 8$ or $x = -16$. Because the base cannot be negative, use $x = 8$. So, the base is 8 feet and the height is 16 feet.   **65.** Because the solutions are $-\frac{b}{a}$ and $\frac{b}{a}$, $a \neq 0$ and b is any real number.
67. Samantha; sample answer: She rewrote the equation to have zero on one side. Then she factored and used the Zero Product Property. Ignatio should have subtracted 12 from each side before factoring. Then he could have used the Zero Product Property.   **69.** Sample answer: $x^2 - 3x + \frac{9}{4} = 0$; $\frac{3}{2}$

Lesson 12-5

1. 169 **3.** $\frac{361}{4}$ **5.** $\frac{25}{4}$ **7.** 121 **9.** 144
11. $-2.9, 4.9$ **13.** $-3.3, 0.3$ **15.** $-1.4, 3.4$
17. $\varnothing$ **19.** $1, 1.5$ **21.** $-3.5, -1.1$
23. vertex form: $y = (x - 6)^2 - 20$; axis of symmetry: $x = 6$; extrema: minimum at $(6, -20)$; zeros: 1.5, 10.5

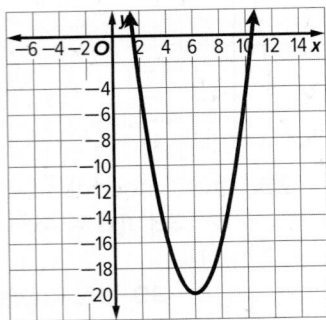

25. vertex form: $y = (x - 4)^2 - 6$; axis of symmetry: $x = 4$; extrema: minimum at $(4, -6)$; zeros: 1.6, 6.4

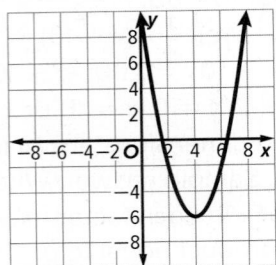

27. about 5.2 ft; 5 ft **29.** The area of the garden will be 4800 ft² for a width of 40 ft or a width of 60 ft. $(x)(200 - 2x) = 4800, -2x^2 + 200x - 4800 = 0, x = 40$ or $x = 60$.
31. $4x(x + 3) = 352; 4x^2 + 12x = 352;$
$x^2 + 3x = 88; \left(x + \frac{3}{2}\right)^2 = 88 + \frac{9}{4}; x + \frac{3}{2} = \pm\frac{19}{2};$
$x = -11$ or 8; Because the dimensions must be positive, use $x = 8$. So, the dimensions are 32 inches by 11 inches. **33a.** Sample answer: $4x^2 + 28x = 226$ or $4x^2 + 28x - 226 = 0$
33b. Sample answer: $4x^2 + 28x + c = 226 + c$. $\left(\frac{28}{2}\right)^2 = 4c$, so $c = 49$. $4x^2 + 28x + 49 = 226 + 49$. $(2x + 7)^2 = 275$. Take the square root and solve the two resulting equations: $2x + 7 \approx 16.58; x \approx 4.79$. $2x + 7 \approx -16.58; x \approx -11.79$.
33c. Sample answer: $(4)(-226) = -904$, there are no factor pairs of -904 that sum to 28. So, the equation cannot be factored. Because one

solution is approximately 4.79, then the equation has one solution that is between 4 and 5.
35. $y = ax^2 + bx + c$
$y = a\left(x^2 + \frac{b}{a}x\right) + c$
$y = \left[a\,x^2 + \frac{b}{a}x + \left(\frac{b}{2a}\right)^2\right] + c - a\left(\frac{b}{2a}\right)^2$
$y = a\left[x - \left(-\frac{b}{2a}\right)\right]^2 + \frac{4ac - b^2}{4a}$
If $h = \frac{b}{2a}$ and $k = \frac{4ac - b^2}{4a}$, then $y = a(x - h)^2 + k$. The axis of symmetry is $x = -\frac{b}{2a}$ or $x = h$.
37. $n^2 + \frac{1}{3}n + \frac{1}{9}$; It is the trinomial that is not a perfect square. **39.** Sample answer: Because the leading coefficient is 1, completing the square is simpler. Graphing the related function with a graphing calculator allows you to estimate the solution. Factoring is not possible.

Lesson 12-6

1. $-7, 7$ **3.** $-4, 9$ **5.** $\varnothing$ **7.** $2.2, -0.6$
9. $-3, -\frac{6}{5}$ **11.** $0.5, -2$ **13.** $0.5, -1.2$ **15.** 3
17. 25.8 seconds **19.** $\left(\frac{60 - x}{4}\right)^2 + \left(\frac{x}{4}\right)^2 = 117;$
24 in. and 36 in. **21.** -135; no real solution
23. 0; one real solution
25. 18.25; two real solutions **27.** -24; no real solutions **29.** 32; two real solutions
31. 89; two real solutions **33.** -64; no real solutions **35.** $-3.7, -0.3$ **37.** $-5.4, -0.6$
39. $\frac{1}{2}, 1$ **41.** $-4\frac{1}{2}, 1$ **43.** $-\frac{2}{3}, 3$ **45.** 0
47. 1 **49.** Sample answer: The value under the radical is $2^2 - 4 \cdot 3q$. For the solutions to be real, that value must be positive. So $2^2 - 4 \cdot 3q \geq 0$. Solve for q; $q < \frac{1}{3}$. The solutions are real when $q < \frac{1}{3}$. For the solutions to be complex and nonreal, $2^2 - 4 \cdot 3q < 0$. Solve for q; $q > \frac{1}{3}$. The solutions are complex and nonreal when $q > \frac{1}{3}$. **51a.** the starting height **51b.** Solve the equation $0 = -16t^2 + 90t + 120$. Find t when the firework is 0 ft off the ground. Use the Quadratic Formula, where $a = -16, b = 90$, and $c = 120$. $t = -1.11$ and 6.74. Given that time cannot be negative, the firework hits the ground 6.74 seconds after launch.
51c. 246.56 feet; 2.81 seconds; Sample answer: Determine the vertex of the parabola. It is on the line of symmetry, which is approximately $t = 2.81$ (midway between the zeros found

in part b). Substitute 2.81 for t in the original equation, $-16(2.81)^2 + 90(2.81) + 120 = 246.56$. **51d.** Because of the -16, the parabola opens down. It has x-intercepts of -1.11 and 6.74. It has a y-intercept of 120. It has a vertex of $(2.81, 246.56)$ **51e.** $-16t^2 + 90t + 120 = 200$ or $-16t^2 + 90t - 80 = 0$. $a = -16$, $b = 90$, and $c = -80$, so

$$t = \frac{-90 \pm \sqrt{90^2 - 4(-16)(-80)}}{2(-16)} =$$

$$\frac{-90 \pm \sqrt{90^2 - 2980}}{-32} = \frac{45 \pm \sqrt{745}}{16}, \text{ or}$$

approximately 1.1 seconds and 4.5 seconds. **53a.** always **53b.** sometimes **55a.** 24.1 feet; The area of the enclosure is $w(w - 20) = w^2 - 20w$. Solve the equation $w^2 - 20w = 100$, or $w^2 - 20w - 100 = 0$. Using the Quadratic Formula, $a = 1$, $b = -20$, and $c = -100$, so

$$w = \frac{-(-20) \pm \sqrt{(-20)^2 - 4(1)(-100)}}{2(1)} = \frac{20 \pm \sqrt{800}}{2} =$$

$10 \pm 10\sqrt{2}$. The solution $10 - 10\sqrt{2}$ is not valid since it is negative. So, the width should be approximately 24.1 feet. **55b.** Solve the equation $w^2 - 20w = R$, or $w^2 - 20w - R = 0$. Using the Quadratic Formula, $a = 1$, $b = -20$, and $c = -R$, so

$$w = \frac{-(-20 \pm \sqrt{(-20)^2 - 4(1)(-R)}}{2(1)} =$$

$$\frac{20 \pm \sqrt{400 + 4R}}{2} = \frac{20 \pm 2\sqrt{100 + R}}{2} =$$

$10 \pm \sqrt{100 + R}$. The solution $10 - \sqrt{100 + R}$ is not valid since it is negative. So, the width must be $10 + \sqrt{100 + R}$. **57.** The St. Louis Arch is 630 feet wide and 630 feet high. Using the points $(0, 0)$, $(315, 630)$, and $(630, 0)$, the equation $y = -\frac{2}{315}x^2 + 4x$ represents the height x feet from one base. The solutions of the equation are $x = 0$ and $x = 630$. This means that the two bases of the St. Louis Arch are 630 feet apart. **59.** one real solution **61.** two real solutions **63.** Sample answer: The polynomial can be factored to get $f(x) = (x - 4)^2$, so the only real zero is 4. The discriminant is 0, so there is 1 real zero. The discriminant can be used to determine how many real zeros there are. Factoring can be used to determine what they are. **65.** Sample answer: Factoring is easy if the polynomial is factorable and complicated if it is not. Not all equations are factorable. Graphing gives only approximate answers, but it is easy to see the number of solutions. Using square roots

is easy when there is no x-term. Completing the square can be used for any quadratic equation and exact solutions can be found, but the leading coefficient has to be 1 and the x^2- and x-terms must be isolated. It is also easier if the coefficient of the x-term is even; if not, the calculation becomes harder when dealing with fractions. The Quadratic Formula will work for any quadratic equation, and exact solutions can be found. This method can be time consuming, especially if an equation is easily factored. **67.** Sample answer: $x^2 + 2x + 1 = 0$; 0; The zero has only one square root. Thus, when the determinant is zero, the Quadratic Formula yields only one value when you simplify the numerator of the formula.

Lesson 12-7

1. $(-1, -3)$ and $(1, -3)$

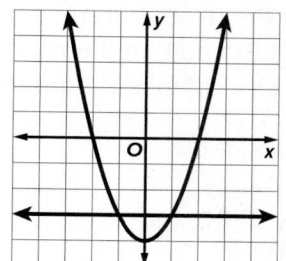

3. $(0, 1)$

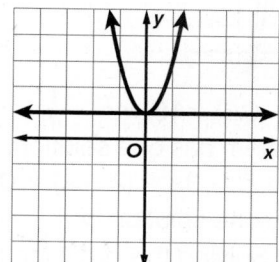

5. $(4, 3)$ and $(-2, 3)$ **7.** $(2, -3)$ and $(3, -4)$
9. $(3, -3)$ and $(-2, 7)$ **11a.** no solution
11b. There is no solution, so the baton will not reach the height of the ceiling. **13.** June
15. $\left(\frac{1}{2}, 1\right)$ and $(5, -17)$
17. $(1, -1)$
19. $(-1, -2)$ and $(2, 4)$
21. $(-4, 0)$ and $(8, 36)$ **23.** $(-3, 6)$ and $(5, 2)$
25. $(3, 3)$ **27.** $\left(\frac{-1 + \sqrt{5}}{2}, \frac{-3 + 3\sqrt{5}}{2}\right)$ and $\left(\frac{-1 - \sqrt{5}}{2}, \frac{-3 - 3\sqrt{5}}{2}\right)$ **29.** no solution **31.** -1

33. no solution   **35a.** one   **35b.** two
35c. none

37a.

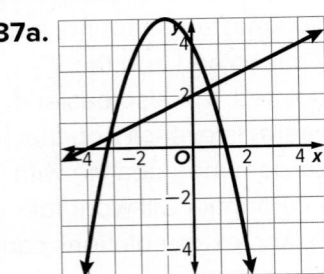

37b. Sample answer: The player will hit the target where the graphs intersect. Based on the graphs, this appears to happen at approximately (−3.1, 0.4) and (0.6, 2.3).
37c. (−3.1, 0.4) and (0.6, 2.3); these match the approximations in part b.   **39.** A graph of a system shows the solutions to the system where the two lines or curves intersect. If the lines or curves intersect twice, there are two solution; if they intersect once, there is one solutions if they do not intersect, there are no solutions.
41. The rock will land in the river. A graph of the functions shows that the path of the rock does not intersect with the cliff face before the river (the x-axis).

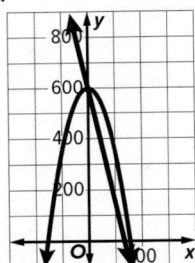

43. (0.8, 0.6) and (−0.8, −0.6)   **45.** $k = -\frac{1}{4}$; Sample answer: If the system has one solution, then $x^2 + 2 = 3x + k$ and $x^2 − 3x + (2 − k) = 0$. This equation has exactly one solution if $b^2 − 4ac = 0$, or $9 − 4(2 − k) = 0$. Solving this equation for k shows that $k = \frac{1}{4}$.   **47.** Sample answer: The maximum number of solutions of a quadratic and linear equation is 2. The linear function cannot change direction, and the quadratic function only changes direction once, so the two can intersect when the quadratic function is increasing and again when it is decreasing. Two linear functions have a maximum of one solution if their graphs intersect once. They have infinitely many solutions if the two functions have the same graph.   **49.** The third system does not belong because it has no solution. The other systems have two solutions.

Lesson 12-8

1. linear; $y = 0.2x − 8.2$ **3.** exponential; $y = 3 \cdot 4^x$ **5.** quadratic; $y = 4.2x^2$
7. exponential; $y = 4 \cdot 0.5^x$
9. quadratic; $y = −3x^2$
11. exponential; $y = 0.5 \cdot 3^x$
13. quadratic; $y = 3x^2$

15a.

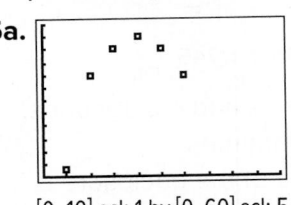

[0, 10] scl: 1 by [0, 60] scl: 5

15b. quadratic   **15c.** $y = −2.857x^2 + 22.429x + 9$; $R^2 = 0.969$; $D = \{x \mid 0 \leq x \leq 8.233\}$; $R = \{y \mid 0 \leq y \leq 53.016\}$
15d. 26 feet   **17.** quadratic; $R^2 \approx 0.980$
19. exponential; $R^2 \approx 0.997$
21. exponential; $y = 1.05 \cdot 2^x$ **23a.** exponential
23b. $y = 20 \cdot 0.5^x$ **23c.** about 0.0098 g
23d. ≈ 701.6 years   **25a.** Linear; the first differences are constant   **25b.** −18 ft/s; the elevator is descending at 18 ft/s.
25c. $y = −18x + 142$; the coefficient of x is −18, which is the value of the first differences in part **a** and the constant rate of change for the function in part **b**.
27. Look up data on the temperature at any place from 6:00 A.M. to 2:00 P.M. and put that data in a scatterplot. Then find the coefficient of determination to determine which model is the best fit. Sample: temperature data from Aug 10 from Iowa City, IA (plot below) shows a quadratic model is best ($R^2 = 0.955$).

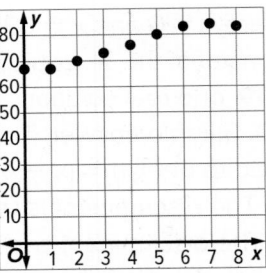

29. Sample answer: Diego is correct. Any function with equal non-zero first differences would have to have second differences of 0. For example if the first differences are all 4, then the second differences are $4 − 4 = 0$ for all terms.   **31.** Linear functions have a constant first difference and quadratic functions have a

constant second difference, so cubic equations would have a constant third difference.
33. Sample answer: If one linear term is ax, the next term is $a(x + 1)$, and the difference between the terms is $a(x + 1) - ax = ax + a - ax$, or a. If one exponential term is a^x, the next term is a^{x+1}, and the ratio of terms is $\frac{a^{x+1}}{a^x}$, or a.

Lesson 12-9

1. $(f + g)(x) = 3x^2 - x - 4$
3. $(f + h)(x) = -5x - 9$
5. $(g - h)(x) = 3x^2 + 4x + 5$
7. $(h - g)(x) = -3x^2 - 4x - 5$
9. $(f \cdot g)(x) = 11x^3 - 66x^2 + 33x$
11. $(g \cdot h)(x) = -x^3 + 10x^2 - 27x + 12$
13. $(f \cdot j)(x) = 11x(2^x) + 33x$ **15a.** $f(t) + g(t) = 7t + 4$ is Jan's speed while walking with the moving walkway.   **15b.** $f(t) - g(t) = t + 2$ is Jan's speed while walking against the moving walkway.
17a. $P(x) = 15 - 0.75x$; $T(x) = 180 + 10x$
17b. $R(x) = -7.5x^2 + 15x + 2700$
17c. $2640   **19.** $\left(\frac{1}{4}\right)^x + x^2$
21. $(p \cdot q)(x) = -2x^3 - 10x$
23. $(q \cdot r)(x) = x^4 - x^3 + 7x^2 - 5x + 10$
25. $(p \cdot t)(x) = -2x\left(\frac{1}{4}\right)^x + 10x$
27a. $\ell(x) = -2x + 20$, $w(x) = -2x + 12$
27b. $(\ell \cdot w)(x) = 4x^2 - 64x + 240$ represents the area of the base of the completed box.
27c. $D = \{x \mid 0 < x < 6\}$; The side lengths of the squares must be positive numbers that are less than half the width of the rectangle to form a box.   **27d.** $(\ell \cdot w)(1.5) = 153$; If the squares are 1.5 inches per side, the area of the base of the box is 153 square inches.
29a. $f(n) = n^2 + 4$; $g(n) = 4n$ **29b.** $h(n)$ is the sum of $f(n)$ and $g(n)$, so $h(n) = n^2 + 4 + 4n$; $h(10) = 144$.   **29c.** The pattern forms a square with sides of length $n + 2$, so the total number of tiles is $(n + 2)^2$. When $n = 10$, this equals 12^2 or 144, so the answer to part **b** is correct.
31a. $A(x) = (12 + 4x)(8 + 2x) - 96$; The area of the gravel border is the product of the function that gives its length, $L(x) = 12 + 4x$, and the function that gives its width, $W(x) = 8 + 2x$, minus the constant function that gives the area of the flower bed, $F(x) = 96$.

31b. Disagree; sample answer: the function may be written as $A(x) = 8x^2 + 56x$, which shows that this is a quadratic function with $a > 0$ and $c = 0$, so the graph is an upward-opening parabola through the origin. As x increases, the area increases.
33. $m(x) = 3^x - 2$ **35.** Sample answer: $f(x) = x^2 + 12x - 7$ and $g(x) = -2x + 5$
37. Sometimes; sample answer: A linear equation with two terms has an x-term and a constant. A quadratic equation with three terms has an x^2-term, an x-term, and a constant. Multiplying the equations gives an x^3-term, an x^2-term, an x-term, and a constant. However, sometimes one or more of the terms will cancel. So, the product may have only 3 terms sometimes.   **39.** Delfina's solution is incorrect. Sample answer: She only multiplied the x-terms by other x-terms and constants by other constants. The correct answer is $(f \cdot g)(x) = 24x^3 + 10x^2 - 170x + 152$.

Module 12 Review

1.

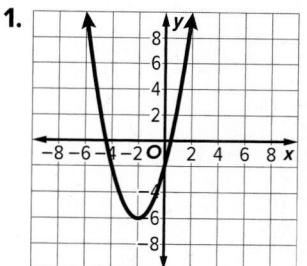

3. B   **5.** -2

7.

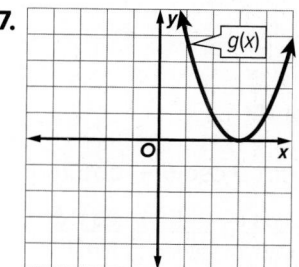

9. C   **11.** C
13. Vertex form: $y = 2(x + 1)^2 + 4$; extrema: $(-1, 4)$; because the x^2-term is positive, the parabola opens upward and the extrema is a minimum.   **15.** 5.7 ft   **17.** No real solutions   **19.** 17   **21.** C

Module 13

Module 13 Opener

1. $-4a(4a - 1)$ **3.** prime **5.** $\frac{\sqrt{2}}{2}$ **7.** $-\frac{\sqrt{3}}{3}$

Lesson 13-1

1. $\frac{4}{5}$ **3.** $\frac{5}{4}$ **5.** $-\frac{5}{4}$ **7.** $\frac{4}{5}$ **9.** $\sqrt{2}$

11. -2.5 **13.** $\frac{\sqrt{7}}{4}$ **15.** $\frac{1}{2}$ **17.** $\cos \theta$ **19.** $\csc \theta$

21. 1 **23.** $-\cot \theta$ **25.** -4 **27.** $\cos \theta$

29. $d = \sec^2 \theta$ **31.** $\frac{1 + \cos \theta}{\sin \theta}$ **33.** $\sec \theta$

35. 2 **37.** $2 \cos^2 \theta$ **39.** -1 **41.** $-\cot^2 \theta$

43. Sample answer: The functions $\cos \theta$ and $\sin \theta$ can be thought of as the lengths of the legs of a right triangle, and the number 1 can be thought of as the measure of the corresponding hypotenuse.

45. Sample answer: $\frac{\sin \theta}{\cos \theta} \cdot \sin \theta$ and $\frac{\sin^2 \theta}{\cos \theta}$.

47. Ebony; Jordan did not use the identity that $\sin^2 \theta + \cos^2 \theta = 1$ and made an error adding rational expressions.

Lesson 13-2

1. $\cos^2 \theta + \tan^2 \theta \cos^2 \theta \overset{?}{=} 1$

$\cos^2 \theta + \frac{\sin^2 \theta}{\cos^2 \theta} \cdot \cos^2 \theta \overset{?}{=} 1$

$\cos^2 \theta + \sin^2 \theta \overset{?}{=} 1$

$1 = 1 \checkmark$

3. $1 + \sec^2 \theta \sin^2 \theta \overset{?}{=} \sec^2 \theta$

$1 + \frac{1}{\cos^2 \theta} \cdot \sin^2 \theta \overset{?}{=} \sec^2 \theta$

$1 + \tan^2 \theta \overset{?}{=} \sec^2 \theta$

$\sec^2 \theta = \sec^2 \theta \checkmark$

5. $\frac{1 - \cos \theta}{1 + \cos \theta} \overset{?}{=} (\csc \theta - \cot \theta)^2$

$\frac{1 - \cos \theta}{1 + \cos \theta} \overset{?}{=} \csc^2 \theta - 2 \cot \theta \csc \theta + \cot^2 \theta$

$\frac{1 - \cos \theta}{1 + \cos \theta} \overset{?}{=} \frac{1}{\sin^2 \theta} - 2 \cdot \frac{\cos \theta}{\sin \theta} \cdot \frac{1}{\sin \theta} + \frac{\cos^2 \theta}{\sin^2 \theta}$

$\frac{1 - \cos \theta}{1 + \cos \theta} \overset{?}{=} \frac{1}{\sin^2 \theta} - \frac{2 \cos \theta}{\sin^2 \theta} + \frac{\cos^2 \theta}{\sin^2 \theta}$

$\frac{1 - \cos \theta}{1 + \cos \theta} \overset{?}{=} \frac{1 - 2 \cos \theta + \cos^2 \theta}{\sin^2 \theta}$

$\frac{1 - \cos \theta}{1 + \cos \theta} \overset{?}{=} \frac{(1 - \cos \theta)(1 - \cos \theta)}{(1 - \cos^2 \theta)}$

$\frac{1 - \cos \theta}{1 + \cos \theta} \overset{?}{=} \frac{(1 - \cos \theta)(1 - \cos \theta)}{(1 - \cos \theta)(1 + \cos \theta)}$

$\frac{1 - \cos \theta}{1 + \cos \theta} = \frac{1 - \cos \theta}{1 + \cos \theta} \checkmark$

7. $(\sin \theta - 1)(\tan \theta + \sec \theta) \overset{?}{=} -\cos \theta$

$\sin \theta \tan \theta + \sin \theta \sec \theta - \tan \theta - \sec \theta \overset{?}{=} -\cos \theta$

$\frac{\sin^2 \theta}{\cos \theta} + \frac{\sin \theta}{\cos \theta} - \frac{\sin \theta}{\cos \theta} - \frac{1}{\cos \theta} \overset{?}{=} -\cos \theta$

$\frac{\sin^2 \theta}{\cos \theta} - \frac{1}{\cos \theta} \overset{?}{=} -\cos \theta$

$\frac{\sin^2 \theta - 1}{\cos \theta} \overset{?}{=} -\cos \theta$

$\frac{-\cos^2 \theta}{\cos \theta} \overset{?}{=} -\cos \theta$

$-\cos \theta = -\cos \theta \checkmark$

9. $\sec^2 \theta - \tan^2 \theta \overset{?}{=} \frac{1 - \sin \theta}{\cos \theta}$

$\frac{1}{\cos \theta} - \frac{\sin \theta}{\cos \theta} \overset{?}{=} \frac{1 - \sin \theta}{\cos \theta}$

$\frac{1 - \sin \theta}{\cos \theta} = \frac{1 - \sin \theta}{\cos \theta} \checkmark$

11. $(\sin \theta + \cos \theta)^2 \overset{?}{=} \frac{2 + \sec \theta \csc \theta}{\sec \theta \csc \theta}$

$(\sin \theta + \cos \theta)^2 \overset{?}{=} \frac{2 + \frac{1}{\cos \theta} \cdot \frac{1}{\sin \theta}}{\frac{1}{\cos \theta} \cdot \frac{1}{\sin \theta}}$

$(\sin \theta + \cos \theta)^2 \overset{?}{=} \left(2 + \frac{1}{\cos \theta \sin \theta}\right) \cdot \frac{\cos \theta \sin \theta}{1}$

$\sin^2 \theta + 2 \sin \theta \cos \theta + \cos^2 \theta \overset{?}{=} 2 \cos \theta \sin \theta + 1$

$2 \cos \theta \sin \theta + \cos^2 \theta + \sin^2 \theta = 2 \cos \theta \sin \theta + \cos^2 \theta + \sin^2 \theta \checkmark$

13. $\csc^2 \theta - 1 \overset{?}{=} \frac{\cot^2 \theta}{\csc \theta + \sin \theta}$

$\cot^2 \theta \frac{1}{\cos \theta - \frac{\sin \theta}{\cos \theta}} \overset{?}{=} \cot^2 \theta \frac{1}{\cos \theta - \frac{\sin \theta}{\cos \theta}}$

$\cot^2 \theta = \cot^2 \theta \checkmark$

15. $\csc^2 \theta \overset{?}{=} \cot^2 \theta + \sin \theta \csc \theta$

$\cot^2 \theta + 1 \overset{?}{=} \cot^2 \theta + \sin \theta \cdot \frac{1}{\sin \theta}$

$\cot^2 \theta + 1 = \cot^2 \theta + 1 \checkmark$

17. $\tan \theta \cos \theta \overset{?}{=} \sin \theta$

$\frac{\sin \theta}{\cos \theta} \cdot \cos \theta \overset{?}{=} \sin \theta$

$\sin \theta = \sin \theta \checkmark$

19. $(\tan \theta)(1 - \sin^2 \theta) \overset{?}{=} \sin \theta \cos \theta$

$\tan \theta \cos^2 \theta \overset{?}{=} \sin \theta \cos \theta$

$\frac{\sin \theta}{\cos \theta} \cdot \cos^2 \theta \overset{?}{=} \sin \theta \cos \theta$

$\sin \theta \cos \theta = \sin \theta \cos \theta \checkmark$

21. $\frac{\sin^2 \theta}{1 - \sin^2 \theta} \overset{?}{=} \tan^2 \theta$

$\frac{\sin^2 \theta}{\cos^2 \theta} \overset{?}{=} \tan^2 \theta$

$\left(\frac{\sin \theta}{\cos \theta}\right)^2 \overset{?}{=} \tan^2 \theta$

$\tan^2 \theta = \tan^2 \theta \checkmark$

23. $n_1 \sin \theta_p = n_2 \sin \theta_r$

$\quad n_1 \sin \theta_p = n_2 \sin (90 - \theta_p)$

$\quad n_1 \sin \theta_p = n_2 \cos \theta_p$

$\quad \dfrac{\sin \theta_p}{\cos \theta_p} = \dfrac{n_2}{n_1}$

$\quad \tan \theta_p = \dfrac{n_2}{n_1}$

25. An identity can be verified by transforming one side of the equation to form of the other side, or by transforming both sides of the equation separately so that both sides form the same expression.
Sample answer by transforming one side of the equation:

$$\frac{1}{1 - \sin^2 \theta} \stackrel{?}{=} \tan^2 \theta + 1$$

$$\frac{1}{1 - (1 - \cos^2 \theta)} \stackrel{?}{=} \tan^2 \theta + 1$$

$$\frac{1}{\cos^2 \theta} \stackrel{?}{=} \tan^2 \theta + 1$$

$$\sec^2 \theta \stackrel{?}{=} \tan^2 \theta + 1$$

$$\tan^2 \theta + 1 = \tan^2 \theta + 1 \checkmark$$

Sample answer by transforming both sides of the equation:

$$\frac{1}{1 - \sin^2 \theta} \stackrel{?}{=} \tan^2 \theta + 1$$

$$\frac{1}{1 - (1 - \cos^2 \theta)} \stackrel{?}{=} \frac{\sin^2 \theta}{\cos^2 \theta} + 1$$

$$\frac{1}{\cos^2 \theta} \stackrel{?}{=} \frac{\sin^2 \theta}{\cos^2 \theta} + \frac{\sin^2 \theta}{\cos^2 \theta}$$

$$\frac{1}{\cos^2 \theta} \stackrel{?}{=} \frac{\sin^2 \theta + \cos^2 \theta}{\cos^2 \theta}$$

$$\frac{1}{\cos^2 \theta} = \frac{1}{\cos^2 \theta} \checkmark$$

27.
$$\frac{1 + \tan \theta}{1 + \cot \theta} \stackrel{?}{=} \tan \theta$$

$$\frac{1 + \tan \theta}{1 + \frac{1}{\tan \theta}} \stackrel{?}{=} \tan \theta$$

$$\left(\frac{\tan \theta}{\tan \theta}\right)\left(\frac{1 + \tan \theta}{1 + \frac{1}{\tan \theta}}\right) \stackrel{?}{=} \tan \theta$$

$$\frac{\tan \theta (1 + \tan \theta)}{\tan \theta + 1} \stackrel{?}{=} \tan \theta$$

$$\tan \theta = \tan \theta \checkmark$$

29. $\dfrac{\sec^2 \theta - \tan^2 \theta}{\cos^2 \theta + \sin^2 \theta} \stackrel{?}{=} 1$

$$\frac{1}{\cos^2 \theta + \sin^2 \theta} \stackrel{?}{=} 1$$

$$\frac{1}{1} \stackrel{?}{=} 1$$

$$1 = 1 \checkmark$$

31. $\cot (-\theta) \cot \left(\dfrac{\pi}{2} - \theta\right) \stackrel{?}{=} 1$

$$\left(\frac{\cos (-\theta)}{\sin (-\theta)}\right)\left(\frac{\cos \left(\frac{\pi}{2} - \theta\right)}{\sin \left(\frac{\pi}{2} - \theta\right)}\right) \stackrel{?}{=} 1$$

$$\left(\frac{\cos \theta}{-\sin \theta}\right)\left(\frac{\sin \theta}{\cos \theta}\right) \stackrel{?}{=} 1$$

$$-1 \neq 1$$

35. No; use the Pythagorean Identity to rewrite the equation as $\sin^2 A + (1 - \sin^2 B) = 1$. This simplifies to $\sin^2 A = \sin^2 B$, which is true if $\sin A = \sin B$ or $\sin A = -\sin B$. Because A and B are both less than 180°, $\sin A > 0$ and $\sin B > 0$. So, the second equation cannot be true. The first equation is true if $A = B$ or if $A = 180° - B$.

37. $\sin^2 \theta - \cos^2 \theta = 2 \sin^2 \theta$; the other three are Pythagorean identities, but this is not.

39. Sample answer: $\sin^2 45° + \cos^2 45° = 1$, but $\sin 45° \neq \sqrt{1 - \cos 45°}$ and $\sin^2 30° + \cos^2 30° = 1$, but $\sin 30° \neq \sqrt{1 - \cos 30°}$.

41. $f(\theta) = \dfrac{1}{2} \sin \theta$

Lesson 13-3

1. $\dfrac{\sqrt{2}}{2}$ **3.** $\dfrac{\sqrt{2} - \sqrt{6}}{4}$ **5.** $2 - \sqrt{3}$ **7a.** $\dfrac{3 + 4\sqrt{3}}{10}$;

$\dfrac{4 - 3\sqrt{3}}{10}$ **7b.** 96.9°   **7c.** no

9.
$$\cos \left(\frac{\pi}{2} + \theta\right) \stackrel{?}{=} -\sin \theta$$

$$\cos \frac{\pi}{2} \cos \theta - \sin \frac{\pi}{2} \sin \theta \stackrel{?}{=} -\sin \theta$$

$$(0)\cos \theta - (1)\sin \theta \stackrel{?}{=} -\sin \theta$$

$$-\sin \theta = -\sin \theta \checkmark$$

11.
$$\cos (180° + \theta) \stackrel{?}{=} -\cos \theta$$

$$\cos 180° \cos \theta - \sin 180 \sin \theta \stackrel{?}{=} -\cos \theta$$

$$-1 \cdot \cos \theta - 0 \cdot \sin \theta \stackrel{?}{=} -\cos \theta$$

$$-\cos \theta = -\cos \theta \checkmark$$

13. $-\dfrac{1}{2}$ **15.** $\dfrac{\sqrt{2}}{2}$ **17.** $-\dfrac{\sqrt{2}}{2}$ **19.** $-\dfrac{\sqrt{2}}{2}$

21. $\dfrac{\sqrt{6} - \sqrt{2}}{4}$ **23.** $-\dfrac{\sqrt{2}}{2}$ **25.** $-\dfrac{\sqrt{3}}{2}$

27.
$$\sin (90° + \theta) \stackrel{?}{=} \cos \theta$$

$$\sin 90° \cos \theta + \cos 90° \sin \theta \stackrel{?}{=} \cos \theta$$

$$1 \cdot \cos \theta + 0 \cdot \sin \theta \stackrel{?}{=} \cos \theta$$

$$\cos \theta = \cos \theta \checkmark$$

29.
$$\cos (270° - \theta) \stackrel{?}{=} -\sin \theta$$

$$\cos 270° \cos \theta + \sin 270° \sin \theta \stackrel{?}{=} -\sin \theta$$

$$0 \cdot \cos \theta + (-1) \cdot \sin \theta \stackrel{?}{=} -\sin \theta$$

$$-\sin \theta = -\sin \theta \checkmark$$

31.
$$\sin\left(\theta - \frac{\pi}{2}\right) \stackrel{?}{=} -\cos \theta$$

$$\sin \theta \cos \frac{\pi}{2} - \cos \theta \sin \frac{\pi}{2} \stackrel{?}{=} -\cos \theta$$

$$(\sin \theta)(0) - (\cos \theta)(1) \stackrel{?}{=} -\cos \theta$$

$$-\cos \theta = -\cos \theta \checkmark$$

33a. $m\angle TPQ = A$; Sample answer: Because $\overline{QT}$ is parallel to the x-axis, $m\angle OQT = A$ since this angle is an alternate interior angle with $\angle QOS$. $\angle OQT$ is complementary to $\angle TQP$ because $\angle OQP$ is a right angle. $\angle TPQ$ is complementary to $\angle TQP$ because these are the acute angles in a right triangle. So, $\angle OQT \cong \angle TPQ$ because these angles are both complementary to the same angle. Therefore, $m\angle TPQ = A$.

33b. In right triangle POQ, $\sin B = \frac{PQ}{OP}$, but $OP = 1$, so $PQ = \sin B$. Also in right triangle POQ, $\cos B = \frac{OQ}{OP}$, and $OP = 1$, so $OQ = \cos B$.

33c. In $\triangle QOS$, $\cos A = \frac{QS}{OQ}$. Therefore, $OS = OQ \cos A$. But from part b, $OQ = \cos B$, so by substitution, $OS = \cos B \cos A$.

33d. In $\triangle TPQ$, $\sin A = \frac{QT}{PQ}$. Therefore, $QT = PQ \sin A$. But from part b, $PQ = \sin B$, so by substitution, $QT = \sin B \sin A$.

33e. In $\triangle POR$, $\cos(A + B) = \frac{OR}{OP} = OR = OS - RS$, by the Segment Addition Postulate. From part c, $OS = \cos B \cos A$. Also, $RS = QT$ because $TQRS$ is a rectangle, and from part d, $QT = \sin B \sin A$. So, $\cos(A + B) = \cos B \cos A - \sin B \sin A = \cos A \cos B - \sin A \sin B$.

35. $\sqrt{2} - \sqrt{6}$ **37.** $-2 + \sqrt{3}$ **39.** $2 - \sqrt{3}$

41. $\cos(A + B) \stackrel{?}{=} \dfrac{1 - \tan A \tan B}{\sec A \sec B}$

$\cos(A + B) \stackrel{?}{=} \dfrac{1 - \frac{\sin A}{\cos A} \cdot \frac{\sin B}{\cos B}}{\frac{1}{\cos A} \cdot \frac{1}{\cos B}}$

$\cos(A + B) \stackrel{?}{=} \dfrac{1 - \frac{\sin A}{\cos A} \cdot \frac{\sin B}{\cos B}}{\frac{1}{\cos A} \cdot \frac{1}{\cos B}} \cdot \dfrac{\cos A \cos B}{\cos A \cos B}$

$\cos(A + B) \stackrel{?}{=} \dfrac{\cos A \cos B - \sin A \sin B}{1}$

$\cos(A + B) = \cos(A + B) ✓$

43. $\sin(A + B)\sin(A - B) \stackrel{?}{=} \sin^2 A - \sin^2 B$

$(\sin A \cos B + \cos A \sin B)(\sin A \cos B - \cos A \sin B) \stackrel{?}{=} \sin^2 A - \sin^2 B$

$\sin^2 A \cos^2 B - \cos^2 A \sin^2 B \stackrel{?}{=} \sin^2 A - \sin^2 B$

$\sin^2 A \cos^2 B + \sin^2 A \sin^2 B - \sin^2 A \sin^2 B - \cos^2 A \sin^2 B \stackrel{?}{=} \sin^2 A - \sin^2 B$

$\sin^2 A(\cos^2 B + \sin^2 B) - \sin^2 B(\sin^2 A + \cos^2 A) \stackrel{?}{=} \sin^2 A - \sin^2 B$

$\sin^2 A(1) - \sin^2 B(1) \stackrel{?}{=} \sin^2 A - \sin^2 B$

$\sin^2 A - \sin^2 B = \sin^2 A - \sin^2 B ✓$

45a. $\tan(A + B) = \dfrac{\sin(A + B)}{\cos(A + B)} =$
$\dfrac{\sin A \cos B + \cos A \sin B}{\cos A \cos B - \sin A \sin B}$; divide the numerator and denominator of this fraction by $\cos A$ $\cos B$. This gives $\dfrac{\frac{\sin A \cos B}{\cos A \cos B} + \frac{\cos A \sin B}{\cos A \cos B}}{\frac{\cos A \cos B}{\cos A \cos B} - \frac{\sin A \sin B}{\sin A \sin B}} =$
$\dfrac{\tan A + \tan B}{1 - \tan A \tan B}.$

45b. $\tan(A - B) = \tan(A + (-B))$
$= \dfrac{\tan A + \tan(-B)}{1 - \tan A \tan(-B)}$, but $\tan(-B) = -\tan B$, so
$\tan(A - B) = \dfrac{\tan A - \tan B}{1 + \tan A \tan B}.$

47. Sample answer: $\sin \dfrac{13\pi}{12} = \sin\left(\dfrac{5\pi}{6} + \dfrac{\pi}{4}\right) =$
$\sin \dfrac{5\pi}{6} \cos \dfrac{\pi}{4} + \cos \dfrac{5\pi}{6} \sin \dfrac{\pi}{4} = \dfrac{1}{2} \cdot \dfrac{\sqrt{2}}{2} +$
$\left(-\dfrac{\sqrt{3}}{2}\right)\left(\dfrac{\sqrt{2}}{2}\right) = \dfrac{\sqrt{2} - \sqrt{6}}{4}$

49. $\theta = \dfrac{\pi}{4}$; by the sum and difference formulas, the equation can be written as $\cos\theta\cos\pi - \sin\theta\sin\pi = \sin\theta\cos(-\pi) - \cos\theta\sin(-\pi)$, or $-\cos\theta = -\sin\theta$, which shows that $\theta = \dfrac{\pi}{4}$.

51. $\sin(19° + 11°)$; $\dfrac{1}{2}$ **53.** $\cos(111° - 21°)$; 0

55. $\tan \dfrac{\pi}{16} + \dfrac{3\pi}{16}$; 1 **57.** $\sin(-2\theta)$

59. $\cot(A + B) = \dfrac{1}{\tan(A + B)}$

$\cot(A + B) = \dfrac{1}{\frac{\tan A + \tan B}{1 - \tan A \tan B}}$

$\cot(A + B) = \dfrac{1 - \tan A \tan B}{\tan A + \tan B}$

$\cot(A + B) = \dfrac{1 - \frac{1}{\cot A} \cdot \frac{1}{\cot B}}{\frac{1}{\cot A} + \frac{1}{\cot B}} \cdot \dfrac{\cot A \cot B}{\cot A \cot B}$

$\cot(A + B) = \dfrac{\cot A \cot B - 1}{\cot A + \cot B}$

61. Sample answer: $A = 35°$, $B = 60°$, $C = 85°$; $0.7002 + 1.7321 + 11.4301 \stackrel{?}{=} (0.7002)(1.7321)$ (11.4301); $13.86 = 13.86 ✓$

63. $\sin(90° - \theta) \stackrel{?}{=} \cos\theta$ Original equation

$\sin 90° \cos\theta - $
$\cos 90° \sin\theta \stackrel{?}{=} \cos\theta$ Difference Identity

$1 \cdot \cos\theta - 0 \cdot \sin\theta \stackrel{?}{=} \cos\theta$ $\cos 90° = 0$;
                         $\sin 90° = 1$

$\cos\theta = \sin\theta ✓$ Simplify.

Lesson 13-4

1. $-\dfrac{4\sqrt{5}}{9}, \dfrac{1}{9}, \sqrt{\dfrac{3 + \sqrt{5}}{6}}, \sqrt{\dfrac{3 - \sqrt{5}}{6}}$

3. $-\dfrac{24}{25}, -\dfrac{7}{25}, \dfrac{\sqrt{5}}{5}, -\dfrac{2\sqrt{5}}{5}$ **5.** $\dfrac{24}{25}, -\dfrac{7}{25}, \dfrac{2\sqrt{5}}{5}, \dfrac{\sqrt{5}}{5}$

7. $\frac{336}{625}, -\frac{527}{625}, \frac{3}{5}, \frac{4}{5}$ **9.** $-\frac{720}{1681}, -\frac{1519}{1681}, \frac{5\sqrt{41}}{41}, \frac{4\sqrt{41}}{41}$

11. $S = 4 \sin\theta \cos\theta$ **13.** $\frac{\sqrt{2+\sqrt{3}}}{3}$

15. $-\frac{\sqrt{2-\sqrt{3}}}{2}$ **17.** $2 + \sqrt{3}$ **19.** $\frac{\sqrt{2+\sqrt{2}}}{2}$

21. $-\frac{\sqrt{2-\sqrt{3}}}{2}$ **23.** $-\frac{\sqrt{2-\sqrt{2}}}{2}$ **25.** $\frac{24}{25}, \frac{7}{25}, \frac{24}{7}$

27. $-\frac{3}{5}, -\frac{4}{5}, \frac{3}{4}$ **29.** $-\frac{4\sqrt{21}}{25}, \frac{17}{25}, -\frac{4\sqrt{21}}{17}$

31a. $\frac{\sqrt{2-\sqrt{2}}}{2}$; $\frac{\sqrt{2+\sqrt{2}}}{2}$ **31b.** $\frac{\sqrt{2}}{2}$; $-\frac{\sqrt{2}}{2}$

33a. Sample answer: sin 240° = sin (180° + 60°)
= sin 180° cos 60° + cos 180° sin 60°
$= -\frac{\sqrt{3}}{2}$

33b. Sample answer: sin 240° = sin (270° − 30°)
= sin 270° cos 30° − cos 270° sin 30°
$= -\frac{\sqrt{3}}{2}$

33c. sin 240° = sin (2 · 120°)
= 2 sin 120° cos 120°
$= -\frac{\sqrt{3}}{2}$

33d. sin 240° = sin $\frac{480°}{2}$
$= -\sqrt{\frac{1-\cos 480°}{2}}$
$= -\frac{\sqrt{3}}{2}$

35. ∠PBD is an inscribed angle that subtends the same arc as the central angle ∠POD, so $m\angle PBD = \frac{1}{2}\theta$. By right triangle trigonometry, $\tan\frac{1}{2}\theta = \frac{PA}{BA} = \frac{PA}{1+OA} = \frac{\sin\theta}{1+\cos\theta}$.

37. sin 2θ = sin(θ + θ)
= sin θ cos θ + cos θ sin θ
= 2 sin θ cos θ
cos 2θ = cos(θ + θ)
= cos θ cos θ − sin θ sin θ
= cos²θ − sin²θ

You can find alternate forms for cos 2θ by making substitutions into the expression cos²θ − sin²θ.

cos²θ − sin²θ = (1 − sin²θ) − sin²θ (Substitute 1 − sin²θ for cos²θ.)
= 1 − 2 sin²θ (Simplify.)
cos²θ − sin²θ = cos²θ − (1 − cos²θ) (Substitute 1 − cos²θ for sin²θ.)
= 2 cos²θ − 1 (Simplify.)

39. Sample answer: Since $d = \frac{v^2 \sin 2\theta}{b}$, d is at a maximum when sin 2θ = 1, that is, when 2θ = 90° or θ = 45°.

Lesson 13-5

1. 60°, 120°, 240°, 300°

3. $\frac{\pi}{6}, \frac{\pi}{2}, \frac{5\pi}{6}, \frac{3\pi}{2}$ **5.** 240°, 300°

7. $\frac{\pi}{3} + 2k\pi, \frac{5\pi}{3} + 2k\pi$

9. $\frac{2\pi}{3} + 2k\pi, \frac{4\pi}{3} + 2k\pi$ **11.** 45 + k · 180°

13. 270° + k · 360° **15.** approximately 7 P.M.

17. [30°, 150°], (180°, 360°)

19. $\frac{\pi}{2}, \pi$

21. 0, $\frac{2\pi}{3}$ **23.** 225° + 360k°, 315° + 360k°

25. 45°, 135° **27.** 225°, 315° **29.** π, $\frac{3\pi}{2}$

31. 0 + 2kπ, $\frac{2\pi}{3} + 2k\pi$, and $\frac{4\pi}{3} + 2k\pi$

33. kπ **35.** $\frac{2\pi}{3} + 2k\pi, \frac{4\pi}{3} + 2k\pi$

37. 210° + k · 360° and 330° + k · 360°

39. 225° + k · 360° and 315° + k · 360°

41. 45° + k · 90°

43. $\frac{\pi}{3} + k\pi$ and $\frac{2\pi}{3} + k\pi$, or 60° + k · 180° and 120° + k · 180°

45. $\frac{\pi}{2} + 2\pi k, \frac{7\pi}{6} + 2\pi k, \frac{11\pi}{6} + 2\pi k$ or 90° + k · 360°, 210° + k · 360°, 330° + k · 360°

47. 30°, 150°, 270°

49. k · 120° or 120° + k · 360° and 240° + k · 360° **51.** 0.0026 s

53a. The maximum value is approximately 15.66, so there are about 15 h 40 min of daylight in Seattle on the longest day of the year.

53b. The graph of y = f(x) intersects the graph of y = 14 at x ≈ 111 and x ≈ 233; this represents the period from April 21 to August 21.

55a. The period is about 11.6 months, which shows that the business has a yearly cycle and it is likely that there is more business during the summer months.

55b. Approximately $366; 6 years is 72 months; Evaluating the function for x = 72 shows that P(72) ≈ 366.

57. $\frac{\pi}{3} < x < \pi$ or $\frac{5\pi}{3} < x < 2\pi$

59. Sample answer: All trigonometric functions are periodic. Therefore, once one or more solutions are found for a certain interval, there will be additional solutions that can be found by adding integral multiples of the period of the function to those solutions.

61. 0, b, or 2b

Module 13 Review

1. $\frac{\sqrt{17}}{8}$ **3.** D   **5.** B, D

7. Step 1: Original equation
    Step 2: Pythagorean Identity
    Step 3: Factor.
    Step 4: Simplify.
    Step 5: Write as two functions.
    Step 6: Divide.
    Step 7: Reciprocal Identity

9. B   **11.** A   **13.** $\frac{\sqrt{2-\sqrt{2}}}{2}$ **15.** A, C

17. A, D, F, H

Glossary

English	Español

A

30°-60°-90° triangle A right triangle with two acute angles that measure 30° and 60°.

45°-45°-90° triangle A right triangle with two acute angles that measure 45°.

absolute value The distance a number is from zero on the number line.

absolute value function A function written as $f(x) = |x|$, in which $f(x) \geq 0$ for all values of x.

accuracy The nearness of a measurement to the true value of the measure.

additive identity Because the sum of any number a and 0 is equal to a, 0 is the additive identity.

additive inverses Two numbers with a sum of 0.

adjacent angles Two angles that lie in the same plane and have a common vertex and a common side but have no common interior points.

adjacent arcs Arcs in a circle that have exactly one point in common.

algebraic expression A mathematical expression that contains at least one variable.

algebraic notation Mathematical notation that describes a set by using algebraic expressions.

alternate exterior angles When two lines are cut by a transversal, nonadjacent exterior angles that lie on opposite sides of the transversal.

alternate interior angles When two lines are cut by a transversal, nonadjacent interior angles that lie on opposite sides of the transversal.

altitude of a parallelogram A perpendicular segment between any two parallel bases.

triángulo 30°-60°-90° Un triángulo rectángulo con dos ángulos agudos que miden 30° y 60°.

triángulo 45°-45°-90° Un triángulo rectángulo con dos ángulos agudos que miden 45°.

valor absoluto La distancia que un número es de cero en la línea numérica.

función del valor absoluto Una función que se escribe $f(x) = |x|$, donde $f(x) \geq 0$, para todos los valores de x.

exactitud La proximidad de una medida al valor verdadero de la medida.

identidad aditiva Debido a que la suma de cualquier número a y 0 es igual a, 0 es la identidad aditiva.

inverso aditivos Dos números con una suma de 0.

ángulos adyacentes Dos ángulos que se encuentran en el mismo plano y tienen un vértice común y un lado común, pero no tienen puntos comunes en el interior.

arcos adyacentes Arcos en un circulo que tienen un solo punto en común.

expresión algebraica Una expresión matemática que contiene al menos una variable.

notación algebraica Notación matemática que describe un conjunto usando expresiones algebraicas.

ángulos alternos externos Cuando dos líneas son cortadas por un ángulo transversal, no adyacente exterior que se encuentran en lados opuestos de la transversal.

ángulos alternos internos Cuando dos líneas son cortadas por un ángulo transversal, no adyacente interior que se encuentran en lados opuestos de la transversal.

altitud de un paralelogramo Un segmento perpendicular entre dos bases paralelas.

altitude of a prism or cylinder A segment perpendicular to the bases that joins the planes of the bases.

altitude of a pyramid or cone A segment perpendicular to the base that has the vertex as one endpoint and a point in the plane of the base as the other endpoint.

altitude of a triangle A segment from a vertex of the triangle to the line containing the opposite side and perpendicular to that side.

ambiguous case When two different triangles could be created or described using the given information.

amplitude For functions of the form $y = a \sin b\theta$ or $y = a \cos b\theta$, the amplitude is $|a|$.

analytic geometry The study of geometry that uses the coordinate system.

angle The intersection of two noncollinear rays at a common endpoint.

angle bisector A ray or segment that divides an angle into two congruent angles.

angle of depression The angle formed by a horizontal line and an observer's line of sight to an object below the horizontal line.

angle of elevation The angle formed by a horizontal line and an observer's line of sight to an object above the horizontal line.

angle of rotation The angle through which a figure rotates.

apothem A perpendicular segment between the center of a regular polygon and a side of the polygon or the length of that line segment.

approximate error The positive difference between an actual measurement and an approximate or estimated measurement.

arc Part of a circle that is defined by two endpoints.

arc length The distance between the endpoints of an arc measured along the arc in linear units.

altitud de un prisma o cilindro Un segmento perpendicular a las bases que une los planos de las bases.

altitud de una pirámide o cono Un segmento perpendicular a la base que tiene el vértice como un punto final y un punto en el plano de la base como el otro punto final.

altitud de triángulo Un segmento de un vértice del triángulo a la línea que contiene el lado opuesto y perpendicular a ese lado.

caso ambiguo Cuando dos triángulos diferentes pueden ser creados o descritos usando la información dada.

amplitud Para funciones de la forma $y = a \, \text{sen} \, b\theta$ o $y = a \cos b\theta$, la amplitud es $|a|$.

geometría analítica El estudio de la geometría que utiliza el sistema de coordenadas.

ángulo La intersección de dos rayos no colineales en un extremo común.

bisectriz de un ángulo Un rayo o segmento que divide un ángulo en dos ángulos congruentes.

ángulo de depresión El ángulo formado por una línea horizontal y la línea de visión de un observador a un objeto por debajo de la línea horizontal.

ángulo de elevación El ángulo formado por una línea horizontal y la línea de visión de un observador a un objeto por encima de la línea horizontal.

ángulo de rotación El ángulo a través del cual gira una figura.

apotema Un segmento perpendicular entre el centro de un polígono regular y un lado del polígono o la longitud de ese segmento de línea.

error aproximado La diferencia positiva entre una medida real y una medida aproximada o estimada.

arco Parte de un círculo que se define por dos puntos finales.

longitude de arco La distancia entre los extremos de un arco medido a lo largo del arco en unidades lineales.

area   The number of square units needed to cover a surface.

arithmetic sequence   A pattern in which each term after the first is found by adding a constant, the common difference d, to the previous term.

asymptote   A line that a graph approaches.

auxiliary line   An extra line or segment drawn in a figure to help analyze geometric relationships.

average rate of change   The change in the value of the dependent variable divided by the change in the value of the independent variable.

axiom   A statement that is accepted as true without proof.

axiomatic system   A set of axioms from which theorems can be derived.

axis of symmetry   A line about which a graph is symmetric.

axis symmetry   If a figure can be mapped onto itself by a rotation between 0° and 360° in a line.

área   El número de unidades cuadradas para cubrir una superficie.

secuencia aritmética   Un patrón en el cual cada término después del primero se encuentra añadiendo una constante, la diferencia común d, al término anterior.

asíntota   Una línea que se aproxima a un gráfico.

línea auxiliar   Una línea o segmento extra dibujado en una figura para ayudar a analizar las relaciones geométricas.

tasa media de cambio   El cambio en el valor de la variable dependiente dividido por el cambio en el valor de la variable independiente.

axioma   Una declaración que se acepta como verdadera sin prueba.

sistema axiomático   Un conjunto de axiomas de los cuales se pueden derivar teoremas.

eje de simetría   Una línea sobre la cual un gráfica es simétrico.

eje simetría   Si una figura puede ser asignada sobre sí misma por una rotación entre 0° y 360° en una línea.

B

bar graph   A graphical display that compares categories of data using bars of different heights.

base   In a power, the number being multiplied by itself.

base angles of a trapezoid   The two angles formed by the bases and legs of a trapezoid.

base angles of an isosceles triangle   The two angles formed by the base and the congruent sides of an isosceles triangle.

base edge   The intersection of a lateral face and a base in a solid figure.

base of a parallelogram   Any side of a parallelogram.

base of a pyramid or cone   The face of the solid opposite the vertex of the solid.

gráfico de barra   Una pantalla gráfica que compara las categorías de datos usando barras de diferentes alturas.

base   En un poder, el número se multiplica por sí mismo.

ángulos de base de un trapecio   Los dos ángulos formados por las bases y patas de un trapecio.

ángulo de la base de un triángulo isosceles   Los dos ángulos formados por la base y los lados congruentes de un triángulo isosceles.

arista de la base   La intersección de una cara lateral y una base en una figura sólida.

base de un paralelogramo   Cualquier lado de un paralelogramo.

base de una pirámide o cono   La cara del sólido opuesta al vértice del sólido.

bases of a prism or cylinder   The two parallel congruent faces of the solid.

bases of a trapezoid   The parallel sides in a trapezoid.

best-fit line   The line that most closely approximates the data in a scatter plot.

betweenness of points   Point C is between A and B if and only if A, B, and C are collinear and $AC + CB = AB$.

bias   An error that results in a misrepresentation of a population.

biconditional statement   The conjunction of a conditional and its converse.

binomial   The sum of two monomials.

bisect   To separate a line segment into two congruent segments.

bivariate data   Data that consists of pairs of values.

boundary   The edge of the graph of an inequality that separates the coordinate plane into regions.

bounded   When the graph of a system of constraints is a polygonal region.

box plot   A graphical representation of the five-number summary of a data set.

bases de un prisma o cilindro   Las dos caras congruentes paralelas de la figura sólida.

bases de un trapecio   Los lados paralelos en un trapecio.

línea de ajuste óptimo   La línea que más se aproxima a los datos en un diagrama de dispersión.

intermediación de puntos   El punto C está entre A y B si y sólo si A, B, y C son colineales y $AC + CB = AB$.

sesgo   Un error que resulta en una tergiversación de una población.

declaración bicondicional   La conjunción de un condicional y su inverso.

binomio   La suma de dos monomios.

bisecar   Separe un segmento de línea en dos segmentos congruentes.

datos bivariate   Datos que constan de pares de valores.

frontera   El borde de la gráfica de una desigualdad que separa el plano de coordenadas en regiones.

acotada   Cuando la gráfica de un sistema de restricciones es una región poligonal.

diagram de caja   Una representación gráfica del resumen de cinco números de un conjunto de datos.

C

categorical data   Data that can be organized into different categories.

causation   When a change in one variable produces a change in another variable.

center of a circle   The point from which all points on a circle are the same distance.

center of a regular polygon   The center of the circle circumscribed about a regular polygon.

center of dilation   The center point from which dilations are performed.

datos categóricos   Datos que pueden organizarse en diferentes categorías.

causalidad   Cuando un cambio en una variable produce un cambio en otra variable.

centro de un círculo   El punto desde el cual todos los puntos de un círculo están a la misma distancia.

centro de un polígono regular   El centro del círculo circunscrito alrededor de un polígono regular.

centro de dilatación   Punto fijo en torno al cual se realizan las homotecias.

center of rotation The fixed point about which a figure rotates.

centro de rotación El punto fijo sobre el que gira una figura.

center of symmetry A point in which a figure can be rotated onto itself.

centro de la simetría Un punto en el que una figura se puede girar sobre sí misma.

central angle of a circle An angle with a vertex at the center of a circle and sides that are radii.

ángulo central de un círculo Un ángulo con un vértice en el centro de un círculo y los lados que son radios.

central angle of a regular polygon An angle with its vertex at the center of a regular polygon and sides that pass through consecutive vertices of the polygon.

ángulo central de un polígono regular Un ángulo con su vértice en el centro de un polígono regular y lados que pasan a través de vértices consecutivos del polígono.

centroid The point of concurrency of the medians of a triangle.

baricentro El punto de intersección de las medianas de un triángulo.

chord of a circle or sphere A segment with endpoints on the circle or sphere.

cuerda de un círculo o esfera Un segmento con extremos en el círculo o esfera.

circle The set of all points in a plane that are the same distance from a given point called the center.

círculo El conjunto de todos los puntos en un plano que están a la misma distancia de un punto dado llamado centro.

circular function A function that describes a point on a circle as the function of an angle defined in radians.

función circular Función que describe un punto en un círculo como la función de un ángulo definido en radianes.

circumcenter The point of concurrency of the perpendicular bisectors of the sides of a triangle.

circuncentro El punto de concurrencia de las bisectrices perpendiculares de los lados de un triángulo.

circumference The distance around a circle.

circunferencia La distancia alrededor de un círculo.

circumscribed angle An angle with sides that are tangent to a circle.

ángulo circunscrito Un ángulo con lados que son tangentes a un círculo.

circumscribed polygon A polygon with vertices outside the circle and sides that are tangent to the circle.

poligono circunscrito Un polígono con vértices fuera del círculo y lados que son tangentes al círculo.

closed If for any members in a set, the result of an operation is also in the set.

cerrado Si para cualquier número en el conjunto, el resultado de la operación es también en el conjunto.

closed half-plane The solution of a linear inequality that includes the boundary line.

semi-plano cerrado La solución de una desigualdad linear que incluye la línea de limite.

codomain The set of all the y-values that could possibly result from the evaluation of the function.

codominar El conjunto de todos los valores y que podrían resultar de la evaluación de la función.

coefficient The numerical factor of a term.

coeficiente El factor numérico de un término.

coefficient of determination An indicator of how well a function fits a set of data.

coeficiente de determinación Un indicador de lo bien que una función se ajusta a un conjunto de datos.

cofunction identities Identities that show the relationships between sine and cosine, tangent and cotangent, and secant and cosecant.

collinear Lying on the same line.

combination A selection of objects in which order is not important.

combined variation When one quantity varies directly and/or inversely as two or more other quantities.

common difference The difference between consecutive terms in an arithmetic sequence.

common logarithms Logarithms of base 10.

common ratio The ratio of consecutive terms of a geometric sequence.

common tangent A line or segment that is tangent to two circles in the same plane.

complement of A All of the outcomes in the sample space that are not included as outcomes of event A.

complementary angles Two angles with measures that have a sum of 90°.

completing the square A process used to make a quadratic expression into a perfect square trinomial.

complex conjugates Two complex numbers of the form $a + bi$ and $a - bi$.

complex fraction A rational expression with a numerator and/or denominator that is also a rational expression.

complex number Any number that can be written in the form $a + bi$, where a and b are real numbers and i is the imaginary unit.

component form A vector written as $<x, y>$, which describes the vector in terms of its horizontal component x and vertical component y.

identidades de cofunción Identidades que muestran las relaciones entre seno y coseno, tangente y cotangente, y secante y cosecante.

colineal Acostado en la misma línea.

combinación Una selección de objetos en los que el orden no es importante.

variación combinada Cuando una cantidad varía directamente y / o inversamente como dos o más cantidades.

diferencia común La diferencia entre términos consecutivos de una secuencia aritmética.

logaritmos comunes Logaritmos de base 10.

razón común El razón de términos consecutivos de una secuencia geométrica.

tangente común Una línea o segmento que es tangente a dos círculos en el mismo plano.

complemento de A Todos los resultados en el espacio muestral que no se incluyen como resultados del evento A.

ángulo complementarios Dos ángulos con medidas que tienen una suma de 90°.

completar el cuadrado Un proceso usado para hacer una expresión cuadrática en un trinomio cuadrado perfecto.

conjugados complejos Dos números complejos de la forma $a + bi$ y $a - bi$.

fracción compleja Una expresión racional con un numerador y / o denominador que también es una expresión racional.

número complejo Cualquier número que se puede escribir en la forma $a + bi$, donde a y b son números reales e i es la unidad imaginaria.

forma de componente Un vector escrito como $<x, y>$, que describe el vector en términos de su componente horizontal x y componente vertical y.

composite figure   A figure that can be separated into regions that are basic figures, such as triangles, rectangles, trapezoids, and circles.

composite solid   A three-dimensional figure that is composed of simpler solids.

composition of functions   An operation that uses the results of one function to evaluate a second function.

composition of transformations   When a transformation is applied to a figure and then another transformation is applied to its image.

compound event   Two or more simple events.

compound inequality   Two or more inequalities that are connected by the words *and* or *or*.

compound interest   Interest calculated on the principal and on the accumulated interest from previous periods.

compound statement   Two or more statements joined by the word *and* or *or*.

concave polygon   A polygon with one or more interior angles with measures greater than 180°.

concentric circles   Coplanar circles that have the same center.

conclusion   The statement that immediately follows the word *then* in a conditional.

concurrent lines   Three or more lines that intersect at a common point.

conditional probability   The probability that an event will occur given that another event has already occurred.

conditional relative frequency   The ratio of the joint frequency to the marginal frequency.

conditional statement   A compound statement that consists of a premise, or hypothesis, and a conclusion, which is false only when its premise is true and its conclusion is false.

figura compuesta   Una figura que se puede separar en regiones que son figuras básicas, tales como triángulos, rectángulos, trapezoides, y círculos.

solido compuesta   Una figura tridimensional que se compone de figuras más simples.

composición de funciones   Operación que utiliza los resultados de una función para evaluar una segunda función.

composición de transformaciones   Cuando una transformación se aplica a una figura y luego se aplica otra transformación a su imagen.

evento compuesto   Dos o más eventos simples.

desigualdad compuesta   Dos o más desigualdades que están unidas por las palabras *y* u *o*.

interés compuesto   Intereses calculados sobre el principal y sobre el interés acumulado de períodos anteriores.

enunciado compuesto   Dos o más declaraciones unidas por la palabra *y* o *o*.

polígono cóncavo   Un polígono con uno o más ángulos interiores con medidas superiores a 180°.

círculos concéntricos   Círculos coplanarios que tienen el mismo centro.

conclusión   La declaración que inmediatamente sigue la palabra *entonces* en un condicional.

líneas concurrentes   Tres o más líneas que se intersecan en un punto común.

probabilidad condicional   La probabilidad de que un evento ocurra dado que otro evento ya ha ocurrido.

frecuencia relativa condicional   La relación entre la frecuencia de la articulación y la frecuencia marginal.

enunciado condicional   Una declaración compuesta que consiste en una premisa, o hipótesis, y una conclusión, que es falsa solo cuando su premisa es verdadera y su conclusión es falsa.

cone   A solid figure with a circular base connected by a curved surface to a single vertex.

confidence interval   An estimate of the population parameter stated as a range with a specific degree of certainty.

congruent   Having the same size and shape.

congruent angles   Two angles that have the same measure.

congruent arcs   Arcs in the same or congruent circles that have the same measure.

congruent polygons   All of the parts of one polygon are congruent to the corresponding parts or matching parts of another polygon.

congruent segments   Line segments that are the same length.

congruent solids   Solid figures that have exactly the same shape, size, and a scale factor of 1:1.

conic sections   Cross sections of a right circular cone.

conjecture   An educated guess based on known information and specific examples.

conjugates   Two expressions, each with two terms, in which the second terms are opposites.

conjunction   A compound statement using the word *and*.

consecutive interior angles   When two lines are cut by a transversal, interior angles that lie on the same side of the transversal.

consistent   A system of equations with at least one ordered pair that satisfies both equations.

constant function   A linear function of the form $y = b$; The function $f(x) = a$, where a is any number.

constant of variation   The constant in a variation function.

cono   Una figura sólida con una base circular conectada por una superficie curvada a un solo vértice.

intervalo de confianza   Una estimación del parámetro de población se indica como un rango con un grado específico de certeza.

congruente   Tener el mismo tamaño y forma.

ángulo congruentes   Dos ángulos que tienen la misma medida.

arcos congruentes   Arcos en los mismos círculos o congruentes que tienen la misma medida.

poligonos congruentes   Todas las partes de un polígono son congruentes con las partes correspondientes o partes coincidentes de otro polígono.

segmentos congruentes   Línea segmentos que son la misma longitud.

sólidos congruentes   Figuras sólidas que tienen exactamente la misma forma, tamaño y un factor de escala de 1:1.

secciones cónicas   Secciones transversales de un cono circular derecho.

conjetura   Una suposición educada basada en información conocida y ejemplos específicos.

conjugados   Dos expresiones, cada una con dos términos, en la que los segundos términos son opuestos.

conjunción   Una declaración compuesta usando la palabra *y*.

ángulos internos consecutivos   Cuando dos líneas se cortan por un ángulo transversal, interior que se encuentran en el mismo lado de la transversal.

consistente   Una sistema de ecuaciones para el cual existe al menos un par ordenado que satisfice ambas ecuaciones.

función constante   Una función lineal de la forma $y = b$; La función $f(x) = a$, donde a es cualquier número.

constante de variación   La constante en una función de variación.

constant term   A term that does not contain a variable.

término constante   Un término que no contiene una variable.

constraint   A condition that a solution must satisfy.

restricción   Una condición que una solución debe satisfacer.

constructions   Methods of creating figures without the use of measuring tools.

construcciones   Métodos de creación de figuras sin el uso de herramientas de medición.

continuous function   A function that can be graphed with a line or an unbroken curve.

función continua   Una función que se puede representar gráficamente con una línea o una curva ininterrumpida.

continuous random variable   The numerical outcome of a random event that can take on any value.

variable aleatoria continua   El resultado numérico de un evento aleatorio que puede tomar cualquier valor.

contrapositive   A statement formed by negating both the hypothesis and the conclusion of the converse of a conditional.

antítesis   Una afirmación formada negando tanto la hipótesis como la conclusión del inverso del condicional.

convenience sample   Members that are readily available or easy to reach are selected.

muestra conveniente   Se seleccionan los miembros que están fácilmente disponibles o de fácil acceso.

converse   A statement formed by exchanging the hypothesis and conclusion of a conditional statement.

recíproco   Una declaración formada por el intercambio de la hipótesis y la conclusión de la declaración condicional.

convex polygon   A polygon with all interior angles measuring less than 180°.

polígono convexo   Un polígono con todos los ángulos interiores que miden menos de 180°.

coordinate proofs   Proofs that use figures in the coordinate plane and algebra to prove geometric concepts.

pruebas de coordenadas   Pruebas que utilizan figuras en el plano de coordenadas y álgebra para probar conceptos geométricos.

coplanar   Lying in the same plane.

coplanar   Acostado en el mismo plano.

corollary   A theorem with a proof that follows as a direct result of another theorem.

corolario   Un teorema con una prueba que sigue como un resultado directo de otro teorema.

correlation coefficient   A measure that shows how well data are modeled by a regression function.

coeficiente de correlación   Una medida que muestra cómo los datos son modelados por una función de regresión.

corresponding angles   When two lines are cut by a transversal, angles that lie on the same side of a transversal and on the same side of the two lines.

ángulos correspondientes   Cuando dos líneas se cortan transversalmente, los ángulos que se encuentran en el mismo lado de una transversal y en el mismo lado de las dos líneas.

corresponding parts   Corresponding angles and corresponding sides of two polygons.

partes correspondientes   Ángulos correspondientes y lados correspondientes.

cosecant   The ratio of the length of a hypotenuse to the length of the leg opposite the angle.

cosecante   Relación entre la longitud de la hipotenusa y la longitud de la pierna opuesta al ángulo.

cosine   The ratio of the length of the leg adjacent to an angle to the length of the hypotenuse.

coseno   Relación entre la longitud de la pierna adyacente a un ángulo y la longitud de la hipotenusa.

cotangent   The ratio of the length of the leg adjacent to an angle to the length of the leg opposite the angle.

cotangente   La relación entre la longitud de la pata adyacente a un ángulo y la longitud de la pata opuesta al ángulo.

coterminal angles   Angles in standard position that have the same terminal side.

ángulos coterminales   Ángulos en posición estándar que tienen el mismo lado terminal.

counterexample   An example that contradicts the conjecture showing that the conjecture is not always true.

contraejemplo   Un ejemplo que contradice la conjetura que muestra que la conjetura no siempre es cierta.

critical values   The z-values corresponding to the most common degrees of certainty.

valores críticos   Los valores z correspondientes a los grados de certeza más comunes.

cross section   The intersection of a solid and a plane.

sección transversal   Intersección de un sólido con un plano.

cube root   One of three equal factors of a number.

raíz cúbica   Uno de los tres factores iguales de un número.

cube root function   A radical function that contains the cube root of a variable expression.

función de la raíz del cubo   Función radical que contiene la raíz cúbica de una expresión variable.

curve fitting   Finding a regression equation for a set of data that is approximated by a function.

ajuste de curvas   Encontrar una ecuación de regresión para un conjunto de datos que es aproximado por una función.

cycle   One complete pattern of a periodic function.

ciclo   Un patron completo de una función periódica.

cylinder   A solid figure with two congruent and parallel circular bases connected by a curved surface.

cilindro   Una figura sólida con dos bases circulares congruentes y paralelas conectadas por una superficie curvada.

D

decay factor   The base of an exponential expression, or $1 - r$.

factor de decaimiento   La base de una expresión exponencial, o $1 - r$.

decomposition   Separating a figure into two or more nonoverlapping parts.

descomposición   Separar una figura en dos o más partes que no se solapan.

decreasing   Where the graph of a function goes down when viewed from left to right.

decreciente   Donde la gráfica de una función disminuye cuando se ve de izquierda a derecha.

deductive argument   An argument that guarantees the truth of the conclusion provided that its premises are true.

argumento deductivo   Un argumento que garantiza la verdad de la conclusión siempre que sus premisas sean verdaderas.

deductive reasoning The process of reaching a specific valid conclusion based on general facts, rules, definitions, or properties.

define a variable To choose a variable to represent an unknown value.

defined term A term that has a definition and can be explained.

definitions An explanation that assigns properties to a mathematical object.

degree The value of the exponent in a power function; $\frac{1}{360}$ of the circular rotation about a point.

degree of a monomial The sum of the exponents of all its variables.

degree of a polynomial The greatest degree of any term in the polynomial.

density A measure of the quantity of some physical property per unit of length, area, or volume.

dependent A consistent system of equations with an infinite number of solutions.

dependent events Two or more events in which the outcome of one event affects the outcome of the other events.

dependent variable The variable in a relation, usually y, with values that depend on x.

depressed polynomial A polynomial resulting from division with a degree one less than the original polynomial.

descriptive modeling A way to mathematically describe real-world situations and the factors that cause them.

descriptive statistics The branch of statistics that focuses on collecting, summarizing, and displaying data.

diagonal A segment that connects any two nonconsecutive vertices within a polygon.

razonamiento deductivo El proceso de alcanzar una conclusión válida específica basada en hechos generales, reglas, definiciones, o propiedades.

definir una variable Para elegir una variable que represente un valor desconocido.

término definido Un término que tiene una definición y se puede explicar.

definiciones Una explicación que asigna propiedades a un objeto matemático.

grado Valor del exponente en una función de potencia. $\frac{1}{360}$ de la rotación circular alrededor de un punto.

grado de un monomio La suma de los exponents de todas sus variables.

grado de un polinomio El grado mayor de cualquier término del polinomio.

densidad Una medida de la cantidad de alguna propiedad física por unidad de longitud, área o volumen.

dependiente Una sistema consistente de ecuaciones con un número infinito de soluciones.

eventos dependientes Dos o más eventos en que el resultado de un evento afecta el resultado de los otros eventos.

variable dependiente La variable de una relación, generalmente y, con los valores que depende de x.

polinomio reducido Un polinomio resultante de la división con un grado uno menos que el polinomio original.

modelado descriptivo Una forma de describir matemáticamente las situaciones del mundo real y los factores que las causan.

estadística descriptiva Rama de la estadística cuyo enfoque es la recopilación, resumen y demostración de los datos.

diagonal Un segmento que conecta cualquier dos vértices no consecutivos dentro de un polígono.

diameter of a circle or sphere   A chord that passes through the center of a circle or sphere.

difference of squares   A binomial in which the first and last terms are perfect squares.

difference of two squares   The square of one quantity minus the square of another quantity.

dilation   A nonrigid motion that enlarges or reduces a geometric figure; A transformation that stretches or compresses the graph of a function.

dimensional analysis   The process of performing operations with units.

direct variation   When one quantity is equal to a constant times another quantity.

directed line segment   A line segment with an initial endpoint and a terminal endpoint.

directrix   An exterior line perpendicular to the line containing the foci of a curve.

discontinuous function   A function that is not continuous.

discrete function   A function in which the points on the graph are not connected.

discrete random variable   The numerical outcome of a random event that is finite and can be counted.

discriminant   In the Quadratic Formula, the expression under the radical sign that provides information about the roots of the quadratic equation.

disjunction   A compound statement using the word *or*.

distance   The length of the line segment between two points.

distribution   A graph or table that shows the theoretical frequency of each possible data value.

domain   The set of the first numbers of the ordered pairs in a relation; The set of *x*-values to be evaluated by a function.

diámetro de un círculo o esfera   Un acorde que pasa por el centro de un círculo o esfera.

diferencia de cuadrados   Un binomio en el que los términos primero y último son cuadrados perfectos.

diferencia de dos cuadrados   El cuadrado de una cantidad menos el cuadrado de otra cantidad.

dilatación   Un movimiento no rígido que agranda o reduce una figura geométrica; Una transformación que estira o comprime el gráfico de una función.

análisis dimensional   El proceso de realizar operaciones con unidades.

variación directa   Cuando una cantidad es igual a una constante multiplicada por otra cantidad.

segment de línea dirigido   Un segmento de línea con un punto final inicial y un punto final terminal.

directriz   Una línea exterior perpendicular a la línea que contiene los focos de una curva.

función discontinua   Una función que no es continua.

función discreta   Una función en la que los puntos del gráfico no están conectados.

variable aleatoria discreta   El resultado numérico de un evento aleatorio que es finito y puede ser contado.

discriminante   En la Fórmula cuadrática, la expresión bajo el signo radical que proporciona información sobre las raíces de la ecuación cuadrática.

disyunción   Una declaración compuesta usando la palabra *o*.

distancia   La longitud del segmento de línea entre dos puntos.

distribución   Un gráfico o una table que muestra la frecuencia teórica de cada valor de datos posible.

dominio   El conjunto de los primeros números de los pares ordenados en una relación; El conjunto de valores *x* para ser evaluados por una función.

dot plot   A diagram that shows the frequency of data on a number line.

double root   Two roots of a quadratic equation that are the same number.

gráfica de puntos   Una diagrama que muestra la frecuencia de los datos en una línea numérica.

raíces dobles   Dos raíces de una función cuadrática que son el mismo número.

E

e   An irrational number that approximately equals 2.7182818....

edge of a polyhedron   A line segment where the faces of the polyhedron intersect.

elimination   A method that involves eliminating a variable by combining the individual equations within a system of equations.

empty set   The set that contains no elements, symbolized by { } or ∅.

end behavior   The behavior of a graph at the positive and negative extremes in its domain.

enlargement   A dilation with a scale factor greater than 1.

equation   A mathematical statement that contains two expressions and an equal sign, =.

equiangular polygon   A polygon with all angles congruent.

equidistant   A point is equidistant from other points if it is the same distance from them.

equidistant lines   Two lines for which the distance between the two lines, measured along a perpendicular line or segment to the two lines, is always the same.

equilateral polygon   A polygon with all sides congruent.

equivalent equations   Two equations with the same solution.

equivalent expressions   Expressions that represent the same value.

evaluate   To find the value of an expression.

e   Un número irracional que es aproximadamente igual a 2.7182818 ....

arista de un poliedro   Un segmento de línea donde las caras del poliedro se cruzan.

eliminación   Un método que consiste en eliminar una variable combinando las ecuaciones individuales dentro de un sistema de ecuaciones.

conjunto vacio   El conjunto que no contiene elementos, simbolizado por { } o ∅.

comportamiento extremo   El comportamiento de un gráfico en los extremos positivo y negativo en su dominio.

ampliación   Una dilatación con un factor de escala mayor que 1.

ecuación   Un enunciado matemático que contiene dos expresiones y un signo igual, =.

polígono equiangular   Un polígono con todos los ángulos congruentes.

equidistante   Un punto es equidistante de otros puntos si está a la misma distancia de ellos.

líneas equidistantes   Dos líneas para las cuales la distancia entre las dos líneas, medida a lo largo de una línea o segmento perpendicular a las dos líneas, es siempre la misma.

polígono equilátero   Un polígono con todos los lados congruentes.

ecuaciones equivalentes   Dos ecuaciones con la misma solución.

expresiones equivalentes   Expresiones que representan el mismo valor.

evaluar   Calcular el valor de una expresión.

even functions Functions that are symmetric in the *y*-axis.

event A subset of the sample space.

excluded values Values for which a function is not defined.

experiment A sample is divided into two groups. The experimental group undergoes a change, while there is no change to the control group. The effects on the groups are then compared; A situation involving chance.

experimental probability Probability calculated by using data from an actual experiment.

exponent When *n* is a positive integer in the expression x^n, *n* indicates the number of times *x* is multiplied by itself.

exponential decay Change that occurs when an initial amount decreases by the same percent over a given period of time.

exponential decay function A function in which the independent variable is an exponent, where $a > 0$ and $0 < b < 1$.

explicit formula A formula that allows you to find any term a_n of a sequence by using a formula written in terms of *n*.

exponential equation An equation in which the independent variable is an exponent.

exponential form When an expression is in the form x^n.

exponential function A function in which the independent variable is an exponent.

exponential growth Change that occurs when an initial amount increases by the same percent over a given period of time.

exponential growth function A function in which the independent variable is an exponent, where $a > 0$ and $b > 1$.

incluso funciones Funciones que son simétricas en el eje *y*.

evento Un subconjunto del espacio de muestra.

valores excluidos Valores para los que no se ha definido una función.

experimento Una muestra se divide en dos grupos. El grupo experimental experimenta un cambio, mientras que no hay cambio en el grupo de control. A continuación se comparan los efectos sobre los grupos; Una situación de riesgo.

probabilidad experimental Probabilidad calculada utilizando datos de un experimento real.

exponente Cuando *n* es un entero positivo en la expresión x^n, *n* indica el número de veces que *x* se multiplica por sí mismo.

desintegración exponencial Cambio que ocurre cuando una cantidad inicial disminuye en el mismo porcentaje durante un período de tiempo dado.

función exponenciales de decaimiento Una ecuación en la que la variable independiente es un exponente, donde $a > 0$ y $0 < b < 1$.

fórmula explícita Una fórmula que le permite encontrar cualquier término a_n de una secuencia usando una fórmula escrita en términos de *n*.

ecuación exponencial Una ecuación en la que la variable independiente es un exponente.

forma exponencial Cuando una expresión está en la forma x^n.

función exponencial Una función en la que la variable independiente es el exponente.

crecimiento exponencial Cambio que ocurre cuando una cantidad inicial aumenta por el mismo porcentaje durante un período de tiempo dado.

función de crecimiento exponencial Una función en la que la variable independiente es el exponente, donde $a > 0$ y $b > 1$.

exponential inequality   An inequality in which the independent variable is an exponent.

desigualdad exponencial   Una desigualdad en la que la variable independiente es un exponente.

exterior angle of a triangle   An angle formed by one side of the triangle and the extension of an adjacent side.

ángulo exterior de un triángulo   Un ángulo formado por un lado del triángulo y la extensión de un lado adyacente.

exterior angles   When two lines are cut by a transversal, any of the four angles that lie outside the region between the two intersected lines.

ángulos externos   Cuando dos líneas son cortadas por una transversal, cualquiera de los cuatro ángulos que se encuentran fuera de la región entre las dos líneas intersectadas.

exterior of an angle   The area outside of the two rays of an angle.

exterior de un ángulo   El área fuera de los dos rayos de un ángulo.

extraneous solution   A solution of a simplified form of an equation that does not satisfy the original equation.

solución extraña   Una solución de una forma simplificada de una ecuación que no satisface la ecuación original.

extrema   Points that are the locations of relatively high or low function values.

extrema   Puntos que son las ubicaciones de valores de función relativamente alta o baja.

extreme values   The least and greatest values in a set of data.

valores extremos   Los valores mínimo y máximo en un conjunto de datos.

F

face of a polyhedron   A flat surface of a polyhedron.

cara de un poliedro   Superficie plana de un poliedro.

factored form   A form of quadratic equation, $0 = a(x - p)(x - q)$, where $a \neq 0$, in which p and q are the x-intercepts of the graph of the related function.

forma factorizada   Una forma de ecuación cuadrática, $0 = a(x - p)(x - q)$, donde $a \neq 0$, en la que p y q son las intercepciones x de la gráfica de la función relacionada.

factorial of n   The product of the positive integers less than or equal to n.

factorial de n   El producto de los enteros positivos inferiores o iguales a n.

factoring   The process of expressing a polynomial as the product of monomials and polynomials.

factorización por agrupamiento   Utilizando la Propiedad distributiva para factorizar polinomios que possen cuatro o más términos.

factoring by grouping   Using the Distributive Property to factor some polynomials having four or more terms.

factorización   El proceso de expresar un polinomio como el producto de monomios y polinomios.

family of graphs   Graphs and equations of graphs that have at least one characteristic in common.

familia de gráficas   Gráficas y ecuaciones de gráficas que tienen al menos una característica común.

feasible region   The intersection of the graphs in a system of constraints.

región factible   La intersección de los gráficos en un sistema de restricciones.

finite sample space   A sample space that contains a countable number of outcomes.

espacio de muestra finito   Un espacio de muestra que contiene un número contable de resultados.

finite sequence A sequence that contains a limited number of terms.

five-number summary The minimum, quartiles, and maximum of a data set.

flow proof A proof that uses boxes and arrows to show the logical progression of an argument.

focus A point inside a parabola having the property that the distances from any point on the parabola to them and to a fixed line have a constant ratio for any points on the parabola.

formula An equation that expresses a relationship between certain quantities.

fractional distance An intermediary point some fraction of the length of a line segment.

frequency The number of cycles in a given unit of time.

function A relation in which each element of the domain is paired with exactly one element of the range.

function notation A way of writing an equation so that $y = f(x)$.

secuencia finita Una secuencia que contiene un número limitado de términos.

resumen de cinco números El mínimo, cuartiles y máximo de un conjunto de datos.

demostración de flujo Una prueba que usa cajas y flechas para mostrar la progresión lógica de un argumento.

foco Un punto dentro de una parábola que tiene la propiedad de que las distancias desde cualquier punto de la parábola a ellos ya una línea fija tienen una relación constante para cualquier punto de la parábola.

fórmula Una ecuación que expresa una relación entre ciertas cantidades.

distancia fraccionaria Un punto intermediario de alguna fracción de la longitud de un segmento de línea.

frecuencia El número de ciclos en una unidad del tiempo dada.

función Una relación en que a cada elemento del dominio de corresponde un único elemento del rango.

notación functional Una forma de escribir una ecuación para que $y = f(x)$.

G

geometric means The terms between two nonconsecutive terms of a geometric sequence; The nth root, where n is the number of elements in a set of numbers, of the product of the numbers.

geometric model A geometric figure that represents a real-life object.

geometric probability Probability that involves a geometric measure such as length or area.

geometric sequence A pattern of numbers that begins with a nonzero term and each term after is found by multiplying the previous term by a nonzero constant r.

geometric series The indicated sum of the terms in a geometric sequence.

medios geométricos Los términos entre dos términos no consecutivos de una secuencia geométrica; La enésima raíz, donde n es el número de elementos de un conjunto de números, del producto de los números.

modelo geométrico Una figura geométrica que representa un objeto de la vida real.

probabilidad geométrica Probabilidad que implica una medida geométrica como longitud o área.

secuencia geométrica Un patrón de números que comienza con un término distinto de cero y cada término después se encuentra multiplicando el término anterior por una constante no nula r.

series geométricas La suma indicada de los términos en una secuencia geométrica.

glide reflection   The composition of a translation followed by a reflection in a line parallel to the translation vector.

greatest integer function   A step function in which $f(x)$ is the greatest integer less than or equal to x.

growth factor   The base of an exponential expression, or $1 + r$.

reflexión del deslizamiento   La composición de una traducción seguida de una reflexión en una línea paralela al vector de traslación.

función entera más grande   Una función del paso en que $f(x)$ es el número más grande menos que o igual a x.

factor de crecimiento   La base de una expresión exponencial, o $1 + r$.

H

half-plane   A region of the graph of an inequality on one side of a boundary.

height of a parallelogram   The length of an altitude of the parallelogram.

height of a solid   The length of the altitude of a solid figure.

height of a trapezoid   The perpendicular distance between the bases of a trapezoid.

histogram   A graphical display that uses bars to display numerical data that have been organized in equal intervals.

horizontal asymptote   A horizontal line that a graph approaches.

hyperbola   The graph of a reciprocal function.

hypothesis   The statement that immediately follows the word *if* in a conditional.

semi-plano   Una región de la gráfica de una desigualdad en un lado de un límite.

altura de un paralelogramo   La longitud de la altitud del paralelogramo.

altura de un sólido   La longitud de la altitud de una figura sólida.

altura de un trapecio   La distancia perpendicular entre las bases de un trapecio.

histograma   Una exhibición gráfica que utiliza barras para exhibir los datos numéricos que se han organizado en intervalos iguales.

asíntota horizontal   Una línea horizontal que se aproxima a un gráfico.

hipérbola   La gráfica de una función recíproca.

hipótesis   La declaración que sigue inmediatamente a la palabra *si* en un condicional.

I

identity   An equation that is true for every value of the variable.

identity function   The function $f(x) = x$.

if-then statement   A compound statement of the form *if p, then q*, where p and q are statements.

image   The new figure in a transformation.

imaginary unit i The principal square root of -1.

incenter   The point of concurrency of the angle bisectors of a triangle.

identidad   Una ecuación que es verdad para cada valor de la variable.

función identidad   La función $f(x) = x$.

enunciado si-entonces   Enunciado compuesto de la forma *si p, entonces q*, donde p y q son enunciados.

imagen   La nueva figura en una transformación.

unidad imaginaria i La raíz cuadrada principal de -1.

incentro   El punto de intersección de las bisectrices interiors de un triángulo.

included angle   The interior angle formed by two adjacent sides of a triangle.

included side   The side of a triangle between two angles.

inconsistent   A system of equations with no ordered pair that satisfies both equations.

increasing   Where the graph of a function goes up when viewed from left to right.

independent   A consistent system of equations with exactly one solution.

independent events   Two or more events in which the outcome of one event does not affect the outcome of the other events.

independent variable   The variable in a relation, usually x, with a value that is subject to choice.

index   In nth roots, the value that indicates to what root the value under the radicand is being taken.

indirect measurement   Using similar figures and proportions to measure an object.

indirect proof   One assumes that the statement to be proven is false and then uses logical reasoning to deduce that a statement contradicts a postulate, theorem, or one of the assumptions.

indirect reasoning   Reasoning that eliminates all possible conclusions but one so that the one remaining conclusion must be true.

inductive reasoning   The process of reaching a conclusion based on a pattern of examples.

inequality   A mathematical sentence that contains $<$, $>$, $\leq$, $\geq$, or $\neq$.

inferential statistics   When the data from a sample is used to make inferences about the corresponding population.

infinite sample space   A sample space with outcomes that cannot be counted.

infinite sequence   A sequence that continues without end.

ángulo incluido   El ángulo interior formado por dos lados adyacentes de un triángulo.

lado incluido   El lado de un triángulo entre dos ángulos.

inconsistente   Una sistema de ecuaciones para el cual no existe par ordenado alguno que satisfaga ambas ecuaciones.

creciente   Donde la gráfica de una función sube cuando se ve de izquierda a derecha.

independiente   Un sistema consistente de ecuaciones con exactamente una solución.

eventos independientes   Dos o más eventos en los que el resultado de un evento no afecta el resultado de los otros eventos.

variable independiente   La variable de una relación, generalmente x, con el valor que sujeta a elección.

índice   En enésimas raíces, el valor que indica a qué raíz está el valor bajo la radicand.

medición indirecta   Usando figuras y proporciones similares para medir un objeto.

demostración indirecta   Se supone que la afirmación a ser probada es falsa y luego utiliza el razonamiento lógico para deducir que una afirmación contradice un postulado, teorema o uno de los supuestos.

razonamiento indirecto   Razonamiento que elimina todas las posibles conclusiones, pero una de manera que la conclusión que queda una debe ser verdad.

razonamiento inductive   El proceso de llegar a una conclusión basada en un patrón de ejemplos.

desigualdad   Una oración matemática que contiene uno o más de $<$, $>$, $\leq$, $\geq$, o $\neq$.

estadísticas inferencial   Cuando los datos de una muestra se utilizan para hacer inferencias sobre la población correspondiente.

espacio de muestra infinito   Un espacio de muestra con resultados que no pueden ser contados.

secuencia infinita   Una secuencia que continúa sin fin.

informal proof   A paragraph that explains why the conjecture for a given situation is true.

prueba informal   Un párrafo que explica por qué la conjetura para una situación dada es verdadera.

initial side   The part of an angle that is fixed on the *x*-axis.

lado inicial   La parte de un ángulo que se fija en el eje *x*.

inscribed angle   An angle with its vertex on a circle and sides that contain chords of the circle.

ángulo inscrito   Un ángulo con su vértice en un círculo y lados que contienen acordes del círculo.

inscribed polygon   A polygon inside a circle in which all of the vertices of the polygon lie on the circle.

polígono inscrito   Un polígono dentro de un círculo en el que todos los vértices del polígono se encuentran en el círculo.

intercept   A point at which the graph of a function intersects an axis.

interceptar   Un punto en el que la gráfica de una función corta un eje.

intercepted arc   The part of a circle that lies between the two lines intersecting it.

arco intersecado   La parte de un círculo que se encuentra entre las dos líneas que se cruzan.

interior angle of a triangle   An angle at the vertex of a triangle.

ángulo interior de un triángulo   Un ángulo en el vértice de un triángulo.

interior angles   When two lines are cut by a transversal, any of the four angles that lie inside the region between the two intersected lines.

ángulos interiores   Cuando dos líneas son cortadas por una transversal, cualquiera de los cuatro ángulos que se encuentran dentro de la región entre las dos líneas intersectadas.

interior of an angle   The area between the two rays of an angle.

interior de un ángulo   El área entre los dos rayos de un ángulo.

interquartile range   The difference between the upper and lower quartiles of a data set.

rango intercuartil   La diferencia entre el cuartil superior *y* el cuartil inferior de un conjunto de datos.

intersection   A set of points common to two or more geometric figures; intersection   The graph of a compound inequality containing *and*.

intersección   Un conjunto de puntos communes a dos o más figuras geométricas; intersección   La gráfica de una desigualdad compuesta que contiene la palabra *y*.

intersection of *A* and *B*   The set of all outcomes in the sample space of event *A* that are also in the sample space of event *B*.

intersección de *A* y *B*   El conjunto de todos los resultados en el espacio muestral del evento *A* que también se encuentran en el espacio muestral del evento *B*.

interval   The distance between two numbers on the scale of a graph.

intervalo   La distancia entre dos números en la escala de un gráfico.

interval notation   Mathematical notation that describes a set by using endpoints with parentheses or brackets.

notación de intervalo   Notación matemática que describe un conjunto utilizando puntos finales con paréntesis o soportes.

inverse   A statement formed by negating both the hypothesis and conclusion of a conditional statement.

inverso   Una declaración formada negando tanto la hipótesis como la conclusión de la declaración condicional.

inverse cosine The ratio of the length of the hypotenuse to the length of the leg adjacent to an angle.

inverse functions Two functions, one of which contains points of the form (a, b) while the other contains points of the form (b, a).

inverse relations Two relations, one of which contains points of the form (a, b) while the other contains points of the form (b, a).

inverse sine The ratio of the length of the hypotenuse to the length of the leg opposite an angle.

inverse tangent The ratio of the length of the leg adjacent to an angle to the length of the leg opposite the angle.

inverse trigonometric functions Arcsine, Arccosine, and Arctangent.

inverse variation When the product of two quantities is equal to a constant k.

isosceles trapezoid A quadrilateral in which two sides are parallel and the legs are congruent.

isosceles triangle A triangle with at least two sides congruent.

inverso del coseno Relación de la longitud de la hipotenusa con la longitud de la pierna adyacente a un ángulo.

funciones inversas Dos funciones, una de las cuales contiene puntos de la forma (a, b) mientras que la otra contiene puntos de la forma (b, a).

relaciones inversas Dos relaciones, una de las cuales contiene puntos de la forma (a, b) mientras que la otra contiene puntos de la forma (b, a).

inverso del seno Relación de la longitud de la hipotenusa con la longitud de la pierna opuesta a un ángulo.

inverso del tangente Relación de la longitud de la pierna adyacente a un ángulo con la longitud de la pierna opuesta a un ángulo.

funciones trigonométricas inversas Arcsine, Arccosine y Arctangent.

variación inversa Cuando el producto de dos cantidades es igual a una constante k.

trapecio isósceles Un cuadrilátero en el que dos lados son paralelos y las patas son congruentes.

triángulo isósceles Un triángulo con al menos dos lados congruentes.

J

joint frequencies Entries in the body of a two-way frequency table. In a two-way frequency table, the frequencies in the interior of the table.

joint variation When one quantity varies directly as the product of two or more other quantities.

frecuencias articulares Entradas en el cuerpo de una tabla de frecuencias de dos vías. En una tabla de frecuencia bidireccional, las frecuencias en el interior de la tabla.

variación conjunta Cuando una cantidad varía directamente como el producto de dos o más cantidades.

K

kite A convex quadrilateral with exactly two distinct pairs of adjacent congruent sides.

cometa Un cuadrilátero convexo con exactamente dos pares distintos de lados congruentes adyacentes.

L

lateral area The sum of the areas of the lateral faces of the figure.

área lateral La suma de las áreas de las caras laterales de la figura.

lateral edges   The intersection of two lateral faces.

lateral faces   The faces that join the bases of a solid.

lateral surface of a cone   The curved surface that joins the base of a cone to the vertex.

lateral surface of a cylinder   The curved surface that joins the bases of a cylinder.

leading coefficient   The coefficient of the first term when a polynomial is in standard form.

legs of a trapezoid   The nonparallel sides in a trapezoid.

legs of an isosceles triangle   The two congruent sides of an isosceles triangle.

like radical expressions   Radicals in which both the index and the radicand are the same.

like terms   Terms with the same variables, with corresponding variables having the same exponent.

line   A line is made up of points, has no thickness or width, and extends indefinitely in both directions.

line of fit   A line used to describe the trend of the data in a scatter plot.

line of reflection   A line midway between a preimage and an image; The line in which a reflection flips the graph of a function.

line of symmetry   An imaginary line that separates a figure into two congruent parts.

line segment   A measurable part of a line that consists of two points, called endpoints, and all of the points between them.

line symmetry   A graph has line symmetry if it can be reflected in a vertical line so that each half of the graph maps exactly to the other half.

linear equation   An equation that can be written in the form $Ax + By = C$ with a graph that is a straight line.

aristas laterales   La intersección de dos caras laterales.

caras laterales   Las caras que unen las bases de un sólido.

superficie lateral de un cono   La superficie curvada que une la base de un cono con el vértice.

superficie lateral de un cilindro   La superficie curvada que une las bases de un cilindro.

coeficiente líder   El coeficiente del primer término cuando un polinomio está en forma estándar.

patas de un trapecio   Los lados no paralelos en un trapezoide.

patas de un triángulo isósceles   Los dos lados congruentes de un triángulo isósceles.

expresiones radicales semejantes   Radicales en los que tanto el índice como el radicand son iguales.

términos semejantes   Términos con las mismas variables, con las variables correspondientes que tienen el mismo exponente.

línea   Una línea está formada por puntos, no tiene espesor ni anchura, y se extiende indefinidamente en ambas direcciones.

línea de ajuste   Una línea usada para describir la tendencia de los datos en un diagrama de dispersión.

línea de reflexión   Una línea a medio camino entre una preimagen y una imagen; La línea en la que una reflexión voltea la gráfica de una función.

línea de simetría   Una línea imaginaria que separa una figura en dos partes congruentes.

segmento de línea   Una parte medible de una línea que consta de dos puntos, llamados extremos, y todos los puntos entre ellos.

simetría de línea   Un gráfico tiene simetría de línea si puede reflejarse en una línea vertical, de modo que cada mitad del gráfico se asigna exactamente a la otra mitad.

ecuación lineal   Una ecuación que puede escribirse de la forma $Ax + By = C$ con un gráfico que es una línea recta.

linear extrapolation   The use of a linear equation to predict values that are outside the range of data.

linear function   A function in which no independent variable is raised to a power greater than 1; A function with a graph that is a line.

linear inequality   A half-plane with a boundary that is a straight line.

linear interpolation   The use of a linear equation to predict values that are inside the range of data.

linear pair   A pair of adjacent angles with noncommon sides that are opposite rays.

linear programming   The process of finding the maximum or minimum values of a function for a region defined by a system of inequalities.

linear regression   An algorithm used to find a precise line of fit for a set of data.

linear transformation   One or more operations performed on a set of data that can be written as a linear function.

literal equation   A formula or equation with several variables.

logarithm   In $x = b^y$, y is called the logarithm, base b, of x.

logarithmic equation   An equation that contains one or more logarithms.

logarithmic function   A function of the form $f(x) = \log$ base b of x, where $b > 0$ and $b \neq 1$.

logically equivalent   Statements with the same truth value.

lower quartile   The median of the lower half of a set of data.

extrapolación lineal   El uso de una ecuación lineal para predecir valores que están fuera del rango de datos.

función lineal   Una función en la que ninguna variable independiente se eleva a una potencia mayor que 1; Una función con un gráfico que es una línea.

desigualdad lineal   Un medio plano con un límite que es una línea recta.

interpolación lineal   El uso de una ecuación lineal para predecir valores que están dentro del rango de datos.

par lineal   Un par de ángulos adyacentes con lados no comunes que son rayos opuestos.

programación lineal   El proceso de encontrar los valores máximos o mínimos de una función para una región definida por un sistema de desigualdades.

regresión lineal   Un algoritmo utilizado para encontrar una línea precisa de ajuste para un conjunto de datos.

transformación lineal   Una o más operaciones realizadas en un conjunto de datos que se pueden escribir como una función lineal.

ecuación literal   Un formula o ecuación con varias variables.

logaritmo   En $x = b^y$, y se denomina logaritmo, base b, de x.

ecuación logarítmica   Una ecuación que contiene uno o más logaritmos.

función logarítmica   Una función de la forma $f(x) =$ base $\log b$ de x, donde $b > 0$ y $b \neq 1$.

lógicamente equivalentes   Declaraciones con el mismo valor de verdad.

cuartil inferior   La mediana de la mitad inferior de un conjunto de datos.

M

magnitude   The length of a vector from the initial point to the terminal point.

magnitud   La longitud de un vector desde el punto inicial hasta el punto terminal.

magnitude of symmetry   The smallest angle through which a figure can be rotated so that it maps onto itself.

magnitud de la simetria   El ángulo más pequeño a través del cual una figura se puede girar para que se cargue sobre sí mismo.

major arc   An arc with measure greater than 180°.

arco mayor   Un arco con una medida superior a 180°.

mapping   An illustration that shows how each element of the domain is paired with an element in the range.

cartografía   Una ilustración que muestra cómo cada elemento del dominio está emparejado con un elemento del rango.

marginal frequencies   In a two-way frequency table, the frequencies in the totals row and column; The totals of each subcategory in a two-way frequency table.

frecuencias marginales   En una tabla de frecuencias de dos vías, las frecuencias en los totales de fila y columna; Los totales de cada subcategoría en una tabla de frecuencia bidireccional.

maximum   The highest point on the graph of a function.

máximo   El punto más alto en la gráfica de una función.

maximum error of the estimate   The maximum difference between the estimate of the population mean and its actual value.

error máximo de la estimación   La diferencia máxima entre la estimación de la media de la población y su valor real.

measurement data   Data that have units and can be measured.

medicion de datos   Datos que tienen unidades y que pueden medirse.

measures of center   Measures of what is average.

medidas del centro   Medidas de lo que es promedio.

measures of spread   Measures of how spread out the data are.

medidas de propagación   Medidas de cómo se extienden los datos son.

median   The beginning of the second quartile that separates the data into upper and lower halves.

mediana   El comienzo del segundo cuartil que separa los datos en mitades superior e inferior.

median of a triangle   A line segment with endpoints that are a vertex of the triangle and the midpoint of the side opposite the vertex.

mediana de un triángulo   Un segmento de línea con extremos que son un vértice del triángulo y el punto medio del lado opuesto al vértice.

metric   A rule for assigning a number to some characteristic or attribute.

métrico   Una regla para asignar un número a alguna caracteristica o atribuye.

midline   The line about which the graph of a function oscillates.

linea media   La línea sobre la cual oscila la gráfica de una función periódica.

midpoint   The point on a line segment halfway between the endpoints of the segment.

punto medio   El punto en un segmento de línea a medio camino entre los extremos del segmento.

midsegment of a trapezoid   The segment that connects the midpoints of the legs of a trapezoid.

segment medio de un trapecio   El segmento que conecta los puntos medios de las patas de un trapecio.

midsegment of a triangle   The segment that connects the midpoints of the legs of a triangle.

segment medio de un triángulo   El segmento que conecta los puntos medios de las patas de un triángulo.

minimum   The lowest point on the graph of a function.

mínimo   El punto más bajo en la gráfica de una función.

minor arc An arc with measure less than 180°.

mixture problems Problems that involve creating a mixture of two or more kinds of things and then determining some quantity of the resulting mixture.

monomial A number, a variable, or a product of a number and one or more variables.

monomial function A function of the form $f(x) = ax^n$, for which a is a nonzero real number and n is a positive integer.

multi-step equation An equation that uses more than one operation to solve it.

multiplicative identity Because the product of any number a and 1 is equal to a, 1 is the multiplicative identity.

multiplicative inverses Two numbers with a product of 1.

multiplicity The number of times a number is a zero for a given polynomial.

mutually exclusive Events that cannot occur at the same time.

arco menor Un arco con una medida inferior a 180°.

problemas de mezcla Problemas que implican crear una mezcla de dos o más tipos de cosas y luego determinar una cierta cantidad de la mezcla resultante.

monomio Un número, una variable, o un producto de un número y una o más variables.

función monomial Una función de la forma $f(x) = ax^n$, para la cual a es un número real no nulo y n es un entero positivo.

ecuaciones de varios pasos Una ecuación que utiliza más de una operación para resolverla.

identidad multiplicativa Dado que el producto de cualquier número a y 1 es igual a, 1 es la identidad multiplicativa.

inversos multiplicativos Dos números con un producto es igual a 1.

multiplicidad El número de veces que un número es cero para un polinomio dado.

mutuamente exclusivos Eventos que no pueden ocurrir al mismo tiempo.

N

natural base exponential function An exponential function with base e, written as $y = e^x$.

natural logarithm The inverse of the natural base exponential function, most often abbreviated as ln x.

negation A statement that has the opposite meaning, as well as the opposite truth value, of an original statement.

negative Where the graph of a function lies below the x-axis.

negative correlation Bivariate data in which y decreases as x increases.

negative exponent An exponent that is a negative number.

función exponencial de base natural Una función exponencial con base e, escrita como $y = e^x$.

logaritmo natural La inversa de la función exponencial de base natural, más a menudo abreviada como ln x.

negación Una declaración que tiene el significado opuesto, así como el valor de verdad opuesto, de una declaración original.

negativo Donde la gráfica de una función se encuentra debajo del eje x.

correlación negativa Datos bivariate en el cual y disminuye a x aumenta.

exponente negativo Un exponente que es un número negativo.

negatively skewed distribution A distribution that typically has a median greater than the mean and less data on the left side of the graph.

net A two-dimensional figure that forms the surfaces of a three-dimensional object when folded.

no correlation Bivariate data in which x and y are not related.

nonlinear function A function in which a set of points cannot all lie on the same line

nonrigid motion A transformation that changes the dimensions of a given figure.

normal distribution A continuous, symmetric, bell-shaped distribution of a random variable.

nth root If $a^n = b$ for a positive integer n, then a is the nth root of b.

nth term of an arithmetic sequence The nth term of an arithmetic sequence with first term a_1 and common difference d is given by $a_n = a_1 + (n-1)d$, where n is a positive integer.

numerical expression A mathematical phrase involving only numbers and mathematical operations.

distribución negativamente sesgada Una distribución que típicamente tiene una mediana mayor que la media y menos datos en el lado izquierdo del gráfico.

red Una figura bidimensional que forma las superficies de un objeto tridimensional cuando se dobla.

sin correlación Datos bivariados en los que x e y no están relacionados.

función no lineal Una función en la que un conjunto de puntos no puede estar en la misma línea

movimiento no rígida Una transformación que cambia las dimensiones de una figura dada.

distribución normal Distribución con forma de campana, simétrica y continua de una variable aleatoria.

raíz enésima Si $a^n = b$ para cualquier entero positivo n, entonces a se llama una raíz enésima de b.

enésimo término de una secuencia aritmética El enésimo término de una secuencia aritmética con el primer término a_1 y la diferencia común d viene dado por $a_n = a_1 + (n-1)d$, donde n es un número entero positivo.

expresión numérica Una frase matemática que implica sólo números y operaciones matemáticas.

O

oblique asymptote An asymptote that is neither horizontal nor vertical.

observational study Members of a sample are measured or observed without being affected by the study.

octant One of the eight divisions of three-dimensional space.

odd functions Functions that are symmetric in the origin.

one-to-one function A function for which each element of the range is paired with exactly one element of the domain.

onto function A function for which the codomain is the same as the range.

asíntota oblicua Una asíntota que no es ni horizontal ni vertical.

estudio de observación Los miembros de una muestra son medidos o observados sin ser afectados por el estudio.

octante Una de las ocho divisiones del espacio tridimensional.

funciones extrañas Funciones que son simétricas en el origen.

función biunívoca Función para la cual cada elemento del rango está emparejado con exactamente un elemento del dominio.

sobre la función Función para la cual el codomain es el mismo que el rango.

open half-plane   The solution of a linear inequality that does not include the boundary line.

opposite rays   Two collinear rays with a common endpoint.

optimization   The process of seeking the optimal value of a function subject to given constraints.

order of symmetry   The number of times a figure maps onto itself.

ordered triple   Three numbers given in a specific order used to locate points in space.

orthocenter   The point of concurrency of the altitudes of a triangle.

orthographic drawing   The two-dimensional views of the top, left, front, and right sides of an object.

oscillation   How much the graph of a function varies between its extreme values as it approaches positive or negative infinity.

outcome   The result of a single event; The result of a single performance or trial of an experiment.

outlier   A value that is more than 1.5 times the interquartile range above the third quartile or below the first quartile.

medio plano abierto   La solución de una desigualdad linear que no incluye la línea de limite.

rayos opuestos   Dos rayos colineales con un punto final común.

optimización   El proceso de buscar el valor óptimo de una función sujeto a restricciones dadas.

orden de la simetría   El número de veces que una figura se asigna a sí misma.

triple ordenado   Tres números dados en un orden específico usado para localizar puntos en el espacio.

ortocentro   El punto de concurrencia de las altitudes de un triángulo.

dibujo ortográfico   Las vistas bidimensionales de los lados superior, izquierdo, frontal y derecho de un objeto.

oscilación   Cuánto la gráfica de una función varía entre sus valores extremos cuando se acerca al infinito positivo o negativo.

resultado   El resultado de un solo evento; El resultado de un solo rendimiento o ensayo de un experimento.

parte aislada   Un valor que es más de 1,5 veces el rango intercuartílico por encima del tercer cuartil o por debajo del primer cuartil.

P

---

parabola   A curved shape that results when a cone is cut at an angle by a plane that intersects the base; The graph of a quadratic function.

paragraph proof   A paragraph that explains why the conjecture for a given situation is true.

parallel lines   Coplanar lines that do not intersect; Nonvertical lines in the same plane that have the same slope.

parallel planes   Planes that do not intersect.

parallelogram   A quadrilateral with both pairs of opposite sides parallel.

parábola   Forma curvada que resulta cuando un cono es cortado en un ángulo por un plano que interseca la base; La gráfica de una función cuadrática.

prueba de párrafo   Un párrafo que explica por qué la conjetura para una situación dada es verdadera.

líneas paralelas   Líneas coplanares que no se intersecan; Líneas no verticales en el mismo plano que tienen pendientes iguales.

planos paralelas   Planos que no se intersecan.

paralelogramo   Un cuadrilátero con ambos pares de lados opuestos paralelos.

parameter   A measure that describes a characteristic of a population; A value in the equation of a function that can be varied to yield a family of functions.

parent function   The simplest of functions in a family.

Pascal's triangle   A triangle of numbers in which a row represents the coefficients of an expanded binomial $(a + b)^n$.

percent rate of change   The percent of increase per time period.

percentile   A measure that tells what percent of the total scores were below a given score.

perfect cube   A rational number with a cube root that is a rational number.

perfect square   A rational number with a square root that is a rational number.

perfect square trinomials   Squares of binomials.

perimeter   The sum of the lengths of the sides of a polygon.

period   The horizontal length of one cycle.

periodic function   A function with y-values that repeat at regular intervals.

permutation   An arrangement of objects in which order is important.

perpendicular   Intersecting at right angles.

perpendicular bisector   Any line, segment, or ray that passes through the midpoint of a segment and is perpendicular to that segment.

perpendicular lines   Nonvertical lines in the same plane for which the product of the slopes is −1.

phase shift   A horizontal translation of the graph of a trigonometric function.

pi   The ratio $\dfrac{\text{cricumference}}{\text{diameter}}$.

parámetro   Una medida que describe una característica de una población; Un valor en la ecuación de una función que se puede variar para producir una familia de funciones.

función basica   La función más fundamental de un familia de funciones.

triángulo de Pascal   Un triángulo de números en el que una fila representa los coeficientes de un binomio expandido $(a + b)^n$.

por ciento tasa de cambio   El porcentaje de aumento por período de tiempo.

percentil   Una medida que indica qué porcentaje de las puntuaciones totales estaban por debajo de una puntuación determinada.

cubo perfecto   Un número racional con un raíz cúbica que es un número racional.

cuadrado perfecto   Un número racional con un raíz cuadrada que es un número racional.

trinomio cuadrado perfecto   Cuadrados de los binomios.

perimetro   La suma de las longitudes de los lados de un polígono.

periodo   La longitud horizontal de un ciclo.

función periódica   Una función con y-valores aquella repetición con regularidad.

permutación   Un arreglo de objetos en el que el orden es importante.

perpendicular   Intersección en ángulo recto.

mediatriz   Cualquier línea, segmento o rayo que pasa por el punto medio de un segmento y es perpendicular a ese segmento.

líneas perpendiculares   Líneas no verticales en el mismo plano para las que el producto de las pendientes es −1.

cambio de fase   Una traducción horizontal de la gráfica de una función trigonométrica.

pi   Relación $\dfrac{\text{circunferencia}}{\text{diámetro}}$

piecewise-defined function A function defined by at least two subfunctions, each of which is defined differently depending on the interval of the domain.

piecewise-linear function A function defined by at least two linear subfunctions, each of which is defined differently depending on the interval of the domain.

plane A flat surface made up of points that has no depth and extends indefinitely in all directions.

plane symmetry When a plane intersects a three-dimensional figure so one half is the reflected image of the other half.

Platonic solid One of five regular polyhedra.

point A location with no size, only position.

point discontinuity An area that appears to be a hole in a graph.

point of concurrency The point of intersection of concurrent lines.

point of symmetry The point about which a figure is rotated.

point of tangency For a line that intersects a circle in one point, the point at which they intersect.

point symmetry A figure or graph has this when a figure is rotated 180° about a point and maps exactly onto the other part.

polygon A closed plane figure with at least three straight sides.

polyhedron A closed three-dimensional figure made up of flat polygonal regions.

polynomial A monomial or the sum of two or more monomials.

polynomial function A continuous function that can be described by a polynomial equation in one variable.

función definida por piezas Una función definida por al menos dos subfunciones, cada una de las cuales se define de manera diferente dependiendo del intervalo del dominio.

función lineal por piezas Una función definida por al menos dos subfunciones lineal, cada una de las cuales se define de manera diferente dependiendo del intervalo del dominio.

plano Una superficie plana compuesta de puntos que no tiene profundidad y se extiende indefinidamente en todas las direcciones.

simetría plana Cuando un plano cruza una figura tridimensional, una mitad es la imagen reflejada de la otra mitad.

sólido platónico Uno de cinco poliedros regulares.

punto Una ubicación sin tamaño, solo posición.

discontinuidad de punto Un área que parece ser un agujero en un gráfico.

punto de concurrencia El punto de intersección de líneas concurrentes.

punto de simetría El punto sobre el que se gira una figura.

punto de tangencia Para una línea que cruza un círculo en un punto, el punto en el que se cruzan.

simetría de punto Una figura o gráfica tiene esto cuando una figura se gira 180° alrededor de un punto y se mapea exactamente sobre la otra parte.

polígono Una figura plana cerrada con al menos tres lados rectos.

poliedros Una figura tridimensional cerrada formada por regiones poligonales planas.

polinomio Un monomio o la suma de dos o más monomios.

función polinómica Función continua que puede describirse mediante una ecuación polinómica en una variable.

polynomial identity   A polynomial equation that is true for any values that are substituted for the variables.

population   All of the members of a group of interest about which data will be collected.

population proportion   The number of members in the population with a particular characteristic divided by the total number of members in the population.

positive   Where the graph of a function lies above the x-axis.

positive correlation   Bivariate data in which y increases as x increases.

positively skewed distribution   A distribution that typically has a mean greater than the median.

postulate   A statement that is accepted as true without proof.

power function   A function of the form $f(x) = ax^n$, where a and n are nonzero real numbers.

precision   The repeatability, or reproducibility, of a measurement.

preimage   The original figure in a transformation.

prime polynomial   A polynomial that cannot be written as a product of two polynomials with integer coefficients.

principal root   The nonnegative root of a number.

principal square root   The nonnegative square root of a number.

principal values   The values in the restricted domains of trigonometric functions.

principle of superposition   Two figures are congruent if and only if there is a rigid motion or series of rigid motions that maps one figure exactly onto the other.

prism   A polyhedron with two parallel congruent bases connected by parallelogram faces.

identidad polinomial   Una ecuación polinómica que es verdadera para cualquier valor que se sustituya por las variables.

población   Todos los miembros de un grupo de interés sobre cuáles datos serán recopilados.

proporción de la población   El número de miembros en la población con una característica particular dividida por el número total de miembros en la población.

positiva   Donde la gráfica de una función se encuentra por encima del eje x.

correlación positiva   Datos bivariate en el cual y aumenta a x disminuye.

distribución positivamente sesgada   Una distribución que típicamente tiene una media mayor que la mediana.

postulado   Una declaración que se acepta como verdadera sin prueba.

función de potencia   Una ecuación polinomial que es verdadera para una función de la forma $f(x) = ax^n$, donde a y n son números reales no nulos.

precisión   La repetibilidad, o reproducibilidad, de una medida.

preimagen   La figura original en una transformación.

polinomio primo   Un polinomio que no puede escribirse como producto de dos polinomios con coeficientes enteros.

raíz principal   La raíz no negativa de un número.

raíz cuadrada principal   La raíz cuadrada no negativa de un número.

valores principales   Valores de los dominios restringidos de las functiones trigonométricas.

principio de superposición   Dos figuras son congruentes si y sólo si hay un movimiento rígido o una serie de movimientos rígidos que traza una figura exactamente sobre la otra.

prisma   Un poliedro con dos bases congruentes paralelas conectadas por caras de paralelogramo.

probability   The number of outcomes in which a specified event occurs to the total number of trials.

probability distribution   A function that maps the sample space to the probabilities of the outcomes in the sample space for a particular random variable.

probability model   A mathematical representation of a random event that consists of the sample space and the probability of each outcome.

projectile motion problems   Problems that involve objects being thrown or dropped.

proof   A logical argument in which each statement is supported by a statement that is accepted as true.

proof by contradiction   One assumes that the statement to be proven is false and then uses logical reasoning to deduce that a statement contradicts a postulate, theorem, or one of the assumptions.

proportion   A statement that two ratios are equivalent.

pure imaginary number   A number of the form bi, where b is a real number and i is the imaginary unit.

pyramid   A polyhedron with a polygonal base and three or more triangular faces that meet at a common vertex.

Pythagorean identities   Identities that express the Pythagorean Theorem in terms of the trigonometric functions.

Pythagorean triple   A set of three nonzero whole numbers that make the Pythagorean Theorem true.

probabilidad   El número de resultados en los que se produce un evento especificado al número total de ensayos.

distribución de probabilidad   Una función que mapea el espacio de muestra a las probabilidades de los resultados en el espacio de muestra para una variable aleatoria particular.

modelo de probabilidad   Una representación matemática de un evento aleatorio que consiste en el espacio muestral y la probabilidad de cada resultado.

problemas de movimiento del proyectil   Problemas que involucran objetos que se lanzan o caen.

prueba   Un argumento lógico en el que cada sentencia está respaldada por una sentencia aceptada como verdadera.

prueba por contradicción   Se supone que la afirmación a ser probada es falsa y luego utiliza el razonamiento lógico para deducir que una afirmación contradice un postulado, teorema o uno de los supuestos.

proporción   Una declaración de que dos proporciones son equivalentes.

número imaginario puro   Un número de la forma bi, donde b es un número real e i es la unidad imaginaria.

pirámide   Poliedro con una base poligonal y tres o más caras triangulares que se encuentran en un vértice común.

identidades pitagóricas   Identidades que expresan el Teorema de Pitágoras en términos de las funciones trigonométricas.

triplete Pitágorico   Un conjunto de tres números enteros distintos de cero que hacen que el Teorema de Pitágoras sea verdadero.

Q

quadrantal angle   An angle in standard position with a terminal side that coincides with one of the axes.

quadratic equation   An equation that includes a quadratic expression.

ángulo de cuadrante   Un ángulo en posición estándar con un lado terminal que coincide con uno de los ejes.

ecuación cuadrática   Una ecuación que incluye una expresión cuadrática.

quadratic expression   An expression in one variable with a degree of 2.

quadratic form   A form of polynomial equation, $au^2 + bu + c$, where u is an algebraic expression in x.

quadratic function   A function with an equation of the form $y = ax^2 + bx + c$, where $a \neq 0$.

quadratic inequality   An inequality that includes a quadratic expression.

quadratic relations   Equations of parabolas with horizontal axes of symmetry that are not functions.

quartic function   A fourth-degree function.

quartiles   Measures of position that divide a data set arranged in ascending order into four groups, each containing about one fourth or 25% of the data.

quintic function   A fifth-degree function.

expresión cuadrática   Una expresión en una variable con un grado de 2.

forma cuadrática   Una forma de ecuación polinomial, $au^2 + bu + c$, donde u es una expresión algebraica en x.

función cuadrática   Una función con una ecuación de la forma $y = ax^2 + bx + c$, donde $a \neq 0$.

desigualdad cuadrática   Una desigualdad que incluye una expresión cuadrática.

relaciones cuadráticas   Ecuaciones de parábolas con ejes horizontales de simetría que no son funciones.

función cuartica   Una función de cuarto grado.

cuartiles   Medidas de posición que dividen un conjunto de datos dispuestos en orden ascendente en cuatro grupos, cada uno de los cuales contiene aproximadamente un cuarto o el 25% de los datos.

función quíntica   Una función de quinto grado.

R

radian   A unit of angular measurement equal to $\frac{180°}{\pi}$ or about 57.296°.

radical equation   An equation with a variable in a radicand.

radical expression   An expression that contains a radical symbol, such as a square root.

radical form   When an expression contains a radical symbol.

radical function   A function that contains radicals with variables in the radicand.

radicand   The expression under a radical sign.

radius of a circle or sphere   A line segment from the center to a point on a circle or sphere.

radius of a regular polygon   The radius of the circle circumscribed about a regular polygon.

radián   Una unidad de medida angular igual o $\frac{180°}{\pi}$ alrededor de 57.296°.

ecuación radical   Una ecuación con una variable en un radicand.

expresión radicales   Una expresión que contiene un símbolo radical, tal como una raíz cuadrada.

forma radical   Cuando una expresión contiene un símbolo radical.

función radical   Función que contiene radicales con variables en el radicand.

radicando   La expresión debajo del signo radical.

radio de un círculo o esfera   Un segmento de línea desde el centro hasta un punto en un círculo o esfera.

radio de un polígono regular   El radio del círculo circunscrito alrededor de un polígono regular.

range   The difference between the greatest and least values in a set of data; The set of second numbers of the ordered pairs in a relation; The set of y-values that actually result from the evaluation of the function.

rate of change   How a quantity is changing with respect to a change in another quantity.

rational equation   An equation that contains at least one rational expression.

rational exponent   An exponent that is expressed as a fraction.

rational expression   A ratio of two polynomial expressions.

rational function   An equation of the form $f(x) = \frac{a(x)}{b(x)}$, where $a(x)$ and $b(x)$ are polynomial expressions and $b(x) \neq 0$.

rational inequality   An inequality that contains at least one rational expression.

rationalizing the denominator   A method used to eliminate radicals from the denominator of a fraction or fractions from a radicand.

ray   Part of a line that starts at a point and extends to infinity.

reciprocal function   An equation of the form $f(x) = \frac{n}{b(x)}$, where n is a real number and $b(x)$ is a linear expression that cannot equal 0.

reciprocal trigonometric functions   Trigonometric functions that are reciprocals of each other.

reciprocals   Two numbers with a product of 1.

rectangle   A parallelogram with four right angles.

recursive formula   A formula that gives the value of the first term in the sequence and then defines the next term by using the preceding term.

reduction   A dilation with a scale factor between 0 and 1.

reference angle   The acute angle formed by the terminal side of an angle and the x-axis.

rango   La diferencia entre los valores de datos más grande or menos en un sistema de datos; El conjunto de los segundos números de los pares ordenados de una relación; El conjunto de valores y que realmente resultan de la evaluación de la función.

tasa de cambio   Cómo cambia una cantidad con respecto a un cambio en otra cantidad.

ecuación racional   Una ecuación que contiene al menos una expresión racional.

exponente racional   Un exponente que se expresa como una fracción.

expresión racional   Una relación de dos expresiones polinomiales.

función racional   Una ecuación de la forma $f(x) = \frac{a(x)}{b(x)}$, donde $a(x)$ y $b(x)$ son expresiones polinomiales y $b(x) \neq 0$.

desigualdad racional   Una desigualdad que contiene al menos una expresión racional.

racionalizando el denominador   Método utilizado para eliminar radicales del denominador de una fracción o fracciones de una radicand.

rayo   Parte de una línea que comienza en un punto y se extiende hasta el infinito.

función recíproca   Una ecuación de la forma $f(x) = \frac{n}{b(x)}$, donde n es un número real y $b(x)$ es una expresión lineal que no puede ser igual a 0.

funciones trigonométricas recíprocas   Funciones trigonométricas que son reciprocales entre sí.

recíprocos   Dos números con un producto de 1.

rectángulo   Un paralelogramo con cuatro ángulos rectos.

formula recursiva   Una fórmula que da el valor del primer término en la secuencia y luego define el siguiente término usando el término anterior.

reducción   Una dilatación con un factor de escala entre 0 y 1.

ángulo de referencia   El ángulo agudo formado por el lado terminal de un ángulo en posición estándar y el eje x.

reflection A function in which the preimage is reflected in the line of reflection; A transformation in which a figure, line, or curve is flipped across a line.

reflexión Función en la que la preimagen se refleja en la línea de reflexión; Una transformación en la que una figura, línea o curva se voltea a través de una línea.

regression function A function generated by an algorithm to find a line or curve that fits a set of data.

función de regresión Función generada por un algoritmo para encontrar una línea o curva que se ajuste a un conjunto de datos.

regular polygon A convex polygon that is both equilateral and equiangular.

polígono regular Un polígono convexo que es a la vez equilátero y equiangular.

regular polyhedron A polyhedron in which all of its faces are regular congruent polygons and all of the edges are congruent.

poliedro regular Un poliedro en el que todas sus caras son polígonos congruentes regulares y todos los bordes son congruentes.

regular pyramid A pyramid with a base that is a regular polygon.

pirámide regular Una pirámide con una base que es un polígono regular.

regular tessellation A tessellation formed by only one type of regular polygon.

teselado regular Un teselado formado por un solo tipo de polígono regular.

relation A set of ordered pairs.

relación Un conjunto de pares ordenados.

relative frequency In a two-way frequency table, the ratios of the number of observations in a category to the total number of observations; The ratio of the number of observations in a category to the total number of observations.

frecuencia relativa En una tabla de frecuencia bidireccional, las relaciones entre el número de observaciones en una categoría y el número total de observaciones; La relación entre el número de observaciones en una categoría y el número total de observaciones.

relative maximum A point on the graph of a function where no other nearby points have a greater y-coordinate.

máximo relativo Un punto en la gráfica de una función donde ningún otro punto cercano tiene una coordenada y mayor.

relative minimum A point on the graph of a function where no other nearby points have a lesser y-coordinate.

mínimo relativo Un punto en la gráfica de una función donde ningún otro punto cercano tiene una coordenada y menor.

remote interior angles Interior angles of a triangle that are not adjacent to an exterior angle.

ángulos internos no adyacentes Ángulos interiores de un triángulo que no están adyacentes a un ángulo exterior.

residual The difference between an observed y-value and its predicted y-value on a regression line.

residual La diferencia entre un valor de y observado y su valor de y predicho en una línea de regresión.

rhombus A parallelogram with all four sides congruent.

rombo Un paralelogramo con los cuatro lados congruentes.

rigid motion A transformation that preserves distance and angle measure.

movimiento rígido Una transformación que preserva la distancia y la medida del ángulo.

root   A solution of an equation.

rotation   A function that moves every point of a preimage through a specified angle and direction about a fixed point.

rotational symmetry   A figure can be rotated less than 360° about a point so that the image and the preimage are indistinguishable.

raíz   Una solución de una ecuación.

rotación   Función que mueve cada punto de una preimagen a través de un ángulo y una dirección especificados alrededor de un punto fijo.

simetría rotacional   Una figura puede girar menos de 360° alrededor de un punto para que la imagen y la preimagen sean indistinguibles.

S

sample   A subset of a population.

sample space   The set of all possible outcomes.

sampling error   The variation between samples taken from the same population.

scale   The distance between tick marks on the *x*- and *y*-axes.

scale factor of a dilation   The ratio of a length on an image to a corresponding length on the preimage.

scatter plot   A graph of bivariate data that consists of ordered pairs on a coordinate plane.

secant   Any line or ray that intersects a circle in exactly two points; The ratio of the length of the hypotenuse to the length of the leg adjacent to the angle.

sector   A region of a circle bounded by a central angle and its intercepted arc.

segment bisector   Any segment, line, plane, or point that intersects a line segment at its midpoint.

self-selected sample   Members volunteer to be included in the sample.

semicircle   An arc that measures exactly 180°.

semiregular tessellation   A tessellation formed by two or more regular polygons.

sequence   A list of numbers in a specific order.

muestra   Un subconjunto de una población.

espacio muestral   El conjunto de todos los resultados posibles.

error de muestreo   La variación entre muestras tomadas de la misma población.

escala   La distancia entre las marcas en los ejes *x* e *y*.

factor de escala de una dilatación   Relación de una longitud en una imagen con una longitud correspondiente en la preimagen.

gráfica de dispersión   Una gráfica de datos bivariados que consiste en pares ordenados en un plano de coordenadas.

secante   Cualquier línea o rayo que cruce un círculo en exactamente dos puntos; Relación entre la longitud de la hipotenusa y la longitud de la pierna adyacente al ángulo.

sector   Una región de un círculo delimitada por un ángulo central y su arco interceptado.

bisectriz del segmento   Cualquier segmento, línea, plano o punto que interseca un segmento de línea en su punto medio.

muestra auto-seleccionada   Los miembros se ofrecen como voluntarios para ser incluidos en la muestra.

semicírculo   Un arco que mide exactamente 180°.

teselado semiregular   Un teselado formado por dos o más polígonos regulares.

secuencia   Una lista de números en un orden específico.

series   The indicated sum of the terms in a sequence.

serie   La suma indicada de los términos en una secuencia.

set-builder notation   Mathematical notation that describes a set by stating the properties that its members must satisfy.

notación de construción de conjuntos   Notación matemática que describe un conjunto al declarar las propiedades que sus miembros deben satisfacer.

sides of an angle   The rays that form an angle.

lados de un ángulo   Los rayos que forman un ángulo.

sigma notation   A notation that uses the Greek uppercase letter S to indicate that a sum should be found.

notación de sigma   Una notación que utiliza la letra mayúscula griega S para indicar que debe encontrarse una suma.

significant figures   The digits of a number that are used to express a measure to an appropriate degree of accuracy.

dígitos significantes   Los dígitos de un número que se utilizan para expresar una medida con un grado apropiado de precisión.

similar polygons   Two figures are similar polygons if one can be obtained from the other by a dilation or a dilation with one or more rigid motions.

polígonos similares   Dos figuras son polígonos similares si uno puede ser obtenido del otro por una dilatación o una dilatación con uno o más movimientos rígidos.

similar solids   Solid figures with the same shape but not necessarily the same size.

sólidos similares   Figuras sólidas con la misma forma pero no necesariamente del mismo tamaño.

similar triangles   Triangles in which all of the corresponding angles are congruent and all of the corresponding sides are proportional.

triángulos similares   Triángulos en los cuales todos los ángulos correspondientes son congruentes y todos los lados correspondientes son proporcionales.

similarity ratio   The scale factor between two similar polygons.

relación de similitud   El factor de escala entre dos polígonos similares.

similarity transformation   A transformation composed of a dilation or a dilation and one or more rigid motions.

transformación de similitud   Una transformación compuesto por una dilatación o una dilatación y uno o más movimientos rígidos.

simple random sample   Each member of the population has an equal chance of being selected as part of the sample.

muestra aleatoria simple   Cada miembro de la población tiene la misma posibilidad de ser seleccionado como parte de la muestra.

simplest form   An expression is in simplest form when it is replaced by an equivalent expression having no like terms or parentheses.

forma reducida   Una expresión está reducida cuando se puede sustituir por una expresión equivalente que no tiene ni términos semejantes ni paréntesis.

simulation   The use of a probability model to imitate a process or situation so it can be studied.

simulación   El uso de un modelo de probabilidad para imitar un proceso o situación para que pueda ser estudiado.

sine   The ratio of the length of the leg opposite an angle to the length of the hypotenuse.

seno   La relación entre la longitud de la pierna opuesta a un ángulo y la longitud de la hipotenusa.

sinusoidal function   A function that can be produced by translating, reflecting, or dilating the sine function.

skew lines   Noncoplanar lines that do not intersect.

slant height of a pyramid or right cone   The length of a segment with one endpoint on the base edge of the figure and the other at the vertex.

slope   The rate of change in the y-coordinates (rise) to the corresponding change in the x-coordinates (run) for points on a line.

slope criteria   Outlines a method for proving the relationship between lines based on a comparison of the slopes of the lines.

solid of revolution   A solid figure obtained by rotating a shape around an axis.

solution   A value that makes an equation true.

solve an equation   The process of finding all values of the variable that make the equation a true statement.

solving a triangle   When you are given measurements to find the unknown angle and side measures of a triangle.

space   A boundless three-dimensional set of all points.

sphere   A set of all points in space equidistant from a given point called the center of the sphere.

square   A parallelogram with all four sides and all four angles congruent.

square root   One of two equal factors of a number.

square root function   A radical function that contains the square root of a variable expression.

square root inequality   An inequality that contains the square root of a variable expression.

standard deviation   A measure that shows how data deviate from the mean.

función sinusoidal   Función que puede producirse traduciendo, reflejando o dilatando la función sinusoidal.

líneas alabeadas   Líneas no coplanares que no se cruzan.

altura inclinada de una pirámide o cono derecho   La longitud de un segmento con un punto final en el borde base de la figura y el otro en el vértice.

pendiente   La tasa de cambio en las coordenadas y (subida) al cambio correspondiente en las coordenadas x (carrera) para puntos en una línea.

criterios de pendiente   Describe un método para probar la relación entre líneas basado en una comparación de las pendientes de las líneas.

sólido de revolución   Una figura sólida obtenida girando una forma alrededor de un eje.

solución   Un valor que hace que una ecuación sea verdadera.

resolver una ecuación   El proceso en que se hallan todos los valores de la variable que hacen verdadera la ecuación.

resolver un triángulo   Cuando se le dan mediciones para encontrar el ángulo desconocido y las medidas laterales de un triángulo.

espacio   Un conjunto tridimensional ilimitado de todos los puntos.

esfera   Un conjunto de todos los puntos del espacio equidistantes de un punto dado llamado centro de la esfera.

cuadrado   Un paralelogramo con los cuatro lados y los cuatro ángulos congruentes.

raíz cuadrada   Uno de dos factores iguales de un número.

función raíz cuadrada   Función radical que contiene la raíz cuadrada de una expresión variable.

square root inequality   Una desigualdad que contiene la raíz cuadrada de una expresión variable.

desviación tipica   Una medida que muestra cómo los datos se desvían de la media.

standard error of the mean The standard deviation of the distribution of sample means taken from a population.

standard form of a linear equation Any linear equation can be written in this form, $Ax + By = C$, where $A \geq 0$, A and B are not both 0, and A, B, and C are integers with a greatest common factor of 1.

standard form of a polynomial A polynomial that is written with the terms in order from greatest degree to least degree.

standard form of a quadratic equation A quadratic equation can be written in the form $ax^2 + bx + c = 0$, where $a \neq 0$ and a, b, and c are integers.

standard normal distribution A normal distribution with a mean of 0 and a standard deviation of 1.

standard position An angle positioned so that the vertex is at the origin and the initial side is on the positive x-axis.

statement Any sentence that is either true or false, but not both.

statistic A measure that describes a characteristic of a sample.

statistics An area of mathematics that deals with collecting, analyzing, and interpreting data.

step function A type of piecewise-linear function with a graph that is a series of horizontal line segments.

straight angle An angle that measures 180°.

stratified sample The population is first divided into similar, nonoverlapping groups. Then members are randomly selected from each group.

substitution A process of solving a system of equations in which one equation is solved for one variable in terms of the other.

supplementary angles Two angles with measures that have a sum of 180°.

error estandar de la media La desviación estándar de la distribución de los medios de muestra se toma de una población.

forma estándar de una ecuación lineal Cualquier ecuación lineal se puede escribir de esta forma, $Ax + By = C$, donde $A \geq 0$, A y B no son ambos 0, y A, B y C son enteros con el mayor factor común de 1.

forma estándar de un polinomio Un polinomio que se escribe con los términos en orden del grado más grande a menos grado.

forma estándar de una ecuación cuadrática Una ecuación cuadrática puede escribirse en la forma $ax^2 + bx + c = 0$, donde $a \neq 0$ y a, b, y c son enteros.

distribución normal estándar Distribución normal con una media de 0 y una desviación estándar de 1.

posición estándar Un ángulo colocado de manera que el vértice está en el origen y el lado inicial está en el eje x positivo.

enunciado Cualquier oración que sea verdadera o falsa, pero no ambas.

estadística Una medida que describe una característica de una muestra.

estadísticas El proceso de recolección, análisis e interpretación de datos.

función escalonada Un tipo de función lineal por piezas con un gráfico que es una serie de segmentos de línea horizontal.

ángulo recto Un ángulo que mide 180°.

muestra estratificada La población se divide primero en grupos similares, sin superposición. A continuación, los miembros se seleccionan aleatoriamente de cada grupo.

sustitución Un proceso de resolución de un sistema de ecuaciones en el que una ecuación se resuelve para una variable en términos de la otra.

ángulos suplementarios Dos ángulos con medidas que tienen una suma de 180°.

surface area The sum of the areas of all faces and side surfaces of a three-dimensional figure.

survey Data are collected from responses given by members of a group regarding their characteristics, behaviors, or opinions.

symmetric distribution A distribution in which the mean and median are approximately equal.

symmetry A figure has this if there exists a rigid motion—reflection, translation, rotation, or glide reflection—that maps the figure onto itself.

synthetic division An alternate method used to divide a polynomial by a binomial of degree 1.

synthetic geometry The study of geometric figures without the use of coordinates.

synthetic substitution The process of using synthetic division to find a value of a polynomial function.

system of equations A set of two or more equations with the same variables.

system of inequalities A set of two or more inequalities with the same variables.

systematic sample Members are selected according to a specified interval from a random starting point.

área de superficie La suma de las áreas de todas las caras y superficies laterales de una figura tridimensional.

encuesta Los datos se recogen de las respuestas dadas por los miembros de un grupo con respecto a sus características, comportamientos u opiniones.

distribución simétrica Un distribución en la que la media y la mediana son aproximadamente iguales.

simetría Una figura tiene esto si existe una reflexión-reflexión, una traducción, una rotación o una reflexión de deslizamiento rígida-que mapea la figura sobre sí misma.

división sintética Un método alternativo utilizado para dividir un polinomio por un binomio de grado 1.

geometría sintética El estudio de figuras geométricas sin el uso de coordenadas.

sustitución sintética El proceso de utilizar la división sintética para encontrar un valor de una función polynomial.

sistema de ecuaciones Un conjunto de dos o más ecuaciones con las mismas variables.

sistema de desigualdades Un conjunto de dos o más desigualdades con las mismas variables.

muestra sistemática Los miembros se seleccionan de acuerdo con un intervalo especificado desde un punto de partida aleatorio.

T

tangent The ratio of the length of the leg opposite an angle to the length of the leg adjacent to the angle.

tangent to a circle A line or segment in the plane of a circle that intersects the circle in exactly one point and does not contain any points in the interior of the circle.

tangent to a sphere A line that intersects the sphere in exactly one point.

term A number, a variable, or a product or quotient of numbers and variables.

tangente La relación entre la longitud de la pata opuesta a un ángulo y la longitud de la pata adyacente al ángulo.

tangente a un círculo Una línea o segmento en el plano de un círculo que interseca el círculo en exactamente un punto y no contiene ningún punto en el interior del círculo.

tangente a una esfera Una línea que interseca la esfera exactamente en un punto.

término Un número, una variable, o un producto o cociente de números y variables.

term of a sequence   A number in a sequence.

terminal side   The part of an angle that rotates about the center.

tessellation   A repeating pattern of one or more figures that covers a plane with no overlapping or empty spaces.

theorem   A statement that can be proven true using undefined terms, definitions, and postulates.

theoretical probability   Probability based on what is expected to happen.

transformation   A function that takes points in the plane as inputs and gives other points as outputs. The movement of a graph on the coordinate plane.

translation   A function in which all of the points of a figure move the same distance in the same direction; A transformation in which a figure is slid from one position to another without being turned.

translation vector   A directed line segment that describes both the magnitude and direction of the slide if the magnitude is the length of the vector from its initial point to its terminal point.

transversal   A line that intersects two or more lines in a plane at different points.

trapezoid   A quadrilateral with exactly one pair of parallel sides.

trend   A general pattern in the data.

trigonometric equation   An equation that includes at least one trigonometric function.

trigonometric function   A function that relates the measure of one nonright angle of a right triangle to the ratios of the lengths of any two sides of the triangle.

trigonometric identity   An equation involving trigonometric functions that is true for all values for which every expression in the equation is defined.

término de una sucesión   Un número en una secuencia.

lado terminal   La parte de un ángulo que gira alrededor de un centro.

teselado   Patrón repetitivo de una o más figuras que cubre un plano sin espacios superpuestos o vacíos.

teorema   Una afirmación o conjetura que se puede probar verdad utilizando términos, definiciones y postulados indefinidos.

probabilidad teórica   Probabilidad basada en lo que se espera que suceda.

transformación   Función que toma puntos en el plano como entradas y da otros puntos como salidas. El movimiento de un gráfico en el plano de coordenadas.

traslación   Función en la que todos los puntos de una figura se mueven en la misma dirección; El movimiento de un gráfico en el plano de coordenadas.

vector de traslación   Un segmento de línea dirigido que describe tanto la magnitud como la dirección de la diapositiva si la magnitud es la longitud del vector desde su punto inicial hasta su punto terminal.

transversal   Una línea que interseca dos o más líneas en un plano en diferentes puntos.

trapecio   Un cuadrilátero con exactamente un par de lados paralelos.

tendencia   Un patrón general en los datos.

ecuación trigonométrica   Una ecuación que incluye al menos una función trigonométrica.

función trigonométrica   Función que relaciona la medida de un ángulo no recto de un triángulo rectángulo con las relaciones de las longitudes de cualquiera de los dos lados del triángulo.

identidad trigonométrica   Una ecuación que implica funciones trigonométricas que es verdadera para todos los valores para los cuales se define cada expresión en la ecuación.

trigonometric ratio   A ratio of the lengths of two sides of a right triangle.

trigonometry   The study of the relationships between the sides and angles of triangles.

trinomial   The sum of three monomials.

truth value   The truth or falsity of a statement.

two-column proof   A proof that contains statements and reasons organized in a two-column format.

two-way frequency table   A table used to show frequencies of data classified according to two categories, with the rows indicating one category and the columns indicating the other.

two-way relative frequency table   A table used to show frequencies of data based on a percentage of the total number of observations.

relación trigonométrica   Una relación de las longitudes de dos lados de un triángulo rectángulo.

trigonometría   El estudio de las relaciones entre los lados y los ángulos de los triángulos.

trinomio   La suma de tres monomios.

valor de verdad   La verdad o la falsedad de una declaración.

prueba de dos columnas   Una prueba que contiene declaraciones y razones organizadas en un formato de dos columnas.

tabla de frecuencia bidireccional   Una tabla utilizada para mostrar las frecuencias de los datos clasificados de acuerdo con dos categorías, con las filas que indican una categoría y las columnas que indican la otra.

tabla de frecuencia relativa bidireccional   Una tabla usada para mostrar las frecuencias de datos basadas en un porcentaje del número total de observaciones.

U

unbounded   When the graph of a system of constraints is open.

undefined terms   Words that are not formally explained by means of more basic words and concepts.

uniform motion problems   Problems that use the formula $d = rt$, where d is the distance, r is the rate, and t is the time.

uniform tessellation   A tessellation that contains the same arrangement of shapes and angles at each vertex.

union   The graph of a compound inequality containing *or*.

union of A and B   The set of all outcomes in the sample space of event A combined with all outcomes in the sample space of event B.

unit circle   A circle with a radius of 1 unit centered at the origin on the coordinate plane.

univariate data   Measurement data in one variable.

no acotado   Cuando la gráfica de un sistema de restricciones está abierta.

términos indefinidos   Palabras que no se explican formalmente mediante palabras y conceptos más básicos.

problemas de movimiento uniforme   Problemas que utilizan la fórmula $d = rt$, donde d es la distancia, r es la velocidad y t es el tiempo.

teselado uniforme   Un teselado que contiene la misma disposición de formas y ángulos en cada vértice.

unión   La gráfica de una desigualdad compuesta que contiene la palabra *o*.

unión de A y B   El conjunto de todos los resultados en el espacio muestral del evento A combinado con todos los resultados en el espacio muestral del evento B.

círculo unitario   Un círculo con un radio de 1 unidad centrado en el origen en el plano de coordenadas.

datos univariate   Datos de medición en una variable.

upper quartile   The median of the upper half of a set of data.

cuartil superior   La mediana de la mitad superior de un conjunto de datos.

valid argument   An argument is valid if it is impossible for all of the premises, or supporting statements, of the argument to be true and its conclusion false.

argumento válido   Un argumento es válido si es imposible que todas las premisas o argumentos de apoyo del argumento sean verdaderos y su conclusión sea falsa.

variable   A letter used to represent an unspecified number or value; Any characteristic, number, or quantity that can be counted or measured.

variable   Una letra utilizada para representar un número o valor no especificado; Cualquier característica, número, o cantidad que pueda ser contada o medida.

variable term   A term that contains a variable.

término variable   Un término que contiene una variable.

variance   The square of the standard deviation.

varianza   El cuadrado de la desviación estándar.

vertex   Either the lowest point or the highest point of a function.

vértice   El punto más bajo o el punto más alto en una función.

vertex angle of an isosceles triangle   The angle between the sides that are the legs of an isosceles triangle.

ángulo del vértice de un triángulo isósceles   El ángulo entre los lados que son las patas de un triángulo isósceles.

vertex form   A quadratic function written in the form $f(x) = a(x - h)^2 + k$.

forma de vértice   Una función cuadrática escribirse de la forma $f(x) = a(x - h)^2 + k$.

vertex of a polyhedron   The intersection of three edges of a polyhedron.

vértice de un polígono   La intersección de tres bordes de un poliedro.

vertex of an angle   The common endpoint of the two rays that form an angle.

vértice de un ángulo   El punto final común de los dos rayos que forman un ángulo.

vertical angles   Two nonadjacent angles formed by two intersecting lines.

ángulos verticales   Dos ángulos no adyacentes formados por dos líneas de intersección.

vertical asymptote   A vertical line that a graph approaches.

asíntota vertical   Una línea vertical que se aproxima a un gráfico.

vertical shift   A vertical translation of the graph of a trigonometric function.

cambio vertical   Una traducción vertical de la gráfica de una función trigonométrica.

volume   The measure of the amount of space enclosed by a three-dimensional figure.

volumen   La medida de la cantidad de espacio encerrada por una figura tridimensional.

W

work problems   Problems that involve two people working at different rates who are trying to complete a single job.

problemas de trabajo   Problemas que involucran a dos personas trabajando a diferentes ritmos que están tratando de completar un solo trabajo.

x-intercept   The *x*-coordinate of a point where a graph crosses the *x*-axis.

intercepción *x*   La coordenada *x* de un punto donde la gráfica corte al eje de *x*.

y-intercept   The *y*-coordinate of a point where a graph crosses the *y*-axis.

intercepción *y*   La coordenada *y* de un punto donde la gráfica corte al eje de *y*.

z-value   The number of standard deviations that a given data value is from the mean.

valor *z*   El número de variaciones estándar que separa un valor dado de la media.

zero   An *x*-intercept of the graph of a function; a value of *x* for which $f(x) = 0$.

cero   Una intercepción *x* de la gráfica de una función; un punto *x* para los que $f(x) = 0$.

Index

Index